PERIODIC CHART OF THE ELEMENTS

1A	2A	3B	4B	5B	6B	7B	8			1B	2B	3A	4A	5A	6A	7A	
																1 H 1.00797 ±0.00001	2 He 4.0026 ±0.00005
3 Li 6.939 ±0.0005	4 Be 9.0122 ±0.00005											5 B 10.811 ±0.003	6 C 12.01115 ±0.00005	7 N 14.0067 ±0.00005	8 O 15.9994 ±0.0001	9 F 18.9984 ±0.00005	10 Ne 20.183 ±0.0005
11 Na 22.9898 ±0.00005	12 Mg 24.312 ±0.0005											13 Al 26.9815 ±0.00005	14 Si 28.086 ±0.001	15 P 30.9738 ±0.00005	16 S 32.064 ±0.003	17 Cl 35.453 ±0.001	18 Ar 39.948 ±0.0005
19 K 39.102 ±0.0005	20 Ca 40.08 ±0.005	21 Sc 44.956 ±0.0005	22 Ti 47.90 ±0.005	23 V 50.942 ±0.0005	24 Cr 51.996 ±0.001	25 Mn 54.9380 ±0.00005	26 Fe 55.847 ±0.003	27 Co 58.9332 ±0.00005	28 Ni 58.71 ±0.005	29 Cu 63.54 ±0.005	30 Zn 65.37 ±0.005	31 Ga 69.72 ±0.005	32 Ge 72.59 ±0.005	33 As 74.9216 ±0.00005	34 Se 78.96 ±0.005	35 Br 79.909 ±0.002	36 Kr 83.80 ±0.005
37 Rb 85.47 ±0.005	38 Sr 87.62 ±0.005	39 Y 88.905 ±0.0005	40 Zr 91.22 ±0.005	41 Nb 92.906 ±0.0005	42 Mo 95.94 ±0.005	43 Tc (99)	44 Ru 101.07 ±0.005	45 Rh 102.905 ±0.0005	46 Pd 106.4 ±0.05	47 Ag 107.870 ±0.003	48 Cd 112.40 ±0.005	49 In 114.82 ±0.005	50 Sn 118.69 ±0.005	51 Sb 121.75 ±0.005	52 Te 127.60 ±0.005	53 I 126.9044 ±0.00005	54 Xe 131.30 ±0.005
55 Cs 132.905 ±0.0005	56 Ba 137.34 ±0.005	57 *La 138.91 ±0.005	72 Hf 178.49 ±0.005	73 Ta 180.948 ±0.0005	74 W 183.85 ±0.005	75 Re 186.2 ±0.05	76 Os 190.2 ±0.05	77 Ir 192.2 ±0.05	78 Pt 195.09 ±0.005	79 Au 196.967 ±0.0005	80 Hg 200.59 ±0.005	81 Tl 204.37 ±0.005	82 Pb 207.19 ±0.005	83 Bi 208.980 ±0.0005	84 Po (210)	85 At (210)	86 Rn (222)
87 Fr (223)	88 Ra (226)	89 †Ac (227)	104 (257)														

*Lanthanum Series

58 Ce 140.12 ±0.005	59 Pr 140.907 ±0.0005	60 Nd 144.24 ±0.005	61 Pm (147)	62 Sm 150.35 ±0.005	63 Eu 151.96 ±0.005	64 Gd 157.25 ±0.005	65 Tb 158.924 ±0.0005	66 Dy 162.50 ±0.005	67 Ho 164.930 ±0.0005	68 Er 167.26 ±0.005	69 Tm 168.934 ±0.0005	70 Yb 173.04 ±0.005	71 Lu 174.97 ±0.005

†Actinium Series

90 Th 232.038 ±0.0005	91 Pa (231)	92 U 238.03 ±0.005	93 Np (237)	94 Pu (242)	95 Am (243)	96 Cm (247)	97 Bk (247)	98 Cf (249)	99 Es (254)	100 Fm (253)	101 Md (256)	102 No (253)	103 Lw (257)

Atomic Weights are based on C^{12}—12.0000 and Conform to the 1961 Values

Printed in U. S. A.

GARTH L. LEE, Ph.D.
Professor and Head, Department of Chemistry,
Utah State University, Logan, Utah

HARRIS O. VAN ORDEN, Ph.D.
Professor of Chemistry,
Utah State University, Logan, Utah

RONALD O. RAGSDALE, Ph.D.
Associate Professor of Chemistry,
University of Utah, Salt Lake City, Utah

ILLUSTRATED BY ALEXIS KELNER
Salt Lake City, Utah

GENERAL AND ORGANIC CHEMISTRY

W. B. SAUNDERS COMPANY • PHILADELPHIA • LONDON • TORONTO • 1971

W. B. Saunders Company: West Washington Square
Philadelphia, Pa. 19105

12 Dyott Street
London, WC1A 1DB

1835 Yonge Street
Toronto 7, Ontario

General and Organic Chemistry SBN 0-7216-5680-3

 Made in the United States of America. Press of W. B. Saunders Company. Library of Congress catalog card number 75-145559.

Print No.: 9 8 7 6 5 4 3 2 1

PREFACE

This text was written to be *read* and understood by the student. It has evolved from our separate experiences in teaching chemistry to those for whom this book is intended: majors in the biological sciences, humanities and arts, agriculture, forestry, nursing, home and family living and education—students who have a secondary interest in the subject. There is a balance among the topics which are presented. Many beginning texts emphasize physical chemistry and principles to the exclusion of all else. This volume presents material in inorganic, nuclear, organic, physical and biochemistry.

Many physical principles and concepts are introduced in the initial portion of the book by use of models and with a leaning toward the historical approach; later these principles are employed in the development of the descriptive material. The descriptive chemistry is related to the periodic relationships of the elements and to the modern concepts of atomic structure and bonding. A considerable amount of descriptive chemistry is included to relate the study to the students' experiences.

This text takes into consideration the limited mathematical background of many of the students who will use it. The necessary mathematics is presented in such a way that the student must "think through" the solution. Memorization and routine application of formulas or patterns for problem solving are discouraged.

Marginal notes are used throughout the text to draw the students' attention to key ideas, new concepts and pertinent questions.

The Additional Reading sections contain references to such popular journals as Scientific American, Chemistry and the Journal of Chemical Education, which present short articles on many subjects of current interest. Reference is also made to chapters in other texts where the student may find alternative treatments of difficult subjects or where he may delve more deeply into topics in which he has a special interest.

Pollution, which is of great concern to the public and the students, is frequently treated in the text, where it is integrated with related descriptive chemistry. Some pollutants considered are DDT, sulfur oxides, mercury, nitrogen oxides, lead and ozone.

The text is beautifully illustrated—but it is illustrated primarily to aid in the discussion of the subject. The second color is used to illustrate atomic and molecular structure, bonding and periodic relationships. The illustrations were developed by Mr. Alexis Kelner, a scientifically oriented artist, in consultation with the authors.

We should like to acknowledge the assistance of our colleagues at Utah State University and at the University of Utah, who have used an earlier text by two of the authors, and thank them for their suggestions and help in making the text more usable. We also thank the production staff at Saunders for their excellent workmanship, and Professor Eugene Rochow, Emeritus Professor of Chemistry, Harvard University, for his careful reading of the manuscript and for his detailed criticism and review. All have helped in making the book a more readable, more teachable text.

GARTH L. LEE

HARRIS O. VAN ORDEN

RONALD O. RAGSDALE

CONTENTS

FIFTEEN • ACTIVE NON-METALS–THE HALOGENS 321

SIXTEEN • THE SULFUR FAMILY 340

SEVENTEEN • RATES OF REACTION AND EQUILIBRIUM 359

TWENTY-FOUR • COLLOIDS 525

TWENTY-FIVE • ORGANIC CHEMISTRY AND THE ALKANES 534

TWENTY-SIX • STABILITY AND REACTIVITY OF ORGANIC COMPOUNDS 571

ONE • INTRODUCTION—CHEMISTRY

1.1 CHEMISTRY DEFINED

Chemistry is a study of the behavior of matter. It is primarily concerned with those **changes in matter** that are **called chemical reactions,** with the heat effects associated with the changes and with the factors such as temperature and pressure that control them.

The chemist may be concerned with the production of a better paint, or the reaction steps in the corrosion of a pipeline or a bridge, or the production of a finer, more durable artificial fiber for fabrics; or he might be studying the side effects of LSD or other hallucinogenic drugs. He may be studying the mechanism by which some plants and microscopic animals fix nitrogen (change the nitrogen of the air into a product useful for plant growth), hoping that his studies will help feed the untold millions who will inhabit our planet if the population explosion continues. On the other hand he might be striving to develop another "pill" to limit that population growth. Or perhaps he is looking for harmful side reactions that might arise from the continued use of such a drug. He might be determining the order of amino acids in an animal or plant protein. He might be studying the composition and reactions of DNA (deoxyribonucleic acid), which serves, like the magnetic tape of a computer, to determine hair color and other genetic characteristics. Or he might be attempting to grow one of the simplest forms of life, a virus, outside a living organism.

the chemist might be searching for a compound which would replace DDT.

Some study chemistry to become chemists, to enter the vast and rapidly expanding chemical industry either as technicians, research scientists, product salesmen, managers or in one of the many other positions available. People in other fields of interest study chemistry because a knowledge of its principles is necessary for an understanding of the subject matter encompassed by their areas. A geologist must learn various techniques of analysis to ascertain the composition of minerals and to understand their formation. Biologists must understand chemistry in order to follow the multitude of biochemical reactions that take place in plants and animals, and to understand their purpose in the various organisms. Since most biological phenomena are based ultimately on chemical and physical processes, the biologist needs an ever increasing knowledge of chemistry and physics. The agricultural specialist who is concerned with animal nutrition, or the use of insecticides, herbicides, or fertilizers deals primarily with chemicals, and he too must understand chemical principles. Manufacturers of fibers, protective coatings, oils and greases, drugs and toys are all concerned with chemical products and must know something of the relationship between molecular structure and the properties of the chemicals with which they are working.

Many study chemistry to better comprehend the nature of the physical world. The student of chemistry soon learns of the similarities that exist in the myriad forms of matter. He learns of the marvelous order of physical things and of the immutability of natural law. All these considerations help man to be at home in the world in which he lives. This is a chemical age. With no conception of chemical changes, one can know little of life processes and other changes in matter that occur continually about us.

1.2 CHEMISTRY IS A SCIENCE

A science is a field of study or investigation in which the scientific method is employed. The scientific method is a system for attacking a problem in which careful, **accurate observation and vivid imagination** are the two major factors. In principle the problem-solving strategy may be separated into four phases:

1. Facts are discovered by **quantitative observation** of the behavior of a system—the system being the collection of chemicals studied—and the facts are recorded.

2. The observed facts are correlated or generalized in the **statement of a natural law.** Quite often the law is mathematical in nature.

3. The question is considered, "What is there about the system that makes it behave as described by the law?" A mental picture, called a **theory,** is evolved to explain the observed behavior.

4. Often a theory not only explains the workings of the law or laws that generated it, but will predict certain other relationships. **The predicted relationships are examined** by experiments and are found to be either true or false. If the expected relationships prove to be false, the theory is false, and this stimulates development of new hypotheses; new experiments then follow. The scientist holds to the best and simplest picture until a better one is found. The scientific approach will be illustrated in later chapters in discussions of the development of fundamental chemical theories.

the four steps may not follow in the order listed in all scientific investigations, but the elements of each step are found in each application of the scientific method.

The scientific method has been extremely fruitful in the physical sciences. Developments arising from its application are in large measure responsible for the great technological advancement of this age. It is proving to be useful in other fields, such as sociology, economics and psychology. A mastery of this method and an ability to use it to solve problems is in itself an excellent reason for studying chemistry.

It is obvious that by its nature science has definite limitations. It is limited to observations of either naturally occurring or staged phenomena. Science thrives when observations are quantitative and exact. Because the theories must be tested by observations, advancement is slow. Scientific truth is not absolute, but it becomes more and more sure as experimental evidence is accumulated.

Chemistry is very closely related to physics; both are studies of matter. But the emphasis in physics is, in general, different from the emphasis in chemistry. Physics is a study of matter in relation to the energy it possesses by virtue of its position or motion—its heat-conducting properties, its electrical properties and its relationship to the transmission of light, sound and heat. Despite the different emphasis, certain areas of study are common to both disciplines and are necessary for an understanding of either. Throughout this text material will be drawn from physics when necessary.

a two-semester course in high school algebra is sufficient.

Mathematics is a necessary tool for expressing the quantitative relationships found in chemistry. Extended studies in chemistry require a more extensive knowledge of mathematics, but in this course a knowledge of **algebra** and a reasonable facility in working so-called "story problems" will suffice.

SUGGESTED READING

Garrett, A. B.: "The Discovery Process and the Creative Mind." J. Chem. Ed., *41*:479, 1964.

Keller, E.: "The Scientific Needs of Urban America." Chemistry, *41*(5):16, 1968.

TWO • MATTER

The physical world about us is composed of matter. The books you read, the food you eat, the records you play, your automobile —all are material. The thoughts expressed in this book, the satisfied feeling you have when you view your dinner, the music that comes from the record and the feeling of power you have when you press on the accelerator are real, but they are not material.

2.1 CHARACTERISTICS OF MATTER

If something is colored, if it has an odor, if it is hard (or soft), if it is hot (or cold), it is composed of matter. Anything that can be detected by any one of the five senses is material. However, not all samples of matter have all the above properties, nor can all samples of matter be detected by every one of the senses. Only two characteristics are common to all portions of matter: **all occupy space** and **all possess mass.** Either of these is a sufficient criterion for the existence of matter, because anything that possesses mass must occupy space, and vice versa. Things not material meet neither requirement.

The **volume** possessed by a sample of matter may be varied, sometimes drastically, but it can never be eliminated entirely. A large volume of air may be compressed into a football or an automobile tire, but an attempt to add an unlimited amount will result in bursting the container.

Two important manifestations of mass are **weight** and **inertia.** The weight of a body is the attraction of the earth due to the force of gravity upon that body. The weight of a body is proportional to its mass and is inversely proportional to the square of the distance of the body from the center of the earth. Thus, the weight varies with the altitude but exists, by definition, everywhere in the earth's

because of the inertia it possesses, it is difficult to cause a speeding car to turn a corner.

gravitational field. Both the balance and the scales (to be discussed in the next section) employ weight to determine the mass of a body.

Inertia is the property of matter that causes it to resist a change in motion or in direction of motion. Because of inertia, it takes energy to start an automobile rolling or to turn it from its straight line path when it is in motion. The greater the mass, the greater the inertia.

2.2 UNITS OF MEASUREMENT—THE DECIMAL SYSTEM

The **space** or **volume** occupied by a body of matter may be measured **quantitatively** in any of several arbitrary units or dimensions. The pint, the cubic foot, the gallon, the teaspoon and the bushel are all units of volume. These are seldom used in courses in chemistry; instead, units of the **decimal (metric) system** are used because they are more easily adapted to computation.

In the decimal system, which is used in many kinds of measurements, a basic unit is chosen of such a size to be convenient for stating common measurements. Other units differ from the basic unit by factors of 10 or 1/10 and are related to the basic unit by universally understood prefixes. The values of the prefixes are listed in Table 2.1, in which the derived units of length are related to the basic unit, the meter. The abbreviations or symbols for the units also appear in the table. The units enclosed in parenthesis in Table 2.1 are seldom used, although their counterparts in other systems find frequent use. Names of the last three units in Table 2.1 do not follow the system. They do differ from the basic unit by powers of 1/10 but are not related to that unit by name. This is unfortunate, but the units are used and must be remembered.

The relationships of the decimeter, millimeter, and the centimeter to the meter are illustrated in Figure 2.1.

Stating a measurement of length given in one unit in terms of another is done by merely moving the decimal point the correct number of places to the right or to the left. Using unit labels on the conversion factors, one can carry out the conversions with little

TABLE 2.1 UNITS OF LENGTH—DECIMAL SYSTEM

Unit	Abbreviation	Size
meter	m	39.37 inches
(*deka*meter)	dkm	10 meters
(*hecto*meter)	hm	100 meters
*kilo*meter	km	1000 meters
(*deci*meter)	dm	1/10 meter
*centi*meter	cm	1/100 meter
*milli*meter	mm	1/1000 meter
micron	μ	10^{-6} meter
millimicron	mμ	10^{-9} meter
angstrom unit	Å	10^{-10} meter

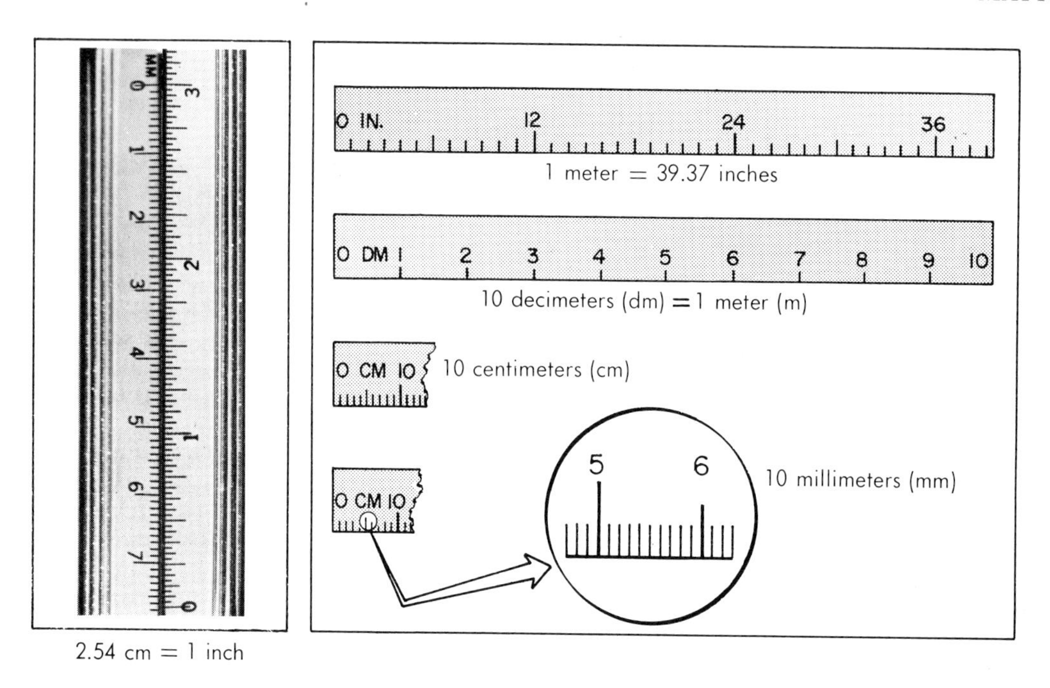

FIGURE 2.1 COMPARISON OF THE ENGLISH AND METRIC SYSTEMS.

chance of error. For example,

$$1.00\,\cancel{\text{cm}} \times \frac{1\,\cancel{\text{m}}}{100\,\cancel{\text{cm}}} \times \frac{1000\text{ mm}}{\cancel{\text{m}}} = 10\text{ mm}$$

Note how units cancel to leave only the unit or units desired. The equalities most commonly used in conversion factors, obtained in this manner, are listed in Table 2.2.

Example: Find 0.023 cm in Å.

Solution:

$$0.023\,\cancel{\text{cm}} \times \frac{10^8\text{ Å}}{\cancel{\text{cm}}} = 0.023 \times 10^8\text{ Å}$$

$$= 2.3 \times 10^{-2} \times 10^8\text{ Å} = 2.3 \times 10^6\text{ Å}$$

TABLE 2.2

1 m	= 100 cm = 1000 mm
1 cm	= 10 mm
1 cm	= 10^8 Å
1 Å	= 10^{-8} cm
1 mm	= 10^{-3} m
1000 m	= 1 km

Units of volume for measurements taken in the laboratory are often derived from units of length. The cubic meter, represented by m^3, is the volume enclosed in a cube 1 m on an edge. This basic unit of volume is too large for use in expressing most volumes used in the laboratory. The cubic centimeter (cm^3) is more convenient. It is the volume of a cube 1 cm on an edge, and is nearly equal to the combined volumn of two nickels. It is one millionth the size of a cubic meter.

$$1\ \text{m}^3 \times 100\ \text{cm/m} \times 100\ \text{cm/m} \times 100\ \text{cm/m} = 10^6\ \text{cm}^3.$$

the capital letter L is used as the abbreviation for liter to avoid confusion of the lower case l with the number 1.

The cubic decimeter (dm^3), being equal to 1000 cm^3, would be a useful unit for the stating of volumes intermediate between the cubic centimeter and the cubic meter. However, it is seldom used. Instead the liter (L), the volume of 1000 g of air-free water, is used. The liter is 1000.027 cm^3. The milliliter (ml), is one thousandth of a liter and has almost exactly the same volume as the cubic centimeter. In this book they will be used interchangeably (usually the cm^3 for solids, the ml for liquids).

Three pieces of glassware commonly used for measuring and dispensing measured volumes of liquids in the laboratory are the graduated cylinder, the buret and pipet, all shown in Figure 2.2. The **graduated cylinder** is the chemist's measuring cup. It is made in various sizes, the total volume varying from 5 ml up to 3 L; it is used in making quick, approximate measurements.

The **buret** is a long glass tube that is mounted vertically, with graduations running from 0 at the top to the total volume in milliliters at the bottom. At the bottom is a stopcock with a glass tip. Liquid can be dispensed by opening the stopcock. The amount dispensed is indicated by the difference in levels before and after dispensing. The buret is used when the chemist desires to add a solution slowly, even a drop at a time, while keeping account of the total amount added.

The **pipet** is a length of glass tubing with a tip on one end and usually a bulb in the center. The pipet delivers one set volume of liquid drawn up into the pipet to the mark just above the bulb. With the forefinger over the end of the tube, one can transfer the contents of the pipet to another container, into which the liquid empties when the forefinger is removed.

The basic unit of **mass** used in scientific work is the **gram.** It appears with other derived units below:

kilogram (kg)	= 1000 gram	≈ 2.2 pounds
gram (g)	≈ 1/454 pound	= 1/28.4 ounces
milligram (mg)	= 0.001 g	
microgram (gamma)	= 0.001 mg	$= 1.0 \times 10^{-6}$ g

Consistent with the above definitions are the following equalities:

$$0.37\ \text{kg} \times 1000\ \text{g/kg} = 370\ \text{g};\ 370\ \text{g} \times 1000\ \text{mg/g} = 3.7 \times 10^5\ \text{mg}$$

$$0.00034\ \text{kg} \times 1000\ \text{g/kg} = 0.34\ \text{g};\ 0.34\ \text{g} \times 1000\ \text{mg/g} = 340\ \text{mg}$$

Again the change from one unit to another is accomplished by

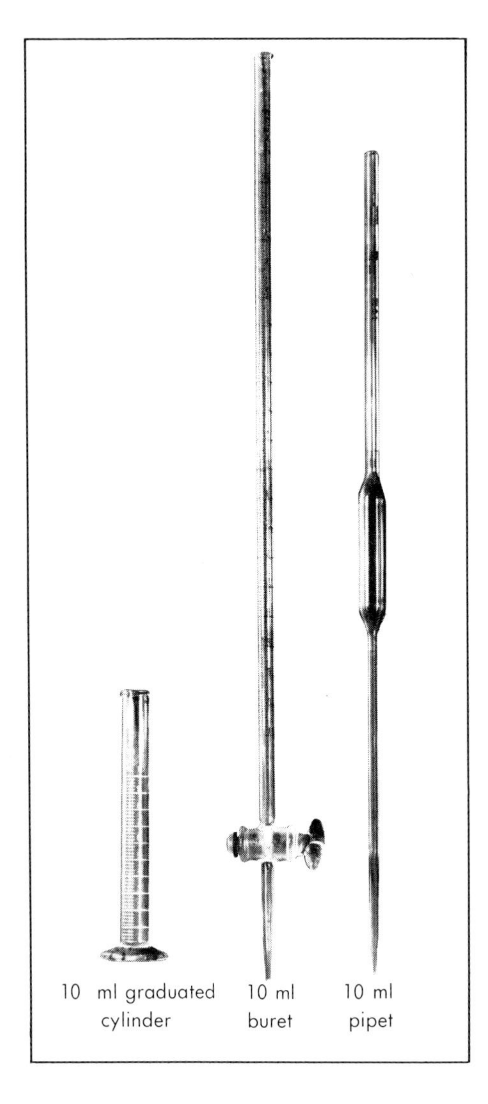

FIGURE 2.2 TYPICAL VOLUMETRIC WARE.

moving the decimal point, the conversions being much simpler than those in the system of ounces, pounds and tons.

All types of scales or balances use the force of gravity in weighing, or finding the mass of objects. A 10-pound bag of potatoes stretches the spring on a grocer's scales twice as far as a 5-pound bag. The pointer on the dial rotates as the spring is lengthened.

A two-pan balance is constructed like a "teeter" or "see-saw." The analytical balance shown in Figure 2.3 is an example. A metal bar is supported or suspended at the center so that it balances. Two pans of equal mass are suspended from or set on the bar, one on either end. The object to be weighed is placed on one pan, tipping the balance arm in that direction. Weights (pieces of brass of known mass) are added, one by one, to the other pan until the bar is restored to its original horizontal position.

When the bar is horizontal, each pan load is pulled with the same force; therefore, each contains the same mass. The mass of the object is found by taking the sum of the masses of the weights on the other pan.

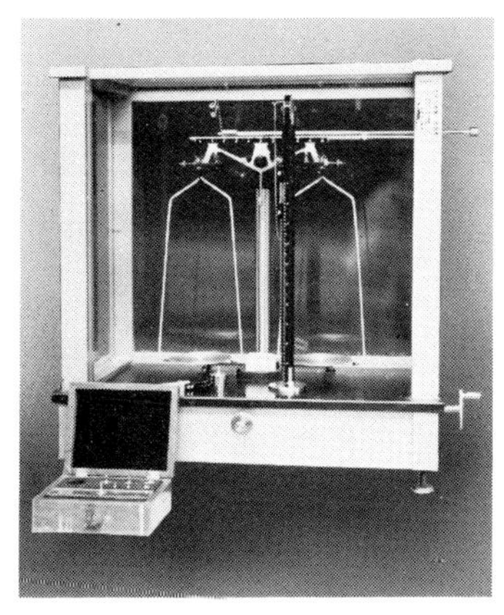

Analytical balance

Triple beam balance

Single pan balance

FIGURE 2.3 CHEMICAL LABORATORY BALANCES.

The sliding weight balance (i.e., triple beam balance; see Figure 2.3) also uses the principle observed in a "see-saw" arrangement. A 10-g weight 10 cm from the center or point of suspension will balance a 5-g weight 20 cm from that point on the other arm of the balance. Thus, by moving one weight back and forth on one arm of the balance, the chemist can balance objects with different masses placed on the pan on the opposite arm. A scale on the bar enables one to read the balancing mass directly.

The modern balance in Figure 2.3 is a one-pan automatic balance that enables one to make rapid weight determination by a substitution technique. An object is weighed by placing it on the pan, and by turning knobs standard weights are removed until their mass is less than and within 1 g of the mass of the object. An optical lever system is then used to measure the weight difference (less than 1 g) between the object and weights removed.

2.3 HETEROGENEOUS AND HOMOGENEOUS MATTER

All bodies of matter, either as they appear in nature or as they occur as a result of man's arranging, may be placed in one of two classes. They must be either homogeneous or heterogeneous. A **heterogeneous body** is an obvious mixture. We can see at once that it is composed of different materials. Concrete, soil, wood, granite and bean soup are mixtures. On the other hand, salt water, a gold ring, some sugar or some air are all examples of **homogeneous matter.** They are composed of only one **phase.** A phase constitutes all parts of a system that have exactly the same physical properties. One crystal of sugar has the same properties as any other crystal of sugar, although the sizes of the crystals may differ. All sugar crystals together constitute a phase. A heterogeneous system contains *more* than one phase. A mixture of iron and sulfur contains two phases; all particles of iron belong to one phase, and all particles of sulfur belong to the other.

The phases of a heterogeneous mixture may be separated in different ways. At times, a solid mixture may be separated into phases with tweezers and a magnifying glass. In a mixture of solid and liquid such as sand in water, one can decant the liquid (pour it off) or filter it. In other instances ingenuity is required. For example, a mixture of sand and salt may easily be separated because they differ greatly in one of their properties: solubility in water. Salt is very soluble, while sand is insoluble. To separate them, one needs merely to add water, stir and filter. The salt dissolves in the water and the sand remains on the sieve or filter paper. The salt may be recovered by evaporating the water. Other methods of separation make use of differences in other physical properties, such as melting point, boiling point, density and so forth.

A phase may be either a **solution** or a **pure substance**. A solution is a homogeneous mixture. Salt water is a solution, a mixture of salt and water. A 14 carat yellow gold ring contains 14 parts of gold to 10 parts of copper by weight and is homogeneous. Air is a homogeneous mixture, a gaseous solution of oxygen and nitrogen and minute amounts of other gases.

2.4 THE THREE STATES OF MATTER

On the basis of differences in certain physical properties, all materials may be divided into three general classes or states. It is apparent that salt, water and air are quite different from each other. Each is an example of one of the states of matter. Salt is a **solid,** air is a **gas,** and water is a **liquid.**

True solids have one distinguishing characteristic. Each solid possesses a certain form or structure. When viewed through a microscope, table salt appears as a mass of tiny cubes. The individual cubes appear in many sizes, but, barring accidental imperfections, all have square corners. If these cubes are crushed into still tinier particles, they are also cubes. Salt is a true solid; it has a definite structure. Attempts to change the structure of individual units are resisted by the unit or crystal to the point of its destruction.

true solids are often defined as crystalline solids.

Liquids differ from solids in that individual units have no definite form or shape. In fact, for this reason one can hardly speak of individual units. Because of this lack of rigid structure, liquids can flow and consequently can assume the form of a container, or of the lower portion of a container. All liquids have some resistance to flow. This resistance is known as viscosity. Some liquids such as water, alcohol and gasoline have very low viscosity, but others (molasses, lubricating oil and tar) present a great deal of resistance. Their high viscosity indicates that there is some transitory or temporary arrangement or structure within the liquid that must be altered as the liquid flows. This is true, but it is not a rigid form and differs greatly from that of solids. One well-known liquid is quite often classed as a solid because of its unusually high viscosity. This liquid is glass, window glass. Investigations have shown that glass has no definite long-range repetitive internal structure.

One property common to both solids and liquids is the small percentage change in their volumes caused by variations in temperature or pressure. Liquids and solids are relatively incompressible.

Gases differ from solids in that they possess no apparent structure. They differ from both solids and liquids in that they possess no special volume. A gas assumes both the form and volume of its container. A gas will fill completely and very rapidly any space into which it is admitted. Since a sample of gas may occupy different volumes, the density of the gas will vary with that volume. In general, the density of gases at room temperature and pressure is many times less than that of solids or liquids. For that reason, densities of solids and liquids are usually recorded as grams per cubic centimeter, whereas the densities of gases are given in grams per liter.

quantitative calculations relating the change in the volume of a gas sample with changes in temperature and pressure will be treated in chapter 6.

2.5 CHANGES IN STATE

In general, under the proper conditions, a particular sample of matter may exist in each of the three states. Consider water, which is a liquid at room temperature. When cooled below 0° centigrade, water freezes into brittle, translucent ice crystals. On warming above that temperature, it is converted again to water. This same water may be converted to vapor, or steam, and recondensed as water. A cold spoon held over the spout of a boiling coffee percolator condenses the steam to drops of water.

When water is distilled (as, for example, during its purification), it is converted to steam and then condensed to water again. A laboratory apparatus used to accomplish this is shown in Figure 2.4. Water is boiled in the distilling flask. Steam leaves the flask

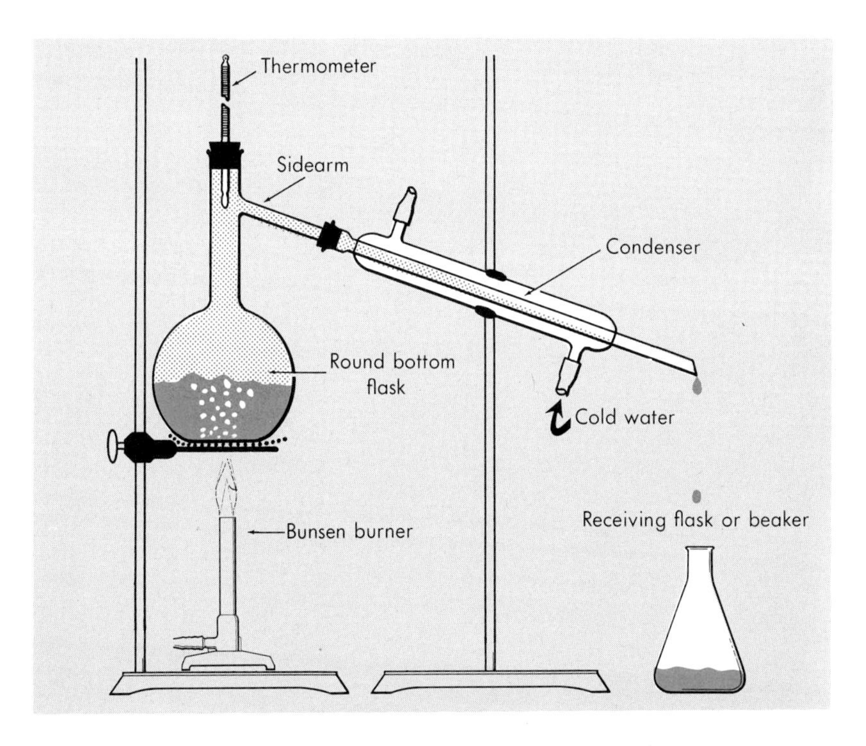

FIGURE 2.4 LABORATORY DISTILLATION APPARATUS.

through the side arm where it is cooled in the water-jacketed condenser and converted (condensed) to water, which drips into the receiving flask.

Some substances once thought to exist in only one or two states can be converted to other states, but only at extremely high or low temperatures. Steel and salt may be melted to liquids and then vaporized, if the temperature is high enough. In contrast, hydrogen gas may be liquefied and finally frozen, but only at temperatures near absolute zero (the coldest possible temperature). However, there are other substances that break down chemically (decompose) when heated to temperatures below their boiling points or even their melting points. Thus not all substances can exist in all three states.

It is often thought that changes in state proceed progressively from solid, to liquid, to gas (vapor) as heat is added, and that the reverse progression occurs in the exact same order. This is not always the case. Dry ice (solid carbon dioxide) changes directly to a gas at room temperature and pressure. A change in state from solid directly to vapor is known as sublimation. Thus we say that solid carbon dioxide ("dry ice") sublimes. At high pressure dry ice will melt, and so it exists as a liquid at 20° C only under such pressure. For this reason, carbon dioxide is shipped as a liquid in tanks. Iodine is purified by sublimation. By one method, iodine is placed near an electric light bulb in a closed container. The iodine is vaporized by the heat of the lighted bulb and travels in the **vapor phase** to cooler surfaces in the container, where it collects as pure iodine crystals. Water vapor from the air is converted directly to a solid (snow) in winter, and is collected on refrigerator pipes as frost.

Changes in state are called physical changes, or physical reactions. The nomenclature is obvious; there is no change in chemical composition, and the material lost when heated is recovered when cooled. When water is frozen to ice, only the physical aspects of the matter are changed. It is still water. Steam is still water and may be condensed to the common liquid.

2.6 PHYSICAL PROPERTIES

A **property is a characteristic** of a type of matter **that aids in describing it or distinguishing it** from other types of matter. Properties are classified, for convenience, into two categories: either physical or chemical. Chemical properties are those observed when a chemical reaction occurs or is occurring. Combustibility, the property of combining with the oxygen in air with a flame, is a chemical property; the combustible substance is changed chemically in the burning. The rate at which one chemical substance reacts with another is also a chemical property. In the chapters in which we discuss the elements and groups of elements, the chemical properties of the elements and their compounds will be the primary consideration.

Physical properties are those properties that can be observed without the occurrence of a chemical reaction. Hardness, electrical conductivity, color, tensile strength and odor are all physical properties. Freezing point (the temperature at which a solid can be in equilibrium with its liquid), boiling point and the degree of solubility of a material in water or another solvent are all physical properties. They are characteristics that are related to so-called **physical** reactions.

2.7 DENSITY

Perhaps the most important physical property of a substance is its density. Like many other properties, density has significance only when it is applied to a homogeneous material, a one-phase system. **Density is defined as the mass per unit volume.** It combines two characteristics of a particular sample of matter into a property of the substance. The three quantities, density, volume and mass, are mutually related:

$$\text{Density} = \frac{\text{Mass}}{\text{Volume}} \text{ or } D = \frac{M}{V}$$

Other forms of the same equation are

$$M = DV \text{ and } V = \frac{M}{D}$$

One merely substitutes the values into the equation to solve for one quantity when the other two are known, as is shown in the following examples. Notice that when the equation is used properly, the answer has the correct units.

Example: Calculate the density of alcohol, knowing that 80 ml weighs (has a mass of) 64 g.

$$\text{Density} = 64 \text{ g}/80 \text{ ml} = 0.80 \text{ g/ml}$$

One can reason that if 64 g is distributed equally among 80 ml, one ml will contain 1/80 of that 64 g. One can check his answer by noting that there are more milliliters than grams, and therefore there will be less than 1 g per ml. Notice that the labels g and ml appear correctly in the answer g/ml.

Example: At 0°C and 1 atmosphere pressure, 22.4 L of oxygen gas has a mass of 32 g. Find the density in g/L.

$$32 \text{ g}/22.4 \text{ L} = 1.43 \text{ g/L}$$

Again, 32 g is distributed equally among 22.4 L, and so each liter contains more than 1 g, or about 1.5 g.

Density is used in several ways. Three of them will be discussed here.

1. Density is a **physical property;** its **numerical value** is characteristic of the material and helps to identify it. When comparing substances it means nothing to say "water weighs 1050 g and iron weighs 6000 g," but it is useful to know that 1 cm^3 of water weighs 1 g and 1 cm^3 of iron weighs 7.86 g. When you say that iron is heavier than water, you are comparing the weights of equal volumes and are speaking of densities. The list below gives the densities of some common substances in g/ml (g/cm^3).

Ethyl alcohol	0.80
Water	1.00
Aluminum	2.71
Bromine	3.4
Iron	7.86
Copper	8.92
Silver	10.5
Lead	11.3
Mercury	13.6
Gold	19.3

2. Using density, one can calculate the *mass* of a certain *volume of liquid*.

units are "cancelled," as one cancels factors in a fraction, to give the correct units in the answer.

Example: Concentrated sulfuric acid has a density of 1.84 g/ml. Calculate the weight of 50. ml of sulfuric acid.

$$50\ \text{ml} \times 1.84\ \text{g/ml} = 92\ \text{g}$$

Again if 1 ml weighs almost 2 g, 50 ml will weigh almost 100 g. Cancellation of units (just as you cancel numbers in fractions) gives you the correct units for your answer.

3. Using density and mass, one can calculate the **volume** of a liquid sample.

Example: How many ml of bromine must one take to get 56 g? The density of bromine is 3.4 g/ml.

$$\frac{56\ \text{g}}{3.4\ \text{g/ml}} = 16\ \text{ml}$$

Not only does the answer have the units of volume, but the answer is reasonable, for if 1 ml contains more than 3 g, then one needs less than 1/3 as many milliliters as grams. The unit g cancels out and ml is left in the numerator as the correct unit for the answer.

The density of a liquid may be **measured directly** with a **hydrometer** (Figure 2.5). The hydrometer utilizes the principle discovered by Archimedes that a floating body settles down in a liquid until it displaces a volume of that liquid having a mass equal to its own. When placed in a liquid, the hydrometer floats partially submerged as shown in Figure 2.6. The lighter the liquid (the lower its density), the greater the amount that must be displaced

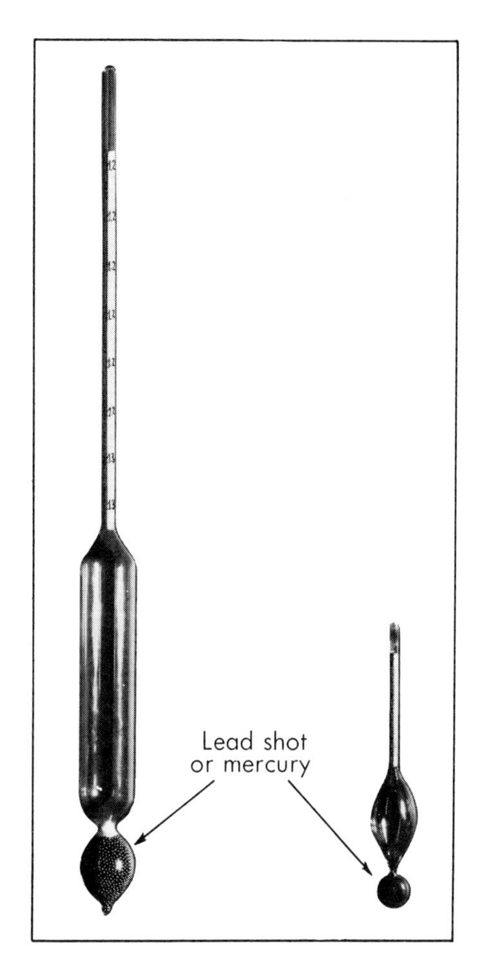

FIGURE 2.5 HYDROMETERS.

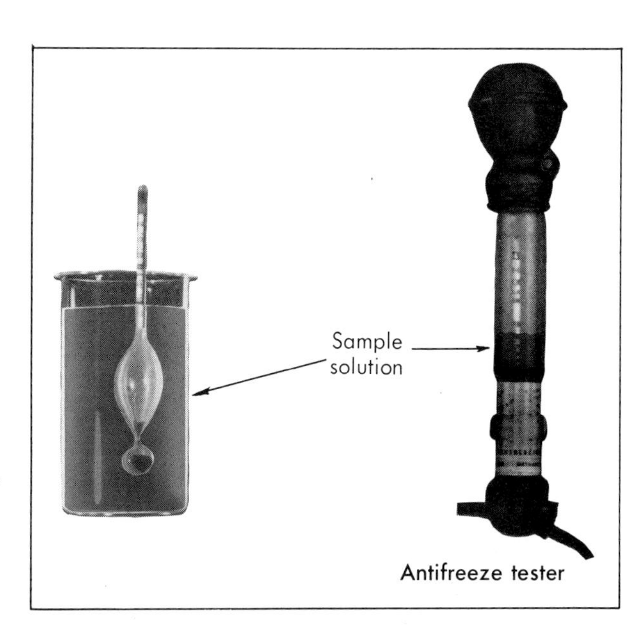

FIGURE 2.6 USING A HYDROMETER.

TABLE 2.3 TABLE OF DENSITIES IN g/ml OR g/cm^3

Substance	Density
Ethyl alcohol	0.806 at 0°C
	0.785 at 25°C
Water	1.000 at 4°C
	0.997 at 25°C
	0.958 at 100°C
Table salt	2.165 at 25°C
Marble	2.7 at 25°C

Metals*			**Alloys***		
Aluminum	(Al)	2.71	Aluminum bronze	(90% Cu, 10% Al)	7.60
Copper	(Cu)	8.92	Bell metal	(78% Cu, 22% Sn)	8.70
Gold	(Au)	19.30	Gold coinage	(90% Au, 10% Cu)	17.17
Lead	(Pb)	11.34	Magnalium	(70% Al, 30% Mg)	2.00
Magnesium	(Mg)	1.74	Solder	(50% Sn, 50% Pb)	8.73
Tin	(Sn)	7.29			

* The densities of metals and alloys are given at 20°C.

by the hydrometer and the greater the extent of submersion. If the liquid is less dense than the average density of the hydrometer, the hydrometer will sink to the bottom. On the other hand, the greater the density, the higher the hydrometer floats. The hydrometer is calibrated in density units, so that one can take the reading of the hydrometer scale at the surface of the water.

Hydrometers are used to test the antifreeze solution in automobile radiators (Fig. 2.6). Ethylene glycol, the major ingredient of most permanent antifreezes, is more dense than water. Addition of ethylene glycol to water gives a solution of density greater than that of water alone. The density of the solution increases, and the freezing point decreases as ethylene glycol is added. A service station attendant merely draws some of the antifreeze solution up into a chamber containing the hydrometer. The higher the hydrometer floats, the greater the percentage of ethylene glycol in the solution, and the lower the freezing point of the solution. Although density is the property determined, the hydrometer is calibrated in terms of the freezing point of the solution having that density.

Table 2.3 contains the densities of some substances. Comparison of these values shows that the densities of different substances exhibit wide variations. Some of the densities will be useful in solving the problems at the end of the chapter.

EXERCISES

2.1 Name the three physical states of matter.

2.2 Describe the appearance of each state in terms that distinguish it from each of the other two.

2.3 Describe in a general way the internal structure of a substance in each of the three states of matter.

2.4 Why is glass not a true solid? Which of the following *are* true solids: coal tar, ice, paraffin, salt, dry ice, wood?

2.5 In which state does a substance exhibit the greatest density? The least density?

2.6 Under what conditions does carbon dioxide exist as a liquid?

2.7 Define viscosity. Name three substances with high viscosity.

2.8 How may salt water be separated into pure substances?

2.9 Which of the following are chemical reactions: the burning of wood, the baking of bread, the drying of sand, the souring of milk, the grinding of salt, the frying of bacon?

2.10 Which of these processes include a chemical reaction?

a. Table salt is heated to 804°C, at which temperature it changes to colorless liquid. When it is cooled, it hardens suddenly in one translucent piece.
b. Blue vitriol (hydrated copper sulfate) is heated to 110°C; the blue crystals change to a white powder as water vapor is lost. The white powder is cooled to room temperature in dry air and remains as such.
c. Sugar is heated in a crucible to 186°C, at which temperature it fumes and leaves a black residue.
d. Pure oxygen is cooled to −183°C; it condenses to a pale blue liquid.
e. A piece of ordinary glass is heated until it softens. It is then bent back and forth at that temperature until it becomes cloudy, solidifies and breaks.

2.11 Fill in this table with equal values in the units indicated.

m	cm	mm	Å
2.71			
	.389		
		3.64	
			24.3

2.12 Write the following numbers in correct exponential form: (a) 2700 (b) 0.0134 (c) 0.0000019 (d) 4.00043 (e) 64×10^{-5}.

2.13 Sand was spilled into 30.6 ml of water in a graduated cylinder until the total volume was 46.7 ml. How many cm^3 of sand were added?

2.14 9.72 g of a metal was added to 26.21 ml of water in a graduated cylinder, increasing the volume to 29.81 ml. Calculate the density of the metal.

2.15 A watch glass was balanced on a two pan balance with one 10 g, one 5 g, two 200 mg and one 50 mg weight. Salt was added until the whole was balanced by one 20 g weight. Find the weight of the salt added.

2.16 25 ml of concentrated hydrochloric acid weighs 34.25 g. Calculate the density of concentrated HCl.

2.17 Find the weight of a bar of aluminum 1.5 cm × 1.5 cm × 15 cm. The density of aluminum is 2.7 g/cm^3.

2.18 A graduated cylinder has an inside diameter of 1 cm. Calculate the distance between the 25 ml and 26 ml marks. The area of a circle is πr^2.

2.19 At 1 atm pressure and 0°C, 18 g of nitrogen occupies 14.4 liters. Calculate the density of nitrogen gas in g/L at these conditions.

2.20 Hydrogen gas has a density of 1/14 that of nitrogen. Find the volume of 8.0 g of H_2 at 1 atm and 0°C.

2.21 At 0°C, 45 ml of ethyl alcohol weighs 36.3 g. Calculate the density of ethyl alcohol at 0°C.

2.22 What is the volume of 36.6 g of ethyl alcohol at 25°C?

Answer: 46.2 ml

2.23 What is the weight of 46.2 ml of ethyl alcohol at 25°C?

2.24 50.0 g of aluminum bronze has a volume of 6.58 cm^3. Calculate the density.

2.25 a. Calculate the volume of 40.0 g of tin.
b. Calculate the volume of 40.0 g of lead.
c. Calculate the volume of 80.0 g of solder.

2.26 Calculate the weight of a bar of gold coinage 1.00 cm × 0.400 cm × 15.0 cm.

2.27 Some gold shot was added to 32.6 ml of water in a graduated cylinder until the final volume was 34.2 ml. Calculate the weight of the gold shot.

2.28 How many cm^3 of copper and how many cm^3 of tin are needed to make 75 cm^3 of bell metal?

2.29 A flask was filled level full with 554 ml of water at 4°C and then warmed to 100°C. How many g of water were lost from the flask?

2.30 A watch glass was balanced on a two pan balance with the following weights: 1 10-g, 1 5-g, 2 2-g, 1 500-mg, and 1 10-mg. Metal filings were then added to the watch glass until it was balanced with the following weights: 2 20-g, 2 200-mg, and 1 50-mg. The metal was poured carefully into a graduated cylinder containing 14.50 ml of water. The final volume was 16.96 ml. Which of the metals in the table above were the filings?

2.31 What is the volume of 20 g of rock salt?

2.32 Calculate the volume of 100 g of a 20 per cent (by weight) solution of salt. Density is 1.15 g/ml.

2.33 Calculate the change in total volume when 20 g of salt is dissolved in 80 g of water.

2.34 a. Calculate the weight of 56 ml of a 20 per cent salt solution.
b. How many grams of salt are contained in the 56 ml of solution?

2.35 What length of an aluminum rod is equal in weight to a 30 cm magnesium rod of the same diameter?

2.36 A cube of zinc 3 cm × 3 cm × 1 cm weighs 64.3 g. Calculate the density of zinc.

2.37 Marble chips dropped into a graduated cylinder containing 32.1 ml of water raised the level to 39.5 ml. How many grams of chips were used?

SUGGESTED READING

Moreau, H.: "The Genesis of the Metric System and the Work of the International Bureau of Weights and Measures." J. Chem. Ed., *30*:3, 1953.

THREE • DALTON'S ATOMIC THEORY—ATOMS, MOLECULES AND MOLES

3.1 ELEMENTS AND COMPOUNDS

Pure substances may be separated into two classes, **elements** and **compounds.** Elements, by far the smaller group, are combined to make compounds; compounds are thus composed of elements. Elements are the substances left when any type of material is reduced to its ultimate chemical constituents. Experimentally it is rather difficult to determine whether a pure substance is a compound or an element. History reveals that many compounds were thought at first to be elements. Substances remain classified as elements only as long as they cannot be shown to be compounds. An element can never be changed to another substance without the addition of a second substance; it cannot be separated into two or more substances, nor can it be the sole product of the chemical reaction of two or more substances.

the elements of Group II of the periodic table are known as alkaline earth elements because their oxides, which are earthy in appearance, were first thought to be elements.

There are 105 chemical elements known, but there are more than one million different compounds composed of these elements, with the number of new compounds growing daily as new substances are isolated or synthesized.

Some of the more common elements are listed with their symbols in Table 3.1. The metals (copper, aluminum, iron and gold) are immediately recognizable. Sulfur, a yellow solid, and carbon (charcoal) are commonplace. The nitrogen and oxygen mixture called air is also well known.

TABLE 3.1 SOME OF THE MORE COMMONLY KNOWN ELEMENTS

Element	Symbol	Element	Symbol
Aluminum	Al	Lead	Pb
Arsenic	As	Mercury	Hg
Calcium	Ca	Nickel	Ni
Carbon	C	Nitrogen	N
Chlorine	Cl	Oxygen	O
Copper	Cu	Platinum	Pt
Gold	Au	Silver	Ag
Hydrogen	H	Sodium	Na
Iodine	I	Sulfur	S
Iron	Fe	Uranium	U

The symbols serve as designations for the names of the elements. They are usually the first one or two letters of the name, with the **first letter in capitals** and the **second letter,** if it is used, a **lower case letter.** The symbol is often derived from the Latin name; for example, Ag for silver, from *argentum*, and Na for sodium, from *natrium*. The student should memorize the symbols of the elements listed.

The identification of a compound as a compound may be a difficult problem. Heterogeneous bodies of matter are obviously not pure substances except when two states of the same substance are simultaneously present, as in a mixture of ice and water at 0°C. A homogeneous material may be either a pure substance, a compound or an element, or it may be a solution (a homogeneous mixture, the subject of Chapter 12). The failure of exhaustive attempts to decompose the matter further will identify a one-phase body of material as an element. It may be still more difficult to distinguish between a solution and a compound. Both are homogeneous and both contain more than one element, but the combining of elements to form a compound must follow strict quantitative rules known as the Laws of Chemical Change, whereas substances can come together to form a **solution** in varying proportions. Consequently, many common substances were not known to be compounds until near the end of the eighteenth century when accurate chemical balances were developed, and chemists began weighing the reactants and products of chemical reactions they had performed.

Scientific laws are statements correlating or summarizing experimental observations. The laws of chemical combination are no exception; their basis is found in the laboratory.

3.2 LAW OF CONSERVATION OF MASS

In a chemical reaction no change can be detected in the total mass of the substance involved, for mass is neither gained nor lost in a chemical reaction. For example, when 11.50 g of sodium metal combines with 17.73 g of chlorine gas, 29.23 g of table salt is produced. The figure 29.23 is the sum of 11.50 g + 17.73 g, and so the mass remains constant. This is true for all reactions studied. The burning of leaves is another example of a chemical reaction. In this process, no mass is lost. If one were to weigh the leaves burned and the oxygen used from the air in the burning, the sum of the two weights would equal the sum of the weights of carbon dioxide and water vapor (that passed off into the air) and of the ash left behind. The Law of Conservation of Mass is accepted because no chemical reaction has yet been observed in which mass was either lost or gained.

The Law of Conservation of Mass applies equally well to physical changes. For example, when salt is dissolved in water, the weight of the salt water is equal to the sum of the weights of the salt and the water weighed separately. The Law of Conservation of Mass does not differentiate between compounds and solutions.

In **nuclear reactions,** in which some of the weight of the nuclei of certain atoms (Chapter 4) is changed into energy, some matter is lost, that is, some mass is converted to energy. Reactions occurring in the atomic bomb are nuclear reactions in which enormous quantities of heat energy are gained by conversion of nuclear mass to energy. The energies involved in chemical reactions are so small, compared with those released in nuclear reactions, that the difference in mass corresponding to the energy of reaction would be too small to be detected, even if the most accurate microbalance were employed.

3.3 LAW OF DEFINITE PROPORTIONS (CONSTANT COMPOSITION)

This law states that when elements react to form a compound, they do so in a definite ratio by weight. A corrollary of this law states that all samples of a given compound contain the same elements in exactly the same weight proportions. When sodium reacts with chlorine gas, 11.50 g of sodium react with 17.73 g of chlorine to form 29.23 g of table salt. If 12.00 g of sodium is allowed to react with 17.73 g of chlorine, some sodium metal will be left over because the sodium and chlorine combine in only one fixed proportion. Also, if 11.50 g of sodium and 18.73 g of chlorine were mixed, 1 g of chlorine gas would be left over. Sodium chloride is a compound; it has a definite composition regardless of the method of preparation, the location of the experiment or the identity of the investigator.

In contrast, the composition of solutions may vary over a wide range. Any amount of table salt up to 36 g may be dissolved in 100 g of water at room temperature to make salt water. The Law of Definite Proportions permits one to distinguish compounds from solutions.

Another criterion of compound formation becomes apparent if we consider the substances in the above example. Sodium is an active metal that reacts violently with water; chlorine is a greenish, poisonous gas. The product of their union is sodium chloride, table salt, very different in properties from either reactant. In general, compounds have properties very different from those of the elements of which they are composed.

sucrose (table sugar), $C_{12}H_{22}O_{11}$, is another example. white, sweet sugar certainly differs in properties from black, flaky carbon, hydrogen gas and oxygen gas.

An apparent contradiction of the Law of Definite Proportions exists in the fact that when carbon combines with oxygen (as when charcoal burns in air), the gaseous product may vary from 57 to 73 per cent oxygen by weight, the balance of the percentage being carbon. The problem is resolved when one realizes that there are **two** common compounds of carbon and oxygen: **carbon dioxide** and **carbon monoxide.** Each compound obeys the Law of Definite Proportions, carbon dioxide being 73 per cent oxygen and carbon monoxide being 57 per cent oxygen. Under differing reaction conditions, one obtains varying proportions of the two gases and therefore a varying percentage of oxygen in the mixed gaseous product. Often the same two elements can form two or more compounds, but in each compound the composition is constant.

3.4 THE LAW OF SIMPLE MULTIPLE PROPORTIONS

the Law of Multiple Proportions refers to two or more compounds, each of which contains the same two elements.

This law states a relationship between the relative masses of one element that may combine with a set mass of another element in a series of their compounds. The generalization applies to any series of compounds, each one of which contains both elements. It may be treated as follows: In a series of compounds containing the same two elements, **the weights of one element which are combined with a fixed weight of the other element bear a simple numerical relationship to each other.**

An example of the application of this law can be made with three compounds, namely, carbon monoxide, carbon dioxide and sodium carbonate (see Table 3.2). Fourteen grams of carbon monoxide contain 6 g of carbon and 8 g of oxygen. Twenty-two grams of carbon dioxide contain 6 g of carbon and 16 g of oxygen. Fifty-three grams of sodium carbonate contain 6 g of carbon, 24 g of oxygen and 23 g of sodium. Each of the samples considered contains 6 g of carbon, but the amounts of oxygen united with the 6 g of carbon are 8, 16 and 24 g respectively, which are in the ratio of 1 to 2 to 3. The Law of Simple Multiple Proportions is a general law and is obeyed by any series of compounds containing the same two elements.

The Law of Definite Proportions and the Law of Simple Multiple Proportions were the inspiration for Dalton's Atomic Theory.

TABLE 3.2 EXAMPLES OF THE LAW OF SIMPLE MULTIPLE PROPORTIONS

Compound	Grams				Ratio of Oxygen
	Total	*C*	*O*	*Na*	
Carbon monoxide	14	6	8	–	8/8 = 1
Carbon dioxide	22	6	16	–	16/8 = 2
Sodium carbonate	53	6	24	23	24/8 = 3

3.5 THE POSTULATES OF DALTON'S ATOMIC THEORY

John Dalton, an English Quaker school teacher, had observed the diffusion of gases into an evacuated space and had observed the interdiffusion of two or more gases to give a homogeneous mixture, such as the mixture of nitrogen, oxygen and water vapor that we call air. He knew of the combination of elements to form compounds and knew the laws of chemical change, which are discussed later in this chapter. In fact, he had assisted in formulating the Law of Simple Multiple Proportions. In an attempt to explain these laws and to reconcile his observations of the behavior of gases, he suggested the **Atomic Theory of Matter.** With few amendments it has remained one of the basic unifying principles of modern chemistry and has gained for its author a place among the greatest chemists who ever lived.

Dalton's theory may be summarized in the following four postulates:

1. **All matter is composed of tiny particles** called **atoms.**
2. **There are many kinds of atoms,** differing both in mass and in other properties. There are as many kinds of atoms as there are elements and no more, because **all atoms of a given element act alike,** but differ from those of other elements in their chemical properties.
3. Atoms can neither be destroyed nor worn out.
4. Atoms may combine to form the smallest units (called molecules) of compounds. Chemical combination occurs between atoms. They may form pairs, groups of three and so forth, but in all cases **each small unit of a compound is identical with all others, that is, each unit contains the same group of atoms** in the same proportions.

3.6 MOLECULES

According to the atomic theory, a **compound** must contain identical pairs, identical triplets, or identical clusters of atoms. For example:

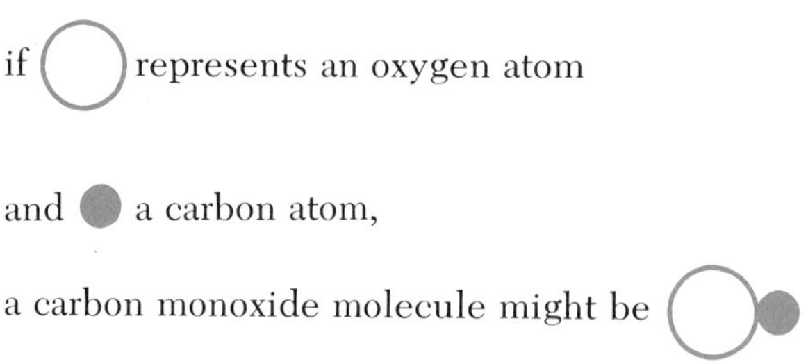

Any measurable sample of carbon **monoxide** gas contains diatomic (two-atom) molecules much too small to be seen by any type of microscope. A larger sample of the gas contains more molecules. On the other hand, carbon **dioxide** also contains molecules of carbon and oxygen, but, as the formula indicates, there are **two** oxygen atoms associated with each carbon atom in the molecule:

the carbon atom and the oxygen atom of the carbon monoxide molecule are bound together in the molecule.

Billions of these triatomic (three-atom) molecules are required to make a measurable sample of carbon dioxide gas. Sodium carbonate, commonly called washing soda, has a more complicated basic unit. It is composed of sodium, carbon and oxygen, and in it two atoms of sodium, one atom of carbon and three atoms of oxygen are joined together.

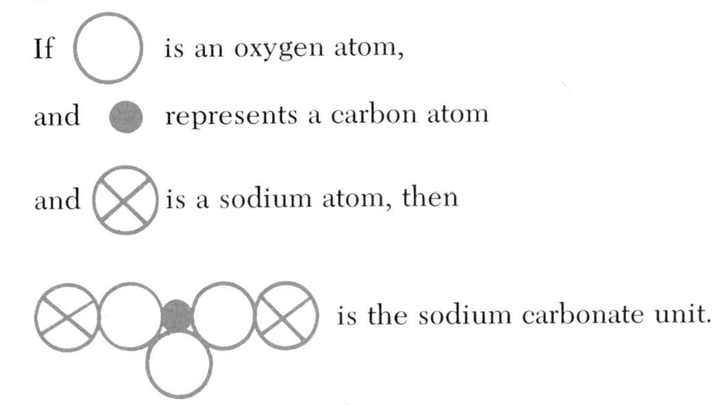

Today Dalton's picture of molecules is accepted with amendments which will be discussed later in this chapter. However, molecular representations such as those shown above are seldom used any more. Instead, they have been replaced by less cumbersome **formulas.** In formulas, elemental symbols are used to designate kinds of atoms. In modern chemical notation, CO represents a carbon monoxide molecule (containing one atom of carbon and one atom of oxygen), and CO_2 represents carbon dioxide. The subscript 2 indicates that two oxygen atoms are joined with one carbon atom in the molecule. Likewise, the formula of sodium carbonate, Na_2CO_3, signifies that two sodium, one carbon and three oxygen atoms are contained in a "molecule" of that compound. The formula may indicate the structure (the order and arrangement of the various atoms in the molecule), but very often it does not.

More complicated molecules exist in nature. The formula for sucrose (table sugar) is $C_{12}H_{22}O_{11}$. A molecule of sucrose contains 12 atoms of carbon, 22 of hydrogen and 11 of oxygen. The formula for calcium nitrate is $Ca(NO_3)_2$. The simplest unit of calcium nitrate contains one calcium atom, two nitrogen atoms and six oxygen atoms. (The subscript following the parentheses indicates that there are two NO_3 groups associated with one calcium atom in the unit.) The formula $BaCl_2{\cdot}2\ H_2O$ might be written $BaCl_2(H_2O)_2$ and indicates that one barium atom, two chlorine atoms, four hydrogen atoms and two oxygen atoms form the smallest unit of crystalline barium chloride.

3.7 CHEMICAL EQUATIONS

According to Dalton, a chemical reaction can be a change only in atomic groupings. Since atoms are not altered in chemical reactions, substances can be changed to other substances only by rearranging the atoms into new molecules (or groups), either by breaking molecules apart or by separating molecules into smaller units and from these remaking new ones.

The burning of coke is a chemical process. Observation has shown that the major reaction is the combining of carbon from the coke with oxygen from air to form carbon dioxide. This reaction is represented by an **equation** which is shown using the formulas of the substances used and formed:

$$C + O_2 \longrightarrow CO_2$$

Carbon and oxygen react to form carbon dioxide.

(Notice that oxygen is diatomic, that is, there are two atoms in a molecule, which is represented as O_2.) The changes in atomic groupings are apparent in the equation. A carbon atom unites with a molecule of oxygen to form a new molecule, that of carbon dioxide. By convention, carbon and oxygen, the substances combining (used), are called **reactants** and appear to the left of the arrow in the equation; carbon dioxide (which is formed) is the **product** and appears to the right.

When carbon is burned in a limited supply of oxygen, carbon monoxide is formed:

$$C + O_2 \longrightarrow CO$$

Carbon Oxygen Carbon monoxide

The symbolic statement is not a true equation because it is not balanced; that is, more atoms are on the left side of the arrow than on the right. Since atoms are neither destroyed nor created in a chemical reaction, it is reasonable that two atoms of carbon must react with one molecule (of two atoms) of oxygen to yield two molecules (each of which are two atom units) of carbon monoxide. **The balanced equation** is therefore

$$2C + O_2 \rightarrow 2CO$$

Another and more complicated reaction occurs when carbon monoxide diffuses into porous iron oxide in a blast furnace. The unbalanced equation for the reaction

$$Fe_2O_3 + CO \rightarrow Fe + CO_2$$

Iron(III) oxide Iron

merely shows the experimental fact that iron(III) oxide and carbon monoxide disappear to form iron (molten at that temperature) and carbon dioxide. The balanced equation

$$Fe_2O_3 + 3CO \rightarrow 2Fe + 3CO_2$$

shows that **three molecules** of carbon monoxide are required to convert **one molecule** of iron(III) oxide into iron. Equations must be balanced; that is, the same total numbers and kinds of atoms must appear both in the reactants (on the left) and in the products (on the right). This equality is achieved by placing the correct coefficients (numbers), determined by inspection, before the chemical formulas. No formulas may be changed in the balancing operation.

3.8 AGREEMENT OF THEORY WITH THE LAWS OF CHEMICAL COMBINATION

Dalton's Atomic Theory is consistent with the three laws of chemical combination. That is, if his picture is correct, these laws describe the observed behavior of matter as it undergoes chemical reaction. The Law of Conservation of Mass is followed, for atoms are merely regrouped during reaction. Since the atoms can neither be destroyed nor created in chemical reaction, mass can neither be lost nor gained. Further, if every submicroscopic molecule (smallest unit) of a compound contains an identical group of atoms, all samples of matter containing them will have a constant composition, and the elements must be combined in *definite proportions.*

Three compounds whose formulas were presented in this chapter contain the same two elements: carbon and oxygen. They are carbon monoxide (CO), carbon dioxide (CO_2) and sodium carbonate (Na_2CO_3). If amounts of each of these compounds are considered such that each contains the same amount of carbon, the amount of oxygen in them will be in the ratio of 1 to 2 to 3. If one molecule of each is taken, each has one carbon atom, and the numbers of oxygen atoms contained are one, two and three, respectively. Any sample of the materials containing an equal number of molecules will contain an equal weight of carbon and will contain weights of oxygen in the above ratio.

3.9 ATOMIC WEIGHTS

"atomic weights" is a misnomer. the atomic weight of an element is the average mass of an atom relative to the masses of the other elements. mass and weight are used interchangeably, but strictly this is incorrect.

Inherent in Dalton's Atomic theory is the idea that all atoms, although very small, possess **mass.** Since the proportion by weight of the elements combined in a compound depends on the **masses** as well as the number of particles in the formula, it is important to know these **masses.**

Historically, the relative weights of atoms were determined indirectly through an abstract analysis of experimental data involving masses and volumes of gases. We shall consider this method after discussing the gas laws in a later chapter. For the moment we shall consider a modern, more direct method by which the atomic weights are found. This method involves the mass spectrograph. We shall consider its principle of operation briefly.

Consider an experiment in which spheres of different densities are dropped from a bridge in the path of a **constant** stiff wind. The spheres might be an iron ball, a baseball and a tennis ball. The wind tends to blow the spheres away from the bridge, having little effect on the iron ball, but causing the lighter spheres to fall a considerable distance downwind. The lighter the sphere, the farther it will be carried.

Figure 3.1 is a diagram showing the curved path taken by the respective balls. One could drop several spheres of known densities, noting the spots where they hit the water. One could then use this scale to find the density of another sphere.

for this analogy to hold, the velocity of the wind must be constant.

The **mass spectrograph** (Fig. 3.2) uses a similar principle. A gas is bombarded with high speed electrons, forming positively charged atoms called **ions.** The ions are hurled into a strong magnetic field and are drawn toward the negative plate. However, the inertia of the particles is such that they cannot turn abruptly, but follow a curved path to strike a photographic plate below. The development of the plate then reveals a black line at the point at which the charged atoms struck it. Lighter ions again are pulled farther from the vertical path, while the paths of heavier ions are bent only slightly.

Knowing the charge on the ion, the strength of the magnetic field, the velocity of the particle and the point at which it hits the plate, one can calculate the mass of the charged atom. Comparison of the positions where hydrogen ions and oxygen ions strike the plate shows the oxygen atom to be almost 16 times as heavy as the hydrogen atom. More careful examination of the photographic plate shows that not all oxygen atoms are the same, but there are a few of greater mass than the most common type. The larger atoms have a mass either 1/16 or 1/8 greater than that of the most common oxygen atom. No atoms have masses in between these values. **Each of these three kinds of oxygen atoms** is called an **isotope.** Similar experiments have shown that most elements as they are found in nature are mixtures of isotopes; that is, they contain two or more kinds of **atoms which differ from each other only in mass.**

Dalton was wrong, it appears, when he postulated that all atoms of an element are identical. Furthermore, it appears that the Law of Definite Proportions is faulty, for a molecule of carbon mon-

FIGURE 3.1 PATHS OF VARIOUS SPHERES FALLING IN THE WIND.

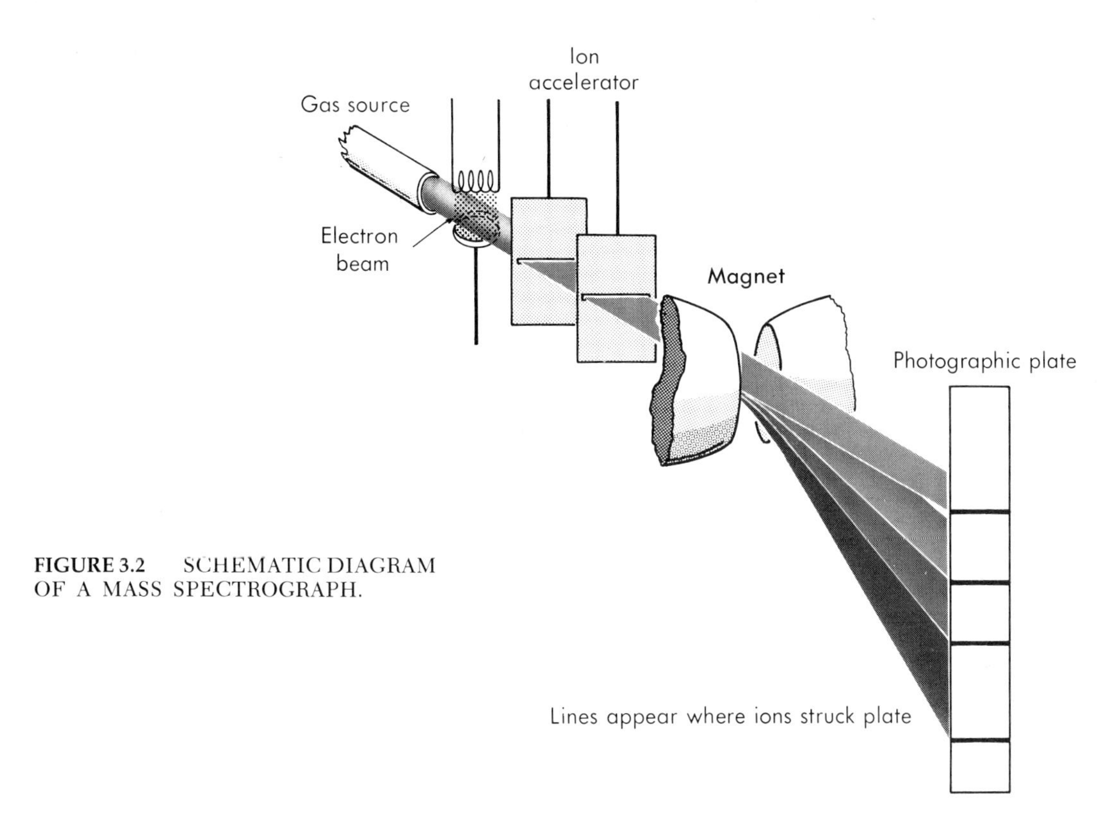

FIGURE 3.2 SCHEMATIC DIAGRAM OF A MASS SPECTROGRAPH.

oxide with the most common oxygen atom will have a larger percentage of carbon **by weight** than a molecule containing a carbon atom and a heavier oxygen atom. However, this ordinarily causes no trouble since nature usually presents us with a nearly *constant* mixture of the isotopes of an element; therefore, we use an average atomic weight.

because Dalton was unaware of the occurrence of isotopes, the atomic theory is not strictly correct.

Development and refinement of the mass spectrograph, an instrument with which scientists measure the masses of gaseous positive ions, has allowed researchers to set relative values for the atomic masses of all the elements in relation to the others. To establish a scale of masses, one need only choose a value for one element (or one isotope of that element) and compare all other elements with it. In 1815, Prout saw that if the mass of a hydrogen atom, the lightest atom, were taken as unity, all other atoms would have comparative masses of nearly whole numbers. Consequently, hydrogen was first given an arbitrary mass of one, and the masses of all other atoms were stated in terms equivalent to the number of hydrogens necessary to equal their masses. Later oxygen was taken as the standard, since oxygen combines with almost all other elements. Average oxygen atoms were said to have a mass of exactly 16, with the masses of all other atoms stated in relation to the mass of the oxygen atom. With the discovery of isotopes, it was noted that the average mass of oxygen varied slightly with different samples taken from air and water and that calculation of a suitable average value was difficult. In 1962, by international agreement, this problem was resolved with the adoption of carbon-12, the common isotope of carbon, as the basis for comparative values; a mass value of **exactly 12.00000** was assigned to this atom. This

TABLE 3.3 ATOMIC WEIGHTS OF SOME COMMON ELEMENTS

Element	Symbol	Atomic Weight
Aluminum	Al	26.9815
Arsenic	As	74.9216
Calcium	Ca	40.08
Carbon	C	12.01115
Chlorine	Cl	35.453
Copper	Cu	63.54
Gold	Au	196.967
Hydrogen	H	1.00797
Iodine	I	126.9044
Iron	Fe	55.847
Lead	Pb	207.19
Mercury	Hg	200.59
Nickel	Ni	58.71
Nitrogen	N	14.0067
Oxygen	O	15.9994
Platinum	Pt	195.09
Silver	Ag	107.870
Sodium	Na	22.9898
Sulfur	S	32.064
Uranium	U	238.03

gave scientists an easily measured standard and changed the existing table of values only slightly. The average mass of an oxygen atom on the new scale is 15.9994; the average mass of a carbon atom is 12.01115. The comparative average masses of the atoms are called the *atomic weights.** The atomic weights of some of the more common elements are shown in Table 3.3. Note the great variance in the comparative masses of atoms of different elements.

An average uranium atom is 238.03/1.008 or nearly 238 times as heavy as the average hydrogen atom, and thus is 238/12 as heavy as the most common carbon atom.

3.10 THE AVOGADRO NUMBER—THE MOLE

From carefully controlled measurements with a mass spectrograph, the average hydrogen atom is found to have a mass of 1.67×10^{-24} gram. Therefore in 1.008 grams of hydrogen, the mass in grams equal to the atomic weight, there are

$$\frac{1.008 \text{ g}}{1.67 \times 10^{-24} \text{ g/atom}} = 6.02 \times 10^{23} \text{ atoms}$$

Because the atomic weights indicate the relative masses of the atoms, in 26.98 grams of aluminum, 16.00 grams of oxygen, or

* Note that the terms *atomic weight* and *atomic mass* are used interchangeably here. This is customary among chemists.

32.064 grams of sulfur there are also 6.02×10^{23} atoms. Each of these quantities of material, having a *weight in grams numerically equal to the atomic weight,* is called the *gram-atomic weight* (or, more simply, a *mole*). The number 6.02×10^{23} is known as the *Avogadro Number,* in honor of Amedeo Avogadro, who predicted long before the masses of individual particles could be determined that the number of grams of any element equal to its atomic weight contained a fixed or constant number of atoms. A **mole** is a number of particles or units of **any kind of matter,** the number being the Avogadro Number.

3.11 FORMULA WEIGHTS (MOLECULAR WEIGHTS)

From the table of atomic weights, it can be seen that the average carbon atom has a mass of 12.0111, and the average mass of the oxygen atom is 15.994. It follows that the carbon monoxide molecule (CO) has a mass equal to the sum of the masses of the two atoms, or $12.0111 + 15.9994 = 28.0105$. The two-atom molecule is 28.0105/12.0111 as heavy as the average carbon atom. The figure 28.010 is called the **molecular weight** or the **formula weight** of carbon monoxide. The atomic weight is the relative weight of an atom; the molecular weight is the relative weight of a molecule, both measured on the same scale. In 28.010 grams of carbon monoxide there are 6.02×10^{23} molecules; that is, 28.010 grams of carbon monoxide is 1 mole, because it contains 6×10^{23} molecules (particles).

units of molecular weight are identical to units of atomic weight.

The molecular weight of water (H_2O) is then $2(1.008) + 15.999 = 18.015$. The water molecule is about 18/16 as heavy as the oxygen atom. In all cases, reference is made to the average atom and the average molecule.

The molecular weight of sodium sulfate (Na_2SO_4) is calculated the same way:

$$\begin{aligned} 2\ Na &= 2 \times 22.990 = 45.980 \\ 1\ S &= 32.064 = 32.064 \\ 4\ O &= 4 \times 15.999 = \underline{63.996} \\ & \qquad 142.040 \end{aligned}$$

The molecular or formula weight is 142.040, the sum of the masses of the seven atoms of the molecule. It also follows that in 142.040 grams of sodium sulfate there are 6.02×10^{23} "molecules" (or formula units), each composed of two sodium atoms—one sulfur and four oxygen atoms. In this 142.040 grams there are $2 \times 6.02 \times 10^{23}$ atoms of sodium, 6.02×10^{23} atoms of sulfur and $4 \times 6.02 \times 10^{23}$ atoms of oxygen. But since each "molecule," or formula group, contains 2 Na, 1 S and 4 O atoms, there are only 6.02×10^{23} groups. One mole (that is, one gram molecular weight) of sodium sulfate, which is 142.040 grams, contains two moles of sodium atoms, one mole of sulfur atoms and four moles of oxygen atoms.

One more example to illustrate the calculation of the formula weight or molecular weight might be instructive. In this example, and in most of the problems to follow, approximate integral values of atomic weights are used.

Example: Calculate the formula weight of potassium alum, $KAl(SO_4)_2 \cdot 12\ H_2O$.

$$
\begin{aligned}
1\ K &= 1 \times 39 = 39 \\
1\ Al &= 1 \times 27 = 27 \\
2\ S &= 2 \times 32 = 64 \\
8\ O &= 8 \times 16 = 128 \\
24\ H &= 24 \times 1 = 24 \\
12\ O &= 12 \times 16 = 192 \\
& \qquad\qquad\quad \overline{474}
\end{aligned}
$$

(Taking integral values of the atomic weights.)

The molecular weight of potassium alum is 474. One of these 48 atom units weighs 474/12 as much as one carbon-12 atom. Furthermore, in one mole, or 474 g, of potassium alum there are 6.02×10^{23} of these 48 atom units.

3.12 CALCULATION OF ELEMENTAL COMPOSITION

If the formula of a compound is known, one can use the atomic weights to calculate the weight of each element in a sample of the compound.

Example: How many grams of oxygen are there in 50 g of water?

$$
\begin{aligned}
2\ H &= 2 \times 1 = 2 \text{ parts H} \\
1\ O &= 1 \times 16 = 16 \text{ parts O} \\
& \qquad \text{in } 18 \text{ total parts}
\end{aligned}
$$

16/18 of any amount of water is oxygen
$16/18 \times 50\ g = 44.5$ g of oxygen in 50 g of water
$2/18 \times 50\ g = 5.5$ g of hydrogen in 50 g of water
and $44.5\ g + 5.5\ g = 50$ g total

Example: How many grams of potassium are there in 68.0 g of $KClO_3$? Adding the atomic weights as found in the periodic table:

$$
\begin{aligned}
K &= 39.1 \\
Cl &= 35.5 \\
3 \times O &= 48.0 \\
& \quad \overline{122.6}
\end{aligned}
$$

In any sample 39.1 of every 122.6 equal parts by weight are potassium, and 39.1/122.6 of 68.0 grams of $KClO_3$ is potassium. Hence

$$39.1/122.6 \times 68.0\ g\ KClO_3 = 21.6\ g\ K \text{ in } 68.0\ g\ KClO_3.$$

3.13 CALCULATION OF PERCENTAGE COMPOSITION

This is similar to the above calculation. One merely takes the weight fraction of 100%.

Example: What is the percentage composition of water?

$$\text{Again } 2\ H = 2 \times 1 = 2$$
$$O = 1 \times 16 = \underline{16}$$
$$18$$

$$2/18 \times 100\% = 11\% \text{ hydrogen}$$
$$16/18 \times 100\% = \underline{89\% \text{ oxygen}}$$
$$100\% \text{ total}$$

Example: Calculate the percentage composition of potassium chlorate, $KClO_3$.

$$K = 39.1 \qquad \frac{39.1}{122.6} \times 100\% = 31.9\%\ K$$
$$Cl = 35.5 \qquad \frac{35.5}{122.6} \times 100\% = 29.0\%\ Cl$$
$$3\ O = \underline{48.0} \qquad \frac{48.0}{122.6} \times 100\% = \underline{39.1\%\ O}$$
$$122.6 \qquad\qquad 100.0\%\ \text{Total}$$

3.14 SIMPLEST FORMULA

remember that the subscripts of a molecular formula give the ratio of the number of moles of the elements in a compound, not the ratio by weight.

The simplest ratio of atoms in a molecule (moles of atoms in 1 mole of a compound) can be found from an analysis of that compound if one uses the atomic weights. These ratios give us the formula. The formula CO_2 tells two facts: There are one carbon and two oxygen atoms per molecule, and the ratio of the number of moles of carbon atoms to the number of moles of oxygen atoms in the compound is 1 to 2. Likewise sodium, sulfur and oxygen are combined in sodium sulfate (Na_2SO_4) in the ratio of 2 moles of sodium atoms to 1 mole of sulfur atoms to 4 moles of oxygen atoms respectively. When one finds that ratio, he has the simplest formula. The example below illustrates the method for finding that ratio.

Example 1: A 100-g sample of a compound of hydrogen and oxygen contains 11.1 g of hydrogen and 88.9 g of oxygen. Calculate the simplest formula.

The first step in finding the ratio of the moles of atoms in the compound is to find the actual moles of each atom in the sample.

For hydrogen there are

$$\frac{11.1 \text{ g}}{1.00 \text{ g/mole of H}} = 11.1 \text{ moles of hydrogen atoms}$$

For oxygen there are

$$\frac{88.9 \text{ g}}{16.00 \text{ g/mole}} = 5.55 \text{ moles of oxygen atoms}$$

The ratio between the moles of atoms can usually be found by dividing the actual number of moles for both atoms by the smaller of the two numbers:

$$\frac{11.1 \text{ moles of H atoms}}{5.55 \text{ moles of O atoms}} = \frac{2.00 \text{ moles of H atoms}}{1.00 \text{ mole of O atoms}}$$

The simplest formula is H_2O_1 or H_2O.

Example 2: A sample of an iron oxide contains 0.252 g of iron and 0.096 g of oxygen. Find the simplest formula consistent with these facts.

Solution: Dividing the mass of each element in the sample of iron oxide by its atomic weight gives the actual number of moles of each atom in that sample.

Fe	0.252 g	O	0.096 g
Fe	$\frac{0.252 \text{ g}}{56.0 \text{ g/mole of Fe}}$	O	$\frac{0.096 \text{ g}}{16.0 \text{ g/mole of O}}$
Fe	0.0045 mole	O	0.0060 mole

Dividing by the smaller number of moles helps in finding the ratio.

Fe	$\frac{0.0045 \text{ mole}}{0.0045 \text{ mole}}$		O	$\frac{0.0060 \text{ mole}}{0.0045 \text{ mole}}$
Fe	1.00	or	O	1.33
Fe	1		O	$1\frac{1}{3}$

Because the number of moles of oxygen atoms is exactly $1\frac{1}{3}$, it can be made a whole number by multiplying by 3. Multiplying both subnumbers by 3 gives

$$Fe_3O_4$$

the smallest whole number ratio.

3.15 SOME ECONOMIC CONSIDERATIONS IN CHEMICAL PROCESSES

It is obvious that quantitative relationships among raw materials and products are of utmost importance in chemical industries. Chemical processors need to know the exact amount of each raw material necessary to produce a given quantity of product. The economic feasibility of a process depends primarily on the amount and cost of raw materials, the amount and price of products and the cost of transforming the raw materials into finished products.

A chemical process may depend upon one or several chemical reactions. If only one chemical reaction is involved, one need simply find relationships between reactants and products. If the process involves a series of chemical reactions, one must solve the quantitative problem for each reaction and finally combine them for the whole process. There are two aspects to the problem of quantities in a chemical reaction: (1) There is a theoretical or a 100 per cent relationship based upon the equation for the reaction (or equations for the reactions), assuming that in each reaction all the reactants will be converted to products. It is the purpose of this section to describe these calculations. (2) Reactions seldom give the theoretical yield. One rarely gains 100 per cent of the product most desired because reactions may stop before the reactants are used up. Often "side-reactions" occur (the chemicals react in another way simultaneously to produce a second, unwanted set of products). Finally, some of the product is lost in purification. A potential processor might carry out a "trial-run" under the conditions used in the process to determine what per cent of the calculated possible yield might be expected. In any case, the yield will not be greater than the theoretical (100 per cent) possible. Great sums of money are spent in industrial research to find better reaction conditions—conditions under which the yield of products will more nearly approach 100 per cent.

3.16 MOLAR RELATIONSHIPS AMONG REACTANTS AND PRODUCTS

To calculate the theoretical relationships among reactants and products, one needs the balanced equation for the reaction. Consider the equation for the thermal (high temperature) decomposition of potassium chlorate in the presence of a trace of manganese dioxide:

$$2\ KClO_3 \rightarrow 2\ KCl + 3O_2$$

Equal numbers of each kind of atom appear in the reactants and in the products in a balanced equation. This agrees with the fact that atoms are neither gained nor lost in a chemical reaction.

The equation may be applied in a quantitative way to the decomposition of any amount of potassium chlorate:

1. In the trivial and hypothetical case of the decomposition of 2 "molecules" of potassium chlorate, the equation indicates that

upon decomposition 2 "molecules" of $KClO_3$ yield two "molecules" of KCl and 3 molecules of O_2.

2. In general terms, potassium chlorate decomposes into potassium chloride and oxygen in the molecular ratio 2 to 2 to 3, respectively.

3. In terms of numbers of gram-molecular weights or moles for the trivial case, the decomposition of 2 moles, the equation reads:

2 moles of $KClO_3$ yields 2 moles of KCl and 3 moles of O_2

4. In general terms, potassium chlorate decomposes into potassium chloride and oxygen in the ratio:

2 moles of $KClO_3$ to 2 moles of KCl to 3 moles of O_2

Thus,

a. When 1 mole of $KClO_3$ decomposes,

1 mole of KCl and 3/2 moles of O_2 are formed.

b. When 1 mole of oxygen is prepared,

2/3 mole $KClO_3$ disappears and 2/3 mole KCl appears.

c. When 0.505 mole $KClO_3$ is decomposed,

0.505 mole KCl and $0.505 \times 3/2$ mole $= 0.757$ mole of O_2 are formed.

The equation for the preparation of hydrogen gas by reaction of white-hot iron with steam may also be used to illustrate molar relationships:

$$3\ Fe + 4\ H_2O \rightarrow Fe_3O_4 + 4\ H_2$$

Example 1. How many moles of Fe_3O_4 can be produced by reaction of 1 mole of Fe?

$$1 \text{ mole Fe} \times \frac{1 \text{ mole } Fe_3O_4}{3 \text{ mole Fe}} = 1/3 \text{ mole } Fe_3O_4$$

Example 2. How many moles of steam (H_2O) react with 0.462 mole Fe?

$$0.462 \text{ mole Fe} \times \frac{4 \text{ moles } H_2O}{3 \text{ moles Fe}} = 0.615 \text{ mole } H_2O$$

Example 3. How many moles of iron are consumed in the preparation of 0.25 mole H_2?

$$0.250 \text{ mole } H_2 \times \frac{3 \text{ moles Fe}}{4 \text{ moles } H_2} = 0.188 \text{ mole Fe}$$

3.17 WEIGHT RELATIONSHIP AMONG REACTANTS AND PRODUCTS

By using atomic and molecular weights, one can readily translate mole relationships into weight relationships. The product of the number of moles of a sample and its molecular weight is the weight in grams. The quotient of weight of a sample divided by the molecular weight gives the number of moles in the sample. Examples of these calculations appear below. They refer to the process for producing hydrogen gas shown in Section 3.16 and to the equation used there:

$$3\ \text{Fe} + 4\ \text{H}_2\text{O} \rightarrow \text{Fe}_3\text{O}_4 + 4\ \text{H}_2$$

Example 1. How many grams of iron must react to produce 0.250 mole H_2? One step added beyond the previous calculations:

$$0.250\ \text{mole H}_2 \times \frac{3\ \text{moles Fe}}{4\ \text{moles H}_2} = 0.188\ \text{mole Fe}$$

the ratio 3/4 is correct in this calculation, because according to the chemical equation fewer moles of iron are required than moles of hydrogen produced.

and 0.188 mole Fe is:

$$0.188\ \text{mole Fe} \times \frac{55.8\ \text{g Fe}}{\text{mole Fe}} = 10.5\ \text{g Fe}$$

The problem set up in full is:

$$0.25\ \text{mole H}_2 \times \frac{3\ \text{moles Fe}}{4\ \text{moles H}_2} \times 56\ \text{g Fe/mole Fe} = 10.5\ \text{g Fe}$$

Example 2. What weight of white-hot iron will react with 4.0 g of steam? The molecular weight of water is $2 \times 1.0 + 16 = 18$. One mole of water is 18 g. Four grams of H_2O is:

$$\frac{4\ \text{g}}{18\ \text{g/mole H}_2\text{O}} = 4/18\ \text{mole H}_2\text{O}$$

Thus, 4/18 mole of H_2O reacts with:

$$4/18\ \text{mole H}_2\text{O} \times \frac{3\ \text{moles Fe}}{4\ \text{moles H}_2\text{O}} = 1/6\ \text{mole Fe}$$

(The fraction $\frac{3}{4}$ is from this portion of the equation.)

$$3\ \text{Fe} + 4\ \text{H}_2\text{O} \rightarrow$$

And 1/6 mole Fe is:

$$\frac{1}{6}\ \text{mole Fe} \times 55.8\ \text{g Fe/mole Fe} = 9.30\ \text{g Fe}$$

The complete setup is:

$$\frac{4\text{ g}}{18\text{ g/mole}} \times \frac{3}{4} \times 55.8\text{ g/mole} = 9.30$$

Example 3. How many g of Fe_3O_4 are produced along with 7.25 g of H_2? There are:

$$\frac{7.25\text{ g}}{2.00\text{ g/mole}} = 3.63\text{ moles of } H_2$$

From the relevant portion of the balanced equation ($\rightarrow Fe_3O_4 + 4\ H_2$), the moles of iron oxide produced are:

$$3.63\text{ moles } H_2 \times \frac{1\text{ mole } Fe_3O_4}{4\text{ moles } H_2} = 0.908\text{ mole } Fe_3O_4$$

Because 1 mole of Fe_3O_4 is $(3 \times 56) + (4 \times 16)$ g = 232,

$$0.908\text{ mole } Fe_3O_4 \times 232\text{ g } Fe_3O_4/\text{mole } Fe_3O_4 = 211\text{ g } Fe_3O_4$$

The setup is:

$$\frac{7.25\text{ g}}{2\text{ g/mole}} \times \frac{1}{4} \times 232\text{ g/mole} = 211\text{ g } Fe_3O_4$$

In summary the following may be noted:

1. The Atomic Theory was a product of the scientific method. Quantitative observations were summarized in the laws of chemical combination. Dalton formulated the theory to explain those laws. The theory has been examined and amended.

2. There are two notable amendments to the theory:

   a. **Not all atoms of an element are identical,** but a natural sample of an element may be a nearly constant mixture of several isotopes which differ by 1 or 2 mass units.

   b. Atoms can be destroyed. Some, such as radium, decompose spontaneously; others only after being bombarded with high-energy particles. However, these **are not** chemical reactions and do not occur under conditions one normally encounters in the chemical laboratory.

3. Chemical reactions consist of a regrouping of atoms. The equation for the reaction denotes equality in number and kinds of atoms in the reactants and in the products. The equation must be a balanced equation.

$$C + O_2 \rightarrow CO$$

is not a balanced equation, but

$$2\ C + O_2 \rightarrow 2\ CO$$

is balanced.

4. Using relative atomic masses (atomic weights of the atoms),
   a. the percentage composition of a compound of known formula can be calculated,
   b. the simplest formula for a compound can be found from an elemental analysis of a pure sample that gives the weight of each element in it and
   c. with the aid of the balanced equation the weight relationship can be determined between the reactants and products of a chemical reaction.

EXERCISES – SET I

3.1 If the Law of Definite Proportions is ever to be disproved, how will it be done?

3.2 In what respect are salt and ice alike; how are they different from window glass, paraffin and tar?

3.3 List five homogeneous foods and five heterogeneous foods.

3.4 List three chemically pure substances, used as such, in human nutrition.

3.5 2.00 g of sulfur was burned, consuming 3.00 g of oxygen, to form 5.00 g of sulfur trioxide. Which law of chemical change is demonstrated by this experiment?

3.6 Write the names of the elements contained in the compounds represented by each of these formulas: (a) $AlPO_4$, (b) $Ca(OH)_2$, (c) $AgNO_3$, (d) H_3AsO_4, (e) $NiCO_3$, (f) $NaCl$, (g) $KClO_3$, (h) $Hg(NO_3)_2$, (i) Na_2SO_4, (j) $KClO_4$.

3.7 Which two substances of those listed in Question 6 can be used together to illustrate the Law of Simple Multiple Proportions?

3.8 Samples of five pure substances were analyzed for the same two elements, X and Y. The analyses are tabulated below:

Compound	*g of X*	*g of Y*
A	2.43	1.42
B	2.43	2.84
C	1.62	1.52
D	1.62	1.42
E	4.86	4.26

Which analysis is incorrect? Why?

3.9 Define: (a) solution, (b) compound, (c) element, (d) heterogeneous, (e) phase.

3.10 Which of the following must contain at least two elements: (a) a solution, (b) a compound, (c) a heterogeneous body, (d) a homogeneous body, (e) a pure substance, (f) an impure substance?

3.11 Using the compounds, NO, NO_2, N_2O_3 and N_2O_5, illustrate the Law of Simple Multiple Proportions.

EXERCISES—SET II

Balance each of the following equations:

3.12 $CaO + HCl \rightarrow CaCl_2 + H_2O$

3.13 $H_2SO_4 + NaOH \rightarrow Na_2SO_4 + H_2O$

3.14 $Fe + H_2O \rightarrow Fe_3O_4 + H_2$

3.15 $Mg + N_2 \rightarrow Mg_3N_2$

3.16 $H_2O + Na \rightarrow H_2 + NaOH$

3.17 $C + H_2O \rightarrow CO + H_2$

3.18 $C_3H_8 + O_2 \rightarrow CO_2 + H_2O$

3.19 $C_3H_8 + H_2O \rightarrow CO_2 + H_2$

3.20 $Al(OH)_3 + H_2SO_4 \rightarrow Al_2(SO_4)_3 + H_2O$

3.21 $PCl_5 + H_2O \rightarrow H_3PO_4 + HCl$

3.22 $Cl_2 + H_2O \rightarrow HCl + O_2$

3.23 $H_2S_2O_7 + H_2O \rightarrow H_2SO_4$

3.24 $H_2O_2 + N_2H_4 \cdot H_2O \rightarrow N_2 + H_2O$

3.25 $PbS + H_2O_2 \rightarrow PbSO_4 + H_2O$

3.26 $Ag_2O + H_2O_2 \rightarrow Ag + H_2O + O_2$

3.27 $Fe_2O_3 + H_2 \rightarrow Fe + H_2O$

3.28 $Fe_2O_3 + C \rightarrow Fe + CO_2$

EXERCISES—SET III

3.29 Write the symbols for the following elements: gold, silver, mercury, barium, tin, lead, iodine, bromine, iron and nickel.

3.30 The atomic weight of sodium is 22.997, of iron 55.85, of oxygen 15.999 and of hydrogen 1.008.

a. Which contains more atoms, 1 gram of hydrogen or 10 grams of oxygen?

b. Will 55.85 pounds of iron have the same number of atoms as 1.008 pounds of hydrogen?
c. Which of the elements listed above has atoms of the greatest mass?
d. What is the molecular weight of Fe_2O_3?
e. How many pounds of Fe_2O_3 have the same number of molecules as there are atoms in 8 pounds of oxygen?

3.31 a. How many sulfur atoms are there in 1 molecule of $Al_2(SO_4)_3$ (aluminum sulfate)?
b. How many oxygen atoms are there in 1 molecule of $Al_2(SO_4)_3$?
c. How many atoms are there in 1 molecule of $Al_2(SO_4)_3$?

3.32 a. How many sulfur atoms are there in 1 mole of $Al_2(SO_4)_3$?
b. How many oxygen atoms are there in 1 mole of $Al_2(SO_4)_3$?
c. How many atoms are there in 1 mole of $Al_2(SO_4)_3$?

EXERCISES – SET IV

3.33 Calculate the molecular weight of:

a. NaCl b. SO_2 c. $Ca(NO_3)_2$ d. Na_2CO_3
e. $Al_2(SO_4)_3 \cdot 18\ H_2O$ f. $BaCl_2 \cdot 2\ H_2O$ g. $Ba(OH)_2$ h. CCl_4

3.34 What fraction of the mass of

a. NaCl is chlorine b. SO_2 is oxygen c. Na_2CO_3 is sodium
d. $Ca(NO_3)_2$ is oxygen e. $Al_2(SO_4)_3 \cdot 18\ H_2O$ is oxygen

3.35 How many grams of oxygen are there in:

a. 60 g of SO_2 b. 60 g of SO_3 c. 75 g of $CaSO_4$
d. 80 g of $Ba(OH)_2$ e. 160 g of $Al_2(SO_4)_3$

3.36 What is the per cent of

a. Oxygen in SO_2 b. sulfur in SO_2 c. sodium in Na_2SO_4
d. sulfur in $Na_2SO_4 \cdot 10\ H_2O$ e. sodium in $NaClO_3$

3.37 Calculate the percentage composition of

a. Na_2CO_3 b. $CaSO_3$ c. $BaCl_2 \cdot 2\ H_2O$

3.38 Calculate the simplest formula for each of the compounds described below. (Use precise values for atomic weights.)

a. 50.00 g of a compound of sodium, phosphorus and oxygen contains 21.04 g Na and 9.45 g P.
b. The oxide that constitutes the most important source of manganese contains 63.19 per cent manganese.
c. An oxide of manganese in which manganese and oxygen are combined in the ratio by weight of 1.717 to 1.000.
d. 74.91 g of arsenic (As) is combined with 3.024 g of hydrogen in arsine.

e. In the oxide of arsenic, white arsenic, 74.91 g of As is combined with 24.00 g of oxygen.
f. A 25 g sample of a compound contains 7.97 g K, 7.23 g Cl and 9.79 g O.
g. 50 g of a compound contains 13.29 g K, 17.68 g Cr and 19.04 g O.

3.39 A compound of hydrogen and nitrogen contains the two elements in the ratio of 3 to 21 g. Calculate the simplest formula. The molecular weight of the compound is 32. What is the molecular formula?

3.40 a. How many atoms are in one formula of $Ca(NO_3)_2 \cdot 6\ H_2O$?

b. Calculate the molecular weight of $Ca(NO_3)_2$.

3.41 How many grams of calcium are there in 25 g of calcium carbonate ($CaCO_3$)?

3.42 Calculate the mass of 16 ml of concentrated hydrochloric acid. Density is 1.37 g/ml.

3.43 How many moles (or what fraction of a mole) is:

a. 100 g $KClO_3$, b. 35 g Fe, c. 10 g CO, d. 74 g PbS, e. 56 g Mn_3O_4, f. 106 g CO_2, g. 25 g CH_4, h. 36 g SO_2?

3.44 What is the weight of one mole of:

a. KCl, b. O_2, c. H_2, d. H_2O, e. Fe, f. H_2O_2, g. MnO_2, h. CH_4, i. CS_2?

3.45
a. How many moles of O_2 are produced from the decomposition of 1 mole of $KClO_3$?
b. How many moles of hydrogen are gained by the reaction of 1 mole of iron with steam? By the reaction of 0.625 mole of Fe?
c. How many moles of water will react with 1 mole of Fe? With 0.625 mole of Fe?
d. How many moles of iron are produced by reaction of 1 mole of CO with Fe_3O_4? By reaction of 0.32 mole of CO?
e. How many moles of H_2O_2 react with 1 mole of PbS? With 0.308 mole?
f. How many moles of MnO_2 are needed to produce 1 mole of Mn_3O_4? To produce 0.244 mole of Mn_3O_4?
g. How many moles of oxygen combine with CH_4 to produce 1 mole of CO_2? To produce 2.41 moles of CO_2?
h. How many moles of O_2 combine with 1 mole of CH_4 to give CO? With 1.56 moles of CH_4?
i. How many moles of oxygen are required to produce 1 mole of SO_2 in the complete combustion of carbon disulfide? To produce 0.563 mole of SO_2?

3.46 a. How many g of KCl are produced upon the decomposition of 100 g of $KClO_3$? How many g of O_2?

b. How many g of hydrogen gas are produced by reaction of 35 g of red hot iron with steam?
c. How many g of iron may be produced by reaction of 10 g of CO with Fe_3O_4?
d. How many g of H_2O_2 will react with 74 g of lead sulfide to produce $PbSO_4$?
e. How many g of MnO_2 must decompose to gain 56 g of Mn_3O_4?
f. How much oxygen is consumed in the complete burning of a sample of methane if 106 g of CO_2 are obtained?
g. How much oxygen is consumed in burning 25 g of CH_4 when the products are CO and H_2O?
h. 36 g of sulfur dioxide were produced by the burning of carbon disulfide. How much oxygen was consumed?

EXERCISES – SET V

Equations for reactions involved in problems 3.45 and 3.46 are:

Preparation of oxygen from potassium chlorate:

$$2\ KClO_3 \rightarrow 2\ KCl + 3\ O_2$$

Reduction of iron oxide with carbon monoxide:

$$Fe_3O_4 + 4\ CO \rightarrow 3\ Fe + 4\ CO_2$$

Preparation of hydrogen by passing steam over hot iron:

$$3\ Fe + 4\ H_2O \rightarrow Fe_3O_4 + 4\ H_2$$

Oxidation of lead sulfate with hydrogen peroxide:

$$PbS + 4\ H_2O_2 \rightarrow PbSO_4 + 4\ H_2O$$

Preparation of oxygen by thermal decomposition of manganese dioxide:

$$3\ MnO_2 \rightarrow Mn_3O_4 + O_2$$

Complete combustion of methane:

$$CH_4 + 2\ O_2 \rightarrow CO_2 + 2\ H_2O$$

Incomplete combustion of methane:

$$2\ CH_4 + 3\ O_2 \rightarrow 2\ CO + 4\ H_2O$$

Complete combustion of carbon disulfide:

$$CS_2 + 3\ O_2 \rightarrow CO_2 + 2\ SO_2$$

SUGGESTED READING

Dinga, G. P.: "The Elements and the Derivation of Their Names and Symbols." Chemistry, *41*(2):20, 1968.

Kieffer, W. F.: *The Mole Concept in Chemistry.* Reinhold Publishing Corp., New York, 1962 (paperback).

Nash, L.: *Stoichiometry.* Addison-Wesley Publishing Co., Reading, Mass., 1966 (paperback).

Quick, F. J.: *Workbook for Introductory College Chemistry.* The Macmillan Co., New York, 1966.

Schaum, D.: *Schaum's Outline of Theory and Problems of College Chemistry,* 5th Ed. Schaum Publishing Co., New York, 1966.

FOUR • ATOMIC STRUCTURE

According to the Atomic Theory, compounds are composed of identical "clusters" or groups of atoms called molecules. Different compounds differ in the number, kinds and arrangement of the atoms in the cluster. Each cluster may be simply a pair of two unlike atoms, or a group of three or four or more atoms, up to a thousand or more. A chemical reaction is a regrouping of atoms. Therefore, it is important to know the nature of the forces that hold the atoms together. A knowledge of the structures of the individual atoms themselves assists in this understanding. Despite the submicroscopic size of the atoms, much is known of the still smaller particles of which they are constructed. We shall discuss these briefly.

4.1 SUBATOMIC PARTICLES, THE "BUILDING BLOCKS" OF ATOMS, THE ELECTRON

The most familiar atomic fragment is the electron, the carrier of electricity. A stream of electrons moving along a metallic conductor constitutes an electric current. Some properties of these electrons can be observed if they are forced to jump across a void. This is done in a **cathode ray** tube.

the terms anode and cathode, which here refer to the positively and negatively charged electrodes of a cathode ray tube, should not be confused with the same terms used in a different sense in discussing electrolytic cells, Chapter 14.

A cathode ray tube (Fig. 4.1) is a cylindrical vacuum tube with an electrode (a metal disk) at either end. The electrode is connected to a wire sealed through the glass wall of the tube. A high voltage is placed across the tube, charging one electrode (the cathode) negatively and the other electrode (the anode) positively. A small but measurable current flows through the "vacuum." Furthermore, at the anode end of the tube the glass tubing beyond the anode glows faintly except where the shadow of the electrode is cast. The glowing of the tube is said to be caused by invisible **cathode rays** that emanate from the cathode and move toward the anode. Those

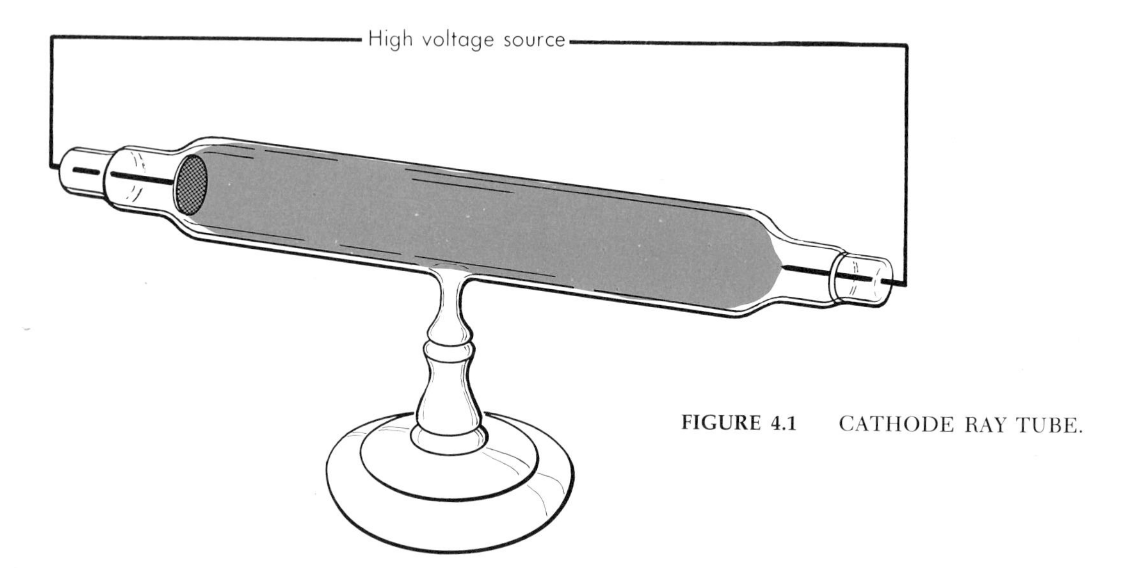

FIGURE 4.1 CATHODE RAY TUBE.

that miss the anode and strike the glass cause it to glow. The operation of this tube suggests that the rays are negative particles—negative because they leave the cathode and are drawn toward the anode, and particles (that is, they possess mass) because they have inertia and do not bend abruptly when passing the anode, but continue on a short distance to strike the glass.

A clearer picture of the electron can be obtained with a modified cathode ray tube (Fig. 4.2). In this tube a large disk is placed in front of the cathode. The disk has a slit across the middle and serves to stop all the rays except those passing through the slit. A thin metal strip, called a screen, treated so that impacts of cathode rays cause it to glow, is placed diagonally across the tube from the end of the slit described above to the opposite end of the tube.

When high voltage is impressed across the tube as before, a straight, green-glowing line traces the image of the slit on the screen (Fig. 4.2b), showing the particles to be traveling in a straight line after being hurled through the slit. When an electric field is placed about the screen with the positive pole above and the negative pole below, the green line is seen to curve upward beginning at the slit near the cathode (Fig. 4.2c). Again the rays bend toward the positive pole (are negatively charged), and the curve is gradual, showing that the momentum of the particles (the "mass in motion") resists the bending force.

usually the bending of the cathode rays is done by bringing the pole of a magnet up to the tube, behind the metal strip. the electrostatic field associated with the magnetic field causes the particles to follow a curved path.

The picture tube of a television set is a modified cathode ray tube. An electron gun (cathode) fires electrons toward the front of the tube. Electron impacts on the treated inner surface of the front face of the tube cause it to glow, lighting up the tube. The glow is seen from the outside of the tube, on the television screen. By an automatic changing of the electrostatic field around the electron stream within the tube, the electrons are made to spray across the front of the tube in strips, similar to the way words are written across a page. The electrons spray across the tube so rapidly that the face of the tube is covered in a fraction of a second and the whole picture glows. Any dark portion of the picture being displayed involves stopping the spray of electrons, leaving that part of the strip dark. Thus, a black and white picture is obtained.

The magnitude of the charge of the electron was determined by Dr. Robert A. Millikan in his Nobel Prize-winning experiment (about 1910), in which an oil drop of known mass surrounding a particle of 1 electron charge was held suspended in an electrostatic field. The experiment will not be discussed here, but the results will be used. The electron charge is 1.6×10^{-19} coulomb.

A current of approximately 1 ampere flows through a lighted 100-watt bulb. An ampere is 1 coulomb per second, or $\frac{1}{1.6 \times 10^{-19}} = 6.3 \times 10^{18} =$ (6,300,000,000,000,000,000) electrons per second. Knowing the charge, one can calculate the mass of an electron from the extent of its bending in an electrostatic field (mass spectrograph principle). Compared to the mass of the carbon-12, the most common isotope of carbon, the electron has a mass of 1/1837 of an atomic weight unit.

The electron is important not only because it is a carrier of electricity but because it is a building block of the atom. The best evidence that electrons constitute a part of the atom comes from a study of the rays emitted in **radioactivity.** Radium, a typical radioactive element, decomposes into other elements by the emission of three types of rays: **alpha, beta and gamma rays.** When these rays are passed through an electrostatic field, alpha rays bend slightly toward the negative pole, beta rays curve more sharply toward the positive pole and gamma rays are unaffected (Fig. 4.3).

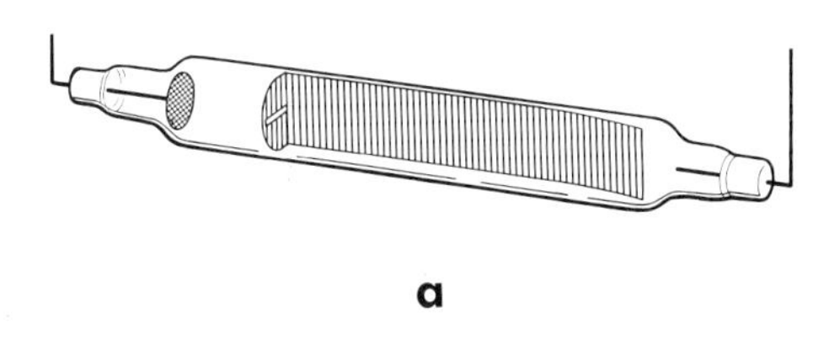

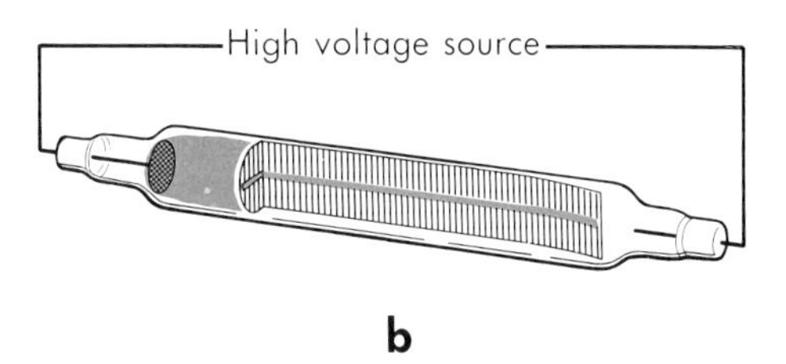

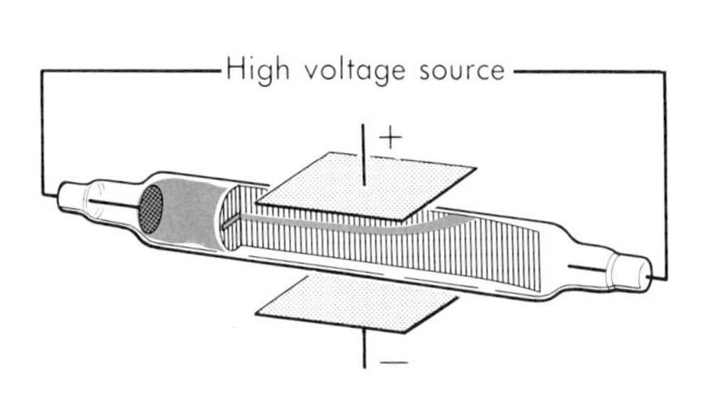

FIGURE 4.2 DEFLECTION OF A CATHODE RAY BEAM.

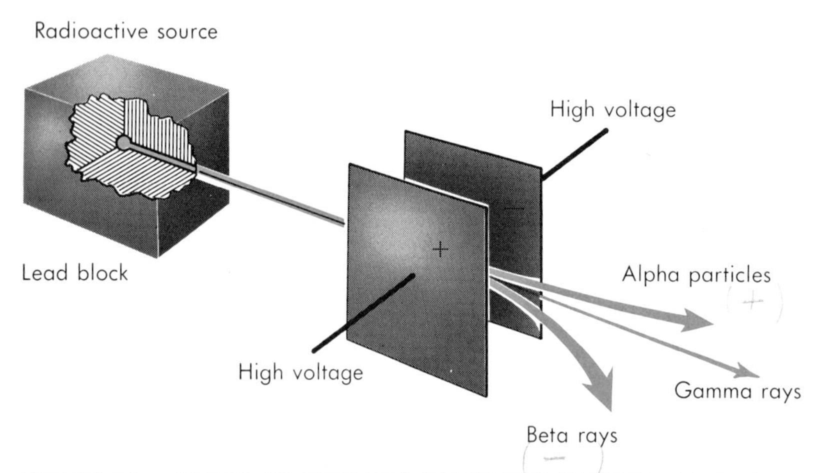

FIGURE 4.3 PATHS OF NUCLEAR PARTICLES PASSING THROUGH AN ELECTRIC FIELD.

The beta ray has been shown to be a stream of electrons hurled from radioactive atoms. The alpha ray is a particle much heavier than an electron, having a mass of approximately 4 atomic weight units.

Its charge is positive and is exactly twice that of the electron. The alpha particle is not considered a building block of atoms, for it in turn appears to be composed of still smaller particles, which will be discussed presently. Upon gaining two electrons, the alpha particle becomes an atom of helium gas.

The gamma ray is a high energy radiation similar to but containing more energy than x-rays. X-rays penetrate flesh but not bones, and therefore they show a profile of bones on a photographic plate activated by the x-rays that pass through a portion of the human body. Gamma rays pass through the bones as well. Both rays destroy body tissue, and both have been used to destroy cancerous cells; at the same time they destroy normal cells. In any case, since gamma rays are light-like radiations, they are not fundamental particles of matter.

4.2 POSITIVE PARTICLES, PROTONS

When a high voltage is impressed across a cathode ray tube containing a gas at low pressure, a bright glow is observed. The color of the light emitted is characteristic of the gas in the tube. If the gas is neon, the glow is bright red; if sodium vapor, intense yellow; if mercury vapor, a light green. These colors are seen in the glow of neon signs, sodium vapor street lights and ultraviolet lamps. The common fluorescent light contains a mixture of gases at low pressure which glow when excited by an electric field. The inner surface of the tube is coated with a phosphorescent material, and the excited gases cause this to glow, too. The gas mixture and the coating are varied in composition to give the desired color.

If the negative electrode of the tube containing the gas is perforated (Fig. 4.4), the fluorescence can be seen to extend in the tube a short distance beyond the negative electrode. This indicates the momentum which carries the particle beyond the electrode is evidence that the particle possesses mass.

that **positive particles** are formed in the gas, through which electrons are passed. These positive particles are drawn toward the negative electrode. Some particles pass through the holes. Because those particles are ultimately responsible for the fluorescence, there is a glow behind the cathode.

Further investigation has shown these positive particles to have a mass very nearly equal that of the atoms of the gas in the tube. The electrons emitted from the cathode and accelerated toward the anode strike the molecules of the gas with sufficient force to dislodge additional electrons, leaving positive ions. For the bombarding of a sodium atom in sodium vapor, the reaction is:

$$Na + e^- \rightarrow Na^+ + 2e^-$$

The positive sodium atom, now called an **ion,** moves toward the cathode (—), and the electrons continue toward the anode. When the sodium ion gains another electron, it emits the characteristic yellow light of sodium.

When bombarded with electrons, hydrogen gas gives the smallest positive particles of all gases, since the hydrogen atom is the smallest of atoms. This positive particle, a hydrogen ion (H^+), is also called the **proton** and is represented by **p^+**. Note here that electrons are removed from atoms with the expenditure of a relatively small quantity of energy, and their loss does not change the atom permanently. Upon capturing another electron, the atom is restored.

The proton can be produced by more energetic means from substances other than hydrogen. When **alpha particles** from radioactively decaying radium are hurled at high velocity from a high energy accelerator into nitrogen gas, particles having the same charge and mass as protons are ejected. Furthermore, these particles become hydrogen gas upon gaining electrons. Oxygen gas

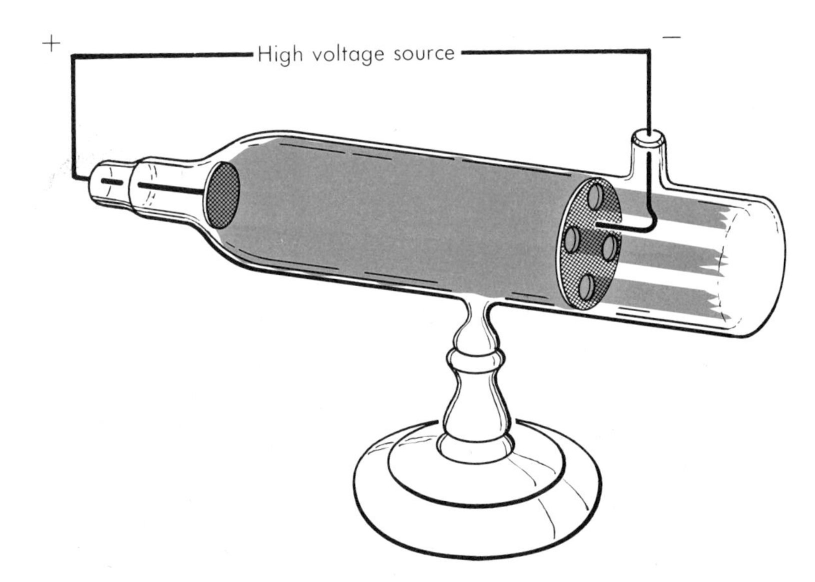

FIGURE 4.4 DETECTING POSITIVE PARTICLES IN A CATHODE RAY TUBE.

is created at the same time by the reaction represented by this equation:

$$^{14}N \quad + \quad ^{4}\alpha \quad \rightarrow \quad ^{1}p \quad + \quad ^{17}O$$

A nitrogen atom with a mass of 14	+	An alpha particle with a mass of 4	→	A proton with a mass of 1	+	An oxygen atom with a mass of 17

Another process yielding protons is:

$$^{27}Al + {}^{4}\alpha \rightarrow {}^{1}p + {}^{30}Si$$

An aluminum atom is changed to a silicon atom by accepting an alpha particle and losing a proton.

Protons may be absorbed into an atom and thus change it into an atom of another element. For example, a proton hurled into a lithium atom changes it to 2 helium atoms:

$$^{7}Li + {}^{1}p \rightarrow 2\ {}^{4}He$$

Many other reactions are known in which a proton is either absorbed or expelled by an atom. In each case the atom is changed into another kind of atom. On the basis of these reactions, the proton is accepted as a **fundamental building unit of the atom.** In summary, the proton has a mass of approximately 1, compared to the carbon-12 atom with 12, and has a charge equal in magnitude to the electron but of opposite sign.

4.3 NEUTRON

Another type of radiation is observed when alpha particles bombard targets of certain light metals such as beryllium and lithium. At first this radiation was classed as very energetic gamma rays, for they were not deflected in an electric field, but they were more penetrating than any gamma ray known. Further experimentation showed that the radiation could dislodge protons from paraffin wax. Paraffin contains hydrogen as one of its elements, but a gamma ray devoid of mass would be unable to dislodge the hydrogen. It was concluded from the velocity of the ejected proton that the radiation was composed of neutral particles with a mass approximately equal to that of the proton. A study of the product of the original bombardment verified this, as shown by the equation:

$$^{9}Be + {}^{4}\alpha \rightarrow {}^{12}C + \text{radiation of mass 1, charge 0}$$

The sum of the mass units of beryllium and the alpha particle is 13; the carbon atom formed from their union is 12. The mass of the particle is approximately 1. In summary, the third building block is the **neutron** (the word is a contraction of neutral-proton), which has no charge and has a mass of about 1, compared to the carbon-12 atom taken as 12.

Typical reactions involving the neutron are the conversion of an oxygen atom by proton bombardment to a fluorine atom and a neutron:

$$^{18}O + {}^{1}p \rightarrow {}^{18}F + {}^{1}n$$

and the conversion of a silver atom to a heavier silver isotope by the capture of a neutron:

$$^{107}Ag + {}^{1}n \rightarrow {}^{108}Ag$$

Much more evidence could be cited for the neutron, but this is perhaps enough to justify its position as an atomic structural unit. The neutron holds an important position in world affairs. It is the particle responsible for the propagation of the chain reaction occurring in an exploding atomic bomb (See Chapter 23).

4.4 OTHER PARTICLES OF MATTER

Thus far, four particles of matter have been discussed. Three, the electron, proton and neutron, are important as building blocks of matter. The fourth, the alpha particle, appears to be an intermediate structural unit composed of two neutrons and two protons. Of the approximately **40 subatomic particles** known, three are mentioned here.

The bombarding of magnesium or aluminum with alpha particles releases (among other particles) a particle having the mass of an electron but a **positive** charge. It is called a **positron.** This particle eluded discovery for some time because of its attraction for the ever present electron. On collision, the positron and electron destroy each other and give rise to a photon (a quantum of radiation), as shown by the following equation:

$$e^{+} + e^{-} \rightarrow \text{photon} + \text{energy}$$

Cosmic ray investigations revealed the existence of another particle, the **meson,** which has an electrical charge equal in magnitude to that of an electron but either positive or negative, and a mass about one-tenth of the proton. The mesons made out of the neutrons and protons in an atom are thought to be necessary to hold the protons of the atom together in the nucleus.

The final subatomic particle we will consider is the **neutrino,** an electrically neutral particle having a mass equal to or less than that of an electron. Mathematical calculations indicated that such a particle exists if momentum and mass are to be conserved in many nuclear reactions, and more recently the neutrino has been observed experimentally. Many other atomic fragments are known, but they are not of importance for our study of chemistry.

4.5 RELATIVE NUMBER OF PARTICLES IN THE ATOM

In the preceeding section, the subatomic particles of matter were presented. Their properties are given in Table 4.1, with the masses of the particles compared on the chemist's atomic weight scale. Three of the particles, the **neutron,** the **proton** and the **electron,** when placed together in various numbers, combinations and

TABLE 4.1 SUBATOMIC PARTICLES

Name	Symbol	Mass (Carbon 12 = 12)	Charge
H atom	H	1.00797 a.m.u.	0
Carbon-12 isotope	^{12}C	12	0
Electron	e^-	1/1837	−1
Proton	p^+	1.0072	+1
Neutron	n	1.0086	0
Positron	e^+	1/1837	+1
Meson	μ	ca. 1/10	±1
Neutrino		very small	0

arrangements, are necessary and sufficient to explain the structure of the atoms. An atom, therefore, may be defined as **an electrically neutral particle composed of electrons, protons and neutrons and having only one center.**

The number of subatomic particles in an atom varies from one atomic species to another. The number and kind of these particles determine the properties of the atom. All atoms having the same number of protons are so nearly identical chemically that they are known by that number, the **atomic number,** and are given the same name. For example, all atoms containing only 1 proton are hydrogen atoms and have an atomic number of 1. All oxygen atoms (atomic number 8) contain 8 protons. Elements are listed in the periodic table in the order of increasing atomic number (proton number). The periodic table is reproduced in Figure 4.5. Memorize the elements with white symbols and their corresponding atomic numbers, noting their location in the table.

The number of electrons in the atom equals the atomic number, because to be neutral the atom must contain an equal number of negative and positive particles. Chlorine has an atomic number of 17 and possesses 17 protons and 17 electrons.

There is no exact relationship between the atomic number and the number of neutrons in an atom, but the number of neutrons can be calculated from the atomic number and the mass number. The mass number is the whole number most nearly equal to the atomic weight. For example, consider the major isotope of aluminum with atomic number 13 and mass 27, represented as $^{27}_{13}Al$. The atom contains 13 p^+, accounting for 13 mass units. It also contains 13 e^-, but their mass is negligible. The remainder of the mass must be accounted for by neutrons.

the subscript is the atomic number of the element whose atom is represented by the symbol. the superscript is the mass number.

$$27 - 13 = 14 \text{ extra mass units.}$$

Each neutron has a mass of approximately 1, so there are 14 neutrons. The number of neutrons in any other atom may always be found by subtracting the atomic number from the mass number. Another example: uranium has an atomic number of 92 and an atomic weight of 238.03 ($^{238}_{92}U$). There are 92 e^-, 92 p^+ and 238 − 92 = 146 n.

Application of this method to all atoms shows that, among lighter elements, the atom contains nearly an equal number of neutrons and electrons. **As the atomic number increases, the ratio of**

1 H																	1 H	2 He
3 Li	4 Be												5 B	6 C	7 N	8 O	9 F	10 Ne
11 Na	12 Mg												13 Al	14 Si	15 P	16 S	17 Cl	18 Ar
19 K	20 Ca	21 Sc		22 Ti	23 V	24 Cr	25 Mn	26 Fe	27 Co	28 Ni	29 Cu	30 Zn	31 Ga	32 Ge	33 As	34 Se	35 Br	36 Kr
37 Rb	38 Sr	39 Y		40 Zr	41 Nb	42 Mo	43 Tc	44 Ru	45 Rh	46 Pd	47 Ag	48 Cd	49 In	50 Sn	51 Sb	52 Te	53 I	54 Xe
55 Cs	56 Ba	57 La	58 Ce ➔ 71 Lu	72 Hf	73 Ta	74 W	75 Re	76 Os	77 Ir	78 Pt	79 Au	80 Hg	81 Tl	82 Pb	83 Bi	84 Po	85 At	86 Rn
87 Fr	88 Ra	89 Ac	90 Th ➔ 103 Lr	104 Ku	105 Ha													

FIGURE 4.5 THE PERIODIC TABLE.

neutrons to protons increases up to about 3 to 2 among the heaviest atoms. One of the heaviest, uranium, has a ratio of 146 to 92, more than 3 to 2.

Isotopes were defined previously as atoms of the same element with different masses. Since they are atoms of the same element, isotopes contain the same number of protons and electrons. The difference in mass is due to a different number of neutrons in the two atoms. Compare the two isotopes of chlorine, one with a mass of 35, the other with a mass of 37:

$^{35}_{17}Cl$	17 p^+ 17 e^- $35 - 17 = 18$ n	$^{37}_{17}Cl$	17 p^+ 17 e^- $37 - 17 = 20$ n

The atomic weight of chlorine, 35.5, is a weight average of the masses of the two isotopes.

4.6 REGIONS WITHIN THE ATOM

In 1911 Ernest Rutherford devised and carried out a simple experiment that gave chemists more insight into atomic structure. He knew that when an alpha particle strikes a metal surface coated with ZnS, a bright flash occurs. Also, if one places a sample of radium opposite a piece of sheet metal coated with zinc sulfide, a spot is seen to glow brightly (Fig. 4.6).

Rutherford placed a thin piece of gold foil between the alpha-emitting radium and the zinc sulfide screen (Fig. 4.7). The gold foil was very thin—4000 Å or 0.00004 cm thick. However, there were over 1000 layers of gold atoms in the foil. Rutherford could see no difference in the intensity of the light at the center of the zinc sulfide screen when the gold foil was in place. Apparently the alpha particles passed right through the gold. Occasionally, however, an isolated flash or scintillation appeared further out towards the edge of the screen. The screen was moved to other positions around and facing the metal foil. Occasional scintillations

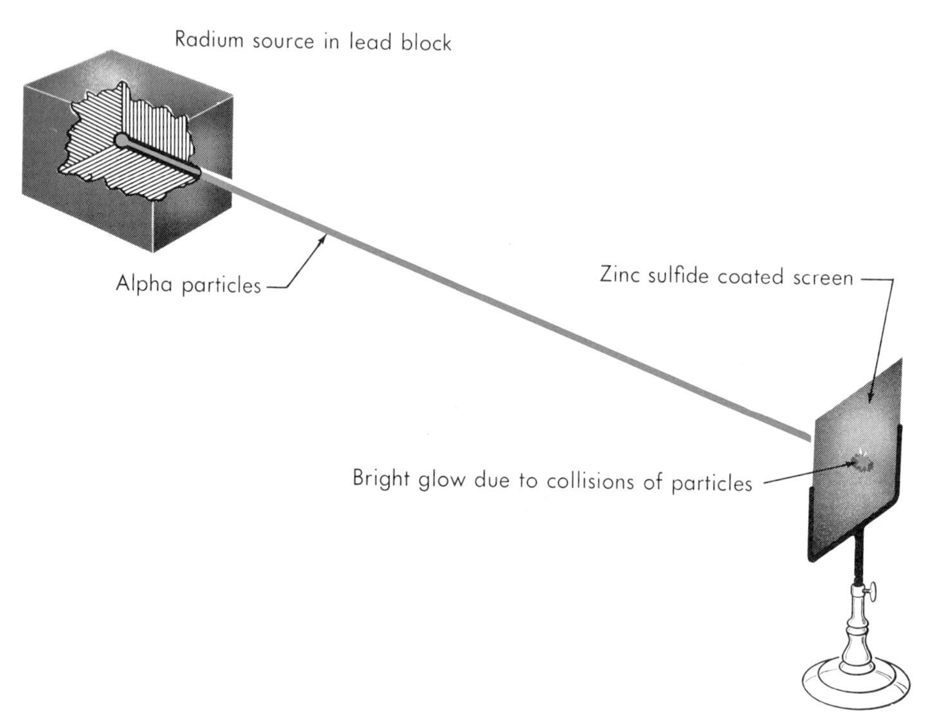

FIGURE 4.6 ALPHA PARTICLES STRIKING A SENSITIZED SCREEN.

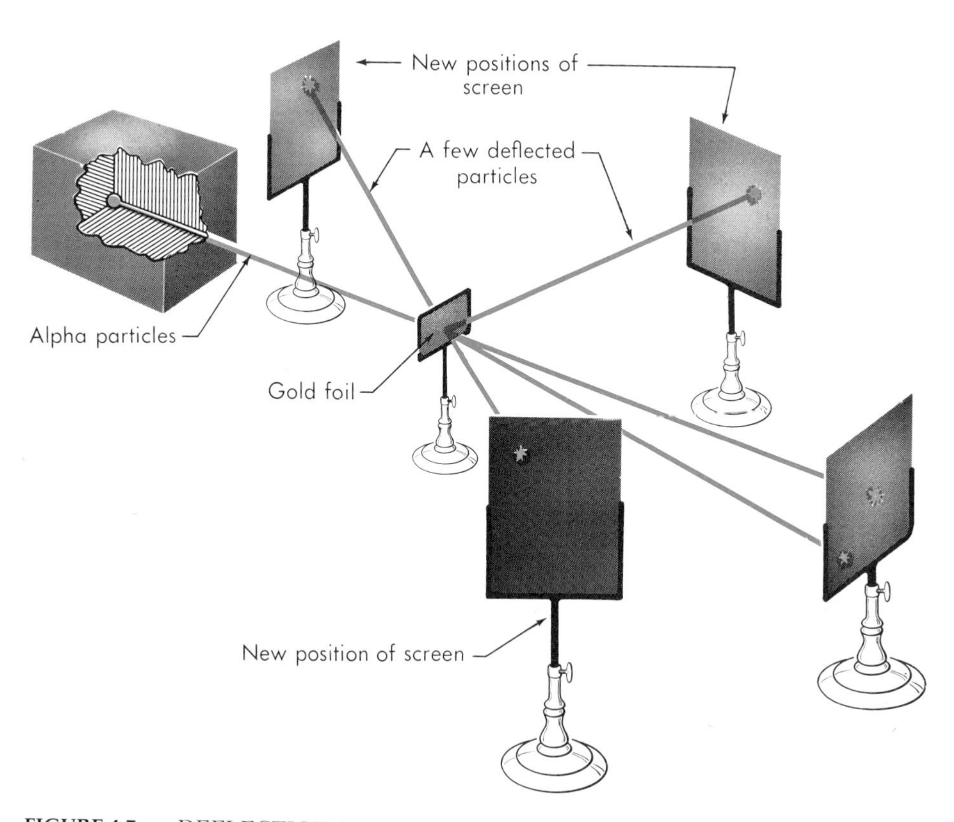

FIGURE 4.7 DEFLECTION OF SOME ALPHA PARTICLES BY GOLD FOIL.

were observed at all angles. Some alpha particles bounced directly back from the foil. Further studies showed that only 1 in 100,000 alpha particles deviated from the straight line. However, those particles that did not hit the center of the screen were found to be reflected from the foil at different angles to the incident radiation.

Rutherford reasoned that the alpha particles which passed straight through the foil could have hit none of the subatomic particles except possibly an electron with negligible mass. Therefore he concluded that an atom is mostly unoccupied space. If only 1 in 100,000 particles were deflected by 1000 layers of atoms, only 1 in 100,000,000 would be deflected by each single layer of gold atoms. He reasoned further that nearly all the total mass of the atom must lie in one small part of the volume—the center of the atom, which he called the *nucleus*—because, when the alpha particle did hit something, it was heavy enough to bounce the particle off in any direction. This would mean that all the protons and neutrons must be in a tiny nucleus, whereas the electrons with their negligible mass might circulate in the space about the nucleus. He thought of the atom as a tiny solar system with the nucleus corresponding to the sun, and the electrons—corresponding to the planets—orbiting the nucleus. The nucleus with its circling electrons, like a spinning top, occupies a volume circumscribed only by orbiting electrons.

if the relative sizes are maintained and the radius of the nucleus were one foot, the atom would have a radius of approximately two miles.

A mathematical analysis of data from this and other experiments shows the nucleus to have a diameter about 1/10,000 the diameter of the whole atom. The diameter of a nucleus is about **10^{-12} cm**; of an atom, approximately **10^{-8} cm**. These diameters vary from atom to atom but are of this order of magnitude.

Rutherford was surprised with the results of his experiments and the picture of the atom that it gave him. He had held the view, then current, that the atom was a homogeneous, spherical cloud or field of mass and positive charge. Electrons were imbedded in the positively charged cloud, or else they floated on its spherical surface. The possible removal of all or some of the electrons of the atoms of a substance (accomplished by rubbing the substance against another) gave a theory for static electricity. Such an atom, with no concentration of either charge or mass, could not account for deflection of the paths of the bombarding alpha particles.

The new picture of the neutrons and protons bound tightly in a small nucleus with the electrons in orbit about it agrees with the relative energies required to remove particles from the atom. Electrons are rather easily removed by electron impact in a cathode ray tube, or by interaction with incident light. Removal of either protons or neutrons requires collision with a high energy alpha particle or with other energetic particles. Further, the ion (atom minus an electron) soon finds another electron and reverts to its former state. Loss of a proton or neutron changes the atom permanently.

4.7 ELECTRONIC STRUCTURE

The nucleus, nuclear reaction and the uses of nuclear energy will be discussed in Chapter 23. The number of protons in the nucleus (the atomic number) sets the number of electrons in the atom; otherwise the protons and neutrons have little to do with the chemical properties of substances. The number and arrangement of the electrons about the nucleus alone hold the key to the reactivity and the bonding characteristics of the atom.

First, one should note that the electron is held in its position about the nucleus by the electrostatic attraction between it and the positively charged nucleus. The questions follow: "Why doesn't the electron fall into the nucleus?" and "What keeps it out in the tremendous space surrounding the small, charged center?" The answers are found in the fact that the particles possess energy. The energy that keeps an electron out of the nucleus as explained by a simple atomic model suggested by Niels Bohr is the **kinetic energy** of the electron as it circles the nucleus.

There is evidence that an atom can exist in any one of a number of energy states; that is, an electron (or electrons) in the Bohr atom can occupy any one of a number of orbits, depending upon the energy it possesses. This is the case for a hydrogen atom, the simplest of all, which has only one electron about its proton nucleus. As an analogy consider the circular swinging of an object on the end of a string (Fig. 4.8).

With no motion and no energy the object hangs straight down. However, as it is given energy in the form of circular motion, the object swings out and up against gravity. In following circle **a** (Fig. 4.8), the weight has little kinetic energy. In circle **b** the weight is moving much faster, has more energy and describes a wider arc. The greater the energy, the greater the diameter of the circle as the object surges out and up.

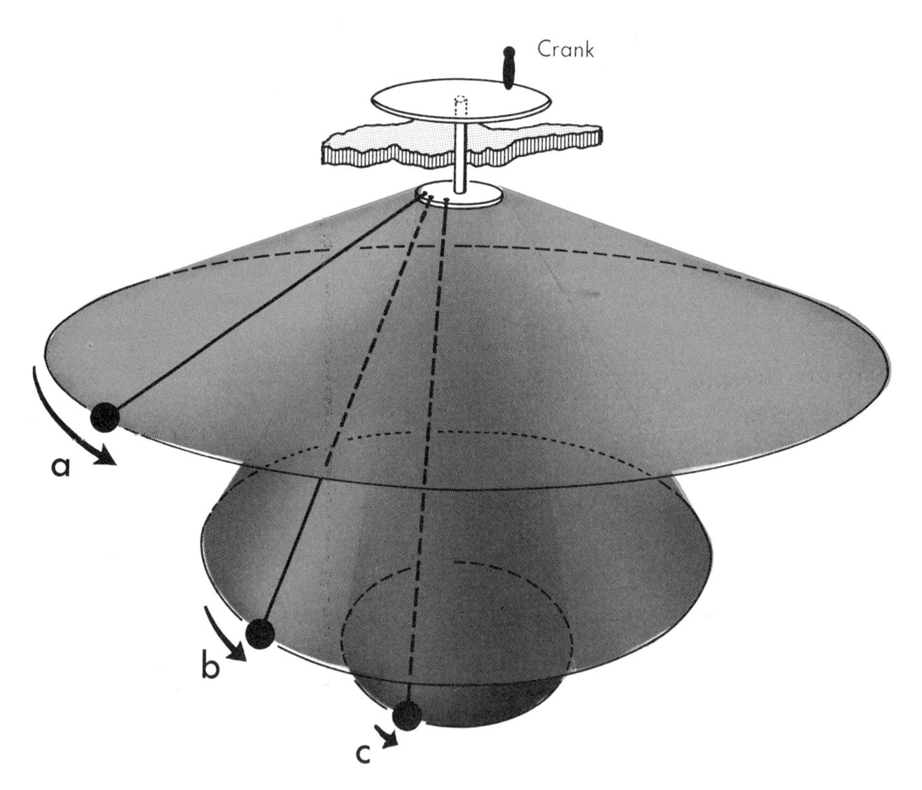

FIGURE 4.8 PATHS OF AN OBJECT AT THE END OF A STRING.

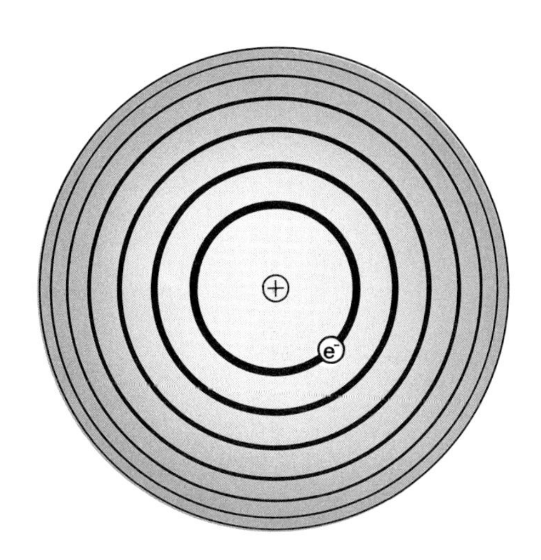

FIGURE 4.9 THE BOHR HYDROGEN ATOM.

According to Bohr, the electron of the atom moves out in a "wider" orbit as more energy is given to it. Here the analogy between the weight on the string and the electron in the Bohr atom ends. In contrast to the path of the weight on the string, which may change gradually from a small to a large diameter, in an atom there are certain well-defined **orbits**, and the electron must receive exactly the right amount of energy to lift it from one to another. The electron must jump from one orbit to another; it cannot remain in between two orbits. The hydrogen atom, according to Bohr, may be represented as in Figure 4.9.

About the nucleus there are many orbits. In this sketch the electron is placed in the innermost orbit, and the atom is said to be in the ground state. By absorption of energy the electron may be lifted into one of the higher orbits.

4.8 SPECTRA: EVIDENCE FOR ORBITS

Experimental evidence that atoms have fixed energy states comes from many sources, perhaps most simply from a careful examination of the light emitted from atoms in fluorescent lamps. It has been stated that gases at low pressure emit light when bombarded with cathode rays. The color of the light is characteristic of the gas. Hydrogen fluorescence is pale blue; sodium, yellow; mercury, green; and neon, red. When a narrow beam of the pale blue light from a hydrogen lamp is passed through a prism, the beam is separated into several color components. The four components of hydrogen radiation appearing in the visible spectrum, together with their respective wavelengths, are: orange (6563 Å), blue (4861 Å), blue-violet (4341 Å) and violet (4101 Å) (see Figure 4.10). When hydrogen radiation is photographed after passing through a prism, lines appear on the developed film at positions corresponding to the wavelength of each color component. Relatively large blank areas appear between the narrow spectral lines on the film.

The line spectra for other excited gases are similar to that described for hydrogen but differ in the color (wavelength) and num-

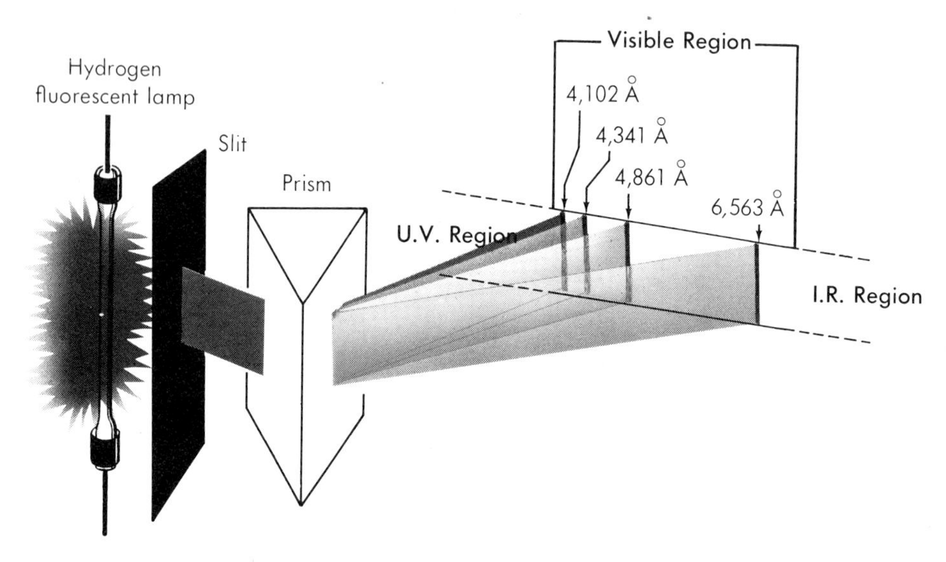

FIGURE 4.10 EMISSION SPECTRUM OF HYDROGEN.

ber of the colored components and may be contrasted with the spectrum of sunlight. When a narrow beam of sunlight is passed through a prism, it gives the complete rainbow (Fig. 4.11). There are only narrow breaks between colors; all varieties of color are present in sunlight. It is known that the energy of the **photons**, which are packets of radiant energy, varies with the color (wavelength) of the photon. Red photons have the least energy; yellow have more; then come green, blue and finally violet, which contain the highest energy photons of visible light. Separating the light colors by a prism arranges the photons according to the energies they possess. While sunlight contains light of many energies, radiation from pale blue light of glowing hydrogen contains only certain energy packets of a definite size. The Bohr model of the hydrogen atom with electron orbits may be used to explain this. When a cathode ray strikes an electron of a hydrogen atom (Fig. 4.12a), the electron may be completely removed from the atom, leaving the ion, which is drawn toward the cathode. Or, as shown in Figure 4.12b, the electron may receive just enough energy to lift it to an orbit of higher energy within the atom.

The bombarding particle is slowed down by the collision, some of its energy being taken up in raising the electron to a higher level. The **excited electron** tends to return to the lowest orbit possible, putting the atom in the ground state. It may do so in one step or in many steps, as shown in paths 1, 2, 3 and 4 in Figure 4.13, but only by giving up energy as photons of light. Path 1 yields one photon; path 2 yields two photons; path 3, two photons; and path 4, three photons. But the energy of the one photon yielded when an electron follows path 1 is equal to the sum of the energies of both photons from path 2, both photons from path 3 and so on. The photon liberated via path 1 has a wavelength of 949.76 Å (ultraviolet); the photons via path 2, 1026 Å (ultraviolet) and 12,818 Å (infrared); the photons via path 3, 1216 Å (ultraviolet) and 4101 Å (violet); and the photons by path 4, 1216 Å (ultraviolet), 6563 Å (orange) and 12,818 Å (infrared). Note that some of the photons are the same

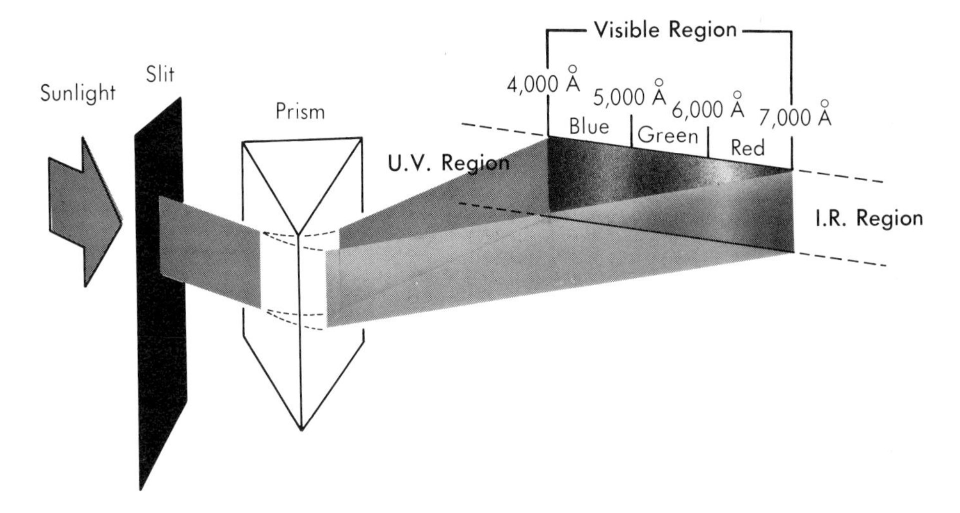

FIGURE 4.11 THE SPECTRUM OF SUNLIGHT.

from steps in two different paths. This occurs because the electron falls between the same two levels in both paths.

In a hydrogen lamp there are billions and billions of atoms "excited" every instant by the impact of electrons; each electron may return by one of the paths shown or by other paths. In any case there will be great numbers of electrons jumping from level 4 to 3, 4 to 2, 4 to 1, 3 to 1 and so forth. For each of these there is a photon of definite wavelength which corresponds to a definite energy and which gives definite thin lines on the photographic film when the spectrum of hydrogen is photographed.

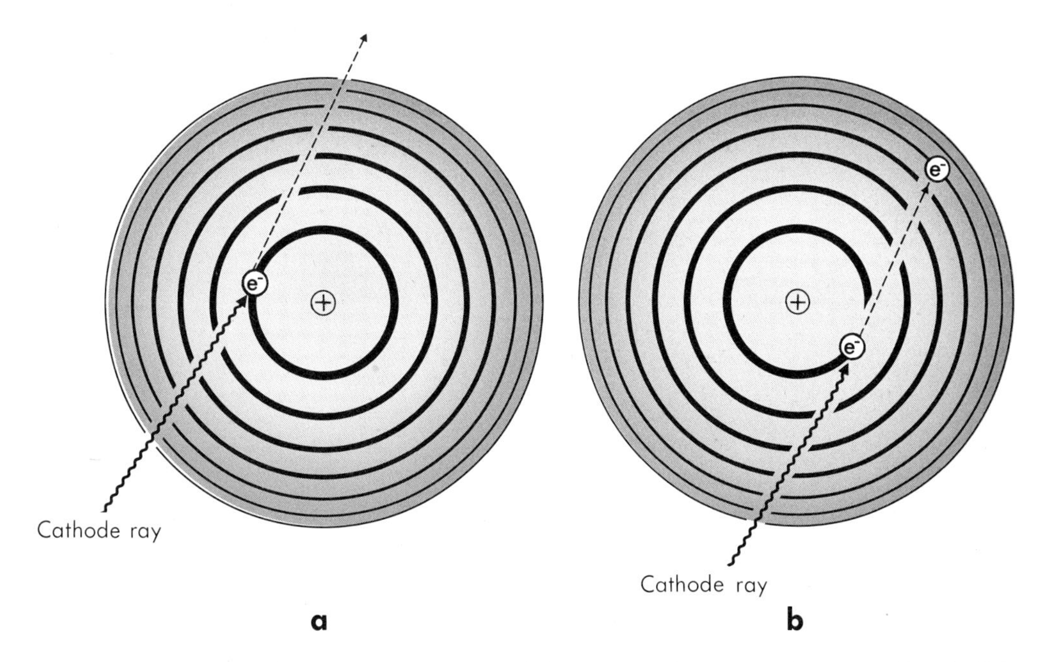

FIGURE 4.12 a. EJECTION OF AN ELECTRON FROM AN ATOM. b. EXCITATION OF AN ELECTRON TO A HIGHER ENERGY LEVEL.

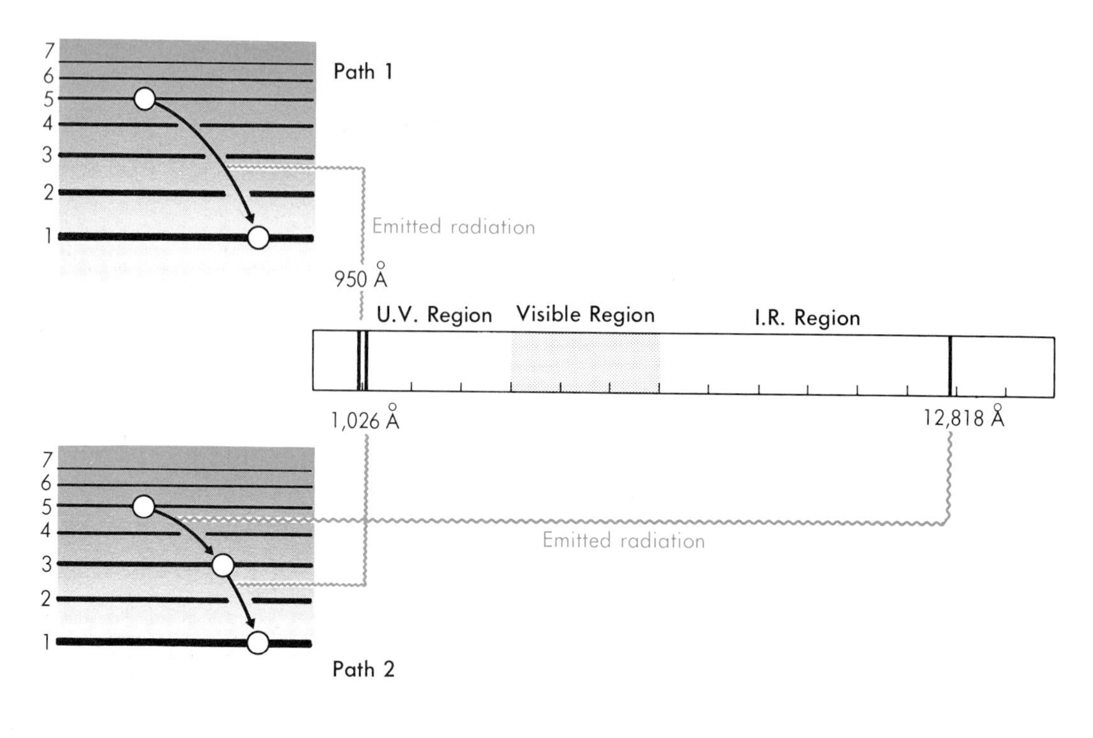

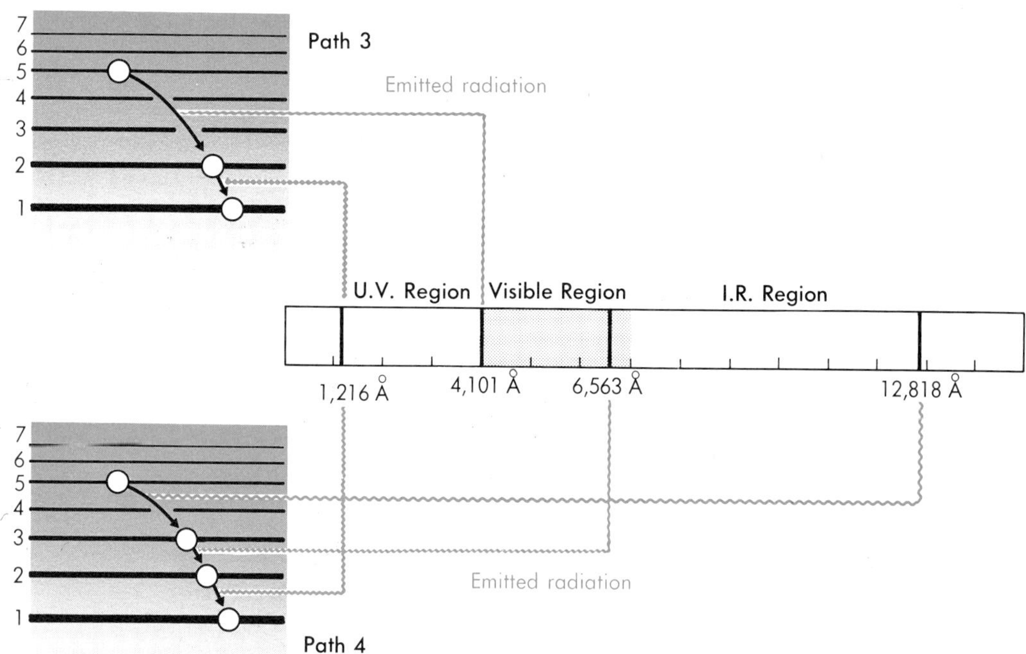

FIGURE 4.13 EMISSION OF LIGHT BY ELECTRONS DROPPING TO LOWER ENERGY LEVELS.

4.9 ORBITAL ARRANGEMENT

Bohr's original picture of the atom has undergone many changes. The electrons in atoms are no longer thought to follow only circular or elliptical orbits. In fact the paths of electrons are neither known nor postulated today. Instead the electrons are assumed to occupy certain regions in space called **orbitals.** Each atom has many possible orbitals in which electrons might be found. Orbitals may differ both in their shapes and in the energy required of an electron that occupies them. The orbital arrangements of all atoms are assumed to be similar, whether or not the orbitals are occupied. That is, the hydrogen atom has the same pattern of orbitals as the uranium atom, although in hydrogen only one orbital (usually the one of lowest energy) contains one electron, while the uranium atom contains 92 electrons in 48 orbitals.

The atom's many orbitals of various sizes and shapes are classified according to the values of a set of three small numbers (**n, ℓ and m**) called **quantum numbers.** The letters n, ℓ and m appear in the Wave Equation, an equation which describes the orbitals. When a **permissible** set of small integral numbers is substituted in the Wave Equation for the letters, the equation may be solved (with other than a permissible set it is insoluble). In solving the equation, one can find the volume in space relative to the nucleus in which an electron is most likely to be located. An envelope around the volume of greatest probability gives the shape of the orbital which has that set of quantum numbers. Each of the three quantum numbers will be discussed below. Rules for determining permissible values will be given and applied, and the shapes of the corresponding orbitals will be indicated.

the wave equation is a complicated differential equation and will not be discussed here.

The **principal quantum number,** n, may have values of 1, 2, 3 and so on. It is also called the **quantum level number**; that is, an orbital for which n = 1 is said to be in the first quantum level. The number of orbitals in a quantum level is equal to the square of the principal quantum number. For example, there is 1^2 or 1 orbital in the first quantum level; 3^2 or 9 in the third. Not only does the number of orbitals per level increase with the number of the level, but also the energy an electron must possess in the orbital and the distance that orbital extends out from the nucleus increases.

The secondary (or sublevel) quantum number, ℓ, may have values (beginning with zero) of 0, 1, 2, 3 and so forth (most often designated by s, p, d, f and so forth), but within a certain quantum level the maximum value of the ℓ number is one less than n. The sublevel number or letter, therefore, indicates a particular sublevel within a quantum level, and the number of possible values of ℓ gives the maximum number of sublevels in a particular quantum level. For example, in the first quantum level, n = 1, the only possible value for ℓ is 0. There is only one sublevel. That is, the level is not divided. There are two sublevels in quantum level 2, for ℓ may have values of both 0 and 1. Application of the restriction that ℓ is equal to or less than $n - 1$ reveals that the number of sublevels in a quantum level is equal to the principal quantum number. The relative energies of the orbitals within a quantum level increases with the value of the secondary quantum number. That is, the d orbitals ($\ell = 2$) of quantum level 3 have electrons of higher energy than the p orbitals ($\ell = 1$), which in turn have a higher energy than s orbitals.

The secondary quantum number (or letter designation) also indicates the **shape** of the orbital. All orbitals with the same secondary quantum number are similar in shape, and orbitals with different secondary quantum numbers have different configurations. An s orbital ($\ell = 0$) is a sphere; that is, the volume in which an electron occupying that orbital is most likely to be found may be circumscribed by a spherical shell about the nucleus as the center (Fig. 4.14a). A p orbital ($\ell = 1$) has two **lobes or probability volumes.** An electron in a p orbital is most likely to be found close to one of two points on opposite sides and equidistant from the nucleus. The probability fades in all directions from each of these points (Fig. 4.14b). A d orbital ($\ell = 2$) has four lobes, something like four spheres in a plane, distorted in such a way that they almost touch at the nucleus; they resemble a four leaf clover (Fig. 4.14c). The electron in one of these orbitals has an equal probability of being within any one of the four lobes. The fifth d orbital has aspects of the four similar d orbitals but will not be described here. An f orbital ($\ell = 3$) has six probability lobes. As can be seen, the orbitals become more complicated as the sublevel number increases.

The **magnetic** (orbital) **quantum number,** m, may have all

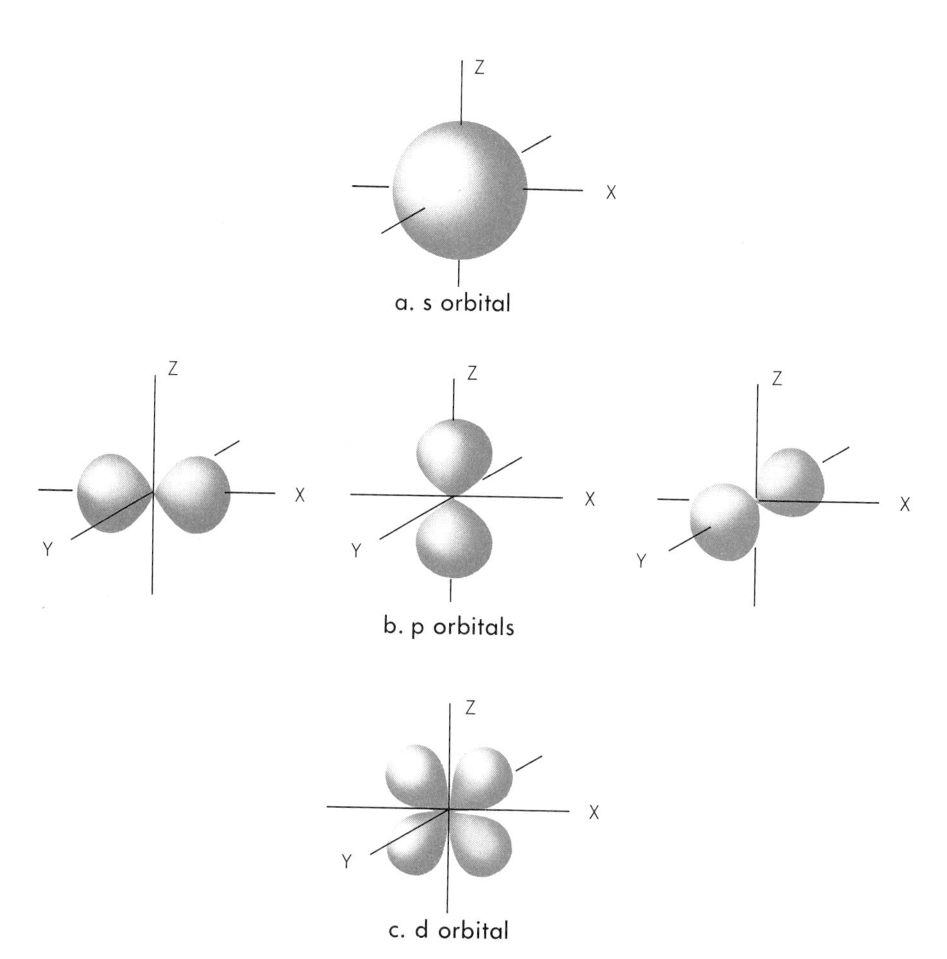

FIGURE 4.14 SOME ATOMIC ORBITALS.

values (positive, negative and zero), equal to or less than the value of ℓ for a particular sublevel. In an s sublevel ($\ell = 0$), the only possible value for m is 0. In a p sublevel ($\ell = 1$), the possible values for m are +1, 0 and −1; in a d sublevel, +2, +1, 0, −1, −2. The orbital quantum number indicates the orientation of the orbital in space. The number of possible magnetic quantum numbers per sublevel gives the number of possible orientations.

Because no two orbitals of the same subgroup can have the same direction in space, the number of orbitals per sublevel is limited and fixed. In agreement with this principle, there is only one orientation for (and only one s orbital in) the first ($\ell = 1$) sublevel. This is in agreement with the fact that a sphere has no direction and the fact that only one sphere can be drawn with the same center and radius. There are three p orbitals, for there are three possible values of m when ℓ is equal to unity. Each of the three orbitals is directed along an axis, X, Y, or Z, of a Cartesian coordinate system (Fig. 4.14b), the three orbitals often are labeled p_x, p_y and p_z. There are **five** d orbitals with five different orientations, and **seven** f orbitals with seven different orientations, but these will not be discussed at this time.

we are not concerned here with the actual values of m but with the number of m values, the multiplicity of m.

As indicated in the discussion of the principal quantum number, orbitals of the same type, such as s, p, d, or f, circumscribe an increasing volume as the principal quantum number increases.

Finally, there is a fourth quantum number, the **spin quantum number,** m_s. The spin quantum number has only two possible values, +1/2 and −1/2. These are related to the so-called spin of an

TABLE 4.2 ELECTRON CAPACITIES OF QUANTUM LEVELS AND SUBLEVELS

Quantum Level n	Orbitals per Sublevel	Orbital Notation	Electrons per Sublevel	Quantum Level Capacity $2n^2$
5	9	5g	18	50
	7	5f	14	
	5	5d	10	
	3	5p	6	
	1	5s	2	
	25			
4	7	4f	14	32
	5	4d	10	
	3	4p	6	
	1	4s	2	
	16			
3	5	3d	10	18
	3	3p	6	
	1	3s	2	
	9			
2	3	2p	6	8
	1	2s	2	
	4			
1	1	1s	2	2

electron about its axis, and of course there can be only two directions of spin. An electron spinning in its orbital creates a magnetic field about the atom. That spin may be either clockwise or counterclockwise, as indicated by the two values of ms, +1/2 and −1/2. In any orbital, regardless of the values of the other three quantum numbers, m_s may assume either of its two possible values.

Because no two electrons in an atom may have an identical set of the four quantum numbers, a maximum of only two electrons can be found in one orbital. In that orbital, the first three quantum numbers are the same, but the spin quantum number has two possible values, so two electrons can occupy an orbital together. The number of sublevels per quantum level, orbitals for the various sublevels, and the electron capacities for the various orbitals, sublevels and quantum levels, are all summarized in Table 4.2. Summing the values in that table shows that the number of orbitals per level is equal to the quantum number for that level squared, n^2, that the electron capacity is equal to $2n^2$.

4.10 ELECTRON CONFIGURATIONS OF THE ELEMENTS

The relative energies of the atomic orbitals are shown in Figure 4.15. As stated in the discussion of the principal quantum number, each succeeding quantum level contains orbitals of higher energy than those in the preceding level. However, as the levels separate into sublevels of differing energies, the **energies of orbitals** in two different quantum levels **overlap** in such a way that the higher orbitals of one quantum level are higher in energy than the lower orbitals of the succeeding quantum level. Knowing the relative energies of the orbitals, one can assign the electron configurations of the atoms of the elements in their usual, lowest energy or ground states. In the assignment the following rules are observed:

1. Electrons are added to the orbitals in the order of increasing energy, beginning with the orbital of lowest energy.
2. Each orbital can hold a maximum of two electrons.
3. Electrons occupy orbitals of any one sublevel singly, unless doing so forces one of a pair into an orbital of higher energy.

The nitrogen atom, with 7 electrons, contains 2 electrons in the first quantum level, 2 electrons in the one orbital of the first sublevel of level 2, and 1 electron in each of the three orbitals of the second sublevel of quantum level 2. The electronic structure of the nitrogen atom is represented in orbital notation as:

$${}_7\text{N}\ 1s^2 2s^2 2p_x^1 2p_y^1 2p_z^1$$

in which the symbols have the meanings indicated below.

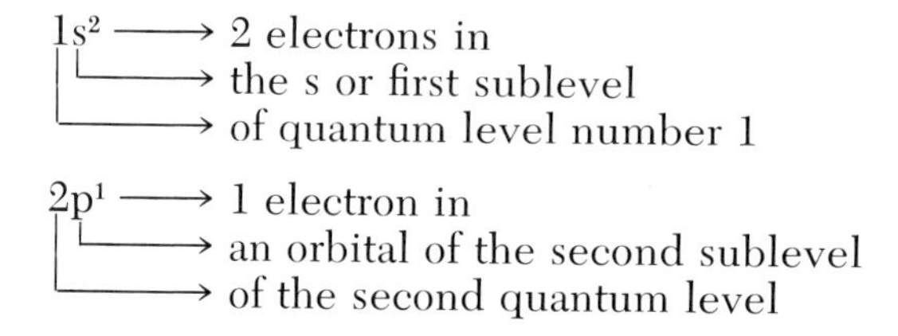

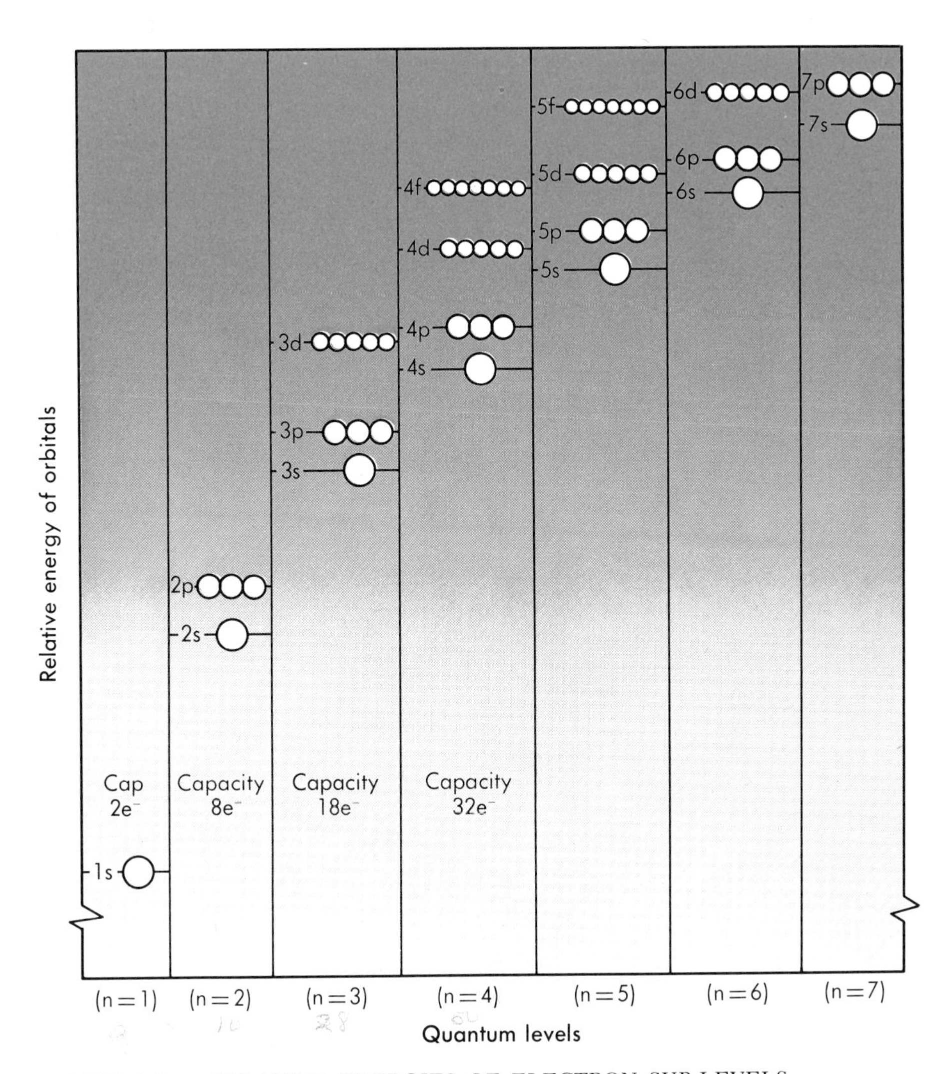

FIGURE 4.15 RELATIVE ENERGIES OF ELECTRON SUB-LEVELS.

The electron structure of the potassium atom, atomic number 19 with 19 electrons, is represented in Figure 4.16. The structure in orbital notation is:

$$_{19}K1s^2 2s^2 2p^6 3s^2 3p^6 4s^1$$

(The total number of electrons in the atom, the subscript, is equal to the sum of all the superscripts in the notation.)

In the potassium atom the first quantum level is filled with 2 electrons ($1s^2$); the second quantum level is filled to capacity with 8 ($2s^2 2p^6$), but the third quantum level contains only 8 electrons ($3s^2 3p^6$) of a total capacity of 18, represented by ($3s^2 3p^6 3d^{10}$). Nevertheless, the 19th electron of potassium enters the 4s sublevel, leaving five d orbitals of the third quantum level vacant. This occurs because electrons fill the orbitals in the order of increasing energy, and the 4s orbital accepts electrons of lower energy than do 3d orbitals. Calcium, with 20 electrons, has the following electron structure:

$$_{20}Ca : 1s^2 2s^2 2p^6 3s^2 3p^6 4s^2$$

Scandium (atomic number 21) has this electron configuration (note the d electron, arrow):

$$_{21}Sc : 1s^2 2s^2 2p^6 3s^2 3p^6 \overset{\downarrow}{3d^1} 4s^2$$

The nine succeeding elements, $_{22}Ti$ to $_{30}Zn$, contain from 2 to 10 electrons in the 3d sublevel. The series of elements from scandium to zinc, in which the "under" quantum level has from 9 to 18 electrons, is called a transition series.

The **valence** electron orbital configurations of the first 20 elements of the periodic table are shown with their symbols in Figure 4.17. The valence electrons are those electrons in the outermost quantum level—those that are involved in chemical bonding (see Chapter 5). In Figure 4.17 the atomic symbols represent the kernels of the atoms, that is, the nucleus and all the electrons except the valence electrons. To complete the atomic symbols, the **outermost** (or valence) electrons are placed as dots about the symbol. In this table paired dots represent electrons which occupy an orbital jointly. However, there is no distinction in the dot formula between s and p orbitals. The notation beneath the symbol for each atom indicates the orbitals in the valence level occupied by electrons. Both the number of valence electrons and the shape of the orbitals occupied by the electrons are important in bonding. The number of electrons in the valence shell corresponds to the group number, the number of the family or group of elements in a vertical column of the periodic table, as shown in Figure 4.18. The group number is shown by a Roman numeral above the group, and the valence orbitals (s, p, d and f) being filled are also designated in Figure 14.18 according to their position in the periodic table.

The electron configurations by sublevel of all the atoms in the ground state are listed in order of increasing atomic number in Table 4.3.

Relative energy of orbitals
1s
2s
2p
3s
3p
3d
4s
4p
4d
5s
5p
Potassium (K)
(n = 1)
(n = 2)
(n = 3)
(n = 4)
(n = 5)
Quantum levels

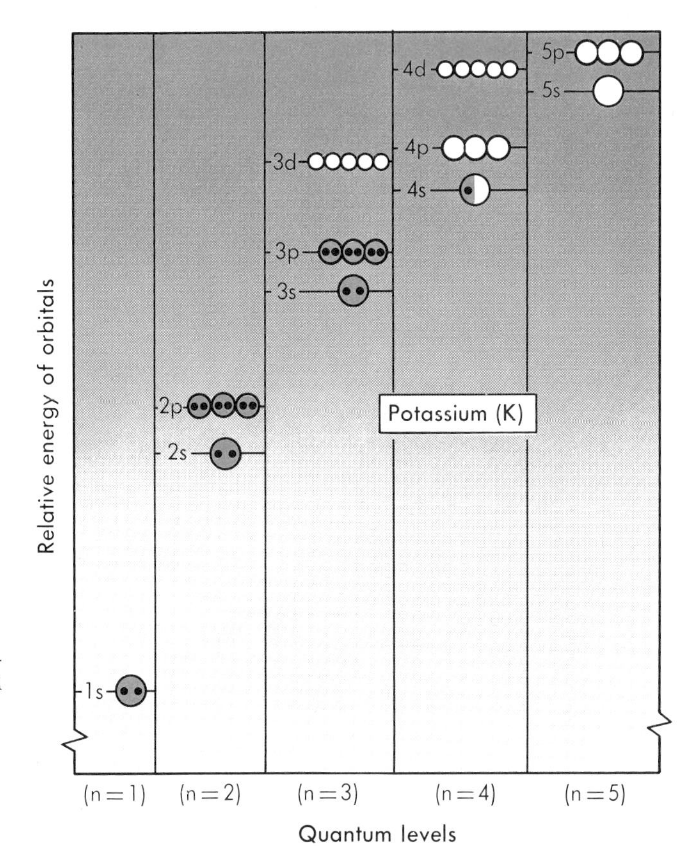

FIGURE 4.16 ELECTRON ORBITAL-ENERGY DIAGRAM FOR POTASSIUM.

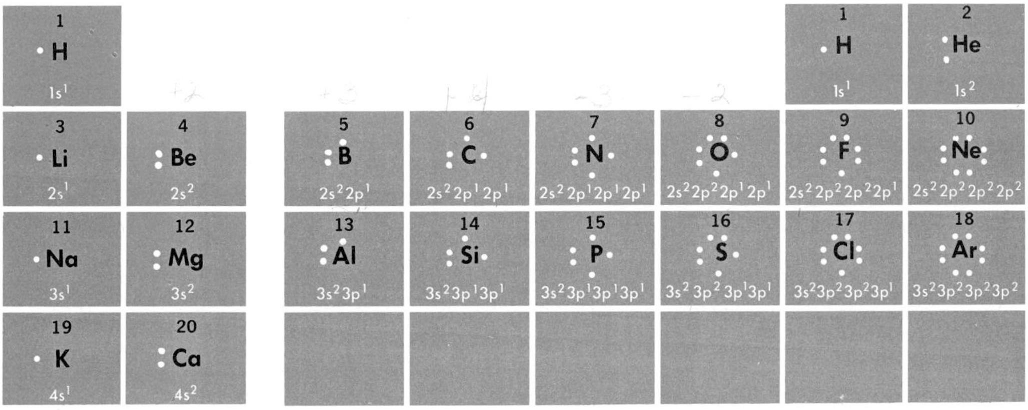

1 H $1s^1$
1 H $1s^1$
2 He $1s^2$
3 Li $2s^1$
4 Be $2s^2$
5 B $2s^2 2p^1$
6 C $2s^2 2p^1 2p^1$
7 N $2s^2 2p^1 2p^1 2p^1$
8 O $2s^2 2p^2 2p^1 2p^1$
9 F $2s^2 2p^2 2p^2 2p^1$
10 Ne $2s^2 2p^2 2p^2 2p^2$
11 Na $3s^1$
12 Mg $3s^2$
13 Al $3s^2 3p^1$
14 Si $3s^2 3p^1 3p^1$
15 P $3s^2 3p^1 3p^1 3p^1$
16 S $3s^2 3p^2 3p^1 3p^1$
17 Cl $3s^2 3p^2 3p^2 3p^1$
18 Ar $3s^2 3p^2 3p^2 3p^2$
19 K $4s^1$
20 Ca $4s^2$

FIGURE 4.17 VALENCE ELECTRON CONFIGURATIONS OF THE FIRST 20 ELEMENTS.

TABLE 4.3 ELECTRON CONFIGURATION OF THE ATOMS

Element	Atomic No.	1	2	3	4	5	6	7
H	1	1						
He	2	2						
Li	3	2	1					
Be	4	2	2					
B	5	2	3					
C	6	2	4					
N	7	2	5					
O	8	2	6					
F	9	2	7					
Ne	10	2	8					
Na	11	2	8	1				
Mg	12	2	8	2				
Al	13	2	8	3				
Si	14	2	8	4				
P	15	2	8	5				
S	16	2	8	6				
Cl	17	2	8	7				
Ar	18	2	8	8				
K	19	2	8	8	1			
Ca	20	2	8	8	2			
Sc	21	2	8	9	2			
Ti	22	2	8	10	2			
V	23	2	8	11	2			
Cr	24	2	8	13	1			
Mn	25	2	8	13	2			
Fe	26	2	8	14	2			
Co	27	2	8	15	2			
Ni	28	2	8	16	2			
Cu	29	2	8	18	1			
Zn	30	2	8	18	2			
Ga	31	2	8	18	3			
Ge	32	2	8	18	4			
As	33	2	8	18	5			
Se	34	2	8	18	6			
Br	35	2	8	18	7			
Kr	36	2	8	18	8			
Rb	37	2	8	18	8	1		
Sr	38	2	8	18	8	2		
Y	39	2	8	18	9	2		
Zr	40	2	8	18	10	2		
Nb	41	2	8	18	12	1		
Mo	42	2	8	18	13	1		
Tc	43	2	8	18	14	1		
Ru	44	2	8	18	15	1		
Rh	45	2	8	18	16	1		
Pd	46	2	8	18	18			
Ag	47	2	8	18	18	1		
Cd	48	2	8	18	18	2		
In	49	2	8	18	18	3		
Sn	50	2	8	18	18	4		
Sb	51	2	8	18	18	5		
Te	52	2	8	18	18	6		
I	53	2	8	18	18	7		
Xe	54	2	8	18	18	8		

TABLE 4.3 ELECTRON CONFIGURATION OF THE ATOMS (CONT'D)

Element	Atomic No.	1	2	3	4	5	6	7
Cs	55	2	8	18	18	8	1	
Ba	56	2	8	18	18	8	2	
La	57	2	8	18	18	9	2	
Ce	58	2	8	18	20	8	2	
Pr	59	2	8	18	21	8	2	
Nd	60	2	8	18	22	8	2	
Pm	61	2	8	18	23	8	2	
Sm	62	2	8	18	24	8	2	
Eu	63	2	8	18	25	8	2	
Gd	64	2	8	18	25	9	2	
Tb	65	2	8	18	26	9	2	
Dy	66	2	8	18	27	9	2	
Ho	67	2	8	18	28	9	2	
Er	68	2	8	18	29	9	2	
Tu	69	2	8	18	31	8	2	
Yb	70	2	8	18	32	8	2	
Lu	71	2	8	18	32	9	2	
Hf	72	2	8	18	32	10	2	
Ta	73	2	8	18	32	11	2	
W	74	2	8	18	32	12	2	
Re	75	2	8	18	32	13	2	
Os	76	2	8	18	32	14	2	
Ir	77	2	8	18	32	15	2	
Pt	78	2	8	18	32	17	1	
Au	79	2	8	18	32	18	1	
Hg	80	2	8	18	32	18	2	
Tl	81	2	8	18	32	18	3	
Pb	82	2	8	18	32	18	4	
Bi	83	2	8	18	32	18	5	
Po	84	2	8	18	32	18	6	
At	85	2	8	18	32	18	7	
Rn	86	2	8	18	32	18	8	
Fr	87	2	8	18	32	18	8	1
Ra	88	2	8	18	32	18	8	2
Ac	89	2	8	18	32	18	9	2
*Th	90	2	8	18	32	19	9	2
*Pa	91	2	8	18	32	20	9	2
*U	92	2	8	18	32	21	9	2
*Np	93	2	8	18	32	22	9	2
*Pu	94	2	8	18	32	23	9	2
*Am	95	2	8	18	32	24	9	2
*Cm	96	2	8	18	32	25	9	2
*Bk	97	2	8	18	32	26	9	2
*Cf	98	2	8	18	32	27	9	2
*Es	99	2	8	18	32	28	9	2
*Fm	100	2	8	18	32	29	9	2
*Md	101	2	8	18	32	30	9	2
*No	102	2	8	18	32	31	9	2
*Lr	103	2	8	18	32	32	9	2
*Ku	104	2	8	18	32	32	10	2
*Ha	105	2	8	18	32	32	11	2

* Probable configurations

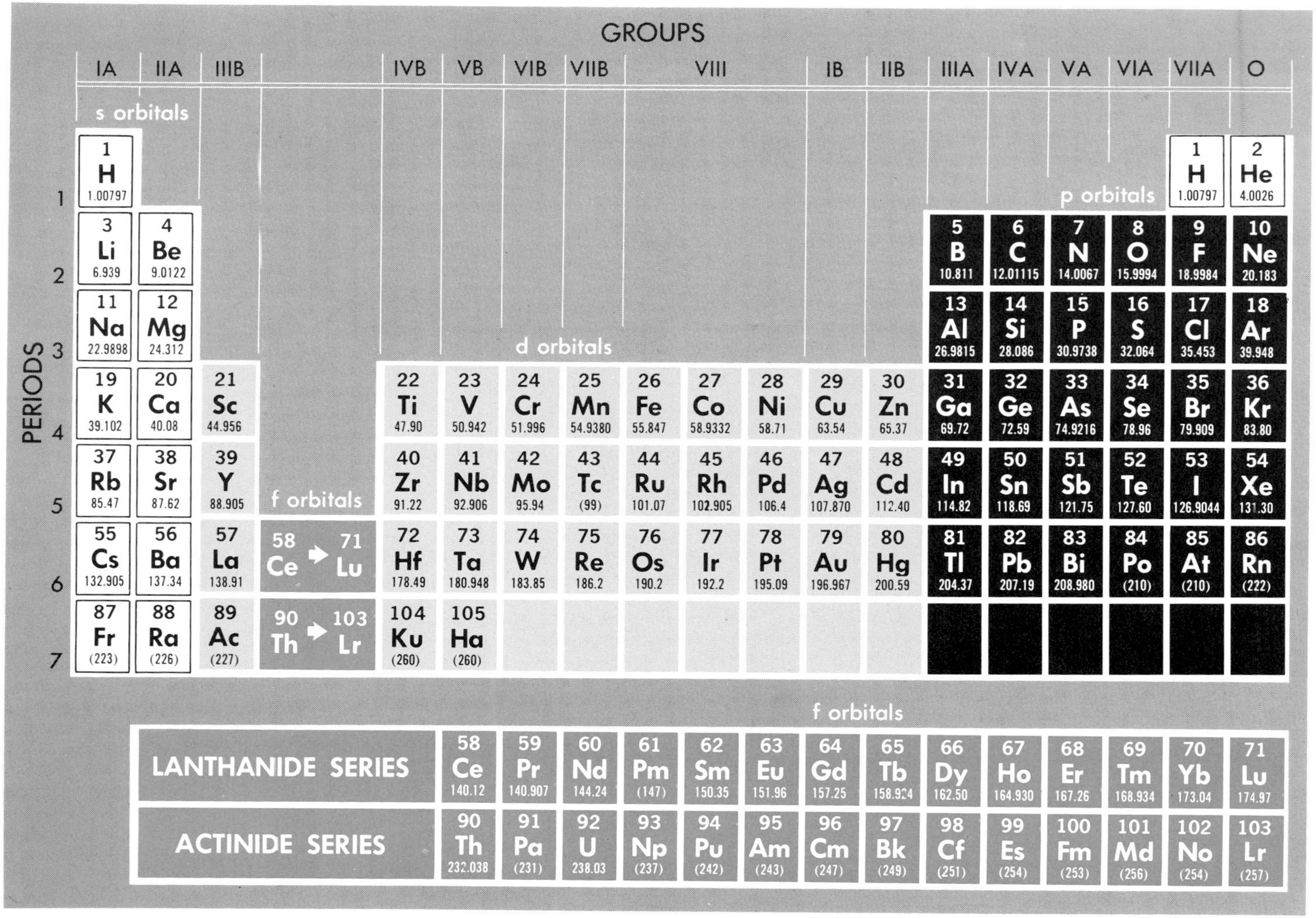

GROUPS

PERIODS	IA	IIA	IIIB		IVB	VB	VIB	VIIB	VIII	VIII	VIII	IB	IIB	IIIA	IVA	VA	VIA	VIIA	O
1	1 H 1.00797																	1 H 1.00797	2 He 4.0026
2	3 Li 6.939	4 Be 9.0122												5 B 10.811	6 C 12.01115	7 N 14.0067	8 O 15.9994	9 F 18.9984	10 Ne 20.183
3	11 Na 22.9898	12 Mg 24.312												13 Al 26.9815	14 Si 28.086	15 P 30.9738	16 S 32.064	17 Cl 35.453	18 Ar 39.948
4	19 K 39.102	20 Ca 40.08	21 Sc 44.956		22 Ti 47.90	23 V 50.942	24 Cr 51.996	25 Mn 54.9380	26 Fe 55.847	27 Co 58.9332	28 Ni 58.71	29 Cu 63.54	30 Zn 65.37	31 Ga 69.72	32 Ge 72.59	33 As 74.9216	34 Se 78.96	35 Br 79.909	36 Kr 83.80
5	37 Rb 85.47	38 Sr 87.62	39 Y 88.905		40 Zr 91.22	41 Nb 92.906	42 Mo 95.94	43 Tc (99)	44 Ru 101.07	45 Rh 102.905	46 Pd 106.4	47 Ag 107.870	48 Cd 112.40	49 In 114.82	50 Sn 118.69	51 Sb 121.75	52 Te 127.60	53 I 126.9044	54 Xe 131.30
6	55 Cs 132.905	56 Ba 137.34	57 La 138.91	58 Ce → 71 Lu	72 Hf 178.49	73 Ta 180.948	74 W 183.85	75 Re 186.2	76 Os 190.2	77 Ir 192.2	78 Pt 195.09	79 Au 196.967	80 Hg 200.59	81 Tl 204.37	82 Pb 207.19	83 Bi 208.980	84 Po (210)	85 At (210)	86 Rn (222)
7	87 Fr (223)	88 Ra (226)	89 Ac (227)	90 Th → 103 Lr	104 Ku (260)	105 Ha (260)													

s orbitals · d orbitals · p orbitals · f orbitals

LANTHANIDE SERIES	58 Ce 140.12	59 Pr 140.907	60 Nd 144.24	61 Pm (147)	62 Sm 150.35	63 Eu 151.96	64 Gd 157.25	65 Tb 158.924	66 Dy 162.50	67 Ho 164.930	68 Er 167.26	69 Tm 168.934	70 Yb 173.04	71 Lu 174.97
ACTINIDE SERIES	90 Th 232.038	91 Pa (231)	92 U 238.03	93 Np (237)	94 Pu (242)	95 Am (243)	96 Cm (247)	97 Bk (249)	98 Cf (251)	99 Es (254)	100 Fm (253)	101 Md (256)	102 No (254)	103 Lr (257)

FIGURE 4.18 ARRANGEMENT OF THE PERIODIC TABLE ACCORDING TO ATOMIC ORBITALS.

4.11 ELECTRONEGATIVITY

electronegativity values are approximate at best. at times the values calculated for the same element vary somewhat when calculated from different compounds. despite the inaccuracies, the concept is very useful.

The type of **bond** to be formed between two atoms (See Chapter 5) is influenced by the **electronegativity** of the two atoms. This is the relative attraction each atom has for the bonding electrons. Through calculations using bond energies, which are taken up in Chapter 5, Linus Pauling has arrived at a scale of relative electronegativities for the elements. Electronegativity values are shown in Figure 4.19, where they are listed below the symbols of the elements in the periodic table. The electronegativities increase from element to element upward through most groups and across each period from left to right in the table. As a consequence of the two trends, the electronegativity increases from a minimum of 0.7 at cesium to a maximum of 4.0 for fluorine. These trends are not observed in the series of transition elements in which the electronegativity varies little from element to element.

An approximate scale of electronegativity can be gained by considering the sizes of atoms. Because the occupied orbitals of highest principal quantum number extend the greatest distance from the nucleus, and because the sizes of those orbitals depend greatly upon the attraction of the nucleus for the electrons in the orbitals, the smallness of the atom radius is a measure of the magnitude of the attraction of the atom for those electrons in its outermost orbitals. That is, the smaller the radius of an atom, the greater the attraction of an atom for the electrons in its valence level which holds the electrons that participate in bonding.

The atomic radii of many of the elements are shown in Figure 4.20 with the atomic symbols arranged in their positions in the periodic table. According to the table, the atomic radii of the atoms increase from top to bottom in most groups of the periodic table. This is to be expected, because each succeeding element in the group has one more filled or partially filled shell of electrons under

1 H 2.1																	1 H 2.1	2 He —
3 Li 1.0	4 Be 1.5												5 B 2.0	6 C 2.5	7 N 3.0	8 O 3.5	9 F 4.0	10 Ne —
11 Na 0.9	12 Mg 1.2												13 Al 1.5	14 Si 1.8	15 P 2.1	16 S 2.5	17 Cl 3.0	18 Ar —
19 K 0.8	20 Ca 1.0	21 Sc 1.3		22 Ti 1.5	23 V 1.6	24 Cr 1.6	25 Mn 1.5	26 Fe 1.8	27 Co 1.8	28 Ni 1.8	29 Cu 1.9	30 Zn 1.6	31 Ga 1.6	32 Ge 1.8	33 As 2.0	34 Se 2.4	35 Br 2.8	36 Kr —
37 Rb 0.8	38 Sr 1.0	39 Y 1.3		40 Zr 1.4	41 Nb 1.6	42 Mo 1.8	43 Tc 1.9	44 Ru 2.2	45 Rh 2.2	46 Pd 2.2	47 Ag 1.9	48 Cd 1.7	49 In 1.7	50 Sn 1.8	51 Sb 1.9	52 Te 2.1	53 I 2.5	54 Xe —
55 Cs 0.7	56 Ba 0.9	57 La 1.1	58 Ce ➔ 71 Lu	72 Hf 1.3	73 Ta 1.5	74 W 1.7	75 Re 1.9	76 Os 2.2	77 Ir 2.2	78 Pt 2.2	79 Au 2.4	80 Hg 1.9	81 Tl 1.8	82 Pb 1.8	83 Bi 1.9	84 Po 2.0	85 At 2.2	86 Rn —
87 Fr 0.7	88 Ra 0.9	89 Ac 1.1	90 Th ➔ 103 Lr	104 Ku	105 Ha													

FIGURE 4.19 ELECTRONEGATIVITY TREND IN THE PERIODIC TABLE (ELEMENTS IN THE DARKER AREA ARE MORE ELECTRONEGATIVE).

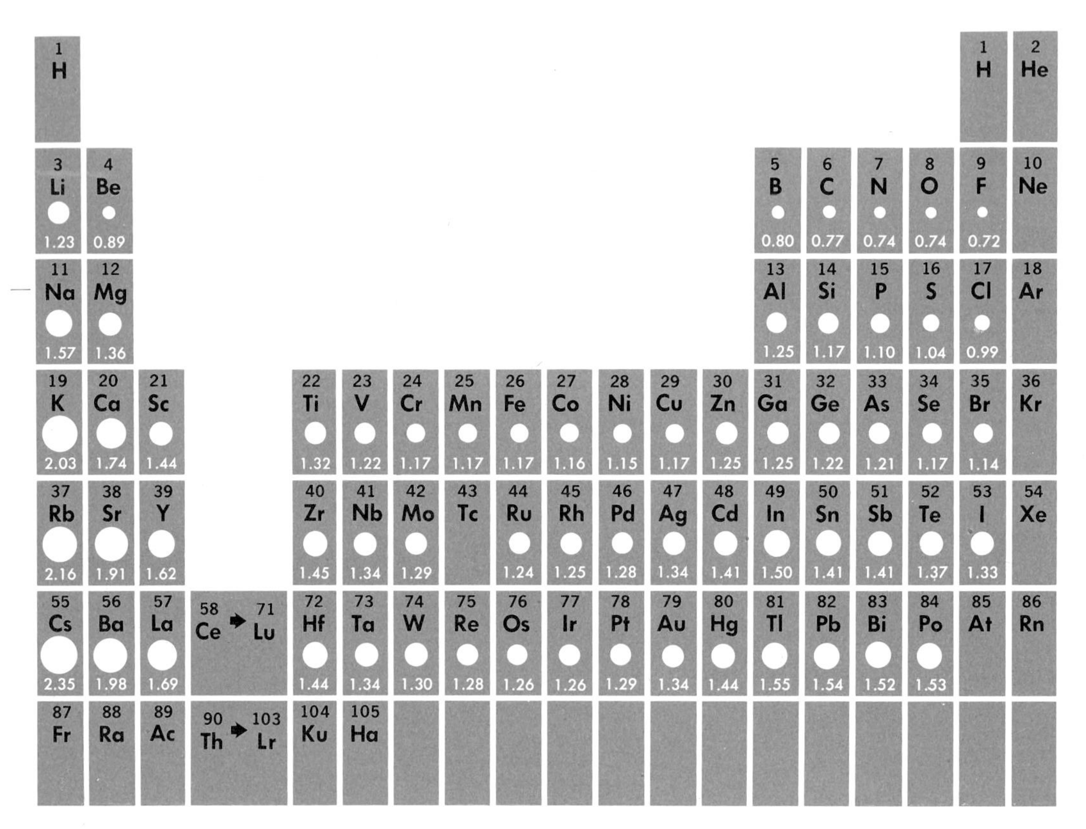

FIGURE 4.20 ATOMIC RADII IN ANGSTROMS.

the electrons of the outermost shell, the valence electrons. Also, as the atomic number increases, the atomic radii decrease from left to right in the periods of eight in the table. For example, the elements from sodium to chlorine inclusive have the following atomic radii, in order: 1.57, 1.36, 1.25, 1.17, 1.10, 1.04 and 0.99 Å. The trend might be explained by the fact that each atom of the period has the same kernel of electrons but that the positive charge on the kernel increases from unity at sodium to seven at chlorine. With the increased charge, the valence electrons are held more tightly and more closely to the nucleus.

Electronegativity increases with the decrease in the size of the atoms. This is generally true in the groups and periods of the periodic table. For as the atomic radii of the elements from lithium to fluorine decrease from 1.23 Å to 0.72 Å, the electronegativity increases from 1.0 to 4.0. Similarly, while the atomic radii of the group of elements from lithium to cesium increase regularly from 1.23 Å to 2.35 Å, the electronegativity decreases from 1.0 to 0.7. This relationship is shown in Figure 4.21, where the atomic radii are plotted against the electronegativity for the elements of Groups I, II and VII and periods 2 and 3 of the periodic table. While within a group or period the electronegativity decreases as the atomic radius increases, there is some difference among elements of different groups or periods. For example, nitrogen and chlorine are both assigned an electronegativity of 3.0, though nitrogen has an atomic radius of 0.74 Å and chlorine a radius of 0.99 Å. Apparently the larger kernels of elements in succeeding periods of the periodic table do not allow the valence electrons to approach the nucleus as closely as do the smaller kernels in preceding periods. Despite the slight disagreement, the size of an atom is useful in indicating the atom's approximate electronegativity.

Variations in the first ionization energy (sometimes called "ionization potential") of the atoms agree generally with the variations in the assigned electronegativity of the elements. That is, as the electronegativity of the elements increases, so does the ionization energy. The ionization energy is defined as the energy required to remove one electron from an atom in the vapor state. The ionization energy is measured by bombarding the vapor of the element at low pressure with electrons at gradually increasing energy, and noting the applied potential at which ionization starts.

Electrons are shot into the vapor from cathode to anode with a gradually increasing voltage while the current through the tube is being monitored. At first, collisions of the cathode rays (electrons) with atoms merely raise the electrons to higher orbitals within the atoms, and the gas glows as the electrons fall back. Eventually, with the increasing voltage, the cathode rays gain sufficient energy to remove electrons completely from the atoms. At this point the dislodged electrons and the positive ions formed assist in carrying the electric current through the gas in the tube, and a sharp

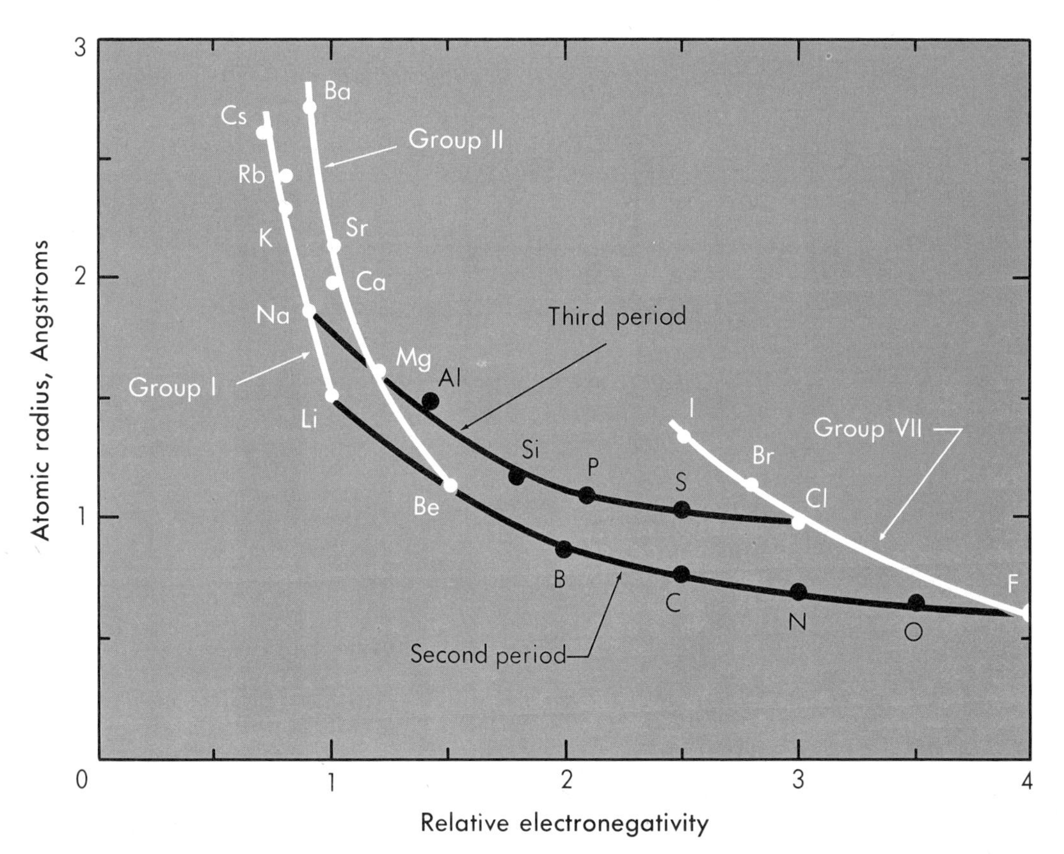

FIGURE 4.21 CORRELATION OF ATOMIC RADII AND ELECTRONEGATIVITIES.

TABLE 4.4 IONIZATION ENERGIES – ELECTRONEGATIVITIES AND ATOMIC RADII FOR ELEMENTS IN PERIODS 2 AND 3

	Li	Be	B	C	N	O	F
Electronegativity	1.0	1.5	2.0	2.5	3.0	3.5	4.0
Radius Å	1.23	0.889	0.80	0.771	0.74	0.74	0.72
Ionization Energies (kcal/mole)	124	215	191	260	336	314	402
	Na	**Mg**	**Al**	**Si**	**P**	**S**	**Cl**
Electronegativity	0.9	1.2	1.5	1.8	2.1	2.5	3.0
Radius Å	1.57	1.36	1.25	1.17	1.10	1.04	0.994
Ionization Energies (kcal/mole)	119	176	138	188	254	239	300

increase in the current is observed. The voltage at which the current increases is called the ionization potential. Because the voltage across the tube is directly proportional to the energy of the electrons, that voltage is listed in electron volts (eV) as the ionization energy. Conversion to kcal per mole is made by multiplying the number of electron volts by the factor 23.1 kcal per mole eV. The first ionization energies (kcal/mole) of the elements in periods 2 and 3 are given in Table 4.4 with the respective atomic radii and electronegativity values. The ionization energies for most of the elements are shown in Figure 4.22. While there are exceptions, the ionization energy increases from left to right in a particular period as the size of the atoms decreases and the electronegativity increases. In a particular group of the table, the ionization energy and electronegativity decrease from top to bottom while the atomic radius increases.

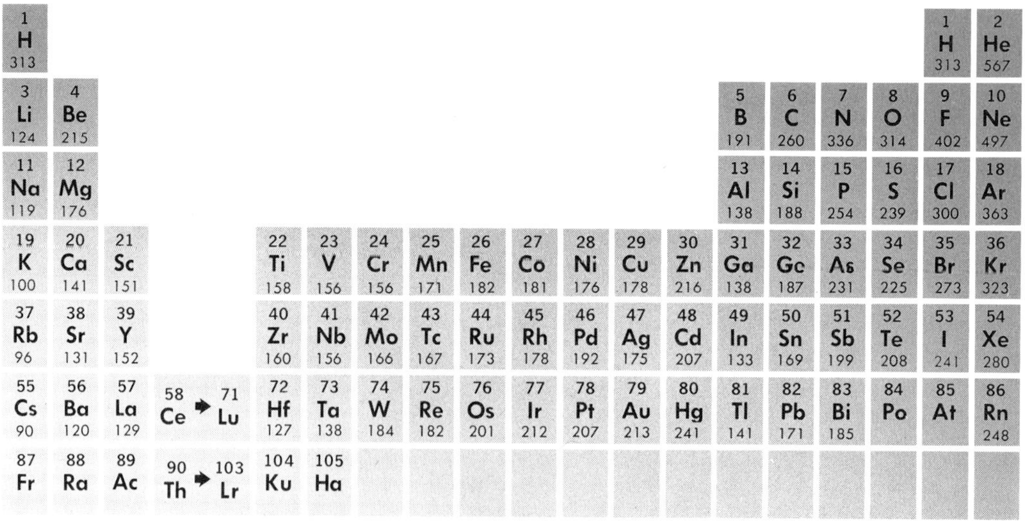

1 H 313																	1 H 313	2 He 567
3 Li 124	4 Be 215												5 B 191	6 C 260	7 N 336	8 O 314	9 F 402	10 Ne 497
11 Na 119	12 Mg 176												13 Al 138	14 Si 188	15 P 254	16 S 239	17 Cl 300	18 Ar 363
19 K 100	20 Ca 141	21 Sc 151		22 Ti 158	23 V 156	24 Cr 156	25 Mn 171	26 Fe 182	27 Co 181	28 Ni 176	29 Cu 178	30 Zn 216	31 Ga 138	32 Ge 187	33 As 231	34 Se 225	35 Br 273	36 Kr 323
37 Rb 96	38 Sr 131	39 Y 152		40 Zr 160	41 Nb 156	42 Mo 166	43 Tc 167	44 Ru 173	45 Rh 178	46 Pd 192	47 Ag 175	48 Cd 207	49 In 133	50 Sn 169	51 Sb 199	52 Te 208	53 I 241	54 Xe 280
55 Cs 90	56 Ba 120	57 La 129	58 Ce → 71 Lu	72 Hf 127	73 Ta 138	74 W 184	75 Re 182	76 Os 201	77 Ir 212	78 Pt 207	79 Au 213	80 Hg 241	81 Tl 141	82 Pb 171	83 Bi 185	84 Po	85 At	86 Rn 248
87 Fr	88 Ra	89 Ac	90 Th → 103 Lr	104 Ku	105 Ha													

FIGURE 4.22 TREND IN IONIZATION ENERGY (ELEMENTS IN DARKER AREAS HAVE THE GREATEST IONIZATION ENERGIES; ENERGIES GIVEN IN KCAL/MOLE OF ATOMS).

4.12 PARAMAGNETISM

In 1921 Stern and Gerlach passed a stream of silver vapor, issuing from boiling silver, between the poles of a strong magnet and observed a split in the stream of atoms. Instead of passing directly through the magnetic field to collect in one spot on a cold metal surface, the stream of atoms split into two streams which separated slightly but definitely to form two spots on the collecting surface, as shown in Figure 4.23. Stern and Gerlach reasoned that separation into two paths occurred because in each silver atom there is an unpaired electron, as seen in the electron notation for silver. Using the symbol of krypton to represent all the electrons in that atom, the electron configuration for silver is:

$$_{47}Ag[Kr]4d^{10}5s^{1}$$

A spinning electron creates a magnetic field, the direction of the magnetic field being dependent upon the direction of the spin, clockwise or counterclockwise. When there are two electrons in an orbital, one must spin clockwise (have a spin quantum number of +1/2) while the other spins counterclockwise (has a spin quantum number of −1/2). With the two electrons in the same orbital spinning in **opposite** directions, their respective fields cancel and there is no resultant magnetic behavior. However, when only one electron is present in an orbital, the electron may spin either clockwise or counterclockwise with equal probability. In passing through a magnetic field, atoms with opposing spins will move equally and oppositely. Any element with atoms containing one or more orbitals occupied by a single electron exhibits a tendency to move to a region of lower magnetic field intensity when passing through a magnetic field (unless the electrons happen to spin oppositely in the case of an even number of single electrons). This tendency is known as paramagnetism.

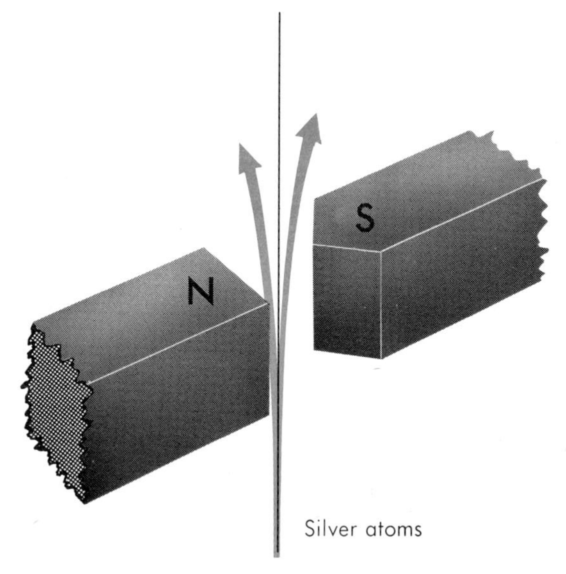

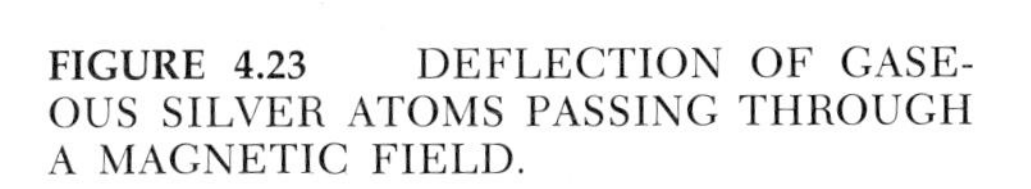
FIGURE 4.23 DEFLECTION OF GASEOUS SILVER ATOMS PASSING THROUGH A MAGNETIC FIELD.

EXERCISES

4.1 What carries the electric current through a cathode ray tube?

4.2 What experimental evidence is there that cathode rays are not light rays?

4.3 What are gamma rays?

4.4 Describe the kinds of rays emitted by radium.

4.5 What happens to the alpha particle lost by radium?

4.6 Describe the principle of operation of the picture tube in a television set.

4.7 What is an ion?

4.8 What evidence is there that positive ions are present in a fluorescent tube?

4.9 How is the mass of the proton determined?

4.10 What evidence is there that the proton is a fundamental unit of all atoms?

4.11 Why should a neutron be more penetrating (approach closer to the nuclei of atoms) than either alpha particles or protons?

4.12 What is the composition of the alpha particle in terms of protons and neutrons?

4.13 Give the reasoning behind the conclusion Rutherford reached from the results of his alpha particle–gold foil experiment.

4.14 What other supporting evidence is there that electrons are outside the nucleus and therefore more loosely bound than protons and neutrons?

4.15 Give the compositions of the following atoms in protons, neutrons and electrons: (a) $^{16}_{8}O$, (b) $^{37}_{17}Cl$, (c) $^{35}_{17}Cl$, (d) the more abundant isotope of calcium, (e) the more abundant isotope of bismuth. See the table of atomic weights on the inside back cover.

4.16 Write the notation for the most abundant isotope of each element below, as is done here for the most abundant isotope of oxygen: $^{16}_{8}O$. (a) helium, (b) sodium, (c) cesium, (d) potassium, (e) vanadium, (f) uranium, (g) radon.

4.17 Define isotope.

4.18 Give the total number of electrons that will fill each of the first 5 electron energy levels of an atom.

4.19 List the visible light colors in the order of increasing energy of the photons.

4.20 Define an orbital. How many electrons may occupy a single orbital at one time?

4.21 Define the term "transition element."

4.22 Why are there never more than 8 electrons in the outermost quantum level of an atom in the lowest energy state?

4.23 Why should magnesium and calcium be more alike in chemical properties than potassium and calcium?

4.24 What similarity in electron structure exists among the following atoms: oxygen, sulfur, tellurium and selenium?

4.25 Write one of the many possible electron configurations for a magnesium atom that might momentarily emit a photon.

4.26 Why does the 19th electron of potassium enter the fourth quantum level when the third is incomplete?

4.27 Radium, with a mass of 226, emits an alpha particle. Show the atomic species formed using this type of notation: ${}^{16}_{8}O$.

4.28 Where does the alpha particle ejected from the radium nucleus gain the necessary electrons to become a helium gas atom?

4.29 A radium atom, ${}^{226}_{88}Ra$, may decay to ${}^{206}_{82}Pb$ by the stepwise emission of these particles in this order: alpha, alpha, alpha, beta, beta, alpha, beta, beta and alpha. Indicate all atomic species that occur in the decay sequence by the type notation used for radium and lead.

4.30 Write the electron subshell configuration for the following atoms: Fe, O, Br, Mg, N, S, F and Sc. Example: ${}_{17}Cl$: $1s^2 2s^2 2p^6 3s^2 3p^5$.

4.31 How does an orbital differ from an orbit?

4.32 Define electronegativity. Using the table, show how electronegativity varies among elements in the periodic table.

SUGGESTED READING

Clark, H. M.: "The Origin of Nuclear Science." Chemistry, *40*(7):8, 1967.

Masterton, W. L. and Slowinski, E. J.: *Chemical Principles*, 2nd Ed. W. B. Saunders Company, Philadelphia, 1969. Chapter 7.

Mulliken, R. S.: "Electrons – What They Are and What They Do." Chemistry, *40*(4):13, 1967.

Sienko, M. J. and Plane, R. A.: *Chemistry*, 3rd Ed. McGraw-Hill Book Co., New York, 1966. Chapter 3.

Sisler, H. H.: *Electronic Structure, Properties, and the Periodic Law*. Reinhold Publishing Corp., New York, 1962 (paperback).

FIVE • CHEMICAL BONDING

5.1 PERIODIC TABLE

In the last chapter the elements of the periodic table were considered in the order and position they occupy in the table. They were seen to be listed according to increasing atomic number in periods of 2, 8, 8 and 18, one below the other, such that elements with the same number of electrons in the outermost level fall in the same vertical column or group.

historical development of the relationship between atomic weights and properties.

To avoid the erroneous assumption that a knowledge of electronic structure led chemists to arrange the elements in a systematic table, it should be noted that electrons were not recognized until more than 30 years after the table was in use. At the time the periodic table was first proposed, the only picture of atomic structure available was one proposed by Prout, who believed that all atomic weights were integral multiples of the atomic weight of hydrogen. In accord with this hypothesis, all atoms were composed of an integral number of hydrogen atoms. This mistaken idea of structure, however, had little to do with the development of the periodic table. Instead the table was conceived solely from studies of the properties of the elements.

In 1829, Dobereiner suggested that like elements occurred in groups of three, which he called "triads." He further noted that within these groups either the elements had very nearly the same atomic weight or that the atomic weight of one element was midway between those of the other two. The atomic weights of the triad *iron, cobalt* and *nickel* are very nearly the same. On the other hand, lithium, sodium and potassium, calcium, strontium and barium, and chlorine, bromine and iodine are similar groups in which the middle element has an atomic weight midway between the other two.

In 1852, Dumas observed that like elements differ in atomic weight by some multiple of 8. For example, consider the atomic weights of lithium (7), sodium (23) and potassium (39). They differ by steps of 16 units.

Thirteen years later, Newlands saw a further relationship between atomic weights and chemical properties. He noted that when the elements are listed in the order of increasing atomic weights, they fall in groups of seven, each group repeating one by one the properties of the first seven. Not considering hydrogen, he listed the elements shown in Figure 5.1.

FIGURE 5.1 NEWLAND'S OCTAVES.

Li	Be	B	C	N	O	F
Na	Mg	Al	Si	P	S	Cl
K	Ca	Cr	Ti	Mn	Fe	Co Ni

In Newlands' table the atomic weights increase through each row, reading left to right, and the properties are repeated; that is, sodium and potassium have properties similar to those of lithium. Since Newlands, another element scandium (Sc) has been found between calcium (Ca) and titanium (Ti). However, even when the elements are moved to the right one space, beginning at titanium, the properties do not match those of the elements above. Newlands' theory does not hold among elements of greater atomic weight than calcium. His successive groups of seven elements, with the eighth repeating properties of the first, were likened to the musical scale and were called "Newlands' Octaves." The noble gases were not known then and so were not included in his table.

Mendeleev and Meyer propose the periodic classification.

Newlands had taken a great step toward constructing the periodic table, but it remained for Dimitri Mendeleev and Lothar Meyer, working independently of each other, to propose the table in much the same form as we know it today. Both men published their tables the same year, 1869, but Mendeleev arrived at his from a study of **chemical properties,** while Meyer used **physical properties.** Meyer's periodic table, which is very similar to Mendeleev's, appears in Table 5.1. Their tables are remarkably similar to the modern version of the periodic table (Figure 5.2), considering the fact that only about 57 of the 105 elements known today had been discovered. Furthermore, they knew nothing of atomic structure and had only very elementary theories of chemical bonding. Mendeleev is usually credited with the development of the table. In April of 1969, the American Chemical Society celebrated the centennial anniversary of the appearance of Mendeleev's table. At first the tables were not accepted, but when Mendeleev predicted the properties for three then undiscovered elements, which proved to be correct upon their discovery, he gained many supporters.

Changes in the table since that time have consisted mainly of additions. However, in some few instances in which the atomic weights do not follow the atomic numbers, the positions of some

GROUPS

PERIODS	IA	IIA	IIIB		IVB	VB	VIB	VIIB	VIII	VIII	VIII	IB	IIB	IIIA	IVA	VA	VIA	VIIA	O
1	1 H 1.00797																	1 H 1.00797	2 He 4.0026
2	3 Li 6.939	4 Be 9.0122												5 B 10.811	6 C 12.01115	7 N 14.0067	8 O 15.9994	9 F 18.9984	10 Ne 20.183
3	11 Na 22.9898	12 Mg 24.312												13 Al 26.9815	14 Si 28.086	15 P 30.9738	16 S 32.064	17 Cl 35.453	18 Ar 39.948
4	19 K 39.102	20 Ca 40.08	21 Sc 44.956		22 Ti 47.90	23 V 50.942	24 Cr 51.996	25 Mn 54.9380	26 Fe 55.847	27 Co 58.9332	28 Ni 58.71	29 Cu 63.54	30 Zn 65.37	31 Ga 69.72	32 Ge 72.59	33 As 74.9216	34 Se 78.96	35 Br 79.909	36 Kr 83.80
5	37 Rb 85.47	38 Sr 87.62	39 Y 88.905		40 Zr 91.22	41 Nb 92.906	42 Mo 95.94	43 Tc (99)	44 Ru 101.07	45 Rh 102.905	46 Pd 106.4	47 Ag 107.870	48 Cd 112.40	49 In 114.82	50 Sn 118.69	51 Sb 121.75	52 Te 127.60	53 I 126.9044	54 Xe 131.30
6	55 Cs 132.905	56 Ba 137.34	57 La 138.91	58 Ce → 71 Lu	72 Hf 178.49	73 Ta 180.948	74 W 183.85	75 Re 186.2	76 Os 190.2	77 Ir 192.2	78 Pt 195.09	79 Au 196.967	80 Hg 200.59	81 Tl 204.37	82 Pb 207.19	83 Bi 208.980	84 Po (210)	85 At (210)	86 Rn (222)
7	87 Fr (223)	88 Ra (226)	89 Ac (227)	90 Th → 103 Lr	104 Ku (260)	105 Ha (260)	106	107											

LANTHANIDE SERIES	58 Ce 140.12	59 Pr 140.907	60 Nd 144.24	61 Pm (147)	62 Sm 150.35	63 Eu 151.96	64 Gd 157.25	65 Tb 158.924	66 Dy 162.50	67 Ho 164.930	68 Er 167.26	69 Tm 168.934	70 Yb 173.04	71 Lu 174.97
ACTINIDE SERIES	90 Th 232.038	91 Pa (231)	92 U 238.03	93 Np (237)	94 Pu (242)	95 Am (243)	96 Cm (247)	97 Bk (249)	98 Cf (251)	99 Es (254)	100 Fm (253)	101 Md (256)	102 No (254)	103 Lr (257)

FIGURE 5.2 THE PERIODIC TABLE.

TABLE 5.1 LOTHAR MEYER'S PERIODIC TABLE

I	II	III	IV	V	VI	VII	VIII	IX
	B 11.0	Al 27.3				?In 113.4		Tl 202.7
	C 11.97	Si 28				Sn 117.8		Pb 207.4
			Ti 48		Zr 89.7			
	N 14.01	P 30.9		As 74.9		Sb 122.1		Bi 207.5
			V 51.2		Nb 93.7		Ta 182.2	
	O 15.96	S 31.98		Se 78		Te 128?		
			Cr 52.4		Mo 95.6		W 183.5	
	F 19.1	Cl 35.38		Br 79.75		I 126.5		
			Mn 54.8		Ru 103.5		Os 198.6	
			Fe 55.9		Rh 104.1		Ir 196.7	
			Co=Ni 58.6		Pd 106.2		Pt 196.7	
Li 7.01	Na 22.99	K 39.04		Rb 85.2		Cs 132.7		
			Cu 63.3		Ag 107.66		Au 196.2	
?Be 9.3	Mg 23.9	Ca 39.9		Sr 87.0		Ba 136.8		
			Zn 64.9		Cd 111.6		Hg 199.8	

modern periodic table arranged in atomic number sequence rather than in atomic weight order.

elements have been changed. In the modern table, elements are arranged in *periods* in the order of increasing atomic number. The properties vary continuously from left to right in those periods. Further, the periods are of such length that similar elements fall in the same vertical column or *group*. The success of the use of the periodic table in classifying an element by its position depends upon the **periodic dependence of properties upon the atomic number.**

5.2 THE NONBONDERS—NOBLE GASES

An interesting paradox exists in the fact that an understanding of the nature of chemical bonding can be aided by studying the particular elements that form few chemical bonds. These are the so-

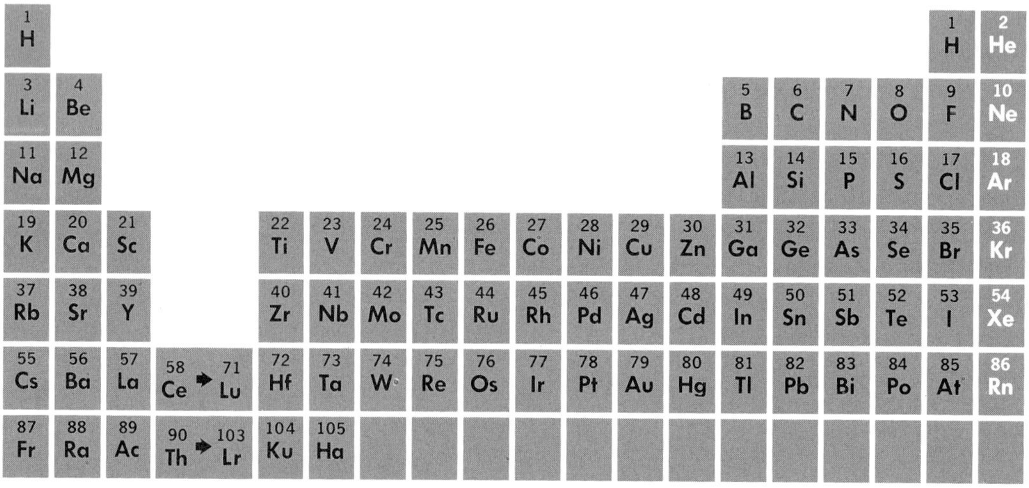

1 H																	1 H	2 He
3 Li	4 Be												5 B	6 C	7 N	8 O	9 F	10 Ne
11 Na	12 Mg												13 Al	14 Si	15 P	16 S	17 Cl	18 Ar
19 K	20 Ca	21 Sc		22 Ti	23 V	24 Cr	25 Mn	26 Fe	27 Co	28 Ni	29 Cu	30 Zn	31 Ga	32 Ge	33 As	34 Se	35 Br	36 Kr
37 Rb	38 Sr	39 Y		40 Zr	41 Nb	42 Mo	43 Tc	44 Ru	45 Rh	46 Pd	47 Ag	48 Cd	49 In	50 Sn	51 Sb	52 Te	53 I	54 Xe
55 Cs	56 Ba	57 La	58 Ce → 71 Lu	72 Hf	73 Ta	74 W	75 Re	76 Os	77 Ir	78 Pt	79 Au	80 Hg	81 Tl	82 Pb	83 Bi	84 Po	85 At	86 Rn
87 Fr	88 Ra	89 Ac	90 Th → 103 Lr	104 Ku	105 Ha													

FIGURE 5.3 THE NOBLE GASES.

called noble gases, the elements listed in the last group of the periodic table and indicated in Figure 5.3.

These elements are "new" elements, neither known nor suspected when the periodic table was first constructed. Argon, helium, neon, krypton and xenon were identified in the last 10 years of the nineteenth century. Radon was discovered during the next 10-year period.

Argon was the first identified. In 1785, Cavendish mixed oxygen with air and passed a spark through it repeatedly until no further reaction took place. The reactions are identical to those occurring when lightning passes through air: First nitrogen(II) oxide is produced by the electric discharge. Subsequent reaction of NO with oxygen produces nitrogen dioxide.

$$N_2 + O_2 \rightarrow 2\,NO$$

$$2NO + O_2 \rightarrow 2NO_2$$

The nitrogen dioxide was then dissolved in potassium hydroxide solution by bubbling the gas through the solution. Excess oxygen was removed by passing the rest of the gas over hot copper.

$$2\,Cu + O_2 \rightarrow 2\,CuO$$

Cavendish reported that a small volume of gas remained after removing the oxygen, amounting to about 1/120th of the volume of his original sample of air. Apparently he never guessed that he had isolated a new element.

In 1894, Lord Rayleigh prepared nitrogen gas by two different methods: by removing oxygen and carbon dioxide from dry air, and by decomposing ammonia, NH_3. He found that nitrogen from air had a density of 1.2572g/L, while nitrogen from ammonia had a density of 1.2506. Rayleigh knew he had not made an error as large as the difference between the two. He reasoned correctly that the atmosphere must contain an unknown gas of density greater than nitrogen. With Sir William Ramsay, he investigated large quantities of air, removing the oxygen by passing the gas over hot copper and removing the nitrogen by passing it over hot magnesium.

$$3\,Mg + N_2 \rightarrow Mg_3N_2$$

Magnesium nitride

Again, as in Cavendish's experiments, a small residue of gas remained. The gas was identified spectroscopically and named argon ("the lazy one"). It is interesting that Cavendish's 1/120th part agrees well with the 0.9 per cent argon in air.

Helium was found in the sun's atmosphere nearly 30 years before it was isolated and identified on earth. In 1868, when a spectroscope was first used during a solar eclipse, the spectrum of sunlight revealed the presence of a series of yellow lines. These lines were different from the yellow spectral lines of any known element. It was concluded that the lines were emitted by some new element glowing in the sun's heat. This unknown element was promptly named helium ("the sun"). The "ium" ending indicates that the new element was thought to be a metal.

In 1891, Hillebrand, an American chemist, observed that certain minerals contained small quantities of an unreactive gas. He supposed that it might be nitrogen and dismissed any further considerations. Ramsay reinvestigated this unreactive gas, and in 1895 showed it to be a new element. He further showed it to have the spectral lines observed for helium in the sun's spectrum in 1868. The new element fell heir to that name.

With the finding of both helium and argon and the placing of them next to hydrogen and chlorine in the periodic table (Figure 5.4), it was apparent to Ramsay that three or four more elements should exist to fill the spaces numbered in the table. With this in view, he conducted an intensive investigation of air, and in 1898 with Travers he found three of the group: neon, krypton and xenon.

Near the turn of the century it was found that not only helium

1 H																	1 H	2 He
3 Li	4 Be												5 B	6 C	7 N	8 O	9 F	1
11 Na	12 Mg												13 Al	14 Si	15 P	16 S	17 Cl	18 Ar 39.948
19 K	20 Ca	21 Sc		22 Ti	23 V	24 Cr	25 Mn	26 Fe	27 Co	28 Ni	29 Cu	30 Zn	31 Ga	32 Ge	33 As	34 Se	35 Br	2
37 Rb	38 Sr	39 Y		40 Zr	41 Nb	42 Mo	43 Tc	44 Ru	45 Rh	46 Pd	47 Ag	48 Cd	49 In	50 Sn	51 Sb	52 Te	53 I	3
55 Cs	56 Ba	57 La	58 Ce → 71 Lu	72 Hf	73 Ta	74 W	75 Re	76 Os	77 Ir	78 Pt	79 Au	80 Hg	81 Tl	82 Pb	83 Bi	84 Po	85 At	4
87 Fr	88 Ra	89 Ac	90 Th → 103 Lr	104 Ku	105 Ha													

FIGURE 5.4 VACANCIES IN THE PERIODIC TABLE.

TABLE 5.2 COMPOSITION OF CLEAN, DRY AIR NEAR SEA LEVEL IN PER CENT BY VOLUME

Nitrogen	78.09	Neon	0.0018
Oxygen	20.95	Helium	0.00053
Argon	0.93	Krypton	0.0001
Carbon dioxide	0.03	Xenon	0.000009

but also a heavier gas accumulated around decaying radium. This gas was named radon.

why was it apparent that other noble gases existed?

$$^{226}_{88}Ra \rightarrow \ ^{4}_{2}He^{2+} + \ ^{222}_{86}Rn + 2e^{-}$$

Radium Alpha particle Radon

Ramsay did not discover radon, but he did much towards determining its properties. A tribute might properly be made to Ramsay, who had a hand in the discovery and identification of each noble gas element. The precision of his work becomes apparent when one considers the small percentage of the individual gases in air (Table 5.2).

5.3 SOURCES, PROPERTIES AND USES OF THE NOBLE GASES

Despite the low percentage of noble gases in air, it is the major source of all the noble gases with the exception of helium and radon, although they also occur in some volcanic gases. Natural gas, used for heating, often contains a substantial proportion of helium (0.5 to 1.5 per cent by volume) and is the commercial source of that element. It is estimated that, if the demand were sufficient, 18 million cubic feet of helium could be produced annually. Radon is obtained in only small amounts from decaying radium, and, because it is radioactive itself, it is used in small vials for treatment of cancer.

All of the noble gases are colorless and odorless. Some of their physical properties are given in Table 5.3.

Helium is unique among all elements in that it will form a true solid only under a pressure of about 25 atmospheres. There is so little attraction between atoms that they will not hold together in the absence of external pressure, even when almost motionless.

TABLE 5.3 PHYSICAL PROPERTIES OF THE NOBLE GASES

	Atomic Weight	Boiling Point	Freezing Point	Solubility in Water ml/L at 20°C	Electrons in the Outermost Level
Helium	4.0026	−268.9	−272.1(?)	13.8	2
Neon	20.183	−246	−248.73	14.7	8
Argon	39.948	−185.8	−189.3	37.9	8
Krypton	83.80	−151.9	−169	73	8
Xenon	131.30	−109.3	−140	110.9	8
Radon	222	−62	−71	–	8

The noble gases are the **most chemically inactive group of elements** in the periodic table. The atoms do not pair to form molecules and they form no compounds among themselves. In fact, until 1962 no *bona fide* compounds of the gases were known. Some helides (helium-containing compounds) had been reported, and some hydrates had been detected, but these were not compounds in the accepted sense. The noble gases were called "inert" gases as an indication of chemical unreactivity toward all other substances.

In 1933, Linus Pauling predicted the stability of some fluorine compounds of the higher atomic weight noble gases, but no one prepared them. Early in 1962, Neil Bartlett identified the compound $O_2^+PtF_6^-$, which occurred by direct combination of oxygen with the very reactive gas, platinum hexafluoride. Knowing that molecular oxygen had an ionization energy approximately that of xenon, he mixed xenon and platinum hexafluoride and obtained a compound of the formula $Xe^+PtF_6^-$. Chemists at the Argonne National Laboratory quickly repeated Bartlett's experiments and confirmed his results. Further, upon decomposing the hexafluoride with heat, they suspected formation of a xenon fluoride compound. They then attempted to form the xenon fluoride compound directly. When a mixture of xenon and fluorine gas was heated for one hour at 400°C, then cooled to dry-ice temperature, a white crystalline solid appeared that proved to be xenon tetrafluoride (XeF_4), a stable crystalline solid with a melting point of about 114°C.

With the confirmed discovery of noble-gas compounds, workers in many laboratories entered the field. Xenon difluoride, xenon hexafluoride, and difluorides of radon and krypton have been prepared. Hydrolysis of xenon fluorides with water has yielded, among other products, an oxyfluoride of xenon ($XeOF_2$) and the explosive trioxide XeO_3.

Although the preparation of noble gas compounds has been hailed as one of the great modern advances in inorganic chemistry, the gases as a group must be regarded as very inactive. Only the most active non-metals have yet been found to combine with the gases. There is little likelihood that many stable compounds will be prepared.

The elements of group zero in the periodic table can be said to have two properties in common: *high volatility* (all are gases at room temperature) and *chemical inactivity*. These properties appear to be related to the structural similarity between the gases. Each gas, except helium, has 8 electrons in the outermost level (helium has a complete first shell of 2). The electrons are held tightly to the atom, leaving little electrical field to attract other noble-gas atoms. Held so tightly, the electrons are not easily lost to other atoms to form ionic bonds, nor are they shared by other than the most electronegative atoms. The stability of the complete electron octet of the noble gases is emphasized in the electron configuration gained by other atoms when they react to form compounds.

why are the noble gases unreactive?

Helium has been used in lighter-than-air craft. It has almost the same lifting power as hydrogen and is non-flammable. It is blown around magnesium parts during welding because magnesium will burn in air, but not in an atmosphere of helium. Helium is mixed with oxygen and used in treating certain respiratory diseases. Its tiny atoms find their way through obstructions and assist the circulation of oxygen. The same mixture is used by deep-sea divers, because helium does not dissolve in blood as much as nitrogen.

Neon is used in neon signs. Red neon signs contain neon gas. Neon signs of other colors (blue, green, etc.) have painted tubes or contain the right mixture of argon or mercury with the neon. Argon is used in ordinary incandescent lamps. The glowing filament of tungsten will not burn in argon and the argon atmosphere prevents the rapid sublimation of the white hot filament onto the glass walls of the bulb. The other three gases, krypton, xenon and radon, are so rare that they have little commercial use.

5.4 ELECTROVALENT OR IONIC BONDING

The study of ionic bonding will be initiated by examining a simple ionic compound, namely, table salt. Table salt is a white solid composed of small cubic crystals having a density of 2.18 g per cm^3. It melts sharply at 804°C, is soluble in water and has a characteristic taste.

Repeated analyses of table salt from different sources show it to be 39.3 per cent sodium and 60.7 per cent chlorine. Therefore, the substance is a compound. It obeys the Law of Definite Proportions. For each gram-atomic weight (23 g) of sodium there are $\frac{23\text{g}}{39.3\text{g}} \times 60.7\text{ g} = 35.5\text{ g}$ (1 gram-atomic weight) of chlorine, and hence the formula is NaCl.

Consider the electrolysis of molten sodium chloride. The salt-filled cell diagrammed in Figure 5.5 is placed in a furnace at a temperature greater than 804°C, the temperature at which NaCl melts. Immersed in the melt, one at each end of the cell, are two graphite

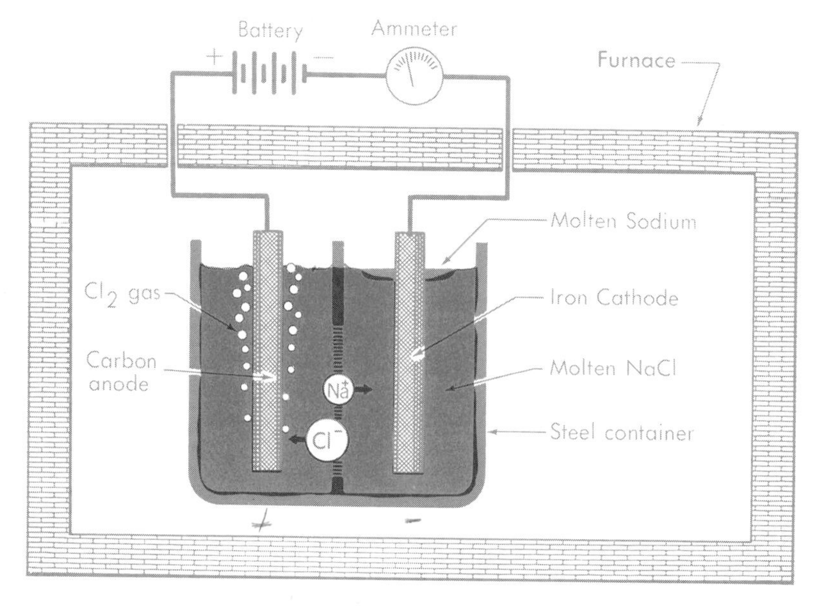

FIGURE 5.5 CELL FOR ELECTROLYSIS OF SODIUM CHLORIDE.

electrodes connected to the opposite poles of a storage battery (similar to the one in an automobile). The electrode connected to the (+) pole of the battery is called the anode; the cathode is connected to the (−) pole. A divider, perforated below the surface, separates the cell into the anode and cathode compartments.

When the circuit is closed, the following events are observed: (1) an electric current flows through the external conductor (the wire) as noted on the ammeter. (2) A greenish yellow gas bubbles up around the anode and escapes. This gas, chlorine, is very poisonous. (3) A liquid, a silvery pool of molten metal, appears and floats on the surface of the molten salt in the cathode compartment. This metal is sodium. (4) The level of the molten salt drops. Salt is being consumed.

These questions arise: (1) Why does sodium accumulate at the cathode? And how is it released from the compound? (2) Why does chlorine appear at the anode? And how is it released? (3) How is the electric circuit completed through the melt?

how does the number of protons compare with the number of electrons in a sodium ion? in a chloride ion?

These questions are answered quite reasonably by the ionic theory, which states that the sodium exists in the compound in some form with a positive charge. It has a positive charge, because in compound formation, one of its electrons is lost to a chlorine atom, which consequently carries a negative charge. When the circuit of the electrolysis cell is closed and the cathode becomes negative, the **positive** sodium **ions** are attracted toward it. Simultaneously the **negative** chloride **ions** migrate toward the positive electrode.

Upon arriving at the cathode, each sodium ion gains an electron, becomes an atom and rises to collect with others as a pool at the surface. Upon arriving at the anode, each chloride ion loses an electron to the anode (becomes an uncharged ion), pairs with another chlorine atom, and escapes as a molecule of chlorine gas. The processes occurring at the electrodes may be represented by the following equations:

Cathode

Na^+	+	e^-	$\rightarrow$	Na
Sodium ion		From the cathode		Sodium atom

Anode

$2\,Cl^-$	$\rightarrow$	Cl_2	+	$2e^-$
2 chloride ions		1 chlorine molecule		To the anode

The reactions permit the existence of an electric current, the flowing of a stream of electrons, from the anode to the cathode through the wire, ammeter and battery. Electrons are taken up from the chloride ions at the anode and flow through the circuit driven by the battery to replace electrons at the cathode, where they are accepted by the sodium ions. The number of electrons given up by the chloride ions must then be equal to the number accepted by the sodium ions.

In the discussion of the theory above, a charge of +1 was assumed for each sodium ion and a charge of −1 for each chloride ion. Careful studies of this electrolysis experiment show that 1 gram-atomic weight of sodium and 1 gram-atomic weight of chlorine are liberated with the passage of 96,500 coulombs (or 6.02×10^{23} electrons) of electricity. This amounts to 1 electron per ion formed or lost.

The preceding experiment was discussed to show why it is assumed that in the melt (and the solid) of NaCl all the sodium exists as sodium ions (Na^+) and all the chlorine exists as chloride ions (Cl^-). In the formation of one formula-weight of NaCl from the atoms, Avogadro's number of electrons is transferred.

$$11^+\ \overline{2})\overline{8})\overline{1} \quad + \quad 17^+\ \overline{2})\overline{8})\overline{7} \quad \rightarrow \quad 11^+\ \overline{2})\overline{8}) \quad + \quad 17^+\ \overline{2})\overline{8})\overline{8})$$

Na, sodium atom **Cl, chlorine atom** **Na^+, sodium ion** **Cl^-, chloride ion**

Na, $1s^22s^22p^63s^1$ + Cl, $1s^22s^22p^63s^23p^5$ → Na^+, $1s^22s^22p^6$ + Cl^-, $1s^22s^22p^63s^23p^6$

In "kernel" form

$$Na\cdot + :\underset{\cdot\cdot}{\ddot{Cl}}\cdot \rightarrow Na^+ + :\underset{\cdot\cdot}{\ddot{Cl}}:^-$$

In combining, sodium has acquired the same electron arrangement as neon (2–8). The chlorine atom, by gaining an electron, acquires the same electron arrangement as argon (2–8–8). A further indication of the stability of the noble gas or octet configuration is found in the fact that sodium is found in nature as Na^+ in compounds. Sodium is never found free (uncombined) in nature. Furthermore, considerable energy is required to take the electron from the chloride ion and transfer it to the sodium. In this instance electrical energy was provided by the battery. Other examples of ionic compounds with their electron arrangements are shown in the following examples. In each case the **"octet" or "inert gas" structure** is attained by the ions in the compound.

Example 1. Write the electronic structure and the formula of magnesium chloride.

$$:\underset{\cdot\cdot}{\ddot{Cl}}\cdot \ + \ \cdot Mg\cdot \ + \ \cdot\underset{\cdot\cdot}{\ddot{Cl}}: \ \rightarrow \ :\underset{\cdot\cdot}{\ddot{Cl}}:^- \ \ Mg^{2+} \ \ :\underset{\cdot\cdot}{\ddot{Cl}}:^-;$$

$MgCl_2$

Example 2. Find the formula of aluminum fluoride.

$$\begin{array}{ccc} :\ddot{F}: & & :\underset{\cdot\cdot}{\ddot{F}}:^- \\ | & & \\ :\underset{\cdot\cdot}{\ddot{F}}\cdot + \cdot \dot{Al}\cdot + \cdot\underset{\cdot\cdot}{\ddot{F}}: & \rightarrow & :\underset{\cdot\cdot}{\ddot{F}}:^- \ \ Al^{3+} \ \ :\underset{\cdot\cdot}{\ddot{F}}:^-; \end{array}$$

AlF_3

Example 3. Find the formula of potassium oxide.

$$K\cdot \ + \ \cdot\underset{\cdot\cdot}{\ddot{O}}\cdot \ + \ \cdot K \ \rightarrow \ K^+ \ \ :\underset{\cdot\cdot}{\ddot{O}}:^{2-} \ \ K^+;$$

K_2O

5.5 METALS AND NON-METALS IN IONIC COMPOUNDS

In the three compounds used as examples in the consideration of ionic bonding, the elements whose atoms lost electrons and be-

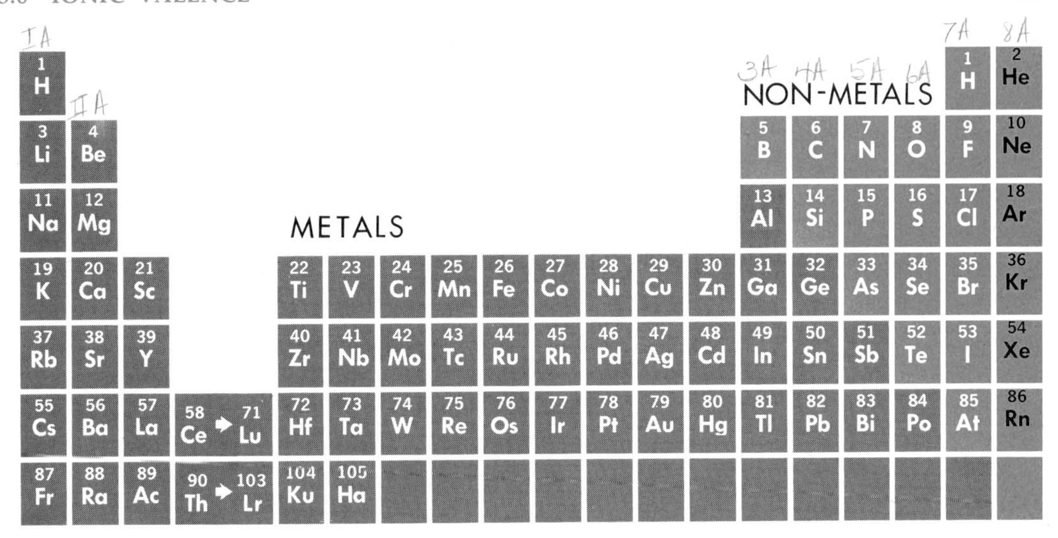

FIGURE 5.6 METALS AND NON-METALS.

metal atoms lose electrons to non-metal atoms when reaction occurs to form an ionic compound.

came positively charged ions (Mg, Al and K) are found on the left side of the periodic table. They are metals, as are all the elements to the left and below the stair-step divider in the Figure 5.6. **Metals lose electrons** in the formation of ionic bonds. Non-metals gain electrons in ionic bond formation. They appear in the upper right corner of the periodic table. Comparison of the periodic table with the divider between metals and non-metals with the chart of electronegativities (Fig. 4.19) shows that the metals are those elements which have atoms with lesser attraction for electrons (smaller electronegativity), while non-metal atoms have a greater attraction. This accounts for the fact that metals generally lose electrons while non-metals gain them.

5.6 IONIC VALENCE

what is valence?

The metal atoms used in illustrating ionic bonding (magnesium, aluminum and potassium) each have a different capacity in combining with the same element. The fluorides of these metals are: aluminum trifluoride, AlF_3; magnesium difluoride, MgF_2; and potassium fluoride, KF. The combining capacities, which are 3 to 2 to 1 for aluminum, magnesium and potassium respectively, are related to a property called **valence, which is the combining capacity** of an element. Potassium has an ionic valence of +1, magnesium of +2 and aluminum of +3. The ionic valence of an element is equal to the charge an atom of that element will assume when it becomes an ion in bond formation. For a metal the valence is positive and is equal to the number of electrons lost by the metal atom. For a non-metal the valence is negative and is equal to the number of electrons gained by the atom in becoming a negative ion. Oxygen has an ionic valence of −2, and chlorine and fluorine of −1. The valence of a non-metal, therefore, is equal to the number of electrons in the outermost shell of the neutral atom minus 8.

what is the valence of sulfur?

One can find the ionic valence of the elements that appear in the right and left extremes of the periodic table by using the group numbers. The group number, the Roman numeral above the vertical columns, is the number of electrons in the outermost shell of the group. For example calcium, which is Group II, has two valence electrons and has an ionic valence of +2. Oxygen is in Group VI, has 6 valence electrons, must gain 2 electrons in ionic combination and so has an ionic valence of −2.

Among the transition elements, the determination of ionic valence is more difficult because of an added complication. Consider iron, atomic number 26, with the electron configuration represented below. Two well-known iron oxides exist: iron(II) oxide (FeO) and iron(III) oxide (Fe_2O_3). In both compounds oxygen has a valence of −2; that is, it requires 2 electrons to complete the noble gas configuration. In iron(II) oxide, which contains an equal number of iron and oxygen atoms, iron has a valence of +2. In iron(III) oxide, in which the atomic ratio is two atoms of iron to three atoms of oxygen, the valence of iron is +3. Consideration of the electronic structure of iron

$$_{26}Fe_{2\text{-}8\text{-}14\text{-}2}$$

indicates that the iron(II) ion formed by the loss of 2 electrons is the ion expected,

$$_{26}Fe^{2+}_{2\text{-}8\text{-}14}$$

but the Fe(III) ion also occurs:

$$_{26}Fe^{3+}_{2\text{-}8\text{-}13}$$

One of the electrons in the next-to-the-outermost level takes part and is lost in the ionic bonding. Again this accents the stability of the octet configuration. The number 14 is not a stable number, and one of the electrons can be lost. In all the transition elements, the number of electrons in the next outermost level is between 9 and 18 inclusive, never 8. One or two of these may be lost at times, giving the element a variable valence. Even the number 18, when it fills the shell, is not as stable as 8; copper (2–8–18–1) has two ions, Cu^+ and Cu^{2+}.

what is the valence of Ag in Ag_2O?

Once electrons are transferred, ions, both positive and negative, are formed and carry the charge specified. They are held together + to − and − to +. In a melt each ion is quite independent of the others, and the ions move around each other and migrate to opposite poles in an electrolysis cell.

In a solid the ions are arranged in an orderly array called a **crystal lattice.** For example, a sodium chloride crystal appears somewhat as represented in Figure 5.7. Pick out any sodium ion (Na^+) in the interior of the crystal. There are four chloride ions (Cl^-) surrounding it in the plane. Also, there is a Cl^- in front and one in back. In other words, around each Na^+ there are six nearest neighbors which are chloride ions. There is no special bond between the sodium ion and any one of the chloride ions. About each Cl^- in the crystal there are also six nearest neighbors, all sodium ions. There is no pairing of atoms. There are no molecules, just the alternating + and − ions, attracted and held next to one another by the attraction of plus and minus charges for each other.

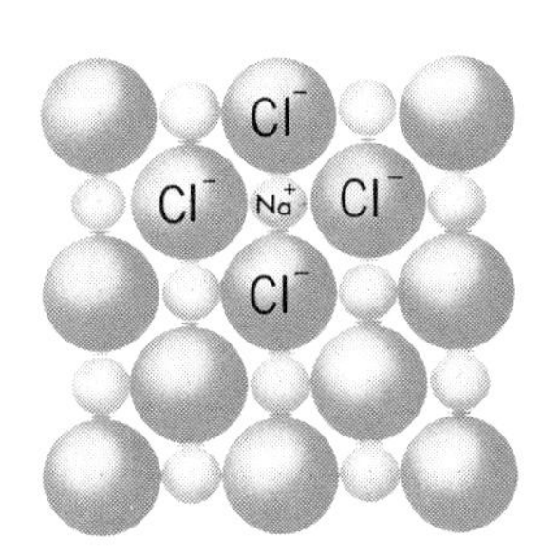

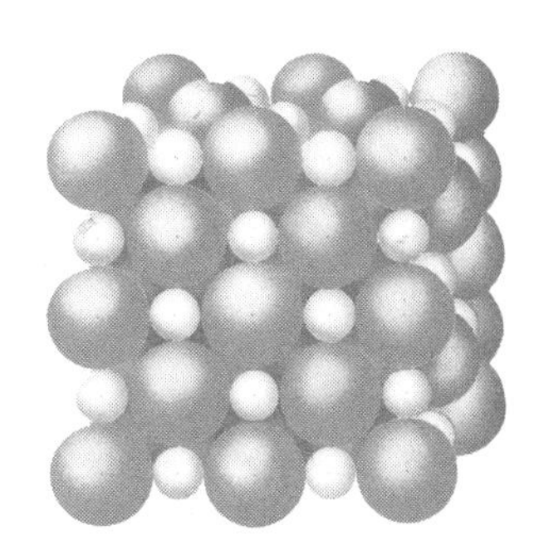

FIGURE 5.7 SODIUM AND CHLORIDE IONS IN A SODIUM CHLORIDE CRYSTAL.

Sharing electrons

5.7 COVALENT BONDING. THE LEWIS THEORY

In 1916, Gilbert N. Lewis at the University of California postulated that atoms in a molecule or a complex ion (both molecules and complex ions will often be referred to as molecules in this section) are bonded atom to atom such that each atom is similar to a link in a chain. Furthermore, he postulated that two atoms of a bond in a molecule are held together by the attraction of their kernels for one or more **pairs** of electrons held by and between them. The **shared electrons,** together with the electrons held exclusively by each atom, complete the electron structures of both atoms such as to give each a **noble gas electron configuration.** In the case of the hydrogen chloride molecule, each atom contributes one electron of an electron pair shared between them, as shown in this equation:

$$\text{H}\cdot + \cdot\ddot{\underset{..}{\text{Cl}}}\text{:} \rightarrow \text{H:}\ddot{\underset{..}{\text{Cl}}}\text{:}$$

Eight electrons surround the chlorine atom in the hydrogen chloride molecule, giving chlorine the electron configuration of argon. Two electrons, the shared pair, give the hydrogen atom the configuration of helium. Two atoms held together by the **cosharing** of one, two or three **pairs of electrons** are said to be bound by a **covalent bond.** One shared pair of electrons is called a single covalent bond. The valence electron formula, the dot formula such as shown for the hydrogen chloride molecule, contains all the valence electrons of the atoms of the molecule, placed in pairs so as to give each hydrogen atom two electrons and all others eight. The dot formula is a convenient bookkeeping device to keep track of all the electrons and to see that all atoms have achieved stable electron structures.

In the methane molecule (CH_4) the carbon atom is bonded to each of four hydrogen atoms by a single covalent bond, to which each atom of each bond contributes one electron of a shared pair:

$$4\text{H}\cdot + \cdot\dot{\underset{\cdot}{\text{C}}}\cdot \rightarrow \begin{array}{c}\text{H}\\ \text{H:}\ddot{\underset{..}{\text{C}}}\text{:H}\\ \text{H}\end{array} \text{ or } \begin{array}{c}\text{H}\\ |\\ \text{H—C—H}\\ |\\ \text{H}\end{array}$$

(an electron pair may be represented by a —).

A nitrogen atom with five valence electrons combines with

three hydrogen atoms to form the ammonia molecule:

$$3\,H\cdot + \cdot\dot{\underset{\cdot}{N}}: \rightarrow \begin{array}{c} H \\ H:\ddot{N}: \\ \ddot{H} \end{array}$$

In each molecule of both gases, methane and ammonia, the central atom is surrounded by eight valence electrons, and all hydrogens have a shared pair.

Two and three pairs of electrons must be shared by two atoms in the molecules of many substances if all atoms represented in the dot formulas are to achieve **noble gas configurations.** With a total of 12 valence electrons, the oxygen atoms of an oxygen gas molecule must share four electrons, that is, they must have a double covalent bond. The dot formula consistent with this picture is :Ö::Ö:. However, magnetic measurements, which will be discussed later, indicate that the molecule contains two unpaired electrons. Consequently, the dot formulas :Ö:Ö: and :Ö÷Ö: may be more correct, although none of the three formulas is completely satisfactory. A reasonable picture is to be found using the Molecular Orbital Theory, which will be introduced shortly.

how many covalent bonds are present in one molecule of H_2O?

If the nitrogen molecule has a triple bond as shown by the dot formula :N:::N:, six of the ten valence electrons must be held jointly by the two atoms if they are to achieve the octet structure. The hydrogen cyanide and carbon monoxide molecules also have ten electrons and, therefore, have a configuration similar to that of the nitrogen molecule, H:C:::N: and :C:::O:.

Dot formulas for several molecules are shown in Table 5.4. In

TABLE 5.4 VALENCE ELECTRON STRUCTURES FOR SOME MOLECULES AND IONS

H—Ö: | H **Water**	:C̈l: :C̈l:C̈:C̈l: :C̈l: **Carbon tetrachloride**	:C̈l—C̈l: **Chlorine**	H:H **Hydrogen**	
H:Ö: :C̈l: **Hypochlorous acid**	H H H:C̈:C̈:H H H **Ethane**	H—C═Ö: | H **Formaldehyde**	:Ö S :Ö: :Ö: **Sulfur trioxide**	
:S̈::Ö: :Ö: **Sulfur dioxide**	:C≡O: **Carbon monoxide**	:Ö═C═Ö: **Carbon dioxide**	H:C:Ö:H :Ö: **Hydrogen formate (formic acid)**	
:Ö: 2− :S̈:S:Ö: :Ö: **Thiosulfate ion**	:Ö 2− C :Ö: :Ö: **Carbonate ion**	:Ö:C̈l:Ö: − :Ö: **Chlorate ion**	H:C:::N: **Hydrogen cyanide**	H+ H:N̈:H H **Ammonium ion**

some cases, dashes are used for electron pairs. Note that the formula for each molecule contains a **total number of electrons equal to the sum of all the valence electrons** of the participating atoms. Where possible, all electrons are arranged in pairs, and, where possible, every atom has, with its shared electrons, the electron configuration of a noble gas atom. Dot formulas for a number of complex ions also appear in the table. The same rules are followed in constructing the dot formula for a complex ion as for a molecule, except that the total number of electrons is increased over that provided by valence electrons of the bonded atoms by the number of the negative charges the ion bears, or is decreased by the number of positive charges shown on the ion. For example, the negative charge of unity on the hydroxide ion (OH^-) indicates that it possesses one more electron than the valence electrons of its two atoms. The dot formula therefore contains $1+6+1=8$ valence electrons, $:\ddot{\underset{\cdot\cdot}{O}}:H^-$. The hydroxide ion cannot occur alone but is always found in a compound such as sodium hydroxide (NaOH) in which the positive ion, Na^+ in the case of this specific compound, is deficient by the electron held in excess by the hydroxide ion.

The sulfate ion (SO_4^{2-}) found in sodium sulfate (Na_2SO_4) may be represented by the dot formula

$$\begin{array}{c} :\ddot{O}:^{2-} \\ :\ddot{\underset{\cdot\cdot}{O}}:\ddot{\underset{\cdot\cdot}{S}}:\ddot{\underset{\cdot\cdot}{O}}: \\ :\underset{\cdot\cdot}{O}: \end{array}$$

There are 32 electrons represented which, when placed correctly, give the ion four single bonds. With the two extra electrons over those provided by the bonded atoms, as indicated by the charge of negative two, plus six electrons from each of five atoms, the total available electrons for formula building is 32. The cyanide ion (CN^-) found in potassium cyanide (KCN) has a triple bond, $:C{\equiv}N:^-$. The extra electron brings the total number of electrons in the CN^- ion to ten, the same number of electrons as found in the nitrogen gas molecule; therefore, the electron configurations of that molecule and the cyanide ion are similar.

As shown by the dot formulas, some molecules contain both single and double or both single and triple bonds or, in some cases, all three bond types. For some molecules, two or three dot formulas may be drawn for the same arrangement of atoms. A case in point is the carbonate ion (CO_3^{2-}) found in sodium carbonate, Na_2CO_3. Each of the formulas,

$$\begin{array}{c} :\ddot{O}\ ^{2-} \\ \| \\ :\ddot{\underset{\cdot\cdot}{O}}:C:\ddot{\underset{\cdot\cdot}{O}}: \end{array} \qquad \begin{array}{c} :\ddot{O}:\ ^{2-} \\ :\ddot{\underset{\cdot\cdot}{O}}:\ddot{C}{=}O: \end{array} \qquad \begin{array}{c} :\ddot{O}:\ ^{2-} \\ :\ddot{O}{=}\ddot{C}:\ddot{\underset{\cdot\cdot}{O}}: \end{array}$$

represent the ion equally well. Each is equivalent but each would predict that the ion would have two single bonds and one double bond. Measurements have shown that atoms bound by a double bond are held more closely together than those bound by a single bond. X-ray studies show that **all three bonds in the carbonate ion are of the same length,** somewhat greater than that expected of a double bond and somewhat shorter than that expected of a single bond. Some molecules have an odd number of valence electrons, which makes it impossible to place all electrons in pairs. The dot formula for the nitrogen(II) oxide molecule ($:\dot{N}::\ddot{O}:$) does not meet one of the criteria of a dot formula, that is, that each atom have a noble gas electron configuration. The possibility of more than one dot formula for a given molecule, the occurrence of molecules with an odd number of electrons and the misleading implications of the dot formula for oxygen all point out deficiences in explaining bonding with the Lewis Octet Theory. These problems will be dealt with more successfully by the Molecular Orbital Theory to be discussed later in the chapter. Despite its limitations, the octet idea is still very useful.

5.8 SALTS: FORMULAS AND NAMES

Many ionic compounds (salts) are encountered in inorganic chemistry. These compounds contain both simple and complex ions. The compound is named by the ions it contains, the **positive ion first**, the negative ion last. For example the compound $CuSO_4$ contains copper(II) ions (Cu^{2+}) and sulfate ions (SO_4^{2-}).

Table 5.5 is a list of the most common ions and their formulas. One should learn the names and formulas (including the charge) of all these ions.

Three simple rules may assist in memorizing the table.

1. **Simple positive ions retain their elemental name** unless there are two or more possible valences. In those cases the ionic change is indicated by a Roman numeral in parentheses following the name. Some have common names which may be encountered in reading, i.e., copper(II) ion, Cu^{2+}, was earlier named cupric ion; copper(I) ion, Cu^{+}, was named cuprous.

2. **Simple negative ions end in *-ide***, i.e., sulfide ion (S^{2-}), chloride ion (Cl^{-}), and so forth. In fact the proper names of all

TABLE 5.5 SOME COMMON IONS†

Name	Formula	Name	Formula
Acetate	$C_2H_3O_2^-$	Iodide	I^-
✔Aluminum	Al^{3+}	Lead(II)	Pb^{2+}
*Ammonium	NH_4^+	Magnesium	Mg^{2+}
✔Barium	Ba^{2+}	Manganese(II)	Mn^{2+}
*Bicarbonate	HCO_3^-	Mercury(I)	Hg_2^{2+}
✔Bromide	Br^-	Mercury(II)	Hg^{2+}
Calcium	Ca^{2+}	✔Nickel(II)	Ni^{2+}
*Carbonate	CO_3^{2-}	*Nitrate	NO_3^-
*Chlorate	ClO_3^-	*Nitrite	NO_2^-
*Chloride	Cl^-	Permanganate	MnO_4^-
*Chlorite	ClO_2^-	*Phosphate (ortho)	PO_4^{3-}
Chromate	CrO_4^{2-}	✔Potassium	K^+
✔Copper(II)	Cu^{2+}	Silver	Ag^+
✔Copper(I)	Cu^+	Sodium	Na^+
*Cyanide	CN^-	*Sulfate	SO_4^{2-}
✔Iron(III)	Fe^{3+}	*Sulfide	S^{2-}
✔Iron(II)	Fe^{2+}	*Sulfite	SO_3^{2-}
Fluoride	F^-	Zinc	Zn^{2+}
*Hydroxide	OH^-		

† The elements marked by checks ✔ and asterisks * are to be used in Problem 7.20 at the end of the chapter.

what is the name assigned to Fe^{3+}?

binary compounds, ionic or covalent, end in *-ide*. The suffix is added to the stem of the name of the more negative element. Examples are: nitrogen(IV) oxide (NO_2) and nitrogen(I) oxide (N_2O).

3. Complex negative ions containing oxygen end in either *-ate* or *-ite*. When the negative complex ions differ only in the number of oxygen atoms they contain, the one with more oxygen atoms ends in *-ate*, the one with fewer in *-ite*.

From the list of names of ions, one can name any compound from its formula, and from the name one can identify the ions present. Furthermore, knowing the charges carried by the ions and knowing that all compounds are neutral one can write the formula from the name. For example, aluminum sulfate contains aluminum ions (Al^{3+}) and sulfate ions (SO_4^{2-}). Because aluminum sulfate must carry no net positive nor negative charge, the ions in any crystal must appear in the ratio of 2 to 3:

$$2 \times (+3) + 3 \times (-2) = 0$$

The formula is $Al_2(SO_4)_3$.

5.9 VALENCE BOND APPROACH

The Valence Bond Theory, proposed by F. London and W. Heitler in 1927, and the Molecular Orbital Theory of Mulliken, introduced at about the same time, refined the postulates of Lewis and placed a mathematical basis under his ideas. Both theories, although quite different in their approaches, apply quantum mechanics to the problem. In the valence bond method, electron probabilities are calculated for the two atoms brought together with their own probability functions. For a single bond, a term evaluates the probability with each electron with its original atom. Then, because the electrons are identical, a term is added in which the electrons are exchanged. The added term (called an exchange integral) predicts an increased electron probability between the two nuclei and predicts a lower energy for the system when the atoms are together (that is, provided the two electrons are of opposite spins). The calculated decrease in energy corresponds to the energy required to break the bond.

quantum mechanical theories of bonding.

For calculations which give a value more nearly equal to the bond energy, two other terms are introduced, one term of which is often of some importance. In these terms the two electrons are considered to be with one atom and then the other. Pictorially the three terms are represented by

$$\underset{(A)}{H^+{:}\ddot{\underset{\cdot\cdot}{Cl}}{:}^-}, \quad \underset{(B)}{H{:}\ddot{\underset{\cdot\cdot}{Cl}}{:}} \text{ and } \underset{(C)}{H{:}^-\ddot{\underset{\cdot\cdot}{Cl}}{:}^+}$$

for the hydrogen chloride molecule. Structure A represents the term for which both shared electrons are on the chlorine atom; structure B, on both atoms; and structure C, on the hydrogen. Structure B represents the exchange integral. Because chlorine has a much greater electronegativity than hydrogen (see Chapter 4),

for HF, would a structure such as A be more or less important than in HCl?

term C for dot formula C is negligible. The actual molecule is considered to be represented by both structures A and B, in this case 83 per cent B and 17 per cent A. That is, the molecule can be represented by neither A nor B alone but has characteristics of both. However, B is a closer representation.

The **valence bond** method deals with molecules or complex ions which have two or more possible dot formulas in much the same way as it handles hydrogen chloride. According to that theory, when two or more equivalent, or nearly equivalent, electron structures, none of which leaves any atom without a noble gas configuration, can be drawn for a molecule without changing the positions of any atom in the unit, no one of the structures represents the molecule correctly. Instead, the molecule is called a **resonance hybrid.** A hybrid is a structure intermediate between all the contributing structures and is more stable than any one of them. The word intermediate refers to every structural feature of the dot formula which may be changed in any two or more of the formulas. The three equivalent structures for the carbonate ion

in the true structure of CO_3^{2-} are the carbon to oxygen bonds all of the same length or do they differ? why?

$$
\begin{array}{ccc}
:\ddot{O}\;^{2-} & :\ddot{O}:\;^{2-} & :\ddot{O}:\;^{2-} \\
\| & | & | \\
:\ddot{\underset{..}{O}}\!-\!C\!-\!\ddot{\underset{..}{O}}: & :\ddot{\underset{..}{O}}\!-\!C\!=\!\ddot{O}: & :\ddot{O}\!=\!C\!-\!\ddot{\underset{..}{O}}:
\end{array}
$$

would then predict that each carbon to oxygen bond should be a $1\frac{1}{3}$ order bond. The nitrate ion (NO_3^-) and the sulfur trioxide molecule are resonance hybrids with bonding similar to the carbonate ion, inasmuch as the electrons in the dot formula are placed identically. Carbon dioxide is a resonance hybrid with these contributing formulas:

$$
\begin{array}{ccc}
:\ddot{\underset{..}{O}}\!-\!C\!\equiv\!O: & :\ddot{O}\!=\!C\!=\!\ddot{O}: & :O\!\equiv\!C\!-\!\ddot{\underset{..}{O}}: \\
(A) & (B) & (C)
\end{array}
$$

with B contributing the most to the structure.

5.10 THE MOLECULAR ORBITAL METHOD

According to the **Molecular Orbital Theory,** all electron orbitals within a molecule or complex ion are associated with the entire molecule. In other words, there is a pattern for orbitals about two or three positive centers of a molecule just as there is a pattern of s, p and d orbitals about the one positive center of an atom. In theory, one determines the pattern of **molecular orbitals** for the molecule, arranges them in the order of increasing energy and adds the electrons for the entire molecule following the same rules used in filling atomic orbitals. That is: (a) orbitals, atomic or molecular, can hold only two electrons, (b) electrons are added to the lowest energy orbital first and (c) if there are not enough electrons to fill orbitals **of equal energy,** they are added one to an orbital before pairing. Actually one need only consider the placing of the valence electrons, those which appear in the dot formula. The great difficulty with this approach is that the mathematical equations to be solved for the molecular orbitals are so difficult that they cannot be solved without making some simplifying assumptions. One method of approximate solution depends upon the assumption that molecular orbitals should be a linear combination of the atomic orbitals of the atoms bonded together, the so-called LCAO approximation. That is, two orbitals (one orbital from each of the two atoms) may be considered to combine to form two orbitals, *both* of which are associated with *both* atoms. One of the molecular orbitals is of lower energy than the average energy of the two atomic orbitals from which it is considered to be composed, has an increased electron probability between the two nuclei and is called a **bonding orbital.** The second molecular orbital of the two is higher in energy than the atomic orbitals, has decreased electron probability between the two nuclei and is called an **antibonding orbital.**

5.11 HOMOPOLAR BONDING IN DIATOMIC MOLECULES

The molecular orbitals that result from a linear combination of the s orbitals of the valence shells of two hydrogen atoms are depicted in Figure 5.8, along with the atomic orbitals. They are arranged such that orbitals are one above the other in the order of increasing energy. The molecular orbitals formed by the combination of the orbitals of the s and p sublevels of two atoms, the s with the s, and the p_x with the p_x, and so on, are shown in Figure 5.9. These molecular orbitals (Fig. 5.9) are available for occupancy by all the valence electrons of a diatomic, homopolar molecule, a molecule in which the two atoms involved present the same set of atomic orbitals for molecular orbital formation.

Two types of molecular orbitals are represented in Figures 5.8 and 5.9, as indicated by the Greek letter prefixes σ (sigma) and π (pi) in the orbital notation. A **sigma orbital** is uniformly symmetrical about the internuclear axis, the line passing through the two nuclei. That is, at all angles of rotation about that axis, the orbital appears to have the same shape. The sigma orbitals in the figure are "formed" by the combination of atomic orbitals s and s, and p_x and p_x, which are, also, uniformly symmetrical about the internuclear axis. A **pi orbital** is **not** uniformly symmetrical but has a twofold symmetry about the internuclear axis. That is, twice in a

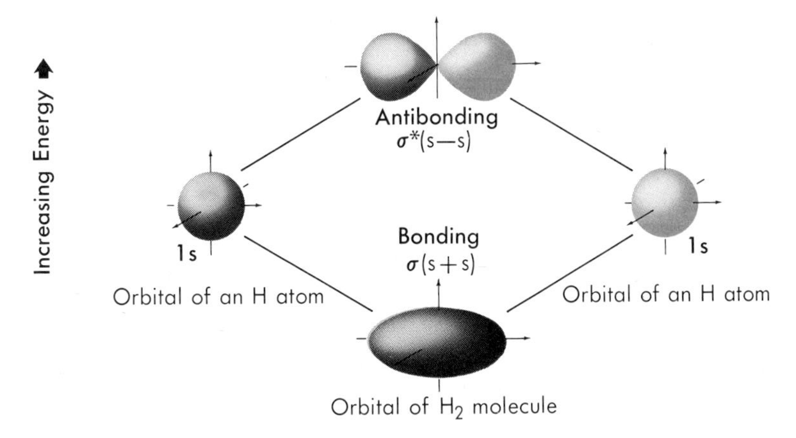

FIGURE 5.8 MOLECULAR ORBITALS OF HYDROGEN.

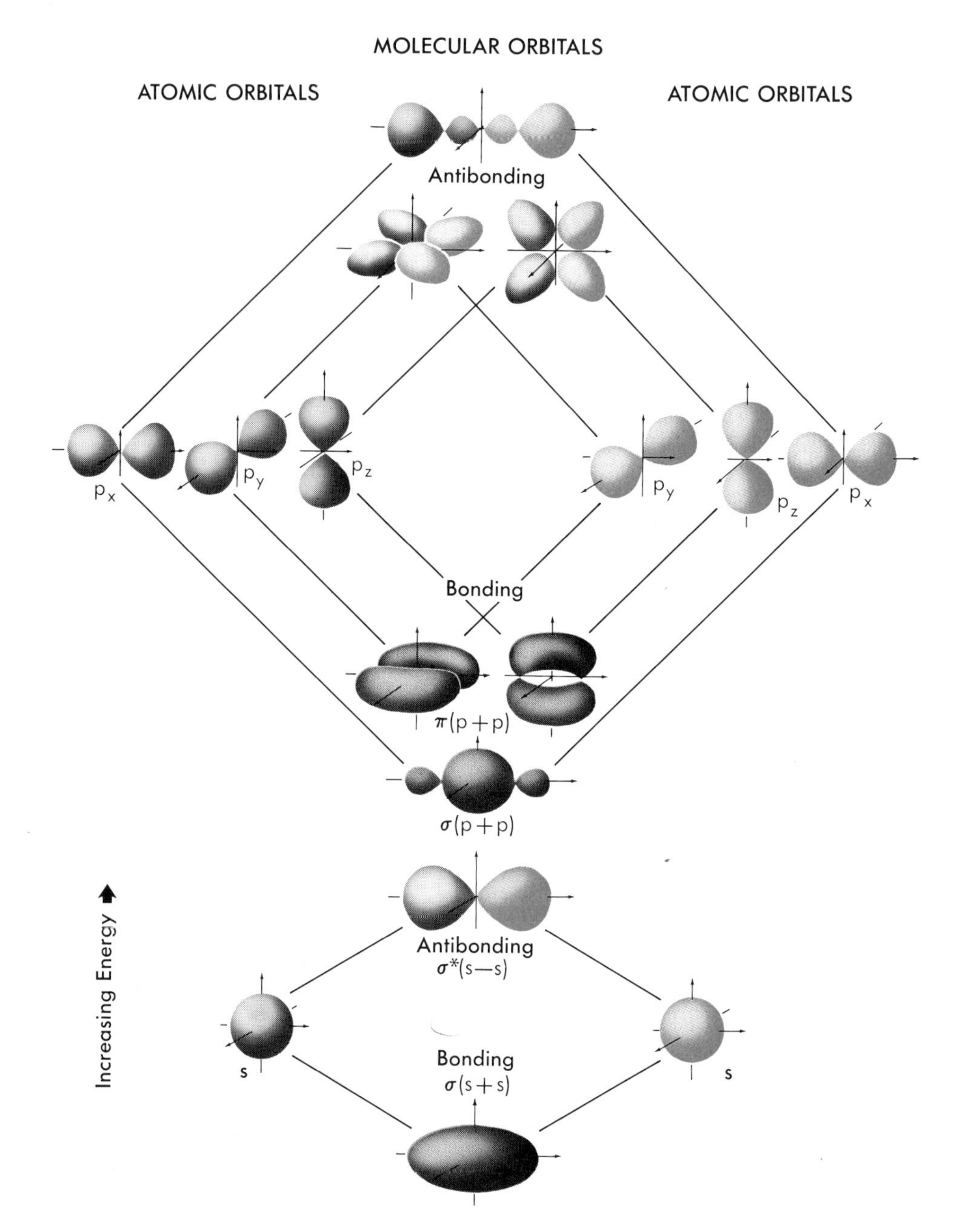

FIGURE 5.9 MOLECULAR ORBITALS FORMED FROM S AND P ATOMIC ORBITALS.

full circle rotation about that axis the orbital appears to be in the same position—at 0° (or 360°) and 180°. The orbital has an upper and lower or a right hand half and a left hand half. Pi orbitals are considered in the LCAO method to be formed of atomic orbitals, a p_y and a p_y, or a p_z and a p_z, orbitals which have a twofold symmetry about the internuclear axis.

the number of covalent bonds between two atoms is the number of bonding electrons minus the number of antibonding electrons divided by 2.

The bonding orbitals represented in Figures 5.8 and 5.9 do have a region or two regions of great electron probability between the two positive kernels, and each antibonding orbital has high electron probability regions positioned so as to pull the kernels apart. Furthermore, the patterns of a bonding orbital and its corresponding antibonding orbital are such that the two together, one overlaying the other, approximate the electron probabilities of the two uncombined atomic orbitals taken together. This is to say, an antibonding orbital filled with two electrons effectively cancels the attractive effect of its corresponding filled bonding orbital. Therefore, in order to have a bond between two atoms of a "molecule," there must be an **excess of filled bonding orbitals** over the filled antibonding orbitals. The number of filled bonding orbitals in excess of the filled antibonding orbitals corresponds to the multiplicity of the bond.

The two electrons of the hydrogen molecule shown in the dot formula, H:H, enter the bonding σ(s+s) orbital. Because there are no electrons to enter the σ^*(s–s) orbital, the bond is composed of the one bonding orbital and contains a single bond. The hypothetical helium molecule, "He_2," has four electrons. The electrons will enter, two each, the σ(s+s) and σ^*(s–s) orbitals. The σ^*(s–s) orbital cancels the attractive effect of the σ(s+s) orbital for the two nuclei, and no He_2 molecule exists.

The lithium molecule (Li_2) contains two valence electrons. They enter the σ(s+s) orbital, forming a single bond. However, the electronegativity (the relative attraction for electrons) of lithium is so small that the bond is weak. The Li_2 molecule is never found in the solid metal but has been detected in lithium vapor. On the other hand, the vapor of beryllium contains no diatomic molecules, as would be expected, inasmuch as the molecule would have one filled bonding and one filled antibonding orbital.

The 14 electrons of the valence shell of chlorine, :C̤̈l:C̤̈l:, fill all the orbitals of Figure 5.9 except the one of highest energy, σ^*(p–p). The following pair of orbitals cancel each other, σ^*(s–s) and σ(s+s), and $2\pi^*$(p–p) and 2π(p+p), leaving only the σ(p+p) orbital. Therefore, the bond in the chlorine molecule is a single bond, a single *sigma* bond, and is not unlike the bond in the boron molecule, B_2 or :B:B:, found in trace amounts in the metal vapor. That bond also is a single sigma bond. However, because of the weak attraction of the boron atom for its valence electron, the bond is much weaker than the chlorine to chlorine bond.

The nitrogen molecule, :N:::N:, has the five molecular orbitals of lowest energy filled in its valence shell as indicated by the following notation: σ(s+s) + σ^*(s–s) + σ(p+p) + π(p+p) + π(p+p). The only filled antibonding orbital, σ^*(s–s), cancels the effect of the σ(s+s) orbital, leaving three filled bonding orbitals in excess. The bond is a triple bond, one sigma plus two pi bonds.

The oxygen molecule, :Ö::Ö:, has two more electrons than the nitrogen molecule. These enter one each into the equal energy π^*(p–p) orbitals. Each of these half-filled antibonding orbitals half cancels the corresponding bonding orbital, leaving the **oxygen molecule** with an unusual **double bond** composed of one sigma and two half pi bonds. This accounts for the **paramagnetic** behavior of oxygen gas, and also accounts for the failure of the dot formula to represent the bond correctly.

Several generalizations can be drawn from a study of homopolar molecules with respect to the molecular orbitals. Any hypothetical diatomic noble gas molecule would have an equal number of filled bonding and antibonding orbitals, and therefore should not exist. Because of this, only the valence electrons (those beyond the previous noble gas molecule configuration) need to be considered in the bonding of diatomic molecules. The limit of bond multiplicity is three, for that is the maximum number of bonding orbitals, namely σ(p+p), π(p+p) and π(p+p) that can be filled in succession. The first dash of a bond in a molecular formula represents a sigma orbital. The second dash of a double bond and the second and third dashes of a triple bond represent pi orbitals. The dot formula is of great help in determining the type of bond in the molecular orbital method.

is CaC_2 an ionic or covalent compound?

Bonding within homopolar complex ions is similar to the bonding in molecules which have the same dot formula and the same number of valence electrons. For example, the bonding in the carbide ion (C_2^{2-}, :C:::C:$^{2-}$) found in calcium carbide (CaC_2) is similar to the bonding in the nitrogen molecule. The carbide ion has a triple bond, i.e., a sigma pi pi bond. The hypochlorite ion (ClO^-, :C̤l:O̤:$^-$) found in sodium hypochlorite (NaOCl) has a single sigma bond similar to the bond in the chlorine molecule.

5.12 HETEROPOLAR BONDING AND POLYATOMIC MOLECULES

The hydrogen chloride molecule, HCl or H:C̤̈l:, contains a heteropolar bond—one heteropolar sigma bond. The participating atoms do not present the exact same set of orbitals for molecular orbital formation, but the s orbital of hydrogen and a p orbital of chlorine combine to form the bonding and antibonding orbitals sketched in Figure 5.10. The two shared electrons (dot formula) enter the bonding orbital; the other six valence electrons enter atomic orbitals of the chlorine atom, all of which are lower in energy than the antibonding molecular orbital. Because the bonding orbital is uniformly symmetrical about the internuclear axis, the bond is a sigma bond. All diatomic molecules in which hydrogen is one but not both of the atoms have the same type of sigma bond.

All polyatomic molecules (molecules having more than two atoms) contain heteropolar bonds. A central atom (an atom bonded to two or more other atoms) cannot present the same set of orbitals as presented by each of the terminal atoms to which it is bonded, because it must divide its orbitals among the two or more bonds that it forms. In polyatomic molecules, as in diatomic molecules, all single bonds are sigma bonds and one portion of all multiple bonds is a filled sigma molecular orbital. Sigma orbitals are formed between atoms by the combining of two atomic orbitals, one of the orbitals coming from each atom such that the molecular orbital is uniformly symmetrical about a line drawn between the two nuclei. Because **atomic orbitals**, except for the s orbital, are **directed in space** with respect to the other orbitals of the atom, the molecular orbitals which they form are also directed in space. Thus the sigma orbitals of the central atom or atoms set the gross geometry of the molecule. The p orbitals not involved in sigma bond formation combine in parallel fashion to give pi orbitals corresponding to the second or the second and third dashes of a multiple bond appearing

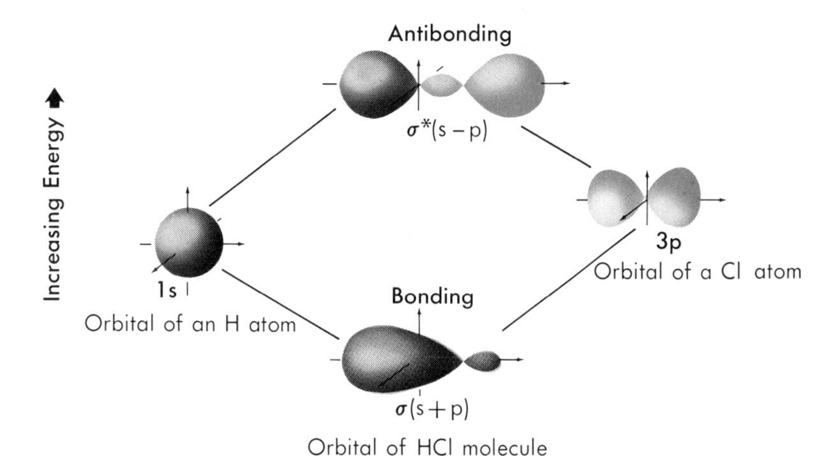

FIGURE 5.10 BONDING IN HYDROGEN CHLORIDE.

in a dot formula and have little to do with the angles between the bonds of an atom.

In the simple picture of the water molecule (Fig. 5.11), each oxygen to hydrogen bond is considered to be similar to the hydrogen to chlorine bond of hydrogen chloride as shown in Figure 5.10. Because the two p orbitals that form sigma bonds with the s orbitals of the hydrogen atoms are mutually perpendicular, the sigma bonds formed are also mutually perpendicular. According to this picture, the water molecule should not be linear but bent. That the molecule is a bent molecule is consistent with its properties, namely, polarity, high boiling point, and so on, which will be discussed later. Furthermore, electron diffraction studies of the molecule, which will not be discussed in this text, also show it to be bent.

FIGURE 5.11 SIMPLE PICTURE OF THE WATER MOLECULE.

5.13 HYBRIDIZATION

The picture of the bonding of the water molecule discussed in the previous section, and its "prediction" that the molecule should be bent, was considered a victory for the molecular orbital method. But the electron diffraction studies which later proved the water molecule to be bent showed the angle between the bonds at the oxygen atom to be 105°, and not 90°, the angle between the p orbitals involved in molecular orbital formation. Because of the difference in angles, the explanation is thought to be incorrect.

The bonding in the methane molecule (CH_4) cannot be explained adequately by a picture of the carbon atom with one s and three p orbitals in its outermost quantum level. The **methane** molecule, whose dot formula is shown in Figure 5.12a, has been found by electron diffraction studies not to be planar as its formula suggests, but instead to have a **tetrahedral shape.** That is, if the carbon atom is thought to be placed in the center of a regular tetrahedron (a three-sided pyramid), the hydrogen atoms fall at the apices (Fig. 5.12b). Also, if the carbon atom is placed in the center of a cube of the proper dimensions, the hydrogen atoms are at opposite corners, as in Figure 5.12c. All carbon to hydrogen bonds in methane are equivalent. The distance between the center of the carbon atom and the center of each hydrogen atom is 1.08 Å, and the angle between a carbon to hydrogen bond and any other carbon to hydrogen bond in the molecule is 109°28′.

If the four sigma bonds were formed, three by the three p orbitals of carbon combining each with an s orbital of hydrogen and one by the combining of the s orbital of carbon with an s orbital of hydrogen, only three of the four bonds would be alike and the angle between any pair of the three equivalent bonds would be 90°. The sigma bond formed by the two s orbitals would be shorter, and the angles between that bond and between each of the others would be greater than 135°. This is not the case, for all bonds are equivalent. To make four equivalent bonds, the s and three p orbitals of the carbon atom in methane are said to *hybridize* (to mix together) and produce a new set of four orbitals.

The valence orbital electron configurations, listed for the first 20 elements of the periodic table in Figure 4.17, are the configurations of the atoms as determined from atomic spectra, where the atoms are uncombined and in the gas phase and far enough from neighboring atoms to act independently. When an atom is in close proximity with other atoms, and especially when it is bonded to other atoms, the energies and shapes of its orbitals may change. The change in the atomic orbitals upon bonding to give the correct

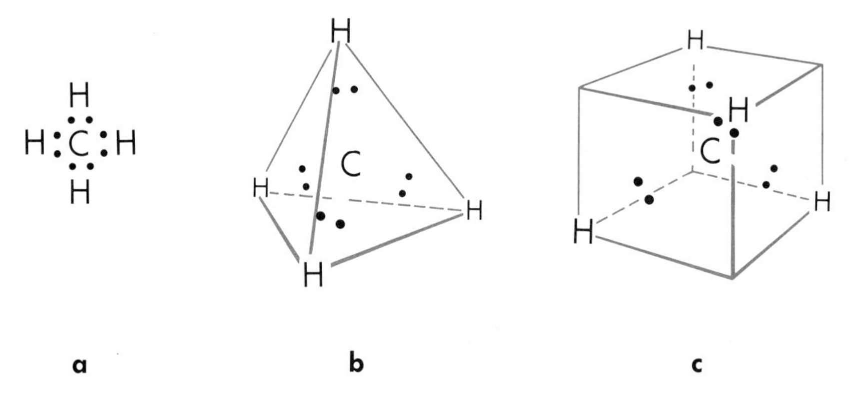

FIGURE 5.12 THREE REPRESENTATIONS OF METHANE.

geometry of the molecules and complex ions being formed is called **hybridization.** Hybridization is thought to occur only at the time of bond formation, and no atom is thought to exist independently with a hybridized configuration. However, a knowledge of these hybridized configurations is necessary in order to place the atoms together in covalent bonding.

In hybridization (Fig. 5.13a) the s orbital of the valence level of an atom combines with one, two or three of the p orbitals in the same level of the same atom to yield two equal, three equal or four equal orbitals respectively. The nature of the hybridization in a hybridized atomic orbital is indicated by its symbol, which contains the symbols of all the orbitals combining in the hybridization. If more than one of the same type of orbital is involved in hybridization, the number is indicated by superscript. For example, **sp hybridization** comes about by the combining of one s and one p orbital to give two sp orbitals; **sp^2 hybridization** comes from the combining of the s and two p orbitals to give three sp^2 orbitals; and **sp^3 hybridization** comes from the combining of the s orbital with

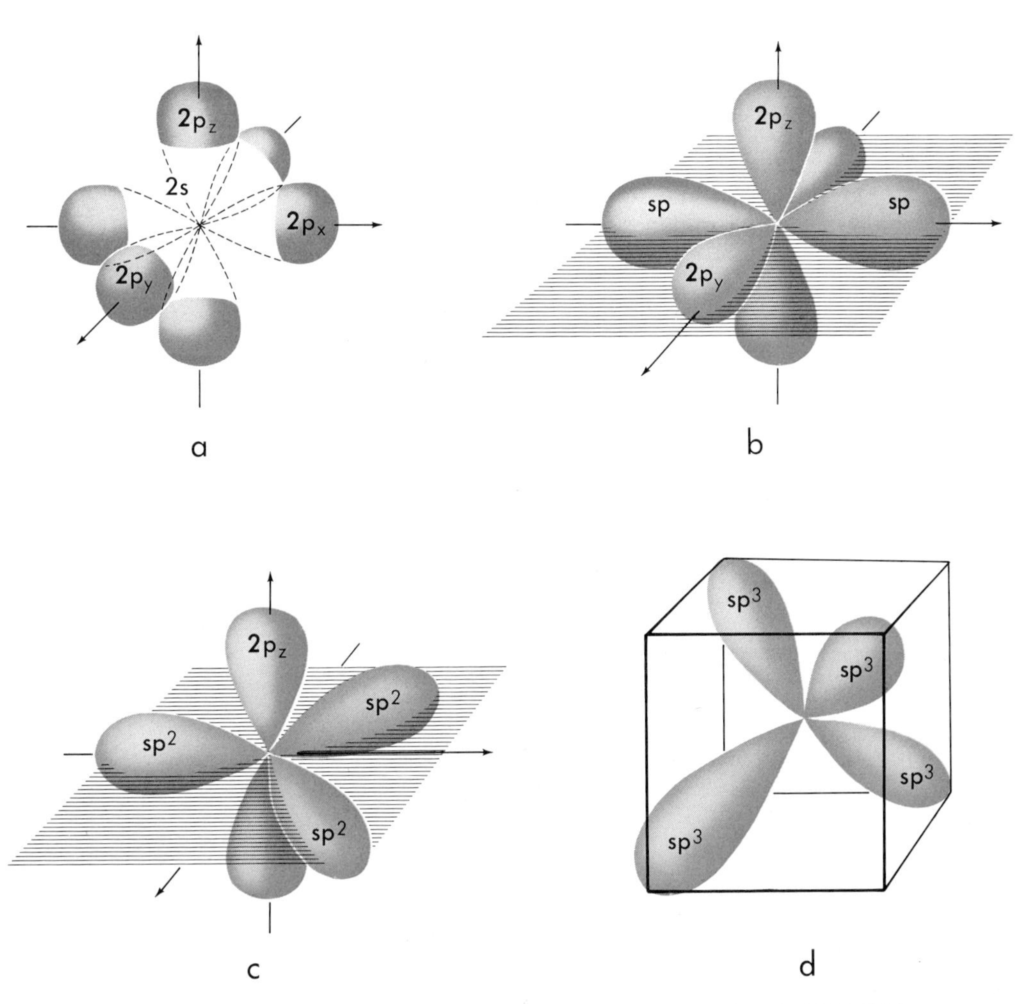

FIGURE 5.13 HYBRIDIZATION OF SOME ATOMIC ORBITALS.

all three p orbitals to form four equivalent sp^3 orbitals (those needed in the methane molecule). At bond formation a *carbon atom* may be thought of as having any one of these valence electron configurations:

1. $s^1p^1p^1p^1$,
2. $(sp)^1(sp)^1p^1p^1$ sp hybridized,
3. $(sp^2)^1(sp^2)^1(sp^2)^1p^1$ sp^2 hybridized (ess-pee-squared), and
4. $(sp^3)^1(sp^3)^1(sp^3)^1(sp^3)^1$ sp^3 hybridized (ess-pee-cubed).

The quantum level number is omitted before each orbital designation to simplify the notation. In each "atom" there are four valence orbitals and four valence electrons as indicated by those superscript numbers outside the parentheses. One electron is placed in each orbital of the hybridized carbon atoms in anticipation of the bonding.

The **two sp orbitals** of an sp hybridized atom **are directed in space at 180°** with respect to each other (point in opposite directions). The two remaining p orbitals are unchanged, are mutually perpendicular and are both perpendicular to the hybridized orbitals (see (Figure 5.13b). The **three sp^2 orbitals** of the sp^2 hybridized atom **are at angles of 120°**; they lie in the same plane and therefore are directed toward the apices of an equilateral triangle. The remaining p orbital is perpendicular to the plane of the hybridized orbitals, as shown in Figure 5.13c. The **four orbitals of the sp^3** hybridized atom are directed toward the apices of a regular tetrahedron or toward the opposite corners of a cube (Fig. 5.13d). The **angle** between each pair of orbitals **is 109°28′.**

Notations and sketches are shown only for orbitals of the one isolated and three hybridized atoms of carbon. Atoms of other elements have similar sets of orbitals, both hybridized and unhybridized, but differ in the number of valence electrons that occupy those orbitals. The number of valence electrons corresponds to the group number of the element in the periodic table.

If a central atom (an atom bonded to two or more other atoms) of a molecule or complex ion is bonded by single bonds to four other atoms, and if the angle between each pair of bonds at the atom is 109°28′ (or nearly so), the central atom will be sp^3 hybridized. The four atoms attached to that central atom will be at the apices of a tetrahedron. If the central atom is bonded to three other atoms and those three atoms define a plane containing the central atom (that is, the molecule is planar), the central atom is sp^2 hybridized. If the central atom is bonded to two other atoms and all three atoms lie in a straight line (the three are linear), the central atom is sp hybridized. The generalized rules for hybridization are stated below. If any two (or more) bond angles of a molecule are

1. equal to, or nearly equal to, 109°28′, the central atom is sp^3 hybridized.
2. equal to, or nearly equal to, 120°, the central atom is sp^2 hybridized.
3. equal to, or nearly equal to, 180°, the central atom is sp hybridized.

Returning to the *methane molecule,* which has four single bonds all on the central atom, the carbon atom is sp^3 hybridized and presents the following orbital-electron structure for bond formation: C $(sp^3)^1(sp^3)^1(sp^3)^1(sp^3)^1$. Each molecular orbital involved in the bonding is a bonding sigma orbital formed by the combining of

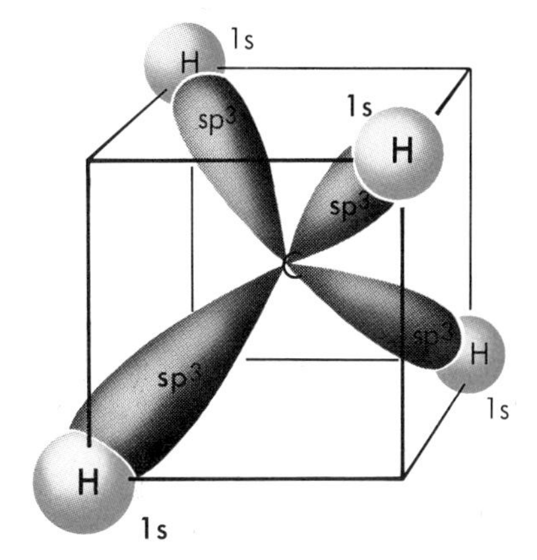

FIGURE 5.14 TETRAHEDRAL CONFIGURATION OF METHANE.

an s orbital of a hydrogen atom and an sp^3 orbital of the carbon atom (Fig. 5.14). Because the sp^3 orbitals are directed toward the apices of a regular tetrahedron, so are the molecular orbitals. The molecule is three dimensional, with the angle between any pair of bonds equal to the tetrahedral angle, 109°28′.

The bonding of an *ammonium ion,* H—N(—H)(—H)—H^+ (N bonded to four H), is very similar to that of methane. Each bond is a sigma bond with two electrons occupying an orbital formed by the combining of an sp^3 orbital of the hybridized nitrogen atom and the s orbital of a hydrogen atom. The molecule is tetrahedral—the angle between any two bonds is 109°28′. The *carbon tetrachloride* molecule, $:\ddot{C}l$—C—$\ddot{C}l:$ (C bonded to four $:\ddot{C}l:$), differs in molecular orbitals from the methane molecule in only one respect: the p orbitals of chlorine—instead of the s orbitals as in hydrogen—combine with the sp^3 orbitals of carbon to form the four bonding orbitals of the molecule. Because the geometry of the molecule is set by the hybridization of the central atom, the carbon tetrachloride molecule also is tetrahedral.

All the molecular orbitals of the ammonia molecule, $:H:\ddot{N}:H$ (with H below N), and of the *hydronium* ion, $H:\ddot{O}:H^+$ (with H below O), are similar, and each is thought to be formed by the combining of the s orbital of a hydrogen atom with an sp^3 orbital of an hybridized central atom. The molecule or ion is a pyramid with the central atom, nitrogen or oxygen, at the apex and the hydrogen atoms at the corners of the triangular base. The unshared electron pair occupies an sp^3 orbital. The hybridized central atom is required because the angle between the bonds of the ammonia molecule, 107°, is nearly the tetrahedral angle. The extended unshared pair of electrons (in an sp^3 orbital) is consistent with the availability of those electrons in acid-base reactions involving either the ammonia molecule or the hydronium ion.

By similar arguments, the two bonding orbitals of the water molecule are thought to be formed by the combining of sp^3 orbitals of oxygen and s orbitals of hydrogen. The measured angle of 105° between the two bonds is nearly the tetrahedral angle (Fig. 5.15a). One of the two unshared pairs of electrons that occupy sp^3 orbitals extending out from the molecule is shared by the proton, H^+, in the formation of the hydronium ion (Fig. 5.15b) in a molecular orbital formed by the combining of the sp^3 orbital with the s orbital of hydrogen.

The carbon atom in the *formaldehyde* molecule, $\mathrm{H{-}\overset{\overset{H}{|}}{C}{=}\ddot{O}\!:}$, is sp^2 hybridized: C $(sp^2)^1(sp^2)^1(sp^2)^1p^1$. Three sigma molecular orbitals formed by the combining of the hybridized atomic orbitals of carbon, two with s orbitals of hydrogen atoms and one with a p orbital of the oxygen atom, hold six electrons and are the skeleton of the molecule. The p orbital of the carbon atom, which is perpendicular to the plane of the molecule, combines with a p orbital of oxygen to form a localized pi orbital. The two electrons that occupy the pi orbital in bonding bring the total number of valence electrons about the carbon atom to eight. The planar molecule has one double and two single bonds, and the angle between adjacent bonds at the carbon atom is about 120°.

The dot formula for the *hydrogen cyanide* molecule, H:C:::N:, is similar to the formula for the nitrogen molecule in that each contains ten electrons and has a triple bond. In addition, however, hydrogen cyanide has a single bond, and this requires a hybridization. Because the molecule is linear, the carbon is sp hybridized: C $(sp)^1(sp)^1p^1p^1$. One sp atomic orbital forms a sigma molecular orbital with the s atomic orbital of the hydrogen atom, making a single C to H bond; the other sp orbital forms the sigma portion of a triple C to N bond by combining with a p orbital of the nitrogen atom. The two p orbitals of the carbon atom and the remaining p orbitals of nitrogen form two pi molecular orbitals. Occupation of the four molecular orbitals each by a pair of electrons gives the single and triple bonds of the molecule. A pair of electrons occupies the s orbital of nitrogen.

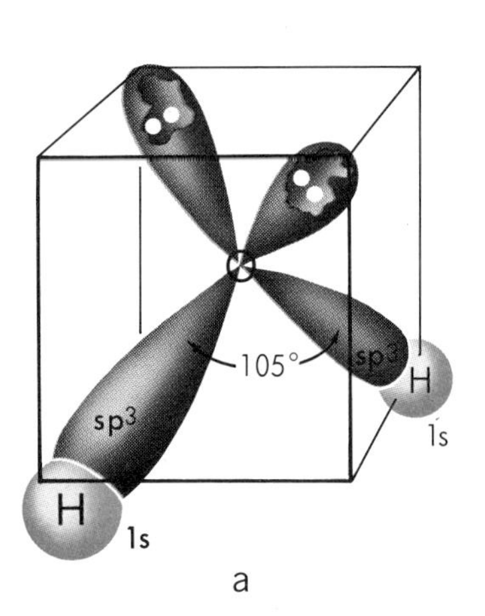

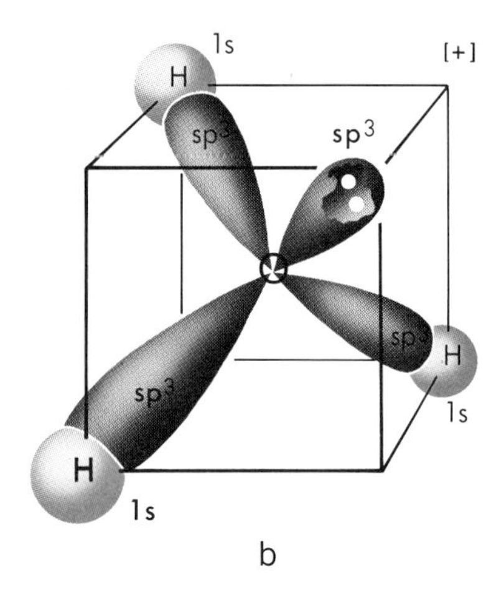

FIGURE 5.15 TETRAHEDRAL ARRANGEMENT OF ORBITALS IN THE WATER MOLECULE AND IN THE HYDRONIUM ION.

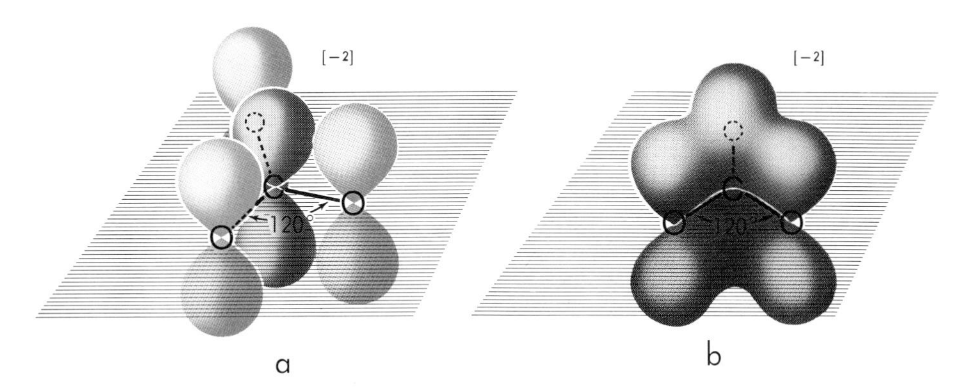

FIGURE 5.16 DELOCALIZED MOLECULAR ORBITAL OF THE CARBONATE ION.

The carbon atom of the *carbonate* ion,

```
    :Ö: 2−
     |
:Ö=C—Ö:
```

(only one of the three equivalent electron formulas is shown) is sp^2 hybridized: C $(sp^2)^1(sp^2)^1(sp^2)^1p^1$. Sigma molecular orbitals formed by the combining of the hybridized atomic orbitals of carbon with p orbitals of the three oxygen atoms gives the skeleton of the ion – a planar ion with bond angles of 120°. Each atom of the ion, as shown in Figure 5.16a, has a p orbital perpendicular to the plane of the molecule. The four p orbitals overlap to form delocalized pi molecular orbitals. In effect, the second pair of electrons represented in the double bond of the dot formula enters the delocalized orbital of lowest energy represented in Figure 5.16b. All three carbon to oxygen bonds are identical, being composed of one sigma bond and one third of a pi bond. Thus, the molecular orbital theory gives a reasonable picture of the resonance hybrid. According to the theory of resonance, each bond of the real ion is an "average" of what is indicated for it by the three dot formulas:

```
     :Ö 2−         :Ö: 2−        :Ö: 2−
      ||            |             |
:Ö—C—Ö:      :Ö=C—Ö:      :Ö—C=Ö:
```

That is, each C to O bond is not a double bond as it is represented in one of the three formulas, nor a single bond as it is shown by the other two dot formulas, but is a one and one third bond. Further, as seen by the orbital picture (Fig. 5.16b), the ion is in the shape of an equilateral triangle.

The *formate* ion of sodium formate, Na^+

```
    :Ö:−
     |
:Ö=C—H
```

, is a resonance hybrid. The structure,

```
    :O:−
     ||
:Ö—C—H
```

, is equivalent to the valence electron structure for the ion in the formula above. In terms of resonance, the carbon to oxygen bonds should be identical; they should be neither single nor double bonds but should be intermediate between them. According to the molecular orbital picture, the carbon atom is sp^2 hybridized. The skeleton of the molecule is three sigma molecular orbitals formed by the combining of the hybridized orbitals, two with p atomic orbitals of the oxygen atoms

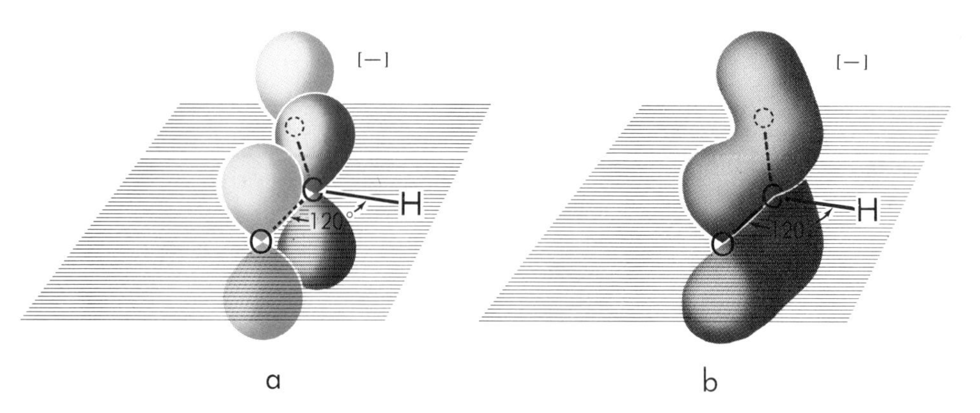

FIGURE 5.17 DELOCALIZED MOLECULAR ORBITAL OF THE FORMATE ION.

and one with the s atomic orbital of the hydrogen. A delocalized pi molecular orbital formed by the overlapping of p orbitals of the carbon atom and the two oxygen atoms is above and below the oxygen to carbon to oxygen portion of the ion (Fig. 5.17a). With the electrons added to the four molecular orbitals, each carbon to oxygen bond is equivalent to one sigma and one half of one pi bond, as shown in Figure 5.17b.

5.14 GENERALIZATIONS

The dot formula (valence electron structure) conveys much information to the chemist who understands molecular orbitals. It gives the general geometry of and indicates the types of molecular orbitals used in forming the molecule or complex ion. Characteristics of molecules with one central atom are related to the types of their dot formulas in Table 5.6. If in the dot formula .

1. **the central atom is bonded to four other atoms** through four single bonds and **the molecule is tetrahedrally shaped,** the central atom is sp^3 hybridized and all the bonds are sigma bonds.

what type of atom (i.e., sp^3, sp^2 or sp) is the middle carbon atom in $CH_3C{\equiv}CH$, $CH_3CH{=}CH$ and $CH_3CH_2CH_3$?

2. the central atom is bonded to three atoms with one double and two single bonds, **the molecule is planar;** the three bonded groups are at the apices of an equilateral, or nearly equilateral, triangle; the hybridization of the central atom is sp^2, and some pi bonding is present.

3. the central atom is bonded to two groups either by two double bonds or by a triple and a single bond, **the molecule is linear;** the central atom is sp hybridized, and there is pi bonding.

4. the central atom is bonded to three groups by single bonds and is surrounded by eight electrons, **the molecule is pyramidal** and the central atom is considered to be sp^3 hybridized with an unshared pair of electrons occupying a hybridized atomic orbital. The bonds are strictly sigma bonds.

5. the central atom is bonded to two groups with two single bonds and is surrounded by eight electrons, the molecule is considered sp^3 hybridized and contains two unshared electron pairs in hybridized orbitals.

TABLE 5.6 DOT FORMULAS, SHAPE AND HYBRIDIZATION OF THE CENTRAL ATOM

Type of Dot Formula	Shape	Hybridization of Central Atom	Example	Rule
—X— (with bonds above and below)	tetrahedral	sp^3	CH_4, NH_4^+	1
—$\ddot{X}$— (with bond below)	pyramidal	sp^3	NH_3, H_3O^+	4
—$\ddot{X}$:	bent	sp^3	H_2O	5
—X: (with single bond above)	bent	sp^2	NO_2^-, SO_2	6
—X— (with double bond above and below)	planar	sp^2	SO_3, SeO_3, CO_3^{2-}	2
=X= , —X≡	linear	sp	HCN, CO_2	3

6. the central atom is bonded to two groups with a single and a double bond and is surrounded by eight electrons, the central atom is considered sp^2 hybridized, and a hybridized orbital of the central atom is occupied by a pair of electrons. Because of the double bond there is pi bonding.

7. If the dot formula is not unique, that is, if more than one formula may be drawn for the molecule, there is delocalized bonding. The bonds which appear as both single and double bonds in the different formulas are involved in the delocalization to form bonds intermediate in character, one and one half, or one and one and one third, and so forth.

These rules do not cover the so-called electron-deficient molecules such as BF_3 (:F̤: / B̤:F̤: / :F̤:) and BBr_3 (:B̤r: / B̤:B̤r: / :B̤r:). The bonding of these molecules will be discussed when the central element is studied. The two molecules cited here are sp^2 hybridized with three single bonds and no unshared electron pairs.

5.15 DELOCALIZATION IN CARBON CHAINS

The study of a molecule, which is basically a chain of atoms, might be approached by considering every atom that is not terminal (at the end of the chain) to be a central atom with the geometry and hybridization predicted by the pattern of the octet of electrons about it. Except for possible delocalization, the total geometry is largely the sum of the geometry of the parts. However, delocalization is dependent upon the bonding of atoms adjacent to the "central" atom under consideration. Consider the compound 1,4-pentadiene (the nomenclature will be discussed in a later chapter), which has the following dot formula:

$$\begin{array}{c} \mathrm{H} \\ \mathrm{H:C::C:\ddot{C}:C::C:H} \\ \ddot{\mathrm{H}}\ \ \ddot{\mathrm{H}}\,\ddot{\mathrm{H}}\,\ddot{\mathrm{H}}\ \ \ddot{\mathrm{H}} \\ 1\ \ \ 2\ \ 3\ \ 4\ \ \ 5 \end{array}$$

Double bonds appear between the first and second and between the fourth and fifth carbon atoms. The first, second, fourth and fifth carbons are sp^2 hybridized; the third is sp^3 hybridized. Inasmuch as the two double bonds are separated by a carbon atom with sp^3 hybridization, and therefore with no p orbitals, the p orbitals perpendicular to the plane of the molecule appear in isolated pairs which combine to form localized pi bonds, as in Figure 5.18a. On the other hand, the electrons are delocalized in 1,3-pentadiene:

$$\begin{array}{c} \mathrm{H} \\ \mathrm{H:C::C:C::C:\ddot{C}:H} \\ \ddot{\mathrm{H}}\ \ \ddot{\mathrm{H}}\,\ddot{\mathrm{H}}\ \ \ddot{\mathrm{H}}\,\ddot{\mathrm{H}} \\ 1\ \ \ 2\ \ 3\ \ \ 4\ \ 5 \end{array} \quad \text{or} \quad \begin{array}{ccccccccccc} & & & & & & & & \mathrm{H} & & \\ & & & & & & & & | & & \\ \mathrm{H} & — & \mathrm{C} & = & \mathrm{C} & — & \mathrm{C} & = & \mathrm{C} & — & \mathrm{C} & — & \mathrm{H} \\ & & | & & | & & | & & | & & | & & \\ & & \mathrm{H} & & \mathrm{H} & & \mathrm{H} & & \mathrm{H} & & \mathrm{H} & & \end{array}$$

The first, second, third and fourth carbon atoms of the pentadiene molecule are sp^2 hybridized; the fifth is sp^3 hybridized. The unhybridized p orbitals, one on each of the first four carbon atoms, are parallel, each being perpendicular to the plane of the molecule. Being on successive atoms of the chain, they combine to form delocalized molecular pi orbitals. The four electrons not involved in sigma bonding enter the delocalized orbitals, one of which is sketched in Figure 5.18b. Because of the delocalization, the double bonds and the single bonds between them are actually neither double nor single bonds but are "one and a fraction" bonds. Electrons occupy orbitals above and below the plane of all carbon atoms, except for the fifth in 1,3-pentadiene. The apparent double bonds in the dot formulas for 1,4-pentadiene and 1,3-pentadiene are said to be **isolated** and **conjugated** respectively. Isolated double bonds are true double bonds, while alternate double and single bonds (conjugated double bonds) are neither double nor single but are fractional bonds with pi bonding running the full length of the conjugation.

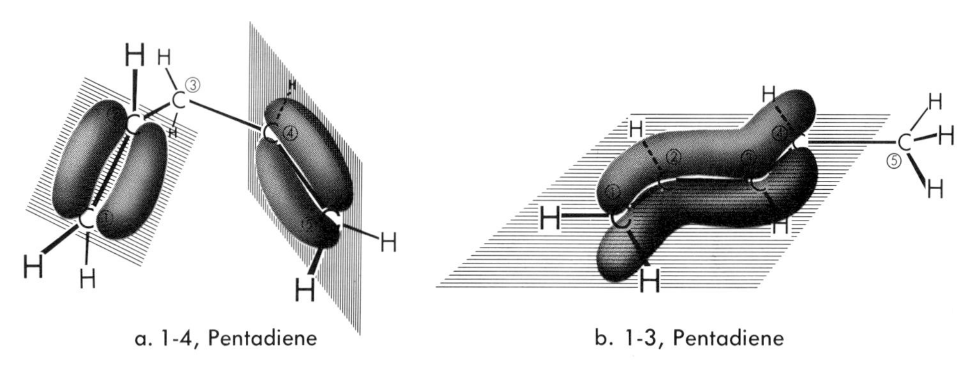

FIGURE 5.18 LOCALIZED AND DELOCALIZED PI BONDING IN PENTADIENES.

The benzene molecule, C_6H_6, contains a completely conjugated ring of carbon atoms:

```
           H                          H
           |                          |
   H       C       H          H       C       H
     \   /   \\   /             \   //  \   /
       C       C                  C       C
       ||      |       and        |       ||
       C       C                  C       C
     /   \   //   \             /   \\  /   \
   H       C       H          H       C       H
           |                          |
           H                          H
```

Because of the conjugation there are no single nor double bonds, but all carbon to carbon bonds are one and one half bonds. Six electrons occupy pi bonding molecular orbitals about the ring. One of the delocalized pi orbitals is shown in Figure 5.19.

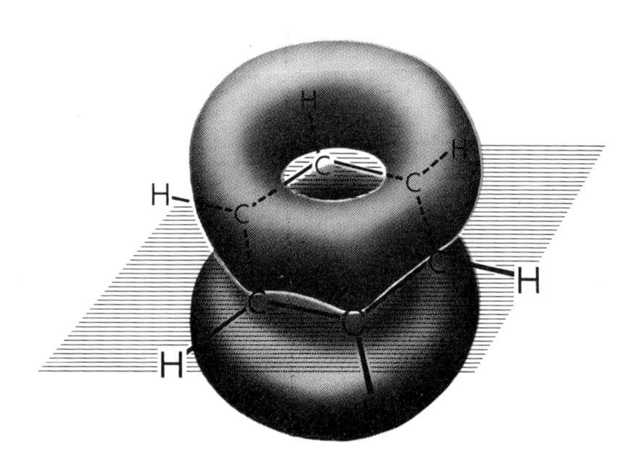

FIGURE 5.19 DELOCALIZED PI ORBITAL IN BENZENE.

5.16 BOND LENGTHS

Evidence for the delocalization of pi electrons in conjugated double bond systems is found in the comparison of the carbon to carbon bond lengths (the distances between the centers of carbon atoms bound together) in various structures (see Table 5.7). **The greater the multiplicity** of the bond, the more tightly the atoms are bound together and **the shorter the bond length.** Regardless of the

TABLE 5.7 CARBON TO CARBON BOND LENGTHS

Bond Type	Compound	Bond Length, Å
Single C to C	Alkanes and derivatives (Chap. 25)	1.54
Benzene C to C	Benzene (Chap. 28)	1.39
Single C to C	1,3-butadiene between carbons 2 and 3 (Chap. 27)	1.46
Double C to C	Alkenes (Chap. 27)	1.34
Triple C to C	Alkynes (Chap. 27)	1.21

hybridization of the two carbon atoms, the single carbon to carbon bond length is about 1.54 Å; the isolated double bond, 1.34 Å; and the triple bond, 1.21 Å. The so-called single bond in 1,3-butadiene indicated by the arrow, $CH_2{=}CH\overset{\downarrow}{-}CH{=}CH_2$, is 1.46 Å in length. Conjugation has shortened it 0.08 Å as compared to an "isolated" single bond. All carbon to carbon distances in the completely conjugated benzene molecule are 1.39 Å. This distance is nearer to the double bond length than to the single bond length. However, a plot of 1.39 Å on a smooth curve connecting the bond lengths of the three integral bonds (Fig. 5.20) shows that the carbon to carbon bond in benzene corresponds to the distance expected of a one and one half bond. This was the conclusion arrived at in the discussion of the molecular orbitals of benzene.

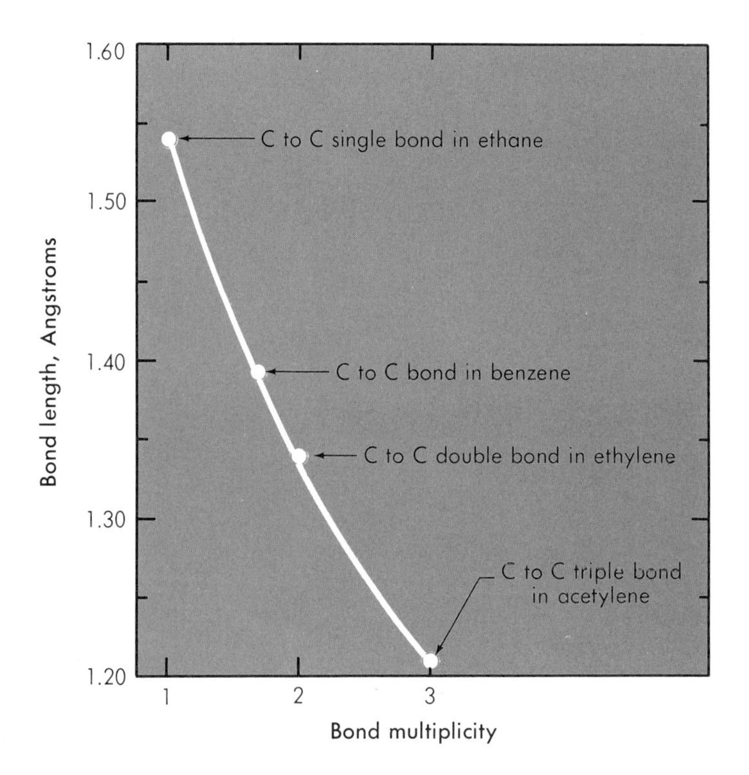

FIGURE 5.20 BOND LENGTHS VERSUS MULTIPLICITY OF CARBON TO CARBON BONDS.

5.17 POLAR MOLECULES (DIPOLES)

Molecules as a whole are neutral, that is, they bear no electrostatic charge. However, a molecule may be polar—one end may be negative or positive with respect to the other due to the unequal sharing of the electrons in bonding orbitals because of a difference in the electronegativity of the bonding atoms. In the bond between a sodium atom and a chlorine atom, the pair of electrons that might be shared in a molecular orbital formed by the combining of the 3s orbital of sodium and the 3p orbital of chlorine is attracted toward chlorine much more than toward sodium because of the greater electronegativity of the chlorine atom (see Table 4.4)—so much more, that the chlorine atom holds the electron pair exclusively and is a negative ion, while the sodium atom bears a plus charge, as shown in Figure 5.21. In the chlorine molecule (Cl_2) the bonding pair of electrons is in the molecular orbital formed by the combining of identical orbitals of identical atoms and therefore is shared equally between them. In the hydrogen chloride molecule a pair of electrons is shared by the two atoms in a molecular orbital formed by the overlapping of the 1s orbital of hydrogen and the 3p orbital of chlorine, but because of the greater electronegativity of chlorine, a difference not great enough to cause a complete transfer of electrons, the electrons are shared unequally. In effect, the electrons are held nearer the chlorine atom, giving it a fractional negative charge (δ^-) and giving the hydrogen end of the molecule an equal positive charge (δ^+). The molecular orbital is distorted in such a way that the electron probability is greater about the chlorine atom.

arrange these compounds in order of polarity: HBr, HF, LiH and F_2.

The **polarity of molecules** can be demonstrated and measured by an electrical apparatus called a capacitor. A capacitor is composed of two parallel metal plates mounted closely together. The two plates are connected to opposite poles of a battery (Fig. 5.22a). When the key is closed, a current flows through the wire, pulling electrons (e^-) from the left plate to the right plate. If the plates are moved closer together, a greater charge can be placed on the plates with the same voltage, because the positive plate attracts electrons across the void. The charge on the plates can also be increased if the air between the capacitor plates is replaced with dry HCl gas. This can be explained if the hydrogen chloride molecules are polar.

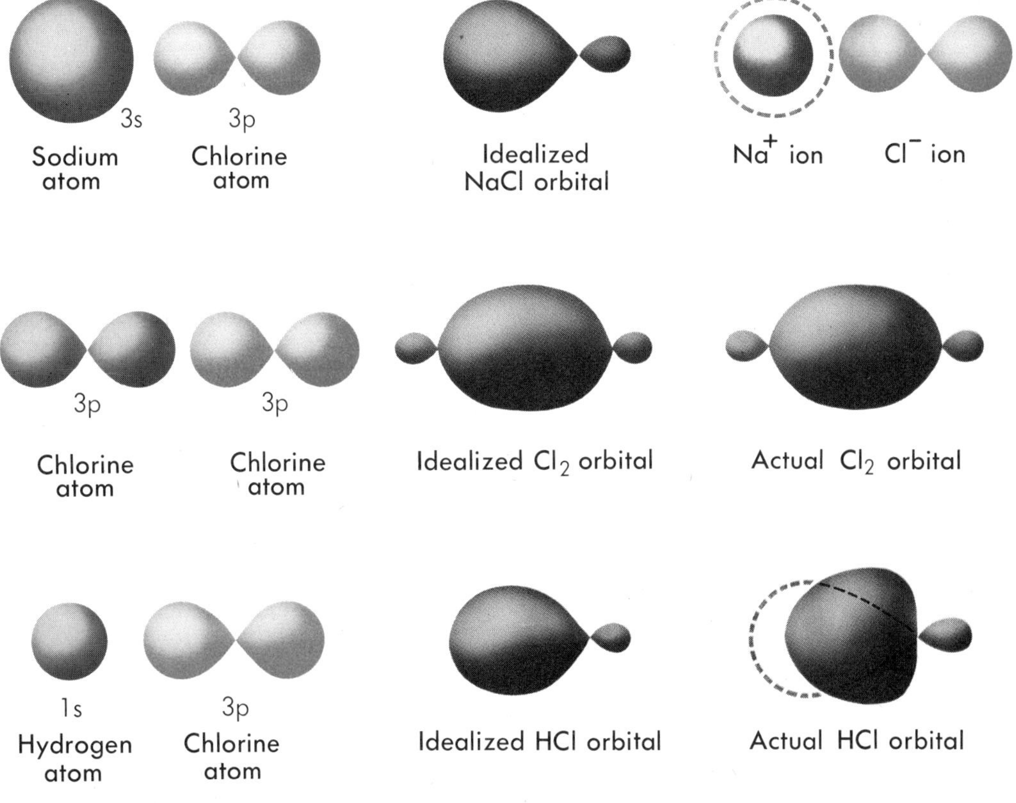

FIGURE 5.21 IONIC, NON-POLAR COVALENT AND POLAR COVALENT BONDS.

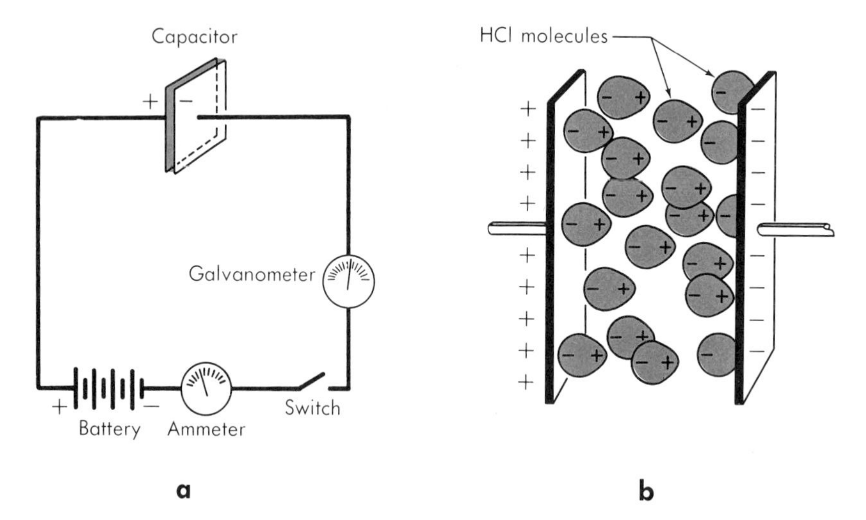

FIGURE 5.22 DIRECTION OF POLARITY.

If so, they would tend to be oriented + to − and − to + in the electrostatic field, as demonstrated in Figure 5.22b. The + ends of the molecules would be nearer the negative plate, attracting the electrons and making it possible to place more there without increasing the voltage. Replacing the hydrogen chloride gas between the plates with air causes a momentary reverse flow of electrons, restoring the charge on the plates to the value held originally with air.

is the molecule CH_2Cl_2 polar or non-polar?

Gases which impart an increased capacitance to a capacitor contain polar molecules. From the magnitude of the increase in capacitance, the strength of the electrostatic field and the pressure of the gas, the dipole moment can be calculated. The **dipole moment** is a measure of the total polarity of the molecule in terms of the product of the charge on the ends of the dipole and the charge separation. Values for several gases appear in Table 5.8. The dependence of the dipole on differences in electronegativity can be seen in the dipole values for the gases of the series hydrogen chloride, hydrogen bromide and hydrogen iodide. A decrease in the dipole moment follows the decreasing difference among the electronegativities of the three elements. The dipole moment of the water is great because the two polar bonds are complementary. The direction of the polarity is along the bisector of the angle towards the negative oxygen atom, as shown in Figure 5.23. Each

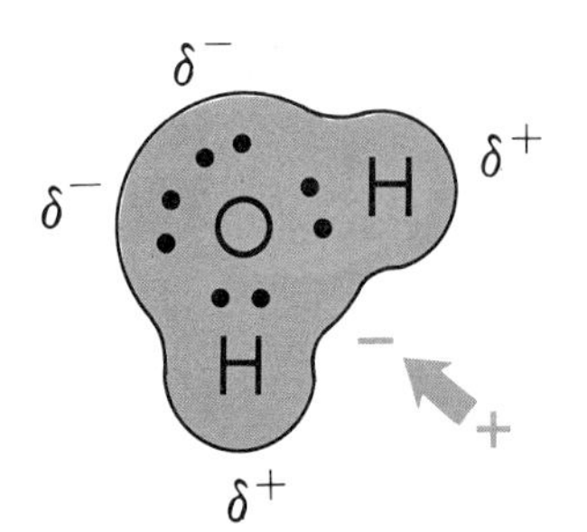

FIGURE 5.23 POLARITY OF THE WATER MOLECULE.

TABLE 5.8 DIPOLE MOMENTS IN DEBYE UNITS

HCl	1.03	NH_3	1.49
HBr	0.78	SO_2	1.61
HI	0.38	CO_2	0.0
H_2O	1.85	CO	0.11
H_2S	0.95		

nitrogen to hydrogen bond is less polar than the oxygen to hydrogen bond of water, yet the ammonia molecule is nearly as polar as water because of the combined polarity of the three bonds directed from the triangular base of the three hydrogens toward the nitrogen.

Even though carbon tetrachloride, sulfur trioxide and carbon dioxide each contain polar bonds, none are dipoles. The polarity of the bonds cancels out in the symmetry of the whole molecule. No one chlorine atom of carbon tetrachloride can be more negative than another, nor can either of the oxygens of carbon dioxide nor any one of the oxygens of sulfur trioxide be any more negative than another. Three types of symmetry are found here: tetrahedral, triangular and linear, each of which rules out any molecular polarity.

5.18 HYDROGEN BONDING

Hydrogen bonds are weak bonds between two highly electronegative atoms of the same molecule or between neighboring molecules through a hydrogen atom provided by one of them. The only **electronegative atoms** that can participate in the bond along with the hydrogen atom are **nitrogen, oxygen** and **fluorine.** The bond may be between two identical or two different atoms. The hydrogen bond is unlike either ionic or covalent bonds, in which electrons are either shared or transferred. Instead, the atoms, now negative because of an octet of electrons about them, are attracted to the positive proton between them. The enthalpy of formation of the bond is about 3 to 9 kilocalories per mole, about 1/10th to 1/20th that of a single covalent bond. Hydrogen bonds are nevertheless important. **Water has a very high boiling point** when compared with other binary hydrogen compounds of the elements of Group VI of the periodic table, as shown in Table 5.9. Ammonia and

TABLE 5.9 BOILING POINTS OF BINARY COMPOUNDS OF HYDROGEN (1 ATMOSPHERE PRESSURE), °C

Group VI Elements		Group VII Elements	
Compound	**Boiling Point**	**Compound**	**Boiling Point**
H_2O	100	HF	20
H_2S	−61	HCl	−85
H_2Se	−41	HBr	−67
H_2Te	−2	HI	−36

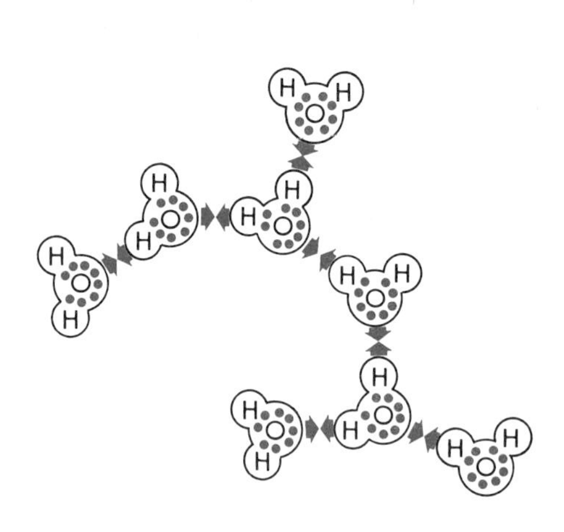

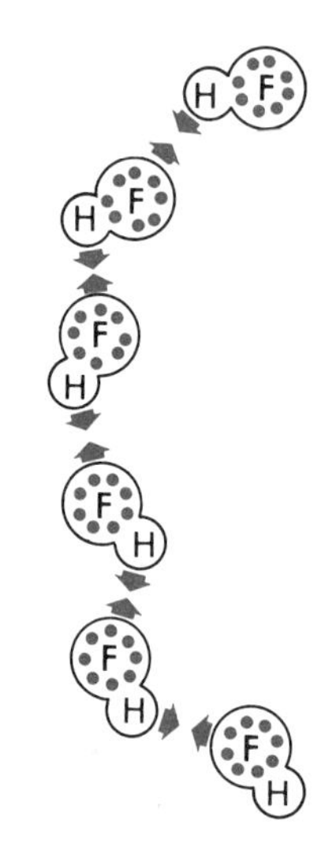

FIGURE 5.24 HYDROGEN BONDING.

hydrogen fluoride also have high boiling points when compared with other binary hydrogen compounds of Groups V and VII respectively. This is due mostly to hydrogen bonding between molecules, with the corresponding unusually high mutual attraction, as shown in Figure 5.24. The role of DNA as a tape in setting the pattern for the synthesis of certain proteins is aided by hydrogen bonding, as will be described in a later chapter.

5.19 PHYSICAL PROPERTIES VERSUS BOND TYPE

Molecules are held together in one unit by the sharing of electrons, but there is no electron sharing between molecules. Unless the molecules are strongly polar, or unless they are united by hydrogen bonding, the forces holding them together are very weak. Non-polar covalent substances have relatively low melting and boiling points, because little energy is required to separate the molecules; the solids are relatively soft and waxy. Highly polar molecules of a polar covalent substance tend to line up positive ends to negative ends, and because of higher intermolecular forces they have melting points higher than non-polar molecular substances. Ionic compounds, as solids set up in a lattice with unlike ions arranged as far as possible in a plus to minus fashion, usually have high melting and boiling points.

Ionic substances tend to be soluble in water and other polar solvents because the ions are strongly attracted to the polar water molecules. **Non-polar substances are soluble in non-polar solvents,** such as carbon tetrachloride, benzene, hexane and toluene, but are not soluble in polar solvents. Polar molecules, such as water, are mutually attracted and in drawing together exclude the non-polar molecules for which they have no special attraction. Polar covalent compounds tend to be soluble in water. Hydrocarbons are insoluble in water because there is very little difference in the electronegativity of carbon and hydrogen and consequently little polarity in the bonds. However, if the compound has some hydroxyl, —OH, amine, $—NH_2$ or other solubilizing groups, groups that can form hydrogen bonds with water, the substance may be miscible with the polar solvent.

EXERCISES

5.1 On what bases were the first periodic tables constructed?

5.2 What structural similarity exists among elements within a group (or family) of the periodic table?

5.3 Name the most important property of the noble gases.

5.4 Write the symbols of the complete electronic configurations of the first 21 elements.

5.5 In what way do the noble gases aid in a study of bonding?

5.6 Who had most to do with the discovery and identification of the noble gas elements?

5.7 What is an interesting item about the discovery, identification and naming of helium?

5.8 State four uses of the noble gases.

5.9 Compare the boiling points and freezing points of the noble gases (Table 5.7). Do they vary as you would expect? Explain.

5.10 What evidence is there that sodium and chloride ions exist in molten salt?

5.11 What does this anode equation mean: $2\ Cl^- \rightarrow Cl_2 + 2e^-$?

5.12 How might one show experimentally that the charges on the sodium and chloride ions are +1 and −1 respectively?

5.13 Define ionic valence.

5.14 How is the group number in the periodic table related to the ionic valence of the elements?

5.15 Explain why transition metals may exhibit more than one ionic valence.

5.16 Write the formulas for the following binary compounds from a knowledge of valences: (a) sodium fluoride, (b) lithium oxide, (c) calcium nitride, (d) sodium hydride, (e) aluminum bromide, (f) barium oxide, (g) magnesium nitride.

5.17 What is the noble gas configuration? Why is it called the noble gas configuration? What is the significance of that electron configuration in bonding?

5.18 Write the electronic configuration and show charges, where they exist, for: (a) the lithium atom and the lithium ion, (b) the fluorine atom and the fluorine ion, (c) Fe; Fe^{2+}, iron(II) ion; Fe^{3+}, iron(III) ion, (d) O^{2-}, F^-, Ne, Na^+, Mg^{2+} and Al^{3+}.

5.19 Write "kernel" electron configurations for the following atoms: Ca, O, N, K, C and S.

5.20 In the table of simple and complex ions (Table 5.15) (a) memorize all ions: name, formula and charge, (b) write the total electronic configuration of all ions checked (✓), (c) write the valence electron structure (dot formula) for all starred (*) ions, (d) be able to write the correct formula for the ionic compound that might be formed from each pair of plus and minus ions.

5.21 Write correct formulas for: (a) aluminum nitrate, (b) sodium phosphate, (c) calcium bicarbonate, (d) silver chromate, (e) nickel sulfate, (f) barium sulfide, (g) potassium bromide, (h) ammonium nitrite, (i) calcium carbonate, (j) mercury(I) chloride.

5.22 Write valence electron structures (dot formulas) for: (a) hydrogen gas, (b) nitrogen gas, (c) ammonia gas, (d) carbon tetrachloride, (e) ammonium ion, (f) hydroxide ion.

5.23 What are sigma bonds, pi bonds and multiple bonds?

5.24 In what respects are the methane molecule (CH_4) and the ammonium ion similar?

5.25 Write the formula for an ion that contains an atom that is hybridized and has an occupied delocalized pi orbital.

5.26 How many electrons can a p orbital hold? An sp^2 orbital? A delocalized orbital?

5.27 Which orbitals are involved in all hybridizations of the carbon atom: 1s, 2s, 2p, 3s, sp or 3d?

5.28 Give the hybridizations for the central atom of a linear molecule, a Y-shaped molecule and a tetrahedral molecule.

5.29 What are the bond angles in methane (CH_4), carbonate ion (CO_3^{2-}) and carbon dioxide?

5.30 Show how you can determine from the dot formula that a molecule or ion (a) contains pi bonding, (b) contains an atom with hybridized orbitals, (c) is linear, (d) is tetrahedral.

5.31 Which contains a delocalized bond: 1,3-pentadiene or 1,4-pentadiene? How do the so-called double bonds differ in the two compounds? What types of hybridized carbon atoms are found in each of the compounds?

5.32 Define resonance and give three examples of molecules stabilized by resonance.

5.33 What experimental evidence is there that the bonds in benzene are neither double nor single but are intermediate between the two?

5.34 Why does water have a much higher boiling point than most covalent compounds with nearly the same molecular weight?

5.35 How is it determined experimentally that molecules of a gas are polar?

5.36 The sulfur trioxide molecule (SO_3) contains polar bonds but is not polar. Explain.

5.37 Define: (a) polar bond, (b) polar molecule, (c) dipole.

5.38 Which of the following contain polar covalent bonds? (a) HCl, (b) H_2O, (c) CO_2, (d) CO, (e) NH_3, (f) CCl_4, (g) CH_4, (h) $HClO_4$, (i) NaCl, (j) Li_2O, (k) N_2, (l) BaS, (m) O_2. Explain why in each case.

5.39 There are three possible positions for hydrogen in the periodic table: (a) above lithium, (b) above fluorine, (c) above and little to the left of carbon. Explain in what respect each position is reasonable.

5.40 Why should methyl alcohol (CH_3OH) be soluble in water?

5.41 Write formulas for each of the following: sodium chloride, potassium chlorate, aluminum phosphate, iron(II) sulfate, copper(I) sulfide, sodium aluminum hexafluoride.

SUGGESTED READING

Barrow, G. M., Kenney, M. E., Lassila, J. D., Little, R. L. and Thompson, W. E.: *Understanding Chemistry*, Vol. II, Chemical Bonding (programmed instruction). W. A. Benjamin, New York, 1966.

Companion, A. L.: *Chemical Bonding*. McGraw-Hill Book Co., New York, 1964.

Ferreira, R.: "The Molecular Orbital Theory," *Chemistry, 41*(6):8, 1968.

Gray, H. B.: *Electrons and Chemical Bonding*. W. A. Benjamin, New York, 1965.

Lee, G. L.: *Principles of Chemistry—A Structural Approach*, Chapter 7. International Textbook, Scranton, Pa., 1970.

Ryschkewitsch, G. E.: *Chemical Bonding and the Geometry of Molecules*. Van Nostrand-Reinhold, New York, 1963.

SIX • GASES, LIQUIDS AND SOLIDS

6.1 THE GASEOUS STATE

Everyone is acquainted with gases, especially the gas mixture called air. In many ways a gas mixture behaves like a pure gas. For this reason air can be used as an example in much of our discussion.

It is common knowledge that air can be compressed into a smaller volume. It is also common knowledge that, as it is compressed, it exerts a force (called pressure) resisting that compression. This principle can be illustrated by consideration of an old-fashioned tire pump. The pump, Figure 6.1a, composed of a narrow, hollow, metal cylinder about two feet in length and two inches in diameter, is closed at one end. The tube is fitted with an air tight piston with a one-way valve; the piston can be raised and lowered by means of a metal rod with a wooden handle. A rubber hose is connected to the cylinder near the bottom or closed end. When the pump is in use, the other end of the hose is connected to the valve stem of a tire. Near the top of the cylinder is a small hole or air vent.

With the hose connected to the tire valve stem, the piston is raised to the top (Fig. 6.1b), and air passes down through the valve in the piston. The piston is then forced down, trapping the air in the cylinder and compressing it as the downward motion continues. As the piston approaches the bottom of the tube, the pressure of the air in the cylinder becomes sufficiently great that it is difficult to finish the stroke. When the pressure of the air in the pump exceeds the pressure in the tire, the air is admitted to the inner tube by the action of a one-way valve. The piston is raised again, letting in more air to be compressed in the cylinder and admitted to the tire. The process is repeated over and over until the tire is inflated. Anyone who has inflated a tire with such a pump knows that a large quantity of air at the pressure at which it enters the pump must be

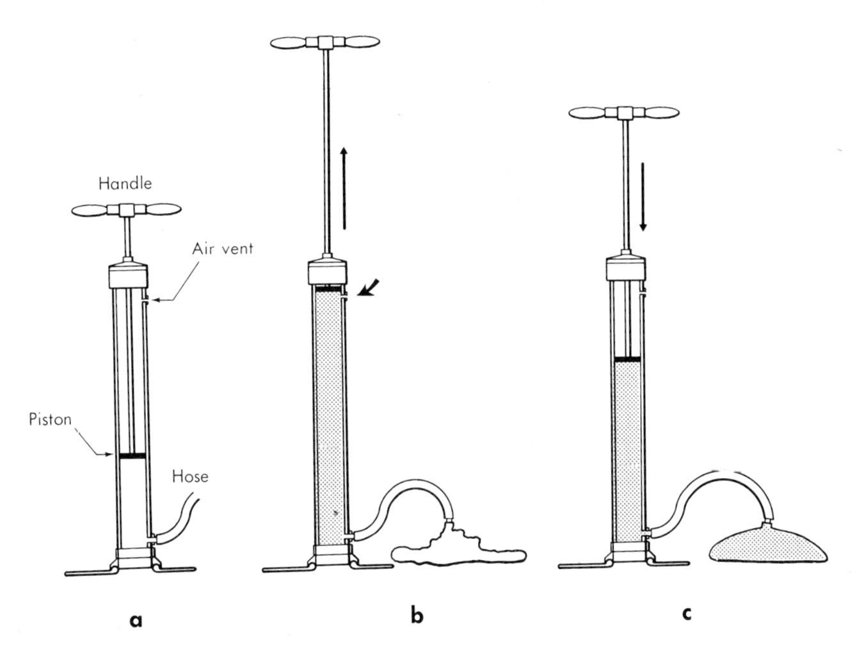

FIGURE 6.1 THE TIRE PUMP.

added to fill the tire with air to a pressure sufficient to support an automobile.

Compressed air keeps a football "tight"; it keeps a deep-sea diver's suit from being crushed by the tons of water over him. It operates jackhammers, paint sprayers, and many water pressure systems.

6.2 BOYLE'S LAW

the gas laws are commonly known by the name of the formulator. it would be well to learn them in that connection.

The quantitative relationship between the volume and the pressure exerted by a sample of gas is given by Boyle's Law. Robert Boyle (1627–1691), one of the founders of modern science, came to the conclusion, after many experiments with gases, that **when the temperature is maintained constant, the volume of a given sample of gas varies inversely with the pressure.** This means that if the pressure is doubled, the volume of the sample will be halved. It also tells us that if the pressure exerted on two liters of air is increased from 14.5 pounds per square inch to 43.5 pounds per square inch, the volume will be reduced to 2/3 liter:

$$2\text{ L} \times \frac{14.5\text{ lbs./in.}^2}{43.5\text{ lbs./in.}^2} = \frac{2}{3}\text{ L}$$

The law may be expressed mathematically as follows:

$$V \propto \frac{1}{P}, \; V = \frac{k}{P} \text{ (where k is a constant)}$$

$$\text{or } VP = k$$

But this is true only when the **temperature** and the **amount of gas** do not change. If any leakage occurs, the law does not apply.

6.3 MANOMETER

In scientific work, pressure is often expressed in "mm of mercury" because a mercury manometer is often used to measure pressure.

The **closed-end** manometer (shown connected to a glass bulb in Figure 6.2) is a slender U-tube mounted vertically, with mercury filling the lower portion. One arm of the U-tube is sealed off. Before sealing, the air was pumped from it, leaving as nearly a complete vacuum as possible. The other arm is connected to a container holding a gas sample.

Figure 6.2a shows the position of the mercury when the gas bulb contains no gas. No gas pressure acts on either mercury surface, so the surfaces are at the same level. In Figure 6.2b a sample of gas has been admitted to the bulb. The pressure in the flask has pushed the mercury down, around and up into the closed arm of the U-tube. The pressure is proportional to the height (h) to which the mercury level rises in the closed arm of the manometer above the level in the other arm.

A **barometer** is a special type of closed-end manometer used to measure the pressure of the gaseous atmosphere around it. The earth is surrounded by a layer of air some 50 miles in depth. The air layer is pulled to the earth by gravity and therefore presses down upon all objects on the earth's surface.

The effects of atmospheric pressure can be observed by attaching a vacuum pump to an empty can. As the air is pumped out and the pressure inside decreases, the weight of the atmosphere crushes the can. Atmospheric pressure is observed when one drinks

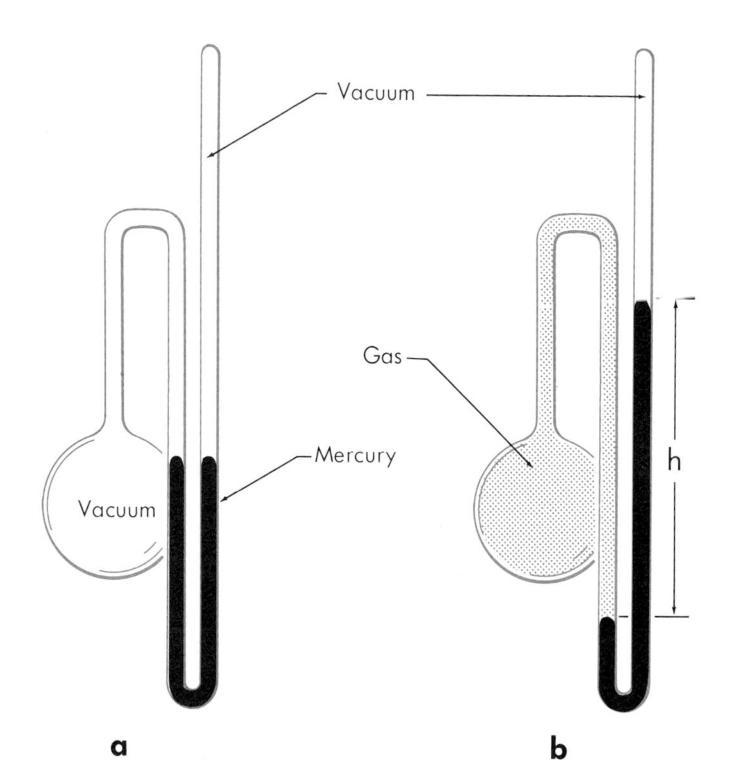

FIGURE 6.2 THE CLOSED-END MANOMETER.

FIGURE 6.3 DRINKING BEVERAGE WITH A STRAW.

a beverage through a straw. One doesn't draw the liquid up the straw; one draws air out of the straw, decreasing the pressure. The air around the straw and above the liquid pushes down on the water level, and the liquid rises in the straw (see Fig. 6.3).

The barometer may be a simple, closed-end manometer open to the air, as in Figure 6.4a, but usually it is a straight, closed-end, inverted tube in a dish of mercury, as in Figure 6.4b. Both types operate on the same principle. The air pushes down on the mercury surface open to the atmosphere, forcing the mercury column up the tube until its weight per square centimeter is equal to the atmospheric pressure.

The surface of the earth is uneven, with low valleys, and high land and mountain ranges penetrating the atmosphere. At high ele-

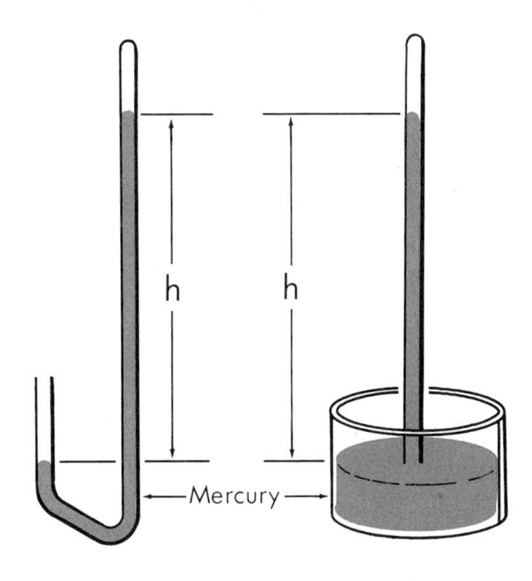

FIGURE 6.4 BAROMETERS.

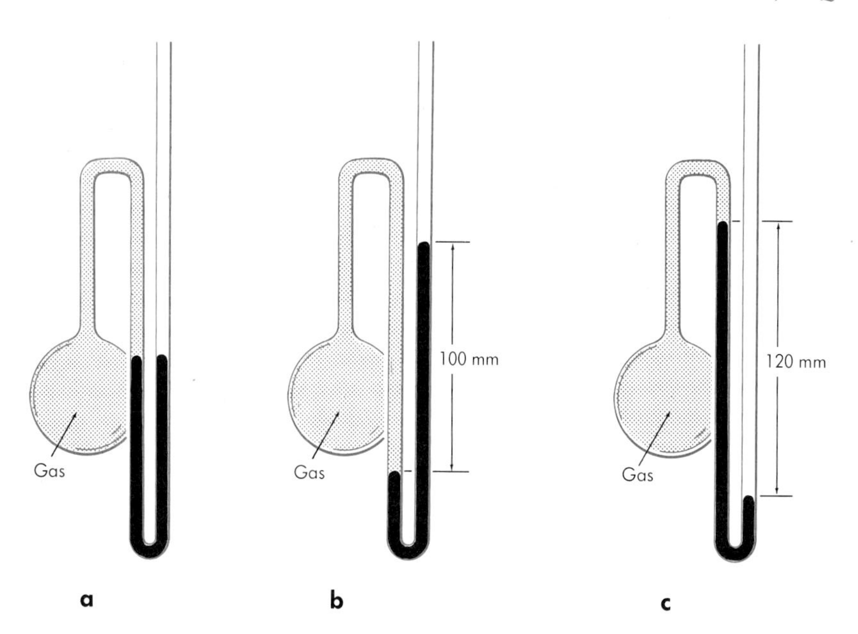

FIGURE 6.5 OPEN-END MANOMETER.

vations the air blanket is not so thick, and the atmospheric pressure is less than at sea level. At sea level the average barometric pressure is 760 mm or 14.7 pounds per square inch. This value has been taken as a standard for comparing gases. When one speaks of a gas at **standard pressure**, he means a **pressure of 760 mm of mercury.** One atmosphere (1 atm) is 760 mm; 2 atm = 2 × 760 mm = 1520 mm. At higher elevations the barometric pressure may be 640 mm of mercury or 640/760 of one atmosphere. Barometer readings are used in weather forecasting. When the humidity (relative moisture content) of the air is high, water vapor takes the place of the heavier air, the pressure decreases and the barometer drops. Should the mass of moisture-laden air at low barometric pressure contact a similar mass of cold air, rain or snow results.

Open-end manometers attached to a gas bulb, as shown in Figure 6.5, are sometimes used to measure gas pressures. In Figure 6.5a the mercury levels are the same, and the pressure on the gas sample is equal to the barometric pressure since the barometric pressure acts on the surface open to the atmosphere. In Figure 6.5b the pressure on the gas sample equals the barometric pressure plus 100 mm. In Figure 6.5c the pressure exerted on the gas sample is 120 mm less than the barometric pressure. If the barometric pressure were 730 mm, the pressure in the bulbs in Figures a, b and c of Figure 6.5 would be equal to 730 mm, 830 mm and 610 mm, respectively.

the prevailing barometric pressure must be known in order to determine pressures with an open-end manometer.

Calculations of changes in gas volume with pressure when the pressure is stated in "mm mercury" are similar to those with the pressure given in pounds per square inch.

Example 1: 150 ml of hydrogen gas at 25°C and 760 mm is expanded until the pressure is 380 mm. Calculate the final volume.

Solution: The volume varies inversely with the pressure. Here the pressure is reduced from 760 mm to 380 mm, and the volume increases by the factor 760/380.

$$150 \text{ ml} \times \frac{760 \text{ mm}}{380 \text{ mm}} = 300 \text{ ml}$$

One may note that the label on the answer is correct; milliliters expresses volume.

Example 2: Find the volume at standard pressure and 18°C of a 2.5 L air sample measured at 18°C and 570 mm.

Solution: As the pressure increases, the volume decreases. The new volume is given by the product of the initial volume and the ratio of the pressures.

$$2.5 \text{ L} \times \frac{570 \text{ mm}}{760 \text{ mm}} = 1.9$$

The pressures may be stated in atmospheres (Example 3).

Example 3: 280 ml of oxygen gas at 25°C and 1.0 atm is compressed to 1.5 atm at the same temperature. Find its final volume.

$$280 \text{ ml} \times \frac{1.0 \text{ atm}}{1.5 \text{ atm}} = 190 \text{ ml}$$

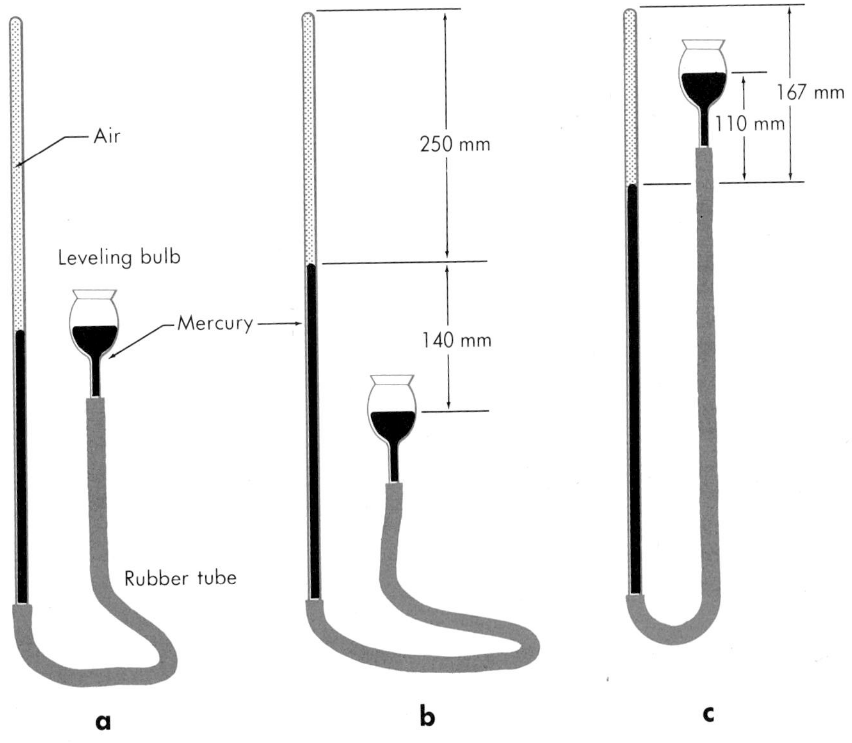

FIGURE 6.6 BOYLE'S LAW APPARATUS.

6.4 BOYLE'S LAW APPARATUS

A simple laboratory apparatus for examining Boyle's Law is shown in Figure 6.6a. A sample of air is trapped in the inverted tube over mercury which fills the lower end of the tube, a connecting hose and a leveling bulb. The inverted tube acts as both the gas container and one arm of the manometer. The leveling bulb and hose constitute the other arm and also a device for varying the pressure on the gas sample. When the leveling bulb is raised, mercury runs into the tube, compressing the sample. The pressure on the sample at any instant is equal to the barometric pressure plus the difference in mercury levels if the level in the bulb is higher, or the barometric pressure minus the difference in levels if the level in the bulb is lower. The length of the gas sample (the distance from the end of the closed tube to the mercury level) is a measure of the volume.

in this case the cross sectional area of the tube containing the gas sample is 1.00 cm^2 such that for each 1-cm change in height of the mercury in the tube, there is a change of 1 ml in volume.

Suppose that at a barometric pressure of 640 mm the leveling bulb is adjusted in such a way that the level of the mercury in it is 140 mm below the mercury level in the tube (Fig. 6.6b), and the volume of the gas sample is 250 ml at 25°C. Now, with no change in temperature and no leakage, the leveling bulb is lifted until the level of mercury in it is 110 mm above the level in the tube (Fig. 6.6c).

From Boyle's Law the new volume may be calculated as follows: 250 ml volume at (640 − 140) mm pressure and 25°C is compressed to a pressure of (640 + 110) mm. What is the resultant volume?

$$250 \text{ ml} \times \frac{500 \text{ mm}}{750 \text{ mm}} = 167 \text{ ml}$$

The measurement shows that the volume of the air sample is 167 ml, as shown in Figure 6.6c, and Boyle's Law is verified.

6.5 TEMPERATURE

Charles' Law is concerned with the **variation of volume with temperature changes.** This requires a definition or discussion of temperature. Some qualitative statements of temperatures are: "It is hot." "It is warm." "It is cold." When a person says, "The beaker is hot," he means that the beaker is of such a temperature that when in contact with his hand, heat flows from the beaker to his hand, giving a burning sensation.

The first law of temperature states that when two objects are brought in contact, heat must flow from the one of higher temperature to the other. Because our skin is so sensitive to heat flow, one can touch an object and sense that its temperature is higher, lower or the same as body temperature. It is difficult, however, to tell by the sense of touch *how much* colder one object is than another. A good conductor, such as iron, copper or another metal, feels colder than the rug or straw it lies upon, because the metal carries the heat away faster. Thus, the sensation of heat flow is not a suitable standard for measuring temperature.

For centuries it has been known that most substances expand upon heating and contract when cooled. Furthermore, with liquids and solids, on which ordinary pressure changes have little effect, the volume is always the same at the same temperature. If calibrated, the volume can be a quantitative measure of the temperature of the object or material that surrounds it. Ordinary thermometers operate upon that principle, measuring the temperature by the extent of expansion of a liquid.

The **mercury thermometer** utilizes the thermal expansion of mercury to indicate temperature. The expansion of mercury with small temperature changes is so slight as to be unobservable if the mercury is contained in an ordinary container. The thermometer is simply a glass bulb sealed to a fine capillary tube and filled with mercury in such a way that the mercury stands part way up the capillary, as in Figure 6.7.

Expansion of the mercury appears as a rising thread in the capillary tube. The thread rises when the mercury is heated and falls when it is cooled. Graduations appear on the capillary tube or on a card mounted adjacent to it where one can read the temperature directly. Two temperature scales are used: the Fahrenheit and the centigrade. Thermometers are known by the scale they carry.

Zero on the **Fahrenheit thermometer scale** was set by placing

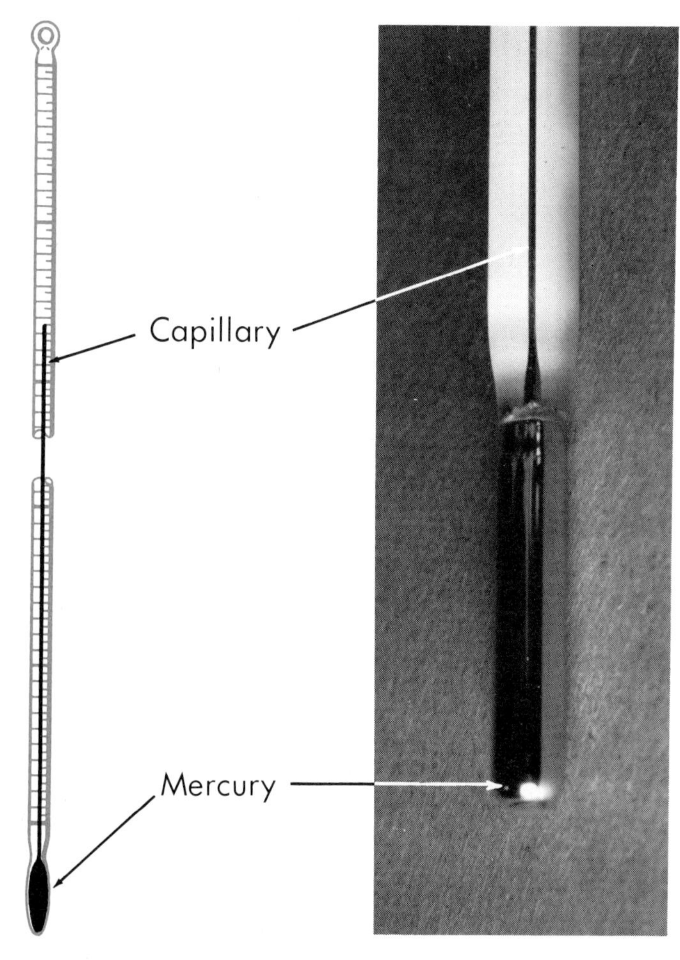

FIGURE 6.7 MERCURY THERMOMETER.

an unmarked thermometer in the coldest mixture of ice, water and table salt that could be made. The position of the mercury was marked on the glass and labeled 0°F. This temperature was assumed to be the lowest temperature that could be attained in a laboratory. Body temperature was then selected to be 96. The thermometer was placed in someone's mouth and the mercury level marked and labeled 96°F. It is interesting that normal body temperature was measured incorrectly by 2.6°. To finish calibrating the thermometer, the distance between the 0° and 96° marks was divided into 96 equal portions with 95 marks. Degrees of the same length are marked off above 96° and below 0°F to complete the scale (Figure 6.8). With this thermometer the freezing point of water is +32° and the boiling point is 212°. There are 180 F° (read Fahrenheit degrees) between the freezing and the boiling points of water. This is the scale used commonly by doctors, weather observers, cooks and most industrial establishments in the USA; all other countries use the centigrade scale.

Physical properties of water are standards for the **centigrade thermometer.** The freezing point of pure water is taken as zero and the boiling point at 760 mm as 100°C. To calibrate the thermometer, one merely places it in a slush of ice and water, makes a mark on the glass even with the mercury and labels it 0°C. One sets the pressure above a vessel of water at 760 mm, heats the vessel until the water boils and holds the bulb just above the water until steam condenses on it and drips steadily back into the vessel. He marks the mercury level on the glass stem of the thermometer and labels it 100°C. The thermometer between 0° and 100° is divided into 100 equal portions. Degrees of equal length are marked above 100°C and below 0°C, as was done with the Fahrenheit thermometer.

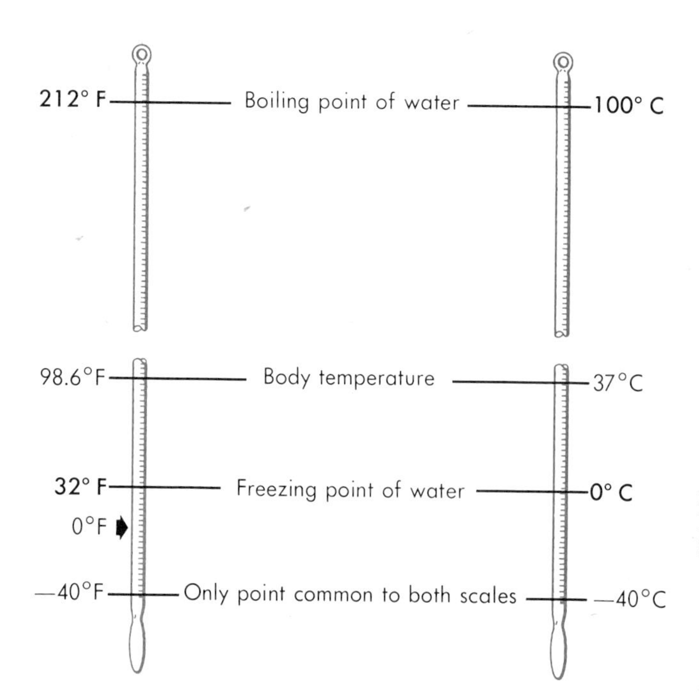

FIGURE 6.8 COMPARISON OF FAHRENHEIT AND CENTIGRADE SCALES.

The centigrade scale is used in scientific work in the U.S. and all the world. The trend is toward its use in industry, too; even in England it is now official. It was found to be much easier to calibrate a thermometer with the centigrade scale than with the Fahrenheit. Normal body temperature varies, and the mixture of ice, salt and water giving the lowest temperature is difficult to prepare.

Both scales may be placed on the same thermometer. The mercury stands the same place at the same conditions regardless of the scale, but the scales are set in relation to each other so that mercury is at 32° on the Fahrenheit scale and 0° on the centigrade scale at freezing, at 98.6°F and 37°C when placed in your mouth, and at 212°F and 100°C in boiling water (Fig. 6.8).

One can convert °F to °C and vice versa if he remembers that there are 212° − 32° = 180F° between freezing and boiling, but only 100° − 0° = 100C°. Thus the centigrade degree is 180/100 = 9/5 as large as the Fahrenheit degree. 0°C starts at 32°F so:

$$\text{degrees C} = (\text{degrees F} - 32) \times \frac{5}{9}$$

Example: How is 88°F expressed on the centigrade scale?

Solution: (88 − 32)5/9 = 31.1°C.

6.6 CHARLES' LAW

When a given quantity of air is heated, it must increase either in volume or in pressure or both. If one has his automobile tires filled to the required pressure in cold weather, he will find on checking his tires after driving for some distance that the pressure has increased several pounds per square inch. He might notice that the tires are hot from the friction on the road surface. When the tires are cooled, the pressure returns to its initial value. Children become victims of this phenomenon when they take air-filled balloons out in the hot sun, only to see them expand and burst. The expansion of air in the balloon with the heat absorbed at the earth's surface and the attendant decrease in density account for the rising of the balloon and its displacement with cold air from above. A wind is just a large scale displacement of warm, rising air with cold air.

The French physicist, **J. A. C. Charles** (1746–1823), studied the effect of temperature changes on the volume of a sample of gas at constant pressure. He and those who followed him made observations such as these:

1. When a gas sample is heated from 0°C to 1°C, the final volume is 274/273 of the initial volume.
2. When a gas sample is heated from 25°C to 30°C, the volume increases to 303/298 of its original value.

That is, a 250 ml sample of air at 25°C and 1 atm is heated to 30°C at 1 atm. Its final volume is 254 ml

$$\text{and } 250 \text{ ml} \times \frac{303}{298} = 254 \text{ ml}$$

3. When a sample of gas is heated from 100°C to 105°C, the volume increases to 378/373 of its original value. That is: 3.40 L of gas at 100°C becomes

$$3.40\text{ L} \times \frac{378}{373} = 3.45\text{ L at }105°\text{C}$$

As expected, an increase in volume always attended a rise in temperature. Furthermore, if Charles added 273 to each temperature, then multiplied the original volume by the ratio of the final temperature to the initial temperature, he could calculate the final volume. Notice in the three examples (a, b and c) given above: 274 = 273 + 1°C, 273 = 273 + 0°C, 303 = 273 + 30°C; 298 = 273 + 25°C; 378 = 273 + 105°C, and so forth.

To illustrate, the calculation for observation 3 is:

$$3.40\text{ L} \times \frac{273 + 105°\text{C}}{273 + 100°\text{C}} = 3.40\text{ L} \times \frac{378}{373} = 3.45\text{ L}$$

Charles further reasoned that since the method applied at all temperatures, **−273°C must be the absolute bottom of temperature** and 25°C was really 273 + 25 = 298° above the bottom or lowest possible temperature. Readings on this new temperature scale are then **absolute.**

$$25°\text{C} = 273 + 25° = 298°\text{A}$$

$$-60°\text{C} = (273 - 60)°\text{A} = 213°\text{A}$$

In all cases, 273 is added to the temperature in °C to find °A. A simple conversion of °C to °F shows that 0°A is also −459°F. The three temperature scales are compared in Table 6.1. The Absolute Temperature Scale used here is identical with the Kelvin Temperature Scale and in every case °K can be substituted for °A.

It is of interest that further research has shown Charles to be correct: −273°C (or more exactly, −273.16°C) is the lowest temperature possible. Thermodynamic studies verify it. Zero degrees absolute has almost been attained in the laboratory, that is, within a few thousandths of a degree.

The formal statement of Charles' Law is then apparent: **With the pressure constant, the volume of a sample of gas varies directly with the absolute temperature.** This law may be stated mathe-

TABLE 6.1 FAHRENHEIT, CENTIGRADE AND ABSOLUTE TEMPERATURE SCALES

°F	°C	°A
392°F	200°C	473°A
212°F	100°C	373°A
32°F	0°C	273°A
−148°F	−100°C	173°A
−328°F	−200°C	73°A
−459°F	−273°C	0°A

matically:

$$V \propto T, V = kT \text{ or } V/T = k$$

(where k is a constant)

$$\text{or } \frac{V_1}{T_1} = \frac{V_2}{T_2}$$

where V_1 and T_1 equal the volume and temperature at one time, and V_2 and T_2 are the volume and temperature of the same gas sample at another volume and temperature. It must be stressed that the law holds only when gas is neither added to nor lost from the sample, and when the pressure remains constant.

In solving problems using Charles' Law, one might substitute into various forms of the equation above or one might apply the corrections to the volume, as Charles found them.

Example 1. A 150 ml sample of oxygen at 1 atm and 25°C is heated to 80°C with the pressure remaining at 1 atm. Calculate the final volume.

$$150 \text{ ml} \times \frac{(273 + 80)°A}{(273 + 25)°A} = 177 \text{ ml}$$

Reasoning it through, one would first change all temperatures to °A. He would say to himself, "The temperature is raised from 298°A to 353°A. The volume increases with a rise in temperature, so multiply the volume by a ratio of the temperatures greater than 1, that is 353/298 and not 298/353." The answer, as anticipated, is a larger final volume (177 ml) than the original (150 ml).

Example 2. 3.5 L of nitrogen at −23°C is heated to 210°C. During the process the pressure remains at 660 mm. To what volume did the nitrogen expand?

The temperatures are

$$273 - 23°C = 250°A \text{ and}$$

$$273 + 210°C = 483°A$$

The volume increases with the temperature:

$$3.5 \text{ L} \times \frac{483°A}{250°A} = 6.75 \text{ L}$$

A **standard temperature** has been chosen to be used with **standard pressure** when gas volumes are compared. Standard temperature is 0°C or 273°A.

Example 3. What is the volume at STP (*S*tandard *T*emperature and *P*ressure) of 2.60 L of oxygen measured at 760 mm and +27°C?

Solution: The absolute temperature decreases by the fraction:

$$\frac{273}{273 + 27} = \frac{273°A}{300°A}$$

and so does the volume:

$$2.60\ L \times \frac{273°A}{300°A} = 2.34\ L$$

6.7 CHARLES' LAW APPARATUS

With additional equipment the apparatus used to test Boyle's Law (Fig. 6.6) may be used to study Charles' Law. This addition is a glass jacket around the sample tube connected through a rubber tube to an Erlenmeyer flask partly filled with water and arranged such that it may be heated at the proper time (see Figure 6.9).

Again the volume of the air sample trapped in the closed-end tube is proportional to the length of the tube it occupies. The pressure is equal to the barometric pressure plus the vertical distance between the mercury level in the bulb and the level of the water in the sample tube.

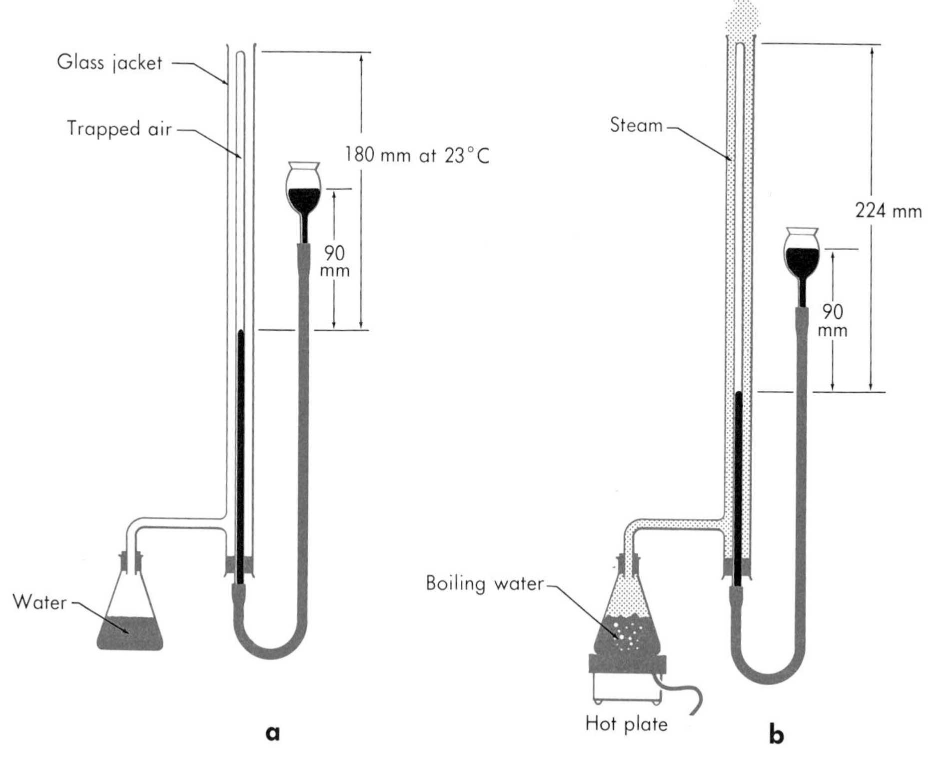

FIGURE 6.9 CHARLES' LAW APPARATUS.

To test Charles' Law, one may measure the "volume" (length) of the sample and the vertical distance between the mercury levels at room temperature, recording both of these measurements along with the temperature. Next the water in the flask is boiled gently until steam rises in the jacket and envelops the tube containing the air sample, raising it to the boiling point of water at the prevailing atmospheric pressure. The gas sample expands, which also changes the vertical distance between the mercury levels. The leveling bulb must then be lowered until the same difference exists between the mercury levels as it did at room temperature. The new volume (length) of the sample is then measured. One may calculate the new volume at boiling temperature from the volume measured at room temperature by Charles' Law and see how closely it agrees with the measured volume.

A sample calculation is shown below. Suppose the volume of the air sample is 180 mm at 23°C, room temperature, with the barometric pressure at 630 mm and the mercury level in the bulb 90 mm above the level in the sample tube. What should the volume be when surrounded with steam with the mercury levels again adjusted to a vertical distance of 90 mm?

Solution: *Initially* there are 180 ml of air in the tube at 23°C (296°A) and 630 mm + 90 mm = 720 mm. Finally, the sample is at 720 mm as before and at 95°C (368°A), the boiling point of water when the barometric pressure is 630 mm. The volume should increase by the factor 368°A/296°A and is then:

$$180 \text{ ml} \times \frac{368°\text{A}}{296°\text{A}} = 224 \text{ ml}$$

Because temperature and pressure have an independent effect on volume changes, one can carry out corrections for both on the same sample of gas and may do so in either order.

Example 1. Find the volume at STP of 180 ml of nitrogen measured at 35°C and 540 mm.

Initially: 180 ml at (35 + 273)°A and 540 mm

Finally: ? ml at 273°A and 760 mm

The volume decreases with the temperature:

$$180 \text{ ml} \times \frac{273°\text{A}}{308°\text{A}}$$

and decreases when the pressure increases:

$$180 \text{ ml} \times \frac{273°\text{A}}{308°\text{A}} \times \frac{540 \text{ mm}}{760 \text{ mm}} = 113 \text{ ml}$$

Example 2. What volume will 265 ml of a gas at −23°C and 1520 mm occupy at 670 mm and 0°C?

Initially: 265 ml at (273 − 23)°A and 1520 mm

Finally: ? ml at 273°A and 670 mm

The volume increases as the temperature increases:

$$265 \text{ ml} \times \frac{273°\text{A}}{250°\text{A}}$$

and increases as the pressure decreases:

$$265 \text{ ml} \times \frac{273°\text{A}}{250°\text{A}} \times \frac{1520 \text{ mm}}{670 \text{ mm}} = 656 \text{ ml}$$

Example 3. What is the volume at STP of 1.54 L of oxygen at 2.40 atm and 90°C?

Initially: 1.54 L at 2.40 atm and (90 + 273)°A

Finally: ? L at 1.0 atm and 273°A

$$1.54 \text{ L} \times \frac{2.4 \text{ atm}}{1 \text{ atm}} \times \frac{273°\text{A}}{363°\text{A}} = 2.78 \text{ L}$$

6.8 DALTON'S LAW OF PARTIAL PRESSURES

The third quantitative law describing the behavior of gases was formulated by John Dalton (1766–1844), the same Dalton whose atomic theory has been discussed. He found from his experience with air and other gas mixtures that **in a mixture of gases, the total pressure exerted by the mixture is equal to the sum of the pressure each gas would exert if alone in the container.** This concerns a mixture and not a compound formed by the combination of the gases in the mixture.

Individual gases of a mixture diffuse through each other and throughout the entire volume of the container. In that sense each gas fills the container, but in so doing, its contribution to the total pressure is only that which it would exert if alone in the container. The total pressure is then the sum of the pressures of each gas in the mixture. Gases in a mixture do not divide the volume among themselves, but divide the pressure.

Some calculations using Dalton's Law are shown below:

Example 1. In a mixture of nitrogen and hydrogen gases with a total pressure of 760 mm mercury, nitrogen exerts a pressure (partial pressure) of 190 mm. Find the pressure of hydrogen in the mixture.

Solution: The total pressure = pressure of nitrogen + pressure of hydrogen

$$P_T = P_{N_2} + P_{H_2}$$

$$\text{and } P_{H_2} = P_T - P_{N_2}$$

$$= 760 \text{ mm} - 190 \text{ mm} = 570 \text{ mm}$$

---

each gas of a mixture fills the container. the gases of the mixture divide the total pressure between themselves.

Example 2. 100 ml of oxygen at 380 mm and 25°C is placed in a 200 ml flask. Carbon dioxide is added to the flask with the oxygen until the total pressure is 690 mm. Calculate the pressure of (a) oxygen and of (b) carbon dioxide in the flask.

Solution: (a) If no CO_2 were present, a "Boyle's Law" calculation would tell us:

$$P_{O_2} = 380 \text{ mm} \times \frac{100 \text{ ml}}{200 \text{ ml}} = 190 \text{ mm}$$

Furthermore, addition of CO_2 does not alter the pressure exerted by oxygen; it remains 190 mm.

(b) The carbon dioxide exerts the balance of the 690 mm:

$$P_{CO_2} = 690 \text{ mm} - 190 \text{ mm} = 500 \text{ mm}$$

Example 3. How many ml of oxygen gas measured at 760 mm and 20°C must be added to 500 ml of neon at 460 mm and 20°C to bring the total pressure to 760 mm?

Solution: The pressure of oxygen in the mixture in a 500 ml container must be: 760 mm − 460 mm = 300 mm.

That is, 500 ml at 300 mm are needed. At 760 mm the volume will be:

$$500 \text{ ml} \times \frac{300 \text{ mm}}{760 \text{ mm}} = 197 \text{ ml}$$

Example 4. 400 ml of oxygen and 100 ml of nitrogen both at STP are allowed to mix in a larger volume at a total pressure of 640 mm, the temperature remaining at 0°C. (a) What is the partial pressure of oxygen? (b) What is the volume of the container?

Solution: (a) 4/5 of the gas by volume is oxygen. It will exert 4/5 of the final pressure.

$$\frac{400 \text{ ml}}{(100 \text{ ml} + 400 \text{ ml})} \times 640 \text{ mm} = \frac{400 \text{ ml}}{500 \text{ ml}} \times 640 \text{ mm}$$

$$= \frac{4}{5} \times 640 \text{ mm} = 512 \text{ mm}$$

(b) The oxygen pressure in the new container is 512 mm (part a). 400 ml of oxygen at 760 mm and 0°C is changed to 512 mm at 0°C. Its volume is then:

$$400 \text{ ml} \times \frac{760 \text{ mm}}{512 \text{ mm}} = 594 \text{ ml}$$

594 ml is the total volume, for all gases in a mixture occupy the total volume and divide the pressure among themselves.

It should be noted in passing that **Charles', Boyle's and Dalton's Laws do not hold precisely.** That is, when one calculates a new volume of a gas sample with changes in temperature or pressure or both, and then measures the change very precisely, he will find a small difference. The error varies with the kind of gas and its temperature and pressure. However, if the initial and final temperatures are well above the condensation temperature of the gas (the boiling point of its liquid), and if the pressures are no more than a few atmospheres, the error is usually small enough to be neglected.

The nature of the error and its source will be discussed in part in the next chapter.

6.9 THE LIQUID STATE

Under the proper conditions any and all gases may be liquefied. Water vapor, a gas, may be liquefied by holding a cold spoon over the spout of a boiling kettle. Rain occurs as a result of a condensation of water vapor. The reverse process, the change of a liquid to a gas, has often been observed. Boiling water decreases in amount as it evaporates. Even at room temperature water evaporates slowly. The liquids alcohol, acetone, gasoline and ether evaporate more readily than water at room temperature.

6.10 VAPOR PRESSURE

A careful study of the phenomenon of evaporation is interesting. Consider a container partially filled with water and attached to a manometer from which the air has just been pumped. At the instant the apparatus is isolated from the pump (by closing the

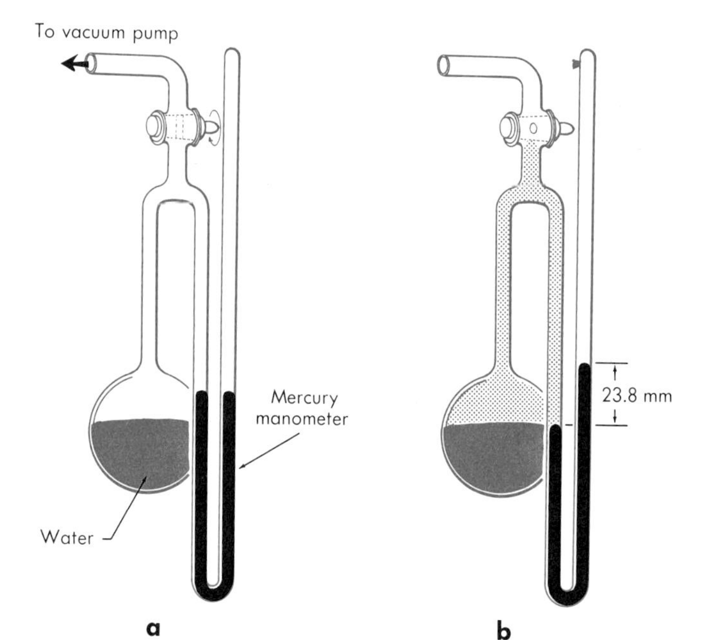

FIGURE 6.10 MEASUREMENT OF VAPOR PRESSURE.

TABLE 6.2 VAPOR PRESSURE TABLE FOR WATER

Temperature °C	Pressure mm	Temperature °C	Pressure mm
0	4.6	50	92.5
5	6.5	60	149.4
10	9.2	70	233.7
15	12.8	80	355.1
20	17.5	90	525.8
25	23.8	100	760.0
30	31.8	110	1074.6
35	42.2		
40	55.3		

stopcock), the mercury levels in the manometer are equal, as shown in Figure 6.10a. However, evaporation rapidly occurs and the pressure of the vapor above the liquid rises rapidly, pushing the mercury level down, and mercury rises up into the closed end of the manometer. If the vapor pressure is relatively great at the temperature measured, the pressure may rise so rapidly that the mercury is shot up into the closed arm of the manometer, sometimes breaking it. Very shortly the pressure ceases changing, and the levels in the mercury manometer remain constant unless the temperature changes. At 25°C the difference in the mercury levels, the pressure, is 23.8 mm (Fig. 6.10b). The pressure, 23.8 mm mercury, is "the vapor pressure" of water at 25°C.

This vapor pressure of water can be changed in only one way: by changing the temperature. Neither adding more water nor admitting gas above the liquid has any measurable effect on the vapor pressure of the water. If the temperature is lowered, the vapor pressure decreases; if it is raised, the vapor pressure increases. The vapor pressures of water at various temperatures are listed in Table 6.2.

6.11 BOILING POINT

At standard pressure, water boils at 100°C. At this temperature further addition of heat causes bubbles to appear in the water, rise to the surface, growing as they rise, and burst, violently agitating the entire liquid. Under these conditions, water disappears very rapidly. At all temperatures below 100°C, when the atmospheric pressure is 760 mm, water cannot boil and the loss of water as vapor is relatively slow. When the vapor pressure exceeds 760 mm by an amount equal to the pressure exerted by the liquid in the container, gas bubbles can grow in the liquid, because the pressure in the vapor sphere is greater than the pressure that rests down upon it. At this point molecules are not only escaping from the surface but are escaping into the vapor bubbles which grow very rapidly as they rise to the surface. Vapor escapes so rapidly, absorbing heat of vaporization, that the temperature of the liquid at the surface never rises above 100°C.

At high altitudes water will boil at temperatures several degrees below 100°C, because the average barometric pressure will be lower than 760 mm. A vapor pressure of 645 mm corresponds to a boiling temperature of 95°C. Of course the vapor pressure of water does not terminate at 760 mm but rises beyond at higher temperatures in a closed container. This principle is employed in the operation of a pressure cooker. The vessel is sealed, and water vapor is retained to give a higher pressure. At a gauge pressure of 5 lbs per square inch, an additional pressure of about 250 mm, the temperature is 108°C. The rate of chemical reactions, among which are cooking reactions, is often doubled by an increase in temperature of 10°C. Therefore, cooking can be speeded very greatly by the higher temperature of reaction.

6.12 CALCULATIONS INVOLVING THE VAPOR PRESSURE IN GAS MIXTURES

Water evaporates continuously from lakes and rivers, entering into the atmosphere until the humidity is 100 per cent. When the humidity is 100 per cent, the partial pressure of water vapor in the air is the vapor pressure. At 25°C the humidity is 100 per cent when the water vapor in the air has a pressure of 23.8 mm. At 0°C, 100 per cent humidity is reached at 4.6 mm. This explains why rain or snow occurs when air with a relatively high humidity is cooled suddenly. If air having a humidity of 75 per cent at 25°C (a partial pressure of water vapor of 0.75×23.8 mm $= 17.4$ mm) is cooled to 0°C, where the **vapor pressure, the maximum possible partial pressure of water vapor,** is 4.6 mm, the excess water vapor must condense into droplets which fall to the earth as rain.

Since gases prepared in the laboratory are often collected over water, a problem arises. Water vapor is contained in the gas, hence the total pressure exerted is not due to the gas alone. Fortunately, when a sample of gas is confined in a small container over water, the vapor pressure is reached in a moment or two, and one may consider the humidity to be 100 per cent. The pressure of the water vapor is then the **vapor pressure** and may be found in vapor pressure tables such as Table 6.2.

the vapor pressure is the maximum value of the pressure of vapor in contact with its liquid at that temperature.

Consider a laboratory exercise in which oxygen is prepared by heating potassium chlorate in the presence of a catalyst in a test tube.

$$2\ KClO_3 \xrightarrow[\text{heat}]{\text{catalyst}} 2\ KCl + 3O_2$$

The oxygen is collected by displacement of water from an inverted bottle in a water bath (Fig. 6.11a). As the oxygen is produced and bubbles up through the water in the bottle, the water level descends until no more oxygen is admitted. To calculate the volume of oxygen at STP, which is often necessary, one must know the pressure in "the mixture" (a mixture because it contains water vapor).

To calculate the volume, the delivery tube is lowered and removed. The bottle is then raised or lowered until the water level inside is level with that in the pan around it (Fig. 6.11b). The pressure of the mixture is then the barometric pressure. At 20°C the

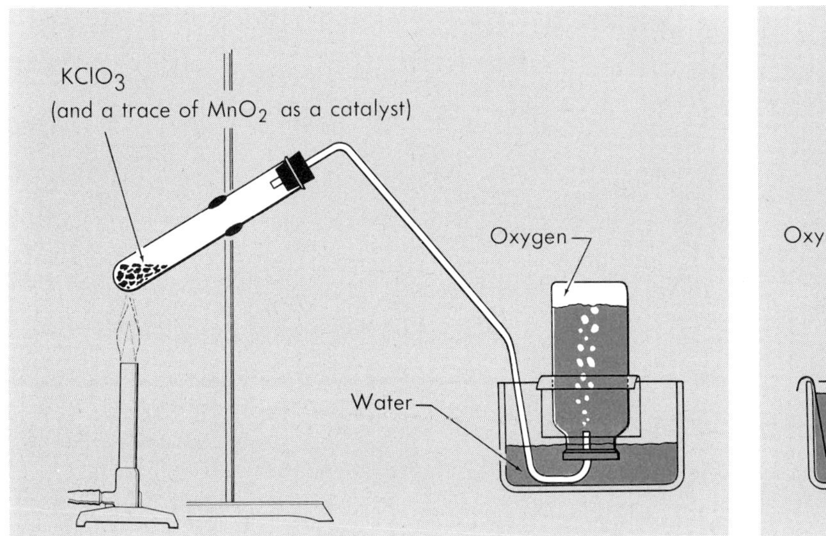

FIGURE 6.11 COLLECTION OF OXYGEN BY DISPLACEMENT OF WATER.

vapor pressure of water (Table 6.2) is 17.5 mm. Applying Dalton's Law, one finds the pressure of oxygen when the barometric pressure is 715 mm mercury is:

$$\begin{aligned} P_{O_2} &= P_T - P_{H_2O} \\ &= 715.0 \text{ mm} - 17.5 \text{ mm} \\ &= 697.5 \text{ mm} \end{aligned}$$

Example 1. What is the pressure of nitrogen which has been collected over water at 25°C and 638 mm?

Solution: The vapor pressure of water at 25°C is 23.8 mm (Table 6.2). The partial pressure of nitrogen is:

$$P_{N_2} = 638.0 - 23.8 \text{ mm} = 614.2 \text{ mm}$$

Example 2. Find the dry volume at STP of 500 ml of nitrogen measured over water at 15°C and 680 mm.

Solution: $P_{N_2} = 680 \text{ mm} - 12.8 = 667 \text{ mm}$. Applying pressure and temperature corrections to the volume:

$$500 \text{ ml} \times \frac{667 \text{ mm}}{760 \text{ mm}} \times \frac{273°\text{A}}{288°\text{A}} = 416 \text{ ml}$$

Example 3. What is the dry volume at STP of oxygen in 400 ml of a mixture of nitrogen and oxygen (80 per cent nitrogen and 20 per cent oxygen by volume) measured over water at 30°C and 720 mm?

Solution: The vapor pressure of water at 30°C is 31.8 mm

$$P_{N_2} + P_{O_2} = 720 \text{ mm} - 31.8 \text{ mm} = 688.2 \text{ mm}$$

$$P_{O_2} = 0.20 \times 688 \text{ mm} = 137.6 \text{ mm}$$

In the gas mixture there is 400 ml of oxygen at 30°C and 137.6 mm. At STP the volume would be:

$$400 \text{ ml} \times \frac{137.7 \text{ mm}}{760 \text{ mm}} \times \frac{273°\text{A}}{(273 + 30)°\text{A}} = 65.3 \text{ ml}$$

6.13 SURFACE TENSION

Liquids exhibit a property called **surface tension.** This property is demonstrated by drops of water on a waxy or oily surface (Fig. 6.12a). The drops tend to form in a spherical shape, as if an invisible membrane held them together. Surface tension is also seen when a drop of water spreads out on a *clean,* smooth glass plate in a thin unbroken film (Fig. 6.12b). Water is not attracted to (is incompatible with) wax, and pulls itself together. Water *is* attracted to a clean glass surface. The drop spreads out, clinging to as much of the attracting surface as possible. Because of surface tension, the water film remains unbroken; the water holds together. Obviously, water molecules exhibit cohesion (attraction for one another).

The rising of a thread of liquid in a glass capillary tube placed in a liquid, as shown in Figure 6.12c, is also due to surface tension. A water film rises up the inside surface of the tube and pulls more liquid up behind it. This results merely in a "rounding up" of the liquid surface on a flat glass surface or inside a large tube, but in a

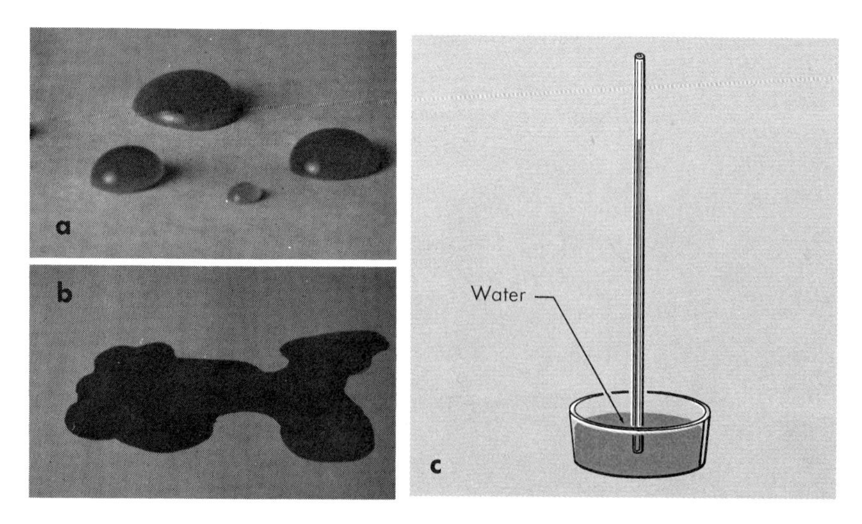

FIGURE 6.12 DEMONSTRATIONS OF SURFACE TENSION.

small capillary the water is pulled from all sides and rises up the tube. The smaller the diameter, the greater the "capillary rise." This phenomenon accounts in part for the lifting of water from roots to leaves in plants and trees.

Surface tension varies from substance to substance and decreases with temperature, disappearing at the critical temperature where gas and liquid become identical.

6.14 VISCOSITY

Another physical property of a liquid is its viscosity, its resistance to flow. Some liquids (water, gasoline, acetone and cleaning fluid) have a very low viscosity. Others (lubricating oil, honey, tar and phosphoric acid) are very viscous. Viscosity generally decreases with a rise in temperature. For this reason lubricating oil of higher viscosity can be used in an automobile in the summertime. Some liquids, however, do increase in viscosity when heated.

6.15 DIFFUSION AND THE BROWNIAN MOVEMENT

Two additional phenomena, **diffusion and Brownian movement**, merit discussion. Diffusion cannot be observed in pure liquids nor in any other homogeneous body of matter, but it can be observed when something is added to change the appearance of a portion of a liquid sample. For example, when a crystal of copper sulfate is placed in a beaker of water, it begins to dissolve, imparting a blue color to the water about the crystal. The color is not confined to the water immediately about the crystal, but spreads (diffuses) slowly until all the liquid is blue.

The second interesting phenomenon is observed when successively smaller clay particles are dropped into a column of water. At first the heavy particles drop straight to the bottom of the vessel. As the particles become smaller, they fall more slowly until eventually very tiny particles seem to stay suspended in the water indefinitely. If one observes a single suspended particle with a

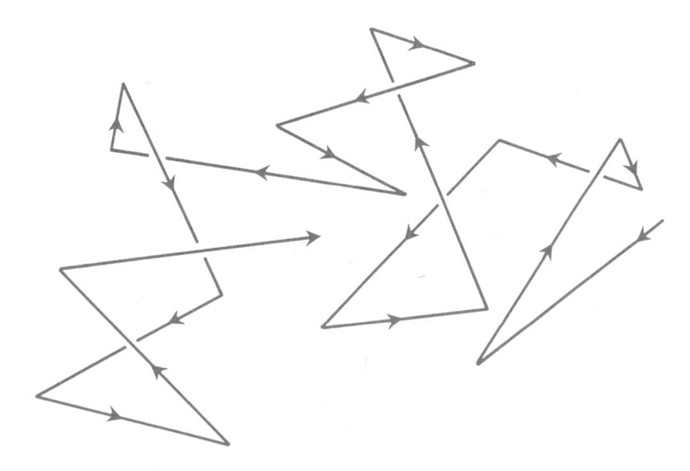

FIGURE 6.13 PATH OF A PARTICLE IN BROWNIAN MOTION.

microscope, one will see that the particle is zigzagging rapidly in the water, never coming to rest. The phenomenon is well known; it is called the **Brownian movement.** The path of particle might be as represented in Figure 6.13.

6.16 THE SOLID STATE–CRYSTAL STRUCTURE

A true solid, as defined by the chemist, must be a pure substance and must exhibit a characteristic crystal pattern. The crystal pattern of a particular solid is often identified by the angles which the faces (flat sides) of the crystal make with one another. The faces of a salt crystal (Fig. 6.14) are perpendicular to one another, and the crystals appear as cubes. Adjacent faces of calcite crystals are not perpendicular to one another, but opposite faces are parallel. Graphite crystals are thin platelets. Each true solid has its own crystal pattern and every crystal of the solid conforms to the pattern. Two other crystals with characteristic patterns are shown in Fig. 6.14.

a

b

c

FIGURE 6.14 CRYSTALS OF NACL, $NH_4H_2PO_4$ AND $CUSO_4$.

6.17 MELTING POINT

A true solid has a **definite and constant melting point** (melting temperature). Each sample of a particular true solid begins to melt at the same temperature and remains at that temperature until completely melted. In contrast, pseudo-solids soften over a range of temperatures. This property can be illustrated by an experimental determination of the melting point of ice, a true solid.

A beaker of ice with a thermometer frozen in it is removed from a freezer at −20°C and allowed to warm to room temperature. During warming, temperature readings are taken every few seconds. A plot of temperature versus time reveals a graph similar to Figure 6.15a. The temperature rises steadily to 0°C. At that point liquid first appears. Melting progresses, but no temperature change occurs until no ice remains. When all the ice is melted, the tempera-

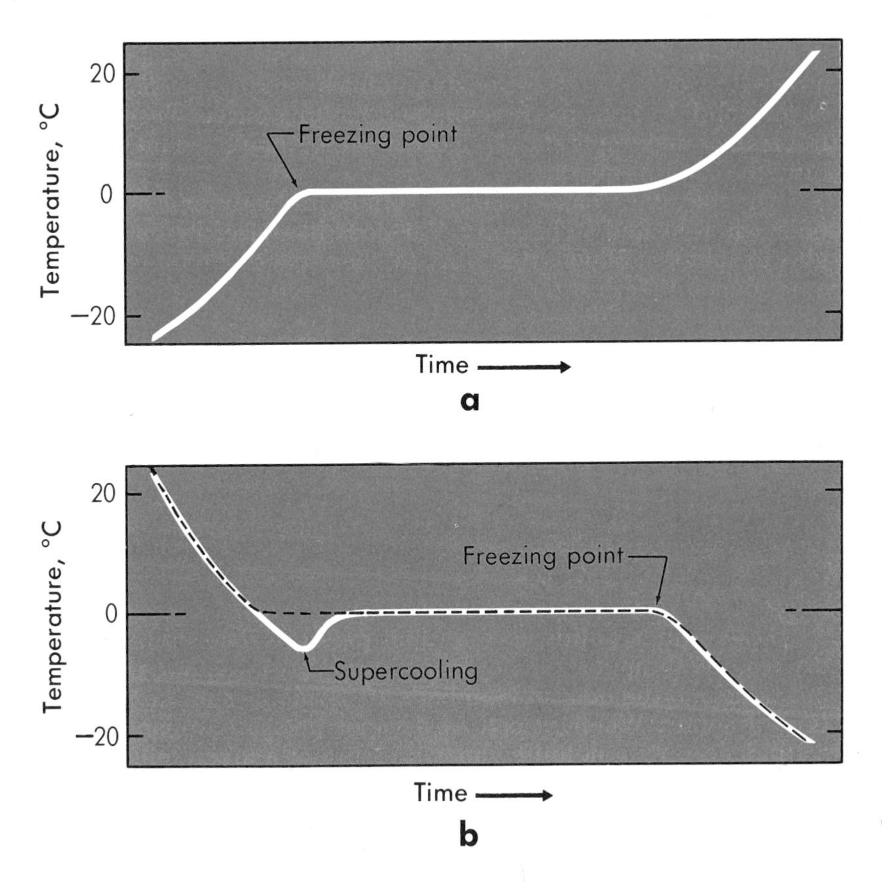

FIGURE 6.15 MELTING AND FREEZING CURVES.

ture of the liquid starts rising and continues to rise until it reaches the temperature of the surroundings.

One can observe the same melting (freezing) point with a thermometer (or thermocouple) by following the temperature of a beaker of water placed in a freezing bath. If temperature readings taken at time intervals are plotted against time, the result *may be* a curve such as the one represented by the dotted line in Figure 6.15b. The temperature falls steadily to 0°C, when ice may appear. The temperature reading remains constant until all the liquid is frozen and then drops steadily to the temperature of the bath. Usually, however, the temperature of the water drops steadily until it is somewhat below 0°C (white line on Figure 6.15b). The water is said to have been supercooled. Suddenly, at no set temperature, ice appears and the temperature rises to 0°C, remains there until all is frozen and then drops steadily as before. Only at 0°C are both water and ice present. This is the melting point. If heat is added in an attempt to raise the temperature of an ice-water mixture, melting will occur, but the temperature will remain at 0°C until all the solid has disappeared. Attempts to cool a mixture of ice and water are frustrating. The water begins to freeze, but the temperature remains at 0°C until all the liquid has disappeared.

The phenomenon noted in the events observed when pure water is cooled below its melting point, namely supercooling, is a general phenomenon. All liquids tend to supercool. The extent of supercooling may often be modified by rapid stirring and can be arrested at or below the freezing point by adding a small crystal of the solid. In both the freezing and melting curves the complete freezing or melting is shown to occur at the same constant temperature, the melting point.

the two characteristics, constant melting point and crystal pattern, are important criteria for true solids.

On the other hand, plastics soften gradually and eventually pour when heated sufficiently. Tar does likewise. Pyrex glass begins to soften at 500°C; it is risky to keep it at that temperature for a prolonged time lest the vessel start to deform. However, it is not so soft that it cannot be kept for a short time at 550°C without losing its shape. It melts and runs at about 700°C. Plastic, tar and glass are not true solids. Not only do they lack a definite melting point, but they have no definite crystal habit. They are liquids that become thicker and thicker with cooling until they are hard and brittle.

6.18 VAPOR PRESSURE

Like liquids, **solids have a vapor pressure.** The sublimation of dry ice, the disappearance of snow in sub-zero weather, the disappearance and the odor of mothballs are evidences of the vaporization of solids. When placed in a vapor pressure apparatus (Fig. 6.16), ice exerts a definite vapor pressure at a given temperature, just as water does. The vapor pressure of ice varies with temperature. At 0°C the vapor pressure is 4.579 mm (the same as for water). Some values are recorded in Table 6.3.

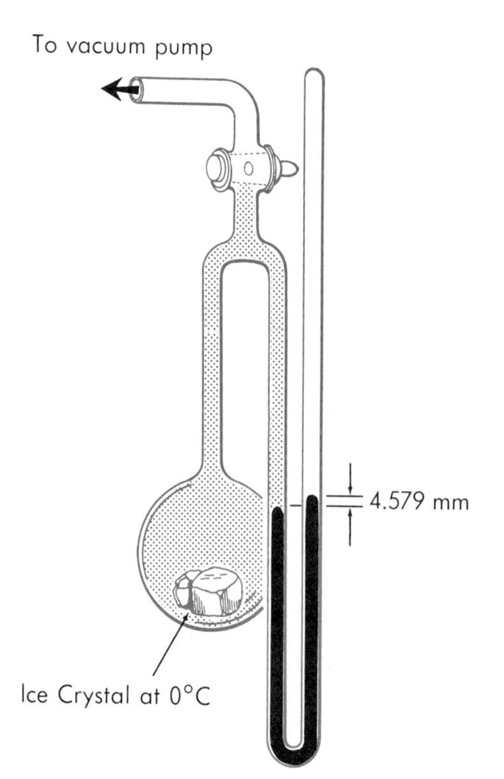

FIGURE 6.16 MEASURING THE VAPOR PRESSURE OF ICE.

TABLE 6.3 VAPOR PRESSURE TABLE OF ICE

Temperature	Vapor Pressure mm
0°C	4.579
−10°C	1.947
−20°C	0.770
−30°C	0.280
−40°C	0.094
−50°C	0.029

6.19 HEAT EFFECTS

The heat required to raise the **temperature of 1 gram of material 1°C** is called **the specific heat.** The specific heat varies somewhat with temperature and varies greatly with changes in state.

The specific heat of water at 15°C is by definition **exactly** 1 calorie. That is, 1 calorie of heat is required to raise 1 gram of water from 14.5°C to 15.5°C. At other temperatures, water has very nearly the same specific heat. The specific heat of ice near 0°C is about 0.5 calorie. And the specific heat of steam near 100°C is approximately 0.5 calorie per gram per degree.

because of hydrogen bonding, water has an unusually large heat of vaporization.

Heat is also exchanged in changing from one state to another. Heat is required to melt or fuse a solid and to vaporize either a solid or a liquid. The heat required to melt 1 gram of a solid is called its **specific heat of fusion.** The specific heat of fusion of water at 0°C is 80 calories. The heat required to vaporize 1 gram of liquid is called the **specific heat of vaporization.** The heat required to vaporize 1 gram of water at the boiling point is 540 calories.

Consider the magnitude of the heat of vaporization of water, 540 calories per gram. Five and four tenths grams of water can be heated all the way from 0°C to 100°C with the same number of calories that are required to vaporize only one gram once 100°C is reached.

One can see why scalding with steam is more serious than with hot water. About nine times more heat is gained from the condensation of the steam at 100°C than is gained in cooling it down from 100°C to body temperature, 37°C.

A refrigerator is kept cool by the continual vaporization of a liquid. A freon gas, a compound of fluorine and carbon, is compressed until liquefied in a motor driven compressor outside the refrigerated compartment. During this process the heat of vaporization is lost. The liquid is then forced into pipes in the refrigerator where the pressure is reduced, allowing the liquid to evaporate. The liquid takes the heat of vaporization from the surroundings, cooling them. The gas is then drawn outside, compressed and condensed to release the heat gained inside the refrigerator. The process of vaporization inside taking up heat and liquefaction outside giving up heat is repeated over and over in the refrigerator cycle. Operation of the refrigerator warms the room while it cools its own interior; it merely pumps heat from inside of the refrigerator to the outside.

EXERCISES

6.1 State each: Boyle's Law, Charles' Law, Dalton's Law.

6.2 Describe two kinds of manometers.

6.3 Explain how a barometer works.

6.4 An open-end manometer and a closed-end manometer are used to measure the pressure of the same gas sample. Why are the vertical distances between mercury levels different?

6.5 What happens to a gas when it is heated in a sealed and rigid container?

6.6 How is a thermometer calibrated in degrees centigrade?

6.7 Which degrees are larger, °F or °C? How much larger?

6.8 In what respect are *degrees absolute* absolute?

6.9 Which of these will fill a 10 liter flask at 25°C: 20 g hydrogen gas, 10 g hydrogen gas, 0.01 g hydrogen gas or 5×10^{-4} g hydrogen gas?

6.10 When a gas sample is collected over water, will its volume be the same as when collected over mercury at the same temperature and pressure? Explain.

6.11 How can the vapor pressure of a liquid be changed?

6.12 Explain how the rising of a liquid in a capillary tube shows surface tension.

6.13 Name five liquids which have a high viscosity.

6.14 Describe Brownian movement.

6.15 Define a true solid.

6.16 What is supercooled water?

6.17 Define: (a) specific heat, (b) heat of fusion, (c) specific heat of vaporization.

PROBLEMS

6.1 36 ml of a gas at 25°C and 1 atm is heated to 85°C and 1 atm. Calculate its final volume.

6.2 What is the volume at STP of 64 ml of a gas measured at 27°C and 760 mm?

6.3 72 L of a gas at 760 mm and 0°C is compressed until the pressure is 1220 mm at 0°C. Calculate the final volume of the gas.

6.4 What is the volume of a sample of gas at 380 mm and 300°A which occupies 26 ml at 640 mm and 300°A?

6.5 What volume will a sample of gas occupy at 1220 mm and −23°C, if it occupies 45 ml at 27°C and 700 mm?

6.6 Calculate the volume at STP of 150 ml of gas measured at 645 mm and 28°C.

6.7 Calculate the temperature in °C of the following Fahrenheit temperatures: 212° (boiling point of water), 98.6° (body temperature), 32° (freezing point of water), 0° and −40°.

6.8 Find the volume at STP of the pure hydrogen in a 10 L mixture of hydrogen, helium and nitrogen at 27°C, in which the partial pressure of hydrogen is 150 mm and the total pressure is 630 mm.

6.9 A tire pump has a cylinder 20 inches in length. How far can the piston be pushed down before air enters the tire, which is at a gauge pressure of 29.4 pounds per square inch? The barometric pressure is 14.7 pounds per square inch.

6.10 A sealed two-quart can is half full of water and has a total pressure of 760 mm at 25°C. It is heated to 100°C without bursting. What is the total internal pressure of the can at that temperature?

6.11 500 ml of oxygen gas saturated with water vapor at 40°C and 680 mm pressure is compressed to 250 ml at the same temperature. Calculate the final pressure.

6.12 100 ml of oxygen gas is collected over water at 25°C and at a barometric pressure of 635 mm. (a) What is the pressure of oxygen? (b) What volume would the dry oxygen occupy at the same barometric pressure and temperature?

6.13 Find the volume at STP of 175 ml of hydrogen measured over water at 28°C and 648 mm.

6.14 100 ml of O_2 at STP is added to 400 ml of N_2 at STP, and the mixture is collected over water at 26°C and 645 mm. (a) What is the oxygen pressure in the wet gas mixture? (b) What is the volume of the wet gas mixture?

6.15 The volume of an airtight irregularly shaped cavity was determined by releasing into it 100 L of carbon dioxide at 25°C and 1 atm. After mixing, the partial pressure of carbon dioxide was 87 mm. Calculate the volume of the cavity.

6.16 How much heat must be absorbed by 15 g of ice at −15°C to convert it to steam at 110°C?

6.17 How many g of steam at 100°C need be added to 600 g of water at 20°C to raise the temperature to 85°C?

6.18 Nitrogen gas is added to an 8-liter container of oxygen gas at 380 mm and 25°C until the total pressure is 640 mm. What is the final pressure of the oxygen gas?

SUGGESTED READING

Conant, J. B. (ed.): "Robert Boyle's Experiments in Pneumatics." In *Case Histories in Experimental Science* (Case 1). Harvard University Press, Cambridge, Mass., 1950.

Etzel, H. W.: "Ionic Crystals." J. Chem. Educ., *38*:225, 1961.

Feifer, N.: "The Relationship Between Avogadro's Principle and the Law of Gay-Lussac." J. Chem. Educ., *43*:411, 1966.

Garrett, A. B., Lippincott, W. T. and Verhoek, F. H.: *Chemistry: A Study of Matter.* Blaisdell Publishing Co., Waltham, Mass., 1968. Chapters 12–15.

Neville, R. G.: "The Discovery of Boyle's Law, 1661–62." J. Chem. Educ., *39*:356, 1962.

SEVEN • THE KINETIC THEORY

7.1 INTRODUCTION

The development of the Kinetic Theory of matter, from its beginning in the seventeenth century, is a fascinating story of a successful application of the scientific method. The theory developed hand in hand with discovery of the phenomena and the formulation of the laws presented in Chapter 6. Newly found facts led to new laws. New laws required reexamination of old theories. Some postulates were rejected, some amended and some verified, until the Kinetic Theory emerged as one of the basic theories of physics and chemistry. The Kinetic Theory holds an exalted position in scientific theory because of its simplicity and because it explains so well the many facts and laws that are known concerning the behavior of matter in its three states. Some of these facts and laws discussed previously in this text are:

1. Matter in all states possesses a heat capacity.
2. Gases are fluid.
3. Gases fill completely any containers to which they are admitted.
4. Gases are compressible.
5. Gases exert a pressure.
6. When the pressure exerted on a gas sample at constant temperature is increased, the volume of the gas sample is reduced. The volume varies inversely with the pressure, as described by Boyle's Law.
7. When the temperature is raised, the volume or the pressure, or both, of a gas sample is increased. If the pressure remains constant, the volume varies with the temperature, as described by Charles' Law.
8. Each gas in a mixture of gases fills the entire volume of its

container and exerts the pressure it would exert if alone in the container. The total pressure is the sum of the pressures each gas would exert if alone in the container, as described by Dalton's Law.

9. Gases follow Boyle's, Charles' and Dalton's Laws with little deviation at temperatures well above the boiling point of the gas and at less than a few atmospheres pressure. Deviations can be very great at low temperatures and high pressures.

10. Liquids often supercool.
11. Liquids are more dense than gases.
12. Liquids are almost imcompressible.
13. Liquids and gases exhibit Brownian movement.
14. Liquids are fluid, but they exhibit viscosity.
15. Liquids exhibit surface tension.
16. Liquids have a set vapor pressure at constant temperature.
17. The vapor pressure of a liquid increases rapidly with temperature.
18. There is a heat of vaporization.
19. Solids possess a rigid, definite structure.
20. Solids exert a vapor pressure.
21. There is a heat of sublimation.
22. Pure solids melt completely at constant temperature.
23. There is a heat of fusion.

The Kinetic Theory must agree with each of these facts and with many others. Failure to agree with any **one fact** invalidates the theory unless it can be amended to agree.

To conserve time and to avoid repetition of details, the development of the Kinetic Theory will not be discussed here. Rather, the postulates of the theory which apply to all three states of matter will be listed first, followed by a list of the postulates that are specific for each phase, gaseous, liquid and solid. The theory will be examined (when it seems best) to see if it explains adequately the observed facts and laws in the list above.

7.2 POSTULATES OF THE KINETIC THEORY

1. **Matter in all states is composed of particles.** This is the central postulate of Dalton's Atomic Theory. These particles may be either molecules, atoms or ions, but the term molecule will be used interchangeably with the term particle in this chapter.

2. **These particles exhibit mutual attraction.** The magnitude of the attraction varies from substance to substance, depending upon ion charge, polarity, size and mass and other factors.

3. **These particles have kinetic energy.** The kinetic energy is distributed at random among the particles in the forms of translational and/or vibrational and/or rotational motion. The idea that particle motion exists in all forms of matter is a great contribution to physical theory. The possible forms of motion (energy) for a diatomic molecule are illustrated in Figure 7.1.

4. **The average kinetic energy increases with temperature and changes with a change in state.** According to the theory, matter may remain in either of the condensed states over a range of temperatures. Addition of heat energy simply causes an increase in energy in the forms already extant. Eventually, however, the energy increase cannot be accommodated without a change of state. There-

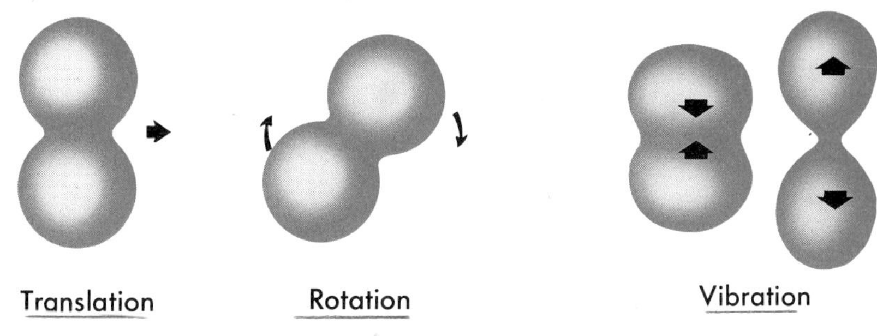

FIGURE 7.1 FORMS OF MOTION FOR A DIATOMIC MOLECULE.

fore, a great and sudden change takes place; the solid melts, the liquid vaporizes. The amount of energy possessed by particles with constant mutual attraction at the same distance of separation determines the state. For a given substance, particle energy increases progressively from solid to liquid to gas.

5. **Collisions between the particles are perfectly elastic.** Energy is transferred from particle to particle by collision, but no energy is lost in the collisions. All the energy is accounted for in the resulting motions of the colliding particles.

The **fact** that **matter in all states possesses a heat capacity**, that is, that energy is required to raise the temperature of all matter (Chapter 6), agrees with the above statements. An increase in temperature is accompanied by an increase in kinetic energy. The energy level is raised by added heat. The 1 calorie of heat that raises 1 gram of water from 14.5°C to 15.5°C merely increases the molecular motion of water by that amount of energy.

7.3 MATTER IN THE GASEOUS STATE

Three of the postulates of the kinetic theory as stated for all states must be specified:

the kinetic theory for gases includes the postulates listed for all matter as well as those listed here.

1. **The average kinetic energy of the particles is very, very great.**

2. As a result, **the effect of the existing mutual attraction is negligible.** Therefore, the particles tend to be very far apart in the available empty space.

3. Not only does the **total kinetic energy increase with temperature**, but the **translational energy is proportional to the absolute temperature.**

If these postulates describe the system correctly, it is apparent that: (a) gases fill the container, (b) gases are fluid and (c) gases interdiffuse. Because the particles of a gas are separated from each other by relatively great distances, gases can be compressed. Compression merely forces the molecules closer together. Expansion allows them to move farther apart, which they will do because of their rapid motion.

Gases exert pressure, which increases with compression and decreases with expansion. One must think of the pressure being exerted not by the gas as a whole, but by the individual molecules colliding with the walls of the container. Consider a target mounted on a spring, as shown by the dotted outline in Figure 7.2. When

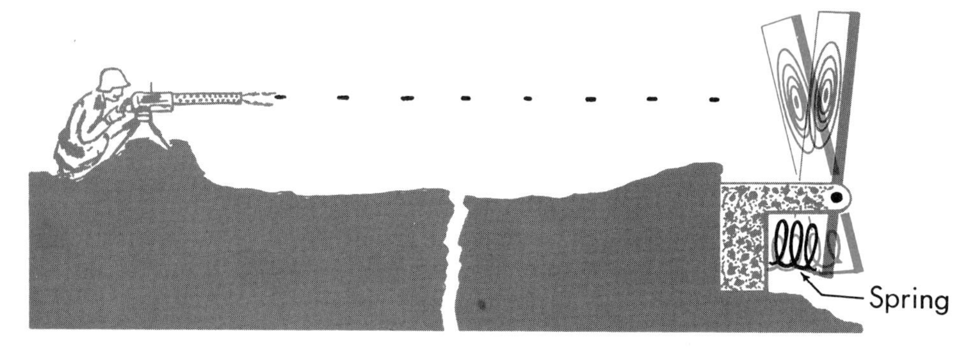

FIGURE 7.2 PRESSURE CAUSED BY BULLETS ON A TARGET.

a rifle bullet strikes the bull's eye, the target is pushed back momentarily, then lurches forward when the force of the bullet embedded in it is spent.

Now suppose the target is fired upon with a machine gun so that the firing rate, which is very slow at first, gradually increases. At first the target lurches backward and forward with each hit. Soon, however, as the firing rate increases, the next bullet hits the target before it has completely recovered from the last. The target vibrates wildly, never quite returning to its original position. As the firing speed increases, the vibration lessens until the target is held quite steadily against the compressed spring. Greater firing speed now merely tends to force it back farther. A steady force is being exerted on the target by the colliding bullets. The distance the target is pushed back against the spring is a measure of the force exerted upon it. The more rapid the firing and the higher the velocity of the bullets, the greater the force.

Pressure is defined as the force exerted per unit area. Each square centimeter of the inner walls of a gas container is a target sustaining a great number of molecular collisions per second. The force on this square centimeter area is the pressure. If the container is non-rigid, the walls are pushed out and stretched until the inner force or pressure is equaled by the tension or force of the walls.

When a gas is compressed, as by a tire pump, more molecules are crowded into a smaller volume. For example, when 2 liters of gas are compressed to 1 liter, twice as many molecules are present in that one liter to collide with the walls of the container, and the pressure doubles (see Figure 7.3). The volume varies inversely with the pressure (Boyle's Law).

The translational energy (the moving energy) of molecules is proportional to the absolute temperature. Therefore, at 10°C or 283°A, the energy of molecules is 283/273 as great as at 0°C. The molecules hit harder and more often with greater velocity, thus increasing the pressure if the container is rigid. If the pressure is to remain the same, the volume must be increased proportionately to the increase in absolute temperature (see Figure 7.4).

A gas exerts the same pressure in a mixture of gases that it would if it were in the container alone. That is, **the pressure of a gas mixture is equal to the sum of the pressures each gas would exert if in the container alone.** As shown in Figure 7.5a, b and c, each gas in a container at constant temperature exerts a pressure

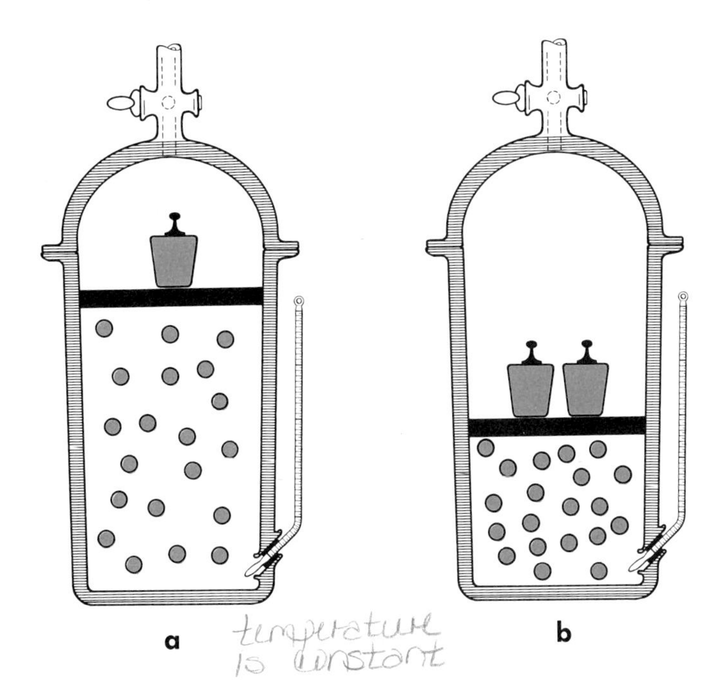

FIGURE 7.3 THE EFFECT OF PRESSURE ON THE VOLUME OF A GAS SAMPLE AT CONSTANT TEMPERATURE.

FIGURE 7.4 THE EFFECT OF INCREASING TEMPERATURE ON THE VOLUME OF A GAS.

depending only on the number of particles present. When the three gases are enclosed together in a container of the same volume, as in Figure 7.5d, all three kinds of molecules are present. All kinds collide with all others and with the sides of the container. All contribute as before to the total pressure, in accordance with Dalton's Law. Furthermore, since Boyle's and Charles' Laws hold for each gas, they hold for the mixture also.

No gas behaves precisely in accordance with Boyle's and Charles' Laws. Only a so-called "perfect" gas does; therefore, these

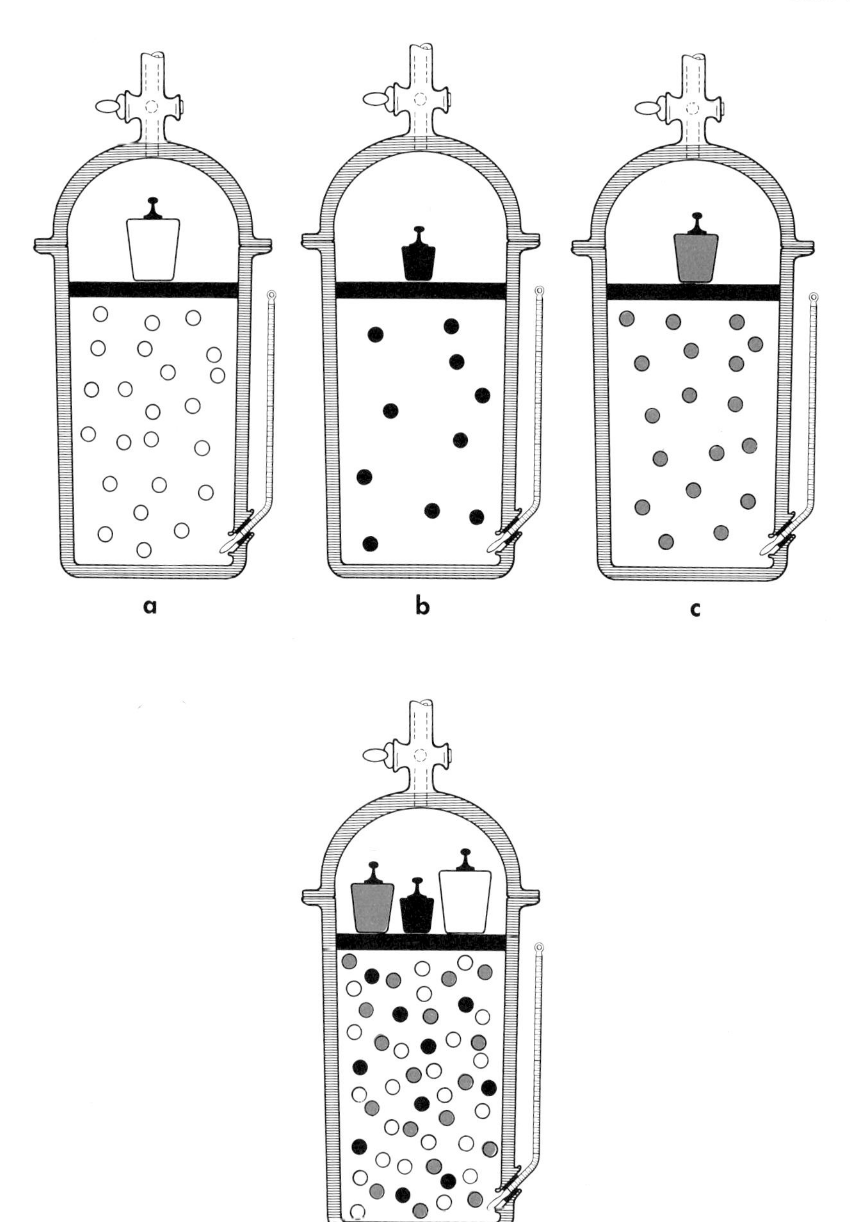

FIGURE 7.5 THE TOTAL PRESSURE COMPARED TO THE PARTIAL PRESSURES OF THE GASES OF A MIXTURE.

laws are called the "perfect gas laws." No real gas is perfect, but most are very nearly so. The deviations are small at less than a few atmospheres pressure and a temperature well above the boiling point of a particular gas. These deviations exist because (1) attractions, although they are small compared to the energy of gas molecules, do exist between the molecules and (2) although the molecules are small in the space that surrounds them, compression cannot decrease the molecular volume as it can the free space between them.

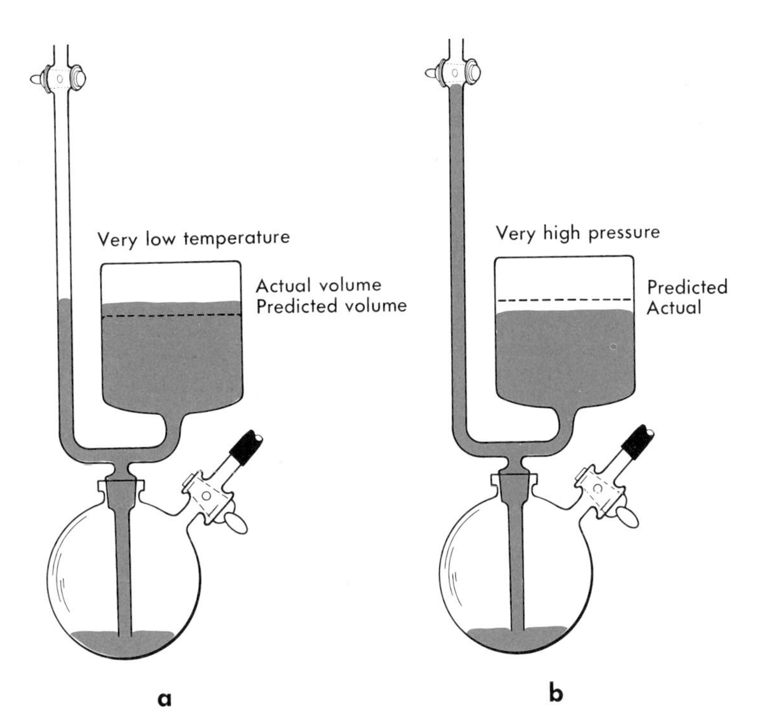

FIGURE 7.6 PREDICTED AND MEASURED VOLUMES OF REAL GASES AT A TEMPERATURE NEAR THE BOILING POINTS AND AT HIGH PRESSURE.

At temperatures near the boiling point of the gas, the kinetic energy is decreased (the molecules slow down) until attractions between them are no longer negligible. They tend to stick together, decreasing the volume, as shown in Figure 7.6a. At high pressure the molecules are no longer a great distance apart but may be very close together. A doubling of pressure under these conditions can result in only a slight decrease in volume, as shown in Figure 7.6b.

at high pressures and temperatures near the boiling point, deviations are great. The vapor may even condense.

There are equations that account for these differences from "perfect gas" behavior, but they are much more complicated than Boyle's and Charles' Laws. Unless one is interested in the behavior of gases at high pressures or low temperatures, the answers obtained by the simple calculation studied here are satisfactory.

7.4 MATTER IN THE LIQUID STATE

Only two postulates of the Kinetic Theory as it applies to all states need to be specified for matter in the liquid state.

1. *The average kinetic energy of the particles is much less than in the gas phase.*

2. *The existent mutual attraction of the particles is sufficient to hold them close together.*

The energy is still great, for as the particles bump into their neighbors, roll and tumble over one another, vibrate, and spin, they may strike minute but visible suspended particles (many, many times bigger than they) with sufficient force that these particles may be seen to follow a wild, zigzag path, i.e., to **exhibit Brownian movement**, as discussed in Chapter 6.

The importance of mutual attraction between particles in liquids is shown by the fact that liquids have a constant volume and assume the overall shape of the lower part of their containers. Mutual attraction is the important factor in **surface tension.** When liquid particles rest upon a surface for which they have little attraction, they tend to pull themselves together into a sphere.

hydrogen bonding constitutes a preferential attraction.

Intermolecular attraction is one of two factors involved in viscosity. Molecules of a liquid must slide past or roll over one another when the liquid is poured. Molecules with strong forces of attraction for one another tend to be more viscous. The configuration or shape of the molecules is another factor influencing viscosity. If the molecules are spheres or are smooth in shape there is little resistance to flow. If the molecules are irregularly shaped, long and zigzag, they experience difficulty in sliding past one another, as shown in Figure 7.7, and they exhibit a high viscosity.

Close examination of the Kinetic Theory shows it to be consistent with the phenomenon of evaporation. Although a molecule in a liquid moves very rapidly, the average energy it possesses over a period of time is insufficient to pull it away from its neighbors. However, not all molecules possess the same energy. In fact, the energy a molecule possesses changes many times per second as it loses or gains energy in repeated collisions. Because the collisions

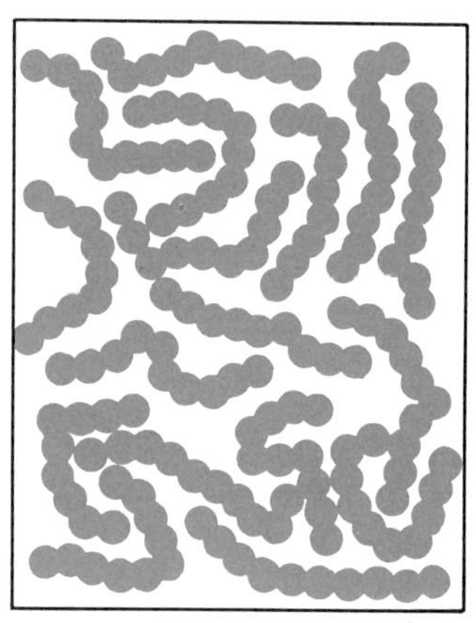

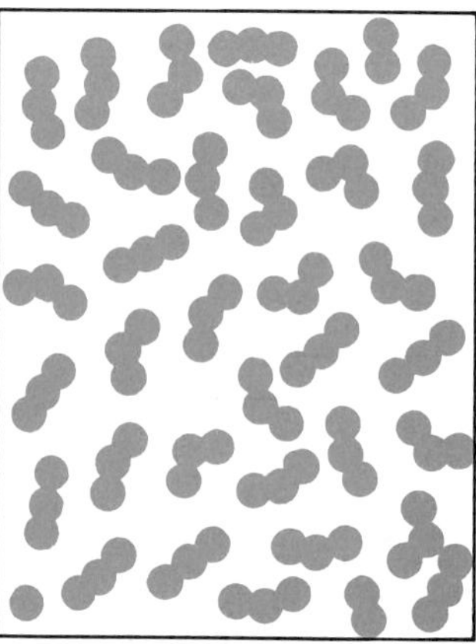

FIGURE 7.7 RELATIVE VISCOSITY VERSUS MOLECULAR STRUCTURE.

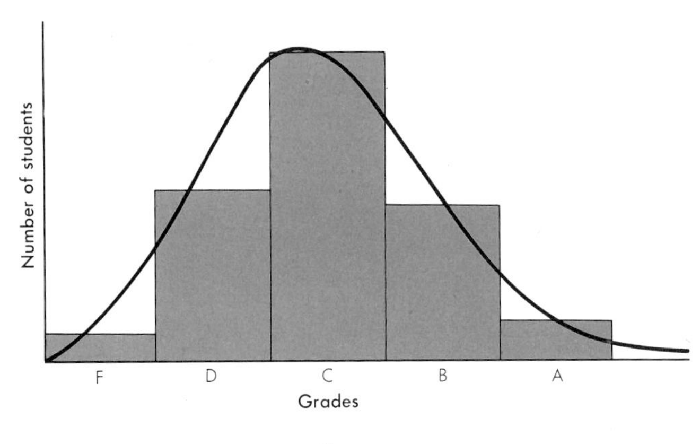

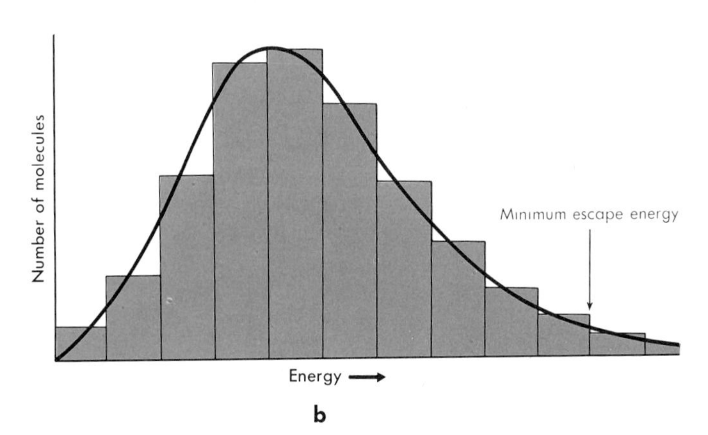

FIGURE 7.8 PROBABILITY CURVES FOR COURSE GRADES AND ENERGIES OF MOLECULES.

between molecules are random, the distribution of energy among the molecules at any one instant is random. If random, the energy distribution follows the so-called "bell-shaped" curve, derived by mathematicians and used by teachers of large classes in evaluating student performance and assigning grades. As shown in Figure 7.8a, there are few students with F's, more with D's, many with an average C grade, fewer again with B grades and few with A grades.

As for the molecules of a liquid (Fig. 7.8b), very few have low energies, many have nearly average energies and very few have very high kinetic energy—energy sufficient to overcome the attraction of their neighbors. Most molecules experience instants of high energy while buried in the liquid, only to lose it to another molecule in a split-second collision.

The very few molecules which gain sufficient energy to escape while at the upper surface of the liquid are responsible for the vapor pressure. Consider water occupying the lower part of a flask, as in Figure 7.9a, from which the air has just been evacuated. High energy molecules are leaving the surface to occupy the space above. The presence of the vapor molecules creates a pressure above the liquid, a pressure that increases as more molecules enter that space. However, the molecular movement is two-directional;

molecules collide with each other and with the container walls, and fall back into the liquid layer, as shown in Figure 7.9b. At first the number returning per second is relatively small, but as the number above the liquid increases, the number returning increases. Eventually molecules return as often as they leave the liquid. At this point (Fig. 7.9c) **dynamic equilibrium** is reached, a condition in which neither the total number of molecules per unit of vapor volume nor the pressure above the liquid changes. This constant pressure is the **vapor pressure.** For water, the vapor pressure at 25°C is 23.8 mm.

The presence of other kinds of molecules above a liquid does not alter its vapor pressure. The rate at which molecules leave the liquid depends only on the temperature; the rate of return depends on the pressure of the vapor just above the liquid as well as on the temperature. A slower rate of diffusion up through a layer of dissimilar molecules in the gas above the liquid may lengthen the time required for the vapor pressure to be established throughout the system, but once equilibrium is attained, the pressure will be the same as if no other gas were present.

Remember that the high energy molecules are the only molecules that can escape from the liquid. Energy in gases and liquids is in the form of particle energy. **The high energy particles are the "hotter" particles.** Continual loss of "hotter" particles from a liquid due to vaporization will lower the temperature of the liquid, unless heat is allowed to flow in. The specific heat of vaporization is the amount of heat that must be added to the liquid to maintain the temperature during the evaporation of 1 gram. It is the energy necessary to maintain average particle energy while losing 1 gram of "hot" particles. The specific heat of vaporization of water is 540 calories per gram at 100°C.

Examination of the vapor pressure table in Chapter 6, part of which is repeated here in Table 7.1 and Figure 7.10, shows that the vapor pressure of water increases with the temperature. This is true of all liquids and would be predicted by the Kinetic Theory. The higher the temperature, the greater the molecular energy, the greater the rate of escape and the greater the vapor pressure. However, the increase in vapor pressure is more than might be expected.

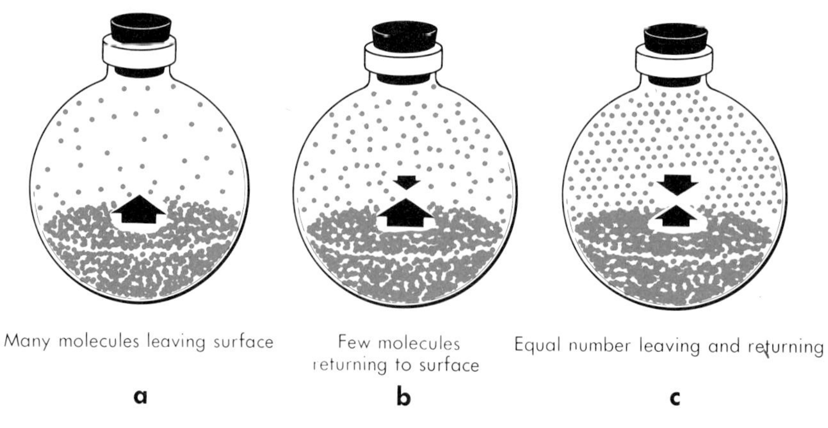

FIGURE 7.9 THE ATTAINMENT OF DYNAMIC EQUILIBRIUM AT THE VAPOR PRESSURE.

TABLE 7.1 VAPOR PRESSURE OF WATER

Temperature		Pressure
°C	°A	*mm Hg*
0	273	4.6
10	283	9.2
20	293	17.5
30	303	31.8
40	313	55.3
60	333	149.4

At 40°C (see Table 7.1) the vapor pressure of water is 55.3 mm; at 60°C it is 149.4 mm. For an increase in absolute temperature of

$$\frac{(333 - 313)}{313} \times 100\% = 6.5\%$$

the change in vapor pressure is:

$$\frac{149.4 - 55.3}{55.3} \times 100\% = 170\%$$

The average molecular energy does not increase 170 per cent with 6.5 per cent increase in absolute temperature. However, the

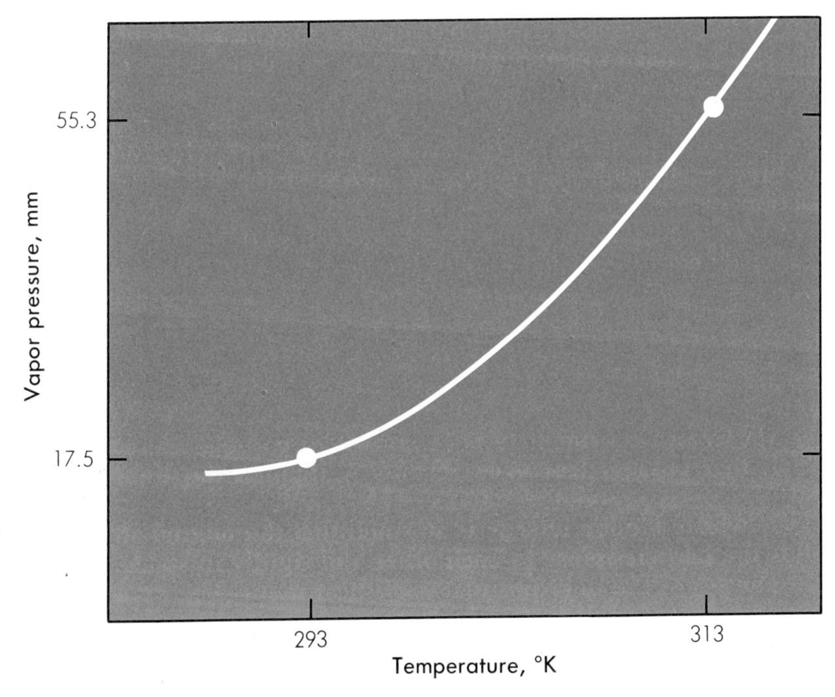

FIGURE 7.10 VARIATION OF THE VAPOR PRESSURE WITH TEMPERATURE.

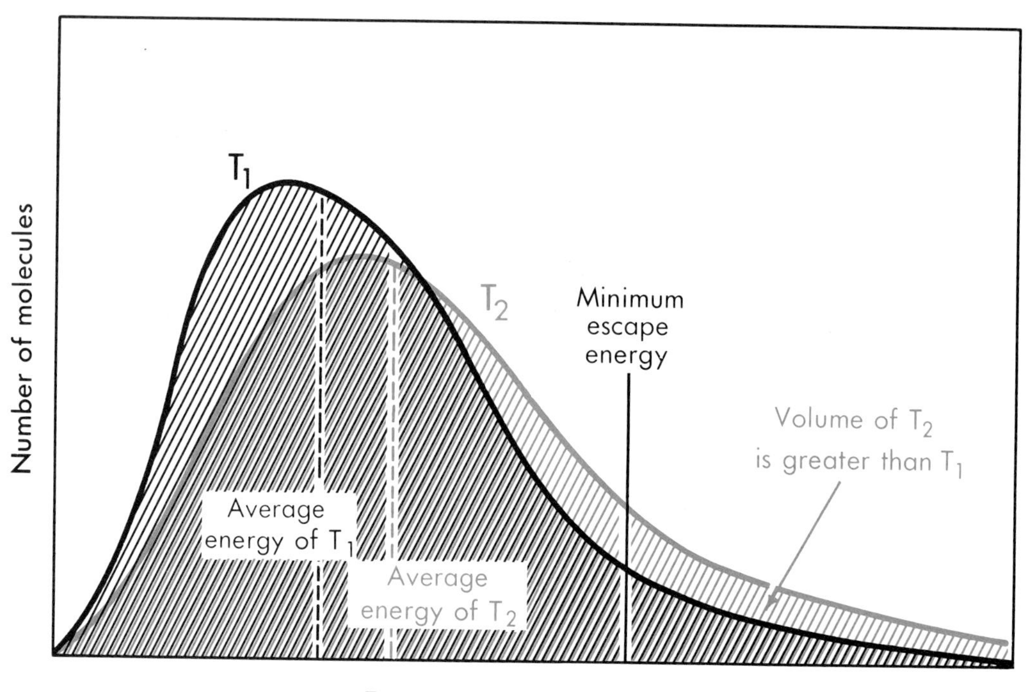

FIGURE 7.11 THE EFFECT OF A TEMPERATURE CHANGE ON THE FRACTION OF THE MOLECULES WITH ENERGY TO ESCAPE.

vapor pressure depends on the number of molecules with an energy sufficient to escape and that number may nearly double with that small change in temperature. The effect of an increase in temperature both on the average energy of molecules in a liquid and on the magnitude of the fraction of those molecules with sufficient energy to escape is shown qualitatively in Figure 7.11. In this figure the average molecular energy is seen to increase by only a small fraction, while the number of molecules with an energy greater than that sufficient to escape has more than doubled. Again, the Kinetic Theory is consistent with the facts.

7.5 MATTER IN THE SOLID STATE

Matter in the solid state is composed of particles

1. **that possess less average energy than in the liquid state** and

2. **whose kinetic energy is so small that the mutual attraction holds the particles close to each other and rigidly in space.**

These are the only postulates of the Kinetic Theory as it applies to all matter that must be specified (limited) to apply only to solids.

The only motion available to a particle in the solid state is **vibration** about its average position. The particle acts as though it were continually agitated while suspended on crossed elastic

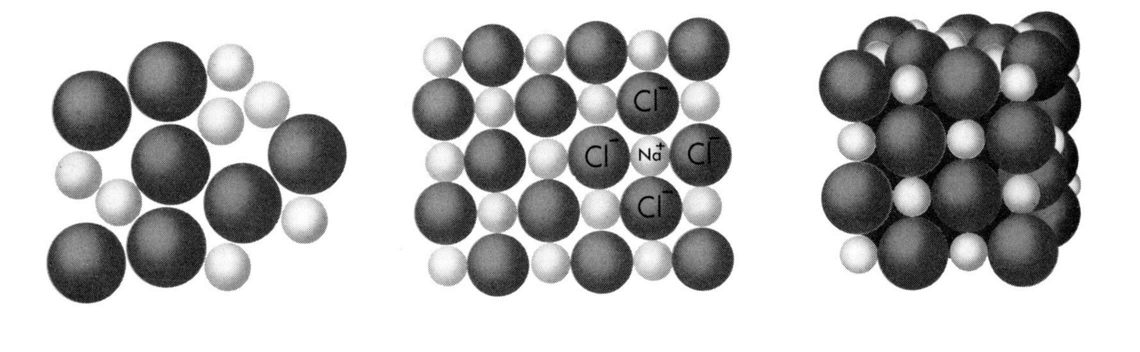

FIGURE 7.12 THE CUBIC LATTICE OF SODIUM CHLORIDE.

bands, as it moves back and forth and up and down, but never leaves its position in the lattice.

Because they are pulled tightly together by mutual attraction and because there are preferential attractions, such as between plus and minus ions and between plus and minus ends of polar molecules, solid particles tend to "pack" together in a **repeating pattern or lattice.** The sodium and chloride ions of solid table salt are not grouped at random, as shown in Figure 7.12a. Equal forces of attraction for each identical ion force them to assume an arrangement in which all distances between the centers of closest ions of opposite charge are equal, as shown in Figures 7.12b and c. This results in a lattice. The same is true for hydrogen chloride. The polar hydrogen chloride molecules are arranged in solid HCl, as shown in Figures 7.13b and c, not as shown in Figure 7.13a. This submicroscopic pattern of particles is continued throughout a large crystal. Molecules or ions are arranged in layers. The crystal breaks between layers; it grows a layer at a time.

The existence of vibrational energy admits an explanation of the loss of individual molecules to *form a vapor about the solid* and *exert vapor pressure*. A molecule in the surface layer of a crystal is attracted to its neighbors in the outer layer and to those behind it. These forces are great. A molecule with an average energy of vibration cannot pull away. However, not all molecules have the same energy of vibration. Energy is passed around by collision of the molecules with their neighbors as they vibrate. A wave of energy may pass through a crystal by being transferred from one neighbor to the next, much as the starting jolt passes from car to car throughout the length of a train.

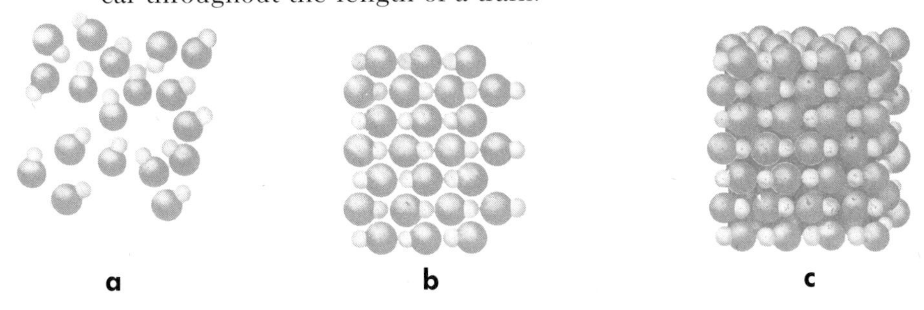

FIGURE 7.13 THE CRYSTAL LATTICE OF SOLID HYDROGEN CHLORIDE.

The vibrational energy possessed by the molecules on the surface of a crystal varies in magnitude as does the translational energy of molecules in a liquid. At any one instant, at a given temperature, a certain fraction of the surface molecules possesses sufficient energy to vibrate free of the solid. These molecules escape and form a vapor phase. Some molecules colliding with the crystal are held. Molecules leave and molecules return. The rate of leaving depends only on the temperature. The rate of return depends on the pressure also. When the pressure is such that molecules return at the same rate they escape, equilibrium is attained and that pressure is the vapor pressure.

Pure liquids freeze completely (until no liquid is left) *at constant temperature*. Pure solids melt completely at this same temperature, which is called either the melting point or the freezing temperature.

At 0°C, water and ice are in dynamic equilibrium. Water molecules are leaving the ice and becoming part of the liquid phase and are doing so very rapidly. At the same time, water molecules collide with the ice crystals and are held in the lattice and thus leave the liquid phase. If no heat is added to nor taken from the system, the rates of leaving and returning are equal at the melting points. Although the shape of the ice mass may change, the amount remains constant. A state of dynamic equilibrium exists in all three equilibrium systems: water and vapor, solid and vapor and solid and liquid. However, temperature alone decides the rate at which molecules leave an area of the surface of either a liquid or a solid. Pressure has a negligible effect, because the states are incompressible. At 0°C, that rate at which molecules leave an area of water surface equals the rate at which they leave the crystal layer, and there is equilibrium. Above 0°C, molecules leave the crystal faster than they return and the ice melts. Below 0°C, molecules leave the liquid phase faster than they return from the crystal, and the water freezes.

Suppose one tries to raise the temperature of a mixture of ice and water by heating. The temperature may rise slightly above 0°C. As explained above, the rate at which particles leave the solid is greater than the rate at which they return, and the ice begins melting. In so doing, it absorbs calories (the heat of fusion) and returns the temperature to 0°C as long as any ice remains. A similar explanation can be given for freezing when both solid and liquid are present. As heat leaves the system, momentarily reducing the temperature, freezing occurs, liberating the heat of fusion to return (or hold) the temperature at 0°C until all the water is frozen.

With ice and vapor (similar to water and vapor) at equilibrium below 0°C, the loss of molecules in the solid equals the return from the vapor. If the temperature is raised 1°C, molecules leave the solid faster as the ice begins to vaporize. As the ice vaporizes, however, the pressure increases, gradually increasing the rate of return until equilibrium is restored. Thus ice and vapor (and also water and vapor) may exist together over a range of temperatures. But as the temperature is changed, the pressure of the vapor must be allowed to change if the coexistence of the two phases is to continue.

Supercooling has been mentioned in the discussion at the beginning of this chapter. It was stated that the water must be stirred or agitated when cooled if it is to begin freezing at 0°C. Often it supercools several degrees below the freezing point before freezing starts. Apparently this is due to the lack of any pattern for starting the lattice. If a tiny crystal of the solid is dropped into a supercooled liquid, it freezes immediately because ions or molecules can then add to the pattern. Scratching or stirring increases the probability of the particles' forming the first unit of the pattern. Once the pattern is there, equilibrium conditions are approached and supercooling is impossible.

EXERCISES

7.1 Why is the theory discussed in this chapter called the Kinetic Theory?

7.2 Write the postulates of the Kinetic Theory as they apply to (a) matter in all states, (b) matter in the gaseous state, (c) matter in the liquid state, (d) matter in the solid state. Certain postulates listed in (a) must be repeated, since they apply to all three states.

7.3 Describe the types of particle motion that may take place in: (a) gases, (b) liquids, (c) solids.

7.4 Which has greatest average particle energy: (a) hydrogen at 0°C and 1 atm, (b) hydrogen at −10° and 1 atm, (c) hydrogen at 0°C and 2 atm, (d) hydrogen at 10°C and 1 atm?

7.5 Explain, using the Kinetic Theory: (a) how gases flow, (b) what happens when a gas is compressed at constant temperature, (c) how each gas of a mixture fills a container completely, (d) why liquids are more dense than gases, (e) why liquids are nearly incompressible, (f) where the heat of fusion (the energy) goes when a solid melts and (g) why liquids supercool.

7.6 Define: (a) perfect gas, (b) dynamic equilibrium, (c) kinetic energy, (d) perfectly elastic collisions, (e) particle (as used in this chapter), (f) theory.

7.7 Which two postulates of the Kinetic Theory of gases become untrue at low temperatures and high pressures?

7.8 Which postulate of the Kinetic Theory does not hold for gases at 0°A?

7.9 What two factors in theory play an important part in determining the viscosity of a liquid?

7.10 At which of these temperatures may an equilibrium exist: (a) −10°C, (b) 0°C, (c) +10°C, (d) 105°C.

a. Between ice and water?
b. Between ice and vapor?
c. Between water and vapor?

7.11 Explain why spilled mercury collects in little spheres.

7.12 Why do pure solids melt completely at one temperature?

7.13 How do the forces of attraction between water molecules vary from state to state?

7.14 What causes pressure on the inner walls of a gas container?

7.15 Why does the vapor pressure of ice increase very rapidly with temperature?

7.16 How many molecules in any state (relative amount) possess nearly the average energy?

7.17 What causes a clay particle to move as it undergoes Brownian movement?

7.18 In what state is diffusion most rapid?

7.19 Which should be **expected** to have the highest boiling point: (a) a solid of small non-polar molecules, (b) a polar substance, (c) an ionic substance? Explain.

7.20 Can a gas exist at absolute zero? Explain.

SUGGESTED READING

Hildebrand, J. H.: *An Introduction to Molecular Kinetic Theory.* Reinhold Publishing Corp., New York, 1964 (paperback).
Lee, G. L.: *Principles of Chemistry—A Structural Approach.* International Textbook Co., Scranton, Pa., 1970. Chapter 4.
Masterton, W. L. and Slowinski, E. J.: *Chemical Principles,* 2nd Ed. W. B. Saunders Co., Philadelphia, 1969. Chapter 5.
Sienko, M. J. and Plane, R. A.: *Chemistry,* 3rd Ed. McGraw-Hill Book Co., New York, 1966. Chapter 6.
Slabaugh, W. H.: "The Kinetic Structure of Gases," J. Chem. Educ., *30*:68, 1953.

EIGHT • THE MOLECULAR THEORY

Joseph Louis Gay-Lussac (1778–1850), one of the pioneers in the study of the gases, took up the work of Jacques Alexandre Charles in 1801. It was Charles who stated, in 1787, that different gases expand or contract to the same extent with the same changes in temperature. It was Gay-Lussac who found the increase in gas volume to be 1/273 of the volume of the gas at 0°C for each degree rise in temperature. He stated the temperature-volume law in the form used today. For this reason Charles' Law is also known as Gay-Lussac's Law. Both men contributed to its formulation, both should share in the honor.

8.1 THE LAW OF COMBINING VOLUMES

Continuing his studies of gases, Gay-Lussac turned his attention to determining the per cent of oxygen in air. This he did by exploding known amounts of hydrogen in known quantities of air. In 1805, he discovered that oxygen and hydrogen combine in the volume ratio of 1 to 2. For example, 500 ml of oxygen combine with 1000 ml of hydrogen (when both are at the same temperature and pressure). This interesting fact led him to investigate volume relationships in other gaseous reactions. He also discovered that:

1. One liter of hydrogen chloride combines with 1 liter of ammonia to yield ammonium chloride.
2. One liter of carbon monoxide reacts with 1/2 liter of oxygen to give 1 liter of carbon dioxide.

3. One liter of oxygen burns with solid sulfur to yield 1 liter of sulfur dioxide.

From these and other similar observations he formulated a relationship, which, once stated, is obvious, called the **Law of Combining Volumes: When two or more gases are involved in a chemical reaction either as reactants or as products, the volumes of the gases produced or consumed are in the ratios of small integers.**

8.2 AVOGADRO'S HYPOTHESIS

The Law of Combining Volumes appeared at approximately the same time as Dalton's Theory. If Dalton's Theory were correct, molecules would combine in the ratios of small whole numbers. It remained for an Italian professor of physics, Amadeo Avogadro (1776–1856), to draw the two ideas together, and in 1811 he presented this postulate: **Equal volumes of gases at the same temperature and pressure contain an equal number of molecules.**

Table 8.1 lists the molecular weights and densities of several gases. The great extremes in molecular masses are pointed out by the inclusion of hydrogen (the lightest gas) and uranium hexafluoride (one of the densest of gases).

According to Avogadro's Hypothesis, there is the same number of uranium hexafluoride molecules in 1 liter of that gas as there are hydrogen molecules in 1 liter of hydrogen gas at the same temperature and pressure, even though the uranium hexafluoride molecule is approximately 175 times heavier than the hydrogen molecule. At first this may appear unreasonable, or even impossible, until one remembers that the molecules in the gases are separated by relatively great distances. This is true even of uranium hexafluoride molecules.

Apparently Avogadro sensed that which was postulated years later, that is, that all gases at the same temperature have the same average molecular translational energy. The translational energy, T.E., depends on both the mass (m) and the velocity (v) of the molecule (T.E. = $1/2\ mv^2$). Uranium hexafluoride molecules move much more slowly than hydrogen molecules and collide with one another and with the sides of the container less often, but they hit much harder. The decrease in frequency of collision is offset exactly by the increased energy of impact; thus, each molecule, regardless of size, contributes equally to pressure and volume.

TABLE 8.1 THE MOLECULAR WEIGHTS AND DENSITIES OF SOME GASES AT STP

Gas	Molecular Weight	Density at STP g/L
H_2	2.016	0.0894
O_2	31.999	1.429
CO_2	44.010	1.964
Cl_2	70.906	3.17
UF_6	352.02	15.71

8.3 MOLAR VOLUME

one mole of nitrogen gas (N_2) weighs 28 grams, occupies 22.4 liters at STP and contains 6.02×10^{23} molecules.

In a previous chapter, it was learned that a gram-molecular weight (1 mole) of any compound contains 6.02×10^{23} molecules (or units), the Avogadro number. In the preceding discussion, it was found that equal volumes of gases at the same temperature and pressure contain an equal number of molecules. It follows that there is a molar volume: that one mole of gas occupies a fixed volume at a fixed temperature and pressure. Using density at STP one can find the molar volume at these conditions.

The volume of 1 mole of oxygen gas is the volume of 32 g of the gas:

1.43 g of O_2 occupies 1 liter at STP (Table 8.1)

$$\frac{32 \text{ g/mole}}{1.43 \text{ g/L}} = 22.4 \text{ L/mole at STP}$$

There are 22.4 of the 1.43 gram liters in 32 g of oxygen. For hydrogen the molar volume is:

$$\frac{2.016 \text{ g/mole}}{0.0894 \text{ g/L at STP}} = 22.4 \text{ L/mole at STP also.}$$

We conclude that the volume of 1 mole of any gas, weighing the gram-molecular weight and containing 6.02×10^{23} molecules, is 22.4 L at STP.

8.4 CALCULATION OF MOLECULAR WEIGHTS

The existence of the molar volume (a set volume for a mole of all gases) affords a method for determining the molecular weight of gases.

Example 1. The density (weight of 1 liter) of a gas composed of carbon and oxygen is 1.964 g/L at STP. Find the molecular weight of the gas.

Solution: One mole of the gas is 22.4 L at STP. One needs to find the weight of 22.4 liters.

$$1.964 \text{ g/L} \times 22.4 \text{ L/mole} = 44 \text{ g/mole}$$

The gas is carbon dioxide (CO_2):

$$12 + (2 \times 16) = 44 \text{ g/mole}$$

Example 2. 2.5 L of a gas at STP composed of carbon and oxygen weighs 3.12 g. Calculate the molecular weight of the gas.

Solution:

$$\frac{3.12\ \text{g}}{2.5\ \text{L}} = 1.25\ \text{g/L (mass of 1 liter)}$$

$$\frac{3.12\ \text{g}}{2.5\ \text{L}} \times 22.4\ \text{L/mole} = 28\ \text{g/mole (mass of 22.4 L)}$$

The gas is carbon monoxide (CO):

$$12 + 16 = 28\ \text{g/mole}$$

Example 3. Two liters of a gas composed of sulfur and oxygen weighs 4.4 g at 640 mm and 25°C. Calculate the molecular weight of the gas.

Solution:

a. Find the volume the 4.4 g could occupy at STP. (Boyle's and Charles' Laws).

$$2\ \text{L} \times \frac{640\ \text{mm}}{760\ \text{mm}} \times \frac{273°\text{A}}{298°\text{A}} = 1.54\ \text{L at STP}$$

b. Find the mass of 1 liter at STP.

$$\frac{4.4\ \text{g}}{1.54\ \text{L}} = 2.86$$

c. Weight of 22.4 L at STP.

$$2.92\ \text{g/L} \times 22.4\ \text{L/mole} = 64\ \text{g/mole}$$

Altogether the setup is:

$$\frac{4.4\ \text{g}}{\left(2\ \text{L} \times \frac{640\ \text{mm}}{760\ \text{mm}} \times \frac{273°\text{A}}{298°\text{A}}\right)} \times 22.4\ \text{L/mole} = 64\ \text{g/mole}$$

The gas is SO_2:

$$32 + (2 \times 16) = 64\ \text{g/mole}$$

Example 4. The density of SO_3 is 3.47 g/L at 820 mm and 30°C. Calculate the molecular weight of SO_3.

Solution: First calculate the volume that the 1 liter would occupy at STP and proceed as in Example 3.

$$\frac{3.47\ \text{g}}{\left(1\ \text{L} \times \frac{820\ \text{mm}}{760\ \text{mm}} \times \frac{273°\text{A}}{303°\text{A}}\right)} \times 22.4\ \text{L/mole} = 80\ \text{g/mole}$$

The gas is SO_3:

$$32 + (3 \times 16) = 80\ \text{g/mole}$$

Example 5. 6.405 g of $KClO_3$ and a trace of MnO_2 catalyst were heated until 350 ml of oxygen collected over water at 640 mm and 24°C was evolved. After cooling, 6.031 g of a $KClO_3$–KCl mixture remained. Calculate the molecular weight of oxygen. The vapor pressure of water at 24°C is 22.1 mm.

Solution:

a. $0.350 \text{ L} \times \frac{273°\text{A}}{297°\text{A}} \times \frac{(640 - 22.1) \text{ mm}}{760 \text{ mm}} = 0.262 \text{ L at STP}$

b. This 0.262 L at STP weighs:

$$6.405 - 6.031 = 0.374 \text{ g}$$

c. One liter at STP weighs:

$$\frac{0.374 \text{ g}}{0.262 \text{ L}} = 1.43 \text{ g}$$

d. 22.4 L at STP (1 mole) weighs:

$$1.43 \text{ g/L} \times 22.4 \text{ L/mole} = 32 \text{ g/mole}$$

Altogether, the setup is:

$$\frac{0.374 \text{ g}}{\left(0.350 \text{ L} \times \frac{273°\text{A}}{297°\text{A}} \times \frac{(640 - 22) \text{ mm}}{760 \text{ mm}}\right)} \times 22.4 \text{ L/mole} = 32 \text{ g/mole}$$

The molecular weight of volatile liquids may also be calculated by the same method. Although the liquid cannot be vaporized at 1 atm and 0°C, the volume it *would have* at STP may be calculated from the volume occupied by the vapor at a temperature above the boiling point.

Example 6. 2.5 L of chloroform vapor at 150°C and 610 mm weighs 6.92 g. Calculate the molecular weight of the liquid.

Solution: Calculated just like a permanent gas.

$$\frac{6.92 \text{ g}}{\left(2.51 \text{ L} \times \frac{273°\text{A}}{423°\text{A}} \times \frac{610 \text{ mm}}{760 \text{ mm}}\right)} \times 22.4 \text{ L/mole} = 119.5 \text{ g/mole}$$

The formula for chloroform is $CHCl_3$:

$$12 + 1.0 + (3 \times 35.5) = 119.5 \text{ g/mole}$$

Example 7. A simple laboratory experiment for determining the molecular weight of liquids with boiling points lower than that of water is described below. Three or four ml of a chlorinated hydrocarbon (for example, a compound containing chlorine, carbon and possibly hydrogen) is added to a previously weighed dry Erlenmeyer flask capped with aluminum foil crimped about the mouth. The Erlenmeyer flask is placed in water, and the water is heated to boiling for about five minutes, as shown in Figure 8.1. The liquid vaporizes and escapes through the pinhole, taking the air out with it. In a short time the flask is filled only with the vapor (of the liquid) at the boiling point of water. The flask is then cooled, which condenses the vapor and draws air back into the flask. The flask is weighed again when cool. The increase in weight is due to the few drops of condensed liquid that filled the flask as vapor when hot.

Suppose the complete volume of the Erlenmeyer flask (level-full) is 139 ml. During the experiment the barometric pressure was 760 mm. The increase in weight (weight of flask + aluminum foil + condensed liquid) minus (weight of flask + aluminum cap) = 0.680 g. Calculate the molecular weight of the liquid.

Solution: 139 ml of vapor at 100°C (boiling point) at 760 mm weighs 0.680 g.

$$\frac{0.680 \text{ g}}{\left(0.139 \text{ L} \times \frac{273°\text{A}}{373°\text{A}}\right)} \times 22.4 \text{ L/mole} = 150 \text{ g/mole}$$

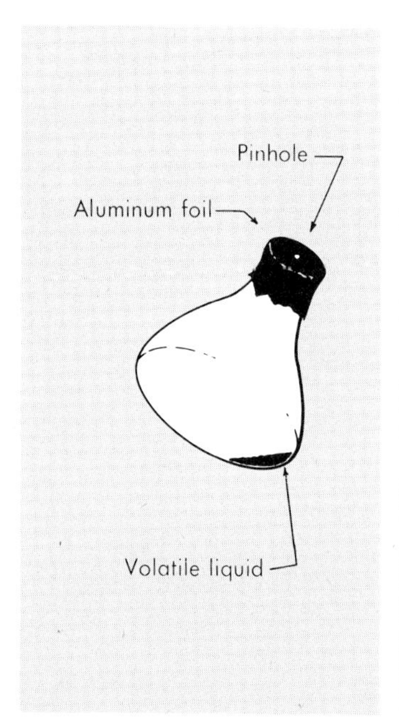

a

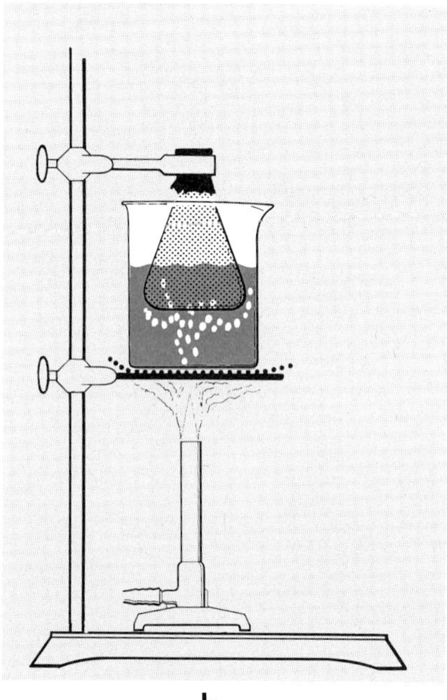

b

FIGURE 8.1 MOLECULAR WEIGHT DETERMINATION FROM VAPOR DENSITY.

The liquid is carbon tetrachloride (CCl_4). It has a molecular weight of $12 + (4 \times 35.5) = 154$ g/mole. The error occurs because carbon tetrachloride has an appreciable vapor pressure at room temperature. When the flask, foil and condensed liquid are weighed, less air was present in the flask (being at 760 mm minus the vapor pressure of CCl_4) than was present at the first weighing when the air pressure was 760 mm.

8.5 CALCULATION OF GAS DENSITIES

When either the molecular weight or the molecular formula (from which the molecular weight can be calculated) is known, one can calculate (using the molar volume at STP) the density of a gas at any temperature and pressure.

Example 1. What is the density in g/L of oxygen gas at STP?

A mole of oxygen gas, O_2, weighs 32.0 g; it occupies 22.4 L at STP. 32.0 g is divided between the 22.4 L.

$$\frac{32.0\ \text{g/mole}}{22.4\ \text{L/mole}} = 1.43\ \text{g/L}$$

Example 2. What is the density in g/L of carbon dioxide at 25°C and 700 mm?

One mole of CO_2 is $12 + (2 \times 16) = 44$ g. It occupies

$$22.4\ \text{L/mole} \times \frac{298}{273} \times \frac{760}{700} = 26.5\ \text{L at 25°C and 760 mm.}$$

One liter weighs

$$\frac{44\ \text{g/mole}}{26.5\ \text{L/mole}} = 1.66\ \text{g/L}$$

The complete setup is:

$$\frac{44\ \text{g/mole}}{\left(22.4\ \text{L/mole} \times \frac{298}{273} \times \frac{760}{700}\right)} = 1.66\ \text{g/L}$$

8.6 THE CANNIZZARO PRINCIPLE AND ATOMIC WEIGHTS

The value of Avogadro's Hypothesis was not fully appreciated for over 40 years. It was in 1858 that the Italian chemist Stanislao Cannizzaro (1826–1910) showed how it might be used to determine atomic weights. Until that time the formula of water was thought to be HO. The atomic weight of oxygen was taken as 8 compared to 1 for hydrogen.

By a simple application of Avogadro's Hypothesis, Cannizzaro showed that **many elemental gases are diatomic** (contain two-atom molecules). Consider the reaction of the gases hydrogen and chlorine. It is shown experimentally that 1 volume of hydrogen combines with 1 volume of chlorine to produce 2 volumes of hydrogen chloride, all measured at the same temperature and pressure. That is:

1 liter hydrogen + 1 liter chlorine → 2 liters hydrogen chloride

According to Avogadro's Hypothesis:

X molecules of hydrogen + X molecules of chlorine $\xrightarrow{\text{become}}$ 2X molecules of hydrogen chloride

The product, hydrogen chloride, is a pure substance; it is composed of identical molecules. Each molecule must contain at least 1 hydrogen atom and at least 1 chlorine atom. But there are twice as many hydrogen chloride molecules formed as there were either chlorine or hydrogen molecules. Therefore, each hydrogen molecule and each chlorine molecule must contain at least 2 atoms.

The reaction is

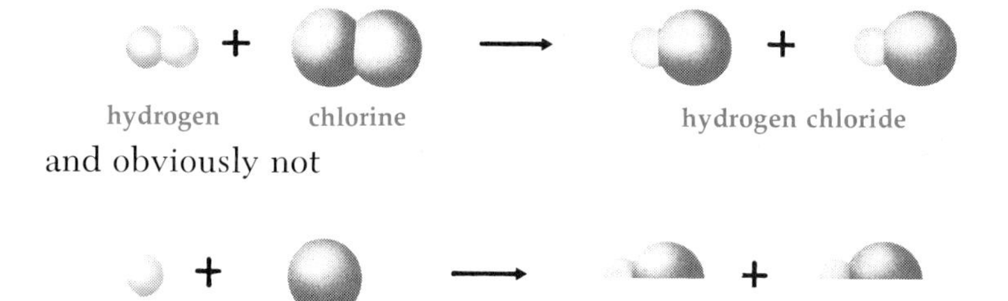

and obviously not

If the explanation is repeated in molar quantities, it is found that when 22.4 L of hydrogen at STP is mixed with 22.4 L of chlorine at STP in a 44.8 L rigid container, exploded with an electric spark, and then cooled to 0°C, the pressure of the 44.8 liters of hydrogen chloride gas is 760 mm.

In equation form,

a. 22.4 L of hydrogen at STP + 22.4 L chlorine at STP → 44.8 liters of hydrogen chloride at STP

b. 1 mole hydrogen + 1 mole chlorine → 2 moles hydrogen chloride

c. 6.02×10^{23} molecules of hydrogen + 6.02×10^{23} molecules of chlorine → 1.204×10^{24} molecules of hydrogen chloride

Hence, the molecular equation is:

$$H_2 + Cl_2 \rightarrow 2\ HCl$$

The experiment above does not rule out the possibility that the hydrogen molecules could be H_4. The equation might then be:

$$H_4 + Cl_2 \rightarrow 2\ H_2Cl$$

However, because 1 mole of hydrogen gas has never been observed to be found in more than 2 moles of a product, its formula is taken as H_2. One mole of hydrogen gas (H_2), 22.4 liters at STP, weighs 2.016 g. The weight of one mole of H atoms is:

$$\frac{2.016\ \text{g/mole H}_2}{2\,\dfrac{\text{gram-atomic weight}}{\text{mole H}_2}} = 1.008\ \text{g/gram-atomic weight}$$

The atomic weight is 1.008. By similar arguments the 22.4 L of chlorine at STP (70.906 g) containes 2 moles of chlorine atoms and the atomic weight is:

$$\frac{70.906}{2} = 35.453$$

In the formation of water from its elements, 2 moles of hydrogen plus 1 mole of oxygen form 2 moles of water. The equation must be

$$2\ H_2 + O_2 \rightarrow 2\ H_2O$$

or

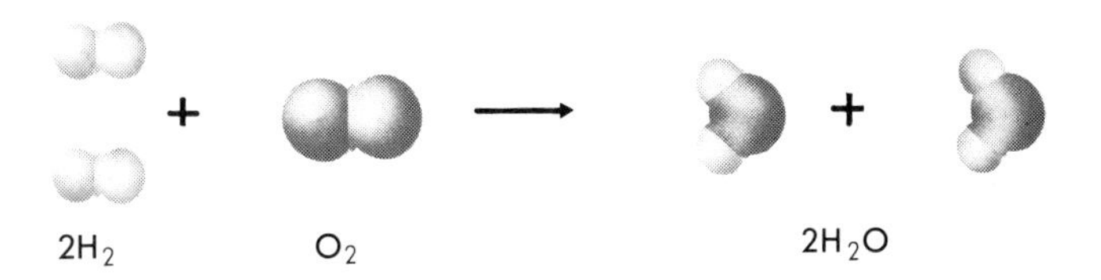

Hydrogen was previously shown to be diatomic (H_2). Oxygen must also be diatomic because after the reaction, oxygen atoms appear in twice as many particles as they did in elemental form. Were the formula of water HO, as it was once thought to be, equal volumes of the diatomic molecules of hydrogen and oxygen must react. This is not the case. The formula of water is H_2O; hydrogen has an atomic weight of 1.008, and the atomic weight of oxygen is $18.016 - 2.016 = 16.00$.

It should be noted that the experimental evidence does not preclude tetratomic oxygen (O_4). That is, a mole of oxygen (32 grams and 22.4 liters) might contain 6.02×10^{23} molecules of O_4. If such were the case, water would be H_2O_2 and the atomic weight of oxygen would be 8. This is improbable. In no reaction has a molecule of oxygen been shown to divide four ways. If it were to do so, the formula H_2O_2 must be accepted. Observations indicate that one mole of oxygen is 2 gram-atomic weights, and the atomic weight of oxygen is 16.00.

Following the same reasoning, the analysis of a series of compounds containing a common element may be used to determine the atomic weight of that element.

Example 1. Below are a series of gases that have been analyzed for an element X. The analyses are given in grams of X per mole.

Gases	Mol. Weight	Gram X/mole	
A	36.5	35.5	35.5
B	71.0	71.0	2×35.5
C	50.5	35.5	35.5
D	85.0	71.0	2×35.5
E	119.5	106.5	3×35.5

Find the atomic weight (maximum atomic weight) of element X.

Solution: The smallest mass of element X per mole of gas is 35.5, and all weights are multiples of 35.5. This is the atomic weight. The element X is chlorine. The gases are HCl, Cl_2, CH_3Cl, CH_2Cl_2 and $CHCl_3$.

Example 2. Find the atomic weight of element Y from the following data.

Gas	Mol. Weight	%Y
A	64	50.0
B	44	72.7
C	80	60.0

Solution: There are

$$0.50 \times 64 = 32 \text{ g of Y/mole of A}$$
$$0.727 \times 44 = 32 \text{ g of Y/mole of B}$$
$$0.60 \times 80 = 48 \text{ g of Y/mole of C}$$

A mole of a compound must contain one atomic weight of each element in grams or a multiple of the atomic weight. The largest number that will divide into both 32 and 48 to give a whole number is 16. The element is oxygen. The gases are SO_2, CO_2 and SO_3.

8.7 VOLUME RELATIONSHIPS AMONG REACTANTS AND PRODUCTS OF A CHEMICAL REACTION

Just as weight relationships among reactants and products of a chemical reaction can be calculated using molecular weights, volume relationships may be calculated by using molecular volumes (volume per mole). For solids and liquids the molar volume differs from substance to substance and may be found by dividing the molecular weight by the density. Calculations of gas volumes (of both reactants and products) are simplified by the fact that all gases have the same molar volume at STP: 22.4 liters.

Example 1. Referring to the process represented by this equation,

$$3\ Fe + 4\ H_2O \rightarrow Fe_3O_4 + 4\ H_2$$

a. How many liters of hydrogen gas at STP may be prepared from reaction of steam with 40 g of Fe? The moles of H_2 produced will be:

$$\frac{40\text{ g Fe}}{56\text{ g Fe/mole Fe}} \times \frac{4\text{ mole }H_2}{3\text{ mole Fe}} = 0.95\text{ mole }H_2$$

the numbers 4 and 3 in the fraction 4/3 are coefficients in the equation for the reaction.

But a mole of any gas is 22.4 L at STP, and the volume of the 0.95 mole H_2 is:

$$0.95\text{ mole} \times 22.4\text{ L/mole} = 21.3\text{ L at STP}$$

The complete setup for the problem is:

$$\frac{40\text{ g}}{56\text{ g/mole}} \times 4/3 \times 22.4\text{ L/mole} = 21.3\text{ L at STP}$$

b. How many g of steam (H_2O) are consumed in producing 10.0 L of hydrogen gas at STP? The moles of hydrogen produced are:

$$\frac{10.0\text{ L}}{22.4\text{ L/mole}} = 0.446\text{ mole }H_2$$

This would then require:

$$\frac{10}{22.4}\text{ mole} \times 4/4 \times 18\text{ g/mole} = 8.03\text{ g water.}$$

The volume at STP must be multiplied by fractions of absolute temperature or of pressures if the volume at other conditions is desired.

c. How many liters of hydrogen gas measured at 150°C and 680 mm may be prepared from the reaction of 40 g of iron with steam? At STP the volume is:

$$\frac{40\text{ g}}{56\text{ g/mole}} \times 4/3 \times 22.4\text{ L/mole} = 21.3\text{ L at STP}$$

At 150°C (423°A) and 680 mm the volume of this gas sample is:

$$21.3\text{ L} \times \frac{423°\text{A}}{273°\text{A}} \times \frac{760\text{ mm}}{680\text{ mm}} = 36.9\text{ L}$$

The complete setup is:

$$\frac{40 \text{ g}}{56 \text{ g/mole}} \times 4/3 \times 22.4 \text{ L/mole} \times \frac{423°\text{A}}{273°\text{A}} \times \frac{760 \text{ mm}}{680 \text{ mm}} = 36.9 \text{ L}$$

Example 2. Volume to volume calculations can be made by the same method. Acetylene burns with oxygen according to the following equation:

$$2\ C_2H_2 + 5\ O_2 \rightarrow 4\ CO_2 + 2\ H_2O$$

How many liters of CO_2 at 640 mm and 25°C are produced when 15 L of acetylene at 25°C and 640 mm are burned?

$$\frac{15 \text{ L} \times \frac{640 \text{ mm}}{760 \text{ mm}} \times \frac{273°\text{A}}{298°\text{A}}}{22.4 \text{ L/mole}} \times 4/2 \times 22.4 \text{ L/mole} \times \frac{760 \text{ mm}}{640 \text{ mm}} \times \frac{298°\text{A}}{273°\text{A}} = 30 \text{ L}$$

Notice that all figures in the mathematical equation will cancel except 4/2, the molar ratio in the equation, and 15 L.

$$15 \text{ L} \times 4/2 = 30 \text{ liter of } CO_2$$

This simplification reminds one of Avogadro's Hypothesis: Equal volumes of gas at the same temperature and pressure contain an equal number of molecules. Therefore, since there is twice the number of CO_2 molecules formed, the volume of CO_2 is double the volume of the acetylene burned.

PROBLEMS

8.1 What fraction of a mole is: (a) 2 liters of gas at STP? (b) 3 liters of oxygen measured at −23°C and 890 mm? (c) 4 liters of nitrogen collected over water at 27°C and 685 mm? (d) What fraction of a mole of water vapor is there in part c?

8.2 How many grams is: (a) 11.2 liters of oxygen at STP? (b) 0.95 L of sulfur dioxide, SO_2 at STP? (c) 2.5 L of ethane (C_2H_6) at 127°C and 640 mm? (d) 45 L of ammonia gas (NH_3) at 500°A and 800 mm?

8.3 How many molecules are there in each gas sample described in (a) to (d) in Problem 8.2?

8.4 Calculate the density of: (a) chlorine gas at STP, (b) water vapor at its boiling point and 1 atm, (c) benzene (C_6H_6) at its boiling point, 80°C, at 1 atm, (d) bromine vapor at 80°C and 480 mm.

8.5 A hydrocarbon gas is 85.7 per cent carbon and 14.3 per cent hydrogen. It has a density of 1.875 g/L at STP. Find the molecular weight and the molecular formula.

8.6 A gas has a density of 1.429 g/L at 819°C and 1520 mm. Calculate the molecular weight of the gas.

8.7 Calculate the volume of 5.6 g of oxygen gas collected over water at 27°C and 640 mm.

8.8 Calculate the density in g/L at STP of the following gases: (a) CO_2, (b) SiF_4, (c) SO_2, (d) NH_3, (e) NF_3, (f) C_2H_2.

8.9 Calculate the molecular weight of each of the following gases: (a) gas A has an STP density of 1.429 g/L; (b) gas B has a density of 2.86 g/L at STP; (c) 2 g of gas C occupies 1.02 L at STP; (d) 2 L of gas D at STP weighs 31.4 g; (e) 3 L of gas E at 640 mm and 27°C weighs 8.19 g; (f) 2.5 g of gas F occupies 0.915 L at 640 mm and 27°C.

8.10 Choose from this list of gases the formula of the gas described in each part: CH_4, NH_3, Ne, CO, NO, O_2, Ar.

a. density of 1.25 g/L at STP.
b. density of 0.714 g/L at STP.
c. density of 4.88 g/L at −123°C and 1520 mm.
d. 3.5 L at 870 mm and 368°A weighs 4.25 g.

8.11 Find the volume at STP of 175 ml of hydrogen measured over water at 28°C and 648 mm. How many g of hydrogen is this?

8.12 7.6 g of O_2 and 34 g of N_2 are mixed and collected over water at 26°C and 645 mm. (a) What is the pressure of oxygen in the wet gas mixture? (b) What is the volume of the wet gas mixture?

8.13 Analysis of samples of three pure gases revealed the following weight of element X per liter at STP: (a) 1.071 g, (b) 2.142 g, (c) 1.607 g. Calculate the probable atomic weight of element X.

8.14 0.500 L of a gas sample at 640 mm and 27°C was found to contain 0.631 g of element X (Problem 8.13). Is the gas sample pure? Supply calculations for evidence.

8.15 Three L of hydrogen gas and 1 liter of nitrogen gas, both measured at 27°C and 640 mm, are passed as a mixture over an impure iron catalyst at high pressure and high temperature. The final gas mixture exerts a pressure of 400 mm at 27°C in a 4 liter container. What fraction of the hydrogen gas originally in the container has combined with nitrogen to form ammonia gas, NH_3?

8.16 What is the volume of (a) 1 mole of O_2 at STP, (b) 0.36 mole of CH_4 at STP, (c) 1 mole of CO_2 at 760 mm and −63°C, (d) 2 moles of CH_4 at 150°C and 600 mm, (e) 0.564 mole H_2 at 3 atm and 273°C?

8.17 How many moles is (a) 3 L of O_2 at STP, (b) 1 L of CO_2 at STP, (c) 14 liters of CH_4 at 760 mm and 25°C, (d) 37 L of CO at 187°C and 2.4 atm, (e) 15.5 L of SO_2 at 25°C and 640 mm?

8.18 How many liters of O_2 at STP may be produced upon the thermal decomposition of 100 g of $KClO_3$? How many liters of oxygen at 2 atm and 150°C?

8.19 What volume of hydrogen at STP is evolved when 35 g of white-hot iron react with steam? What volume at 25°C and 640 mm?

8.20 How many liters of CO_2 at STP are produced when 10 L of CO at STP react with iron oxide? How many liters of CO_2 at 640 mm and 125°C are produced when 10 L of CO at 640 mm and 25°C reacts with iron oxide?

8.21 How many g of Fe are produced when 15 L of CO measured at 890 mm and 27°C reacts with Fe_3O_4? How many liters of CO_2 at STP are produced?

SUGGESTED READING

Feifer, N.: "The Relationship Between Avogadro's Principle and the Law of Gay-Lussac." J. Chem. Educ., *43*:411, 1966.
Hildebrand, J. H.: *An Introduction to Molecular Kinetic Theory.* Reinhold Publishing Corp., New York, 1964 (paperback).
Lee, G. L.:*Principles of Chemistry—A Structural Approach.* International Textbook Co., Scranton, Pa , 1970. Chapter 5.
Masterton, W. L. and Slowinski, E. J.: *Chemical Principles,* 2nd Ed. W. B. Saunders Co., Philadelphia, 1969. Chapter 5.
Neville, R. G.: "The Discovery of Boyle's Law, 1661–62." J. Chem. Educ., *39*:356, 1962.
Reilly, D.: "Robert Boyle and His Background." J. Chem. Educ., *28*:178, 1951.
Sienko, M. J. and Plane, R. A.: *Chemistry,* 3rd Ed. McGraw-Hill Book Co., New York, 1966. Chapter 6.

NINE • OXYGEN AND OXIDATION

9.1 INTRODUCTION

Thus far we have considered concepts of atomic structure and bonding and some basic principles and ideas of physics and chemistry relative to the properties and characteristics of matter. The gaseous, liquid and solid states have been studied. The importance of the electronic configuration of the atoms of an element in determining its properties has been stressed. **The periodic table** has been shown to be an arrangement of the **elements in groups having similar electronic structures** and chemical properties.

These basic ideas and concepts will be used in discussing the physical and chemical properties of the elements in the chapters to follow. Many of the properties of an element and its compounds will be correlated with or predicted, or both, from its position in the periodic table.

9.2 HISTORY OF OXYGEN

Oxygen is the most abundant element on earth and is necessary for life processes. Even though oxygen surrounds us and the atmosphere consists of from 18 to 21 per cent oxygen (by volume, depending upon the humidity), it was not the first element to be discovered. In fact 21 elements, shown in white lettering in Figure 9.1, were discovered before oxygen was isolated and combustion (oxidation) was understood. Many rarer elements, such as gold, silver, copper and mercury were known hundreds of years before oxygen. Carbon (charcoal) and sulfur were known and used in gunpowder in the sixteenth century.

Prior to 1777, the mechanism of combustion or burning was explained by the so-called *phlogiston theory.* According to this theory, combustible materials contain varying amounts of phlo-

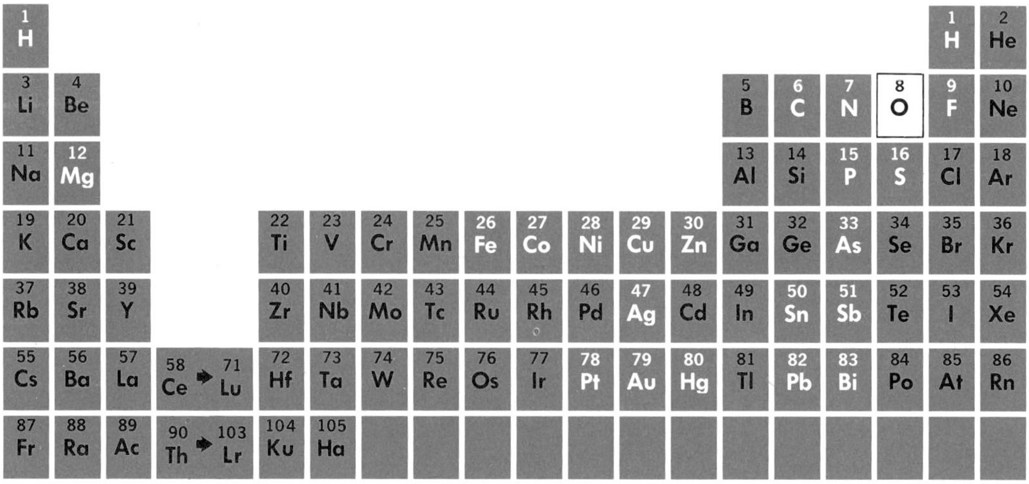

1 H																	1 H	2 He
3 Li	4 Be												5 B	6 C	7 N	8 O	9 F	10 Ne
11 Na	12 Mg												13 Al	14 Si	15 P	16 S	17 Cl	18 Ar
19 K	20 Ca	21 Sc		22 Ti	23 V	24 Cr	25 Mn	26 Fe	27 Co	28 Ni	29 Cu	30 Zn	31 Ga	32 Ge	33 As	34 Se	35 Br	36 Kr
37 Rb	38 Sr	39 Y		40 Zr	41 Nb	42 Mo	43 Tc	44 Ru	45 Rh	46 Pd	47 Ag	48 Cd	49 In	50 Sn	51 Sb	52 Te	53 I	54 Xe
55 Cs	56 Ba	57 La	58 Ce → 71 Lu	72 Hf	73 Ta	74 W	75 Re	76 Os	77 Ir	78 Pt	79 Au	80 Hg	81 Tl	82 Pb	83 Bi	84 Po	85 At	86 Rn
87 Fr	88 Ra	89 Ac	90 Th → 103 Lr	104 Ku	105 Ha													

FIGURE 9.1 ELEMENTS KNOWN BEFORE OXYGEN.

giston. When the material burns, phlogiston is released into the atmosphere. The greater its phlogiston content, the more completely the substance can burn. However, air can accept only a limited amount of phlogiston. When either the phlogiston content of the burning substance is depleted or the air becomes saturated with phlogiston and can accept no more, the reaction ceases. The reasoning is clear. In changing to ashes a burning material loses something, and combustible substances may not burn completely in a limited supply of air. The phlogiston theory remained secure until the quantitative experiments of Lavoisier (La-vwah-zee-ay′) showed that the ash could contain more mass than the substance burned.

how was oxygen discovered?

Two men are credited with the discovery of oxygen, a Swedish chemist, Karl Wilhelm Scheele, and an Englishman, Joseph Priestley. Scheele prepared oxygen in 1772 by heating a number of salts (silver carbonate, mercury(II) carbonate, mercury(II) oxide, magnesium nitrate and manganese(IV) dioxide). The preparation from manganese dioxide is depicted by the equation:

$$3MnO_2 \xrightarrow{heat} Mn_3O_4 + O_2$$

When Priestley first prepared oxygen in 1774, he was unaware of Scheele's findings. Scheele's experiments were sent to his publisher in 1775 and printed in his book, "Fire and Air," which was not published until 1777. In August, 1776, Scheele, who was aggravated over the delay in publishing his book, wrote to a friend, saying, "I have thought for some time back, and I am now more convinced, that the greater number of my laborious experiments on fire will be repeated, possibly in a somewhat different manner, by others, and that their work will be published sooner than my own, which is concerned also with air. It will then be said that my experiments are taken, it may be in a slightly altered form, from their writings." Priestley published his work periodically and soon gained wide acclaim for his discovery.

Priestley was one of the first to see the advantage of collecting gases over mercury instead of water. Substances in the presence of mercury can be heated to well over 100°C, and very few gases are soluble in mercury. In 1774, he experimented with various "kinds of air." He heated all available substances over mercury. If a gas was obtained, he tested it by admitting a burning candle and by placing mice in an atmosphere of the gas. In one of these experiments oxygen was evolved. The details of this experiment are interesting, and a representation of it is given in Figure 9.2. As usual, he inverted a bottle full of mercury in a dish of mercury in such a way that no air entered the inverted bottle. He then pushed mercury(II) oxide under the edge of the bottle and let it rise to the top of the mercury. He heated the oxide by focusing the sun's rays on it with a large magnifying glass. Soon the mercury level descended as oxygen formed over it in the bottle and the quantity of the red oxide diminished. The equation for the reaction is:

$$2HgO \xrightarrow{\text{heat}} 2Hg + O_2$$

Priestley discovered that a candle burned in the "air" with a remarkably vigorous flame. His description of the testing of "dephlogisticated air" (O_2) is intriguing:

> My reader will not wonder that, after having ascertained the superior goodness of dephlogisticated air by mice living in it, and the other tests above mentioned, I should have the curiosity to taste it myself. I have gratified that curiosity by breathing it, drawing it through a glass syphon, and by this means I reduced a large jar full of it to the standard of common air. The feeling of it to my lungs was not sensibly different from that of common air, but I fancied that my breast felt peculiarly light and easy for some time afterwards. Who can tell but that, in time, this pure air may become a fashionable article in luxury? Hitherto only two mice and myself have had the privilege of breathing it.

Neither Priestley nor Scheele realized that he had discovered a new element. They both considered the gas to be very pure air. It remained for the great French scientist, Antoine Laurent Lavoisier, to demonstrate that oxygen was a pure substance, a component of air and a reactant consumed when substances were burned.

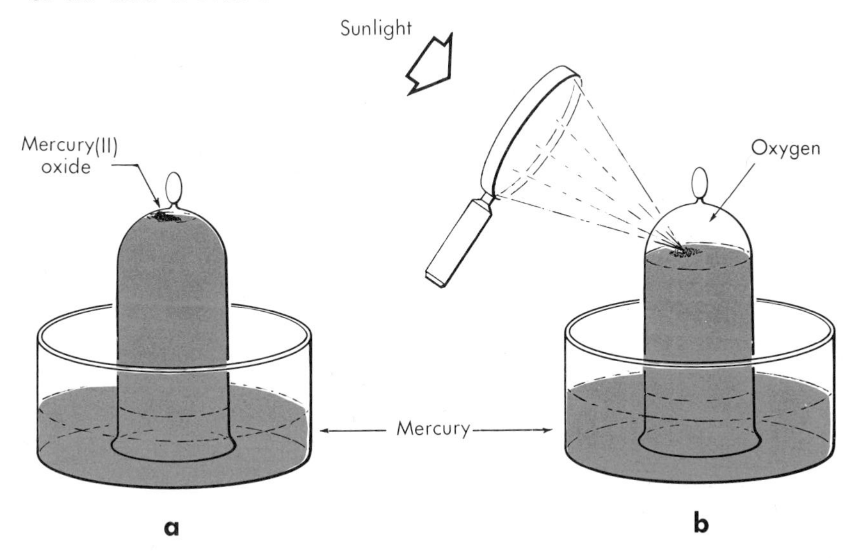

FIGURE 9.2 PREPARATION OF OXYGEN BY HEATING MERCURY(II) OXIDE.

Lavoisier's conclusions were based on a number of experiments. In one of these he placed a piece of tin metal in a flask of air and sealed it. He heated the flask, weighed the apparatus when cool and found there was no change in weight. (The tin present, however, had changed in appearance and properties.) On breaking the flask, air rushed in, showing that the amount of air in the flask had diminished. Furthermore, the solid present had a greater mass than the tin from which it was formed, the weight equaling the weight of the original tin plus the loss in weight in the air. Clearly a part of the air was transferred to the tin. He found that only a certain portion of the air would react with heated tin.

can you show that oxygen is an element?

Later the same year, 1777, he heated mercury mildly in air and noted a decrease in the volume or amount of air as mercury(II) oxide formed. He next heated the oxide to a higher temperature, decomposing it as Priestley had done in preparing oxygen. He recovered the same volume of gas as a result of the decomposition of the oxide as had been consumed in its formation. These experiments were followed by other oxidation experiments, in which burning was shown to be a combination with oxygen. Lavoisier's work not only established oxygen as a new and important element, and clarified the nature of burning or combustion, it also verified the **Law of Conservation of Mass** by showing that combustion reactions are in accord with that law.

9.3 THE OCCURRENCE OF OXYGEN

The relative abundances of the principal elements in the earth's crust (which includes a layer of the crust 10 miles thick, the

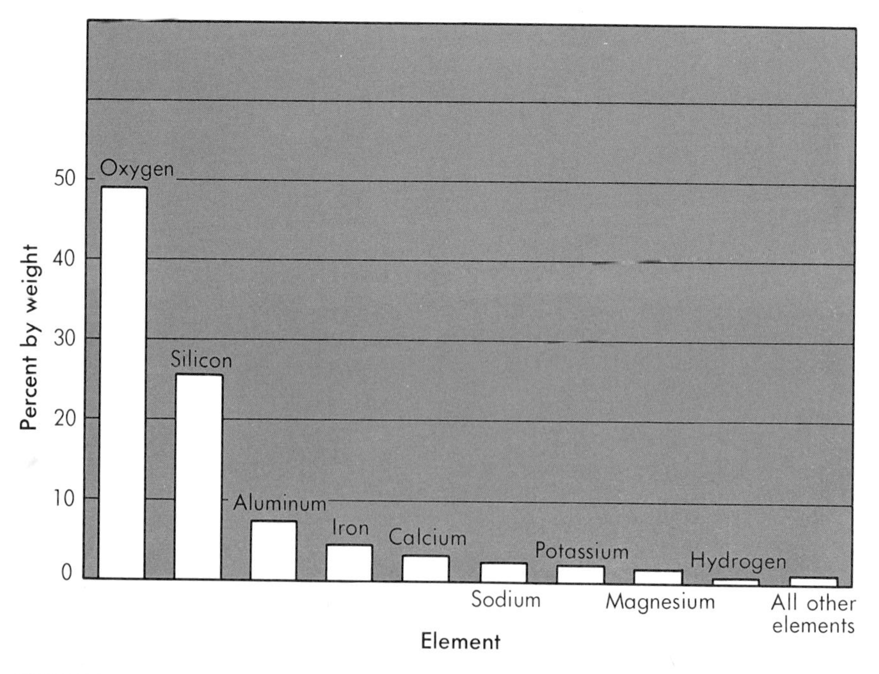

FIGURE 9.3 THE RELATIVE ABUNDANCE OF SOME ELEMENTS IN THE EARTH'S CRUST.

hydrosphere and the atmosphere) are shown in Figure 9.3. It can be seen that nine elements comprise about 99 per cent of the earth's crust; oxygen and silicon make up three fourths of the mass, and oxygen alone contributes 49 per cent. Dry air is 21 per cent oxygen by volume, with nearly all of the oxygen in elemental form as O_2. A small amount of oxygen occurs in carbon dioxide and in oxide gases of other elements such as nitrogen and sulfur. **These oxide gases are notable contributors to air pollution.**

oxygen and silicon are the most abundant elements.

Due to the great abundance of oxygen and because, as will be shown later, most of the elements combine with it, it is not surprising that oxygen was very early chosen as a standard when the relative masses of atoms (atomic weights) were being established.

9.4 PROPERTIES OF OXYGEN

Oxygen is element number 8. It appears in the upper right corner of the periodic table between nitrogen and fluorine and above sulfur. Situated in this position, it is similar to the elements around it in the table in that it has a great electronegativity. Among the active elements (all the elements excluding the noble gases, which form very few compounds), oxygen has an electronegativity second only to fluorine. This means that compared with any element with which it combines, except for fluorine, oxygen has a greater attraction for the bonding electrons.

Oxygen falls among the elements that are gases at room temperature and 1 atmosphere pressure. In Figure 9.4 the physical states of the elements are compared relative to their positions in the periodic table. Approximately 5 per cent of the elements are liquids, 10 per cent are gases and the others are solids. The gases are concentrated as a group in the upper right portion of the table. Except for oxygen, which may be triatomic under special conditions, and the noble gases which are monatomic, all the gases are

Cs, Fr and Ga have melting points near room temperature.

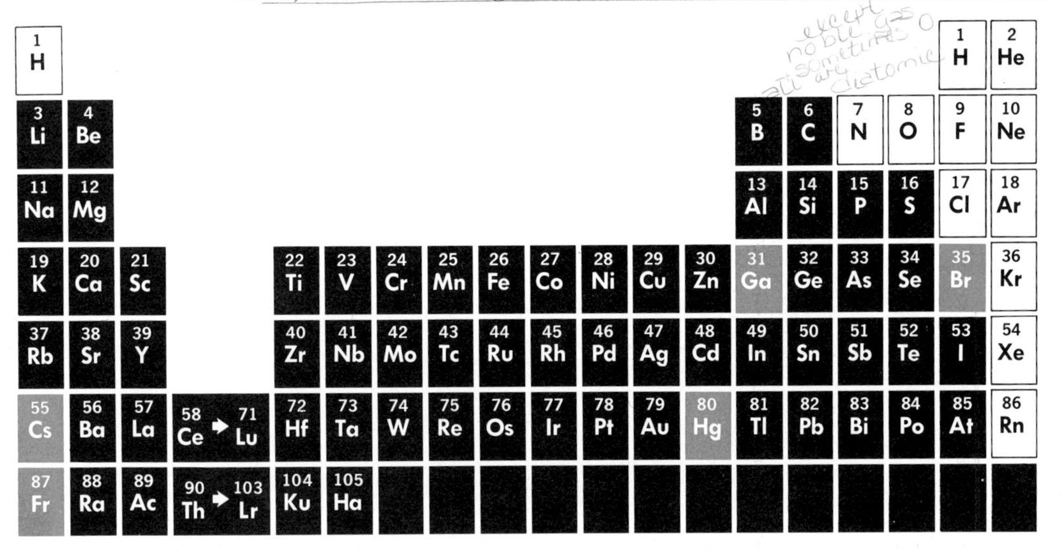

FIGURE 9.4 PHYSICAL STATES OF THE ELEMENTS AT ROOM TEMPERATURE AND 1 ATMOSPHERE PRESSURE (SOLIDS, BLACK; LIQUIDS, ORANGE; GASES, WHITE).

TABLE 9.1

	Molecular weight	Density	Boiling point	Freezing point
Oxygen	32.00	1.429g/l	−183°C	−219°C
Ozone	48.00	2.144g/l	−112°C	−245°C

diatomic. All the diatomic molecules consist of atoms that have a great electronegativity. Consequently they hold their electrons, shared and unshared, closely about them.

Triatomic oxygen (O_3), called **ozone, is an allotrope** of diatomic oxygen; ozone may occur under special conditions. The physical properties of ozone and of oxygen are compared in Table 9.1. Notice that ozone has a greater density and has higher melting and boiling points than diatomic oxygen. The triatomic molecules are less stable than the more common two-atom molecules, and O_3 is both more reactive and a stronger oxidizing agent than O_2. Consistent with its stronger oxidizing ability is the fact that energy is required to make ozone from oxygen:

an allotrope is a different form of the same element.

$$3O_2 + \text{energy} \rightarrow 2O_3$$

The energy may be supplied either by illuminating the gas with ultraviolet light, passing oxygen gas over a hot platinum wire or passing it through an electric discharge, as shown in Figure 9.5.

In the stratosphere, the region of space 7 to 30 miles above the earth's surface, ozone is approximately 300 times more abundant than it is in the troposphere (the space from 0 to 7 miles above the earth's surface). The presence of **ozone** in the stratosphere is very important to our existance, since the gas **absorbs ultraviolet radiation** and greatly reduces the intensity of ultraviolet rays reaching the earth. The absorption of ultraviolet light by ozone results from the following reaction

$$O_3 \xrightarrow{h\nu} O_2 + O$$

where $h\nu$ represents light energy, which in this case must be of wavelengths between 1200 and 2900 Å.

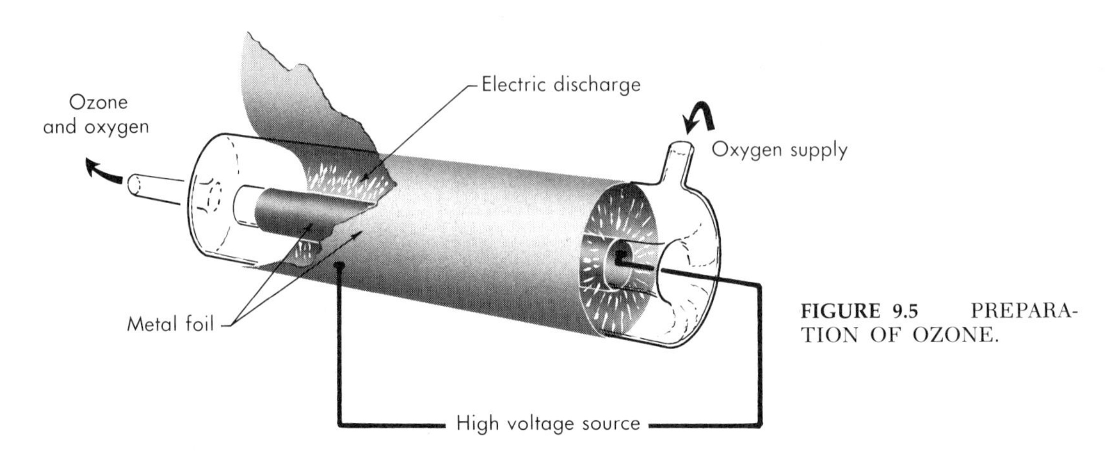

FIGURE 9.5 PREPARATION OF OZONE.

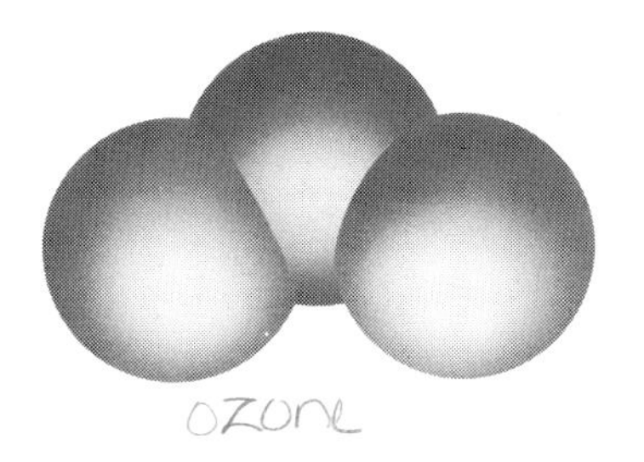

FIGURE 9.6 OZONE AND OXYGEN MOLECULES.

On the other hand, ozone in the troposphere is a harmful **air pollutant.** In Los Angeles County two million students through the high school level are excused from "strenuous indoor and outdoor activities . . . when the forecast concentration of ozone in the atmosphere reaches 0.35 parts per million . . . because smog is an increasing health hazard which may seriously affect the lungs of young people." The smog for which Los Angeles has received a reputation is not due to initial air pollutants but to the chemicals formed when these compounds react in the presence of radiation from the sun and is, therefore, referred to as a **photochemical smog.** When hydrocarbons (most of which come from automobile exhausts) are exposed to sunlight in the presence of NO and NO_2, a complex series of chemical reactions occurs. One of the irritating and toxic products is ozone.

Oxygen, with symmetrical, non-polar molecules, is only sparingly soluble in water. Models representing oxygen and ozone are shown in Figure 9.6. At 18°C and 1 atmosphere, only 32.3 ml (43.2 milligrams) may dissolve in 1 liter of water. Nevertheless, this small quantity of dissolved oxygen is essential for marine life. Fish will die in newly boiled and cooled water, since most of the oxygen has been carried off with the escaping steam as the solubility of the gas decreases with an increase in temperature. **Thermal pollution** is one of the concerns of modern society. Great quantities of water warmed in cooling the condensers of all power plants, including atomic reactors, are returned to lakes or streams from which they were drawn, increasing the temperature of the entire body of water. The oxygen content of the water is less at the higher temperature. Consequently, thermal pollution has a detrimental effect on the ecology (plant and animal life) of rivers and lakes.

9.5 THE BONDING OF OXYGEN

Many of the compounds which oxygen forms may be predicted from the position of the element in the periodic table. In Group VI, containing elements with six valence elections, an oxygen atom must gain two electrons by either transfer or sharing to achieve an octet configuration. The ionic valence of the element is −2. Oxygen may share electrons with its neighbors, which also have a great electronegativity. In covalent bonding, the oxygen atom may contribute one electron to each of two single bonds, or it may contribute two electrons to a double bond, or it may accept two electrons contributed by a less electronegative element for sharing. When already attached to two other atoms by single bonds, oxygen may contribute 2 electrons for sharing with the third atom. When the

oxygen atom is bonded to either two or three atoms, it **is usually sp^3 hybridized,** as described in Chapter 5 in the discussion of the water molecule. Because the angle between the two O to H bonds of the water molecule (104.5°) is nearer the tetrahedral angle (109.5°) than to the right angle, the orbitals of oxygen involved in the bonding are thought to be hybridized and not simple p orbitals. The remaining two sp^3 orbitals of the oxygen atom in water each possess an electron pair, called in this situation an unshared pair. The angles between the bonds on the oxygen atom are listed for several compounds in Table 9.2. All the angles quoted in the table, with the exception of the O—O—H bond in hydrogen peroxide and the O—O—O bond in ozone, approach the tetrahedral angle. The oxygen atom in each is thought to be sp^3 hybridized.

In the ozone molecule, in which the O—O—O bond differs from all the other bonds in that the oxygen participates in both a single and a double bond, :Ö—O=Ö:, (more correctly two $1\frac{1}{2}$ bonds by resonance), the angle is more nearly the 120° angle of an sp^2 hybrid, and the oxygen atom is said to be sp^2 hybridized.

The formulas for normal oxides of metals (those with the simple O^{2-} ion) can be predicted using the ionic valences of the metals. For example, iron(II) oxide has the formula of FeO; magnesium oxide, with the magnesium ion Mg^{2+}, is MgO; and lithium oxide, with the lithium ion Li^+, is Li_2O. Because of the many possibilities in covalent bonding, the formulas of the non-metal oxides cannot be predicted. The normal oxides of many of the elements, those in which the other element has or is assigned a positive valence equal to the group number of the element in the periodic table, are shown in Figure 9.7.

The number of metal oxides is increased and the predictability of metallic oxide formulas confused by the capability for oxygen atoms to share electrons among themselves. The metallic oxides whose formulas were "predicted" from valences are the so-called **normal oxides**. Just as there are normal oxides, K_2O, BaO, Al_2O_3, and so on, there are also **peroxides,** Na_2O_2, BaO_2, K_2O_2, and so on, in which oxygen would appear to have a charge of −1, and the **superoxides,** KO_2, RbO_2 and so on, in which the oxygen appears to have a charge of $-\frac{1}{2}$. These apparent charges of −1 and $-\frac{1}{2}$ for oxygen are oxidation states or oxidation numbers (to be discussed later in the chapter). They are not real charges on individual atoms but are half of the charge borne by two oxygen atoms bound together by a covalent bond. The peroxide ion may be represented by the electron configuration, $[:\ddot{O}:\ddot{O}:]^{2-}$, in which each atom has an

TABLE 9.2 BOND ANGLES IN OXYGEN COMPOUNDS

Angle	Compound	Degrees
O—O—H	H_2O_2	97
H—O—H	H_2O	104.5
F—O—F	OF_2	101.5
Cl—O—Cl	Cl_2O	110.8
C—O—C	$(CH_3)_2O$	111
C—O—H	CH_3OH	107.9
O—O—O	O_3	116.8

GROUPS

IA	IIA	IIIB		IVB	VB	VIB	VIIB	VIII	VIII	VIII	IB	IIB	IIIA	IVA	VA	VIA	VIIA	O
E_2O																	E_2O_7	EO_4
1 H H_2O	EO												E_2O_3	EO_2	E_2O_5	EO_3	1 H	2 He
3 Li Li_2O	4 Be BeO												5 B B_2O_3	6 C CO_2	7 N N_2O_5	8 O	9 F	10 Ne
11 Na Na_2O	12 Mg MgO	E_2O_3		EO_2	E_2O_5	EO_3	E_2O_7		EO_4		E_2O	EO	13 Al Al_2O_3	14 Si SiO_2	15 P 'P_2O_5'	16 S SO_3	17 Cl Cl_2O_7	18 Ar
19 K K_2O	20 Ca CaO	21 Sc Sc_2O_3		22 Ti TiO_2	23 V V_2O_5	24 Cr CrO_3	25 Mn Mn_2O_7	26 Fe	27 Co	28 Ni	29 Cu Cu_2O	30 Zn ZnO	31 Ga Ga_2O_3	32 Ge GeO_2	33 As As_2O_5	34 Se SeO_3	35 Br	36 Kr
37 Rb Rb_2O	38 Sr SrO	39 Y Y_2O_3		40 Zr ZrO_2	41 Nb Nb_2O_5	42 Mo MoO_3	43 Tc Tc_2O_7	44 Ru RuO_4	45 Rh	46 Pd	47 Ag Ag_2O	48 Cd CdO	49 In In_2O_3	50 Sn SnO_2	51 Sb Sb_2O_5	52 Te TeO_3	53 I	54 Xe XeO_4
55 Cs Cs_2O	56 Ba BaO	57 La La_2O_3	58 Ce → 71 Lu	72 Hf HfO_2	73 Ta Ta_2O_5	74 W WO_3	75 Re Re_2O_7	76 Os OsO_4	77 Ir	78 Pt	79 Au Au_2O	80 Hg HgO	81 Tl Tl_2O_3	82 Pb PbO_2	83 Bi Bi_2O_5	84 Po	85 At	86 Rn
87 Fr Fr_2O	88 Ra RaO	89 Ac Ac_2O_3	90 Th → 103 Lr	104 Ku	105 Ha													

FIGURE 9.7 FORMULAS FOR OXIDES IN WHICH THE POSITIVE ELEMENT HAS A VALENCE CORRESPONDING TO ITS GROUP NUMBER.

octet of valence electrons. The superoxide ion has one less electron than the peroxide ion. It is inadequately described by either of the electron structures $:\ddot{O}\cdot\ddot{O}:^{-}$, and $:\ddot{O}::\cdot\ddot{O}:^{-}$, but may be understood in terms of bonding and antibonding orbitals of the molecular orbital theory of bonding.

The Molecular Orbital Theory not only describes the bonding in the oxygen molecule and in the diatomic negative oxygen ions, peroxide and superoxide, but also describes the bonding in the dioxygenyl ion (O_2^+) found in the compound O_2PtF_6, which led Niel Bartlett to the preparation of the first noble gas-containing compound, $XePtF_6$. The molecular orbitals occupied by the valence electrons of the diatomic molecule and ions are shown in Table 9.3. In the series O_2^+, O_2, O_2^- and O_2^{2-}, **the bonding order decreases** successively by half-steps **from $2\frac{1}{2}$ to 1** as each succeeding electron is added to an antibonding orbital. The determination of the bonding order of dioxygenyl ion can be made by listing the molecular

TABLE 9.3 MOLECULAR ORBITALS OCCUPIED BY ELECTRONS BY THE DIATOMIC IONS AND MOLECULE OF OXYGEN

Molecular Orbitals	**Electrons per Molecular Orbital**			
	O_2^+	O_2	O_2^-	O_2^{2-}
σ(p−p)				
π^*(p−p) π^*(p−p)	1 0	1 1	2 1	2 2
π(p+p) π(p+p)	2 2	2 2	2 2	2 2
σ(p+p)	2	2	2	2
σ^*(s−s)	2	2	2	2
σ(s+s)	2	2	2	2

orbitals and canceling the bonding orbitals by the corresponding antibonding orbitals:

$$\begin{matrix} 2e^- & 2e^- & 2e^- & 2e^- & 2e^- & 1e^- \\ \sigma(s{+}s) + & \sigma^*(s{-}s) + & \sigma(p{+}p) + & \pi(p{+}p) + & \pi(p{+}p) + & \pi/2^*(p{-}p) \end{matrix}$$

$$\text{Bond order} \quad = \quad \begin{matrix} \sigma(p{+}p) + \pi(p{+}p) + \pi/2(p{+}p) \\ 1 \quad + \quad 1 \quad + \quad \tfrac{1}{2} \end{matrix} \quad = \quad 2\tfrac{1}{2}$$

bonding scheme for O_2^+, where * represents an antibonding molecular orbital.

In a similar fashion the bond order of O_2, O_2^- and O_2^{2-} can be found from Table 9.3.

Because of the many possible covalent oxides and the normal, super- and peroxides, the actual formula of the oxides must be determined either by gas density measurements and analyses of percentage compositions by weight or by other structural studies performed in the laboratory.

Actually bonds within the oxides, and in other series of compounds, are neither entirely ionic nor entirely covalent but vary from completely ionic to partially ionic–partially covalent to completely covalent. The nature of the bond depends upon the relative electronegativity of the elements of the atoms in the bond. Because **oxygen** is highly electronegative, its **bonds with other highly electronegative atoms are highly covalent,** while its **bonds with atoms of low electronegativity are largely ionic.** The variations in the ionic and covalent character of the bonds in oxides will be illustrated by the oxides of the elements of the third period in the periodic table, the elements sodium through chlorine. The electronegativity increases from left to right from sodium to chlorine, as seen in the regular decrease in the atomic radius (Table 9.4);

TABLE 9.4 PHYSICAL AND CHEMICAL PROPERTIES OF BINARY OXIDES

Third Period Elements							
Property	Na	Mg	Al	Si	P	S	Cl
Atomic radius, Å	1.57	1.36	1.25	1.17	1.10	1.04	0.99
Ionization energy (kcal/mole) $M_{(g)} \rightarrow M^+_{(g)} + e^-$	118.5	176.3	138.0	187.9	254.0	238.9	300.0
Electronegativity differences (oxygen–third period element)	2.6	2.3	2.0	1.7	1.4	1.0	0.5
Third Period Binary Oxides							
	Na_2O	MgO	Al_2O_3	SiO_2	P_4O_{10}	SO_3	Cl_2O_7
M.p. (°C)	920	3802	2027	1700	360 (sublimes)	17 (α form)	−81.5
Acid-base character of the oxides	strongly basic	basic	amphoteric	weakly acidic	moderately acidic	strongly acidic	very strongly acidic

basic or acidic depending on what it reacts with

the radius decreases as the valence electrons are pulled with greater force toward the nucleus. That the ionization energy generally increases as the radius decreases is further evidence of the greater electron attractiveness of the elements at the extreme right of the row. In line with the above discussion, the oxides of the elements on the left side of the row should be highly ionic, and covalent character should be at a maximum in the oxides on the right, as the electronegativity difference decreases to only 0.5 between oxygen and chlorine, as shown in Table 9.4. The type of bonding in a compound may be reflected in the melting point of the crystalline solid. Substances composed of small covalent molecules have low melting points because a relatively low level of energy is required to separate the individual molecules. The melting temperatures of polar covalent substances are greater than for corresponding non-polar covalent substances. Polarity is due to partial ionic character. The melting points of ionic compounds are high because positive and negative ions must be separated somewhat in the melting. Compare the melting points of the series of oxides appearing in Table 9.4. The melting points of the elements to the left in the table are high, indicating ionic character; those on the right are low, indicating increasing covalent character. The melting point of silicon dioxide (SiO_2) is an anomaly. The solid is not composed of discrete, three-atom molecules but is a three-dimensional network of silicon and oxygen atoms. Because covalent bonds must be broken in melting, the melting point of SiO_2 is very high.

carbon in the form of diamond, BN, SiC, and AlN are other examples of giant covalent molecules which consist of an indefinitely extended network of atoms and have high melting points.

The nature of the bonding in a binary oxygen containing compound is reflected in the acidity or basicity of the oxide as seen in the reaction that occurs when the oxide is dissolved in water. In Chapter 13 an acid is defined as a substance that donates protons, H^+, and a base as a substance that accepts protons in reaction. If the bonding is largely ionic, the oxide ion accepts an H^+ from water in dissolving:

$$Na_2O + H_2O \rightarrow 2\ Na^+ + OH^- + OH^-$$

The two hydroxide ions are shown separated from the sodium ion in this equation to indicate the separation of the ions when the product remains in solution. The hydroxide ions formed in the reaction are basic. The reaction is actually one of the oxide ion with water:

$$O^{2-} + H_2O \rightarrow 2\ OH^-$$

If the bonding within an oxide is covalent, the oxygen usually remains with the atom to which it is bonded when the compound is added to water. The atom to which it is bonded accepts a hydroxide ion from water, freeing a proton (H^+), which in turn is donated to a water molecule to form the hydronium ion, H_3O^+. The process is illustrated by the equation for the reaction of covalent sulfur trioxide with water:

$$SO_3 + 2\ H_2O \rightleftarrows HSO_4^- + H_3O^+$$

In hypothetical steps the reaction is

$$\begin{array}{c} O \\ | \\ O{-}S \\ \| \\ O \end{array} + \begin{array}{c} O{-}H \\ | \\ H \end{array} \rightarrow \begin{array}{c} O \\ | \\ O{-}S{-}OH^- \\ | \\ O \end{array} + H^+$$

$$H^+ + \begin{array}{c} O{-}H \\ | \\ H \end{array} \rightarrow \begin{array}{c} H{-}O{-}H^+ \\ | \\ H \end{array}$$

Oxides with a high degree of both ionic and covalent character are called **amphoteric oxides** and may act under different conditions as either acids or bases. With an acid stronger than water they act as a base, and with a base stronger than water they act as an acid. Aluminum oxide is an amphoteric oxide. It will accept protons from the hydronium ions in an acid solution in water:

an oxide that is capable of reacting with an acid and a base is amphoteric.

$$Al_2O_{3(s)} + 6\,H_3O^+ \rightarrow 2\,Al^{+3}_{(aq)} + 3\,H_2O + 6\,H_2O$$

The aluminum to oxygen bond is cleaved in the reaction, freeing the aluminum ion and releasing water. In a basic solution, one which contains excess hydroxide ions (OH^-), the oxide reacts with water to produce hydroxide ions.

The acidic character of aluminum oxide is illustrated in the reaction of Al_2O_3 with a strong base such as sodium hydroxide:

$$Al_2O_{3(s)} + 2OH^-_{(aq)} + 3\,H_2O \rightarrow 2\,Al(OH)^-_{4(aq)}$$

The acidic or basic nature of the oxides of the third period of the periodic table are given in Table 9.4.

9.6 OXIDATION STATES

Oxidation state (number) is related to the ionic valence. In fact, in binary ionic compounds the oxidation state of each ion is its valence. For any simple monatomic ion, the oxidation state is the ionic valence, i.e., the actual charge. The oxidation states of some simple ions in compounds are listed in Table 9.5. Considering the ions in magnesium chloride ($MgCl_2$), the second entry in the table, for example, the magnesium has lost its two valence electrons, and therefore has an oxidation state of +2. To balance the positive oxidation state of the magnesium atom, each chlorine atom as the chloride in magnesium chloride must have an oxidation state of −1. The oxidation state (number) is equal in magnitude to the excess number of electrons held by each atom.

for simple monatomic ions the oxidation state is the net charge on the ion.

The oxidation states of atoms in molecules and in complex ions are assigned in a more arbitrary fashion. In these covalent units the oxidation states are assigned as though the more electronegative atom of two atoms sharing a pair of electrons holds the shared electrons completely. In the hydrogen chloride molecule, a hydrogen and a chlorine atom share an electron pair between them, H:Ċ̤l: . Chlorine, being more electronegative than hydrogen, is assigned

TABLE 9.5 OXIDATION STATES OF SIMPLE IONS

Compound	Oxidation States of Simple Ions
KCl	K^+, +1; Cl^-, −1
$MgCl_2$	Mg^{2+}, +2; Cl^-, −1
NaOH	Na^+, +1
$NaAl(SO_4)_2$	Na^+, +1; Al^{3+}, +3
$MgCO_3$	Mg^{2+}, +2
NH_4Cl	Cl^-, −1
Li_2O	Li^+, +1; O^{2-}, −2
Na_2O_2	Na^+, +1

the pair of electrons when determining oxidation states. Its oxidation state is −1, whereas that of hydrogen is +1.

Potassium perchlorate is an ionic compound. The perchlorate ion, with a charge of −1, has this configuration:

```
   ..−
  :O:
..  ..  ..
:O:Cl:O:
..  ..  ..
  :O:
   ..
```

When all the electrons shared by chlorine and oxygen are assigned to oxygen, the more electronegative atom of each bond, oxygen has an oxidation state of −2, and chlorine has a state of +7. (One of the electrons among the shared pairs is the extra electron that gives the complex ion its charge.)

Because it is difficult to remember the relative electronegativity of each element participating in covalent bonding, a simple set of rules is given in Table 9.6 for the assignment of oxidation states for each element in all compounds.

TABLE 9.6 RULES FOR THE ASSIGNMENT OF OXIDATION STATES

1. An element in the free state has an oxidation state of zero.
2. The oxidation state of a simple ion is the ionic charge.
3. Oxidation states are assigned in accordance with the rules listed below insofar as possible so that
   a. in compounds the sum of all oxidation states is zero;
   b. in complex ions the sum of all oxidation states is the ionic charge.
   When application of two rules would violate condition a or condition b, the rule listed first takes precedence.
4. Some common oxidation states are
   a. Alkali metals +1
   b. Alkaline earth metals +2
   c. Fluorine −1
   d. Hydrogen +1 (except in metal hydrides, e.g., NaH, −1)
   e. Oxygen −2 (except in peroxides, e.g., Na_2O_2, −1)
5. Using the above rules, one can assign to other elements such oxidation states as satisfy conditions a and b.

Examples for the assignment of oxidation states to all elements of a compound or complex ion are listed below:

1. $KClO_3$:
   K(+1), O(−2), Cl(+5) for +1 +5 +3(−2) = 0
2. Na_2O:
   Na(+1), O(−2) for 2(+1)+(−2) = 0
3. Na_2O_2:
   Na(+1), O(−1) for 2(+1)+2(−1) = 0
4. $HClO_2$:
   H(+1), O(−2), Cl(+3) for +1 +3 +2(−2) = 0
5. CaH_2:
   Ca(+2) and H(−1) for +2 +2(−1) = 0
6. SO_4^{2-}:
   O(−2) and S(+6) for +6 +4(−2) = −2
7. $S_2O_3^{2-}$:
   O(−2) and S(+2) for 2(+2) +3(−2) = −2
8. MnO_4^-:
   O(−2) and Mn(+7) for +7 +4(−2) = −1
9. CH_2O:
   H(+1), O(−2) and C(0) for 0 +2(+1) −2 = 0
10. Fe_3O_4:
   O(−2) and Fe(+8/3) for 3 × 8/3 +4(−2) = 0

note that the sum of all oxidation numbers in a neutral compound is zero.

Only in the case of simple ions does the oxidation state correspond to an electrical charge on the atom, and only in the case of a simple ion is the charge equal to the electron excess or deficit of the ion over that of the neutral atom. At times the oxidation state is the total number of electrons contributed by the atom in sharing. In the nitrate ion, nitrogen provides five electrons for sharing, and its oxidation state is +5:

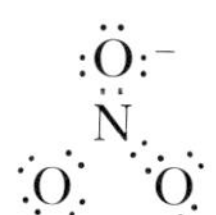

However, this relationship does not hold in all cases. The carbon in formaldehyde (CH_2O) (oxidation state of 0) contributes four electrons to sharing. Each iron atom in Fe_3O_4 does not contribute 8/3 electrons for sharing in Fe_3O_4.

Although there is no good physical representation for all oxidation states, oxidation states are useful. They assist in classifying compounds, in identifying reactions and in balancing equations which involve oxidation and reduction.

9.7 OXIDATION AND REDUCTION

Approximately 13 million tons of oxygen are produced annually in the United States. Most of that oxygen is used in industry as a reactant in many chemical processes. Reactions with oxygen are called oxidation reactions. If the reaction is accompanied by a flame, as many are, it is also called a combustion reaction. Consider the reaction for the complete combustion of carbon in burning coke:

$$C + O_2 \rightarrow CO_2$$

in the combustion of carbon, oxygen is the oxidizing agent.

According to the rules, carbon has an oxidation state of 0 in its free or uncombined state and an oxidation state of +4 in carbon dioxide. In the reaction the oxidation state of carbon has been made more positive, from 0 to +4. In the oxidation of lithium metal, the oxidation state of lithium is changed from 0 to +1:

$$\overset{(0)}{4\,Li} + O_2 \rightarrow 2\,\overset{(+1)}{Li_2O}$$

In many reactions not involving oxygen, the oxidation state of an element is made more positive. **The term oxidation** has been adopted to **cover** not only the **reactions** of elemental oxygen but all reactions **in which the oxidation state of a substance becomes more positive.** Oxidation is said to occur in these reactions, and the element whose oxidation state becomes more positive is said to have been oxidized. Consider the burning of hydrogen gas in chlorine gas to form hydrogen chloride,

$$\overset{(0)}{H_2} + Cl_2 \rightarrow 2\,\overset{(+1)}{HCl}$$

The oxidation state of hydrogen is changed from 0 to +1, and thus has become more positive; therefore, hydrogen has been oxidized. The oxidizing agent in this case is not oxygen but chlorine. Copper metal is dissolved in hot sulfuric acid according to this equation:

$$\overset{(0)}{Cu} + 2\,H_2SO_4 \rightarrow SO_2 + \overset{(+2)}{CuSO_4} + 2\,H_2O$$

The oxidation states 0, written above Cu, and +2, written above Cu in copper(II) sulfate, show that oxidation has occurred. Again, in this example oxygen is not the oxidizing agent, since oxygen gas is not present. Hot concentrated sulfuric acid is the oxidizing agent.

Reduction, the opposite of oxidation, always occurs along with oxidation. In reduction the oxidation state of an element becomes less positive. During the complete combustion of carbon (first equation in this section), in which carbon was oxidized to carbon dioxide, the oxidation state of oxygen becomes less positive, changing from 0 to −2. Oxygen is **reduced** and carbon is the reducing agent. Sulfur in the sulfuric acid is reduced in oxidizing copper, the oxidation state of sulfur changing from +6 to +4; hence, copper is the reducing agent. Notice that in all oxidation-reduction equations **the substance oxidized is the reducing agent, while the substance reduced is the oxidizing agent.** Oxidation of one substance is always accompanied by the reduction of another.

9.8 PREPARATION OF OXYGEN

Air is the most important commercial source of oxygen. Air is the cheapest source because the removal of oxygen from it involves only the separation of the gas from a mixture; no chemical bonds are broken during the separation. The separation involves first the liquefaction of the air by applying pressure and lowering the tem-

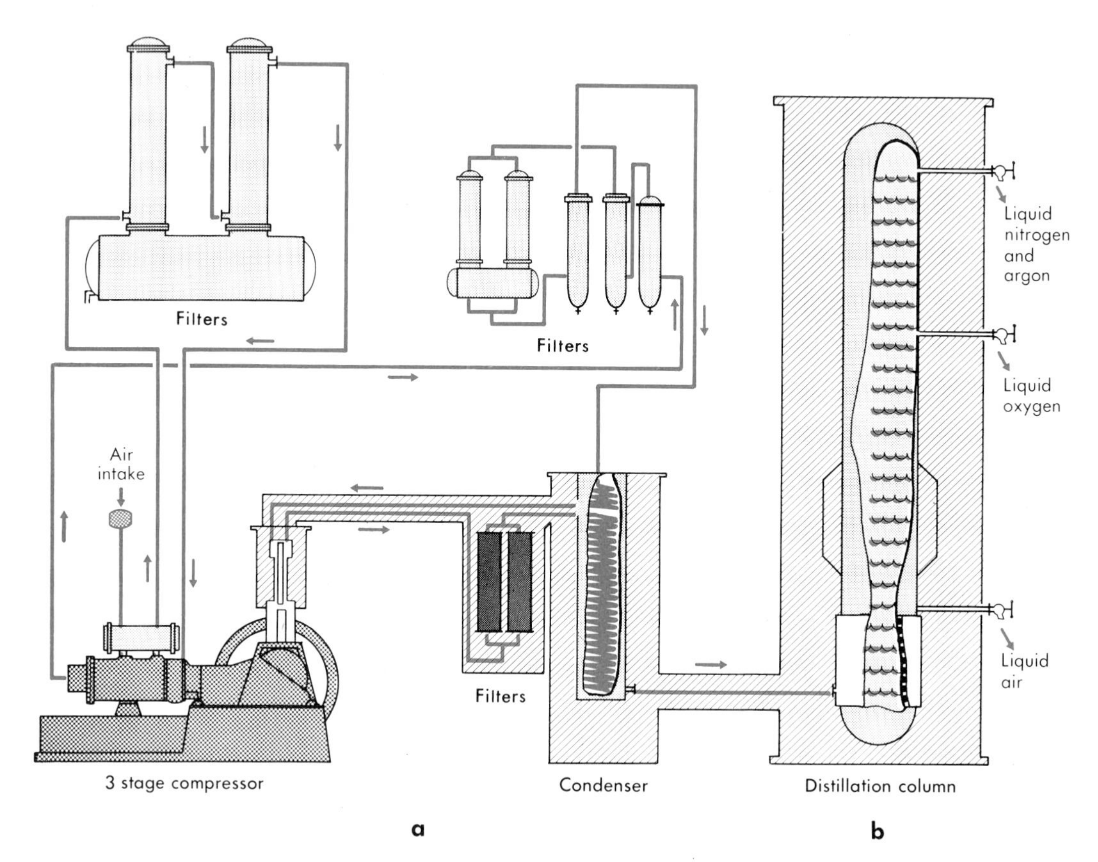

FIGURE 9.8 SEPARATION OF OXYGEN FROM LIQUID AIR.

perature; this is followed by the fractional distillation of the liquid air. Figure 9.8a is a diagram depicting the steps involved in the liquefaction of air. The distillation of liquid air is shown in Figure 9.8b. Nitrogen boils at a lower temperature (−196°C) than oxygen (−183°C), so nitrogen, being more volatile, boils off first during the distillation.

Preparation of oxygen in the laboratory involves the decomposition of oxygen-containing compounds. These compounds are usually decomposed by heating. Some examples represented by equations are

$$2\,HgO \xrightarrow{\Delta} 2\,Hg + O_2$$

$$2\,PbO_2 \xrightarrow{\Delta} 2\,PbO + O_2$$

$$2\,KClO_3 \xrightarrow{\Delta} 2\,KCl + 3\,O_2$$

the symbol Δ represents heat.

The oxygen is usually collected by water displacement, as shown in Figure 9.9. In this process, manganese dioxide (MnO_2) is used as a catalyst to speed the thermal decomposition of the potassium chlorate. When heated alone, potassium chlorate decomposes slowly at 400°C, and even more slowly, or not at all, at lower temperatures. At 400°C, potassium chlorate may also decompose slowly by another process, forming potassium perchlorate:

$$4\,KClO_3 \xrightarrow{heat} 3\,KClO_4 + KCl$$

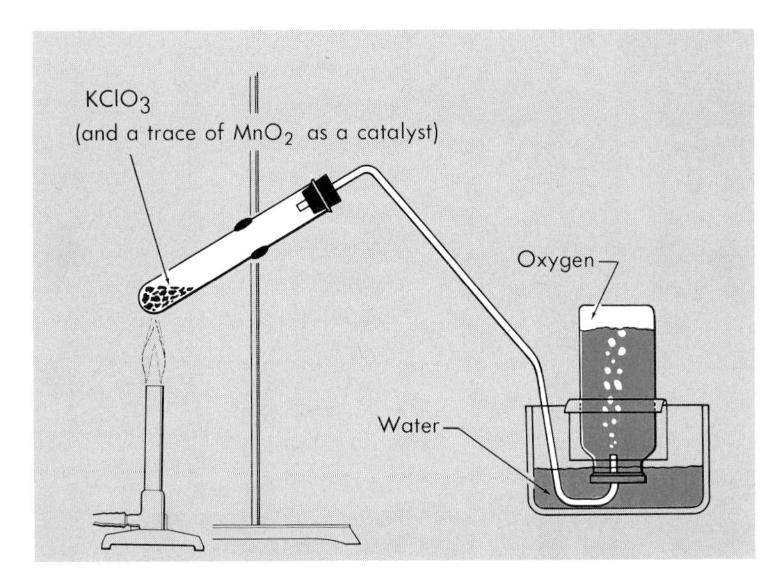

FIGURE 9.9 CHEMICAL PREPARATION OF OXYGEN.

However, when a trace of manganese dioxide is added, decomposition to give oxygen occurs rapidly at 200°C, much below the temperature required to form the perchlorate. At 200°C, manganese dioxide contributes no oxygen and is found unchanged mixed with the potassium chloride left when all the potassium chlorate has been decomposed. The unusual capability of certain materials, called **catalysts,** to **hasten chemical reactions** without being consumed is often not understood; nevertheless, they are used extensively in industry. Much research time is spent in the search for new and useful catalysts.

Very pure oxygen may be prepared by the electrolysis of water. Hydrogen also is obtained. A trace of either a base or an acid must be added to the water to aid the conduction of the electric current through the water, because water is an extremely poor conductor. Water is decomposed by electrical energy into its components, hydrogen and oxygen, in accord with the equation

$$2\,H_2O \xrightarrow[\text{trace acid}]{\text{electrolysis}} 2\,H_2 + O_2$$

Figure 9.10 shows a demonstration cell for the electrolysis of water. The apparatus is filled with a very dilute solution of sulfuric acid. Leads from platinum electrodes are sealed through the glass walls of the electrolysis cell and are connected to the poles of a storage battery. When the circuit is closed, bubbles rise from both electrodes, and gas collects in both tubes. The gas collects above the cathode at exactly double the rate that it forms in the other. This gas burns in the air, forming water; it is hydrogen. The gas collected above the anode supports combustion; it is oxygen. Figure 9.10a shows the apparatus before closing the circuit, whereas Figure 9.10b shows the apparatus after the circuit has been closed for a few minutes. Some oxygen and hydrogen are produced electrolytically today at isolated power plants when electric power is plentiful and cheap.

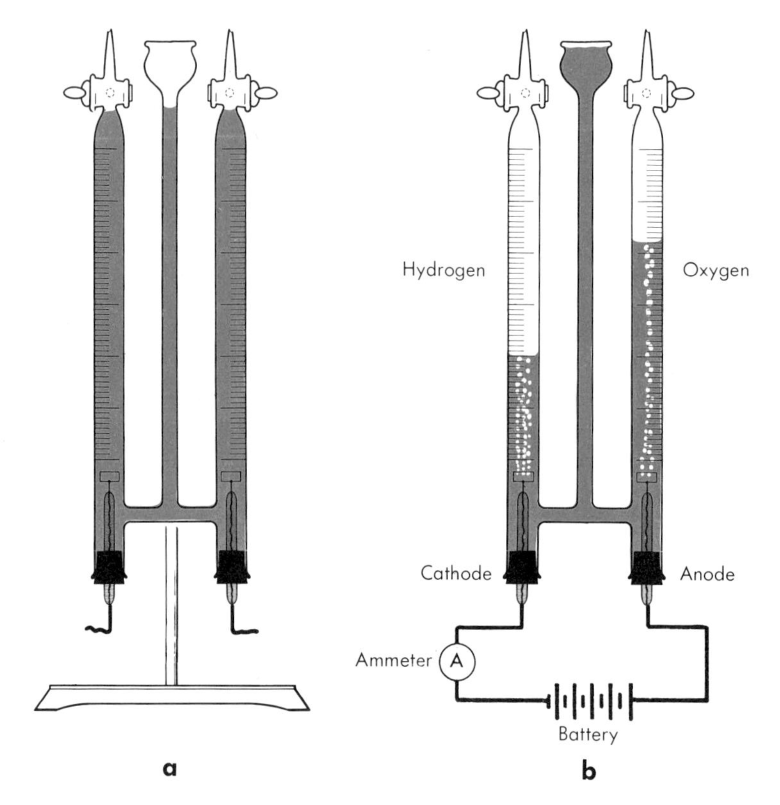

FIGURE 9.10 ELECTROLYSIS OF WATER TO PRODUCE HYDROGEN AND OXYGEN.

9.9 SOME REACTIONS AND USES OF OXYGEN

The many uses of oxygen depend upon the ease with which oxygen combines with many materials. Rapid oxidation accompanied by a flame is known as burning or combustion. Slower combination is known simply as oxidation. When coal burns, the major reaction is the combustion of carbon, to form carbon dioxide:

one kcal equals 1000 calories.

$$\underset{\text{From coal}}{C} + \underset{\text{From air}}{O_2} \rightarrow \underset{\text{Carbon dioxide}}{CO_2} + 94 \text{ kcal}$$

Or, if insufficient oxygen is present, carbon monoxide is formed:

$$2\,C + O_2 \rightarrow 2\,CO + 52.8 \text{ kcal}$$

Natural gas, mostly methane (CH_4), burns with oxygen with three possible sets of products, depending on the amount of oxygen present. The three reactions may take place simultaneously. Carbon dioxide and water are formed when an excess of oxygen is present:

$$CH_4 + 2\,O_2 \rightarrow CO_2 + 2\,H_2O + 212 \text{ kcal}$$

When less oxygen is present, carbon monoxide is the carbon containing product:

$$2\,CH_4 + 3\,O_2 \rightarrow 2\,CO + 4\,H_2O + 289\text{ kcal}$$

With still less oxygen, carbon (soot) accumulates:

$$CH_4 + O_2 \rightarrow C + 2\,H_2O + 118\text{ kcal}$$

A luminous flame occurs when little oxygen is present, because the flakes of carbon present among the gaseous products are heated to incandescence in the flame. A sooty flame is yellow, because the carbon particles glow with a yellow light in the heat of the flame.

When 1 mole of carbon (12 grams) is completely oxidized to CO_2, 94,000 calories of heat are liberated. When 2 moles (24 grams) of carbon are oxidized to CO, 52,800 calories (= 26,400 calories per 12 grams) are released. It is important economically for a furnace to have a sufficient draft (supply of air) such that the carbon be completely oxidized to CO_2. Similarly, when 1 mole of CH_4 is completely oxidized to CO_2 and water, 212 kilocalories of heat are liberated, in contrast to 289 kilocalories for 2 moles (or 144.5 kcal per 1 mole) when CO and water are formed. Even less heat results when carbon and water are produced (118 kcal per 1 mole).

Reactions that **give off heat** are called **exothermic** reactions. Certain reactions are **endothermic**, that is, **heat is absorbed** during the reaction. All furnaces use oxygen from the air to combine with the fuel (coal, gas, or fuel oil), liberating carbon oxides, water vapor and heat. In an automobile engine, air provides the oxygen for the explosion (combustion) of gasoline vapors in the cylinders to push the piston.

Oxygen of inhaled air is trapped by hemoglobin in the lungs and carried to the cells, where it is used to oxidize carbohydrates for energy and body heat. Oxidation of sugar in the body cells yields carbon dioxide and water in accord with the equation

$$\underset{\text{Table sugar}}{C_{12}H_{22}O_{11}} + 12\,O_2 \rightarrow 12\,CO_2 + 11\,H_2O + 1349\text{ kcal}$$

The same products are obtained when sugar is completely burned in an open flame.

Various torches premix oxygen and a combustible gas in order to produce very hot flames. A rationale for premixing the gases will be given shortly. Oxygen and natural gas are used in a glass blower's torch:

$$2\,O_2 + CH_4 \rightarrow CO_2 + 2\,H_2O + \text{heat}$$

A hotter flame is obtained with an oxygen-hydrogen mixture:

$$2\,H_2 + O_2 \rightarrow 2\,H_2O + \text{heat}$$

The oxyacetylene torch mixes oxygen and acetylene prior to burning and gives a flame reaching 2800°C, which is sufficiently hot to melt steel and to patch it by melting steel welding rods into breaks. The reaction is represented by this equation:

$$2\,C_2H_2 + 5\,O_2 \rightarrow 4\,CO_2 + 2\,H_2O + 312\text{ kcal}$$

Since the most important uses of oxygen involve combustion and since many tragedies result from uncontrolled combustion, the factors that affect the combustion of oxygen merit discussion.

1. Proper *mixing of the reactants* (oxygen and the combustible material), intimately and in the right proportions, aids combustion. For this reason the gaseous fuel and oxygen are mixed in an acetylene torch. Reactions cannot occur unless the reactants contact each other, molecule for molecule. Air and gasoline vapor are mixed in the carburator to give fast, even burning in the automobile cylinder. Explosions in flour mills result only when powdered flour is diffused or dispersed through the air. Even **aluminum dust** will **explode** if blown into (mixed with) air in the right proportions and ignited.

2. An *increase in concentration* of either reactant, oxygen or the combustible material, with the exception of inert materials, aids combustion. An oxygen-acetylene mixture burns much more rapidly and hotter than an air-acetylene mixture because the gases are in closer contact and there is no unreactive nitrogen present to be heated. Rocket fuel mixtures use liquid oxygen, the most highly concentrated form of oxygen available.

3. Combustion proceeds more rapidly at *higher temperatures*. Most substances will not burn in oxygen at room temperature but must be heated to or above the kindling temperature for that substance with a match or a spark to initiate the combination. The flame spreads as the gases are heated. **Spontaneous combustion** illustrates the importance of temperature. If an easily oxidized substance such as an oil is absorbed in a pile of rags, it begins to oxidize slowly in air, producing heat. If the heat escapes too slowly through the rags, the temperature rises. The oxidation proceeds faster as the temperature increases, giving more heat until the kindling temperature is reached and the rags burst into flame. Many a haystack has burned because damp hay began to rot (oxidize), producing sufficient heat to raise a portion of the hay to a temperature above its kindling temperature.

4. *Oxidation (not combustion)* is aided in the body by catalysts. Enzymes in the body catalyze the oxidation of foodstuffs so that they can be oxidized rapidly at body temperature, which is far below the kindling temperature. We have much to learn about these body catalysts and should recognize that the reactions by which foods are oxidized are quite different from simple burning in oxygen gas, even though the ultimate products, carbon dioxide and water, may be the same.

EXERCISES

9.1 Define (a) oxidation, (b) earth's crust (as used here), (c) troposphere, (d) stratosphere, (e) reduction, (f) a peroxide, (g) combustion, (h) phlogiston, (i) catalyst, (j) enzyme, (k) atmosphere, (l) triatomic, (m) pollutant, (n) photochemical reaction, (o) thermal pollution, (p) an acid, (q) a base, (r) oxidation states, (s) exothermic, (t) spontaneous combustion.

9.2 Judging from the position of oxygen in the periodic table:

a. Is it a metal or a non-metal?
b. What is its ionic valence?
c. What is its electronegativity as compared with that of its neighbors?
d. Is it more reactive than sulfur?

9.3 Name all the elements that are gases at room temperature. In what general portion of the periodic table are they found?

9.4 Why is chlorine a gas at room temperature and bromine not? Why is radon a gas and bromine not?

9.5 Describe the phlogiston theory of combustion. What experimental facts did Lavoisier use to dispel the theory?

9.6 Compare the densities, melting points, boiling points and chemical reactivities of oxygen (O_2) and ozone (O_3).

9.7 In what respect is ozone in the stratosphere beneficial?

9.8 Why are some smogs called photochemical smogs?

9.9 Why should oxygen gas be less soluble in water than hydrogen chloride gas?

9.10 List in order the five most abundant elements in the earth's crust.

9.11 Which bonds are largely covalent: the O to H bond in NaOH; the Na to Cl bond in NaCl; the S to O bond in $BaSO_4$?

9.12 Describe the reaction conditions for the laboratory preparation of oxygen from potassium chlorate.

9.13 What is a catalyst? Give an example of catalysis.

9.14 Describe the major commercial method for the preparation of oxygen.

9.15 What might one do to speed up the rate of combustion of two reactants?

9.16 Explain how spontaneous combustion occurs.

9.17 Why should the flame from a oxy-hydrogen torch be hotter than that of hydrogen burning in air?

9.18 How many kilocalories of heat are evolved when 34.0 grams of methane (CH_4) burn completely to carbon dioxide and water? How-many calories are evolved when the same amount of methane burns to carbon monoxide and water?

9.19 Write valence electron structures (dot formulas) for oxygen, sodium peroxide, sodium oxide, ozone and sulfur dioxide.

9.20 Determine the oxidation state of nitrogen in each of these known compounds:

(a) $NaNO_2$
(b) N_2O
(c) HNO_3
(d) NH_3
(e) NH_2OH
(f) N_2H_4
(g) NO
(h) N_2O_3
(i) N_2O_5
(j) HNO_2

9.21 Determine the oxidation states of each element in each of the following compounds:

(a) Na_2SO_4
(b) NH_4Cl
(c) $Ca(HCO_3)_2$
(d) $SOCl_2$
(e) $Ba(NO_3)_2$
(f) Hg_2Cl_2
(g) Al_2O_3
(h) $HClO_4$
(i) $ZnCO_3$
(j) Mg_3N_2

SUGGESTED READING

Cloud, P., and Gibor, A.: "The Oxygen Cycle," *Scient. Amer., 223* (3): 110, 1970.
Forber, E.: "Oxygen—The Element with Two Faces," *Chemistry, 39* (5): 17, 1966.
Gale, G.: "Phlogiston Revisited—Explanatory Models and Conceptual Change." *Chemistry, 41:* (4): 16, 1968.
Jolly, W. L.: *The Chemistry of the Non-Metals.* Prentice-Hall, Inc., Englewood Cliffs, New Jersey, 1966 (paperback).
Plane, R. A., and Hester, R. E.: *Elements of Inorganic Chemistry,* W. A. Benjamin Co., New York, 1965.
Stone, J. K.: "Oxygen and Steelmaking," *Scient. Amer., 218* (4): 24, 1968.

TEN • HYDROGEN AND WATER

10.1 HISTORY OF HYDROGEN

References concerning the preparation of hydrogen appear early in scientific history. In the beginning of the Revival of Learning following the Dark Ages, when chemistry was being born from alchemy, Paracelsus (1493–1541), a "practicer of medicine," is believed to have observed a gas that evolved when metals were treated with acids. Van Helmont (1577–1644), a Belgian physician of the next century, apparently knew of the same gas, and in the seventeenth century Robert Boyle, whose gas law we have already studied, prepared and isolated hydrogen.

However, it remained for the shy, eccentric Englishman, **Henry Cavendish** (1731–1810), to characterize the gas and thus gain recognition for its discovery. Figure 10.1 shows the dates of the discovery of hydrogen and the other elements of the periodic table. Hydrogen was the nineteenth element to be discovered and characterized. In 1766, Cavendish published a paper in which he described his preparation of "inflammable air" and discussed its properties. He had prepared the gas in the usual way—by treating a metal with acid. Because hydrogen would not support combustion, he identified it with phlogiston (air minus O_2). In 1783, 11 years after Scheele's discovery of oxygen and 6 years after Lavoisier announced the modern theory of combustion, Cavendish combined hydrogen and oxygen and obtained water. Using this experiment as a basis, Lavoisier named the new gas *"hydro-gen,"* or *water-former*. Cavendish clung to the phlogiston theory and never recognized that in burning, hydrogen had combined with oxygen.

10.2 OCCURRENCE OF HYDROGEN

Hydrogen is very widely distributed in nature, being found in water and in almost all compounds that are or were constituents of living things. It is estimated that 16 per cent of all the atoms of the

1 H 1766																	1 H	2 He 1895
3 Li 1817	4 Be 1798												5 B 1808	6 C 1600	7 N 1772	8 O 1772	9 F 1771	10 Ne 1898
11 Na 1807	12 Mg 1755												13 Al 1827	14 Si 1823	15 P 1669	16 S 1600	17 Cl 1774	18 Ar 1894
19 K 1807	20 Ca 1808	21 Sc 1879		22 Ti 1791	23 V 1830	24 Cr 1797	25 Mn 1774	26 Fe 1600	27 Co 1735	28 Ni 1751	29 Cu 1600	30 Zn 1746	31 Ga 1875	32 Ge 1886	33 As 1250	34 Se 1817	35 Br 1826	36 Kr 1898
37 Rb 1861	38 Sr 1790	39 Y 1794		40 Zr 1789	41 Nb 1801	42 Mo 1778	43 Tc 1937	44 Ru 1844	45 Rh 1803	46 Pd 1803	47 Ag 1600	48 Cd 1817	49 In 1863	50 Sn 1600	51 Sb 1600	52 Te 1782	53 I 1811	54 Xe 1898
55 Cs 1860	56 Ba 1808	57 La 1839	58 Ce → 71 Lu	72 Hf 1923	73 Ta 1802	74 W 1781	75 Re 1925	76 Os 1803	77 Ir 1803	78 Pt 1735	79 Au 1600	80 Hg 1600	81 Tl 1861	82 Pb 1600	83 Bi 1600	84 Po 1898	85 At 1940	86 Rn 1900
87 Fr 1939	88 Ra 1898	89 Ac 1899	90 Th → 103 Lr	104 Ku 1965	105 Ha 1970													

LANTHANIDE SERIES	58 Ce 1803	59 Pr 1885	60 Nd 1843	61 Pm 1947	62 Sm 1879	63 Eu 1896	64 Gd 1880	65 Tb 1843	66 Dy 1886	67 Ho 1879	68 Er 1843	69 Tm 1879	70 Yb 1878	71 Lu 1907
ACTINIDE SERIES	90 Th 1828	91 Pa 1917	92 U 1789	93 Np 1940	94 Pu 1940	95 Am 1945	96 Cm 1944	97 Bk 1950	98 Cf 1950	99 Es 1952	100 Fm 1953	101 Md 1955	102 No 1958	103 Lr 1961

FIGURE 10.1 DATES OF DISCOVERY OF THE ELEMENTS.

earth's crust (i.e., the atmosphere, plants, animals, oceans and the 10-mile shell) are hydrogen atoms. But because the hydrogen atom is so light (the lightest of all elements), less than 1 per cent of the earth's crust (by weight) is hydrogen. See Figure 9.3 for a comparison of the abundance of hydrogen with the eight most abundant elements.

Hydrogen is probably the **most abundant element in the universe,** since the stars and sun consist mainly of hydrogen. **Elemental (or free) hydrogen** is very uncommon on the earth, because the earth's gravitational attraction is too small to hold hydrogen molecules. However, uncombined hydrogen is sometimes found in gases issuing from volcanos.

10.3 PHYSICAL PROPERTIES OF HYDROGEN

The electronic structure of hydrogen and the bonding in diatomic H_2 were discussed in Chapter 5. The bonding involves the sharing of a pair of electrons in the bonding $\sigma(s+s)$ molecular orbital.

Hydrogen is element number 1. The atom consists of 1 electron and 1 proton and is the smallest of all the atoms. Hydrogen is the least dense of all gases. At STP the density of hydrogen is equal to (2.016 g/22.4 L =) 0.0898 g per L. Next in density comes helium, a monatomic gas having a density of (4.003 g/22.4 L =) 0.178 g per L.

Being much lighter than air, hydrogen will rise in the atmosphere. Hydrogen issuing from a generator can be collected with **a test tube held upside down.** The hydrogen rises in the tube, forcing the air down and out of the mouth. A balloon filled with hydrogen rises in air. Additional weight can be added, and the balloon will rise if the sum of the weights of the balloon, the added mass and the hydrogen is less than the weight of the volume of air the balloon displaces.

Air has an average molecular weight of 29 (being 4/5 nitrogen and 1/5 oxygen) and a density of 1.299 g per L at STP. One liter of hydrogen will lift, in addition to its own weight, (1.299 g/l − 0.0898 g/L =) 1.21 grams. By the same reasoning, 1 liter of helium will lift (1.299 g/L − 0.178 g/L =) 1.12 grams.

Hydrogen diffuses faster than any other gas, since its mass is less than that of any other gas and according to the kinetic theory all gases at the same temperature have equal translational energy. The rapid diffusion of hydrogen can be demonstrated by placing an inverted beaker of hydrogen over a porous clay cup, as shown in Figure 10.2. Hydrogen diffuses inward through the walls of the clay cup faster than the air within the cup diffuses outward. The resulting pressure inside the cup forces the water in the bottle out through the jet.

Hydrogen is a **colorless, odorless, and tasteless** gas. The gas is only slightly soluble in water because the non-polar molecules are not attracted to the dipoles of water. The diatomic, non-polar molecules have little attraction among themselves, and the liquid boils at a very low temperature, −253°C. It is a flammable gas, burning quietly when emitted from a jet in air, but exploding with violence when ignited in a mixture with air or oxygen.

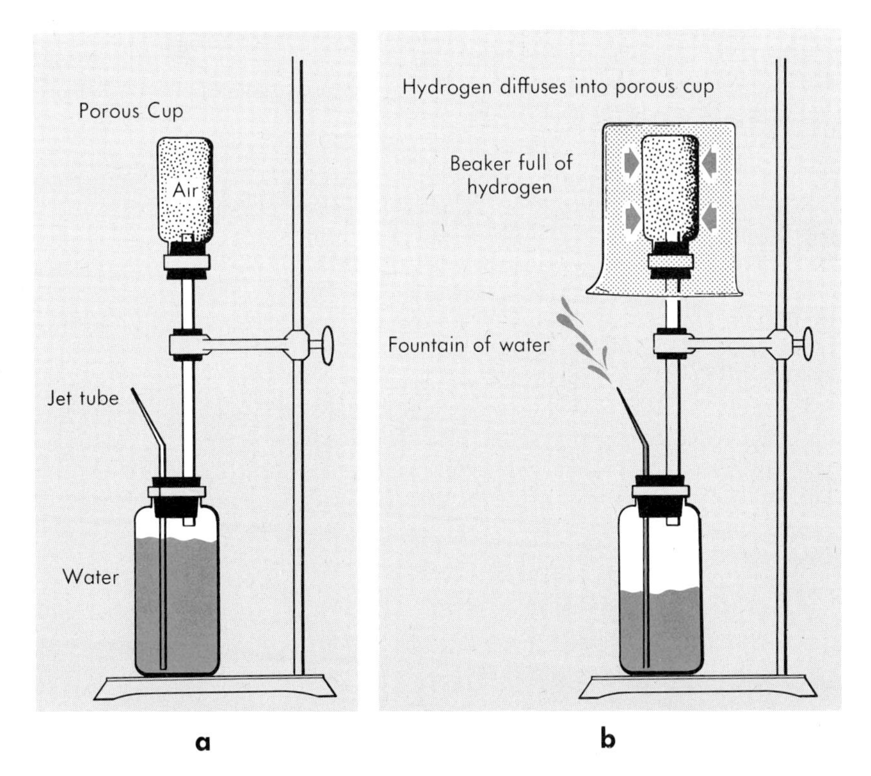

FIGURE 10.2 HYDROGEN DIFFUSION EXPERIMENT.

10.4 ISOTOPES OF HYDROGEN

Hydrogen atoms come in three sizes (masses). The three—protium, (common hydrogen), deuterium, and tritium—are listed in Table 10.1 with a summary of properties, names and occurrences. Atoms of all three types of hydrogen possess one proton but the deuterium atom has one neutron and the tritium atom has two neutrons.

there are three hydrogen isotopes.

TABLE 10.1 HYDROGEN ISOTOPES

	Mass	% of Natural Hydrogen	Boiling Point	Freezing Point
Protium ^1_1H (light hydrogen)	1.008	99.9844	−253°C	−259°C
Deuterium ^2_1H (heavy hydrogen)	2.0146	0.0156	−249°C	−254.6°C
Tritium ^3_1H	3.017	10^{-15} to 10^{-16}		

The per cent of deuterium in hydrogen is so small that it makes no perceptible difference in the atomic weight, although the isotope is nearly twice as heavy as light hydrogen. Nevertheless, considerable amounts of deuterium are obtained for use in experimental nuclear reactors by the repeated electrolysis of water, which has one part of heavy water (D_2O) in 6400 parts of ordinary water. Hydrogen is the only element having isotopes with a mass ratio as great as 2 to 1. Consequently, there is a greater difference between the properties of hydrogen and deuterium than between isotopes of any other element. The freezing and boiling points of heavy hydrogen are approximately four degrees greater than for ordinary hydrogen. The isotopic hydrogens can be separated from a mixture by repeated distillations.

In general, deuterium reacts with the same substances as does ordinary hydrogen, forming similar compounds. Sometimes compounds containing heavy hydrogen or heavy water itself are used to identify the role of hydrogen or water in a chemical reaction. Because it can be traced (followed through a reaction), deuterium serves as a tracer in the study of chemical and physiological changes. However, deuterium usually reacts more slowly than ordinary hydrogen. Tritium is produced only by nuclear transformations. It is formed from the action of cosmic rays on some atomic nuclei. It is radioactive and gradually undergoes decay; hence, only traces are found in nature. Deuterium and tritium are involved in the fusion process which occurs in the so-called hydrogen bomb and in the reactions within the sun that furnish life-giving heat and light.

10.5 HYDROGEN AND THE PERIODIC TABLE

Hydrogen and helium are found in period 1 of the periodic table. Helium is a noble gas and belongs specifically to Group 0. In the periodic table we have shown hydrogen as a member of both the alkali metal family (Group I) and the halogen family (Group

GROUPS

IA	IIA	IIIB		IVB	VB	VIB	VIIB	VIII	VIII	VIII	IB	IIB	IIIA	IVA	VA	VIA	VIIA	O
EH																	HE	
1 H	EH$_2$												EH$_3$	EH$_4$	H$_3$E	H$_2$E	1 H	2 He
3 Li LiH	4 Be BeH_2												5 B 'BH_3'	6 C CH_4	7 N H_3N	8 O H_2O	9 F HF	10 Ne
		EH$_3$		EH$_4$	EH$_5$	EH$_6$	EH$_7$		EH$_8$		EH	EH$_2$						
11 Na NaH	12 Mg MgH_2												13 Al AlH_3	14 Si SiH_4	15 P H_3P	16 S H_2S	17 Cl HCl	18 Ar
19 K KH	20 Ca CaH_2	21 Sc		22 Ti	23 V	24 Cr	25 Mn	26 Fe	27 Co	28 Ni	29 Cu CuH	30 Zn ZnH_2	31 Ga 'GaH_3'	32 Ge GeH_4	33 As H_3As	34 Se H_2Se	35 Br HBr	36 Kr
37 Rb RbH	38 Sr SrH_2	39 Y		40 Zr	41 Nb	42 Mo	43 Tc	44 Ru	45 Rh	46 Pd	47 Ag	48 Cd CdH_2	49 In InH_3	50 Sn SnH_4	51 Sb H_3Sb	52 Te H_2Te	53 I HI	54 Xe
55 Cs CsH	56 Ba BaH_2	57 La	58 Ce ➡ 71 Lu	72 Hf	73 Ta	74 W	75 Re	76 Os	77 Ir	78 Pt	79 Au	80 Hg HgH_2	81 Tl	82 Pb PbH_4	83 Bi	84 Po	85 At	86 Rn
87 Fr	88 Ra RaH_2	89 Ac	90 Th ➡ 103 Lr	104 Ku	105 Ha													

FIGURE 10.3 SOME BINARY COMPOUNDS OF HYDROGEN.

VII). Because the hydrogen atom has one electron, an s electron in its outermost and only electron shell, it resembles the alkali metals. Because it needs only one electron, either by sharing or transfer, to gain a noble gas configuration, the element is similar to the halogens.

the simplest compounds are represented in figure 10.3. for example, besides CH_4 there are thousands of compounds of the general formulas C_nH_{2n+2} and C_nH_{2n} where n can have numerous values.

Certain known binary hydrogen compounds are arranged in Figure 10.3 in such a way that the second element is placed in its position in the periodic table. One can predict the formulas listed for compounds of the elements of Groups V, VI and VII if hydrogen is assumed to be like the alkali metals. Likewise, the formulas for the simple compounds of hydrogen with members of Groups I, II, III, and IV (as listed in Fig. 10.3) can be predicted if hydrogen is assumed to be a halogen (member of Group VII). However, the **nature of the bonding** between hydrogen and the elements of Groups V, VI and VII in binary compounds **is very different** from the bonding between a metal of Group I and the corresponding non-metal atoms (Groups V, VI and VII). For example, compare the bonding in hydrogen fluoride gas formed by the combination of hydrogen and fluorine,

$$H_{2(g)} + F_{2(g)} \rightarrow 2\ HF_{(g)} + \text{heat}^*$$

and the bonding in sodium fluoride, NaF, a solid, formed by the spontaneous reaction of sodium and fluorine,

$$2\ Na_{(s)} + F_2 \rightarrow 2\ NaF_{(s)} + \text{heat}$$

* A number of reactions of hydrogen with non-metals (such as O_2 and Cl_2) occur with violence or an explosion. For an explosion to occur, reactions must be highly exothermic, proceed rapidly and produce gaseous products which, as a consequence of the sharp rise in temperature, expand rapidly.

Hydrogen fluoride is bound by a single, sigma covalent bond – the sharing of a pair of electrons – a bond similar to the single bond in hydrogen and fluorine molecules:

$$\mathrm{H:\ddot{\underset{\cdot\cdot}{F}}:}$$

Sodium fluoride is an ionic compound, a salt, in which the oppositely charged ions are held together in a lattice in crystals of the solid. The valence electron of the Group I metals has been transferred to the fluorine atom as shown by the dot formula,

$$\mathrm{Na^{+}} \quad \mathrm{:\ddot{\underset{\cdot\cdot}{F}}:^{-}}$$

Not only do the binary compounds of hydrogen with the non-metals of Groups V, VI and VII, which appear in Figure 10.3, have the formulas that would be predicted were hydrogen an alkali metal, but the formula is that required to give the non-metals an octet configuration. **The compounds** of Groups V, VI and VII **have the formulas H_3X, H_2X and HX** (where X represents the non-metals of Groups V, VI and VII respectively in the series). Dot formulas for a compound from each group show the achievement of the noble gas electron configuration:

$$\begin{array}{c}\mathrm{H}\\ \mathrm{:\ddot{N}:H}\\ \mathrm{\ddot{H}}\end{array};\quad \begin{array}{c}\\ \mathrm{H:\ddot{O}:}\\ \mathrm{\ddot{H}}\end{array};\ \text{and}\ \mathrm{H:\ddot{\underset{\cdot\cdot}{F}}:}$$

Also the simple binary hydrides of carbon and silicon agree with the octet principle:

$$\begin{array}{c}\mathrm{H}\\ \mathrm{H:\ddot{\underset{\cdot\cdot}{C}}:H}\\ \mathrm{H}\\ \text{methane}\end{array} \qquad \begin{array}{c}\mathrm{H}\\ \mathrm{H:\ddot{\underset{\cdot\cdot}{Si}}:H}\\ \mathrm{H}\\ \text{silane}\end{array}$$

The binary compounds of hydrogen with the non-metals as listed in Figure 10.2 are not the only binary compounds of hydrogen with those elements. The number of known compounds of hydrogen with the elements of the second period are listed in Table 10.2. In these compounds two or more non-metal atoms are bound together and then surrounded with sufficient hydrogen atoms to achieve the octet configuration. Some examples are shown in Table 10.3. The great number of carbon hydrogen compounds (commonly called **hydrocarbons**) is due to the fact the carbon atom can share electrons with other carbon atoms, as well as with hydrogen, and because it has only four valence electrons, which permits formation of branched chains of atoms and also multiple bonds. A complete branch of chemistry called **organic chemistry** has developed around these hydrocarbon compounds and compounds derived from them by adding various other elements. The later chapters of this text are devoted to the chemistry of these compounds.

Boron is a metalloid. It too forms binary compounds with hydrogen, but the electron arrangement is different here than in hydrocarbons. The simplest of the boron hydrides is diborane,

TABLE 10.2 NUMBER OF BINARY HYDRIDES OF THE SECOND PERIOD ELEMENTS

Element	Li	Be	B	C	N	O	F	Ne
Number of Hydrides	1	1	15	hundreds of thousands	4	2	1	0

TABLE 10.3 ELECTRON DOT STRUCTURES OF SOME BINARY COMPOUNDS OF HYDROGEN

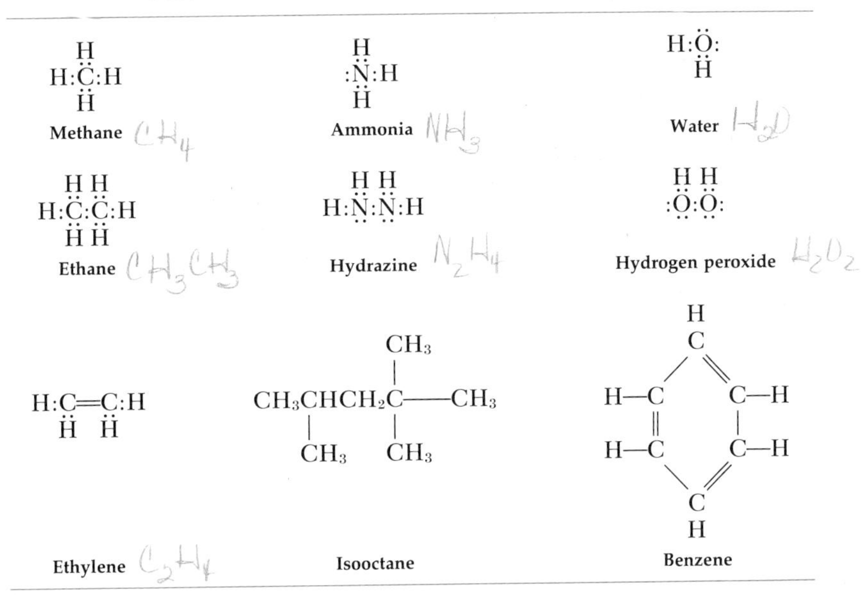

B_2H_6, which has a total of 12 valence electrons. When these electrons are distributed around (or among) the atoms, which are placed as experimental evidence suggests,

H H H
B B
H H H

the boron hydrides are electron-deficient compounds.

there appears to be too few electrons to give the octet configuration about the two boron atoms. Consequently, such compounds are referred to as **electron-deficient compounds.**

Often the covalent, binary hydrogen compounds (compounds of hydrogen with non-metals) are designated by common (unsystematic) names. Some of these are: ammonia (NH_3), methane (CH_4:), ethane (CH_3CH_3), ethylene ($CH_2{=}CH_2$), phosphine (PH_3) and water (H_2O). Others are more frequently known by systematic names: hydrogen sulfide (H_2S), hydrogen fluoride (HF) and hydrogen chloride (HCl).

Compounds of hydrogen with the metals in Groups I and II of the periodic table (Fig. 10.3) are ionic compounds in which the hydrogen atom bears a single negative charge and is called **the hydride ion.** The metal hydrides are similar in many respects to the metal halides, which are binary compounds of the halogens with the active metals in which the halides exhibit their normal ionic valence, −1. The formation of sodium chloride and sodium hydride from the elements illustrates the similar behavior of chlorine and hydrogen in the reactions

$$2\,Na_{(s)} + Cl_{2(g)} \rightarrow 2\,Na^{+}Cl^{-}{}_{(s)} + \text{heat}$$

$$2\,Na_{(s)} + H_{2(g)} \rightarrow 2\,Na^{+}H^{-}{}_{(s)} + \text{heat}$$

The existence of **the hydride ion with a charge of minus one** is demonstrated by the electrolysis of molten anhydrous sodium hydride, NaH, as shown in Figure 10.4. Hydrogen gas is liberated at the anode. This would not be possible were the hydride ion positively charged; but, being negative, the ion is attracted to the anode where it releases an electron to that electrode to become a neutral atom:

$$:H^{-} \rightarrow \cdot H + e^{-}$$

Two hydrogen atoms unite to form a stable electron configuration by sharing two electrons, and a hydrogen molecule is formed:

$$2\,H\cdot \rightarrow H:H$$

As the hydrogen gas is released at the positive electrode, positive sodium ions are drawn to the cathode where they acquire electrons, one per each ion, to become atoms and float to the surface of the molten hydride.

The great difference between the binary compounds of hydrogen with the alkali metals (the ionic hydrides) and the binary com-

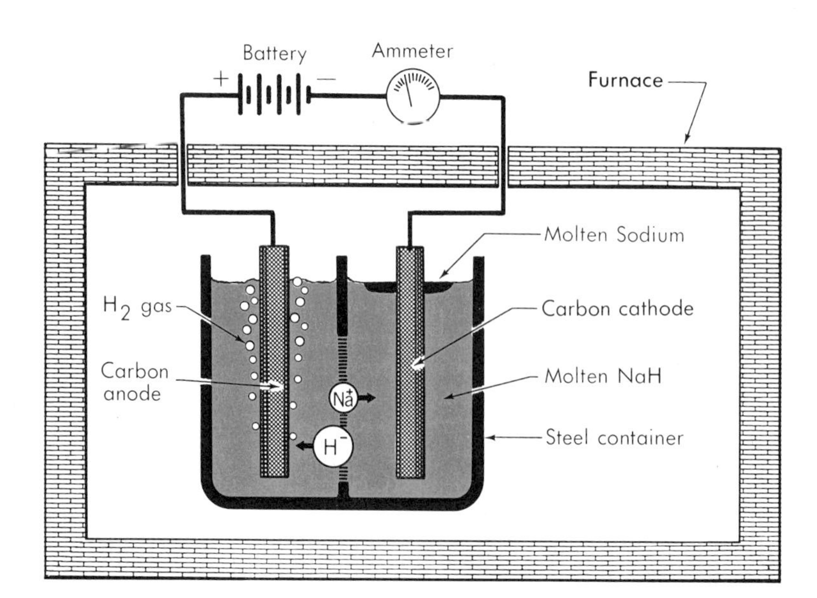

FIGURE 10.4 ELECTROLYSIS OF SODIUM HYDRIDE.

pounds of hydrogen with the non-metals (the covalent hydrogen compounds) can be seen when a comparison of the reactions of sodium hydride and hydrogen chloride with water as illustrated by the equations is made.

$$Na^+H^-_{(s)} + H_2O \rightarrow Na^+ + OH^- + H_{2(g)}$$

$$H:\underset{\cdot\cdot}{\ddot{Cl}}:_{(g)} + H_2O \rightarrow H_3O^+ + :\underset{\cdot\cdot}{\ddot{Cl}}:^-$$

The hydride ion, being a very strong base, **is a proton acceptor.** Upon addition to water, the hydride ion abstracts a proton from water, leaving a hydroxide ion

$$H:^- + H:\underset{H}{\ddot{O}}: \rightarrow H:H + :\underset{\cdot\cdot}{\ddot{O}}:H^-$$

Hydrogen chloride, in contrast to ionic NaH, is a **strong acid, a proton donor.** When the gas is dissolved in water, hydrogen chloride releases an H^+ to the water molecule,

$$:\underset{\cdot\cdot}{\ddot{Cl}}:H + :\underset{H}{\ddot{O}}:H \rightarrow :\underset{\cdot\cdot}{\ddot{Cl}}:^- + H:\underset{H}{\ddot{O}}:H^+$$

the symbol H_3O^+ refers to the hydrogen ion in water, although it is probably more complex, perhaps being $H_9O_4^+$. the symbol H^+ is used in some texts to represent H_3O^+, but H^+ does not exist in water.

to form the chloride ion and the **hydronium ion.**

During war-time, calcium hydride and lithium hydride pills are used as a portable source of hydrogen gas for inflating observation and weather balloons. The pills are removed from specially sealed water-tight packages and placed in a container open to the mouth of the collapsed balloon. On addition of water, the hydrogen gas is released to fill the balloon, which rises into the atmosphere.

A number of the transition metals absorb and transmit hydrogen at elevated temperatures. In fact, a heated palladium thimble attached to a glass system may be used to purify hydrogen. The gas passes through the thimble at about 300°C, leaving impurities behind. Metals saturated with the gas are often incorrectly called compounds. They are named interstitial hydrides because the hydrogen atoms occupy the interstices (the holes between the metal atoms in the crystal). A constant hydrogen-to-metal ratio is not found in these hydrides, and the law of definite proportions is not obeyed.

10.6 PREPARATION

One method of preparing hydrogen is that employed by Cavendish and those who preceded him: *the reaction of a metal with an acid.* Zinc reacts with dilute sulfuric acid as represented by the equation:

$$Zn + \underset{\text{sulfuric acid}}{H_2SO_4} \rightarrow \underset{\text{zinc sulfate in solution}}{ZnSO_4} + H_2$$

A laboratory apparatus for the preparation of hydrogen is shown in Figure 10.5. **All connections must be airtight** to prevent loss of

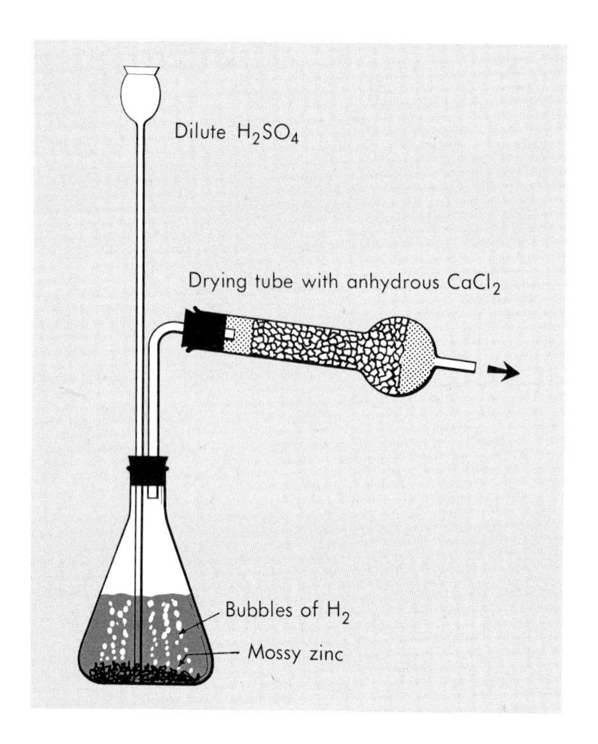

FIGURE 10.5 APPARATUS FOR PREPARATION OF HYDROGEN.

the gas and to minimize possibility of a fire. Mossy zinc, used because of its large surface area, is placed in the Erlenmeyer flask. After replacing the stopper, dilute sulfuric acid is added through the thistle tube. The end of the thistle tube is immersed in the acid to prevent the escape of hydrogen. The hydrogen passes through a loose plug of glass wool, on which any acid spray is collected, and then through granular anhydrous calcium chloride, where water vapor is removed by the reaction:

$$CaCl_2 + 2\ H_2O \rightarrow CaCl_2 \cdot 2\ H_2O$$

Many anhydrous substances have a great attraction for water. Anhydrous calcium chloride combines with the water to form a dihydrate, a compound with two water molecules per formula of salt. After all the air has been expelled from the reaction flask and the drying tube by the escaping gas, the stream of dry hydrogen issuing from the tip may be lighted and observed to burn.* Alternatively, the evolved hydrogen can be collected by water displacement in the manner illustrated for the preparation of oxygen in Figure 9.9.

do not light the hydrogen until all the air has been expelled.

An equation which more fully depicts the reaction species in the preparation of hydrogen is:

$$Zn + 2\ H_3O^+ + 2\ HSO_4^- \rightarrow H_2 + Zn^{2+} + 2\ HSO_4^- + 2\ H_2O$$

* Since the experimenter does not *know* when "all the air has been expelled from the reaction flask," he may try to light the stream of hydrogen too soon, producing a violent explosion and a spray of glass. The following method should be used: Collect a sample of the issuing gas in an inverted test tube, carry it to a flame 14 feet away, and ignite the gas. If the mixture is explosive, there will be a "pop," and no harm is done. When the issuing gas is all hydrogen, it will ignite at the mouth of the test tube and burn quietly. The flame may then be carried back *in the test tube* to light the gas at the tip. This method is absolutely safe.

Sulfuric acid is a strong acid and in aqueous solution it is present as the hydronium ion (H_3O^+) and the hydrogen sulfate ion (HSO_4^-). Zinc hydrogen sulfate (a salt) is, also, dissociated in water. The nature of acids and salts in aqueous solution is discussed in Chapter 13. Let it suffice here to write the above equation as

$$Zn + 2\,H_3O^+ \rightarrow H_2 + Zn^{+2} + 2\,H_2O$$

Since the hydrogen sulfate ion appears on both sides of the equation (i.e., it is unchanged in the reaction), it is not included in the equation. This preparation of hydrogen involves an oxidation-reduction reaction. Zinc loses electrons (is oxidized) and, because it accepts electrons from the substance reduced, is the reducing agent.

$$Zn \rightarrow Zn^{+2} + 2\,e^-$$

Hydronium ion is reduced, that is, gains electrons, and because it takes the electrons from the substance oxidized, is the oxidizing agent.

$$2\,H_3O^+ + 2\,e^- \rightarrow H_2 + 2\,H_2O$$

the piece of sodium should be about one-half the size of an eraser on a pencil. dropping in a large piece of sodium can result in an explosion. the experiment demonstrated here should be done at arm's length with the experimenter wearing safety goggles.

Hydrogen can be prepared by this method using any metal that will reduce (donate electrons to) the hydrogen ion and any acid which will provide the hydrogen ion in sufficient concentration. **The reactivities of metals** with acid solutions **vary greatly.** Some metals cannot reduce the hydrogen ion and some are such powerful reducing agents that they are capable of reducing water itself (freeing hydrogen from water).

One metal that reacts with water to produce hydrogen gas is sodium. When a small piece of sodium is dropped into a beaker containing a small quantity of water, as in Figure 10.6a and b, it

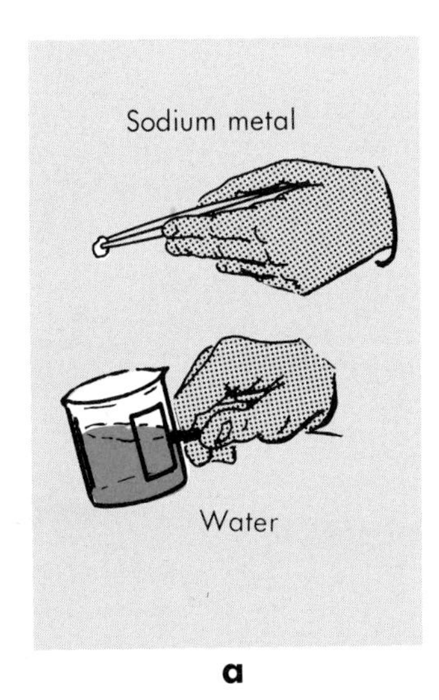

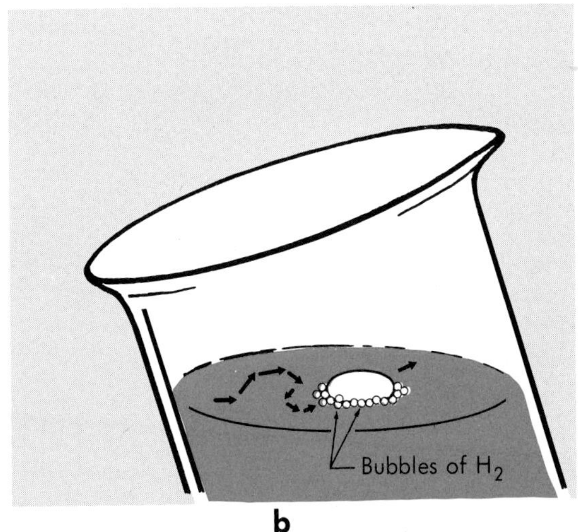

FIGURE 10.6 SODIUM REACTING WITH WATER.

moves rapidly at random on the water surface as a molten sphere, melted by the heat of the reaction. Observation reveals that gas bubbles are forming and bursting at a rapid rate at the metal-water interface, causing the metal to dart about.

The gas liberated in the bubbling is flammable and explodes in air when ignited. A solution of sodium hydroxide remains in the test tube. The net ionic equation for this reaction is

$$2\,Na + 2\,H_2O \rightarrow 2\,Na^+ + 2\,OH^- + H_2$$

From a series of experimental observations using different metals, it is possible to arrange the metals in an order of decreasing activity or **ability to produce hydrogen** from acid solutions, giving a so-called *activity series.* A partial activity series (one including only some of the known metals) appears in Table 10.4.

The **relative activity** of all the metals with water and acid is indicated by their position in the periodic table (see Figure 10.7). Notice that those metals sufficiently active to react with cold water are in the extreme left-hand side of the periodic table. Those that do not react with acids are among or near the transition elements at the bottom center of the periodic table.

Even though a metal does not react with the hydrogen ion to yield hydrogen gas, this does not mean that the metal will not react with an acid in another way. For example, copper reacts with an oxidizing acid such as nitric acid:

$$Cu + 4\,H_3O^+ + 4\,NO_3^- \rightarrow Cu^{+2} + 2\,NO_3^- + 6\,H_2O + 2\,NO_2$$

In this reaction the H_3O^+ ions are not reduced (H_2 is not formed), but the nitrate ion (NO_3^-) is the oxidizing agent in the acidic solu-

TABLE 10.4 PARTIAL ACTIVITY SERIES

Metal	Reaction with water	Reaction with acids
K Ba Ca Na	Will replace H_2 from cold water: $Ba + 2H_2O \rightarrow Ba(OH)_2 + H_2$	React explosively with acids: $Ca + 2HCl \rightarrow CaCl_2 + H_2$
Mg	Will react with boiling water: $Mg + 2H_2O \rightarrow Mg(OH)_2 + H_2$	
Al Mn Zn Fe Cd Ni	The red-hot metal will react with steam: $3Fe + 4H_2O \rightarrow Fe_3O_4 + 4H_2$	React vigorously with acids to give H_2: $Mg + 2HCl \rightarrow MgCl_2 + H_2$
Sn Pb	No reaction with either water or steam	Hardly react with acids
Cu Hg Ag Au Pt	No reaction with water	Will not react with nonoxidizing acids

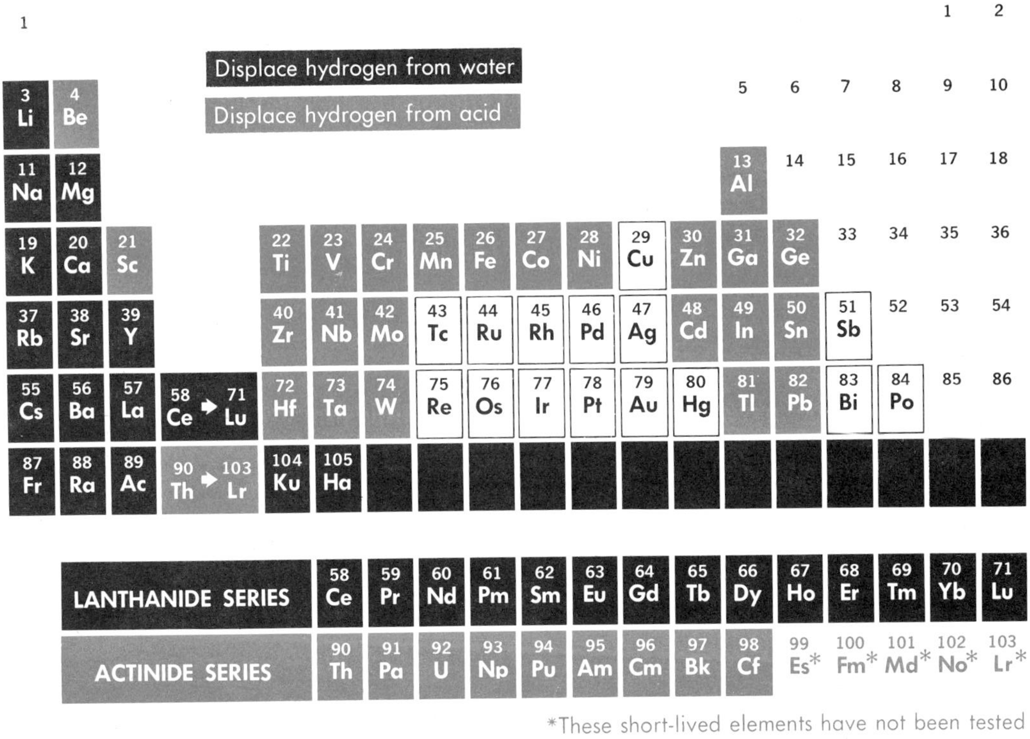

1 2

Displace hydrogen from water

Displace hydrogen from acid

3 Li, 4 Be, 5, 6, 7, 8, 9, 10

11 Na, 12 Mg, 13 Al, 14, 15, 16, 17, 18

19 K, 20 Ca, 21 Sc, 22 Ti, 23 V, 24 Cr, 25 Mn, 26 Fe, 27 Co, 28 Ni, 29 Cu, 30 Zn, 31 Ga, 32 Ge, 33, 34, 35, 36

37 Rb, 38 Sr, 39 Y, 40 Zr, 41 Nb, 42 Mo, 43 Tc, 44 Ru, 45 Rh, 46 Pd, 47 Ag, 48 Cd, 49 In, 50 Sn, 51 Sb, 52, 53, 54

55 Cs, 56 Ba, 57 La, 58 Ce → 71 Lu, 72 Hf, 73 Ta, 74 W, 75 Re, 76 Os, 77 Ir, 78 Pt, 79 Au, 80 Hg, 81 Tl, 82 Pb, 83 Bi, 84 Po, 85, 86

87 Fr, 88 Ra, 89 Ac, 90 Th → 103 Lr, 104 Ku, 105 Ha

LANTHANIDE SERIES: 58 Ce, 59 Pr, 60 Nd, 61 Pm, 62 Sm, 63 Eu, 64 Gd, 65 Tb, 66 Dy, 67 Ho, 68 Er, 69 Tm, 70 Yb, 71 Lu

ACTINIDE SERIES: 90 Th, 91 Pa, 92 U, 93 Np, 94 Pu, 95 Am, 96 Cm, 97 Bk, 98 Cf, 99 Es*, 100 Fm*, 101 Md*, 102 No*, 103 Lr*

*These short-lived elements have not been tested

FIGURE 10.7 RELATIVE REACTIVITY OF METALS WITH WATER AND ACIDS.

tion. The nitrate ion is reduced to nitrogen dioxide gas (NO_2), and copper is oxidized to Cu^{+2} ions.

10.7 COMMERCIAL PRODUCTION OF HYDROGEN

Burning coke reacts with steam to produce a commercial fuel, a mixture of hydrogen and carbon monoxide called water gas.

$$\underset{\text{coke}}{C} + \underset{\text{steam}}{H_2O} \rightarrow \underset{\text{carbon monoxide}}{CO} + \underset{\text{hydrogen}}{H_2}$$

Since carbon monoxide is so poisonous, water gas is seldom used in the home. Both carbon monoxide and hydrogen burn, and the mixture is an efficient fuel. In preparing water gas, preheated air is blown over burning coke to raise the temperature to white heat. Steam is then passed over the burning coke, producing water gas. The steam gradually quenches the burning coke. Just before total quenching, air is again blown over the coke, regenerating the burning for another blast of steam.

Water gas may be passed over a heated catalyst with more steam to convert the carbon monoxide to carbon dioxide, thereby producing more hydrogen if pure (carbon monoxide-free) hydrogen is desired:

$$\underset{\text{carbon monoxide}}{CO} + \underset{\text{steam}}{H_2O} \xrightarrow[\text{heat}]{\text{catalyst}} CO_2 + H_2$$

Carbon dioxide may be removed (scrubbed) from the mixture by passing it through lime water (a solution of calcium hydroxide) or through a solution of sodium hydroxide.

One of the chief industrial sources of hydrogen is the catalytic reaction of hydrocarbons with steam. When propane, for example, is passed with steam over a nickel catalyst at elevated temperatures, carbon dioxide and hydrogen are formed.

$$\underset{\text{propane}}{C_3H_8} + 6\ H_2O \rightarrow 3\ CO_2 + 10\ H_2$$

Carbon dioxide is removed as before. Other hydrocarbons, such as methane (CH_4), ethane (C_2H_6), ethylene (C_2H_4), and so on, may also be used.

Hydrogen, like oxygen, is a product of the electrolysis of water. A certain amount is produced for commercial purposes by this method in areas where electric power is cheap.

$$2\ H_2O \xrightarrow{\text{electrolysis}} 2\ H_2 + O_2$$

10.8 USE OF HYDROGEN TO PREPARE OTHER CHEMICALS

Approximately 8 million tons of hydrogen were used in the United States in 1969. Most of the hydrogen that is produced is for use in hydrogenation (addition of hydrogen to an element or compound), as a fuel or as a reducing agent (for the removal of oxygen from a compound). A great deal of hydrogen has been used as fuel in the space program.

The most important commercial use of hydrogen is in the production of other valuable chemicals by hydrogenation. Some examples follow.

1. *Ammonia is produced by hydrogenating nitrogen gas in the presence of a catalyst,* under pressure:

$$\underset{\text{hydrogen}}{3\ H_2} + \underset{\text{nitrogen}}{N_2} \xrightarrow[450^\circ]{\text{catalyst}} \underset{\text{ammonia}}{2\ NH_3}$$

what are some of the uses of hydrogen?

Ammonia is used as a base in chemical processes and as a liquid fertilizer. It is a major raw material for the production of nitric acid, explosives and many fertilizers.

2. *Methyl alcohol (wood alcohol) is produced by combining carbon monoxide and hydrogen* under pressure:

$$\underset{\text{carbon monoxide}}{CO} + 2\ H_2 \xrightarrow[350^\circ]{\text{catalyst}} \underset{\text{methyl alcohol}}{CH_3OH}$$

Methyl alcohol is used in the commercial synthesis of many organic chemicals.

3. *Coal releases hydrocarbons when hydrogen is passed over it in the presence of a catalyst.* This is a possible source of gasoline in the future.

4. *Vegetable oils are hardened by combination with hydrogen* to make margarine and solid shortening.

10.9 HYDROGEN AS A FUEL

1. In some engines the propulsion comes from burning hydrogen with oxygen, which is a highly exothermic reaction. The water vapor, expanded at the very high temperature caused by the heat of the reaction, is blasted from the rear of the rocket, thrusting it forward.

$$2\,H_2 + O_2 \rightarrow 2\,H_2O + 116{,}000 \text{ cal}$$

2. In the *hydrogen-oxygen torch*, the two gases are premixed in proper proportions and burned together at a high temperature as they emerge from the tip, as shown in Figure 10.8. Small objects may be heated in an oxy-hydrogen flame to about 2800°C.

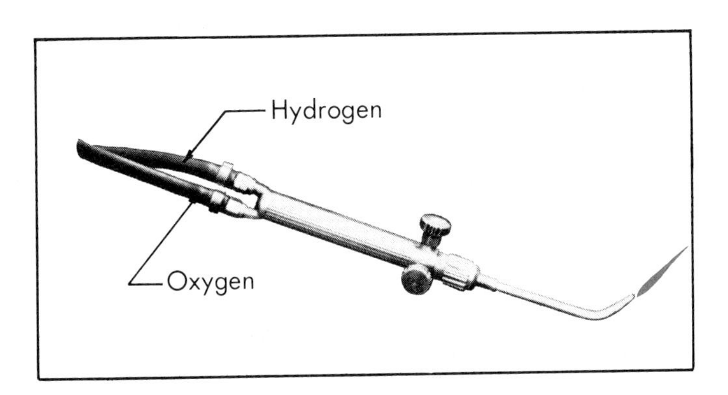

FIGURE 10.8 THE HYDROGEN-OXYGEN TORCH.

3. An interesting method of obtaining high temperatures is by use of an atomic hydrogen torch. A great quantity of energy is required to split a hydrogen molecule into atoms:

$$\underset{\text{molecule}}{H_2} + \underset{\text{103,000 cal/mole}}{\text{energy}} \rightarrow \underset{\text{atoms}}{2\,H}$$

Upon recombination, the energy is released. This recombination is catalyzed by a metal surface. In the atomic hydrogen torch, hy-

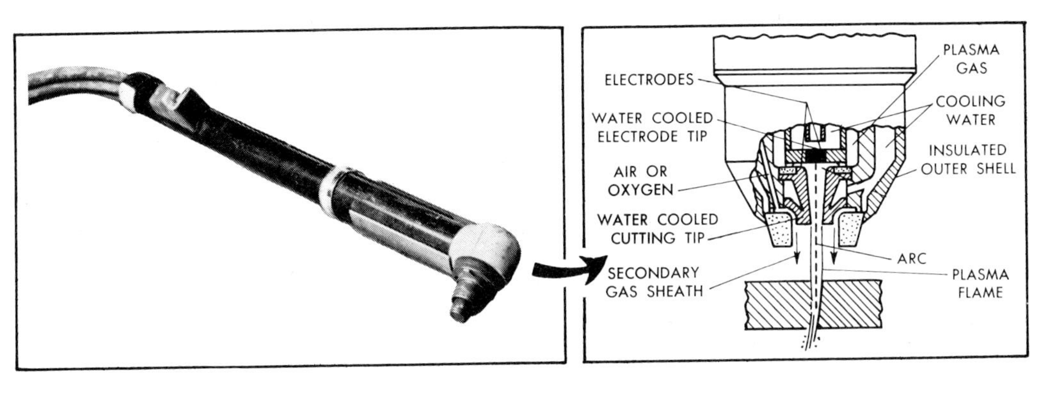

FIGURE 10.9 THE PLASMA TORCH.

drogen gas is passed between two sparking electrodes. Electrical energy from the arc splits the molecules into atoms. When the stream of atoms hits the metal surface, they pair together, liberating that large quantity of heat absorbed in the separation. The hydrogen then burns with air, liberating more calories. The temperature of the flame reaches 5000°C and higher and is sufficient to melt all metals. Further refinements of the atomic hydrogen torch has resulted in the development of the plasma torch (Fig. 10.9). A number of gases including hydrogen can be used to produce high temperatures.

10.10 HYDROGEN IN THE COMMERCIAL PRODUCTION OF SOME METALS

Figure 10.10 illustrates an experiment in which hot copper(II) oxide is reduced to copper metal in a stream of hydrogen. *Hydrogen will reduce the hot oxides of metals of intermediate activity.* In general all oxides of the metals between iron and copper in the activity series (Table 10.6) can be reduced by hydrogen when heated. Oxides of the metals below copper can be reduced merely by heating, while oxides of those above iron are stable at high temperature even in the presence of hydrogen. Two metals, tungsten and molybdenum, are prepared commercially by reduction with hydrogen. The equation for producing tungsten is

$$\underset{\text{tungsten(VI) oxide}}{WO_3} + 3\,H_2 \xrightarrow{\text{heat}} \underset{\text{tungsten}}{W} + 3\,H_2O$$

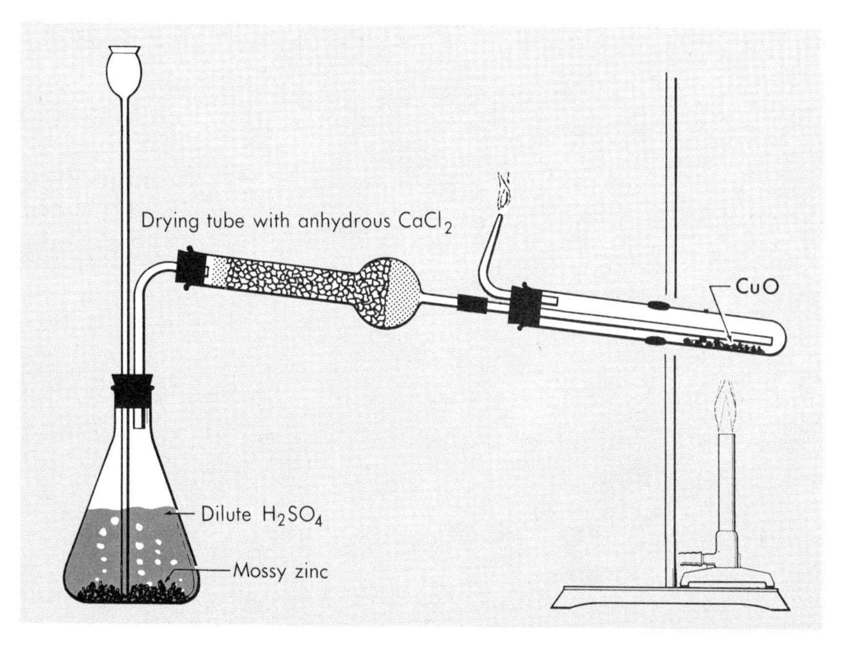

FIGURE 10.10 THE REDUCTION OF COPPER OXIDE WITH HYDROGEN.

10.11 HYDROGEN WAS USED IN LIGHTER-THAN-AIR CRAFT

The lifting power of hydrogen gas in air has been discussed in another section. Hydrogen was used to fill weather balloons, dirigibles and balloons, especially in Germany between the two World Wars. Several tragedies resulted because of the extreme flammability of hydrogen in air. An example is shown in Figure 10.11, the explosion of the dirigible Hindenburg. Because of the danger of explosion, unreactive helium, with nearly as much lifting power, has replaced hydrogen in lighter-than-air craft.

FIGURE 10.11 THE HINDENBURG EXPLOSION.

10.12 WATER

Water is the most important hydrogen containing compound, and probably the most important of all compounds. Its role as the major component of the human body, and the solvent in which all of the body chemistry takes place, makes it of utmost importance. Its utility is related to its physical and chemical properties, which are, in turn, dependent upon the structure of the water molecule.

The water molecule is not linear; it is bent with the angle between the OH bonds at the oxygen atom being 104.5°, as shown in Fig. 10.12a. The oxygen atom in the water molecule is thought to be sp^3 hybridized; two of the eight electrons of the molecule are in each of the two sigma bonds shared by and bonding a hydrogen atom to the oxygen. The remaining four valence electrons occupy the sp^3 orbitals, two in each. The molecule may be thought of as

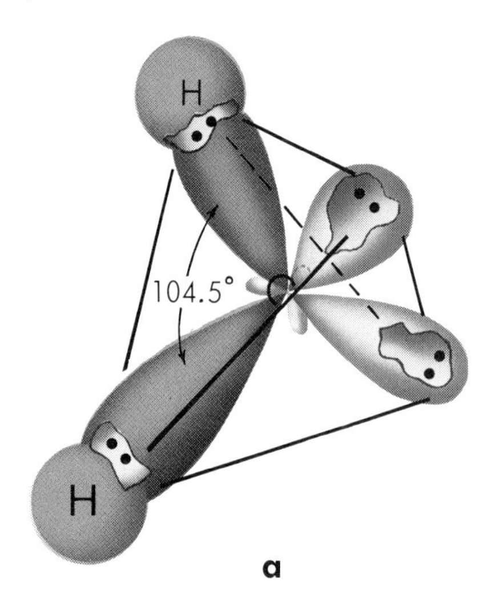

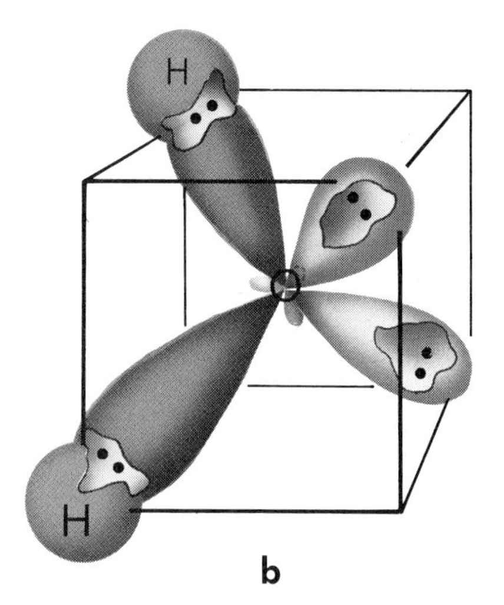

FIGURE 10.12 THE WATER MOLECULE.

occupying a tetrahedron with the oxygen in the center of the pyramid, the two hydrogen atoms at apices and the sp^3 orbitals holding the unshared electron pairs pointing in space toward the other two apices. Alternately the molecule can be thought of as occupying a section of a cube, as shown in Figure 10.12b. The figure is a distorted cube because the angle between the OH bonds is less than the tetrahedral angle, 109.5°; the angle between the orbitals of the unshared pairs is greater than 109.5°. It is argued that the repulsion between the negative unshared electron pairs in the sp^3 orbitals opens that angle and consequently closes the angle slightly between the bonds where the protons (H^+) embedded in the bonding orbitals partially neutralize the negative charge.

10.13 HYDROGEN BONDING IN WATER

Because the water molecule is bent and because the oxygen atom has a much greater electronegativity than hydrogen, the water molecule is highly polar, with the polarity directed along the bisector of the bond angle toward the oxygen atom as shown in Figure 10.13.

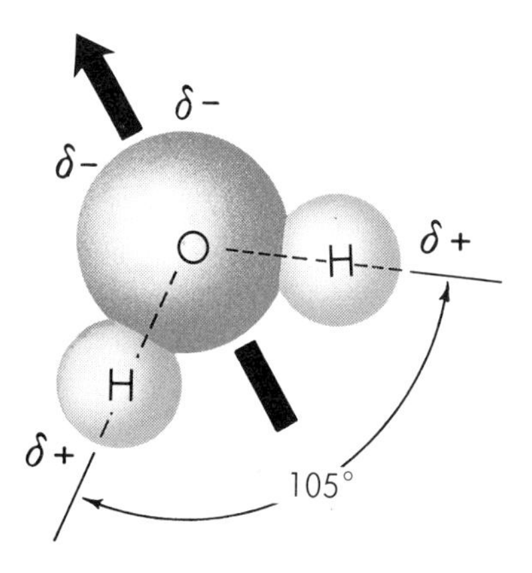

FIGURE 10.13 THE POLAR WATER MOLECULE.

Water molecules form **hydrogen bonds** among themselves and between other molecules having structures with that capability. As discussed in Chapter 5, a hydrogen bond is a relatively weak attraction between two highly electronegative atoms through a hydrogen atom which is attached to one of them by a single covalent bond. The oxygen atom of the water molecule is an electronegative atom. It bonds readily through one or both of its hydrogen atoms to the oxygen of another molecule or other molecules as illustrated in Figure 10.14. The hydrogen bond thus formed is considered essentially electrostatic since the hydrogen atom has the capacity to share only two electrons. As stated in Chapter 5, hydrogen bonding occurs between small electronegative atoms such as fluorine, oxygen and nitrogen through a hydrogen attached to one of them. Only 5 to 10 per cent of the energy required to separate a normal covalent bond is needed to separate a hydrogen bond.

Evidence used to support the existence of hydrogen bonds is found when the boiling points of the binary hydrogen containing compounds of a family of non-metals are compared. Consider the normal boiling points of the other simple hydrogen compounds of Group VI: H_2Te, hydrogen telluride; H_2Se, hydrogen selenide; H_2S, hydrogen sulfide, which are −1.8°C, −41.5°C and −60.8°C, respectively. As the identity of the Group VI element changes from the heavy members of the group to the lighter members, the boiling points become more negative. Were this trend to continue, water should have a boiling point of approximately −90°C rather than the observed value of 100°C. Similarly the boiling points of hydrogen fluoride (HF) and ammonia (NH_3) do not follow the trends established by the hydrogen compounds of the elements below them in their respective groups in the periodic table. Instead

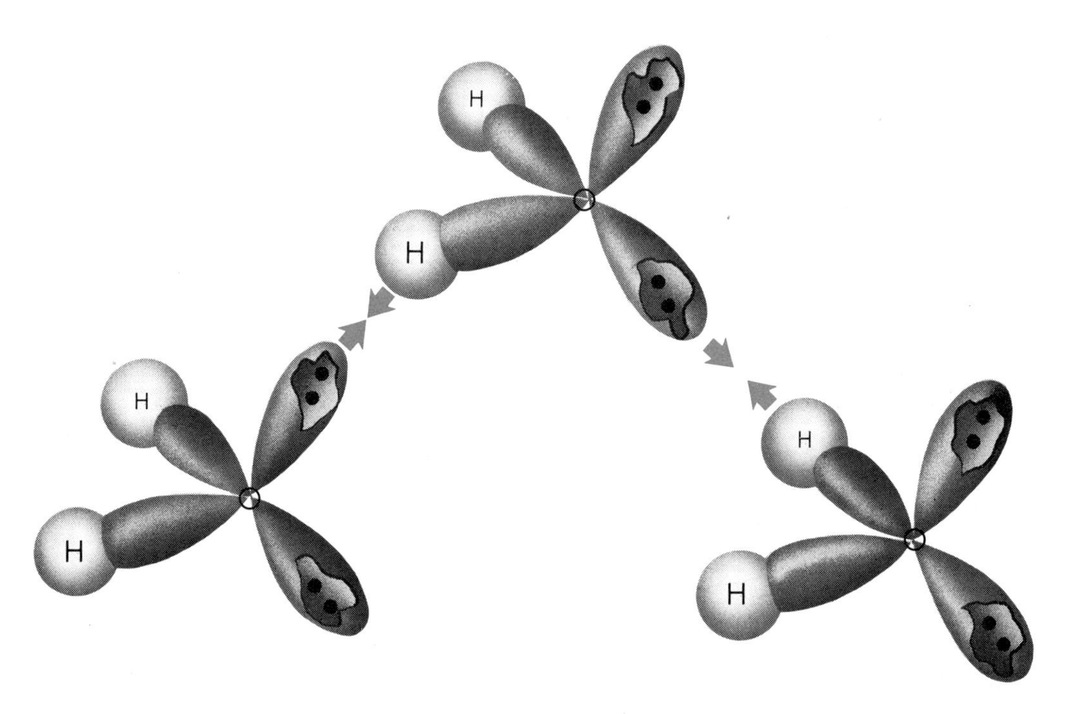

FIGURE 10.14 HYDROGEN BONDING IN WATER.

their boiling points are abnormally high. The boiling points of the Group IVA and VIA series of hydrogen containing compounds are shown in Figure 10.15 with those of the noble gases. It can be seen in the table that the boiling points of the noble gases and the covalent hydrogen compounds of Group IV follow regular trends; that is, the boiling point of the hydrogen compound of the first member of that family is not unusually high. There is no hydrogen bonding in methane (CH_4).

The decreasing contraction of water with cooling, which eventually results in an expansion below 4°C, and the expansion of the liquid upon freezing are attributed to hydrogen bonding. As the temperature of the water decreases, more water molecules are held in hydrogen bonding. When ice is formed, hydrogen bonding is nearly 100 per cent complete (except for crystal imperfections). Without hydrogen bonding and aided by polarity, water molecules can pack together tightly, with the V-shaped molecules folding

expansion of water upon freezing can be attributed to hydrogen bonding.

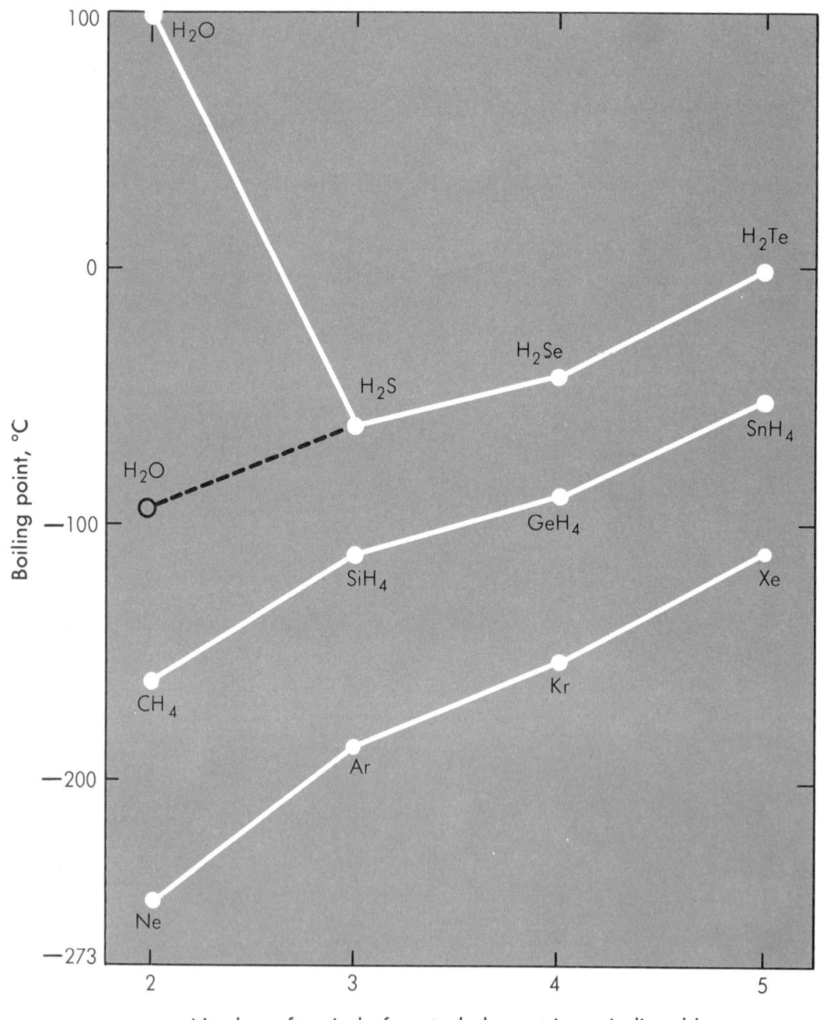

FIGURE 10.15 ABNORMAL BOILING POINT OF WATER.

TABLE 11.1 ELECTRON CONFIGURATIONS OF THE ALKALI AND ALKALINE EARTH ELEMENTS

Alkali Metals*		Alkaline Earth Metals*		
$_{3}$Li	$1s^2$ | $2s^1$	$_{4}$Be	He	$2s^2$
$_{11}$Na	$1s^2$ $2s^2$ $2p^6$ | $3s^1$	$_{12}$Mg	Ne	$3s^2$
$_{19}$K	$1s^2$ $2s^2$ $2p^6$ $3s^2$ $3p^6$ | $4s^1$	$_{20}$Ca	Ar	$4s^2$
$_{37}$Rb	$1s^2$ $2s^2$ $2p^6$ $3s^2$ $3p^6$ $3d^{10}$ $4s^2$ $4p^6$ | $5s^1$	$_{38}$Sr	Kr	$5s^2$
$_{55}$Cs	$1s^2$ $2s^2$ $2p^6$ $3s^2$ $3p^6$ $3d^{10}$ $4s^2$ $4p^6$ $4d^{10}$ $5s^2$ $5p^6$ | $6s^1$	$_{56}$Ba	Xe	$6s^2$
$_{87}$Fr	$1s^2$ $2s^2$ $2p^6$ $3s^2$ $3p^6$ $3d^{10}$ $4s^2$ $4p^6$ $4d^{10}$ $4f^{14}$ $5s^2$ $5p^6$ $5d^{10}$ $6s^2$ $6p^6$ | $7s^1$	$_{88}$Ra	Rn	$7s^2$

* Colored area represents the electron configuration of the noble gas preceding the given alkali or alkaline earth element. The clear boxes show the valence electrons for the respective Group IA and IIA metals.

The electron configurations of the alkali and alkaline earth elements are shown in Table 11.1. The alkali metals have one valence electron; the alkaline earths, two. In chemical reaction, the atoms of the elements lose the valence electrons, giving the alkali metals ions a charge of +1; the alkaline earths, +2. Reason for these valencies, +1 and +2, can be seen in the comparative ionization energies of the elements, given in Table 11.2.

The first ionization energy, the **energy required to remove the first electron** from the neutral gaseous atom, is lower for the Group I metals than for the Group II metals. Within each group the ionization energies decrease from top to bottom in the table. The first ionization energy of the alkali metal varies from approximately one

TABLE 11.2 IONIZATION ENERGIES IN KCAL PER MOLE FOR THE ALKALI AND ALKALINE EARTH ELEMENTS

Alkali Element	I_1*	I_2	Alkaline Earth Element	I_1	I_2	I_3
Li	124.3	1744	Be	214.9	419.9	3548
Na	118.5	1091	Mg	176.3	346.6	1848
K	100.1	733.6	Ca	140.9	273.8	1181
Rb	96.31	634	Sr	131.3	254.3	‡
Cs	89.78	579	Ba	120.2	230.7	‡
Fr	†	†	Ra	121.7	234.0	‡

* I_1 represents the energy in kcal per mole that is required to remove an electron from the gaseous atom. I_2 and I_3 represent the energy necessary to remove the second and third electrons.

† Francium has been prepared synthetically in very small amounts and the ionization energies have not yet been measured.

‡ Values have not been measured.

half to two thirds of the value for the corresponding alkaline earth element. In contrast, the second ionization energy for each Group I element (the energy required to take an electron away from the atom after one has already been removed) is greater than for the alkaline earth metal of the same period. The second ionization energy of each alkali metal is 6 to 12 times greater than the first ionization energy, while the second ionization energy for each alkaline earth metal is only about double the first. In the alkaline earth elements the great increase in ionization energy for each atom comes with the removal of the third electron. That is, the third ionization energy for the Group II elements is from 4 to 9 times greater than the second. The comparisons of ionization energies indicate that the M^+ ion is formed rather easily from the alkali metal atoms, and the M^{+2} ion from the alkaline earth atoms, while the formation of M^{+2} alkali ions and M^{+3} alkaline earth ions would be very diffi-

why does sodium not form Na^{2+} ion?

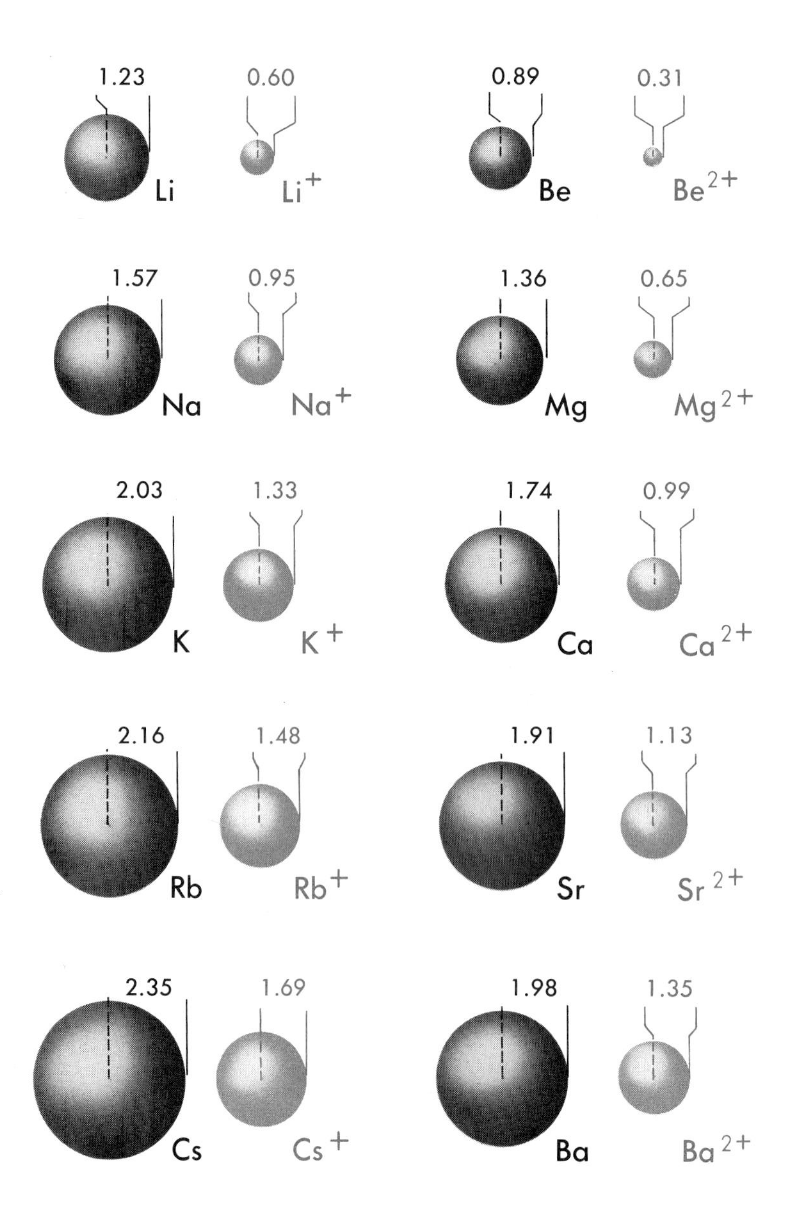

FIGURE 11.1 RELATIVE SIZES OF SOME ATOMS AND IONS.

cult. Those comparisons also confirm the stability of the noble gas valence electron configuration, since the taking of two electrons from a Group I atom and of three from a Group II atom would break into that electron arrangement.

The trends in the ionization energies listed in Table 11.2 can be understood by considering the relative sizes of the atoms and ions of the first two families of the periodic table (Fig. 11.1). Atomic size increases from top to bottom in either family or group, from lithium to cesium in Group I and from beryllium to barium in Group II. An increase in size is expected as the number of filled electronic shells increases with the atomic weight. The first ionization energy decreases within a group or family as the atomic radii increase. The decrease in ionization energy accompanying an increase in size is expected, because the valence electron is further removed from the nucleus which attracts it.

the radius of Be^{2+} is much smaller than the radius of the beryllium atom.

The fact that the first ionization energies of the alkali metals are less than the first ionization energies of the corresponding alkaline earth elements is consistent with the relative size of the atoms of the elements. The radius of an atom of each Group II element is less than the radius of the atom of the Group I element that immediately precedes it in the table. The corresponding Group III atom is smaller than the atom of Group II, and the trend of decreasing size extends from left to right in the rows of eight across the periodic table. There is a corresponding trend toward increasing ionization energy. As a result, the elements of lowest ionization energy are those at the extreme left and bottom of the table.

The very large values of I_2 (the second ionization energy) for the elements of Group I, and of I_3 (the third ionization energy) for elements of Group II, are reasonable when one compares the sizes of the atoms with the sizes of the ions (Fig. 11.1). It is evident that the size of the atom or ion is not the only factor in determining the ionization energy, for sodium and barium have nearly the same ionization energies (118.5 and 120.2 kcal per mole, respectively) but their atomic radii are 1.57 and 1.98 Å, respectively. On the other hand, the second ionization energy for lithium is almost equal to the third for magnesium, and the two ions Na^+ and Ca^{2+} have nearly equal radii. From our argument, the singly charged ions of the Group II elements must have radii much more nearly the same as those of the atoms than of the doubly charged ions. Otherwise the great increase in the third ionization energy over the second could not be explained.

The sizes of the atoms of the Group I and II metals also **are reflected in their relative densities.** If the atoms could pack together in the solid in such a way that there were no holes (interstices) between the atoms, the density would be inversely proportional to the size of the atoms, because density is equal to the ratio of the atomic weight and the atomic volume. However, the atoms tend to be spherical in shape and they pack together in the metal crystals in different patterns such that the percentage of the total volume which is interstices varies somewhat from metal to metal. The relative densities of the alkali and alkaline earth metals are compared to some common metals in Figure 11.2, in which the metals are arranged in the periodic table. The magnitudes of the densities are proportional to the intensity of the color in Figure 11.2. A comparison of the densities with the sizes of the atoms, as

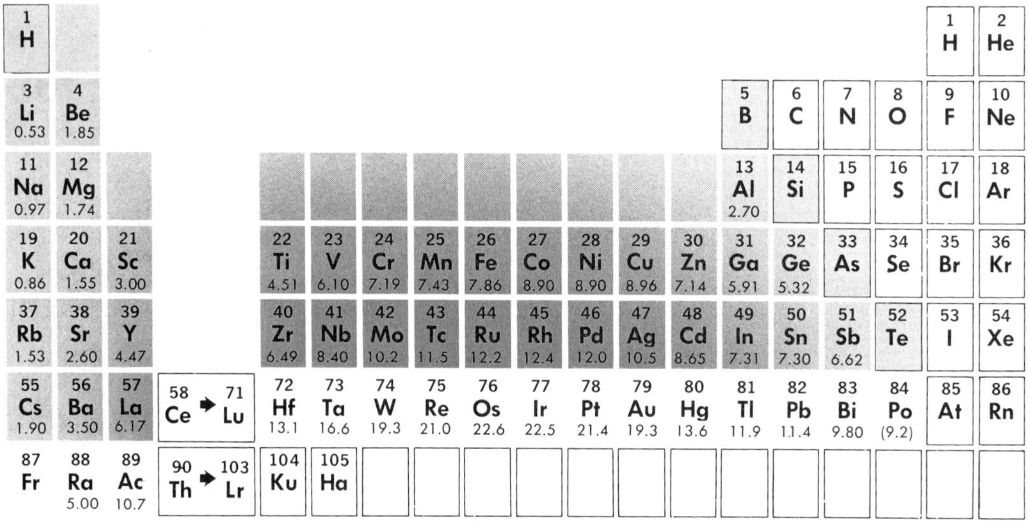

1 H																	1 H	2 He
3 Li 0.53	4 Be 1.85												5 B	6 C	7 N	8 O	9 F	10 Ne
11 Na 0.97	12 Mg 1.74												13 Al 2.70	14 Si	15 P	16 S	17 Cl	18 Ar
19 K 0.86	20 Ca 1.55	21 Sc 3.00		22 Ti 4.51	23 V 6.10	24 Cr 7.19	25 Mn 7.43	26 Fe 7.86	27 Co 8.90	28 Ni 8.90	29 Cu 8.96	30 Zn 7.14	31 Ga 5.91	32 Ge 5.32	33 As	34 Se	35 Br	36 Kr
37 Rb 1.53	38 Sr 2.60	39 Y 4.47		40 Zr 6.49	41 Nb 8.40	42 Mo 10.2	43 Tc 11.5	44 Ru 12.2	45 Rh 12.4	46 Pd 12.0	47 Ag 10.5	48 Cd 8.65	49 In 7.31	50 Sn 7.30	51 Sb 6.62	52 Te	53 I	54 Xe
55 Cs 1.90	56 Ba 3.50	57 La 6.17	58 Ce → 71 Lu	72 Hf 13.1	73 Ta 16.6	74 W 19.3	75 Re 21.0	76 Os 22.6	77 Ir 22.5	78 Pt 21.4	79 Au 19.3	80 Hg 13.6	81 Tl 11.9	82 Pb 11.4	83 Bi 9.80	84 Po (9.2)	85 At	86 Rn
87 Fr	88 Ra 5.00	89 Ac 10.7	90 Th → 103 Lr	104 Ku	105 Ha													

FIGURE 11.2 RELATIVE DENSITIES OF METALS.

in Figure 11.1, indicates that a large size results in low density as explained above. The densities, along with other properties, are given for all the active metals in Table 11.3.

The active metals have a metallic luster resembling shiny silver. Often this is not apparent, for even when kept in oil to protect them from the oxygen and water vapor in the air, the **metals become coated** with a dull gray oxide. When the metals are freshly cut, the shiny surface is visible, but it turns gray almost immediately. The **alkali metals are soft,** and can be cut with a knife very easily. The alkaline earth metals in the group have somewhat the same softness as lead. All are good conductors of heat and electricity. Sodium is a good conductor of both heat and electricity, but its softness and its reactivity restrict its use as a conductor.

The low ionization energy might account for electrical conduction. The outermost electrons are held very loosely in the interstices of the metal and can easily be pushed from atom to atom by an electrical potential. The softness is also related to a relatively small attraction for the valence electrons. If atoms (ions in the crystal) had a great attraction for electrons, they would be held together in a rigid crystalline structure, but the attraction is small and the crystals are soft. The melting points are low (especially for the alkali metals) for the same reason. Cesium, for example, is a liquid above 29°C.

From a consideration of the ionization energies it was argued that Group I elements will only have the +1 oxidation state or

TABLE 11.3 PHYSICAL PROPERTIES OF THE ACTIVE METALS

	Li	Na	K	Rb	Cs	Be	Mg	Ca	Sr	Ba
Atomic weight	6.939	22.990	39.102	85.47	132.905	9.0122	24.312	40.08	87.62	137.34
Density at 20°C g/cm³	0.535	0.971	0.862	1.532	1.90	1.85	1.738	1.55	2.54	3.5
Melting point, °C	179	97.9	63.5	39.0	28.5	1277	651	850	770	725
Boiling point, °C	1336	883	758	700	670	2970	1107	1487	1380	1140

TABLE 11.4 SOME TYPICAL REACTIONS OF THE ALKALI AND ALKALINE EARTH METALS

Alkali Metals		Alkaline Earth Metals
	With hydrogen	
$2\ M + H_2 \rightarrow 2\ MH$		$M + H_2 \rightarrow MH_2$
	With the halogens ($X = F_2$, Cl_2, Br and I_2)	
$2\ M + X_2 \rightarrow 2\ MX$		$M + X_2 \rightarrow MX_2$
	With sulfur	
$16\ M + S_8 \rightarrow 8\ M_2S$		$8\ M + S_8 \rightarrow 8\ MS$

valence in chemical compounds and that Group II elements will have only an oxidation state of +2. Some typical reactions of the active elements in which the predicted oxidation state is attained are shown in Table 11.4.

The reactions of oxygen with the active metals are not shown in Table 11.4 since oxides, peroxides and superoxides may be obtained. The formation of peroxides and superoxides is due to the high reactivity of the active metals.

11.4 THE ACTIVITY SERIES

In Chapter 10 the relative reactivities of some metals in displacing hydrogen were given. Now we shall consider quantitatively the relative abilities of the active metals and other metals to displace hydrogen from a solution containing 1 mole of hydronium ion per liter. These data are given in the *electromotive series* or *activity series,* where the metals are listed in the order of decreasing reactivity with acid solutions (Table 11.5).

The half reactions in Table 11.5 are written as reductions, and the reactions **that proceed to the right more readily are those assigned a larger positive voltage.** Therefore, the oxidizing strength (ability to accept electrons) increases (for the species on the left of the arrow in the equations) as one goes down the table. The strongest reducing agents (the metal in the reduced state) are at the top of the table on the right side of the arrow.

The rates of reaction with water or acids follow rather closely, in this case, the strength of the reducing agent. Metals with a high negative potential (−3.02 to −2.7) evolve hydrogen with explosive violence. Metals with intermediate values (−2.34 to −0.44) replace hydrogen at a reasonable rate. Metals with a still lower negative value react very slowly with acids, if at all. Metals with a positive value do not replace hydrogen from acids.

The order of the alkali elements in the activity series—Cs, Li, Rb, K and Na—is much as expected. Only lithium appears to be out of order. On the basis of its larger ionization energy, lithium should follow sodium. The apparent contradiction disappears when one realizes that the process of oxidation in water solutions involves more than the loss of electrons. The reaction of lithium metal with

TABLE 11.5 ELECTROMOTIVE SERIES FOR SOME METALS*

Oxidizing Agent (Oxidized State)		Reducing Agent (Reduced State)	Standard Electrode Reduction Potential (E°)
$Cs^+ + e^-$	=	Cs	−3.02
$Li^+ + e^-$	=	Li	−3.02
$Rb^+ + e^-$	=	Rb	−2.99
$K^+ + e^-$	=	K	−2.92
$Ba^{+2} + 2e^-$	=	Ba	−2.90
$Sr^{+2} + 2e^-$	=	Sr	−2.89
$Ca^{+2} + 2e^-$	=	Ca	−2.87
$Na^+ + e^-$	=	Na	−2.71
$Mg^{+2} + 2e^-$	=	Mg	−2.34
$Be^{+2} + 2e^-$	=	Be	−1.70
$Al^{+3} + 3e^-$	=	Al	−1.67
$Zn^{+2} + 2e^-$	=	Zn	−0.76
$Cr^{+3} + 3e^-$	=	Cr	−0.71
$Fe^{+2} + 2e^-$	=	Fe	−0.44
$Ni^{+2} + 2e^-$	=	Ni	−0.25
$Sn^{+2} + 2e^-$	=	Sn	−0.14
$Pb^{+2} + 2e^-$	=	Pb	−0.13
$2H^+ + 2e^-$	=	H_2	0.00 (Definition)
$Cu^{+2} + 2e^-$	=	Cu	+0.34
$Hg_2^{+2} + 2e^-$	=	2 Hg	+0.80
$Ag^+ + e^-$	=	Ag	+0.80
$Au^+ + e^-$	=	Au	+1.70

(Arrow down along left side: Increasing Strength as Oxidizing Agents; arrow up along reducing agent column: Increasing Strength as Reducing Agents)

* The equations are written as half reactions (for a whole reaction both an oxidizing agent and reducing agent are required).

dilute hydrochloric acid may be represented by the equation:

$$2\,Li + 2\,HCl \rightarrow 2\,LiCl + H_2$$

Lithium and hydrochloric acid react in the ratio of 2 gram-atomic weights (moles) to 2 moles, producing 1 mole of hydrogen gas for each 2 moles of lithium consumed. The ionic equation,

$$2\,Li + 2\,H_3O^+ + 2\,Cl^- \rightarrow 2\,Li^+ + 2\,Cl^- + 2\,H_2O + H_2$$

showing that hydrochloric acid and lithium chloride are ionic in solution, reduces to the net ionic equation:

$$2\,Li + 2\,H_3O^+ \rightarrow 2\,Li^+ + H_2 + 2\,H_2O$$

The equation shows that lithium loses an electron to the hydronium ion. Upon accepting two electrons, two hydronium ions are changed to two water moles and one molecule of hydrogen:

$$2\,H_3O^+ + 2\,e^- \rightarrow 2\,H_2O + H_2$$

The equation does not show that lithium is a solid as a reactant nor that lithium ions, a product, are hydrated. The process of oxidation which the metal undergoes acting as a reducing agent and which is represented by the equation $M_{(s)} = M^+_{(aq)} + e$ can be divided into three steps represented by these equations:

what properties of a metal affect its reducing ability?

$$M_{(s)} + \text{Sublimation energy} \rightarrow M_{(gas)}$$

$$M_{(gas)} + \text{Ionization energy} \rightarrow M^+_{(gas)} + e^-$$

$$M^+_{(gas)} + X\,H_2O \rightarrow M(H_2O)^+_X + \text{Hydration energy}$$

where $M(H_2O)^+_X$ represents $M^+_{(aq)}$. The three equations indicate that large **sublimation energies,** large **ionization energies** and small **hydration energies** cause metals to be weak reducing agents. Conversely, small ionization energies, small sublimation energies and large hydration energies cause metals to be strong reducing agents. Small ions have larger hydration energies than large ions of the same charge because the negative dipole of the water molecule can approach a small positive ion more closely. Thus lithium, with a small ionic radius and a correspondingly great hydration energy, is a stronger reducing agent than sodium despite its relatively larger ionization energy.

11.5 OCCURRENCE OF THE ACTIVE METALS

Only four of the active metals, magnesium, calcium, sodium and potassium, occur abundantly in nature (see Figure 9.3). The average abundances of Groups I and II elements in the igneous rocks of the earth are given in Table 11.6. Because of their great chemical reactivity, the elements are never found free in nature, but occur as the positive ion in deposits of salts of the metals, in silicate minerals and in sea water. Large deposits of sodium chloride are found in the remains of ancient inland seas. The percentage of sodium chloride in ocean water is small, but constitutes up to 25 to 27 per cent of isolated inland seas such as the Great Salt Lake. Salty seas and salt deposits are the major sources of sodium. Good, usable deposits of potassium are rare. Potassium salts are often mixed with chloride or sulfate salts of magnesium, which are difficult to separate. This accounts for the high price and scarcity of potassium compounds. Calcium is often found as the carbonate ($CaCO_3$) in marble, chalk, limestone, dolomite, coral and Iceland spar. It also occurs as the sulfate. Magnesium is found chiefly as the carbonate ($MgCO_3$) in mineral deposits, in dolomite, and as $MgCl_2$ in salt brine and in sea water.

TABLE 11.6 AVERAGE ABUNDANCE OF GROUPS IA AND IIA ELEMENTS OF IGNEOUS ROCK

Alkali metals	Per cent	Alkaline earth metals	Per cent
Li	0.0065	Be	0.0006
Na	2.83	Mg	2.09
K	2.59	Ca	3.63
Rb	0.031	Sr	0.03
Cs	0.0007	Ba	0.025
Fr	–	Ra	0.00000000013

11.6 PREPARATION OF THE ACTIVE METALS

Since an active metal always occurs in nature in ionic compounds, the preparation of the free metal involves the conversion of the metal ion into a metal atom by reduction. The electron or electrons originally lost in compound formation must be returned to the ion. Since these elements have the least affinity for electrons of all the elements, and cannot therefore readily take electrons from other elements, they are usually **reduced from their ions electrolytically.** Electrical energy is used to force the return of the electrons to the metal ion.

Sodium is usually prepared by the electrolysis of fused sodium chloride. The electrolysis of salt was mentioned and diagrammed in Chapter 5 in the discussion of ionic bonding. The Downs cell is commonly used for the commercial production of sodium. Fused sodium chloride (with calcium chloride added to lower the melting temperature of the solid from 800°C to 590°C) is electrolyzed in an iron pot lined with fire brick (Fig. 11.3).

A positive carbon electrode stands in the bottom, surrounded by a negative ring-shaped iron electrode. The electrodes are separated by a perforated steel divider extending from a solid steel canopy over the anode. The steel divider serves to separate the products, chlorine gas and sodium, preventing their explosive recombination. Sodium metal is formed at the surface of the cathode by this half cell reaction:

$$Na^+ + e^- \rightarrow Na\ (metal)$$

The molten metal, being less dense than the chloride melt, rises

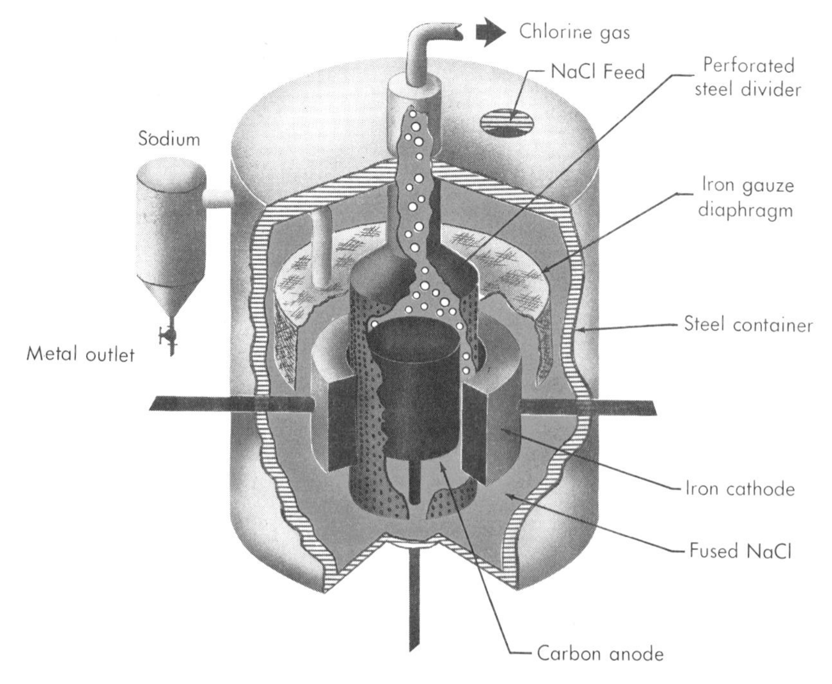

FIGURE 11.3 THE DOWNS CELL

under the iron diaphram to fill the outlet pipe, which exits to the left, and spills over into a side container. Chlorine gas collects above the anode and is guided out through the solid steel canopy and piped off to pressure tanks for shipping.

There is little demand for potassium as a metal. There are but few instances where cheaper sodium cannot be used instead of potassium. When wanted, potassium can be prepared in the same manner as sodium—by electrolysis of the molten chloride. Potassium also is obtained by heating a mixture of potassium fluoride with calcium carbide to about 1000°C in a steel reaction vessel:

$$2\,KF + CaC_2 \xrightarrow{\text{heat}} 2\,K + CaF_2 + 2C$$

Potassium is vaporized at this temperature and is collected by condensing the potassium vapor in hot paraffin.

Calcium is prepared by electrolyzing fused calcium chloride at 800°C, using a graphite anode and iron cathode. The electrolysis is started with the cathode just touching the surface of the liquid. As the calcium collects on the electrode, it is raised slowly. A metallic rod of calcium forms as an extension of the electrode as the electrode is slowly moved away from the surface of the melt.

11.7 USES OF SODIUM

Sodium is the only active metal used extensively in industry. The uses, which follow, depend upon either its chemical reactivity or physical properties.

1. A large proportion of the sodium produced (more than 20,000 tons annually) is mixed with lead to give a chemically reactive alloy of the two metals which is formed (molded) into bricks. The bricks are used in making tetraethyl lead, a gasoline additive.

2. Sodium is used as a reductant and as a catalyst in the production of numerous organic compounds, such as synthetic detergents.

sodium is used in the preparation of many organic compounds.

3. Some sodium metal is used for **heat conduction** (to abstract heat from a nuclear reactor core, for example). Molten sodium is sealed into hollow valve stems of valves to be used in airplane and truck engines. The sodium, molten at operating temperature, conducts the heat rapidly away from the valve head to aid in preventing warping.

4. Sodium and other alkali metals are used in photoelectric tubes. The light energy that is absorbed causes electrons to be released from the metal, thus causing a flow of current as the tube is irradiated.

An example of a photoelectric cell is shown in Figure 11.4. One side of the inner surface of a vacuum tube is coated with a fine film or mirror of sodium metal. The tube is so arranged in the meter that light may shine through the opposite side of the tube upon the sodium surface. The positive pole of the battery is connected through a galvanometer to a wire inside the vacuum tube. The sodium mirror is connected to the negative pole of the battery. When no light falls upon the sodium surface (in the dark), no current flows through the galvanometer. When light shines on the sodium

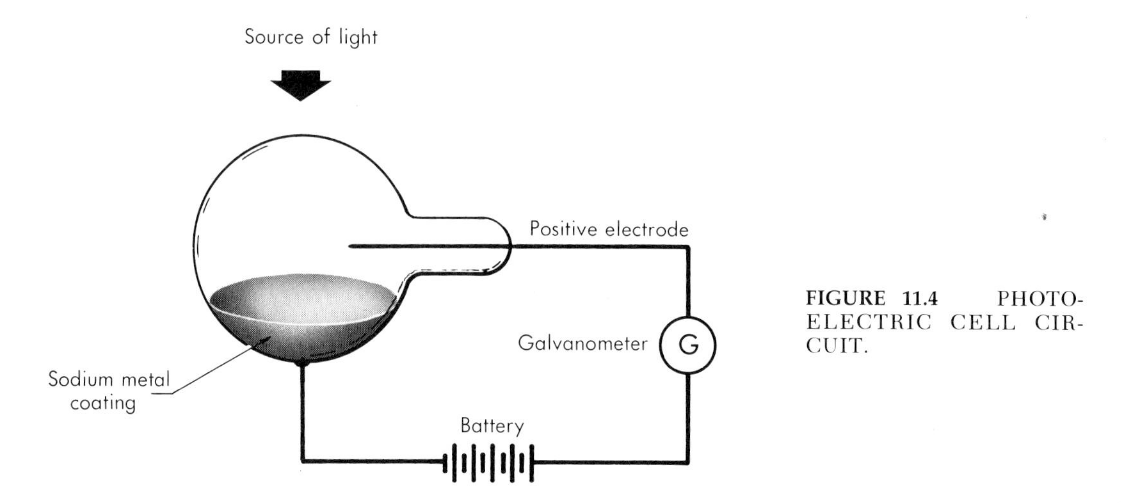

FIGURE 11.4 PHOTOELECTRIC CELL CIRCUIT.

surface, energy is absorbed, ejecting electrons. They are attracted to the positive electrode wire in the tube, completing the circuit and causing a current to flow. The number of electrons liberated, and therefore the current, is proportional to the light intensity.

5. A small quantity of sodium is used in sodium vapor lamps. The characteristic yellow light is visible at greater distances on foggy nights, and the yield of light (in lumens per watt or per dollar) is greatly increased over that of incandescent lamps.

11.8 SOME COMPOUNDS OF THE ACTIVE METALS

1. *Potassium Compounds:* These are not used in great amounts. In almost all cases, the corresponding sodium compounds are just as suitable and are much cheaper. Potassium compounds are important in agriculture. Not less than 0.1 per cent **potassium** in the soil **is necessary for** the healthy growth of the **plants.** The supply of potassium is continually being removed by the harvesting of crops and must be replenished. Some may be returned by adding plant and organic waste or ashes to the soil, but more may be added in chemical fertilizers in the form of potassium sulfate, potassium chloride or potassium nitrate. Potassium nitrate is found in deposits as a residue of the decay of animal matter.

2. *Sodium Compounds:* Several compounds of sodium are important commercially: sodium chloride ($NaCl$), sodium hydroxide ($NaOH$), sodium carbonate (Na_2CO_3), sodium bicarbonate ($NaHCO_3$) and sodium sulfate (Na_2SO_4). *Sodium chloride* is important as a seasoning for foods. Because of its availability in nature, it is the source of other sodium chemicals and for sodium itself, for hydrochloric acid and for chlorine. The electrolysis of salt under various conditions illustrates both the use of salt and the preparation of some of these compounds. (a) When anhydrous salt is melted and electrolyzed, sodium metal and chlorine gas are produced (Downs cell; Fig. 11.3), the reactions occurring at the electrodes

and the overall reaction being:

$$\begin{array}{ll} \text{Cathode:} & 2(Na^+ + e^- \rightarrow Na) \\ \text{Anode:} & 2Cl^- \rightarrow Cl_2 + 2e^- \\ \hline \text{Overall:} & 2Na^+Cl^- \rightarrow 2Na + Cl_2 \end{array}$$

(b) When water is added, making a brine solution, other products result because of a change in the cathode reaction. Sodium metal cannot be prepared in the presence of water. If it were formed, sodium would react violently with water to give hydrogen gas and sodium hydroxide. Instead, however, water reacts at the cathode giving hydrogen and the hydroxide ion directly:

$$2\,H_2O + 2e^- \rightarrow 2OH^- + H_2$$

or, including the sodium ion:

$$2Na^+ + 2H_2O + 2e^- \rightarrow 2NaOH + H_2$$

The reaction at the anode is unchanged by the addition of water. The two electrode reactions are added here to give the overall reaction.

$$\begin{array}{r} 2Na^+ + 2H_2O + 2e^- \rightarrow 2NaOH + H_2 \\ 2Cl^- \rightarrow Cl_2 + 2e^- \\ \hline 2Na^+ + 2Cl^- + 2H_2O \rightarrow 2NaOH + Cl_2 + H_2 \end{array}$$

All three products are marketed. **Sodium hydroxide** is used in large quantities in industry and in the production of various chemicals, including soap. It is recovered from the solution. **Hydrogen** and **chlorine** are collected separately, compressed in tanks, and sold. Several million tons of chlorine are produced this way every year, along with hydrogen gas for which there is a great demand in industry.

Various types of cells have been designed to produce sodium hydroxide, chlorine gas and hydrogen gas from brine. In Chapter 15 a cell that uses a mercury cathode is discussed in connection with the mercury pollution problem. In another type of electrolytic cell, the Nelson cell, shown in Figure 11.5, carbon dioxide is added, giving another product. In operation CO_2 may be bubbled into the cell during the electrolysis of brine. The carbon dioxide reacts with the sodium hydroxide formed to give $NaHCO_3$, sodium bicarbonate:

$$\begin{array}{l} 2Na^+Cl^- + 2H_2O \xrightarrow{\text{electrolysis}} 2NaOH + Cl_2 + H_2 \\ 2NaOH + 2CO_2 \longrightarrow 2NaHCO_3 \end{array}$$

The overall reaction is:

$$2NaCl + 2H_2O + 2CO_2 \xrightarrow{\text{electrolysis}} 2NaHCO_3 + Cl_2 + H_2$$

Sodium bicarbonate is baking soda. It is used medicinally and in baking powder, but the demand for it is not great. Most of the white

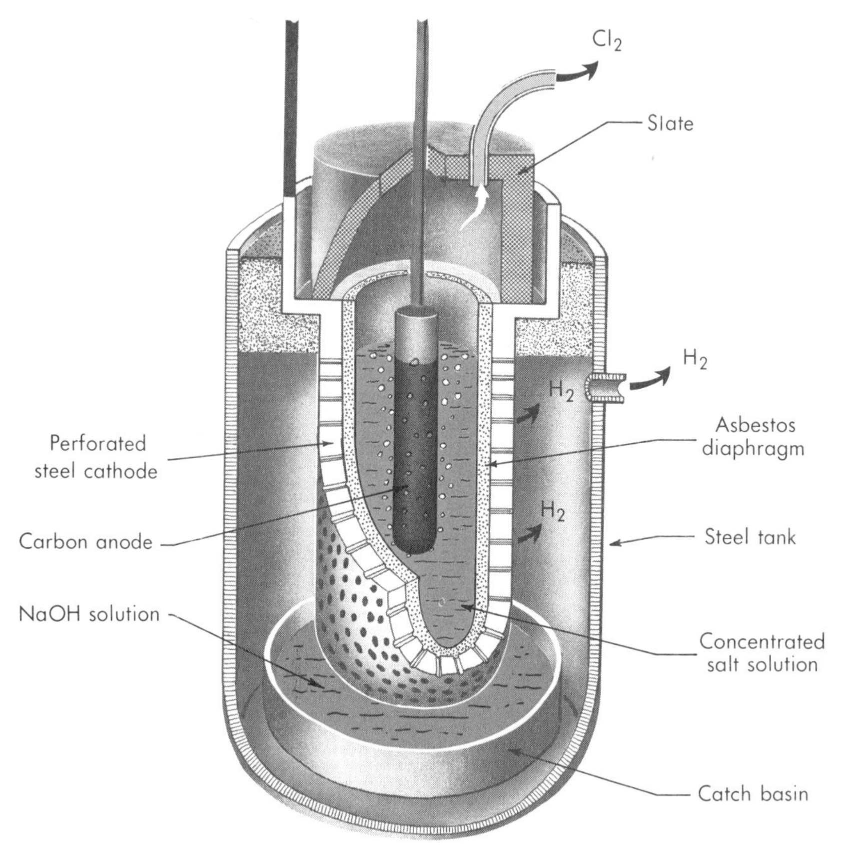

FIGURE 11.5 THE NELSON CELL.

salt is heated to form sodium carbonate:

$$2NaHCO_3 \xrightarrow{\text{heat}} \underset{\text{Sodium carbonate}}{Na_2CO_3} + H_2O + CO_2$$

The carbon dioxide liberated in the heating is added to the carbon dioxide bubbled into the electrolysis cell, decreasing waste in the process. **Sodium carbonate is washing soda.** It is used in many cleaning preparations and also in the manufacture of glass, paper, soap and many other chemicals. More than five million tons of sodium carbonate are used per year.

There are other processes for producing these compounds, although the electrolytic methods are the most important. Sodium bicarbonate and sodium carbonate are produced by the Solvay process. In this process, carbon dioxide and ammonia are bubbled into brine (concentrated salt solution):

$$NaCl + NH_3 + CO_2 + H_2O \rightarrow NaHCO_{3(s)} + NH_4Cl$$

Sodium bicarbonate precipitates because of its slight solubility in water. The crystals are collected, washed and dried. Again sodium carbonate is prepared by roasting sodium bicarbonate. The solution left behind contains ammonium chloride; it is treated with a slurry of $Ca(OH)_2$, driving off ammonia:

$$2NH_4Cl + Ca(OH)_2 \rightarrow 2NH_3 + CaCl_2 + 2H_2O$$

The ammonia is used in the original step. The $CaCl_2$ solution is dried to gain calcium chloride, a salt used on icy roads in winter time.

3. *Calcium Compounds:* The most important compound of calcium is $CaCO_3$. Calcium carbonate occurs in vast deposits as limestone. It is used in that form as a building material. It contributes to the mortar that holds the blocks together. Limestone is heated to 1100°C in a lime kiln to make quicklime by the elimination of carbon dioxide

$$\underset{\text{Limestone (calcium carbonate)}}{CaCO_3} \xrightarrow{\text{heat}} \underset{\text{Quicklime (calcium oxide)}}{CaO} + CO_2$$

(This is the source of the carbon dioxide used in the Solvay process and in the electrolytic production of sodium carbonate.)

Calcium oxide is "slaked" by the addition of water, with a great evolution of heat:

$$CaO + H_2O \rightarrow \underset{\text{Slaked lime (calcium hydroxide)}}{Ca(OH)_2} + \text{heat}$$

Mortar is made by adding 1 part **calcium hydroxide** and 3 parts **sand** with sufficient water to make a paste. Bricks or stones are laid with mortar between them. At first the mortar merely dries, and calcium hydroxide bonds the sand particles and bricks together. In time, carbon dioxide in the air diffuses into the mortar, which is quite porous, and reacts with the hydroxide, changing it to calcium carbonate. Water escapes as vapor:

$$Ca(OH)_2 + CO_2 \rightarrow CaCO_3 + H_2O$$

Calcium carbonate is harder than calcium hydroxide. The mortar becomes stronger with time. But heat will change either calcium carbonate or calcium hydroxide in the mortar to calcium oxide, which will not hold the sand particles together. Therefore, a brick wall is weakened by fire and must be destroyed.

$$Ca(OH)_2 \xrightarrow{\text{heat}} CaO + H_2O$$

$$CaCO_3 \xrightarrow{\text{heat}} CaO + CO_2$$

Gypsum is a hard shiny white crystalline material with the formula $CaSO_4 \cdot 2H_2O$. When heated, gypsum loses some of its water and becomes a white powder called "plaster of Paris":

$$\underset{\text{Gypsum}}{2CaSO_4 \cdot 2H_2O} \xrightarrow{\text{heat}} \underset{\text{Plaster of Paris}}{(CaSO_4)_2 \cdot H_2O} + 3H_2O$$

When water is added, plaster of Paris takes up water, again forming gypsum, the hard, rocky material. Plaster of Paris cannot harden without water. This is true of Portland cement, a very complicated mixture of compounds that contains calcium among other elements.

EXERCISES

11.1 List five general properties of metals.

11.2 Where are the metals in the periodic table?

11.3 Tell what is meant by malleability, ductility and elasticity.

11.4 Describe the bonding that exists between atoms in metals.

11.5 Why are metals better electrical conductors than are the solid nonmetals?

11.6 What chemical property applies to all metals?

11.7 Based upon the values in the electromotive series, are beryllium and magnesium active metals?

11.8 Are francium and radium active metals? Why were they not discussed?

11.9 Predict the melting point of francium.

11.10 List the alkali metals and the alkaline earth metals.

11.11 How many valence electrons does each alkali metal atom have?

11.12 Write the formulas for two superoxides, two peroxides and two normal oxides.

11.13 Name three metals that are less dense than water.

11.14 Write an equation for the reaction of sodium with water.

11.15 Describe how sodium metal is produced.

11.16 Compare the ionization energies of the elements of Groups I and II in the periodic table. Show how that energy is related to the size of the atoms.

11.17 Why is lithium a better reducing agent in aqueous solution than sodium?

11.18 State three uses for sodium metal.

11.19 Write equations for:

(a) the electrolysis of molten salt.
(b) the electrolysis of brine.
(c) the electrolysis of brine with carbon dioxide.

11.20 Describe the Solvay process.

11.21 Describe the preparation of mortar, starting with limestone. Describe the setting-up of mortar.

11.22 Why are elemental beryllium crystals much harder than the solid forms of the other elements of Group II?

11.23 Why should cesium have a lower melting point than sodium?

11.24 What two factors about a metal ion should affect the melting point of that metal?

11.25 How many (a) grams of sodium hydroxide, (b) liters of chlorine at STP and (c) liters of hydrogen at 10 atm and 60°C are gained by the electrolysis of 88 grams of sodium chloride in brine?

SUGGESTED READING

Ashcroft, N. W.: "Liquid Metals." Scient. Amer., *221* (1): 72, 1969.
Fern, W. O.: "Potassium," Scient. Amer., *181* (2): 16, 1949.
Gilman, H., and Eisch, J. J.: "Lithium." Scient. Amer., *208* (1): 88, 1963.
Heilbrunn, L. V.: "Calcium and Life." Scient. Amer., *184* (7): 60, 1951.
Schechter, W. H. and Kleinberg, J.: "Oxides of the Alkali and Alkaline Earth Metals." J. Chem. Educ., *24:* 302, 1947.
Schubert, J.: "Beryllium and Berylliois," Scient. Amer., *199* (2): 27, 1958.

TWELVE • SOLUTIONS

12.1 GENERAL ASPECTS

most liquids encountered in nature are solutions.

Solutions were discussed briefly in Chapter 2, in which elements and compounds were identified and described. A solution was described as a homogeneous sample of matter that is neither element nor compound. The distinction is correct, but a positive definition is needed.

A solution is a homogeneous mixture. That is, a sample taken from any part of it exhibits the same chemical and physical properties as any other sample. Salt water is a solution. All samples taken from various portions of the liquid have the same intensity of taste, the same density, the same index of refraction and the same percentage composition.

A solution is a **mixture of two or more pure substances.** A compound contains two or more elements but is not a mixture. Sometimes is is a very difficult laboratory problem to demonstrate that a solution is not a compound and vice versa. However, two general rules are obeyed by mixtures: (1) a mixture retains, to a large extent, the properties of the substances in the mixture, and (2) a slight variation in composition causes a correspondingly slight, continuous change in properties.

Salt and water mix in particular and restricted proportions to form solutions (mixtures) of salt water. An 8 per cent salt solution (by mass) is quite salty to the taste and has a density of 1.0559 g/ml. A 9 per cent solution is a bit more salty and has a density of 1.0596 g/ml. Both these solutions retain the characteristic taste of salt, and both are very similar in appearance and density to water. The density and saltiness increase steadily, with no sharp changes as more salt is added, until the limit is reached and no more salt dissolves.

12.2 TYPES OF SOLUTIONS

There are many types of solutions. Solutions may be liquid, solid or gaseous. All gas mixtures are solutions, and all gases dissolve in each other in all proportions. A gas may also dissolve in a liquid. Oxygen is slightly soluble in water and will dissolve to form a liquid solution. Two liquids often mix to form a solution, as when alcohol is poured into water. A solid may dissolve in a liquid to form a liquid solution. Solid solutions are not as well known but many occur. A gold ring is a solution of gold and copper. A silver dental filling is a solution of silver and mercury. At 300°C, hydrogen gas dissolves in palladium metal to form a solid solution.

Only **binary solutions** are discussed above. A binary solution is a homogeneous mixture of **two** substances. A solution may contain many substances, but general chemistry is concerned mostly with binary solutions, specifically binary liquid solutions, in which one of the substances is water.

The terms **solute** and **solvent** are often used in referring to the components of a binary solution. The distinction between the two is not always clear, but solvent usually refers to the substance present in greater amount or to the substance that the resultant solution resembles more closely. The solute is often considered to be dissolved in the solvent. In most cases in which water is one of the substances, *it* is termed the solvent.

12.3 CONCENTRATION

Because the proportion of solute to solvent may be varied, terms must be used in describing the solution to express that proportion. In colloquial conversation a solution is said to be "strong" or "weak." A strong solution is one containing a higher proportion of solute than usual. A weak solution has less than the usual amount of solute. Neither of these terms is precise, and each has other meanings in chemistry. In chemical language the terms that express this ratio between solute and solvent are called **concentration** terms. If a solution has more than the usual amount of solute, it is said to be **concentrated** or to have a high concentration. Solutions with a small or less than the usual proportion of solute are said to be **dilute** or of low concentration.

There are many quantitative terms to express concentration. The most convenient expression is usually chosen for each case. When a distinction is made between solvent and solute, the concentration is expressed in terms of the solute. Several concentration terms will be presented here, along with a convenient method of preparation and a description of the apparatus used in working with solutions.

1. **Volume Ratio.** This term is often used when both solvent and solute are liquids. It is also used in cooking. A syrup containing three cups of sugar and two cups of water has a volume ratio of 3 to 2. Only volume-measuring apparatus is needed to make solutions of a definite volume ratio.

2. **Weight Per Cent and Mole Per Cent.** The concentrations of many solutions are stated in **weight per cent.** An 8 per cent solution of salt has already been referred to in this chapter. An 8 per cent salt solution contains salt and water in the proportions of 8

the concentration term, mole per cent, is important in chemistry.

parts of salt to 92 parts of water by weight. Of a total of 100 parts, 8 parts are sodium chloride. A balance is used in making solutions of certain weight percentages. The weight of solvent and solute present in the solution must be known. When the concentration is stated in per cent, it is percentage by weight unless stated otherwise. Weights of solute and solvent also are needed to calculate the mole per cent of a component of a solution. A 5 mole per cent solution of sugar is one in which 5 per cent of the moles present are moles of sugar. A small quantity of a 5 mole per cent sugar solution may be made by mixing 0.05 mole of sugar and 0.95 mole of water:

$$\begin{aligned} 0.05\text{ mole} \times 342\text{ g/mole} &= 17.1\text{ g sugar} \\ 0.95\text{ mole} \times 18\text{ g/mole} &= \underline{17.1\text{ g water}} \\ & 34.2\text{ g solution} \end{aligned}$$

This solution happens to be 50 per cent by weight, but it is 5 mole per cent sugar.

An 8 per cent (by weight) salt (NaCl) solution has a mole per cent of:

$$\frac{8\text{ g}/58.5\text{ g/mole}}{8\text{ g}/58.5\text{ g/mole} + 92\text{ g}/18\text{ g/mole}} \times 100\% = 2.61\text{ mole per cent NaCl}$$

3. **Grams per Liter and Moles per Liter—Molarity.** In chemical work the concentration of many solutions is expressed in **grams of solute per liter of total solution.** A solution of sodium sulfate containing 10 g/L is of such a concentration that in 1 liter of the solution there are 10 g of Na_2SO_4. Consequently, there is not exactly 1 liter of water in 1 liter of solution but just that amount required to make the whole equal to one liter. One uses both a balance and

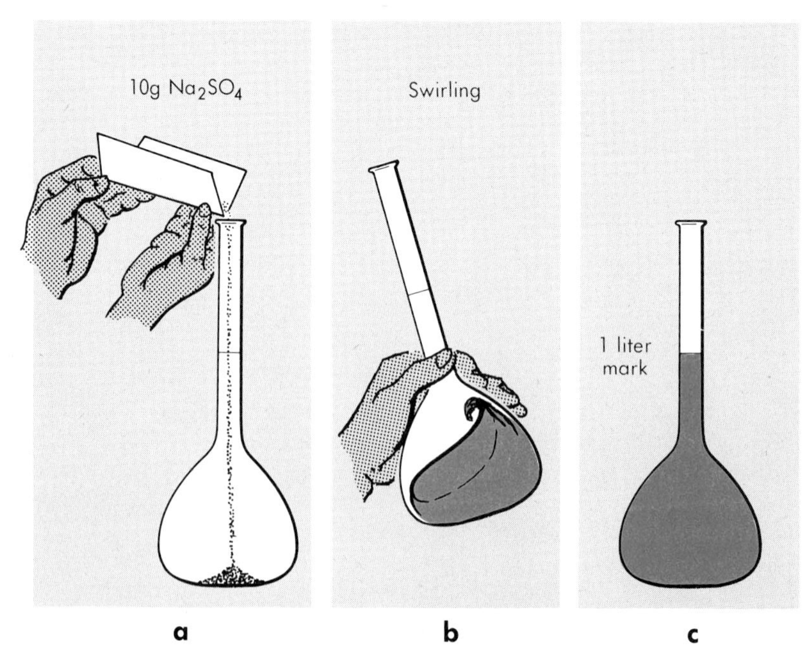

FIGURE 12.1 PREPARING A SOLUTION OF KNOWN CONCENTRATION.

a volume measure to make this solution. In making one liter of the solution, 10 g of Na_2SO_4 are weighed and added to a container capable of holding and measuring exactly 1 liter (see Figure 12.1). For precise work a volumetric flask is used. Water is added and the mixture is swirled to dissolve the salt; then more water is added until exactly 1 liter of solution is obtained. This 1 liter of solution contains 10 g of Na_2SO_4. Every portion of this solution has a concentration of 10 g/L.

The concentration term used most commonly in analytical chemistry is **molarity.** The molarity is the concentration of a solution expressed in terms of the **number of moles of solute in 1 liter of total solution.** A 5 molar solution of hydrochloric acid (5 M HCl) is a solution of such a concentration that 1 liter of it contains 5 moles of dissolved hydrogen chloride.

Again, equipment to measure both mass and volume is required to make a solution of known molarity. Suppose one were to make 1 liter of a 5 molar solution of sodium nitrate (5 M $NaNO_3$). First, 5 moles of sodium nitrate

$$5 \text{ moles} \times 85 \text{ g/mole} = 425 \text{ g}$$

are weighed on a balance. Second, the 425 g of $NaNO_3$ is transferred to a 1 liter flask, as depicted in Figure 12.2. Third, water is added to dissolve the salt, and then the flask is filled to the 1 liter mark.

Five moles of sodium nitrate are contained in 1 liter of the resulting solution; the solution is said to be 5 molar. A solution of the same concentration can be made by dissolving 2.5 moles of $NaNO_3$ in enough water to make 500 ml of solution. Molarity does not tell "how much" solution there is, **but how concentrated** it is.

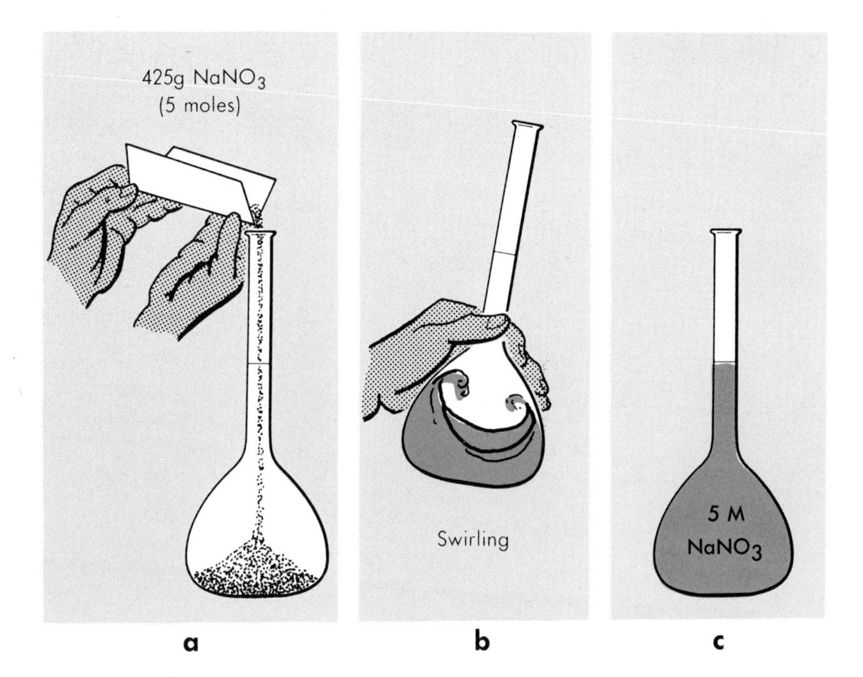

FIGURE 12.2 MAKING A 5 M $NaNO_3$ SOLUTION.

If 1 liter of solution contains 5 moles of solute, one-half liter contains 2.5 moles; one-tenth liter, 0.5 moles; but all are at the same concentration. Herein lies the value of the terminology. If 2 moles of solute are required, and there is available only a 5 molar solution, one must take 2/5 of a liter, or

$$\frac{2 \text{ moles}}{5 \text{ moles/L}} \times 1000 \text{ ml/L} = 400 \text{ ml}$$

All acids and bases used as common reagents in the laboratory are kept in an aqueous (water) solution and are used in solution. The total amount of solution added is not important, but the number of moles of solute is. When concentration is given in moles per liter, one can easily determine the number of moles present in a certain volume, as illustrated above. The concentration term *molarity* will be used extensively throughout the balance of this book.

Often dilute solutions are made from more concentrated solutions by the addition of water. For example, suppose 150 ml of 1.8 M HCl solution is required, and only 6.0 M HCl is available. The new solution will be

$$\frac{1.8}{6.0} = 0.3 \text{ as concentrated as the 6 molar solution}$$

and therefore will contain in 10 volumes the amount of HCl contained in 3 volumes of the more concentrated solution.

$$\frac{3}{10} \times 150 \text{ ml} = 45 \text{ ml}$$

$$\frac{1.8}{6.0} \times 150 \text{ ml} = 45 \text{ ml}$$

Therefore, to 45 ml of 6 M HCl is added sufficient water to make 150 ml of solution.

12.4 SOLUBILITY

Consider the change in density when salt is added to water at 20°C. The density increases gradually, as shown in Table 12.1 and Figure 12.3. The gradual change in this physical property with

TABLE 12.1 DENSITY OF SALT SOLUTIONS

Per Cent	Density g/ml
0	0.9982
1	1.0053
4	1.0268
8	1.0559
16	1.1162
24	1.1804
26	1.1972

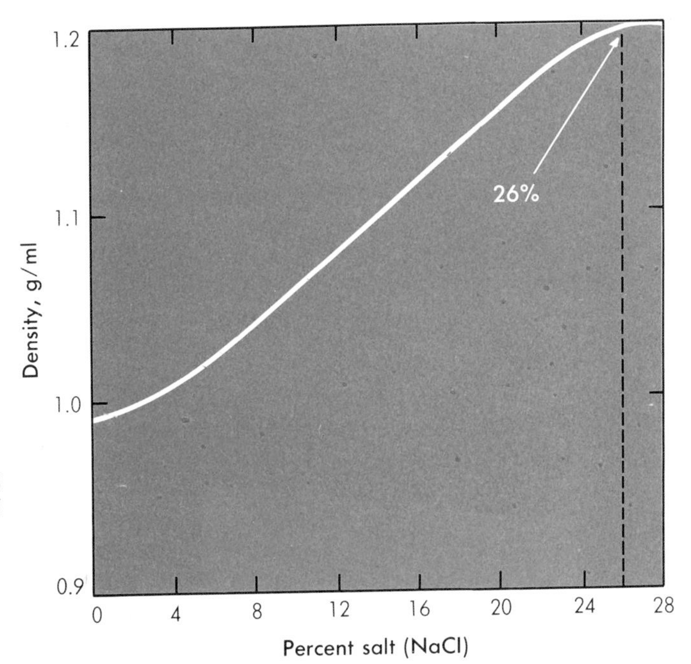

FIGURE 12.3 DENSITY VERSUS PERCENTAGE COMPOSITION FOR NaCl SOLUTIONS.

changing composition occurs as expected for a solution. However, addition of more salt beyond 26 per cent causes no change in the density of the solution. The reason is obvious to the operator, because no more salt dissolves. Neither time nor stirring can cause the excess salt to dissolve. The solution is said to be **saturated** with respect to salt. The concentration of the solute in the solution at saturation is its **solubility.** The concentration of table salt in the water is 26 per cent, or in terms of molarity the total weight of 1 liter is

$$1.197 \text{ g/ml} \times 1000 \text{ ml/L} = 1197 \text{ g/L},$$

the weight of salt in the 1 liter is

$$1197 \text{ g/L} \times 0.26 = 311.5 \text{ g/L},$$

and the number of moles per liter is

$$\frac{1197 \text{ g/L} \times 0.26}{58.5 \text{ g/mole}} = 5.4 \text{ moles/L} = 5.4 \text{ M}$$

To repeat, no more than 35 g of table salt can be dissolved in 100 g of water at 20°C. Many substances have a limited solubility in water, and in other solvents as well. Some are less soluble than salt and some are more soluble, but most of them have a limit. To understand the reason for this, one must consider the mechanism of dissolving. Suppose a crystal of sodium chloride is placed in water. It begins to dissolve immediately; ions (both positive and negative) disappear from the crystal faces into the solution. Oppositely charged ions on all sides and behind in the crystal lattice hold ions of average energy securely on the crystal face, but the energy of ions in a crystal follows a random distribution. Most ions vibrate with an energy near the average, but a few ions have much less than average energy and a few much more. The energy passes very rapidly from ion to ion by internal collisions. If an ion on the crystal surface gains sufficient energy and vibrates far enough from its position to be surrounded by polar water molecules, it may be attracted by them from the crystal. The dissolved ion is attracted by the polar water molecules that are around it, and these become oriented in such a way that opposite charges come together as shown in Figure 12.4. By this procedure ions escape into the solution continuously to become **solvated** by the solvent molecules. Ions never cease leaving the crystal, and dissolving would appear never to stop if it were not true that ions can and do return to the crystal. When a solvated ion moving about in the solution strikes the crystal face with sufficient force, the water molecules are pushed aside. If the ion rests in an exact spot left by a like ion, it will remain. Despite the "if's," this happens at a rapid and an increasing rate, as the concentration of ions in the solution increases with dissolving. Eventually the rate of ion return (**crystallization**) equals the rate of dissolving, and the concentration of ions in the solution remains constant. A **dynamic equilibrium** exists at **saturation.**

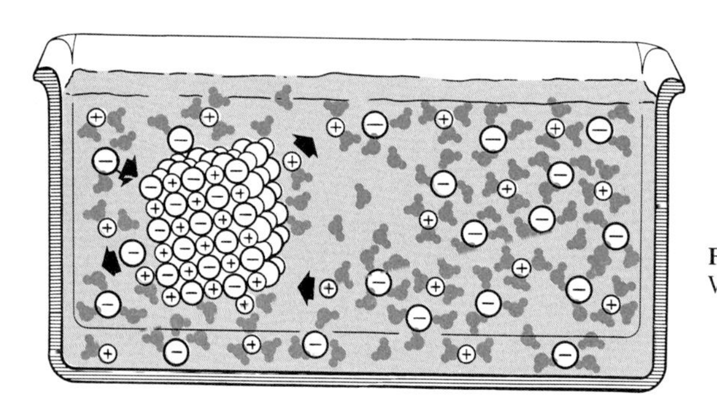

FIGURE 12.4 SOLVATION OF IONS WITH WATER.

There is experimental evidence that ions are leaving and returning to the faces of a crystal in a saturated solution. On standing, small crystals in a saturated solution grow larger and consolidate, while the total mass of the crystals remains constant.

12.5 TEMPERATURE EFFECT ON SOLUBILITY

Neither stirring nor powdering a solute will increase its solubility. Both aid in speedy attainment of saturation but cannot increase the total amount of solute that dissolves. Stirring brings the crystal faces in contact with a less concentrated portion of the solution and thus reduces the amount of recrystallization occurring during dissolving. Powdering the solute increases the dissolving area, but likewise increases the recrystallization area.

A rise in temperature increases both the vibration of ions in the crystal and the movement of solvated ions in the solution, and therefore it increases both the rate of dissolving and the rate of crystallization. However, the two rates are seldom increased to the

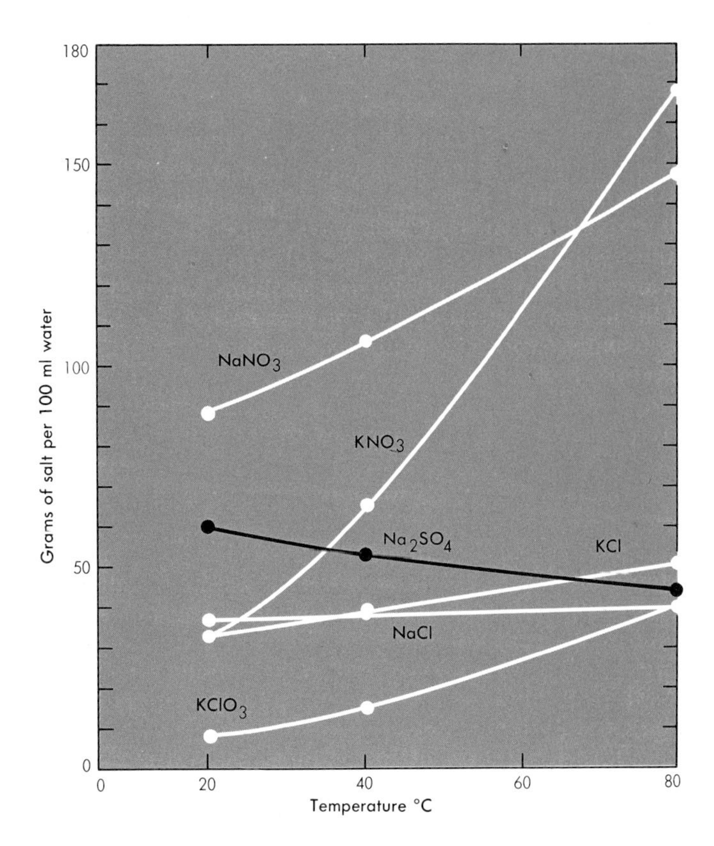

FIGURE 12.5 TEMPERATURE-SOLUBILITY CURVES FOR VARIOUS SALTS IN WATER.

same extent. If the rate of dissolving is increased more by the increase in temperature, the solubility increases. If the rate of crystallization is increased more, the solubility is decreased.

The solubility of a group of salts in grams of solute per 100 grams of water at several temperatures is given in Table 12.2 and is shown graphically in Figure 12.5.

In general the solubility of a salt increases with an increase in temperature, but there are notable exceptions. Five of the six salts listed in Table 12.2 show an increasing solubility with increasing temperature. The solubility of table salt (NaCl) changes very little. The solubility of sodium sulfate (Na_2SO_4) decreases.

TABLE 12.2 SOLUBILITIES OF SEVERAL SALTS AT SEVERAL TEMPERATURES IN g/100 ml OF WATER AND THEIR RESPECTIVE HEATS OF SOLUTION

Salt	Solubility at Temperature °C 20	40	80	Heat of Solution cal/gram
KCl	33	39	50	+60
KNO_3	33	65	168	+84
$KClO_3$	8	15	40	+83
NaCl	37	38	39	+22
$NaNO_3$	88	106	147	+59
Na_2SO_4	*60	53	44	− 3.88

* Metastable

12.6 HEATS OF SOLUTION

Heat effects often accompany the preparation of a solution. When 20 g of sodium thiosulfate crystals are dissolved in 50 g of water with stirring, the temperature of the 70 g of solution drops 10 to 12C°. In contrast, when 20 g of sodium hydroxide pellets are dissolved in 50 g of water with agitation, the temperature rises about 50C°. Heat is absorbed when sodium thiosulfate is added to water. Heat is liberated when sodium hydroxide is added to water. The heat absorbed or liberated is called the **heat of solution,** and is usually expressed as the *molar* heat of solution (the heat change involved when 1 mole of solute is dissolved). The heat of solution varies from solute to solute, changes with the concentration of the solution, and is proportional to the amount of solute dissolved. The molar heat of solution of sodium hydroxide in water is −10,300 calories (exothermic); for sodium thiosulfate it is +11,370 calories (endothermic).

The dissolving of most salts is endothermic. This is understandable because ions in a crystal are held tightly by the opposite charges of adjacent ions. Each ion that escapes must gain sufficient energy to overcome the forces binding it in the lattice. As discussed previously, only the highest energy molecules are able to escape. As they escape, the average energy per particle (the temperature of the crystal) decreases.

The separation of ions from a crystal is not the only process occurring when a salt dissolves. Indeed the solid would not dissolve without **solvation.** The surrounding of the ion by strongly attracting polar water molecules aids in the dissolution of the salt. Just as energy is absorbed when ions are pulled from attracting particles in a lattice, energy is liberated when an ion becomes bound to a group of water molecules. The heat of solution is the difference between the binding energy of the lattice and the binding energy of the solvent molecules. If the energy required to overcome the lattice forces is greater than that regained in solvation, the heat of solution is positive (endothermic). If the energy regained in solvation is greater, the dissolving is exothermic.

An interesting relationship exists between the heats of solution and the change of solubility with temperature. A glance at Table 12.2 reveals that those **salts** whose **solubility increases with temperature have a positive heat of solution,** that is, the dissolving is endothermic. The one salt, sodium sulfate, whose solubility decreases with temperature, has a negative (exothermic) heat of solution. Furthermore, the greater the heat of solution, the greater the change of solubility with temperature. This relationship holds for the group of salts in Table 12.2.

if a solute that has a negative heat of solution were more soluble at higher temperatures, there would be no limit to its solubility. dissolving would raise the temperature; this would dissolve more solute, which would raise the temperature still higher, and so on.

12.7 SUPERSATURATION

Observe in Table 12.2 that potassium nitrate (KNO_3) is one of the salts with an increased solubility at higher temperatures. The solubility at various temperatures is

33 g in 100 g water at 20°C
64 g in 100 g water at 40°C
167 g in 100 g water at 80°C

If 64 g of potassium nitrate are added to 100 g of water at 80°C, it will dissolve quickly and completely. Now suppose the solution is at room temperature or is allowed to cool to room temperature. One would expect that when the temperature reached and then dropped just below 40°C, crystals would appear. However, often this does not happen. If the cooling solution is kept very still, the temperature may drop to near room temperature before **any** crystals appear. Below 40°C, the solution is more than saturated; it is **supersaturated.** Agitation of the solution or scratching the inside of the container usually initiates crystallization; adding a crystal of the solute invariably does. When crystallization does start, it proceeds rapidly, almost instantaneously, until the concentration of the solution is reduced to the solubility limit. The supersaturated solution then becomes a saturated solution.

Saturation exists when there is a dynamic equilibrium between the solution and the solid crystals with respect to the transfer of

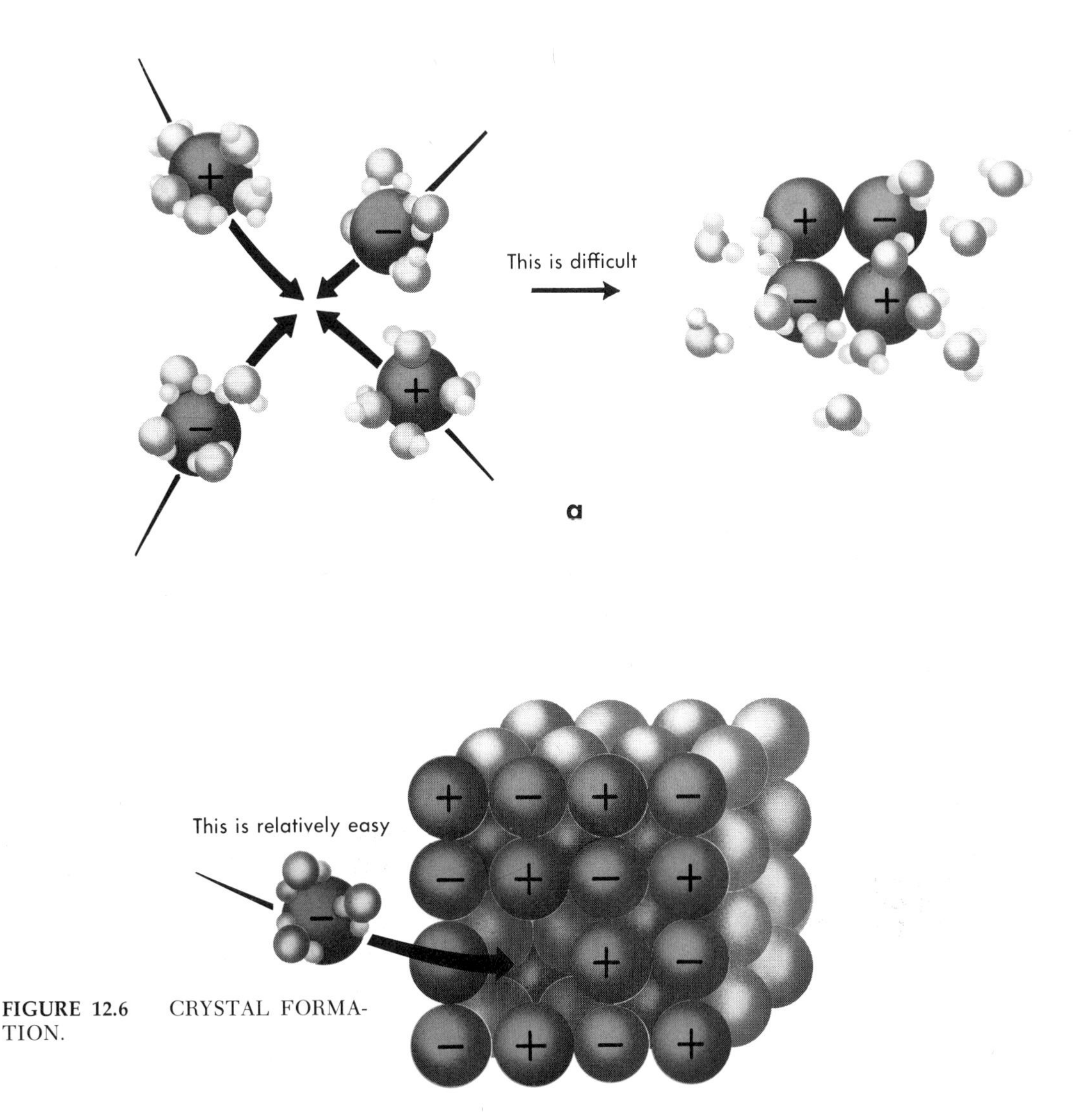

FIGURE 12.6 CRYSTAL FORMATION.

solute ions or molecules from one to the other. In a supersaturated solution, with not one crystal of solute present, no such equilibrium is possible. It appears that the solvated ions experience difficulty in forming the first closely packed unit of the crystal, as shown in Figure 12.6a. Once the lattice does appear, either by chance, by agitation, or by the addition of a small crystal, crystallization can and does occur. Figure 12.6b illustrates the relative ease with which ions may add to the highly attracting rigid crystal face. Equilibrium is soon attained, and the solution is reduced to saturation.

12.8 COLLIGATIVE PROPERTIES OF SOLUTIONS —LOWERING OF THE VAPOR PRESSURE

The mere presence of **any solute** in a solution modifies the physical properties of the solvent in the solution. The effects are of practical and theoretical interest. In this chapter we shall consider solutions in which the solute is a non-volatile non-electrolyte. Those properties discussed below that depend upon the number of particles of solute in a given mass of solvent are called **colligative** properties.

a non-electrolyte is a substance that does not separate into ions when dissolved in water.

The vapor pressure of water above a solution is always less than the vapor pressure of pure water at the same temperature. This may be demonstrated by placing two beakers, one containing water and one containing a sugar solution, in a vacuum desiccator (Fig. 12.7a). After the beakers stand for a time, the volume of pure solvent in the beaker on the left has decreased, while the volume of solution in the beaker on the right has increased by an equal amount. Eventually all the solvent will be transferred to the solution (Fig. 12.7b). The vapor pressure of a solution can be measured with a manometer in the same manner that the vapor pressure of a pure liquid is measured. In many solutions of non-volatile solutes the decrease in vapor pressure is found to be proportional to the concentration of the solute.

The vapor pressure of a solution containing 1 mole of sugar in 1000 g of water has a vapor pressure about 1/55.5 less than that of pure water at the same temperature. A solution containing 2

a

b

FIGURE 12.7 TRANSFER OF SOLVENT TO A SOLUTION.

moles of sugar in 1000 g of water has a vapor pressure about 2/55.5 less than water, and so on.

12.9 DECREASE IN VAPOR PRESSURE AND ELEVATION OF THE BOILING POINT

The boiling point is defined as that temperature at which the vapor pressure is 760 mm. It is obvious that if a solution has a lower vapor pressure than the pure solvent at the same temperature, it will have a higher boiling point. A sugar solution has a lower vapor pressure than pure water at 100°C and must be heated to a temperature higher than 100°C to exert a vapor pressure of 760 mm. The increase in the boiling point by addition of a solute is proportional to the concentration of solute in the solution. A solution containing 1 mole of a non-volatile solute in 1000 g of water boils at 100.52°C. The **elevation** in **boiling point** therefore is **0.52C°.** A solution containing only 1/2 mole of solute per 1000 g of water boils 0.26C° above normal, or at 100.26°C.

12.10 DEPRESSION OF THE FREEZING POINT

It is common knowledge that ice melts when salt is thrown upon it, even though the temperature remains below 0°C. It also is known that a mixture of milk, cream, sugar and flavoring (ice cream) will not freeze in ice water but will freeze when surrounded by a mixture of ice, salt and water. Similarly the addition of a solute known as "antifreeze" can lower the freezing point of water as much as 60F° (33C°). All of these are examples of a depression of the freezing point of water by the addition of a solute. Again, the freezing point depression is proportional to the concentration of the solution. A solution in water of such a concentration that **1 mole of solute** is contained in **1000 g** of water freezes at **−1.86°C**; the freezing point is depressed 1.86C°. One half mole of sugar in 1000 g of water forms a solution which freezes at −0.93°C.

12.11 OSMOTIC PRESSURE

A simple experiment may be devised to observe the phenomenon of osmosis (Fig. 12.8). A carrot is hollowed out from the top and filled with a concentrated sugar syrup. A cork bearing a glass tube is plugged tightly into the hole. The carrot is then placed in water. Presently the solution can be seen rising in the glass tube, while the water level in the beaker falls slowly. The solution will rise several meters in the glass tube. This passage of water from the solvent through a **semipermeable membrane** into a solution is known as **osmosis.** The carrot is a semipermeable membrane. It is semipermeable or partly permeable, because water can permeate it, but the solute (sugar) cannot. In osmosis, solvent passes through a semipermeable membrane from a solvent to a solution, or from one solution to another solution of greater concentration.

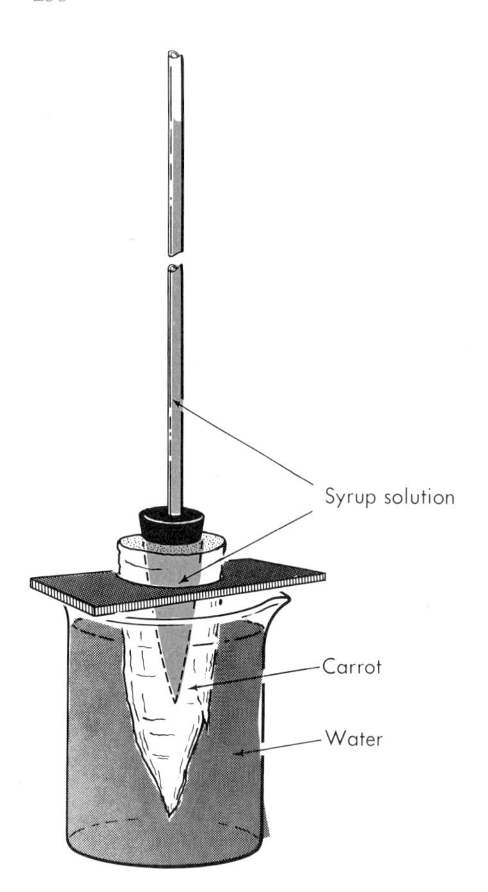

FIGURE 12.8 AN OSMOSIS EXPERIMENT.

Osmosis and osmotic pressure might be more easily visualized by considering the cell shown in Figure 12.9. A solution of solute in water is separated from water (solvent) by a thin membrane that water alone can pass through. Osmosis occurs immediately, and water flows through the membrane into the solution. The change in volumes may be arrested by forcing a piston down on the solution. At a certain force per square inch (pressure) the flow stops. If the pressure is increased, the flow of solvent is reversed. The pressure that is required above the solution to stop the osmosis is called the osmotic pressure.

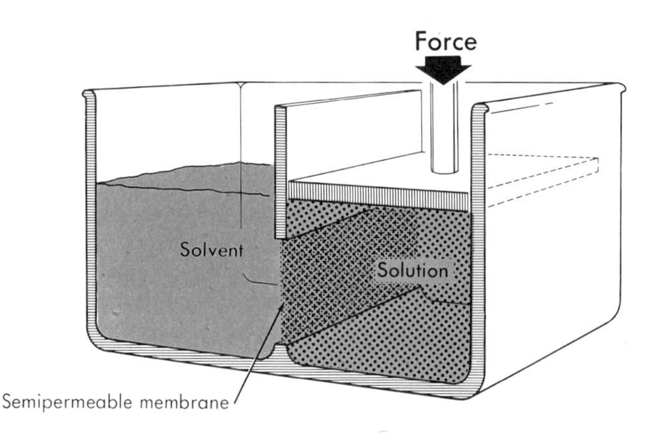

FIGURE 12.9 ILLUSTRATION OF OSMOTIC PRESSURE.

A solution containing **1 mole of sugar,** or another non-ionizing solute, in **1000 g of water** has an osmotic pressure of **24.4 atm at 25°C.** A solution containing only 1/2 that amount of solute in 1000 g of water has an osmotic pressure of 12.2 atm at 25°C. Osmotic pressure increases with temperature, but for simplicity it will be calculated in this text only at 25°C. Plant roots gain moisture from the soil by osmosis, and osmosis is involved in the distribution of liquid in the animal body.

Osmotic pressures as high as 250 atmospheres have been measured. The membrane in this case is a specially prepared porous ceramic cup, similar to a flower pot, and the solution is very concentrated because the osmotic pressure of a solution is proportional to the concentration.

12.12 THE INTERRELATEDNESS OF THESE FOUR COLLIGATIVE PROPERTIES

colligate is defined "to relate by some reasonable explanation." the colligative properties are related through their dependence upon the reduced escaping tendency of solvent molecules.

Lowering of the vapor pressure and elevation of the boiling point are, of course, directly related. Vapor pressure depression, freezing point depression and osmotic pressure **depend upon a decreased tendency for molecules of a solvent to leave** the liquid when a solute is present.

The vapor pressure of a solvent is lowered when a solute is added because non-volatile solute molecules decrease the rate of evaporation by occupying positions at the surface of a solution from which solvent molecules could be escaping. Correspondingly, a **smaller rate of return** of molecules to the liquid is required to establish equilibrium, as illustrated in Figure 12.10a, b.

The freezing point of a solvent or solution is that temperature at which there is an equilibrium between the passage of molecules from solid to liquid and vice versa. Figure 12.11 shows the condition that exists between a solvent and its solid at the freezing point. When a solute is present, the rate of molecules leaving the solution

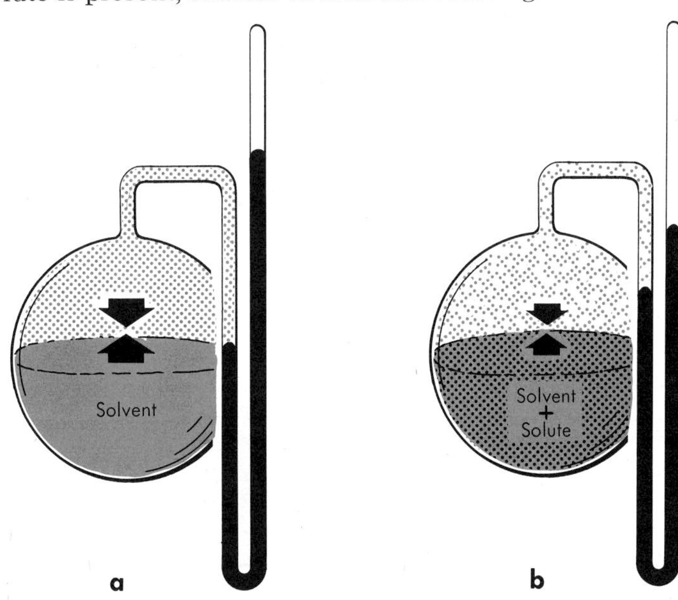

FIGURE 12.10 A COMPARISON OF THE VAPOR PRESSURES OF A SOLVENT AND ITS SOLUTION.

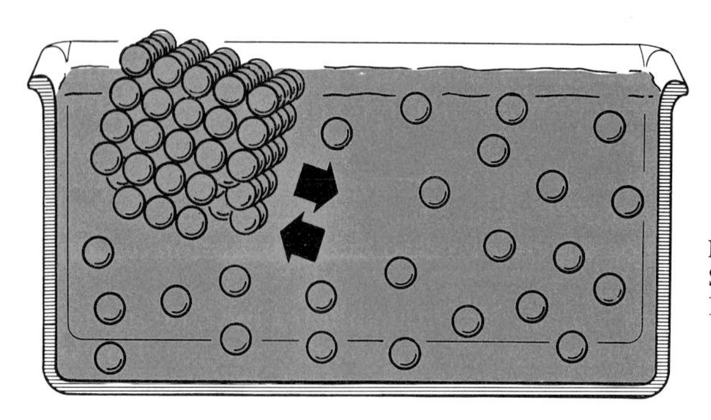

FIGURE 12.11 EQUILIBRIUM BETWEEN SOLID AND LIQUID AT THE FREEZING POINT.

to return to the solvent crystal is decreased because fewer solvent molecules are in the liquid surface about the crystal. In order that equilibrium be reestablished, the temperature must be decreased, which decreases the energy of vibration and decreases the number of water molecules with enough energy to leave the solid.

Osmosis occurs, and osmotic pressure exists, only because molecules of solvent leave the pure solvent more rapidly than they leave a solution. In Figure 12.12a, a semipermeable membrane is

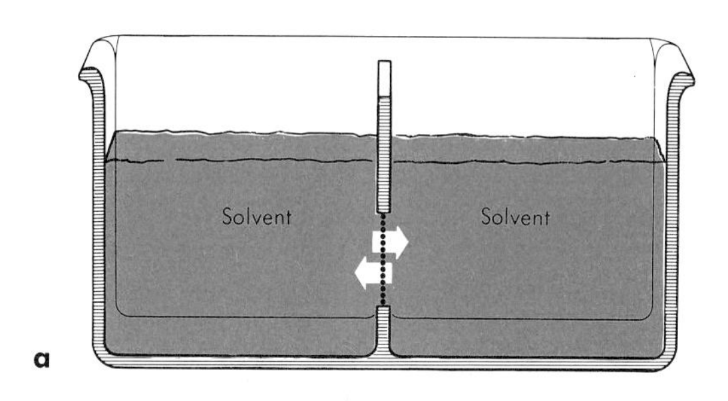

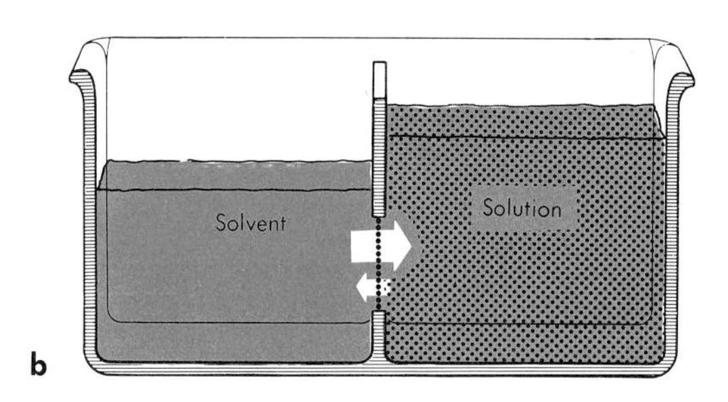

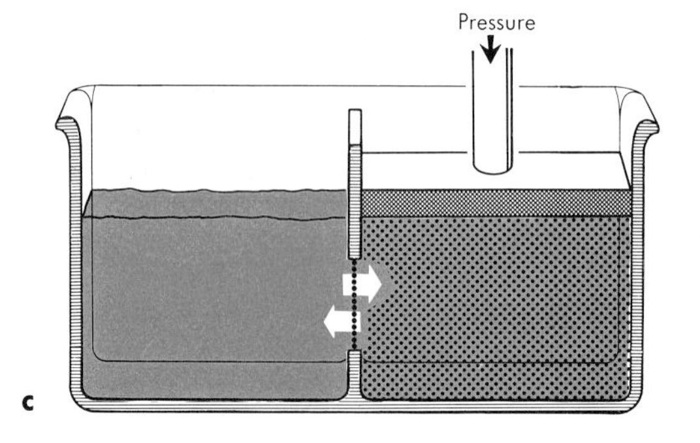

FIGURE 12.12 OSMOTIC PRESSURE RESTORES EQUILIBRIUM.

shown separating two compartments of pure solvent. Solvent molecules move through the membrane at an equal rate in both directions, and osmosis is not observed. In Figure 12.12b, the membrane separates solvent from solution. Fewer solvent molecules occupy positions against the membrane in the solution side, and therefore fewer molecules pass from solution to solvent than from solvent to solution. Osmosis results. As shown in Figure 12.12c, sufficient pressure has been applied to the solution to increase the rate at which molecules leave the solution to the rate at which they return. This pressure is the osmotic pressure.

From this pictorial representation of theory, one can see that all these colligative properties should be proportional to the concentration of the solute. Further, the magnitude of the changes depends solely on the proportion of solute particles present. One mole of any non-volatile, non-ionizing solute should be as effective as a mole of any other in a fixed volume of solvent.

12.13 CALCULATIONS

Two common calculations are associated with the colligative properties of solutions:

1. From the molecular weight of the solute and weight proportion of solute to solvent one can **calculate the anticipated change in properties.**

2. From the weight proportions of solute and solvent and the actual measured property one can **calculate the molecular weight of the solute.**

The value of each colligative property is proportional to the concentration. For example, a solute of such concentration that 1 mole of solute is contained in 1000 g of water (1) will have a vapor pressure 1/55.5 less than that of the solvent at the same temperature, (2) will boil at 100.52°C (at 1 atm pressure), (3) will freeze at −1.86°C and (4) has an osmotic pressure of 24.4 atm at 25°C.

Example 1. Calculate the four colligative properties of a solution containing 27 g of glucose ($C_6H_{12}O_6$) in 100 g of water.

Solution: There are 27 g glucose per 100 g water. This corresponds to

$$\frac{27\ \text{g}}{100\ \text{g water}} \times 1000\ \text{g} = 270\ \text{g glucose in 1000 g water}$$

This 270 g is:

$$\frac{270\ \text{g}}{180\ \text{g/mole}} = 1.5\ \text{moles of glucose}$$

a. The *vapor pressure* at 25°C is:

$$23.8\ \text{mm} - 1.5\ \text{mole} \times \frac{1}{55.5\ \text{mole}} \times 23.8\ \text{mm} = 23.2\ \text{mm}$$

b. The *boiling point* is:

$$100°C + 0.52°C/mole \times 1.5\ mole = 100.78°C$$

c. The *freezing point* is:

$$0°C - 1.5\ mole \times 1.86°C/mole = -2.79°C$$

d. The *osmotic pressure* at 25°C is:

$$1.5\ mole \times 24.4\ atm/mole = 36.6\ atm$$

Example 2. A solution containing 6.0 g of a non-volatile non-electrolyte in 150 g of water freezes at −1.00°C. Calculate the molecular weight of the solute.

Solution: With 1000 g of water there would be:

$$6.0 \times \frac{1000\ g}{150\ g} = 40\ g\ of\ solute$$

But this 40 g lowers the freezing point by only 1.00C°, so

$$40\ g \times \frac{1.86C°/mole}{1.00C°} = 74.4\ g/mole$$

would lower the freezing point to −1.86°C and is therefore 1 mole.

12.14 SOLUTIONS OF LIQUIDS IN LIQUIDS, MISCIBILITY AND IMMISCIBILITY

Alcohol and water together in any proportion enter one phase and are said to be completely miscible. Other pairs of liquids such as oil and water are immiscible; that is, they are almost totally insoluble in each other and form two phases.

Oils and greases are soluble in carbon tetrachloride, benzene and gasoline. For this reason, these solvents have been used as dry cleaning fluids (non-water fluids) for removing oily and greasy spots. Because of the toxicity of carbon tetrachloride and because of the flammability of benzene and gasoline, special non-toxic and non-flammable solvents now are used for this purpose. A dry cleaning agent must be miscible with oils and must be volatile so that the cleaning fluid remaining in the fabric after cleaning evaporates readily, leaving the fabric "dry."

ethyl alcohol is completely miscible in water; some alcohols are not.

Laboratory experience has shown that gasoline, carbon tetrachloride and benzene (Group I) are immiscible with water. Methyl alcohol, ethyl alcohol and acetone (Group II) are all completely soluble. **The liquids within each group and often liquids from both**

groups are miscible. Liquids may be classed in three groups. The liquids listed above represent Groups I and II; water may be called "Group III." All the liquids in Group I contain non-polar molecules. Molecules of Group II are moderately polar. Water molecules are highly polar. Non-polar liquids tend to dissolve in non-polar liquids; polar liquids dissolve in polar liquids. Non-polar liquids and highly polar liquids are immiscible. If the attraction between like and unlike molecules is nearly the same, the two liquids will mix. If attractions are very different, solubility is limited. Liquids of Group II are sufficiently polar to dissolve in water, but not polar enough to be immiscible with non-polar liquids.

Consider water and carbon tetrachloride. When poured together, the denser carbon tetrachloride settles in a layer at the bottom. Even after vigorous shaking, the liquids separate completely. The highly polar water molecules tend to orient themselves positive end to negative end and to draw themselves tightly together, excluding the carbon tetrachloride molecules.

In summary, if the molecules of both liquids are polar, or if the molecules of both liquids are non-polar, they are likely to be miscible. However, if one liquid contains polar molecules and the other non-polar molecules, they are likely to be immiscible. Figure 12.13a represents polar molecules of two liquids. The molecules of the liquids attract each other and therefore remain together. Figure 12.13b represents non-polar molecules of two liquids; there is no distinction between the attraction of the molecules, and they remain together. In Figure 12.13c polar molecules are drawn to-

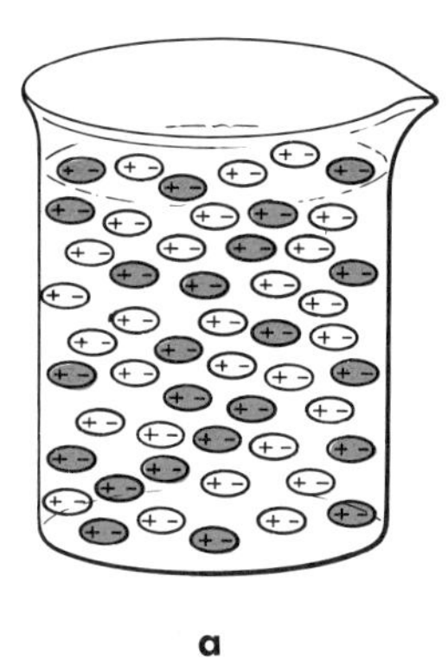

a

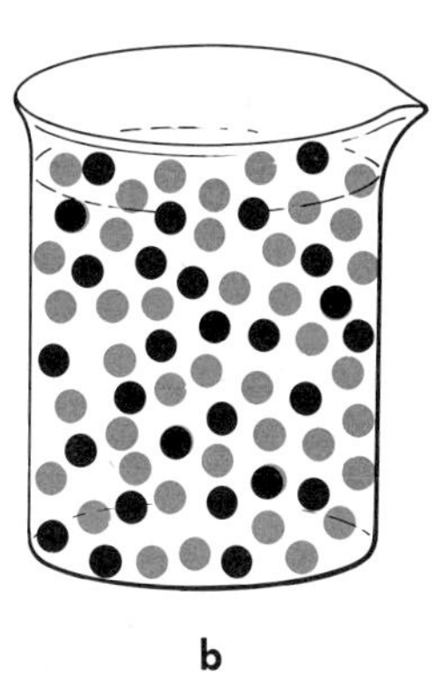

b

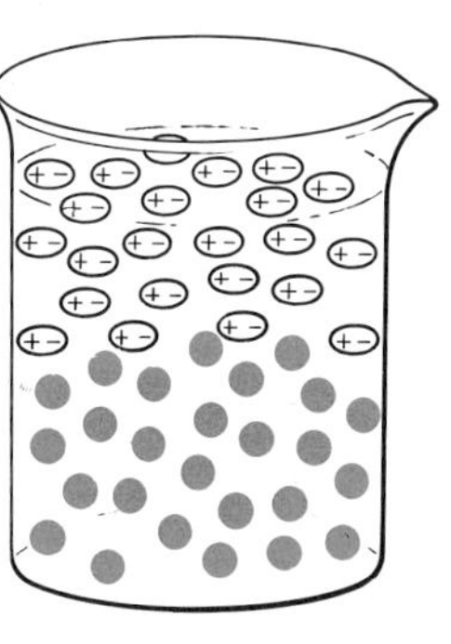

c

FIGURE 12.13 POLARITY OF MOLECULES VERSUS MISCIBILITY.

gether by their attraction for one another, "squeezing out" the nonpolar molecules. The two liquids remain separate, the less dense liquid forming a layer above the other.

12.15 COLLIGATIVE PROPERTIES

The vapor pressure of a solution of two liquids might be predicted if one were to consider each of them to be the solvent, with the normal capability of lowering vapor pressure, and then add the two vapor pressures together. Consider a 50 mole per cent solution of ethyl alcohol in water. Molecules in the solution might be represented as in Figure 12.14 (even though real molecules are not lined up so neatly). One half the molecules near the surface are water molecules. There is one half the number of water molecules in position to escape that there would be if the liquid were pure water. It follows that the vapor pressure of water above the alcohol-water solution should be one half that of pure water. This agrees with Raoult's Law, which states that the partial pressure of a liquid vapor over a solution is equal to the product of the mole fraction of the liquid in the solution and the vapor pressure of the pure liquid. By the same argument, the pressure of ethyl alcohol vapor above the solution should be one half the vapor pressure of pure alcohol.

As a result, the total vapor pressure of the solution should be intermediate between the vapor pressures of the two pure liquids. This should be reflected in the boiling point of the solution, which should be intermediate between the boiling points of the two liquids. Water boils at 100°C and ethyl alcohol at 78.5°C, both at 1 atm pressure. A solution containing water and alcohol in the weight ratio of 18 to 46 respectively (a 50 mole per cent solution) has a boiling point of 79.8°C. The boiling point of the solution lies between those of the two liquids, but it is nearer the boiling point of alcohol.

It should be noted that since the alcohol has a higher vapor pressure above the 50 mole per cent water solution than does water, the vapor will contain a higher proportion of alcohol. While the weight per cent of alcohol in the solution is 72, the vapor is 84

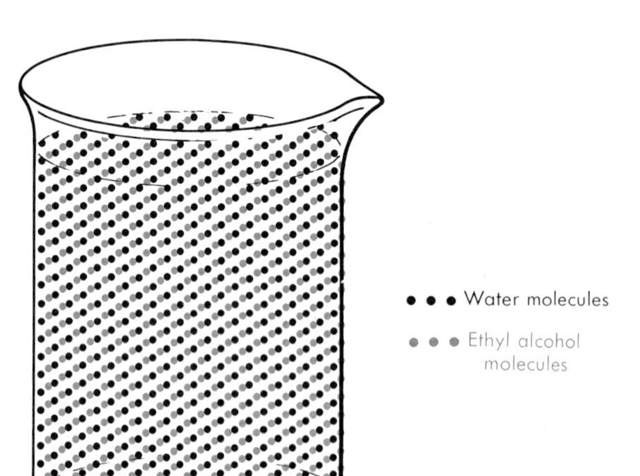

FIGURE 12.14 DISTRIBUTION OF ALCOHOL WITH WATER IN A 50 MOLE PER CENT SOLUTION.

a 95 per cent by weight solution of ethyl alcohol in water actually boils at a temperature below that of pure alcohol because of an unusually great attraction between molecules of the two liquids.

per cent alcohol. Since the vapor is more concentrated in the more volatile component than in the liquid, a more concentrated alcohol solution can be gained by vaporizing a portion of the alcohol-water solution and then condensing that portion. The process can be repeated with the condensate, achieving more separation. This is done in fractionating columns in the alcohol industry. Repeated distillations take place as the vapor moves up the column and the aqueous condensate moves down. Gasoline is separated from less volatile petroleum products by the same process.

Liquid solutes have the same effect on the freezing points of liquid solvents as do solid solutes. Methyl alcohol and ethylene glycol (permanent antifreeze) reduce the freezing point of water to the same extent, mole for mole, as sugar.

12.16 SOLUTIONS OF GASES IN LIQUIDS

Solutions of gases in liquids are fully as important as liquid-liquid solutions. All aquatic animal life depends on oxygen dissolved in water. Fish gain oxygen for metabolism from water. Water in an aquarium must be continually replaced with fresh water, or air must be bubbled into the water to maintain the oxygen supply. Many cold beverages contain carbon dioxide dissolved in water, which imparts a crisp, tart taste. Carbon dioxide bubbles from the drink when the cap is removed.

A solution of a gas in water is easily prepared; the gas needs merely to come in contact with the liquid. The dissolving of oxygen in water can be hastened by increasing the contact of the two by spraying water in air.

Both temperature and pressure affect the solubility of gases. Raising the temperature increases the motion of gas molecules and tends to expel them. Air can be effectively removed from water by boiling. The escaping steam carries the air molecules from the solution.

The amount of gas that can be dissolved in a liquid is proportional to the pressure of the gas above the liquid. This is a statement of Henry's Law. The greater the pressure, the greater the amount of gas that will dissolve. Carbon dioxide is dissolved in cold drinks, and the bottles are capped at about 5 atm pressure. When the pressure of carbon dioxide above the liquid is reduced to 1 atm by removing the cap, the gas escapes by bubbling.

These rules do not apply when a chemical reaction or special chemical bonding occurs between the solvent and the gas. Hydro-

TABLE 12.3 SOLUBILITY OF GASES IN WATER IN MOLES PER LITER OF SOLUTION AT 1 atm AND 0°C

Gas	Solubility
Nitrogen	1.05×10^{-3}
Oxygen	1.3×10^{-4}
Hydrogen	9.6×10^{-4}
Carbon Monoxide	1.57×10^{-3}
Hydrogen Sulfide	0.208
Ammonia	11
Hydrogen Chloride	12

chloric acid and an ammonia solution are both solutions of gas in water. Here bonds are formed by chemical reaction, or hydration. The solubilities of ammonia and hydrogen chloride are several moles per liter, as compared to a small fraction of a mole per liter for many other gases, as shown in Table 12.3.

EXERCISES

12.1 Define (a) homogeneous, (b) mixture, (c) solution, (d) solute, (e) solvent.

12.2 Distinguish between a compound and a solution on the basis of their properties and properties of their components.

12.3 Describe two examples of each: (a) gaseous solution, (b) liquid solution, (c) solid solution.

12.4 What is a binary solution?

12.5 What is concentration?

12.6 Consider a 3 M solution of NaCl.

a. Calculate the grams of solute per liter of solution.
b. Describe steps in making 1 liter of the solution.
c. Describe how 500 ml of the solution are made.

12.7 Water is added to 200 ml of a solution of a certain molarity until the total volume is 600 ml. What is the molarity of the final solution as compared to the molarity of the original solution?

12.8 Define (a) saturated solution, (b) solubility, (c) dynamic equilibrium, (d) solvated molecule.

12.9 How does a change in temperature affect the solubility of salts? How is this effect related to the heat of solution?

12.10 Describe a method for preparing a supersaturated solution.

12.11 List four colligative properties.

12.12 Name at least two ways in which all four colligative properties are related.

12.13 Describe osmosis.

12.14 A solution of sugar in water has a vapor pressure 1/55.5 less than that of pure water at the same temperature. What is the (a) boiling point, (b) freezing point and (c) osmotic pressure of the solution at 25°C?

PROBLEMS

12.15 How many moles

a. of oxalic acid ($H_2C_2O_4$) is 54 g?
b. of sodium chlorate ($NaClO_3$) is 100 g?
c. of sulfuric acid are contained in 238 g of a 20 per cent solution?
d. of $CuSO_4 \cdot 5\ H_2O$ is 500 g?

12.16 How many moles of potassium chloride are there in 255 ml of a 6.5 M solution? How many grams?

12.17 How many ml of 0.75 M $Ba(NO_3)_2$ solution may be made (a) with 1.35 moles of $Ba(NO_3)_2$? (b) with 450 g of $Ba(NO_3)_2$?

12.18 How many grams of H_3PO_4 are there in 360 ml of 3.7 M H_3PO_4?

12.19 How many ml of 0.75 M $BaCl_2$ solution must be taken to obtain 50 g of $BaCl_2$?

12.20 Which one of these solutions contains the most $CaCl_2$?

a. 1000 ml of 0.35 M $CaCl_2$
b. 300 ml of 1.2 M $CaCl_2$
c. 300 g of 10 per cent solution of $CaCl_2$

12.21 Which solution of H_2SO_4 is more concentrated, a 1 M solution or one containing 10 g of H_2SO_4 per liter?

12.22 200 ml of 3.4 M HCl is added to 100 ml of 1 M HCl. Calculate the concentration (in moles per liter) of the final solution.

12.23 Of the three solutions, made as follows, which solution is most dilute?

a. 150 g NaCl is dissolved in sufficient water to make 1 L of solution.
b. 1 mole NaCl dissolved in sufficient water to make 300 ml of solution.
c. 300 ml of 3.5 M NaCl and sufficient water to make 400 ml of solution.

12.24 35 g of HCl is added to 500 ml of 3.5 M HCl. Assume that the volume remains 500 ml. What is the final concentration?

12.25 What is the vapor pressure at 25°C, the osmotic pressure at 25°C, the freezing point and the boiling point for each of the solutions described below? In each case the solute is non-volatile and a non-electrolyte. (The water vapor pressure table is on page 137.)

a. 1000 g of water + 1 mole of solute
b. 500 g of water + 1 mole of solute
c. 326 g of water + 1/2 mole of solute
d. 3420 g of water + 3.21 moles of solute

12.26 Urea is a solid that does not ionize. The molecular weight of urea is 60. Calculate the four properties listed in Problem 12.25 for the following solutions.

a. 500 g of water + 60 g urea
b. 100 g of water + 3.0 g urea
c. 320 g water + 7.8 g urea

12.27 A solution containing 500 g of water and 25 g of solute boils at 100.52°C at 1 atm.

a. What is the apparent molecular weight of the solute?
b. What is the vapor pressure of the solution at 100°C?
c. What is the freezing point of the solution?
d. What osmotic pressure will the solution exert at 25°C?

12.28 What is the molecular weight of the solute in each of the following cases (assume a non-volatile, non-electrolyte):

a. 24 g in 360 g of water freezes at −1.06°C
b. 13.5 g in 500 g of water boils at 100.20°C
c. 15 g in 750 g of water has a vapor pressure of 750 mm at 100°C
d. 18.4 g in 520 g of water has an osmotic pressure of 16 atm.

12.29 Calculate the mole per cent of alcohol in a solution made by adding 100 ml of ethyl alcohol (C_2H_5OH) (density = 0.7 g/ml) to 100 ml of water.

12.30 Calculate the freezing point of a solution containing 22.5 g of glucose ($C_6H_{12}O_6$) in 500 g of water.

12.31 Which of these aqueous glucose ($C_6H_{12}O_6$) solutions begins to freeze first when all are cooled together: (a) 50 per cent by weight, (b) 9.1 mole per cent, (c) 2.78 M?

SUGGESTED READING

Bailar, J. C., Jr., Moeller, T. and Kleinberg, J.: *University Chemistry.* D. C. Heath Co., Boston, 1965. Chapter 10.
Sienko, M. J. and Plane, R. A.: *Chemistry,* 3rd Ed. McGraw-Hill Book Co., New York, 1966. Chapter 10.
Sisler, H. H.: *Chemistry in Non-aqueous Solvents.* Reinhold Publishing Corp., New York, 1961 (paperback).
Snyder, A. E.: "Desalting Water by Freezing." Scient. Amer., *207* (6):41, 1962.

THIRTEEN • IONS IN SOLUTION: ACIDS, BASES AND SALTS

13.1 THE THEORY OF ELECTROLYTIC DISSOCIATION

The structure of atoms has already been discussed (Chapter 4). One who understands the electrical nature of atomic structure can visualize the electrical imbalance that occurs in atoms when electrons are transferred from atom to atom in a chemical reaction. Ions (charged atoms) exist in pure substances. It might be assumed that ions exist in water solutions of the same substances.

Arrhenius's Theory of Electrolytic Dissociation states "Upon addition to water, certain solutes are separated into positive and negative fragments." This is an elementary concept, but one must remember that Arrhenius suggested his theory in 1887 before much was known of atomic structure and before either the electron or the proton were known.

a theory depends upon agreement of observations and laws derived from observations.

The quantitative observations that led to the Theory of Electrolytic Dissociation are fourfold. Solutions of certain substances exhibit all four of the effects to be discussed in Sections 13.2 to 13.5.

13.2 ABNORMAL COLLIGATIVE PROPERTIES

Some of the first experiments with colligative properties were carried out by Cavendish (the discoverer of hydrogen), who studied the freezing points of solutions and found the depression to be proportional to the concentration of the solute. Little more was done until F. M. Raoult, in an extensive series of experiments in 1881, discovered the laws of freezing point depression, vapor pressure depression and boiling point elevation that were introduced in the previous chapter to calculate the magnitude of the changes in properties.

However, Raoult found numerous exceptions to his laws. While many solutes behaved as predicted, many others had a much greater effect on the freezing and boiling points than he had calculated.

For example, a solution containing table salt in the ratio of 1 mole to 1000 grams of water froze at −3.72°C and boiled at 101.04°C, the change in these properties being just double that calculated. It happened that all deviations from expected behavior were positive; a solute either changed the properties as predicted or changed them more than predicted. For other solutes the changes often were twice, triple or quadruple the value calculated.

13.3 ELECTRICAL CONDUCTIVITY

The same solutes that imparted an enhanced effect to the colligative properties of water were found to convert water into an electrolytic conductor. Electrolysis experiments had been carried out with solutions of these solutes since about 1800. At first, interest centered on the fact that when wires from the two poles of a battery were placed in a solution of one of these solutes, water was decomposed and hydrogen and oxygen gases were liberated (Fig. 13.1). Later, questions arose as to the means by which the current was carried through the solution. Because their solutions carried an electric current, these solutes were called **electrolytes.** This name will be used throughout the remainder of the chapter.

13.4 RAPID REACTION RATES

It was observed that reactions between electrolytes in solution occurred instantaneously. Often two electrolytes react slowly (if

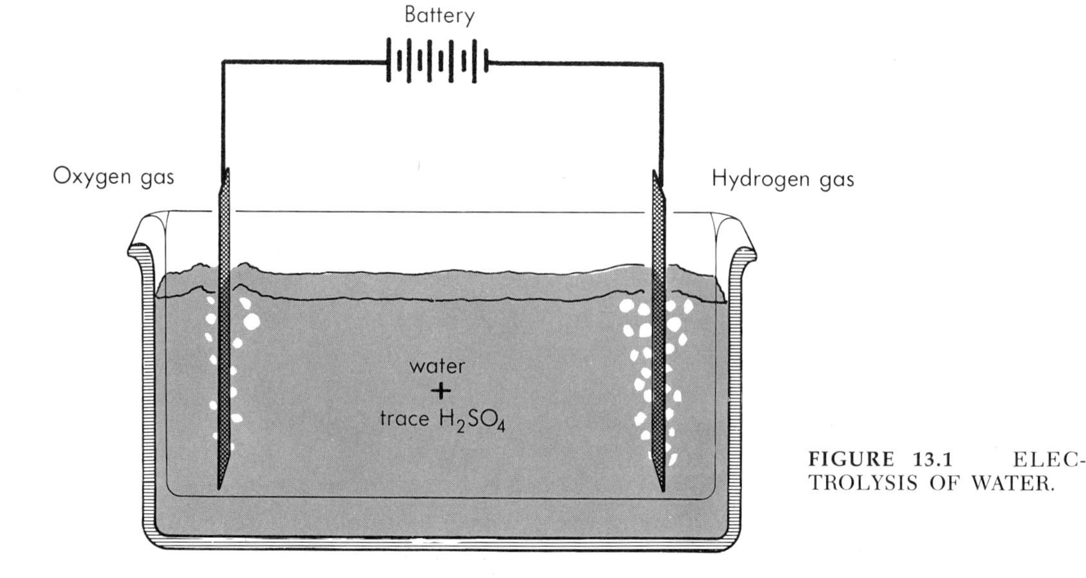

FIGURE 13.1 ELECTROLYSIS OF WATER.

at all) when mixed in the solid state, but immediately if the electrolytes are first dissolved in water. This is contrasted with reactions of solutions of non-electrolytes, which might progress for hours before the reaction is complete. For example, silver nitrate crystals and sodium chloride crystals may be placed together. If they are kept perfectly dry, one can separate the crystals after an indefinite length of time with tweezers and magnifying glass. The solids do not react. However, when salt water is added to a silver nitrate solution, a white precipitate of silver chloride forms immediately, almost completely depleting one or both of the reactants:

$$\underset{\text{In solution}}{AgNO_3} + \underset{\text{In solution}}{NaCl} \rightarrow \underset{\text{White precipitate}}{AgCl} + \underset{\text{In solution}}{NaNO_3}$$

Solutions of hydrochloric acid and sodium bicarbonate react immediately to release carbon dioxide gas:

$$\underset{\text{In solution}}{HCl} + \underset{\text{In solution}}{NaHCO_3} \rightarrow \underset{\text{Gas}}{CO_2} + H_2O + \underset{\text{In solution}}{NaCl}$$

Bubbling begins upon mixing and ends abruptly at the completion of the reaction.

Consider the rate of reactions of solutions of a non-electrolyte, ethyl alcohol, and a weak electrolyte, acetic acid, to produce ethyl acetate:

$$\underset{\text{Acetic acid}}{HC_2H_3O_2} + \underset{\text{Ethyl alcohol}}{C_2H_6O} \rightarrow \underset{\text{Ethyl acetate}}{C_4H_8O_2} + H_2O$$

Analyses performed at various time intervals after the two solutions are mixed reveal that several weeks elapse before the maximum amount of ethyl acetate is formed without a catalyst.

13.5 OTHER PROPERTIES CHARACTERISTIC OF FRAGMENTS OF THE ELECTROLYTE

Solutions of the electrolytes copper(II) sulfate, copper(II) chloride, copper(II) nitrate and copper(II) bromide are shades of blue, bluish green, or green, the color depending somewhat on the concentration. Solutions of sodium sulfate, sodium chloride, sodium acetate, sodium oxalate and sodium nitrate are colorless, as are solutions of sodium chloride, ammonium chloride, potassium chloride and calcium chloride. In Table 13.1 these and other solutions are placed in groups according to similarities in color or some other distinguishing property.

The table indicates that copper salts, in solution, are blue or green, while chloride ion is colorless, chromate compounds are yellow, acids are sour and hydroxides are slick to the touch and taste bitter. The properties are associated with only one part of the solute. For example, in a copper(II) sulfate solution, the copper ion alone is responsible for the blue color.

The fact that solutions containing copper compounds are blue

TABLE 13.1 GROUPING OF SOME ELECTROLYTES BY A COMMON PROPERTY

$CuSO_4$ $Cu(NO_3)_2$ $CuCl_2$ $CuBr_2$	Blue or green	Na_2SO_4 $NaCl$ $NaOAc$ $Na_2C_2O_4$	Colorless	$NaCl$ NH_4Cl KCl $CaCl_2$	Colorless
$(NH_4)_2CrO_4$ K_2CrO_4 Na_2CrO_4	Yellow	HCl HNO_3 H_2SO_4 H_3PO_4	Sour taste	$NaOH$ $Ca(OH)_2$ $Ba(OH)_2$	Slippery feel, bitter taste

many chemicals are poisonous; do not taste chemicals.

and that the blue color is contributed by copper enables the student to interpret an observation concerning conduction by solutions of electrolytes.

When a crystal of hydrated copper(II) sulfate ($CuSO_4 \cdot 5\ H_2O$) is placed in water, the blue color diffuses slowly throughout the solution. If a container with 2 or 3 liters of solution is used, the diffusion may take hours, as indicated in Figure 13.2a. When wires from a battery are placed in the solution containing the crystal, the positive wire near the crystal and the negative wire on the opposite side of the container, a circuit is formed and a current flows. While conduction occurs, the blue color of copper diffuses more rapidly toward the negative electrode, as shown in Figure 13.2b.

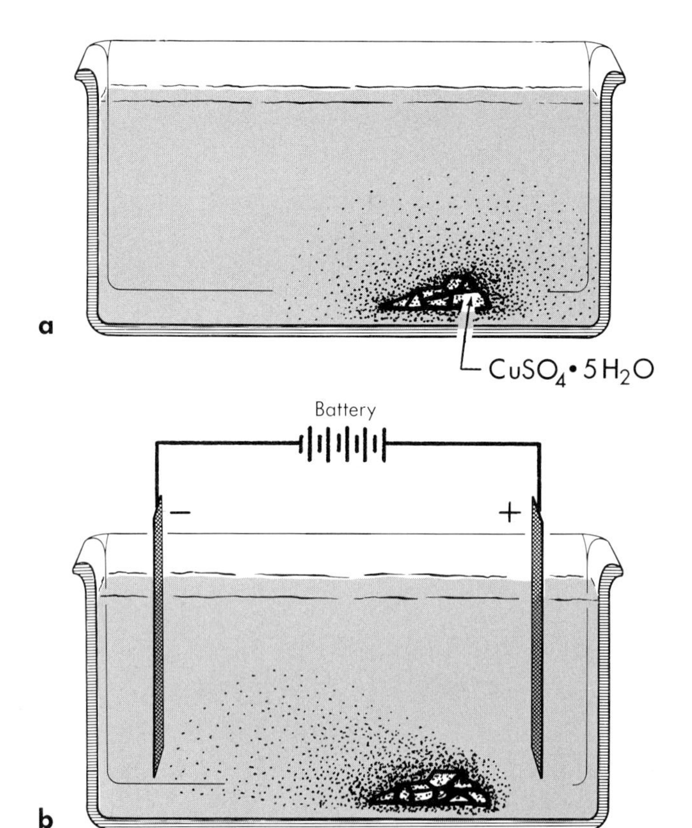

FIGURE 13.2 SIMPLE DIFFUSION AND ELECTRODIFFUSION OF COPPER(II) SULFATE IN WATER.

It should be emphasized that the **electrolytes,** the solutes which convert water into a conductor of electricity, are: (1) the **same** substances whose solutions have abnormal colligative properties, (2) the **same** substances which act so rapidly in solution and (3) the **compounds** that contribute properties to the solution characteristic of parts of their molecules. Solutions of non-electrolytes (1) do not conduct an electric current, (2) freeze at the temperature predicted, (3) may react slowly and (4) act as a whole unit in contributing properties to a solution.

13.6 AGREEMENT OF THEORY AND FACT

Svante Arrhenius, as a college student, visualized a model that explained the facts to him. This picture (theory) is embodied in the statement already quoted, "Upon addition to water, electrolytes are separated into positive and negative fragments." These fragments are now called ions.

A. If the theory were true, an **electric current might be carried** by the movement of the charged fragments through the solution. Positively charged ions can be drawn toward the negative electrode and negatively charged ions toward the positive electrode. The movement of the blue in copper(II) sulfate solution toward the negative electrode might be explained if the copper fragment is the positive ion.

B. If electrolytes are separated into positive and negative ions upon addition to water, the **colligative properties** of the **resulting solution should be more pronounced** than those of a solution of a non-electrolyte of the same concentration. The colligative properties of solutions of non-electrolytes were found to depend on the concentration of molecules, regardless of their size. A solution containing 60 g of urea in 1000 g of water and a solution containing 342 g of sugar in 1000 g of water both have the same osmotic pressure, the same freezing point, the same boiling point and the same vapor pressure. Assuming that charged particles are neither more nor less effective than molecules, one mole (58.5 g) of the simple electrolyte sodium chloride should change the colligative properties of 1000 g of water just twice as much as 342 g of sugar, because there are twice as many particles (1 mole of sodium ions and 1 mole of chloride ions).

C. If electrolytes are separated into ions when added to water, they should **react very rapidly in solution.** Reaction between electrolytes may be merely a pairing of oppositely charged ions. Consider the reaction of silver nitrate and sodium chloride in solution.

$$AgNO_3 + NaCl \rightarrow AgCl_{(s)} + NaNO_3$$

Silver ions and chloride ions join and precipitate. Such ions must bear opposite charges. Assuming that the charge on the silver ion is positive, the equation becomes:

$$Ag^+ + Cl^- \rightarrow AgCl_{(s)}$$

Nothing happens to the sodium and nitrate fragments. They were ions in solution before reaction and they remain as such afterwards. No compound is decomposed. Oppositely charged silver and chloride ions need only find one another as they move rapidly about in the solution, which is very rich in both of them initially.

D. If an electrolyte is separated into ions when dissolved in water, **properties of the solution will be characteristic of portions of the electrolyte.** A copper(II) nitrate solution, a copper(II) sulfate solution and a copper(II) chloride solution all contain the same copper fragment, the same copper(II) ion (Cu^{2+}), and all have nearly the same blue color. All acid solutions contain hydrogen ion, all are sour and all conduct electricity. The sour taste of the solution must be a property of the hydrogen fragment. If electrolytes are separated into fragments in solution, the solution should have the properties of those fragments.

E. If agreement with this fourfold evidence were insufficient to justify acceptance of the theory of electrolytic dissociation, one might consider the **reactions that occur at the electrodes in electrolytic conduction.** Brine, a solution of table salt, conducts an electric current. A cell for the electrolysis of a brine solution is diagrammed in Figure 13.3. Hydrogen gas bubbles form around the negative electrode. Chlorine gas bubbles up around the positive electrode, while a solution of sodium hydroxide is concentrated about the negative electrode.

Assuming that the electrolyte, sodium chloride, is divided into positive sodium ions and negative chloride ions dispersed among water molecules, chloride ions are drawn to the positive electrode to lose electrons to it and become neutral chlorine gas molecules.

$$2\,Cl^- \rightarrow Cl_2 + 2\,e^-$$

Sodium ions are drawn to the negative electrode, however, where electrons are accepted by water molecules:

$$2\,H_2O + 2\,e^- \rightarrow H_2 + 2\,OH^-$$

leaving hydroxide ions in solution with the sodium ions. Electrons released by chloride ions at one electrode and accepted by water molecules at the other electrode must flow through the external

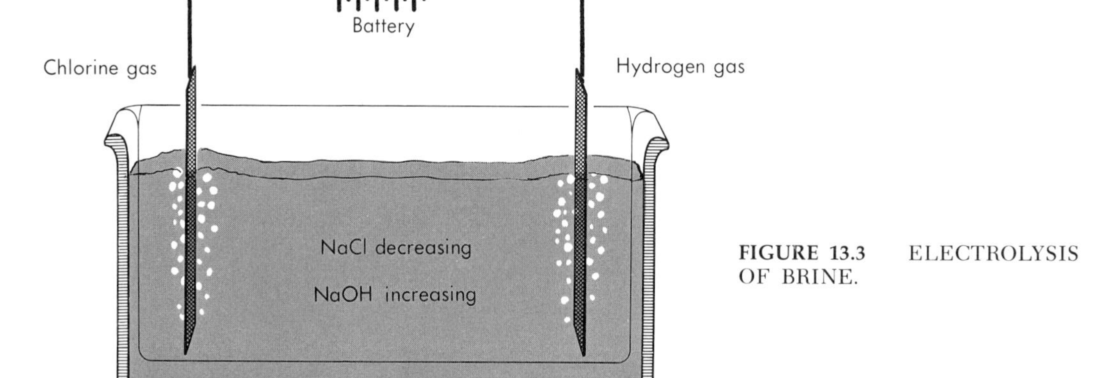

FIGURE 13.3 ELECTROLYSIS OF BRINE.

circuit, making an electric current. How could one explain the electrolytic process without ions?

The ionic theory explains the phenomena discussed above and many others. In fact, as it has been amended and expanded, it is in agreement with all known evidence.

One **minor amendment of Arrhenius's Theory** should be noted. Arrhenius suggested that electrolytes are "separated into ions" by the action of the water into which they are dissolved. This is true of some strong electrolytes, namely the acids, among which is hydrochloric acid. In a dilute solution, hydrogen chloride reacts with water according to this equation:

$$HCl_{(g)} + H_2O \xrightarrow{100\%} H_3O^+ + Cl^-$$

Neither hydronium ions nor chloride ions existed until the two molecular substances were brought together. Hydrogen chloride is **ionized literally** when dissolved. **All strong acids** exhibit covalent bonding when pure and are **ionized when dissolved in water.**

not all strong electrolytes can be made to form an aqueous solution that is a good conductor of electricity. some are not soluble enough. regardless, all the strong electrolyte in solution is in the form of ions.

Another class of strong electrolytes is the salts, compounds that are ionic in the pure state. Table salt is one of these. Sodium ions and chloride ions are associated in an orderly arrangement in a salt crystal. When table salt is dissolved in water, the ions are separated, the orderly association is broken and salt is **dissociated.** A crystalline salt is already ionized and therefore cannot be ionized when dissolved.

13.7 WEAK ELECTROLYTES

The preceding discussion dealt with strong electrolytes, substances which are completely ionized in solution. Hydrogen chloride, sodium chloride and ammonium sulfate are strong electrolytes. The first two compounds have a double effect and the third compound has a triple effect on the colligative properties of their solutions as compared with solutions of non-electrolytes.

There are other substances whose solutions have colligative properties of a magnitude just a **little** greater than those of non-electrolytes, and which conduct an electric current poorly. These substances are called weak electrolytes. They are covalent substances when pure and are not completely separated into ions by water; only a small fraction is ionized at one time. Some substances slightly ionized in solution are acetic acid ($HC_2H_3O_2$), ammonia (NH_3), oxalic acid ($H_2C_2O_4$) and hydrogen sulfide (H_2S). Some of the most common are listed in Table 13.2.

13.8 TYPES OF IONIC REACTIONS

A great many reactions of electrolytes in solution are known. These reactions may be divided conveniently, though somewhat arbitrarily, into five classes.

A. Some reactions consist merely of the **"pairing" of ions to**

TABLE 13.2 NET IONIC EQUATIONS

In Net Ionic Equations

1. Strong electrolytes in solution are written as ions.
2. Precipitates are written as molecules.
3. Covalent substances (including gases) are written as molecules.
4. Weak electrolytes are written as molecules.
5. Complex ions are written as such.
6. Ions undergoing no change, appearing exactly the same on the left and the right side of the arrow, are removed from the equation.

Solubilities of Ionic Compounds

1. Most alkali metal and ammonium compounds are soluble ($KClO_4$ is slightly soluble).
2. All nitrates are soluble.
3. All acetates are soluble ($AgC_2H_3O_2$ is slightly soluble).
4. All sulfates are soluble except $BaSO_4$, $PbSO_4$ and Hg_2SO_4. $CaSO_4$ and $SrSO_4$ are slightly soluble.
5. All chlorides, bromides and iodides are soluble except those of Ag^+, Pb^{2+}, Hg_2^{2+}, and Tl^+.
6. All hydroxides, phosphates, carbonates and sulfides are insoluble except those of the alkali metals. $Ba(OH)_2$ is moderately soluble.

Strong Electrolytes Include

1. All ionic substances except the halides and cyanides of Hg, Cd and Zn.
2. The strong acids are $HClO_4$, HNO_3, HBr, HI, H_2SO_4, HCl.

Weak Electrolytes Include

1. The weak acids H_2O, HCN, H_2CO_3, $H_2C_2O_4$, HF, HNO_2, HOAc, H_2S, HOCl, H_2SO_3 and H_3BO_3.
2. The weak base, NH_3.

form a precipitate. The silver nitrate–sodium chloride reaction has already been cited as one of these. The reaction occurring when barium chloride solution is mixed with a solution of washing soda (Na_2CO_3) is another example. The overall equation for the reaction is:

$$BaCl_2 + Na_2CO_3 \rightarrow BaCO_{3(s)} + 2\ NaCl$$

If dissolved, strong electrolyte substances are written as ions and the equation becomes:

$$Ba^{2+} + 2\ Cl^- + 2\ Na^+ + CO_3^{2-} \rightarrow BaCO_{3(s)} + 2\ Na^+ + 2\ Cl^-$$

All that has occurred is:

$$^*(i)\ Ba^{2+} + CO_3^{2-} \rightarrow BaCO_3$$

The sodium and chloride ions remain in solution.

* Equations marked with small Roman numerals are discussed more fully later in this chapter.

To some the equation may indicate that neutral pairs of barium and carbonate ions are formed or perhaps that neutral molecules result. This is not true. Actually, barium and carbonate ions are adding one-by-one to ionic crystals from a very supersaturated solution of the salt formed when the solutions containing the ions are mixed.

B. Some ions in **solution combine to form,** among other products, **a gas,** which escapes.

1. Sodium bicarbonate solution reacts with a hydrochloric acid solution to form a gas:

$$HCl + NaHCO_3 \rightarrow NaCl + H_2O + CO_{2(g)}$$

Again writing ions in solution as ions, the equation is:

$$H_3O^+ + Cl^- + Na^+ + HCO_3^- \rightarrow Na^+ + Cl^- + 2\ H_2O + CO_{2(g)}$$

When non-participating ions are eliminated, the equation becomes:

$$(ii)\ H_3O^+ + HCO_3^- \rightarrow 2\ H_2O + CO_{2(g)}$$

2. Ammonia gas may be prepared by mixing solutions of ammonium chloride and sodium hydroxide.

$$NH_4Cl + NaOH \rightarrow NaCl + NH_3 + H_2O$$

Writing ions in solution as ions, the equation becomes:

$$NH_4^+ + Cl^- + Na^+ + OH^- \rightarrow Na^+ + Cl^- + NH_3 + H_2O$$

All that occurs may be represented by:

$$(iii)\ NH_4^+ + OH^- \rightarrow NH_3 + H_2O$$

C. Some ions in solution pair to form a **non-ionized or a weakly ionized substance or substances.**

1. Consider the reaction of hydrochloric acid and sodium hydroxide in solution:

$$HCl + NaOH \rightarrow NaCl + H_2O$$

When ions in solution are written as ions:

$$H_3O^+ + Cl^- + Na^+ + OH^- \rightarrow Na^+ + Cl^- + 2\ H_2O$$

the only thing that occurs is:

$$(iv)\ H_3O^+ + OH^- \rightarrow 2H_2O$$

Water is a very weak electrolyte, for all practical purposes a non-ionized substance.

2. Another reaction is that of sodium acetate and hydrochloric acid. Sodium acetate ($NaC_2H_3O_2$) is often represented

by the formula NaOAc. When treated with a strong acid, NaOAc forms slightly ionized acetic acid:

$$NaOAc + HCl \rightarrow \underset{\text{Acetic acid}}{HOAc} + NaCl$$

In ionic form,

$$H_3O^+ + Cl^- + Na^+ + OAc^- \rightarrow HOAc + Na^+ + Cl^- + H_2O$$

The net process which occurs is:

$$(v)\ H_3O^+ + OAc^- \rightarrow H_2O + HOAc$$

Two slightly ionized substances, water and acetic acid, are formed.

3. Acetic acid reacts with sodium hydroxide solution:

$$HOAc + NaOH \rightarrow NaOAc + H_2O$$

Again writing ions in solution as ions:

$$HOAc + Na^+ + OH^- \rightarrow Na^+ + OAc^- + H_2O$$

All that occurs is:

$$(vi)\ HOAc + OH^- \rightarrow OAc^- + H_2O$$

Here a slightly ionized substance, HOAc, reacts with a hydroxide ion to give the *very* slightly ionized substance, water.

D. In some reactions involving ions in solution, **complex ions are formed.**

1. When ammonia solution is added to a solution of copper sulfate, the pale blue color is changed to an intense dark blue. This is due to the formation of the complex ion, $Cu(NH_3)_4^{2+}$. The overall equation is

$$CuSO_4 + 4\,NH_3 \rightarrow Cu(NH_3)_4SO_4$$

Writing ions as ions,

$$Cu^{2+} + SO_4^{2-} + 4\,NH_3 \rightarrow Cu(NH_3)_4^{2+} + SO_4^{2-}$$

All that occurs may be represented by

$$(vii)\ Cu^{2+} + 4\,NH_3 \rightarrow Cu(NH_3)_4^{2+}$$

2. Silver chloride, a white salt insoluble in water, is dissolved by the addition of ammonia:

$$AgCl_{(s)} + 2\,NH_3 \rightarrow Ag(NH_3)_2Cl$$

Writing ions in solution as ions

$$(viii)\ AgCl + 2\,NH_3 \rightarrow Ag(NH_3)_2^+ + Cl^-$$

The equation represents the process which occurs on mixing.

3. Gold will not dissolve in concentrated nitric acid, but it does dissolve in a mixture of concentrated nitric acid and concentrated hydrochloric acid by formation of a complex ion:

$$Au + 4\,HCl + HNO_3 \rightarrow HAuCl_4 + NO + 2\,H_2O$$

Writing ions as ions,

$$Au + 4\,H_3O^+ + 4\,Cl^- + H_3O^+ + NO_3^- \rightarrow H_3O^+ + AuCl_4^- + NO + 6\,H_2O$$

All that occurs is represented by

$$(ix)\ \ Au + 4\,H_3O^+ + 4\,Cl^- + NO_3^- \rightarrow AuCl_4^- + NO + 6\,H_2O$$

E. Some reactions are accomplished by the **transfer of electrons to** or from or between ions.

1. When chlorine water is added to a solution of potassium iodide, molecular iodine is formed, as evidenced by the appearance of a brown coloration. The reaction is

$$Cl_2 + 2\,KI \rightarrow 2\,KCl + I_2$$

Writing ions in solution as ions,

$$Cl_2 + 2\,K^+ + 2\,I^- \rightarrow 2\,K^+ + 2\,Cl^- + I_2$$

All that occurs is

$$(x)\ \ Cl_2 + 2\,I^- \rightarrow 2\,Cl^- + I_2$$

2. Zinc metal dissolves in a solution of copper(II) sulfate, while copper precipitates out on the zinc surface. The reaction is:

$$Zn + CuSO_4 \rightarrow ZnSO_4 + Cu$$

Writing free ions as ions,

$$Zn + Cu^{2+} + SO_4^{2-} \rightarrow Zn^{2+} + SO_4^{2-} + Cu$$

All that occurs is represented by

$$(xi)\ \ Zn + Cu^{2+} \rightarrow Cu + Zn^{2+}$$

This ionic reaction is relatively slow because it depends on migration of copper(II) ions to the zinc metal surface.

13.9 NET IONIC EQUATIONS

Those equations in Section 13.8 marked with small Roman numerals are called **net ionic equations.** They are reproduced below:

(i) $Ba^{2+} + CO_3^{2-} \rightarrow BaCO_3$
(ii) $H_3O^+ + HCO_3^- \rightarrow 2\,H_2O + CO_2$
(iii) $NH_4^+ + OH^- \rightarrow NH_3 + H_2O$
(iv) $H_3O^+ + OH^- \rightarrow 2\,H_2O$
(v) $H_3O^+ + OAc^- \rightarrow H_2O + HOAc$
(vi) $HOAc + OH^- \rightarrow OAc^- + H_2O$
(vii) $Cu^{2+} + 4\,NH_3 \rightarrow Cu(NH_3)_4^{2+}$
(viii) $AgCl + 2\,NH_3 \rightarrow Ag(NH_3)_2^+ + Cl^-$
(ix) $Au + 4\,H_3O^+ + 4\,Cl^- + NO_3^- \rightarrow AuCl_4^- + NO + 6\,H_2O$
(x) $Cl_2 + 2\,I^- \rightarrow 2\,Cl^- + I_2$
(xi) $Zn + Cu^{2+} \rightarrow Zn^{2+} + Cu$

Each of these net ionic equations was determined by the same general steps: (1) The balanced overall equation was written. (2) All strong electrolytes in solution were separated into ions. Ions in solution were written as separate ions. This gives the total ionic equation. It is an attempt to show the substances in the form in which they exist at reaction time. (3) Ions which appear free and unchanged on both sides of the equation are eliminated; obviously they take no part in the reaction. They are there because the reacting ion cannot appear alone. They are often called "spectator" ions. The net ionic equation includes only the participants.

Step (2) is, possibly, the only difficult step of the three. One must know which are "ions in solution." This is a twofold problem. (a) Are they soluble (in solution) and (b) are they ions? Table 13.2 includes a statement of rules for writing net ionic equations, a list of solubilities of salts and lists of common strong and weak electrolytes.

Both **overall** equations and net ionic equations are useful. However, each serves a different function. The overall equation:

$$Na_2CO_3 + BaCl_2 \rightarrow BaCO_3 + 2\,NaCl$$

may be used to calculate the weight proportions of the reactants, sodium carbonate and barium chloride. It can be used to determine the maximum possible yields of the precipitate, barium carbonate, and of the salt in solution, sodium chloride, when certain weights of reactants are used. The overall equation gives the **material balance.**

The **net ionic** equation:

$$Ba^{2+} + CO_3^{2-} \rightarrow BaCO_{3(s)}$$

states that the only thing that happens when the two water solutions of sodium carbonate and barium chloride are mixed is the precipitation of barium carbonate. In fact, one could **predict this reaction** if he knew that of the salts sodium carbonate, barium chloride, barium carbonate and sodium chloride, only barium carbonate is insoluble in water.

Net ionic equations show the change or changes that occur when ions in solution participate in chemical reactions. The change may be the formation or disappearance of (a) a precipitate, (b) a gas, (c) a covalent substance or a weak electrolyte, (d) a complex ion or (e) an element.

The following two examples illustrate the writing of balanced net ionic equations.

Example 1. Write the balanced net ionic equation for the reaction occurring when a solution of nitric acid (HNO_3) is added carefully and slowly to a solution of sodium hydroxide (NaOH).

Solution: Table 13.2 includes HNO_3 with the strong electrolytes and states that alkali metal hydroxides are soluble:

$$H_3O^+ + NO_3^- + Na^+ + OH^- \rightarrow \text{products}$$

"Trading" of ions would give $NaNO_3$, a soluble ionic substance, and H_2O, a very weak electrolyte.

The equations are:

Overall: $HNO_3 + NaOH \rightarrow NaNO_3 + H_2O$
Total Ionic: $H_3O^+ + NO_3^- + Na^+ + OH^- \rightarrow Na^+ + NO_3^- + H_2O$
Net ionic (minus spectator ions): $H_3O^+ + OH^- \rightarrow 2\,H_2O$

Example 2. Acetic acid ($HC_2H_3O_2$, often abbreviated HOAc) is added to a solution of potassium hydroxide. Write the net ionic equation for the reaction that occurs.

Solution: Exchange of ions indicates that this overall equation represents the reaction:

$$HOAc + KOH \rightarrow KOAc + H_2O$$

The net ionic equation is:

$$HOAc + OH^- \rightarrow OAc^- + H_2O$$

13.10 ACIDS, BASES AND SALTS

The traditional theory of acids and bases was developed throughout the nineteenth century, reaching its full stature with the publication of Arrhenius's Theory of Electrolytic Dissociation in 1887. During this period, a number of criteria for acids and for bases were determined. An acid is a substance whose solution:

A. conducts an electric current,

B. has a sour taste,

C. dissolves or destroys certain metals and their oxides (that is, is corrosive),

D. changes the color of certain natural or synthetic dyes, whose color is different in acidic and basic solution. (One such dye is litmus, which is often absorbed on paper strips and used for testing acidity. Litmus is red in acid solutions; blue in basic solutions.)

E. reacts with solutions of bases and in so doing loses its characteristic properties,

F. reacts with zinc and similar metals to produce hydrogen gas.

A base is a substance whose solution:

A. conducts an electric current,

B. has a bitter taste and a soapy feel,

C. will dissolve or corrode certain materials,

D. changes the color of certain dyes,

E. reacts with acids, and in so doing loses its characteristic properties.

It will be seen later that many acids and many bases do not possess all the above characteristics. Because they do not, newer concepts of acids and bases have been formulated to include them. Consider the following examples of acid-base reactions.

Example 1. **Hydrochloric acid,** a solution of hydrogen chloride in water, agrees with all criteria for acids. A **sodium hydroxide** solution has all the listed characteristics of bases. Consider the reaction between the solutions, the reaction itself being one criterion:

$$HCl + NaOH \rightarrow NaCl + H_2O$$

Writing ions in solution as ions:

$$H_3O^+ + Cl^- + Na^+ + OH^- \rightarrow Na^+ + Cl^- + 2\ H_2O$$

All that has occurred is represented by:

$$H_3O^+ + OH^- \rightarrow 2\ H_2O$$

In the process water is formed. The resulting solution has properties of neither an acid nor a base.

Certain conclusions (not all correct) might be made from the results of this reaction:

A. The so-called acid properties of a solution are properties of the hydronium ion, H_3O^+.

B. Basic properties are properties of the hydroxide ion, OH^-.

C. Acids and bases, or H_3O^+ and OH^-, are, in a sense, opposites. They are said to neutralize each other.

D. Reaction of an acid and a base always produces water and an ionic compound called a salt.

These conclusions are not completely general but hold for reactions of strong acids and strong bases in water solution.

Example 2. The reaction of another strong acid and strong base is represented below:

Base		Acid		Salt		Water
KOH	+	HNO_3	$\rightarrow$	KNO_3	+	H_2O
Potassium hydroxide		Nitric acid		Potassium nitrate		

Writing ions as ions:

$$K^+ + OH^- + H_3O^+ + NO_3^- \rightarrow K^+ + NO_3^- + 2\,H_2O$$

Eliminating spectator ions:

$$\underset{\text{Base}}{OH^-} + \underset{\text{Acid}}{H_3O^+} \rightarrow \underset{\text{Water}}{2\,H_2O}$$

Again, all that occurs is the formation of water from the acid, H_3O^+, and the base, OH^-.

Example 3. Another substance always called an acid is acetic acid, $HC_2H_3O_2$ (HOAc). However, an acetic acid solution does not act in accordance with all the criteria. It is a poor conductor of electricity and will not react with zinc to produce hydrogen gas. However, the solution will react with a more reactive metal such as sodium.

Because of these limitations, acetic acid is called a weak acid. It reacts with bases:

$$HOAc + NaOH \rightarrow NaOAc + H_2O$$

Writing ions in solution as ions:

$$HOAc + Na^+ + OH^- \rightarrow Na^+ + OAc^- + H_2O$$

Eliminating spectator ions:

$$HOAc + OH^- \rightarrow OAc^- + H_2O$$

A small fraction of acetic acid exists in ionic form in solution because it is a conductor, although a poor one, of electricity:

$$HOAc + H_2O \leftrightarrows H_3O^+ + OAc^-$$

This is an equilibrium reaction. If the H_3O^+ present is used up, more will be formed. However, there is no reason to believe that the H^+ must be passed to a water molecule to form a H_3O^+, then passed on to the OH^- of the base. The acid properties of a solution of HOAc may not be solely those of H_3O^+. The net ionic equation

$$HOAc + OH^- \rightarrow OAc^- + H_2O$$

may well portray the major course of the reaction. Again a salt solution and water are formed. The reaction of a weak acid and a base is commonly called neutralization. This point will be discussed later.

Example 4. Ammonia solution is considered a base. Again, it is a poor conductor. Ammonia solution reacts with acids:

$$HCl + NH_3 \rightarrow NH_4^+ + Cl^-$$

Writing ions in solution as ions:

$$H_3O^+ + Cl^- + NH_3 \rightarrow NH_4^+ + H_2O + Cl^-$$

And the net ionic equation is:

$$H_3O^+ + NH_3 \rightarrow NH_4^+ + H_2O$$

Here the only water formed is that liberated when H^+ is taken from the hydronium ion. Hydroxide ions exist in an ammonia solution

$$NH_3 + H_2O \leftrightarrows NH_4^+ + OH^-$$

but only a very low concentration of OH^- exists. There is no reason to believe that all ammonium ions are prepared by ionization of ammonia, releasing an OH^- to form water with the hydronium ion. The formation may be direct, as shown by the net ionic equation. If so, OH^- is not the sole contributor of basic properties.

Ammonia solution and acetic acid solution can react together by neutralization:

$$HOAc + NH_3 \rightarrow NH_4OAc$$

With ions in solution written as ions:

$$HOAc + NH_3 \rightarrow NH_4^+ + OAc^-$$

There are no spectator ions. Only ions are formed.

From the preceding discussion, it is apparent that any acid, strong or weak, may be neutralized by any base, strong or weak. It is obvious that **all the acids discussed above** have the same capacity to react with bases, that one mole of one of these acids in solution will neutralize the same amount of base as one mole of another. Likewise any of the bases mentioned will neutralize the same amount of acid as one mole of the others. Further, 1 mole of any **acid mentioned** will neutralize one of any **base mentioned.** This is true because each contributes either 1 mole of H^+ or accepts 1 mole of H^+.

In summary:

A. One mole of HCl gives 1 mole of H_3O^+ in solution ready for reaction, as does 1 mole of HNO_3.

B. One mole of HOAc contributes 1 mole of H^+ for reaction with a base.

C. One mole of KOH and 1 mole of NaOH each furnishes 1 mole of OH^- to accept 1 mole of H^+.

D. One mole of NH_3 accepts 1 mole of H^+.

13.11 POLYPROTIC ACIDS AND POLYHYDROXY BASES

Certain other acids have a greater reacting capacity than those already discussed and may contribute 2 or 3 moles of hydrogen ions per mole for reaction with a base. One of these is sulfuric acid, a solution of hydrogen sulfate in water. On addition to water, hydrogen sulfate is ionized:

$$H_2SO_4 + H_2O \xrightarrow{100\%} H_3O^+ + HSO_4^-$$

The first hydrogen atom is lost from each molecule to form hydronium and monohydrogen sulfate ions. Sulfuric acid is a strong electrolyte. Further ionization occurs, but only to a limited extent, because HSO_4^- is a weaker acid:

$$HSO_4^- + H_2O \rightleftarrows H_3O^+ + SO_4^{2-}$$

but H^+ may be gained from both the H_3O^+ and the weak acid, HSO_4^- by a stronger base than water. One mole of H_2SO_4 has **twice** the capacity of 1 mole of HCl in neutralization:

$$H_2SO_4 + 2\ NaOH \rightarrow Na_2SO_4 + 2\ H_2O$$

In total ionic form:

$$H_3O^+ + HSO_4^- + 2\ Na^+ + 2\ OH^- \rightarrow 2\ Na^+ + SO_4^{2-} + 3\ H_2O$$

The net ionic equation represents the reaction occurring:

$$H_3O^+ + HSO_4^- + 2\ OH^- \rightarrow SO_4^{2-} + 3\ H_2O$$

Other polyprotic acids are sulfurous acid (H_2SO_3), phosphoric acid (H_3PO_4), hydrogen sulfide (H_2S) and oxalic acid ($H_2C_2O_4$). However, these are all weak acids; only a small fraction of even the first hydrogens are transferred to water molecules, and a correspondingly smaller fraction of the second hydrogens are lost. The complete neutralization of phosphoric acid is represented by the equation:

$$H_3PO_4 + 3\ NaOH \rightarrow Na_3PO_4 + 3\ H_2O$$

In net ionic form:

$$H_3PO_4 + 3\ OH^- \rightarrow PO_4^{3-} + 3\ H_2O$$

Some bases have a greater capacity than others for combining with acids. One mole of some bases may combine with 2 or 3 moles of an acid such as HCl. Barium hydroxide, $Ba(OH)_2$, will accept

2 moles of protons per mole:

$$Ba(OH)_2 + 2\,HCl \rightarrow BaCl_2 + 2\,H_2O$$

Writing ions as ions:

$$Ba^{2+} + 2\,OH^- + 2\,H_3O^+ + 2\,Cl^- \rightarrow Ba^{2+} + 2\,Cl^- + 4\,H_2O$$

In net ionic form (after dividing all coefficients by 2):

$$OH^- + H_3O^+ \rightarrow 2\,H_2O$$

All strong bases release hydroxide ions when dissolved, for, like other salts, a strong base is completely ionized in the solid, and its ions are separated upon dissolving.

13.12 EQUIVALENT WEIGHTS OF ACIDS AND BASES

Acids may contribute 1, 2 or 3 moles of H^+ per mole, and bases may accept 1, 2 or 3 moles of H^+ per mole. Suppose weights of various acids and bases are chosen so as to be equivalent to 1 mole of hydrochloric acid in reaction capacity—equivalent in the sense that they, the weights, will replace or neutralize the 1 mole of HCl. Table 13.3 lists the **equivalent weights,** as chosen above, for some of the common acids and bases.

TABLE 13.3 FORMULAS, MOLECULAR WEIGHTS AND EQUIVALENT WEIGHTS OF SOME COMMON ACIDS AND BASES

Name	Formula	Molecular Weight	Equivalent Weight
Acids			
Hydrochloric acid	HCl	36.5	36.5
Nitric acid	HNO_3	63	63
Sulfuric acid	H_2SO_4	98	49
Acetic acid	$HC_2H_3O_2$	60	60
Hydrocyanic acid	HCN	27	27
Sulfurous acid	H_2SO_3	82	41
Phosphoric acid	H_3PO_4	98	32.6
Oxalic acid	$H_2C_2O_4$	90	45
Bases			
Sodium hydroxide	$NaOH$	40	40
Potassium hydroxide	KOH	56	56
Barium hydroxide	$Ba(OH)_2$	171	85.5
Calcium hydroxide	$Ca(OH)_2$	74	37
Aluminum hydroxide	$Al(OH)_3$	78	26
Ammonia	NH_3	17	17

The equivalent weight may be defined as the weight of an acid or a base that will contribute or combine with 1 mole of H^+ in chemical reaction.

Normality. In reactions of acids and bases, the equivalent weight is more fundamental than the molecular weight, because 1 equivalent weight of anything will either react with or replace any other equivalent weight. For this reason, the concentration of solutions of acids and bases is often stated in terms of the number of equivalents per liter; the **concentration in terms of equivalents per liter is the normality.** A 6 normal solution of hydrochloric acid, represented by 6 N HCl, is of such a concentration that 1 liter of the solution contains 6 equivalent weights, or 6 × 36.5 g = 219.0 g, of HCl. One-fourth liter of the same solution contains

$$\frac{6}{4}\text{ moles or }\frac{219}{4}\text{ g of HCl}$$

Consideration of a few elementary calculations may assist in understanding normalities.

Example 1. How many grams of H_3PO_4 are contained in 300 ml of 2 N H_3PO_4?

Solution: The equivalent weight is 1/3 the molecular weight, i.e., $\frac{98\text{ g/mole}}{3\text{ eq/mole}}$.

There are 2 equivalents per liter and there is

$$\frac{300\text{ ml}}{1000\text{ ml/L}} = 0.3\text{ L}$$

Hence

$$\frac{98}{3}\text{ g/eq} \times 2\text{ eq/L} \times \frac{300}{1000}\text{ L} = 19.6\text{ g}$$

Example 2. How many liters of 0.025 N $Ca(OH)_2$ may be made from 50 g of $Ca(OH)_2$?

Solution: Needed for 1 liter:

$$0.025\text{ eq/L} \times \frac{74}{2}\text{ g/eq} = 0.925\text{ g/L}$$

There are 50 g, which is more than enough to prepare 50 liters:

$$\frac{50\text{ g}}{0.925\text{ g/L}} = 54\text{ L}$$

Example 3. Water is added to 60 ml of 3.5 N H_2SO_4 until the volume is 180 ml. What is the normality of the final solution?

Solution: The same amount of solute is present in 180 ml as was present in 60 ml. The final solution is more dilute by the fraction 60 ml/180 ml.

$$3.5 \text{ eq/L} \times \frac{60 \text{ ml}}{180 \text{ ml}} = 1.2 \text{ eq/L} = 1.2 \text{ N}$$

Note the similarity between calculations of molarities and calculations of normalities.

13.13 TITRATIONS

Consider an experiment in which the course of a reaction between a solution of sulfuric acid and a solution of barium hydroxide is followed by a conductivity apparatus. Ten milliliters of a dilute solution of sulfuric acid is transferred from a stock solution by a pipette to a 100 ml beaker. The electrodes of a conductivity apparatus are dipped into the solution, as shown in Figure 13.4a, and the light globe glows brightly. A 0.12 N $Ba(OH)_2$ solution is added in portions from a burette with stirring. With the first addition of the hydroxide solution, a white precipitate appears, which becomes heavier as more base is added. Simultaneously the light dims (Fig. 13.4b). When 19.22 ml are added, the light is extinguished (Fig. 13.4c). Addition of one more drop of barium hydroxide solution causes the light to glow dimly. The intensity of the light increases steadily as more base is added.

The reaction of the two solutions may be represented by these equations

Overall:

$$H_2SO_4 + Ba(OH)_2 \rightarrow BaSO_4 + 2\,H_2O$$

Net ionic:

$$H_3O^+ + HSO_4^- + Ba^{2+} + 2\,OH^- \rightarrow BaSO_{4(s)} + 3\,H_2O$$

At the equivalence point, that is, when an equivalent amount of base has been added to the acid, all has been converted to insoluble barium sulfate and covalent water, and practically no ions are present in solution. We find that 19.22 ml of 0.12 N $Ba(OH)_2$ is equivalent to (accepts the protons contributed by) 10 ml of the acid solution. The acid solution contains 19.22/10 times as many equivalents per liter as the basic solution; hence, its concentration is

$$\frac{19.22}{10} \times 0.12 \text{ eq/L} = 0.23 \text{ eq/L} = 0.23 \text{ N}$$

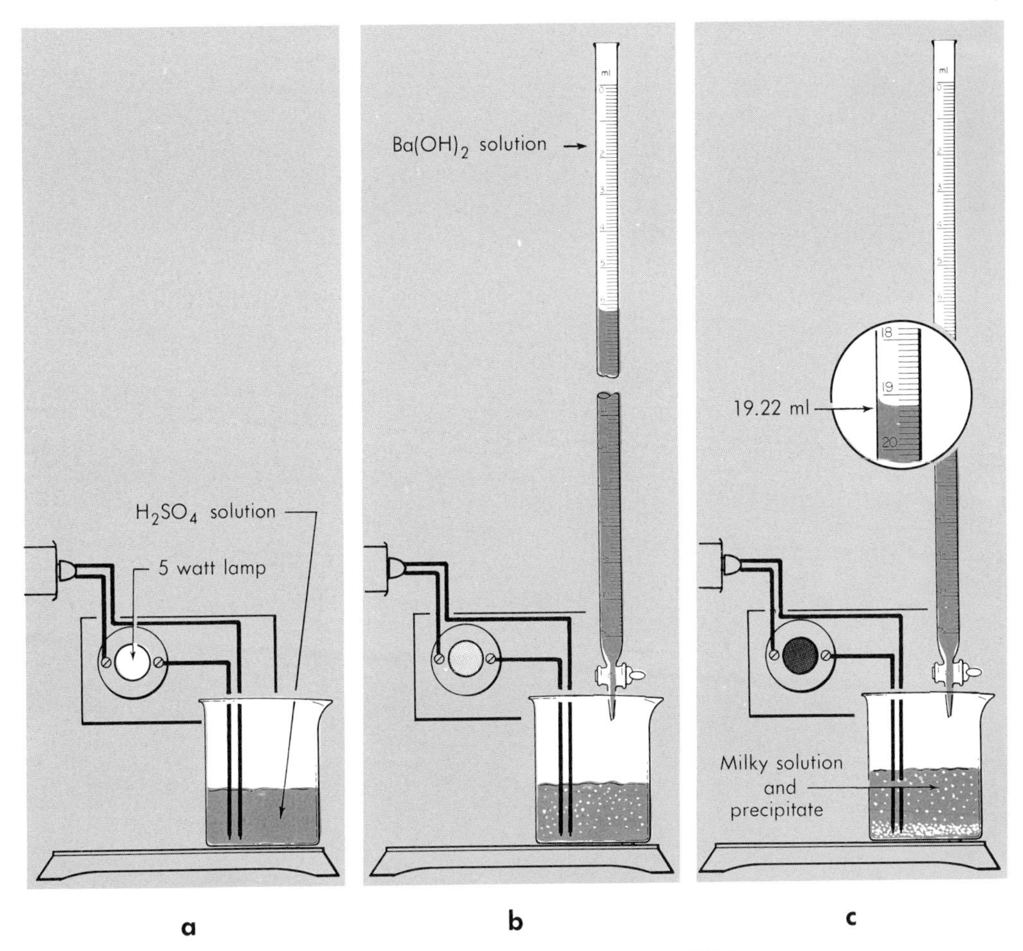

FIGURE 13.4 CONDUCTOMETRIC TITRATION.

The process described above is called a conductometric titration. Barium hydroxide solution was titrated with the sulfuric acid solution to learn the concentration of the acid. It is a conductometric titration because conductivity measurements were used to indicate the neutralization point of the two solutions. However, this method can be used only when the salt formed is insoluble; otherwise the conductance of the solution does not drop to zero. Certain dyes called indicators are used to indicate the point of neutralization when an acid and a base are titrated. A commonly used indicator is phenolphthalein, an organic compound whose structure is as complex as the spelling of its name. The indicator solution is made by dissolving 1 gram of phenolphthalein in 1 liter of a 90 per cent alcohol solution. Two drops of this solution (about 0.10 ml) are sufficient to turn 200 ml of water containing one drop of 0.10 N NaOH solution a definite pink. Any other base listed previously in this chapter will turn phenolphthalein pink. On the other hand, 200 ml of water, or 200 ml of any colorless acid solution, will remain colorless when the indicator is added.

If a sodium hydroxide solution is added slowly to a solution of an acid containing a drop or two of phenolphthalein, no color will appear until the acid has been consumed by the reaction and one more drop of base has been added. Titrations make use of this fact.

A representative titration may be carried out in this manner. Exactly 10 ml of 0.201 N HCl is transferred from a stock solution with a pipette to a 125 ml Erlenmeyer flask. About 50 ml of distilled water and two drops of indicator solution are added to give sufficient volume, as shown in Figure 13.5a. A 50-ml burette is rinsed with distilled water and with a sodium hydroxide solution of unknown concentration, and it is finally filled with that solution to the zero mark. Readings are made at the bottom of the meniscus (Fig. 13.5b). Sodium hydroxide solution is then added to the acid solution with swirling or mixing. Near the end point (neutralization point) the solution surrounding the drop of base just added turns pink and trails around with swirling until it disappears. When the transient color appears, the base is added dropwise, swirling after each drop until a pink color is obtained that remains for at least 30 seconds (Fig. 13.5c). (The solution will slowly become acidic by dissolving CO_2 from the air if it is allowed to stand and the pink color will slowly disappear.)

Suppose the first permanent pink is obtained when 17.31 ml of NaOH has been added. At this point an excess of base has been added, but it is a negligible excess. One can now calculate the normality of the base.

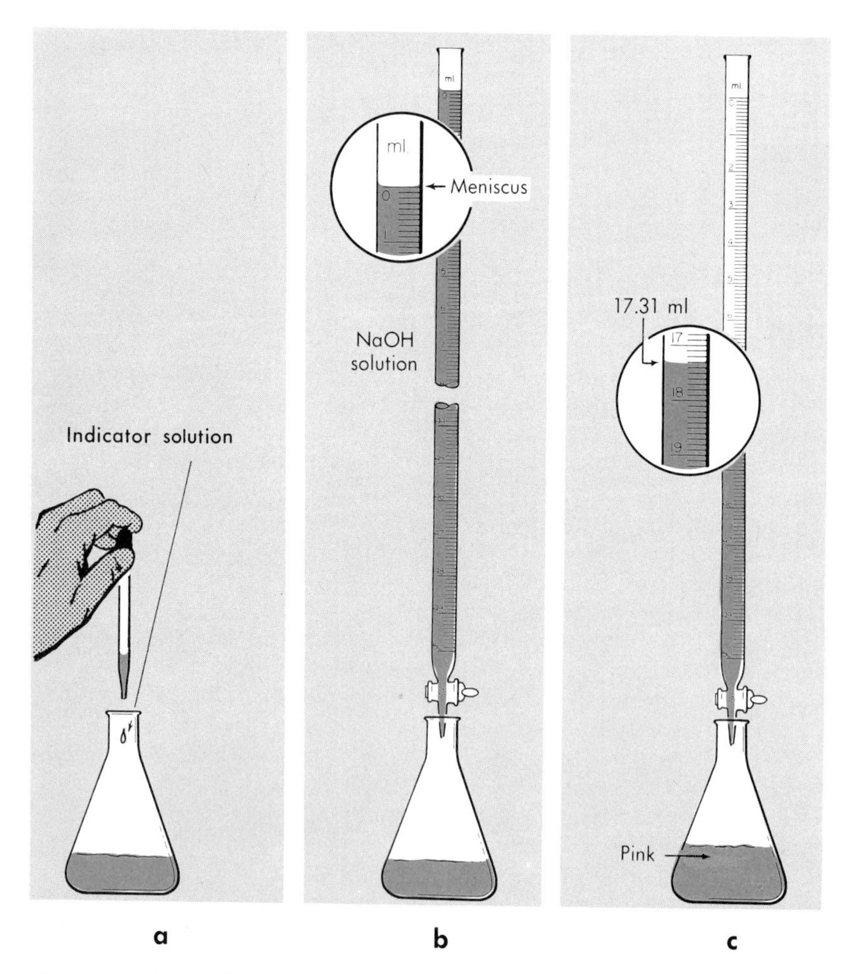

FIGURE 13.5 ACID-BASE TITRATION WITH END-POINT INDICATOR.

Example 1. What is the concentration of a solution of sodium hydroxide if 17.31 ml is required to neutralize 10 ml of 0.201 N HCl?

Solution: The sodium hydroxide solution is more dilute than the acid solution. The concentration is

$$0.201\ \text{N} \times \frac{10\ \text{ml}}{17.31\ \text{ml}} = 0.116\ \text{N}$$

Notice that one could determine the concentration of the acid solution if the concentration of the base to be added were known. With phenolphthalein indicator, the base is added to the acid solution, because the change of colorless to pink is more easily detected than the reverse.

Example 2. In a titration, 7.41 ml of 0.351 N NaOH were required to neutralize 10 ml of an H_2SO_4 solution. Find the molarity of the sulfuric acid.

Solution: The sulfuric acid solution is more dilute than the base and has a normality of

$$\frac{7.41\ \text{ml}}{10\ \text{ml}} \times 0.351\ \text{N} = 0.260\ \text{N}\ H_2SO_4$$

There are 1/2 as many moles as there are equivalents per liter of solution, because a mole of H_2SO_4 is 2 equivalents. Hence, the molarity is

$$0.260\ \text{eq/L} \times 1/2\ \text{mole/eq} = 0.130\ \text{mole/L} = 0.130\ \text{M}$$

13.14 ACID-BASE CONCEPTS

We will consider three concepts of acids and bases. Each succeeding concept is more inclusive than the preceding one. The concept is revealed by the respective definitions of acid, base, salt and neutralization.

1. *The Arrhenius concept of acids and bases* is the traditional idea. According to it:

A. An **acid** is an electrolyte which furnishes H^+ ions in solution. In more modern interpretations the hydronium ion (H_3O^+) is furnished.

B. A **base** is an electrolyte that furnishes hydroxide ions (OH^-) in solution.

C. **Neutralization** is the combining of the H_3O^+ of the acid and the OH^- of the base to form water.

D. A **salt** is an electrolyte that is neither acid nor base, that furnishes neither a H_3O^+ nor an OH^- and that may be formed, along with water, in neutralization.

Consider the reaction of hydrochloric acid and sodium hydroxide:

$$\underset{\text{From the acid HCl}}{H_3O^+} + \underset{\text{From the base NaOH}}{OH^-} \rightarrow \underset{\text{Water}}{2\,H_2O}$$

The reaction is a neutralization. When the solution is evaporated, the salt, NaCl, remains. This concept includes only strong acids and bases, unless one assumes that the few H_3O^+ and OH^- ions formed (as shown for acetic acid and ammonia)

$$HOAc + H_2O \leftrightharpoons H_3O^+ + OAc^-$$

and

$$NH_3 + H_2O \leftrightharpoons NH_4^+ + OH^-$$

pair together

$$H_3O^+ + OH^- \rightarrow 2\,H_2O$$

and as they are used up, more ions form to keep the process going. This may not be correct; much of the reaction may be direct:

$$HOAc + NH_3 \rightarrow NH_4OAc$$

2. *The Brønsted-Lowry Concept,* proposed independently by J. N. Brønsted in Denmark and T. M. Lowry in England in 1922–1923, is more inclusive than the Arrhenius Theory, covering both the so-called hydrolysis reactions and reactions taking place in solutions other than water. According to this theory:

the Brønsted-Lowry Concept of acids and bases will be used throughout the balance of the text.

A. An **acid** is an ion or a molecule that can give up a proton (H^+) in chemical reaction.

B. A **base** is a substance or an ion of a substance that can accept a proton.

C. **Neutralization** is the reaction occurring when a proton (H^+) is transferred from an acid to a base.

Reactions of strong acids and bases agree with this definition, because an H^+ is transferred from the hydronium ion to the hydroxide ion.

Consider a series of different reactions:

(i) Heat is evolved when hydrogen chloride gas, a covalent substance, is dissolved in water and ions are formed. The equation for the reaction is

$$HCl_{(g)} + H_2O \rightarrow H_3O^+ + Cl^- + \text{heat}$$

(ii) Acetic acid can be prepared by treating sodium acetate

with a solution of a strong acid:

$$H_3O^+ + OAc^- \rightarrow HOAc + H_2O$$

(iii) Acetic acid will react with ammonia solution:

$$HOAc + NH_3 \rightarrow NH_4^+ + OAc^-$$

(iv) Ammonia may be prepared by adding sodium hydroxide to an ammonium chloride solution.

$$NH_4^+ + OH^- \rightarrow NH_3 + H_2O$$

Notice that reactions (i) to (iv) are all neutralization reactions in the Brønsted sense. In each reaction, a proton is transferred. Select the acid, the proton donor, in each equation.

Acids

(i) $HCl + H_2O \rightarrow H_3O^+ + Cl^-$

(ii) $H_3O^+ + OAc^- \rightarrow HOAc + H_2O$

(iii) $HOAc + NH_3 \rightarrow NH_4^+ + OAc^-$

(iv) $NH_4^+ + OH^- \rightarrow H_2O + NH_3$

Note that the acid in each succeeding equation is a product of the equation above. It is the product that results when the base accepts the proton. Obviously the acid strength decreases from the very strong acid $HCl_{(g)}$ through the series H_3O^+, HOAc and NH_4^+ to H_2O. Hydrogen chloride will react with all the bases listed; H_3O^+ will react with those of (ii), (iii) and (iv); HOAc will react with the bases in (iii) and (iv); while NH_4^+ reacts with only OH^-. H_2O will lose a proton to none of these bases.

Now select the base, the substance that accepts the proton:

Bases

(i) $HCl + H_2O \rightarrow H_3O^+ + Cl^-$

(ii) $H_3O^+ + OAc^- \rightarrow HOAc + H_2O$

(iii) $HOAc + NH_3 \rightarrow NH_4^+ + OAc^-$

(iv) $NH_4^+ + OH^- \rightarrow H_2O + NH_3$

The base in each equation is a product of the succeeding equation. When the bases are listed as below, the basic strength decreases through the series:

$$OH^- > NH_3 > OAc^- > H_2O > Cl^-$$

When the acids are listed in decreasing strength and the bases in increasing strength, an interesting relationship is found:

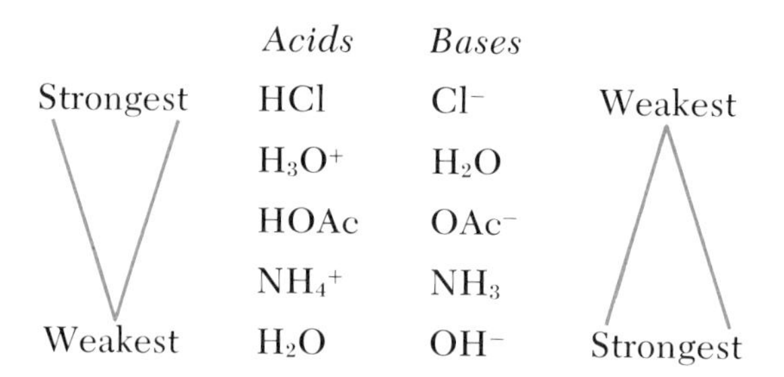

	Acids	*Bases*	
Strongest	HCl	Cl^-	Weakest
	H_3O^+	H_2O	
	HOAc	OAc^-	
	NH_4^+	NH_3	
Weakest	H_2O	OH^-	Strongest

In Brønsted terminology, Cl^- is the **conjugate** base of the acid HCl, and HCl is the conjugate acid of the base Cl^-. Because HCl is a strong acid, and loses H^+ readily, Cl^- is a weak base, that is, a poor acceptor of H^+. Of the bases, OH^- is the strongest one listed (it accepts a proton most readily), and its conjugate acid H_2O is a weak acid (a poor proton donor). The acid in hydrochloric acid solution is H_3O^+. The base in a solution of barium hydroxide is OH^-.

the stronger an acid, the weaker its conjugate base.

Hydrogen chloride and ammonia react in the air to form a white smoke which settles on nearby surfaces as a white precipitate of NH_4Cl. Notice that the "dustiest" reagent bottles are those containing concentrated ammonia water and concentrated hydrochloric acid; the HCl bottle is dirtiest because the ammonia gas diffuses more rapidly. The reaction is:

$$NH_{3(g)} + HCl_{(g)} \rightarrow NH_4^+Cl^-_{(s)}$$

(Charges are shown to indicate that the solid is ionic.) According to Arrhenius, this is not a neutralization equation because neither H_3O^+ nor OH^- is involved. Because an (H^+) is transferred, it is a "Brønsted-Lowry" neutralization.

3. The Brønsted-Lowry Concept limits the term "acid" to substances that possess hydrogen. The **Lewis Concept,** introduced by G. N. Lewis the same year that Brønsted presented his idea, recognizes that other compounds act similarly to acids when water is not present and that these might properly be called acids also. For example, boron trichloride reacts with ammonia, just as hydrogen chloride does:

(i) $:\ddot{Cl}{-}B(\ddot{Cl}:)_2 + :NH_3 \rightarrow :\ddot{Cl}{-}B(\ddot{Cl}:)_2:NH_3$

Lewis acid **Lewis base**

(ii) $:\ddot{Cl}:H + :\ddot{N}H_3 \rightarrow [:\ddot{Cl}:]^- [H:\ddot{N}:H_3]^+$

In both reactions, the base contributes electrons for sharing with the acid or a fragment of an acid.

Lewis's definitions are:

A. An **acid** is an ion or molecule that can accept electrons for covalent bond formation.

B. A **base** is an ion or molecule that can donate a pair of electrons for covalent bond formation.

a coordinate covalent bond is one in which one atom furnishes both electrons of the bond.

C. **Neutralization** is the formation of the coordinate covalent bond.

These definitions include a reaction long considered an acid-base reaction by geologists and metallurgists, but one not included in the other theories because the reactants are not the hydroxide and hydronium ions and no proton is transferred. This reaction is acid-base by all theories:

$$Ca(OH)_2 + H_2SO_4 \rightarrow CaSO_4 + 2\ H_2O$$

but the reaction of the anhydrides is acid-base only according to the Lewis Theory:

$$CaO + SO_3 \rightarrow CaSO_4$$

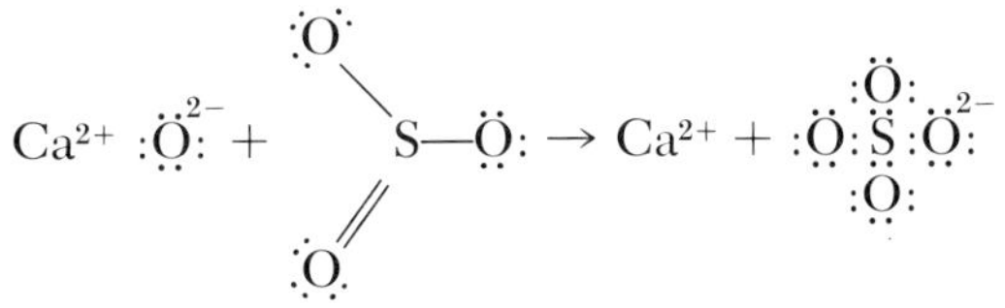

$$Ca^{2+}\ :\ddot{\underset{..}{O}}:^{2-} + \ (O{=})(O{-})S{-}\ddot{\underset{..}{O}}: \rightarrow Ca^{2+} + \left[:\ddot{\underset{..}{O}}:\ddot{\underset{..}{S}}:\ddot{\underset{..}{O}}: \text{ with } :\ddot{O}: \text{ above and } :\underset{..}{O}: \text{ below S}\right]^{2-}$$

The oxide is the base, because the oxide ion contributes the 2 electrons for the covalent bond. Sulfur trioxide is the acid, because it accepts the pair of electrons for sharing.

The Lewis Concept is more inclusive than the Brønsted-Lowry Theory. Because most acid-base reactions appearing in this text take place in water solutions, for which the Brønsted-Lowry definition is sufficient, they will be discussed in terms of that concept.

EXERCISES – Set I

13.1 State Arrhenius's Theory of Electrolytic Dissociation.

13.2 Which types of electrolytes are ionized when dissolved in water? Which types are simply dissociated?

13.3 Three different solutions containing equal numbers of moles of solute in the same weight of water freeze at these temperatures: −0.42, −0.50 and −0.84°C. What kinds of substances might these solutes be?

13.4 What ions conduct current when a dilute solution of sulfuric acid is electrolyzed?

13.5 Name four properties of solutions of electrolytes that distinguish them from solutions of non-electrolytes.

13.6 Why is the precipitation of silver metal on a copper wire that is immersed in a silver nitrate solution a slower reaction than the precipitation of silver chloride accomplished by mixing solutions of silver nitrate and sodium chloride?

13.7 Name two tests one might use to distinguish a weak electrolyte from a strong electrolyte.

13.8 Name five different types of ionic reactions. Write a net ionic equation to represent each type.

13.9 What does a net ionic equation show?

13.10 Why are overall equations useful?

13.11 What is a spectator ion?

13.12 How are each of the following written when they appear in net ionic equations: (a) dilute sulfuric acid, (b) dilute hydrochloric acid, (c) acetic acid, (d) sulfurous acid, (e) ammonia solution, (f) oxalic acid, (g) calcium hydroxide solution, (h) sodium chloride solution, (i) precipitate of barium sulfate, (j) copper metal, (k) elemental sulfur?

13.13 What color is litmus paper in an acid solution? What color is phenolphthalein in acid solution?

13.14 Why is the base usually titrated into the acid solution when phenolphthalein is the indicator?

13.15 Why cannot the conductometric apparatus be used to indicate the neutralization point (end point) when ammonia solution is titrated into hydrochloric acid solution?

13.16 In what sense is 1/2 mole of H_2SO_4 equivalent to 1 mole of HCl? to one mole of ammonia?

13.17 Name three polyprotic acids.

13.18 Write three definitions of a base.

13.19 Why is the Brønsted-Lowry Concept more general than the Arrhenius Concept?

13.20 Hydrocyanic acid is a weaker acid than acetic acid. How do the acetate ion (OAc^-) and the cyanide ion (CN^-) compare in strength as bases?

EXERCISES—Set II

13.21 Write each equation in balanced net ionic form:

a. $NaOH + HCl \rightarrow NaCl + H_2O$
b. $Ca(OH)_2 + HCl \rightarrow CaCl_2 + H_2O$
c. $HOAc + Ca(OH)_2 \rightarrow Ca(OAc)_2 + H_2O$

d. $HOAc + NaOH \rightarrow NaOAc + H_2O$
e. $H_2S + NH_3 \rightarrow (NH_4)_2S$
f. $H_2S + NaOH \rightarrow Na_2S + H_2O$
g. $HCl + NH_3 \rightarrow NH_4Cl$
h. $AgNO_3 + NaCl \rightarrow AgCl_{(s)} + NaNO_3$
i. $BaCl_2 + Na_2SO_4 \rightarrow BaSO_{4(s)} + NaCl$
j. $Cu(NO_3)_2 + H_2S \rightarrow CuS_{(s)} + HNO_3$
k. $Ca(HCO_3)_2 \rightarrow CaCO_{3(s)} + H_2O + CO_2$
l. $Cl_2 + NaBr \rightarrow Br_2 + NaCl$
m. $AgCl_{(s)} + NH_3 \rightarrow Ag(NH_3)_2Cl$
n. $Br_2 + NH_4I \rightarrow NH_4Br + I_2$
o. $F_2 + H_2O \rightarrow HF + O_2$
p. $Na + H_2O \rightarrow NaOH + H_2$
q. $Zn + Cu(NO_3)_2 \rightarrow Zn(NO_3)_2 + Cu$
r. $Al + CuCl_2 \rightarrow AlCl_3 + Cu$
s. $Ba(OII)_2 \mid Na_2SO_4 \rightarrow BaSO_4 + NaOH$
t. $HNO_3 + Ba(OH)_2 \rightarrow Ba(NO_3)_2 + H_2O$

PROBLEMS

13.22 22.4 L of a pure gas at STP is dissolved in 2000 g of water. The freezing point of the solution is $-1.86°C$. Is the solute gas a non-electrolyte or an electrolyte?

13.23 7.1 g of Na_2SO_4 is dissolved in 200 g of water. Calculate:

a. the freezing point of the solution,
b. the boiling point of the solution,
c. the vapor pressure at 80°C,
d. the osmotic pressure at 25°C.

13.24 A quantity of salt (NaCl) was spilled into 200 ml of water. The resulting solution boiled at 100.30°C. How much salt was spilled?

13.25 Calculate the molecular and equivalent weights of each of the following acids and bases:

a. HCl
b. HOAc (CH_3COOH)
c. H_3PO_4
d. $H_2C_2O_4$
e. NaOH
f. $Ba(OH)_2$
g. $Ca(OH)_2$
h. $Al(OH)_3$
i. NH_3

13.26 How many grams of solute are contained in 1 liter of:

a. 1 M HCl
b. 1 N HCl
c. 1.5 M H_2SO_4
d. 1.5 N H_2SO_4
e. 0.023 N $Ca(OH)_2$
f. 0.129 M $Ba(OH)_2$
g. 0.63 N H_3PO_4

13.27 How many grams of solute are there in:

a. 2500 ml of 1.8 N H_2SO_4
b. 280 ml of 0.15 N $Ba(OH)_2$
c. 954 ml of 0.89 N H_3PO_4
d. 3.2 L of 0.8 N NaOH
e. 0.75 L of 6.4 N HCl

13.28 How many milliliters of the following solutions can be made from 100 g of the solute:

a. 1 M $H_2C_2O_4$
b. 3 M $H_2C_2O_4$
c. 2.4 N $H_2C_2O_4$
d. 0.012 M $Ca(OH)_2$
e. 0.012 N $Ca(OH)_2$

13.29 What is the molarity of each of the following solutions:

a. 2.4 N HCl
b. 0.38 N H_2SO_4
c. 12.1 N HNO_3
d. 6.0 N HOAc
e. 0.03 N $H_2C_2O_4$

13.30 What is the normality of each of the following solutions:

a. 3.6 M HNO_3
b. 0.29 M H_2SO_4
c. 0.36 M $H_2C_2O_4$
d. 2.4 M H_3PO_4
e. 0.58 M HCN

13.31 Write a general rule relating the normality to the molarity of solutions.

13.32 Calculate the normality of the solutions made as indicated below:

a. 100 ml of 1.8 N HCl is diluted with water to 200 ml.
b. 100 ml of 2.9 N HCl is diluted to 250 ml.
c. 150 ml of 0.54 N H_2SO_4 is diluted to 450 ml.
d. 320 ml of 1.6 M H_2SO_4 is diluted to 684 ml.
e. 1.2 L of 0.32 M $H_2C_2O_4$ is diluted to 4.8 L.
f. 200 ml of 2 M H_2SO_4 is added to 200 ml of 4 N H_2SO_4.
g. 300 ml of 2.4 M H_2SO_4 is added to 250 ml of 2.4 N H_2SO_4.

13.33 Calculate the normality of each of the following solutions for which no concentration is given:

a. 20 ml of NaOH is neutralized by 30 ml of 2 N HCl.
b. 13,400 ml of $Ca(OH)_2$ is neutralized by 10.0 ml of 1.2 N H_2SO_4.
c. 25 ml of 3.0 M H_2SO_4 is neutralized with 35 ml of NaOH.
d. 11.2 ml of 2.6 N HOAc is neutralized with 341 ml of $Ba(OH)_2$.
e. 290 ml of HOAc solution neutralized 20 g of NaOH dissolved in 100 ml of water.
f. 10 ml of 0.261 N H_2SO_4 is dissolved in 100 ml of water and then neutralized with 13.5 ml of NaOH.

SUGGESTED READING

Luder, W. F.: "Contemporary Acid-Base Theory." J. Chem. Educ., *25:* 555, 1948.
Mogul, P. H., and Schmuckler, J. S.: "Dilute Solutions of Strong Acids: The Effect of Water on pH." Chemistry, *42* (9): 14, 1969.
Sienko, M. J., and Plane, R. A.: *Chemistry*, 3rd Ed. McGraw-Hill Book Co., New York, 1961, Chapter 10.
Sisler, H. H.: *Chemistry in Non-aqueous Solvents.* Reinhold Publishing Corp., New York, 1961 (paperback).
Vanderwerf, C. A.: *Acids, Bases, and the Chemistry of the Covalent Bond.* Reinhold Publishing Corp., New York, 1961 (paperback).

FOURTEEN • ELECTRICAL ENERGY AND OXIDATION-REDUCTION REACTIONS

14.1 INTRODUCTION

Certain **spontaneous chemical reactions** may be harnessed in such a way that part of the energy normally converted to heat is converted to electrical energy. The conversion of chemical energy to electrical energy occurs in the lead storage battery of an automobile and in a flashlight battery.

Similar spontaneous reactions are observed in a general chemistry laboratory. Among these is the reaction between copper metal and the solution of a silver salt. When a copper wire is placed in a silver sulfate solution, as in Figure 14.1, it immediately turns silver colored. Within moments a silver "fur" covers the wire and, in time, may grow until the tiny silver branches are more than a half inch long. When the silver is rubbed from the copper wire, it is seen to be thin and rough. During the process the colorless silver sulfate solution has turned blue. Chemical analyses show that the solution contains copper(II) sulfate. Silver ion has left the solution; copper(II) ion has replaced it. The overall equation is:

$$Cu + Ag_2SO_4 \rightarrow CuSO_4 + 2\ Ag$$

and the net ionic equation is:

$$Cu + 2\ Ag^+ \rightarrow Cu^{2+} + 2\ Ag$$

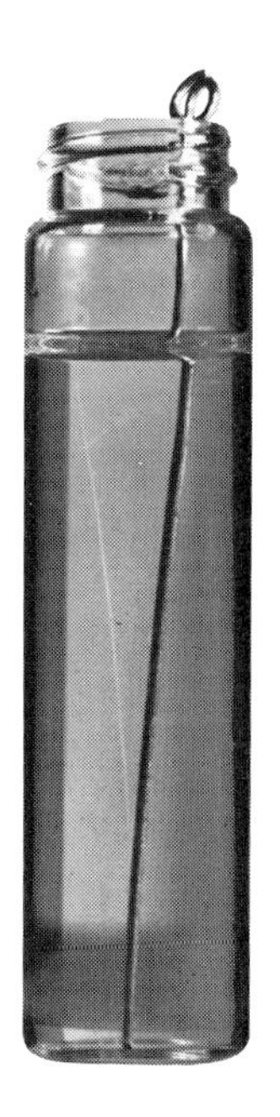
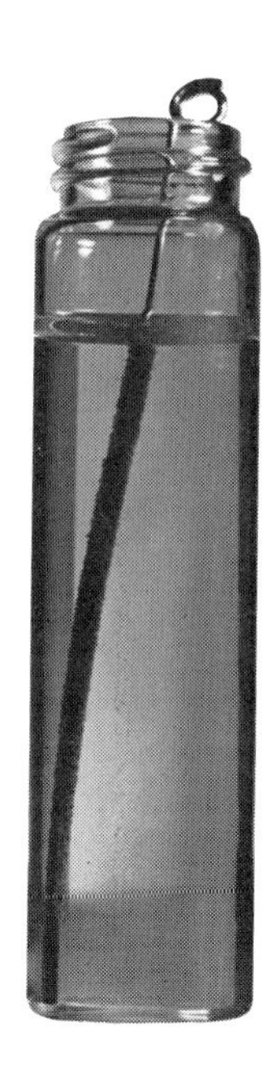

FIGURE 14.1 GROWTH OF SILVER ON A COPPER WIRE.

Sulfate ions remain unchanged in the solution throughout the reaction. The replacement of silver ion in the solution with copper(II) ion is an **oxidation-reduction reaction.** The oxidation-reduction reaction can be separated into two parts called **half-cell equations.** In one half-cell reaction, oxidation occurs; that is, electrons are lost. In the other half-cell reaction, reduction occurs; that is, electrons are taken up by the reactants. For this particular reaction the half-cell reactions are

$$Cu \rightarrow Cu^{2+} + 2\ e^- \text{ (oxidation)}$$

and

$$2\ Ag^+ + 2\ e^- \rightarrow 2\ Ag \text{ (reduction)}$$

The reaction consists of the **transfer of electrons** from copper atoms to silver ions. The copper atoms become ions and the silver ions become atoms. There are many other reactions in which more active metals replace the ions of less active metals in solution. Zinc metal dissolves in copper(II) sulfate solution to precipitate copper, as do iron and nickel. The reaction of zinc with strong acid solutions to produce hydrogen gas is a similar reaction.

a spontaneous oxidation-reduction reaction which can be separated such that oxidation and reduction take place at different sites may be used to produce electrical energy.

14.2 THE SILVER-COPPER PRIMARY CELL

It might appear from the equation that electrons are passed directly from the copper atom to the silver ions when silver deposits on copper, that silver ions collide with copper atoms on the surface of the wire and while in contact exchange electrons. This happens at first perhaps, until a layer of silver atoms covers the copper surface. However, silver atoms can precipitate on one another, growing out from the wire in branches, as represented by the photograph and diagram in Figure 14.2. When a silver atom adds to the end of a branch that is but a fraction of an inch long, it is precipitating at a distance of perhaps 1,000,000 atoms from the nearest copper atom which might lose the electron. But silver metal is an excellent conductor of electricity. Two electrons travel from each dissolving copper atom on the wire, out through the silver branch, to be accepted by two silver ions at the silver surface.

A primary cell may be constructed for this reaction by separating the dissolving and precipitating sites by a greater distance and by connecting them by a different metallic conductor above the solution, as in Figure 14.3. In this cell the copper metal and the silver salt solution are not allowed to mix but are in separate compartments divided by a porous plate. In one compartment a copper bar is suspended in a solution of copper(II) nitrate. A silver bar rests in a silver nitrate solution in the second compartment. The porous divider allows the solutions to come into contact but retards their mixing.

When the two metal bars are connected by an electrical conductor through a galvanometer, the galvanometer needle is deflected; an electrical current is flowing. The electron stream flows from the copper bar to the silver. At the same time, the copper bar dissolves, the silver bar thickens and the copper(II) nitrate solution becomes a deeper blue. It is clear that the same reaction is occurring that occurred when the copper wire was placed in the solution of silver sulfate. The overall reaction is the same

$$Cu + 2\,Ag^+ \rightarrow Cu^{2+} + 2\,Ag$$

with the half-cell reactions taking place in different halves of the apparatus.

oxidation takes place at the anode; reduction, at the cathode.

This apparatus is called a **primary electrical cell.** The copper and silver bars are called electrodes. Oxidation occurs at the copper

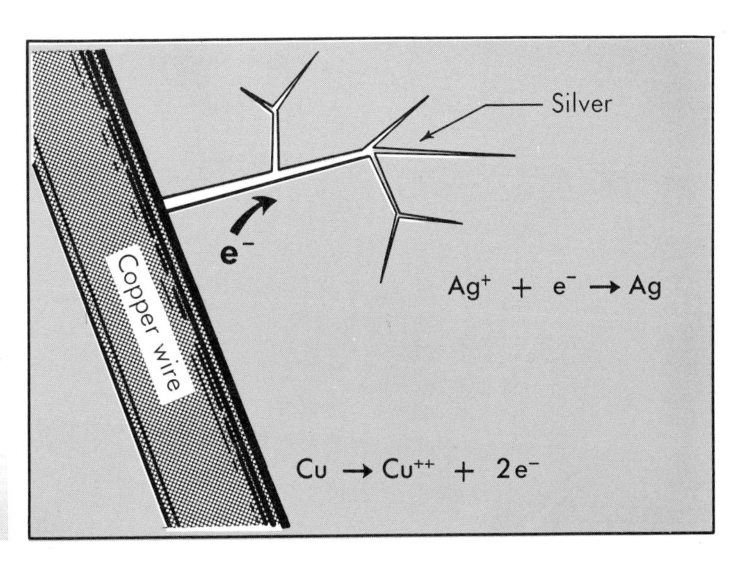

FIGURE 14.2 CRYSTALS OF SILVER ON A COPPER WIRE.

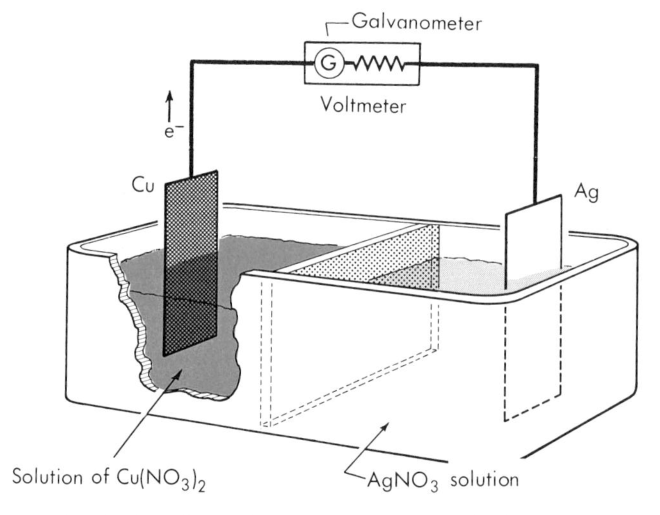

FIGURE 14.3 PRIMARY CELL.

electrode, which is called the **anode** because oxidation occurs at its surface:

$$Cu \rightarrow Cu^{2+} + 2\ e^- \text{ (oxidation)}$$

Reduction occurs at the silver electrode, which is called the **cathode** because reduction occurs there:

$$Ag^+ + e^- \rightarrow Ag \text{ (reduction)}$$

The cell is a primary cell because it operates spontaneously. A secondary cell, or an electrolysis cell, requires another cell or another source of electricity to keep it running.

14.3 OXIDATION-REDUCTION

Because the reaction occurring in every electrolytic cell is an oxidation-reduction reaction, the terms should be reviewed and applied to several cases. The definitions of the terms and the method for balancing the equations apply as well as to the many oxidation reactions which occur outside electrolytic cells.

The reaction taking place in the **Daniell cell** is oxidation-reduction:

$$Zn + CuSO_4 \rightarrow ZnSO_4 + Cu$$

$$\text{i. } Zn + Cu^{2+} \rightarrow Zn^{2+} + Cu$$

Oxidation-reduction occurs in the cell represented by:

$$Pt|H_2,\ H_2SO_4\|CuSO_4|Cu$$

$$H_2 + CuSO_4 \rightarrow H_2SO_4 + Cu$$

$$\text{ii. } H_2O + H_2 + Cu^{2+} + SO_4^{2-} \rightarrow HSO_4^- + Cu + H_3O^+$$

the double vertical line in cell notation is used to separate the cell into half cells; the single vertical line separates the electrode from substances that react with it or at its surface.

The reaction occurring in the **lead storage cell** is oxidation-reduction:

$$Pb + PbO_2 + 2\ H_2SO_4 \rightarrow 2\ PbSO_4 + 2\ H_2O$$

$$\text{iii. } Pb + PbO_2 + 2\ HSO_4^- + 2\ H_3O^+ \rightarrow 2\ PbSO_4 + 4\ H_2O$$

Oxidation has occurred, because in each reaction an element has **increased in oxidation state.** (The small Roman numerals refer to the equation of the same number above):

$$\text{i. } \overset{0}{Zn} \text{ to } \overset{+2}{Zn^{2+}}$$

$$\text{ii. } \overset{0}{H_2} \text{ to } \overset{+1}{H_3O^+}$$

$$\text{iii. } \overset{0}{Pb} \text{ to } \overset{+2}{PbSO_4}$$

Reduction has occurred, because in each reaction an element has **decreased in oxidation state:**

$$\text{i. } \overset{+2}{Cu^{2+}} \text{ to } \overset{0}{Cu}$$

$$\text{ii. } \overset{+2}{Cu^{2+}} \text{ to } \overset{0}{Cu}$$

$$\text{iii. } \overset{+4}{PbO_2} \text{ to } \overset{+2}{PbSO_4}$$

These reactions need not occur only in electrical cells. When occurring in cells, if they are spontaneous and if the oxidation and reduction portions can be made to occur at separate sites, one can gain **energy** from them. Reaction (i) occurs when a zinc bar is immersed in a copper sulfate solution.

The reaction occurring when concentrated hydrochloric acid is heated with manganese dioxide to produce chlorine gas is oxidation-reduction:

$$4\ HCl + MnO_2 \rightarrow MnCl_2 + Cl_2 + 2\ H_2O$$

$$\text{iv. } 4\ H_3O^+ + 2\ Cl^- + MnO_2 \rightarrow Mn^{2+} + Cl_2 + 6\ H_2O$$

Oxidation-reduction occurs when a basic potassium chlorate solution is mixed with a solution of potassium iodide to release iodine:

$$KClO_3 + 3\ H_2O + 6\ KI \rightarrow KCl + 3\ I_2 + 6\ KOH$$

$$\text{v. } ClO_3^- + 6\ I^- + 3\ H_2O \rightarrow Cl^- + 3\ I_2 + 6\ OH^-$$

An element undergoes an increase in oxidation state (is oxidized) in each of these reactions:

$$\text{iv. } \overset{-1}{Cl^-} \text{ to } \overset{0}{Cl_2}$$

$$\text{v. } \overset{-1}{I^-} \text{ to } \overset{0}{I_2}$$

While another element in each reaction suffers a decrease in oxidation state (is reduced):

$$\text{iv. } \overset{+4}{MnO_2} \text{ to } \overset{+2}{Mn^{2+}}$$

$$\text{v. } \overset{+5}{ClO_3^-} \text{ to } \overset{-1}{Cl^-}$$

Notice that oxidation occurs simultaneously with reduction. This is always the case, because oxidation is a loss of electrons and reduction is a gain of electrons. The loss must equal the gain. Observe the electron loss and gain in the five equations discussed above, as given in Table 14.1.

These equations showing oxidation or reduction are named half-cell equations because they indicate reactions that occur or might occur at one electrode, or one-half the reaction of an electrolytic cell.

Notice further that the half-cell equations are combined together in the net ionic equation in such a way that there is no net gain nor loss in electrons. For example, in equation (v) six electrons are needed in the reduction:

$$ClO_3^- + 3\ H_2O + 6\ e^- \rightarrow Cl^- + 6\ OH^-$$

When iodide ions change to an iodine molecule, only two electrons are freed:

$$2\ I^- \rightarrow I_2 + 2\ e^-$$

TABLE 14.1 HALF-CELL EQUATIONS SHOWING GAIN AND LOSS OF ELECTRONS

Oxidation	Reduction
(i) $Zn: \rightarrow Zn^{2+} + 2\ e^-$	$Cu^{2+} + 2\ e^- \rightarrow Cu:$
(ii) $H_2O + H_2 + SO_4^{2-} \rightarrow HSO_4^- + H_3O^+ + 2\ e^-$	$Cu^{2+} + 2\ e^- \rightarrow Cu:$
(H:Ö: + H:H + :Ö:S:Ö:$^{2-}$ → H:Ö:S:Ö:$^-$ + H:Ö:H$^+$ + 2 e$^-$; $8\ e^- + 2\ e^- + 32\ e^- = 32\ e^- + 8\ e^- + 2\ e^-$)	
(iii) $Pb + HSO_4^- + H_2O \rightarrow PbSO_4 + H_3O^+ + 2\ e^-$	$PbO_2 + HSO_4^- + 3\ H_3O^+ + 2\ e^- \rightarrow PbSO_4 + 5\ H_2O$
(Pb: + H:Ö:S:Ö:$^-$ + H:Ö: → Pb^{2+}:Ö:S:Ö:$^{2-}$ + H:Ö:H$^+$ + 2 e$^-$; $4\ e^- + 32\ e^- + 8\ e^- = 2\ e^- + 32\ e^- + 8\ e^- + 2\ e^-$)	(:Ö:Pb:Ö: + H:Ö:S:Ö:$^-$ + 3 H:Ö:H$^+$ + 2 e$^-$ → Pb^{2+}:Ö:S:Ö:$^{2-}$ + 5 H:Ö: ; $16\ e^- + 32\ e^- + 3 \times 8\ e^- + 2\ e^- = 2\ e^- + 32\ e^- + 5 \times 8\ e^-$)
(iv) $2\ Cl^- \rightarrow Cl_2 + 2\ e^-$	$MnO_2 + 4\ H_3O^+ + 2\ e^- \rightarrow Mn^{2+} + 6\ H_2O$
(2:C̈l:$^-$ → :C̈l:C̈l: + 2 e$^-$; $2 \times 8\ e^- = 14\ e^- + 2\ e^-$)	(:Ö:Mn:Ö: + 4 H:Ö:H$^+$ + 2 e$^-$ → Mn^{2+} + 6 H:Ö: ; $16\ e^- + 4 \times 8\ e^- + 2\ e^- = 2\ e^- + 6 \times 8\ e^-$)
(v) $2\ I^- \rightarrow I_2 + 2\ e^-$	$ClO_3^- + 3\ H_2O + 6\ e^- \rightarrow Cl^- + 6\ OH^-$
(2:Ï:$^-$ → :Ï:Ï: + 2 e$^-$; $2 \times 8\ e^- = 14\ e^- + 2\ e^-$)	(:Ö:C̈l:Ö:$^-$ + 3 H:Ö: + 6 e$^-$ → :C̈l:$^-$ + 6:Ö:H$^-$; $26\ e^- + 3 \times 8\ e^- + 6\ e^- = 8\ e^- + 6 \times 8\ e^-$)

Therefore, 6 iodide ions must change to 3 iodine molecules while 1 chlorate ion changes to 1 chloride ion to give the 6 electrons needed for the reduction

$$ClO_3^- + 3\ H_2O + 6\ e^- \rightarrow Cl^- + 6\ OH^-$$
$$3(2\ I^- \rightarrow I_2 + 2\ e^-)$$
$$\overline{ClO_3^- + 6\ I^- + 3\ H_2O \rightarrow 3\ I_2 + Cl^- + 6\ OH^-}$$

This is the **balanced** net ionic equation. Were one to use the equation to determine the relative amounts of reagent potassium chlorate required to oxidize a given amount of potassium iodide, one would have to add the necessary spectator ions to the equation to obtain the overall equation:

$$KClO_3 + 6\ KI + 3\ H_2O \rightarrow 3\ I_2 + KCl + 6\ KOH$$

The substances, potassium chlorate and potassium iodide, are needed in the molar ratio of 1 to 6 respectively.

14.4 ION-ELECTRON METHOD OF BALANCING OXIDATION-REDUCTION EQUATIONS

There are several different methods for balancing oxidation-reduction equations. The method described here is called the **ion-electron method.** For the purpose of simplicity the hydronium ion will be represented as H^+ in this chapter. One must remember that each H^+ is attached to a water molecule not shown. Should one desire to obtain the balanced equation with hydronium ions, one must change all H^+ in the final equation to H_3O^+ and then add one H_2O to the opposite side of the equation for every hydronium ion.

An oxidation-reduction equation can be balanced by following these steps:

1. Write the equation in the net ionic form.
2. Separate the equation into two parts by pairing ions or compounds containing the same element in changed oxidation states.
3. Balance each part as follows:
   A. In an acid solution:
      (1) Balance the elements by using H_2O and H^+ as needed.
      (2) Balance electrically by adding electrons to get the half-cell equation.
   B. In a basic solution:
      (1) Balance the elements by adding H_2O and OH^- as needed. Note: $2\ OH^- \rightarrow H_2O + O$ (an oxygen contained in a compound or ion)
      (2) Balance electrically by adding electrons to get the half-cell equation.
4. Multiply the half-cell equations by appropriate factors such that the number of electrons is equal in both equations, and then combine them.
5. Reduce the coefficients of the equation to the lowest possible integers.

Example 1. Potassium permanganate reacts with concentrated hydrochloric acid to produce chlorine gas and manganese(II) ion. Balance the equation for the reaction.

Overall equation:

$$HCl + KMnO_4 \rightarrow Cl_2 + KCl + MnCl_2 + H_2O$$

Complete ionic equation (writing H_3O^+ as H^+):

$$H^+ + Cl^- + K^+ + MnO_4^- \rightarrow Cl_2 + K^+ + Cl^- + Mn^{2+} + Cl^- + H_2O$$

Net ionic equation (unbalanced):

$$MnO_4^- + Cl^- + H^+ \rightarrow Cl_2 + Mn^{2+} + H_2O \text{ (Step 1)}$$

One half-cell reaction (Step 2):

$$\overset{+7}{MnO_4^-} \rightarrow \overset{+2}{Mn^{2+}} \text{ (reduction)}$$

$$MnO_4^- + 8\ H^+ \rightarrow Mn^{2+} + 4\ H_2O \text{ (Step 3A, 1)}$$

(Add H_2O to balance O's first, then add H^+ to balance H's).

$$MnO_4^- + 8\ H^+ + 5\ e^- \rightarrow Mn^{2+} + 4\ H_2O \text{ (Step 3A, 2)}$$

The other half-cell reaction (Step 2):

$$\overset{-1}{Cl^-} \rightarrow \overset{0}{Cl_2} \text{ (oxidation)}$$

$$2\ Cl^- \rightarrow Cl_2 \text{ (Step 3A, 1)}$$

$$2\ Cl^- \rightarrow Cl_2 + 2\ e^- \text{ (Step 3A, 2)}$$

Combining half-cell reactions (Step 4):

$$\begin{array}{r} 2(MnO_4^- + 8\ H^+ + 5\ e^- \rightarrow Mn^{2+} + 4\ H_2O) \\ 5(2\ Cl^- \rightarrow Cl_2 + 2\ e^-) \\ \hline 2\ MnO_4^- + 10\ Cl^- + 16\ H^+ \rightarrow 2\ Mn^{2+} + 5\ Cl_2 + 8\ H_2O \end{array}$$

Step 5 does not apply here since the equation already has the smallest coefficients.

Notice that all products and reactants must be known, except the hydronium ion (H^+) and water. These two are added as required in the balancing.

The balanced equation (Step iv) may be converted to the overall equation using molecular formulas by "combining" ions and adding the necessary spectator ions:

$$\underset{(2\ K^+ \text{ and } 6\ Cl^- \text{ added})}{(2\ K^+ + 2\ MnO_4^-) + (16\ H^+ + 16\ Cl^-)} \rightarrow \underset{(4\ Cl^-,\ 2\ K^+ \text{ and } 2\ Cl^- \text{ added})}{(2\ Mn^{2+} + 4\ Cl^-) + (2\ K^+ + 2\ Cl^-)} + 5\ Cl_2 + 8\ H_2O$$

Equal numbers of each kind of ion were added to both sides of the equation to give a balanced total ionic equation. The chemicals are required in the molar ratios shown in this equation:

$$2\ KMnO_4 + 16\ HCl \rightarrow 2\ MnCl_2 + 2\ KCl + 5\ Cl_2 + 8\ H_2O$$

Two moles of $KMnO_4$ react with 16 moles of HCl to yield 5 moles of chlorine gas.

The balanced equation may be used to determine the relative quantities of reactants and products involved. Using the equation above, the number of milliliters of 6.0 M HCl required to reduce 25 g of $KMnO_4$ is:

$$\frac{25\ \text{g}}{158\ \text{g/mole}} \times \frac{16}{2} \times \frac{1}{6\ \text{moles/L}} \times 1000\ \text{ml/L} = 211\ \text{ml of}$$

6.0 M HCl to yield

$$\frac{25\ \text{g}}{158\ \text{g/mole}} \times \frac{5}{2} \times 22.4\ \text{L/mole} = 8.86\ \text{L of}\ Cl_2\ \text{at STP}$$

Example 2. Balance the half-cell equation for the reduction of dilute nitric acid to nitrogen(II) oxide in an acid solution:

$$\underset{\textbf{Dilute}}{HNO_3} \rightarrow NO_{(g)}$$

In net ionic form:

$$NO_3^- \rightarrow NO$$

This is all of the half-cell equation that is needed to begin balancing it. The nitrate ion is consumed and nitrogen(II) oxide gas is formed. But NO_3^- cannot change to NO as the only product; two oxygen atoms are not accounted for. The only possible additional product is water:

$$NO_3^- \rightarrow NO + 2\ H_2O$$

The hydrogen atoms needed on the left side of the half-cell equation must be in the form of hydronium ions:

$$4\ H^+ + NO_3^- \rightarrow NO + 2\ H_2O$$

The equation is balanced elementally (the atoms balance) but not electrically. Each of 4 hydronium ions is minus 1 electron, and 1 nitrate ion has a charge of minus 1, but the products are neutral. For four H^+ ions and one NO_3^- ion to become one NO and two H_2O molecules, three electrons must be taken up:

$$4\ H^+ + NO_3^- + 3\ e^- \rightarrow NO + 2\ H_2O$$

$$4^+ + 1^- + 3^- = 0 \qquad\qquad 0 + 0 = 0$$

The half-cell equation is balanced. **Electrons are gained** by the reactants; **reduction has occurred.**

Example 3. Zinc metal reacts with the nitrate ion in an acid solution to give the zinc ion and ammonium ion. Obtain the balanced net ionic equation:

$$Zn + NO_3^- \xrightarrow{\text{acid solution}} Zn^{2+} + NH_4^+$$

Oxidation half-cell equation:

$$Zn \rightarrow Zn^{2+} \text{ (elementally balanced)}$$

$$Zn \rightarrow Zn^{2+} + 2\ e^- \text{ (electrically balanced)}$$

Reduction half-cell equation:

$$NO_3^- \rightarrow NH_4^+$$

$$NO_3 \xrightarrow{\text{acid solution}} NH_4^+ + 3\ H_2O$$

$$NO_3^- + 10\ H^+ \rightarrow NH_4^+ + 3\ H_2O \text{ (balanced elementally)}$$

$$NO_3^- + 10\ H^+ + 8\ e^- \rightarrow NH_4^+ + 3\ H_2O \text{ (balanced electrically)}$$

$$1^- + 10^+ + 8^- = 1^+ + 0$$

Adding equations together so that electrons drop out:

$$4(Zn \rightarrow Zn^{2+} + 2\ e^-)$$

$$\underline{NO_3^- + 10\ H^+ + 8\ e^- \rightarrow NH_4^+ + H_2O}$$

$$4\ Zn + NO_3^- + 10\ H^+ \rightarrow 4\ Zn^{2+} + NH_4^+ + 3\ H_2O$$

This is the balanced net ionic equation.

Equations of reactions occurring in **basic** solutions are not encountered so frequently, and the equations are more difficult to balance. One example will be discussed.

Example 4. As stated before in this chapter, the chlorate ion in a basic solution oxidizes the iodide ion to I_2. The only other product is the chloride ion. Balance the equation:

$$ClO_3^- + I^- \xrightarrow{\text{basic solution}} Cl^- + I_2$$

Choosing the simplest half-cell equation (oxidation):

$$I^- \rightarrow I_2$$

$$2\ I^- \rightarrow I_2 \text{ (balanced elementally)}$$

$$2\ I^- \rightarrow I_2 + 2\ e^- \text{ (balanced electrically)}$$

The reduction half-cell equation is

$$ClO_3^- \rightarrow Cl^-$$

One must balance the elements in the equation by adding only water (H_2O) and the hydroxide ion (OH^-). To keep the hydrogens balanced, one must add H_2O to one side and 2 OH^- to the other.

one H_2O molecule takes one O atom from a ClO_3^- ion to form two OH^- ions.

$ClO_3^- + 3\ H_2O \rightarrow Cl^- + 6\ OH^-$ (balanced elementally)

$-1 + 0 = -1 \qquad -1 + -6 = -7$

$ClO_3^- + 2\ H_2O + 6\ e^- \rightarrow Cl^- + 6\ OH^-$ (balanced electrically)

Combining half-cell equations:

$$3(2\ I^- \rightarrow I_2 + 2\ e^-)$$
$$ClO_3^- + 3\ H_2O + 6\ e^- \rightarrow Cl^- + 6\ OH^-$$
$$6\ I^- + ClO_3^- + 3\ H_2O \rightarrow 3\ I_2 + Cl^- + 6\ OH^-$$

In this reaction iodide ions lose electrons; iodine increases in oxidation state; the iodide ion is oxidized. How is oxidation accomplished? The chlorate ion and water in a basic solution are the **oxidizing agents.**

In the same reaction the chlorate ion and water gain electrons in becoming the chloride ion and the hydroxide ion; the oxidation state of chlorine decreases from +5 to −1; chlorate and water are reduced. The iodide ion gives the chlorate ion and water the electrons, and, therefore, the iodide ion is the **reducing agent.**

The iodide ion loses electrons. However, it is not correct to say that the chlorate ion alone gains electrons. The chlorate ion and water gained electrons to become hydroxide ions and chloride ions. The chlorate ion alone is not the oxidizing agent. Further, since a change in the concentration of any one of the ions changes the oxidizing strength, the water and the chlorate, the chloride and the hydroxide ions together are the oxidizing agents.

14.5 THE SOURCE OF ELECTROMOTIVE FORCE

When electrons flow from one point to another they are moved by **electromotive force.** Electromotive force consists in the potential difference in electric charge between two points. The greater the difference in electric charge, the greater the force with which electrons are drawn to the more positive of the two points. The volt is the unit of electromotive force.

Electrons flow from the copper anode to the silver cathode, therefore there is an electromotive force between them. To understand this one must first analyze the situation existing at the surface of each electrode before the external conductor, the wire, is connected between them. Copper atoms have a slight tendency to lose the two valence electrons and to enter the solution as ions; a small proportion of the surface atoms do this. At the same time copper(II) ions collide with the electrode surface to gain electrons and remain there. These two opposite reactions become equal in rate, and an equilibrium exists with respect to this reaction between the electrode surface and the solution:

$$Cu \rightleftarrows Cu^{2+} + 2\ e^-$$

A small concentration of electrons resides in the copper electrode, the magnitude of the charge depending upon the relative ionization energy of copper, the energy of hydration of the copper(II) ions and the bonding forces within copper metal. A like situation exists at the silver electrode. Atoms leave at a very slow rate, leaving their electrons behind, and ions return to the surface to take up electrons. This equilibrium exists at the electrode surface:

$$Ag \rightleftarrows Ag^+ + e^-$$

However, silver has a higher ionization energy than copper and holds its valence electron more tightly than the copper atom holds its electrons. Therefore, the electron concentration at the silver electrode is less than at the copper. There exists a difference between the charges on the two electrodes. When the two electrodes are connected with a material that allows a free passage of electrons, the electrons flow from the region of greater concentration (the copper electrode) to the region of lower concentration (the silver electrode).

Unless the two electrode solutions come in contact, no current flows. The excess of positive copper(II) ions about the copper electrode would attract the excess electrons to the copper surface, and the nitrate ions remaining about the silver electrode would soon repel electrons at that point. The porous plate allows nitrate ions to pass from the silver compartment to the anode compartment so that they equal twice the number of copper(II) ions coming into the solution. It also allows the copper(II) ions to migrate into the cathode compartment.

One can see from the two equilibria existing at the electrodes before connecting them together that the concentration of the two ions in their respective solutions affects the magnitude of the electron charge on the electrodes before the circuit is closed, and therefore affects the voltage of the cell. The higher the copper(II) ion concentration, the lower the electron concentration at the anode and the lower the voltage. The greater the silver ion concentration, the lower the electron concentration at the cathode and the greater the cell voltage.

14.6 CELL POTENTIALS

Two simply constructed primary cells are sketched in Figures 14.4 and 14.5; namely the Daniell cell and the copper-silver cell discussed in the previous sections.

The walls of the porous cup replace the porous plate. The electrode reactions are those represented beneath the sketches, with oxidation occurring at the anode and reduction at the cathode. Electrons move from left to right in the external conductor of both cells as sketched. During operation zinc dissolves and the copper plates out (precipitates) in the Daniell cell, while copper dissolves and the silver precipitates in the copper-silver cell.

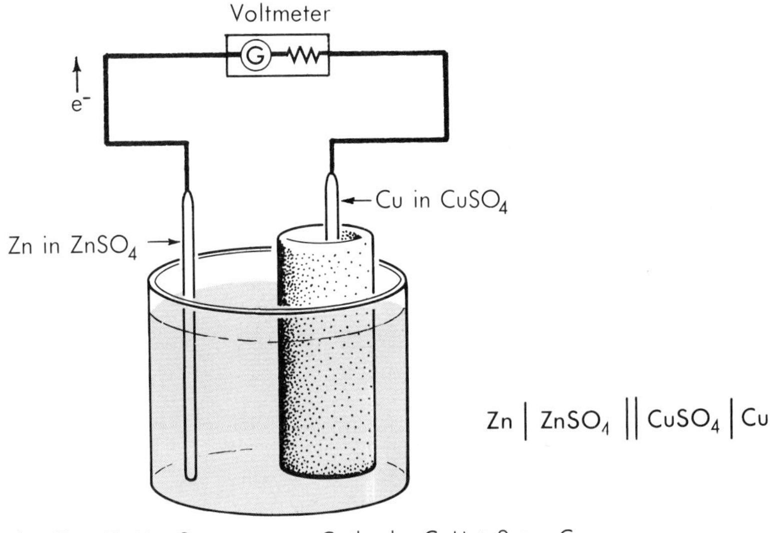

FIGURE 14.4 DANIELL CELL.

The voltage of the Daniell cell is 1.10 volts, as compared to a voltage of 0.46 volts for the copper-silver cell. This is true only when the concentrations of the ions involved in the net ionic equations of the cell reactions are 1 molar. No attempt will be made to define voltage in fundamental terms. It suffices to say that if the same resistance is connected in series with a galvanometer across the electrodes of either cell, the current will be 1.10/0.46 as great from the Daniell cell as from the copper-silver cell. A voltmeter is a galvanometer with a high resistance connected in series with it to limit the current. It should be obvious that if one primary cell sends a current of more than twice as many amperes through a resistance than another cell does, it must have a potential or voltage more than twice as great.

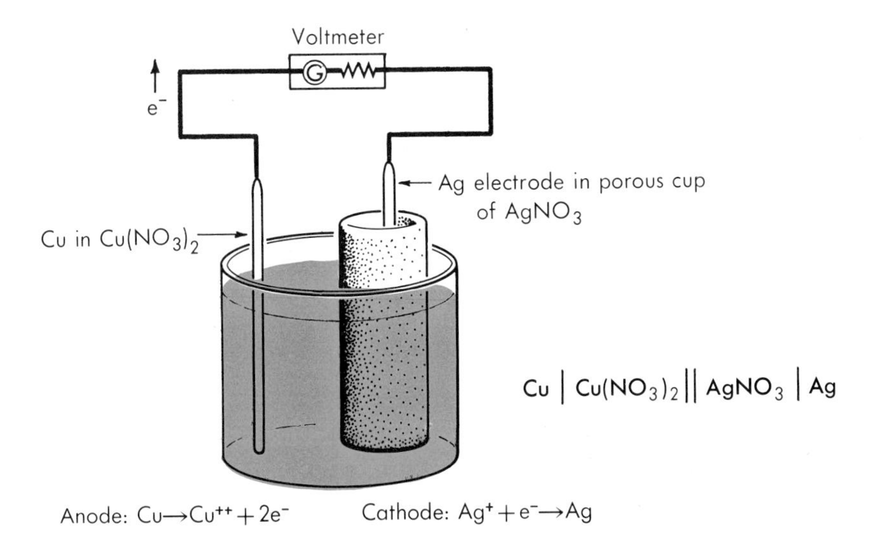

FIGURE 14.5 SILVER-COPPER CELL.

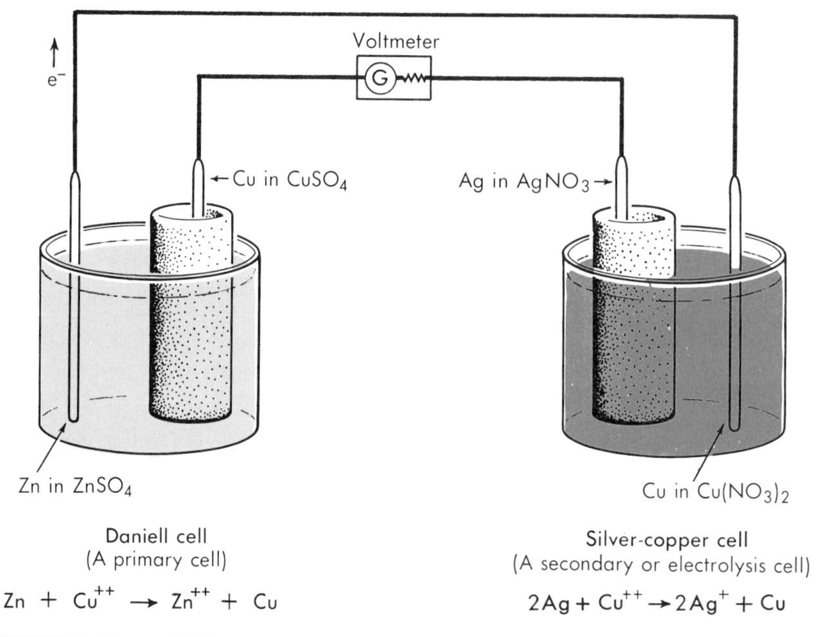

FIGURE 14.6 PRIMARY-SECONDARY CELL SYSTEM.

An interesting experiment may be performed to demonstrate that the Daniell cell has a greater electromotive force than the copper-silver cell. One can connect the two anodes and the two cathodes together as in Figure 14.6. When this is done, the reactions in the Daniell cell are the same as before, but the reactions in the copper-silver cell are reversed. Silver dissolves and copper precipitates. The latter cell has become a **secondary cell** or an **electrolysis cell.** Its motivation comes from the cell of higher voltage, which drives a current through the cell of lower voltage against its natural tendency. The preparation of chlorine gas, sodium hydroxide solution and sodium metal and the purification of aluminum and copper all take place in secondary cells where the electrons move as forced by an external electromotive force. The silver electrode, which was the cathode when the cell acted as a primary cell, is now the anode because oxidation is occurring there. The copper electrode has become the cathode because reduction of copper(II) ions occurs there. However, the charge in the solution about each electrode, now determined by the Daniell cell, is the same as before. The voltmeter in the circuit reads $1.10 - 0.46 = 0.64$ volts.

14.7 TABLE OF ELECTRODE POTENTIALS (VOLTAGES)

In the cells discussed, a metal was either dissolved or precipitated at each electrode. Very often this is not the case. The cell represented by this notation:

$$Zn|ZnSO_4||H_2SO_4,\ H_2|Pt$$

is sketched in Figure 14.7. This cell contains the zinc electrode of the Daniell cell and the so-called hydrogen electrode. The **hydrogen electrode** consists of a platinum strip immersed in a solution of hydronium ions over which hydrogen gas is bubbled. The platinum metal provides a surface to which hydrogen gas and hydronium ions may come, the hydrogen to give up electrons when the electrode acts as an anode, and the hydronium ions to gain electrons when the electrode is the cathode. The platinum also serves as a conductor to carry the electrons to or from the reaction site. Platinum is an inert metal and takes no other part in the reaction. When the zinc ion concentration and the hydronium ion concentration are both 1 molar and the hydrogen gas pressure is 1 atmosphere, zinc acts as the anode. Dissolving zinc liberates electrons which travel through the external conductor to the hydrogen electrode, where they are absorbed in the reduction of hydronium ions, as shown by the electrode equations in Figure 14.7. The voltage of this cell is 0.76.

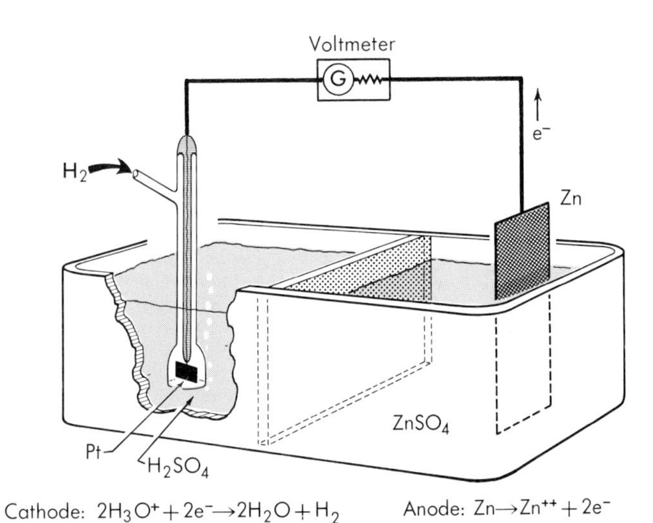

FIGURE 14.7 HYDROGEN ELECTRODE IN A PRIMARY CELL WITH ZINC.

A primary cell may be made of a copper electrode and the hydrogen electrode, as in Figure 14.8. The hydrogen electrode is the anode in this cell. Oxidation of hydrogen and the plating of copper occurs as represented by the half-cell reactions. Electrons move from the hydrogen electrode to the copper electrode with a measured voltage of 0.34. Hence the copper electrode is 0.34 volts more positive than the hydrogen electrode. The hydrogen electrode is 0.76 volts more positive than the zinc electrode. Therefore the copper electrode should be $+0.34 - (-.76) = 1.10$ volts more positive than the zinc electrode. This is found to be true because the Daniell cell joins these two electrodes and has 1.10 volts. In the silver-copper cell, electrons flow to the silver electrode with a voltage of 0.46. The silver electrode should be $0.34 + 0.46 = 0.80$ volts more positive than the hydrogen electrode. Hence the hydrogen-silver cell has a potential of 0.80 volts.

When one knows the voltages of the two cells with a common electrode, one can calculate the voltage of the cell composed of the two dissimilar electrodes, as was done above.

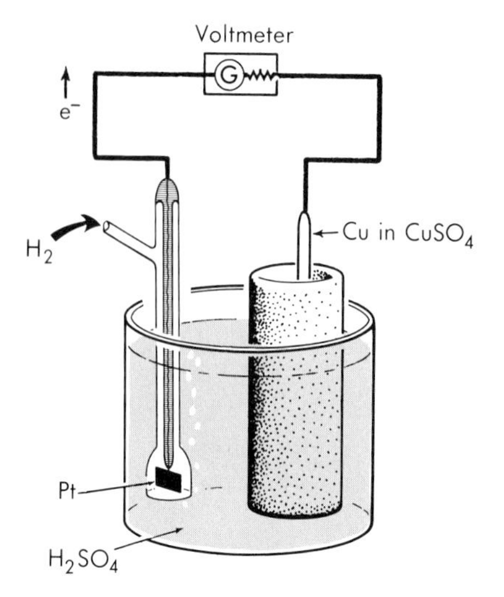

Anode: $H_2 + 2H_2O \rightarrow 2H_3O^+ + 2e^-$ Cathode: $Cu^{++} + 2e^- \rightarrow Cu$

$$Pt \mid H_2;\ H_2SO_4 \parallel CuSO_4 \mid Cu$$

FIGURE 14.8 HYDROGEN ELECTRODE IN A PRIMARY CELL WITH COPPER.

TABLE 14.2 ELECTROMOTIVE SERIES

Oxidized form			Reduced form	Voltage
Cs^+	$+ e^-$	⇄	Cs	−3.02
Li^+	$+ e^-$	⇄	Li	−3.02
Rb^+	$+ e^-$	⇄	Rb	−2.99
K^+	$+ e^-$	⇄	K	−2.92
Ba^{2+}	$+ 2 e^-$	⇄	Ba	−2.90
Sr^{2+}	$+ 2 e^-$	⇄	Sr	−2.89
Ca^{2+}	$+ 2 e^-$	⇄	Ca	−2.87
Na^+	$+ e^-$	⇄	Na	−2.71
Mg^{2+}	$+ 2 e^-$	⇄	Mg	−2.34
Be^{2+}	$+ 2 e^-$	⇄	Be	−1.70
Al^{3+}	$+ 3 e^-$	⇄	Al	−1.67
Mn^{2+}	$+ 2 e^-$	⇄	Mn	−1.05
Zn^{2+}	$+ 2 e^-$	⇄	Zn	−0.76
Cr^{2+}	$+ 2 e^-$	⇄	Cr	−0.71
S	$+ 2 e^-$	⇄	S^{2-}	−0.51
Fe^{2+}	$+ 2 e^-$	⇄	Fe	−0.44
Cd^{2+}	$+ 2 e^-$	⇄	Ce	−0.40
Co^{2+}	$+ 2 e^-$	⇄	Co	−0.28
Ni^{2+}	$+ 2 e^-$	⇄	Ni	−0.25
Sn^{2+}	$+ 2 e^-$	⇄	Sn	−0.14
Pb^{2+}	$+ 2 e^-$	⇄	Pb	−0.13
$2\ H_3O^+$	$+ 2 e^-$	⇄	$H_2 + 2\ H_2O$	0.00
Bi^{3+}	$+ 3 e^-$	⇄	Bi	+0.20
As^{3+}	$+ 3 e^-$	⇄	As	+0.30
Cu^{2+}	$+ 2 e^-$	⇄	Cu	+0.34
I_2	$+ 2 e^-$	⇄	$2\ I^-$	+0.53
Fe^{3+}	$+ e^-$	⇄	Fe^{2+}	+0.75
Hg_2^{2+}	$+ 2 e^-$	⇄	2 Hg	+0.80
Ag^+	$+ e$	⇄	Ag	+0.80
Br_2	$+ 2 e^-$	⇄	$2\ Br^-$	+1.06
Cl_2	$+ 2 e^-$	⇄	$2\ Cl^-$	+1.36
Au^+	$+ e^-$	⇄	Au	+1.68
F_2	$+ 2 e^-$	⇄	$2\ F^-$	+2.85

The activity series, or **electromotive series** (Table 14.2), lists for each electrode the voltage of a cell joining that electrode and the hydrogen electrode, with the sign being **negative** if the hydrogen electrode is the cathode and **positive** if the hydrogen electrode is the anode. To find the voltage of a cell composed of any two electrodes listed, subtract the more negative value from the more positive. The cathode will be the electrode with the more positive voltage in the table. Electrons will flow toward the electrode with the more positive number. For example, a cell of this notation: $Pb|Pb(NO_3)_2||AgNO_3|Ag$ has a voltage of $0.080-(-0.13)=0.93$ volts. A cell represented by this notation: $Zn|Zn(NO_3)_2||Pb(NO_3)_2|Pb$ has a voltage of $-0.13-(-0.76)=0.63$ volts; zinc is the anode. These particular voltages are obtained only when the concentrations of all the ions participating in the net ionic reaction are 1 molar.

the cell potential found by subtracting the standard electrode potential of the anode from the standard electrode potential of the cathode is valid only when the concentrations of all ions appearing in the net-ionic equation are 1 molar.

14.8 SOME COMMON PRIMARY CELLS—THE GRAVITY CELL

The primary cell that provided electricity for the old transcontinental telegraph was the gravity cell (Fig. 14.9a), a variation of the Daniell cell. This cell is interesting because an ingenious arrangement makes it unnecessary to divide it into two compartments. The copper electrode, in the form of a "crowfoot," rests on the bottom of a cylindrical glass vessel, surrounded by a copper(II) sulfate solution. At the top of the cell is a zinc crowfoot electrode in a solution containing zinc sulfate and sodium sulfate. Because the copper(II) sulfate solution is more dense than the zinc sulfate–sodium sulfate solution, it tends to remain on the bottom under the influence of gravity. Hence the name gravity cell.

Operation of the cell tends to keep the copper(II) ions down

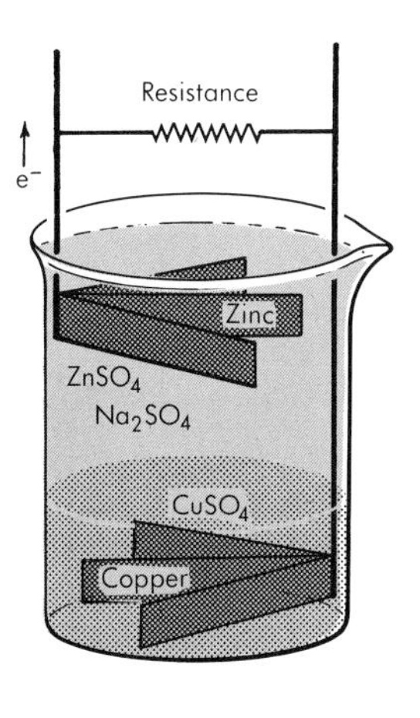

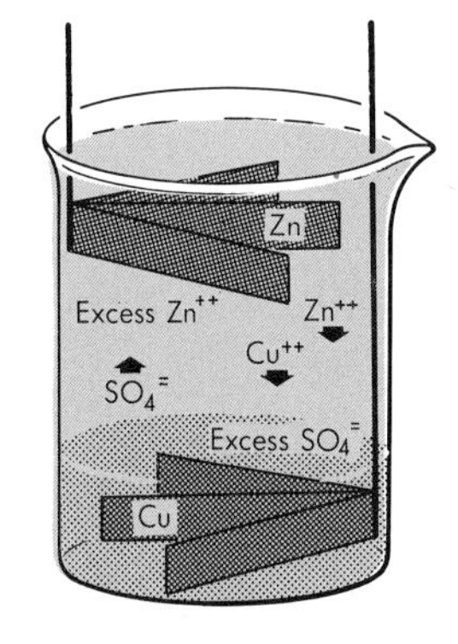

FIGURE 14.9 GRAVITY CELL.

about the cathode. For this reason, when the cell is not in use, the electrodes are connected through a large resistance. A small but continuous current flows through the cell. When current is drawn from the cell, copper(II) ions leave the solution in the neighborhood of the cathode, to be precipitated as atoms on the cathode. Sulfate ions remain behind and give a negative charge to the area surrounding the copper (Fig. 14.9b). Copper(II) ions are attracted to the negative area and remain about the cathode. At the same time zinc ions enter the solution from the anode. The existence of more positive ions than negative ions about the anode gives the area a positive charge, and the sulfate ions are attracted to the anode. Zinc ions tend to move from the more positive solution about the anode to the cathode and the copper(II) ions. This presents no difficulty, because zinc ions will not react with copper. One needs only to keep copper(II) ions away from the zinc electrode.

14.9 THE DRY CELL

The dry cell is the most familiar of all primary cells. Flashlight batteries and transistor radio batteries are dry cells. The dry cell is not completely dry, but it contains a stiff paste of water, ammonium chloride, zinc chloride and manganese dioxide surrounding a graphite electrode, all enclosed in a zinc container, as shown in Figure 14.10. At the inner surface of the zinc container this reaction occurs:

with the solution in a semi-dry paste, the flashlight cell can be carried in any position.

$$Zn \rightarrow Zn^{2+} + 2\ e^-$$

This is an oxidation reaction, and the zinc is the anode. The graphite rod serves as an inert electrode, much as the platinum foil does in the hydrogen electrode. At the rod's surface manganese dioxide and ammonium ions rearrange, with the absorption of electrons, to form new substances:

$$2\ MnO_2 + 2\ NH_4^+ + 2\ e^- \rightarrow 2\ NH_3 + H_2O + Mn_2O_3$$

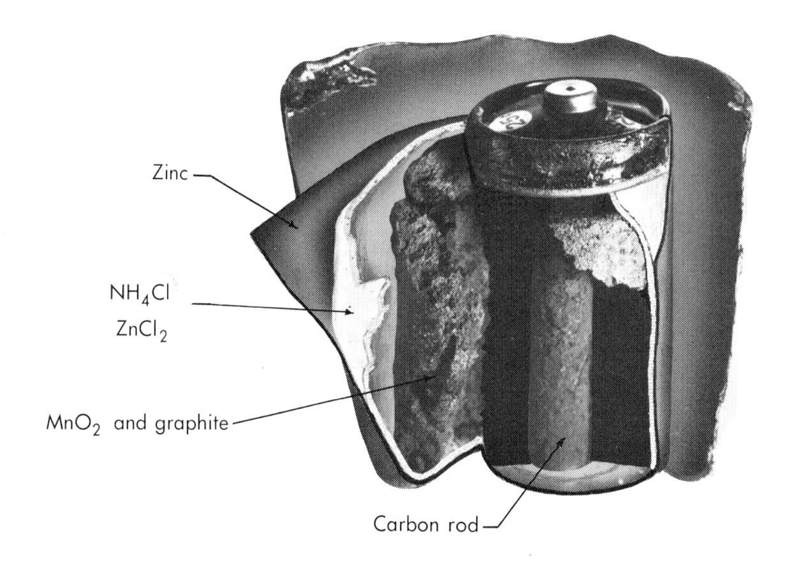

FIGURE 14.10 THE DRY CELL.

Because electrons are taken up by the reactants near it, the graphite rod is the cathode. Electrons flow from zinc to graphite through an external conductor. Zinc ions form a complex ion, $Zn(NH_3)_4^{2+}$, with the ammonia formed in the cathode reaction so that no odor of ammonia is detectable. The overall reaction is:

$$2\ Zn + 4\ MnO_2 + 4\ NH_4^+ \rightarrow Zn(NH_3)_4^{2+} + Zn^{2+} + 2\ H_2O + 2\ Mn_2O_3$$

The cell runs down (fails to produce more electricity) when either the ammonium chloride or the manganese dioxide in the vicinity of the cathode is depleted or when the zinc container is eaten through and lets the liquid out. If it stands overnight, a dead cell may have a temporary revival. Some ammonium chloride solution may diffuse to the cathode. Important characteristics of dry cells are their lightness and the absence of either acids or bases in the fluid.

14.10 THE LEAD STORAGE CELL

The source of electricity to start the automobile engine and to ignite the gasoline-air mixture in the cylinders is a battery of either three or six lead storage cells. The anode of a lead storage cell is a lead grid with spongy lead pressed into the grid work. The cathode is a lead grid with lead oxide in the grid, as in Figure 14.11. The electrodes are suspended in a solution of sulfuric acid. The lead grids take no part in the cell reactions but carry the current to the lead battery terminals. When the cell is in operation, spongy lead on the anode dissolves, but the lead ions which are formed precipitate lead sulfate directly on the plate:

$$Pb + HSO_4^- + H_2O \rightarrow PbSO_4 + H_3O^+ + 2\ e^-$$

Lead(IV) oxide dissolves at the cathode to become lead(II) ions, which also precipitate as lead sulfate:

$$PbO_2 + 3\ H_3O^+ + HSO_4^- + 2\ e^- \rightarrow PbSO_{4(s)} + 5\ H_2O$$

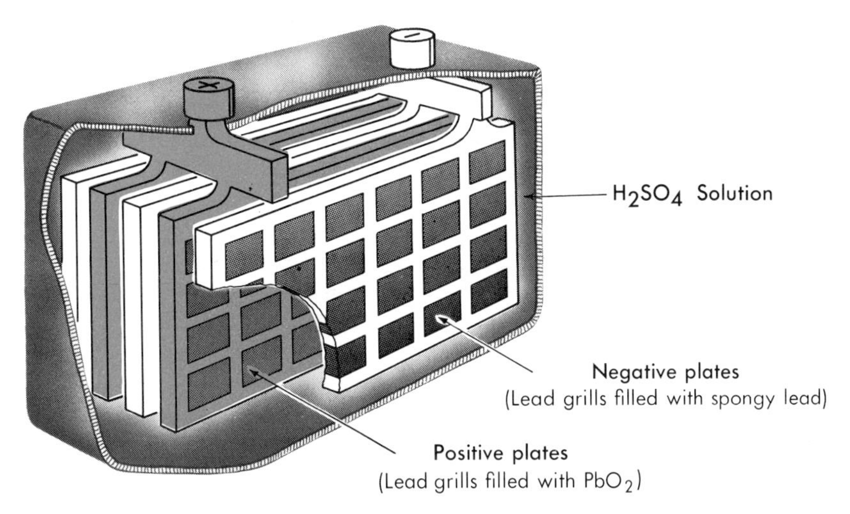

FIGURE 14.11 THE LEAD STORAGE BATTERY.

The overall reaction is:

$$Pb_{(s)} + PbO_{2(s)} + 2\ H_2SO_4 \rightarrow 2\ PbSO_{4(s)} + 2\ H_2O$$

the small s in parentheses after a formula in an equation indicates a solid.

Lead sulfate forms at both electrodes; sulfuric acid is consumed and water is formed. The cell is discharged when the spongy lead and lead(IV) oxide are almost completely converted to lead sulfate and when the sulfuric acid cell solution is reduced to a very low concentration. The sulfuric acid concentration is easily checked to determine the "charge" of the cell. Sulfuric acid has a density about twice that of water. When the cell is discharged, the density of the solution approaches that of water. The density is tested with a hydrometer.

A lead storage cell can be **recharged**; that is, the lead oxide and the spongy lead can be regenerated in place, and the sulfuric acid can be restored by reversing the flow of electrons. This is done constantly by the alternator while the engine is operating. The alternator impresses a higher voltage than the cell delivers, and changes it (during charging) to a secondary cell. The lead storage cell supports many, many cycles of charging and discharging.

Each cell provides 2 volts. Three cells or 6 cells are connected anode to cathode to give either 6 or 12 volts. Each storage cell has a cathode and an anode of several plates. These larger electrodes are capable of greater power (a greater current at the same voltage) or of a longer life, for a greater weight of reactants is present. Electrical energy depends upon both the voltage and the quantity of electricity that flows.

14.11 CORROSION

The corrosion (oxidation) of metals in the atmosphere is often related to the reactions that occur in primary cells. Consider a drop of water on an iron surface, as in Figure 14.12. This constitutes an electrical cell. The edge of the drop on the iron surface, where air, water and iron come together, is the cathode. By the absorption of atmospheric oxygen and electrons,

$$O_2 + 2\ H_2O + 4\ e^- \rightarrow 4\ OH^-$$

hydroxide ions are formed. Hydroxide ions diffuse throughout the

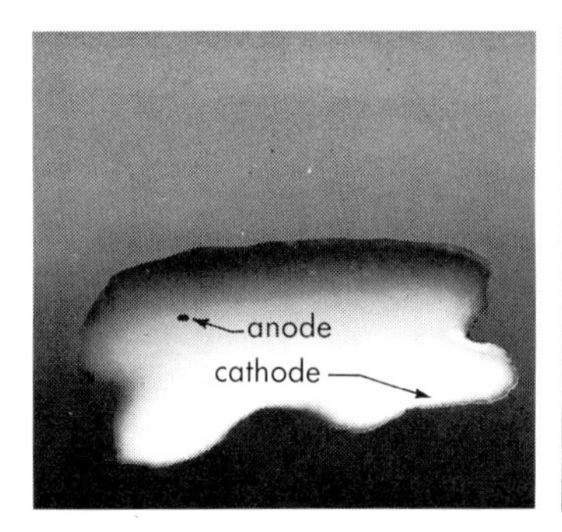

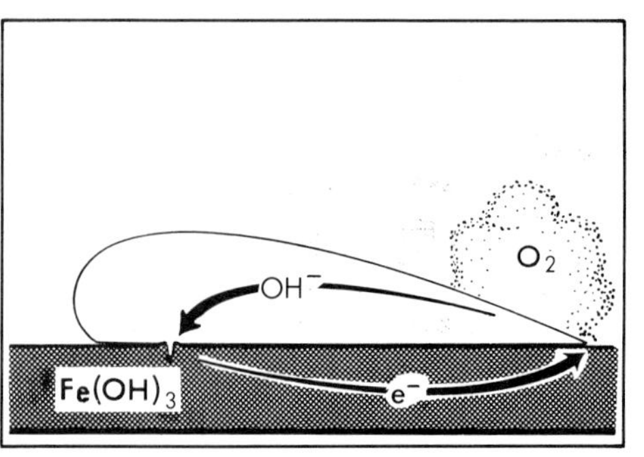

FIGURE 14.12 CORROSION ON A MOIST IRON SURFACE.

droplet. Electrons are produced at the site of the anode reaction, which is under the drop on the iron surface:

$$Fe + 3\ OH^- \rightarrow \underset{\textbf{Rust}}{Fe(OH)_3} + 3\ e^-$$

The brown precipitate dries to iron(III) oxide. Electrons produced at the water-metal interface are carried through the metal to the reduction site at the edge of the droplet. The overall balanced reaction is:

$$4\ Fe + 3\ O_2 + 6\ H_2O \rightarrow 4\ Fe(OH)_3$$

Iron is often coated with less active metals (such as tin) to prevent corrosion. Oxidation of tin is very slow and the iron is protected until the tin corrodes through or is broken in one spot. The tin then becomes a liability, because it offers a better surface for the reaction with oxygen than does iron. Iron oxidizes at the break more rapidly than if no tin were present, since iron is the anode and tin is the cathode.

Iron is often coated with zinc to protect it from atmospheric oxidation. Galvanized ware is zinc-coated iron. Zinc, more reactive than iron, becomes the anode and iron, the cathode. It might appear that oxidation would be hastened, but the presence of zinc retards the oxidation of iron even when a break occurs in the coating. When a drop of water covers the hole, more active zinc is dissolved in preference to iron. Often the zinc coating almost completely disappears before a hole penetrates the iron.

14.12 OTHER OXIDATION-REDUCTION EQUATIONS

Some oxidation-reduction equations do **not occur in water** solution, do not involve ions and cannot be balanced by the method described in this chapter. One of these reactions might well have been the first reaction called oxidation. One such reaction is the burning of coal:

$$\overset{0}{C} + O_2 \rightarrow \overset{+4}{C}O_2$$

The oxidation state of carbon is raised from 0 to +4 as indicated. The oxidation state of oxygen is decreased from 0 to −2:

$$C + \overset{0}{O_2} \rightarrow C\overset{-2}{O_2}$$

No electron change is obvious. Neutral carbon and an oxygen molecule with its shared electrons merely join and share their electrons. Initially it was thought that carbon dioxide was ionic, that oxide ions carried a charge of −2 and that electrons were exchanged in the reaction, but this is not so.

Reduction may well have been named for the reaction of coke

at high temperatures with iron(III) oxide (Fe_2O_3) in an ore which is reduced in weight and composition to iron.

$$\overset{+3 \;\longrightarrow\; 0}{2\ Fe_2O_3 + 3\ C \xrightarrow[\text{heat}]{} 4\ Fe + 3\ CO_2}$$
$$0 \longleftarrow +4$$

Again, although oxidation states change, no clear case of electron transfer is apparent.

Perhaps the method for balancing oxidation-reduction equations should be expanded to include this type of reaction. However, most reactions studied in elementary chemistry occur in water solutions and involve ions. Those that do not are generally easily balanced by inspection.

EXERCISES – Set I

14.1 Define oxidation, reduction, oxidation state, oxidizing agent and reducing agent.

14.2 Write the oxidation states (numbers) of all elements in the following formulas:

(a) KCl, (b) $KClO_3$, (c) ClO_4^-, (d) BrO^-, (e) PO_4^{3-}, (f) HSO_4^-, (g) MnO_2, (h) $S_2O_3^{2-}$, (j) $Na_2S_2O_3$, (k) $Cr_2O_7^{2-}$

14.3 In this balanced half-cell equation:

$$MnO_4^- + 2\ H_2O + 3\ e^- \rightarrow MnO_2 + 4\ OH^-$$

a. What element changes oxidation state?
b. Is it oxidized or reduced?
c. What has gained electrons in this reaction?

14.4 Mercury(II) sulfide is dissolved in aqua regia (a mixture of concentrated HCl and HNO_3) to form sulfur, the tetrachloromercurate(II) ion, $HgCl_4^{2-}$, and nitrogen dioxide gas, NO_2.

a. Write a balanced net ionic equation.
b. What is the oxidizing agent?
c. What is oxidized?

14.5 Name three metals that will replace silver in solution.

14.6 What is meant by a spontaneous reaction?

14.7 Write three oxidation-reduction reactions.

14.8 Write the half-cell equations for the dissolving of zinc in copper(II) sulfate solution.

14.9 What charge does the anode bear in a primary cell?

14.10 Define anode. The definition should hold for both primary and secondary cells.

14.11 Write the overall reaction, the net ionic equation and the electrode reactions for the copper-silver cell, the zinc-hydrogen cell and the copper-hydrogen cell.

14.12 Describe how one might obtain the electrode voltage of the tin-tin ion electrode.

14.13 Could one use zinc in place of platinum for the hydrogen electrode? Explain.

14.14 What is the purpose of the porous plate?

14.15 Write the reactions occurring at the electrodes while the lead storage cell is being charged.

EXERCISES – Set II

14.16 Obtain the balanced net ionic equation for each of these reactions by the ion-electron method.

a. $Cu^{2+} + I^- \rightarrow Cu^+ + I_2$
b. $CeO_2 + I^- \rightarrow Ce^{3+} + I_2$ (in acid)
c. $Cu + H^+ + NO_3^- \rightarrow Cu^{2+} + NO_2 + H_2O$ (conc. acid)
d. $Cu + H^+ + NO_3^- \rightarrow Cu^{2+} + NO + H_2O$ (dil. acid)
e. $Zn + H^+ + NO_3^- \rightarrow Zn^{2+} + NH_4^+ + H_2$) (dil. acid)
f. $C + NO_3^- \rightarrow CO_2 + NO_2 + H_2O$ (in acid)
g. $Sn + HNO_2 \rightarrow SnO_2 + N_2 + H_2O$ (dil. acid)
h. $Cr_2O_7^{2-} + Fe^{2+} + H^+ \rightarrow Fe^{3+} + Cr^{3+} + H_2O$
i. $CrO_4^{2-} + I^- + H^+ \rightarrow Cr^{3+} + I_2 + H_2O$
j. $S + H^+ + HSO_4^- \rightarrow SO_2 + H_2O$
k. $Co^{2+} + I^- + IO_3^- + H_2O \rightarrow Co(OH)_2 + I_2$
l. $Sb^{3+} + MnO_4^- \rightarrow H_3SbO_4 + Mn^{2+}$ (in acid)
m. $I^- + MnO_4^- \rightarrow I_2 + Mn^{2+}$ (in acid)
n. $H_2S + H^+ + NO_3^- \rightarrow S + NO$
o. $NH_3 + O_2 \rightarrow NO + H_2O$
p. $MnO_2 + C_2O_4^{2-} + H^+ \rightarrow Mn^{2+} + CO_2$
q. $MnO_4^- + AsO_2^- \rightarrow MnO_2 + AsO_4^{3-} + OH^-$
r. $Ca(OCl)_2 + I^- \rightarrow I_2 + Ca^{2+} + Cl^-$ (in acid)
s. $Mn^{2+} + S_2O_8^{2-} \rightarrow MnO_2 + HSO_4^- + H^+$
t. $TeO_3^{2-} + I^- \rightarrow Te + I_2$ (in acid)
u. $U^{2+} + MnO_4^- \rightarrow Mn^{2+} + UO_2^{2+}$ (in acid)
v. $As + H^+ + NO_3^- \rightarrow H_3AsO_4 + NO$
w. $ClO_3^- \rightarrow ClO_4^- + ClO_2^-$
x. $Pd + HSO_4^- \rightarrow Pd^2 + SO_2$

PROBLEMS

14.17 How many liters of NO_2 at STP are formed by the reaction of 24.0 g of copper metal with an excess of concentrated nitric acid?

14.18 How many grams of potassium dichromate, $K_2Cr_2O_7$, react to oxidize 100 ml of 2.0 M Fe^{2+} to iron(III) ion?

14.19 How many grams of MnO_2 are required to produce 2.60 mole of electrons when reduced to manganese(II) ion in an acid solution?

14.20 How many ml of hot concentrated nitric acid (15 M) react to dissolve a half dollar? A half dollar weighs 11.6 g and is 40 per cent silver.

14.21
a. Sketch this cell: $Mg|Mg^{2+}(1\ M)||Ag^{+}(1\ M)|Ag$.
b. Identify anode and cathode.
c. Indicate direction of electron flow.
d. Calculate the voltage of the cell.

14.22 Consider the two cells connected as indicated:

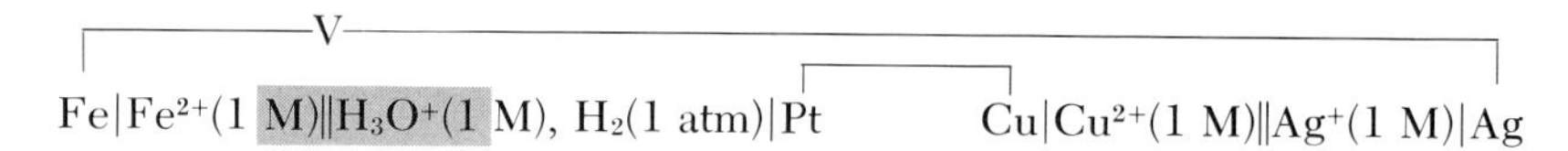

a. What is the voltage of the cell?
b. Indicate the direction of electron flow in both external conductors.
c. Write both oxidation half-cell equations.
d. Write both reduction half-cell equations.
e. What substances are dissolved? What substances are formed?

14.23 The cell, $Mg|Mg^{2+}(1\ M)||Ag^{+}(1\ M)|Ag$, is allowed to discharge until the concentration of magnesium ion is increased to 1.25 M. What is the silver ion concentration? Has the voltage changed during the discharging? If so, has it increased or decreased?

SUGGESTED READING

Austin, L. G.: "Fuel Cells." Scient. Amer., *201*, (4):72, 1959.

Bailar, J. C., Jr., Moeller, T. and Kleinberg, J.: *University Chemistry*, D. C. Heath and Co., Boston, 1965. Chapter 12.

Brockman, C. J.: "Primary Cells – A Brief Historical Sketch." J. Chem. Educ., *4*:770, 1927.

Garrett, A. B.: "Nuclear Batteries." J. Chem. Educ., *33*:446, 1956.

Weissbart, J.: "Fuel Cells – Electrochemical Converters of Chemical to Electrical Energy." J. Chem. Educ., *38*:267, 1961.

FIFTEEN • ACTIVE NON-METALS —THE HALOGENS

15.1 INTRODUCTION

Three elements in Group VII of the periodic table were identified in the early part of the nineteenth century (see Figure 9.3). Chlorine was prepared in 1774 by the Swedish chemist Carl Wilhelm Scheele, who also is celebrated for his discovery of oxygen two years earlier. But just as he had called oxygen "pure air" and had not appreciated the import of his discovery, he supposed chlorine to be a compound containing oxygen. Thirty-six years later **Sir Humphry Davy,** finding it impossible to decompose the gas, recognized it to be an element and named it chlorine (from a Greek word meaning greenish-yellow).

Iodine was prepared in 1811 by Courtois, a manufacturer of saltpeter (KNO_3), which is a component of black gunpowder. Courtois was preparing potassium from the ash of kelp and other seaweeds. Treatment of the ash with hot sulfuric acid released a dense violet vapor, which collected in glistening black crystals on a cool surface. Gay-Lussac proved the crystalline substance to be an element and named it after the color of its vapor. (The word iodine also derives from the Greek.)

In 1826, bromine was prepared by passing chlorine into the bitter solution left when sodium chloride is crystallized from sea water. Balard, its discoverer, named the element from a Greek word meaning "stench."

Although compounds of fluorine had been known for years, pure fluorine was first prepared in 1886 by Moissan. (His method will be described later in the chapter.)

Astatine (discovered in 1940), the fifth element of the group, is radioactive. It occurs in nature in very minute amounts; it is a product of decay of actinium. The element has been prepared artificially by nuclear reactions induced by the bombardment of bismuth with alpha particles.

15.2 OCCURRENCE

The halogen family is the most **active non-metal group.** The halogens ("salt formers") are electron acceptors and are commonly found in nature in salts as the halide ions, F^-, Cl^-, Br^- and I^-. The elements are quite rare, the estimated percentages in the earth's crust being 0.1 per cent F, 0.2 per cent Cl, 0.001 per cent Br and 0.001 per cent I. However, there are readily available concentrated sources of each. Fluorine occurs in nature as fluorspar (CaF_2) and cryolite (Na_3AlF_6). Most naturally occurring salts of the other three elements are very soluble and have been washed or are being washed by rain and seepage into the ocean and into land-locked seas. The salts of these elements, especially sodium chloride, have been well known for centuries.

About 60 elements have been detected in sea water. More elements will probably be found there in the future as more sensitive analytical methods are developed. Ocean water contains about 3.5 per cent solid residue, and a cubic mile of sea water holds about 166 million tons of solids. The concentrations in terms of tons per cubic mile of 19 of the most abundant elements in the ocean, along with silver and gold, are listed in Table 15.1.

Notwithstanding the large quantities of residue in sea water, only four elements have been extracted in any large amounts.

TABLE 15.1 CONCENTRATION, AMOUNTS AND PRINCIPAL SPECIES OF SOME ELEMENTS IN SEA WATER*

Element	Concentration (mg/L)	Amount of Element in Sea Water (tons/cubic mile)	Principal Species
Chlorine	19,000	89.5×10^6	Cl^-
Sodium	10,500	49.5×10^6	Na^+
Magnesium	1350	6.4×10^6	Mg^{2+}; $MgSO_4$
Sulfur	885	4.2×10^6	SO_4^{2-}
Calcium	400	1.9×10^6	Ca^{2+}, $CaSO_4$
Potassium	380	1.8×10^6	K^+
Bromine	65	306,000	Br^-
Carbon	28	132,000	HCO_3^-; CO_2; CO_3^{2-}; organic compounds
Strontium	8	38,000	Sr^{2+}, $SrSO_4$
Boron	4.6	23,000	$B(OH)_3$, $B(OH)_2O^-$
Silicon	3	14,000	$Si(OH)_4$, $Si(OH)_3O^-$
Fluorine	1.3	6100	F^-
Argon	0.6	2800	Ar
Nitrogen	0.5	2400	NO_3^-; NO_2^-; NH_4^+, N_2; organic compounds
Lithium	0.17	800	Li^+
Rubidium	0.12	570	Rb^+
Phosphorus	0.07	330	HPO_4^{2-}; $H_2PO_4^-$; PO_4^{3-}; H_3PO_4
Iodine	0.06	280	IO_3^-, I^-
Barium	0.03	140	Ba^{2+}, $BaSO_4$
Silver	0.003	1	$AgCl_2^-$; $AgCl_3^{2-}$
Gold	0.000004	0.02	$AuCl_4^-$

* These data were obtained from *The Sea*, M. N. Hill, editor, Vol. 2, 1963, pp. 4–5, Interscience Publishers, New York, and *The Mineral Resources of the Sea* by J. L. Mero, 1965, pp. 26–27, Elsevier Publishing Co., New York.

at the present time the ocean is the commercial source for only four elements.

Sodium and **chlorine** in the form of table salt (NaCl), **magnesium** and some of its compounds and elementary **bromine** are produced from ocean water. Extraction of other minerals from sea water have not yet proved successful. However, patents have been granted covering the extraction of iodine, potassium, calcium sulfate, gold and silver. German scientists tried to extract gold after World War I because Fritz Haber, winner of a Nobel Prize for the fixation of nitrogen, thought that this would be a way to pay off war debts. In the future the ocean will probably be the commercial source for other minerals.

Water covers approximately three fourths of the earth's surface; there are 320,000,000 cubic miles of ocean water. This limitless source of salt was tapped by the ancients. The Romans dammed off pools of sea water and allowed the water to evaporate, leaving the salt. This important substance, so necessary for nutrition, was carried throughout the Roman Empire. Soldiers were often paid with salt *(sal)*, hence the word salary. Salt is still prepared by solar evaporation.

All inland seas contain a high percentage of sodium chloride, but vary in the percentage of other salts, depending upon the soluble salts present in the area draining into the lake. The Dead Sea is 25 per cent salt; it is especially high in magnesium chloride and iodine. Great Salt Lake in Utah is saturated with salts, but contains more sulfates than does ocean water. Owen's Lake, California, is relatively high in carbonates. The Stassfurt deposits (remains of an inland sea) in Germany are high in potassium salts. Sodium chloride is found in underground deposits where it is either mined or dissolved in water and pumped up as brine.

Iodine occurs in nature as an iodate. The chief compound is **sodium iodate** ($NaIO_3$), an impurity in saltpeter ($NaNO_3$) **which is found in Chile.**

15.3 PHYSICAL PROPERTIES

The properties of the Group VIIA elements are those which might be expected from their position in the periodic table. The magnitudes of the properties listed in Table 15.2 illustrate trends in properties in the halogen family. For example the boiling points of the elements of the group are low, varying from −188°C for fluorine to +184°C for iodine. Furthermore, the boiling points increase regularly from fluorine to iodine. These boiling points are expected. Non-polar molecules of non-metal atoms are held together only by **van der Waal's forces,** rather weak forces compared to those between ions or between polar molecules. Nevertheless, these are the forces that cause the behavior of the vapors to deviate from Boyle's and Charles' gas laws. They are the forces that hold the molecules together in soft crystals below their boiling points.

what are van der Waal's forces?

One theory suggests that van der Waal's forces are fundamentally electrostatic, as are forces between polar molecules and among ions. By some means, according to the theory, the cloud of electrons surrounding one molecule in a liquid or solid becomes distorted to one side of the molecule. This creates temporary positive and negative ends on the molecule. Under the influence of the first

TABLE 15.2 PROPERTIES OF THE HALOGENS

Property	Fluorine (F)	Chlorine (Cl)	Bromine (Br)	Iodine (I)
Molecular formula	F_2	Cl_2	Br_2	I_2
Molecular weight	38.00	70.91	159.8	253.8
Appearance at room temperature	Light yellow gas	Greenish-yellow gas	Reddish-brown liquid	Purple-black crystals
Melting point	−220	−101	−7.2	114
Boiling point	−188	−35	59	184
Valence electrons	$2s^22p^5$	$3s^23p^5$	$4s^24p^5$	$5s^25p^5$
Ionic radius of X^-, Å	1.36	1.81	1.95	2.16
Covalent radius of X in X_2, Å	0.72	0.99	1.14	1.33
Ionization energy for $X_{(g)} \rightarrow X^+_{(g)} + e^-$ (kcal/mole)	401.5	300	273	241
E° in volts for $X_{2(aq)} + 2e^- \rightarrow 2X^-_{(aq)}$	2.87	1.36	1.06	0.53
Electronegativity	4.0	3.0	2.8	2.5

molecule, a polarity is induced in neighboring molecules as the electron clouds of those molecules draw away from the distorted electron cloud of the first. The non-polar molecules, then under the influence of one another, become temporary dipoles, and the attraction between positive and negative ends of the temporarily polar molecules holds them together. The greater the number of electrons in a molecule, the greater the size of the molecules (the size of the electron clouds), and the more loosely the electrons are held the greater the charge that can be induced by the process described here, and, therefore, the greater the attraction between molecules. The attractions should increase from fluorine to iodine as the size and number of electrons increase and as the ionization energy decreases. With increased mutual attraction the heavier molecules of the group require more energy and a higher temperature to separate them in vaporization.

iodine is purified by sublimation.

Iodine is an unusual solid in that it has a considerable vapor pressure at its melting point (90 mm pressure at 113.5°C). When heated slowly the solid is converted, without melting, to dense violet fumes. The vapor condenses to a solid on cold surfaces in shiny black flakes, which are, except for their color, similar to hoarfrost. This process, which can be demonstrated in the laboratory as shown in Figure 15.1, is called **sublimation.** Iodine crystals heated in the evaporating dish vaporize and recrystallize on the underside of a cold watch glass which covers the dish. Iodine is purified on a commercial scale by sublimation, because the vaporizing iodine leaves nonvolatile impurities as a residue.

In Table 9.2 the chlorine atom was compared in size with atoms of the other third period elements. The trend in decreasing size as one goes from left to right across the periodic table was emphasized. The relatively small sizes of the halogen atoms cause these elements to have large ionization energies and a great attraction for electrons. Except for the noble gases, elements of the halogen group have the highest ionization energies of any family of elements. This great attraction for electrons is reflected in high electronegativity values. Fluorine is the most electronegative of the halogens; iodine is the least. Hence fluorine is the most reactive,

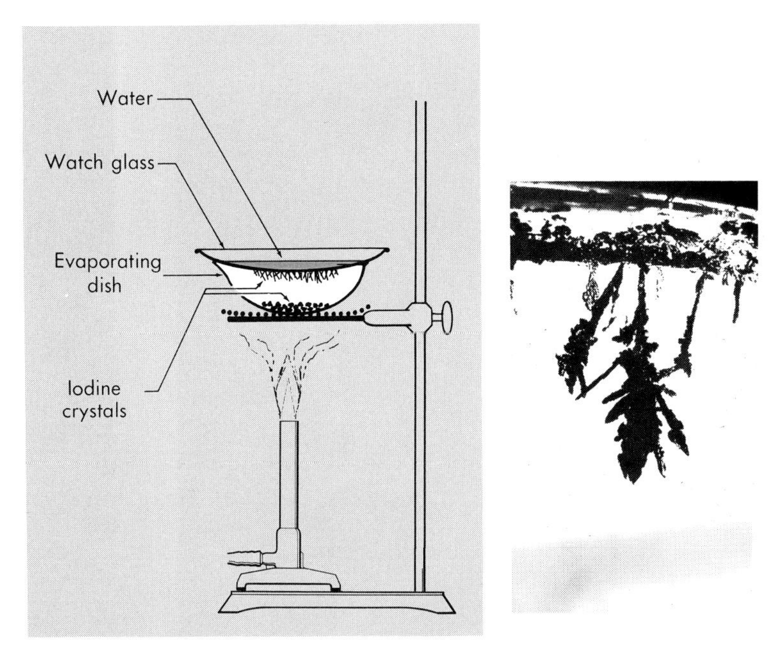

FIGURE 15.1 SUBLIMATION OF IODINE.

with the activity decreasing successively from member to member down the group to iodine. This trend is illustrated by the **reduction potentials** (E° values in Table 15.2) where **fluorine is the strongest oxidizing agent** and iodine is the weakest of the group. The relative reactivities will be emphasized in the discussion of the preparations of the elements and their compounds. The Group VII elements are all toxic and are dangerous to handle because of their high reactivity. The halogen vapors are extremely irritating to the nose and throat, and contact with the skin can cause severe burns.

fluoride ions are larger than fluorine atoms.

The relative sizes of the atoms and positive ions of the active metals were shown in Figure 11.1. This figure may be compared with the sizes of the halogen atoms and the halide ions (−1 oxidation state) shown in Figure 15.2. The covalent radii (half the distance between the nuclei of the diatomic molecule) increase as the atomic number becomes greater. As expected, the ionic radii are larger than both the atomic radii and the covalent radii, because each ion has one electron in excess of the positive charge in the nucleus.

Since fluorine is the most electronegative element, it is considered to have a −1 oxidation state in all its compounds. In Figure 15.3 are listed known binary fluorides in which the element bonded with fluorine has an oxidation state corresponding to its respective group number. The number of fluoride ions in the compound is equal to the group number of the second element. Blanks in the table of fluorides do not necessarily indicate that fluorine does not react with that element, but only the absence of a compound EF_n, where n corresponds to the group number. For example, nitrogen forms three binary fluorides, N_2F_2, N_2F_4 and NF_3, but not NF_5. Chlorine forms the compounds ClF, ClF_3 and ClF_5 with fluorine, but not ClF_7. In addition, fluorine reacts directly with the noble gases xenon, krypton, and radon to yield the compounds XeF_2, XeF_4, XeF_6, KrF_2 and RnF_2.

Tables similar to Figure 15.3 can be assembled for the other binary halides: chlorides, bromides and iodides. Because of the

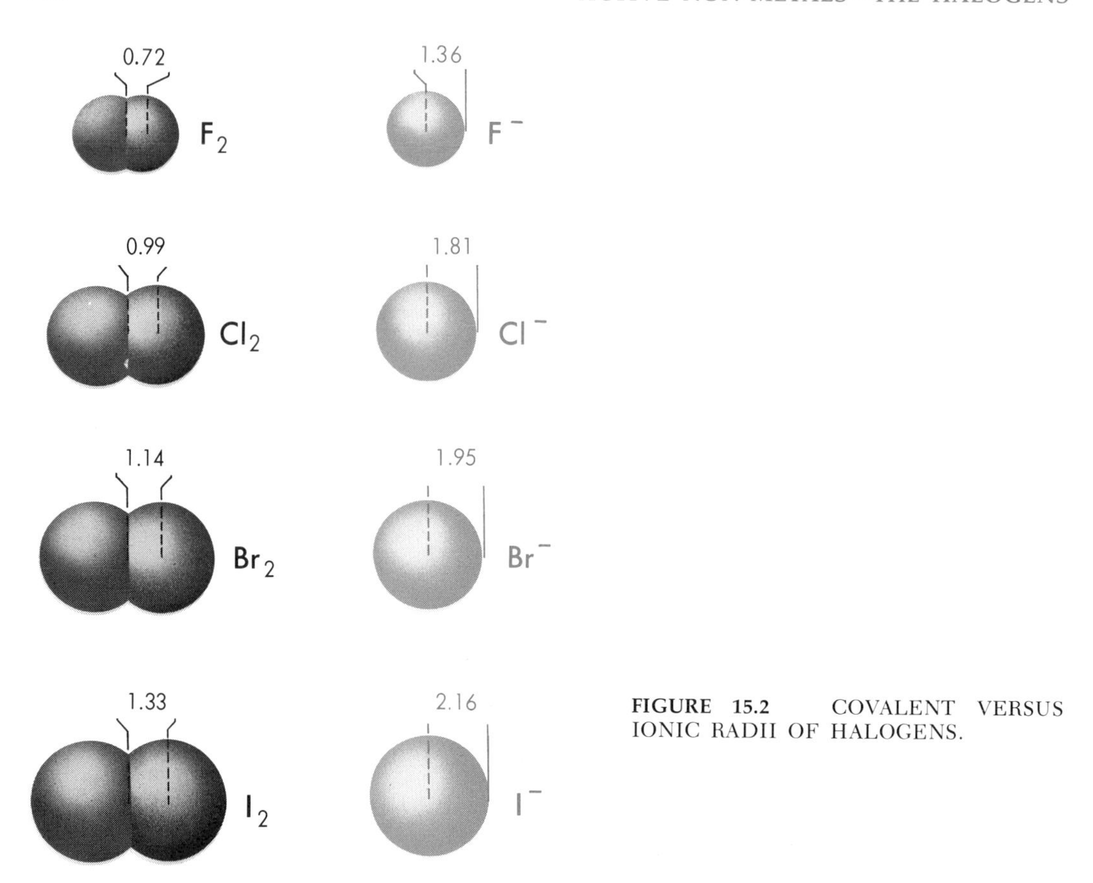

FIGURE 15.2 COVALENT VERSUS IONIC RADII OF HALOGENS.

GROUPS

IA	IIA	IIIB		IVB	VB	VIB	VIIB	VIII			IB	IIB	IIIA	IVA	VA	VIA	VIIA	O
EF	EF_2	EF_3		EF_4	EF_5	EF_6	EF_7	EF_8			EF	EF_2	EF_3	EF_4	EF_5	EF_6	EF_7	EF_8
1 H HF																	1 H	2 He
3 Li LiF	4 Be BeF_2												5 B BF_3	6 C CF_4	7 N	8 O	9 F	10 Ne
11 Na NaF	12 Mg MgF_2												13 Al AlF_3	14 Si SiF_4	15 P PF_5	16 S SF_6	17 Cl	18 Ar
19 K KF	20 Ca CaF_2	21 Sc ScF_3		22 Ti TiF_4	23 V VF_5	24 Cr CrF_6	25 Mn	26 Fe	27 Co	28 Ni	29 Cu CuF	30 Zn ZnF_2	31 Ga GaF_3	32 Ge GeF_4	33 As AsF_5	34 Se SeF_6	35 Br	36 Kr
37 Rb RbF	38 Sr SrF_2	39 Y YF_3		40 Zr ZrF_4	41 Nb NbF_5	42 Mo MoF_6	43 Tc	44 Ru	45 Rh	46 Pd	47 Ag AgF	48 Cd CdF_2	49 In InF_3	50 Sn SnF_4	51 Sb SbF_5	52 Te TeF_6	53 I IF_7	54 Xe
55 Cs CsF	56 Ba BaF_2	57 La LaF_3	58 Ce → 71 Lu	72 Hf HfF_4	73 Ta TaF_5	74 W WF_6	75 Re ReF_7	76 Os	77 Ir	78 Pt	79 Au	80 Hg HgF_2	81 Tl TlF_3	82 Pb PbF_4	83 Bi BiF_5	84 Po	85 At	86 Rn
87 Fr	88 Ra	89 Ac AcF_3	90 Th → 103 Lr	104 Ku	105 Ha													

FIGURE 15.3 SOME BINARY FLUORIDES OF THE ELEMENTS.

TABLE 15.3 MELTING POINTS OF SOME FLUORIDES OF THE THIRD PERIOD ELEMENTS

NaF	MgF_2	AlF_3	SiF_4	PF_5	SF_6	ClF_3
988	1266	1291	−90.2	−83	−50.5	−83
Primarily ionic bonding			Primarily covalent bonding			

decreased reactivity of these halogens, fewer compounds are known in which the element has an oxidation state corresponding to the respective group member in the periodic table as compared to the number of compounds formed with fluorine.

In Table 15.3 it can be seen that the melting points of the binary halides vary with the position of the central (other) element in the periodic table. Data are given for fluorides of the third period elements. The high melting point of NaF, 988°C as compared to −83°C for ClF_3, is explained by the difference in the nature of the chemical bonds in the two compounds. **Sodium fluoride is ionic**; it has a crystalline lattice that consists of a three-dimensional geometrical array of positive and negative ions, as shown in Figure 15.4. Each sodium ion inside the crystal is surrounded by six fluoride ions each at an equal distance from the sodium ion. Conversely each fluoride ion is surrounded by six sodium ions each at an equal distance from the fluoride ion. The electrostatic forces between the ions are relatively strong, and in melting the crystal many of these ionic bonds must be broken such that ions or groups of ions may move with respect to one another. As a result, high temperatures are necessary to melt this ionic crystal. The low melting points of the covalent compounds are the result of weak intermolecular forces (forces of attraction among the individual molecules) called van der Waals forces.

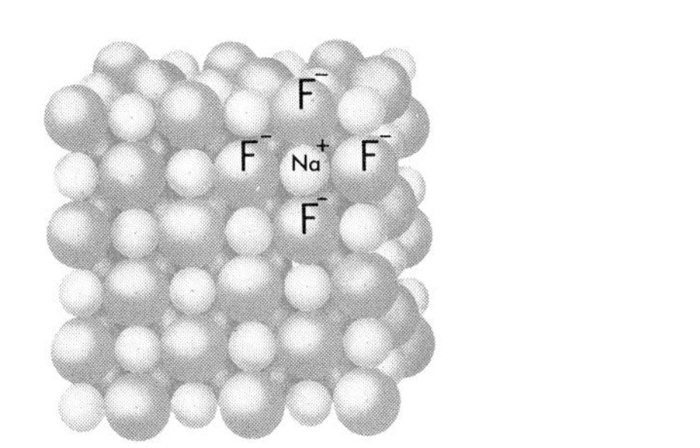

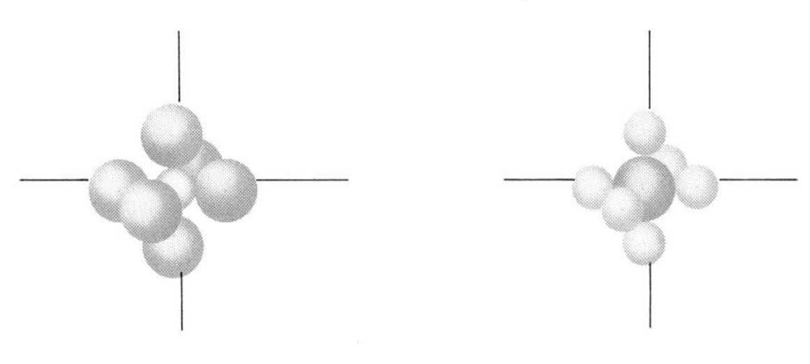

FIGURE 15.4 SODIUM FLUORIDE CRYSTAL LATTICE.

15.4 PREPARATION OF THE HALOGENS

Because of high reactivity, **halogens** are never found free (uncombined) in nature, but usually **occur as halides** (see Table 15.1). Consequently their preparation involves oxidation of the halide. As already noted in Chapters 9 and 14, oxidation involves the loss of electrons. The halides (X^-, −1 oxidation state) are very stable, as indicated by the high positive reduction potential values in

Table 15.2. These values are given for the half reaction:

$$X_{2(aq)} + 2e^- \rightarrow 2\,X^-_{(aq)}$$

where X_2 represents a halogen molecule and X^-, a halide ion. The positive value indicates that the reaction proceeds most readily "to the right," that is, to form product from reactant as represented by the equation. Half-reactions that go readily to the right are difficult to reverse to the left. The oxidation of fluoride ion is the most difficult. The possible methods for preparing each halogen are determined largely by the relative ease or difficulty of oxidation.

15.5 PREPARATION OF FLUORINE

Since the fluoride ion is so stable (that is, fluorine has such an attraction for its valence electrons), fluorine cannot be prepared from fluorides by oxidation with other chemicals. Instead, the fluoride ion is oxidized by electrolysis, i.e., electrical energy is needed to remove the electron. The electrolysis cell, constructed of special fluoride-coated metals, is very similar operationally to the Downs cell (Fig. 11.3) which is used to prepare sodium metal and chlorine gas. Hydrogen fluoride is electrolyzed at 100°C from a liquid mixture of HF and KF. Hydrogen is evolved at the surface of the cathode; fluorine is evolved at the anode. To avoid an explosion, the gases must be kept apart:

$$H_2 + F_2 \rightarrow 2\,HF + 128{,}400 \text{ cal}$$

15.6 PREPARATION OF CHLORINE

1. Chlorine may be produced by the same method as fluorine, that is, by the electrolysis of a liquid chloride. (The electrolysis of sodium chloride was described in Chapter 11.)

2. Unlike fluorine, chlorine can be prepared in the presence of water. The electrolysis of a concentrated solution of sodium chloride yields chlorine by a process to be discussed presently. Fluorine cannot be prepared by electrolysis of a halide salt in a water solution, because the fluorine formed reacts rapidly with water, oxidizing and replacing the oxygen:

fluorine can oxidize water.

$$2F_2 + 2H_2O \rightarrow 4HF + O_2$$

3. Certain chemicals (strong oxidizing agents) have sufficient attraction for electrons to prepare chlorine from chlorides by chemical oxidation. Some chemicals that will oxidize chlorides in the presence of an acid are potassium permanganate ($KMnO_4$), manganese dioxide (MnO_2) and nitric acid (HNO_3). The equation is shown for MnO_2 reacting with hydrochloric acid:

$$MnO_2 + 4\,HCl \rightarrow MnCl_2 + Cl_2 + 2\,H_2O$$

The preparation is illustrated in Figure 15.5. Dilute hydrochloric

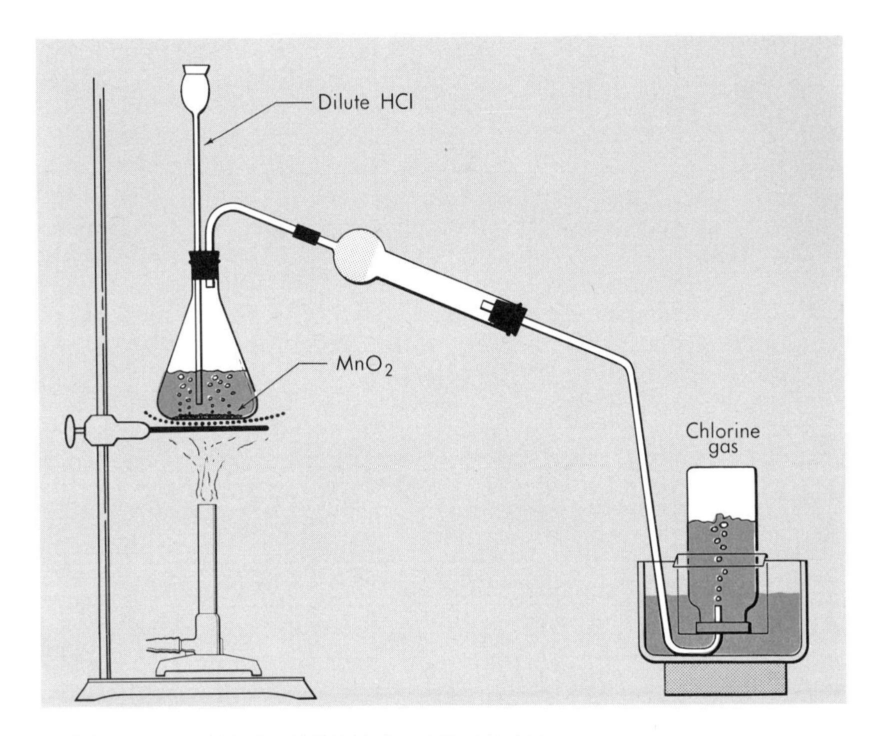

FIGURE 15.5 PREPARATION OF CHLORINE.

acid is poured through the thistle tube over powdered manganese dioxide. The chlorine gas is collected over water after passing through glass wool which removes acid droplets.

The bulk of the chlorine (about 10 million tons per year) produced in the United States is made by the electrolytic oxidation of chloride ion in aqueous sodium chloride. **Chlorine ranks fifth** of all chemicals in the amount produced. With the usual graphite electrodes, the products of electrolysis are hydrogen, chlorine and a solution of sodium hydroxide.

Anode reaction	$2Cl^- \rightarrow Cl_2 + 2e^-$
Cathode reaction	$2H_2O + 2e^- \rightarrow H_2 + 2OH^-$
Cell reaction	$2Cl^- + 2H_2O \rightarrow Cl_2 + H_2 + 2OH^-$

However, when both electrodes are graphite, pure sodium hydroxide is difficult to obtain. Most chlorine is produced in an electrolysis cell with a mercury cathode. Using mercury, sodium is first obtained at the cathode instead of hydrogen because the sodium dissolves in the mercury to form a mixture called an amalgam. Upon reaction with water, the sodium from the amalgam produces hydrogen and sodium hydroxide:

$$2Na(Hg)_x + 2H_2O \rightarrow H_2 + 2Na^+ + 2OH^- + XHg$$

Approximately 1.6 million pounds of mercury were purchased by the chemical industry in 1969 for the production of chlorine. Unfortunately about the same amount of mercury was lost from the electrolytic cell and found its way into the waste water. The escape of mercury has become a serious pollution problem, since it tends to settle in the sediment of lakes and rivers. Small organisms on the lake bottom ingest the mercury, bottom-feeding fish eat the small organisms and large game fish eat the bottom-feeding fish. The mercury becomes more concentrated with each successive step. Because of mercury pollution, fishing is banned at many points on Lake Champlain and in the St. Clair River-Lake Erie water system, and people are warned against eating the fish. Some mercury-cell chlorine plants have been closed by court order. The problem of escaping mercury can be solved (if sufficient pressure is brought to bear on the chemical industry) by using care and good equipment. Some plants are currently operating with negligible mercury discharge in the waste water.

mercury insecticides are also pollutants, and game birds (pheasants) are the victims.

15.7 PREPARATION OF BROMINE

Bromine may be prepared by any of the general methods used to prepare chlorine; bromides may be oxidized by even milder oxidizing agents than are used for chlorine. In fact, the commercial preparation of bromine depends upon the use of chlorine as an oxidizing agent.

$$Cl_2 + NaBr \rightarrow 2\,NaCl + Br_2$$

As noted earlier, the major source of bromine is bromide in ocean water. The flow diagram in Figure 15.6 illustrates a process for the extraction of bromine from sea water, a process which in-

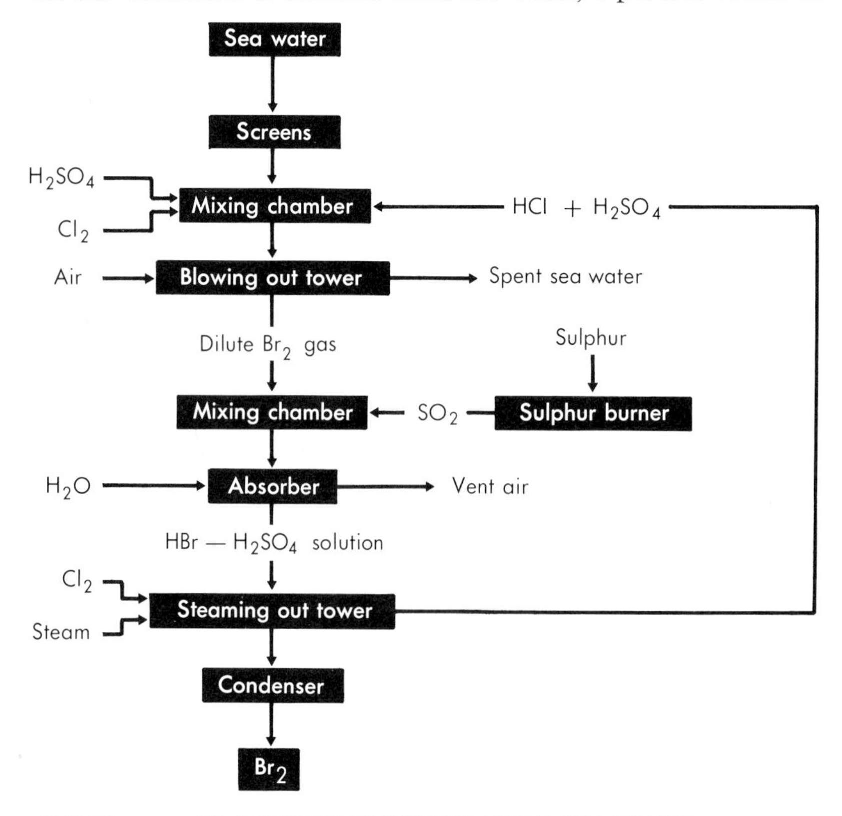

FIGURE 15.6 BROMINE PRODUCTION FROM SEA WATER.

volves mixing sulfuric acid and chlorine with sea water. The acid prevents the hydrolysis (reaction with water) of chlorine. The chlorine oxidizes the bromide ion to elemental bromine. Elemental bromine has now been prepared but in such a large amount of water that the solution is not colored. To concentrate the bromine into a smaller volume, air is blown through the water carrying the bromine with it as a vapor. Passage through a sodium carbonate solution "strips" the air of the bromine by the following reaction:

chlorine is an important oxidizing agent.

$$3Br_2 + 6Na_2CO_3 + 3H_2O \rightarrow 5NaBr + NaBrO_3 + 6NaHCO_3$$

The bromine as sodium bromide and sodium bromate is now in a small volume. Acidification with sulfuric acid reverses the reaction

$$5NaBr + NaBrO_3 + 3H_2SO_4 \rightarrow 3Br_2 + 3Na_2SO_4 + 3H_2O$$

restoring the bromine, which is distilled from the solution.

15.8 PREPARATION OF IODINE

Iodine can be prepared from the halide salt by any method used to prepare fluorine, chlorine or bromine. Even **milder oxidizing agents can be used to oxidize iodides** than those used for the oxidation of bromides. Bromine can be used as the oxidizing agent, although chlorine, because of its greater availability, is the usual oxidizing agent. Iodine was first discovered by Courtois, who oxidized iodide with hot concentrated sulfuric acid:

$$5H_2SO_4 + 8KI \xrightarrow{\Delta} 4I_2 + 4K_2SO_4 + 4H_2O + H_2S$$

The iodide Courtois used was obtained from seaweed ash. Growing seaweed requires iodine, which is stored in the stems and stalks.

Most of the iodine produced commercially is prepared from sodium iodate, an impurity in Chilean saltpeter (KNO_3). The iodate-containing solution left when the saltpeter is purified is treated with sodium hydrogen sulfite:

$$\underset{\text{Sodium iodate}}{2NaIO_3} + \underset{\text{Sodium hydrogen sulfite}}{7NaHSO_3} \rightarrow \underset{\text{Sodium hydrogen sulfate}}{5NaHSO_4} + 2Na_2SO_3 + I_2 + H_2O$$

The precipitated iodine is filtered, dried and separated by sublimation from solid impurities.

15.9 GENERAL CONSIDERATIONS

The ***decrease* in activity of the halogens** through the series fluorine to iodine should be apparent. From this fact it is clear that each succeeding halogen can be prepared by all methods for the preceding one plus one or two more. The activity trend is further highlighted in that each halogen will replace in solution halide ions of any halogen that follows it, as illustrated with chlorine and bromine in the following reactions:

can I_2 oxidize any halides?

$$Cl_2 + 2\ NaBr \rightarrow 2\ NaCl + Br_2$$

$$Cl_2 + 2\ NaI \rightarrow 2\ NaCl + I_2$$

$$Br_2 + 2\ KI \rightarrow 2\ KBr + I_2$$

Of course, it is not practical to replace chlorine or any of the other halogens with fluorine, because fluorine reacts very rapidly with water to produce oxygen as well.

15.10 USES OF THE HALOGENS

Some of the important uses of the halogens are summarized in Table 15.4. The chemistry of fluorine has been developed mostly since World War II, because it was not until then that methods were found to handle the very reactive gas, and then only because large amounts of fluorine were used in the production of the first atomic bombs. One method for the separation of the isotope ^{235}U from the more abundant ^{238}U made use of the different rates of diffusion of the uranium hexafluoride gases, $^{235}UF_6$ and $^{238}UF_6$. Great quantities of fluorine were needed for the separation. Fluorocarbons are unaffected by rather high temperatures and are inert to most chemical reagents. **A polymeric fluorocarbon, Teflon,** is used on the inside surface of cooking utensils, such as non-stick frying pans. Other fluorocarbons are used for lubricants, gaskets, tubing and refrigerants.

Chlorine is a so-called "heavy" or large volume chemical. Large amounts, approximately two thirds of the total production,

TABLE 15.4 IMPORTANT USES OF THE HALOGENS

Halogen	Use
Fluorine	Nuclear energy field (preparation of UF_6) Teflon $(CF_2CF_2)_x$ Refrigerant gases (such as CCl_2F_2, CCl_3F) Propellents in spray cans (same gases) NaF (fluoridation of water and as an insecticide) HF (etching of glass, catalyst in petroleum refining)
Chlorine	Commercial preparation of Br_2 and I_2 Cl_2 purifying water, insecticides, plastics, solvents HCl (industrially important acid) NaOCl, $CaOCl_2$, bleaching ClO_3^- (chlorates), matches
Bromine	$C_2H_4Br_2$, gasoline additive AgBr, photography
Iodine	AgI, photography NaI, iodized salt for thyroid I_2, antiseptic

are used for bleaching textiles and for bleaching wood pulp for paper. Traces of chlorine are added to drinking water sources to kill bacteria and oxidize coloring matter. One of the large uses of chlorine has been in the preparation of chlorine-carbon compounds for use as pesticides. One notable example is DDT, 1,1-bis (*p*-chlorophenyl)-2,2,-trichloroethane. It now appears that DDT has a Dr. Jekyll and Mr. Hyde complex. A few years ago, in a period of one year, DDT brought about a 40 per cent decrease in the mortality rate in Malay by controlling the mosquitoes. The main problem with DDT is that it decomposes very slowly in the soil or in animal or plant residues and as a result remains in our environment for long periods of time. It is insoluble in water, but dissolves in the fatty tissue of animals where it accumulates in dangerous amounts. It has been blamed for altering the reproductive mechanism in birds and fish. Here is another challenge for the chemist: to develop new insecticides that will not have the undesirable properties found in the chlorinated hydrocarbons. A number of countries and states are now banning the use of DDT in an attempt to prevent a "silent spring."

new insecticides are required for mosquito control.

a few years ago Rachel Carson in her book, *Silent spring*, decried the use of DDT.

Two of the main uses of bromine are in photography and in gasoline additives. Ethylene bromide (1,2-dibromoethane) is added to "leaded" gasoline, so that the tetraethyl lead which is used as an antiknock agent will be changed to lead bromide instead of lead oxide which deposits in the cylinders. Lead bromide is volatile and escapes with the exhaust gases. To limit pollution of the atmosphere by toxic lead compounds, automobile engines are being designed to use unleaded fuels. One result of this change could be that the demand for bromine will decrease.

Iodine has few uses outside of the chemical laboratory. It is best known as tincture of iodine (an alcohol solution of iodine), an antiseptic used to treat wounds and abrasions of the skin. Iodine and its compounds are also used medically in treating thyroid disorders and occasionally for other diseases.

15.11 CLASSIFICATION OF HALOGEN-CONTAINING COMPOUNDS

The halogens are found in many types of compounds. Compounds of the halogens with carbon and hydrogen, in which all the bonding is covalent, are discussed as organo-halides later in the text. The salts and acids are summarized here, with emphasis upon the electronic structure of the negative ion. It is convenient to classify these compounds in accord with the oxidation states of the halogens. Table 15.5 lists some chlorine compounds. With few exceptions, there are corresponding compounds of bromine and iodine. The systematic nomenclature for these compounds employs the prefixes and suffixes that are italicized in the names of the acids and salts. Notice that the oxidation states of chlorine in the compounds are odd numbers and that they extend from −1 to +7. In each case, as illustrated by the electronic structures in Table 15.5, an 8-electron configuration of electrons is about chlorine in each complex ion.

some examples of halogen compounds that do not follow the octet rule are ClF_3, ClF_5, BrF_5 and IF_7.

TABLE 15.5 OXIDATION STATES OF CHLORINE

Oxidation State of Chlorine	Acid	Salts	Anion and Electronic Structure
−1	HCl, *hydro*chlor*ic* acid	KCl, potassium chlor*ide* $MgCl_2$ magnesium chlor*ide* AgCl, silver chlor*ide*	Cl^- $\quad :\underset{..}{\overset{..}{Cl}}:^-$
0	Cl_2, chlorine (neither acid nor salt)		
+1	HOCl, *hypo*chlor*ous* acid	$Ca(OCl)_2$, calcium *hypo*chlor*ite* NaOCl, sodium *hypo*chlor*ite*	OCl^- $\quad :\underset{..}{\overset{..}{O}}:\underset{..}{\overset{..}{Cl}}:^-$
+3	$HClO_2$, chlor*ous* acid	$NaClO_2$, sodium chlor*ite* $Mg(ClO_2)_2$, magnesium chlor*ite*	ClO_2^- $\quad :\underset{..}{\overset{..}{O}}:\underset{..}{\overset{..}{Cl}}:^-$ / $:\underset{..}{\overset{..}{O}}:$
+5	$HClO_3$, chlor*ic* acid	$KClO_3$, potassium chlor*ate* $Ba(ClO_3)_2$, barium chlor*ate*	ClO_3^- $\quad :\underset{..}{\overset{..}{O}}:\overset{..}{Cl}:\underset{..}{\overset{..}{O}}:^-$ / $:\underset{..}{\overset{..}{O}}:$
+7	$HClO_4$, *per*chlor*ic* acid	$KClO_4$ potassium *per*chlor*ate* $Ca(ClO_4)_2$ calcium *per*chlor*ate*	ClO_4^- $\quad :\overset{..}{O}:$ / $:\underset{..}{\overset{..}{O}}:\overset{..}{Cl}:\underset{..}{\overset{..}{O}}:^-$ / $:\underset{..}{\overset{..}{O}}:$

15.12 HYDROGEN HALIDES

Further periodic relationships can be observed from a consideration of the formation of the hydrogen halides. The hydrogen halides may be prepared directly from the elements. Hydrogen and fluorine will react even at −253°C, at which temperature hydrogen is a liquid and fluorine a solid. When the two gases are burned together in a torch, the flame is so hot it melts rock and concrete. Hydrogen and chlorine burn smoothly together in a torch but react explosively if premixed and exposed to sunlight. Hydrogen and bromine may be made to combine at 200°C in the presence of a platinum catalyst. The partial reaction of hydrogen and iodine, even at a high temperature, is too slow to be of primary importance. If the hydrogen halides were prepared from the elements at room temperature, the **reaction with fluorine would be the most exothermic,** whereas the reaction with iodine would be endothermic. These results are shown quantitatively in the following equations:

$$H_2 + F_2 \rightarrow 2HF + 128{,}400 \text{ cal}$$

$$H_2 + Cl_2 \rightarrow 2HCl + 22{,}060 \text{ cal}$$

$$H_2 + Br_2 \rightarrow 2HBr + 8660 \text{ cal}$$

$$H_2 + I_2 + 6200 \text{ cal} \rightarrow 2HI$$

Hydrogen fluoride and hydrogen chloride are prepared commercially by heating their salts with concentrated sulfuric acid:

$$CaF_2 + H_2SO_4 \xrightarrow{\Delta} CaSO_4 + 2HF_{(g)}$$

$$2NaCl + H_2SO_4 \xrightarrow{\Delta} Na_2SO_4 + 2HCl_{(g)}$$

Since HF and HCl are more volatile than H_2SO_4, the hydrogen halides escape as volatile gases. Neither HBr nor HI may be prepared successfully (without other products) by this method because they are oxidized to the halogen by hot, concentrated sulfuric acid. All the hydrogen halides can be prepared by using a non-volatile, non-oxidizing acid on a soluble halide, as demonstrated for HI:

$$NaI + H_3PO_4 \xrightarrow{\Delta} NaH_2PO_4 + HI_{(g)}$$

Some of the properties of the hydrogen halides are given in Table 15.6. **Each hydrogen halide is a gas at room temperature and atmospheric pressure.** Each is very soluble in water. Hydrogen chloride dissolves in water up to 37 per cent, or up to 12 moles of acid per liter of solution, to produce concentrated hydrochloric acid. Hydrochloric acid is more important than either hydrobromic acid or hydroiodic acid simply because it is produced more cheaply. It is one of the three strong acids produced and used in the largest amounts in industry.* About 2 million tons of hydrochloric acid are produced per year in the United States. Great amounts of the acid are pumped down oil wells and forced out through the porous, oil-bearing rock. It dissolves the rock, opening the pores and increasing the oil flow by this reaction:

$$2\,HCl + MgCO_3 \rightarrow MgCl_2 + H_2O + CO_2$$

Hydrochloric acid is used in the recovery of some metals from their ores and is used in cleaning metals that are to be galvanized or

* The other two are sulfuric acid, H_2SO_4, the largest volume chemical, (about 30 million tons per year) and nitric acid, HNO_3, (about 6 million tons per year).

TABLE 15.6 SOME PROPERTIES OF THE HYDROGEN HALIDES

	HF	HCl	HBr	HI
Color	Colorless	Colorless	Pale yellow	Pale yellow
Molecular weight	20.01	36.46	80.92	127.91
Melting point (°C)	−83.1	−114.8	−88.5	−50.8
Boiling point (°C)	19.54	−84.9	−67.0	−35.38
Solubility in water (g/100g of water at 20°C)	35.3	42	49	57
$HX_{(g)} \rightarrow H_{(g)} + X_{(g)}$ (kcal/mole)	134.6	103.2	87.5	71.4
Internuclear distance in H—X (Å)	0.92	1.28	1.41	1.62

tinned. It is used in converting corn starch to dextrins and in the manufacture of soaps, glues and dyes. Aqueous hydrogen chloride is the acid found in gastric juices. The percentage varies from 0.2 to 0.4 per cent.

One would predict that the boiling points and melting points of the hydrogen halides would increase with an increase in molecular weight, since there would be an increase in the van der Waals forces between the covalent molecules. From a consideration of the trend in the boiling points of HCl, HBr and HI in Figure 15.7 we would predict a boiling point of −110°C for HF. With a boiling point of 19.54°C, **hydrogen fluoride is unique** among the hydrogen halides. The boiling point is much greater for HF because of hydrogen bonding, which causes greater association of the molecules. Hydrogen bonds are only formed with very electronegative atoms like fluorine, oxygen and nitrogen, as was discussed earlier in Chapter 9.

All the hydrohalic acids except hydrofluoric acid dissociate or ionize completely according to the equation

$$HX + H_2O \xrightarrow{100\%} H_3O^+ + X^-$$

whereas the ionization of hydrofluoric acid can be represented as follows:

$$HF + H_2O \rightleftarrows H_3O^+ + F^-$$

$$F^- + HF \rightleftarrows HF_2^-$$

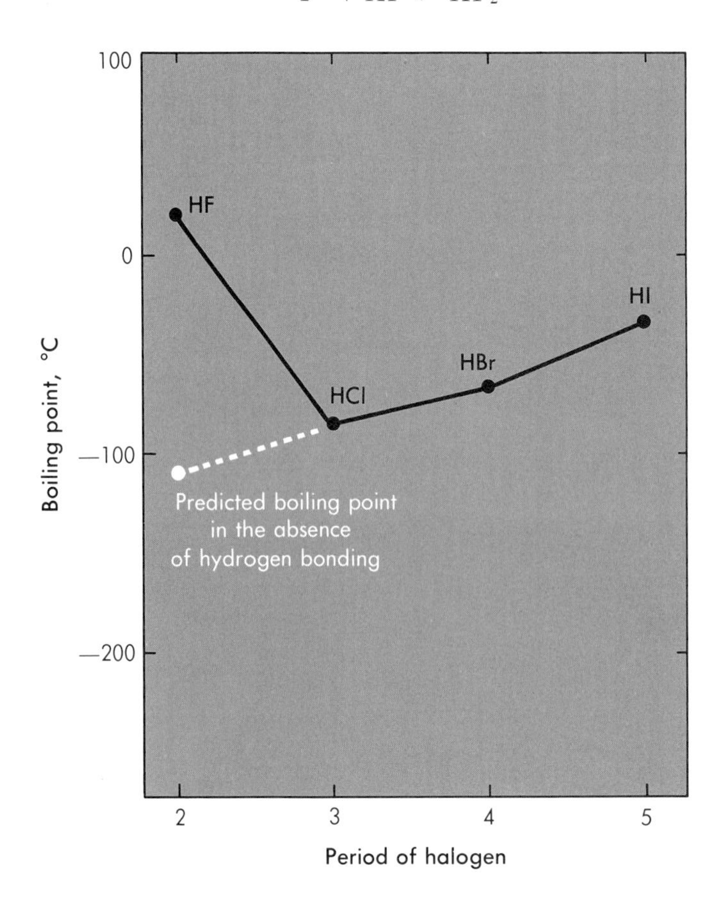

FIGURE 15.7 BOILING POINTS OF THE HYDROGEN HALIDES.

or $$2HF + H_2O \rightleftarrows H_3O^+ + HF_2^-$$

The degree of ionization of HF is quite small. Since the ionization of HCl, HBr and HI is complete, they are classified as strong acids, but HF is a weak acid because of limited ionization. The relative strength of HF compared to the other HX acids can be explained on the basis of the H—X bond lengths and the energy required to break these bonds (Table 15.6). The length of the H—F bond is about one half that of the H—I bond. As noted in Chapter 5, the shorter the bond length, the greater the strength of the bond. Hydrofluoric acid has a specific property (which no other acid has) of dissolving or etching glass. Because of this property it is commonly thought that HF is the strongest acid, but this property does not depend upon the per cent dissociation. Dry hydrogen fluoride dissolves glass equally as well as the acid solution. To illustrate the reaction, calcium silicate, one of the compounds in the complex silicate mixture of glass, is used to represent all such silicates:

hydrofluoric acid dissolves glass, but it is not a strong acid. why?

$$\underset{\text{Calcium silicate}}{CaSiO_3} + 6HF \rightarrow CaF_2 + \underset{\text{Silicon tetrafluoride}}{SiF_{4(g)}} + 3H_2O$$

To etch designs on glass, the glass is first covered with a thin film of paraffin, by melting the wax on the glass. The design is then scratched through the wax film with a sharp instrument. When the acid is placed on the glass, it dissolves the glass only where the waxy film is broken. When sufficient glass is dissolved, the hydrofluoric acid is washed away. **Etching with hydrofluoric acid** is important, because glass scored mechanically is weakened and breaks easily at the scratch. Etched glass retains its original strength.

Hydrofluoric acid cannot be stored in glass vessels, so plastic containers are used. Extreme caution should be exercised in handling hydrofluoric acid (such as using rubber or plastic gloves), for it causes very painful, slow-healing burns.

15.13 SOME OXYGEN-HALOGEN COMPOUNDS

The oxyhalogen acids and salts are the most common compounds with the halogens in positive oxidation states. Some examples for chlorine compounds were given in Table 15.5. There is no evidence for the existence of oxyfluorine acids or oxyfluorine salts, since this would require fluorine to have a positive oxidation state. The main interest in the **oxyhalogen compounds** is in their use as **oxidizing agents.** These compounds are all very reactive and thermally unstable. Extreme care and caution must be exercised in working with them.

Hypochlorous acid (HOCl) is prepared by the reaction of chlorine with water:

$$Cl_2 + H_2O \rightleftarrows HCl + HOCl$$

The extent to which the reaction proceeds to the right can be enhanced by the addition of a base which neutralizes the acids that are formed:

$$Cl_2 + 2\,NaOH \rightarrow \underset{\text{Sodium hypochlorite}}{NaClO} + NaCl + H_2O$$

The hypochlorite ion attacks and oxidizes many dyes and coloring materials, making them colorless. Clorox and other bleaching solutions contain hypochlorites.

An example of a halogen oxide is chlorine dioxide (ClO_2). This reddish-yellow gas is explosive, decomposing to chlorine and oxygen. Despite its explosiveness, chlorine dioxide is used extensively to bleach flour and paper. The strong and rapidly acting oxidizing agent penetrates the material to oxidize the coloring matter. One way of preparing the gas is by reducing a chlorate solution with sulfur dioxide:

$$2\,KClO_3 + SO_{2(g)} \rightarrow 2\,ClO_{2(g)} + KHSO_4 + KCl$$

EXERCISES

15.1 Name the four common halogens.

15.2 State the colors and physical states of the halogens at room temperature and pressure.

15.3 Give the source of the names chlorine, iodine and bromine.

15.4 Where and in what form are the halogens found in nature?

15.5 Why should the elemental halogens be more soluble in carbon tetrachloride than in water?

15.6 Which halogen is produced primarily from ocean water?

15.7 Explain why the color of iodine solutions in carbon tetrachloride is violet (as is iodine vapor), whereas solutions of iodine in water are brown.

15.8 Why is chlorine more soluble in sodium hydroxide solution than in water?

15.9 Which is the most active non-metal of all?

15.10 Why should bromine have a higher boiling point than chlorine?

15.11 Which halogen has the greatest attraction for electrons? Why?

15.12 Why are fluorides not as abundant in ocean water as are chlorides and bromides?

15.13 Why is it impossible to produce fluorine by the electrolysis of a solution of calcium fluoride?

15.14 List the steps in the preparation of bromine from sea water.

15.15 How can one separate iodine from non-volatile impurities?

15.16 Cite four sets of experimental evidence that the chemical activity of the halogens increases successively through the series iodine to fluorine.

15.17 Why do the fluorides of Group IA elements have high melting points and those of Groups VIA and VIIA low melting points?

15.18 Define oxidizing agent.

15.19 Which of the halogens is the best oxidizing agent?

15.20 Why is the industrial preparation of sodium hydroxide a source of mercury pollution?

15.21 List six industrial uses for the halogens.

15.22 Determine the oxidation state of the halogen in each compound:

a. $HBrO$ b. $CaCl_2$ c. $NaIO_3$ d. KIO_4 e. $KBrO_3$ f. Cl_2O g. $CHCl_3$ h. CCl_4

15.23 Determine the oxidation states of each element in each compound:

a. $KAl(SO_4)_2$ b. $NaHCO_3$ c. CO_2 d. Na_2O_2 e. CH_2O f. Na_3AlF_6 g. Fe_2O_3 h. $H_2Cr_2O_7$ i. $PbCrO_4$ j. $Co(NH_3)_6Cl_3$ k. NH_4Cl l. H_2SO_4

15.24 What acid is found in the stomach?

15.25 Which is the weakest acid and why?

a. HF b. HCl c. HBr d. HI

15.26 What is the color of a solution of bromine in carbon tetrachloride? of potassium bromide in water?

15.27 Why is the heating of sodium iodide with sulfuric acid a poor way to prepare hydrogen iodide?

SUGGESTED READING

Carson, R.: *Silent Spring*. Houghton Mifflin Co., New York, 1962.
Johnson, R. C.: *Introductory Descriptive Chemistry*. W. A. Benjamin, Inc., New York, 1966 (paperback).
Jolly, W. L.: *The Chemistry of the Non-metals*. Prentice-Hall, Inc., Englewood Cliffs, N. J., 1966 (paperback).
Keller, E.: "The DDT Story." *Chemistry*, *43* (2): 8, 1970.
Navratil, J. D.: "Fluorine—A Hostile Element." Chemistry, *42* (2): 11, 1969.
Peakall, D. B.: "Pesticides and the Reproduction of Birds." Scient. Amer. *222* (4): 72, 1970.
Sherwood, M.: "Sea of Chemicals." Chemistry, *43* (7): 34, 1970.
Wenk, E., Jr.: "The Physical Resources of the Ocean." Scient. Amer. *221* (3): 166, 1969.
Wooster, W. S.: "The Ocean and Man." Scient. Amer. *221* (3): 218, 1969.

SIXTEEN • THE SULFUR FAMILY

16.1 HISTORY

Sulfur has been known for at least 4000 years, but it has been used widely for less than 200 years. For many years the island of Sicily was the chief source of sulfur. The people of Egypt used sulfur for bleaching cotton as early as 2000 B.C. Egyptian paintings dating from 1600 B.C. contain a pigment made of bright yellow sulfur. In Homeric times, sulfur was used as a fumigant and a purifier. Romans used sulfur in finishing cast bronze, and also in medicines as a fumigant and in finishing cloth. Records indicate that the early Britains were acquainted with the element. It was probably brought to them by their Roman conquerors, from whom came the common name "brimstone," a variation of "bryn stone" ("the stone that burns"). However, sulfur was not identified as an element until 1810, several years after one of its compounds, sulfuric acid, became a recognized commercial product. Only 575,000 tons of sulfur were produced in 1900, but by 1969, some 10 million tons were used in the United States. This is about 100 pounds per year for every person in this country, and yet most people very seldom see the element. In 1900, Sicily was the major producer; the United States produced not more than 3000 tons. Today the United States and Canada produce most of the world's supply of sulfur.

16.2 OCCURRENCE

Sulfur occurs free in nature and also as sulfides and sulfates. The more common sulfides are pyrite (FeS_2); chalcocite (Cu_2S), a major copper ore; chalcopyrite ($CuFeS_2$); sphalerite (ZnS), a zinc ore; galena (PbS), a lead ore, and bornite (Cu_5FeS_4). The commonly occurring sulfates are gypsum ($CaSO_4 \cdot 2\ H_2O$), anhydrite ($CaSO_4$), kieserite ($MgSO_4 \cdot H_2O$) and anglesite ($PbSO_4$). Elemental and combined sulfur occur in coal and petroleum. One important commercial source, which yields about 50 per cent of the world's sulfur, is the **hydrogen sulfide** that is present as a noxious impurity in "sour" **natural gas.** The hydrogen sulfide must be removed before the natural gas is burned. The by-product, sulfur, can be obtained at a very low cost compared to that obtained by the Frasch process, which is discussed later in this chapter.

Sulfur is the fifteenth element in abundance in the earth's crust, comprising an estimated 0.06 per cent of the crust's total mass. It is found in plant and animal tissue and is the eighth most abundant element (0.25 per cent by weight) in the human body.

Native or elemental sulfur occurs in many places, often in volcanic regions, encrusted on the walls and vents of craters. Volcanic areas containing sulfur are found in Sicily, Japan, Chile, Mexico, Iceland, Norway and Spain.

An interesting story concerns the presence of sulfur deposits on volcanic crater walls. While Cortez was conquering Mexico in the middle of the sixteenth century, his army ran short of gun powder. Black powder may be made of charcoal, saltpeter and sulfur. Supplies of the first two were available, but sulfur was not to be had. One soldier, Francisco Montano, had himself lowered in a basket 400 to 500 feet into the crater of Popocatepetl, where he scraped sufficient sulfur from the walls for their needs.

Native sulfur deposits, more important than those in volcanic areas, accompany gypsum and anhydrous calcium sulfate accumulations. The major deposits are found in the Texas-Louisiana-Mexico area, in Sicily and in Poland. Fifty per cent of all sulfur produced is obtained from these deposits.

16.3 PRODUCTION OF SULFUR – THE FRASCH PROCESS

In the Texas-Louisiana area, sulfur, mixed with calcium sulfate rock, lies between 500 and 3000 feet below the earth's surface in salt domes thrust up by underlying columns of salt (Fig. 16.1). The first salt-sulfur dome was discovered in 1869 during an oil well-drilling operation. Various attempts were made to sink shafts to the sulfur mine by conventional methods, but due to noxious fumes, all failed. An engineer, D. Herman Frasch, conceived and developed, after ten years of experimentation, a plan to melt the sulfur underground and force it up as a liquid. The process which bears his name has made the United States the major supplier of sulfur.

In the Frasch process the sulfur is melted by superheated water and raised with an air lift. Sulfur melts at about 113°C (which is above the normal boiling point of water). However, by placing the water under a pressure of 100 pounds per square inch (nearly 7 atm), the temperature can be raised to 170°C. The installation is a well consisting of three concentric pipes (Fig. 16.2). The outermost pipe carries the hot water and reaches to the bottom of the sulfur-bearing stratum. It is perforated at the sulfur layer so that hot water may pass out into the stratum. The melted sulfur enters and stands in the second pipe. Compressed air, forced down the

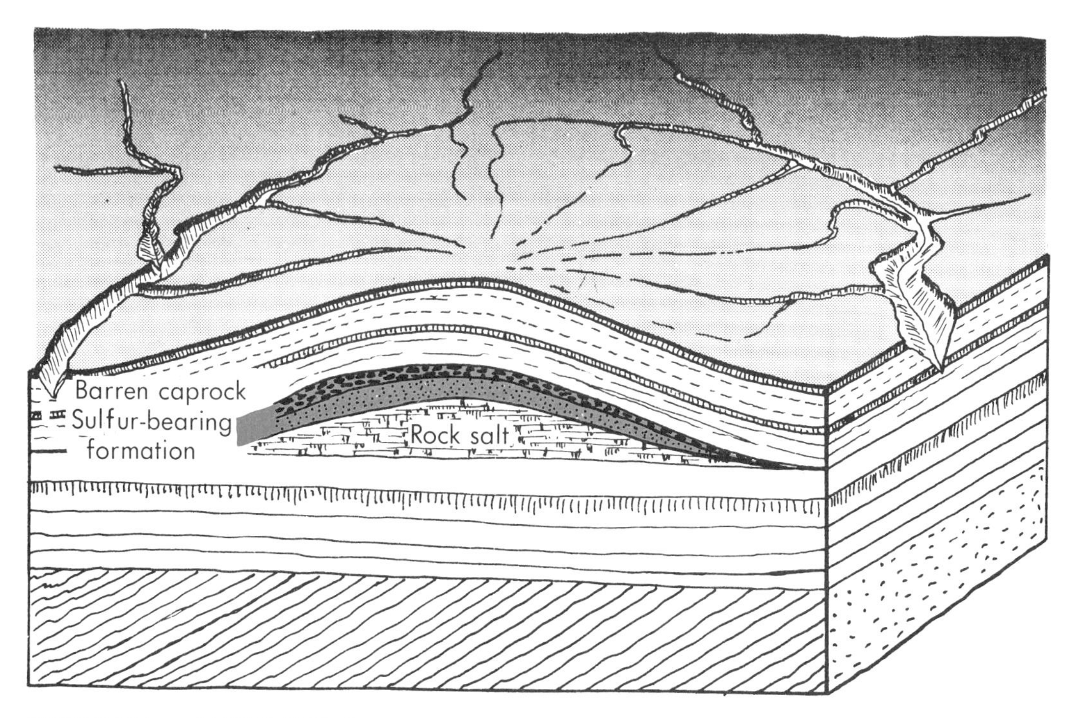

FIGURE 16.1 SALT DOME CONTAINING ELEMENTAL SULFUR.

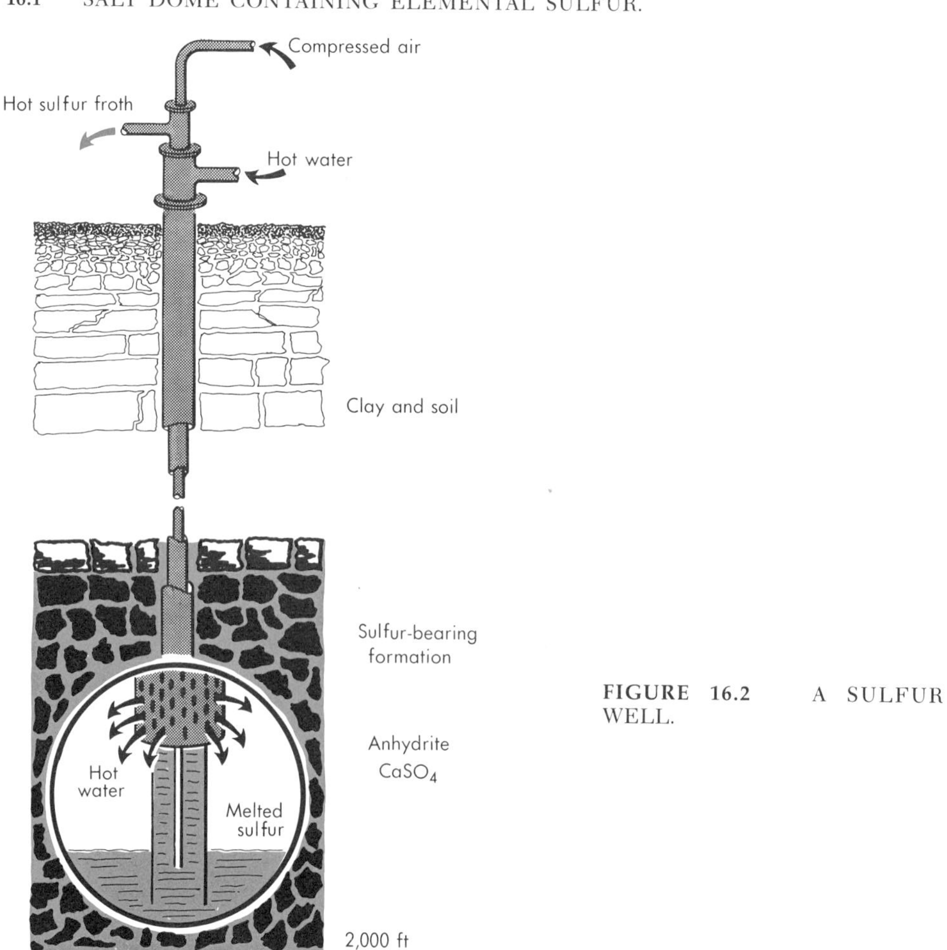

FIGURE 16.2 A SULFUR WELL.

innermost pipe, which does not quite reach the bottom of the well, rises up through the second pipe carrying a mixture of melted sulfur, water and froth. At the surface the liquid mixture is de-aerated and flooded into huge vats formed by temporary nesting walls. The sulfur cools, hardens and dries. As the mound of hardened sulfur rises, the nesting walls are raised until mountains about 375 feet long, 200 feet wide and 30 feet high are formed. When needed, the sulfur is blasted from the face of the mound and loaded by power shovel into railroad cars. The sulfur obtained by the Frasch process is 99.05 per cent pure, probably the purest product mined, except gold dust. Over 9,000,000 tons of sulfur were extracted from the first salt dome worked by the Frasch process.

16.4 PRODUCTION OF SULFUR—NATURAL GAS

since sulfur is readily available from the Frasch process and natural gas, the metal industry has not had an incentive to remove all the sulfur dioxide from smelter smoke.

The increased demand for fuels and the national concern for a cleaner environment are demanding a greater removal of sulfur from natural gas and fuel oil. As a result, sulfur output is now greater than consumption. In the first step the hydrogen sulfide and natural gas are separated from the liquid hydrocarbons. The hydrogen sulfide, an acid, is absorbed in a basic organic amine solution. This separates the H_2S from the natural gas, which is then carried off and sold. The hydrogen sulfide is then removed from the organic amine by a distillation process and is partially burned:

$$2\ H_2S + 3\ O_2 \rightarrow 2\ H_2O + 2\ SO_2$$

The moist mixture of unchanged reactant, H_2S, and the product, SO_2, is cooled and, with catalysis, reacts to form sulfur:

$$2\ H_2S + SO_2 \xrightarrow{\text{catalyst}} 2\ H_2O + 3\ S_{(g)} + \text{heat}$$

The sulfur vapor is condensed and collected as molten sulfur.

16.5 PHYSICAL PROPERTIES OF SULFUR

Some physical properties of sulfur and its related elements are shown in Table 16.1. As would be expected from periodic relationships, the melting points, and boiling points and ionic and covalent radii of these elements increase from oxygen to tellurium. Also the ionization energy and electronegativity values decrease from member to member down the periodic family.

In the solid state **sulfur and selenium exist in different allotropic forms.** This is the phenomenon (polymorphism) in which there are different forms of the same substance in the same physical state. We have considered the allotropes of oxygen, O_2 and O_3, in Chapter 9, and are acquainted with the two forms of carbon, diamond and graphite. The two important forms of solid sulfur, rhombic and monoclinic, are illustrated in Figure 16.3. The rhombic form is stable at room temperature, whereas the monoclinic form becomes stable at 95.5°C. The two crystalline forms contain S_8

TABLE 16.1 SOME PHYSICAL PROPERTIES OF THE SULFUR FAMILY ELEMENTS

	Oxygen	Sulfur	Selenium	Tellurium
Formula of molecule	O_2	S_8	Se_8	Te
Atomic weight	15.9994	32.064	78.96	127.60
Melting point (°C)	−218.4	112.8 (rhombic)	217 (gray)	452
Boiling point (°C)	−183.0	444.6	685 (gray)	1390
Valence electrons	$2s^22p^4$	$3s^23p^4$	$4s^24p^4$	$5s^25p^4$
Ionic radius of X^{-2} (Å)	1.40	1.84	1.98	2.21
Covalent radius (Å)	0.73	1.02	1.16	1.34
Ionization energy, $X_{(g)} \rightarrow X^+_{(g)} + e^-$ (kcal/mole)	314	239	225	208
Electronegativity	3.5	2.5	2.4	2.1

"ring" molecules (Fig. 16.4) in which the sulfur atoms are linked in a closed chain. The bonds between sulfur atoms are single covalent bonds, as represented in the structure in Figure 16.4 (each line represents an electron pair). Because two mutually perpendicular p orbitals of each sulfur are involved in bonding, the sulfur atoms are positioned alternately up and down in a "puckered" ring.

The changes that occur when sulfur is melted and heated are quite complicated. Sulfur exhibits a "double" melting point; when the solid is heated, it melts at 112.8°C. Then at any point below 119°C it may suddenly solidify again and then remelt at 119°C. The unusual changes can be explained in part on the basis of structure. **Rhombic sulfur** is stable up to 95.5°C; **monoclinic sulfur** is stable between 95.5°C and its melting point of 119.25°C. However, the change from one crystal habit to the other is very slow because of the restricted movement of the S_8 molecules in the solid. Therefore, when rhombic sulfur is heated rapidly, the change to the monoclinic structure does not occur, but the unstable rhombic sul-

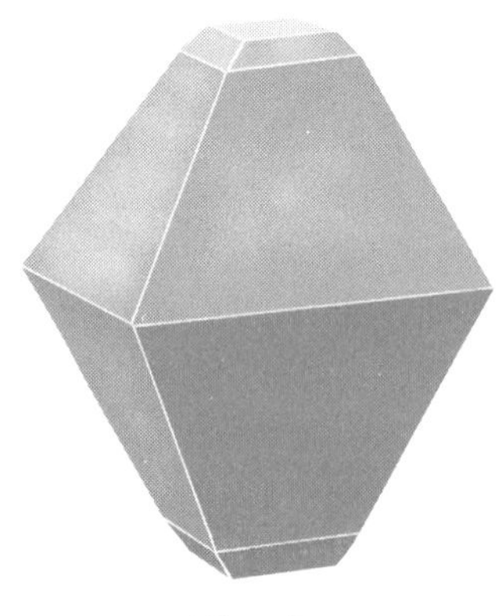

FIGURE 16.3 CRYSTALLINE FORMS OF SULFUR.

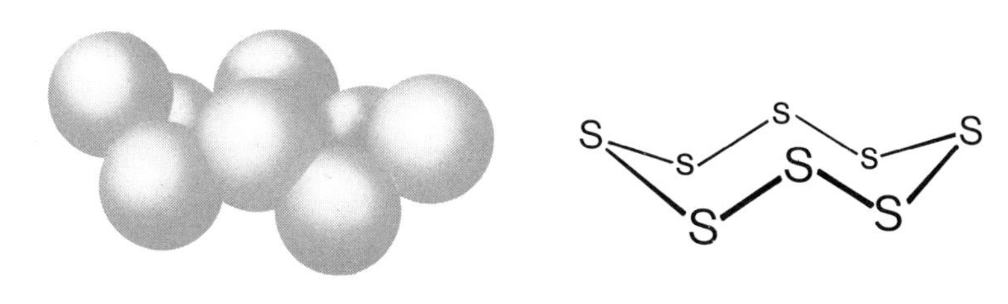

FIGURE 16.4 THE SULFUR MOLECULE.

fur melts at its own melting point, 112.8°C, only to freeze sometimes to the stable monoclinic sulfur and melt again at 119°C. When liquid sulfur, which is a free-flowing, straw colored liquid at the melting point, is heated from the melting point to 160°C, there is little change in the viscosity or other properties of the liquid; but beginning at approximately 160° and continuing to 200°C the sulfur becomes extremely viscous and darkens in color to brown and almost black. A change in molecular structure accompanies the unexpected increase in viscosity. The S_8 rings break open and join together to form long chains, some as long as 10,000 atoms. The change from nearly spherical S_8 molecules, which roll easily over one another, to coiled, tangled chains accounts for the great increase in viscosity. As the temperature rises above 200°C, the natural increase in vibrational movement allows the chains to slide more easily past one another, and the viscosity decreases as expected. When cooled rapidly, the chains do not break rapidly enough, no orderly pattern can be assumed by the odd-sized molecules and so plastic sulfur, a soft rubbery material, results from the fast cooling (Fig. 16.5). In time the plastic sulfur reverts to the stable rhombic form. In plastic sulfur the chains of sulfur atoms are in the form of coils and the rubbery character is due to coiling and uncoiling of the chains under stress.

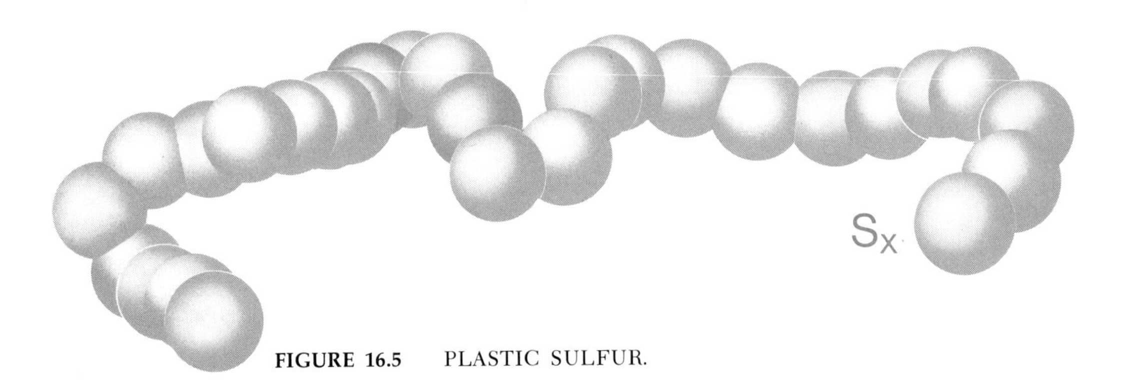

FIGURE 16.5 PLASTIC SULFUR.

16.6 CHEMICAL PROPERTIES OF SULFUR

Members of Group VIA of the periodic table are not as closely related to each other as are the elements in the halogen family. Note the differences in the physical properties listed in Table 16.1. Chemical properties vary from those of the active nonmetals, oxygen and sulfur, to those of metallic tellurium. Oxygen is more reactive than sulfur, but the reactions of sulfur and selenium (and often tellurium) are similar. The members of Group VI have a strong tendency to attain a noble gas electronic configuration in chemical reaction (Table 16.2). Each element has six valence electrons and therefore can either gain two electrons or share some or all of its valence electrons in covalent bonding. The gaining of two electrons is represented by the reaction of calcium and sulfur atoms in Figure 16.6. The noble gas electronic configuration can also be attained by the formation of covalent bonds. Some examples are:

how many valence electrons does sulfur have?

$$\mathrm{H{:}\ddot{\underset{\cdot\cdot}{O}}{:}} \qquad \mathrm{H{:}\ddot{\underset{\cdot\cdot}{S}}{:}} \qquad \mathrm{H{:}\ddot{\underset{\cdot\cdot}{Se}}{:}} \qquad \mathrm{H{:}\ddot{\underset{\cdot\cdot}{Te}}{:}}$$
$$\mathrm{H} \qquad\quad \mathrm{H} \qquad\quad \mathrm{H} \qquad\quad \mathrm{H}$$

TABLE 16.2 THE ELECTRONIC CONFIGURATIONS OF THE OXYGEN FAMILY ELEMENTS*

Element			Ion	
$_{8}O$	He	$2s^2 2p^4$	$_{8}O^{2-}$	Ne
$_{16}S$	Ne	$3s^2 3p^4$	$_{16}S^{2-}$	Ar
$_{34}Se$	Ar	$3d^{10} 4s^2 4p^4$	$_{34}Se^{2-}$	Kr
$_{52}Te$	Kr	$4d^{10} 5s^2 5p^4$	$_{52}Te^{2-}$	Xe
$_{84}Po$	Xe	$4f^{14} 5d^{10} 6s^2 6p^4$	$_{84}Po^{2-}$	Rn

* The colored area represents either the electron configuration of the noble gas preceding the given element in the oxygen family or a noble gas electronic configuration.

Elements of the oxygen family would be expected to form compounds similar to those listed for oxygen in Table 9.5, but the reactivity of the Group VI elements as non-metals decreases going down the family, and there are fewer compounds of tellurium than there are of the other members of the family. Two series of parallel compounds of oxygen and sulfur are listed in Table 16.3.

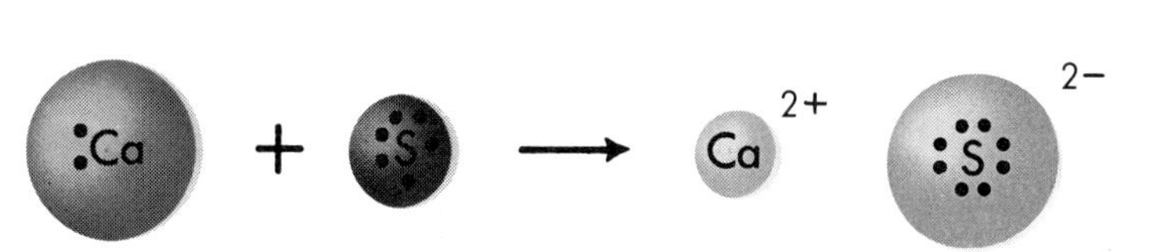

FIGURE 16.6 ELECTRON EXCHANGE TO ACQUIRE THE OCTET CONFIGURATION.

TABLE 16.3 SOME SULFUR AND OXYGEN COMPOUNDS

Oxygen	H_2O	CO_2	FeO	ZnO	Na_2O	$NaOH$	As_2O_3
	Water	Carbon dioxide	Iron(II) oxide	Zinc oxide	Sodium oxide	Sodium hydroxide	Arsenic(III) oxide

Sulfur	H_2S	CS_2	FeS	ZnS	Na_2S	$NaHS$	As_2S_3
	Hydrogen sulfide	Carbon disulfide	Iron(II) sulfide	Zinc sulfide	Sodium sulfide	Sodium hydrosul-fide	Arsenic(III) sulfide

16.7 COMPOUNDS OF SULFUR

Sulfur and its related elements, selenium and tellurium, exist in four common oxidation states: −2, 0, +4 and +6. Some compounds of the elements in these oxidation states are listed in Table 16.4. The compounds will be studied in the order of increasing oxidation states of the Group VI elements with the major emphasis on sulfur compounds.

TABLE 16.4 COMMON OXIDATION STATES OF SULFUR, SELENIUM AND TELLURIUM

Oxidation State	Formula	Name
−2	H_2S	Hydrogen sulfide
	CuS	Copper(II) sulfide
	H_2Se	Hydrogen selenide
	$FeSe$	Iron(II) selenide
	H_2Te	Hydrogen telluride
	$ZnTe$	Zinc telluride
0	S_8	Sulfur
	Se_8	Selenium
	Te	Tellurium
+4	SO_2	Sulfur dioxide
	Na_2SO_3	Sodium sulfite
	SF_4	Sulfur tetrafluoride
	SeO_2	Selenium dioxide
	K_2SeO_3	Potassium selenite
	TeO_2	Tellurium dioxide
	$BaTeO_3$	Barium tellurite
+6	SO_3	Sulfur trioxide
	H_2SO_4	Hydrogen sulfate, sulfuric acid
	$CaSO_4$	Calcium sulfate
	SF_6	Sulfur hexafluoride
	SeO_3	Selenium trioxide
	$FeSeO_4$	Iron(II) selenate
	TeO_3	Tellurium trioxide
	$MgTeO_4$	Magnesium tellurate

16.8 OXIDATION STATE OF −2

Binary compounds with hydrogen and with metals (Table 16.4) illustrate the −2 oxidation state of sulfur, selenium, and tellurium. The binary hydrides are all colorless gases with unpleasant odors. They are highly poisonous, but because of their intense odors, they can be detected much below the lethal concentrations. The boiling points of the gases increase from H_2S (−60°C) to H_2Te (−1.8°C), as one would expect with increasing numbers of electrons (greater van der Waals forces). As already discussed in Chapter 9, the boiling point of water is "abnormally" high, compared to hydrides in this family, because of hydrogen bonding.

bond angles in H_2S, H_2Se and H_2Te are ~90°, in contrast to 104.5° for H_2O. the larger sizes and lower electronegativities of S, Se and Te enables these elements to use p orbitals for bonding because of the decrease in the repulsion between bonding electron pairs.

The binary hydrides can be pictured as being formed by the overlap of two p orbitals, each with a half-filled s orbital of separate hydrogen atoms. This is shown for H_2S in Figure 16.7. Consequently the resulting molecule is bent, with a bond angle of about 90°.

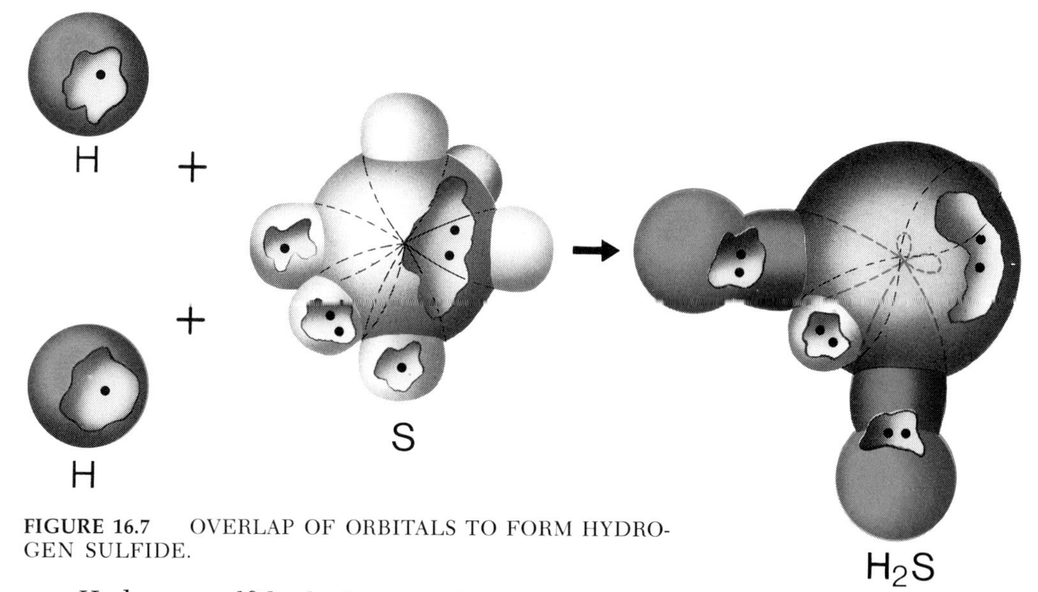

FIGURE 16.7 OVERLAP OF ORBITALS TO FORM HYDROGEN SULFIDE.

Hydrogen sulfide, hydrogen selenide and hydrogen telluride can be prepared easily by reaction of dilute non-oxidizing acids on metal sulfides, selenides and tellurides. This reaction is demonstrated for hydrogen sulfide. Iron(II) sulfide reacts with dilute hydrochloric acid:

$$FeS_{(s)} + 2\ HCl \rightarrow H_2S_{(g)} + FeCl_2$$

The net ionic equation for the reaction is:

$$FeS_{(s)} + 2\ H_3O^+ \rightarrow H_2S_{(g)} + Fe^{2+} + 2\ H_2O$$

This reaction can be carried out in a Kipp generator, an old but ingenious apparatus (Fig. 16.8). Pieces of iron sulfide are placed on the tray in the lower portion of the generator. With the stopcock closed, dilute hydrochloric acid is poured down the funnel. The pressure keeps the acid separated from the iron sulfide, and no reaction occurs. To obtain H_2S gas, one opens the stopcock, releasing the pressure in the lower chamber and allowing the acid to rise up to cover the iron sulfide. The H_2S which is formed leaves through the stopcock. To stop the reaction, the stopcock is closed.

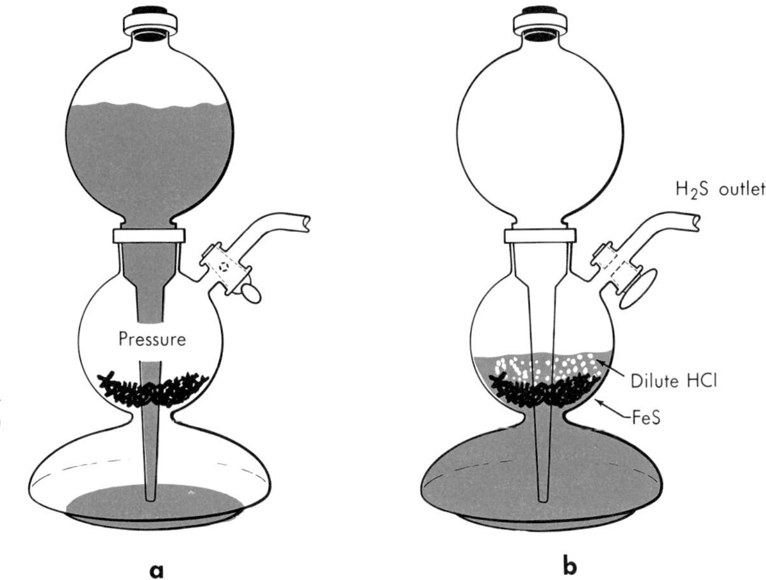

FIGURE 16.8 HYDROGEN SULFIDE (KIPP) GENERATOR.

H_2S is a weak acid.

The generated gas builds up pressure, forcing the acid level down below the sulfide again. When more H_2S gas is needed, one merely opens the stopcock and repeats the procedure. Being able to turn off H_2S gas is almost as important as being able to prepare it.

The hydrogen compounds are all moderately soluble in water and are weak diprotic acids in aqueous solution. The trend in acid strength parallels that of the hydrogen halides. Thus H_2S is the weakest acid of the group, and H_2Te is the strongest. The ionization occurs in two steps, as shown for hydrogen sulfide:

$$H_2S + H_2O \rightleftarrows H_3O^+ + HS^-$$

$$HS^- + H_2O \rightleftarrows H_3O^+ + S^{2-}$$

Because of the two available hydrogen atoms, the acid, even though it is very weak, has just as great a capacity per mole in neutralization as does sulfuric acid.

16.9 SOME REACTIONS OF HYDROGEN SULFIDE

1. Hydrogen sulfide burns readily in air to give sulfur dioxide and water:

$$2\ H_2S + 3\ O_2 \rightarrow 2\ SO_2 + 2\ H_2O$$

2. Hydrogen sulfide reacts with sulfur dioxide on a moist surface to produce sulfur:

$$2\ H_2S + SO_2 \xrightarrow[\text{surface}]{\text{moist}} 3\ S + 2\ H_2O$$

Reactions 1 and 2 are those used to convert hydrogen sulfide gas from sour natural gas to sulfur. Equation 2 represents the reaction occurring on the walls of cooling volcanos. Two of the escaping gases, hydrogen sulfide and sulfur dioxide, react as they come together on the moist walls.

3. Several metals have very insoluble sulfides. H_2S may be used as a source of sulfide ions to precipitate many metal ions from solution for identification. One of these metals is copper. When hydrogen sulfide is bubbled into a solution of copper sulfate, the following reaction occurs:

$$H_2S + CuSO_4 \rightarrow CuS_{(s)} + H_2SO_4$$

The net ionic equation is:

$$2\ H_2O + H_2S + Cu^{2+} \rightarrow CuS_{(s)} + 2\ H_3O^+$$

Because the **metal sulfides show a wide variation in solubility,** it is possible to separate the metal ions into groups of similar solubilities. For example, FeS, MnS and ZnS are insoluble in water but soluble in dilute hydrochloric acid, whereas CuS, PbS and HgS are insoluble in water and in dilute, non-oxidizing acids. In contrast, the sulfides Na_2S, K_2S and CaS are all soluble in water.

16.10 OXIDATION STATE OF +4

The most important compound of sulfur in the +4 oxidation state is sulfur dioxide (SO_2). Analogous compounds are known for selenium and tellurium, but they will not be discussed here since they have no great industrial importance. Because sulfur in sulfur dioxide has an oxidation state between its maximum and minimum values, it can behave either as an oxidizing agent or a reducing agent. The importance of sulfur dioxide lies primarily in the fact that it is an intermediate in the preparation of sulfuric acid. Other examples of its use include those in which it acts as a reducing agent, as in the bleaching of silk, wool and paper. It is also used as a preservative for many dried fruits.

why can SO_2 react as either an oxidizing agent or a reducing agent?

Sulfur dioxide is a colorless gas having a suffocating, irritating odor; it is somewhat poisonous, especially to plants. It is classed by some pollution experts as one of the most potent air pollutants. It appears to attack the human respiratory system and damages plant leaves. It also attacks certain metal and stone structures, particularly those made of limestone. Most of the sulfur dioxide emitted to the atmosphere comes from the burning of coal, which contains from 1 to 3.5 per cent sulfur. In fact, about 50 per cent of the SO_2 in the air is emitted from combustion of fuel for electric power generation. If all the sulfur dioxide from this source were recovered, it would meet the nation's sulfuric acid requirements. **Another major source of sulfur dioxide pollution** is the **roasting of sulfide ores** such as ZnS, FeS_2 and Cu_2S as the first step in the processing of these metals. This step is illustrated for Cu_2S:

$$2\ Cu_2S + 3\ O_2 \rightarrow 2\ Cu_2O + 2\ SO_2$$

Much of the sulfur dioxide produced in these processes is scrubbed from the smoke stack gases and converted to sulfuric acid. Still a typical smelter may emit 1000 tons of sulfur dioxide to the atmosphere per day. The sulfur dioxide pollution problem can be solved, but it will require more research. The subject of air pollution has become an "in thing" in our society, but we as citizens need to demand more action from both industry and government, and we must be prepared to meet increased costs which will be required to clean up the air.

Most of the sulfur dioxide used for the production of sulfuric acid is produced by burning of sulfur in air:

$$S + O_2 \rightarrow SO_2$$

The molecules of sulfur dioxide are bent, and the structure of the compound may be represented by the following resonance structures (see Chapter 5):

Because of its bent structure, SO_2 is a polar molecule, and this property is reflected in the ease with which sulfur dioxide may be liquefied. Its boiling point at 1 atm pressure is $-10°C$, and at room temperature a pressure of about 4 atm will liquefy the gas.

Sulfur dioxide is a non-metal oxide, and therefore has acidic properties (see Chapter 9). It is moderately soluble in water, producing solutions of sulfurous acid, H_2SO_3, a weak diprotic acid.

16.11 OXIDATION STATE OF +6

Sulfur trioxide (SO_3) is a colorless liquid that freezes at 15°C, boils at 46°C and has a choking odor. As the anhydride of sulfuric acid, it reacts violently with water, liberating heat to produce sulfuric acid:

$$SO_3 + H_2O \rightarrow H_2SO_4 + 40{,}500 \text{ calories}$$

The SO_3 molecule is planar, with the sulfur at the center of an equilateral triangle. The O—S—O bond angles are 120°, and the sulfur is sp^2 hybridized. As with SO_2, no single electronic configuration can be written for SO_3, but it may be represented by the following resonance structures:

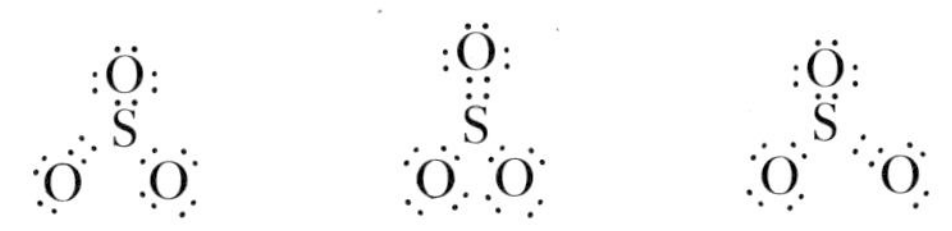

Upon combination with water to produce H_2SO_4, the planar sulfur trioxide is converted to a tetrahedral molecule. The structures of

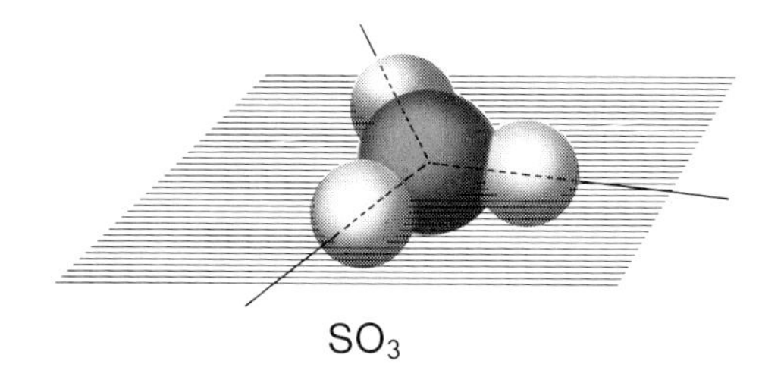

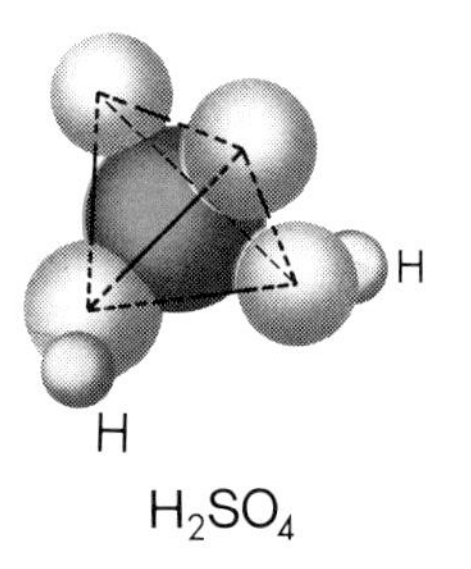

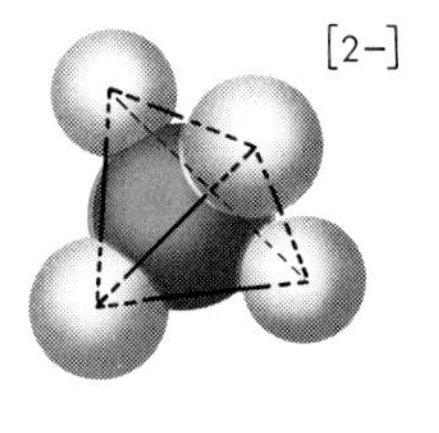

FIGURE 16.9 STRUCTURES OF SO_3, SO_4^{2-} AND H_2SO_4.

the SO_3, H_2SO_4 molecules and the SO_4^{2-} ion are depicted in Figure 16.9.

Sulfur trioxide is produced by the reaction of sulfur dioxide with air.

$$2\ SO_2 + O_2 \rightleftarrows 2\ SO_3 + 47{,}000 \text{ cal}$$

Since the reaction is very slow at ordinary temperatures, the commercial preparation is carried out at elevated temperatures, ranging from 400°C to 700°C, and in the presence of a catalyst, which increases the rate of the reaction. The conditions of the process are set as a compromise between a fast rate of reaction and a high conversion of SO_2 to SO_3. At low temperatures conversion is 100 per cent, but the reaction is too slow to be economically feasible. At very high temperatures the rate of reaction is great, but the per cent conversion is low. It is predicted by Le Chatelier's Principle that the equilibrium for the above reaction is far to the left at high temperatures, since the reaction is exothermic.

There are two major methods of manufacture for sulfuric acid: the contact process and the lead chamber process.

1. **Contact Process.** Preparation of sulfuric acid by the contact process usually begins with crude sulfur (99.05 per cent pure) from the Frasch process.

A. Sulfur is burned in air to give sulfur dioxide:

$$S + O_2 \rightarrow SO_2$$

or the sulfur dioxide may be recovered from smelter gases and purified.

B. The most important step is the catalytic oxidation of sulfur dioxide to sulfur trioxide by air. The gas mixed with air is passed over a vanadium pentoxide catalyst at about 400 to 450°C. About 97 per cent of the sulfur dioxide is converted to sulfur trioxide:

$$2\ SO_2 + O_2 \xrightarrow[400°C]{V_2O_5} 2\ SO_3$$

C. Since sulfur trioxide is the anhydride of sulfuric acid, it would seem logical that it should merely be added to water. However, mechanical problems arise. **The sulfur trioxide vapor** (at the temperature of addition) **forms fog droplets with water vapor.** The droplets diffuse slowly and the reaction comes almost to a standstill. Consequently, the sulfur trioxide vapor is dissolved in concentrated sulfuric acid to form fuming sulfuric acid:

$$H_2SO_4 + SO_3 \rightarrow H_2S_2O_7$$

D. Water is then added, converting the fuming sulfuric acid to sulfuric acid, two molecules for one:

$$H_2S_2O_7 + H_2O \rightarrow 2\ H_2SO_4$$

Steps C and D occur simultaneously, with water being metered into the concentrated acid as the sulfur trioxide dissolves. Pure concentrated sulfuric acid is formed in this manner, the solution being 98 per cent H_2SO_4. The acid is pure because the sulfur dioxide from which it is made is pure. Unless pure sulfur dioxide is passed over the catalyst, it is soon poisoned and is no longer effective.

2. **Lead Chamber Process.** About 30 per cent of the sulfuric acid used today is still produced by the old lead chamber process. In this process, the sulfur dioxide gas need not be pure enough to be used in the contact process. It is mixed with air and the oxides of nitrogen (NO and NO_2), then sprayed with steam into huge lead-lined chambers. Sulfuric acid collects on the floor of the chamber from the reaction

$$2\ SO_2 + O_2 + 2\ H_2O \xrightarrow{NO,\ NO_2} 2\ H_2SO_4$$

The nitrogen oxides are catalysts. The exact role of the catalyst mixture is not known, but in some way it acts as a carrier of oxygen. The solution gained from this process is never more than 60 or 70 per cent sulfuric acid. Furthermore, it is quite impure. However, some industries, such as the fertilizer industry, can use the acid as it is. It is not economical to purify or concentrate it.

16.12 PROPERTIES AND USES OF SULFURIC ACID

why is sulfuric acid important?

Sulfuric acid is the chemical produced in the largest volume in the United States. Such large amounts are used that it is a fairly accurate index of the economic prosperity of a country. In the United States about 85 per cent of the sulfur consumed is used to make sulfuric acid. The uses of sulfuric acid are summarized in Figure 16.10.

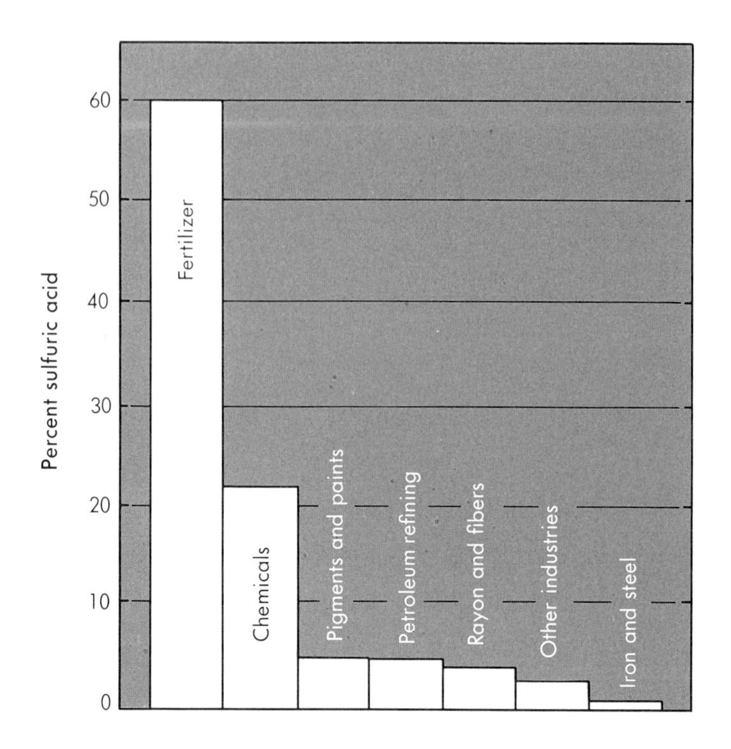

FIGURE 16.10 UTILIZATION OF SULFURIC ACID.

Concentrated sulfuric acid, as sold commercially and as obtained directly from the contact process, **is a thick, syrupy, colorless liquid** with a boiling point of 338°C. It fumes slightly in moist air because of slowly escaping sulfur trioxide vapor. Concentrated sulfuric acid has a density (1.84 g/ml) almost double that of water and is 98.3 per cent H_2SO_4. The solution is 18.4 molar in acid, as shown by the following calculation:

$$1.84 \text{ g/ml} \times 1000 \text{ ml/L} = 1840 \text{ g of } H_2SO_4 \text{ and } H_2O \text{ per liter}$$

$$1840 \text{ g/L} \times 0.983 = 1812 \text{ g of } H_2SO_4\text{/L}$$

$$\frac{1812 \text{ g } H_2SO_4\text{/L}}{98 \text{ g } H_2SO_4\text{/mole}} = 18.4 \text{ moles/L} = 18.4 \text{ M}$$

The complete setup is:

$$\frac{1.84 \text{ g/ml} \times 1000 \text{ ml/L} \times 0.983}{98 \text{ g/mole}} = 18.4 \text{ M}$$

Sulfuric acid is a strong acid. The first ionization occurs completely when sufficient water is added:

$$H_2SO_4 + H_2O \xrightarrow{100\%} H_3O^+ + HSO_4^-$$

The second ionization is comparatively small:

$$HSO_4^- + H_2O \leftrightarrows H_3O^+ + SO_4^{2-}$$

although bisulfate ion (HSO_4^-) ionizes to a greater extent than acetic acid. Because it is a strong acid, and because the boiling point is high, concentrated sulfuric acid is used for the production of three other mineral acids, namely, hydrochloric acid, hydrofluoric acid, and nitric acid:

$$H_2SO_4 + NaCl \xrightarrow{\Delta} Na^+HSO_4^- + HCl_{(g)}$$

$$H_2SO_4 + NaF \xrightarrow{\Delta} Na^+HSO_4^- + HF_{(g)}$$

$$H_2SO_4 + NaNO_3 \xrightarrow{\Delta} Na^+HSO_4^- + HNO_{3(g)}$$

The gases produced are bubbled into water. This is an important method for the production of HCl and HF.

As shown in Figure 16.10, the largest use of sulfuric acid is in the production of fertilizers. One fertilizer, ammonium sulfate, is made simply by the acid-base reaction of crude sulfuric acid with ammonia:

$$2\ NH_3 + H_2SO_4 \rightarrow (NH_4)_2SO_4$$

Ammonium sulfate is a water soluble plant source of nitrogen and sulfate. As a source of phosphorus, insoluble phosphate rock can be rendered soluble and available to plants with sulfuric acid:

$$\underset{\text{Insoluble}}{Ca_3(PO_4)_2} + \underset{\text{Concentrated}}{2\ H_2SO_4} \rightarrow \underset{\text{Calcium dihydrogen phosphate}}{Ca(H_2PO_4)_2} + 2\ CaSO_4$$

The dried mixture of the two salts is called superphosphate.

Steel is usually "pickled" (cleaned of iron rust) by immersion in a bath of sulfuric acid before it is coated with zinc or tin. Sulfuric acid is also used as an electrolyte in lead storage batteries and in the production of many chemicals and drugs. **Hot concentrated sulfuric acid is an oxidizing agent.** Sulfuric acid is used in the petroleum industry to oxidize tars and organic sulfides for removal.

Copper metal is an unreactive metal and will not reduce the hydrogen ion to hydrogen. However, copper will dissolve in hot concentrated sulfuric acid, with the H_2SO_4 acting as an oxidizing agent:

$$Cu + \underset{\text{Concentrated}}{2\ H_2SO_4} \xrightarrow{\Delta} CuSO_4 + SO_2 + 2\ H_2O$$

The acid also will oxidize carbon:

$$C + \underset{\text{Concentrated}}{2\ H_2SO_4} \xrightarrow{\Delta} CO_2 + SO_2 + 2\ H_2O$$

Sulfuric acid is a strong dehydrating agent. Not only will concentrated sulfuric acid remove water from air bubbled through it, but it also will subtract the elements of water from some compounds, decomposing them. Wood is charred by sulfuric acid, largely by loss of water. Sugar is decomposed by a concentrated solution of sulfuric acid.

$$C_{12}H_{22}O_{11} \xrightarrow{H_2SO_4} 12\ C + 11\ H_2O$$

FIGURE 16.11 DEHYDRATION OF SUGAR WITH SULFURIC ACID.

Water is tied up and removed from the sugar by the acid, and a fluffy black residue is left behind (Fig. 16.11). Toluene and concentrated nitric acid are converted to trinitrotoluene by concentrated sulfuric acid and heat.

$$\underset{}{C_7H_8} + \underset{\substack{\text{Nitric}\\\text{acid}\\\text{(concentrated)}}}{3\ HONO_2} \xrightarrow{H_2SO_4} \underset{\text{T.N.T.}}{C_7H_5{\cdot}(NO_2)_3} + 3\ H_2O$$

EXERCISES

16.1 How long has sulfur been known?

16.2 What is the origin of the term "brimstone"?

16.3 What nation is the world's chief producer of sulfur?

16.4 Name four compounds of sulfur that occur in natural deposits.

16.5 How is sulfur obtained from its ore in Sicily?

16.6 Describe the Frasch process.

16.7 What are the two major commercial sources of sulfur?

16.8 Why must the hot water used in the Frasch process be under pressure?

16.9 Is the sulfur rising from a sulfur well dissolved in water?

16.10 Describe how sulfur is obtained from the hydrogen sulfide found in sour natural gas.

16.11 Discuss the "double" melting point of sulfur.

16.12 How does monoclinic sulfur differ from rhombic sulfur?

16.13 How might one account for the darkening in color and the increase in viscosity observed when liquid sulfur is heated above 160°C?

16.14 What is the chief use of sulfur?

16.15 What is the ionic valence of sulfur? What are its usual oxidation states?

16.16 Why are not more people poisoned by the deadly poison, hydrogen sulfide?

16.17 Write the equation that indicates the source of sulfur deposits on volcanic crater walls.

16.18 Why are several million tons of sulfur dioxide produced each year?

16.19 What is the source of SO_2 gas in an iron smelter?

16.20 Why is sulfur trioxide vapor not dissolved directly in water to prepare sulfuric acid?

16.21 What is the oxidation state of sulfur in each of the following:

(a) S_2Cl_2, (b) SO_2, (c) SO_4^{2-}, (d) SO_3^{2-}, (e) SO_3, (f) $S_2O_3^{2-}$, (g) $S_2O_4^{2-}$, (h) $H_2S_2O_7$, (i) $KHSO_4$

16.22 Name three outstanding chemical properties of concentrated sulfuric acid.

16.23 Name four uses of sulfuric acid.

16.24 Write equations for the preparation of sulfuric acid, starting with sulfur and continuing through the contact process.

PROBLEMS

16.25 The density of sulfur vapor at 444.6°C is 3.65 g/L at 1 atm pressure. Show that most of the sulfur vapor at that temperature is in the form of S_8 molecules. Assume that all molecules are either S_2, S_6 or S_8.

16.26 Calculate the molecular weight of sulfur vapor at 1000°C and 1 atm. The density at those conditions is 0.614 g/L.

16.27 A perfect cube of rhombic sulfur, 5 cm on an edge, is burned to sulfur dioxide and then converted by the contact process to sulfuric acid. How many grams of concentrated sulfuric acid solution can be prepared?

16.28 How many liters of oxygen at 700 mm and 25°C are required to burn 50 g of sulfur to sulfur dioxide?

16.29 How many g of sulfur may be formed by reaction of 50 L of SO_2 and 50 L of H_2S, both at STP?

SUGGESTED READING

Bailar, J. C., Jr., Moeller, T. and Kleinberg, J.: *University Chemistry.* D. C. Heath and Co., Boston, 1965. Chapter 16.

Jolly, W. L.: *The Chemistry of the Non-metals.* Prentice-Hall, Inc., Englewood Cliffs, New Jersey, 1966 (paperback).

Pratt, C. J.: "Sulfur," Scient. Amer., *222* (5):62, 1970.

Schaeffer, H. F. and Palmer, G. D.: "Plastic and Allotropic Forms of Sulfur." J. Chem. Educ., *17*:473, 1940.

Wartburg, A. F. and Lodge, J. P., Jr.: "Estimating Concentration of Air Pollutants (Sulfur Dioxide)." Chemistry, *41* (2):29, 1968.

SEVENTEEN • RATES OF REACTION AND EQUILIBRIUM

17.1 INTRODUCTION

Two questions of utmost importance in chemical processing are, "How much?" and "How long?"

The first question may be stated more explicitly. "What proportions of the reactants (the raw materials) may be converted into salable products?" Although the best purification processes concentrate most of the impurities to be thrown away, with a minimum of product, there are always losses incurred in the purification of products.

However, even if none of the products were lost in handling and in purification, one cannot usually expect to obtain a 100 per cent yield of products, as calculated using the equation, because many reactions, by their nature, do not "**go to completion.**" That is, the reaction proceeds only until a certain percentage of the reactants are converted to products; it attains an "**equilibrium,**" and no further increase in product occurs, and as long as the conditions are not changed, no more product is produced. For example, it has been noted that in the preparation of ammonia, only 60 per cent of the nitrogen and hydrogen, when mixed in the correct proportions (1:3) and subjected to a catalyst at 500°C and 1000 atm pressure, can be converted to ammonia. Furthermore, this percentage is the best yield obtainable for the process and was attained only by understanding equilibrium and its dependence on reaction conditions.

The question "How long?" may also be expanded: "How long will it take for the reactants, or a certain percentage of the reactants, to be converted into the desired products?" Time is at a premium in our economy. To prepare hydrogen in the laboratory, zinc metal and hydrochloric acid would be used instead of magnesium and hot water. Both reactions produce pure hydrogen in very good yield, but the first process yields in minutes the amount of gas gained from the second process in hours. Time is of particular significance when reactions are to be carried out at elevated temperatures because fuel is required to maintain the temperature. For reactions that take hours, and not seconds, the fuel costs could be prohibitive.

Often there are competing reactions. Both ethylene, a hydrocarbon gas, and diethyl ether, an anaesthetic, are made by heating ethyl alcohol with the dehydrating agent sulfuric acid:
by heating ethyl alcohol with the dehydrating agent sulfuric acid:

a. $$C_2H_5OH \xrightarrow[\text{concentrated } H_2SO_4]{\text{heat}} \underset{\text{Ethylene}}{C_2H_4} + H_2O$$

b. $$C_2H_5OH + HOC_2H_5 \xrightarrow[\text{concentrated } H_2SO_4]{\text{heat}} \underset{\text{Ether}}{C_2H_5\text{—}O\text{—}C_2H_5} + H_2O$$

The product of reaction (b), the ether, can be gained in good yield by properly controlling conditions, such as by keeping reaction (b) going at a faster rate than reaction (a). Temperature regulates rate and determines which of the products is formed faster. At 140°C, ethyl ether is the principal product; at 180°C, ethylene is obtained.

The two questions, "How much?" and "How long?", are related, since the condition of equilibrium depends on the rate.

17.2 THE RATE OF A REACTION

A chemical reaction has previously been defined as a change in atomic groupings: ions may be formed or changed to atoms or molecules; molecules may be broken down or formed. Regardless of the exact nature of the atomic grouping in a chemical reaction, a substance (or substances) with definite physical properties disappears and a substance (or substances) of different definite properties appears.

The course of a reaction can be followed either by watching how fast a reactant disappears or by watching how rapidly a product appears and increases in amount. The product (or reactant) whose amount of increase (or decrease) is easiest to observe is chosen. If a gas is evolved from liquids, one can watch the bubbling to see how fast the reaction is proceeding, or one can collect and measure the gas above the reaction and note the rate of increase in volume. If a colored product is formed, the increase in color intensity can be monitored. If heat is evolved (that is, if the reaction is **exothermic**), the reaction can be followed by measuring the heat liberated.

for the reaction $A + B \rightarrow 2C$, C is formed at exactly twice the rate at which A is used up.

The rate of a reaction may be expressed in different terms. The chemist may be concerned with the moles of reactant disappearing or with the moles of product appearing per second or he may be concerned with the change in concentration in solution (that is, the change in moles per liter per second). A clearer distinction between the terms will be made later in the chapter, and various important conditions that change the rate of a reaction will be discussed from both experimental and theoretical standpoints.

17.3 FACTORS THAT INFLUENCE THE RATE OF A REACTION

Many factors affect the rate of chemical reactions. Among the most important of these are **temperature,** the **concentrations** of reactants and the presence of **catalysts** and **light.**

It is common knowledge that changes in temperature affect the rate of chemical reactions, for an increase in rate almost invariably accompanies an increase in temperature. By placing food on a burner or in a hot oven, the housewife initiates many complicated chemical reactions that do not occur at room temperatures. If certain foods do not cook satisfactorily at the temperature of boiling water, the housewife may place them and the water in a pressure cooker, in which higher temperatures can be attained. The housewife slows some chemical reactions when she places food in the refrigerator, and she stops them almost completely when she puts food in a freezer.

if the rates of decomposition reactions follow this general rule, how much longer should it take for food to decompose in a refrigerator at 5°C than at room temperature (25°C)?

As a general rule, the rate of most chemical reactions is **doubled** by a temperature increase of **only 10C°.** The term "doubled" is used loosely here. Some reaction rates are indeed doubled, whereas others are tripled and still others are increased only by half. Since a small change in temperature has such a great effect on the reaction rate, kitchen ovens and other apparatus involved with chemical reactions must have reasonably precise temperature controls.

An interesting example of the change in reaction rate with temperature is found in the rate of development and metabolism of so-called "cold-blooded" animals. (Cold-blooded animals are those that assume the temperature of their surroundings.) A tadpole develops two to three times faster in water at 20°C than he does in water at 10°C. The metabolic processes of reptiles are more than doubled with a rise in temperature of 10C°. Warm-blooded animals maintain a constant body temperature, so that external changes in temperature have little or no effect on reaction rates within the body.

An increase in the concentration of one or all the reactants in a reaction mixture increases the rate of reaction. Advantage is taken of this fact when the mixture of air and gasoline vapor is compressed in an automobile engine cylinder to give a higher concentration of the vapors present. Steel wool burns only slowly in the heat of a bunsen flame in air, but it burns with a brilliant white flame in pure oxygen, in which the oxygen concentration is five times greater than that of air. A 1 molar solution of hydrochloric acid reacts much faster with zinc than 1 molar solution of acetic acid, because the concentration of the active component of the solutions, the hydronium ion, is almost 1000 times greater. Silver precipitates on a copper wire much faster from a 1 molar solution of silver nitrate than from a one-half molar solution.

Catalysts are of great importance in chemical processing. Many processes become practical only as a result of the addition of a catalyst, which speeds up the critical reaction:

1. Pure potassium chlorate decomposes very slowly into potassium chloride and oxygen at 400°C in accord with the equation

$$2\ KClO_3 \xrightarrow{\text{heat}} 2\ KCl + 3\ O_2 \quad \text{(see Chapter 9)}$$

At the same temperature, however, the salt is changed more rapidly into potassium perchlorate so that $KClO_4$ becomes the principal product at this temperature:

$$4\,KClO_3 \xrightarrow{\text{heat}} 3\,KClO_4 + KCl$$

When a trace of manganese dioxide is added, decomposition to give oxygen occurs at a much more rapid rate, and even at 200°C $KClO_3$ is almost completely transformed into KCl and O_2, with the formation of very little $KClO_4$.

2. Addition of a few drops of copper(II) sulfate solution to a slowly reacting mixture of zinc metal and sulfuric acid causes the solution to bubble vigorously as hydrogen escapes at a much more rapid rate.

3. Ammonia burns very poorly, if at all, in air, but on the hot surface of a platinum wire ammonia and oxygen of the air unite almost instantaneously. This is the important step—the making of nitrogen(II) oxide—in the Ostwald process for the manufacture of nitric acid.

4. Liquid vegetable oil may be shaken for days with hydrogen gas without any apparent change. But when finely divided nickel or platinum is added, hydrogen is taken up rapidly to convert the oil into a solid fat that is useful for making shortening or butter substitutes (the nickel or platinum is separated afterward!)

5. When a pinch of manganese dioxide is added to a solution of hydrogen peroxide, bubbling occurs as oxygen is rapidly released:

$$2\,H_2O_2 \rightarrow 2\,H_2O + O_2$$

The same result is obtained by adding a drop of blood, which contains a chemical called catalase.

The extra substances added in the examples cited above—copper(II) sulfate solution, platinum wire, finely divided nickel, manganese dioxide and catalase—are catalysts. They all have these characteristics in common: (a) Each increased the rate of a reaction. (b) Each could be recovered unreacted at the termination of the reaction. (c) Each was rather specific for a certain reaction; that is, the catalyst did not increase the rates of all the reactions concerned but increased only the rate of one special reaction. For example, addition of manganese dioxide speeds the decomposition of hot potassium chlorate to form oxygen, but it has no effect on the rate of formation of potassium perchlorate. Catalase in blood differs from the other catalysts cited in that it is a biological material formed in body cells. It is called an **enzyme.**

A mixture of hydrogen and chlorine at room temperature and in the dark fails to react, just as does a mixture of hydrogen and oxygen in a container at room temperature in the light. At high temperature, a mixture of hydrogen and chlorine will combine in the dark, or if ignited in the dark at room temperature with a spark, they unite with an explosion. Likewise, if exposed to visible light, a rapid (explosive) reaction occurs. The reaction can be controlled by varying the intensity of the light, since it goes faster when the intensity is increased. Experiments reveal that **blue-green** light with a wavelength of about 4000 Å is being absorbed by the gas mixture.

Hydrogen and bromine combine to form hydrogen bromide when irradiated with ultraviolet light. The standard method for preparing alkyl halides (organic compounds to be discussed in a later chapter) involves ultraviolet radiation. For example, the bromination of propane, $CH_3CH_2CH_3$,

$$CH_3CH_2CH_3 + Br_2 \rightarrow CH_3CH_2CH_2Br + HBr$$

(one of two possible products)

is accomplished by bubbling the propane gas through a solution of bromine dissolved in carbon tetrachloride (solvent), while the bubbling solution is illuminated with ultraviolet light. The CCl_4 is used to permit control of the concentration.

In other **photochemical** reactions nitrogen dioxide (NO_2) is decomposed to nitrogen(II) oxide, NO and oxygen; acetylene (C_2H_2) is polymerized; carbon monoxide and chlorine combine to give phosgene ($COCl_2$), and acetaldehyde (CH_3CHO) decomposes into ethylene (C_2H_4) and carbon monoxide, each reaction being induced by the absorption of light.

The classic example of a reaction requiring light is **natural photosynthesis** in plants, through which carbon dioxide and water vapor are combined in sunlight to produce glucose and liberate oxygen:

$$6\,CO_2 + 6\,H_2O \rightarrow C_6H_{12}O_6 + 6\,O_2$$

While much work has been carried out to determine the steps in this reaction, and although much is known of the reactions involved, there is still much to be learned.

17.4 ELEMENTARY REACTIONS

The Collision Theory has been proposed to explain the effects of the various factors upon the rates of reaction. According to the theory, a chemical reaction, as represented by the chemical equation, is accomplished by one or several steps called **elementary reactions.** The set of elementary reactions which take the reactants through intermediate stages to the final products are called collectively the **mechanism** for the reaction. The role of temperature, concentrations, catalysts, light and other factors, as they affect the overall rate of a reaction, can be understood only by determining how each of these factors affects the rate of each of the individual steps of the mechanism.

each molecule reacts by itself in a unimolecular reaction.

Most of the elementary reactions that take place in a homogeneous system, in which reactants and products are present in the same phase such as a gas mixture or a solution, are either unimolecular or bimolecular. In a **unimolecular** reaction, individual molecules, twisting, stretching or bending in vibrational motion, rearrange within themselves or fall apart (decompose). In bimolecular reactions, two molecules collide, break bonds in the collision, re-form new bonds and separate as product molecules. A certain minimum energy is required of a certain twisting, stretching molecule to fall apart, and a certain minimum energy is required of two specific colliding molecules to overcome the mutual repulsion of their negative electron clouds and to hit hard enough to rupture bonds. The minimum energy required of the one molecule in the unimolecular reaction and of the two molecules in a bimolecular reaction to react is called the **activation energy** for the reaction.

Some unimolecular reactions occur because of the absorption of light, the important step in a photochemical reaction. The molecule absorbs a photon of light, which raises it to a higher level of energy so that it vibrates and comes apart. In many of these reactions one bond is broken for each photon of light absorbed.

Elementary reactions in heterogeneous systems, such as the dissolving of zinc in hydrochloric acid, involve other elementary steps that will be discussed when reactions in those systems are considered.

The rate of a unimolecular reaction, in terms of the number of moles of reactant disappearing per liter per second, depends upon the number of molecules per unit volume having sufficient vibrational energy to react. The rate of a bimolecular reaction depends upon three factors: (1) the number of collisions per unit volume per second, (2) the fraction of those collisions in which the molecules are oriented properly to react and (3) the fraction of the collisions of correctly oriented molecules that are sufficiently energetic to cause reaction.

17.5 THE EFFECT OF TEMPERATURE CHANGES UPON THE RATE OF ELEMENTARY REACTIONS

To understand the effect of temperature on the rate of a bimolecular reaction, consider the simple reaction of two gases, A_2 and B_2, to form AB:

$$A_2 + B_2 \rightarrow 2\,AB$$

That is, two molecules, one of each reactant, collide and then separate into molecules of product. The reaction is visualized in Figure 17.1.

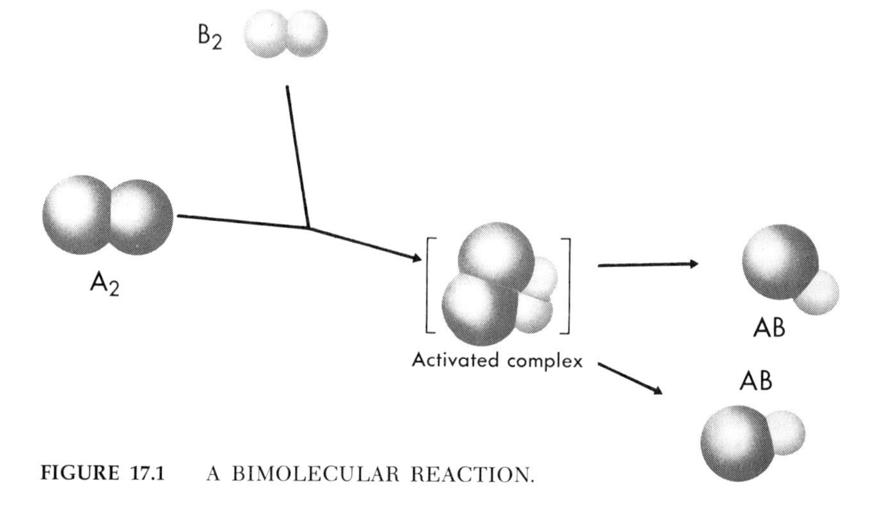

FIGURE 17.1 A BIMOLECULAR REACTION.

The two stable gas molecules, A_2 and B_2, must collide at just the right angle and with sufficient force to upset electron orbitals completely. The "double molecule" in this condition is called the **activated complex.** When orbitals re-form, unlike atoms may be paired together and the product AB formed. At suitable reaction temperatures only a very, very small fraction of the molecular collisions yields products. Certain conditions must be met:

1. The molecules must hit broadside. The unlike atoms must be close enough to form new molecules.

2. The collisions must be between very rapidly moving molecules. Molecules with negative electron shells repel one another. To react, molecules must have sufficient energy (activation energy) to overcome the repulsion, to collide and to break the existent covalent bonds.

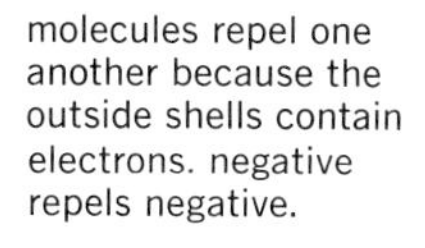
molecules repel one another because the outside shells contain electrons. negative repels negative.

When two molecules approach on a collision course, the repulsion energy increases very rapidly as they come together, as illustrated in Figure 17.2. When the nuclei of atoms of the separate molecules are about the same distance from each other as are the nuclei of the same molecule, the energy is highest and corresponds to the activation energy, because at this point the bonding is indefinite. When the activation complex separates into product molecules, and as these move apart, the repulsion decreases, falling to a level **below** the **energy of the reactants** if the reaction is **exothermic** or remaining **above** if the reaction is **endothermic.**

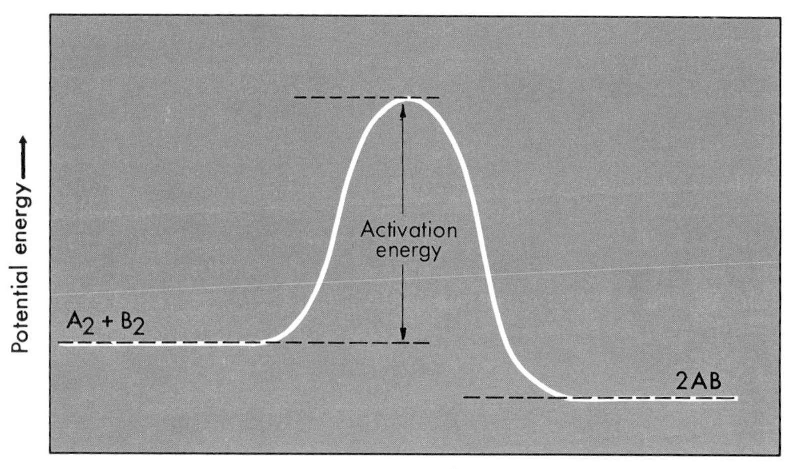

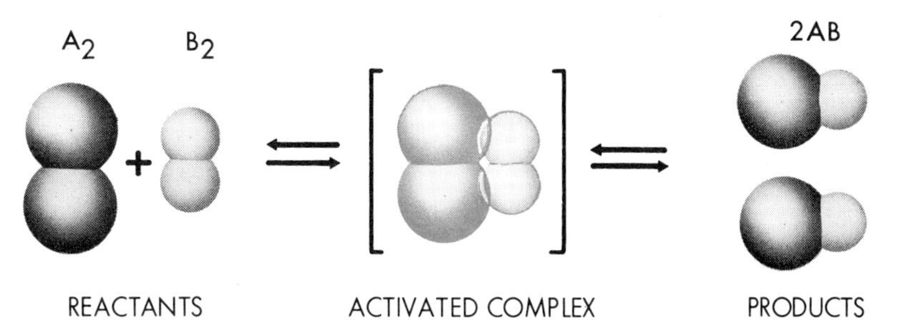

FIGURE 17.2 POTENTIAL ENERGY DIAGRAM FOR THE REACTION $A_2 + B_2 \rightarrow 2AB$.

The activated complex may also regain stability by re-forming the original stable molecule, or molecules.

Remember that not all molecules of a gas have the same translational energy. At any one instant, due to random collisions with one another, a few molecules have very low energy (velocity), many have a velocity very nearly average and another few have very high energies, as shown in Figure 17.3. Only molecules with a combined energy greater than the activation energy for the reaction (represented by the dark bars to the far right on the graph) can collide with sufficient force to react. *If* a molecule of A_2 with sufficiently high energy collides at just the right angle with a molecule of B_2, it also having a high energy, two molecules of AB can result.

Only collision of **high energy molecules** results in chemical reaction; molecules of low or near average velocity cannot react. Two high energy molecules (one of A_2 and one of B_2) must find one another before they collide with other molecules and are thereby (as a result of energy transfer) reduced to low energy molecules. The temperature of gases determines the fraction of high energy molecules present. According to the Kinetic Theory the average translational energy is proportional to the absolute temperature. Likewise, at any given temperature, and at any instant, only a given fraction of the molecules are sufficiently energetic to react. In view of this, why should the rate of a bimolecular reaction double or triple when the temperature is raised from 300 to 310°A, since the energy due to the velocity of the molecules is increased by only 1/30, an increase which cannot double the rate? The 10°C increase in temperature provides a few more collisions per second, but this cannot double the rate of reaction. The answer lies in the fact that the **number of high energy molecules is doubled,** or tripled, as shown in Figure 17.4. At the higher temperature there are still only a few molecules with energies equal to the activation energy, but this few is twice or three times as many as before.

unless there are more molecules without sufficient energy to react, raising the temperature does not increase the rate of a reaction.

The rate of a unimolecular reaction is increased markedly with a rise in temperature for much the same reason as given for a bimolecular reaction. A slight increase in temperature causes only a slight increase in the average vibrational energy. However, the

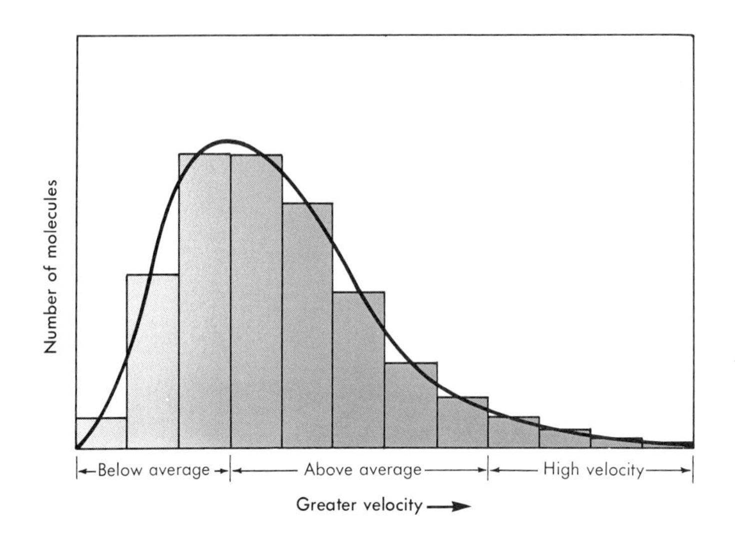

FIGURE 17.3 VELOCITY DISTRIBUTION CURVE (OF MOLECULES).

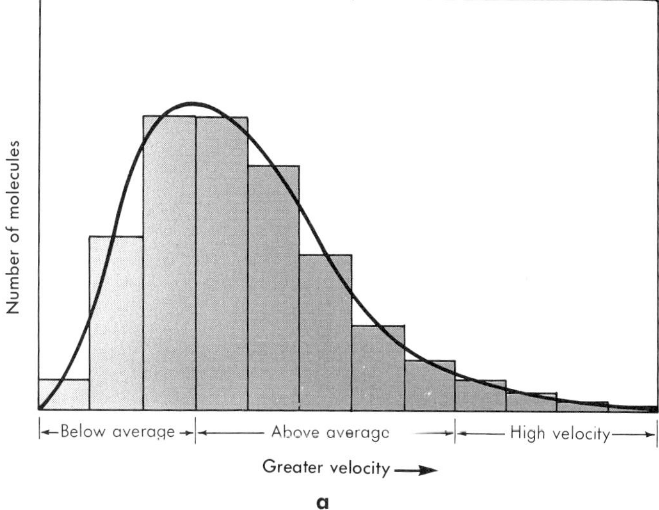

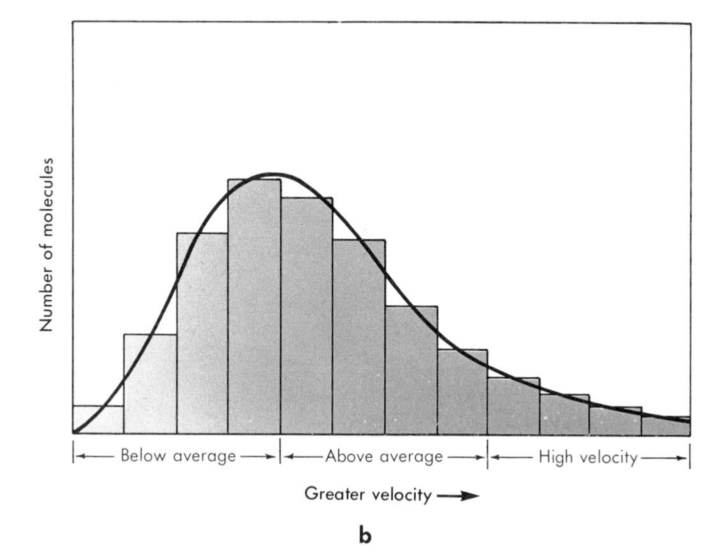

FIGURE 17.4 VELOCITY DISTRIBUTION AT TWO DIFFERENT TEMPERATURES.

number of molecules vibrating with sufficient energy to fall apart is increased markedly.

A photochemical reaction, in which molecules absorb a photon of light, thus converting the energy of the photon into a vibrational energy that is greater than the energy of the bond such that it falls apart, is not affected greatly by a rise in temperature. Since the absorbed energy is more than sufficient to break the bond, little is gained by raising the vibrational energy of the molecule.

17.6 MOLECULAR ORIENTATION AND THE RATES OF CHEMICAL REACTIONS

The importance of the orientation of reacting molecules in a bimolecular collision was noted in the case of the hypothetical reaction

$$A_2 + B_2 \rightarrow 2AB$$

It was stated that unless the molecules met broadside, as shown in Figure 17.1, they could not possibly separate, without additional reaction steps, as two molecules of AB. In most cases the orientations of the molecules at collision are random, and consequently only a very small fraction of the collisions in which the molecules have an energy equal to or greater than the activation energy are successful in forming products. In some reactions the nature of the molecules themselves aids in giving the proper orientations. Two examples will be discussed here.

The reaction between ammonia and hydrochloric acid in solution is a Brønsted acid-base reaction in which the proton is transferred from the hydronium ion to the ammonia molecule:

$$H_3O^+ + NH_3 \rightarrow NH_4^+ + H_2O$$

The polarity of the ammonia molecule contributes to the reactivity of that molecule with hydronium ion because ammonia molecules in the vicinity of positive hydronium ions will tend to be oriented in such a way that the negative end of the dipole, the nitrogen atom at the apex of the pyramid shown in Figure 17.5, is **directed toward** the ion. Thus, the ammonia molecule approaches a collision with the hydronium ion with the proper orientation—the electron pair toward the ion. The hydronium ion is affected to a lesser degree by the presence of the negative charge of the polar molecule, but it will **tend** to be turned in the **right direction** so that upon collision in the solution a hydrogen atom of the hydronium ion collides with the nitrogen atom in such a way that the unshared pair of electrons is between them. The molecule and ion in this configuration constitute the activated complex. Upon separation, the proton may leave with the ammonia:

$$\mathrm{H{-}\underset{H}{\overset{H}{N}}:} + \mathrm{H{-}\underset{H}{\overset{H}{O}}:^{+}} \rightarrow \left[\mathrm{H{-}\underset{H}{\overset{H}{N}}:\cdots H\cdots :\ddot{O}\begin{matrix}\diagup H\\ \diagdown H\end{matrix}}\right]^{+} \rightarrow \mathrm{H{-}\underset{H}{\overset{H}{N}}{-}H}^{+} + \mathrm{:\ddot{O}{-}\underset{H}{H}}$$

A reaction similar to the Brønsted acid-base reaction is the conversion of methyl bromide (CH_3Br) to methanol (CH_3OH) by reaction with a strong base in water:

$$OH^- + CH_3Br \rightarrow CH_3OH + Br^-$$

Because a methyl ion (CH_3^+) is transferred instead of a proton, it is not an acid-base reaction in the Brønsted sense but is a Lewis base

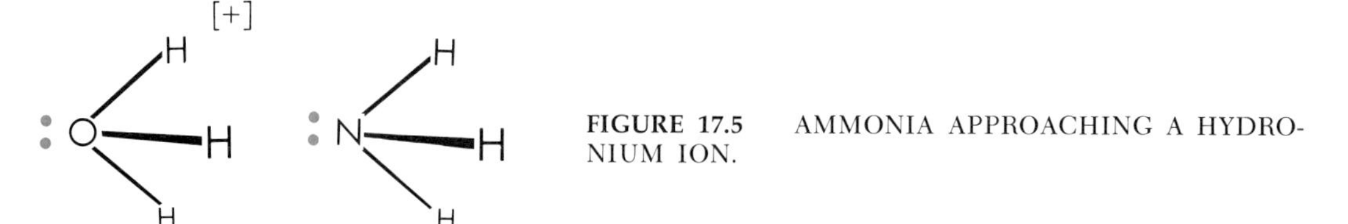

FIGURE 17.5 AMMONIA APPROACHING A HYDRONIUM ION.

displacement. The base, OH^-, presents a pair of electrons to share with the carbon of the methyl ion, which in turn leaves a pair of electrons with the departing base, the bromide ion:

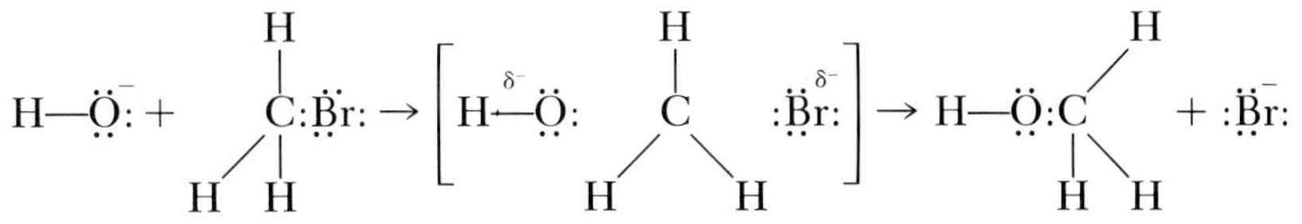

$$H-\ddot{\underset{..}{O}}:^{-} + \underset{H\;\;H}{\overset{H}{C}}:\ddot{\underset{..}{Br}}: \rightarrow \left[H\overset{\delta-}{-}\ddot{\underset{..}{O}}: \quad \underset{H\quad H}{\overset{H}{C}} \quad :\ddot{\underset{..}{Br}}:^{\delta-} \right] \rightarrow H-\ddot{\underset{..}{O}}:\underset{H\;\;H}{\overset{H}{C}} + :\ddot{\underset{..}{Br}}:^{-}$$

The reaction is thought to proceed through a one-step process in which the polarity of the methyl bromide plays an important role in the reactivity. The negative hydroxide ion approaches the methyl bromide molecule opposite the negative bromine of the dipole and presents two electrons to share with the carbon atom. At the same time the bromine atom, taking the shared pair of electrons, leaves the carbon atom and becomes a bromide ion. The molecule is inverted in the process—the carbon to hydrogen bonds bend back toward the departing bromide ion and away from the approaching hydroxide ion. The orientation of the methyl bromide dipole relative to the approaching negative hydroxide ion (positive end to negative ion) places it at the correct angle for reaction.

17.7 CONCENTRATION AND THE RATES OF CHEMICAL REACTIONS

The dependence of the rate of a unimolecular reaction on the concentration of the reactant is simple. The rate is proportional to the concentration of the reactant in moles per liter, for the more molecules there are per liter, the more molecules will fall apart. The rate is related to the concentration in this manner:

$$\text{Rate} = k_1[A]$$

for the unimolecular reaction

$$A \rightarrow B + C$$

a unimolecular reaction is always first order. a first order reaction is not always unimolecular. explain.

where k_1 is a constant for the specific reaction and [A] represents the concentration of A in moles per liter. A reaction in which the rate varies directly with one concentration is said to be of **first order.** A unimolecular reaction is a **first order** reaction.

The rate dependency of a bimolecular reaction upon the concentrations of reactants is a little more complicated. A bimolecular reaction is **second order** with respect to the concentrations. For the hypothetical bimolecular reaction

$$A_2 + B_2 \rightarrow 2AB$$

(used to illustrate the effect of temperature changes on the rate of reactions in the preceding section) the rate equation is

$$\text{Rate} = k_2[A_2][B_2]$$

where k_2 is the specific reaction rate constant for this reaction.

In simple terms this equation says that the rate of reaction is proportional both to the moles per liter of A_2 and to the moles per liter of B_2. If either concentration increases, the reaction goes faster. In fact, if one concentration is doubled, the rate doubles; if both concentrations are doubled, the reaction proceeds four times as fast. **This is true only if the temperature does not change.**

square brackets, [ ], will be used in many instances to indicate moles per liter.

A simple analogy might be helpful in understanding the variation of reaction rate with concentration. Suppose 100 pairs of boys are walking about at random on the university campus. At the same time 100 pairs of girls are walking at random on the same campus. It is springtime. Each time a pair of boys meets a pair of girls, a reaction occurs and the grouping changes. Suppose that in 20 minutes, 10 pairs of boys meet 10 pairs of girls and 20 couples are formed.

Now suppose that 200 pairs of boys and 100 pairs of girls start walking at random on the campus. How long will it take for 20 couples to be formed? Obviously with twice as many boys looking, it should take half the time, or 10 minutes.

But if there are 200 pairs of boys and 200 pairs of girls at the beginning, each one of twice as many boys will be looking for any one of twice as many girls, and only 5 minutes will be required for the first 20 couples to be formed.

Notice that only the initial rate of reaction was considered (the first 20 couples); no mention was made of the time for **all** couples to be paired off. As a reaction proceeds, the concentration of reactants decreases, and the reaction slows down accordingly. This agrees with the rate equation.

a bimolecular reaction is always second order. a second order reaction is not necessarily bimolecular. explain.

The bimolecular reaction of the type

$$2A \rightarrow B + C$$

is also second order with respect to concentrations. However, in this case it is the same concentration. That is,

$$\text{Rate} = k_3[A]^2$$

where k_3 is the specific rate constant for this reaction. The subscripts 1, 2 and 3 are used on the k's to indicate that all specific reaction rate constants are different. According to the equation, the rate at which A decreases in moles per liter per second varies directly with the concentration of A to the second power.

The rate of the photochemical reaction in which a molecule is decomposed (a bond is ruptured) by the absorption of light, such as

$$A \xrightarrow{\text{U.V.}} B + C$$

(where U.V. over the arrow indicates that the reaction proceeds when illuminated by ultraviolet light) may vary with the conditions of reaction in its dependency on concentration. If the gas is at such a high pressure (concentration) that all the light of the particular wavelength is absorbed, the rate of the reaction is zero order with respect to the concentration. Under those conditions the rate depends only upon the intensity of the illumination and not upon the concentration. If the pressure (concentration) of gas A is so low that only a small fraction of the activating light is absorbed, and if the absorption is proportional to the pressure of A, the rate will be first order.

Notice that the concentration of products enters none of the rate equations for elementary reactions. The amount of product has no effect upon the rate of an elementary reaction.

17.8 THE MECHANISM OF A REACTION

Mechanisms for two reactions were described in the discussion of orientation effects, the reaction of ammonia and hydronium ion and the replacement of the bromide ion in methyl bromide with the hydroxide ion. The two mechanisms were very simple, each having only one step. Several complex mechanisms will now be discussed.

The equation for the reaction of tertiary butyl bromide ($(CH_3)_3CBr$) with the hydroxide ion is similar to the equation for the reaction of methyl bromide with hydroxide ion. Bromide ion is displaced by the hydroxide ion:

$$(CH_3)_3CBr + OH^- \rightarrow (CH_3)_3COH + Br^-$$

Tertiary butyl bromide and methyl bromide are equally unstable in a basic solution, but the larger alkyl bromide is much less reactive with the reagent OH^-. The negative hydroxide ion approaching the tertiary butyl bromide molecule opposite the bromine group cannot pass between the large methyl groups to donate and share electrons with the carbon atom. Instead, the tertiary butyl bromide molecule must first lose the bromide ion by ionization, a more energetic process. The hydroxide ion then takes the place occupied previously by the bromide. Because of the larger groups about the central carbon atom, the process changes from a bimolecular reaction (as in the methyl bromide reaction) to a *two-step* mechanism in which a unimolecular reaction is followed by a bimolecular combination:

$$\text{(i)}\quad CH_3{-}\underset{\displaystyle CH_3}{\overset{\displaystyle CH_3}{\underset{|}{\overset{|}{C}}}}{-}Br \rightarrow CH_3{-}\underset{\displaystyle CH_3}{\overset{\displaystyle CH_3}{\underset{|}{\overset{|\,+}{C}}}} + :\ddot{\underset{\cdot\cdot}{Br}}:^- \quad \text{(slow, high activation energy)}$$

$$\text{(ii)}\quad CH_3{-}\underset{\displaystyle CH_3}{\overset{\displaystyle CH_3}{\underset{|}{\overset{|\,+}{C}}}} + :\ddot{\underset{\cdot\cdot}{O}}:H^- \rightarrow (CH_3)_3C{-}OH \quad \text{(fast)}$$

The high activation energy of the unimolecular reaction in the first step, where positive and negative ions must be separated, makes this the slow step. The slow step in a series of consecutive reactions is the so-called **rate determining step** because the overall reaction can not proceed faster than the slowest step.

The reaction represented by the equation

$$ROH + Br^- \rightarrow RBr + OH^-$$

(where R is a hydrocarbon group such as isopropyl, $(CH_3)_2CH{-}$)

is a Lewis base-displacement reaction. In fact, it is the reverse of the reaction discussed above. The reaction ordinarily does not go, but in an acidic solution, an additional reactant, H_3O^+, is present, and one of the products is changed from hydroxide ion to water. Those products, the alkyl bromide and water, are more stable than the reactants (the alcohol, hydronium ion and bromide ion), and therefore the replacement of the hydroxyl group with bromide in an acid solution occurs:

$$ROH + H_3O^+ + Br^- \rightarrow RBr + 2\,H_2O$$

Because it enters into the reaction and gives more stable products, the hydronium ion is not actually a catalyst but a reactant. However, the hydronium ion does react with the alcohol, changing the molecule in such a way that the slow step in the reaction expected without the acid (too slow to be observed) is speeded up. In that respect the hydronium ion functions as a catalyst.

Without the hydronium ion, were the replacement a two step process, the hydroxyl group would have to leave the alcohol as a hydroxide ion (OH^-) to make way for the addition of the bromide ion. The separation of the negative hydroxide ion and the positive alkyl ion (R^+), in opposition to their strong mutual attraction, is an infinitely slow reaction requiring great energy:

$$ROH \rightarrow R^+ + OH^-$$

With hydronium ion present, the Lewis base (ROH) replaces the previous base (H_2O), attached to the hydrogen ion, to form the ion ROH_2^+,

$$\text{(i)}\quad R:\ddot{\underset{..}{O}}:H + H:\underset{\ddot{H}}{\ddot{O}}:H^+ \rightarrow R:\underset{\ddot{H}}{\ddot{O}}:H^+ + H_2O$$

A neutral water molecule then leaves, forming the positive alkyl ion:

$$\text{(ii)}\quad R:\underset{\ddot{H}}{\ddot{O}}:H^+ \rightarrow R^+ + H_2O$$

The negative bromide ion and positive alkyl ion combine in the final step to yield the alkyl bromide product:

$$\text{(iii)}\quad R^+ + Br^- \rightarrow RBr$$

In this three-step mechanism, the second step (the separation of the ROH_2^+ ion into water and R^+) is the slow, or rate determining, step.

Sometimes there is internal competition in a mechanism. That is, one elementary reaction that **opposes** the formation of final products competes with an elementary reaction necessary for product formation. Nitrogen(II) oxide reacts with oxygen spontaneously to yield nitrogen dioxide according to the overall reaction

$$2\,NO + O_2 \rightarrow 2\,NO_2$$

The proposed mechanism comprises three steps:

$$\text{(i)}\ O_2 + NO \rightarrow OONO \quad \text{(slow)}$$
$$\text{(ii)}\ OONO \rightarrow O_2 + NO \quad \text{(fast)}$$
$$\text{(iii)}\ OONO + NO \rightarrow 2\ NO_2 \quad \text{(slow)}$$

The unimolecular reaction (step ii) reverses the formation of products but is a fast reaction. This mechanism provides an explanation for an unusual effect observed for this reaction: The rate of the overall reaction actually **decreases** with an increase in temperature. According to the explanation all three elementary reactions in the mechanism are increased in rate with the rise in temperature, but the rate of step ii increases much faster than that of step iii, so that OONO decomposes so fast to NO and O_2 that reaction iii cannot compete, and products are formed more slowly.

In the lead chamber process for making H_2SO_4, sulfur dioxide, water vapor and air are mixed in the presence of a mixture of nitrogen oxides at high temperature:

$$2\ SO_2 + O_2 + 2\ H_2O \xrightarrow[\Delta]{NO,\ NO_2} 2\ H_2SO_4$$

The mechanism might be

$$\text{(i)}\ NO_2 + SO_2 \rightarrow NO + SO_3$$
$$\text{(ii)}\ SO_3 + H_2O \rightarrow H_2SO_4$$
$$\text{(iii)}\ 2\ NO + O_2 \rightarrow 2\ NO_2$$

Because the reaction progresses satisfactorily when the nitrogen oxides are present but not without them, and because the nitrogen dioxide consumed in the first step is produced in the final step, the nitrogen oxide mixture is a catalyst. Obviously the activation energies of each of the two steps (i and iii) are less than the activation energy for the direct reaction of sulfur dioxide and oxygen, as shown in Figure 17.6, and the reaction proceeds.

The apparently simple reaction between hydrogen gas and

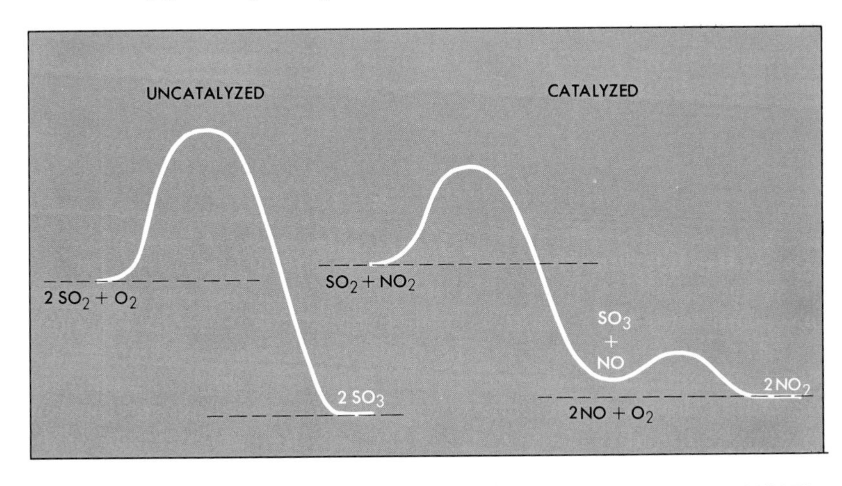

FIGURE 17.6 ACTIVATION ENERGIES OF CATALYZED AND UNCATALYZED REACTIONS.

bromine vapor to form hydrogen bromide

$$H_2 + Br_2 \rightarrow 2\,HBr$$

when the mixture is irradiated with ultraviolet light, is thought to proceed by **five** elementary reactions. The proposed mechanism is:

$$\text{(i)}\quad Br_2 \xrightarrow{\text{U.V. light}} 2\ :\ddot{\underset{..}{Br}}\cdot \qquad \text{(chain initiating step)}$$

$$\left.\begin{array}{ll}\text{(ii)} & :\ddot{\underset{..}{Br}}\cdot + H_2 \rightarrow HBr + H\cdot \\ \text{(iii)} & H\cdot + Br_2 \rightarrow HBr + :\ddot{\underset{..}{Br}}\cdot \end{array}\right\} \quad \text{(chain reaction)}$$

$$\text{(iv)}\quad :\ddot{\underset{..}{Br}}\cdot + HBr \rightarrow Br_2 + H\cdot$$

$$\text{(v)}\quad :\ddot{\underset{..}{Br}}\cdot + :\ddot{\underset{..}{Br}}\cdot \rightarrow Br_2 \qquad \text{(chain stopping step)}$$

The series of reactions is called a free-radical mechanism because entities, atoms in this case, with an odd number of electrons are called free radicals. The reaction is said to have a chain mechanism, with reactions ii and iii constituting the chain. A chain reaction is a two-step portion of a mechanism in which each step produces a necessary reactant for the other, in such a way that each step can follow the other over and over while at the same time forming a final product. As can be seen in the mechanism, steps ii and iii can follow each other over and over, producing hydrogen bromide in each step, as long as both reactants (bromine and hydrogen) are present. Because of the chain, a relatively small number of photons absorbed in the chain initiating step (step i), where the absorption of the photon separates the bromine molecule into atoms, results in the production of many times more molecules of hydrogen bromide. Step iv slows down the process and step v removes free radicals necessary for the propagation of the chain and causes the reaction to cease shortly after the irradiating light is extinguished.

The reaction between hydrogen and bromine will occur at high temperatures in the absence of ultraviolet light. At high temperature there is a sufficiently large number of collisions of sufficiently high energy to break the bonds of bromine molecules and produce the atoms necessary to start and maintain the chain reaction. Once the bromine atoms are formed, the reaction proceeds along much the same path as when free radicals are formed by the absorption of light.

17.9 HETEROGENEOUS REACTIONS

A reaction in which two or more phases are involved is a heterogeneous reaction. Heterogeneous reactions are very common and tend to be more complex than homogeneous reactions.

The burning of coal is a heterogeneous reaction. Carbon and other combustible materials in the solid coal combine with the oxygen of the air. As in many heterogeneous reactions, one reactant is depleted readily at the point of the reaction, and the rate of the reaction is limited by the rate of diffusion to the reaction site. Oxygen is depleted rapidly at the surface of the burning coal and more must diffuse to that site in order for the reaction

$$C + O_2 \rightarrow CO_2$$

to proceed.

The rate of the reaction can be speeded up by a wind or an artificial air current that will bring new oxygen-containing air to the burning surface. The reaction rate can also be increased remarkably by grinding the coal to a powder and then spraying the coal dust in air. A coal dust-air mixture in the proper proportions, when ignited with a match or a spark, may explode. The heat of the reaction, once it is started, brings the dust to burning temperature, and the required oxygen surrounding each particle of coal dust allows the reaction to proceed rapidly.

A lighted match dropped into an open pan of gasoline can be extinguished by the liquid. (Do not try it!!) But a spray of the liquid (tiny droplets in air) explodes with the slightest spark.

Silver precipitates from a silver nitrate solution much more rapidly on fine copper metal shreds than on one large copper wire of the same weight:

$$2\,Ag^{+} + Cu \rightarrow Cu^{2+} + 2\,Ag$$

Again, a much greater area of copper is in contact with the silver-bearing solution. The rate equation for this reaction at a set temperature is of this form:

$$\text{Rate} = k \times \text{area} \times [Ag^{+}]$$

The rate increases directly with the concentration of the silver ion and also directly with the surface area of copper in contact with the solution. The rates of this reaction, as with other heterogeneous reactions, is affected spectacularly by changes in the state of subdivision of the reactants.

17.10 CATALYSIS

The five examples of catalysis cited in Section 17.3 were chosen for variety. There are many more catalyzed reactions and many more kinds of catalysts. The functioning of many catalysts is not understood; certainly they do not all behave alike. Reasonable explanations for the operation of some catalysts have been postulated.

The possible roles of the hydronium ion as a catalyst in the conversion of tertiary butyl bromide to butyl alcohol and the role of nitrogen oxides in the lead chamber process for producing sulfuric acid were both described in the discussion of mechanisms. Explanations for two more examples of catalysis will be given.

When copper sulfate solution is added as a catalyst to a mixture of mossy zinc and sulfuric acid, in which the following reaction is slowly taking place

$$Zn + 2\,H_3O^{+} \rightarrow Zn^{2+} + 2\,H_2O + H_2$$

copper, being less active than either zinc or the hydronium ion, "plates out" in spots on the zinc metal surface. Tiny voltaic cells are then set up between the dissimilar metals in solution. The zinc surface in the solution acts as the reducing agent, and zinc atoms dissolve from the surface, leaving electrons behind:

$$Zn \rightarrow Zn^{2+} + 2e^-$$

The electrons are conducted through the zinc to a copper spot where the hydronium ions can accept electrons more easily and where hydrogen is formed:

$$2\,H_3O^+ + 2\,e^- \rightarrow 2\,H_2O + H_2$$

Cell reactions of this type account for the fact that aluminum that is 99 per cent pure corrodes in sea water, whereas very pure electrolytically refined aluminum does not.

A mixture of the two gases ethylene ($CH_2{=}CH_2$) and hydrogen will remain indefinitely without reacting at temperatures below 200°C. However, when finely divided platinum, nickel or palladium is added to the mixture, the addition reaction

$$CH_2{=}CH_2 + H_2 \rightarrow CH_3CH_3$$

proceeds rapidly at room temperature. This is an example of heterogeneous catalysis—the gases react when they come together at the metallic surface. As evidence, platinum is much more efficient when used in finely divided form, precipitated in a thin layer on the surface of pumice or other inert material, than is a smooth piece of platinum wire. Little is known of the mechanism of this reaction, but some facts about the surface of the metal and the nature of the molecules aid in a partial explanation. Hydrogen is soluble to some extent in metallic platinum at high temperatures. X-ray analyses of the metal reveal that the interstices are large enough to admit hydrogen atoms but not hydrogen molecules. The hydrogen, in passing into the metal, separates into atoms at the surface. The atoms pass from interstice to interstice as the vibration of the atoms within the metal opens passageways between the holes. Furthermore, platinum atoms are arranged in planes of the crystal, such that the centers of the metal atoms are about $1\frac{1}{3}$ Å apart. The distance between the carbon atoms of the double bond in ethylene is 1.34 Å.

It is possible that ethylene molecules collide with the platinum surface, which is already saturated with hydrogen atoms that are separated at distances almost equal to the length of the carbon to carbon double bond. In those collisions in which the carbon atoms each fall directly above an adsorbed hydrogen atom, the pi electrons of the double bond split and pair with the electrons of the two hydrogen atoms to form two single bonds. When the molecule leaves the surface of the metal, the bond has reacted with two hydrogen atoms and the composition is $H_3C{-}CH_3$. As is true of all catalysts, the platinum metal does not make the reactants any more or less stable, but it does provide a pathway that allows the reaction to take place much more easily and rapidly.

17.11 A CHEMICAL EQUILIBRIUM

The existence of equilibrium might be best understood by considering a series of experiments that might be carried out in the laboratory, with the results given below.

Experiment 1. A 1-liter quartz glass vessel containing 0.01 mole H_2 and 0.01 mole I_2 is placed in a furnace at 425°C. After several hours, the flask is removed, "chilled" to room temperature to stop all reaction and its content analyzed. It is found to contain

0.00214 mole $H_2 = 2.14 \times 10^{-3}$ mole

2.14×10^{-3} mole I_2

and 0.0157 mole HI

Experiment 2. A 1-liter vessel containing 0.02 mole of HI gas is placed in the same furnace at the same temperature. After several hours, and after chilling, an analysis shows it to contain

2.14×10^{-3} mole H_2

2.14×10^{-3} mole I_2

and 0.0157 mole HI

Experiment 3. A 1-liter flask containing either 0.01 mole of H_2 and 0.01 mole I_2, or 0.02 mole HI, is placed in a furnace at 420°C for several hours, chilled and analyzed. Analysis reveals the presence of slightly *more than:*

0.0157 mole of HI

and *less than*

2.14×10^{-3} mole of both H_2 and I_2

Experiment 4. A 1-liter flask containing either 0.01 mole H_2 and 0.01 mole I_2, or 0.02 mole HI, is placed in a furnace at 430°C for several hours, chilled and analyzed. Analysis reveals the presence of slightly *less than*

0.0157 mole of HI

and *more than*

2.14×10^{-3} mole of both H_2 and I_2

None of these four experiments reveals any rates of reaction. In each case the reaction has progressed until all changes in concentration have come to a standstill.

When iodine vapor and hydrogen gas **only** are placed in a container at 425°C (Experiment 1), hydrogen iodide is formed. This reaction must occur:

$$H_{2(g)} + I_{2(g)} \rightarrow 2\,HI_{(g)}$$

When hydrogen iodide is placed alone in a container at 425°C for several hours (Experiment 2), analyses show that hydrogen and iodine are both present, as well as HI. So this reaction must occur:

$$2\ HI_{(g)} \rightarrow H_{2(g)} + I_{2(g)}$$

But these reactions are opposites. It might be supposed that if one reaction occurs, the other cannot. However, notice that regardless of the initial amounts, the system attains the same final state: 2.14×10^{-3} mole of H_2, 2.14×10^{-3} mole of I_2, and 1.57×10^{-2} mole of HI. Furthermore, one can alter the percentages of gases in the sealed container by lowering or raising the temperature (Experiments 3 and 4).

A reaction that can proceed in either direction is written with double arrows in the following manner:

$$H_{2(g)} + I_{2(g)} \rightleftarrows 2\ HI_{(g)}$$

and is said to be **reversible.** When gases in the system described in Experiments 1 and 2 reach a constant state with no further changes in concentration, the system is said to be in **equilibrium with respect to the reversible reaction.**

Because equilibrium can be attained from either set of reactants and because a change in temperature changes the equilibrium concentration, it seems obvious that the **state of equilibrium is not static but dynamic.** In other words, 0.01 mole of hydrogen and 0.01 mole iodine vapor do not react in a liter vessel until 2.14×10^{-3}

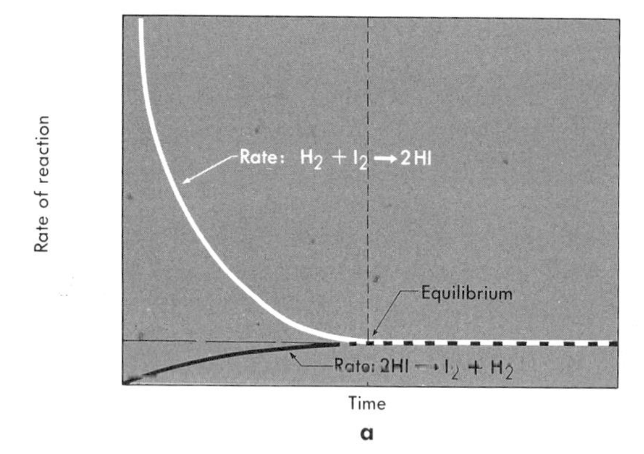

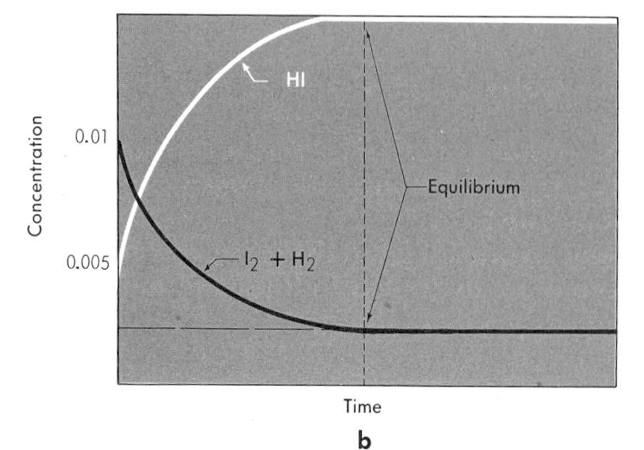

FIGURE 17.7 TIME VERSUS RATE AND TIME VERSUS CONCENTRATION CURVES FOR HI FORMATION AND DECOMPOSITION.

mole of each remain and **then suddenly stop.** Nor does 0.02 mole of hydrogen iodide decompose until only 0.0157 mole remains, and then **suddenly stop.** In both cases the reaction continues as long as the system is kept at a sufficient temperature for an appreciable reaction to occur. The reaction slows as the reactant concentration decreases, but it appears to cease changing or reacting at equilibrium only because the product, or products, react together to produce reactants. When both reaction rates are the same, the concentrations undergo no further change. The diagrams in Figure 17.7 show how the rates of the forward and reverse reaction change with time and how the concentration of reactants and products varies in Experiment 1, in which 0.01 mole of the H_2 and 0.01 mole of I_2 vapor were placed in the 1-liter flask at 425°C. The forward reaction, the combining of hydrogen gas and iodine vapor, goes most rapidly at the beginning but falls off as the reactants are used up in forming the product, hydrogen iodide. There was no reverse reaction rate initially because no hydrogen iodide was present in the flask. With the initial formation of hydrogen iodide the reverse reaction begins and then increases as the forward reaction decreases. Eventually the two reaction rates become equal. That is, hydrogen iodide is being decomposed as fast as it is formed, so the concentrations of all three gases remain constant and equilibrium is attained.

chemical equilibrium is a dynamic equilibrium. explain.

17.12 THE EXPRESSION FOR EQUILIBRIUM CONSTANT

In studying many systems in which equilibrium is obtained with respect to a chemical reaction, chemists have found that a relationship exists between the concentrations of reactants and products; this relationship will have a constant value for that reaction at a given temperature. The expression for that relationship is an equation which gives us a numerical **equilibrium constant.**

The expression for the equilibrium constant contains the product of the equilibrium concentrations of the products of the reaction, raised to the power of their coefficients in the equation for the reaction, divided by the products of the concentrations of the reactants, raised to the power of their proper coefficients. For the reaction

$$H_{2(g)} + I_{2(g)} \xrightarrow{425°C} 2\,HI_{(g)}$$

the equilibrium constant expression is

$$K = \frac{[HI]^2}{[H_2][I_2]}$$

Using the equilibrium concentration values from Experiment 1, one can evaluate K at 425°C.

$$K = \frac{(0.0157)^2}{(2.14 \times 10^{-3})^2} = 53.9$$

The constant, K, is a **number** that changes only when temperature is changed. Regardless of the total pressure of gases in the vessel and regardless of initial gaseous concentrations, the value of the constant does not change. A second determination of K may help to emphasize that fact.

Experiment 5. One-tenth mole of H_2 and 1.24 moles of HI are admitted to a previously evacuated 10-liter flask, placed in a furnace at 425°C, left for several hours, then chilled and analyzed. The analysis reveals the presence of the following amounts of contents: 0.20 mole H_2, 0.10 mole I_2 and 1.04 moles HI, all contained in the 10 liters. Substituting the individual concentrations in moles per liter in the equilibrium constant expression, one obtains

$$\frac{[HI]^2}{[H_2][I_2]} = \frac{(0.104)^2}{0.020 \times 0.010} = 54.0$$

The value of K (54.0) agrees well with 53.9 determined above.

Even though the solvent in which a reaction takes place may also be either a reactant or product, the concentration term for it usually does not appear in the expression for the equilibrium constant because the solvent is in such a great concentration, as compared to other reactants and products, that any changes are usually negligible. No concentration terms are used for solids because while the amounts may change during reaction, the concentrations in the solution do not change. For example, the constant for the dissolving of silver chloride in an ammonia solution, according to the equation

$$AgCl_{(s)} + 2\,NH_3 \rightleftarrows Ag(NH_3)_2^+ + Cl^-$$

is given by

$$K = \frac{[Ag(NH_3)_2^+][Cl^-]}{[NH_3]^2}$$

As another example, sulfur dioxide reacts with oxygen gas in the presence of a vanadium pentoxide catalyst at 450°C to yield an equilibrium mixture of sulfur dioxide, oxygen and sulfur trioxide vapor:

$$2\,SO_{2(g)} + O_{2(g)} \rightleftarrows 2\,SO_{3(g)}$$

The constant for this reaction is given by

$$K = \frac{[SO_3]^2}{[SO_2]^2[O_2]}$$

17.13 DERIVATION OF THE EXPRESSION FOR EQUILIBRIUM CONSTANT

The equilibrium constant can be derived for a particular reaction if the rate equations are known for both the forward and reverse reactions. The rate equations must be found in the laboratory, unless the reaction is assumed to be an elementary reaction for which the rate equation comes directly from the equation for the overall reaction.

Consider the elementary reaction

$$A_{2(g)} + B_{2(g)} \rightleftarrows 2\ AB_{(g)}$$

occurring at any set temperature. The rate of the bimolecular reaction is given by

$$\text{Rate}\ (\rightarrow) = k_1[A_2][B_2]$$

The reverse reaction

$$2\ AB \rightarrow A_2 + B_2$$

is elementary and, therefore, bimolecular. The rate equation for that reaction is

$$\text{Rate}\ (\leftarrow) = k_2[AB]^2$$

The specific reaction rate constants are different and cannot change. Therefore, the concentrations change, [A] and [B] increasing as [AB] decreases and vice versa. When the concentrations adjust themselves so that

$$\text{Rate}\ (\rightarrow) = \text{Rate}\ (\leftarrow)$$

gas AB is formed and decomposed at the same rate. The concentrations cease to change, and the chemical system is in equilibrium with respect to the reversible reaction.

If the rate of the forward reaction is equal to the rate of the reverse reaction, then

$$k_1[A_2][B_2] = k_2[AB]^2$$

Rearranging terms,

$$\frac{k_1}{k_2} = \frac{[AB]^2}{[A_2][B_2]}$$

Because k_1 and k_2 are constants, so also is their ratio, k_1/k_2, which is defined as K, the equilibrium constant. Thus for the reaction

$$A_2 + B_2 \rightleftarrows 2\ AB$$

it is true that

$$K = \frac{[AB]^2}{[A_2][B_2]}$$

This derivation is only for this specific reaction. Expressions for equilibrium constants for reactions with other rate equations must be derived individually. Furthermore, the equilibrium constant is not evaluated in the derivation; the numerical value must be found by the substitution in the equation of a set of concentrations determined by experiment for a system in equilibrium.

17.14 TESTS FOR EQUILIBRIUM

The equilibrium constant, once it has been determined for a particular reaction at a given temperature, can be used to determine whether a reaction system has reached equilibrium. Unless the constant determined for the set of measured concentrations is equal to the equilibrium constant, the system is not at equilibrium.

Example. A mixture of hydrogen, hydrogen iodide and iodine at 425°C is chilled and analyzed and is found to have the following concentrations at 425°C.

$$[H_2] = 0.040 \text{ moles/liter}$$

$$[HI] = 0.15 \text{ moles/liter}$$

and

$$[I_2] = 0.25 \text{ moles/liter}$$

Is the mixture at equilibrium?

Solution:

$$\frac{[HI]^2}{[H_2][I_2]} = \frac{0.15^2}{0.040 \times 0.025} = 22.5$$

The system is not at equilibrium for

$$22.5 \neq 54.0$$

Le Chatelier's Principle (discussed later) can be used to test for equilibrium even though nothing is known about the value of the equilibrium constant. Furthermore, one is not limited to concentrations in using this principle; one may also use temperatures. Le Chatelier's Principle may be applied to the results of Experiments 1 to 4 in Section 17.11. The great French chemist, Henri Louis Le Chatelier (1850–1936) said, "When a stress is applied to a system in equilibrium, the system tends to rearrange itself so as to nullify the stress." The stress is applied to a factor or factors that influence equilibrium, such as temperature or concentration. A stress is placed upon the temperature of a system by heating or cooling it. A stress is applied to a concentration by adding more or extracting some reactant or product from the reaction mixture.

The utility of Le Chatelier's Principle can be seen by applying it to a convenient system, an equilibrium mixture of hydrogen, hydrogen iodide and iodine at 425°C. The overall reaction, in which there is equilibrium, can be represented by the equation with double arrows:

$$H_{2(g)} + I_{2(g)} \rightleftarrows 2\,HI + 3000 \text{ calories}$$

According to Le Chatelier, when the temperature of a system at equilibrium is raised, the system will react in such a way as to lower it. This it can do by using up calories, which is to say by reacting in the endothermic direction. Thus, hydrogen iodide decomposes partially into its elements, thereby using up heat. The concentration of HI decreases slightly; the concentrations of H_2 and I_2 increase. This agrees with the result observed in Experiment 4 (Section 17.11) and does not conflict with the rate theory, which states that the rates of **both** the forward and the reverse reactions must increase with temperature. It is just that the rate in the endothermic direction increases more.

When hydrogen is added to a mixture of the three gases at equilibrium, a stress is placed upon the hydrogen concentration—it is too high. In an effort to nullify the increase, some H_2 and I_2 are converted to HI, increasing the concentration of the HI and decreasing the concentrations of the two elements. Again this is in agreement with the rate theory. Although the forward and reverse reactions are not elementary, the forward rate (combining of the elements) increases as the concentration of H_2 is increased, while the reverse rate is unaffected until the concentration of HI is increased because of the increased forward reaction.

The equilibrium constant expression predicts the same result upon addition of more hydrogen. If before addition

$$\frac{[HI]^2}{[H_2][I_2]} = K = 53.9 \text{ at } 425°C$$

then after the addition of more hydrogen

$$\frac{[HI]^2}{[H_2][I_2]} < K$$

and so $[H_2]$ and $[I_2]$ must decrease while [HI] increases to restore K.

If additional HI is added,

$$\frac{[HI]^2}{[H_2][I_2]} > K$$

Hence HI must be partially converted to H_2 and I_2, reducing its concentration and increasing the concentration of the other two until the constant K is restored. The net result of adding additional HI was an increase in the concentrations of H_2 and I_2.

If additional I_2 vapor is added,

$$\frac{[HI]^2}{[H_2][I_2]} < K$$

so H_2 and I_2 must be partially converted to HI until K is restored. The net effect is an increase in [HI] and a decrease in $[H_2]$.

Finally, applying Le Chatelier's Principle, when HI is added to an equilibrium mixture of the three gases, a "stress" is applied to the [HI]. The only relief that can be obtained results from a decomposition of HI. This the system tends to do. The net result

suppose that Le Chatelier's Principle were not true, and instead when a stress was applied to a system in equilibrium, the system tended to rearrange itself such as to exaggerate the stress; then all equilibrium systems would be potentially explosive.

is an increase in $[H_2]$ and $[I_2]$. Following the same reasoning, one can determine the effect of adding any of the three gases on the concentrations of an equilibrium mixture.

17.15 EFFECT OF CATALYSIS AND STATE OF SUBDIVISION ON EQUILIBRIUM CONCENTRATIONS

A catalyst can increase or decrease the rate of reaction. In the hydrogenation of oils, the platinum surface holds hydrogen atoms firmly in position for collision and addition to the molecules of oil. But this is a two-way situation. Just as the platinum holds the hydrogen so that it is accepted by the oil more easily, it receives it back from the oil just as easily. Both the forward and the reverse reactions are aided. It has been found that a catalyst changes the rate of both forward and reverse reactions **equally.** Hence, the presence of a catalyst does not change equilibrium concentrations.

The effect of the state of subdivision of a catalyst or a reactant on the rate of a reaction is exemplified by the platinum catalyzed hydrogenation. Because the hydrogenation reaction occurs on the platinum surface, the catalyst should be finely subdivided; the finer the platinum particles, the greater the surface area. Again the reverse reaction is benefited as greatly as the forward reaction by the increased area, and the equilibrium concentrations remain unchanged.

17.16 CALCULATIONS USING THE EQUILIBRIUM CONSTANT EXPRESSION

Not only is the expression of the equilibrium constant useful in telling qualitatively how the addition of a reactant will "shift" the equilibrium concentrations, but it may also be used to determine one or two of the actual concentrations when the others are known, and even to give a quantitative measure of the "shift" in equilibrium concentrations.

Example 1. Five grams of hydrogen iodide are placed in a 2.5-liter flask at 425°C. What will be the concentrations of the three gases when equilibrium is attained?

Solution: The initial concentration of HI is

$$\frac{5.0\ \text{g}}{2.5\ \text{L}} \times \frac{1}{128\ \text{g/mole}} = 0.0156\ \text{mole/liter}$$

Let $X = [H_2]$ = the equilibrium concentration of H_2

then $X = [I_2]$ = the equilibrium concentration of I_2 vapor

and $0.0156 - 2X$ = the equilibrium concentration of HI

$$\begin{array}{llll} & X & X & 0.156 - 2\,X \\ \text{for} & H_2 + & I_2 \rightleftarrows & 2\ HI \end{array}$$

In the equilibrium constant expression

$$\frac{[HI]^2}{[H_2][I_2]} = \frac{(0.0156 - 2\,X)^2}{X^2} = 53.9$$

Taking the square roots of both sides of the equation, we obtain

$$\frac{0.0156 - 2\,X}{X} = (53.9)^{1/2} = \sqrt{53.9} = 7.34$$

$$0.0156 - 2\,X = 7.34\,X$$

$$9.34\,X = 0.0156$$

$$X = 0.00167$$

and the equilibrium concentrations are:

$$[H_2] = [I_2] = 1.67 \times 10^{-3} \text{ mole/liter}$$

Example 2. Five hundredths of a gram of hydrogen is added to the 2.5-liter flask in Example 1, and the contents are allowed to return to equilibrium conditions. What is the concentration of iodine vapor?

Solution:

Let $X = [I_2]$

then $[HI] = (0.0156 - 2\,X)$ mole/liter

and $[H_2] = \dfrac{0.05 \text{ g}}{2 \text{ g/mole}} \times \dfrac{1}{2.5 \text{ liters}} + X = (0.01 + X)$ mole/liter

Substituting in the equilibrium constant expression

$$\frac{(0.0156 - 2X)^2}{(0.01 + X)X} = 53.9$$

Solving this equation by means of the quadratic formula gives

$$X = 4.5 \times 10^{-4} \text{ mole/liter} = [I_2]$$

Notice that the addition of 0.01 mole/liter of H_2 reduced the equilibrium concentration of I_2 from 1.67×10^{-3} mole/liter to 4.52×10^{-4} mole/liter. These two calculations are sufficient here. Other similar calculations will be made in the discussion of weak electrolytes.

17.17 APPLICATION OF THE PRINCIPLES OF RATES AND EQUILIBRIUM TO CHEMICAL PROCESSES

A number of processes using catalysts have been described, and many more could be noted. The role of the catalyst is often not understood, but it is apparent that many processes would be impossible and many would be much more expensive without catalysts. The value of a large surface area was discussed in relation to explosions, or uncontrolled reactions. Large surface area is of more value in the controlled reactions of chemical processes, some of which will be mentioned later in the text. In many processes, solids are finely powdered and liquids are stirred violently to increase the area and velocity of reaction.

Temperature is often varied to control the rate of reaction. This has already been discussed in relation to cooking. Temperature might also be varied to increase the equilibrium percentage of a desired product. One simple commercial process, the calcining of lime to produce quicklime, employs both temperature and concentration factors:

$$CaCO_{3(s)} \rightarrow CaO_{(s)} + CO_2$$

Calcium carbonate (limestone) → Calcium oxide (quicklime) + Carbon dioxide

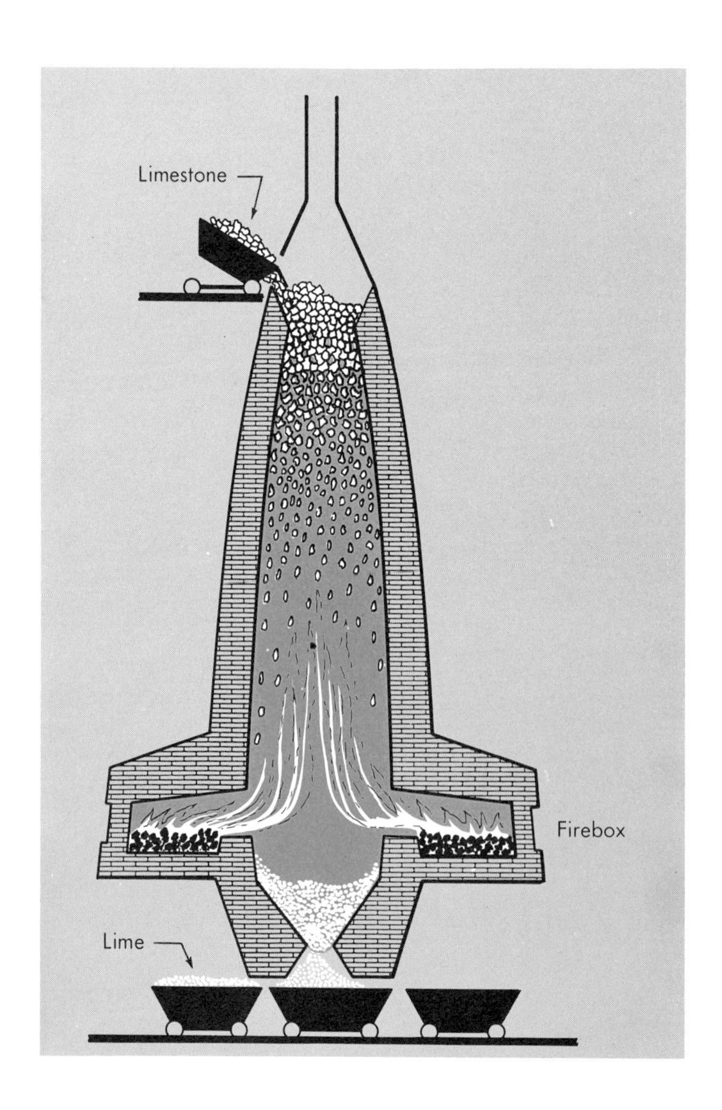

FIGURE 17.8 LIME KILN.

The reaction takes place in a limekiln, a chimney-like furnace, by a blast of hot gases from the bottom, as shown in Figure 17.8. The reaction occurs at about 850°C as the gases move up through the continuously descending limestone in the kiln. High temperature is necessary to raise the equilibrium pressure of the escaping gas to a sufficiently high pressure, or concentration. Notice on the decomposition pressure curve (Figure 17.9) that the decomposition pressure is 760 mm at 912°C. But another restriction is imposed: the temperature must be kept lower than 912°C to prevent the melting of silicates present. A satisfactory compromise is reached at 850°C, at which the equilibrium pressure is necessarily less than 1 atm in the blast of gases. The escaping CO_2 mixes with the gases so that its partial pressure is much less than atmospheric pressure, and the limestone is decomposed rapidly.

A careful choice of reaction conditions is used in the Haber process, in which a mixture of nitrogen and hydrogen is passed over a specially prepared iron catalyst at 500°C and 1000 atm pressure to produce ammonia. The equation for the reaction is

$$N_2 + 3\,H_2 \rightarrow 2\,NH_3 + 32{,}880 \text{ calories}$$

The special conditions are high temperature, high pressure and a catalyst. The catalyst's only role is to increase the rate of reaction, so that equilibrium concentrations are reached rapidly.

The high pressure results in a high concentration of reactants initially and a high concentration of both products and reactants at and near equilibrium. The high concentration of reactants increases the rate of reaction. The effect on equilibrium concentrations is

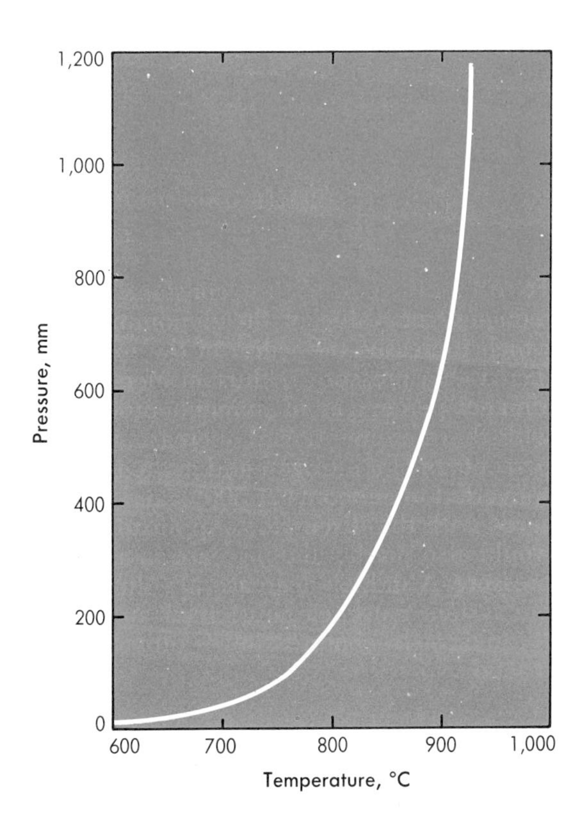

FIGURE 17.9 CARBON DIOXIDE PRESSURE VERSUS TEMPERATURE CURVE FOR $CaCO_3$ DECOMPOSITION.

more subtle, but it can be understood by considering the equilibrium constant expression for the reaction:

$$N_2 + 3\,H_2 \rightleftarrows 2\,NH_3$$

$$\frac{[NH_3]^2}{[N_2][H_2]^3} = K$$

(By convention, the concentrations of products appear in the numerator; the concentration of reactants in the denominator. Remember that each concentration is raised to the power of its coefficient in the balanced equation.)

Now consider a mixture of the three gases at equilibrium. All the concentrations will be equilibrium concentrations and

$$\frac{[NH_3]_1^2}{[N_2]_1[H_2]_1^3} = K$$

(where the subscript indicates concentrations in situation I).

Now suppose the gases are compressed by means of a piston into one half the volume. Instantaneously all concentrations in the container are doubled. At that instant

$$\frac{[NH_3]_2^2}{[N_2]_2[H_2]_2^3} \neq K_{eq}$$

but because

$$[NH_3]_2 = 2[NH_3]_1$$
$$[N_2]_2 = 2[N_2]_1$$
$$[H_2]_2 = 2[H_2]_1$$

then

$$\frac{(2[NH_3]_1)^2}{(2[N_2]_1)(2[H_2]_1)^3} = \frac{2^2}{2 \times 2^3} \times K = \frac{1}{4}\,K$$

To restore the equilibrium constant, some N_2 and H_2 will combine, reducing the value of the denominator and forming some NH_3, which increases the value of the numerator. The net effect of an increase in pressure is thus an increase in the concentrations and a higher percentage of ammonia produced at equilibrium, coupled with the aforementioned increase in rate. Table 17.1 shows the percentage of a nitrogen and hydrogen mixture (in a ratio of 1 to 3) converted to ammonia at 500°C and various pressures. The results agree with the discussion. Application of Le Chatelier's Principle would also predict the increase in ammonia percentage with high pressure. A sudden decrease in volume places a stress on pressure; the system must rearrange itself in order to reduce the pressure. It can do so by compressing 4 volumes to 2 volumes through further reaction:

$$\underset{\text{1 volume}}{N_2} + \underset{\text{3 volumes}}{3\,H_2} \rightarrow \underset{\text{2 volumes}}{2\,NH_3}$$

$$\text{4 volumes} \rightarrow \text{2 volumes}$$

The result is a decrease in N_2 and H_2 concentrations and an increase in the concentration of NH_3.

TABLE 17.1 PER CENT AMMONIA AT EQUILIBRIUM AT 500°C

Pressure Atm.	% NH_3
50	5.56
100	10.61
300	26.44
600	42.15
1000	57.47

High temperature does not favor a high equilibrium concentration of ammonia. The reaction producing ammonia is exothermic. The reverse, the endothermic reaction, is favored. This is shown in Table 17.2, in which the per cent conversion of N_2 and H_2 in mixtures containing 1 volume of nitrogen to 3 volumes of hydrogen at 1000 atm is recorded for several temperatures.

TABLE 17.2 PER CENT CONVERSION OF NITROGEN AND HYDROGEN TO AMMONIA AT 1000 ATM.

Temp. °C	% NH_3
300	92
400	80
500	57
600	31
700	13

At a temperature of 500°C, a maximum of 57 per cent N_2 and H_2 can be converted to NH_3. At 400°C, 80 per cent can be converted. However, the time required at 400°C is much greater than that needed to obtain the 57 per cent yield possible at 500°C. In practice the gases are passed over the catalyst at such a rate that equilibrium is not reached; only about 40 per cent ammonia is gained. The gases are either cooled to liquefy the ammonia or are bubbled through water to "strip" the ammonia. The unreacted elemental gases are passed over the catalyst again with the next mixture of gases. By this recycling procedure, substantially all the nitrogen and hydrogen is converted to ammonia.

EXERCISES

17.1 State two reasons why 100 per cent of the reactants are not converted to products in a chemical reaction.

17.2 Why is time significant in chemical reactions carried out at elevated temperatures?

17.3 Define:

a. rate of reaction
b. catalyst
c. exothermic
d. reversible reaction
e. chemical equilibrium
f. equilibrium constant
g. activation energy
h. mechanism
i. elementary reaction
j. chain mechanism
k. unimolecular reaction
l. bimolecular reaction

17.4 What methods might be used to measure the rate of a reaction?

17.5 What is a general rule relating a rise in temperature to the rate of increase in a reaction rate?

17.6 Name four instances in which the temperature is controlled to control the rate of a chemical reaction.

17.7 Why do many reactions proceed twice as fast with a 10° rise in temperature?

17.8 Will all exothermic reactions proceed at room temperature?

17.9 Explain why the reaction of H_2 and I_2 gases has an activation energy.

17.10 What is k_1 in the equation:

$$\text{Rate} = k_1[A][B]?$$

17.11 What does this mean: $[NH_3]$?

17.12 Why does a newly emptied gasoline container explode more readily when a match is introduced than a full container?

17.13 Describe two ways in which catalysts may take part in a chemical reaction.

17.14 How might one determine that a mixture of gases is at chemical equilibrium?

17.15 Does changing the temperature ever change the value of the equilibrium constant? Explain.

17.16 Does changing the pressure change the value of the equilibrium constant?

17.17 Does a catalyst change the value of the equilibrium constant?

17.18

$$N_2 + O_2 + \text{heat} \rightleftarrows 2\,NO$$

The equation above illustrates an endothermic reversible reaction. What happens to the nitrogen(II) oxide concentration in an equilibrium mixture of the gases when the temperature is raised?

17.19 Explain why a catalyst, high pressure and high temperature are used for the Haber process.

17.20 State Le Chatelier's Principle.

17.21 Use Le Chatelier's Principle to show how the equilibrium pressure of ammonia (in a mixture with hydrogen and nitrogen) varies with both temperature and pressure.

17.22 The important reaction of the contact process is represented by the equation below:

$$2\ SO_{2(g)} + O_{2(g)} \rightleftarrows 2\ SO_{3(g)} + 43{,}800 \text{ cal at } 600°C$$

a. Will one obtain a greater percentage of SO_3 at 650°C?
b. Will one obtain a greater percentage of SO_3 at high pressure?

Explain each answer using Le Chatelier's Principle.

17.23 Write the equilibrium constant expression for each of the following reactions:

a. $C_2H_{6(g)} \rightarrow C_2H_{4(g)} + H_{2(g)}$
b. $2\ NO_{(g)} + O_{2(g)} \rightarrow 2\ NO_{2(g)}$
c. $NO_{2(g)} + SO_{2(g)} \rightarrow SO_{3(g)} + NO_{(g)}$
d. $3\ O_{2(g)} \rightarrow 2\ O_{3(g)}$

17.24 Calculate the concentration of hydrogen iodide in moles per liter if the reaction vessel, Experiment 5, had been reduced in volume to 0.5 liter at 425°C. Use K = 54 at 425°C.

SUGGESTED READING

Benson, S. W.: "Some Aspects of Chemical Kinetics for Elementary Chemistry." J. Chem. Educ., *39:* 321, 1962.
Campbell, J. A.: "Why Do Chemical Reactions Occur?" Prentice-Hall, Englewood Cliffs, New Jersey, 1965 (paperback).
Carmody, W. R.: "Dynamic Equilibrium." J. Chem. Educ., *37:* 312, 1960.
Cornell, P. H., and Reusch, R. N.: *Molecular Equilibrium.* W. B. Saunders Co., Philadelphia, 1963 (a programmed paperback manual).
Harris, G. M.: *Chemical Kinetics.* D. C. Heath and Co., Boston, 1966 (paperback).
King, E. L.: "How Chemical Reactions Occur." W. A. Benjamin Co., New York, 1963 (paperback).
Miller, A. J.: "Le Chatelier's Principle and the Equilibrium Constant." J. Chem. Educ., *31:* 455, 1954.

EIGHTEEN • NITROGEN AND PHOSPHORUS

18.1 PROPERTIES OF GROUP V ELEMENTS

The elements in the nitrogen family, Group VA, have a wider range of properties than is shown by either the elements of the oxygen family or the elements of the halogen family. Properties vary from those of nitrogen and phosphorus, which are decidedly non-metals, to those of arsenic and antimony, which are metalloids, and to bismuth, which is decidedly a metal. The physical properties of the members of Group V are given in Table 18.1.

The electronic configurations of the nitrogen family elements are shown in Table 18.2. As seen from this table, all the elements

TABLE 18.1 SOME PHYSICAL PROPERTIES OF ELEMENTS OF GROUP V

	Nitrogen	**Phosphorus**	**Arsenic**	**Antimony**	**Bismuth**
Molecular formula	N_2	P_4 (white) P_n (red)	As_n (metallic) As_4 (yellow)	Sb (metallic)	Bi (metallic)
Melting point (°C)	−209.9	44.1 (white)	814 (36 atm)	630	271
Boiling point (°C)	−195.8	280 (white)	633 (sublimes)	1380	1560
Atomic radius (Å)	0.92	1.28	1.39	1.59	1.70
First ionization energy (kcal/mole)	336	254	231	199	185
Electronegativity	3.0	2.1	2.0	1.9	1.9

TABLE 18.2 ELECTRONIC CONFIGURATIONS OF THE ELEMENTS OF THE NITROGEN FAMILY

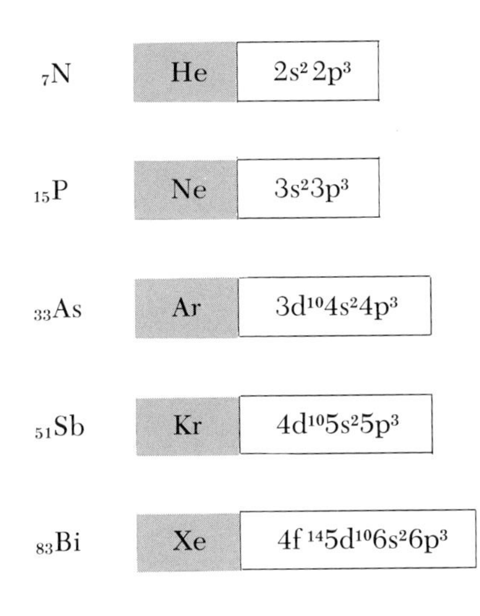

Element	Core	Configuration
$_{7}N$	He	$2s^2 2p^3$
$_{15}P$	Ne	$3s^2 3p^3$
$_{33}As$	Ar	$3d^{10} 4s^2 4p^3$
$_{51}Sb$	Kr	$4d^{10} 5s^2 5p^3$
$_{83}Bi$	Xe	$4f^{14} 5d^{10} 6s^2 6p^3$

have five electrons in the valence shell. Each element has three electrons less than the noble gas of its period, but only the definite non-metals, nitrogen and phosphorus (the elements with higher electronegativities), can acquire an ionic charge of -3. For example,

$$3\ Mg: + 2\ \cdot\dot{\underset{\cdot\cdot}{N}}: \rightarrow 3\ Mg^{2+} + 2\ :\dot{\underset{\cdot\cdot}{N}}:^{3-}$$

None of the members of group VA lose all five electrons to form positive ions with a $+5$ charge, although the elements can be assigned the $+5$ oxidation state in some compounds and complex ions in which there is covalent bonding. Antimony and bismuth (the more metallic elements) can form ions with a $+3$ charge by loss of the three p electrons from the valence shell. These positive ions only exist when combined with anions which are highly electronegative groups. Consequently most of the chemistry of the Group V elements will involve sharing of electrons (covalent bonding), rather than ionic bonds.

Since members of the nitrogen family form compounds with covalent bonding, they exhibit multiple oxidation states between -3 and $+5$. In fact, compounds are known in which nitrogen has all integral oxidation states between -3 and $+5$. Table 18.3 lists some compounds in which the elements of group VA are present in the various oxidation states characteristic of the elements of the nitrogen family. From the table one can see that the most important oxidation states are $+5$, $+3$, 0 and -3.

The **acidity of the oxides** of the group VA elements **decreases as the metallic character (atomic number) of the element increases.** For oxides with the elements in the $+3$ oxidation state, N_2O_3, P_4O_6 and As_4O_6 are acidic oxides, Sb_4O_6 is amphoteric, and Bi_2O_3 is a basic oxide. For the $+5$ oxidation state all the oxides are acidic, but the acidity decreases markedly from N_2O_5 to Bi_2O_5.

TABLE 18.3 COMPOUNDS OF THE NITROGEN FAMILY WHICH ILLUSTRATE VARIOUS OXIDATION STATES

Oxidation State	Nitrogen	Phosphorus	Arsenic	Antimony	Bismuth
+5	N_2O_5	P_4O_{10}	As_2O_5	Sb_2O_5	Bi_2O_5
+4	NO_2	–	–	–	–
+3	N_2O_3	P_4O_6	As_4O_6	Sb_4O_6	Bi_2O_3
+2	NO	–	–	–	–
+1	N_2O	–	–	–	–
0	N_2	P_4	As_4	Sb	Bi
−1	NH_2OH	–	–	–	–
−2	H_2NNH_2	H_2PPH_2	–	–	–
−3	NH_3	PH_3	AsH_3	SbH_3	BiH_3

the acidity of the nitrogen oxides decreases as the oxidation state of nitrogen decreases.

18.2 COMPARISON OF NITROGEN AND PHOSPHORUS

Nitrogen is a non-poisonous, colorless, relatively inactive gas. It occurs in the elemental state in air.

Phosphorus is an active chemical, as indicated by the following properties:

1. It is never found uncombined in nature, but always occurs as a phosphate.

2. Waxy white phosphorus quickly turns yellow in air and fumes as oxidation occurs on the surface. If it is finely divided, phosphorus bursts into flame when it comes in contact with air. As a safety measure, white phosphorus is stored under water.

3. White phosphorus is a deadly poison. It attacks the bones in the jaw and nose and causes them to crumble. Skin burns from phosphorus are painful and heal very slowly.

4. White phosphorus is soluble in organic solvents, especially in carbon disulfide, and is very reactive chemically in such solutions.

molecular structure effects chemical reactivity.

The difference in **chemical reactivity** between nitrogen and phosphorus **can be related to** the differences in **molecular structure.** Nitrogen molecules are diatomic (N_2) and have the electronic structure shown in Figure 18.1. Each nitrogen atom ($\cdot\dot{\underset{\cdot}{N}}\cdot$), lacking three electrons of a completed octet, acquires the octet configuration by forming a triple bond consisting of three shared electron pairs. It will be recalled from a description of molecular orbitals in Chapter 5 that the nitrogen atoms are held together by a σ bond and two π bonds. The bond energy is very high, as 225,000 calories are needed to separate one mole of N_2 into nitrogen atoms. Any

:N⋮⋮N:

:N≡N:

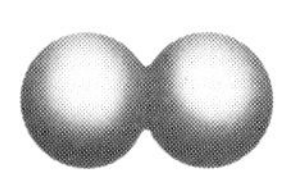

Nitrogen molecule

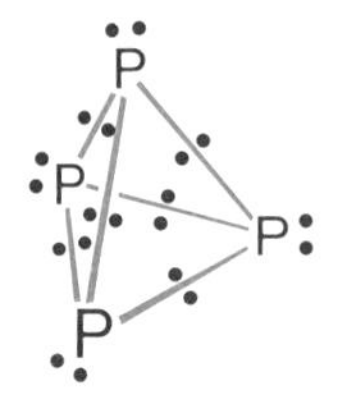

White phosphorus

FIGURE 18.1 MODELS OF THE NITROGEN AND WHITE PHOSPHORUS MOLECULES.

reaction in which the nitrogen atoms are separated must provide this great amount of energy, and as a result few reactions occur with nitrogen and it is relatively unreactive. Given enough electrical or thermal excitation energy, the stable triple bond in the nitrogen molecule is broken, and nitrogen combines with oxygen to give nitrogen (II) oxide by an endothermic reaction:

$$N_2 + O_2 \xrightarrow[\text{discharge}]{\text{electric}} 2\ NO$$

the phosphorus molecule has a tetrahedral structure.

Elemental white phosphorus consists of P_4 molecules. The phosphorus atom ($\cdot\ddot{P}\cdot$), like the nitrogen atom, lacks three electrons of a completed octet. Each phosphorus atom completes its octet by forming a single covalent bond with each of the other three phosphorus atoms of the P_4 molecule. As shown in Figure 18.1, each phosphorus atom has an unshared pair of electrons. When white phosphorus burns, one single bond is broken at a time, and in a limited supply of air the main product is P_4O_6. With excess oxygen, P_4O_{10} is formed. The structures of P_4O_6 and P_4P_{10} are based on P_4 tetrahedra, as shown in Figure 18.2.

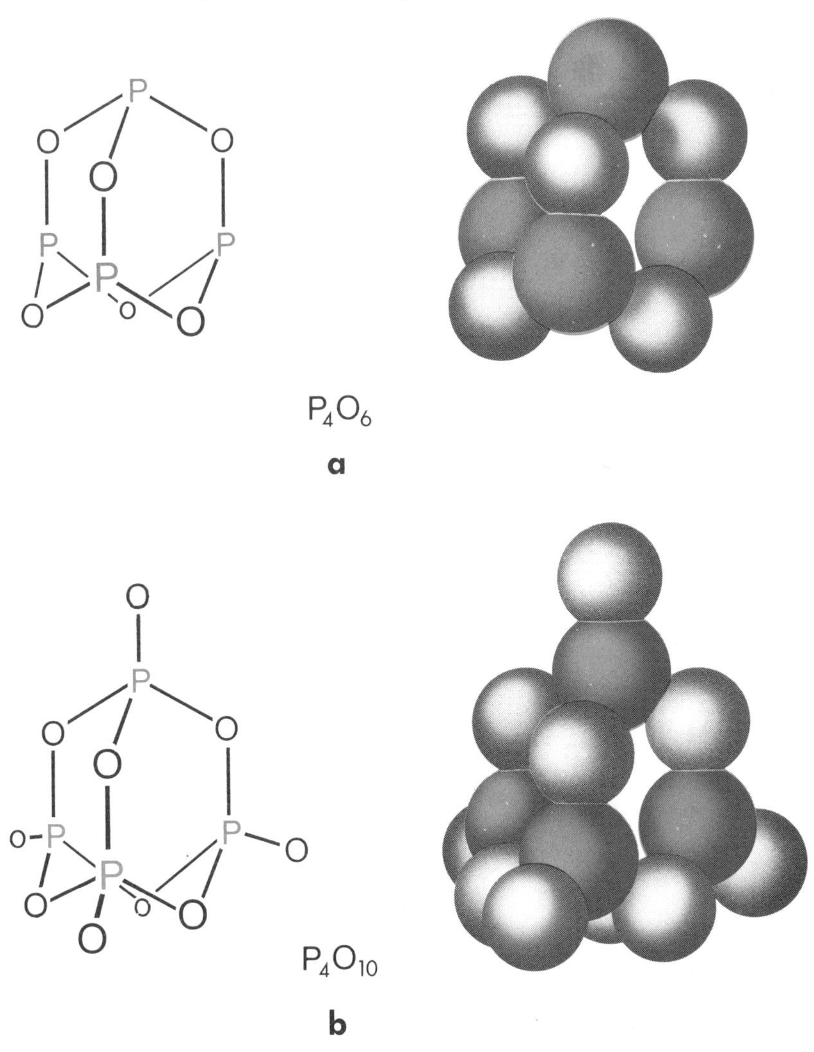

FIGURE 18.2 STRUCTURES OF SOME OXIDES OF PHOSPHORUS.

Further confirmation that the reversal in reactivity of nitrogen and phosphorus is due to molecular structure is found in the existence of other forms of phosphorus. White phosphorus is formed when the vapor is collected under water. When warmed, especially with a trace of iodine catalyst, white phosphorus is converted to *red phosphorus*, which is a polymeric material. *Black phosphorus* can be obtained in crystalline form by heating white phosphorus under very high pressure. The allotropic forms show considerable difference in chemical reactivity; the white form is by far the most reactive. The red and black forms are stable in air, and black phosphorus will not burn in air unless heated above 200°C.

18.3 OCCURRENCE

The elements of Group VA are rather well known despite the fact that only phosphorus is abundant in nature; it comprises 0.188 per cent of the earth's crust, as compared with 0.0046 per cent for nitrogen. Most of the nitrogen in nature is in the atmosphere. Nitrogen is an indispensable element of the biosphere, the shell of living material around the earth. All living matter obtains nitrogen in many complicated compounds in its food, and much work in agriculture is concerned with supplying these nitrogen-bearing foods. As mentioned before, nitrogen is reluctant to combine with other elements to form ions that dissolve in water for plant food. It is fortunate that nitrogen is not more reactive; if it were, much of the oxygen in the air would combine with nitrogen.

Because of the difficulty in obtaining nitrogen compounds, nature passes its combined nitrogen from one living form to another in a complicated cycle (Fig. 18.3). Nitrogen is absorbed as nitrates and ammonium salts into plant roots, and in the plants the

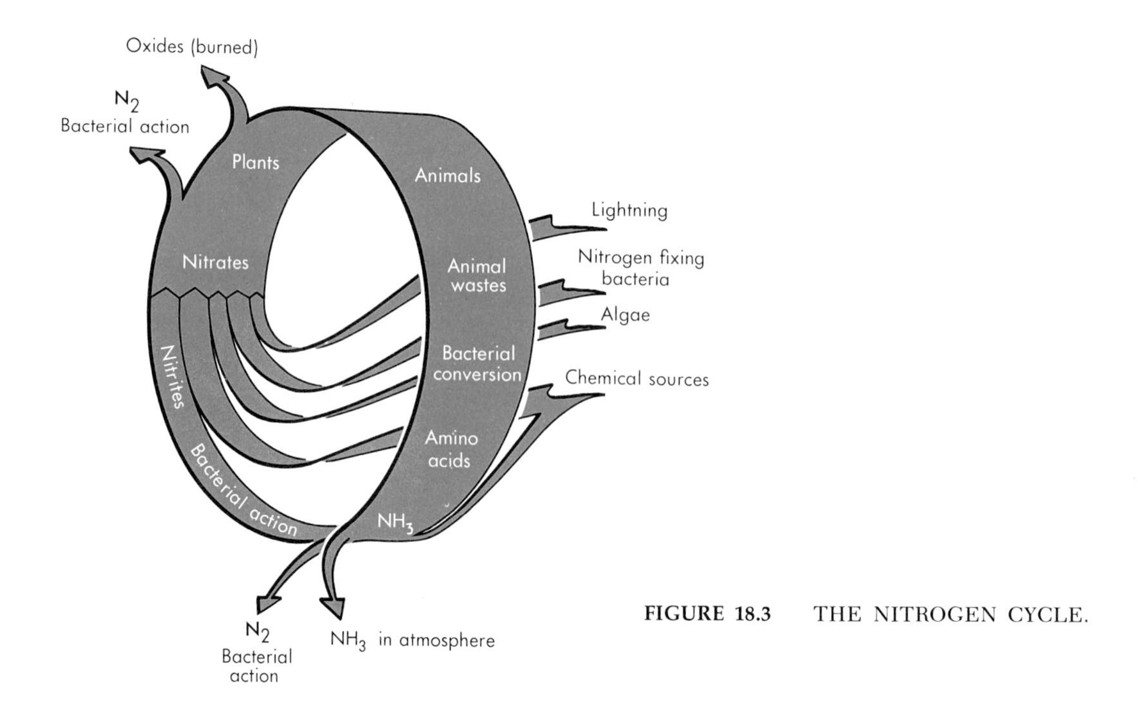

FIGURE 18.3 THE NITROGEN CYCLE.

nitrogen is incorporated into amino acids and proteins. Animals eating the plants gain the proteins for themselves. As one animal eats another, nitrogen-containing proteins are distributed throughout the biosystem. Dead and decaying plant and animal tissues and wastes are decomposed by bacteria to ammonia, amino acids and urea, which in turn are decomposed to nitrites, then to nitrates, which are absorbed by plants to begin the cycle again. However, the cycle is inefficient. Much of the ammonia from decaying proteins is lost to the atmosphere where it is oxidized to N_2. Certain bacteria convert the nitrogen in amino acids to nitrogen gas; this also is lost to the atmosphere. Plant wastes are burned and animal wastes are accumulated in areas where they are not needed.

The cycle is depleted to a much greater extent in cultivated areas than in wild, uninhabited areas, because nitrogen is absorbed by the crops which are then removed from the land.

A process called **nitrogen fixation** supplements the cycle. Nitrogen fixation may occur in nature in three ways:

1. Certain symbiotic bacteria that grow on the roots of **leguminous plants can absorb nitrogen** from the air and combine it chemically for their own use. Furthermore, they store up more of the "fixed" nitrogen than they need and thus build up the nitrogen content of the soil.

why is nitrogen fixation so important?

2. Certain microscopic plants, called **algae,** which occur in the oceans, swamps and rice paddies, **can fix nitrogen** from the air and thus extend the fertility of the soil.

3. Lightning imparts sufficient energy to combine the nitrogen and oxygen molecules in its path. The resulting nitrogen oxides are dissolved by rain and absorbed by the soil. It is estimated that between 7 and 10 pounds of nitrogen in the form of nitric acid are added to each acre of soil per year by lightning.

18.4 COMPOUNDS OF NITROGEN

Nitrogen and its compounds are studied here in the light of man's quest to fix nitrogen to supplement the cycle and to supply his need for industrial compounds of nitrogen without robbing the cycle.

The most obvious source of nitrogen is air. **Nitrogen is obtained commercially by the fractional distillation of air,** as described in Chapter 9. A small amount of very pure nitrogen may be prepared in the laboratory by mixing and heating gently a solution of sodium nitrite and ammonium chloride:

$$NH_4^+ + NO_2^- \rightarrow N_2 + 2\ H_2O$$

Because solid ammonium nitrite is explosive, the solution is not allowed to evaporate to dryness. The reaction involves an **auto-oxidation–reduction.** Referring to the equation above and writing oxidation states of nitrogen:

NH_4^+ is a reducing agent; NO_2^- is an oxidizing agent.

$$\overset{-3}{NH_4^+} + \overset{+3}{NO_2^-} \rightarrow \overset{0}{N_2} + 2\ H_2O$$

The oxidation state of nitrogen in NO_2^- decreases; the nitrogen in NH_4^+ increases in oxidation state. Ammonium nitrite is its own oxidizing and reducing agent. Auto-oxidation–reduction occurs in many high explosives.

In the late nineteenth century there was great concern about famine, since the world's major supply of fixed nitrogen (Chile saltpeter, $NaNO_3$) was rapidly being depleted. Much of the Chile saltpeter was used for fertilizer, and the inevitable exhaustion of sodium nitrite would bring a resurgence of chronic famine. Although there was limited possibility for expanding the production of ammonia from coal, it was necessary to develop other new processes for fixing atmospheric nitrogen.

18.5 PREPARATION OF AMMONIA

1. About 1.4 per cent by weight of coal is nitrogen. When coal is "coked," that is, heated in the absence of air, about 20 per cent of its nitrogen is converted to ammonia. At one time this was the major source of ammonia. The need for ammonia, mostly for fertilizers, greatly exceeded the supply.

2. In the early 1900's **Fritz Haber,** a German chemist, developed a process that involved a continuous recirculation of the elements of **ammonia** at high pressure over a metallic catalyst, where they combined to form ammonia. In the process (discussed relative to chemical equilibrium in Chapter 17), nitrogen from the distillation of liquid air and hydrogen from the catalytic reaction of steam with hydrocarbons or with coal are passed over an impure iron catalyst at 500°C and about 1000 atm:

Fritz Haber received a Nobel Prize for his ammonia process.

$$N_2 + 3\ H_2 \xrightarrow[500°C,\ 1000\ \text{atm}]{\text{iron}} 2\ NH_3 + 32{,}880\ \text{calories}$$

Not more than 60 per cent of the nitrogen and hydrogen can be converted to ammonia under these conditions. The ammonia is liquefied and removed, and the unreacted elemental gases are passed over the catalyst again along with incoming hydrogen and nitrogen. The catalyst is prepared by an elaborate process; it contains at least 95 per cent iron with traces of potassium, aluminum, magnesium and silicon oxides.

The development of this process was of world wide historical importance in World War I because Germany, until that time, depended almost entirely on Chilean saltpeter as its source of fixed nitrogen for explosives. Since her supply of saltpeter was cut off because the Allied Powers controlled the seas, Germany could not have waged war for any extended period. But the Haber process, which utilized the nitrogen from the air, made Germany independent of outside sources. This enabled Germany to extend the war for a long period and helped exhaust her resources and embitter her enemies; it perhaps also contributed to conditions which led to the rise of Hitler in Germany.

The development of **the Haber process has helped remove the threat of famine.** Current United States production of ammonia is about *13 million tons* per year. This is an example of how science has developed a product that can be used either for the benefit or for the destruction of man. Today more than ever before the public is asking the scientist to make ethical judgments concerning the areas in which he will do research. Recently we have seen the removal of secret research from some university campuses, notably secret projects in the area of biological and chemical warfare. Even if scientists were to make good judgments about the areas in which scientific knowledge was to be advanced, most developments can be exploited either for "good" or for "evil." Consequently, we can not logically consider the expansion of science to be a main source of our current problems. *Everyone* must assume a moral responsibility for the use that is made of scientific knowledge, or indeed of any kind of knowledge.

3. Ammonia is prepared in the laboratory by an acid-base reaction. A strong basic solution is added to an ammonium salt, or to a solution of that salt:

$$NH_4Cl + OH^- \rightarrow NH_{3(g)} + H_2O + Cl^-$$

An especially interesting preparation of ammonia involves the use of nitrogen to support combustion. Magnesium is sufficiently active to burn in nitrogen, with a flame, to form a nitride:

$$3\ Mg + N_2 \longrightarrow \underset{\text{Magnesium nitride}}{Mg_3N_2}$$

Addition of water releases ammonia:

$$Mg_3N_2 + 6\ H_2O \rightarrow 3\ Mg(OH)_2 + 2\ NH_{3(g)}$$

In this reaction water is an acid and the nitride ion is a base.

18.6 PROPERTIES AND USES OF AMMONIA

Ammonia is a colorless gas with a strong, characteristic odor. It freezes at −77°C and boils at −33.3°C at a pressure of 1 atmosphere. At greater pressure, it is easily liquefied at room temperature. For this reason it is used in refrigeration. In the refrigerator cycle ammonia* is compressed outside the cooling compartment until it is liquefied, giving up the latent heat of vaporization to the surroundings (Fig. 18.4). The ammonia, now a liquid, is admitted to a container inside the refrigerator where the pressure is released and the liquid vaporizes, taking the heat of vaporization from the surroundings (the inside of the refrigerator). The cycle continues indefinitely, liquefying with pressure outside the refrigerator to release the heat of condensation outside, and vaporizing at reduced pressure inside to take up the heat of vaporization.

liquid ammonia is used as a solvent for some chemical reactions.

Liquid ammonia (with no water present) is a good solvent and is used as the reacting medium for certain substances that are unstable in water. As a solvent, liquid ammonia resembles water. The ammonia molecules are polar, but not as polar as those of water. The self-ionization of ammonia in the liquid is analogous to

* Although ammonia is widely used as the refrigerant in large commercial installations and in cooperative food-storage lockers, in which a very low temperature is required, household freezers and refrigerators usually use an odorless, nontoxic fluorocarbon such as CCl_2F_2. The principle, however, is the same.

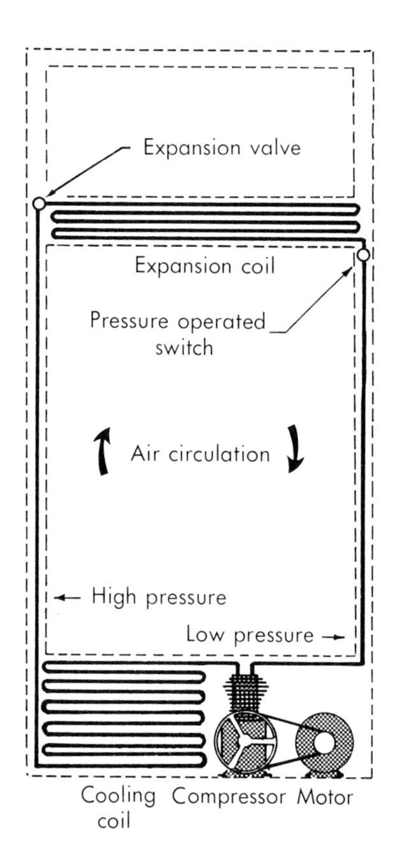

FIGURE 18.4 THE REFRIGERATION CYCLE.

the ionization of water:

$$2\ NH_3 \rightleftarrows NH_4^+ + NH_2^-$$

$$2\ H_2O \rightleftarrows H_3O^+ + OH^-$$

By analogy, NH_2^- is the strongest basic ion in liquid ammonia and NH_4^+ is the acid ion. Neutralization transfers a proton to NH_2^-. Solutions of NH_4Cl and $NaNH_2$ (sodium amide) in ammonia can be titrated with phenolphthalein as an indicator.

Ammonia is very soluble in water. At room conditions, over 300 L of ammonia will dissolve in 1 L of water. The rapid solubility can be demonstrated by an ammonia fountain (Fig. 18.5). A glass tube from an inverted flask of ammonia is immersed in water. A medicine dropper containing water is placed through the stopper so that a few drops of water may be admitted to the flask. As soon as the drops of water enter the flask, ammonia dissolves in it, pulling water up into the flask like a fountain. When ammonia dissolves in water, a weakly basic solution results. Spectroscopic data suggest that it is a hydrated ammonia solution and that only a small concentration of ammonium ions and hydroxide ions are present from this reaction:

$$NH_3 + H_2O \rightleftarrows NH_4^+ + OH^-$$

When an ammonia solution is neutralized by an acid, an ammonium

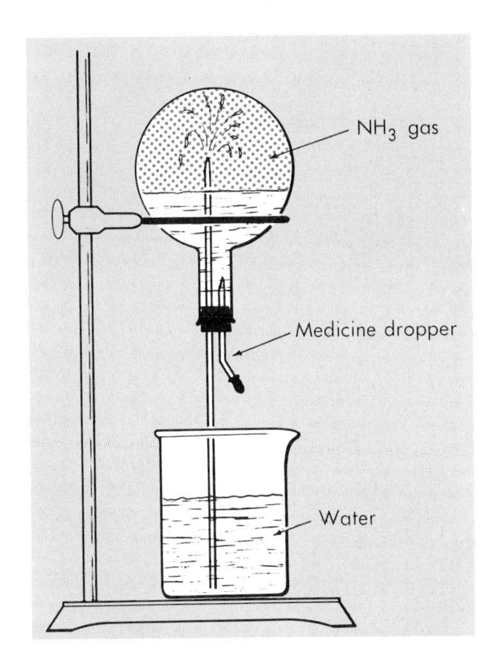

FIGURE 18.5 THE AMMONIA FOUNTAIN.

salt is left in solution. When ammonia and hydrochloric acid are neutralized and the solution has evaporated, ammonium chloride crystals appear, which have the formula $NH_4^+Cl^-$. Ammonium chloride is an ionic compound, a salt, as represented by the assigned charges. The ammonium ion is present in many salts, all of which are ionic, behaving very much like corresponding positive ions of the alkali metals.

Ammonia forms stable complex ions with certain metal ions. A silver ammonia complex results when ammonia solution is added to a silver nitrate solution:

$$Ag^+ + 2\ NH_3 \longrightarrow \underset{\text{Silver ammonia complex ion}}{Ag(NH_3)_2^+}$$

The complex ion is sufficiently stable so that ammonia dissolves a precipitate of insoluble silver chloride

$$AgCl_{(s)} + 2\ NH_3 \rightarrow Ag(NH_3)_2^+ + Cl^-$$

Addition of ammonia to a pale blue solution of copper(II) ion changes the color to an intense dark blue, with the formation of a complex ion:

$$Cu^{2+} + 4\ NH_3 \rightarrow Cu(NH_3)_4^{2+}$$

Ammonia is oxidized by air in the presence of a platinum catalyst as an intermediate step in the preparation of nitric acid, a major use of ammonia.

18.7 PREPARATION OF NITRIC ACID

For many years the only method of preparing nitric acid was by distilling the vapor from a mixture of sulfuric acid and Chilean saltpeter and collecting it in water:

$$NaNO_3 + H_2SO_4 \rightarrow HNO_{3(g)} + NaHSO_4$$

ammonia is an important raw material in the production of nitric acid.

This process is still used but has become relatively unimportant compared to the preparation of nitric acid from NH_3, in view of the availability of large quantities of ammonia.

Ammonia can be burned with air in the presence of a metal catalyst according to any one of three reactions:

$$4\ NH_3 + 5\ O_2 \rightarrow 4\ NO + 6\ H_2O$$

$$4\ NH_3 + 3\ O_2 \rightarrow 2\ N_2 + 6\ H_2O$$

$$2\ NH_3 + 2\ O_2 \rightarrow N_2O + 3\ H_2O$$

Only the first is of any commercial value. When ammonia mixed with oxygen or air is passed over a platinum gauze catalyst at 1000°C, the combustion of nitrogen(II) oxide (NO) occurs almost quantitatively, with the time of contact with the catalyst being less than 0.01 second. This is the first step in the **Ostwald Process** for making nitric acid from ammonia. The catalytic oxidation of ammonia with air can be demonstrated by placing a hot coil of platinum above a concentrated ammonia solution in an Erlenmeyer flask (Fig. 18.6). Oxygen in air diffuses into the mouth of the flask to unite with ammonia vapors on the surface of the platinum wire, where the heat of reaction keeps the wire glowing at red heat.

To convert nitrogen(II) oxide to nitric acid, the oxide is mixed with air and passed upwards through a series of towers down which water is trickling. Two reactions occur in the tower. The first,

$$2\ NO + O_2 \longrightarrow \underset{\text{Nitrogen dioxide}}{2\ NO_2}$$

is a slow reaction and no catalyst is known to speed it. The brown gas (NO_2) then reacts with water.

$$3\ NO_2 + H_2O \rightleftarrows 2\ HNO_3 + NO$$

The nitrogen(II) oxide that is released reacts with air to form more nitrogen dioxide. Nitric acid with a concentration of about 55 per cent is prepared by this method.

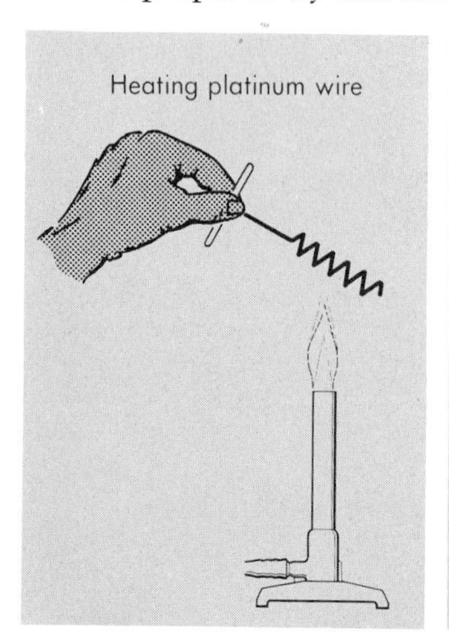

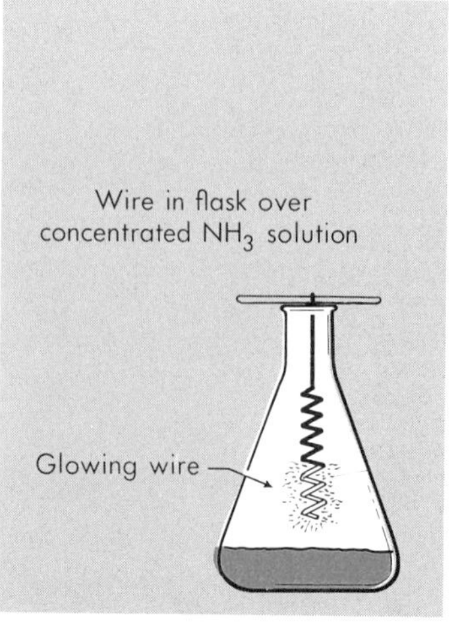

FIGURE 18.6 CATALYTIC OXIDATION OF AMMONIA.

As noted earlier, nitrogen(II) oxide may be prepared directly from its elements. When a mixture of air enriched with oxygen is passed through a high-temperature electric arc, about 2.5 per cent of the oxide is gained:

$$N_2 + O_2 \xrightarrow{\text{energy}} 2\ NO$$

This is the reaction which occurs in the path of lightning. In the moist air the gas is converted to nitrogen dioxide, then to nitric acid. Prior to the development of the Ostwald process, an electric arc was employed in localities where electric power was cheap.

Nitrogen(II) oxide and nitrogen dioxide are chemicals that are involved in the **photochemical smog** for which Los Angeles is noted. A small amount of nitrogen(II) oxide is produced in the combustion of gasoline in the automobile engine. The oxygen comes from the air. Some nitrogen and oxygen of the air are combined in the heat of the exploding mixture of air and hydrocarbons in the cylinder. The resulting oxides of nitrogen oxidize released hydrocarbons (especially in sunlight) to irritating aldehydes and organic nitrites. Nitrogen(II) oxide first reacts with oxygen to produce NO_2 as in the Ostwald process. Nitrogen dioxide undergoes a photochemical reaction in sunlight to produce oxygen atoms.

$$NO_2 \xrightarrow{h\nu} NO + O$$

Oxygen atoms produced in this reaction are extremely reactive and react with oxygen to produce ozone. This reacts with unburned hydrocarbons from the gasoline to produce the aldehydes

$$O + O_2 \rightarrow O_3$$

$$O + \text{hydrocarbons} \rightarrow RCHO^* + RCO_2{}^*$$

motor vehicles are the main source of nitrogen oxides in the atmosphere.

The nitrogen oxides, ozone and aldehydes are all components of smog; they are all very harmful. Motor vehicles are the main source of photochemical smog, since they are the main source of nitrogen oxides and hydrocarbons (Table 18.4).

TABLE 18.4 SOURCES OF SELECTED AIR POLLUTANTS IN MILLIONS OF TONS PER YEAR*

Source	Carbon Monoxide	Sulfur Oxides	Nitrogen Oxides	Hydro-carbons	Particulate Matter	Totals
Motor vehicles	66	1	6	12	1	86
Industry	2	9	2	4	6	23
Power plants	1	12	3	1	3	20
Space heating	2	3	1	1	1	8
Refuse disposal	1	1	1	1	1	5
Totals	72	26	13	19	12	142

* United States Public Health Service, 1966.

Air pollution is important because many **pollutants are injurious to our health,** as noted in a report to Congress by the Department of Health, Education and Welfare in 1968:

* R represents hydrocarbon chains of varying lengths.

The main thrust of the evidence is clear and conclusive—the types or levels of air pollution which are now commonplace in American communities are an important factor in the occurrence and worsening of chronic respiratory diseases, and may even be a factor in producing heightened human susceptibility to upper respiratory infections including the common cold.

18.8 PROPERTIES AND USES OF NITRIC ACID

Concentrated nitric acid (about 68 per cent), which is obtained by distilling an aqueous solution of the acid until it boils at a constant temperature, has a density of 1.42 g/ml. Its molarity is:

$$\frac{1000 \text{ ml/L} \times 1.42 \text{ g/ml} \times 0.68}{63 \text{ g/mole}} = 15.3 \text{ moles/liter} = 15.3 \text{ M}$$

Nitric acid is a strong acid and a good oxidizing and nitrating agent. It ionizes completely in water:

$$HNO_{3(g)} + H_2O \xrightarrow{100\%} H_3O^+ + NO_3^-$$

Noble metals such as silver and copper will dissolve in nitric acid but not in hydrochloric or sulfuric acid. The dissolving is accomplished by an oxidation of the metal:

$$\underset{\text{Conc.}}{Cu + 4\ HNO_3} \rightarrow Cu(NO_3)_2 + 2\ NO_2 + 2\ H_2O$$

$$\overset{0}{Cu} + 2\ \overset{+5}{NO_3^-} + 4\ H_3O^+ \rightarrow \overset{+2}{Cu^{2+}} + 2\ \overset{+4}{NO_2} + 6\ H_2O$$

The use of nitric acid with concentrated sulfuric acid as a nitrating agent is an important commercial process. Nitric acid also is used extensively in the manufacture of explosives, fertilizers, dyes and many other products.

18.9 BONDING AND STRUCTURE OF SOME NITROGEN AND PHOSPHORUS COMPOUNDS

As already noted, nitrogen is found in compounds in which the oxidation state varies from −3 to +5. Among other things, the change in oxidation state indicates a change in the number of electrons shared in covalent bonding. Nitrogen has been found to exhibit all three types of hybridization studied in Chapter 5: sp, sp^2, sp^3. Furthermore, nitrogen is the "central atom" in many molecules and complex ions which have resonance, that is, which have delocalized pi orbitals. Some examples of molecules and ions that exhibit this great variety in bonding are listed in Table 18.5.

Of the molecules and ions listed in Table 18.5, nitrogen(I) oxide (N_2O or NNO) is a linear molecule. With a bond angle of 180° it must be sp hybridized (Chapter 5). The bond angle of the nitrite ion (NO_2^-) is 115°. With the angle that close to 120°, the

TABLE 18.5 SHAPE AND HYBRIDIZATION OF SOME NITROGEN AND PHOSPHORUS MOLECULES AND IONS

Molecule or Ion	Shape	Hybridization of Central Atom
N_2O	linear	sp
NO_2^-	bent	sp^2
NO_3^-	planar	sp^2
NH_3	pyrimidal	sp^3
NH_4^+, PH_4^+, PCl_4^+	tetrahedral	sp^3
PO_4^{-3}	tetrahedral	sp^3
PF_5, PCl_5	trigonal-bipyramidal	sp^3d
PF_6^-, PCl_6^-	octahedral	sp^3d^2

nitrogen atom probably is sp^2 hybridized. One of the three sp^2 orbitals forms no bond, as there are only two oxygen atoms, but is directed outward from the molecule. It holds two electrons that are called **an unshared pair.** The nitrate ion (NO_3^-) is a planar ion with the oxygen atoms at the apices of an equilateral triangle. The bond angles are 120° and the nitrogen atom is sp^2 hybridized, as it was in the nitrite ion. The nitrate ion is very similar in electron structure to the carbonate ion, which was discussed in Chapter 5. There are two equivalent dot formulas for the nitrite ion and three equivalent dot formulas for the nitrate ion, as shown in Figure 18.7. Therefore, the dot formulas are **resonance structures** – the actual structure is intermediate between the participating resonance structures. That means both N to O bonds in the nitrite ion are the same, the three N to O bonds in the nitrate ion are identical and all N to O bonds are shorter than single bonds but longer than double N to O bonds. The delocalized pi orbitals of the nitrite ion are over and under all three atoms; of the nitrate ion, all four atoms. The ammonium ion (NH_4^+), with four N to H bonds, and the ammonia molecule (NH_3) with bond angles of 107°, are sp^3 hybridized. The ammonia molecule has an unshared electron pair.

The phosphorus atom tends not to hybridize in the sp or sp^2 orbital arrangement, but rather, in molecules and ions with one center it tends to form single bonds in an ordinary tetrahedral arrangement. However, phosphorus forms other molecules and ions, with other hybridizations, that make its bonding as varied and interesting as that of the nitrogen atom. The elements of Group V combine with the halogens to form tri- and pentahalides, such as phosphorus trichloride (PCl_3) and **phosphorus pentachloride** (PCl_5). In fact, phosphorus forms tri- and pentahalides by combining directly with all the halogens. The only exception is phosphorus pentaiodine, which is not known to exist. **Nitrogen,** in contrast to all the other members of its group, **forms no pentahalides,** although it forms trihalides.

nitrogen has only four valence orbitals.

The reasons for the inability of nitrogen to form pentahalides can be seen in an analysis of the nature of phosphorus pentachloride and a comparison of the available orbitals in the valence shells of the Group V atoms. In the vapor and liquid phases, the pentachloride of phosphorus is composed of PCl_5 molecules. The phosphorus atom is bonded to each of the chlorine atoms, which are arranged about the phosphorus as if it were at the center of a trigonal bipyramid and they were at the corners, as shown in Figure 18.8. In the solid state the pentachloride has been shown by x-ray studies to be composed of PCl_4^+ and PCl_6^- ions. In effect a Cl^- has

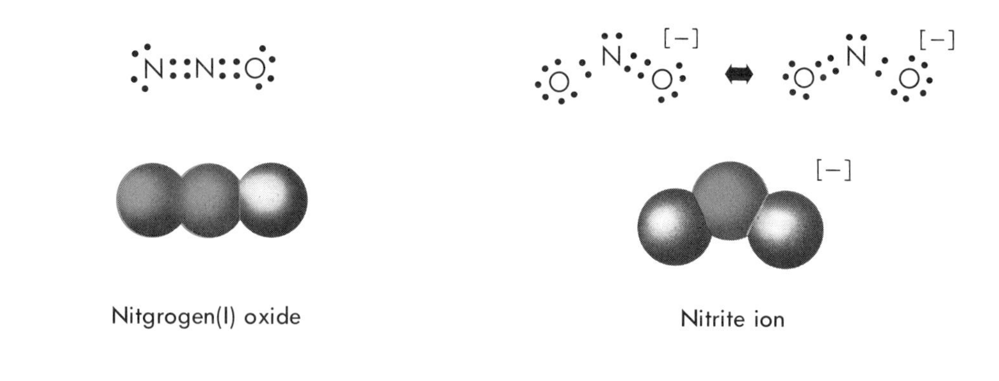

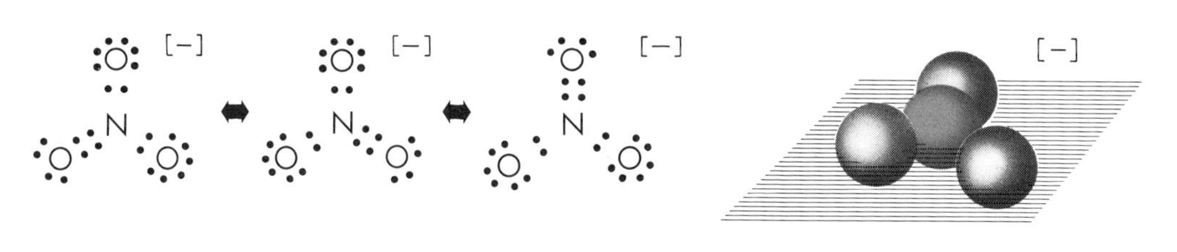

FIGURE 18.7 LEWIS AND SPACE FILLING STRUCTURES FOR SOME NITROGEN COMPOUNDS AND IONS.

been transferred from one PCl_5 molecule to another:

$$PCl_5 + PCl_5 \rightarrow PCl_4^+ + PCl_6^-$$

The PCl_4^+ ion is tetrahedral in shape, similar to the ammonium ion (NH_4^+); the PCl_6^- ion is octahedral. That is, the chlorine atoms may be considered as placed at the apices of a regular octahedron, each bonded to the phosphorus atom at the center, as illustrated in Figure 18.8.

By sp^3 hybridization the isolated phosphorus atom

$$P{:}1s^22s^22p^12p^12p^1$$

could be considered to be hybridized as

$$P{:}1s^22(sp^3)^22(sp^3)^12(sp^3)^12(sp^3)^1$$

When P loses an electron to gain a charge of +1, each sp^3 orbital

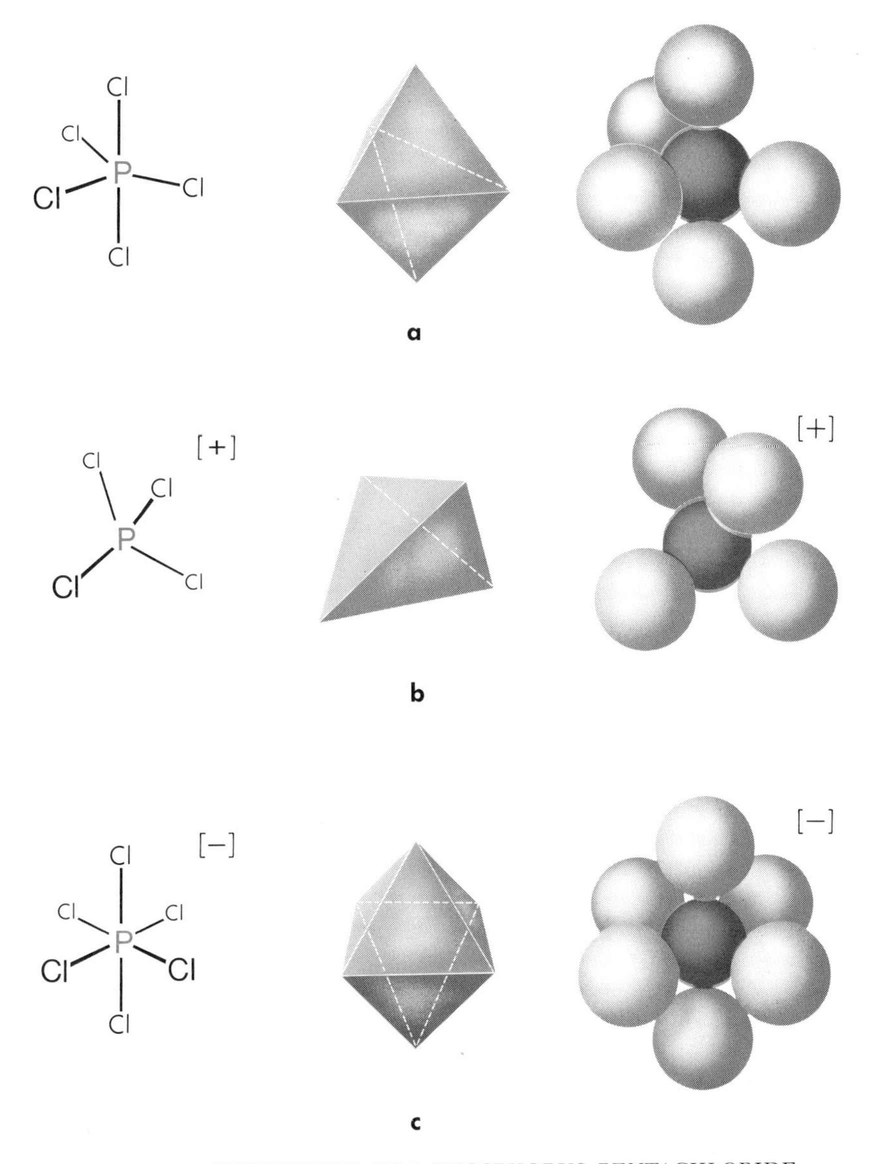

FIGURE 18.8 STRUCTURES FOR PHOSPHORUS PENTACHLORIDE.

has one electron to pair with an electron of a chlorine atom in a $\sigma(sp^3 + p)$ orbital to form PCl_4^+. Whereas all the valence orbitals of nitrogen, an s and three p orbitals, are involved in the hybridization, which gives only four orbitals and limits nitrogen to four bonds, the phosphorus atom has *d orbitals* in its valence shell, as do the atoms of the remaining elements of Group V:

phosphorus has more than four valence orbitals.

$$P{:}3s^2 3p^1 3p^1 3p^1 3d^0 3d^0 3d^0 3d^0 3d^0$$

In this representation of the phosphorus atom, all electrons below the valence level have been omitted, and orbitals with no electrons have the superscript zero. Two of the hybridizations possible with the available set of orbitals in the phosphorus atom are

$$P{:}3(sp^3d)^1 3(sp^3d)^1 3(sp^3d)^1 (sp^3d)^1 3(sp^3d)^1 3d^0 3d^0 3d^0 3d^0$$

and

$$P: 3(sp^3d^2)^1 3(sp^3d^2)^1 3(sp^3d^2)^1 3(sp^3d^2)^1 3(sp^3d^2)^1 3(sp^3d^2)^0 3d^0 3d^0 3d^0$$

With sp^3d hybridization the orbitals are directed with respect to each other so as to form a trigonal bipyramid. Each hybridized orbital combines with a p orbital of a chlorine atom and holds one electron from each atom to form five bonds. Each chlorine atom in the molecule has an octet of electrons; the phosphorus atom has ten. With the **phosphorus atom sp^3d^2 hybridized** (second representation above), there are six identical orbitals directed from the phosphorus atom at the center toward the corners of a regular octahedron. Since the phosphorus atom in the octahedral PCl_6^- ion has a negative charge, it can be considered to be P^-. The extra electron can be placed in the empty sp^3d^2 orbital in the second representation shown above. Each sp^3d^2 orbital combines with a p orbital of a chlorine atom to form a molecular orbital, which holds a pair of electrons, one from each atom. Each chlorine atom obtains a completed octet by sharing the pair of electrons; the phosphorus atom has 12 electrons about it. With the phosphorus pentachloride molecule and the hexachloride ion we have a "violation" of the octet principle. This is quite common among the molecules and complex ions of the non-metals beneath the second period elements in the periodic table; in each case it is due to the availability of d electrons for chemical bonding.

18.10 PHOSPHORUS

As mentioned before, phosphorus occurs in nature as phosphates. One mineral, fluoroapatite, has the formula $3Ca_3(PO_4)_2 \cdot CaF_2$. Some calcium phosphate deposits are of prehistoric animal origin. Calcium phosphate is one of the two minerals present in bones and teeth; the other is the aragonite form of calcium carbonate. Phosphorus also is contained in living cells. The body of an adult human being contains about 3.4 pounds of phosphorus. Animals obtain phosphorus from plants, and plants acquire it from the soil. In many soils, phosphorus is depleted and must be added in fertilizers. One of the major uses of concentrated sulfuric acid is to convert insoluble phosphate rock to soluble dihydrogen phosphate for fertilizer:

$$2\ H_2SO_4 + Ca_3(PO_4)_2 \rightarrow Ca(H_2PO_4)_2 + 2\ CaSO_4$$

Elementary phosphorus is produced by heating a mixture of phosphate rock, sand and coke in an electric furnace:

$$\underset{\text{Phosphate rock}}{2\ Ca_3(PO_4)_2} + \underset{\text{Sand}}{6\ SiO_2} + \underset{\text{Coke}}{10\ C} \rightarrow \underset{\substack{\text{Calcium silicate}\\ \text{(slag)}}}{6\ CaSiO_3} + 10\ CO + P_4$$

The calcium silicate, called slag, melts and runs to the furnace floor. Phosphorus is vaporized, and the vapor is collected in water, where it condenses to a liquid. The liquid is poured into molds to solidify as white phosphorus. Alternatively, the vapor may be mixed with oxygen in a water-cooled, graphite-lined chamber to form P_4O_{10} directly. Phosphorus(V) oxide is frequently represented by its empirical formula P_2O_5 and called phosphorus pentoxide.

Most of the phosphorus produced is converted into compounds, principally the various sodium phosphates used in detergents. Large quantities are used in making insecticides, rat poison and phosphor bronze. Once contained in matches, poisonous white phosphorus has been replaced by phosphorus trisulfide (P_4S_3), which is in the tip of the match above a mixture of potassium chlorate, sulfur and glue. Friction on the tip ignites the P_4S_3, which, in turn, ignites the sulfur–potassium chlorate mixture. The chlorate furnishes oxygen to burn the sulfur. Safety matches in effect have the phosphorus ignitor on the box and the oxidizing-reducing agent mixture on the stick.

Phosphorus(V) oxide is used as **a drying agent** since it has a great affinity for water. Phosphoric acid is prepared from phosphorus(V) oxide by addition of water:

$$P_4O_{10} + 6\ H_2O \rightarrow 4\ H_3PO_4$$

which is the strongest acid, H_3PO_4, H_2SO_4 or $HClO_4$? why?

It can also be obtained from the hydrolysis of phosphorus(V) chloride:

$$PCl_5 + 4\ H_2O \rightarrow H_3PO_4 + 5\ HCl$$

A comparison of the relative acidities of orthophosphoric, sulfuric and perchloric acids shows that H_3PO_4 is weaker than H_2SO_4, which in turn is weaker than $HClO_4$. This trend in acid strengths can be explained in terms of the electronegativity of the central atom, because the electronegativity increases in the same order ($P < S < Cl$) as the acid strength.

Phosphorus(III) oxide (P_4O_6) combines with water to form phosphor*ous* acid (H_3PO_3):

$$P_4O_6 + 6\ H_2O \rightarrow 4\ H_3PO_3$$

Phosphorous acid can also be prepared from the hydrolysis of phosphorus trichloride:

$$PCl_3 + 3\ H_2O \rightarrow H_3PO_3 + 3\ HCl$$

The electronic dot structures of phosphoric and phosphorous acids are

```
       H                  H
       ..                 ..
      :O:                :O:
   .. .. ..              .. ..
 H:O:P:O:H            H:P:O:H
   .. .. ..              .. ..
      :O:                :O:
       ..                 ..
```

Phosphoric acid   Phosphorous acid

In each acid the phosphorous atom is bonded to four other atoms in a tetrahedral configuration. The third hydrogen atom in phosphorous acid is bonded directly to phosphorus and is not acidic. Consequently, H_3PO_3 is a diprotic acid. Two protons (H^+) can be removed from H_3PO_3 by titration; three can be removed from H_3PO_4:

$$2\ H_3PO_4 + 3\ Ca(OH)_2 \rightarrow Ca_3(PO_4)_2 + 6\ H_2O$$

$$H_3PO_3 + Ca(OH)_2 \rightarrow CaHPO_3 + 2\ H_2O$$

Three series of salts can be formed with phosphoric acid, as illustrated by the formulas and names of the sodium salts:

NaH_2PO_4	Sodium dihydrogen phosphate
Na_2HPO_4	Disodium hydrogen phosphate
Na_3PO_4	Trisodium phosphate

The anion of the monosodium salt (NaH_2PO_4) is weakly acidic:

$$H_2PO_4^- + H_2O \rightleftarrows H_3O^+ + HPO_4^{2-};$$

That of the disodium salt is weakly basic. This reaction:

$$HPO_4^{2-} + H_2O \rightarrow H_2PO_4^- + OH^-$$

occurs to a greater extent than does this reaction:

$$HPO_4^{3-} + H_2O \rightarrow H_3O^+ + PO_4^{3-}$$

The phosphate ion is quite basic because the reaction

$$PO_4^{3-} + H_2O \rightarrow HPO_4^{2-} + OH^-$$

occurs to a fairly large extent in solution, and this may account for the detergent action (or detergent-aiding action) of sodium phosphates.

EXERCISES

18.1 Which element of Group V occurs most abundantly in nature?

18.2 Which element of Group V is a stable diatomic molecule at room temperature?

18.3 Which element of Group V is a metal? Which elements are definitely non-metals?

18.4 Compare the chemical reactivity of white phosphorus, red or violet phosphorus and nitrogen.

18.5 Explain why white phosphorus is more active to air oxidation than nitrogen and more active than red or violet phosphorus.

18.6 Why should white and red phosphorus differ more in their properties than do rhombic and monoclinic sulfur?

18.7 Describe "active nitrogen."

18.8 a. Write the ionic valence for each member of Group V.
b. Write the symbols for the possible hybridizations of nitrogen.
c. Write the formula for an ion or molecule having nitrogen in each of the hybridizations.

18.9 Which elements of Group V are necessary for plant and animal life?

18.10 What is the nitrogen cycle?

18.11 Explain why the element nitrogen is an example of "poverty in the midst of plenty."

18.12 Describe five ways by which "fixed" nitrogen may be returned or added to the soil.

18.13 Write equations showing how nitric acid is formed by lightning.

18.14 How did the development of the Haber process affect world history?

18.15 Name the four most abundant gases in dry air in the order of decreasing abundance.

18.16 Explain auto-oxidation reduction.

18.17 What is the oxidation state of nitrogen in each of the following compounds: (a) HNO_2, (b) N_2O_5, (c) NO, (d) N_2O, (e) NO_2, (f) N_2O_4, (g) HNO_3?

18.18 Describe the Haber process.

18.19 Write equations (in two steps) for the preparation of ammonia gas using magnesium.

18.20 How does liquid ammonia resemble water as a solvent?

18.21 Explain the cycle occurring in the cooling system of a refrigerator.

18.22 Describe the Ostwald process for producing nitric acid.

18.23 Explain the role of nitrogen(II) oxide in air pollution.

18.24 How is nitric acid made from Chilean saltpeter?

18.25 What is the major use of ammonia?

18.26 What two important properties does sulfuric acid possess that nitric acid does not possess? What property does nitric acid have that is not exhibited by cold dilute sulfuric acid?

18.27 How is the element phosphorus produced?

18.28 List the uses of phosphorus.

18.29 Name the anhydride of the phosphoric acid.

18.30 Write symbols for the hybridization of phosphorus in five complex ions or molecules in tetrahedral, trigonal bipyramidal and octahedral configurations.

18.31 Explain why phosphorus can have octahedral and trigonal bipyramidal hybridizations while nitrogen cannot.

18.32 Explain how some of the electronic structures of phosphorus in some of its compounds do not conform to the "octet" principle.

SUGGESTED READING

Delwiche, C. C.: "The Nitrogen Cycle." Sci. Amer. *223*(3)136(1970).
Johnson, R. C.: *Introductory Descriptive Chemistry.* W. A. Benjamin, Inc., New York, 1966.
Jolly, W. L.: *The Chemistry of the Non-metals.* Prentice-Hall, Inc., Englewood Cliffs, New Jersey, 1966.
Lagowski, J. J.: "Liquid Ammonia—a Unique Solvent." Chemistry, *41*:4, 1968.
Research Reporter: "Ammonia Synthesis at Room Temperature and Pressure." Chemistry, *42*:7, 1969.

NINETEEN • IONIC EQUILIBRIA

19.1 INTRODUCTION

Many chemical reactions take place in water solutions. Soil chemistry is water solution chemistry. Tree sap and other plant fluids are water solutions. The fluid base of blood is water. Gastric juices that decompose foodstuffs in the stomach are water solutions. In most reactions occurring in the biosphere, water is present and plays an important role.

Many industrial processes also involve water solutions. The treatment of wood pulp, the dissolving of silver from its ore and the preparation and action of fertilizers are processes in which solutions play a leading role.

In all these natural and industrial processes, and in most reactions in which water takes part, ions are involved. In many processes the ionic concentrations are maintained at a very low level but are still very important. There is only about 0.01 mole of hydronium ion per liter of stomach fluid, but that acid concentration is sufficient to digest food. Were the solution more concentrated in the acid ion, the stomach lining would dissolve. Soil solutions must contain some iron in solution. The amount is described in parts per million, but without this concentration the plant leaves turn yellow, a condition called **chlorosis.**

These are but a few examples of the critical importance of ions and ionic concentrations. These concentrations are set and controlled by chemical equilibria that in many ways are similar to those discussed in Chapter 17. To understand the processes described above, or to devise others, one must understand ionic equilibria.

This chapter is divided into four sections; each section is concerned with equilibria involving ions. The first part is concerned with homogeneous reactions, in which all molecules and ions participating are in one phase, that phase being a water solution. Ionization characterizes the reactions, principally the ionization of weak electrolyte acids and bases. The second portion is a study of homogeneous equilibria of soluble strong electrolytes: weak acids and bases and complex ions. Heterogeneous equilibria, a study of dissociation in the dissolving of salts and the dissolving of metals in solution, is the subject of the third section. Part four is an introduction to systems in which more than one equilibrium is involved.

19.2 HOMOGENEOUS EQUILIBRIA–WEAK ELECTROLYTES

Weak electrolytes, considered in Chapter 14, are pure substances (1) whose solutions freeze at a slightly lower temperature than water, boil at a slightly higher temperature and exhibit a slightly greater osmotic pressure than do solutions of non-electrolytes of the same concentration in water, (2) whose solutions conduct an electric current but are poor conductors, (3) that are written as molecules in an ionic equation and (4) whose formation is the motive force in certain ionic reactions.

Weak electrolytes may be contrasted with **strong electrolytes,** which are substances (1) whose solutions exhibit a freezing point, a boiling point and an osmotic pressure two, three or four times greater than a solution of a non-electrolyte of the same concentration, (2) whose solutions are excellent electrolytic conductors, (3) that are written as ions in an ionic equation (when in solution) and (4) whose formation may be the motive force of a reaction only if the strong electrolyte is insoluble in water.

note the differences between strong and weak electrolytes.

Further understanding of weak electrolytes can be obtained by a more detailed comparison with strong electrolytes.

19.3 pH

Before discussing weak electrolytes, which, whether they are acids or bases, are concerned with hydronium ion concentration, an alternate and commonly used method for indicating the hydronium ion concentration should be discussed. In natural systems–soils, plant fluids and animal fluids–the hydronium ion concentration varies well within the limits of 1 M H_3O^+ and 1×10^{-14} M H_3O^+.

Consider the hydronium ion concentrations of solutions of hydrochloric acid. Hydrochloric acid is a strong electrolyte and is completely ionized in water:

$$HCl + H_2O \xrightarrow{100\%} H_3O^+ + Cl^-$$

The concentration of non-ionized HCl in 1 M HCl is zero, so the concentration of H_3O^+ is 1 M, and that of Cl^- is also 1 M. Table 19.1 lists the H_3O^+ concentrations in moles per liter of various HCl solutions.

TABLE 19.1

Molarity of HCl	$[H_3O^+]$	Molarity of HCl	$[H_3O^+]$
0.1 M HCl	$0.1 = 10^{-1}$	0.004 M HCl	4×10^{-3}
0.02 M HCl	2×10^{-2}	0.001 M HCl	1×10^{-3}
0.01 M HCl	1×10^{-2}	0.0006 M HCl	6×10^{-4}

To avoid using very small numbers in indicating hydronium ion concentrations, the chemist uses a term called pH, which bears the relationship to $[H_3O^+]$ as shown below:

as the solution becomes more acidic, the pH becomes less positive.

1 M H_3O^+ has a pH of 0
0.1 M H_3O^+ has a pH of 1
0.01 M H_3O^+ has a pH of 2
0.001 M H_3O^+ has a pH of 3
0.0001 M H_3O^+ has a pH of 4

The pH of a solution increases in magnitude as the acidity decreases. When the acidity $[H_3O^+]$ changes by a factor of 10, the pH changes by 1. The pH is related to the logarithm of the hydronium concentration:

$$0.01 = 1/100 = 1/10^2 = 10^{-2} \quad (-2 \text{ is the log of } 0.01)$$
$$0.0001 = 1/10000 = 1/10^4 = 10^{-4} \quad (-4 \text{ is the log of } 0.0001)$$

The pH is defined by this equation:

$$pH = -\log[H_3O^+]$$

and may be applied as follows:

Example 1. Find the pH of 0.001 M HCl = −3.

Solution: The $[H_3O^+]$ of 0.001 M HCl is 0.001

$$\log[H_3O^+] = \log 0.001 = -3$$

$$pH = -\log[H_3O^+] = -\log 0.001 = 3$$

Example 2. Find the pH of 0.003 M HCl.

Solution: The pH of 0.01 M HCl is:

$$pH = -\log[H_3O^+] = -\log 0.01 = -(-2) = 2$$

The pH of 0.001 M HCl is:

$$pH = -\log[H_3O^+] = -\log 0.001 = -(-3) = 3$$

0.003 M H_3O^+ lies between 0.01 M H_3O^+ and 0.001 M H_3O^+. The pH will be greater than 2 and less than 3. It happens that 0.003 may be written $10^{-2.523}$

$$pH = -\log[H_3O^+] = -\log 10^{-2.523} = -(-2.523) = 2.523$$

One senses that the pH of 0.003 M HCl (2.523) is intermediate between the pH of 0.01 M HCl (2) and the pH of 0.001 M HCl (3), but one may not know the exact value without access to a table of logarithms or a slide rule. Table 19.2 shows logs of two figure numbers. To determine the logarithm of a number, first write it in the correct exponential form. For example find the log of 0.0042.

TABLE 19.2 TABLE OF LOGARITHMS

	.0	.1	.2	.3	.4	.5	.6	.7	.8	.9
1	000	041	079	114	146	176	204	230	255	279
2	301	322	342	362	380	398	415	431	447	462
3	477	491	505	519	532	544	556	568	580	591
4	602	613	623	634	644	653	663	672	681	690
5	699	708	716	724	732	740	748	756	763	771
6	778	785	792	799	806	813	820	826	833	839
7	845	851	857	863	869	875	881	887	892	898
8	903	909	914	919	924	929	935	940	945	949
9	954	959	964	969	973	978	982	987	991	996

$$0.0042 = 4.2 \times 10^{-3}$$

Find the logs of the two parts and add. The fractional logs, logs for numbers between 1 and 10, are positive and must be read off the table. Log 4.2 is found in the table as follows:

	.0	.1	.2	.3	.4	.5	.6	.7	.8	.9
1										
2										
3										
4 ————			0.623							
5										
6										
7										
8										
9										

The log of a power of 10 is that power, in this case −3.

$$\text{Log } 4.2 \times 10^{-3} = +0.623 + (-3)$$

$$\begin{array}{r} -3.000 \\ +0.623 \\ \hline -2.377 \end{array}$$

The pH of a 0.0042 M HCl is then

$$\text{pH} = -\log[H_3O^+] = -\log(4.2 \times 10^{-3}) = -(0.623 - 3.000)$$
$$= -(-2.377) = 2.377$$

19.4 WEAK ACIDS

The fact that hydrochloric acid is a strong acid, as was assumed in the discussion of pH, may be shown with an experiment on conductivity:

Experiment 1. A 0.10 M HCl solution is added to both compartments of a divided glass battery jar, as shown in Figure 19.1, containing two pairs of graphite electrodes. When immersed in a conducting solution, each pair of electrodes completes a circuit through a light bulb. Each pair of electrodes reaches to the bottom of the jar. The pair on the left is uncovered the entire depth of

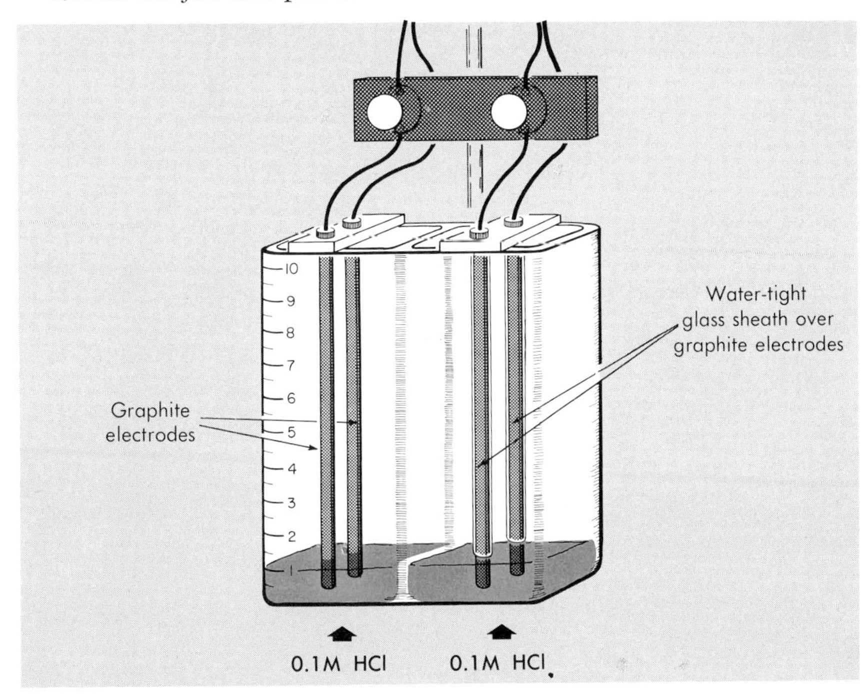

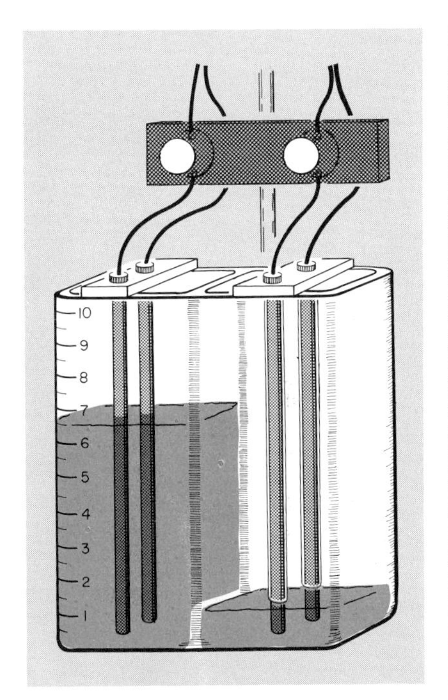

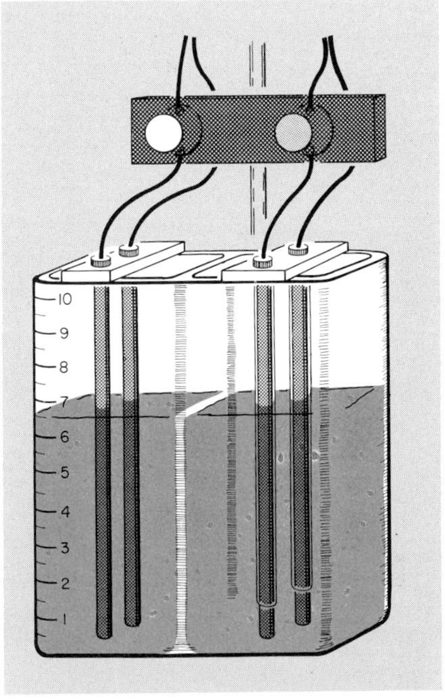

FIGURE 19.1 CONDUCTIVITY OF HCl SOLUTIONS OF DIFFERENT CONCENTRATIONS.

the container, whereas the pair on the right is covered tightly by glass tubes down to 1 cm from the bottom. In effect these electrodes are 1 cm in height. The 0.10 M HCl is added to the depth of 1 cm in each cell, covering each pair of electrodes to that depth. The exposed section of the electrodes on the right is completely covered. The two light bulbs glow with equal intensity. Distilled water is now added to the compartment on the left. As the incoming water mixes and the solution level rises, the bulb on the left retains the same intensity.

The explanation for this is simple. The 0.1 molar hydrochloric acid was completely ionized. Dilution did not change the number of ions. The same number of conducting ions remained between the electrodes, and the intensity of light did not change. Upon addition of distilled water to the compartment on the right, that light dimmed considerably, because, although the same number of ions existed in solution, fewer lay in the 1 cm layer between the 1 cm electrodes.

Experiment 2. After pouring out the hydrochloric acid solutions and washing and drying the battery jar, one fills each compartment with 1 molar acetic acid to the height of 1 cm, as shown in Figure 19.2. Although the molar concentration of this solution is 10 times that of the 0.1 M HCl previously in the jar, the light bulbs glow much less brightly than before. This happens because acetic acid is a weak electrolyte and is only partially ionized in water. Additional evidence of this fact is obtained by adding distilled water to the compartment on the left. As the water is added, the light becomes brighter and brighter. *Apparently, the addition of*

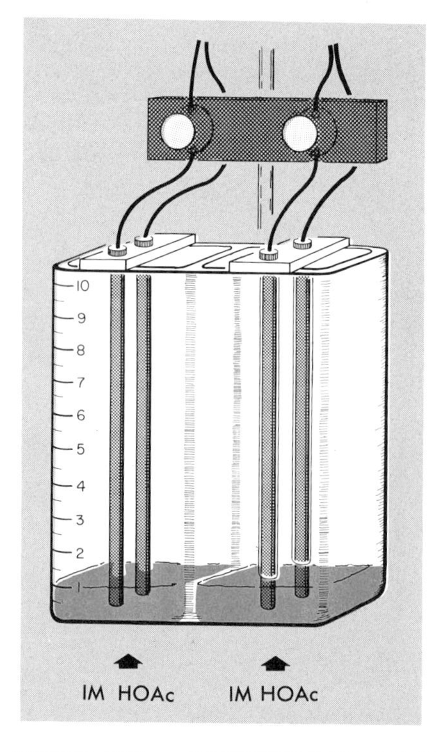

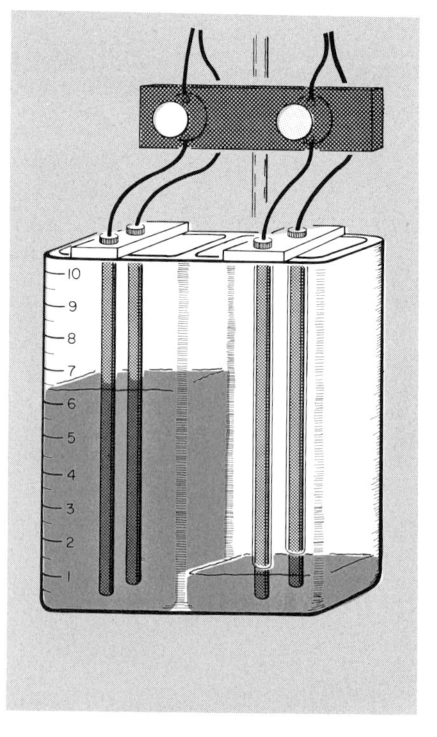

FIGURE 19.2 CONDUCTIVITY OF HOAc SOLUTIONS OF DIFFERENT CONCENTRATIONS.

a 1.0 molar solution of hydrochloric acid has twice as many ions per unit volume as a 0.5 molar solution. a 1.0 molar solution of acetic acid has twice as many ions per unit volume as a 0.24 molar solution of acetic acid.

water to the left compartment has created more ions. If more hydronium and acetate ions were formed by dilution (adding water), acetic acid is not 100 per cent ionized in a 1 molar solution; it is capable of reacting with more water to produce more ions.

The increase in the number of ions can be measured quantitatively by first filling the left compartment to a depth of 10 cm, then removing solution until both light bulbs glow with the same intensity. This happens when the more dilute solution (now 0.1 M HOAc*) is removed to a height of 3.1 cm. It follows that the number of ions was increased by

$$\frac{10}{3.1} = 3.2 \text{ times}$$

with the tenfold dilution.

When distilled water is added to the compartment on the right, the light dims, but not so rapidly as with the solution of hydrochloric acid. More than one-tenth the *original* number of ions are in the 1 cm layer between the 1 cm electrodes when the diluted solution stands 10 cm deep in the cell.

The fact that any solution of acetic acid will produce more H_3O^+ and OAc^- ions when diluted, plus the fact that 1.0 M HOAc has the same capacity to neutralize a base as does 1.0 M HCl, indicates that the small percentage of ionization of acetic acid is governed by an equilibrium reaction:

$$HOAc + H_2O \rightleftarrows H_3O^+ + OAc^-$$

It appears that the equilibrium is dynamic, that the forward and reverse rates are dependent upon concentrations, that the rates change with changing concentrations from dilution and, therefore, change the percentage ionized at equilibrium.

19.5 DERIVATION OF K_a

A consideration of the rate equations indicates how the equilibrium ion concentrations should change with dilution. Following the reasoning of Chapter 17, the rate of the forward reaction for the ionization of acetic acid (a bimolecular reaction) at room temperature,

$$HOAc + H_2O \rightarrow H_3O^+ + OAc^-,$$

may be given by this expression:

$$\text{Rate } (\rightarrow) = k[HOAc][H_2O],$$

---

* Acetic acid is an organic compound with the formula $CH_3{-}\overset{\overset{\large O}{\|}}{C}{-}OH$. Since only the right-hand hydrogen atom is released as H_3O^+ by reaction with water, the acid can be written more simply as HOAc where "Ac" stands for the acetyl group CH_3CO.

where k is the specific reaction rate constant and is equal to the rate when $[HOAc] = [H_2O] = 1$ mole per liter.

However, the concentration of water changes by a very small percentage when a weak electrolyte ionizes in it. Water is at a very high concentration in the solution. Pure water is about 55.5 M water:

$$\frac{1000\ g/L}{18\ g/mole} = 55.5\ moles/L = 55.5\ M$$

Again, nearly 1000 g of water is added to 60 g of acetic acid to make one liter of 1 M HOAc. The concentration drops from 55.5 M to perhaps 55.0 M. Less than 1 per cent of the acetic acid is ionized at equilibrium, and the concentration of water decreases negligibly to 54.99 M. Because the concentration of water is essentially constant, the rate equation may become:

$$\text{Rate}\ (\rightarrow) = k_1[HOAc],$$

these reactions are bimolecular and are, therefore, second order.

where $k_1 = k[H_2O]$, the product of two constants.

The rate equation for the reverse reaction,

$$HOAc + H_2O \leftarrow H_3O^+ + OAc^-,$$

is given by:

$$\text{Rate}\ (\leftarrow) = k_2[H_3O^+][OAc^-]$$

At equilibrium, Rate $(\rightarrow)$ = Rate $(\leftarrow)$

$$k_1[HOAc] = k_2[H_3O^+][OAc^-]$$

In this reaction, k_2 is much greater than k_1, and only very low concentrations of H_3O^+ and OAc^- (compared to the concentration of the non-ionized HOAc) are required to make the rates equal. (This will be shown experimentally later.) Suppose now that to a 100 ml sample of 1 M HOAc, in which the equilibrium concentration of H_3O^+ and OAc^- and undissociated HOAc exists, one adds an additional 100 ml of water. All participating concentrations (except that of water, which does not appear in the rate equations) are one-half the initial value.

Rate $(\rightarrow)$ will have been "halved" $= k_1 \times 1/2 \times [HOAc]$.

Rate $(\leftarrow)$ will have been reduced to "1/4" $= 1/2[H_3O^+] \times 1/2[OAc^-]$

Rate $(\rightarrow) = 2$ Rate $(\leftarrow)$

and H_3O^+ ions are being formed faster than they are being consumed. This inequality in rate persists until the ionic concentrations become great enough and the HOAc concentration small enough to restore equilibrium. Dilution increased the total number of ions, although the number of ions in 100 ml was decreased. Again, in 200 ml of 0.5 M HOAc there are more ions than in 100 ml of 1.0 M HOAc, but fewer ions are present in 100 ml of 0.5 M HOAc than in 100 ml of 1.0 M HOAc.

Rearrangement of the equated rate expressions gives the Mass Law Expression:

$$K_a = \frac{k_1}{k_2} = \frac{[H_3O^+][OAc^-]}{[HOAc]}$$

K_a is the equilibrium constant for the ionization of a weak acid. From this equation it is readily apparent that a simultaneous, proportionate reduction in all three concentrations would cause the expression to be less than K_a until some of the HOAc ionized to decrease the value of the denominator and increase the fraction.

19.6 EVALUATION OF K_a

In the derivation of the expression for the equilibrium constant, one merely shows that if rates depend on concentrations at equilibrium, as they appear to, then the expression

$$\frac{[H_3O^+][OAc^-]}{[HOAc]}$$

is equal to a **number** (at a certain temperature). The equation cannot be used quantitatively until the number has been determined, using experimental data. An experiment may be performed, using the apparatus shown in Figure 19.3, to determine the value of K_a at room temperature.

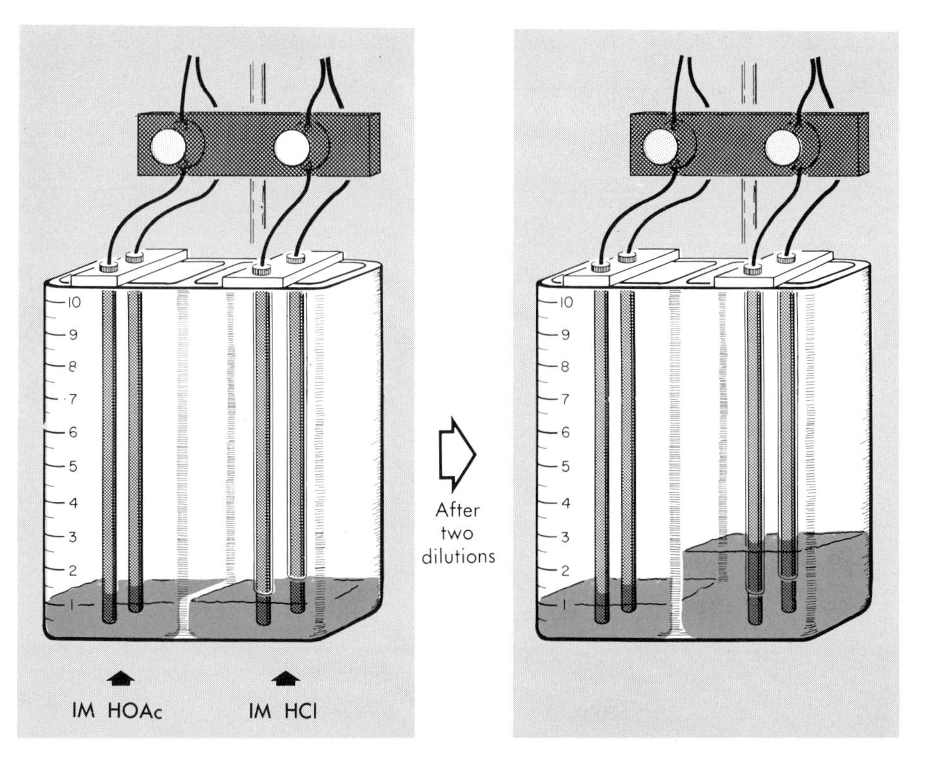

FIGURE 19.3 DETERMINATION OF H_3O^+ CONCENTRATION OF 1M HOAc.

Experiment 3. A 1.0 M HOAc solution is added to the left compartment until the electrode is covered to the height of 1 cm. A 1.0 M HCl solution is added to the right compartment to the same height, just covering the 1 cm electrodes. The light on the right glows much more brightly than the one on the left, because more ions are present.

Distilled water is added with stirring to the HCl in the compartment on the right until the level stands at the 10 cm mark. During the addition the light grows dimmer, but it is still brighter than the light on the left. The solution is then siphoned from the same (right side) compartment down to the 1 cm level. The light intensity remains the same, because the layer of solution between the electrodes was not disturbed.

The HCl solution in the right compartment then is diluted a second time with distilled water to the 10 cm mark. The light intensity is still greater than that on the left. Again the solution is siphoned down to the 1 cm level, and once again distilled water is added, more slowly this time because the light on the right is not much brighter than the light on the left.

When the height of the dilute HCl solution reaches about the 2.5 cm level, both lights glow with the same intensity. It may be assumed that an equal number of ions lies between the 1 cm electrodes in the two vessels. This is not exactly true, for H_3O^+ and OAc^- ions are present in the left hand container, while H_3O^+ and Cl^- ions are present in the right. Unless the Cl^- ion conducts equally as well as the OAc^- ion, the ionic concentrations need not be the same. However, the hydronium ion carries a major share of the current, making the difference between the conductivity of OAc^- and Cl^- negligible.

The final HCl solution has a concentration of:

$$1.0 \text{ mole/liter} \times \frac{1}{10} \times \frac{1}{10} \times \frac{1}{2.5} = 0.0040 \text{ mole/liter}$$

(first, second and third dilutions).

Because all the HCl is ionized, the concentration of H_3O^+ and Cl^- in the HCl solution is 0.0040 M. It follows that the H_3O^+ and OAc^- concentrations in 1 M HOAc are 0.0040 M.

This is not the only method for determining the concentration of hydronium ions in 1.0 M HOAc. Figure 19.4 indicates that 0.0042 M HCl and 1.0 M HOAc have (1) the same sourness of taste, (2) the same conductivity, (3) the same color in indicator solutions and (4) the same reading on a pH meter, which measures precisely the hydronium concentration by its effect on the voltage of a certain primary cell.

To complete the determination of K_a:

the $[H_3O^+] = 0.0042$ mole/liter (measured)

the $[OAc^-] = 0.0042$ mole/liter (appears ion for ion with H_3O^+)

and the non-ionized HOAc concentration:

$$[HOAc] = 1 - 0.0042 \text{ mole/liter} = 0.996 \text{ mole/liter}.$$

FIGURE 19.4 COMMON PROPERTIES OF SOLUTIONS OF THE SAME HYDRONIUM ION CONCENTRATION.

Substituting these values in the equation, one obtains:

$$K_a = \frac{[H_3O^+][OAc^-]}{[HOAc]} = \frac{0.0042 \times 0.0042}{0.996} = 1.75 \times 10^{-5}$$

Using only significant figures, $K_a = 1.8 \times 10^{-5}$. One might as well have used 1 mole per liter for HOAc as 0.996 mole per liter.

The value of the ionization constant (K_a) may be determined using solutions of acetic acid of concentrations other than 1.0 M. One need merely find the hydronium ion concentration in another acetic acid solution in a manner discussed for 1.0 M HOAc, and substitute this figure in the Mass Law Expression. Table 19.3 shows the measured hydronium ion concentrations for HOAc solutions of various molarities, with the value of the ionization constant calculated using those values. The values of the ionization constant calculated at different concentrations are nearly the same and are close to 1.8×10^{-5}.

TABLE 19.3 THE $[H_3O^+]$ AND VALUE OF K_a FOR ACETIC ACID FOR SOLUTIONS OF DIFFERENT CONCENTRATIONS

M HOAc	H_3O^+	K_a
0.100	0.00135	1.82×10^{-5}
0.010	0.00042	1.76×10^{-5}
0.001	0.000124	1.75×10^{-5}

19.7 CALCULATION OF THE H_3O^+ CONCENTRATION IN SOLUTIONS OF WEAK ACIDS

note the great variation in the value of the equilibrium constants for weak acids.

Once the equilibrium constant K_a has been calculated, the constant expression can be used to calculate the H_3O^+ concentration of solutions of other concentrations.

Example 3. What is the hydronium ion concentration of 0.05 M HOAc at room temperature? $K_a = 1.8 \times 10^{-5}$

Solution: Let $X = [H_3O^+]$

And since H_3O^+ was produced only by this reaction

$$\begin{array}{c} 0.05 - X \qquad\qquad X \qquad\quad X \\ HOAc + H_2O \rightarrow H_3O^+ + OAc^-, \end{array}$$

$$[OAc^-] = X$$

$$\text{and } [HOAc] = 0.05 - X$$

Then

$$\frac{[H_3O^+][OAc^-]}{[HOAc]} = \frac{X^2}{0.05 - X} = 1.8 \times 10^{-5}$$

One may solve the quadratic equation:

$$X^2 = (0.05 - X)(1.8 \times 10^{-5})$$

$$X^2 + 1.8 \times 10^{-5}X - 9 \times 10^{-7} = 0$$

$$X = \frac{-1.8 \times 10^{-5} \pm \sqrt{(1.8 \times 10^{-5})^2 + 4 \times 9 \times 10^{-7}}}{2}$$

$$= \frac{-1.8 \times 10^{-5} \pm 1.9 \times 10^{-3}}{2}$$

$$= 9.4 \times 10^{-4} \text{ mole/liter}$$

or one may assume that X is very small and that $0.05 - X \approx^{*} 0.05$

$$\frac{X^2}{0.05} = 1.8 \times 10^{-5}$$

$$X^2 = 90 \times 10^{-8}$$

$$X = 9.5 \times 10^{-4} \text{ mole/liter}$$

In this case 0.05 could be assumed equal to $0.05 - X$ without a great error, for $X = 9.5 \times 10^{-4}$, and so $0.05 - X = 0.04905$. Since the answers are correct to only two significant figures, 0.05 was acceptable. In general, when the equilibrium constant is 5×10^{-4} or smaller, the X may be neglected without serious error.

According to the calculations:

$$[H_3O^+] = 9.4 \times 10^{-4} \text{ mole/liter}$$

$$[OAc^-] = 9.4 \times 10^{-4} \text{ mole/liter}$$

$$\text{and } [HOAc] = \text{a bit less than } 0.05 \text{ mole/liter}$$

The pH may be calculated:

$$pH \quad 4.000 - \log 9.4 = 4.000 - 0.973 = 3.03$$

---

* $\approx$ means approximately equal to.

Example 4. Calculate the hydronium ion concentration and pH of a solution containing 1 mole of HOAc and 0.6 mole of NaOAc (sodium acetate) per liter, given that $K_a = 1.8 \times 10^{-5}$.

Solution: NaOAc, a strong electrolyte, is completely ionized in solution, and the solution contains 0.6 M Na^+ and 0.6 M OAc^-. The hydronium ion present comes from partial ionization of acetic acid,

$$HOAc + H_2O \rightarrow H_3O^+ + OAc^-$$

If $[H_3O^+] = X$

then $[OAc^-] = 0.6 + X$

and $[HOAc] = 1 - X$

Also, $[Na^+] = 0.6$ (but is of no consequence in this problem for it does not appear in the equilibrium expression).

$$\text{Hence, } K_a = \frac{[H_3O^+][OAc^-]}{[HOAc]} = \frac{X(0.6 + X)}{1 - X}$$

but $0.6 + X \approx 0.6$

and $1 - X \approx 1.0$

$$\frac{X \times 0.6}{1} = 1.8 \times 10^{-5}$$

$X = 3 \times 10^{-5}$ mole/liter $= [H_3O^+]$

$pH = 5.00 - \log 3 = 5.00 - 0.477 = 4.52$

the addition of a base, OAc^- in this case, should reduce the acidity.

The addition of the conjugate base (OAc^-) of the acid HOAc suppresses the hydronium ion concentration. This phenomenon is known as the *common ion effect.*

TABLE 19.4 IONIZATION CONSTANTS OF WEAK ACIDS

Acid	Reactions	Constant
Acetic	$H_2O + CH_3COOH \rightarrow H_3O^+ + CH_3COO^-$	1.8×10^{-5}
Benzoic	$H_2O + C_6H_5COOH \rightarrow H_3O^+ + C_6H_5COO^-$	6.6×10^{-5}
Carbonic	$2\ H_2O + CO_2 \rightarrow H_3O^+ + HCO_3^-$	4.2×10^{-7}
Formic	$H_2O + HCOOH \rightarrow H_3O^+ + HCOO^-$	2.1×10^{-4}
Hydrocyanic	$H_2O + HCN \rightarrow H_3O^+ + CN^-$	4.0×10^{-10}
Hydrofluoric	$H_2O + 2\ HF \rightarrow H_3O^+ + HF_2^-$	6.9×10^{-4}
Hydrogen sulfide	$H_2O + H_2S \rightarrow H_3O^+ + HS^-$	1.0×10^{-7}
Nitrous	$H_2O + HNO_2 \rightarrow H_3O^+ + NO_2^-$	4.5×10^{-4}
Oxalic	$H_2O + H_2C_2O_4 \rightarrow H_3O^+ + HC_2O_4^-$	3.8×10^{-2}
Phenol	$H_2O + C_6H_5OH \rightarrow H_3O^+ + C_6H_5O^-$	1.0×10^{-10}
Phosphoric	$H_2O + H_3PO_4 \rightarrow H_3O^+ + H_2PO_4^{2-}$	7.5×10^{-3}
Propionic	$H_2O + C_2H_5COOH \rightarrow H_3O^+ + C_2H_5COO^-$	1.4×10^{-5}

Example 5. There are other weak acids besides HOAc. One, benzoic acid (represented as HOBz), has an equilibrium constant of 6.6×10^{-5}. Find the $[H_3O^+]$ and the pH of 0.02 M HOBz.

Solution:

Let $X = [H_3O^+]$

then $[OBz^-] = X$

and $[HOBz] = 0.02 - X \approx 0.02$

$$K_a = \frac{[H_3O^+][OBz^-]}{[HOBz]} = \frac{X^2}{0.02} = 6.6 \times 10^{-5}$$

$X^2 = 1.32 \times 10^{-6}$

$X = 1.15 \times 10^{-3}$ M H_3O^+

$pH = 3.00 - \log 1.15 = 3.00 - 0.060 = 2.94$

19.8 WEAK ELECTROLYTE BASE

The "case of the weak base" parallels the case of the weak acid. A weak base, such as ammonia, ionizes in water to produce hydroxide ions:

$$NH_3 + H_2O \rightarrow NH_4^+ + OH^-.$$

Conductivity experiments comparing solutions of ammonia with solutions of the strong base NaOH show that only a small fraction of the NH_3 molecules are ionized at one time. The ionization is a reversible reaction and quickly comes to equilibrium:

$$NH_3 + H_2O \rightleftarrows NH_4^+ + OH^-$$

The rate of ionization at a set temperature is a function of concentration, and because it is a bimolecular reaction,

$$\text{Rate } (\rightarrow) = k[NH_3][H_2O]$$

However, because $[H_2O]$ changes but little, a simpler equation is used:

$$\text{Rate } (\rightarrow) = k_1[NH_3]$$

where $k_1 = k[H_2O]$. (The k's in these rate equations have a different magnitude than those for other reversible reactions.)

The reverse rate, the rate of recombination of ions, is a bimolecular reaction and depends upon the ionic concentrations:

$$\text{Rate } (\leftarrow) = k_2[NH_4^+][OH^-]$$

At equilibrium the concentrations have adjusted themselves so that

$$\text{Rate} (\rightarrow) = \text{Rate} (\leftarrow)$$

$$k_1[NH_3] = k_2[NH_4^+][OH^-]$$

and the equilibrium constant for the weak base is

$$K_b = \frac{k_1}{k_2} = \frac{[NH_4^+][OH^-]}{[NH_3]}$$

(The equation is valid only when the concentrations are equilibrium concentrations.)

The expression for K_b is similar to the one for K_a. Just as dilution of a weak acid causes a greater percentage of ionization, so will dilution of a weak base. Again, with the addition of water, each concentration in the numerator is reduced by the same proportion as the one concentration in the denominator. To restore the value of the constant K_b, the $[NH_3]$ must decrease with the consequent and equal increase in $[NH_4^+]$ and $[OH^-]$.

The equilibrium constant expression, the Mass Law Equation, is useful for quantitative calculations when the value of the constant K_b is known. This must be determined using *experimental data.*

It happens that 0.1 M NH_3 has a hydroxide concentration of 0.00135 M. That is, it has the same conductance and same bitter taste, has the same color as acid-base indicators and gives the same reading on the pH meter as 0.00135 M NaOH.

In 0.1 M NH_3:

$[OH^-] = 0.00135$ mole/liter

$[NH_4^+] = 0.00135$ mole/liter (OH^- and NH_4^+ are formed together)

The concentration of non-ionized ammonia is

$$[NH_3] = 0.1 - 0.00135 \approx 0.1$$

Substituting the set of equilibrium concentrations in the expression,

$$K_b = \frac{[NH_4^+][OH^-]}{[NH_3]} = \frac{0.00135 \times 0.00135}{0.1} = 1.82 \times 10^{-5}$$

By chance, K_b for ammonia and K_a for acetic acid have the same numerical value. One may now use the expression to find the concentration of hydroxide ion and the pOH (analogous to pH) of various solutions containing ammonia.

Example 6. Find the concentration of hydroxide ion and the pOH of 0.30 M NH_3.

Solution:

Let $[OH^-] = X$

then $[NH_4^+] = X$ (both come from NH_3, ion per ion)

the non-ionized ammonia concentration:

$$[NH_3] = 0.3 - X \approx 0.3$$

Substituting in the equation:

$$K_b = \frac{[NH_4^+][OH^-]}{[NH_3]} = \frac{X \cdot X}{0.3} = 1.8 \times 10^{-5}$$

$$X^2 = 5.4 \times 10^{-6}$$

$$X = 2.3 \times 10^{-3} \text{ mole/liter} = [OH^-]$$

The pOH of the solution is found as follows:

$$\begin{aligned} pOH &= 3 - \log 2.3 = 3.000 - 0.362 \\ &= 2.64 \end{aligned}$$

Example 7. Find the hydroxide concentration in a 500 ml solution containing 0.20 mole NH_3 and 0.15 mole NH_4Cl.

Solution: In moles per liter the concentrations are 0.40 M NH_3 and 0.30 M NH_4Cl. But NH_3 is slightly ionized (X moles),

$$\underset{0.40 - X}{NH_3} + H_2O \rightarrow \underset{X}{NH_4^+} + \underset{X}{OH^-}$$

and NH_4Cl is in the form of ions in the solution. A 0.30 M NH_4Cl solution contains 0.30 mole NH_4^+ per liter and 0.30 mole Cl^- per liter and no un-ionized NH_4Cl.

The ionic concentrations from both substances are

$$[OH^-] = X \text{ mole/liter}$$

$$[NH_4^+] = 0.30 + X \approx 0.30 \text{ mole/liter}$$

$$[NH_3] = 0.40 \text{ mole/liter}$$

$[Cl^-] = 0.30$, but is of no consequence in the problem.

$$K_b = \frac{[NH_4^+][OH^-]}{[NH_3]} = \frac{0.30\, X}{0.40} = 1.8 \times 10^{-5}$$

$$X = \frac{0.4}{0.3} \times 1.8 \times 10^{-5} = 2.4 \times 10^{-5} \text{ mole/liter}$$

$$[OH^-] = 2.4 \times 10^{-5} \text{ mole/liter}$$

$$\text{The pOH} = 5 - \log 2.4 = 5.00 - 0.382 = 4.618$$

The addition of NH_4^+, the conjugate acid of the base NH_3, suppressed the concentration of hydroxide ion. This is another example of common ion effect.

19.9 FINDING SQUARE ROOTS

In several of the examples discussed, and in several to follow, the square roots of numbers must be determined. The numbers are small fractions, often small enough that it is convenient to write them in exponential form. One thing to remember is that the product of the square roots of two numbers is the square root of their product. As an illustration:

$4 \times 9 = 36$

$2 \times 3 = 6$ Each number on the second line is the square root of the number above it.

This is useful in finding the square roots of numbers in exponential form. Thus the square root of 4×10^{-6} is 2×10^{-3}. Should the coefficient of ten be an uneven number, such as 3.6×10^{-7}, the decimal point must be moved, since the coefficient must be divisible by 2.

$$3.6 \times 10^{-7} = 36 \times 10^{-8}.$$

The square root is:

$$6 \times 10^{-4}.$$

Unfortunately, not all numerical factors of exponential numbers are perfect squares. The square root of a number may be found on a slide rule, or, if one cannot read a slide rule, the answer may be determined using logarithms. The square root of a number is the number whose logarithm is one-half the logarithm of the given number. For example:

a. the log of 9 is 0.954; of 3 is 0.477
b. the log of 4 is 0.602; of 2 is 0.301
c. the log of 16 is 1.204; of 4 is 0.602

Example 8. Find the square root of 7.1.

Solution:

The log of 7.1 is 0.851 (Table 19.2)

$$\frac{0.851}{2} = 0.426$$

The number 0.426 is nearest the logarithm of 2.7 (Table 19.2), hence the square root is 2.7.

Note that 2.7 is not the exact square root of 7.1, but it is near enough for our purposes.

Example 9. Find the square root of 2.4×10^{-4}.

Solution:

The square root $= \sqrt{2.4} \times 10^{-2}$

logarithm of $2.4 = 0.380$ (Table 19.2)

$$\frac{0.380}{2} = 0.190$$

The number whose logarithm is 0.190 is 1.6 (Table 19.2).

$$\sqrt{2.4} \times 10^{-2} = 1.6 \times 10^{-2}$$

Example 10. Find the square root of 8.2×10^{-11}.

Solution:

$$8.2 \times 10^{-11} = 82 \times 10^{-12}$$

$$\text{Log } 82 = 1.914 \text{ (table)}$$

$$\frac{\text{Log } 82}{2} = \frac{1.914}{2} = 0.957$$

The number whose log is 0.957 is 9.1, so

$$\sqrt{82 \times 10^{-12}} = 9.1 \times 10^{-6}$$

19.10 THE CONSTANT K_w

Water is an interesting substance, for not only is it the solvent in water solutions, but it is also both a weak acid and a weak base.

Water is often described as a non-conductor of electricity, but pure water has a very small but measurable conductance because of the ionization

$$H_2O + H_2O \rightarrow H_3O^+ + OH^-$$

This ionization is an equilibrium ionization. The rate of ionization is a function of the concentration of water:

$$\text{Rate } (\rightarrow) = k[H_2O][H_2O]$$

But again, because the water concentration suffers such a small change in dilute solutions of acids and bases, the rate is nearly constant:

$$\text{Rate} = k_1$$

$$\text{where } k_1 = k[H_2O]^2$$

The reverse rate is a function of the ionic concentrations.

$$\text{Rate } (\leftarrow) = k_2[H_3O^+][OH^-]$$

At equilibrium:

$$\text{Rate } (\rightarrow) = \text{Rate } (\leftarrow)$$

$$k_1 = k_2[H_3O^+][OH^-]$$

$$\frac{k_1}{k_2} = [H_3O^+][OH^-] = K_w$$

There is no denominator in the equilibrium constant for water. This differs from all other weak acids in water, for water is also the solvent.

use the Avogadro Number to determine how many hydronium ions there are per milliliter of pure water.

In the above rate equations $k_2 > k_1$; that is, the concentrations of H_3O^+ and OH^- are very low in water compared to the water concentration. In very pure water triple distilled in platinum ware, at 25°C the concentration of hydronium ion is equal to 1.0×10^{-7} mole per liter. The hydroxide ion concentration is also 1.0×10^{-7} mole per liter. Of the 55.5 moles of water present in 1 liter of pure water at or near room temperature, only 1×10^{-7} mole is ionized. This is justification for calling water a covalent substance and a non-electrolyte. Actually, some few ions are present.

If $[H_3O^+] = [OH^-] = 1 \times 10^{-7}$ in pure water at 25°C,

$$K_w = [H_3O^+][OH^-] = 1 \times 10^{-7} \times 1 \times 10^{-7} = 1 \times 10^{-14}$$

If this be true, neither $[H_3O^+]$ nor $[OH^-]$ can be zero in a solution in water. If one is greater than 1.0×10^{-7}, the other must be less. In an acid solution the $[H_3O^+]$ is greater than 1×10^{-7}; therefore, the $[OH^-]$ is less. Furthermore, when either $[H_3O^+]$ or $[OH^-]$ is known, the other can be determined, because their product is known.

Example 11. Find the $[OH^-]$ in a 0.0024 M HCl solution.

Solution: HCl is a strong electrolyte, therefore:

$$[H_3O^+] = 2.4 \times 10^{-3} \text{ moles/liter}$$

Substituting in K_w:

$$K_w = [H_3O^+][OH^-] = 2.4 \times 10^{-3} \times [OH^-] = 1.0 \times 10^{-14}$$

$$[OH^-] = \frac{1.0 \times 10^{-14}}{2.4 \times 10^{-3}} = \frac{10 \times 10^{-15}}{2.4 \times 10^{-3}} = 4.2 \times 10^{-12}$$

19.11 pH VERSUS pOH

Because the product of the concentration of hydronium ion and the concentration of hydroxide ion is always the same in a water solution, an interesting relationship exists between the pH and the pOH. Since logarithms are added in multiplying,

if $[H_3O^+][OH^-] = 1.0 \times 10^{-14}$

then $\log [H_3O^+] + \log [OH^-] = -14$

and $-\log [H_3O^+] - \log [OH^-] = 14$

or $pH + pOH = 14$

Rearranging the equations:

$$pH = 14 - pOH$$

and $$pOH = 14 - pH$$

Example 12. Calculate the hydronium ion concentration, the hydroxide ion concentration, the pH and the pOH of 0.3 M HCN. $K_{HCN} = 4.0 \times 10^{-10}$.

Solution:

a. To find $[H_3O^+]$

let $X = [H_3O^+]$

then $[CN^-] = X$

and $[HCN] = 0.3 - X \approx 0.3$

Substituting;

$$K_{HCN} = \frac{[H_3O^+][CN^-]}{[HCN]} = \frac{X^2}{0.3} = 4.0 \times 10^{-10}$$

$$X^2 = 1.2 \times 10^{-10}$$

$$X = \sqrt{1.2} \times 10^{-5}$$

$$\frac{\log 1.2}{2} = \frac{0.0792 \text{ (table)}}{2} = 0.0396$$

The number whose log is 0.0396 is 1.1:

$$X = 1.1 \times 10^{-5} \text{ mole/liter} = [H_3O^+]$$

b. to find $[OH^-]$:

$$[H_3O^+] = 1.1 \times 10^{-5} \times [OH^-] = 1.0 \times 10^{-14}$$

$$[OH^-] = \frac{1.0 \times 10^{-14}}{1.1 \times 10^{-5}} = \frac{10 \times 10^{-15}}{1.1 \times 10^{-5}} = 9.0 \times 10^{-10} \text{ mole/liter}$$

c. to find the pH:

$$[H_3O^+] = 1.1 \times 10^{-5} \text{ mole/liter}$$

$$pH = 5 - \log 1.1 = 5.00 - 0.040 = 4.96$$

d. to find pOH:

$$[OH^-] = 9.0 \times 10^{-10} \text{ mole/liter}$$

$$pOH = 10 - \log 9 = 10.00 - 0.954 = 9.046$$

or $$pH = 4.96$$

$$pOH = 14.00 - 4.96 = 9.04$$

Pure water is **neutral**; it is the standard for neutrality. Any solution with the same hydronium ion concentration, the same pH, the same hydroxide ion concentration or the same pOH as water is neutral.

In pure water the hydronium ion concentration is the same as the hydroxide ion concentration.

$$[H_3O^+] = [OH^-] = 1.0 \times 10^{-7} \text{ mole/liter}$$

and the

$$pH = 7 - \log 1 = 7.00$$

$$POH = 7 - \log 1 = 7.00$$

The solution is neutral if the pH = 7 and pOH = 7.

The solution is acidic if the pH $<$ 7 and pOH $>$ 7.

The solution is basic if the pH $>$ 7 and pOH $<$ 7.

19.12 HOMOGENEOUS EQUILIBRIA OF STRONG ELECTROLYTES—HYDROGEN-SALT ACIDS

The hydrogen sulfate ion (HSO_4^-), often called bisulfate ion, is an acid. It is obtained in solution by dissolving in water a hydrogen sulfate salt, such as a sodium hydrogen sulfate ($NaHSO_4$). The salt dissociates completely in solution. The sodium ions make no contribution to the acidity of the solution. Hydrogen sulfate ion is an acid, for it reacts with water:

$$HSO_4^- + H_2O \rightarrow H_3O^+ + SO_4^{2-}$$

to produce hydronium ion. The hydrogen sulfate ion contributes a proton to water to form the hydronium ion. But hydrogen sulfate is only a moderately strong (and moderately weak) acid, for in a 1 molar solution of sodium hydrogen sulfate, only about 1/10 of the bisulfate ions decomposes to form hydronium ions. Experiments with dilution and titration show the reaction to be reversible.

hydrogen sulfate ion acts as a weak Bronsted acid in water because a small proportion of the ions lose (H^+) to water.

$$HSO_4^- + H_2O \rightleftharpoons H_3O^+ + SO_4^{2-}$$

The equilibrium constant expression for the equation is:

$$K_a = \frac{[H_3O^+][SO_4^{2-}]}{[HSO_4^-]}$$

As always, the concentration of water does not appear in the equation. For the hydrogen sulfate ion the constant is 1.26×10^{-2}. One can use this constant to find the hydronium ion concentration and the pH of solutions containing the ion. However, the constant is so large that the quadratic equation must be employed.

Many hydrogen-salt acids exist. Four appear in Table 19.5 with their constants. The hydrogen sulfate ion is present in sulfuric acid

TABLE 19.5

Acid	Reactions	Constant
Bicarbonate ion	$H_2O + HCO_3^- \rightarrow H_3O^+ + CO_3^{2-}$	4.8×10^{-11}
Bisulfate ion	$H_2O + HSO_4^- \rightarrow H_3O^+ + SO_4^{2-}$	1.26×10^{-2}
Bisulfide ion	$H_2O + HS^- \rightarrow H_3O^+ + S^{2-}$	1.3×10^{-13}
Dihydrogen phosphate ion	$H_2O + H_2PO_4^- \rightarrow H_3O^+ + HPO_3^{2-}$	6.2×10^{-8}

solutions. Sulfuric acid is called a **diprotic acid** because it may donate two protons per molecule. The second proton is that of the hydrogen sulfate ion. The acidic properties of HSO_4^- are not so important in a solution of H_2SO_4, because the contribution to the concentration of hydronium ion is negligible compared to that of the first ionization of H_2SO_4,

$$H_2SO_4 + H_2O \xrightarrow{100\%} H_3O^+ + HSO_4^-$$

Nevertheless,

$$HSO_4^- + H_2O \rightleftarrows H_3O^+ + SO_4^{2-}$$

is an equilibrium reaction producing a small concentration of hydronium ions. In general, the second and third ionizations of diprotic and triprotic acids contribute little to the acidity of solutions of the acids but do govern the acid concentration if present alone as a hydrogen salt.

19.13 WEAK ACIDS AND BASES—HYDROLYSIS

The ammonium ion is a weak acid. A solution of an ammonium salt, such as ammonium chloride (NH_4Cl), is only slightly acidic. The solution is acidic and the ammonium ion is an acid since this reaction occurs:

$$NH_4^+ + H_2O \rightarrow NH_3 + H_3O^+$$

The ammonium ions donate protons to water molecules. However, it is a weak acid since the fraction of ammonium ions decomposed at any one time in an ammonium chloride solution is very small and changes with the concentration.

$$NH_4^+ + H_2O \rightleftarrows NH_3 + H_3O^+$$

The equilibrium expression for this reaction may be derived as before, and it is

$$K_h = \frac{[NH_3][H_3O^+]}{[NH_4^+]}$$

where K_h represents the constant for a hydrolysis reaction. This type of reaction is called **hydrolysis** (hydro-lysis, or water decomposition). The distinction is not clear, for the ionizations of all acids and all weak bases are water decompositions.

The equilibrium constant (K_h) for this reaction might be evaluated as any other by the the substitution of a set of equilibrium concentrations, but a mathematical trick makes this unnecessary:

$$K_h = \frac{[NH_3][H_3O^+]}{[NH_4^+]} = \frac{[NH_3][H_3O^+][OH^-]}{[NH_4^+][OH^-]}$$

The fraction is not changed by multiplying both the numerator and denominator by the same quantity, the concentration of hydroxide ion.

But
$$[H_3O^+][OH^-] = K_w$$

and
$$\frac{[NH_3]}{[NH_4^+][OH^-]} = \frac{1}{K_b \text{ of } NH_3}$$

so
$$K_h = \frac{K_w}{K_b}$$

The hydrolysis constant of the ammonium ion is equal to the equilibrium constant for water divided by the ionization constant of its conjugate base, ammonia. A second look at the equation for the hydrolysis of ammonia suggests that K_w and K_b of ammonia are involved:

$$NH_4^+ + H_2O \rightleftarrows NH_3 + H_3O^+$$

The equilibrium exists because H_2O on the left and NH_3 on the right are competing for the proton. The equilibrium concentrations depend on the relative attraction of these two molecules for the proton in question.

The acetate ion is a weak base. Solutions of acetate salts, such as sodium acetate and NaOAc, are basic since this reaction occurs:

$$H_2O + OAc^- \rightleftarrows OH^- + HOAc$$

and because the acetate ions accept protons from the water molecules. They are **weakly** basic, because only a small fraction of the acetate ions are converted to acetic acid molecules.

The equilibrium reaction may be represented by this equation:

$$H_2O + OAc^- \rightleftarrows OH^- + HOAc$$

The equilibrium expression is:

$$K_h = \frac{[HOAc][OH^-]}{[OAc^-]}$$

Again the reaction is called hydrolysis, and again the constant for the acetate ion is related to the constant for water and the constant for the conjugate acid.

$$K_h = \frac{K_w}{K_a \text{ of HOAc}}$$

This is reasonable, because a glance at the equation reveals that OAc^- (on the left) competes with OH^- (on the right) for the proton to form HOAc or H_2O. Equilibrium concentrations are set by this competition. Just as the acetate ion, OAc^- (the conjugate base of acetic acid, HOAc), is a base, so also are the hydrogen carbonate ion, HCO_3^- (the conjugate base of carbonic acid, H_2CO_3); the cyanide ion, CN^- (the conjugate base of hydrocyanic acid, HCN); and the sulfate ion, SO_4^{2-} (the conjugate base of the acid hydrogen sulfate ion, HSO_4^-).

The aluminum ion is also a weak acid. A solution of an aluminum salt, such as aluminum sulfate ($Al_2(SO_4)_3$), is acidic because this equilibrium reaction occurs:

$$Al^{3+} + 2\ H_2O \rightarrow AlOH^{2+} + H_3O^+ \text{ or}$$

$$Al(H_2O)_6^{+3} + H_2O \rightarrow Al(H_2O)_5OH^{2+} + H_3O^+$$

note the aluminum ion, Al^{3+}, is the acid, not $Al_2(SO_4)_3$.

19.14 BUFFER SOLUTIONS

In certain chemical systems, it is important to maintain a nearly constant pH. The rates of reactions occurring in solutions are often dependent upon the hydronium ion concentration. The pH can affect the equilibrium concentration if the hydronium ion is a reactant. A change in pH may even alter the products of a reaction. Death results if the pH of the blood strays outside the pH limits of 6.8 to 7.8 (account is taken of this in intravenous feeding), but the blood of normal individuals contains a **buffer** that keeps the pH within the limits of 7.3 to 7.45. A buffer is a set of solution components that holds the pH nearly constant despite shocks to the system by relatively large additions of acids or bases.

To understand buffer action, consider the addition of acid and base to three different solutions with the same pH:

Solution A. A dilute solution of a strong acid, 1.8×10^{-5} M HCl, has a pH of 4.75, because $[H_3O^+] = 1.8 \times 10^{-5}$

$$pH = 5.00 - \log 1.8 = 5.00 - 0.25 = 4.75$$

Solution B. A fairly concentrated solution of a weak acid, 0.8 M HCN ($K_a = 4 \times 10^{-10}$), has a pH of 4.75:

If $$X = [H_3O^+] = [CN^-]$$

$$K_a = \frac{[H_3O^+][CN^-]}{[HCN]} = \frac{X^2}{0.8} = 4 \times 10^{-10}$$

$$X = [H_3O^+] = 1.8 \times 10^{-5}$$

$$pH = 4.75$$

Solution C. A solution containing equal concentrations of a certain weak acid (stronger than HCN) and a salt of that weak acid, 1.00 M HOAc and 1.00 M NaOAc, has a pH of 4.75.

$$K_{HOAc} = 1.8 \times 10^{-5}$$

for if

$$X = [H_3O^+]$$

$$[OAc^-] = [1.00 + X] \text{ mole/liter} \approx 1$$

$$[HOAc] = (1.00 - X) \text{ mole/liter} \approx 1$$

$$K_a = \frac{X \cdot 1}{1} = 1.8 \times 10^{-5}$$

$$[H_3O^+] = X = 1.8 \times 10^{-5} \text{ mole/liter}$$

and $pH = 4.75$

Suppose one adds to separate 1-liter volumes of each solution above: (a) 0.20 mole HCl and (b) 0.20 mole NaOH.

Solution A. To 1 liter of 1.8×10^{-5} M HCl with pH of 4.75 is added:

a. 0.20 mole HCl

$$0.20 \text{ mole/liter} + 0.000018 \text{ mole/liter} \approx 0.2 \text{ mole/liter}$$

$$[H_3O^+] = 2 \times 10^{-1}$$

$$pH = 1 - \log 2 = 1.000 - 0.301 = 0.70$$

b. 0.20 mole NaOH

1.8×10^{-5} mole of NaOH is used in neutralizing the HCl present, leaving 0.20 mole/liter − 0.000018 mole/liter ≈ 0.20 mole NaOH

$$[OH^-] = 2 \times 10^{-1} \text{ mole/liter}$$

$$pOH = 1.000 - \log 2 = 1.000 - 0.301 = 0.699$$

$$pH = 14.00 - 0.699 = 13.3$$

Solution B. To 1 liter of 0.80 M HCN with a pH of 4.75 is added:

a. 0.20 mole HCl

The presence of more $[H_3O^+]$ will shift the equilibrium until just a negligible amount of HCN is ionized.

Even without this equilibrium shift, the $[H_3O^+]$ from HCN is negligible.

$$0.20 \text{ mole/liter} + 1.8 \times 10^{-5} \text{ mole/liter} \approx 0.20 \text{ mole/liter}$$

As before the pH is 0.70.

b. 0.20 mole NaOH. The 0.20 mole of NaOH neutralizes part of the acid present.

$$\underset{HCN}{0.80 - 0.20} + \underset{OH^-}{0.20 - 0.20} \rightarrow \overset{0.20}{CN^-} + H_2O$$

The concentrations are:

$[H_3O^+] = X$ mole/liter

$[CN^-] = 0.2 + X \approx 0.2$ mole/liter

$[HCN] = 0.6 - X \approx 0.6$ mole/liter

$$K_a = \frac{X \times 0.2}{0.6} = 4.0 \times 10^{-10}$$

$$[H_3O^+] = X = \frac{0.6}{0.2} \times 4.0 \times 10^{-10} = 1.2 \times 10^{-9} \text{ mole/liter}$$

$pH = 9 - \log 1.2 = 9.000 - 0.079 = 8.92$

Solution C. To 1 liter of a solution containing 1.0 mole HOAc and 1.0 mole NaOAc per liter with a pH of 4.75 is added:

a. 0.2 mole HCl. The HCl reacts with the acetate ion of the sodium acetate and is consumed, yielding 0.20 mole of HOAc.

$$\underset{0.2-0.2}{H_3O^+} + \underset{1.0-0.2}{OAc^-} \rightarrow \underset{1.0+0.2}{HOAc} + H_2O,$$

If $[H_3O^+]$ comes from the acetic acid:

$[H_3O^+] = X$ mole/liter

$[OAc^-] = X + 0.8 \approx 0.8$ mole/liter

$[HOAc] = 1.2 - X \approx 1.2$ moles/liter

$$K_a = \frac{X \times 0.8}{1.2} = 1.8 \times 10^{-5}$$

$$[H_3O^+] = X = \frac{1.2}{0.8} \times 1.8 \times 10^{-5} = 2.7 \times 10^{-5} \text{ mole/liter}$$

$pH = 5 - \log 2.7 = 5.00 - 0.43 = 4.57$

b. 0.20 mole NaOH. The OH^- from sodium hydroxide neutralizes some of the acetic acid present, producing more acetate ions:

$$\underset{1.0-0.2}{HOAc} + \underset{0.2-0.2}{OH^-} \rightarrow \underset{1.0+0.2}{OAc^-} + H_2O$$

If $[H_3O^+]$ comes from the acetic acid:

$[H_3O^+] = X$

$[OAc^-] = 1.2 + X \approx 1.2$ moles/liter

$[HOAc] = 0.8 - X \approx 0.8$ mole/liter

$$K_a = \frac{X \times 1.2}{0.8} = 1.8 \times 10^{-5}$$

$$[H_3O^+] = X = \frac{0.8}{1.2} \times 1.8 \times 10^{-5} = 1.2 \times 10^{-5} \text{ moles/liter}$$

$pH = 5 - \log 1.2 = 5.00 - 0.079 = 4.92$

The resultant effects on the pH of the addition of acid and base to separate portions of the three solutions is summarized in Table 19.6.

TABLE 19.6 pH OF ORIGINAL SOLUTIONS AND pH WITH ACID AND BASE ADDED

	pH		
Original Solution	*Original Solution (1.0 liter)*	*Original* +0.2 *mole* H_3O^+	*Original* +0.2 *mole* OH^-
(A) 1.8×10^{-5} M HCl	4.75	0.70	13.30
(B) 0.8 M HCN	4.75	0.70	8.92
(C) 1 M HOAc, 1 M NaOAc	4.75	0.57	4.92

From the table one can see that the pH of a dilute solution of a strong acid (solution A) varied widely: 0.70 to 4.75 to 13.30. The solution of a very weak acid (solution B) varied widely also: 0.70 to 4.75 to 8.92, although the pH did not rise as much with the addition of base as it did in solution A. Solution C, a solution of a moderately weak acid and its salt, is a buffer solution; the pH changed very little compared to the changes in solution A and solution B: 4.57 to 4.75 to 4.92. The additions of acid and base were relatively large but illustrate the buffer effect.

Acetic acid and the acetate ion are called a buffer-pair. A solution containing large and equal amounts of the two will have a hydronium ion concentration equal to the constant, and will tend to keep the hydronium concentration near that value, because the HOAc concentration and OAc^- concentration control the H_3O^+ concentration.

see how the relative concentrations of the acetate ion and undissociated acetic acid control the acidity of a solution.

$$\frac{[H_3O^+][OAc^-]}{[HOAc]} = 1.8 \times 10^{-5}$$

Observe the equation for the equilibrium constant. When both the acid and the anion concentrations are known, the $[H_3O^+]$ is set. When both are equal they divide out, and $[H_3O^+]$ equals the constant:

$$\frac{[H_3O^+] \times \text{Equal}}{\text{Equal}} = 1.8 \times 10^{-5}$$

When [HOAc] and $[OAc^-]$ are *large* and equal, they tend to stay nearly equal, because the small amount of base added converts some HOAc to OAc^-,

$$\underset{\text{small decrease}}{HOAc} + \underset{\text{small}}{OH^-} \rightarrow \underset{\text{small increase}}{OAc^-} + H_2O$$

and small additions of acid convert some OAc^- to HOAc

$$\underset{\text{small decrease}}{OAc^-} + \underset{\text{small}}{H_3O^+} \rightarrow \underset{\text{small increase}}{HOAc} + H_2O$$

A solution containing any weak acid and its conjugate base in large, equal concentrations is a buffer. **Any weak acid and its conjugate base are a buffer-pair.** Each buffer tends to keep the H_3O^+ concentration near the value of the particular equilibrium constant of the acid. Likewise any solution containing a weak base and its conjugate acid in relative large and equal concentrations is a buffer solution. *The weak base and its conjugate acid are a buffer-pair.* A solution 1 M in NH_3 and 1 M in NH_4Cl tends to keep the $[OH^-]$ constant, the pOH constant and consequently the pH constant, because the ammonia concentration and ammonium ion concentration control the hydroxide ion concentration, as shown in the equilibrium constant equation:

$$\frac{[NH_4^+][OH^-]}{[NH_3]} = 1.8 \times 10^{-5}$$

If they are equal,

$$\frac{\text{Equal} \times [OH^-]}{\text{Equal}} = 1.8 \times 10^{-5}, \text{ and hence}$$

they divide out and the $[OH^-]$ equals the constant. If they are large, they tend to stay nearly equal, for a small addition of base converts a little ammonium ion to ammonia:

$$\underset{\text{small decrease}}{NH_4^+} + \underset{\text{small}}{OH^-} \rightarrow \underset{\text{small increase}}{NH_3} + H_2O$$

And a small addition of acid converts a small amount of ammonia to ammonium ion.

$$\underset{\text{small decrease}}{NH_3} + \underset{\text{small}}{H_3O^+} \rightarrow \underset{\text{small increase}}{NH_4^+} + H_2O$$

Table 19.7 lists a series of buffer-pairs, the constants of the acid or base, and the pH of a solution containing equal concentrations of each of the pairs.

TABLE 19.7 THE pH OF CERTAIN BUFFER PAIRS

Pair	Constant	pH
Acetic acid + acetate ion	$K_a = 1.8 \times 10^{-5}$	4.75
Carbonic acid + carbonate ion	$K_a = 4.2 \times 10^{-7}$	6.38
Hydrogen cyanide + cyanide ion	$K_a = 4.0 \times 10^{-10}$	9.40
Oxalic acid + oxalate ion	$K_a = 3.8 \times 10^{-2}$	1.42
Ammonia + ammonium ion	$K_b = 1.8 \times 10^{-5}$	9.25
Bisulfate + sulfate ion	$K_a = 1.26 \times 10^{-2}$	1.89

At this point, as a conclusion to the discussion of weak acids and bases, certain relationships among the hydronium ion concentration, the hydroxide ion concentration, pH and pOH of certain solutions are summarized in graphic form in Table 19.8.

TABLE 19.8 pH AND pOH

$[H_3O^+]$	pH	Electrolytes		Common Ion	Hydrolysis	pOH	$[OH^-]$
		Strong	*Weak*				
10	−1	10 M HCl				15	10^{-15}
1	0	1 M HCl				14	10^{-14}
0.1	1	0.1 M HCl	1 M HSO_4^-			13	10^{-13}
0.01	2	0.01 M HCl	1 M HOAc 0.1 M HOAc	1 M HSO_4^- + 1 M SO_4^{2-}		12	10^{-12}
0.001	3	0.001 M HCl				11	10^{-11}
10^{-4}	4	1×10^{-4} M HCl				10	10^{-10}
10^{-5}	5	1×10^{-5} M HCl	1 M HCN	1 M HOAc + 1 M OAc^-		9	10^{-9}
10^{-6}	6				1 M NH_4^+	8	10^{-8}
10^{-7}	7		Water		1 M NaCl 1 M SO_4^{2-}	7	10^{-7}
10^{-8}	8				1 M OAc^-	6	10^{-6}
10^{-9}	9	1×10^{-5} M NaOH		1 M NH_3 + 1 M NH_4^+ 1 M HCN + 1 M CN^-		5	10^{-5}
10^{-10}	10	1×10^{-4} M NaOH				4	10^{-4}
10^{-11}	11	0.001 M NaOH	0.1 M NH_3 1 M NH_3		1 M CN^-	3	0.001
10^{-12}	12	0.01 M NaOH				2	0.01
10^{-13}	13	0.1 M NaOH				1	0.1
10^{-14}	14	1 M NaOH				0	1
10^{-15}	15	10 M NaOH				−1	10

19.15 EQUILIBRIUM FORMATION AND DECOMPOSITION OF COMPLEX IONS

Certain metal ions in solution, especially those of the transition metals, tend to be surrounded by groups of molecules or other ions to form large complex ions. Metal ions are hydrated in water solutions: in water the copper(II) ion is $Cu(H_2O)_4^{2+}$; the aluminum ion, $Al(H_2O)_6^{3+}$; and the iron(III) ion, $Fe(H_2O)_6^{3+}$. For simplicity they are written as simply Cu^{2+}, Al^{3+} and Fe^{3+}, but no such bare ions exist in water. Some commonly known complex ions are $Ag(NH_3)_2^+$, the diammine silver ion; $Cu(NH_3)_4^{2+}$, the tetraammine-copper(II) ion; $Fe(CN)_6^{3-}$, the hexacyanoferrate(III) ion; $Ag(S_2O_3)_2^{3-}$, the dithiosulfatoargentate(I) ion; and $FeSCN^{2+}$, the thiocyanato-iron(III) ion. The prefixes di-, tri-, hexa- etc., indicate the number of groups (called ligands) about the metal ion center. The roman numeral indicates the oxidation state of the metal ion center.

the nature of the bonding within these complex ions will be discussed in Chapter 21.

Often metal-centered complex ions impart a characteristic color to a solution; $Cu(NH_3)_4^{2+}$ is an intense blue; $Cu(H_2O)_4^{2+}$, a powder blue; $FeSCN^{2+}$, an intense red; $Fe(CN)_6^{2-}$, an amber. The nature of the bonding within a complex ion will be discussed later, along with a theory for the occurrence of color.

A complex ion may be formed by a reversible and stepwise exchange of ligands about the metal ion center until a stable maximum number is attained. The stable maximum number is characteristic of the metal center, although it can vary somewhat with the nature of the ligand. The formulas listed above indicate by subnumber the stable maximum ligand number, the so-called **coordination number** for the metal ion in the particular complex.

The reversible nature of the addition of ligands can be observed in a solution of copper(II) nitrate and ammonia. Gradual addition of ammonia in excess of that needed for complex formation causes a gradual deepening of the blue cast. Dilution of a tetraamminecopper(II) solution causes the color to fade more rapidly than the volume increases. In any solution with copper ions (Cu^{2+}, actually $Cu(H_2O)_4{}^{2+}$) and ammonia, there are also these ions: $Cu(NH_3)_4{}^{2+}$, $Cu(NH_3)_3(H_2O)^{2+}$, $Cu(NH_3)_2(H_2O)_2{}^{2+}$ and $Cu(NH_3)(H_2O)_3{}^{2+}$. These are related to one another by the equilibria

$$Cu(NH_3)_4{}^{2+} + H_2O \rightleftarrows Cu(NH_3)_3(H_2O)^{2+} + NH_3$$

$$Cu(NH_3)_3H_2O^{2+} + H_2O \rightleftarrows Cu(NH_3)_2(H_2O)_2{}^{2+} + NH_3$$

$$Cu(NH_3)_2(H_2O)_2{}^{2+} + H_2O \rightleftarrows Cu(NH_3)(H_2O)_3{}^{2+} + NH_3$$

$$Cu(NH_3)(H_2O)_3{}^{2+} + H_2O \rightleftarrows Cu(H_2O)_4{}^{2+} + NH_3$$

A great excess of ammonia will shift all four equilibria (to the left as the equations are written) and cause most of the copper ion to appear as the tetraamminecopper(II) ion. With a smaller ammonia concentration all complex ions and copper(II) ions may appear in concentrations of the same order.

The diamminesilver(I) complex may be formed by adding ammonia to a solution of silver nitrate:

$$Ag(H_2O)_2{}^+ + NH_3 \rightleftarrows Ag(NH_3)H_2O^+ + H_2O$$

$$Ag(NH_3)H_2O^+ + NH_3 \rightleftarrows Ag(NH_3)_2{}^+ + H_2O$$

Both of the equilibria represented by the above equations have an equilibrium constant called the **formation constant.**

$$\frac{[AgNH_3{}^+]}{[Ag^+][NH_3{}^+]} = 2.1 \times 10^3$$

and

$$\frac{[Ag(NH_3)_2{}^+]}{[AgNH_3{}^+][NH_3]} = 1.6 \times 10^3$$

The complexes are written without the waters of hydration, which is incorrect but customary. Water being the solvent, its concentration does not appear in the Mass Law Equation. The constants can be used to find the equilibrium concentrations and to determine quantitatively the change in ionic concentrations with addition of common ions and with dilution. But for this course a qualitative understanding will suffice.

19.16 HETEROGENEOUS EQUILIBRIA – THE SOLUBILITY PRODUCT CONSTANT

The mechanism of the dissolving of ionic substances was discussed in Chapter 13. Ions vibrating on the surface of a crystal in water are occasionally displaced sufficiently to be surrounded by water molecules with their strong dipoles, and carried away into solution. If there were no return of ions to the crystal, all salts would be soluble in water until all the water molecules were used in hydration. Ions in the solution occasionally collide with the crystal face with sufficient force to push aside the surrounding water dipoles. If the ions rest at the right places on the lattice "checker-board," they are held and remain. Saturation occurs when the concentration of ions in solution becomes sufficiently great that the rate of return of ions to the crystal or crystals equals the rate of their leaving. There is, then, no net change in the mass of the crystals. The solution is called a saturated solution. The concentration of the solute in the saturated solution is the solubility.

Consider the rate equations for both the dissolving and precipitating (crystallizing) of a salt, silver chloride. Figure 19.5 illustrates the two rates. The rate at which a solid dissolves is dependent only upon the temperature, which is constant (room temperature), and on the surface area of the crystal or crystals. That is,

$$\text{Rate}\ (\rightarrow) = k_1 \times \text{area}$$

The rate at which ions return to the crystal (precipitation) is proportional to the surface area (the greater the surface area, the greater the number of collisions with the crystal), to the silver ion concentration, and to the chloride ion concentration:

$$\text{Rate}\ (\leftarrow) = k_2 \times \text{area} \times [Ag^+][Cl^-]$$

Unless the solution is saturated, the rates are unequal. At saturation the ionic concentrations will have adjusted themselves so that a dynamic equilibrium exists and

$$\text{Rate}\ (\rightarrow) = \text{Rate}\ (\leftarrow)$$

$$k_1 \times \text{area} = k_2 \times \text{area} \times [Ag^+][Cl^-]$$

The crystal area appears in both rate expressions (because a greater area for leaving means a larger target to return to), and may be divided into the equation:

$$k_1 = k_2[Ag^+][Cl^-]$$

$$K_{sp} = \frac{k_1}{k_2} = [Ag^+][Cl^-]$$

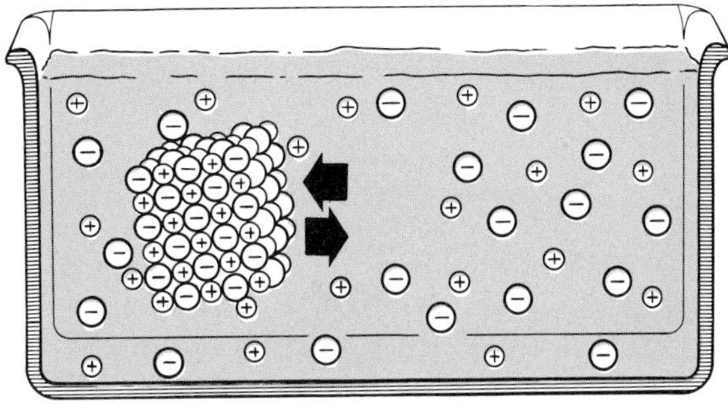

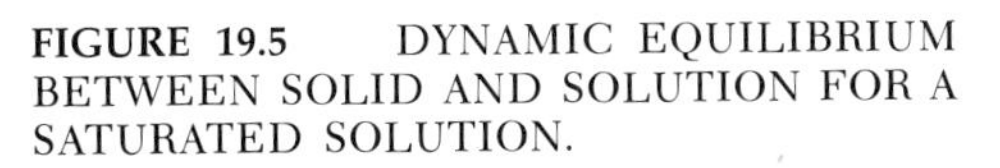

FIGURE 19.5 DYNAMIC EQUILIBRIUM BETWEEN SOLID AND SOLUTION FOR A SATURATED SOLUTION.

K_{sp} is the solubility product constant, the product of the solubilities (concentrations of the ions at saturation):

$$K_{sp} = [Ag^+][Cl^-]$$

This expression indicates that the solubility of a salt depends only on the concentration of its ions in solution. Further, it indicates that one can decrease the concentration of one ion in solution by adding the other. Dilution reduces *both* ionic concentrations and therefore causes more salt to dissolve.

A variation appears in the solubility product expression when the positive and negative ions are not in one-to-one proportion in the crystal. To illustrate, the K_{sp} for lead chloride is thus derived:

$$PbCl_2 \rightleftarrows Pb^{2+} + 2\ Cl^-$$

The rate of dissolving of lead chloride is proportional to the area of the crystal or crystals at a set temperature and is not a function of any ion concentrations.

$$\text{Rate} (\rightarrow) = k_1 \times \text{area}$$

The rate of crystallization of the salt is proportional to the area, the concentration of lead ion and the concentration of chloride ion squared, because 2 chloride ions must precipitate with every lead ion:

$$\text{Rate} (\leftarrow) = k_2 \times \text{area} \times [Pb^{2+}][Cl^-]^2$$

At equilibrium (saturation):

$$\text{Rate} (\rightarrow) = \text{Rate} (\leftarrow)$$

$$k_1 \times \text{area} = k_2 \times \text{area} \times [Pb^{2+}][Cl^-]^2$$

and

$$K_{sp} = \frac{k_1}{k_2} = [Pb^{2+}][Cl^-]^2$$

in contrast to the equilibrium expression for the ionization of a weak acid, the solubility product constant expression has no denominator.

In the Mass Law Equation for equilibrium, each substance with a variable concentration appears to the power of its coefficient in the equation for dissolving, the products customarily being put in the numerator and the reactants in the denominator. The concentration of lead chloride ($PbCl_2$) as such is very small; it is constant as long as any solid is present, and it does not appear in the expression.

All that can be said about the derivation of a solubility product constant is this: If the rate of dissolving and the rate of crystallization are both proportional to the area of the crystals present at equilibrium, and if the rate of precipitation is proportional to the concentrations of the ions raised to the power of the coefficients in the balanced equation, the products of the solubilities of the ions of a salt raised to the power of the coefficients must equal a number which does not change unless the temperature is changed. The K_{sp} equation cannot be used quantitatively unless the value of the constant is known. This can be determined by using a set of equilibrium concentration values.

The solubility of silver chloride in pure water is 1.67×10^{-5} mole per liter at room temperature. That is, solid silver chloride does not dissolve in a solution containing each of its ions at a concentration above 1.67×10^{-5} mole per liter.

Using these equilibrium values,

$$[Ag^+] = 1.67 \times 10^{-5} \text{ mole/liter}$$

$$[Cl^-] = 1.67 \times 10^{-5} \text{ mole/liter}$$

$$K_{sp} = [Ag^+][Cl^-] = (1.67 \times 10^{-5})^2 = 2.8 \times 10^{-10}$$

1.59×10^{-2} mole of lead chloride may be dissolved in 1 liter of pure water at room temperature. At that concentration the solution is saturated. Because the lead chloride liberates its ions upon dissolving according to this equation:

$$PbCl_2(s) \rightarrow Pb^{2+} + 2\ Cl^-$$

dissolving lead chloride yields twice as many chloride ions as lead ions according to the subscripts in the formula.

then $[Pb^{2+}] = 1.59 \times 10^{-2}$ mole/liter

and $[Cl^-] = 2 \times 1.59 \times 10^{-2} \text{ mole/liter} = 3.18 \times 10^{-2}$ mole/liter

Substituting into the K_{sp} equation,

$$K_{sp} = [Pb^{2+}][Cl^-]^2 = (1.59 \times 10^{-2})(3.18 \times 10^{-2})^2 = 1.6 \times 10^{-5}$$

Once the K_{sp} is known, one can carry out the reverse of the calculations in which the K_{sp} is determined from the solubility; that is, can calculate the *solubility*.

Example 13. Find the solubility of barium sulfate ($BaSO_4$) in pure water. The K_{sp} of barium sulfate is 1.0×10^{-10}.

Solution: At saturation:

$$[Ba^{2+}][SO_4^{2-}] = K_{sp} = 1.0 \times 10^{-10}$$

Let X equal the moles of barium sulfate dissolved per liter of solution. Barium sulfate dissociates according to this equation:

$$BaSO_{4(s)} \rightarrow Ba^{2+} + SO_4^{2-}$$

$$[Ba^{2+}] = [SO_4^{2-}] = X$$

$$X^2 = 1.0 \times 10^{-10}$$

$$X = 1.0 \times 10^{-5} \text{ mole of } BaSO_4\text{/liter}$$

If the K_{sp} is known, one can find the solubility of a salt in a solution containing a *common ion*.

Example 14. How many moles of $BaSO_4$ will dissolve in 1 liter of 2 M Na_2SO_4?

$$K_{sp} = 1.0 \times 10^{-10}$$

Solution: Let X = moles of $BaSO_4$ to dissolve per liter.

Then $[Ba^{2+}] = X$

$[SO_4^{2-}] = 2 + X \approx 2$

$K_{sp} = [Ba^{2+}][SO_4^{2-}] = X \times 2 = 1.0 \times 10^{-10}$

$X = 5 \times 10^{-11}$. The solubility of $BaSO_4$ in the solution is 5×10^{-11} mole/liter.

Compare the answers in Examples 13 and 14. Notice how much the presence of the sulfate ion has suppressed the solubility of barium sulfate. Table 19.9 gives the solubility product constants for several common salts at room temperature.

TABLE 19.9 SOLUBILITY PRODUCT CONSTANTS AT ROOM TEMPERATURE

Substance	Equilibrium	Solubility Product Constant
Silver bromide	$AgBr_{(s)} = Ag^+ + Br^-$	3.3×10^{-13}
Calcium carbonate	$CaCO_{3(s)} = Ca^{2+} + CO_3^{2-}$	6.9×10^{-9}
Lead(II) chloride	$PbCl_{2(s)} = Pb^{2+} + 2\ Cl^-$	1.6×10^{-5}
Mercury(I) chloride	$Hg_2Cl_{2(s)} = Hg_2^{2+} + 2\ Cl^-$	1.1×10^{-18}
Silver chloride	$AgCl_{(s)} = Ag^+ + Cl^-$	2.8×10^{-10}
Lead chromate	$PbCrO_{4(s)} = Pb^{2+} + CrO_4^{2-}$	2.0×10^{-16}
Silver chromate	$Ag_2CrO_{4(s)} = 2\ Ag^+ + CrO_4^{2-}$	1.9×10^{-12}
Calcium fluoride	$CaF_{2(s)} = Ca^{2+} + 2\ F^-$	3.6×10^{-11}
Aluminum hydroxide	$Al(OH)_{3(s)} = Al^{3+} + 3\ OH^-$	5×10^{-33}
Copper(II) hydroxide	$Cu(OH)_{2(s)} = Cu^{2+} + 2\ OH^-$	1.6×10^{-19}
Iron(III) hydroxide	$Fe(OH)_{3(s)} = Fe^{3+} + 3\ OH^-$	6×10^{-38}
Silver iodide	$AgI_{(s)} = Ag^+ + I^-$	8.5×10^{-17}
Calcium oxalate	$CaC_2O_{4(s)} = Ca^{2+} + C_2O_4^{2-}$	1.3×10^{-9}
Barium sulfate	$BaSO_{4(s)} = Ba^{2+} + SO_4^{2-}$	1.0×10^{-10}
Lead sulfate	$PbSO_{4(s)} = Pb^{2+} + SO_4^{2-}$	1.3×10^{-8}
Copper(II) sulfide	$CuS_{(s)} = Cu^{2+} + S^{2-}$	4×10^{-36}
Iron(II) sulfide	$FeS_{(s)} = Fe^{2+} + S^{2-}$	4×10^{-17}
Lead sulfide	$PbS_{(s)} = Pb^{2+} + S^{2-}$	4×10^{-26}
Mercury(I) sulfide	$Hg_2S_{(s)} = Hg_2^{2+} + S^{2-}$	1×10^{-45}
Silver sulfide	$Ag_2S_{(s)} = 2\ Ag^+ + S^{2-}$	1×10^{-50}

19.17 REPLACEMENT EQUILIBRIA

When a pure copper wire is placed in 3 M $AgNO_3$, silver precipitates on the copper and copper dissolves in the solution:

$$Cu + 2\ Ag^+ \rightarrow 2\ Ag + Cu^{2+}$$

However, not all the silver is taken from the solution; the reaction is not quite completed. When the concentration of silver ion has been reduced to 2.4×10^{-8} molar, the concentration of copper nitrate having risen to almost 1.5 molar, the reaction stops. Equilibrium has been attained.

The equilibrium constant expression for the equation (as long as some pure copper and pure silver are in contact with the solution) is:

$$K = \frac{[Cu^{2+}]}{[Ag^+]^2} = \frac{1.5}{(2.4 \times 10^{-8})^2} = 2.6 \times 10^{15}$$

A cell may be constructed as shown in Figure 19.6. A copper electrode is placed in water containing a trace of $Cu(NO_3)_2$. The solution is separated by a porous plate from 3 M $AgNO_3$ solution, in which a silver electrode is dipped. A relatively high voltage electric current flows when the electrodes are connected through the voltmeter. The voltage decreases slowly as the reaction proceeds, raising the copper(II) ion concentration and reducing the silver ion concentration. When the concentration of copper(II) ion reaches 1 molar and the silver ion concentration has decreased to 1 molar, the voltage of the cell is 0.46 volt. The value found by subtracting the "voltage" values in the electromotive series (Table 14.2) is: $-0.34 - (-0.80) = 0.46$ volt. As the reaction proceeds and the cell "runs down," the voltage decreases to zero at a silver concentration of 2.4×10^{-8} moles per liter. The system is at equilibrium; no force exists to change the concentrations.

The potential of 0.46 volts for the silver-copper cell when both silver ion and copper(II) ion are 1 molar is a measure of the "unequilibrium." It is the reverse voltage that must be imposed upon the cell to keep the reaction from occurring.

When chlorine is bubbled into a 2 molar sodium bromide solution, chlorine dissolves, replacing bromide ions. Bromine collects on the bottom of the container.

$$Cl_2 + 2\ Br^- \rightarrow 2\ Cl^- + Br_2$$

The bromide ion is not completely replaced by the chlorine. If the

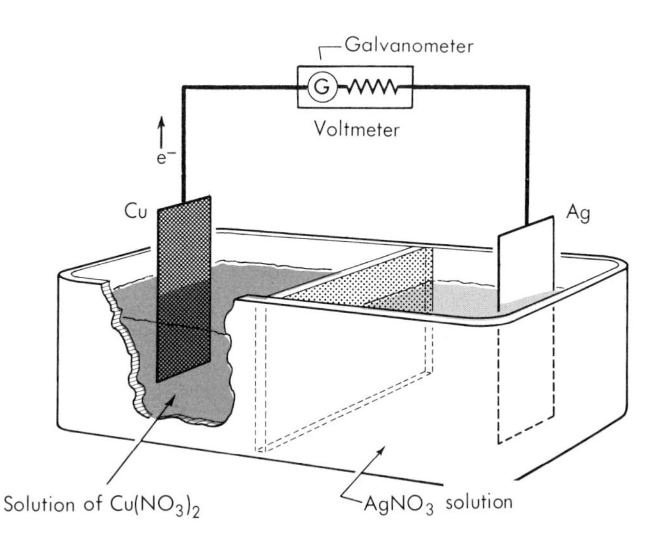

FIGURE 19.6 Cu-Ag PRIMARY CELL.

atmospheric pressure (and therefore the pressure of the chlorine) is 1 atmosphere, the reaction stops when the concentration of bromide ion is reduced to 1.7×10^{-5} mole per liter and the concentration of chloride ion is nearly 2 molar.

The expression for the equilibrium constant may be written in this fashion:

$$K_{eq} = \frac{[Cl^-]^2}{P_{Cl_2}[Br^-]^2} = \frac{(2\ \text{moles/liter})^2}{1\ \text{atm}(1.7 \times 10^{-5}\ \text{mole/liter})^2} = 1.37 \times 10^{10}$$

This reaction may be carried out in a cell. Chlorine at 1 atmosphere pressure is bubbled over a platinum electrode in 1 M NaCl solution, separated by a porous plate from a 1 M NaBr solution with a drop of bromine liquid in the bottom and a platinum electrode in it. Current flows when the electrodes are connected; chlorine dissolves and the bromide ion is oxidized to bromine. The cell voltage is 0.30 and is a measure of the "unequilibrium." If the cell solutions were changed to 1.7×10^{-5} M NaBr and 2 M NaCl, no current would flow. In fact, if the electrodes are left connected, the reaction will proceed until these concentrations are reached.

In summary, displacement (oxidation-reduction) reactions are often reversible reactions. The cell voltage obtained from voltages in the electromotive force series are the measured voltages when all concentrations of participating ions are 1 molar. The voltage is a measure of the "unequilibrium" of the system.

19.18 COMPLEX EQUILIBRIA

It might be thought that the equilibria already considered are sufficiently complex. As a matter of fact, many of the equilibria may be much more complex than the equilibrium expressions indicate. In this section processes will be discussed in which more than one reversible reaction is involved.

A. In many qualitative analysis schemes the silver ion (along with the lead ion and the mercury(I) ion, Hg_2^{2+}) is separated from other ions by the addition of hydrochloric acid. The silver ion precipitates as the chloride:

$$Ag^+ + Cl^- \rightarrow AgCl$$

The lead and mercury(I) ions do likewise:

$$Pb^{2+} + 2\ Cl^- \rightarrow PbCl_2$$

$$Hg_2^{2+} + 2\ Cl^- \rightarrow Hg_2Cl_2$$

Lead chloride may be separated from the filtered precipitate of chlorides by the addition of hot water. (The solubility of lead chloride increases markedly with temperature.) Finally, silver chloride is separated from mercury(I) chloride by adding ammonia solution, in which it is dissolved by formation of a complex ion:

$$AgCl_{(s)} + NH_3 \rightleftarrows AgNH_3^+ + Cl^-$$

$$AgNH_3^+ + NH_3 \rightleftarrows Ag(NH_3)_2^+$$

(The water molecule is not shown in the complex ion $Ag(NH_3)^+$, which is more properly written $Ag(NH_3)(H_2O^+)$. Before ammonia was added, an equilibrium existed between solid AgCl and the ions in solution:

$$AgCl_{(s)} \rightleftarrows Ag^+ + Cl^-$$

Because the K_{sp} of AgCl is small (2.8×10^{-10}), very low concentrations of Ag^+ and Cl^- ions remain in solution at equilibrium. However, when ammonia is added, the concentration of silver ion is greater than can exist in equilibrium with silver ammonia complex ion:

$$AgNH_3^+ \rightleftarrows Ag^+ + NH_3$$

Ammonia ties up the silver ion and decreases the concentration to its equilibrium value. Silver chloride dissolves in an attempt to restore the silver ion concentration. If there is sufficient ammonia, and if the chloride ion concentration does not become too great, all the AgCl dissolves:

$$\begin{array}{lll} & NH_3 & \\ & + & \\ AgCl_{(s)} \rightarrow & Ag^+ + Cl^- & \\ & \downarrow & \\ & AgNH_3^+ \xrightarrow{NH_3} & Ag(NH_3)_2^+ \end{array}$$

One can restore the AgCl precipitate in two ways: by increasing the chloride ion concentration or by decreasing the ammonia concentration. Both are accomplished by adding hydrochloric acid.

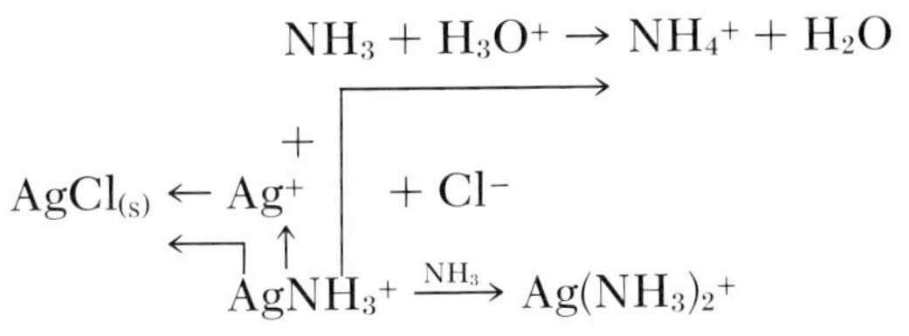

$$\begin{array}{lll} & NH_3 + H_3O^+ \rightarrow NH_4^+ + H_2O & \\ & + & \\ AgCl_{(s)} \leftarrow & Ag^+ \quad + Cl^- & \\ & \uparrow & \\ & AgNH_3^+ \xrightarrow{NH_3} & Ag(NH_3)_2^+ \end{array}$$

B. The separation of copper(II) and iron(II) ions depends on the fact that, while both sulfides, CuS and FeS, will precipitate when H_2S is added to a basic solution containing the two ions Fe^{2+} and Cu^{2+}, only copper(II) sulfide precipitates from an acid solution.

Both sulfides are very insoluble: CuS is less soluble. The K_{sp} for FeS is 4×10^{-17}; for CuS, 4×10^{-36}. In a 1 M Cu^{2+} solution the concentration of sulfide ion cannot exceed 4×10^{-36} mole per liter. In 1 M Fe^{2+} the concentration of sulfide ion cannot exceed 4×10^{-17} mole per liter. A basic solution of hydrogen sulfide precipitates both ions, because it is no longer a solution of hydrogen sulfide but a solution of sulfide ions:

$$H_2S + 2\ OH^- \rightarrow 2\ H_2O + S^{2-}$$

Hydrogen sulfide in solution might contribute sulfide ions for precipitation. It is a weak electrolyte and is partially ionized in two steps.

$$H_2S + H_2O \rightleftarrows H_2O^+ + HS^-$$

$$HS^- + H_2O \rightleftarrows H_3O^+ + S^{2-}$$

hydrogen sulfide is a diprotic acid.

The first step occurs to a slight extent; the second to a much lesser degree, giving a sulfide ion concentration of 1×10^{-13} mole per liter in a neutral solution saturated with H_2S. As low as the sulfide concentration is, it is much greater than needed to precipitate either CuS or FeS from 1 molar solutions. In a 3 M HCl solution, a solution 3 molar in H_3O^+, the two ionization reactions of H_2S are forced sufficiently to the left so that the sulfide ion concentration drops below the 4×10^{-17} mole per liter needed to precipitate FeS, and CuS is precipitated alone.

C. The precipitation of silver ion on copper wire from a 1 M $AgNO_3$ solution may be stopped by adding ammonia solution. When the concentration of ammonia is above 2 molar, the ammonia complex ties up the silver ions, reducing the concentration in solution below that needed for replacement with copper:

$$Ag(NH_3)_2{}^+ \rightleftarrows Ag^+ + 2\ NH_3$$

This is shown graphically in the "crossed" equation below:

$$\begin{array}{ccc} & 2\ NH_3 & \\ & + & \\ Cu + & 2\ Ag^+ & \rightleftarrows Cu^{2+} + 2\ Ag \\ & \downarrow & \\ & 2\ AgNH_3{}^+ & \end{array}$$

EXERCISES

19.1 Define: (a) homogeneous, (b) strong electrolyte, (c) equilibrium, (d) pH.

19.2 How might one show a solution to be a dilute solution of a strong electrolyte and not a more concentrated solution of a weak electrolyte?

19.3 What is the pH of: (a) water, (b) 0.001 M HCl, (c) 2.4×10^{-3} M HNO_3?

19.4 A solution of a pure substance in water conducts an electric current. A solution of the same substance of twice the concentration is not twice as good a conductor as the first. Is the solution that of a strong electrolyte? Explain.

19.5 Derive the equilibrium constant expression for the ionization of propionic acid (HOPr).

19.6 Why are there no terms in the denominator of the expression for K_w?

19.7 Why is a solution of sodium cyanide not neutral?

19.8 Which of these substances in solution is weakly basic, but is a good electrolytic conductor: (a) 1×10^{-5} M HCl, (b) 1 M NH_3, (c) NH_4Cl, (d) NaOAc?

19.9 Derive the equilibrium constant expression for the ionization of the hydrogen sulfate ion (HSO_4^-).

19.10 Define hydrolysis. Write a hydrolysis reaction.

19.11 Define buffer solution. Show why a solution containing 1 mole of formic acid and 1 mole of sodium formate per liter has a pH of 3.68.

19.12 Write the decomposition equation for $Cu(NH_3)_4^{2+}$.

19.13 Derive the K_{sp} expression for the slightly soluble salt, silver chromate (Ag_2CrO_4).

19.14 Why does the voltage of a cell decrease in value as it discharges?

19.15 Addition of which of the following will cause more silver chloride to dissolve in a saturated solution: (a) water, (b) 1 M HCl, (c) 1 M silver nitrate, (d) 6 M NH_3, (e) a zinc wire?

19.16 Explain why H_2S gas bubbled into a solution containing 1 mole of $CuCl_2$, 1 mole of $FeCl_2$ and 3 moles of HCl per liter will precipitate CuS and no FeS.

PROBLEMS

19.17 Calculate the ionization constant for germanic acid (represented as HGer). A 0.60 M solution has a hydronium concentration of 1.34×10^{-5}.

19.18 Calculate the ionization constant for nitrous acid (HNO_2), which in a 0.050 M solution has a hydronium ion concentration of 4.7×10^{-4} mole per liter.

19.19 Calculate K_a for hyponitrous acid ($H_2N_2O_2$). A 0.30 M solution has a hydronium ion concentration of 1.7×10^{-4} mole per liter.

19.20
a. Write the ionic equation for the first ionization of silicic acid (H_4SiO_4).
b. Write the equilibrium constant expression.
c. Calculate the ionization constant from the fact that a 0.20 M solution has a $[H_3O^+]$ of 2.0×10^{-3}.

19.21 Calculate the ionization constant of NH_3. A liter of solution containing 0.10 mole of NH_3 and 0.10 mole of NH_4Cl has a hydroxide ion concentration of 1.8×10^{-5}.

19.22 Calculate the $[H_3O^+]$ of a 0.20 M HBrO solution. $K_a = 2 \times 10^{-9}$.

19.23 Calculate the $[H_3O^+]$ of a solution containing 2.0 moles of HNO_2 and 0.40 mole of $NaNO_2$ per liter. $K_a = 4.5 \times 10^{-4}$.

19.24 Calculate the hydronium ion concentration in moles per liter in a solution containing 0.10 mole of propionic acid (HOPr) and 0.060 mole of sodium propionate (NaOPr) per liter. $K_a = 1.4 \times 10^{-5}$.

19.25 Calculate the $[OH^-]$ of 5.5 M NH_3.

19.26 Calculate the $[H_3O^+]$ of 4.0 M HCN. $K_a = 4 \times 10^{-10}$.

19.27 Calculate the hydronium ion concentration of 3.2 M HOAc.

19.28 For a 2.0 M NH_3 solution, calculate: (a) $[OH^-]$, (b) pOH, (c) $[H_3O^+]$, (d) pH.

19.29 Calculate the $[OH^-]$ of a solution containing 0.20 mole of NH_3 and 0.40 mole of NH_4Cl per liter.

19.30 Calculate the pH of all solutions described in Problems 19.17 to 19.27.

19.31 A saturated solution of silver bromide has a concentration of 5.75 $\times 10^{-7}$ mole per liter. Calculate the solubility product constant.

19.32 A saturated solution of silver iodide is 9.2×10^{-9} M. Calculate the K_{sp}.

19.33 Calculate K_{sp} for calcium fluoride. Calcium fluoride has a solubility of 2.04×10^{-4} mole per liter.

19.34 Calculate the K_{sp} for Ag_2S, knowing that it has a solubility of 1.58 $\times 10^{-24}$ mole per liter in a 1.0×10^{-3} molar solution of sodium sulfide.

19.35 Calculate the K_{sp} for Ag_2S. The solubility of Ag_2S is 1.35×10^{-17} mole per liter.

19.36 Calculate the solubility of $BaSO_4$ in a 2 M solution of $BaCl_2$. $K_{sp} = 1 \times 10^{-10}$.

19.37 Calculate the solubility of $BaSO_4$ in moles per liter.

19.38 Calculate the concentration of sulfide ion necessary to just start precipitation of FeS from a 0.10 M solution of $FeCl_2$. K_{sp} of FeS = 4×10^{-17}.

19.39 The $[H_3O^+]$ of 0.050 M HCN is 4.5×10^{-6}. Calculate the equilibrium constant, K_a, for HCN.

19.40 Calculate K_a for HNO_2. A 0.050 M solution has a $[H_3O^+]$ of 4.7×10^{-3} (0.0047) mole per liter.

19.41 Calculate the ionization constant, K_b, of NH_3 if a 0.10 M solution has an $[OH^-]$ of 1.34×10^{-3} mole per liter.

19.42 Calculate the $[H_3O^+]$ of 0.010 M in HCN. K_a for HCN is 4.0×10^{-10}.

19.43 Calculate the $[H_3O^+]$ of a solution 0.10 M in HCN and 0.10 M in NaCN.

19.44 For a solution containing 2.4 moles of NH_4Cl and 1.2 moles NH_3 per liter calculate: (a) $[OH^-]$, (b) pOH, (c) $[H_3O^+]$, (d) pH.

SUGGESTED READING

Lee, G. L.: *Principles of Chemistry—A Structural Approach,* Chapter 16. International Textbook, Scranton, Pa., 1970.

Sienko, M. and Plane, R.: *Chemistry.* 3rd Ed., Chapter 17, McGraw-Hill Book Co., New York, 1966.

TWENTY • METALLURGY

20.1 INTRODUCTION

Metals were defined in Chapter 5 as those elements that tend to lose electrons in chemical reactions. They are shown in Figure 20.1 to occupy 80 per cent of the periodic table. The general properties of metals were discussed in Chapter 11, which is concerned with the group of metals known as the "active metals." The physical properties of **hardness, ductility, malleability, tensile strength, conductivity**, etc. discussed in that chapter apply more generally to the metals considered here than to the active metals.

This chapter discusses the industrial methods of obtaining and treating some common metals. Metals generally occur in nature as **minerals**, which are naturally occurring metal-containing compounds. These compounds are mixed with rock and often buried beneath the earth's surface. An **ore** is a mineral deposit of sufficient mineral content and of such a physical state that its metal can be economically extracted.

The main topics to be discussed here are: (1) the mining of the ore, (2) the separation of inert rock material from the ore, (3) the reduction of the compound to yield the free metal, (4) the refining, or purification, of the metal and (5) the addition of other metals or substances to prepare alloys, with emphasis on the chemistry involved in the reduction and refining processes.

These five metallurgical steps are accomplished in a variety of ways, and the order of the steps may be changed in the processing of different metals. In the processing of some metal ores, the steps occur simultaneously. In other processes, each step is composed of several smaller steps. Variations in the process are fully as numerous as the number of metals, hence it is impossible to describe the metallurgy of all the common metals. As a compromise, the metallurgy of three metals will be discussed quite thoroughly. These metals—aluminum, copper and iron—are chosen because they present three important different metallurgical schemes, and they are important individual metals.

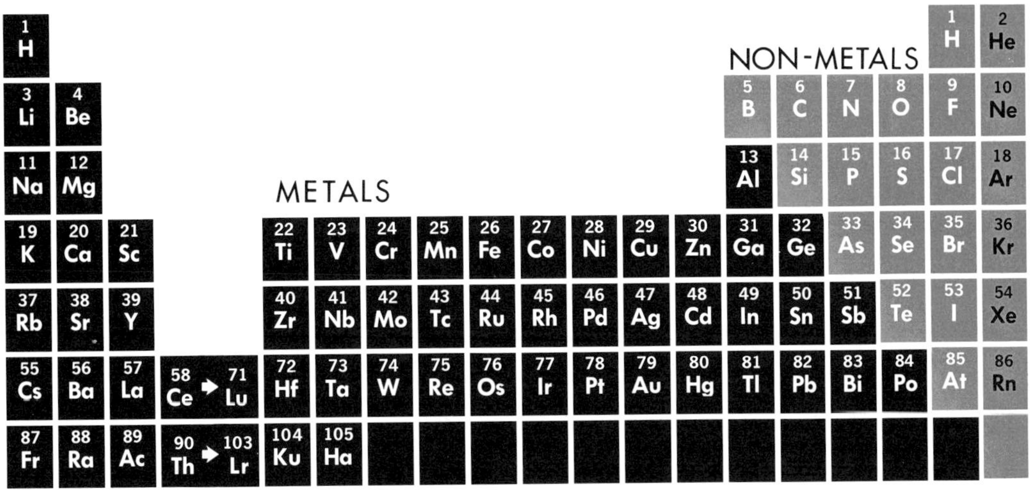

NON-METALS

METALS

1 H																	1 H	2 He
3 Li	4 Be												5 B	6 C	7 N	8 O	9 F	10 Ne
11 Na	12 Mg												13 Al	14 Si	15 P	16 S	17 Cl	18 Ar
19 K	20 Ca	21 Sc		22 Ti	23 V	24 Cr	25 Mn	26 Fe	27 Co	28 Ni	29 Cu	30 Zn	31 Ga	32 Ge	33 As	34 Se	35 Br	36 Kr
37 Rb	38 Sr	39 Y		40 Zr	41 Nb	42 Mo	43 Tc	44 Ru	45 Rh	46 Pd	47 Ag	48 Cd	49 In	50 Sn	51 Sb	52 Te	53 I	54 Xe
55 Cs	56 Ba	57 La	58 Ce → 71 Lu	72 Hf	73 Ta	74 W	75 Re	76 Os	77 Ir	78 Pt	79 Au	80 Hg	81 Tl	82 Pb	83 Bi	84 Po	85 At	86 Rn
87 Fr	88 Ra	89 Ac	90 Th → 103 Lr	104 Ku	105 Ha													

FIGURE 20.1 THE METALS.

20.2 ALUMINUM

Aluminum is a common metal now, but 100 years ago it was more costly than gold, although aluminum is the most abundant metal in the earth's crust. Despite its abundance, there would be little use for aluminum had not an economical process been developed for its recovery. Such a process, in which anhydrous aluminum oxide is electrolyzed in solution in a fused salt (Step 4 below), was developed independently in 1866 by an American, Charles M. Hall, and a Frenchman, P. T. Heroult. Today the production of aluminum is second only to the production of iron.

Aluminum occurs in many clays and rocks, but the sole source of commercial aluminum is the mineral **bauxite,** found in Georgia, Alabama, Tennessee, and Arkansas and in Guyana and Dutch Guiana. Bauxite is almost entirely hydrated aluminum oxide ($Al_2O_3 \cdot XH_2O$ where X is greater than 1 and less than 3) which contains impurities of iron oxide, aluminum silicates and titanium dioxide.

The extraction and purification of aluminum follow the steps outlined in the flow sheet (Figure 20.2):

(1) The bauxite ore is first crushed and ground to a very fine powder. It is then treated with a hot solution (about 169°C, under pressure) of sodium hydroxide made by adding sodium carbonate to calcium hydroxide. In solution, sodium hydroxide is formed and insoluble calcium carbonate precipitates:

$$Na_2CO_3 + Ca(OH)_2 \rightarrow CaCO_3 + 2\ NaOH$$

The basic solution dissolves the hydrated alumina from the inert material and other impurities:

$$Al_2O_3 \cdot 3\ H_2O + 2\ OH^- \rightarrow 2\ Al(OH)_4^-$$

(Here bauxite is written as the trihydrate.) — aluminate ion

(2) The clear solution containing sodium and aluminate ions

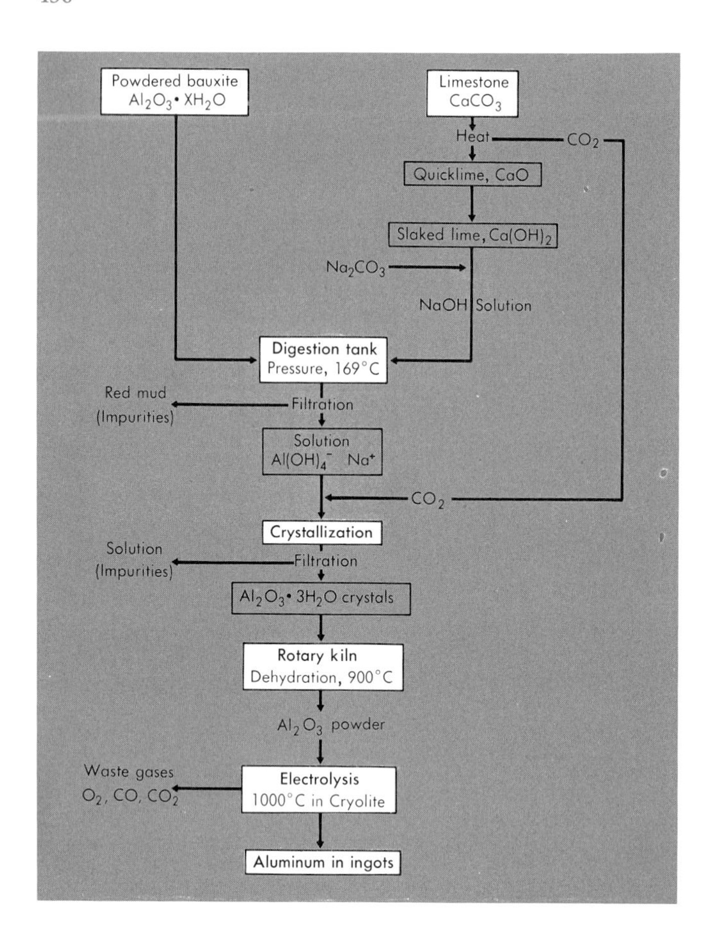

FIGURE 20.2 FLOW SHEET FOR THE EXTRACTION AND PURIFICATION OF ALUMINUM.

is filtered from a "red mud" containing the impurities, and is allowed to cool in tanks into which carbon dioxide is bubbled. The carbon dioxide converts the aluminate ion back to Al_2O_3:

$$2\ Al(OH)_4^- + 2\ CO_2 \rightarrow 2\ HCO_3^- + Al_2O_3 + 3\ H_2O$$

Lowering the temperature decreases the solubility of alumina greatly, and the trihydrate ($Al_2O_3 \cdot 3\ H_2O$) precipitates as large translucent crystals.

(3) The crystals of hydrated alumina are washed, filtered and then dehydrated as they pass through a large rotary kiln at 1000°C.

(4) The white anhydrous powder is fed continuously into molten cryolite (Na_3AlF_6) at 1000°C, where it is electrolyzed. The electrolytic cell is a huge, carbon-lined steel container (about 18′ × 8′ × 3′) (Fig. 20.3).

The bottom lining is the cathode; aluminum is reduced at the cathode and collects as a molten pool to be drawn off periodically. Oxygen gas is liberated at the graphite anodes where it reacts with the graphite, gradually consuming it to form carbon monoxide. The high temperature of the cell is maintained by the electrical resistance of the melt. Aluminum from the Hall process has a purity of about 99 per cent. A great amount of electrical energy is required for this process. Each cell operates at about 50,000 amperes and at 6 volts. Aluminum plants must be located near cheap sources of electricity. If aluminum of higher purity is desired, it may be refined electrolytically.

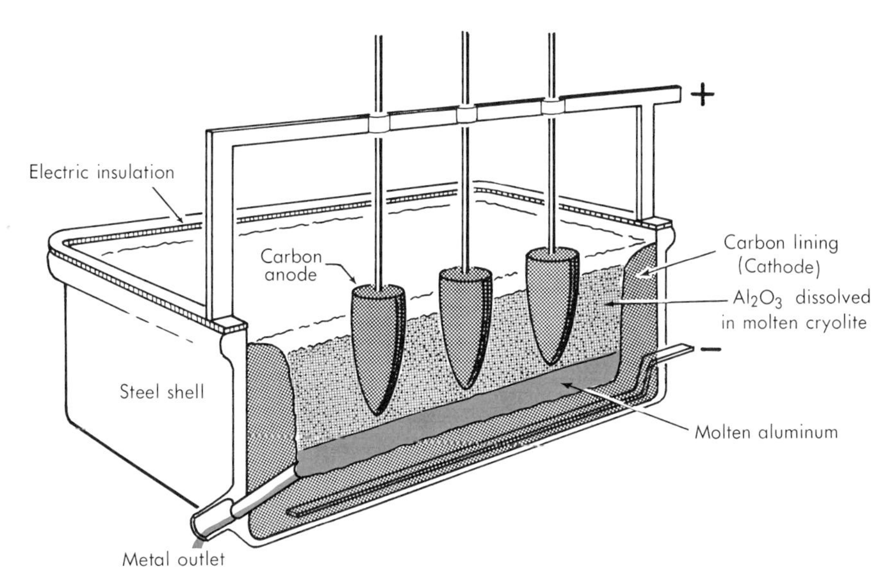

FIGURE 20.3 THE ELECTROLYSIS OF ALUMINUM OXIDE IN MOLTEN CRYOLITE.

Aluminum is a fairly active metal, as shown by its position in the electromotive force series. Powdered aluminum will explode when blown in air and ignited. Impure aluminum becomes pitted when in contact with the atmosphere. Very pure aluminum is sufficiently resistant to corrosion that hulls of ocean-going vessels are made of aluminum covered with a coat of the very pure, electrolytically refined metal. Aluminum's remarkable resistance to oxidation is deceiving. Actually the silver-white surface of newly cut aluminum "tarnishes" almost immediately on contact with air to form a thin, almost transparent oxide film:

$$4\ Al + 3\ O_2 \rightarrow 2\ Al_2O_3$$

pure aluminum is resistant to corrosion.

The film adheres tightly to the aluminum surface; no "buckling" occurs because the increase in weight is nearly compensated by an increase in density of the oxide. The unbroken film prevents further oxidation. The higher the purity of the aluminum surface, the more closely the film adheres and the more resistant the metal is to corrosion. If the metal surface is altered so that the oxide does not adhere, oxidation occurs. If mercury is rubbed on an aluminum surface to form a thin amalgam layer, the oxide does not adhere, and corrosion occurs rapidly.

The use of aluminum depends primarily upon its low density (2.70 g/cm^3). Only magnesium of the structural metals is less dense. The soft, pure metal is not easily machined, but its alloys, made with small amounts of iron, magnesium, zinc, copper, silicon or manganese, are harder and can be machined. The tough alloys are used extensively in aircraft engines and bodies. Aluminum has a high heat conductivity and is therefore used in cooking ware. The metal has good electrical conductivity. A copper wire has less electrical resistance than an aluminum wire of the same diameter, but an aluminum wire with the same weight per 100 feet is a better conductor than copper wire. Aluminum cable with a steel core for greater tensile strength is often used for electrical transmission lines. Aluminum roofing and so-called "tinfoil" candy wrappers are two of many other uses.

Aluminum, because of its resistance to corrosion and because of the "buy-throw away" and "do not return" trends established for the convenience of the consumer, **is a pollution problem.** Discarded aluminum beer and soft drink cans do not rust rapidly like tin cans, which are really tin-coated iron, so they never disappear and are left to clutter the landscape. Some aluminum companies are paying for empties, which they reprocess at a loss in attempts to alleviate the problem. Unfortunately, small scraps such as the "flip-top" openers for soft drink, beer and juice cans are too small to be gathered easily and remain on lawns and park premises forever unless time is taken to gather them. Eventually Americans must realize that natural resources are not limitless. Then there will be fewer throw-aways and non-returnables, which will serve not only to keep the landscape clean but also to conserve our natural resources.

20.3 COPPER

The name "copper" is taken from the Latin word *cuprum,* which refers to the production of the metal on the island of Cyprus by the Romans. Copper has been produced from before the time of Christ. This is due in part to the fact that copper is of sufficiently low reactivity to be found free or uncombined in many areas (the Lake Superior region, Arizona and New Mexico and the Mediterranean area). Rock containing native copper need merely be heated to the melting point of copper to separate the two. Copper occurs in extensive deposits as oxides or sulfides. The minerals are also easily reduced to copper, because of its low reactivity.

Although some high-grade copper ore deposits are being mined and smelted, much of the world's copper is gained from low-grade copper sulfide ores containing from 0.5 to 3 per cent copper. The vastness of easily accessible deposits and a unique concentrating process combine to make the extraction economical.

1. **Mining.** Figure 20.4 illustrates the major steps in the extraction of copper from low-grade ore. (The shaft mine is included to indicate that some copper ore is mined in that manner.) In the process to be described, the low-grade copper ore is blasted from a mountainside in levels. The copper mine in Bingham Canyon, Utah, is a tremendous open pit hollowed out of the walls and the bottom of a canyon. Ore is being taken from both the bottom and the sides of the pit. The walls are cut into steplike levels; each step is about 60 feet high and has a railroad track. Huge ore masses of about 50,000 tons are blasted with ammonium nitrate from a 500-foot strip along the edge of a level. Ore running 0.4 per cent copper or more is loaded into trains to be taken to the mill. Lower grade ore is loaded into other trains to be dumped elsewhere as waste. Blasting moves down from level to level, for blasting off one level widens the level below, until the 33 levels of the 2210-foot side of the pit have been moved back. When the circle at the mine bottom becomes of sufficient diameter for a track, the center is excavated and one more level joins the family of levels above. Approximately two-thirds of the materials blasted is waste and is dumped.

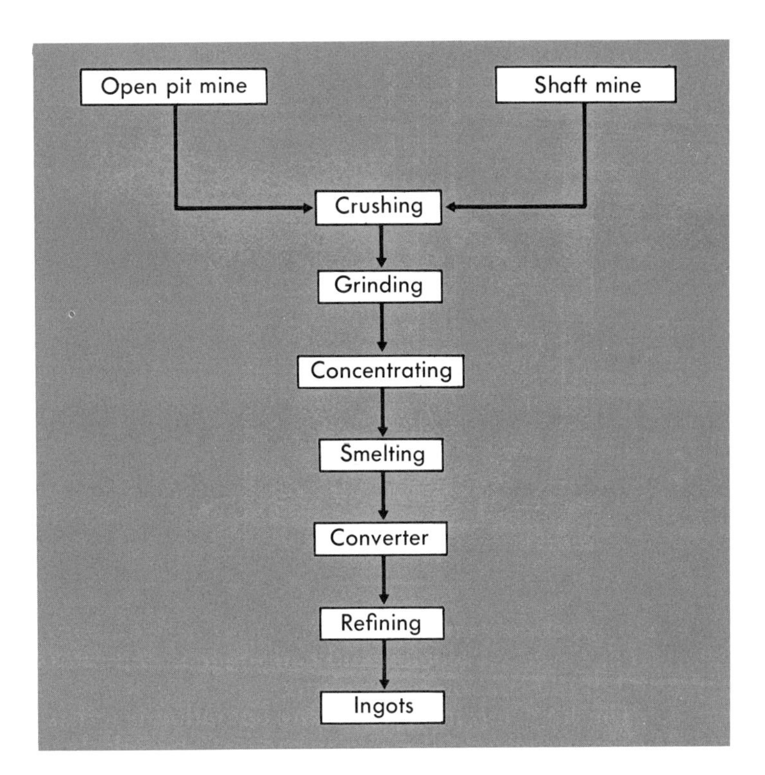

FIGURE 20.4 STEPS IN THE EXTRACTION OF COPPER.

2. **Concentration.** At the mills the ore is dumped into a giant gyratory crusher which can reduce granite boulders 3 feet in diameter to pieces not greater than 6 inches across. The ore is further crushed and then ground in water in ball mills, where steel balls in a rolling steel barrel tumble over each other, pulverizing the ore that chances to be between, until 80 per cent of it will pass through a sieve with 10,000 openings per square inch.

3. The fine ore in a water slurry is passed through a series of flotation cells (Figure 20.5). Here oil and a detergent are added, and the mixture is stirred while air is blown in. The detergent is added to make bubbles with the oil. Oil is added to wet the mineral-bearing particles so that they rise with the bubbles as a froth on the water surface to spill over the side of the cell and collect in a large settling tank. There the bubbles burst in a spray of water to let the mineral settle to the bottom. The rocky material containing no mineral sinks and remains on the bottom of the flotation cell. After passing through several flotation cells, the rock materials are washed out as tailings on a huge dump. Only 4 pounds of each 100 pounds of ore entering the flotation cell are collected in the froth as concentrate. Ninety-six pounds are discarded as tailings.

The product from the flotation process contains about 30 per cent copper. It may contain traces of other metals. At the Kennecott Copper Company mills in Magna, Utah, the concentrate contains 1.5 per cent molybdenite, and also small traces of gold and silver. The molybdenite is worth separating. At those mills an additional flotation is carried out in the same type cells, but with a wetting agent or oil that wets the molybdenum-bearing mineral but not the copper mineral. Molybdenite is taken off as the concentrate and the copper concentrate remains as tailings.

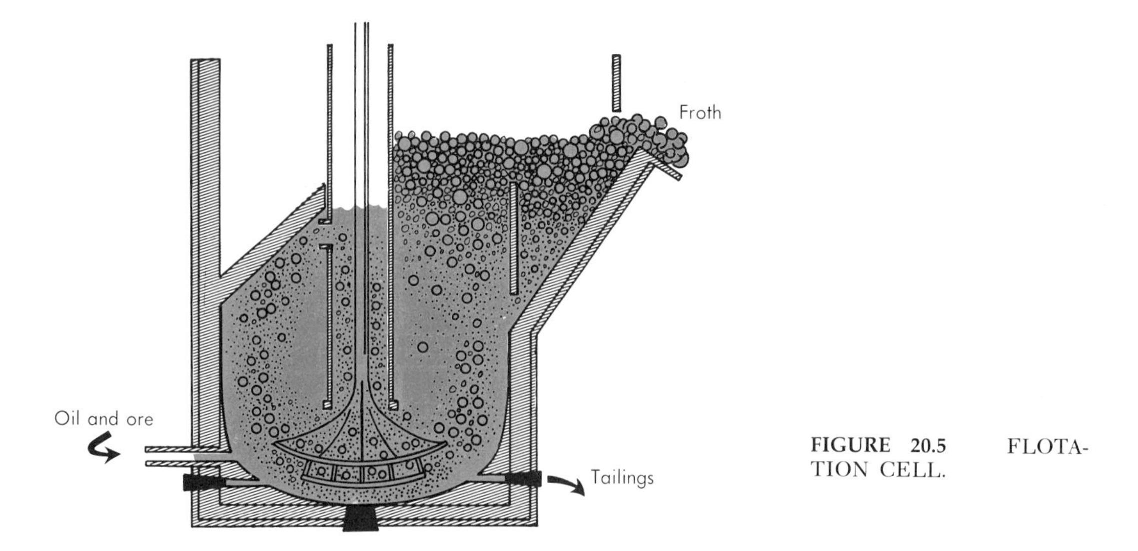

FIGURE 20.5 FLOTATION CELL.

The **copper concentrate** containing 30 per cent copper in the compounds Cu_2S (chalcocite) and $CuFeS_2$ (chalcopyrite) is filtered and dried in preparation for smelting.

4. The first step in the **smelting operation** is a roasting process. The ore is heated intensely in the air. The sulfide is partially converted to oxide. For copper(I) sulfide the reaction is represented by this equation:

$$2\ Cu_2S + 3\ O_2 \rightarrow 2\ Cu_2O + 2\ SO_2$$

The sulfur dioxide is converted to sulfuric acid by either the lead chamber or the contact process.

The roasted ore is then smelted to remove the rocky waste material called gangue, represented as SiO_2. A flux, limestone ($CaCO_3$), is added, which combines with the gangue to form slag, a low melting, glasslike compound:

$$SiO_2 + CaCO_3 \rightarrow CaSiO_3 + CO_2$$

During the process both the slag and the copper sulfide-copper oxide mixture are melted and run to the bottom of the furnace. The heavier copper-containing melt rests on the bottom. Periodically, as the molten slag and sulfide-oxide accumulate, they are run out; the slag is dumped, and the metal sulfide-oxide is poured into forms in which it cools as ingots of **copper matte.**

The copper matte is finally reduced to copper in a converter, in which air is blown through the molten mass. The equations below represent the reactions taking place:

$$Cu_2S + 2\ Cu_2O \rightarrow 6\ Cu + SO_2$$

and

$$O_2 + Cu_2S \rightarrow 2\ Cu + SO_2$$

When conversion to copper is complete, the copper is cast into 700-pound slabs. The escaping of sulfur dioxide gas as the metal cools gives it a blistered appearance and the name "blister copper." Blister copper is about 99.5 per cent copper.

5. Most **copper is purified electrolytically** and used for electrical wiring and electrical equipment. Impurities greatly reduce the conductivity. The presence of 0.03 per cent arsenic reduces the conductivity by about 14 per cent.

The refining takes place in huge electrolytic cells. Seven-hundred-pound blister copper anodes, about 3 feet square and 2 inches thick, are suspended in a copper sulfate solution between thin pure sheets of copper (cathodes) (Figure 20.6). In the operation copper is dissolved from the impure copper anodes:

$$Cu \rightarrow Cu^{2+} + 2\ e^-$$

while at the cathode the copper(II) ion is reduced to copper:

$$Cu^{2+} + 2\ e^- \rightarrow Cu$$

Copper is dissolved at the anode, transported as the copper(II) ion through the electrolyte and deposited on the cathode. As the anode grows thinner, the cathode grows thicker. Purification is accom-

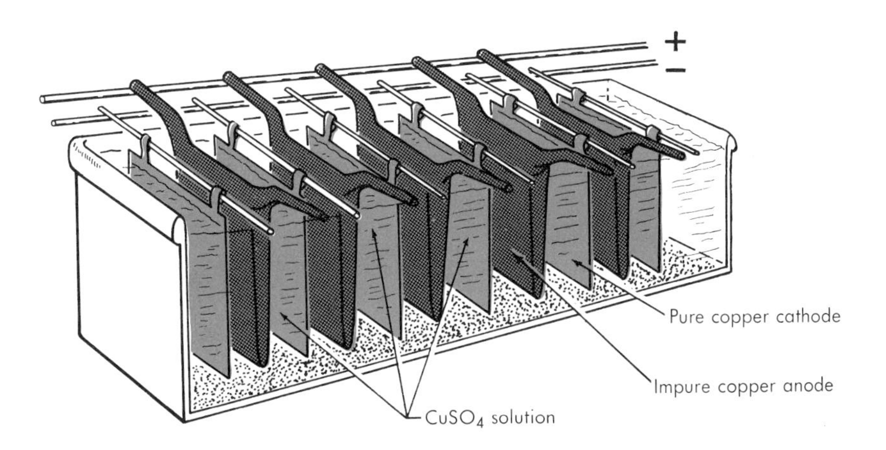

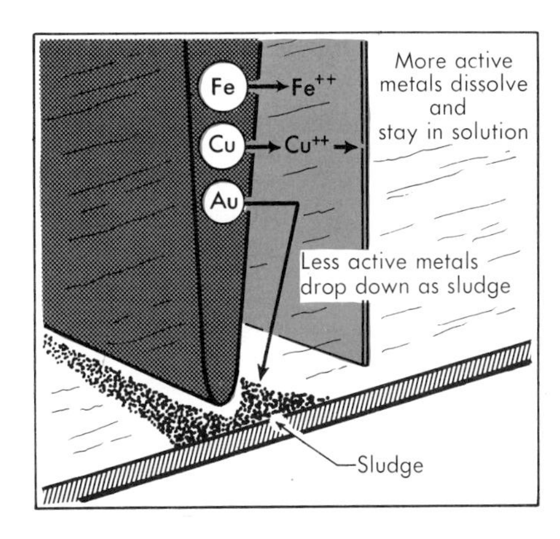

FIGURE 20.6 ELECTROLYTIC PURIFICATION OF COPPER.

plished because: (a) inert material and less reactive metals, such as gold and silver, are not oxidized as easily as copper and are undermined by the dissolving copper until they drop to the bottom of the cell as sludge, (b) more active metals present, such as iron, are ionized more readily than copper and dissolve in the solution, but are less readily reduced than copper ion and therefore remain in solution. The solution must be changed periodically.

The electrodes are removed from the cells every two weeks. One 700-pound block of blister copper makes two 300-pound blocks of pure metal. The scrap that remains is melted and recast for purification. The purified copper blocks (99.9 per cent copper) are melted in electric arc furnaces and cast in ingots of the shapes and sizes desired by manufacturers of copper wire and copper wares.

The sludge in the refining cells is collected to be treated for the extraction of silver and gold. These are contained in the original ore in the proportions of 0.104 ounce and 0.017 ounce per ton, respectively. Because silver and gold stay with the copper mineral, the complete metallurgical process serves to concentrate the precious metal in the sludge. At a price of approximately $35.00 an ounce, the value of the gold recovered is considerable.

Copper has a reddish brown color. (Copper and gold are the only common metals that have neither a silvery nor gray appearance.) It has a density of 8.9 and a melting point of 1083°C. It is second only to silver in electrical conductivity, is resistant to corrosion (after formation of a thin oxide layer) and is much cheaper than silver. For these reasons, its chief use is in electrical wiring and in electrical appliances. It is a ready conductor of heat and consequently is used in and on cooking ware. It is also used in making coins, roofing and water pipes.

Copper is widely used in alloys. Some of these are bronze (copper, tin and zinc), brass (copper and zinc), aluminum bronze (aluminum and copper), and coinage silver (silver and copper).

20.4 IRON

Iron is the most important of all metals. Iron and its alloy with carbon, called steel, are literally the skeleton of the industrial structure. The reasons for iron's importance will be seen in the discussion to follow.

iron has an industrial importance which exceeds that of any other element.

Iron was known in ancient times. The various ancient names meant "metal from heaven," indicating that iron was known as the free metal found in meteors. About 4.7 per cent of the earth's crust is iron. It is second only to aluminum in its abundance. It is believed that the center of the earth is a mixture of molten iron and nickel. Like aluminum, iron is present everywhere in the earth's crust, but also like aluminum, at only relatively few spots is the extraction economical.

20.5 PRODUCTION OF IRON

Many factors are involved in choosing a site for a steel mill. Of prime importance is the availability of good iron ore. Iron does occur free in nature, but only to an extremely limited extent. Minerals containing iron include hematite (Fe_2O_3), magnetite (Fe_3O_4), siderite ($FeCO_3$), pyrite (FeS_2) and chalcopyrite ($FeCuS_2$). The two oxides are the principal minerals from which iron is obtained. Good deposits of siderite are not available, and sulfide ores are not greatly used, because the presence of even small traces of sulfur in iron or steel affects the properties adversely.

The physical state in which the mineral occurs affects its availability. The amount and kind of inert material (gangue) in the ore are important. The amount of gangue must be kept at a minimum and must be of such a composition as to be easily fused with the addition of a melting agent called flux. The expense involved in mining and the expense of transportation are other vital factors.

Not only are the state, availability and transportation of the iron ore important, but a source of good coking coal is necessary, because coke is required in large quantities in the reduction process. Limestone must also be available, because limestone is the flux added to silica-bearing ore to convert the silica gangue to low-melting calcium silicate slag. Water is not needed to produce iron, but it is required in great tonnages in the rolling of steel. All these must be available if a steel mill is to be successful. Approximately 2 tons of high-grade iron ore, 0.5 tons of limestone, 1 ton of coal and 250 tons of water (used over and over) are required to produce 1 ton of steel.

The major steel-producing area of the United States is outlined on the map in Figure 20.7. Steel-making centers are located largely in the Great Lakes area along the southern shores of Lake Michigan, Lake Erie and Lake Ontario. Economic conditions are met there.

The greatest iron deposits in the world are located in the Mesabi Range in northern Minnesota and in adjacent Canada. The

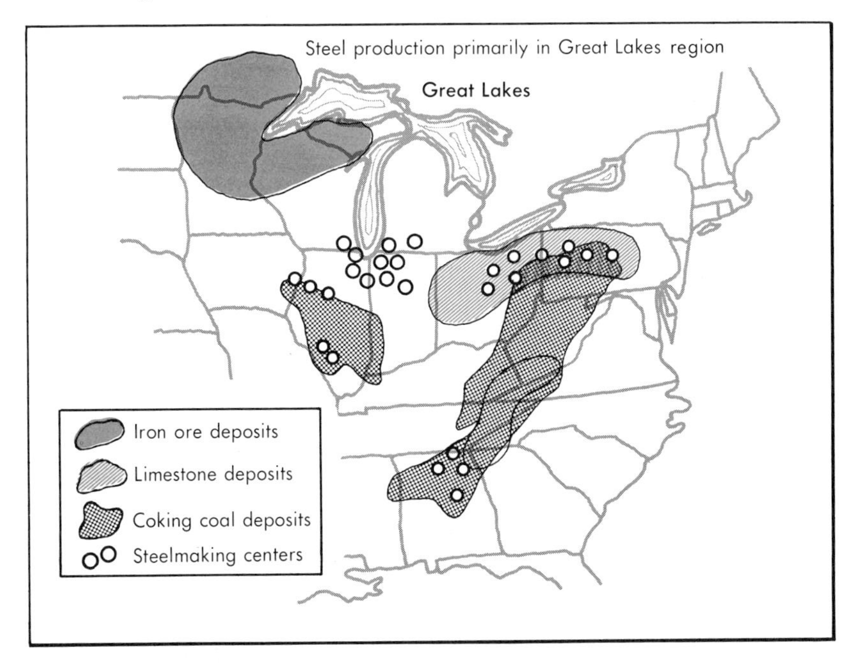

FIGURE 20.7 MAJOR STEEL PRODUCING AREAS OF THE UNITED STATES.

deposits are oxides, hematite and magnetite. Furthermore, the deposits have been raised during the geological history of the area; the covering has been washed off until the rich deposits lie very near the surface. Thin layers of earth have been removed to reveal great deposits of high-grade ore. The ore is mined in the open and loaded into railroad cars. About 2 billion tons of high-grade iron ore have been removed. Approximately 1 billion tons still remain. A much greater supply of low-grade ore will remain when the high-grade ore is gone. The iron ore deposits are many miles from the steel mills, but this disadvantage is offset by cheap water transportation. Lake steamers carry the ore to the mills. It was thought once that the steel mills might shift to other areas. This appears unlikely now, because with the opening of the St. Lawrence Seaway, the rich ores of Labrador and Venezuela may now be transported cheaply.

Great stores of anthracite coal lie in the mountains of Pennsylvania and West Virginia just south of the great steel areas. Limestone deposits occur at the earth's surface in Ohio and Pennsylvania. The lakes and streams nearby contribute water.

Many iron ores are improved before smelting by one or more processes called **beneficiation.** Beneficiation may constitute crushing and grinding the ore to allow the reducing gas to penetrate it more readily in the blast furnace, or it may consist of roasting the ore in air to remove sulfur by oxidation to form sulfur dioxide or convert the ore more completely to an oxide. One beneficiation process, mainly a concentration process, is important now that the depletion of high-grade ores is a possibility.

In this process, low-grade hematite ore is crushed, ground, mixed with powdered coke and roasted. The roasting converts hematite to magnetite, a magnetic material, as the name suggests. The powder of coke and magnetite is then poured down through a magnetic field. The particles containing magnetite are deflected in the magnetic field and fall into a bin separate from the nonmagnetic rock. The powdered concentrate may then be dumped into balling drums that roll powder into moist balls the size of walnuts. Finally the moist balls are hardened in a furnace for later handling.

Iron is produced in the **blast furnace,** a hollow steel giant lined with fire brick (Figure 20.8) and measuring about 100 feet high and 25 feet in diameter. The furnace operates continuously. Iron ore, coke and limestone are fed into the furnace at the top. Preheated air from huge stoves is blown in the bottom, up through the mixture of solids. Molten iron and molten slag are withdrawn periodically as they accumulate at the base of the furnace. The molten slag is dumped as waste; the molten iron is run into interconnected forms to cool.

The reactions that occur to accomplish the conversion of ore to iron are probably many and complex and are not completely known. A plausible set of reactions is represented by the equations below:

1. Coke burns in the incoming air to form carbon monoxide:

$$2\ C + O_2 \rightarrow 2\ CO + 53{,}000 \text{ calories}$$

This reaction contributes the heat to maintain the furnace at the necessary temperature (about 1600°C at the hottest point).

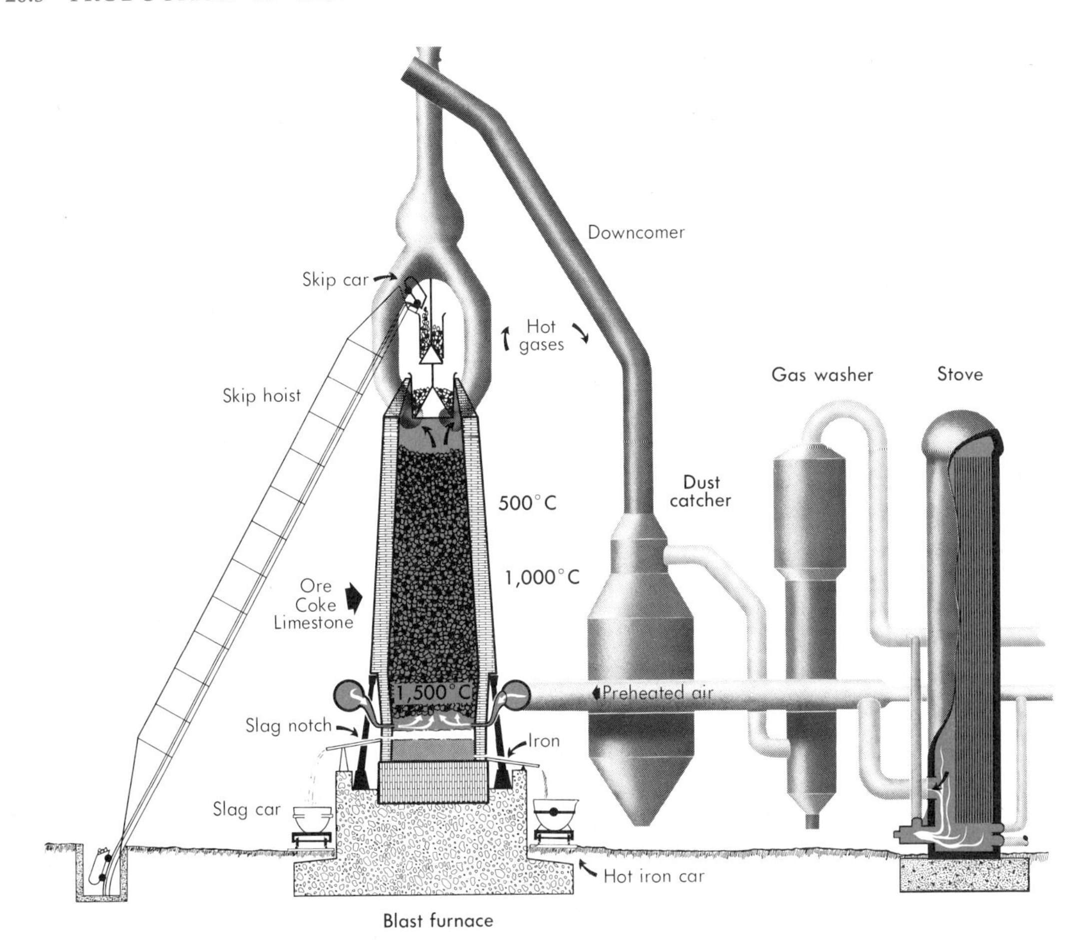

FIGURE 20.8 THE BLAST FURNACE.

2. Carbon monoxide (produced above) is the reducing agent. As a gas, it can diffuse in and through the porous ore, where it reacts to produce iron.

carbon monoxide reduces the iron oxides to iron.

$$Fe_2O_3 + 3\ CO \rightarrow 2\ Fe + 3\ CO_2$$

3. The iron reacts with unburned coke to form iron carbide (which is 6.6 per cent carbon).

$$3\ Fe + C \rightarrow Fe_3C$$

This reaction serves a purpose because iron melts at 1550°C whereas the melting point of iron carbide is 1150°C. Molten iron carbide forms at a considerable distance up inside the furnace, and then runs down and collects in the base.

4. Removal of the inert rock material (gangue) is made easier by this reaction:

$$\underset{\text{Limestone (flux)}}{CaCO_3} + \underset{\text{Gangue}}{SiO_2} \rightarrow \underset{\text{Slag}}{CaSiO_3} + CO_2$$

Not only does slag have a lower melting point than gangue, but it runs nicely down through the charge, whereas molten silica tends to form a sticky layer over the walls and the charge.

5. Coke and limestone assist in removing other impurities combined with the iron, such as sulfur:

$$CaCO_3 + FeS + 2\ C \rightarrow CaS + Fe + 3\ CO$$

The calcium sulfide melts and runs to the furnace base, where it is dissolved in the slag layer above the molten iron carbide.

After cooling, the impure iron (mostly iron carbide) is dumped from the interconnected mold and broken apart. These chunks of metal were thought by some to look like pigs, hence the product from the blast furnace is called **pig iron.** While the metal pigs cool in the forms, a thin layer of slag rises to form over them and protect them from oxidation.

It takes about 12 hours for material added at the top to pass through the blast furnace and come out as iron and slag. Nevertheless, one blast furnace often produces 1,000 tons of pig iron per day.

Pig iron is 92 to 94 per cent iron, with carbon, slag and other impurities. It may be used as it is for some purposes. Pig iron that is poured into a mold and allowed to cool rapidly forms a very hard shiny gray material called white cast iron, largely a solid solution of carbon in iron. When a casting of pig iron is allowed to cool slowly, a hard black material is formed, called black cast iron. Black cast iron is largely a mass of iron carbide crystals and flakes of carbon imbedded in iron. Both forms of cast iron are very brittle. In the past, soil pipes, old iron stoves and certain locomotive parts were made of cast iron. Now tougher steel has almost entirely replaced cast iron.

Wrought iron was made by **puddling** pig iron and iron oxide. Puddling amounts to prolonged stirring of the molten mixture. Carbon is burned from the cast iron:

$$3\ Fe_3C + 2\ Fe_2O_3 \rightarrow 13\ Fe + 3\ CO_2$$

until the iron contains 0.01 to 0.02 per cent carbon and slag fibers. Wrought iron is soft and malleable. The slag prevents oxidation. Anvils, chains, anchors and wire were once made of wrought iron; now they are made of low carbon steel.

20.6 STEEL

Almost all iron used in industry is in the form of steel, which is an alloy. Ordinary steel is iron alloyed with from 0.2 to 1.5 per cent carbon. So-called alloy steels contain iron with a small percentage of such elements as cobalt, vanadium, chromium, tungsten and silicon.

steel is iron alloyed with carbon.

Steel is made from pig iron by removing impurities, adding definite amounts of either carbon or another element or elements and administering such heat and rolling treatments as are needed for the properties desired.

The steel rails that spanned the continent in 1869 were of steel

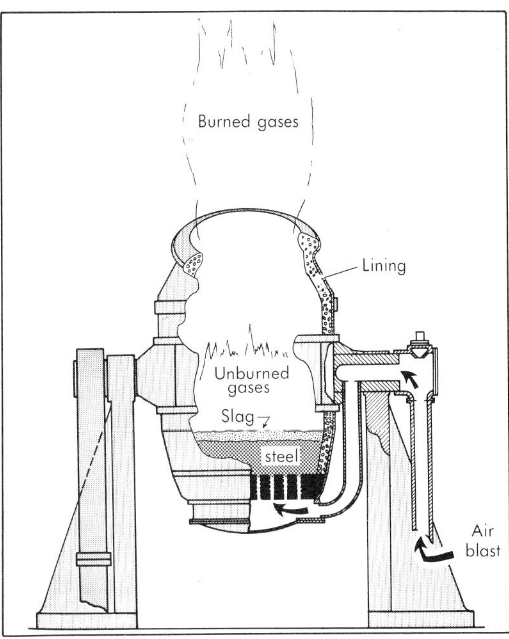

FIGURE 20.9 THE BESSEMER CONVERTER.

made in a Bessemer converter. The Bessemer converter of today is a huge, oval-shaped barrel (Figure 20.9) that may be tipped to accept molten raw iron from the blast furnace and to dump the steel when it is finished. The converter is loaded with 15 to 20 tons of molten pig iron. A blast of air or oxygen forced up through holes in a false bottom burns out impurities:

$$S + O_2 \rightarrow SO_2$$

$$Si + O_2 \rightarrow SiO_2$$

$$Mn + O_2 \rightarrow MnO_2$$

$$C + O_2 \rightarrow CO_2$$

The last reaction is the major reaction. The gases formed escape in the flame. Solid oxides combine with the converter lining to make slag, which rises to the top. The blast of air (with a great flame) continues for 10 to 15 minutes. When the impurities have been burned, which is determined by the color of the flame, the air blast is turned off. Either carbon (usually as spiegeleisen, a high-carbon steel) or another alloying agent is added, and the contents are dumped into a ladle to be cast into ingots for the required further treatment. However, this method, which is the quickest method of making steel, has an inherent weakness. Analysis by flame color is not thoroughly reliable, and there is insufficient time for chemical analysis. Some batches of steel are excellent; others are poor.

Less than ten per cent of the steel produced today is produced in the Bessemer converter. Ninety per cent is prepared in the open hearth furnace. The open hearth furnace is a great steel tray with a thick lining of calcium oxide fire brick (Figure 20.10). Pig iron (raw iron) from the blast furnace, plus rusted scrap iron, pure hematite ore (Fe_2O_3), coke and a small proportion of limestone are placed in the tray, which may hold 150 tons. The whole is heated by burning a preheated mixture of air and natural gas.

Some of the reactions occurring in the melt are represented by these equations:

$$Fe-C + Fe_2O_3 \rightarrow Fe + CO_2$$
$$Fe-Si + Fe_2O_3 \rightarrow Fe + SiO_2$$
$$Fe-P + Fe_2O_3 \rightarrow Fe + P_4O_{10}$$
$$CaCO_3 + SiO_2 \rightarrow CaSiO_3 + CO_2$$
$$6CaCO_3 + P_4O_{10} \rightarrow 2Ca_3(PO_4)_2 + 6CO_2$$

Fe–C, Fe–Si and Fe–P are used to represent alloys of indefinite composition; because the proportions are indefinite, the equations are not balanced.

Samples are taken from the seething white-hot mass (about 1600°C) at intervals and analyzed. When the composition is just right for the steel desired, usually after 9 to 12 hours, the molten steel is discharged into ladles to be cast into ingots. During the heating in the furnace, materials may be added to the charge as required. Just before dumping, a trace of magnesium metal or other active metal is added as a **scavenger.** A scavenger merely combines with any dissolved oxygen that escapes during cooling, and it remains in the steel as an oxide, preventing formation of oxygen bubbles in the steel. Figure 20.10 shows how the burning gases that have passed over the hearth charge are passed through extensive brick-

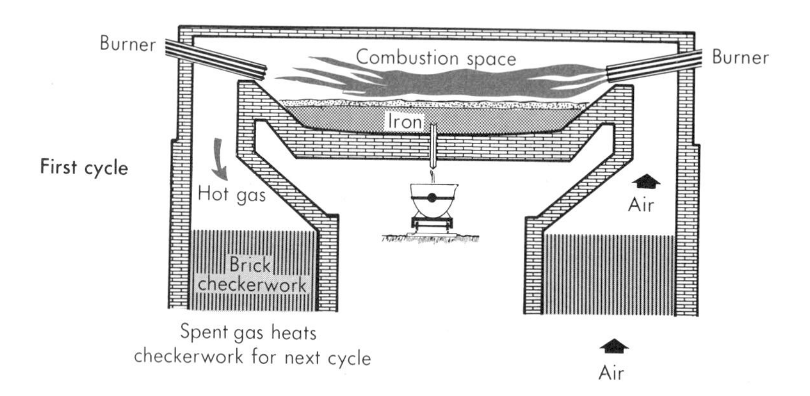

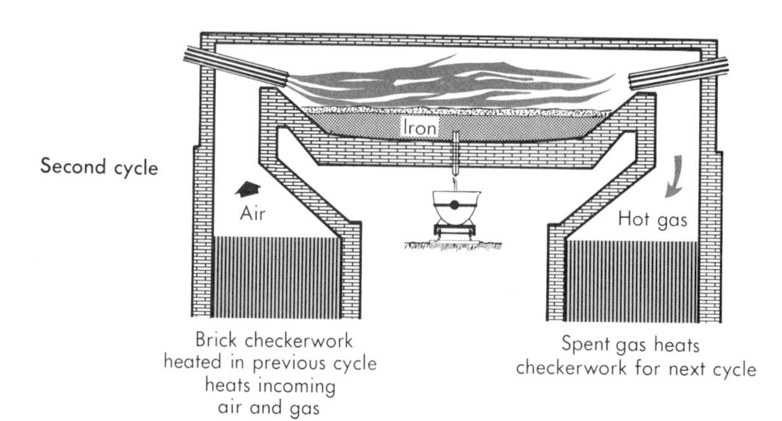

FIGURE 20.10 THE OPEN HEARTH FURNACE.

work, heating it to a high temperature. The flow of gas and air is then reversed, and the gas mixture derives heat from the bricks on the opposite side of the brickwork. Much heat is conserved by this arrangement.

For many structural purposes, the ingots stripped from their molds are carried to soaking pits, which are great gas-heated furnaces maintained at about 1200°C, somewhat below the melting point of the steel. The ingot is left there for the necessary time (perhaps 24 hours) until heated all the way through to the pit temperature. The ingot is then transferred to rolling stands, where it passes between huge, water-cooled rollers and is reduced to slabs, plate, steel sheet or rods. Two hundred and fifty tons of water pass over each ton of steel in rolling. Not only does the rolling reduce the metal to the desired thickness, but it toughens it. The major steps in the production of steel are illustrated in Figure 20.11.

Certain steels receive special treatment to gain desired properties. Hardness and brittleness increase with the increase of carbon. Following are the approximate percentages of carbon in three grades of steel:

0.2%C or less	mild steel (wire, chains, anvils, etc.)
0.2 to 0.6%C	medium steel (structural steel)
0.6 to 1.5%C	high carbon steel (cutting tools, razor blades)

Heat treatment affects the properties of steel to a remarkable degree. This is possible because there are two allotropic forms of iron. Alpha iron is stable between the melting point (1535°C) and 1400°C and also below 906°C. Gamma iron is stable between 906°C and 1400°C. Furthermore, carbon and iron form compounds with different crystal forms and may also form solid solutions. Very rapid cooling from high temperature leaves the steel mostly in the more brittle gamma form, for changes from one form to another in the solid are slow at high temperature and essentially nil at room temperature. Cooling more slowly to 800°C and then quenching leaves the metal partially in the alpha form, whereas very slow cooling gives a still higher percentage of alpha iron.

A wide range of properties may be given **quenched steel** (steel that is cooled very rapidly by immersing it in water, oil or even liquid mercury) by further heat treatment. The metal may be heated merely to a temperature much lower than the melting point and then allowed to cool slowly. Table 20.1 illustrates the change in use occasioned by change in properties of quenched high carbon steel that is reheated to various temperatures. The data are relative, but they do indicate that the higher the temperature is raised

TABLE 20.1 TEMPERATURE OF QUENCHING AND USES FOR "QUENCHED" STEEL

230°C	Razor blades
250°C	Pocket knives
260°C	Scissors and chisels
270°C	Butcher knives
290°C	Watch springs
320°C	Saws

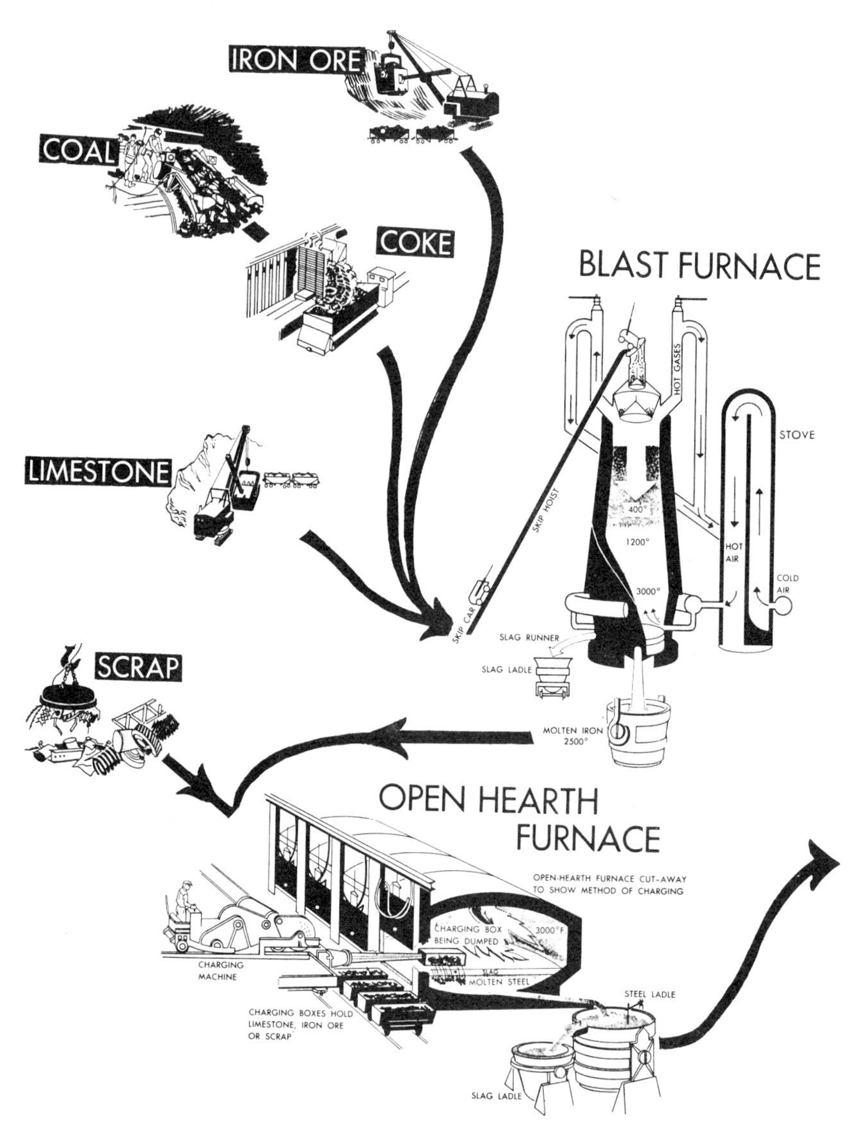

Figure 25.11a. Metallurgy of steel. *

* All temperatures are in °F.

FIGURE 20.11 STEPS IN THE PRODUCTION OF STEEL.

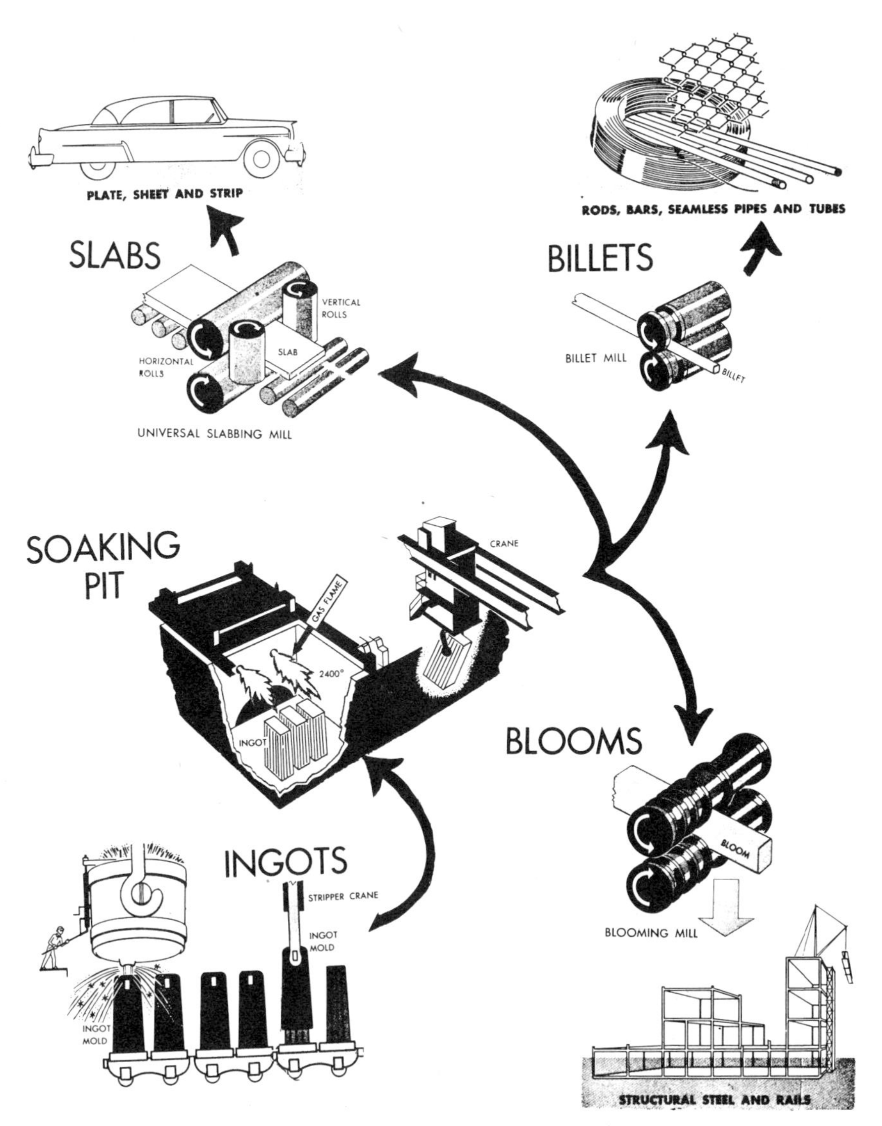

Figure 25.11b. Metallurgy of steel. *

*All temperatures are in °F.

(Courtesy of United States Steel Corporation)

FIGURE 20.11 (CONTINUED)

(even though it is much lower than the melting point) and the more slowly the piece is allowed to cool, the more steel is converted to the softer forms.

A steel that is both tough and hard—tough like medium carbon steel and hard like high carbon steel—is sometimes sought. An axle needs these properties; it must be hard for a bearing surface but tough so it will not break with a sudden jolt. These qualities may be achieved by placing a medium steel rod surrounded by powdered carbon in a furnace and then heating it rapidly. When the outside of the rod becomes hot, it reacts with the carbon. The colder and less accessible metal inside does not. Upon cooling, a sheath of high-carbon steel surrounds medium steel. The process is called casehardening.

So-called alloy steels are made by adding other elements besides (or in place of) carbon. A wide variety of properties can be given steel by changing the alloying components. Some interesting alloys with approximate properties and compositions are listed below:

1. Stainless steel (18 per cent chromium, 8 per cent nickel, 74 per cent iron) is very resistant to corrosion and so remains bright and shiny.
2. Cobalt steel (a small per cent of cobalt) is hard, resists corrosion and maintains its physical properties at high temperatures. It is used for high-speed cutting tools and surgical instruments.
3. Manganese steel (12 to 14 per cent manganese) is extraordinarily hard. Crushing machines and safes are made of it.
4. Chronium-vanadium steel (5 to 10 per cent chromium; 0.15 per cent vanadium) is tough and elastic. It is used for automobile axles and frames.
5. Tungsten steel is used in armor plate and metal-piercing shells. Tools containing as much as 20 per cent tungsten hold a cutting edge even at red heat.
6. Silicon steel (1 to 2 per cent silicon) is hard and strong; steel containing 12 to 15 per cent silicon is resistant to corrosion.
7. Molybdenum steel (6 to 7 per cent molybdenum) retains its temper when heated.
8. Lead steel (0.02 per cent lead) is easily machined.
9. Titanium steel (0.1 per cent titanium) is very hard.

These are only a few of the elements and a few of the percentages used in steel alloys. Other elements used are copper, niobium, zirconium, hafnium, tantalum, nitrogen, phosphorus and sulfur.

The above is but a brief and non-technical discussion of steel. It is estimated that an average of 30 pounds of other elements are found in each ton of steel. A steel may contain as many as five or six different elements. By judicious use of alloying agents, heat treatment and handling (such as hot or cold rolling), steels with an infinite variation of properties are made. For this reason and also because of its accessibility, iron is the most important of all industrial metals. More iron (steel) is produced than all other metals combined.

Industry will see the gradual further displacement of steel by aluminum and magnesium in areas where light weight is at a premium. This substitution will be more pronounced as available high-grade iron ore supplies dwindle. Eventually plastics may displace steel for many uses.

EXERCISES

20.1 List 10 metals.

20.2 Where are the metals in the periodic table?

20.3 Define metallurgy.

20.4 How is aluminum separated from the rock and iron oxide in the ore?

20.5 Describe the Hall process. What purpose does cryolite serve?

20.6 What protects the active metal aluminum from rapid atmospheric corrosion?

20.7 In what industry does aluminum find its major use?

20.8 How does aluminum compare with copper as an electrical conductor?

20.9 Which of these metals occurs free in nature: iron, copper or aluminum? Why?

20.10 Describe the concentration (flotation) process for low-grade copper ores.

20.11 How is molybdenite separated from the copper concentrate?

20.12 Describe the three processes in copper metallurgy for which heat is required. Write the equations. Define the terms: blister copper, copper matte, flux, slag and gangue.

20.13 How is blister copper purified? What happens to the more active metals present? What happens to the less active metals?

20.14 List four uses of copper metal.

20.15 What raw materials are needed to produce steel?

20.16 Which four "common" metals are most abundant in the earth's crust? (See Figure 9.3.)

20.17 Name the two most useful iron ores and write their formulas.

20.18 Write five equations for processes occurring in the blast furnace.

20.19 Why is the steel center of America located on the shores of the Great Lakes?

20.20 Describe one concentration process for low-grade iron ore.

20.21 What is pig iron? Where is it produced? What is its composition?

20.22 Describe three functions of coke in the blast furnace.

20.23 What is cast iron?

20.24 Define steel.

20.25 What are the three major operations in converting pig iron to steel?

20.26 Why is rust scrap iron added to the open hearth charge?

20.27 How may one vary the properties of steel with a set percentage of carbon?

20.28 Write the approximate composition of stainless steel.

20.29 Which common metal is the best electrical conductor?

SUGGESTED READING

Battista, O. A.: "Copper: A Metal Known to Ancient Man." Chemistry, *42* (8):6 (1969).

Battista, O. A.: "Aluminum – Featherweight Champion of Metals." Chemistry, *42* (3):14 (1969).

Burns, R. M.: "Chemical Reactions in the Corrosion of Metals." J. Chem. Educ., *30,* 318 (1953).

Sienko, M. and Plane, R.: *Chemistry.* 3rd Ed., Chapters 21 and 22, McGraw-Hill Book Co., New York, 1966.

TWENTY-ONE • THE TRANSITION ELEMENTS

21.1 INTRODUCTION

The chemistry of the transition elements is currently an area of great research interest. The transition elements are typical metals and lose electrons in reaction to form positive ions. As cited in Chapter 5, the ions seldom achieve the octet configuration of the noble gases and often exhibit two or more valences. For this reason the positive ions they form cannot simply be called "copper ion" or "cobalt ion" but must be labelled as cobalt(III) ion, Co^{3+}, and cobalt(II) ion, Co^{2+}, or as copper(I) ion, Cu^{+}, and copper(II) ion, Cu^{2+}. The Roman numeral following the name in parentheses indicates the charge of the simple ion. Not only do the transition metals lose electrons in chemical reaction and are therefore found as positive ions in compounds, but they bond with negative ions or molecules, often with several at a time, to form metal-centered complex ions or molecules usually referred to simply as complex ions, complex compounds or coordination complexes.

Simple ions, complex negative ions and molecules which surround the centrally positioned metal atom of these complex ions are called **ligands.** Ligands which are or may be bonded to the metal atom center through two or more sites on the ligand, permitting the ligand to surround or partially surround the center atom, are called chelates. Structural formulas for two notable examples of coordination complexes, both containing chelates, are **hemin** (a part of hemoglobin) and **chlorophyll** (the green coloring material in plants). These are shown in Figure 21.1. Metal-centered complex ions are important in analytical chemistry as precipitating reagents, as indicators in titrations, as catalysts in biochemistry and bioinorganic chemistry, in ion exchange reactions and in many other roles.

complex metal ions are extremely important in nature.

Before studying the nature of the bonding and the applications of complex ions, the general properties of the elements themselves will be considered. Attention will be focused upon the first transi-

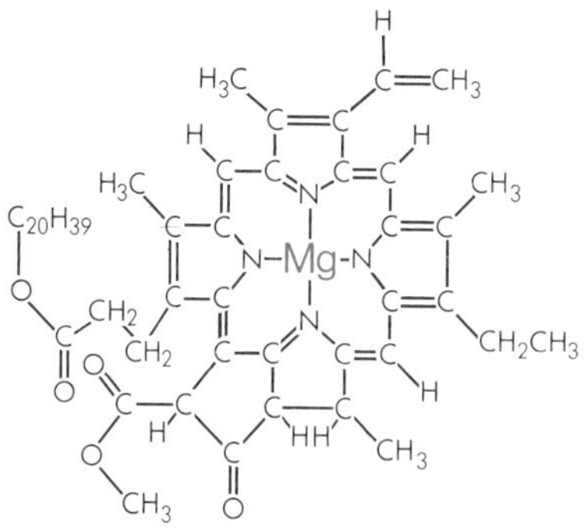

Chlorophyll A

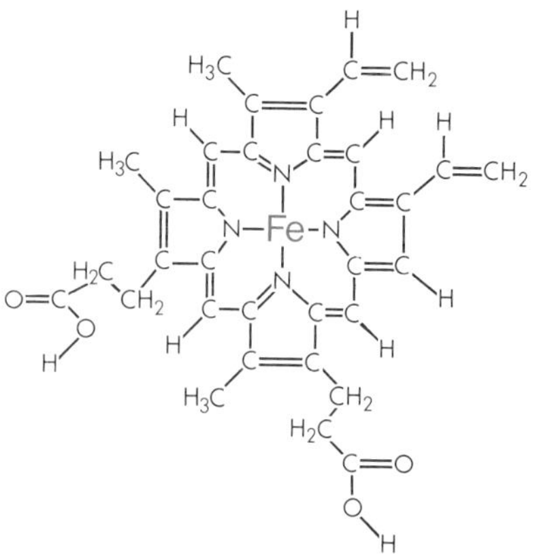

Hemin

FIGURE 21.1 BIOLOGICAL COORDINATION COMPOUNDS: CHLOROPHYLL AND HEMIN.

TABLE 21.1 PROPERTIES OF THE ELEMENTS OF THE FIRST TRANSITION SERIES

Valence	Sc	Ti	V	Cr	Mn	Fe	Co	Ni	Cu	Zn
Electronic Configuration	$3d^14s^2$	$3d^24s^2$	$3d^34s^2$	$3d^54s^1$	$3d^54s^2$	$3d^64s^2$	$3d^74s^2$	$3d^84s^2$	$3d^{10}4s^1$	$3d^{10}4s^2$
Ionization Energy, kcal	151	158	155	156	171	182	181	176	178	217
Atomic radius, Å	1.44	1.32	1.22	1.17	1.17	1.16	1.16	1.15	1.17	1.25
Melting point, °C	1539	1675	1890	1890	1244	1535	1495	1435	1083	419.4
Boiling point, °C	2727	3260	~3000	2482	2097	3000	2900	2732	2595	907
Density, g/cc	2.99	4.54	6.11	7.18	7.3	7.87	8.9	8.90	8.96	7.13

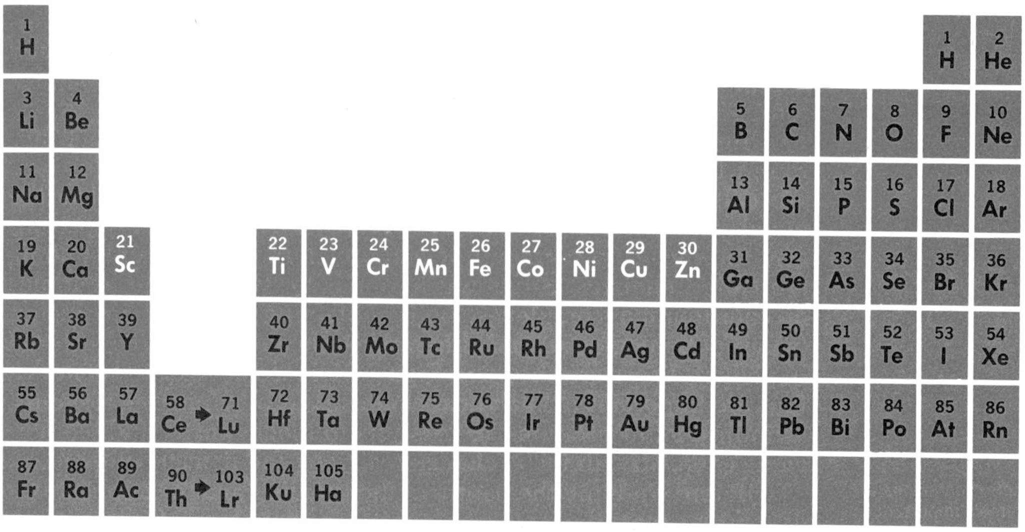

FIGURE 21.2 THE FIRST TRANSITION SERIES (IN WHITE).

tion series, the fourth period elements in the middle of the table from scandium to zinc, as shown in the periodic table in Figure 21.2. Except for copper and zinc, the atoms of all elements in the series have an incompletely filled set of d orbitals. That is, they have from one to nine electrons inclusively in the d orbitals. The ground state electron configurations of the outermost electron level and of the underlying d sublevel of atoms of the series are given in Table 21.1.

21.2 GENERAL PROPERTIES OF THE TRANSITION METALS

The physical properties of the transition elements (melting points, boiling points, density and first ionization energies) are very similar and do not vary a great deal throughout the series (See Table 21.1), and they are generally higher than the same properties of the active metals of Group IA and IIA (Tables 11.2 and 11.3). The greater densities and ionization energies are due primarily to the small size of the atoms as compared to atoms of the active metals, as shown in Figure 21.3.

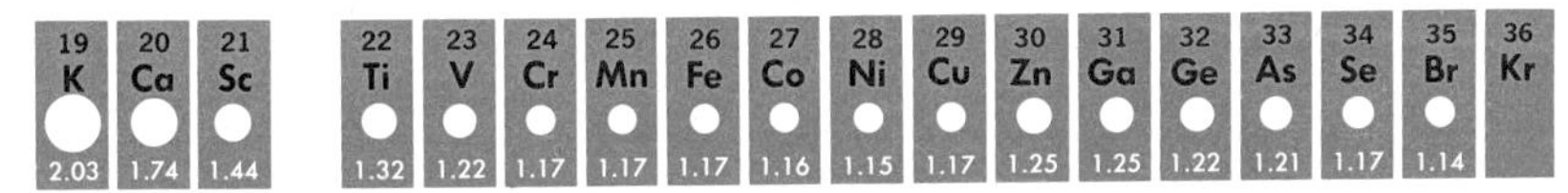

FIGURE 21.3 ATOMIC RADII OF THE FIRST LONG SERIES OF ELEMENTS.

The smaller atoms pack together into a smaller volume, and the smaller atoms have their valence electrons closer to the positive nucleus. The higher melting and boiling points they exhibit are largely due to the fact that the d orbitals of the transition metal atoms overlap between atoms in the bulk metal to form molecular orbitals which, when occupied by the electrons that partially fill the d orbitals of the atoms separately, give rise to partial covalent bonding. The covalent bonding via d orbitals just described for transition elements is added to the metallic bonding of valence electrons in delocalized orbitals in the interstices of metal crystals of all metals, as described in Chapter 11. The additional covalent bonding necessitates more energy and a higher temperature to separate atoms of these metals than is required for the softer and lower melting alkali and alkaline earth metals.

The radii of atoms of elements of the first transition series, as shown in Table 21.1, decrease slightly as one proceeds from left to right in the periodic table. Notice that as one moves from element to element from left to right in the series, an electron is added to an underlying orbital and not to the valence level which determines the radius of the atom. Each additional d electron does not quite cancel the effect of the additional attraction of the increased positive charge on the nucleus to the valence electrons, and the atoms are only slightly decreased in size.

Most transition metals are **paramagnetic** (are attracted to a magnetic field). The existence of paramagnetism suggests that there are unpaired electrons. The unpaired electrons are placed in the incompletely filled sublevel of d orbitals, singly from Sc to Mn and then progressively in pairs.

Not only does a transition metal often show more than one valence in its ionic compounds, as mentioned earlier in the chapter, but it also exhibits **various oxidation states** in its metal-centered complexes. This is possible because varying numbers of the d electrons (as well as the s electrons) can be involved in the chemical bonding. Some common oxidation states for the first series of transition elements are given in Table 21.2. The table shows oxidation states of +2, +3, +4, and +5 for vanadium. Were the valence electrons lost in the bonding, the electron configurations for vanadium for the oxidation states listed would be $4s^0 3d^3$, $4s^0 3d^2$, $4s^0 3d^1$ and $4s^0 3d^0$, respectively.

TABLE 21.2 OXIDATION STATES OF THE FIRST TRANSITION ELEMENTS

Sc	Ti	V	Cr	Mn	Fe	Co	Ni	Cu	Zn
								1	
	2	2	2	2	2	2	2	2	2
3	3	3	3	3	3	3	3		
	4	4		4					
		5							
			6	6	6				
				7					

Most compounds of the transition elements are colored. **The color is related to the presence of 3d electrons.** The 3d electrons absorb light of a particular energy, and in the process a 3d electron is raised to a higher energy level. The light absorbed is of an energy such that it is in the visible spectrum. Upon absorption of a specific band, the remaining light lacks some of its color components and no longer appears white.

Finally, the transition elements have the ability to form many complexes. This property is due to the relatively small sizes of the atoms and to the availability of d orbitals for bonding.

21.3 NATURE OF COORDINATION COMPOUNDS

Insight can be gained into the nature of coordination compounds by considering three distinct compounds having different colors but the same formula: $CrCl_3 \cdot 6H_2O$. One mole of violet $CrCl_3 \cdot 6H_2O$ reacts with excess $AgNO_3$ in solution to form a precipitate consisting of three moles of AgCl. One mole of the light green compound reacts with $AgNO_3$ to precipitate two moles of AgCl. With the dark green compound only one mole of AgCl is obtained. One mole of each of the compounds placed in a dry atmosphere in the presence of a drying agent such as calcium chloride or sulfuric acid results in the loss of zero, one, and two moles of water, respectively. Finally, electrical conductance measurements of solutions of the three compounds indicate that they liberate four, three, and two ions per formula, respectively. The observations cited above and summarized in Table 21.3 show that the three coordination

TABLE 21.3 SUMMARY OF DATA FOR THREE DIFFERENT COMPOUNDS WITH THE FORMULA $CrCl_3 \cdot 6H_2O$

Compound	Number of Cl^- ions precipitated	Number of H_2O molecules removed by a drying agent	Number of ions in complex	Actual formula
Violet $CrCl_3 \cdot 6H_2O$	3	0	4	$[Cr(H_2O)_6]Cl_3$
Light green $CrCl_3 \cdot 6H_2O$	2	1	3	$[Cr(H_2O)_5Cl]Cl_2 \cdot H_2O$
Dark green $CrCl_3 \cdot 6H_2O$	1	2	2	$[Cr(H_2O)_4Cl_2]Cl \cdot 2H_2O$

compounds of the same formula are not identical, and are therefore isomers. On the basis of the data, the formulas of the violet, light green and dark green compounds are best written as: $[Cr(H_2O)_6]Cl_3$, $[Cr(H_2O)_5Cl]Cl_2 \cdot H_2O$ and $[Cr(H_2O)_4Cl_2]Cl \cdot 2H_2O$, respectively.

The species, molecules or ions (water and chloride ion in this case) within the brackets are considered to be arrayed about and strongly bonded to the central metal ion and are called **ligands.** The species outside the brackets occupy positions in the lattice of the solid, either as ions or as water of hydration, and these are less strongly bonded and are separate from the complex with its center and ligands when the compound is dissolved in water. Hence the violet compound furnishes three chloride ions for precipitation with silver ion, or a total of four ions in water solution; the light green compound furnishes two chloride ions, or a total of three ions; and the dark green compound furnishes one chloride ion, or a total of two ions. All six ligands in the violet compound are water molecules. In the light green compound one chloride ion replaces a water molecule as a ligand. In the dark green compound there are four water molecules and two chloride ions as ligands. The water molecules outside the brackets, those that are not ligands, are simply lost among the molecules of the solvent when the compounds containing them dissolve.

Because each compound is neutral and contains three negative chloride ions either within or outside the coordination sphere, the oxidation state of the central chromium atom is +3. The formal names of these compounds, in the order mentioned above, are hexaaquochromium(III) chloride, chloropentaaquochromium(III) chloride monohydrate and dichlorotetraaquochromium(III) chloride dihydrate. The nomenclature of complex compounds is discussed in the last section of this chapter. For now it is sufficient to say that di-, tri-, tetra-, penta- and hexa- refer to the number of a certain ligand present; aquo represents water; and the Roman numeral, three in this case, specifies the oxidation state of the metal center.

the coordination number corresponds to the number of ligand attachments to a metal ion.

The number of ligands grouped about the metal ion in a complex with simple ligands, or the number of points of attachment of ligands to the metal center if some ligands are chelates, is called the **coordination number** of the metal center. Chromium(III) has a coordination number of six in each of the three isomeric compounds just discussed. Complexes with coordinate numbers of all numbers from two to nine are known, but those of four and six are most common for ions of the elements of the first transition series. The ligands appear to group around the metal ion center in a symmetrical array when possible. When there are six ligands, the coordination number is six; the ligands are arranged about the center in such a way that when lines are drawn connecting adjacent ligands, a regular octahedron is formed, as in Figure 21.4. The complex is said to be octahedral. A complex with four ligands is either tetrahedral or square planar; that is, the ligands will either be at the apices of a regular tetrahedron about the metal ion as a center, or they will be at the four corners of a flat square. When the coordination number is two, the complex is linear (Figure 21.4). The geometries for various coordination numbers are given in Table 21.4.

The metal centers found as central atoms or ions in complex ions are most generally transition metals. Other small metal ions at the top of their respective groups in the periodic table, such as lithium, aluminum and beryllium, may appear in complexes. De-

$Cu(CN)_2^-$ [−] [−]

Linear

$CoBr_4^{2-}$ [2−] [2−]

Tetrahedral

$Cu(H_2O)_4^{2+}$ [2+] [2+]

Square planar

$Ni(NH_3)_6^{2+}$ [2+]

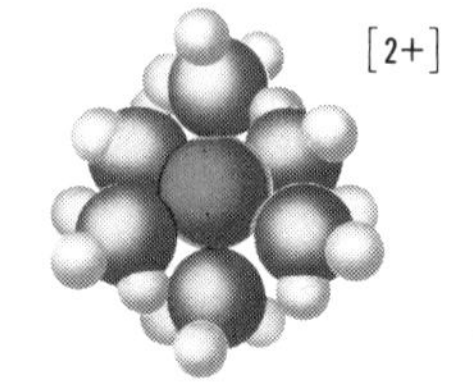

Octahedral

FIGURE 21.4 THE GEOMETRIES OF REPRESENTATIVE METAL-CENTERED COMPLEXES.

TABLE 21.4 COORDINATION NUMBER, GEOMETRY AND HYBRIDIZATION OF SOME COMPLEX IONS

Coordination No.	Geometry	Hybridization	Examples
2	Linear	sp	$Ag(NH_3)_2^+$, $Ag(CN)_2^-$
4	Tetrahedral	sp^3	$Zn(OH)_4^{2-}$, $CoCl_4^{2-}$
4	Square planar	dsp^2	$PtCl_4^{2-}$, $Cu(NH_3)_4^{2+}$
5	Trigonal bipyramid	dsp^3	$Fe(CO)_5$
6	Octahedral	d^2sp^3	$Ni(H_2O)_6^{2+}$, $Cr(NH_3)_6^{3+}$

TABLE 21.5 ELECTRONIC STRUCTURES OF SOME MONO- AND MULTIDENTATE LIGANDS

Name	Ligand	Number of donor sites
Ammonia (ammine)	$H-\ddot{N}H_2$ (H—N: with two H)	1
Water (aquo)	$H-\ddot{O}-H$	1
Chloride (chloro)	$:\ddot{Cl}:^-$	1
Hydroxide (hydroxo)	$H-\ddot{O}:^-$	1
Pyridine	$C_5H_5\ddot{N}$ (ring)	1
Oxalate ion (oxalato)	$^-:\ddot{O}(O=)C-C(=O)\ddot{O}:^-$	2
Ethylenediamine	$H-\ddot{N}H-CH_2-CH_2-\ddot{N}H-H$	2
Diethylenetriamine	$H_2\ddot{N}CH_2CH_2\ddot{N}HCH_2CH_2\ddot{N}H_2$	3
Ethylenediamine tetraacetate ion (ethylenediamine-tetracetato)	$(^-:\ddot{O}-C(=\ddot{O}:)CH_2)_2\ddot{N}CH_2CH_2\ddot{N}(CH_2C(=\ddot{O}:)-\ddot{O}:^-)_2$	

pending on the metal concerned, the center can have oxidation states varying from zero to +5 or +6.

Ligands are always Lewis bases; that is, they must have a pair of electrons extended toward the metal center for sharing. Available electron pairs for water and ammonia, and also for hydroxide and chloride ions, are shown in Table 21.5. As with the ligands found in hemin and chlorophyll, other ligands may bond through two or more coordination positions. At each of these positions the ligand must be a Lewis base and must have an available electron pair. The ligand, **ethylenediamine** (Table 21.5), has two complexing positions and is called a **bidentate ligand,** a name derived from Latin meaning *two-toothed.* Complexes containing multidentate ligands (also called chelates) are often more stable than those with simple ligands. The hexadentate ligand, ethylenediaminetetraacetate ion, forms a complex with calcium ion, a metal ion which is somewhat larger than transition metal ions and has little tendency to form complexes with most ligands. Removal of calcium ions by ethylenediaminetetraacetate ion (EDTA) is one method for softening water. A representation of the $[Ca(EDTA)]^{2-}$ complex appears in Figure 21.5.

ligands are electron pair donors.

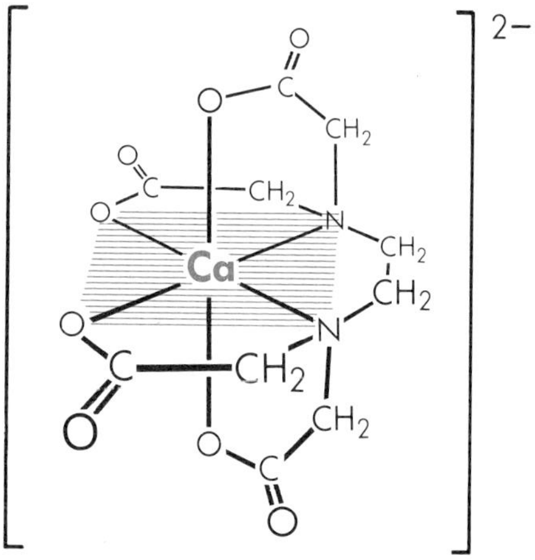

FIGURE 21.5 CHELATE COMPLEX OF CALCIUM.

21.4 BONDING IN COMPLEX IONS

The theories of bonding discussed in Chapter 5, especially the concept of hybridization, which was extended to explain the configuration of the phosphorus pentachloride molecule and the phosphorus hexachloride ion, can be used to account for the linear, tetrahedral, square planar and octahedral structures of metal-centered complexes. According to the theory to be discussed, the available vacant orbitals of the central metal ion or atom hybridize. Each of the hybridized orbitals of the metal center overlaps with the orbital of a ligand containing a pair of electrons. The molecular orbital formed by the overlapping is filled with the electrons furnished by the ligand. A covalent bond in which both electrons are supplied by one of the atoms is called a **coordinate bond.** This

a coordinate bond is formed when a ligand donates a pair of electrons to the central metal ion.

explains the use of the term **coordination compound** for the complexes.

In practice, the bonding of a complex is "predicted" by pairing the electrons of the metal atom or ion center in the lowest energy orbitals possible and then hybridizing the lowest vacant orbitals of the atom in a pattern that becomes apparent in the application. A hybridized orbital of the central atom forms a sigma bond with the orbital and the pair of electrons from the ligand. For example, the ion diamminecopper(I), $Cu(NH_3)_2^+$, contains the copper(I) ion. Copper as an atom has the following electron configuration in the outermost quantum level and underlying d sublevel:

vertical arrows, ↑ and ↓, are used to represent electrons. the arrows pointing up and down indicate the two electrons in the same orbital are of opposite spins.

	3d	4s	4p
Cu	↑↓ ↑↓ ↑↓ ↑↓ ↑↓	↑	☐ ☐ ☐

the copper(I) ion has the configuration,

	3d	4s	4p
Cu^+ or Cu(I)	↑↓ ↑↓ ↑↓ ↑↓ ↑↓	☐	☐ ☐ ☐

The next orbitals available are the s and p orbitals. In this case the s and one p orbital of the metal ion center hybridize.

	3d	4sp	4p
$Cu(NH_3)_2^+$	↑↓ ↑↓ ↑↓ ↑↓ ↑↓	↑↓ ↑↓	☐ ☐
		sp hybridization	

The sp orbitals are at 180 degrees. When they combine with the orbitals of the two ammonia ligands and accept the electron pair from each, the complex will be linear. That bonding electron configuration is represented above where the hybridized orbitals are grouped together.

The fact that the tetracyanonickelate(II) ion is square planar can be explained similarly. A d orbital is available and hybridizes with the s and two p orbitals to give a set of dsp^2 orbitals. These orbitals are in the form of a square, making the whole complex square planar.

	3d	4s	4p
Ni	↑↓ ↑↓ ↑↓ ↑ ↑	↑↓	☐ ☐ ☐
Ni^{+2}	↑↓ ↑↓ ↑↓ ↑ ↑	☐	☐ ☐ ☐
Ni(II)	↑↓ ↑↓ ↑↓ ↑↓ ☐	☐	☐ ☐ ☐

	3d	dsp^2	4p
$Ni(CN)_4^{2-}$	↑↓ ↑↓ ↑↓ ↑↓	↑↓ ↑↓ ↑↓ ↑↓	☐
		dsp^2 hybridization	

Similarly the iron(II) in the hexacyanoferrate(II) ion, $Fe(CN)_6^{4-}$, is d^2sp^3 hybridized. As seen in Chapter 18 for the phosphorus atom in the phosphorus hexachloride ion, the arrangement about a d^2sp^3 hybridized central atom is octahedral.

	3d	4s	4p
Fe	↑↓ ↑ ↑ ↑ ↑	↑↓	☐ ☐ ☐
Fe^{2+}	↑↓ ↑ ↑ ↑ ↑	☐	☐ ☐ ☐
Fe(II)	↑↓ ↑↓ ↑↓ ☐ ☐	☐	☐ ☐ ☐

	3d	d^2sp^3
$Fe(CN)_6^{4-}$	↑↓ ↑↓ ↑↓	↑↓ ↑↓ ↑↓ ↑↓ ↑↓ ↑↓

In the tetrachlorocobaltate(II) ion, $CoCl_4^{2-}$, a 3d orbital should be the next orbital available in the Co^{2+} atom,

	3d	4s	4p
Co	↑↓ ↑↓ ↑ ↑ ↑	↑↓	☐ ☐ ☐
Co(II) (expected)	↑↓ ↑↓ ↑↓ ↑ ☐	☐	☐ ☐ ☐

One would expect that the d orbital made vacant by the maximum pairing of electrons on the d orbitals would hybridize with the s and two p orbitals to give the square planar configuration of tetracyanonickelate(II) ion. Instead, the s and three p orbitals apparently hybridize to give an sp^3, tetrahedral structure, leaving the electrons to occupy three of the d orbitals singly.

	3d	sp^3
$CoCl_4^{2-}$	↑↓ ↑↓ ↑ ↑ ↑	↑↓ ↑↓ ↑↓ ↑↓

sp^3 hybridized

Of the four complexes studied, $Cu(NH_3)_2^+$, $Ni(CN)_4^{2-}$, $Fe(CN)_6^{4-}$ and $CoCl_4^{2-}$, only $CoCl_4^{2-}$ has any unpaired electrons in its d orbitals. Paramagnetism occurs in atoms, molecules or ions only when there are unpaired electrons. Furthermore, the magnitude of the paramagnetism increases with the number of unpaired electrons. While the sp^3 structure found for $CoCl_4^{2-}$ by x-ray studies is not that expected by our first approach, at least the complex has been found to be paramagnetic, and the paramagnetism is of such a magnitude as to suggest three unpaired electrons. The other three complexes are not paramagnetic.

Although the bonding approach used above is very useful in explaining and correlating geometries and electronic structures of many complex ions, it is lacking in other aspects. In particular, it cannot explain the wide variation of colors of the many coordination compounds. We will not go into the bonding theories which can account for the color of transition metal complexes, but the ligands which approach the metal ion in bonding will have coulombic interaction with the d electrons of the metal. This results in **different energy levels for the d electrons.** Consequently, a different frequency of light is needed to promote a d electron to a higher energy level. Each type of ligand will have slightly different interactions with the d electrons, and as a result there will be a different color.

21.5 COORDINATION COMPOUNDS IN NATURE

chlorophyll is a coordination compound.

According to the structure of chlorophyll A as shown in Figure 21.1, the molecule which envelops the magnesium ion is planar. This is not expected, because the next available orbitals in the magnesium ion, Mg^{2+}, are s and p orbitals. One would expect them to hybridize to give a tetrahedral sp^3 configuration. Apparently the rigid framework of the chelate holds the complexing nitrogen atoms in a plane, regardless of anything else. One p orbital should then be left unhybridized and unoccupied with electrons as a d orbital is brought into the hybridization. The hybridization is then sp^2d, which is shaped the same as a dsp^2 hybrid.

The role of **chlorophyll** in nature is related to the absorption of energy from sunlight and the subsequent transfer of this energy in a chemically useful form. Because of the **importance of the photosynthesis process,** chlorophyll has been the subject of an enormous amount of research.

hemoglobin is a coordination compound of iron.

The iron-centered complex which is of most importance to man is **hemoglobin.** This is the compound by which atmospheric oxygen is held and transported through the blood. The structure of hemin, a part of hemoglobin, along with the structure of chlorophyll A (Figure 21.1), emphasizes the similarity of the molecular skeletons of the two complexes about the central metal ions. Oxygen is transported from the lungs to the tissue cells by formation of a complex between the iron atom and an oxygen molecule. The bonding between oxygen and iron is very weak, and oxygen is readily released to the cells. The oxygen complex is bright red, the characteristic color of arterial blood. Release of oxygen changes the color of the complex and makes the blood purplish red, the color of venous blood.

Other donor molecules or ions react to form complexes with the iron atom of hemoglobin. Carbon monoxide forms a more stable complex than that formed with oxygen, so stable that carbon monoxide does not leave the complex readily. This can greatly decrease the number of sites available for transporting oxygen. As a result, when an organism has carbon monoxide poisoning, the cells starve for lack of oxygen. If caught in time, carbon monoxide poisoning can be reversed by increasing the ratio of oxygen to carbon monoxide in the air taken into the lungs. **Carbon monoxide is a dangerous atmospheric pollutant** precisely because of its ability to render the hemoglobin molecule incapable of performing its oxygen-carrying task. In Table 18.4 it was shown that gasoline engines contribute about 90 per cent of the carbon monoxide in the atmosphere. City officials recently banned automobiles from four of the largest shopping areas in Tokyo, Japan for one day. In the Ginza district the carbon monoxide content of the air dropped from 10.5 to 2.3 parts per million within one hour after the streets were closed. Incomplete combustion of gasoline in the automobile engine, producing CO, is becoming a big problem primarily because of the large number of cars in the U. S. In the future we may see more areas banning use of the automobile. Officials of Yosemite National Park are planning to keep automobiles from the valley floor.

21.6 SOME APPLICATIONS OF COMPLEX ION FORMATION

A metal ion in a complex is often more stable with respect to reduction than is the uncomplexed metal ion. Hence the presence of a ligand which can immediately form a complex with any metal favors oxidation of (loss of electrons by) a metal atom. This principle is applied in the commercial production of gold and silver, since the +1 oxidation state of these metals is stabilized by coordination with cyanide ion. Air oxidation of native gold and silver does not occur at a measurable rate. When carried out in the presence of cyanide ion, under conditions which permit complex formation in accordance with the equations which follow, the air oxidation proceeds with formation of the stabilized ion:

$$4\ Ag_{(s)} + O_2 + 8\ CN^- + 2\ H_2O \rightarrow 4\ Ag(CN)_2^- + 4\ OH^-$$

$$4\ Au_{(s)} + O_2 + 8\ CN^- + 2\ H_2O \rightarrow 4\ Au(CN)_2^- + 4\ OH^-$$

In the absence of cyanide ion, oxygen is not a strong enough oxidizing agent to oxidize gold, but in the cyanide solution the indicated complexes are formed. The precious metals may then be obtained by adding zinc, which displaces them in the complex ion:

$$2\ Ag(CN)_2^- + Zn_{(s)} \rightarrow 2\ Ag_{(s)} + Zn(CN)_4^{2-}$$

and

$$2\ Au(CN)_2^- + Zn_{(s)} \rightarrow 2\ Au_{(s)} + Zn(CN)_4^{2-}$$

The oxidation is accomplished by adding a solution of sodium cyanide to the pulverized noble metal ore held in large vats. Air is passed through the mixture for several days. The solution is then filtered, and zinc is added to the solution of the cyanide complexes to free the noble metals from their complex ions.

A qualitative test for the presence of silver ion in solution is accomplished by the addition of hydrochloric acid to the solution. The appearance of a white precipitate confirms the presence of silver ion.

$$Ag^+ + Cl^- \rightarrow AgCl_{(s)}$$

Other ions that also form precipitates with chloride ion, such as Pb^{2+}, may be tested for, using the complexing ability of ammonia. Silver chloride dissolves upon the addition of an ammonia solution.

$$AgCl_{(s)} + 2\ NH_3 \rightarrow Ag(NH_3)_2^+ + Cl^-$$

Lead chloride, if present, does not dissolve and can be observed.

Developing photographs is a chemical process that involves the formation of a silver complex. A photographic film is a plastic strip coated with an emulsion (a gel) in which are imbedded minute silver bromide crystals (Figure 21.6a). When the film is exposed (the picture is taken), light reflected from light-colored and white objects passes through the lens and falls on the emulsion. The incidence of light breaks down the physical structure of the silver bromide crystals so that they are more susceptible to chemical attack. The emulsion must be kept in the dark except for the exposure. After exposure, the film is removed from the camera and in a dark room is developed in a solution of a reducing agent (Figure 21.6b). The agent reduces the silver ion in the light-struck silver bromide crystals to silver, which in a finely divided state is black. The imperfect (exposed) particles of silver bromide are reduced much faster than the exposed particles and in proportion to the degree of exposure. (However, all the silver bromide particles would be reduced if left in the developing solution too long.) The darkening, or developing, is stopped by washing the film in water and then in a "fixing" solution (Figure 21.6c). The fixing solution, sometimes called "hypo," contains sodium thiosulfate ($Na_2S_2O_3$). Washing in sodium thiosulfate removes the unreduced silver bromide, which is dissolved by the formation of a very stable com-

what coordination compound is formed in the fixing process?

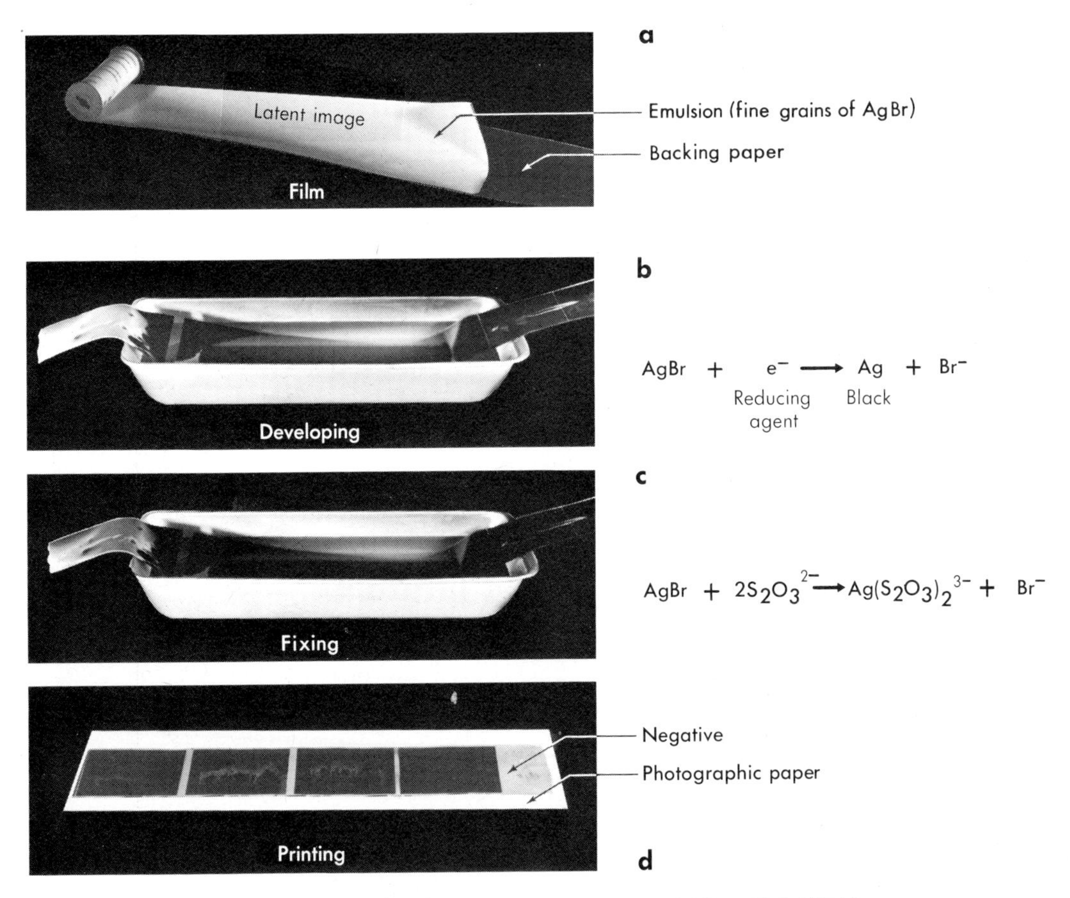

FIGURE 21.6 DEVELOPING A PHOTOGRAPHIC FILM.

plex ion:

$$AgBr + 2\ S_2O_3^{2-} \rightarrow Ag(S_2O_3)_2^{3-} + Br^-$$

The ions in solution are washed from the film. When dried, the treated film is a finished negative, on which white objects appear black, black appears transparent, and gray remains gray, for where light struck the film, black silver remains. The making of the positive (the photographic print) follows the same steps of exposure, developing and fixing. Photographic paper (paper containing a silver bromide emulsion) is exposed to the light through the film (Figure 21.6d). When developed and fixed, a positive is obtained, appearing white where the negative was dark and black where the negative was transparent.

The action of **ion exchange resins** depends upon the formation of complexes. The resin in a synthetic ionic exchanger used to purify water removes metal ions because of the tendency of the ions to form a stable complex ion. A schematic equation for the removal of copper ions from solution appears below.

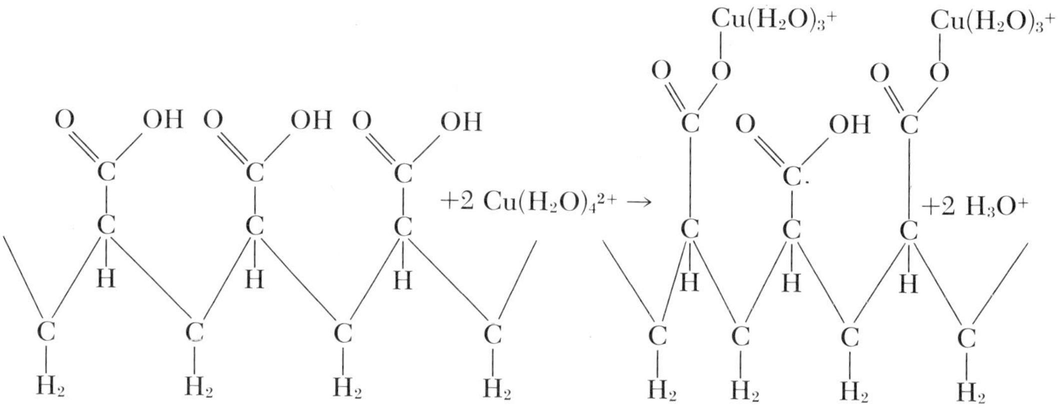

```
                                                          Cu(H2O)3+          Cu(H2O)3+
                                                             |                   |
                                                          O  O                O  O
 O    OH  O    OH  O    OH                                 \\/                 \\/
  \\ /     \\ /     \\ /                                    C   O      OH      C
   C        C        C                                      |    \\   /        |
   |        |        |   +2 Cu(H2O)4 2+ -->                 |      C.          |  +2 H3O+
   C        C        C                                      C      |           C
  /|\      /|\      /|\                                    /|\     C          /|\
 / H \    / H \    / H \                                  / H \   /|\        / H \
C     C        C        C                               C     C    H    C         C
|     |        |        |                               |     |         |         |
H2    H2       H2       H2                              H2    H2        H2        H2
```

21.7 HYDROLYSIS OF METAL IONS AND AMPHOTERISM

"Simple" metal ions in solution are actually complex ions. The water molecule was listed among complexing groups in Table 21.5, and the structure of the hydrated copper(II) ion appeared in Figure 21.4. For example, the ions, Fe^{3+}, Al^{3+}, Ni^{2+} and others are hydrated in solution and in many of their crystalline compounds, and these hydrated ions should be written $Fe(H_2O)_6^{3+}$, $Al(H_2O)_6^{3+}$ and $Ni(H_2O)_6^{2+}$. Just as the proton does not exist in solution but is hydrated to form the hydronium ion, so are these **metal ions hydrated.**

Solutions of metal ions are slightly acidic. That is, in such solutions more protons are complexed with water to form hydronium ions than is the case in water alone. Consider one branch of the

a proton in $Cu(H_2O)_4^{2+}$ ion is more acidic than a proton in a free H_2O molecule.

$Cu(H_2O)_4^{2+}$ ion:

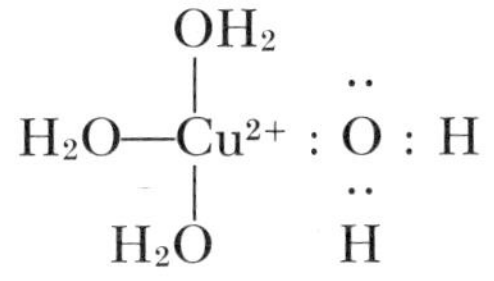

```
        OH2
         |       ..
H2O—Cu2+ : O : H
         |       ..
       H2O       H
```

Electrons are attracted by the doubly-charged Cu^{2+} ion. The pairs of electrons shared between the oxygen and hydrogen atoms are pulled by the charged copper ion toward oxygen, making the attached water molecule more polar and leaving a smaller share of the electrons for each hydrogen atom. The reaction

$$Cu(H_2O)_4^{2+} + H_2O \rightleftarrows Cu(H_2O)_3OH^+ + H_3O^+$$

occurs to only a small extent, but to a greater extent than the ionization of water:

$$2\,H_2O \rightarrow H_3O^+ + OH^-$$

Hence the pH of the solution is less than 7.

Aluminum ion is more acidic than copper(II) ion by the same type of equilibrium reaction:

$$Al(H_2O)_6^{3+} + H_2O \rightleftarrows Al(H_2O)_5OH^{2+} + H_3O^+$$

In a very dilute solution where the hydronium ion concentration remains almost 10^{-7}, the reaction may proceed two steps farther:

$$Al(H_2O)_5OH^{2+} + H_2O \rightleftarrows Al(H_2O)_4(OH)_2^+ + H_3O^+$$

and $$Al(H_2O)_4(OH)_2^+ + H_2O \rightleftarrows Al(H_2O)_3(OH)_{3(s)} + H_3O^+$$

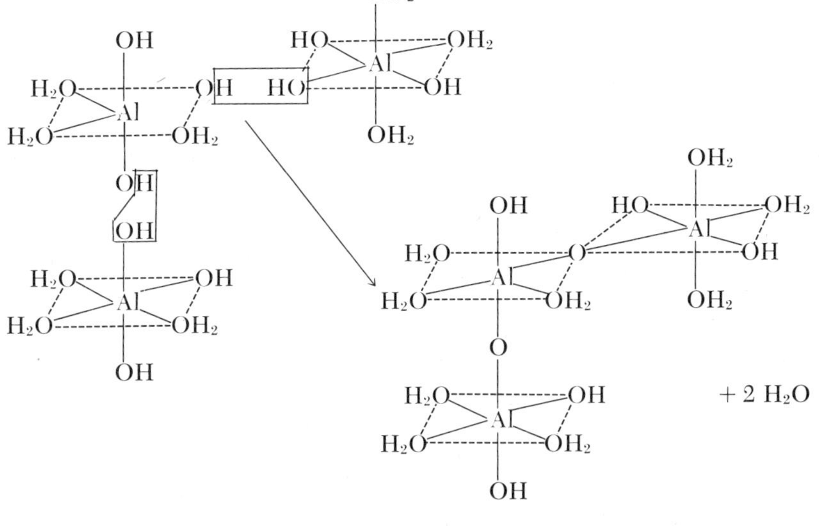

```
         OH                      OH2
         |                HO-----+-----OH2
H2O------+------OH  HO           Al
     Al                 HO-------+-----OH
H2O------+------OH2              |
         |                      OH2
         OH
         OH
H2O------+------OH
     Al
H2O------+------OH2
         |
         OH
```

```
                                            OH2
            OH                     HO-------+-------OH2
            |                               Al
H2O---------+---------O---------------------+-------OH
       Al                                   |
H2O---------+---------OH2                  OH2
            |
            O
            |
H2O---------+---------OH                           + 2 H2O
       Al
H2O---------+---------OH2
            |
            OH
```

As shown in the equations, when three protons are lost from a hydrated aluminum ion, a molecule (a neutral group of atoms) is formed. Not being charged, the molecule has little attraction for solvent dipoles, and it precipitates. The precipitate is often called hydrous aluminum oxide and is written $Al(OH)_3$. The more correct formula is $Al(H_2O)_3(OH)_3$. But even this formula is not entirely correct, because as soon as the molecules are formed, they proceed to join together, with a loss of water as shown at the bottom of the preceding page. An extensive network of molecules soon exists as a filmy white precipitate with the formula of $Al(H_2O)_3(OH)_xO_y$, where $x + 2y = 3$.

Aluminum sulfate is used at times to purify water. In the dilute solution obtained by adding a few crystals of $Al_2(SO_4)_3$ to a pool of water, the hydrated aluminum ion hydrolyzes to hydrous aluminum oxide. The netlike precipitate settles slowly to the bottom, taking dust particles, bacteria and even color from the water.

The aluminum ion is sufficiently acidic to be titrated with a solution of a strong base, such as sodium hydroxide. The addition of three times as much base (in mole quantities) precipitates the aluminum as the hydrous hydroxide:

$$Al(H_2O)_6^{3+} + 3\,OH^- \rightarrow Al(H_2O)_3(OH)_{3(s)} + 3\,H_2O$$

Aluminum hydroxide is a base; it dissolves in an acidic solution, reversing the three-step equation above:

$$Al(H_2O)_3(OH)_3 + 3\,H_3O^+ \rightarrow Al(H_2O)_6^{3+} + 3\,H_2O$$

The hydrous hydroxide is also an acid. Hydroxide ions may take protons from the remaining tightly held water molecules, effectively exchanging them for hydroxide groups. The first step in the reaction occurs when more base is added. It is represented by this equation:

$$Al(H_2O)_3(OH)_3 + OH^- \rightarrow \underset{\text{Aluminate ion}}{Al(H_2O)_2(OH)_4^-} + H_2O$$

The aluminate ion ($Al(H_2O)_2(OH)_4^-$) is commonly written as $Al(OH)_4^-$ or AlO_2^-. This is consistent with the practice of ignoring water of hydration or even "combined" water. "Bookkeeping" shows that

$$Al(OH)_4^- + 2\,H_2O \text{ is equivalent to } Al(H_2O)_2(OH)_4^-$$

and

$$AlO_2^- + 4\,H_2O \text{ is equivalent to } Al(H_2O)_2(OH)_4^-$$

The simpler formulas are often used in preference to the more correct but more cumbersome notation used in this section.

$Al(OH)_3$ and $Zn(OH)_2$ are amphoteric.

Because it is both an acid and a base, aluminum hydroxide is said to be **amphoteric.** Zinc hydroxide also is amphoteric, as is shown by these equations:

$$Zn(H_2O)_2(OH)_2 + H_3O^+ \rightarrow Zn(H_2O)_3(OH)^+ + H_2O$$
acting as a base

$$Zn(H_2O)_2(OH)_2 + OH^- \rightarrow Zn(H_2O)(OH)_3^- + H_2O$$
acting as an acid

Other amphoteric hydroxides (written without water of hydration) are $Be(OH)_2$, $Cr(OH)_3$, $Zn(OH)_2$, $Sn(OH)_4$, $Pb(OH)_2$, $As(OH)_3$ and $Sb(OH)_3$. Notice that the metals in the formulas fall near the "staircase" line separating metals and non-metals in the periodic table. They are among the most electronegative of the metals and therefore hold complexing groups through the shared pair of electrons most tightly. The hydroxide and water groups are held so securely that aluminum hydroxide and zinc hydroxide are covalent and not ionic compounds. This makes the acidic behavior possible. Protons could not be pulled away unless the water molecules in the hydroxide were held tightly by the metal center.

Hydroxides of active metals (electropositive metals), such as barium hydroxide, $Ba(OH)_2$, are basic but not acidic. $Ba(H_2O)_xOH^+$ does not exist as a unit in solutions. The large barium ion surrounded by an indefinite and changing number of water molecules remains free of the hydroxide ions present in solution. Water molecules about the large barium ion are not held sufficiently secure that a proton might be pulled away. Transition metal hydroxides tend to be covalent and show slight amphoteric properties, but the effect becomes significant only for hydroxides of metal having sufficient electronegativity to form very stable complexes with oxygen.

21.8 NOMENCLATURE

Because there are thousands of transition metal complexes known, because more are being found and synthesized every year and because they are complex, often having several different groups in the same complex, the systematic method developed by the International Union of Pure and Applied Chemistry (IUPAC) for naming all inorganic compounds and ions should be used.

Rules for that system adapted especially for complexes are summarized here.

1. As with all ionic compounds, the positive ion is named first; the negative ion is named second with a space between the names. This follows even though one or both of the ions may be metal-centered complexes.

2. Within the name of the complex, ion or molecule, the ligands are named before the metal center—the negative ligands first, then the neutral ligands. The number of identical ligands are indicated by the Greek prefixes, di, tri, tetra, penta, hexa, and so forth. If the name of the ligand itself comtains a prefix, the prefixes, bis, tris, tetrakis, and so forth, are used. That is, the presence of two ethylenediamine ligands on the same metal center is indicated by the name fragment, bis(ethylenediamine).

3. Negative ligands are given the ending -o; neutral ligands are named as molecules:

Cl^-	chloro	$NH_2CH_2CH_2NH_2$	ethylenediamine
OH^-	hydroxo	Cl_3P	trichlorophosphine
CN^-	cyano	H_2NNH_2	hydrazine
F^-	fluoro		

Some exceptions to this rule are:

H_2O aquo NH_3 ammine   CO carbonyl

4. The central metal atom of the complex is named after the ligands, and the oxidation state of the metal in the complex is noted by a Roman numeral set off by parentheses and placed immediately after the metal name, concluding the name of the complex.

5. If the complex is a cation or neutral molecule, the ending of the name is not changed but remains the normal ending of the metal center. If the complex ion is a negative ion, the end of its name, which is the end of the name of the metal center, is changed to -ate. The ending -ate comes before the Roman numeral.

6. There is no spacing between the parts of the name of a complex ion or molecule. Examples of the application of the naming system are illustrated in Table 21.6.

TABLE 21.6 EXAMPLES OF THE NOMENCLATURE OF COORDINATION COMPOUNDS

Formula	Name
$[PtCl_2(NH_3)_2]$	Dichlorodiammineplatinum(II)
$[Co(NO_2)NH_3)_5]Br_2$	Nitropentaamminecobalt(III) bromide
$Na_3[Fe(CN)_6]$	Sodium hexacyanoferrate(III)
$Na_2[Fe(CN)_6]$	Sodium hexacyanoferrate(II)
$[Cr(H_2O)_5Cl]Cl_2 \cdot H_2O$	Chloropentaaquochromium(III) chloride monohydrate
$[Co(NH_2CH_2CH_2NH_2)_3]_2(SO_4)_3$	Tris(ethylenediamine)cobalt(III) sulfate
$[Ag(NH_3)_2]Cl$	Diamminesilver(I) chloride

EXERCISES

21.1 Define and give an example of each:

a. metal (chemical sense)
b. active metal
c. transition element
d. complex ion
e. metal-centered complex ion

21.2 Why do transition elements usually exhibit a number of oxidation states while the active metals do not?

21.3 What characteristics do transition metal ions have that help promote the formation of complex ions?

21.4 What is the coordination number of Co(III) in the complex $[Co(H_2O)_2(NH_3)_2Cl_2]Cl$?

21.5 What is a ligand?

21.6 Ammonia, carbon monoxide, ethylenediamine and water are all ligands, whereas the ammonium ion is not. Explain.

21.7 Explain the difference between a simple covalent bond and a coordinate-covalent bond.

21.8 Why are transition metal complexes sometimes called coordination compounds?

21.9 Why is a metal ion sometimes called a Lewis acid?

21.10 What is a tridentate ligand?

21.11 Explain why Ti^{+3} ion in aqueous solution is colored, whereas Ti^{+4} ion in aqueous solution is colorless.

21.12 Name two transition metal complexes which are industrially important and state their uses.

21.13 What is the hybridization of a square planar complex? Of a linear complex ion? Of an octahedral complex?

21.14 Why is there little difference in the size of the atoms of the fourth period transition elements?

21.15 Name the following compounds: $[Cr(H_2O)_4(Cl)_2]Cl$ and $[Co(NH_5)_5H_2O]SO_4$.

21.16 Write two chemical reactions which illustrate the amphoterism of $[Al(H_2O)_3(OH)_3]$.

21.17 Why would an aqueous solution containing the $Ti(H_2O)_6^{4+}$ complex be more acidic than a solution containing $Ti(H_2O)_6^{3+}$?

21.18 Which of the following compounds would one expect to be colored?

a. $[Cu(H_2O)_4]SO_4 \cdot H_2O$
b. $Al(H_2O)_3(OH)_3$
c. $[Ag(NH_3)_2]Cl$
d. KCl
e. $K_3[Fe(CN)_6]$
f. $[Ni(NH_3)_6]Br_2$
g. $[Zn(NH_3)_4]Cl_2$
h. $[Cr(H_2O)_4Cl_2]Cl \cdot 2H_2O$

SUGGESTED READING

Larsen, E. M.: *Transitional Elements.* W. A. Benjamin, New York, 1965 (paperback).

Martin, D. F., and Martin, B. B.: *Coordination Compounds.* McGraw-Hill, New York, 1964 (paperback).

Masterton, W. L., and Slowinski, E. J.: *Chemical Principles.* 2nd Edition, Chapter 19, W. B. Saunders Co., Philadelphia, 1969.

Research Reporter: "Transfer of Oxygen from Red Blood Cells to Tissues." Chemistry, *40* (5): 35, (1967).

Schubert, J.: "Chelation in Medicine." Sci. Amer., *214* (5): 40 (1966).

TWENTY-TWO • CARBON, SILICON AND BORON

22.1 INTRODUCTION

Atoms of the elements of Groups IIIA and IVA in the periodic table have three and four valence electrons respectively, having the electron structures ns^2np^1 and $ns^2np^1np^1$. From their positions in the middle of the table and from the fact that the atoms have about half the number of electrons of the noble gas configuration, one might assume that each element of the groups might be both metallic and non-metallic. However, because electronegativity increases generally from left to right and from bottom to top in the periodic table, the uppermost elements of the groups have greater attraction for electrons and tend to be more non-metallic, while the lower members of the groups are metallic. This is true in Groups IIIA and IVA, where the stair-step dividing line in the table that separates metals from non-metals passes beneath boron and silicon, placing boron, carbon and silicon among the non-metals. Boron and silicon have some properties intermediate between metals and non metals and are sometimes called metalloids (meaning "like metals").

metalloids are elements that have some metal-like properties.

The remaining members of the groups are generally metallic. This is especially true of tin and lead, which, along with aluminum (which has been discussed in Chapter 20 from the point of view of its metallurgy), are the most important of the lower members of the groups. Being metals with several valence electrons and relatively great attraction for electrons, tin and lead behave similarly to the transition metals, having more than one ionic valence and sharing electrons of ligands to form many metal-centered complex ions. Some properties of the Group IIIA and IVA elements are given in Tables 22.1 and 22.2, in which breaks in the trends from member to member illustrate the change from non-metals to metals. Because chemistry of the type exhibited by the metals of the two

TABLE 22.1 SOME PROPERTIES OF THE GROUP IIIA ELEMENTS

	Boron	Aluminum	Galium	Indium	Thallium
Electronic configuration	$2s^22p^1$	$3s^23p^1$	$4s^24p^1$	$5s^25p^1$	$6s^26p^1$
Melting point (°C)	2300	660.2	29.78	156.6	303.5
Boiling point (°C)	2550 (sublimes)	2467	2403	2000	1460
Atomic radius (Å)	0.80	1.25	1.25	1.50	1.55
Ionization energy (kcal/mole)	191	138	138	133	141
Electronegativity	2.0	1.5	1.6	1.7	1.8
Density (g/cc)	2.34	2.699	5.91	7.31	11.85

TABLE 22.2 SOME PROPERTIES OF THE GROUP IVA ELEMENTS

	Carbon	Silicon	Germanium	Tin	Lead
Electronic configuration	$2s^22p^2$	$3s^33p^2$	$4s^24p^2$	$5s^25p^2$	$6s^26p^2$
Melting point (°C)	3550	1410	937.4	231.9	327.5
Boiling point (°C)	4827	2355	2830	2270	1744
Atomic radius (Å)	0.77	1.17	1.22	1.41	1.54
Ionization energy (kcal/mole)	260	188	182	169	171
Electronegativity	2.5	1.8	1.8	1.8	1.8
Density (g/cc)	3.513	2.33	5.323	7.31	11.35

groups has been discussed elsewhere, boron, carbon and silicon will be studied in this chapter. The two members of Group IVA will be discussed first.

22.2 CARBON AND SILICON

The chemistry of carbon and silicon is concerned primarily with covalent bonding. Usually these atoms do not form simple ions. To do so, the atom must either lose four electrons to achieve a charge of positive four, or it must gain four electrons to gain a negative charge of four. Both elements are always found in complex ions or in molecules in which the carbon or silicon shares electrons with other atoms and attains, with the sharing, a completed octet of valence electrons.

Carbon differs from silicon, and from all other elements in its ability to form thousands of compounds in which many **carbon atoms are bonded to each other** in continuous chains, branched chains and closed chains or rings. This property was illustrated by the listings in Table 10.2 of the known hydrides (compounds with hydrogen) of the elements of the first row of eight in the periodic table. According to that table there are "hundreds of thousands" of compounds containing only the two elements, carbon and hydrogen. The field of organic chemistry is devoted entirely to these compounds of carbon and hydrogen and to those in which the hydrogen is replaced by one or more other elements. Boron and silicon atoms do exhibit a slight ability to combine in chains. There are 15 known borohydrides (Table 10.2), and silicon hydrides are known up to Si_6H_{14}.

Stannane (SnH_4) and plumbane (PbH_4) are the only hydrides prepared with the heavier carbon family elements. The lack of many hydrides with Si—Si bonds, and the absence of any with Sn—Sn and Pb—Pb bonds, is a function of the weakness of those bonds and is related to size of the atom. A comparison of the energies of the single bonds

C—C	83.1 kcal/mole
Si—Si	42.2 kcal/mole
Ge—Ge	37.6 kcal/mole
Sn—Sn	34.2 kcal/mole
C—C	83.1 kcal/mole
C—H	98.8 kcal/mole
C—O	84.0 kcal/mole
C—Cl	78.5 kcal/mole
Si—Si	42.2 kcal/mole
Si—H	70.4 kcal/mole
Si—O	88.2 kcal/mole
Si—Cl	75.7 kcal/mole

shows that the strength of the C—C bond is of the same magnitude as the energies of the bonds between carbon and other elements. **The Si—Si bond,** on the other hand, is much weaker than the bonds that silicon usually forms with other elements.

Another difference between silicon and carbon is the ability of the carbon atom to form multiple bonds, which involve overlap of p orbitals, both with other carbon atoms and with atoms of other elements. Silicon does not participate in this type of pi bonding. This allows for great differences in the properties of their compounds. Consider the melting points of carbon dioxide and silicon dioxide, −56.5°C (at 5.11 atm) and 1710°C, respectively. With double bonding between carbon and each oxygen atom, carbon dioxide is a linear molecule, $:\ddot{O}{=}C{=}\ddot{O}:$. There are discrete three-atom molecules of CO_2, whereas each crystal of **SiO_2 is a giant molecule** in which each silicon is tetrahedrally bonded to four oxygen atoms, and each oxygen atom is bonded to two silicon atoms. The structure of SiO_2 is discussed in more detail later.

carbon can form multiple bonds.

With its capability to form multiple bonds, the carbon atom in its compounds can participate in one triple or a triple and a single bond, two double bonds, or a double and two single bonds; it can be sp, sp^2 and sp^3 hybridized. Furthermore, one or two p orbitals of the carbon atom can be involved in forming a delocalized bonding orbital. This means that complex ions containing the carbon atom may be resonance structures. The hybridization of the carbon atom and the delocalization of the same ions containing the carbon atom were described in detail in Chapter 5. The bonding versatility of carbon contributes to the great variety of compounds classified in organic chemistry.

22.3 OCCURRENCE OF CARBON

Carbon is a relatively rare element, constituting approximately 0.03 per cent in the earth's crust, but it is found in many places. The common substances petroleum, natural gas and limestone contain carbon. All plants and animals contain carbon in an endless variety of compounds. Carbon also occurs in nature in the free state. Coal is largely carbon with volatile impurities; graphite is mined in Ceylon and Madagascar; diamond is found mainly in the form of rough stones in deposits in the Union of South Africa, the Congo, Tanzania, Ghana and other African countries. About 23,000,000 carats (nearly 5 tons) of diamonds are mined annually. A carat weighs 0.200 gram.

22.4 ALLOTROPES OF CARBON (GRAPHITE AND DIAMOND)

The **allotropy** of carbon is most interesting. Few substances differ so greatly in physical properties (Table 22.3) as diamond and graphite, both of which are pure crystalline carbon.

The uses of these two elemental forms of carbon further emphasize their contrasting properties. The diamond may be a gem in an engagement ring or in a pendant, or it may be the stylus of a record player. Millions of dollars worth of diamonds are crushed per year for industrial uses. Small fragments are mounted in the

TABLE 22.3 PHYSICAL PROPERTIES OF PURE CARBON

Property	Diamond	Graphite
Configuration	Octahedral irregular sharp fragments	Flakes and large crystals
Color	Colorless, yellow, brown, black	Black
Light transmission	Transparent, but a high index of refraction and high dispersion	Opaque
Density (g/cm^3)	3.51	2.22
Hardness	Hardest natural material known	Very soft
Conductivity	A non-conductor of electricity	A conductor (approximately 0.001 that of copper)

tips of steel cutting tools, and on the surface of grinding wheels for cutting hard metal, stones and glass. Diamond dust is used for polishing fine bearing surfaces and for polishing other diamonds. Diamond is the hardest abrasive known. The uses of diamond listed above depend mainly upon the remarkable hardness of the substance and its resultant resistance to wear or scratching. Diamond is an almost perfect electrical insulator.

industrial uses of diamond depend upon its hardness.

Pencil lead is a mixture of graphite and clay. The soft graphite-clay mixture wears away and leaves a black line on the paper against which it is rubbed. Graphite is used as a bearing lubricant and may be dry or suspended in water or oil. It forms a slippery film which reduces the friction between two sliding surfaces. **Graphite** with a binder is pressed into **electrodes for electrolysis cells**; its conductivity and its resistance to electrical corrosion make it useful in this capacity.

The remarkable differences in the properties of the hard, glittering diamond and black, soft, flaky graphite suggest that there are great differences in the "internal" structure of the two substances. These differences can be seen if one considers the two structures as revealed by X-ray analysis. In the diamond (Fig. 22.1) each carbon atom is sp^3 hybridized and bonded to four other atoms by single sigma bonds in a three-dimensional lattice. Each of the four valence electrons of each carbon atom is paired with an electron from another carbon atom to form 4 non-polar covalent bonds. The angles between any two bonds radiating from a single carbon atom are 109°28′. The four carbon atoms to which one carbon is attached are at apices of a regular tetrahedron, and each carbon to carbon bond is 1.54 Å in length.

Every atom in a diamond gem stone, except those on the surface, are bonded to four more atoms. Because all atoms are bonded together, the stone is sometimes called a macromolecule (giant molecule). One can see that this three-dimensional, covalently bonded macromolecule should be very rigid and have a high melting point (Table 22.1). A fracture of the crystal in any direction must break many carbon-carbon bonds. However, there are certain

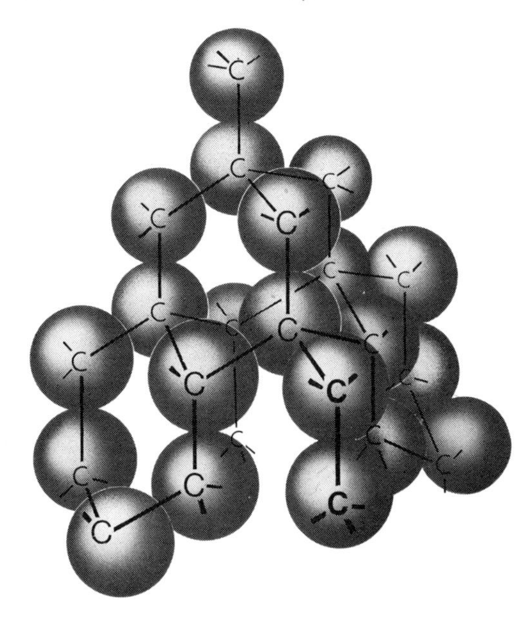

FIGURE 22.1 THE ARRANGEMENT OF C ATOMS IN DIAMOND.

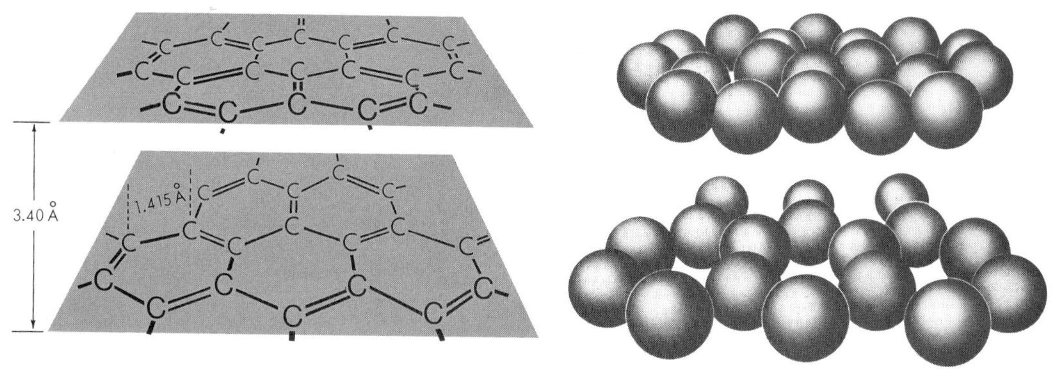

FIGURE 22.2 CARBON ATOMS IN GRAPHITE.

planes along which fewer bonds need be broken. To prevent shattering, a diamond cutter must determine these planes before he cleaves an uncut diamond. Since all the valence electrons are involved in bonding between carbon atoms, a diamond is a very poor conductor of electricity.

Graphite is not bonded in a three-dimensional network as is diamond, but each atom is bonded to three others in layers of repeating hexagons (Fig. 22.2). Each diamond may be considered a molecule, but each layer of graphite is a molecule. The distance between adjacent atoms in a hexagon is 1.415 Å. No covalent bonds exist between the layers, which are only held together by weak van der Waals forces. All the covalent bonds between atoms in each layer are equivalent. The carbon to carbon distance in graphite (1.415 Å), compared to that in diamond (1.54 Å), suggests that **some multiple bonding** must exist **in graphite** to account for the bond shortening. Since each carbon atom is in a plane and bonded to three other carbon atoms, it is sp^2 hybridized. This leaves each carbon atom with the fourth electron in a p orbital, which is perpendicular to the plane of the molecule (Fig. 22.3a). The p orbitals overlap to form delocalized orbitals that extend over the complete layer (Fig. 22.3b). The electrons pair to occupy the orbitals which are lowest in energy. The graphite layer may be represented by resonance structures in which each bond is a $1\frac{1}{3}$ bond (Fig. 22.3c). The electrons in these molecular orbitals belong to the molecule as a whole and are comparatively free to move throughout the entire layer. These mobile electrons account for the **electrical conductivity** and for the weak bonding between graphite layers.

There are several well-known examples of so-called "amorphous" carbon. Among these are charcoal (both from wood and bone), coke and carbon black. **Charcoal** is made by heating wood, or bones, at high temperature in the absence of air. **Coke** is the residue from a high temperature distillation of coal. **Carbon black** is an incomplete combustion product of natural gas formed by impinging the flame on a metal surface. Upon close investigation these forms of carbon appear to be microcrystals of the graphite variety, containing numerous imperfections and impurities. Charcoal, because of its great surface area and the holes in it, has a great ability to absorb matter. **Activated charcoal** (charcoal heated in a steam atmosphere) is added to colored solutions such as cane sugar syrup to absorb the coloring material; it is also used dry in gas-absorbing masks.

activated charcoal is used as a decolorizer.

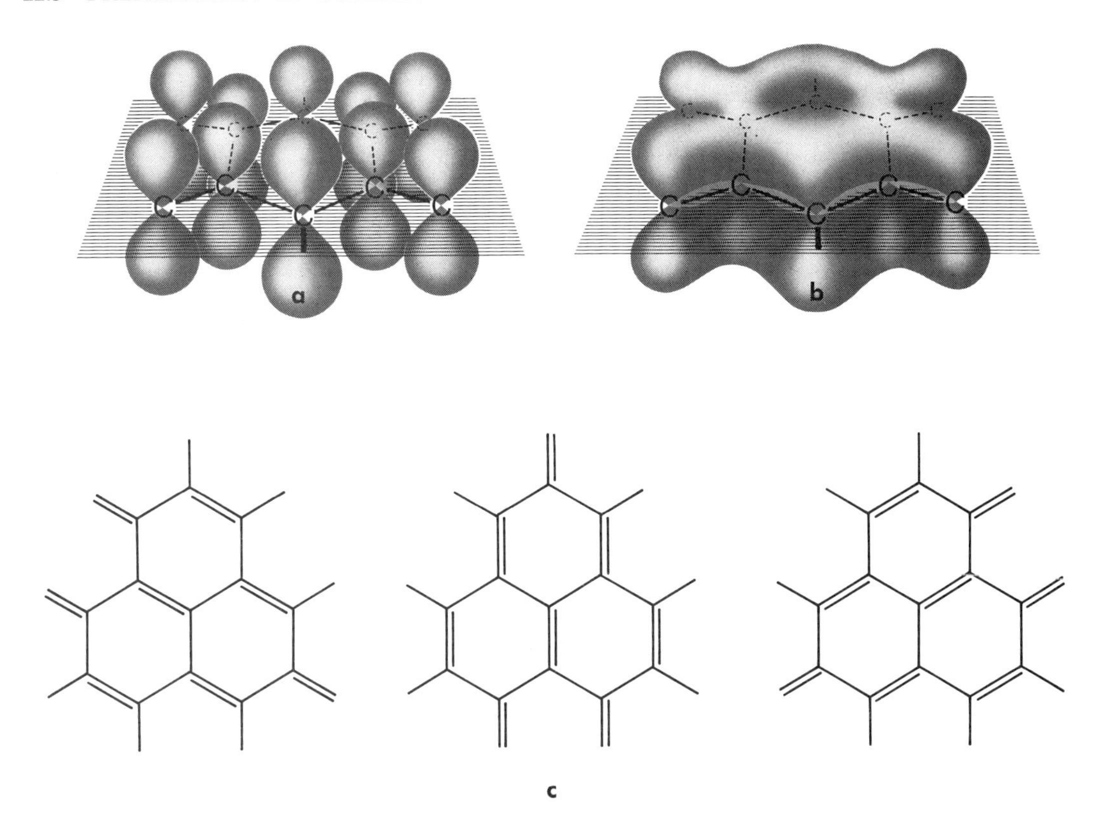

FIGURE 22.3 BONDING IN GRAPHITE.

22.5 PREPARATION OF DIAMOND

Free elemental carbon of the amorphous form is prepared rather easily by the methods mentioned above. Graphite may be prepared by simply heating hard coal to 2000°C in the absence of air in an electric furnace. The manufacture of diamonds is much more difficult. Man has sought methods for making diamonds for many years. In 1880, J. B. Hannay heated various mixtures containing graphite in iron tubes until they exploded, but it is not known for sure if any diamonds were synthesized in this manner. Henri Moissan dissolved carbon in molten iron, then plunged the liquid metal into cold water. The tremendous pressure of the iron as it cooled from the outside and contracted was thought sufficient to press the crystallizing carbon into diamond, but only graphite flakes were found in the steel. In 1940, P. W. Bridgman placed graphite under a pressure of 6,000,000 pounds per square inch, but to no avail. In 1954, a team of scientists with General Electric Company subjected graphite (with other ingredients) to both extremely high temperature (1800°C) and extremely high pressure (2,000,000 psi) and produced a cluster of tiny diamonds. Diamonds sufficiently large for use in cutting tools are being made by this process, and research continues to make them larger and more economically.

Recently a 20 carat diamond was produced. In less than three years from the successful production of man-made diamonds more than 100,000 carats were produced. Man-made diamonds now supply a great many of the requirements for industrial diamonds.

22.6 SOME CARBON-CONTAINING COMPOUNDS

Some inorganic compounds of carbon such as sodium carbonate (Na_2CO_3) and sodium hydrogen carbonate ($NaHCO_3$) were considered briefly in earlier chapters. Other important inorganic compounds of carbon will be discussed here.

1. **Calcium carbide.** When calcium oxide is heated with carbon to about 3000°C, calcium carbide is formed:

$$CaO + 3\,C \xrightarrow{\text{Heat}} \underset{\text{Calcium carbide}}{CaC_2} + CO$$

Calcium carbide is very reactive with water, yielding acetylene gas. Calcium carbide was the fuel for the old miner's lamp. Water dripping down upon the solid carbide released acetylene, which burned with a luminous flame:

$$CaC_2 + 2\,H_2O \rightarrow Ca(OH)_2 + \underset{\text{Acetylene}}{C_2H_2}$$

The C_2^{2-} ion has ten electrons. Like the nitrogen molecule, it contains a triple bond.

$$:C:::C:^{2-}$$

Sodium, potassium, boron, aluminum, silicon and many other elements also form solid carbides.

2. **Carbon dioxide.** Carbon dioxide, used to make sodium carbonate and sodium bicarbonate and liberated in the roasting of limestone, coral, oystershells and other forms of $CaCO_3$, is also found in the respiratory gases of animals. Oxygen is carried from the lungs in the blood and **sugars are oxidized** in the cells, as represented by this overall reaction:

$$C_6H_{12}O_6 + 6\,O_2 \rightarrow 6\,CO_2 + 6\,H_2O + \text{energy}$$

Muscular energy and heat are released in the process. The carbon dioxide formed is carried in turn by the blood to the lungs where it is exhaled. Carbon dioxide is not poisonous. However, its presence in air decreases the percentage of oxygen and necessitates an increased rate of breathing. Animals can tolerate a 50 per cent mixture of carbon dioxide in air for only a short time. However, the amount of carbon dioxide even in a closed, crowded room seldom becomes greater than 2 or 3 per cent. The high humidity that develops in such a closed space is actually more harmful.

Green plants in sunlight **absorb carbon dioxide** from the air and combine it with water to form carbohydrates. The overall equation for the formation of a simple sugar is the reverse of the equation for carbohydrate oxidation in the animal cell:

$$6\,CO_2 + 6\,H_2O \xrightarrow{\text{Sunlight}} C_6H_{12}O_6 + 6\,O_2$$

This reaction, and others similar to it, are important because energy is taken from the sun to maintain a food-energy source, and oxygen is produced for the respiration of animals. **Green plants and animals aid each other** in this way. A balance must exist between them to maintain the present percentages of carbon dioxide and oxygen, but because of the population explosion it may not always be possible to maintain this balance.

The energy requirements of the world's population are extremely large. Most of the energy comes from the burning of the fossil fuels (petroleum, coal and natural gas). From the oxidation of these fuels about ten billion tons of carbon dioxide are dumped into the air each year. As a result, carbon dioxide is becoming a pollutant. Besides its other objectionable features, carbon dioxide causes a "greenhouse effect." Carbon dioxide is transparent to the visible radiation of the sun, but it does absorb infrared radiation that is radiated out from the earth. Consequently, atmospheric CO_2 traps radiated energy from the earth and prevents it from escaping. Glass in greenhouses performs a similar function, trapping the energy that is emitted by the objects in the greenhouse. It appears that the temperature of the earth's surface has increased 0.2°C and that the temperature of the stratosphere has increased 2°C as a result of increasing carbon dioxide in the atmosphere. Further increases could cause melting of glaciers with a resulting rise in the sea level. This may not become much of a problem in time because of the rapid depletion of fossil fuels. Nuclear fuels do not produce carbon dioxide.

what is the "greenhouse effect"?

Some fire extinguishers contain liquid carbon dioxide under pressure. The liquid is vaporized upon leaving the extinguisher to form the gas. The non-flammable gas, which is more dense than air, covers the burning material, excluding oxygen. Other extinguishers contain sulfuric acid and sodium hydrogen carbonate in separate, uncapped compartments. When turned over, the contents mix to form a carbon dioxide foam that covers the burning area with an inert blanket. Carbon dioxide, a product of the complete combustion of carbon-containing materials, cannot be further oxidized.

Carbon dioxide is a colorless, odorless gas with a slightly sour taste. It is very soluble in water and forms the weakly acidic solution:

$$2\,H_2O + CO_2 \rightleftarrows H_3O^+ + HCO_3^-$$

The crisp, tart taste is detected in carbonated beverages, which are saturated with carbon dioxide gas at several atmospheres pressure. When the bottle cap is removed, some gas escapes with bubbling, but enough remains to flavor the drink.

3. **Carbon monoxide.** Carbon monoxide, a product of the incomplete combustion of carbon-containing materials, can be oxidized very easily to carbon dioxide:

$$2\,CO + O_2 \rightarrow 2\,CO_2 + 135{,}300 \text{ calories}$$

As mentioned previously, it is a commercial fuel. Further, it is an extremely important reducing agent. One example of its utility referred to in Chapter 20 was in the reduction of iron oxide in the blast furnace. Carbon monoxide has a great affinity for the iron in hemoglobin, the oxygen carrier in blood. It fills the sites normally occupied by oxygen. Without oxygen, cell metabolism ceases and death results.

why is carbon monoxide such an important reducing agent?

Carbon monoxide will react with several metals (under the proper conditions), forming such carbonyl complexes as $Fe(CO)_5$, $Ni(CO)_4$ and $Co_2(CO)_8$. One of these carbonyls, $Ni(CO)_4$ (tetracarbonylnickel(0)), is very important commercially. In the Mond process for producing nickel, carbon monoxide is passed over a mixture of nickel, copper and iron sulfides at 50°C. Tetracarbonylnickel(0), a liquid that boils at 43°C, is carried away in the vapor. At a still higher temperature (180°C) the carbonyl compound is decomposed; the carbon monoxide is restored and may be used again. Only purified nickel remains.

22.7 SILICON

Silicon is the second most abundant element in the earth's crust, which is estimated to be 27.7 per cent silicon, 49.1 per cent oxygen and 8.1 per cent aluminum. Silicon and oxygen are the principal elements in earth and rocks. The continents rest on great slabs (20 miles thick) of granite, which is largely silicon and oxygen. Sand is silicon dioxide. Kaolin (clay) is an impure, hydrated aluminum silicate with the approximate composition $Al_2(Si_2O_3)(OH)_4$. The three most abundant elements are combined together in kaolin. Other aluminosilicates are feldspar, soapstone and mica. Silicates constitute 87 per cent of the earth's crust.

22.8 SILANES

A metal-silicon compound, magnesium silicide (Mg_2Si), which is prepared by heating magnesium and silicon in the absence of air, reacts with dilute hydrochloric acid in the absence of oxygen

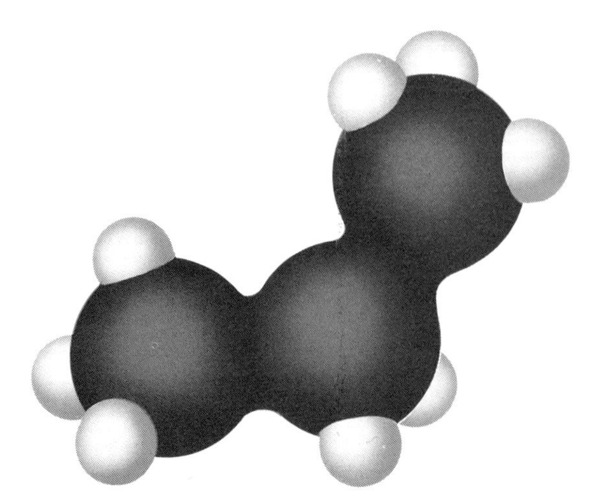

FIGURE 22.4 A MOLECULE OF Si_3H_8.

to produce several members of a family of compounds called **silanes:** SiH_4, Si_2H_6, Si_3H_8, Si_4H_{10} and traces of silicon-hydrogen compounds of greater molecular weight. A model of Si_3H_8 (Fig. 22.4) shows that each silicon atom in the molecule has a configuration similar to that of the carbon atom in the diamond crystal. Four covalent bonds surround each silicon atom, which is sp^3 hybridized; all angles between bonds are about equal to the tetrahedral angle (in diamond), 109°28′. The structure of the silicon atom implies the existence of many larger silicon-hydrogen compounds. However, silanes are not stable because the silicon to silicon bonds are so weak. They burn spontaneously in air to give silicon dioxide and water:

$$SiH_4 + 2\,O_2 \rightarrow SiO_2 + 2\,H_2O$$

They react with weak alkali solutions to produce silicates and hydrogen.

22.9 SILICON DIOXIDE (SILICA)

Sand, quartz and crystobalite are composed of silicon and oxygen in the atomic ratios of 1 to 2. The formula generally used, SiO_2, indicates that there are discrete 3-atom molecules like those of CO_2. This is incorrect; a better formula would be $(SiO_2)_n$, where n is a very large variable integer. Each particle of sand and each quartz crystal may be considered a **macromolecule,** for the silicon and oxygen are covalently bonded together throughout the whole substance. Each silicon atom is bonded to four oxygen atoms; each oxygen atom is bonded to two silicon atoms. In effect, silicon atoms are bonded to one another through oxygen atoms (Fig. 22.5). This makes sand a hard substance, but not so hard as diamond, because with only two bonds per oxygen there are fewer covalent bonds per unit volume in the crystal. As might be expected, the angles between the silicon bonds extending to the oxygen atoms are equal to the tetrahedral angle. The angle between the oxygen bonds is slightly less.

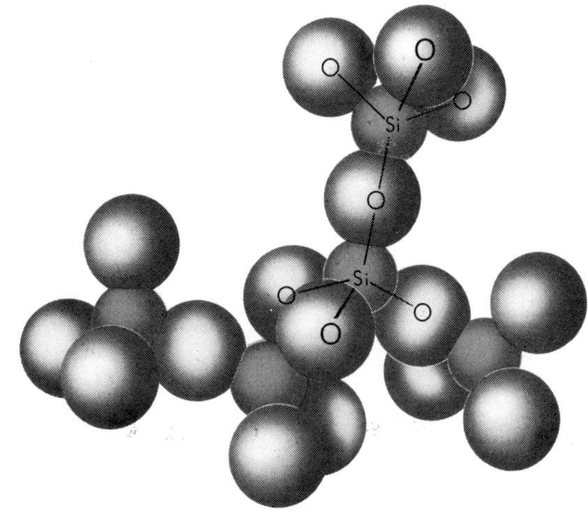

FIGURE 22.5 THE STRUCTURE OF SAND (SiO_2).

Silica can be fused at about 1700°C to a clear, colorless liquid. Orderly formation of the covalent bonds is difficult, and silica supercools readily, hardening to quartz glass, which contains no quartz at all but is simply vitreous silica. Quartz glass is interesting because of its high melting point and because of its permeability to ultraviolet light, which is excluded almost entirely by ordinary glass.

22.10 IGNEOUS ROCKS

The same structural unit, **silicon atoms surrounded by four oxygen atoms** in chains or rings, is the basic unit in igneous rock and in clay soil produced from the rock by weathering. Granite, the most abundant rock, is 70 per cent SiO_2 combined with the elements aluminum, magnesium, calcium, potassium, hydrogen and others. The internal structure of muscovite, a typical mineral found in granite, is illustrated in Figure 22.6. Two silicon-oxygen layers are covalently bonded together through aluminum atoms. The double layers are bonded in turn by potassium ions. The ionic bonding is less rigid, making muscovite a flaky material. In clay the ions in the small particles of muscovite are hydrated with water molecules that wedge in between the double silicate layers, causing the clay to expand and become spongy.

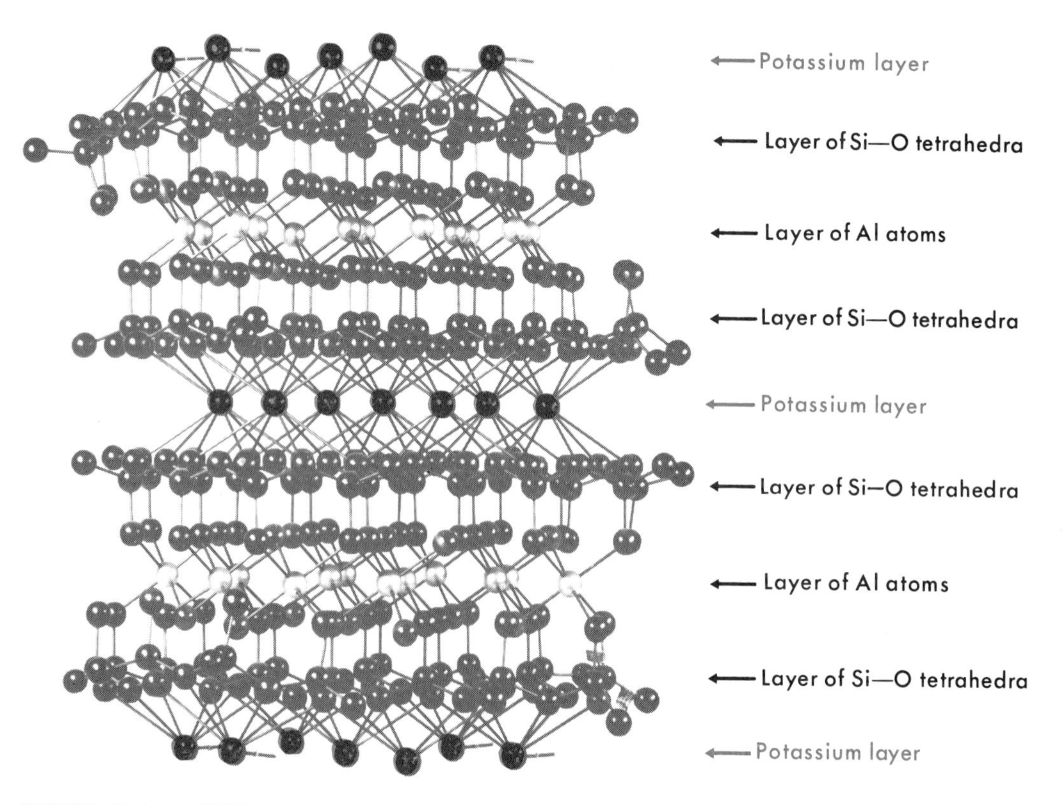

FIGURE 22.6 MUSCOVITE.

22.11 GLASS

Common window glass, called soft glass, is made by melting a mixture of sand, sodium carbonate and calcium carbonate. The melt of 70 to 80 per cent SiO_2, 12 to 17 per cent Na_2O, 5 to 15 per cent CaO and 1 to 4 per cent Al_2O_3 is not a pure compound. It supercools to a hard brittle state with no definite crystal structure.

Pyrex glass contains a higher percentage of silicon dioxide (about 80 per cent), and also 12 per cent B_2O_3, the balance being sodium and aluminum oxides. The boron-containing glass is harder and has a higher melting point than soft glass. In addition, it has a much lower coefficient of expansion; that is, it expands much less when heated. As a result, it stands sudden changes in temperature much better than soft glass. Common glass breaks when one portion is raised to a higher temperature than another because it expands away from the cooler portion. Because of its low coefficient of expansion, pyrex glass is often used for cooking ware.

Traces of transition-metal oxides added to the melt color glass by forming coordination compounds in it. The color is somewhat dependent on the particle size of the oxide impurity and upon the rate of cooling of the glass. Chromium oxide and copper oxide impart a greenish color to glass, cobalt glass is blue, and glass containing dispersed gold is ruby colored.

22.12 SILICONES

The silicones are an interesting family of compounds containing silicon, oxygen, carbon and hydrogen. Silicon atoms are bonded to one another through oxygen atoms in long chains as in silicate rocks. However, the silicon bonds not only with oxygen but also with carbon, as shown here diagrammatically:

```
   CH3     CH3     CH3
   |       |       |
 —Si—O—Si—O—Si—O—
   |       |       |
   CH3     CH3     CH3
```

A characteristic silicone structure

If the chains of silicon and oxygen atoms are short, the substances are liquid and oily; if the chains are long, the substances are rubbery. **The silicones tend to maintain their elasticity and fluidity** over a wide range of temperatures. They are quite unreactive with most chemicals. They are used as a unique form of "inorganic" rubber (they have no C—C bonds in the chain), as lubricants, and for gaskets and electrical insulation when extreme temperature changes are encountered. The boots that made the first footprints on the moon were made of silicone rubber because no other rubbery polymer would stand the high temperature of the moon's surface during the lunar day.

why are the silicones important?

22.13 BORON

As mentioned in the introduction to this chapter, boron is a non-metal, whereas the other elements in the family are metallic in character. The predominant covalent character of boron can be illustrated by the melting points of the fluorides of the Group III elements:

BF_3	−126.7°C
AlF_3	1291°C (sublimes)
GaF_3	800°C (sublimes)
InF_3	1170°C
TlF_3	550°C (decomposes)

The oxides and hydroxides of the boron family exhibit the usual trend of decreasing acidic behavior as one goes down the group. Boric acid, $B(OH)_3$, is weakly acidic, which indicates that boron is a non-metal. Both **$Al(OH)_3$ and $Ga(OH)_3$ are amphoteric,** and $In(OH)_3$ and $Tl(OH)_3$ are basic.

what is an amphoteric compound?

Boron does not occur free in nature. It is present principally in ores as borates such as borax, $Na_2B_4O_7 \cdot 10H_2O$. Only about 0.001 per cent of the earth's crust is composed of boron. Boron is obtained by heating the oxide (B_2O_3) with a large excess of magnesium, which acts as a reducing agent and combines with the oxygen:

$$B_2O_3 + 3\ Mg \rightarrow 2B + 3\ MgO$$

Pure crystalline boron is obtained by converting impure boron to boron tribromide (BBr_3), which is then passed over a hot (1400°C) tungsten wire in the presence of hydrogen:

$$2\ BBr_3 + 3H_2 \rightarrow 2B + 6\ HBr$$

Boron has the electronic configuration $2s^2 2p^1_x 2p_y^0 2p_z^0$, which gives it three valence electrons and four valence orbitals. Consequently, it does not have enough valence electrons to use all of its valence orbitals nor to attain the octet valence electron configuration of the noble gases in forming covalent bonds. Boron in BF_3 is sp^2 hybridized; it is a planar molecule with the electronic structure

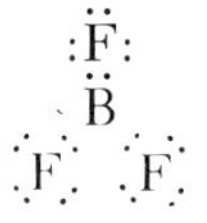

Since the boron atom does not have the octet of electrons in boron trifluoride, it is called an electron-deficient compound. Electron-deficient compounds such as the boron halides and aluminum halides are able to act as Lewis acids by accepting pairs of electrons from some Lewis bases. Some examples are

$$:\ddot{F}:^- \quad + \quad BF_3 \quad \rightarrow \quad [BF_4]^-$$

and

$$H_3N: \quad + \quad BF_3 \quad \rightarrow \quad H_3N:BF_3$$

BF₃ is a Lewis acid.

In the formation of the acid-base complexes, boron becomes sp^3 hybridized. Electron-deficient compounds such as BF_3 are acid catalysts because of their acidic behavior, in the Lewis sense.

Boron forms several hydrides of the general formulas B_nH_{n+4}, where n = 2, 5, 6 and 10, and B_nH_{n+6}, where n = 4, 5, 9 and 10. The simplest boron hydride would be "BH_3," but it does not exist. Instead, the dimer of BH_3 is the simplest hydride isolated.

Diborane is also an **electron-deficient** molecule. The arrangement of the atoms in the molecule is shown in Figure 22.7a. When two "BH_3" groups are brought together so that the atoms are placed in proper relationship to each other

```
H :  . H     . H
    B   :   B
 :     H  :   :
H              H
```

it can be seen that there are only 12 valence electrons to bond 8 atoms together. To accomplish the bonding, each of two electron pairs enters a delocalized molecular orbital involving the two boron atoms and one of the hydrogen atoms between them. The two *boron to hydrogen to boron* bonds are called three-center bonds. A molecular orbital representation is illustrated in Figure 22.7b. The higher boranes have three-center B—H—B bonds and three-center B—B—B bonds as well as bonds with a larger number of centers.

The boranes are very reactive compounds and are spontaneously flammable in air. A great deal of research has been carried out with them in anticipation of their use as rocket fuels. They have a higher heat of combustion per gram than do hydrocarbons.

Borohydride ion (BH_4^-), which is tetrahedral, is also known. It can be prepared by the reaction of lithium hydride and diborane in ether as a solvent:

$$2\ LiH + B_2H_6 \rightarrow 2\ LiBH_4$$

Higher borohydrides have also been prepared. The alkali **boro-**

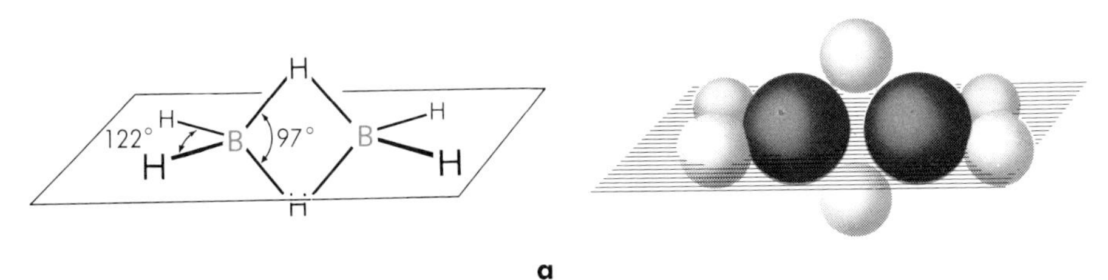

a

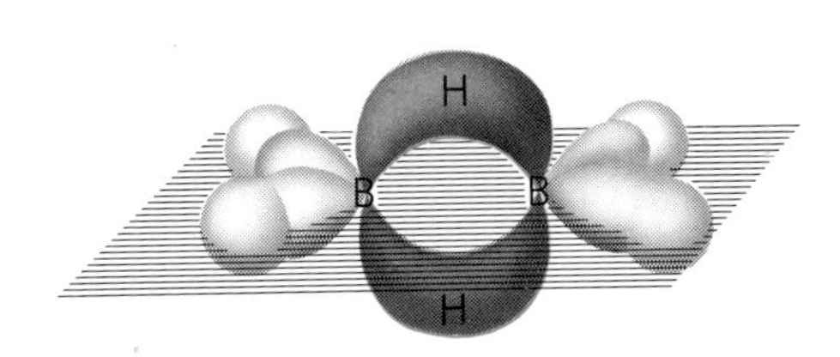

b

FIGURE 22.7 DIBORANE.

hydrides are very important reducing agents in organic chemistry.

Diborane reacts with ammonia to form the ionic compound

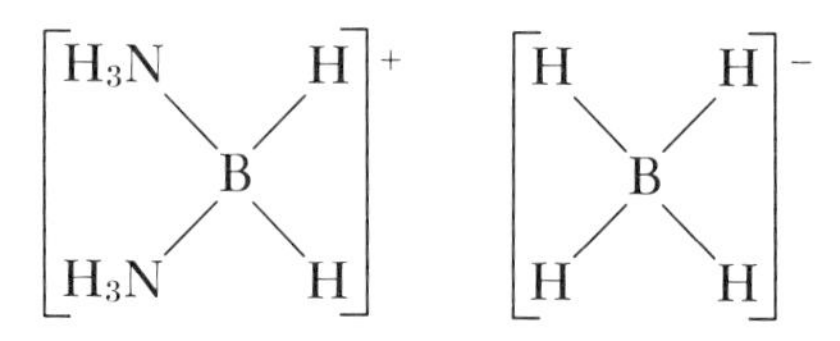

Heating the diammoniate of diborane to 200°C produces $B_3N_3H_6$, known as borazine. Upon further heating a boron nitride is produced that has a structure similar to that of graphite:

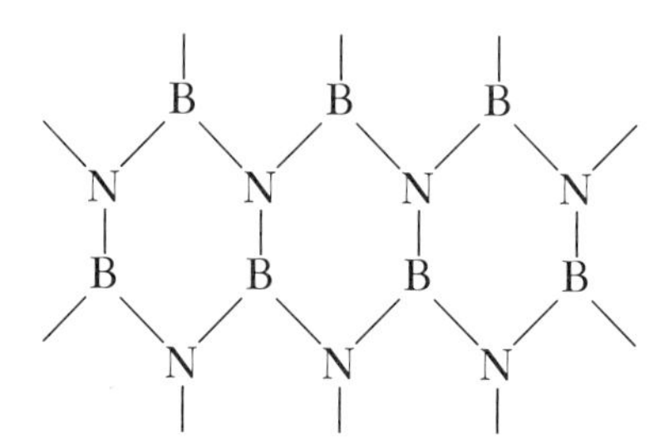

Since each nitrogen atom has five valence electrons and each boron has three electrons, and since the nitrogen and boron atoms are in a 1 to 1 ratio, the structure is isoelectronic with graphite, where each carbon atom supplies four valence electrons. Boron nitride is an unreactive, insoluble, refractory material. The layers are arranged such that a boron atom lies directly under and directly over nitrogen atoms in the layers which it separates. As in graphite, the layers are held together by van der Waals forces. At high temperatures and pressures boron nitride can be converted to a diamond-like structure. The resulting material is known as borazon; it is one of the hardest substances known.

EXERCISES

22.1 Of the elements B, Al, C, Si, Sn and Pb, which are considered metals?

22.2 In the order of their abundance, list the five most abundant elements in the earth's crust.

22.3 List five compounds in which the two most abundant elements appear together.

22.4 Explain the reason for the difference in the physical properties of the two forms of elemental carbon.

22.5 What two elements of bonding do carbon compounds possess that are not possessed by compounds of silicon?

22.6 What is the hybridization of (a) C in CO_2, (b) C in graphite, (c) C in CO_3^{2-}, (d) Si in SiO_2, (e) C in CH_4, (f) B in BF_3, (g) B in BF_4^-, (h) C in diamond?

22.7 Which of these pure elemental solids has the highest melting point: C, Si, Ge, Sn or Pb? Why?

22.8 Why is a diamond called a macromolecule? Why are the properties of borazon so similar to those of diamond? Why are the properties of boron nitride so different?

22.9 Why is there such a great difference between the densities of diamond and graphite?

22.10 Why should the bond angles of carbon in diamond be 109°28′? Why is this angle called the tetrahedral angle?

22.11 What are the bond angles in graphite?

22.12 Why is graphite a better electrical conductor than diamond?

22.13 In what way must a balance exist between the respiration of plants and animals?

22.14 What is the major commercial use of carbon dioxide?

22.15 Describe the effect of carbon dioxide on the temperature of the atmosphere at the earth's surface?

22.16 Describe the action of a sulfuric acid-sodium carbonate fire extinguisher.

22.17 Why should sand be less hard than diamond?

22.18 What three elements are most abundant in clay soils?

22.19 Why does the volume of clay shrink when it sets?

22.20 What is the prominent molecular difference between the gas CO_2 and the solid SiO_2?

22.21 Why is it that boron glass can stand much greater temperature changes than soft glass?

22.22 What is the common structural unit in silica and in silicate materials?

22.23 Describe the action of carbon monoxide as a poison.

22.24 Give two bits of experimental evidence that boron is a non-metal.

22.25 Describe both steps of the preparation of pure boron from sodium borate.

SUGGESTED READING

Battista, O. A.: "Black Magic with Carbon Black." Chemistry, *41* (4):8 (1968).

Bolin, B.: "The Carbon Cycles." Scientific American, *223* (3):124 (1970).
Hutchinson, G. E.: "The Biosphere." Scientific American, *223* (3):44 (1970).
Mellon, E. K.: "Synthetic Borane Chemistry: A Challenge to Chemical Theory." Chemistry, *41* (10):8 (1968).
Research Reporter: "Hannay's Diamonds." Chemistry, *42* (6):20 (1969).
Rochow, E. G.: "*The Metalloids.*" D. C. Heath and Co., Chicago, 1966 (paperback).
Woodwell, G. M.: "The Energy Cycle of the Biosphere." Scientific American, *223* (3):64 (1970).

TWENTY-THREE • THE NUCLEUS AND CHEMISTRY

23.1 INTRODUCTION

A large volume of literature has accumulated from the many scientific investigations of the nuclei of atoms, and hence much is known about these tiny, though massive, centers of atoms. The composition of nuclei is expressed in terms of the kinds and numbers of nucleons (protons and neutrons) they possess. In every instance, except for the lightest isotope of hydrogen, all atomic nuclei are composed of protons and neutrons. Nuclei having certain numbers of neutrons appear to be more prevalent. Structurally, it appears that there are energy levels for nucleons in the nucleus just as there are energy levels for extranuclear electrons that surround the atom. Within nuclear energy levels there are also sublevels which, when they are filled with nucleons, lend a stability to the nucleus not unlike the stability of the filled s, p, d and f orbitals in the electronic structures of atoms and ions.

the nature of the nucleus is studied in nuclear physics.

Theories have been developed to explain the nature of the forces necessary to hold comparatively massive particles (protons), all bearing the same charge, in such a tiny nuclear volume. Studies of these fascinating topics, however, must be left for classes in nuclear physics. This chapter will be devoted to other equally fascinating aspects of the subject: gross nuclear changes, the change in chemical identity associated with the nuclear reactions, the energies involved in nuclear reactions and the use made of nuclear changes and of the products of those changes. The discussion will be divided into five parts: (a) the discussion of nuclear composition as it relates to the size of atoms and their position in the periodic table (some facts presented in Chapter 4 will be repeated here), (b) mass and energy relationships, (c) the radioactive decay of larger nuclei (with the formation of new atoms), (d) nuclear fission and

(e) atomic fusion (union of smaller nuclei). Applications will be interwoven into the discussion.

23.2 COMMON FACTS ABOUT NUCLEI

The nucleus may be assumed to be composed of neutrons and protons only. About 40 subatomic particles have been discovered (mesons, neutrino, and so on). Some may well exist in the nucleus but, if so, they are there at the expense of the nucleons. The remaining type of particle found in the atom is the electron. The electrons occupy the volume about the nucleus.

The composition of the nucleus of an atom is shown by a subscript and a superscript preceeding the symbol. Thus $^{16}_{8}O$ represents the common oxygen nucleus. It contains 16 nucleons, 8 of which are protons and 8 neutrons. $^{238}_{92}U$ is the uranium nucleus, containing 92 protons and a total of 238 nucleons, or $238 - 92 = 146$ neutrons. The number 238, the nucleon total, is called the mass number.

Atoms are numbered and named according to the number of protons in the nucleus. The proton number is the so-called atomic number. All atoms of the same element have the same atomic number. Atoms are arranged in the periodic table in order of increasing atomic number. Most elements have isotopes, which are atoms of the same element with different masses; they are, therefore, atoms having the same number of protons but possessing a different number of neutrons. Usually, stable isotopes of the same element differ by only one or two neutrons, but there may be 2, 3, 4 or more such isotopes. For tin, ten naturally occurring stable isotopes have been found having the mass numbers 112, 114, 115, 116, 117, 118, 119, 120, 122 and 124.

the two isotopes of chlorine, 35 and 37, give it an atomic weight of approximately 35.5.

Among the lighter naturally occurring atoms the number of protons and neutrons is about equal. However, as the number of protons increases, the ratio of the number of neutrons to the number of protons increases. Figure 23.1 shows a plot of the number

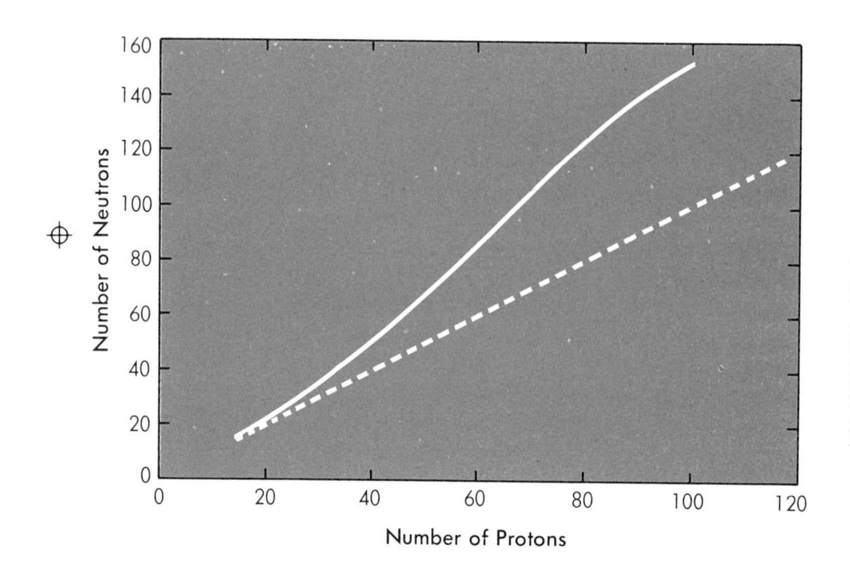

FIGURE 23.1 PLOT OF NEUTRONS/PROTONS FOR SOME TYPICAL NUCLEI (THE SOLID LINE REPRESENTS OBSERVED RATIOS AND THE DASHED LINE 1 TO 1 RATIOS).

of neutrons versus the number of protons for a typical group of atoms. Apparently this increasing number of excess neutrons is necessary for the stability of the larger nuclei. In the heavier nuclei the ratio of neutrons to protons slightly exceeds 1.5 to 1.

23.3 NUCLEAR MASS AND ENERGY

Of great importance is the fact that the mass of the nucleus of an atom is not equal to but less than the combined masses of its separate nucleons. For example, the nucleus of the common isotope of carbon ($^{12}_{6}C$) has a mass of 12.0000 atomic mass units (amu). It contains 6 protons, each with a mass of 1.00814 amu, and 6 neutrons, each one having a mass of 1.00897. The apparent loss of mass due to the combination of nucleons is called the **mass defect.**

Mass defect = sum of individual mass of nucleons present minus the observed mass.

$$\text{Mass defect} = (6 \times 1.00814) + (6 \times 1.00897) - 12.0038 = 0.0989 \text{ amu}$$

Obviously, mass is not conserved when 6 neutrons and 6 protons combine to form a $^{12}_{6}C$ nucleus. Instead, a mass of 0.0989 amu disappears. It is converted to energy. Using Einstein's equation for mass energy equivalence, $E = mc^2$, one finds that 2.14×10^{12} calories of heat would be liberated when the necessary neutrons and protons are brought together to form 1 mole of carbon-12 nuclei. This is equivalent to the energy from 6×10^5 lbs. of coal. Dividing the mass defect by 12 and expressing the value in terms of an energy unit, mev (million electron volts), one arrives at a quantity called the **binding energy per nucleon.** This is, in fact, the binding energy, since it is the energy which must be invested for each of the 12 nucleons of carbon-12 to separate the nucleus into its particles.

The binding energy curve for the atoms appears in Figure 23.2. The binding energy per nucleon for each atom is plotted against the mass number (the number of nucleons). One can see that atoms of a total mass of approximately 60 have the greatest binding energy

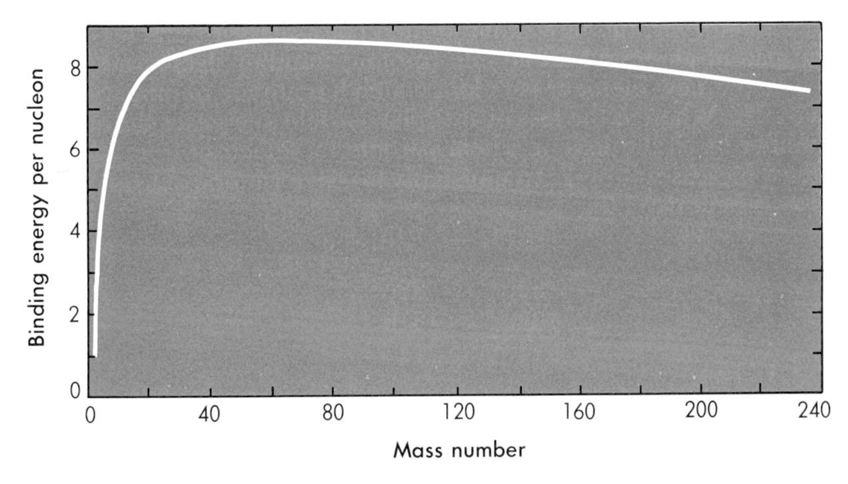

FIGURE 23.2 PLOT OF BINDING ENERGY PER NUCLEON.

per nucleon. The binding energies per nucleon gradually decrease for atoms with a mass number greater than 60 and decrease steeply from this maximum for atoms of smaller mass number. Obviously, if larger nuclei are separated into smaller fragments (of about mass number 60), energy must be released, and if tiny nuclei are combined into larger nuclei, still greater energies must be liberated.

23.4 RADIOACTIVE DECAY

All atoms with an atomic number greater than 83, and a few others which are lighter, are radioactive. That is, they undergo a spontaneous loss of alpha particles, beta particles or gamma radiation, or all three. The nature of some types of radiation was considered in Chapter 4, where it was noted that alpha rays bend very gradually toward the negative electrode of an electric field, that beta particles bend sharply toward the positive electrode and that gamma rays follow a straight path through a transverse electric field. Quantitative tests of the degree of bending show the alpha ray to be a particle, a helium nucleus (He^{2+}) in motion, and the beta particle to be an electron (e^-). The gamma ray is a high-energy light ray similar to an x-ray.

The spontaneous loss of tiny particles from the nucleus is known as **radioactive decay,** for in losing the particles (or radiation) the atom decreases in mass and changes from an atom of one element into an atom of another element in sequence until, in the series beginning with radium, a stable lead atom results. One can follow the pathway of the changes and note the kind of atom remaining as each ray is lost. The first step in the decay of radium is the loss of an alpha particle:

$$^{226}_{88}Ra \rightarrow \ ^{222}_{86}Rn + \ ^{4}_{2}He$$

$$\begin{array}{lll} 88p^+ & 86p^+ & 2p^+ \\ \underline{138n} & \underline{136n} & \underline{2n} \\ 226 & 222 & 4 \end{array}$$

Because the alpha particle (a helium nucleus) contains 2 protons and 2 neutrons, the new atom formed has a mass 4 units less than the parent atom and an atomic (proton) number decreased by 2; hence the new element is radon, which has atomic number 86 and mass 222. Whereas the loss of an alpha particle decreases the atomic number by 2, the loss of a beta particle (an electron from the nucleus) increases it by 1. An unusually massive isotope of lead, $^{214}_{82}Pb$, loses a beta particle:

$$^{214}_{82}Pb \rightarrow \ ^{214}_{83}Bi + e^-$$

$$\begin{array}{ll} 82p^+ & 83p^+ \\ \underline{132n} & \underline{131n} \\ 214 & 214 \end{array}$$

In effect, in the nucleus a neutron is converted to a proton (assumes a +1 charge) with the loss of an electron. Gamma radiation, on the other hand, merely reduces the internal energy of an atom but does not change its identity.

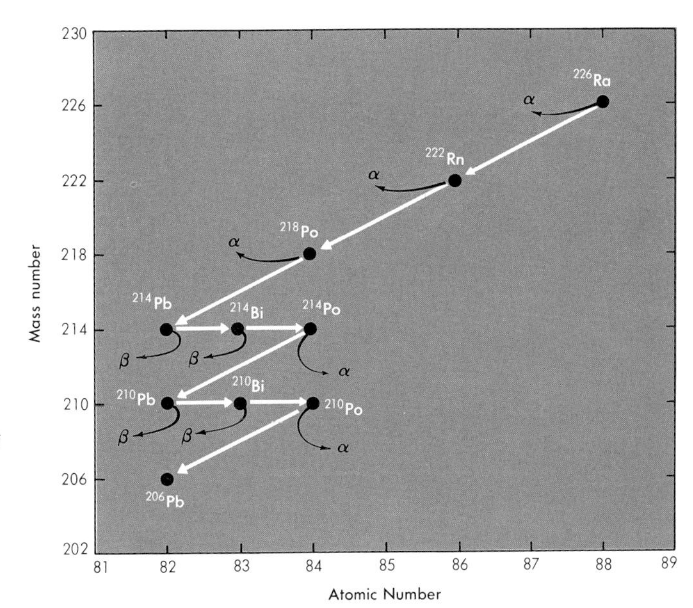

FIGURE 23.3 RADIUM DECAY SERIES.

The radium decay series, which shows the atoms obtained after each radiation or particle loss, is shown in diagrammatic form in Figure 23.3. An increase in atomic number is shown by a move to the right and a decrease by a move to the left.

Uranium, protactinium and thorium, elements 92, 91 and 90, are the largest naturally occurring atoms. They are all radioactive and decay by loss of beta particles and alpha particles in a stepwise fashion through radium to lead.

23.5 DETECTION OF RADIOACTIVE DECAY

There is a variety of methods to detect radioactive decay.

1. Gamma radiation affects *photographic film* in much the same way as does light, but exposure by gamma radiation may be accomplished even through the wrapping. The exposure provides a quantitative measure of gamma radiation.

the so-called radium watch dials which glow in the dark depend upon alpha particle-activated fluorescence.

2. Zinc sulfide and some other minerals fluoresce when bombarded with alpha particles, and the amount of light (or even the number of individual scintillations) can be measured to evaluate the alpha radiation.

3. When gamma rays pass through a gas at low pressure between charged electrodes, they ionize the gas. The ions that are formed are drawn to the charged electrodes, causing a sudden surge of current, which causes the flick of a galvanometer needle or a click in a counter. The tube containing the charged electrodes is called a **Geiger tube** and is the essential feature of the instrument called a Geiger counter (see Figure 23.4). Connected to a counter, a Geiger tube will measure radiation intensity. Geiger counters are carried over the desert in searches for deposits of uranium, and they are widely used for monitoring natural radiation.

Some radioactive material and a Geiger counter can be used to carry out very interesting studies. For example, one can add a radio-

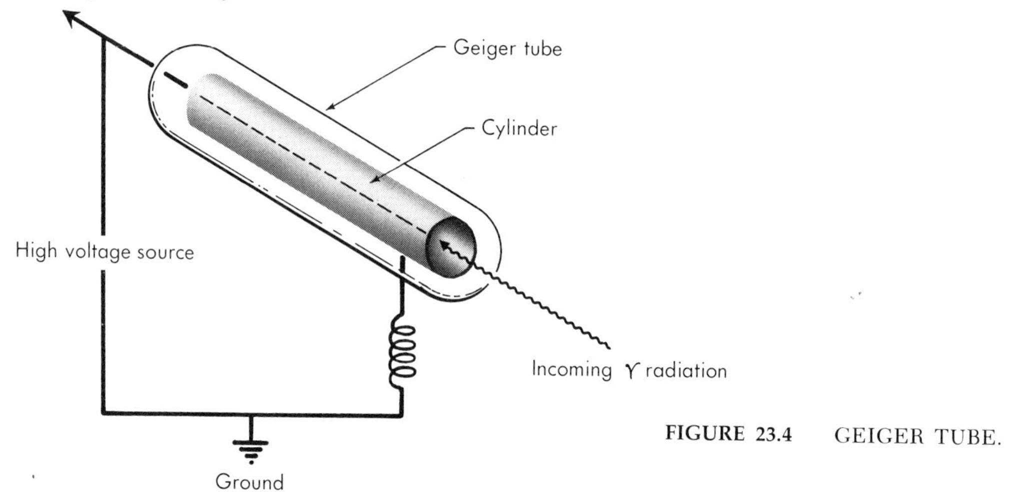

FIGURE 23.4 GEIGER TUBE.

active solute to a body of water, then by monitoring springs below with a Geiger counter, can determine which are fed by the reservoir above. One can determine the volume of blood in the body by admitting a small measured sample of blood (of the same type) containing radioactive material with a known radiation count. After the blood is completely mixed, a like sample of blood will exhibit a count reduced by the dilution factor. One can follow a radioactive chemical from a nutrient solution up the roots and stem of a plant to see where that chemical accumulates. Radioactive iodine is used to determine the activity of the thyroid gland.

4. Possibly the most accurate measure of the quantity of radiation is through a direct analysis of radiating material. Alpha and beta radiations change the identity of the atoms.

A. An alpha-emitting sample may be enclosed in a gastight evacuated container. Helium gas accumulates, one atom per alpha particle. The pressure of the helium is a measure of the amount of radiation.

how are the alpha particle and the helium gas atom related?

B. One can perform a mass analysis. If the radiating sample is a gas, or if it can be vaporized, it can be admitted to the ionization chamber of a mass spectrograph, where fragments of molecules, parent atoms and products are separated because of the difference in their masses.

C. Chemical analysis reveals the amount of accumulated radiation products and is a measure of the radiation time and intensity.

23.6 RATES OF RADIOACTIVE DECAY

Radioactive decay is a **first-order process,** that is, it changes at a rate which depends only upon the concentration or amount of decaying material present. A certain fraction of the unstable atoms present, regardless of the total number, decays in a certain period of time. The rates of radioactive decay reactions are compared by half-lives. The half-life is the time required for one half of the atoms to decay. This time is always the same for the same material regardless of the mass of the sample.

Different radioactive species have different half-lives and therefore decay at different rates. The half-life may vary from a microsecond to thousands of years. Half-lives of some radioactive elements are shown in Table 23.1.

TABLE 23.1 HALF-LIVES FOR CERTAIN DECAY REACTIONS

Reaction			Half-life
$^{3}_{1}H$	⟶	$^{3}_{2}He + \beta^-$	12.4 years
$^{14}_{6}C$	⟶	$^{14}_{7}N + \beta^-$	5568 years
$^{32}_{15}P$	⟶	$^{32}_{16}S + \beta^-$	14.3 days
$^{35}_{16}S$	⟶	$^{35}_{17}Cl + \beta^-$	87.1 days
$^{131}_{53}I$	⟶	$^{131}_{54}Xe + \beta^-$	8.07 days
$^{90}_{38}Sr$	⟶	$^{90}_{39}Y + \beta^-$	27.7 years
$^{221}_{88}Ra$	⟶	$^{217}_{86}Rn + \alpha$	30 seconds
$^{234}_{92}U$	⟶	$^{230}_{90}Th + \alpha$	2.48×10^5 years
$^{257}_{103}Lr$	⟶	$^{253}_{101}Md + \alpha$	8 seconds
$^{60}_{27}Co$	⟶	$^{56}_{25}Mn + \alpha$	5.26 years

α = Alpha particle
β^- = Beta ray (electron)

No conditions of pressure or temperature short of the temperatures existent on the stars or in the center of an atomic explosion have been found to alter the rate of radioactive decay. This principle is used when using radioactivity to determine the age of the earth. If one assumes that deposits of radioactive material were pure parent material at the creation of the earth (and the universe), from the half-life of the process and the relative amounts of parent and daughter material present one can calculate the age of the earth. Estimates of the earth's age by this method are about 5.5 billion years.

write the notation for the common isotope of carbon.

Determination of age by carbon dating assumes that while the tree or plant lived, from which a manuscript or a wooden object was made, it contained the normal percentage of radioactive carbon ($^{14}_{6}C$) in its fibers. When the tree died, no more radioactive carbon could enter by respiration of carbon dioxide in the leaves. The radioactive carbon in the wood then decayed at the usual rate, and its percentage and radioactivity (count) decreased with time. The less radioactivity (the slower the count) found, the older the manuscript or object.

23.7 ATOMIC SYNTHESIS

Larger atomic nuclei, already radioactive, will often absorb another particle hurled into them. The resultant nuclei are also usually radioactive. In 1939, Enrico Fermi proposed that atoms of atomic number greater than 92 (uranium then being the largest atom known) could be made by bombarding uranium nuclei with neutrons.

In the period of time since 1940, atoms of **12 new elements,** up to and including the element of atomic number 105, have been made. All have been synthesized by a similar process—particles are hurled into the nucleus. Capture of the particle by the nucleus is followed by the radioactive emission of one or more other particles.

In 1940, the common uranium isotope of mass 238 was irradiated with neutrons. The following reactions occurred:

A. Capture of a neutron gave uranium 239:

$$^{238}_{92}U + ^{1}_{0}n \rightarrow ^{239}_{92}U$$

B. $^{239}_{92}U$ then emitted a beta particle to yield an atom of a new element having atomic number 93. This element was named neptunium.

$$^{239}_{92}U \rightarrow ^{239}_{93}Np + \beta^-$$

C. Subsequent emission of a second beta particle gave an atom of another new element, which was named plutonium.

$$^{239}_{93}Np \rightarrow ^{239}_{94}Pu + \beta^-$$

D. Americium was made by capture of colliding alpha particles by uranium 238:

$$^{238}_{92}U + ^{4}_{2}He \rightarrow ^{242}_{94}Pu$$

followed by beta emission

$$^{242}_{94}Pu \rightarrow ^{242}_{95}Am + \beta^-$$

The larger transuranic elements, from atomic number 97 through 105, have been prepared since 1954 by bombardment of heavy nuclei with charged nuclei of small atoms using the cyclotron. For example, californium was prepared as shown below:

$$^{238}_{92}U + ^{12}_{6}C \rightarrow ^{244}_{98}Cf + 6n$$

23.8 NUCLEAR FISSION

A radically different type of nuclear reaction can occur when heavy atomic nuclei absorb neutrons. Absorption of slow neutrons with the eventual elimination of a beta particle converted uranium-238 into neptunium, a new element discussed in the preceding section. However, when the uranium-238 atom is bombarded with fast neutrons with an energy exceeding one million electron volts,

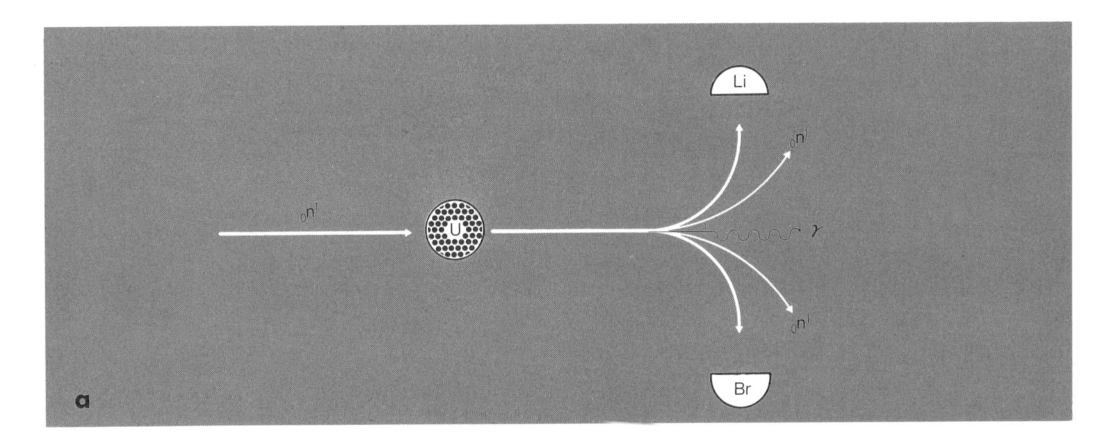

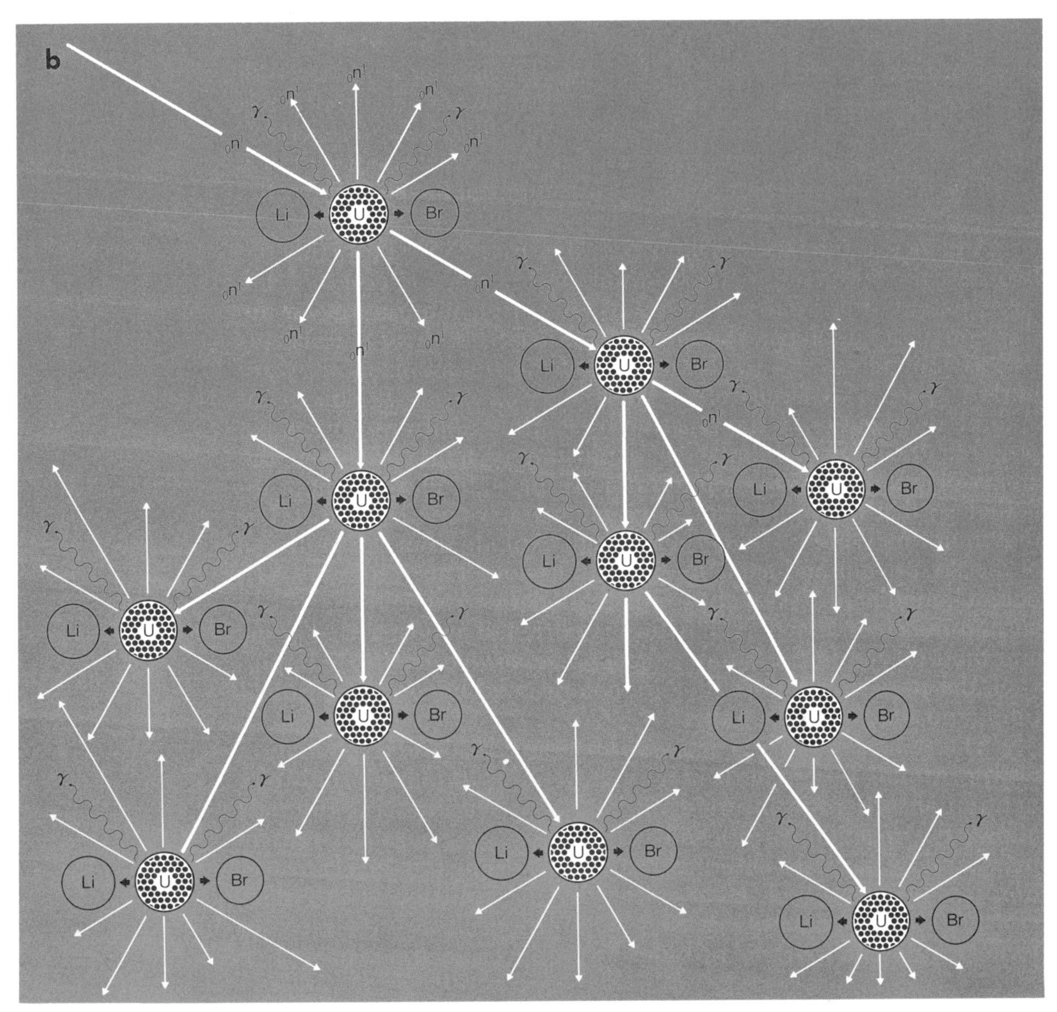

FIGURE 23.5 NUCLEAR FISSION OF URANIUM.

the neutron is absorbed and fission results. The huge atom is separated into two smaller atoms and a shower of subnuclear particles.

Uranium-235, the rarer of the two isotopes of uranium (0.7 per cent in natural deposits), suffers fission after absorbing only thermal (slow) neutrons. One cannot predict which of a variety of possible nuclear products will be obtained. One possible reaction sequence could yield:

$$^{235}_{92}U + ^{1}_{0}n \rightarrow ^{236}_{92}U$$

$$^{236}_{92}U \rightarrow ^{152}_{63}Eu + ^{80}_{35}Br + 4n + 6\beta^-$$

Because the neutron to proton ratio is greater among heavier atoms, as shown in Figure 23.1, the two large fission products cannot be the only fission products. Either neutrons must be expelled, or beta particles must be emitted (converting neutrons in the nucleus to protons). Actually, both processes occur. Gamma radiation accompanies nuclear fission, but the number of rays emitted bears no direct relationship to the composition of the two nuclei formed. Many other pairs of nuclei are possible, each accompanied by neutrons or beta rays or both. Because neutrons are both a product of the reaction and an initiating reactant, a so-called "chain reaction" can occur. Once the reaction is initiated, neutrons from fission can be slowed down and then absorbed by other nuclei, sustaining the reaction (see Figure 23.5).

A glance at the binding energy curve (Fig. 23.2) shows that the two new nuclei have greater binding energy per nucleon than the large parent atom. Mass is lost and converted into energy in the division of the nucleus. The energy is released both as gamma rays and as kinetic energy in the motion of the fragments as they are hurled apart. The large release of energy per each fission event and the propagation of the reaction by emitted neutrons both make possible the atomic bomb. An explosion is achieved when the atoms of uranium-235 or plutonium are brought close enough together so that neutrons from a first chance fission are not lost, but that more than one nucleus is caused to explode from the neutrons of that first event and each succeeding fission. This requires a sufficient amount of fissionable fuel (called the "critical mass") to sustain the chain reaction. The fuel is enriched almost to that point and then placed in the bomb *in several separated parts*. By implosion, the *portions of* atom fuel are compressed beyond the critical limit, and the bomb explodes.

The same principle is active in an atomic reactor for generating power. Atomic fuel is built in a "honeycomb" structure. Rods of a neutron absorber are inserted in the holes. Withdrawing the rods allows the reaction to proceed very slowly, just over the **critical mass,** and by partially withdrawing or reinserting the rods the reaction can be controlled. Heat caused by the collision of the high-velocity fission fragments is used, with proper heat exchangers, to form steam, which drives turbines to produce electric power.

23.9 NUCLEAR FUSION

At particle energies corresponding to the temperature on the sun, two very small nuclei, upon collision, may join together (fuse) to form a larger nucleus. Such reactions are thought to occur on the sun:

$$^1_1H + ^1_1H \rightarrow ^2_1H + \beta^+ + \nu$$

$$^2_1H + ^1_1H \rightarrow ^3_2He + \nu$$

$$^3_2He + ^3_2He \rightarrow ^4_2He + 2^1_1H$$

The result of this three-step process is the conversion of four hydrogen nuclei to one helium nucleus, a neutrino (ν), a positron (e^+) and a tremendous amount of energy (see Figure 23.2). The temperature of an atomic fission bomb is sufficient to initiate reactions similar to these. In the so-called hydrogen bomb, fusion reactions are triggered by atomic fission. Reactions similar to those shown above are called **thermonuclear reactions.**

23.10 BIOLOGICAL EFFECTS

Living organisms are damaged by the radiation resulting from radioactive decay. The radiation is capable of breaking covalent chemical bonds which can cause irreparable damage, leading to sickness and even death. Genetic damage, referred to as **mutation,** can result from radiation. Much is being done in the area of bacterial radiation to study these effects.

Radioactive chemicals are used to treat cancer. They are useful because of their ability to destroy tissue. Just as they can destroy normal cells, the radiation also destroys the abnormal cells in cancerous tissue, and even more readily. One hopes for destruction of the cancerous cells with a minimum destruction of healthy tissue.

EXERCISES

23.1 What justification might there be in the assumption that smaller nuclei are more stable than larger atomic nuclei, just because a very great number of all naturally occurring atoms have an atomic number less than 30?

23.2 Why should a large nucleus be less stable than a smaller nucleus?

23.3 Why must a smaller nucleus (with more than one proton) have approximately the same number of protons as neutrons, while a large nucleus contains more neutrons than protons?

23.4 How does the absorption of an alpha particle by the nucleus change the atomic number and the mass number of an atom? Will it change the chemical properties?

23.5 Two identical atoms lose an identical particle, one a beta particle, the other a valence electron. How do the resulting "atoms" differ?

23.6 An unstable atom has a greater neutron to proton ratio than stable atoms. Describe two possible decay processes to restore a stable ratio.

23.7 Describe the following nuclear processes:

(a) fission
(b) fusion
(c) decay
(d) absorption

23.8 Why must fission of a large nucleus always be accompanied by or be followed by other radiations?

23.9 Why is "mass defect" coupled with "binding energy"?

23.10 A polonium-218 atom loses (in order) an alpha particle a beta, a beta, an alpha, and a beta. Show in standard notation, such as $^{218}_{84}Po$, all atoms formed in the decay series.

SUGGESTED READING

Choppin, G. R.: "Nuclear Fission." Chemistry, *42* (7):25 (1969).
Johnson, R. H.: "Radiation Chemistry." Chemistry, *42* (7):31 (1969).
Keller, E.: "Early Days of Radioactivity in Industry," Part II. Chemistry, *42* (5):16 (1969).
Research Reporter: "Forty Years Ago in the World of Physics." Chemistry, *41* (6):26 (1968).
Research Reporter: "The Uranium Story." Chemistry, *40* (5):30 (1967).
Research Reporter: "Solidifying Radioactive Wastes." Chemistry, *40* (5):33 (1967).
Research Reporter: "A New Chapter on Element 104." Chemistry, *42* (5): 18 (1969).
Seaborg, G.: *Man-made Transuranium Elements.* Prentice-Hall, Inc., Englewood Cliffs, N. J., 1963 (paperback).
Wall, F. E.: "Early Days of Radioactivity in Industry," Part I. Chemistry, *42* (4):17 (1969).

TWENTY-FOUR • COLLOIDS

24.1 INTRODUCTION

A colloid is a "near solution" composed of at least two substances that are insoluble in each other. Tiny particles of one substance are dispersed throughout the other in much the same way as particles of a solute are dispersed throughout the solvent in a solution. To the unaided eye both the solution and the colloid appear as homogeneous substances. In a true solution, the solute particles are individual atoms, molecules or ions. The solute, being soluble in the solvent, separates into these ultimate particles. In contrast, the particles in a colloid are much larger and generally consist of groups or clusters of many molecules or ions. Few exceptions to this exist, except for gigantic polymer molecules such as may be found among proteins or polysaccharides and which may constitute a single particle. A colloid thus contains two phases: the dispersed phase and a continuous phase, in which the particles of dispersed phase are distributed. For reasons to be discussed later the tiny particles of a dispersed phase do not clump together (coalesce) and settle out.

small molecules have a "diameter" of 10^{-8} to 10^{-7} centimeters.

As stated previously, particles of the dispersed phase of a colloid are normally invisible. They range in diameter from 10^{-4} to 10^{-7} cm. The presence of particles is evident when a path of light through the mixture is observed at right angles. A cloudy cylinder appears, known as the Tyndall effect (see Figure 24.1). The individual particles that present a cloudy appearance can be detected by an **ultramicroscope.** (An ultramicroscope is merely a microscope trained on an individual particle in the Tyndall column.) The reflected light from the particle, when viewed in the field of the ultramicroscope, moves at random, showing the particle to be in a state of Brownian motion.

The unique properties of a colloid may be attributed to the **minute size** of the dispersed particles (which, because of their tiny size, do exhibit unusual properties) and to the **wide dispersion** of the particles in the supporting phase. The unusual properties of small particles will be discussed first. The types of colloids

the supporting phase is, also, the continuous phase.

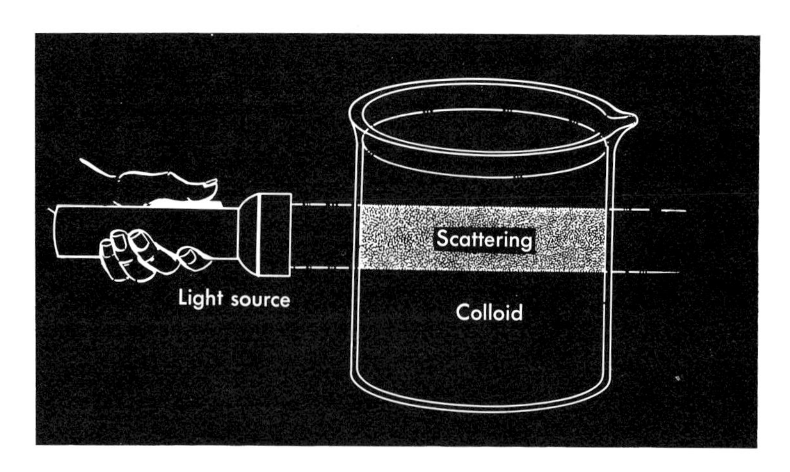

FIGURE 24.1 TYNDALL EFFECT IN COLLOIDS.

which are formed from various particles that are dispersed in different media will then be considered.

24.2 PROPERTIES OF SMALL PARTICLES

One might assume that small particles of a pure substance should have properties identical with those of larger clusters. This is not true of individual ions or molecules, compared to the larger aggregates. Silver nitrate in a solution, when mixed with a solution of sodium chloride, reacts instantaneously to precipitate silver chloride. However, dry crystals of the two solids can remain indefinitely in intimate contact without reacting. Though not as marked as this, differences do exist between colloidal-sized particles and still larger aggregates. Properties influenced by particle size include color, solubility, vapor pressure and ability to absorb other substances.

When powdered, a substance may change color. Finely divided metals appear jet black. Platinum powder precipitated from a solution for use as a catalyst is called platinum black. Mercury(II) sulfide precipitated from a mercury(II) nitrate solution with hydrogen sulfide may appear black, green or blue, depending upon particle size. A colorless salt crystal, when powdered, appears white. Undoubtedly, the scattering of light by the multitude of randomly oriented particle faces has much to do with the apparent change in color.

Tiny particles exhibit a slightly greater solubility than larger particles. One takes advantage of this property when a precipitate is allowed to "digest" before filtering. Often tiny colloidal particles, small enough to pass through the pores of a filter, will, upon standing (or better, upon standing at a higher temperature), disappear as larger particles form and grow larger.

Very tiny particles, on the order of a few molecules or ions, must have a predominance of convex surfaces, as shown in Figure 24.2. Ions or molecules at this surface have fewer attracting neighbors than those in a plane or a nearly planar surface (those on the convex surface are held less tightly and leave more readily, and hence at a greater rate). These particles, therefore, can exist in equilibrium with a more concentrated solution than larger particles.

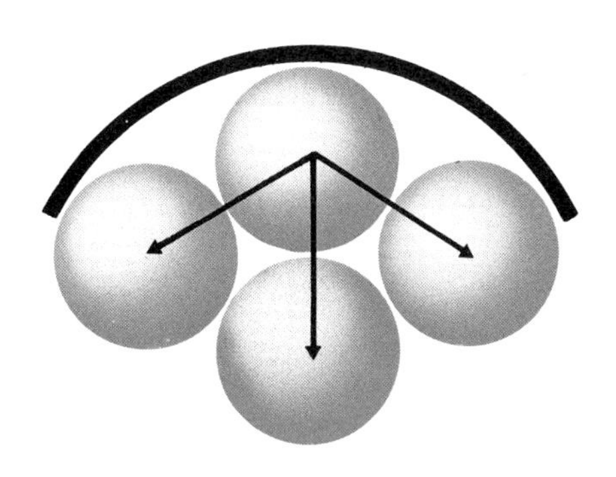

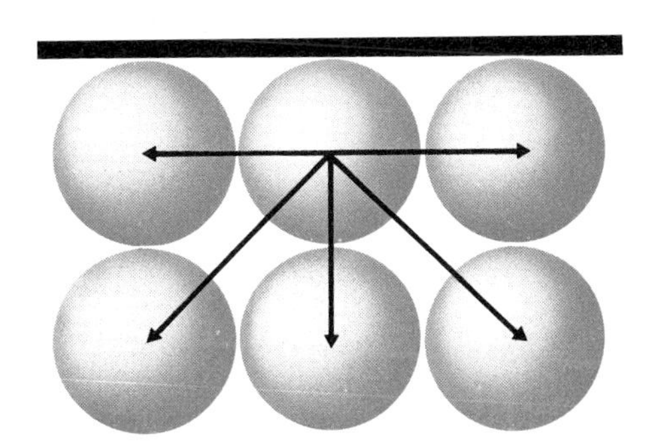

FIGURE 24.2 FEWER NEIGHBORS ON A CONVEX SURFACE.

Tiny particles or tiny droplets have a slightly greater vapor pressure than larger bodies of the same substance. The convex surfaces at which each molecule has fewer neighbors account for the increase in vapor pressure. One problem in laying out "poison gas" in wartime (most poison gases are liquids and are laid down as mists) is to perfect a spray that will deliver tiny droplets all of the same size. Such a mist, under the right conditions, could remain for weeks or longer in a little valley or pocket of undisturbed air. The liquid vaporizes and condenses at the same rate from each drop, keeping them at constant size.

Fine particles often have a great degree of absorbency. A large volume of carbon monoxide can be absorbed on a few cubic centimeters of activated charcoal. Activated charcoal is a very porous material, an aggregate of tiny particles. The property of great absorbency depends upon a large surface area. One cubic centimeter of solid absorbent material has 6 cm^2 area of absorbent surface. The same cube divided into cubes 0.001 mm on an edge has a total surface area of 60,000 cm^2 and should therefore absorb approximately 10,000 times the material absorbed by the single cube. Cornstarch powder dusted on a grease spot on fabric will often absorb the spot and remove it when brushed out because of the large absorbent surface of the tiny starch particles.

activated charcoal is an absorbent used in gas masks.

24.3 TYPES AND BEHAVIOR OF COLLOIDS—AEROSOLS

One type of colloid, more common than first supposed, is the **aerosol.** An aerosol is composed of tiny particles, solid or liquid,

FIGURE 24.3 PARTICLES IN AIR ILLUMINATED BY SHAFTS OF LIGHT.

dispersed in a gas phase. The most common continuous phase is air. Ordinary air is a dilute aerosol. The existence of tiny particles in the air can be seen in shafts of sunlight in a darkened building. Mists, fogs and clouds are aerosols of tiny water droplets in air. A cloud may exist for hours or days and yet not condense to rain or snow.

One reason for the great stability of clouds is the fact that each particle (each droplet) may bear a charge. As a cloud brushes over the surface of the earth, polar points in clay particles may attract electrons from the droplets. The positive droplets formed then repel one another and cannot come together. If the droplets are of constant size with equal vapor pressures, all will remain, losing and gaining equally, and hence the aerosol is stable and remains (Fig. 24.4). However, when the particle charge is neutralized by lightning (electrons traveling up from the earth's surface), the neutralized droplets can collect and fall.

Cloud seeding is based upon the principle that the large negative iodide ion of silver iodide will attract and collect the positive droplets around it. Becoming larger, the droplet formed about the ion falls, collecting others on the way down. (The falling droplets give a conducting path for the neutralization from the earth's surface, and the rain falls.)

Fine dust from parched, over-grazed or over-tilled land may be swept up from the earth to form a huge aerosol (dust cloud). In the sweeping-up process, electrons are lost to the earth's surface.

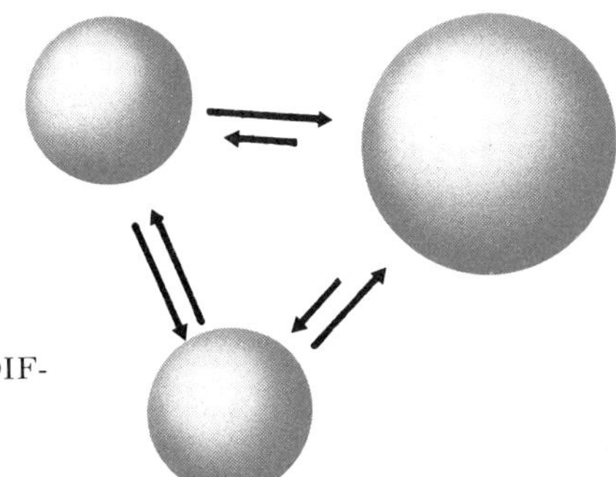

FIGURE 24.4 RATES OF EVAPORATION BETWEEN DIFFERENT SIZE DROPLETS.

The charged particles repel one another. Such clouds have been known to remain in the air for days and to travel for hundreds of miles with the prevailing winds. Ships at sea have been deluged with muddy rains when a dust cloud, neutralized in an electrical storm, comes down with the rain.

in coal dust or flour dust explosions, the dispersion of the tiny solid particles allows the reactants oxygen and dust to come together in combining proportions.

Because air contains oxygen, aerosols of combustible material may be explosive. Flour mills have been blown to bits by ignition of flour dust in air. Coal dust remaining after blasting in a coal mine may be ignited by a spark, causing an explosion. Explosion danger exists in these cases because the combustible material and oxygen are both present. Furthermore the tiny isolated colloid particles can be heated to the kindling temperature by even the smallest spark. Once ignited, the rapidly burning particles heat more and more neighboring particles to the ignition temperature, and the explosion spreads.

24.4 SOLS

aerosols are one class of sols.

The term **sol** is used in the general sense to designate a dispersion of a solid in a solid, gas or liquid. Of these types, the class in which tiny solid particles are dispersed in liquid is by far the most important. The discussion which follows, unless otherwise stated, refers to this type of sol.

A sol exhibits the Tyndall effect. A cloudy path can be seen when a beam of light is passed through the sol in a darkened room. Each particle of the cloudy path can be seen to execute Brownian motion when viewed with an ultramicroscope. A sol can be prepared by shaking finely powdered clay in water. Minute particles remain suspended. At times, when attempts are made to precipitate solids in the laboratory, the solid forms as a sol having particles so small as to pass through the pores of a filter paper. Digesting the precipitate (mentioned above) often helps. Often addition of a salt such as ammonium nitrate helps, by neutralizing the charge on the particles, to collect sol particles into filterable-sized aggregates. Sol particles may be charged by either having lost ions to, or having absorbed ions from, the solution. The charged particles sometimes collect about the ions of opposite charge.

Protein solutions form one important type of sol that has been investigated extensively. Plant or animal tissue can be homogenized in a blender, and the protein then extracted from the homog-

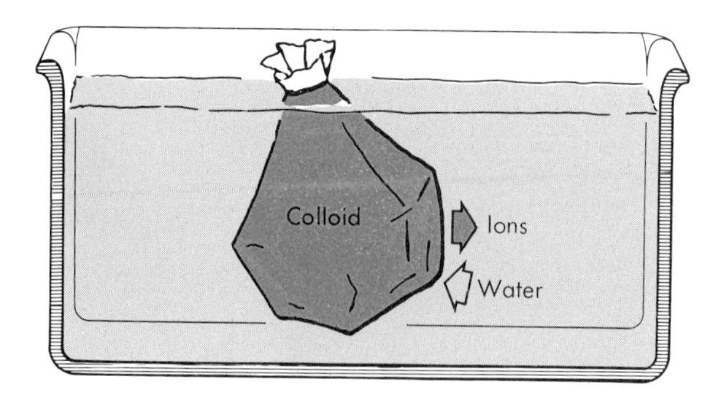

FIGURE 24.5 DIALYSIS.

enate with water. The protein molecules are often of sol-particle size. To study the proteins, one must separate the protein from ions in the solution and then separate the sol mixtures into nearly homogeneous fractions of proteins without taking them out of the water medium or causing them to become denaturated (permanently changed). An **ultracentrifuge** will precipitate colloid particles, and because they precipitate in order of decreasing mass, the centrifuge permits their separation and classification according to size, leaving the ions behind in solution. One can also remove ions by **dialysis.** In dialysis the colloidal solution is placed in a bag having a membrane permeable to ions but not to sol particles. The bag is placed in distilled water. Ions move through the membrane until the concentration is nearly equal inside and out, as shown in Figure 24.5. This reduces the ion concentration inside the bag. A series of successive dialyses, each in distilled water, reduces the concentration of ions greatly.

24.5 GELS

A gel contains solid particles dispersed in a liquid. A gel also exhibits the Tyndall effect. Gel particles, however, are not near-spherical particles as are most sols, but are tiny, interconnected fibers. In a sense the solid is a continuous phase, like a "honeycomb" in a beehive. The network is three-dimensional, with the fibers so close together that the liquid becomes trapped between them. The trapping of the liquid greatly increases the viscosity, and the liquid is not free to flow.

Jelly and **gelatin** deserts are gels and possess the semirigid form characteristic of gels. The continuous liquid phase is water, and the fibers are composed of gelatin extracted from the joints of animal hoofs, purified and dried. Gels often appear in chemical processes where molecules combine in branched chains and networks. The solid itself is, of course, only a very small fraction by weight of the complete gel.

24.6 EMULSIONS

An emulsion is composed of two immiscible liquids. Droplets of one liquid are dispersed throughout the other. The proportion of the dispersed liquid and continuous liquid can vary greatly.

Emulsions with a high proportion of dispersed liquids often appear creamy or milky. Most emulsions contain water as one liquid. The second liquid must be immiscible and is often called the oily liquid or the oil.

Milk is both a sol and an emulsion. Protein and other solids form the sol part, and droplets of fat, which collect and rise eventually as cream, constitute the oily phase of an emulsion.

An emulsion may contain either oil droplets in water or water droplets in oil; that is, either phase may be the continuous phase, as shown in Figure 24.6. It is not always possible to determine which is the continuous phase from the appearance of the emulsion. However, if dissolved salts are present, one can determine the type experimentally. If an emulsion containing ions conducts an electric current, the continuous phase is water. If it does not, the ions are in isolated droplets and cannot pass freely through the solution, as indicated in Figure 24.6.

when water is the continuous phase, there is a continuous path for ions around the oily drops and throughout the emulsion.

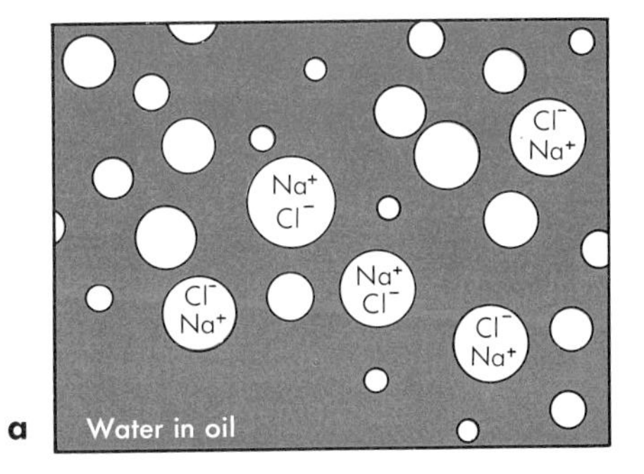

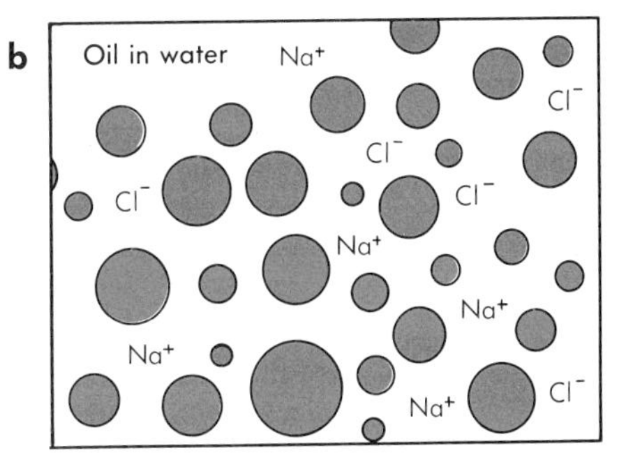

FIGURE 24.6 WATER IN OIL AND OIL IN WATER EMULSIONS.

Salad dressings and cream hair oil are emulsions. Cream hair oil is easily spread on the hair because it is an oil in water emulsion. Water can enter the continuous water layer and thin it out indefinitely. Oil alone cannot be wet and will not mix with the water.

Preparation of a permanent emulsion may be rather difficult. One can shake two liquids together vigorously until small droplets are mixed together. But as soon as agitation ceases, the liquids often separate into layers of the two liquids. The colloid can often be stabilized by adding an **emulsifying agent.** An emulsifying agent has a dual aspect, one water and the other oil. Soap and synthetic detergents are emulsifying agents. As described in Chapter 32, the soap molecule has a long oily end and a short ionic end. The oily end dissolves in oil, the ionic end in water. When the two phases, oil and water, are shaken together with soap, the soap molecules line up around the droplets in the surface between the two phases. The layer of protective colloid keeps the droplets separated so they will not coalesce.

Salad dressing and cream hair oil contain emulsifying agents. Modern automobile engine oil contains a detergent, so that any water in the crankcase is taken up in the form of colloidal particles and removed when the oil is changed. So-called photographic emulsions are really sols of silver halide crystallites in a gel matrix.

EXERCISES

24.1 Define a colloid.

24.2 Define: (a) a gel, (b) a sol, (c) an aerosol, (d) an emulsion.

24.3 What stabilizes a dust cloud (an aerosol) so that the dust may remain aloft for days?

24.4 Why is a dust cloud often broken up in an electrical storm?

24.5 Is a snowflake a collection of colloid particles, or does it occur as an enlargement of one colloidal particle? Explain.

24.6 Fine amber-colored glass is a sol of gold dispersed in a supercooled liquid. What accounts for the color?

24.7 Why should aluminum particles blown in air explode when ignited?

24.8 Why will ducks sink in a pool of water containing detergent?

24.9 Name four emulsions.

24.10 Why is a sol of equal-sized particles in a saturated solution more stable than one with a gradation in sizes?

24.11 Why is a gel more viscous than a sol?

24.12 Why should a small droplet of water have a higher vapor pressure than a large drop?

24.13 A large drop of water, a large drop of salt solution and several small drops of water on a glass plate are placed in an airtight container saturated with water vapor. What happens eventually?

24.14 Why should milk containers be covered when placed in a refrigerator?

24.15 Milk is homogenized by passing it under pressure through a microsieve. What does this process accomplish?

SUGGESTED READING

Bailar, Jr., J. C., Moeller, T. and Kleinberg, J.: *University Chemistry*, Chapter 10, D. C. Heath and Co., Chicago, 1965.

Hauser, E. A.: "Importance of the Implications of Colloid Chemistry in Science Courses." J. Chem. Educ., 26:566 (1949).

Hauser, E. A.: "The History of Colloid Science." J. Chem. Educ., 27:271 (1950).

Nevill, W. A.: *General Chemistry*, Chapter 17, McGraw-Hill, New York, 1967.

TWENTY-FIVE • ORGANIC CHEMISTRY AND THE ALKANES

In Chapters 5 and 22 the structural characteristics of the carbon atom in its elemental form and in some of the compounds in which carbon is found in nature were discussed. Many additional compounds of carbon are known. In fact about 90 per cent of all known chemical compounds contain the element carbon. This fact alone makes an intensive study of these compounds necessary, and the importance of carbon compounds in our daily lives is added reason for learning about their behavior. Most foods, most medicines and most fabrics for clothes contain carbon as an essential element. Detergents, nail polish, nylon, DDT, perfumes, plastic bottles, shaving lotions, soaps, gasolines, tranquilizers, antibiotics, LSD, barbiturates and aspirin are compounds of carbon. All contribute in one way or another to the health, comfort and psychic experience of our lives.

Of particular importance is our knowledge of the changes substances undergo in cells and tissues of the animal body. Much of this knowledge of biological processes has been obtained through the study of carbon compounds, which are the principal constituents of living organisms. Historically, the sources of most carbon compounds were animals and plants. The early association of the study of carbon compounds with living organisms led to the use of the term "organic chemistry" for the field of study concerned with them. The chemicals produced by and isolated from microorganisms, plants and animals were once thought to differ greatly from other chemicals in that some unusual property of living material, "a vital (living) force," was required in their formation. Today, many compounds not found in organisms are studied in organic chemistry, as well as many that are, in fact, poisonous to them. Nevertheless, many toxic substances have structural characteristics similar to those of the compounds which are used, transformed by or produced by living organisms.

Modern organic chemistry is more inclusive than suggested above but is tied closely to the chemistry of carbon compounds, based on the fact that carbon is the key element in most of the structures studied. The role of carbon as the backbone or framework of organic substances is determined by the structure of the carbon atom and the nature of its bonding with other carbon atoms and with atoms of other elements. To seek an understanding of this role we will begin by examining some of the simplest of these compounds, the hydrocarbons.

25.1 THE ALKANES

what is the formula of the alkane having eight carbons?

Hydrocarbons are compounds that contain only the elements carbon and hydrogen. The simplest of the hydrocarbons are the alkanes. The term alkane refers to all compounds with the formula C_nH_{2n+2}, where n is any number. Compounds of this series or class have both common names and official IUPAC* names. Official names have the ending **-ane.** The first member of the series is methane.

25.2 METHANE

refer to Chapter 5 for a review of hybridization.

Methane, the major component of natural gas, is piped from wells to be used as a fuel for domestic and industrial purposes. Its simplest formula, calculated from analysis, is CH_4. The molecular weight as determined from gas densities is 16. Therefore, the simplest formula is also the molecular formula. The molecule appears as shown in Figure 25.1a. The "stick model" illustrated in Figure 25.1b shows the bonds connecting the centers of the hydrogen atoms to the carbon atom. In this compound the carbon atom is sp^3 hybridized. Each electron present in an sp^3 orbital of carbon is paired with a single electron in an s orbital of hydrogen to give a molecule often represented as

$$\begin{array}{c} H \\ \cdot\cdot \\ H:C:H \\ \cdot\cdot \\ H \end{array}$$

The molecule is also designated by the formula

$$\begin{array}{c} H \\ | \\ H—C—H \\ | \\ H \end{array}$$

where each line in the formula corresponds to a pair of electrons, i.e., a single, covalent, sigma type bond. Regardless of the formula used it should be remembered

* IUPAC = International Union of Pure and Applied Chemistry.

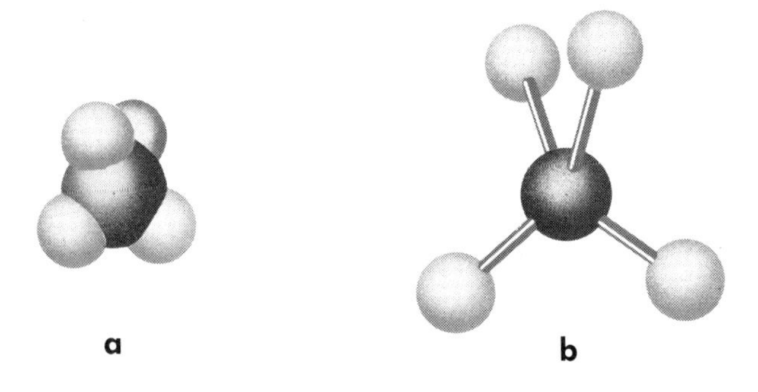

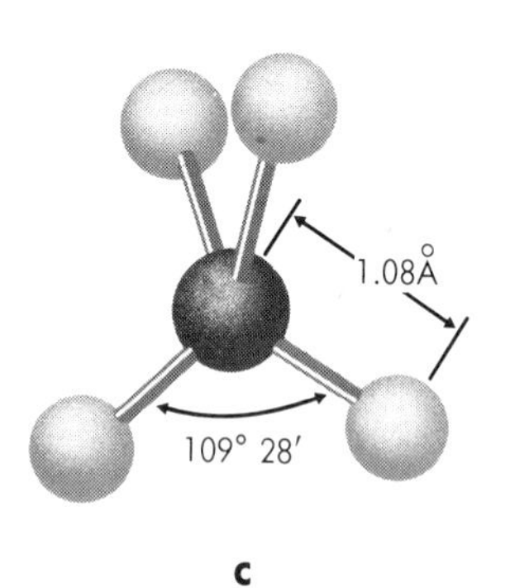

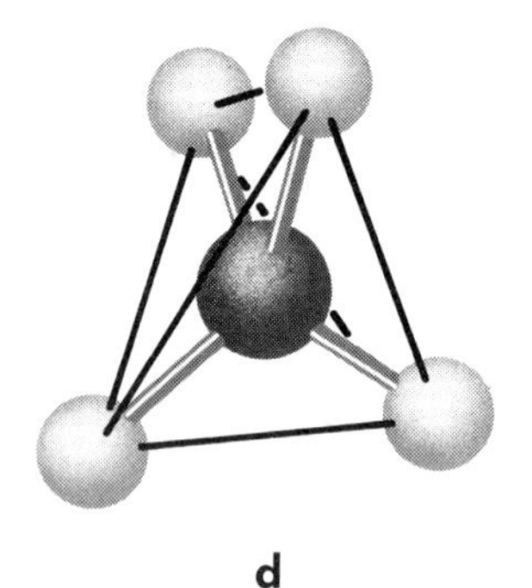

FIGURE 25.1 REPRESENTATIONS OF THE METHANE MOLECULE.

that the molecule is three dimensional with an angle of 109°28′, **the tetrahedral angle** (Fig. 25.1c) between any pair of the four bonds. The length of the carbon to hydrogen bond, i.e., the distance between the center of the carbon atom and the center of any one of the hydrogen atoms bonded to it is 1.08 Å. Lines connecting all pairs of hydrogen atoms form a regular tetrahedron. A plane passed through two of the hydrogens and the carbon atom is perpendicular to a plane passed through the carbon atom and the other two hydrogens.

In methane all hydrogen atoms are bound to carbon in exactly the same way. All C to H bonds are identical in length and are identically related to the other three C to H bonds. Because all H atoms of methane are identical, replacement of any one hydrogen with another atom or group of atoms leads to the same new compound as would replacement of any other hydrogen by the same group. For example, the replacement of a hydrogen atom of methane with a chlorine atom gives one and only one compound having the formula CH_3Cl.

25.3 ETHANE

The second member of the alkane series is ethane, C_2H_6. This molecular formula is not completely satisfactory. It does not indicate the arrangement of the atoms in the molecule. A better formula is $CH_3{-}CH_3$. Still more representative formulas are

$$\begin{array}{ccccc} & H & & H & \\ & | & & | & \\ H- & C & - & C & -H \\ & | & & | & \\ & H & & H & \end{array} \quad \text{and} \quad \begin{array}{c} H\,H \\ \cdot\cdot\;\cdot\cdot \\ H{:}C{:}C{:}H \\ \cdot\cdot\;\cdot\cdot \\ H\,H \end{array}$$

The ethane molecule is shown in Figure 25.2. Like methane, it is three dimensional. The angles between any two bonds extending from carbon are 109°28′. The carbon to hydrogen bond lengths are 1.08 Å as before. The distance between the centers of bonded carbon atoms is 1.54 Å. Also like methane, ethane is a gas. It occurs in trace quantities in natural gas.

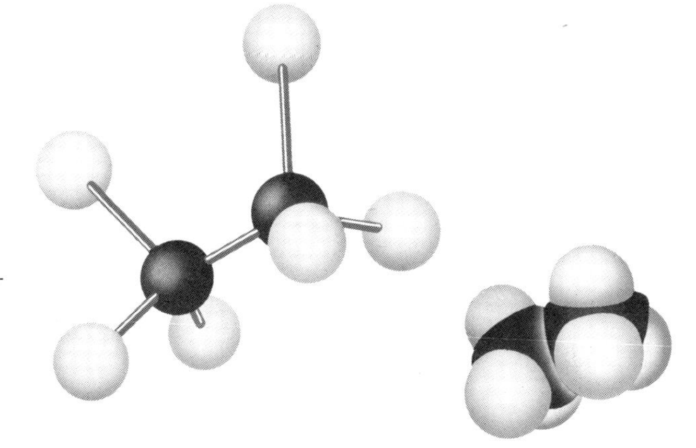

FIGURE 25.2 THE ETHANE MOLECULE.

In contrast with the molecular formulas, CH_4 and C_2H_6, which account for only the numbers and kinds of atoms in a molecule, the formulas

$$\begin{array}{ccc} & H & \\ & | & \\ H- & C & -H, \\ & | & \\ & H & \end{array} \quad \begin{array}{c} H \\ \cdot\cdot \\ H{:}C{:}H, \\ \cdot\cdot \\ H \end{array} \quad CH_3{-}CH_3, \quad \begin{array}{ccccc} & H & & H & \\ & | & & | & \\ H- & C & - & C & -H \\ & | & & | & \\ & H & & H & \end{array} \text{ and } \begin{array}{c} H\,H \\ \cdot\cdot\;\cdot\cdot \\ H{:}C{:}C{:}H \\ \cdot\cdot\;\cdot\cdot \\ H\,H \end{array}$$

detail the structure more completely, and are called **structural formulas.** Structural formulas show the number, kinds and arrangement (sequence) of the atoms in a molecule.

Ethane may be pictured as a substituted CH_4 in which one hydrogen is replaced with the CH_3 (methyl) group of atoms. As expected, one and only one ethane is known. Just as in the case of methane, all hydrogen atoms attached to carbon in ethane occupy identical positions. Replacement of any one hydrogen with another atom (or group) leads to the same new compound. Replacement of a hydrogen atom of ethane with a chlorine atom yields one and only one compound having the formula

how many arrangements of two C and six H atoms can you deduce if only single bonds between atoms are permitted?

$$C_2H_5Cl, \quad \begin{array}{ccccc} & H & & H & \\ & | & & | & \\ H- & C & - & C & -Cl, \\ & | & & | & \\ & H & & H & \end{array} \quad \begin{array}{ccccc} & Cl & & H & \\ & | & & | & \\ H- & C & - & C & -H \\ & | & & | & \\ & H & & H & \end{array} \text{ or } \begin{array}{ccccc} & H & & H & \\ & | & & | & \\ H- & C & - & C & -Cl \\ & | & & | & \\ & H & & H & \end{array}$$

All formulas are for the same compound. More than one monochloro compound would be expected if the hydrogen atoms were not identical.

25.4 PROPANE

The third member of the alkane series is propane, C_3H_8, in which n is three in the general formula, C_nH_{2n+2}. The molecule can be represented by any one of these structural formulas:

$$\mathrm{CH_3CH_2CH_3} \qquad \begin{array}{ccccccc} & \mathrm{H} & & \mathrm{H} & & \mathrm{H} & \\ & | & & | & & | & \\ \mathrm{H}- & \mathrm{C} & - & \mathrm{C} & - & \mathrm{C} & -\mathrm{H} \\ & | & & | & & | & \\ & \mathrm{H} & & \mathrm{H} & & \mathrm{H} & \end{array} \qquad \begin{array}{c} \mathrm{H\,H\,H} \\ \cdot\cdot\ \cdot\cdot\ \cdot\cdot \\ \mathrm{H{:}C{:}C{:}C{:}H} \\ \cdot\cdot\ \cdot\cdot\ \cdot\cdot \\ \mathrm{H\,H\,H} \end{array} \qquad \begin{array}{ccc} \mathrm{CH_3} & - & \mathrm{CH_2} \\ & & | \\ & & \mathrm{CH_3} \end{array} \qquad \begin{array}{ccccc} & \mathrm{H} & & & \\ & | & & & \\ \mathrm{H}- & \mathrm{C} & -\mathrm{H} & \mathrm{H} & \\ & | & & | & \\ \mathrm{H}- & \mathrm{C} & \text{———} & \mathrm{C} & -\mathrm{H} \\ & | & & | & \\ & \mathrm{H} & & \mathrm{H} & \end{array}$$

The angles between two bonds extending from the same carbon are neither 180° nor 90° as they appear in the printed structural formulas but are all 109°28′, and the molecule better as represented in Figure 25.3, where the three carbons are placed at the apices of an isosceles triangle. The carbon to carbon bond distances are 1.54 Å; the carbon to hydrogen bond lengths, 1.08 Å.

FIGURE 25.3 PROPANE.

With respect to position in the molecule, there are two kinds of carbon atoms in propane. The central atom is attached to two other carbon atoms and is called a **secondary carbon atom.** Each of the end or terminal carbons is attached to only one other carbon and is, therefore, called a **primary carbon atom.** Because hydrogen atoms are attached to both the secondary and the primary carbon atoms, there are, structure-wise, two non-equivalent kinds of hydrogen atoms. Named in accord with the type of carbon to which they are bound, propane has six identical **primary** and two identical **secondary hydrogen atoms,** per molecule.

Replacement of one hydrogen in propane with another atom or group of atoms can yield two compounds, depending upon the type of hydrogen replaced. Exchange of a primary hydrogen with chlorine yields 1-chloropropane,

$$\begin{array}{ccccccc} & \mathrm{H} & & \mathrm{H} & & \mathrm{H} & \\ & | & & | & & | & \\ \mathrm{H}- & \mathrm{C} & - & \mathrm{C} & - & \mathrm{C} & -\mathrm{Cl} \\ & | & & | & & | & \\ & \mathrm{H} & & \mathrm{H} & & \mathrm{H} & \end{array} \quad \text{or} \quad \begin{array}{ccccccc} & \mathrm{H} & & \mathrm{H} & & \mathrm{H} & \\ & | & & | & & | & \\ \mathrm{H}- & \mathrm{C} & - & \mathrm{C} & - & \mathrm{C} & -\mathrm{H} \\ & | & & | & & | & \\ & \mathrm{H} & & \mathrm{H} & & \mathrm{Cl} & \end{array} \quad \text{or} \quad \mathrm{CH_3CH_2CH_2Cl};$$

how many nonequivalent kinds of H atoms are there in 1-chloropropane? in 2-chloropropane?

exchange of a secondary hydrogen yields 2-chloropropane,

$$\begin{array}{c} \text{H}\,\text{H}\,\text{H} \\ \cdot\cdot\;\cdot\cdot\;\cdot\cdot \\ \text{H:C:C:C:H} \\ \cdot\cdot\;\cdot\cdot\;\cdot\cdot \\ \text{H}\,\text{Cl}\,\text{H} \end{array} \quad \text{or} \quad \begin{array}{c} CH_3 \\ | \\ Cl-C-CH_3 \\ | \\ H \end{array} \quad \text{or} \quad CH_3CHClCH_3$$

Two or more different compounds with the same molecular formula are **isomers** of one another. 2-Chloropropane and 1-chloropropane are isomers because each has the formula, C_3H_7Cl. Isomeric substances have the same chemical composition but may differ only slightly or very greatly in physical and sometimes chemical properties. Because the differences in the compounds are structural, structural formulas are necessary to distinguish between isomers. **Isomerism** is very important in organic chemistry. There are many possible arrangements of atoms in large molecules.

25.5 BUTANES

The plural title is used for this section because there are two compounds, two isomers, with the molecular formula, C_4H_{10}. Notice that just as there are two different places for the substitution of a chlorine atom into propane as discussed previously, so there are two different places for introducing an additional carbon atom. The introduction of a methyl group into propane in place of hydrogen therefore gives rise to two different compounds. Structural formulas are needed to distinguish them.

Both n-butane and isobutane may be considered as having been obtained through replacement of one hydrogen of propane with the CH_3 group. Replacement of a primary hydrogen of propane with CH_3 yields normal butane; replacement of a secondary hydrogen yields isobutane. Since only two kinds of hydrogen are present in propane only two new compounds are possible.

Structural formulas for one isomer, normal butane (Fig. 25.4a) are

note that no one of the structures written is more correct than the others.

$$CH_3-CH_2-CH_2-CH_3 \quad \begin{array}{c} \quad\quad\quad CH_3 \\ \quad\quad\quad | \\ CH_3CH_2CH_2 \end{array} \quad \text{and} \quad \begin{array}{c} \;\;H\;\;\;H\;\;\;H\;\;\;H \\ \;\;|\;\;\;\;|\;\;\;\;|\;\;\;\;| \\ H-C-C-C-C-H \\ \;\;|\;\;\;\;|\;\;\;\;|\;\;\;\;| \\ \;\;H\;\;\;H\;\;\;H\;\;\;H \end{array}$$

Structural formulas for isobutane (Fig. 25.4b) are

$$\begin{array}{c} CH_3 \\ | \\ CH_3-CH-CH_3 \end{array} \quad \begin{array}{c} \;\;H\;\;\;\;CH_3\;\;H \\ \;\;|\;\;\;\;\;\;|\;\;\;\;\;\;| \\ H-C-C-C-H \\ \;\;|\;\;\;\;\;\;|\;\;\;\;\;\;| \\ \;\;H\;\;\;\;\;\;H\;\;\;\;\;\;H \end{array} \quad \text{and} \quad \begin{array}{c} \quad\quad\quad CH_3 \\ \quad\quad / \\ CH_3-C-H \\ \quad\quad \backslash \\ \quad\quad\quad CH_3 \end{array}$$

"Normal" and "iso" are used in conjunction with the common names of these hydrocarbons. "**Normal**" refers to continuous chain hydrocarbons. "**Iso**" is prefixed to designate "branched" hydrocarbon chains and is used specifically for hydrocarbons with a CH_3 branch off the second carbon.

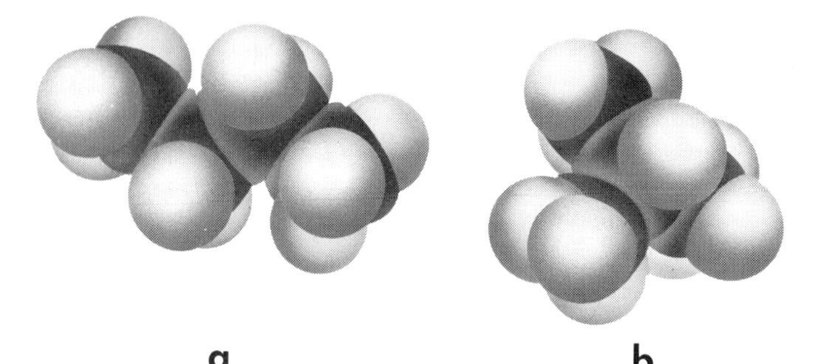

FIGURE 25.4 THE ISOMERIC BUTANES.

Normal butane, often written n-butane, and isobutane, since they differ only in the chain arrangement of the carbon atoms, are **chain isomers** of each other, and are distinctly different compounds with different properties. Like propane, the butanes are used as bottled gas for fuel.

In the normal butane molecule, there are two primary and two secondary carbon atoms, connected in a continuous chain in the order: primary, secondary, secondary, and primary. Attached to the carbons are six identical primary and four identical secondary hydrogen atoms. Isobutane has no secondary carbon atom but contains three primary and one tertiary carbon atoms per molecule. The **tertiary carbon** is the central carbon atom—tertiary because it is attached to three other carbons. Isobutane, therefore, has nine primary and one tertiary hydrogen atoms, i.e., each primary hydrogen atom occupies a position identical in all respects to that occupied by every other primary hydrogen atom in the molecule.

how many primary H's are present in a molecule having five primary C's?

It is useful to distinguish between the types of hydrogen atoms as to position of the carbon to which they are attached, because the capability for a reaction at the hydrogen, the rate at which the reaction might occur and the nature of the product or products depend upon the type of hydrogens in the reacting molecule.

25.6 PENTANES

There are three isomeric pentanes:

normal pentane $CH_3CH_2CH_2CH_2CH_3$

isopentane
$$\begin{array}{c} \quad CH_3 \\ \quad | \\ CH_3CHCH_2CH_3 \end{array}$$

neopentane
$$\begin{array}{c} CH_3 \\ | \\ CH_3—C—CH_3 \\ | \\ CH_3 \end{array}$$

All have the same molecular formula (C_5H_{12}), but each has a different arrangement of the carbon atoms. They are **chain isomers** of each other. Each student should count for each isomer the number of primary, secondary and tertiary carbon atoms it contains and should count the number of each type of hydrogen present in each structure.

Examination of n-pentane shows that all secondary hydrogen atoms present are not identical. Similarly, in isopentane, all primary hydrogens are not identical. In both cases, hydrogens of the same classification are located differently with respect to the other atoms in the molecule and thus are non-equivalent. Because n-pentane contains three kinds of hydrogens, each different in some way from the others, three different products are possible from n-pentane when a single hydrogen atom is replaced with another atom. For example, with chlorine the three products are $CH_3CH_2CH_2CH_2CH_2Cl$, $CH_3CH_2CH_2CHClCH_3$ and CH_3CH_2CH-$ClCH_2CH_3$. Isopentane, because it contains four kinds of non-equivalent hydrogens, should give rise to four products by replacement of a single hydrogen atom. Neopentane contains a type of carbon atom not previously encountered. The central carbon is completely surrounded by and attached to four other carbon atoms. A carbon of this kind is designated as a **quaternary carbon** and names of hydrocarbons in which such a carbon atom is found may have the prefix *neo-*. A quaternary hydrogen is impossible.

note that neohexane (draw its structure) has three kinds of hydrogen atoms.

The pentanes are very volatile (low boiling) liquids obtained from petroleum. They are used as fuels and solvents.

Of the hydrocarbons introduced up to this point, ethane, CH_3—CH_3; propane, CH_3—CH_2—CH_3; n-butane, CH_3—CH_2—CH_2—CH_3, and n-pentane, $CH_3CH_2CH_2CH_2CH_3$, are classified as **continuous chain hydrocarbons** with each carbon atom a link of the chain. The chains of carbon atoms are not actually straight as they appear in the written formulas but they bend through the tetrahedral angle at each carbon. At best they zigzag and may even assume almost circular arrangements (Fig. 25.5) which change freely

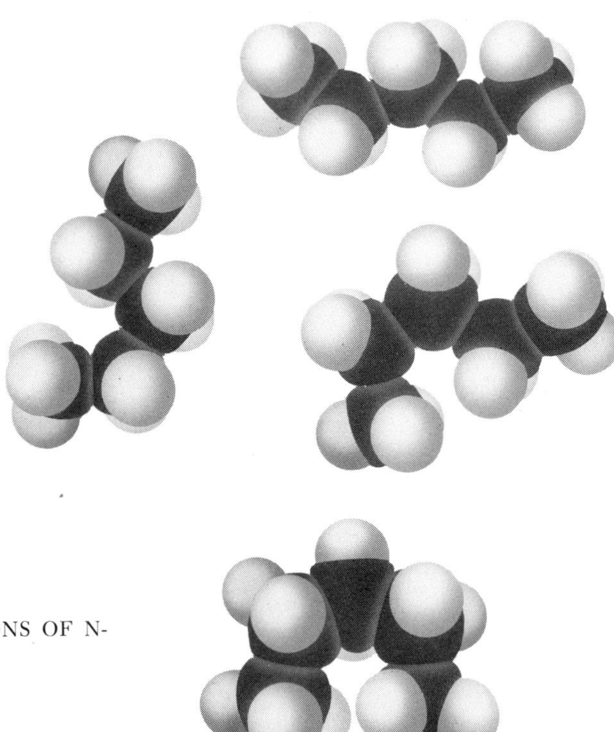

FIGURE 25.5 SOME CONFORMATIONS OF N-PENTANE.

from one to another by rotation about single bonds. These arrangements, since the molecule is not restricted to any one of them, are called **conformations** to distinguish them from **isomers** in which the two or more arrangements that exist do not freely change from one into another. In contrast to the continuous chain hydrocarbons, the compounds

isobutane $CH_3{-}\underset{}{\overset{\overset{\displaystyle CH_3}{|}}{C}}HCH_3$

isopentane $CH_3\overset{\overset{\displaystyle CH_3}{|}}{C}HCH_2CH_3$

and neopentane $CH_3{-}\underset{\underset{\displaystyle CH_3}{|}}{\overset{\overset{\displaystyle CH_3}{|}}{C}}{-}CH_3$

are called **branched chain hydrocarbons.** The first two have but one branch; neopentane has two branches.

25.7 HEXANES

There are five isomeric hexanes:

1. Normal hexane $CH_3CH_2CH_2CH_2CH_2CH_3$

2. Isohexane (2-methylpentane) $CH_3\overset{\overset{\displaystyle CH_3}{|}}{C}HCH_2CH_2CH_3$

3. Neohexane (2,2-dimethylbutane) $CH_3\underset{\underset{\displaystyle CH_3}{|}}{\overset{\overset{\displaystyle CH_3}{|}}{C}}CH_2CH_3$

4. 2,3-Dimethylbutane $CH_3{-}\overset{\overset{\displaystyle CH_3}{|}}{C}H{-}\overset{\overset{\displaystyle CH_3}{|}}{C}H{-}CH_3$

5. 3-Methylpentane $CH_3CH_2\underset{\underset{\displaystyle H}{|}}{\overset{\overset{\displaystyle CH_3}{|}}{C}}CH_2CH_3$

which hexane contains five different kinds of hydrogen atoms?

All are volatile liquids present as a mixture in the fraction of petroleum which distills in the temperature range 60 to 100°C.

25.8 NUMBERS OF ISOMERS

The number of isomers of each alkane increases rapidly with the number of carbon atoms per molecule. Table 25.1 illustrates this point. When four or more carbon atoms are present, isomeric hydrocarbons exist because several different chain arrangements are possible. The number of possible arrangements increases rapidly with each additional carbon atom (Table 25.1).

TABLE 25.1 NUMBER OF ISOMERS

Molecular Formulas	No. of Isomers	Molecular Formulas	No. of Isomers
CH_4	1	C_8H_{18}	18
C_2H_6	1	C_9H_{20}	35
C_3H_8	1	$C_{10}H_{22}$	75
C_4H_{10}	2	$C_{14}H_{30}$	1,858
C_5H_{12}	3	$C_{20}H_{42}$	336,319
C_6H_{14}	5	$C_{30}H_{62}$	4,111,846,763
C_7H_{16}	9		

if it takes one day to make each isomer, how many years would it take to make all the isomers having 14 carbons?

Relatively few of the more than four billion "isomers" of $C_{30}H_{62}$ have been isolated or prepared. This number represents the possible arrangements of 30 carbons connected to one another in chains and branches such that each carbon is connected to at least one but not more than four other carbons in a continuous network. Although some highly branched isomers would be too crowded about the center of the molecule, most, if they were needed, could be either isolated from petroleum or synthesized.

25.9 NOMENCLATURE

With the great number of possible isomers, it would be futile to have common unrelated names for each compound. For this reason the International Union of Pure and Applied Chemistry has developed a systematic method by means of which, in principle, all possible organic compounds can be named. Rules for naming alkanes by the IUPAC system are here listed and applied.

1. All normal continuous chain alkanes have the name ending *-ane*. The *stem names* assigned to the first four are the common names. The stems for the names beyond the fourth member of the series are derived from Greek.

CH_4	methane	n-C_8H_{18}	octane
C_2H_6	ethane	n-C_9H_{20}	nonane
C_3H_8	propane	n-$C_{10}H_{22}$	decane
n-C_4H_{10}	butane	n-$C_{11}H_{24}$	undecane
n-C_5H_{12}	pentane	n-$C_{12}H_{26}$	dodecane
n-C_6H_{14}	hexane	n-$C_{13}H_{28}$	tridecane
n-C_7H_{16}	heptane	n-$C_{14}H_{30}$	tetradecane

2. Groups or chains that do not appear alone but are attached to longer alkane chains are given the *same stem name as the alkane* having the same number of carbons, with *endings changed to -yl.*

Those appearing most frequently are:

how many carbons are there in the n-butyl group? sec-butyl group? t-butyl group?

CH_3—, methyl, related to methane, CH_4
CH_3—CH_2—, ethyl, related to ethane, C_2H_6
CH_3—CH_2—CH_2—, n-propyl
CH_3—CH(CH_3)—, isopropyl
(both related to propane and differing only in the point of attachment)
CH_3—CH_2—CH_2—CH_2—, n-butyl
CH_3—CH(CH_3)—CH_2—, isobutyl
CH_3—CH_2—CH(CH_3)—, secondary butyl (sec-butyl or s-butyl)
CH_3—C(CH_3)(CH_3)—, tertiary butyl (t-butyl)

3. Branched alkanes are named as "branched" alkanes. The "trunk," the longest continuous carbon chain, is named as a normal alkane. Groups that are attached are listed *alphabetically* as modifiers of the trunk name. If a group appears twice, the prefix *di* is used; if three times, *tri* is used; as are *tetra, penta* and so on.

Compound A:

CH_3—CH_2—CH(CH_3)—CH(CH_2—CH_3)—CH_2—CH_3

is an "ethylmethylhexane." The "trunk" is underlined and the attached groups are circled. The compound is an isomer of n-nonane.

Compound B:

CH_3—CH_2—CH(CH_3)——CH(CH_3)—CH_2—CH_3

note that the circled groups are considered to be substituted in place of H's of the main chain.

is a dimethylhexane, as is compound C.

Compound C:

CH_3—C(CH_3)(CH_3)—CH_2—CH_2—CH_2—CH_3

how many carbons per molecule in 2-methyldecane?

4. *The main (longest) carbon chain is numbered from the most branched end.* The position of each group on the chain is indicated by a number preceding its name, which is the number of the carbon to which the group is connected. When two or more identical

groups are attached, the prefixes di, tri, tetra and so on are used and a number for each group precedes the collective name.

Notice the application of these rules in compounds A, B and C:

Compound A:

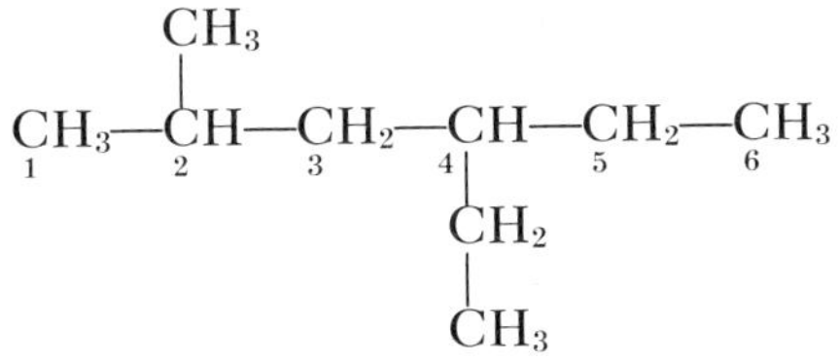

```
        CH3
        |
 CH3—CH—CH2—CH—CH2—CH3
  1   2   3  4|  5    6
              CH2
              |
              CH3
```

4-ethyl-2-methylhexane

Compound B,

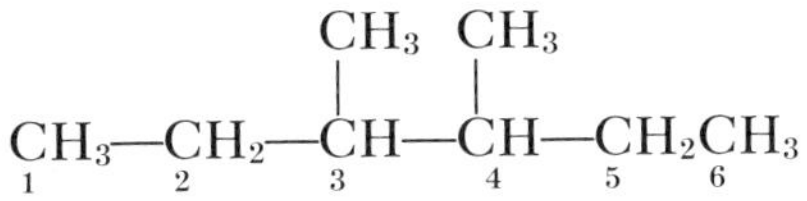

```
            CH3  CH3
            |    |
 CH3—CH2—CH—CH—CH2CH3
  1   2   3   4   5  6
```

3,4-dimethylhexane

Compound C,

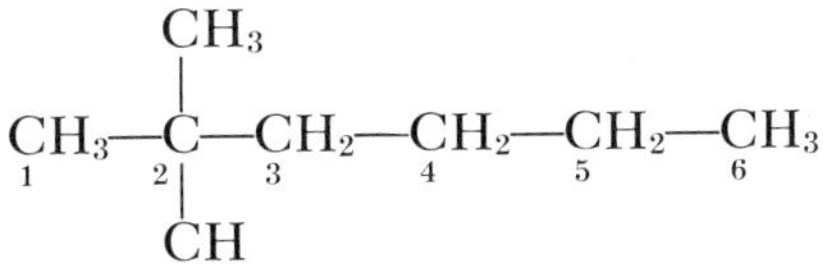

```
       CH3
       |
 CH3—C—CH2—CH2—CH2—CH3
  1  2|  3    4    5    6
       CH
```

2,2-dimethylhexane (not 5,5-dimethylhexane)

Another compound, compound D,

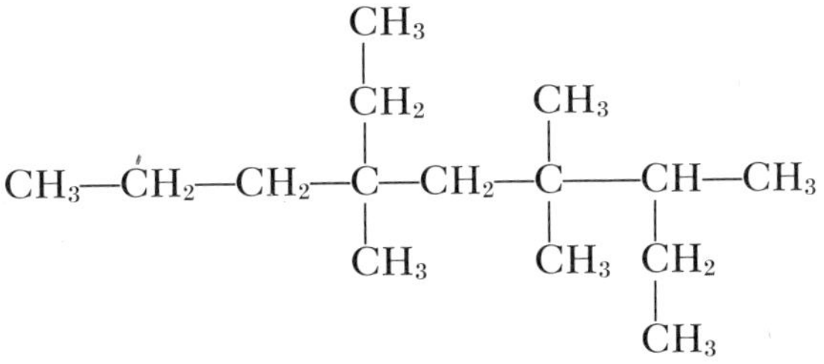

```
                  CH3
                  |
                  CH2       CH3
                  |         |
 CH3—CH2—CH2—C—CH2—C——CH—CH3
                  |         |    |
                  CH3       CH3  CH2
                                 |
                                 CH3
```

6-ethyl-3,4,4,6-tetramethylnonane (not 4-ethyl-4,6,6,7-tetramethylnonane)

The numbering of each group is indicated below.

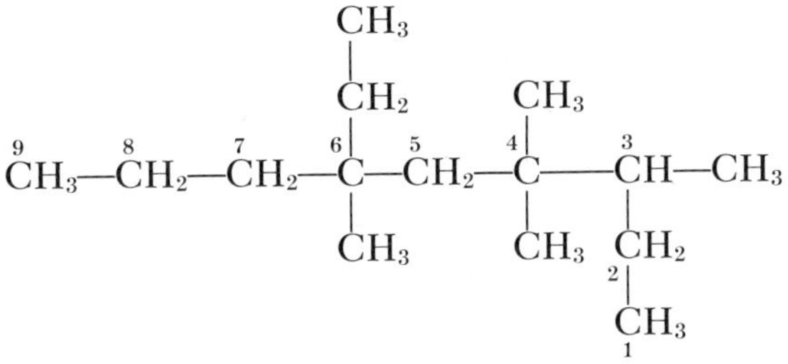

```
                  CH3
                  |
                  CH2       CH3
  9    8    7    6|    5    4|    3
 CH3—CH2—CH2—C—CH2—C——CH—CH3
                  |         |    |
                  CH3       CH3  CH2
                                2|
                                 CH3
                                 1
```

what is the fewest number of carbons in a compound having the prefix 3-methyl as part of its name?

The chain is numbered (from the right) because a branch or substituent appears on the carbon nearer to the terminal carbon on the right than to the end carbon on the left. The chain is numbered from the end nearest the first branching group or substituent. Numbering in this manner gives the set of smallest numbers to the side chains (attached groups). This principle is applied irrespective of the nature of the group or substituent attached.

Compound E,

```
       CH3      CH3
        |        |
CH3—CH—CH—C——CH—CH2—CH3
            |    |    |
      CH3—CH  CH2  CH—CH3
            |    |    |
           CH3  CH3  CH3
```

```
       Methyl          Methyl
        CH3             CH3
         |               |
or CH3—CH——CH————C————CH——————————CH2—CH3
    1    2    3|      4|    5|             Ethyl
               |       |     |
          CH3—CH      CH2   CH————CH3
               |       |    6|    Methyl
              CH3     CH3   CH3
          Isopropyl   Ethyl   7
```

may be named either 4,5-diethyl-3-isopropyl-2,4,6-trimethylheptane or **3,4-diethyl-5-isopropyl-2,4,6-trimethylheptane.**

(The numbering of the carbons in the main chain was reversed to form the second name. The second name, however, is the only acceptable name according to an IUPAC rule that states that if one or more side chains are in equivalent positions the one to be assigned the lowest number is that cited first in the correct name.) Notice that the terms with their number prefixes appear in the name in alphabetical order. Also notice that there are two different seven carbon chains which might have been selected for assigning a name to this molecule. First, the seven carbon chain chosen in assigning the name given above and second, the seven carbon chain which would include the circled ethyl group given on the right as part of the main chain. Where two chains of equal length are present, that chain having the most branches is chosen as the main chain for purposes of assigning a name to the compound. This was done in naming compound E above.

Reconstruction of the formula from the name may assist in understanding the system of nomenclature.

3,4-diethyl-5-isopropyl-2,4,6-trimethyl*heptane*.

a. is a heptane:

```
C—C—C—C—C—C—C
1 2 3 4 5 6 7
```

b. with two ethyl groups, one attached to carbon 3 and one to carbon 4:

```
C——C——C——C——C——C——C
1  2  3|  4|  5  6  7
      CH2 CH2
       |   |
      CH3 CH3
```

c. with an isopropyl group on carbon number 5:

```
C——C——C——C———C——C——C
1  2  3|  4|  5|  6  7
      CH2 CH2   |
       |   |  HC—CH3
      CH2 CH2   |
               CH3
```

d. and with three methyl groups, one each on carbons 2, 4 and 6:

to what does a hydrogen "group," as used here, refer?

```
     CH3      CH3          CH3
      |        |            |
C——C——C——C———C——C——C
1  2  3|  4|  5|  6  7
      CH2 CH2 HC—CH3
       |   |   |
      CH3 CH3 CH3
```

e. Hydrogen "groups" are added until each carbon has 4 bonds

```
      CH3            CH3            CH3
       |              |              |
CH3——CH——CH——C———CH——CH——CH3
              |      |       |
             CH2    CH2  HC—CH3
              |      |       |
             CH3    CH3     CH3
```

One should recognize that this formula and the formula for compound E represent identical compounds.

Additional rules are available and necessary for assigning correct IUPAC names to more complex structures.

25.10 PHYSICAL PROPERTIES OF THE ALKANES

Properties of substances are determined in varying degree by a number of factors. The composition, the bond type, the polarity or non-polarity, the presence or absence of hydrogen bonding, the molecular weight, the size and the shape or configuration of a molecule are among the most important.

The alkanes, being composed solely of carbon and hydrogen (elements of nearly the same electronegativity), with the hydrogens being arranged in a more or less symmetrical way about the carbons, are essentially non-polar and cannot participate in hydrogen bonding. Therefore, they are insoluble in water because the water molecules cling together and do not permit the alkane molecules to come between. On the other hand, alkanes are soluble in non-polar solvents such as benzene, ether, carbon tetrachloride and other petroleum products.

The density of substances depends mostly on the kinds and ratios of atoms in the molecule. Each kind of atom has almost the same size in all compounds and possesses constant mass. The smaller members of the alkane series have a higher ratio of the lighter hydrogen atoms (in methane, 4 to 1) and are therefore less dense. The hydrogen to carbon ratio decreases to a limiting value of almost 2 to 1 with higher molecular weight alkanes, and the density rises to an almost constant value of 0.777 g/ml (Table 25.2). Because they have a density less than that of water, alkanes form a film, or "oil slick," on the surface of water.

TABLE 25.2 PHYSICAL PROPERTIES OF SELECTED PARAFFIN HYDROCARBONS

Name	Formula	Melting Point, °C	Boiling Point, °C	Density at 20°C
Methane	CH_4	−182.5	−161.5	0.424 (at bp)
Ethane	C_2H_6	−183.2	−89	0.546 (at bp)
Propane	C_3H_8	−187.7	−42.1	0.582 (at bp)
Butane	C_4H_{10}	−138.3	−0.55	0.5787
Isobutane	C_4H_{10}	−159.4	−11.7	0.5572
Pentane	C_5H_{12}	−129.7	36.0	0.6262
2-Methylbutane	C_5H_{12}	−159.9	27.9	0.6197
2,2-Dimethylpropane	C_5H_{12}	−15.9	9.5	0.5904
Hexane	C_6H_{14}	−95.4	68.6	0.6594
Heptane	C_7H_{16}	−90.6	98.4	0.6838
Octane	C_8H_{18}	−56.8	125.6	0.7028
Nonane	C_9H_{20}	−53.7	150.7	0.7181
Decane	$C_{10}H_{22}$	−29.7	174.0	0.7299
Undecane	$C_{11}H_{24}$	−25.6	195.8	0.7403
Dodecane	$C_{12}H_{26}$	−9.6	216.3	0.7488
Tridecane	$C_{13}H_{28}$	−6.2	235.5	0.7467
Tetradecane	$C_{14}H_{30}$	5.5	252.5	0.744 (at mp)
Pentadecane	$C_{15}H_{32}$	10	270.5	0.776 (at mp)
Hexadecane	$C_{16}H_{34}$	18	287.5	0.775 (at mp)
Octadecane	$C_{18}H_{38}$	28	317	0.777 (at mp)
Nonadecane	$C_{19}H_{40}$	32	330	0.777 (at mp)

As with any series of similar substances, the boiling point increases regularly with increasing molecular weight. However, as expected, the boiling points are much lower for these non-polar compounds than for polar compounds with corresponding molecular weights. For example, compare the boiling point of methane (−161.5°C) with that of water (100°C). Table 25.2 shows the regular increase in boiling point as the mass of the molecule increases with successive additions of $—CH_2—$ units in the normal alkanes, a so-called **homologous series.**

Each member of the series differs by one $—CH_2—$ unit in the carbon chain from the preceding or following member, and all members of the series may be represented by the formula

predict the boiling point of the normal chain alkane, $C_{20}H_{42}$ (eicosane).

CH_3—$(CH_2)_n$—H, where n is zero or any number. Neighboring compounds are **homologs** of each other. The increase in the boiling point from compound to compound is larger for the smaller members, but it becomes almost constant per —CH_2— added with the higher members.

The boiling point is also affected by changes in structure. Isomeric alkanes exhibit differences in boiling points. In the n-alkane series the boiling point increases with each added —CH_2— group, not only because of an increase in molecular weight, but because longer molecules can wrap around and become entangled with one another and are thus more difficult to separate. In Table 25.3 the boiling point of n-hexane, for example, may be compared with its more branched and more compact isomers, 3-methylpentane, 2-methylpentane, 2,3-dimethylbutane and 2,2-dimethylbutane (Fig. 25.6).

The basic reason for the higher boiling points exhibited by continuous chain over those of isomeric branched chain hydrocarbons lies in the greater intermolecular attraction which exists between adjacent molecules of unbranched over branched chains when closely packed in the liquid. Branching which increases the distance between adjacent chains decreases the intermolecular attrac-

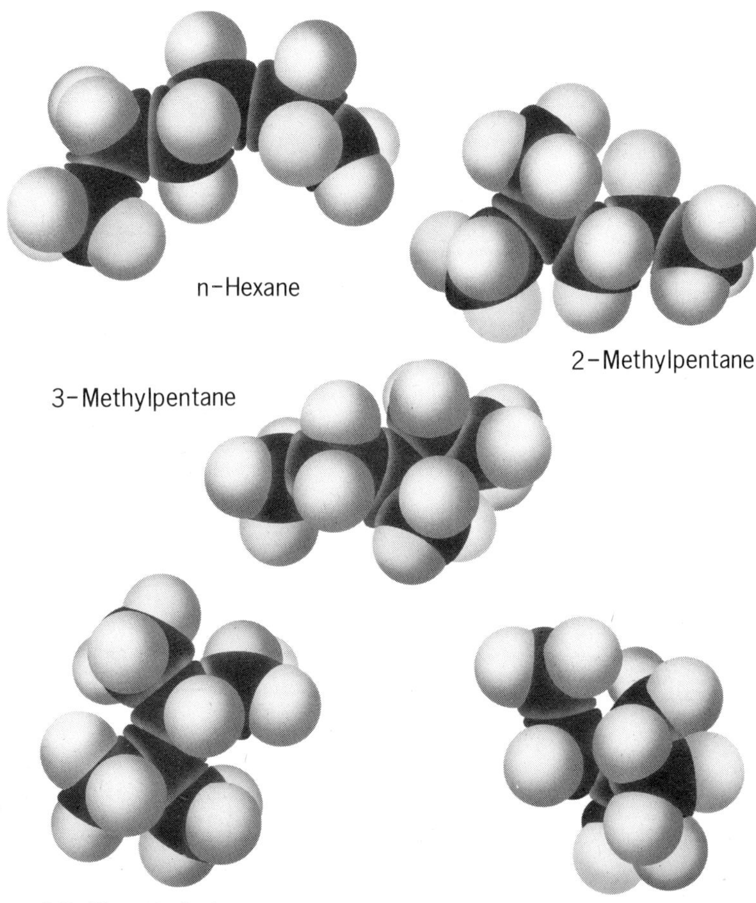

FIGURE 25.6 THE ISOMERIC HEXANES.

TABLE 25.3 BOILING POINTS OF THE ISOMERIC HEXANES, °C

n-hexane	68.7 (a in Fig. 25.6)
3-methylpentane	63.3 (c in Fig. 25.6)
2-methylpentane	60.3 (b in Fig. 25.6)
2,3-dimethylbutane	58.0 (d in Fig. 25.6)
2,2-dimethylbutane	49.7 (e in Fig. 25.6)

tion between neighbors and permits an easier escape of molecules from the liquid. The result is lower boiling points for the branched chain isomers.

The **viscosity,** a measure of the difficulty with which molecules slide by one another, increases with chain length so that the members of the series, excepting the first three members which are gases at room conditions, vary from free flowing liquids to thick oils. The liquids tend to be oily to the touch. The high molecular "solids" are waxy solids. All are non-conductors of electricity.

25.11 CHEMICAL PROPERTIES. GENERAL ASPECTS

The alkanes are singularly unreactive organic compounds. Because they are unaffected by most bases, acids and mild oxidizing agents, they have been named **paraffins** (that is, substances having little affinity). The name paraffin is employed specifically for a waxy mixture of high molecular weight alkanes, paraffin wax. The relative unreactivity of alkanes is apparently due to the stable carbon to carbon *or* carbon to hydrogen bonds which must be broken first before further reaction can occur. When sufficient energy is provided alkanes do undergo reactions. For example, alkanes burn in either pure oxygen or in the air when ignited.

$$CH_3CH_2CH_3 + 5\,O_2 \rightarrow 3\,CO_2 + 4\,H_2O$$

The ignition provides sufficient energy to achieve bond cleavage in a few molecules of the hydrocarbon. The fragments produced then react rapidly with oxygen to form carbon dioxide and water, as shown by the overall reaction. Once the reaction is initiated the energy released by reaction of a few molecules activates (cleaves) other molecules and keeps the reaction going.

Alkanes react with halogens in either sunlight or at high temperatures.

$$CH_3CH_3 + Cl_2 \rightarrow CH_3CH_2Cl + HCl$$

Because the product contains a halogen, the reaction is called **halogenation.** The sunlight or high temperature provides the energy necessary to split the Cl_2 apart and initiate this reaction. Again, once initiated, the reaction proceeds rapidly. The halogenation reaction will be discussed in detail in the next section. Alkanes undergo other transformations. For example, when alkanes are heated to a sufficiently high temperature in the absence of oxygen, they break into smaller molecules. With methane the products are hydrogen and carbon black:

$$CH_4 \rightarrow C + 2\,H_2$$

The high temperature provides the energy required to break the carbon to hydrogen bonds. This type of reaction is referred to as **pyrolysis.** When used in the petroleum industry for breaking large molecular weight hydrocarbons into smaller molecular weight compounds, it is commonly called "**cracking.**" The decomposition of methane is the simplest example of the pyrolysis of an alkane. With compounds containing two or more carbon atoms cleavage of the carbon to carbon bonds also occurs.

25.12 HALOGENATION

Alkanes react with the halogens (chlorine, bromine and iodine) when a mixture of the alkane and a halogen is either placed in sunlight, illuminated with artificially produced ultraviolet light or subjected to a high temperature. The equation for the reaction in which one hydrogen of methane is replaced with chlorine to form CH_3Cl

$$CH_4 + Cl_2 \xrightarrow[\text{light}]{\text{heat}} CH_3Cl + HCl$$

shows the substances reacting and the products formed but does not indicate the role of the sunlight or ultraviolet light. The manner in which each reactant changes and the role of the light is suggested by the so-called "mechanism" of the reaction. A mechanism outlines in detail each type of change which might occur and a possible role of each ingredient of the reaction mixture. When methane and halogen are mixed in sunlight, absorption of the ultraviolet light by the halogen molecule initiates the reaction by giving the halogen molecule sufficient vibrational energy to split. With chlorine this reaction is as represented:

note that each chlorine atom retains all its electrons and is uncharged.

$$\text{a. } Cl_2 + \text{photon} \rightarrow 2\,Cl\cdot$$

The active chlorine atom shown with a dot representing the unpaired electron in a p orbital (called a **free radical**) next attacks a methane molecule and takes a hydrogen atom. The three additional pairs of valence electrons of chlorine will not be shown.

neutral atoms are reactive because they seek to achieve an octet noble gas configuration.

$$\text{b. } Cl\cdot + CH_4 \rightarrow \cdot CH_3 + HCl$$

The methyl free radical, $\cdot CH_3$, then collides with another chlorine molecule,

$$\text{c. } \cdot CH_3 + Cl_2 \rightarrow CH_3Cl + Cl\cdot$$

to form methyl chloride (CH_3Cl) and another chlorine free radical ($Cl\cdot$). This chlorine atom may react again as in b, yielding another methyl radical to react as in c. Reactions b and c follow one another over and over again, yielding products at each step. Reactions of this character are called *chain reactions*. They are alternate links, one following another in a chain. Because of the reaction chain, a small amount of absorbed light may cause extensive reaction.

Although the free radicals are present in much lower concentrations than the molecular species, occasionally chlorine atoms collide and combine.

d. $Cl\cdot + Cl\cdot \rightarrow Cl_2$

Methyl radicals may combine,

e. $CH_3\cdot + CH_3\cdot \rightarrow CH_3CH_3$ (yielding another product, ethane)

and chlorine atoms and methyl radicals unite.

f. $CH_3\cdot + Cl\cdot \rightarrow CH_3Cl$

Because these reactions destroy radicals necessary for the propagation of the chain, they are called **chain termination** (stopping) steps. When the free radicals disappear from the mixture or when either of the reactants is consumed, the reaction ceases.

All chain reactions follow a similar sequence of events, continuing until a reactant is used up completely or until chain termination reactions halt further change. Reaction a is commonly referred to as the **chain initiation step.** Reactions b and c are the **chain propagation** reactions which keep the sequence going. Reactions d, e and f constitute the so-called termination steps of a chain mechanism. Chain reactions proceed in a stepwise manner similar to that outlined previously.

chain propagation requires that one new radical be formed for each one used.

This type of reaction, $CH_4 + Cl_2 \rightarrow CH_3Cl + HCl$, is called **substitution.** Regardless of the intermediate steps (mechanism) involved, the final result is the replacement or substitution of a chlorine atom for a hydrogen in the methane molecule.

Further substitution of hydrogens by halogen can and does occur. The atomic halogen formed by absorption of light, heat or by step c in the chain reaction may attack a CH_3Cl molecule with the following result:

g. $Cl\cdot + CH_3Cl \rightarrow HCl + \cdot CH_2Cl$

h. $\cdot CH_2Cl + Cl_2 \rightarrow CH_2Cl_2 + Cl\cdot$

In this manner 4 hydrogen atoms of methane may be replaced by 4 chlorine atoms. The equations for individual reactions are as follows:

$$CH_4 + Cl_2 \rightarrow CH_3Cl + HCl$$

Chloromethane

$$CH_3Cl + Cl_2 \rightarrow CH_2Cl_2 + HCl$$

Dichloromethane
(methylene chloride)

$$CH_2Cl_2 + Cl_2 \rightarrow CHCl_3 + HCl$$

Trichloromethane
(chloroform)

$$CHCl_3 + Cl_2 \rightarrow CCl_4 + HCl$$

Tetrachloromethane
(carbon tetrachloride)

If an excess of methane gas is used, a mixture of the mono-, di-, tri- and tetrahalogenated methanes is obtained when methane is chlorinated. If the reaction is permitted to take place for an extended period of time with an excess of chlorine gas, the completely substituted product, tetrachloromethane, may be obtained as the only substituted product.

Halogenation of other alkanes occurs in an analogous manner but a greater variety of products is possible. Consider the bromination of propane. Propane is bubbled through a solution of bromine in carbon tetrachloride during illumination with ultraviolet light. In contrast to methane, propane ($CH_3CH_2CH_3$) contains both primary and secondary hydrogens. Two different monobromopropanes can be expected. 1-Bromopropane (I) results from the replacement of a primary hydrogen and 2-bromopropane (II) from replacement of a secondary H as follows:

$$CH_3CH_2CH_3 + Br_2(\text{in } CCl_4) \rightarrow CH_3CH_2CH_2Br \quad (I) + HBr$$

$$CH_3CH_2CH_3 + Br_2(\text{in } CCl_4) \rightarrow \underset{\displaystyle Br}{CH_3CHCH_3} \quad (II) + HBr$$

Because the propane molecule has six primary hydrogens and two secondary hydrogens, and because the probability should be about equal for the bromine atom to collide with and combine with any one of the hydrogens, the monohalogenated product might be expected to be a mixture of I and II in the mole ratio of three to one respectively. Actually almost equal amounts of the two isomers are obtained (48 per cent $CH_3CH_2CH_2Br$ and 52 per cent CH_3CH-$BrCH_3$). This indicates that a secondary hydrogen is abstracted by and eventually replaced by a bromine atom more rapidly than a primary hydrogen.

The reaction (second step of the chain mechanism for halogenation) which removes a secondary H atom

$$\text{i.}\ Br\cdot + CH_3CH_2CH_3 \rightarrow CH_3\dot{C}HCH_3 + HBr$$

proceeds more rapidly than

$$\text{j.}\ Br\cdot + CH_3CH_2CH_3 \rightarrow CH_3CH_2CH_2\cdot + HBr$$

in which a primary hydrogen is captured. Experimental studies have shown that the rate of formation of a free radical generally parallels its stability, that is, the more stable the free radical the faster it is formed from its precursor. Since the rate of reaction i exceeds that of reaction j, the conclusion is reached that the secondary free radical formed in i is more stable than the primary free radical formed in j. This more rapid rate of reaction of secondary over primary hydrogen atoms accounts for the larger than statistically expected yield of 2-bromopropane. As a general observation, secondary free radicals are more stable than primary free radicals.

Further reaction of the bromopropanes with halogen produces a greater variety of polysubstitution products than is possible with methane or ethane

$$CH_3CH_2CH_2Br + Br_2 \text{ (in } CCl_4) \xrightarrow{\text{U.V.}} \left[\begin{array}{l} CH_2BrCH_2CH_2Br \\ \text{1,3-Dibromopropane} \\ CH_3CHBrCH_2Br \\ \text{1,2-Dibromopropane} \\ CH_3CH_2CHBr_2 \\ \text{1,1-Dibromopropane} \end{array}\right] + HBr$$

how many isomers of $C_3H_5Br_3$ are possible?

$$CH_3CHBrCH_3 + Br_2 \text{ (in } CCl_4) \xrightarrow{\text{U.V.}} \left[\begin{array}{l} CH_3CBr_2CH_3 \\ \text{2,2-Dibromopropane} \\ CH_3CHBrCH_2Br \\ \text{1,2-Dibromopropane} \end{array}\right] + HBr$$

Four isomers with the formula $C_3H_6Br_2$ result when propane reacts with bromine. The use of an excess of bromine yields a very complicated mixture of products, including octabromopropane, mixtures which may be separated by fractional distillation.

In the bromination of isobutane, replacement of primary and tertiary hydrogens occurs.

$$CH_3-\overset{\displaystyle CH_3}{\underset{\displaystyle CH_3}{\overset{|}{\underset{|}{C}}}}-H + Br_2 \text{ (in } CCl_4) \rightarrow CH_3-\overset{\displaystyle CH_3}{\underset{\displaystyle CH_2Br}{\overset{|}{\underset{|}{C}}}}-H + HBr$$

1-Bromo-2-methylpropane

$$CH_3-\overset{\displaystyle CH_3}{\underset{\displaystyle CH_3}{\overset{|}{\underset{|}{C}}}}-H + Br_2 \text{ (in } CCl_4) \rightarrow CH_3-\overset{\displaystyle CH_3}{\underset{\displaystyle CH_3}{\overset{|}{\underset{|}{C}}}}-Br + HBr$$

2-Bromo-2-methylpropane

The isobutane molecule has nine primary and one tertiary hydrogen atoms. Assuming that all hydrogen atoms react at the same rate, one would expect nine times as much 1-bromo-2-methylpropane as 2-bromo-2-methylpropane. The actual yield obtained contains twice as much 1-bromo-2-methylpropane as 2-bromo-2-methylpropane. This shows that the tertiary hydrogen of isobutane reacts considerably faster with bromine than does a primary hydrogen. A similar experimental comparison between the rates of displacement of secondary and tertiary hydrogens in a compound having both has shown that tertiary hydrogens react faster with bromine. The results therefore show that this reaction,

$$\text{k. } Br\cdot + CH_3-\overset{\displaystyle CH_3}{\underset{\displaystyle CH_3}{\overset{|}{\underset{|}{C}}}}-H \rightarrow CH_3-\overset{\displaystyle CH_3}{\underset{\displaystyle CH_3}{\overset{|}{\underset{|}{C}}}}\cdot + HBr$$

proceeds at a faster rate than a corresponding reaction in which either a primary or secondary hydrogen atom is abstracted from an alkane. Since the tertiary free radical, $(CH_3)_3C\cdot$, is formed faster than either a primary or secondary free radical it is concluded that it is more stable than both.

Based on the experimental results discussed for the bromination of propane and isobutane, we may now set up the order of ease (rate) of formation of free radicals (which is also the order of stability) as follows:

Least rapidly formed free radical	Primary < Secondary < Tertiary	Most rapidly formed free radical

Many industrially important reactions are free radical reactions (chain reactions). Certain reactions leading to the formation of polymers and the cracking of petroleum are examples. At high temperature the formation of many isomeric free radicals of similar energies leads to very complicated mixtures of products. Large scale separation procedures may permit economic isolation of useful products from these mixtures.

25.13 NITRATION

Another substitution reaction of alkanes (again under exceptional conditions) is nitration. Here an NO_2 group is introduced in place of a hydrogen atom. At 400°C nitric acid vapor reacts with alkanes. As a vapor, nitric acid does not act as an acid and is not ionized into hydrogen ions and nitrate ions; the $HONO_2$ molecule appears to split in a different way under these conditions to form free radicals. The overall reaction with methane is

$$CH_4 + HONO_2 \rightarrow CH_3NO_2 + H_2O$$

Higher alkanes react in much the same manner, but the reaction due to formation of several kinds of free radicals, some of which induce chain cleavage through carbon to carbon bond rupture, gives a more complex mixture of products. Propane is nitrated on a commercial scale. The products are 1-nitropropane and 2-nitropropane (via substitution of an NO_2 group for a primary or a secondary hydrogen atom); nitroethane and nitromethane (via carbon to carbon cleavage of the carbon chain). The products are separated by distillation. Although nitration is complicated by side reactions, it is introduced here to emphasize substitution as a typical chemical reaction of alkanes.

25.14 ADDITIONAL NOMENCLATURE

In the last two sections several substituted alkane products were named with no attendant explanation. One such compound was the **bromosubstituted alkane:**

$$CH_3—CHBr—CH_2Br$$

1,2-Dibromopropane

One may think of the compound as propane with bromine atoms substituted for hydrogens. The IUPAC name, 1,2-dibromopropane, is a logical extension of the naming system used for branched alkanes. The group is named and its position on the chain (the carbon to which it is attached) noted by a number preceding the group name. To extend the system to include alkanes with substituted groups, one needs to know the group names. As new groups are introduced they will be named. The non-alkyl groups encountered thus far are the halogens,

—Cl chloro-

—Br bromo-

—I iodo-

(any one of these may be represented by —X or halo- [for halogen]) and the nitrogen dioxide group

$—NO_2$ nitro-

The correct IUPAC name for the substituted alkane:

$$
\begin{array}{ccccccc}
 & & CH_3 & & & & NO_2 \\
 & & | & & & & | \\
CH_3 & — & CH & — & CHCl & — & C — CH_3 \\
 & & & & & & | \\
 & & & & & & CH_2 \\
 & & & & & & | \\
 & & & & & & CH_3
\end{array}
$$

is 3-chloro-2,4-dimethyl-4-nitrohexane. The compounds

$$CH_3CH_2Br,\ CH_3CH_2CH_2Br \quad \text{and} \quad CH_3CHBrCH_3$$

are properly called bromoethane, 1-bromopropane and 2-bromopropane, respectively, but they are more often labeled simply ethyl bromide, n-propyl bromide and isopropyl bromide.

25.15 PETROLEUM

Most hydrocarbons used today originate from coal, **petroleum** or **natural gas.** The latter two are the most important natural sources of alkanes. Natural gas is largely methane (80 to 85 per cent) accompanied by progressively smaller amounts of ethane, propane, butane and the higher alkanes (3 per cent). Petroleum, the largest source of hydrocarbons, is a complex mixture of alkanes accompanied by varying amounts of **cyclic saturated** (Section 25.17) and **aromatic hydrocarbons** (Chapter 28) along with smaller quantities of oxygen-, sulfur- and nitrogen-containing substances.

Petroleum is a viscous liquid obtained from underground deposits found in the upper strata of the earth's crust. Natural gas and petroleum are trapped in the interstices of porous rocks, such as limestone or sandstone, in reservoirs capped by dome-shaped overlying layers of non-porous rocks. Drilling a well provides an outlet through this impervious cap from which natural gas and petroleum may escape or be removed. Hydrostatic pressure of the underlying liquid present when a well is first drilled forces the natural gas and crude oil through the opening until the excess pressure is released, after which the oil is recovered by pumping.

Petroleum is separated into fractions of differing boiling range by distillation. This is called **refining.** The fractionation is accomplished in refineries located near the oil fields or in localities to which the crude oil is moved by pipeline or other readily available transportation. Distillation of the crude oil depends on progressive separations as the boiling temperature is increased. Various fractions, the quantity and composition of which differ widely with the origin of the petroleum, are collected. The usual fractions obtained as distillate are listed in Table 25.4.

TABLE 25.4 PETROLEUM DISTILLATE

Fraction Name	Boiling Temperature, °C	Carbon Chain Length
Gas	Below 20°	C_1 to C_4
Petroleum ether	20–60°	C_5 to C_6
Ligroin	60–100°	C_6 to C_7
Gasoline	40–205°	C_6 to C_{12}
Kerosene	175–235°	C_{12} to C_{20}
Gas oil	Above 275°	
Lubricating oil	Nonvolatile liquids	
Asphalt or petroleum coke	Residue	

Petroleum from Pennsylvania and other eastern states is composed of alkanes and yields paraffin as the end product of distillation. Oils of this kind are said to have a paraffin base. Petroleums from California, Texas and Mexico, which yield a black tarlike residue when distilled, are referred to as asphalt base petroleums. Midcontinent oils are a composite of both types.

In addition to the naturally occurring petroleums, synthetic petroleum (less complex than natural petroleum) is available through the **Fischer-Tropsch synthesis.**

By this process coal, steam and hydrogen can be converted to a mixture of alkanes. Coal and steam are converted to carbon monoxide and hydrogen by the water-gas reaction

$$C + H_2O \rightarrow CO + H_2$$

The water-gas mixture is enriched with more hydrogen and passed over a cobalt-thoria catalyst at about 250°C to give a mixture of alkanes. For the preparation of octane, the reaction is:

$$8\,CO + 17\,H_2 \xrightarrow[\text{Catalyst}]{} C_8H_{18} + 8\,H_2O$$

For countries with inadequate petroleum resources, and perhaps for all countries as present oil supplies diminish, this product could become the best source of alkanes.

Petroleum is the most important source of motor fuel, which is the chief use of the volatile fractions. Correspondingly, as the demand for gasoline has mushroomed over the years, methods have been developed for converting larger quantities of the less volatile components of petroleum into smaller molecules more useful as fuels. Simultaneously the development of modern engines, with their more exacting fuel requirements, stimulated the discovery of reactions for producing fuels of greater efficiency by combining the small molecules, formerly lost during distillation as gases, into highly branched hydrocarbons having boiling points suitable for gasoline use.

Pyrolysis or "cracking" of alkanes accomplishes the cleavage of large molecular weight alkanes (non-volatile components) into smaller molecules.

$$CH_3(CH_2)_6CH_2CH_2CH_2(CH_2)_5CH_3 \xrightarrow{\Delta} \underset{\text{1-nonene}}{CH_3(CH_2)_6CH{=}CH_2} + \underset{\text{n-heptane}}{CH_3(CH_2)_5CH_3}$$

In this equation hexadecane ($C_{16}H_{34}$) is shown to cleave between the seventh and eighth carbon. However, the cleavage may occur at any point, always yielding an alkane and a hydrogen deficient (unsaturated) hydrocarbon, called an **alkene** (Chapter 27). Suitable catalysts reduce the cracking time considerably. Cracking carried out in the presence of catalysts is known as **catalytic cracking.** Almost all refineries have at least one catalytic cracker.

Because pyrolysis yields a mixture of alkanes and alkenes (Chapter 27) and since alkenes improve the quality of gasoline by causing it to burn more smoothly, the process increases the quantity of gasoline obtainable from petroleum while also giving a superior product. Gasoline made by cracking requires protection to prevent auto-oxidation, which initiates polymer and gum formation. **Antioxidants** (oxidation inhibitors) of various types are added to some gasolines to prevent deterioration.

The presence of alkanes having branched chains also improves the quality of gasoline. The conversion of a continuous chain hydrocarbon into a branched chain hydrocarbon of the same composition yields a better gasoline. Such a conversion is referred to as **isomerization.** The continuous chain hydrocarbon is transformed into its more highly branched isomer. Despite the relative inertness of alkanes, they, in the presence of certain catalysts, react readily in this way. As an example, n-butane when treated with aluminum bromide and HBr is partially converted into isobutane to give an equilibrium mixture containing only 20 per cent n-butane and 80 per cent isobutane:

$$CH_3CH_2CH_2CH_3 \underset{}{\overset{AlBr_3,\ HBr}{\rightleftarrows}} CH_3\overset{\overset{\displaystyle CH_3}{|}}{C}HCH_3$$

Treated with this same catalyst, pure isobutane is converted into the same mixture, 20 per cent n-butane and 80 per cent isobutane; hence the double arrow.

The same type of reaction occurs with higher molecular weight alkanes. Chain fragments break off and reattach on the chain. The point of attachment is not predetermined and an equilibrium mixture (often very complex) is usually obtained. The effect is to increase the percentage of highly branched isomers and thus to provide better fuel.

The quality of gasoline as a fuel is expressed by its **octane number.** This is a figure obtained by comparing the performance of a standard one cylinder test engine when using the gasoline concerned as a fuel with the performance of this same test engine when operated using a blend of two synthetic fuels, "isooctane" (2,2,4-

trimethylpentane) and n-heptane. Isooctane is assigned an octane rating of 100, n-heptane a rating of 0. The **octane rating** is the per cent of the octane present in a fuel mixture with n-heptane which exhibits the same knocking characteristics as the gasoline being examined. Knock refers to the ping that occurs when an automobile is accelerated rapidly. The existence of knock or ping indicates that conditions are unsuitable for efficient performance of the engine with the fuel being used. A gasoline whose performance corresponds to that of a mixture containing 90 per cent isooctane and 10 per cent n-heptane would have an octane rating of 90.

Highly branched alkanes, unsaturated (hydrogen deficient) hydrocarbons (alkenes) and aromatic hydrocarbons (Chapter 28), when present, materially improve the antiknock quality of gasoline. Manufacturing procedures which increase the percentage of molecules of these types in gasolines are widely used. Cracking and aromatization or reforming (discussed in Chapter 28) are two others. **Alkylation** is the addition of branched chain alkanes to alkenes at atmospheric pressure at a relatively low temperature in the presence of a suitable catalyst. This provides a direct route to highly branched alkanes. **Aromatization or reforming** refers to the catalytic process used to convert cycloalkanes and related hydrocarbons into aromatic hydrocarbons (Chapter 28). Aromatization yields a highly improved gasoline since most aromatic hydrocarbons have octane numbers greater than 100.

Substances not obtained from petroleum may be added to gasoline to improve its quality. Of these, the metal alkyls are the most effective. The best known of these is tetraethyl lead, $(C_2H_5)_4Pb$. This compound is added to gasoline together with ethylene bromide and ethylene chloride. The latter two compounds prevent deposition of lead oxide in the engine by converting the lead oxide formed from tetraethyl lead into volatile lead halides which exit in the exhaust.

About 95 per cent of all American gasolines contain added tetraethyl lead and are known as ethyl gasolines.

25.16 COMBUSTION

The most widespread use made of alkanes obtained from petroleum is as a fuel. This follows because of the energy released when alkanes burn in pure oxygen or in air. As shown earlier (Section 25.11) alkanes burn in oxygen only after "ignition." A small volume of the alkane vapor-air mixture is heated locally with an electric spark or flame. Because of this locally high temperature, some of the activated (or *hot*) molecules collide with sufficient velocity to break bonds and start the reaction. The heat liberated then raises other molecules to reaction temperature and the combustion continues. As indicated, the reaction of an alkane with an ample supply of oxygen yields carbon dioxide and water:

$$\underset{\text{Methane}}{CH_4} + 2\,O_2 \rightarrow CO_2 + 2\,H_2O + 211\text{ kcal}$$

why does CO appear in automobile exhaust?

$$2\,\underset{\substack{\text{n-octane}\\(C_8H_{18})}}{CH_3(CH_2)_6CH_3} + 25\,O_2 \rightarrow 16\,CO_2 + 18\,H_2O + 2610\text{ kcal}$$

With insufficient oxygen the following reactions occur simultaneously with those preceding:

$$2\,CH_4 + 3\,O_2 \rightarrow 2\,CO + 4\,H_2O + 290\text{ kcal (145 kcal/mole)}$$

$$2\,C_8H_{18} + 17\,O_2 \rightarrow 16\,CO + 18\,H_2O + \text{heat}$$

and

$$CH_4 + O_2 \rightarrow C + 2\,H_2O + 118\text{ kcal}$$

$$2\,C_8H_{18} + 9\,O_2 \rightarrow 16\,C + 18\,H_2O + \text{heat}$$

It is evident that burning with insufficient oxygen will cause the formation of either poisonous carbon monoxide gas or flakes of soot (carbon) which when heated to incandescence in a flame emit a yellow light. Such side reactions release many fewer calories per molecule of hydrocarbon burned. Complete combustion of one gram-molecular weight (mole) of methane yields 211 kcal, incomplete combustion to carbon monoxide gives 145 kcal, and combustion under conditions which give carbon yields only 118 kcal.

TABLE 25.5 HEATS OF COMBUSTION PER MOLE IN n-ALKANES

	Kilocalories per Mole		Heat per Added	Average Increase
	Gases	Liquids	—CH_2— Group	per Added —CH_2—
CH_4	210.8			
C_2H_6	368.4		157.6	
C_3H_8	526.3		157.9	
n-C_5H_{12}	838.3	833.4	156.0	
n-C_6H_{14}		989.8	156.4	156.4 kcal/mole
n-C_7H_{16}		1149.9	160.0	
n-C_8H_{18}		1305.0	155.1	
n-$C_{10}H_{22}$		1610.7	153.0	

estimate the kcal of heat formed when n-$C_{20}H_{42}$ is burned completely to CO_2 and H_2O.

The heats of combustion for members of the homologous n-alkane series are listed in Table 25.5. The table shows that, as expected, each additional —CH_2— group present in a hydrocarbon chain increases the heat of combustion by nearly a constant factor (156 kcal per mole).

25.17 SYNTHESIS OF PURE ALKANES

Petroleum is largely a mixture of alkanes, as are the various fractions separated from it upon distillation. These fractions are commonly used as mixtures without further separation. Carefully performed fractional distillation, although time consuming and difficult, will permit separation of the pure hydrocarbons. It is often less time consuming, easier and less expensive to synthesize (make) an individual alkane from other readily available chemicals than to attempt its separation as a pure compound from petroleum. Two relatively simple and useful reactions for synthesizing pure alkanes are the **Wurtz reaction** and a reaction between water and a compound having the formula, RMgX, called a **Grignard reagent.**

The Wurtz Reaction. A reaction which forms an alkane takes place when dry **alkyl halide** (halogen-substituted alkane) is added slowly to small pieces of sodium metal. The reaction using bromoethane follows.

how could you make CH_3CH_2Br?

$$CH_3CH_2Br + 2\,Na + BrCH_2CH_3 \rightarrow CH_3CH_2CH_2CH_3 + 2\,NaBr$$

The net result of the reaction is the combination of the two ethyl groups to form n-butane. This reaction is most useful when only a single alkyl halide (here bromoethane) is used such that two identical hydrocarbon groups (CH_3CH_2—) are joined. It is possible to make **unsymmetrical alkanes,** that is, alkanes which cannot be separated into two identical halves. For example, ethyl bromide and n-propyl bromide (1-bromopropane) can combine with elimination of bromine in the presence of sodium to form n-pentane:

$$CH_3CH_2Br + 2\,Na + BrCH_2CH_2CH_3 \rightarrow CH_3CH_2CH_2CH_2CH_3 + 2\,NaBr$$

As expected the yield is low because both n-butane and n-hexane are also formed and in amounts almost equal to that of n-pentane. This reaction is to be expected since the probability of sodium's inducing the combination of two ethyl groups to yield n-butane or n-propyl groups to form n-hexane equals the probability of its inducing the combination of one ethyl with one n-propyl group to form n-pentane. Whenever a mixture of two different alkyl bromides is used, a mixture of products must be separated.

Reaction of Water with a Grignard Reagent. A dry halide when added slowly to small pieces of magnesium suspended in dry ethyl ether (Chapter 29) undergoes reaction to form a type of compound referred to as a Grignard reagent. It has the formula RMgX or ArMgX, where R may be any alkyl and Ar any aryl group (Chapter 5) of atoms. The reaction with 1-bromobutane is as follows:

$$CH_3CH_2CH_2CH_2Br + Mg \rightarrow CH_3CH_2CH_2CH_2MgBr$$

The product ($CH_3CH_2CH_2CH_2MgBr$) is called a Grignard reagent after Nobel Prize recipient Victor Grignard, who first discovered this type of reaction. This compound is very reactive with water and several types of organic compounds to be encountered later in the text. The reactions of Grignard reagents have great utility in organic chemistry. The reaction with water forms a hydrocarbon:

$$CH_3CH_2CH_2CH_2MgBr + H_2O \rightarrow CH_3CH_2CH_2CH_3 + Mg(Br)OH$$

The hydrocarbon is formed in essentially 100 per cent purity and almost 100 per cent of the Grignard reagent is converted into hydrocarbon. Because many long chain alkyl halides are more available in pure form than are the corresponding pure long chain hydrocarbons, conversion of these halides into RMgX can serve as a useful procedure for preparing the corresponding alkane. For example, octadecane can be prepared in the following way:

$$CH_3(CH_2)_{16}CH_2Br^* + Mg \rightarrow CH_3(CH_2)_{16}CH_2MgBr$$

$$CH_3(CH_2)_{16}CH_2MgBr + H_2O \rightarrow CH_3(CH_2)_{16}CH_3 + Mg(Br)OH$$

25.18 CYCLOALKANES

In alkanes the carbon atoms are attached to one another to form continuous or branched chains; these are called *open chain* compounds. Other compounds have closed chains and are classified as cyclic compounds.

One series of such compounds are the *cycloalkanes*. Each of these hydrocarbons may be represented by the formula C_nH_{2n}, where n is any number from 3 to about 30 for known compounds. Each cycloalkane has 2 fewer hydrogen atoms than the open chain alkane with the same number of carbons.

what mixture of products would result if CH_3Br and CH_3CH_2Br were used?

what is the simplest formula for the cycloalkane having 15 carbons?

The simplest structural formula for the cycloalkanes is $(—CH_2—)_n$, where n has the same values as in the molecular formula, C_nH_{2n}. There are no terminal groups (end of the chain groups, $—CH_3$). This fact accounts for the two fewer hydrogens and predicts the general configuration of the compounds. Since all groups are non-terminal $—CH_2—$ groups, the compounds have no end. They are closed chain or cyclic ("circle") hydrocarbons. The term cycloalkane is used because of their similarity in properties to the open chain alkanes. Compare cyclohexane and n-hexane:

```
      CH2                 CH2
     /   \               /   \
  CH2     CH2         CH2     CH3
   |       |           |
  CH2     CH2         CH2     CH3
     \   /               \   /
      CH2                 CH2
```

	Cyclohexane $(—CH_2—)_6$	n-Hexane $CH_3(CH_2)_4CH_3$
b.p.	81.4°C	68.6°C
m.p.	6.5°C	−94.3°C

how many primary H's in cylcohexane? why?

Both are volatile (low boiling) liquids with similar chemical properties. Note that cyclohexane contains only secondary hydrogens.

Cyclopentane, $(CH_2)_5$, and cyclohexane, $(CH_2)_6$, occur naturally. Both are volatile liquids found in crude petroleum. All cycloalkanes can be formed in varying yields by treating the proper open chain dihaloalkanes with zinc dust.. Cyclopropane is prepared from 1-bromo-3-chloropropane.

* This alkyl bromide (1-bromooctadecane) is obtainable through a reaction between HBr and pure $CH_3(CH_2)_{16}CH_2OH$, which may be purchased from chemical supply companies.

$$\begin{array}{l} CH_2{-}Br \\ / \\ CH_2 \qquad\qquad + Zn \rightarrow \\ \backslash \\ CH_2{-}Cl \end{array} \quad \begin{array}{c} CH_2 \\ / \,| \\ CH_2 \;\; | \\ \backslash \,| \\ CH_2 \end{array} \quad + ZnBrCl$$

1-Bromo-3-chloropropane — Cyclopropane

These are **intramolecular** (within the molecule) reactions. The yield is decreased greatly because **intermolecular** (between molecules) reaction also occurs:

$$BrCH_2CH_2CH_2Cl + Zn + BrCH_2CH_2CH_2Cl \rightarrow BrCH_2CH_2CH_2CH_2CH_2CH_2Cl + ZnBrCl$$

In practice the intermolecular reaction is minimized by using a very dilute solution of the dihalide in some inert solvent. In this situation the two ends of the same molecule have more time to meet and react before colliding with ends of other molecules. Other products such as cyclohexane are obtained as by-products.

Cyclopropane, like normal propane, is a gas. It is widely used as a general anesthetic in place of ether. Cyclobutane is also a gas, whereas cyclopentane and cyclohexane are liquids as already pointed out.

All the carbons in these compounds are "saturated," i.e., they have 4 single bonds. With the exception of cyclopropane and, to a lesser degree, cyclobutane, the $—CH_2—$ groups react just like $—CH_2—$ groups in open chain alkanes. The compounds are relatively inert, being limited in reactions to halogenation with sunlight, vapor phase nitration, and combustion.

In contrast, cyclopropane reacts readily with sulfuric acid, hydrobromic acid and bromine, the latter in the dark. The reactions involve opening of the chain through carbon to carbon bond rupture followed by attachement of the halves of the reagent to the carbon ends of the opened chain. The reactions are *additions* and *not substitution.* With bromine in carbon tetrachloride the reaction is:

HBr is not formed in this reaction. why?

$$\begin{array}{c} CH_2 \\ / \,| \\ CH_2 \;\; | \\ \backslash \,| \\ CH_2 \end{array} + Br_2 \text{ (in } CCl_4) \rightarrow \begin{array}{l} \quad CH_2Br \\ \;\;/ \\ CH_2 \\ \;\;\backslash \\ \quad CH_2Br \end{array}$$

1,3-Dibromopropane

When cyclopropane is bubbled into the colored bromine solution, the solution becomes progressively lighter in color. The colorless dibromo compound which is formed remains in the solvent to be separated by fractional distillation.

Furthermore, cyclopropane can be hydrogenated (combined with hydrogen) when mixed with hydrogen under pressure in the presence of finely divided platinum:

$$\begin{array}{l} CH_2 \\ | \quad \backslash \\ | \quad CH_2 + H_2 \xrightarrow[\text{Room Temperature}]{Pt} CH_3CH_2CH_3 \\ | \quad / \\ CH_2 \end{array}$$

Other cycloalkanes can be hydrogenated at higher temperatures; cyclobutane at 120°C and cyclopentane and cyclohexane at 200°C. In all cases the product is a normal alkane.

The greater reactivity of cyclopropane (and cyclobutane) with the addition of reagents has been ascribed to a straining of the bonds in the molecule. Johann Friedrich von Baeyer (1835–1917) reasoned that since the normal carbon bond angle is the tetrahedral angle, 109°28′, the bonds in cyclopropane and cyclobutane are strained. The bond angles in a planar equilateral triangle are 60°, in a planar square, 90° (Fig. 25.7a and b). Therefore, according to the Baeyer strain theory there is an overall strain of 109°28′ − 60° = 49°28′ and 109°28′ − 90° = 19°28′, respectively, in cyclopropane and cyclobutane. These two molecules tend to open up. On the other hand, the angles in a planar pentagon are 108° (very little strain). If the cyclohexane ring is planar (all carbons lie in one plane) the angles would be 120°, but because the ring can "buckle," the angles are 109°28′, and there is no strain. Cyclobutane and cyclopentane are somewhat "buckled" and thus are in reality non-planar.

The heat of combustion of cycloalkanes verifies the higher energy state of cyclopropane and cyclobutane when compared with larger ring cycloalkanes (see Table 25.6). For open chain alkanes each methylene group, $—CH_2—$, contributes very close to 157.4 kcal/mole to the heat of combustion. For cyclopropane the heat of combustion per $—CH_2—$ group is 167 kcal/mole. This is about 10 kcal per methylene higher than for alkanes. Likewise, cyclobutane shows a value of 164 kcal/mole. Again this is higher by 7 kcal/methylene than for the open chain alkanes. If cyclopropane and cyclobutane evolve more energy per $—CH_2—$ than an open chain compound, it can mean only that they contain more energy per $—CH_2—$ group. The greater energy makes it reasonable to suppose, in accordance with the Baeyer strain theory, that the tendency to undergo ring opening reactions is related to this lesser stability of the compounds. Cyclopentane and higher cyclic alkanes have

TABLE 25.6 HEATS OF COMBUSTION OF CYCLOALKANES

Ring Size	Heat of Combustion per CH_2, Kcal/mole
3	166.6
4	164.0
5	157.7
6	157.4

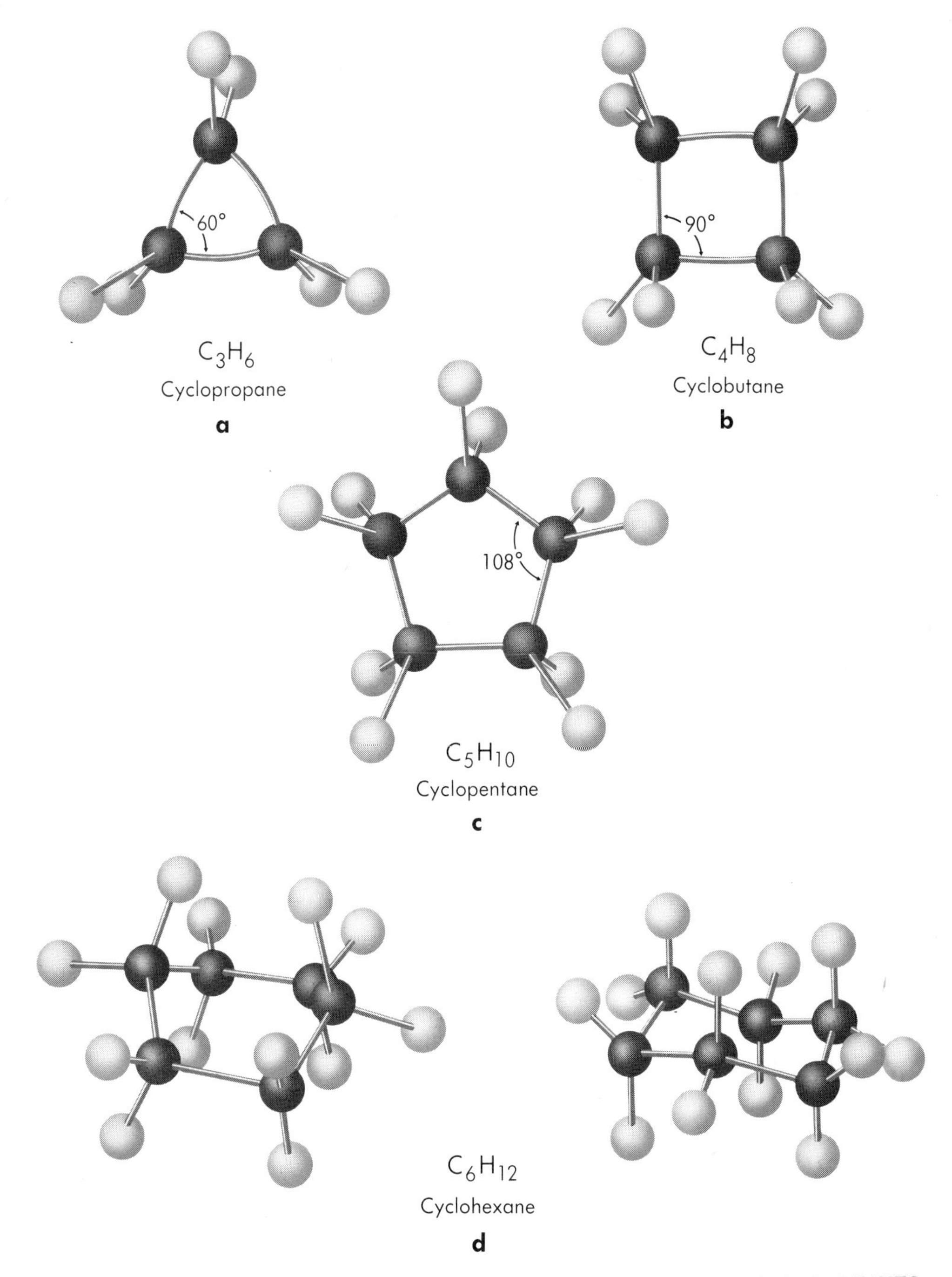

FIGURE 25.7 THE FIRST FOUR MEMBERS OF THE CYCLOALKANES.

heats of combustion in accordance with those of the open chain alkanes and correspondingly in most reactions behave much the same.

EXERCISES

25.1 Define each of the following terms and illustrate each *italicized* term with an example:

a. organic chemistry
b. *organic chemical (compound)*
c. *hydrocarbon*
d. *alkane*
e. *simplest formula* (use isobutane for an example)
f. *molecular formula* (use isobutane for an example)
g. *structural formula* (use isobutane for an example)
h. the tetrahedral angle
i. *continuous chain hydrocarbon*
j. covalent bond
k. *branched hydrocarbon*
l. *isomers*
m. *alkyl group*
n. *polar compound*
o. boiling point
p. *homologous series*
q. *homolog*
r. *n-alkane*
s. *isoalkane*
t. *paraffin*
u. kcal
v. *substitution reaction*
w. *haloalkane*
x. pyrolysis
y. isomerization
z. Baeyer strain theory
aa. cycloalkane
ab. planar
ac. Grignard reagent
ad. Wurtz reaction

25.2 What is the distance between (a) centers of adjacent carbons (C—C) in an alkane and (b) centers of a hydrogen atom and the carbon to which it is attached (C—H)?

25.3 Which formula represents the structure of propane best?

```
                                          H                 CH3
                                          |                 |
CH3—CH2—CH3     CH3—CH2            CH3—C—H         CH3—CH2
                      |                   |
                      CH3                 CH3
```

25.4 Write structural formulas for each of the isomeric hexanes and name each one with the correct IUPAC name.

25.5 Name each of these alkyl groups:

```
CH3—     CH3CH2—     CH3—CH2—CH2—     CH3—CH—
                                           |
                                           CH3
```

25.6 Name each group attached to the main chain of this compound (list them in alphabetical order):

```
       CH3               NO2  I
       |                 |    |
CH3—C——CH—CH2—C——C—CH3
       |      |          |    |
       Br     CH—CH3    Cl   CH2
              |               |
              CH3             CH3
```

25.7 Which of the compounds represented by the following formulas are branched alkanes?

```
CH3(CH2)3CH3    (CH3)4C     (CH3)2CH—CH3    CH3(CH2)3—CH(CH3)2
                  CH3          CH3     H
                  |            |       |
CH3—CH2       CH3—C—CH3    H—C—CH2—C—H
     |            |            |       |
     CH3          CH3          H       CH3
                              H       H
                              |       |
CH3         CH2        CH3   H—C——————C—H  H        (CH3—CH2)2
   \       /   \       /       |       |    |
    CH2          CH2           H    H—C——————C—H
                                       |    |
                                       H    H
```

25.8 How many (different) isomers of n-hexane are represented here?

```
          CH3
          |
a.  CH3—CH—CH2—CH2—CH3
b.  (CH3)2CH—CH2—CH2—CH3
c.  (CH3CH2)2CH—CH3
d.  CH3—CH2—CH2—CH—CH3
                     |
                     CH3
               CH3
               |
               CH2
               |
e.  CH3—CH2—CH—CH3
f.  CH3—CH—CH—CH3
         |    |
         CH3  CH3
g.  CH3—CH2—CH—CH2—CH3
               |
               CH3
h.  (CH3)3C—CH2—CH3
i.  CH3—CH2—CHCH3
               |
               CH2
               |
               CH3
j.  CH3—CH2
         |
         CH3
         CH3
         |
k.  CH3—C—CH2—CH3
         |
         CH3
l.  (CH3)2CHCH(CH3)2
m.  CH3(CH2)4CH3
```

25.9 How is the insolubility of alkanes in water related to their relatively low boiling points?

25.10 Write balanced equations for the complete combustion of propane, n-butane and n-heptane.

25.11 Chlorine gas is permitted to react with propane in the presence of ultraviolet light. Write structural formulas for and name each possible compound formed.

25.12 Write equations illustrating each of the following (include reaction conditions over the arrow):

a. halogenation of methane
b. nitration of ethane
c. cracking of tetradecane
d. isomerization of n-octane
e. complete combustion of methane
f. incomplete combustion of ethane

25.13 Sodium metal is dropped into a mixture of bromomethane, bromoethane, 1-bromopropane and 2-bromopropane. Draw the structures and name all the possible products, using the IUPAC system.

25.14 Why is cyclopropane more reactive with halogens than propane? Why is the reaction of cyclopropane with bromine in the dark called addition?

25.15 Write the correct IUPAC name for each compound whose structure is given:

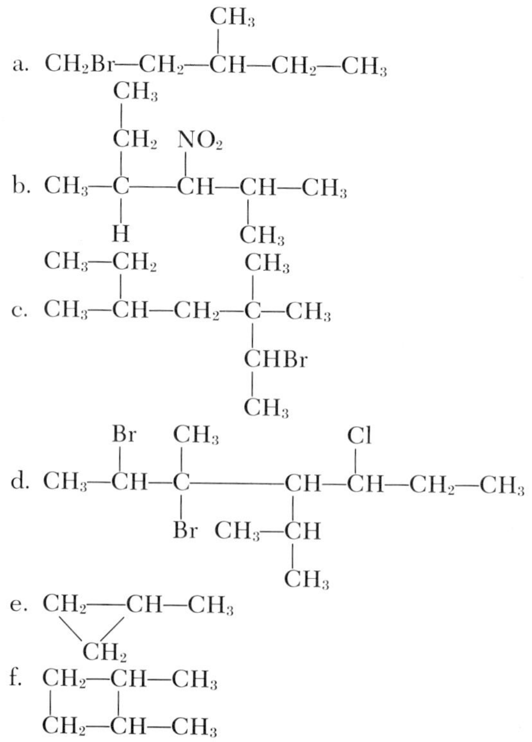

```
                        CH3
                        |
a. CH2Br—CH2—CH—CH2—CH3

           CH3
           |
           CH2  NO2
           |    |
b. CH3—C——CH—CH—CH3
           |         |
           H         CH3

   CH3—CH2        CH3
         |         |
c. CH3—CH—CH2—C—CH3
                   |
                   CHBr
                   |
                   CH3

         Br   CH3             Cl
         |    |               |
d. CH3—CH—C————CH—CH—CH2—CH3
              |        |
              Br  CH3—CH
                       |
                       CH3

e. CH2——CH—CH3
     \  /
     CH2

f. CH2—CH—CH3
   |    |
   CH2—CH—CH3
```

25.16 Name three homologs of n-propane.

25.17 How many different (non-equivalent) kinds of hydrogen atoms are present in n-hexane? isohexane? neohexane? 2,3-dimethylbutane?

25.18 How many monochloro substitution products are obtainable by chlorination of n-hexane? neohexane? 3-methylpentane?

25.19 Write equations for the preparation of the following from suitable haloalkanes of fewer carbon atoms:

a. n-hexane
b. 2,3-dimethylbutane
c. 2,5-dimethylhexane
d. n-pentane
e. Which of the these would be formed in lowest percentage of theoretical yield? Explain.

25.20 Write equations for the preparation of an alkane from 1-chloro-2,2-dimethylpropane using the Grignard reaction.

25.21 Point out the error made in applying the IUPAC rules of nomenclature when the names given below were selected for the compounds represented by these names. Write a correct IUPAC name for each compound.

a. 2-ethylpropane
b. 4-methylpentane
c. 3,3,6-trimethyl-5-ethylheptane
d. 3-bromo-2-ethylbutane
e. 1,6-dibromocyclohexane
f. 1,1,1,1-tetrachloromethane
g. 1-iodo-2-methylethane

25.22 What simple laboratory test is available for distinguishing cyclopropane from n-propane? Describe what would be observed in the test and write equations for any reaction (or reactions) which occur.

25.23 A sample of a hydrocarbon having the formula C_4H_{10} was chlorinated and four different monochloroalkanes were isolated from the reaction mixture. Did this sample of hydrocarbon contain only one kind of molecule or was it a mixture of isomers? Explain.

25.24 A sample of an open-chain pure hydrocarbon having the formula C_5H_{12} was chlorinated and yielded only one monochloroalkane. How many different types of hydrogen atoms are present in this hydrocarbon? What must be the structure of this compound?

25.25 The Wurtz synthesis was carried out using a mixture of two different alkyl halides, RI and R′I. R′I was used in largest amount. Butane, pentane and hexane were obtained as products with hexane being obtained in largest amount. What are the formulas of the alkyl halides, RI and R′I, used in this synthesis?

25.26 You have been given a container filled with a single alkane as a gas. It is known to contain 5, 6, 7 or 8 carbons per molecule. When subjected to combustion it is observed that a 20 ml sample of the gas reacted with oxygen to produce 340 ml of a gaseous mixture containing carbon dioxide and water. All gas volumes are measured at the same temperature and pressure. How many carbons are present in this alkane? If it gives one monobromo substitution product when subjected to bromination, what is a structure which it could have?

25.27 A hydrocarbon was found to contain 83.33 percent carbon and 16.17 percent hydrogen. One liter of the vapor of the compound measured at 40°C and 636 mm of Hg pressure weighed 2.73 g. Careful separation of the products obtained when the hydrocarbon was brominated yielded four monobromo substitution products. Write an acceptable formula (both empirical and structural) for this hydrocarbon.

25.28 Ten ml of a mixture of methane and propane required 41 ml of oxygen for complete combustion. Calculate the number of ml of methane and the number of ml of propane in the sample of gas examined. All gas volumes are measured at the same temperature and pressure.

25.29 An analysis and a molecular weight determination demonstrated that a certain liquid had a molecular formula of C_6H_{14}. What structure would you assign to this hydrocarbon if when subjected to chlorination it gave (a) 3 monochlorosubstitution products, (b) five monochlorosubstitution products or (c) two monochlorosubstitution products?

25.30 A vehicle is powered by a combustion engine fitted with a condenser such that the steam formed when the fuel is burned is condensed to liquid water and the carbon dioxide formed is permitted to escape into the atmosphere. Will the vehicle lose or gain weight as it operates if the fuel used is n-octane? What will be the gain or loss in weight when 5 liters of this fuel is burned?

SUGGESTED READING

Brown, H. C.: "Foundations of Structural Theory." J. Chem. Ed., *36*: 104, 1959.

Ferguson, L. N.: "Alicyclic Chemistry: the Playground of the Organic Chemists." J. Chem. Ed., *46*: 404, 1969.

Hendrickson, J. B., Cram, D. J. and Hammond, G. S.: *Organic Chemistry*, Third Edition. McGraw-Hill Book Company, New York, 1970. Chapters 3, 4 and 20.

Hurd, C. D.: "The General Philosophy of Organic Nomenclature." J. Chem. Ed., *38*: 43, 1961.

Morrison, R. T. and Boyd, R. N.: *Organic Chemistry*, Second Edition. Allyn and Bacon, Inc., Boston, 1966. Chapters 1, 2 and 4.

Nelson, T. W.: "The Origin of Petroleum." J. Chem. Ed., *31*: 399, 1954.

Noller, C. R.: *Chemistry of Organic Compounds*, Third Edition. W. B. Saunders Company, Philadelphia, 1965. Chapters 1, 2, 4 and 5.

Pryor, W. A.: *Introduction to Free Radical Chemistry*. Prentice-Hall, Englewood Cliffs, N. J., 1966 (paperback).

TWENTY-SIX • STABILITY AND REACTIVITY OF ORGANIC COMPOUNDS

26.1 INTRODUCTION

refer to Chapter 17 for a review of reaction rates and equilibrium.

Whether a chemical reaction can take place or not depends upon two things: the relative stability of the products of the reaction when compared with that of the reactants and the availability of a reaction pathway which permits conversion of reactants into products at a reasonable rate.

The stability factor, whose magnitude is expressed by the difference in **free energy** (ΔG) between reactants and products, controls the nature of the product formed when reaction is permitted to proceed until no further change occurs (attains equilibrium) and also determines the maximum yield which might be attained. The second factor, the rate or **kinetic factor,** the availability of a suitably rapid reaction pathway, determines the rate at which product formation takes place and controls the time period required for completion of the reaction.

Frequently in chemical processes the same reactants may undergo either of two (or more) separate transformations to yield different products. For example, when potassium chlorate ($KClO_3$) is heated at 400°C, it may yield either potassium chloride and oxygen gas or potassium chloride and potassium perchlorate or all three products simultaneously.

$$4\,KClO_3 \rightarrow 4\,KCl + 6\,O_2$$
$$4\,KClO_3 \rightarrow 3\,KClO_4 + KCl$$

Similarly, when naphthalene reacts with concentrated sulfuric acid to form naphthalene sulfonic acid, two reactions take place, each of which yields a different product:

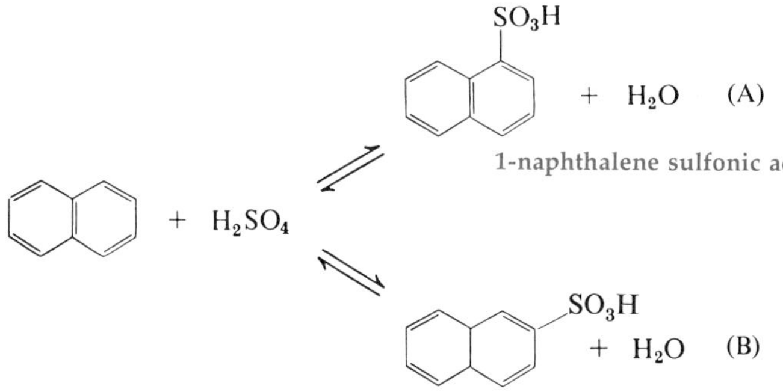

$$C_{10}H_8 + H_2SO_4 \rightleftharpoons C_{10}H_7SO_3H + H_2O \quad (A)$$

1-naphthalene sulfonic acid

$$C_{10}H_8 + H_2SO_4 \rightleftharpoons C_{10}H_7SO_3H + H_2O \quad (B)$$

2-naphthalene sulfonic acid

The symbol (naphthalene ring symbol) refers to a hydrocarbon with the structure

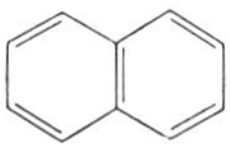

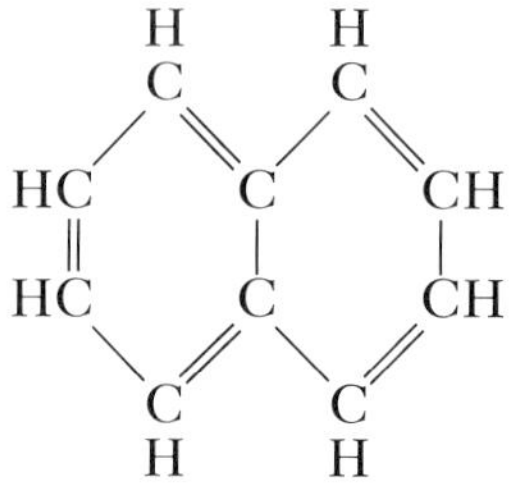

and in this reaction the $—SO_3H$ group is substituted for an H atom located at the position it occupies in the product, 1- or 2-naphthalene sulfonic acid. Each reaction has its own rate and each yields products which have a certain stability with respect to the reactants from which both are formed. Reaction A yields a product at a faster rate than does reaction B. 2-Naphthalene sulfonic acid, the product of reaction B, is more stable than 1-naphthalene sulfonic acid, the product of reaction A. Since all reactants and products are completely interconvertible in this system, when naphthalene and sulfuric acid are allowed to react until no further change occurs, the more stable 2-naphthalene sulfonic acid is the principal product. Under these circumstances product formation is said to be under **thermodynamic (stability) control,** that is, the products obtained are determined by the relative stabilities of the products which might result as compared to that of the reactants used. When this reaction is performed at 165°C the principal organic product is 2-naphthalene sulfonic acid, with negligible amounts of the 1-naphthalene sulfonic acid. Extending the reaction time at this temperature shows no change in yield or nature of the product formed.

If the reaction of naphthalene with sulfuric acid is carried out at 80°C and then terminated after the same reaction time permitted for the reaction at 165°C, the principal organic product obtained is

1-naphthalene sulfonic acid, accompanied by a small yield of 2-naphthalene sulfonic acid. Since both reactions A and B are taking place and since all substances present are interconvertible, product formation here must be determined by the faster rate of formation of the less stable 1-naphthalene sulfonic acid as compared to that of the more stable 2-naphthalene sulfonic acid. When the product obtained is determined by the relative rates of formation of the possible products rather than by their stabilities, the product obtained is said to be under **kinetic (or rate) control.** For reactions in which different products can be produced from the same reactants the product obtained may be determined (controlled) by either its relative rate of formation or by its relative stability. When a product other than the most stable product is obtained, product formation is rate or kinetically controlled. If one product is both more stable and also formed more rapidly under all conditions, the same product will be obtained irrespective of the conditions used. Stability and rate as factors in determining the course of organic reactions are considered in the next two sections.

26.2 STABILITY

Chemical systems tend to "gravitate" to the state of lowest energy—of lowest free energy. A system may arrive at the state by one of two arbitrarily defined classes of changes: either physical reactions or chemical reactions. Physical reactions include changes in state, such as the melting of a solid, the condensing of a vapor and the formation of a solution. Chemical reactions include combustion (combination with oxygen), oxidation-reduction (exchange of electrons), decomposition of chemicals—and in fact, all processes in which atomic groups or groupings are changed.

The change in free energy for a process is a function of three factors: the heat of the reaction as it occurs at constant temperature in an open container (ΔH), the temperature (T), and the change in total **entropy** occasioned by the reaction (ΔS). The heat of reaction has been determined for many reactions and can be calculated for many others using these values. In an **endothermic reaction,** heat is absorbed by the chemicals and the heat of reaction is said to be positive. An **exothermic reaction** in which heat is released has a negative heat of reaction. Changes in entropy have been calculated for many reactions, but the calculation and the quantitative use of entropy are beyond the scope of this text. Qualitatively, the entropy of a system is a reflection or indication of the disorder or randomness of the system. A solid at low temperature with its polar molecules in exact alignment is highly organized and has little entropy. A gas at high temperature with its molecules distributed at random has a large entropy. The sublimation of a solid is accompanied by a great increase in entropy.

refer to Chapter 25 for combustion reactions which are exothermic.

The exact relationship between changes in free energy (ΔG), heats of reaction (ΔH), the temperature (T), and changes in entropy (ΔS) is given by the equation:

$$\Delta G = \Delta H - T\Delta S$$

where ΔG is the change in free energy,

ΔH is the heat of reaction or the change in enthalpy (heat content)
T is the temperature in degrees Kelvin
and ΔS is the change in entropy.

A reaction, physical or chemical, may occur if ΔG is negative, i.e., if there is a decrease in free energy. A decrease will exist with any of the following conditions:

1. the reaction is exothermic and the entropy increases (change in entropy, ΔS, is positive),
2. the reaction is exothermic and the heat of reaction exceeds $T\Delta S$, the product of the absolute temperature and whatever *decrease* in entropy occurs, or
3. the reaction is endothermic, but the product of the temperature and the *increase* in entropy ($T\Delta S$), is greater than the positive heat of reaction (i.e., $T\Delta S > \Delta H$).

The higher the temperature, the more a reaction in which there is an *increase* in entropy is favored, because an increase in T increases the negative term $T\Delta S$. The vaporization of water is an endothermic process (ΔH is positive), but the system increases in entropy when the liquid turns to vapor. At 25°C and 1 atmosphere pressure the positive heat of reaction exceeds the product of the temperature and the change in entropy; therefore ΔG is positive and the reaction does not occur. Instead, the free energy change for the reverse reaction is negative, and therefore water vapor condenses at 25°C at a partial pressure of 1 atmosphere. At 101°C the product of temperature and the change in entropy does exceed the heat of reaction, and hence water vaporizes.

ice has less entropy than liquid water, and liquid water in turn has less than water vapor.

The heat of reaction varies little with the pressures of gases involved in reactions, but the entropy of a system does vary with the pressure. The entropy increases with the expansion of a gas or gases and increases with the dilution of a solute in a solution. This means that a reaction at equilibrium when the concentrations of reactants and products are one mole per liter ($\Delta G = 0$) may not be in equilibrium when the concentrations are all 0.1 molar. A change in concentrations might change ΔS and cause a reaction (a shift in equilibrium).

when the pressure of a gas is changed from 2 atmospheres to 1 atmosphere the entropy of the gas increases.

Although free energy changes are equal to the difference between terms involving both the heat of reaction and the change in entropy, very often in reactions of organic chemicals the changes in entropy are very small as compared to the heats of reaction—so small that they often have a negligible effect on changes in free energy (ΔG) and the heat of reaction may be used as the criterion for the possible occurrence of a reaction. That is, exothermic reactions may occur and endothermic reactions may not. Care must be exercised in applying this test, for the appearance of a gaseous product from reacting solids or liquids, the occurrence of more molecules in the product than among the reactants or a high temperature may invalidate the criterion.

The reaction

$$CH_4 + 2\ O_2 \rightarrow CO_2 + 2\ H_2O + 211\ \text{kcal}$$

is highly exothermic; 211 kcal are liberated for the mole quantities

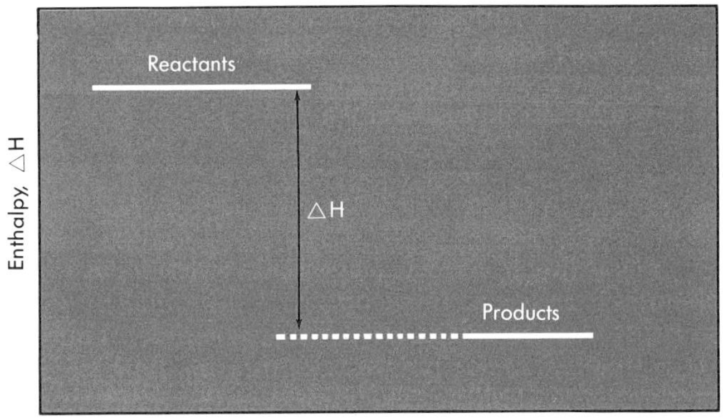

FIGURE 26.1 ENTHALPY CONTENT OF REACTANTS AND PRODUCTS.

in this reaction 3 molecules of reactants produce 3 molecules of products; thus, very little change in entropy results.

indicated by the coefficients in the equation as written. The free energy decreases this amount, since there is little change in entropy in this case. A diagram showing the decrease in enthalpy (Fig. 26.1) is very similar to a diagram showing the decrease in free energy. The decrease in enthalpy which corresponds to the decrease in free energy indicates that the reaction is thermodynamically possible. In common terms the reaction is said to be spontaneous, and, for the specific example, a mixture of methane and oxygen is less stable than the carbon dioxide and water vapor formed which are more stable (i.e., of lower energy content). Methane is unstable with respect to combustion.

26.3 REACTIVITY

fast reactions may give either poor yields or good yields. Rates and yield are independent of each other.

Thermodynamic instability is a necessary but not sufficient condition for product formation. A mixture of methane and oxygen is unstable with respect to formation of carbon dioxide and water. This is demonstrated by the *enthalpy decrease* (which corresponds to the decrease in free energy in this case) observed to accompany the reaction (Section 26.2). This means that product formation can be expected if the proper conditions for reaction are found. The magnitude of the enthalpy decrease is indicative of the yield of product which might be expected. A large enthalpy change (decrease) indicates that a relatively large yield of final product can be expected. A small decrease corresponds to a much smaller yield of product.

Although the reaction of methane and oxygen is a transformation which, when it occurs, gives a relatively large yield of product, it is observed that a mixture of methane and oxygen can stand indefinitely (as can a mixture of hydrogen and oxygen) without any appreciable change or reaction. The rate at which methane reacts with oxygen at room temperature is infinitesimally slow. On the other hand, when the same mixture is provided with additional energy through ignition with a spark or by a flame, the reaction begins and proceeds with explosive rapidity. At room temperature, because the reaction is very slow, methane is said to exhibit low reactivity toward oxygen. At higher temperatures, because it reacts very rapidly, methane is said to exhibit high reactivity toward oxygen. **Reactivity** in conformity with this usage is commonly employed to designate differences in rates of reaction of chemical substances. A substance which undergoes a very rapid reaction is highly reactive. A substance which reacts only very slowly with another substance under the conditions provided exhibits a low reactivity with that substance.

26.4 THEORY OF REACTION RATES (A REVIEW)

According to theory, a chemical reaction, as represented by the chemical equation, is accomplished by one or several successive steps, which are called **elementary reactions.** The set of elementary steps that take reactants through intermediate stages on to the final products is called the mechanism of the reaction. The rate of a reaction then is controlled by the rates of its elementary reactions or, specifically, depends upon the rate of the slowest or so-called **rate determining step** of the mechanism. The rate of a chemical reaction may be changed by altering the rate of the rate determining step or by fashioning a faster pathway which avoids it.

refer to Chapter 17 for a more extended discussion of this topic.

Most elementary reactions are either unimolecular or bimolecular. In a **unimolecular reaction,** individual molecules or ions, twisting, stretching or bending in vibrational motion, rearrange within themselves or fall apart (decompose). In bimolecular reactions two molecules or ions collide together to rupture bonds, and re-form others to form a new or several new molecules. A certain minimum of energy is required to enable any twisting, stretching molecule to fall apart, and a certain minimum of energy is required of two specific colliding molecules, after overcoming the mutual repulsion of their electron clouds, to hit hard enough to break bonds. That minimum energy needed for a specific elementary reaction to occur is called the **activation energy** for that reaction. The activation energies for the successive steps of a mechanism are the energy barriers for the overall process. Each of the three "bumps" on the energy diagram, Figure 26.2, represent the activation energy of an elementary reaction in the mechanism:

i. $A + B \rightarrow C$
ii. $C \rightarrow D$
iii. $D + A \rightarrow E$

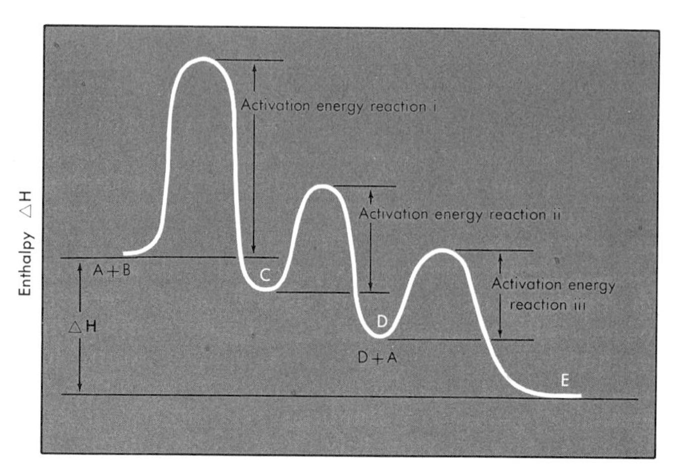

FIGURE 26.2 POTENTIAL ENERGY (HEAT CONTENT, H) DIAGRAM.

for the following hypothetical reaction which is the sum of reactions i, ii and iii,

$$2\,A + B \rightarrow E$$

The rate of a unimolecular reaction, in terms of the number of moles of reactant disappearing per liter per second, depends upon the number of molecules per unit volume having sufficient vibrational energy to react and is directly proportional to concentration. The rate of a bimolecular reaction depends upon three factors: (1) the number of collisions per unit volume per second, (2) the fraction of those collisions in which the molecules are oriented properly to react, and (3) the fraction of the collisions of correctly oriented molecules that are sufficiently energetic to cause reaction. An increase in the concentration of a reactant or reactants increases bimolecular reactions, for it increases the number of collisions. Raising the temperature increases the rates for both unimolecular and bimolecular reactions, for the number of high energy molecules is increased greatly with relatively modest increases in temperature. Not only are there more collisions per second at higher temperatures but the collisions are more energetic. The polarity of neighboring molecules affects their relative orientations and, thus, may change the fraction of the collisions in which the molecules are oriented correctly.

At times the addition in very small amounts of a substance other than one of the reactants increases the rate of a reaction greatly. Such a substance is called a **catalyst.** It does not alter the stability of the system, i.e., change the free energy of the process, but it changes the reactivity. A catalyst must either change the activation energy of the rate determining step, affect the orientation of the molecules in collision, or make possible another pathway or mechanism with a lower activation energy.

To acquaint the student with factors that affect the reactivity of organic compounds, mechanisms that are characteristic of those used for explaining organic reactions will be discussed. Most of the examples are drawn from examples discussed earlier in the text (see Chapter 17) and some from material discussed in the previous chapter. It is hoped that the discussion of these examples by way of review will help the student to visualize other mechanisms to be encountered in subsequent chapters.

26.5 A FREE RADICAL REACTION

The reaction between chlorine and methane in the presence of ultraviolet light at room temperature as discussed in the last chapter proceeds by the proposed mechanism:

$$\begin{array}{rlll}
\text{i.} & Cl_2 & \xrightarrow{\text{Ultraviolet light}} & 2Cl\cdot \\
\text{ii.} & Cl\cdot + CH_4 & \rightarrow & CH_3\cdot + HCl \\
\text{iii.} & CH_3\cdot + Cl_2 & \rightarrow & CH_3Cl + Cl\cdot \\
\text{iv.} & Cl\cdot + Cl\cdot & \rightarrow & Cl_2 \\
\text{v.} & Cl\cdot + CH_3\cdot & \rightarrow & CH_3Cl \\
\text{vi.} & CH_3\cdot + CH_3\cdot & \rightarrow & CH_3CH_3
\end{array}$$

This series of reactions (discussed more fully in Chapter 25) is called a free radical mechanism, because the entities, atoms (Cl·) and groups of atoms (CH_3·), with an odd number of electrons, are called free radicals. The reaction is said to follow a chain mechanism, with reactions ii and iii constituting the chain. A chain reaction is a two step portion of a mechanism in which each step produces a necessary reactant for the other, such that each step can follow the other over and over while at the same time forming a final product, here CH_3Cl. As can be seen in the mechanism steps ii and iii can follow one another over and over, producing CH_3Cl in each step, as long as both reactants (chlorine and methane) are present. Because of the chain, a relatively small number of photons absorbed in the chain-initiating step, step i, results in the production of many times more molecules of methyl chloride. Steps iv, v and vi remove free radicals necessary for the propagation of the chain, causing the reaction to cease shortly after the irradiating light is extinguished.

The reaction between chlorine and methane will occur at high temperatures in the absence of ultraviolet light. At high temperature there are a sufficiently great number of collisions of sufficiently high energy to break the bonds of chlorine molecules to produce atoms necessary to start and maintain the chain reaction. Once the chlorine atoms are formed, the reaction proceeds along the same path as when free radicals are formed by the absorption of light.

26.6 REACTION OF A BRØNSTED ACID AND BASE

An **acid** in the **Brønsted** sense is a molecule or ion that donates a proton, H^+, in reaction. A **base** is a molecule or ion that accepts a proton in a reaction. The acid-base reaction, commonly called **neutralization,** is the transfer of a proton from an acid to a base. The reaction of ammonia in an aqueous solution with added hydrochloric acid is an example of an acid-base reaction. The reacting species are the acid (hydronium ion, H_3O^+) and the base, the ammonia molecule NH_3. The transfer of the proton is evident in the equation for the reaction.

for application of this principle, see Chapters 29, 31 and 33.

$$H_3O^+ + NH_3 \rightarrow NH_4^+ + H_2O$$

The reaction is exothermic and there is little change in entropy. Thus there is a decrease in free energy. Hydronium ions and ammonia together in water are unstable. The neutralization reaction proceeds rapidly when solutions of the two reactants are mixed until very little ammonia and hydronium ion remains, i.e., the system is unstable as predicted by the decrease in free energy.

The reaction of ammonia and hydronium ion is a base displacement reaction in terms of the **Lewis acid-base concept. Lewis** defined an **acid** as an electron pair acceptor and a **base** as an electron pair donor. In the reaction under discussion, a proton of the hydronium ion, upon collision with an ammonia molecule, leaves the

$AlCl_3$, BF_3 and a H^+ are Lewis acids; that is, they are electron acceptors.

OH^-, NH_3 and H_2O are Lewis bases; that is, they are electron donors.

pair of electrons shared with the oxygen of the water molecule and joins to nitrogen, using the previously unshared electron pair on the ammonia molecule (Fig. 26.3). The weaker base, water, is displaced by the stronger Lewis base ammonia.

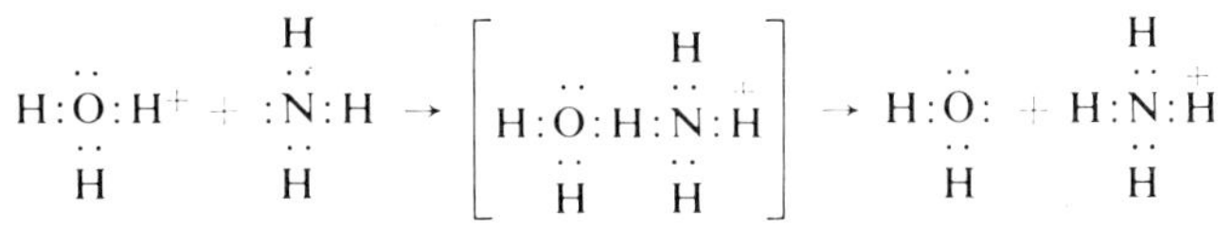

$$\begin{matrix} & \\ H\!:\!\ddot{O}\!:\!H^+ \\ H \end{matrix} + \begin{matrix} H \\ :\!\ddot{N}\!:\!H \\ H \end{matrix} \rightarrow \left[\begin{matrix} & H \\ H\!:\!\ddot{O}\!:\!H\!:\!\ddot{N}\!:\!H^+ \\ H & H \end{matrix}\right] \rightarrow \begin{matrix} & \\ H\!:\!\ddot{O}\!: \\ H \end{matrix} + \begin{matrix} H \\ H\!:\!\ddot{N}\!:\!H^+ \\ H \end{matrix}$$

FIGURE 26.3 THE LEWIS BASE DISPLACEMENT OF H_2O BY NH_3.

Because the electron pair on the nitrogen atom of the ammonia molecule is more available for sharing than an electron pair on the oxygen of the water molecule, the ammonium ion is stable in an acidic solution. The stability of the ammonium ion in the solution contributes to the decrease in free energy—the force that drives the reaction to completion.

The polarity of the ammonia molecule contributes to the reactivity of that molecule with hydronium ion because ammonia molecules in the near vicinity of positive hydronium ions will tend to be oriented so that the negative end of the dipole, the nitrogen atom at the apex of the pyramid (Fig. 26.4), is directed toward the ion. Thus, the ammonia molecule approaches a collision with the

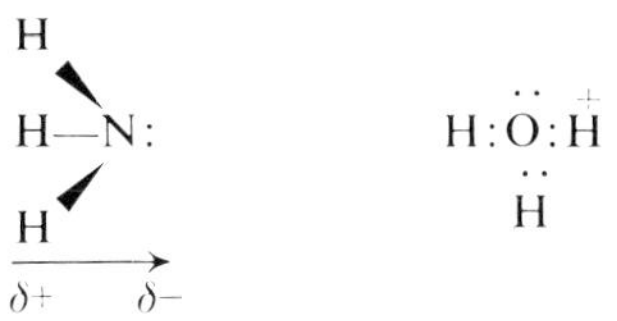

$$\begin{matrix} H \\ H\!-\!N\!: \\ H \\ \delta+ \longrightarrow \delta- \end{matrix} \qquad\qquad \begin{matrix} \\ H\!:\!\ddot{O}\!:\!H^+ \\ H \end{matrix}$$

FIGURE 26.4 ORIENTATION OF AMMONIA NEAR THE POSITIVE ION.

hydronium ion with the proper orientation, i.e., with the electron pair toward the ion. The hydronium ion is affected to a lesser degree by the presence of the negative charge of the polar molecule but will *tend* to be turned in the right direction. Because of the orientation effects of the charges and partial charges, more of the sufficiently energetic collisions are successful than if the colliding particles were oriented at random.

26.7 ANOTHER BASE DISPLACEMENT REACTION

The conversion of methyl bromide, CH_3Br, to methanol, CH_3OH, by reaction with a solution of a strong base in water,

$$OH^- + CH_3Br \rightarrow CH_3OH + Br^-$$

is a Lewis base replacement similar to the reaction of hydronium ion and ammonia. However, it is not an acid-base reaction in the Brønsted sense because no proton is involved. The Lewis base replacement character of the reaction is apparent in the equation:

$$^{-}\!:\ddot{\underset{..}{O}}:H + H:\overset{H}{\underset{H}{\ddot{\underset{..}{C}}}}:\ddot{\underset{..}{Br}}: \rightarrow H:\overset{H}{\underset{H}{\ddot{\underset{..}{C}}}}:\ddot{\underset{..}{O}}:H + :\ddot{\underset{..}{Br}}:^{-}$$

the strong Lewis base, OH^-, takes the CH_3^+ away from another but weaker Lewis base, Br^-

Methanol, more stable in a basic solution than methyl bromide, is formed. The reaction is thought to proceed through a one step concerted process, in which the polarity of the methyl bromide plays an important role in the reactivity. The negative hydroxide ion approaches the methyl bromide molecule opposite the negative bromine of the dipole and presents two electrons to share with the carbon atom. At the same time the bromine atom, taking the shared pair of electrons, leaves the carbon atom and becomes a bromide ion (Fig. 26.5). The molecule is inverted in the process—the carbon to hydrogen bonds bend back towards the leaving bromide ion and away from the approaching hydroxide ion. The orientation of the methyl bromide dipole relative to the approaching negative hydroxide ion (positive end to negative ion) places it at the correct angle for reaction.

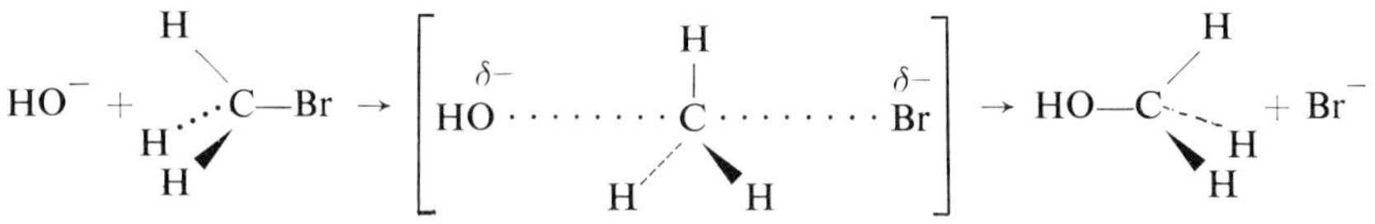

$$HO^- + H_3C\text{—}Br \rightarrow \left[\overset{\delta-}{HO} \cdots\cdots \overset{H}{\underset{H\ \ H}{C}} \cdots\cdots \overset{\delta-}{Br} \right] \rightarrow HO\text{—}CH_3 + Br^-$$

FIGURE 26.5 ACID-BASE DISPLACEMENT OF Br^- WITH OH^-.

The equation for the reaction of tertiary butyl bromide $(CH_3)_3CBr$, with the hydroxide ion is similar to the equation for the reaction of methyl bromide with the same base. Bromide ion is displaced by the hydroxide ion.

$$(CH_3)_3CBr + OH^- \rightarrow (CH_3)_3COH + Br^-$$

Tertiary butyl bromide is more unstable than methyl bromide in basic solution. However, the negative hydroxide ion approaching the tertiary butyl bromide molecule opposite the bromine group does not approach between the large methyl groups to donate and share electrons with carbon. Instead the tertiary butyl bromide molecule must first lose the bromide ion by ionization. The hydroxide ion then takes the place previously occupied by the bromine. Because of the larger groups about the central carbon atom, the process changes from a concerted bimolecular reaction to a two step mechanism in which a unimolecular reaction is followed by a bimolecular combination.

$$\text{i. } CH_3\text{—}\overset{CH_3}{\underset{CH_3}{\overset{|}{\underset{|}{C}}}}\text{—}Br \rightarrow CH_3\text{—}\overset{CH_3}{\underset{CH_3}{\overset{|}{\underset{|}{C}}}}{}^{+} + :\ddot{\underset{..}{Br}}:^{-} \text{ (slow, high activation energy)}$$

because of the difficulty the OH^- has in reacting with the central carbon, the ionization must occur before reaction proceeds.

$$\text{ii. } CH_3\text{—}\overset{\displaystyle CH_3}{\underset{\displaystyle CH_3}{\overset{|}{\underset{|}{C^+}}}} + {}^-\!:\ddot{\underset{..}{O}}\!:\!H \rightarrow CH_3\text{—}\overset{\displaystyle CH_3}{\underset{\displaystyle CH_3}{\overset{|}{\underset{|}{C}}}}\!:\!\ddot{\underset{..}{O}}\!:\!H \text{ (fast)}$$

The high activation energy of the unimolecular reaction, first step, makes it the rate determining step. When t-butyl bromide is allowed to react with OH^-, the replacement product, $(CH_3)_3COH$, is not the product obtained in largest yield. Elimination (loss of HBr) to form isobutylene (see Chapter 27) takes place at a faster rate and more completely than does Br replacement. When base displacement does occur, however, it takes place as outlined.

26.8 ACID "CATALYZED" REPLACEMENT OF A HYDROXYL GROUP WITH BROMINE

The reaction represented by the equation

$$ROH + Br^- \rightarrow RBr + HO^-$$

the weak Lewis base Br^- cannot take the CH_3^+ away from the strong Lewis base OH^-.

where R— is a hydrocarbon group such as isopropyl, $(CH_3)_2CH$—, is a Lewis base replacement reaction similar to the reaction discussed in (Section 26.7). In fact, it is the reverse of that reaction. The reaction does not go, because an alkyl bromide is unstable in a basic solution (Section 26.6) and the free energy change is positive for the reaction represented by the equation. In an acidic solution, an additional reactant, H_3O^+, is present and a product is changed from hydroxide ion to water. Those products, the alkyl bromide and water, are more stable than the reactants: the alcohol, hydronium ion and bromide ion and the replacement of the hydroxyl group with bromine in an acid solution

$$ROH + H_3O^+ + Br^- \rightarrow RBr + 2\ H_2O$$

is accompanied by a decrease in free energy, and the reaction occurs. Because it enters into the reaction giving more stable products and is not regenerated, the hydronium ion is not actually a catalyst but a reactant—hence the quotation marks in the title to this section. However, the hydronium ion does react with the alcohol, changing the molecule such that the slow step in the hypothetical reaction without the acid is speeded up. In that respect the hydronium ion acts similarly to a catalyst.

Without the hydronium ion, were the replacement a two step process, the hydroxyl group must leave the alcohol as an hydroxide ion, OH^-, to make way for the addition of the bromide ion. The separation of the negative hydroxide ion and the positive alkyl ion, R^+, in opposition to their strong mutual attraction, is a slow reaction requiring great energy.

$$ROH \rightarrow R^+ + OH^-$$

With hydronium ion present, the Lewis base, ROH, replaces the base, H_2O, to form the ion, ROH_2^+.

$$\text{i. } R{:}\ddot{\underset{\cdot\cdot}{O}}{:}H + H{:}\ddot{\underset{\ddot{H}}{O}}{:}H^+ \rightarrow R{:}\ddot{\underset{\ddot{H}}{O}}{:}H^+ + H_2O$$

A neutral water molecule then leaves the positive alkyl ion

conversion of the —OH attached to R into OH_2 permits formation of R^+.

$$\text{ii. } R{:}\ddot{\underset{\ddot{H}}{O}}{:}H^+ \rightarrow R^+ + H_2O$$

The negative bromide ion and positive alkyl ion combine in the final step to yield the alkyl bromide product.

$$\text{iii. } R^+ + Br^- \rightarrow RBr$$

In this process, the hydronium serves a dual role, i.e., it decreases the stability of the reactants and increases their reactivity.

26.9 OXIDATION OF A MULTIPLE CARBON TO CARBON BOND

One test for the presence of a multiple carbon to carbon bond in a hydrocarbon is based upon the ready oxidation of that bond with a solution of potassium permanganate. In the oxidation process permanganate ion, purple in solution, is destroyed, leaving the solution decolorized. If a solution of potassium permangate is decolorized when added to a hydrocarbon, the hydrocarbon molecules contain one or more double or triple bonds. Permanganate ion in solution oxidizes the hydrocarbon at the multiple bond and in return is reduced to either manganous ion or to a precipitate of dark brown manganese dioxide, depending upon the conditions of the reaction. In either case the purple color of permanganate ion disappears. If the temperature is elevated, or if either acid or base is added to the permanganate solution, the hydrocarbon chain is cleaved at the multiple bond. For compounds having double bonds, in a dilute solution at room temperature, the process may be arrested at the diol stage, a step preceding that of rupture of the bond. The mild oxidation of hydrocarbon with a double bond, R—CH=CH—R′ (where R— and R′ represent hydrocarbon groups) by the permanganate ion, MnO_4^-, may be represented by a balanced equation using the half-cell reactions 1 and 2 which follow:

$$\text{1. } 4\,H_2O + RCH{=}CHR' \rightarrow \underset{OH}{\underset{|}{RCH}}{-}\underset{OH}{\underset{|}{CHR'}} + 2\,H_3O^+ + 2\,e^- \text{ (oxidation)}$$

$$\text{2. } MnO_4^- + 2\,H_3O^+ + 2\,e^- \rightarrow MnO_3^- + 3\,H_2O \text{ (reduction)}$$

$$H_2O + RCH{=}CHR' + MnO_4^- \rightarrow \underset{OH}{\underset{|}{RCH}}{-}\underset{OH}{\underset{|}{CHR''}} + MnO_3^-$$

The MnO_3^- ion is reduced by reaction with hydrocarbon to either manganous ion or manganese dioxide.

The course of the reaction is not understood completely, but the first step in the mechanism is thought to be the bimolecular combination of the hydrocarbon molecule and the permanganate ion to form an unstable complex ion.

$$
\mathrm{RCH{=}CHR'} + \begin{array}{ccc} O & & O \\ & \diagdown\!\!\diagdown\;\diagup\!\!\diagup & \\ & Mn & \\ & \diagup\!\!\diagup\;\diagdown & \\ O & & O^- \end{array} \rightarrow \begin{array}{ccccc} & H & & H & \\ & | & & | & \\ R- & C & - & C & -R \\ & | & & | & \\ & O & & O & \\ & & \diagdown\;\diagup & & \\ & & Mn & & \\ & & \diagup\!\!\diagup\;\diagdown & & \\ & O & & O^- & \end{array}
$$

The complex is then hydrolyzed by water in what must be more than one step to yield the final products.

$$
\begin{array}{ccccc} & H & & H & \\ & | & & | & \\ R- & C & - & C & -R' \\ & | & & | & \\ & O & & O & \\ & & \diagdown\;\diagup & & \\ & & Mn & & \\ & O & & O & \end{array} + H_2O \rightarrow \begin{array}{ccccc} & H & & H & \\ & | & & | & \\ R- & C & - & C & -R' \\ & | & & | & \\ & O & & O & \\ & | & & | & \\ & H & & H & \end{array} + MnO_3^-
$$

Many organic reactions proceed through the formation of **unstable and reactive intermediates** such as the complex represented here.

26.10 CATALYTIC HYDROGENATION OF ETHYLENE

A mixture of the gases, ethylene, $CH_2{=}CH_2$, and hydrogen, will remain indefinitely without reaction at temperatures below 200°C. However, when finely divided platinum, nickel or palladium is added to the mixture, the addition reaction

$$CH_2{=}CH_2 + H_2 \rightarrow CH_3CH_3$$

proceeds rapidly at room temperature. This is a case of heterogeneous catalysis—the gases react when they come together at the metallic surface. The reaction is exothermic, the heat of reaction being about −30 kcal. per mole of reacting ethylene.

Little is known of the mechanism of the reaction but some facts about the surface of the metal and the nature of the molecules aid in a partial explanation. Hydrogen gas is soluble in metallic platinum at high temperatures. One method for purifying small amounts of the gas is to pass the impure hydrogen at 300°C over the outside of a platinum thimble sealed to a vacuum system. Hydrogen and only hydrogen passes into the outside wall of the thimble, through the metal and into the apparatus. X-ray analyses of the metal reveal that the interstices (holes between the metal atoms) are large enough to admit hydrogen atoms but not hydrogen molecules. The hydrogen gas, in passing through the tube, separates into atoms at the surface and the atoms pass from interstice to interstice as the vibrations of the atoms within the metal open passageways between the holes. Furthermore, platinum atoms in the crystal are arranged in planes of the crystal, such that the centers of the metal atoms are about $1\frac{1}{3}$ Å apart. (The distance between the carbons of the double bond in ethene is 1.34 Å).

It is possible that ethylene molecules collide with the platinum surface already saturated with hydrogen atoms which are separated at distances almost equal to the length of the carbon to carbon double bond. In those collisions in which the carbon atoms fall each directly above an absorbed hydrogen atom, the pi electrons of the double bond split to pair, each one, with the electron of hydrogen to form two single bonds. When the molecule leaves the surface of the metal, the bond is saturated with hydrogens. As is true of all catalysts, the platinum metal does not make the reactants any more or less stable, but it provides a pathway along which the reacting chemicals are more reactive.

EXERCISES

26.1 Define the following:

a. enthalpy
b. entropy
c. mechanism
d. elementary reaction
e. activation energy
f. thermodynamic stability
g. spontaneous reaction
h. reactive
i. thermodynamically unstable
j. Brønsted acid
k. Lewis acid
l. catalyst
m. chain mechanism
n. exothermic
o. reactant
p. product
q. unimolecular reaction
r. bimolecular reaction

26.2 What two conditions must be met for a reaction to take place? Discuss.

26.3 Why is a reaction with a great increase in entropy favored by increasing the temperature of the system?

26.4 What is meant by the statement that the formation of a product is rate controlled?

26.5 Does the entropy of the system increase or decrease when 10 grams of water is melted? Explain.

26.6 How might these factors affect the rate of a reaction?

a. a rise in temperature of the reactants.
b. an increase in the pressure of a reacting gas mixture.
c. the addition of a catalyst.

26.7 A mixture of reactants is thermodynamically stable. Should you add a catalyst to get the reaction started? Explain.

26.8 The reaction of HBr with $CH_2{=}CH{-}CH{=}CH_2$ gives two products, $BrCH_2{-}CHBr{-}CH{=}CH_2$(I) and $BrCH_2{-}CH{=}CH{-}CH_2Br$(II). When the reaction is carried out at a temperature of −80°C, a product mixture containing 20 percent of II and 80 percent of I is obtained. When allowed to react at 40°C for the same time period, a mixture of 80 percent of II and 20 percent of I is obtained. By referring to the reaction between naphthalene and sulfuric acid discussed in Section 26.1, suggest a reason for the difference in yield observed at the two temperatures for the two products. Which compound, I or II, is formed faster at 40°C? At −80°C? Which compound, I or II, is more stable at 40°C? At −80°C?

26.9 A reaction was carried out and shown to have a ΔS of −10.6 cal/mole-degree and a ΔH of −57.8 kcal/mole at 25°C. Calculate the ΔG for this reaction. How does the thermodynamic stability of the reactants compare with that of the products of this reaction?

26.10 A reaction for the conversion of glyceric acid (A) into pyruvic acid (B) and water shows a ΔG of −12,950 calories/mole and a ΔS of 21.7 calories/mole-degree at 25°C. What is the ΔH for this reaction at 25°C? Which are more thermodynamically stable, reactants or products? Is this a fast or slow reaction? Explain.

SUGGESTED READING

Breslow, R.: *Organic Reaction Mechanisms.* W. A. Benjamin, Inc., New York, 1965 (paperback).
Campbell, J. A.: *Why Do Chemical Reactions Occur?* Prentice-Hall, Inc., Englewood Cliffs, N. J., 1965 (paperback).
Caserio, M. C.: "Reaction Mechanisms in Organic Chemistry." J. Chem. Ed., *42*:570, 1965.
Edwards, J. O.: "Bimolecular Nucleophilic Displacement Reactions." J. Chem. Ed., *45*:386, 1968.
Gould, E. S.: *Mechanisms and Structure in Organic Chemistry.* Holt, Rinehart and Winston, Inc., New York, 1959. Chapters 4, 5 and 6.
Hendrickson, J. B., Cram, D. J. and Hammond, G. S.: *Organic Chemistry,* Third Edition. McGraw-Hill Book Company, New York, 1970. Chapters 2, 3, 8 and 9.
Morrison, R. T. and Boyd, R. N.: *Organic Chemistry,* Second Edition. Allyn and Bacon, Inc., Boston, 1966. Chapters 1, 2 and 14.
Noller, C. R.: *Chemistry of Organic Compounds,* Third Edition. W. B. Saunders Company, Philadelphia, 1965. Chapters 2 and 3.
Stewart, R.: *The Investigation of Organic Reactions.* Prentice-Hall, Inc., Englewood Cliffs, N. J., 1966 (paperback).

TWENTY-SEVEN • MULTIPLE CARBON TO CARBON BONDS

27.1 INTRODUCTION

A structural feature consisting of an atom, group of atoms or a variation in bonding which gives to or confers upon an alkane certain recognizable properties is called a **functional group.** The functional group, one of the most important concepts in organic chemistry, provides the basis for relating and remembering properties of organic compounds and for separating them into classes or families of similar characteristics. For example, all hydrocarbons containing one carbon to carbon double bond in addition to "alkane" carbons possess a similar set of properties different from those characteristic of alkanes. Likewise, all compounds containing the carbon to carbon triple bond have a set of unique properties. Both the carbon to carbon double bond and the carbon to carbon triple bond are functional groups.

Each functional group endows a compound with a set of properties distinct and different from those of any other functional group. Each functional group, thus, gives rise to a recognizable class of organic compounds. Hydrocarbons which contain one carbon to carbon double bond are named alkenes; those having the triple bond, alkynes. In addition to the carbon to carbon multiple bonds there are a number of other important functional groups. The most important of these groups, along with the names of the classes of compounds which contain them, are given in Table 27.1.

TABLE 27.1 FUNCTIONAL GROUPS AND RELATED CLASSES OF COMPOUNDS

Functional Group		Class of Compound	
Structure	Name	General Formula	Name
—CH=CH—	Double bond	RCH=CHR	Alkene (olefin)
—C≡C—	Triple bond	RC≡CR	Alkyne (acetylenes)
—OH	Hydroxyl	ROH	Alcohols
		ArOH	Phenols
—OR	Alkoxy	ROR; ArOAr	Ethers
—CH=O	Formyl	RCH=O; ArCH=O	Aldehydes
—C(=O)—	Carbonyl	RC(=O)R; ArC(=O)Ar	Ketones
—COOH	Carboxyl group	RCOOH; ArCOOH	Carboxylic acids
—CN	Cyano	RCN, ArCN	Nitriles
$—NO_2$	Nitro	RNO_2; $ArNO_2$	Nitro compounds
—X	Halo (chloro, bromo, iodo)	RX; ArX	Halo compounds
—COOR	Carbalkoxy (or alkoxy-carbonyl)	RCOOR; ArCOOR	Esters
$—NH_2$	Amino	RNH_2; $ArNH_2$	Amines
$—CONH_2$	Carboxamide	$RCONH_2$; $ArCONH_2$	Amides
R—	Alkyl	RH	Alkane
Ar—	Aryl	ArH	Aromatic hydrocarbon

Substances which have more than one functional group exhibit two or more sets of properties, a set characteristic of each different functional group present. Examples of compounds with several functional groups, such as the amino acids (Chapter 35), will be discussed in later chapters. Subsequent chapters deal with the chemistry of the different functional groups and compounds which contain them. This chapter considers two functional groups, the double and triple bonds, along with the compounds which contain them.

27.2 NATURE OF MULTIPLE BONDS: THE DOUBLE BOND

Ethylene with structural formulas

```
H      H     H         H
 ˙.  .˙       \       /
  C : : C ,    C=C          and    CH2=CH2
 .˙  ˙.       /       \
H      H     H         H
```

is the simplest hydrocarbon containing a double bond. Each atom of the molecule by electron sharing has attained the noble gas configuration. Models of the ethylene molecule are pictured in Figure 27.1. In Figure 27.1a the positions and relative sizes of the atoms are shown. The sketches in Figure 27.1b and c are attempts to describe the bonding involved. In each, both carbons form single electron pair bonds with two hydrogens. These C—H bonds are

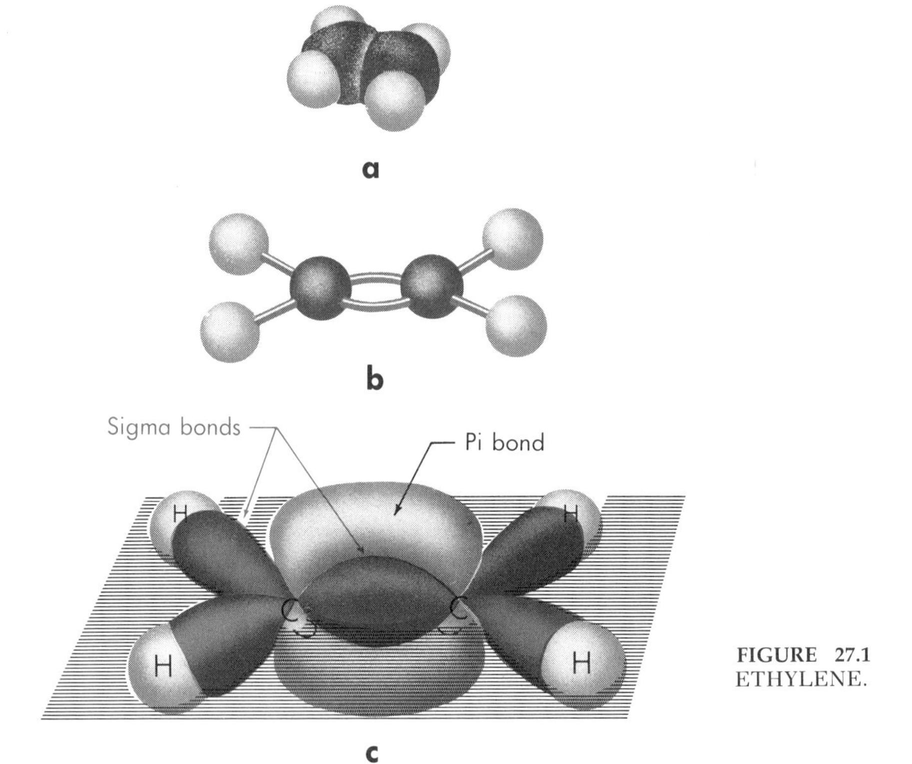

FIGURE 27.1 MODELS OF ETHYLENE.

similar to the C—H bond in alkanes. In b, two of the bonds of each carbon are bent together from the tetrahedral angle between them to make a double bond of two parallel single bonds. By this picture, both portions of the double bond are identical and both pairs of electrons occupy similar orbitals. Recent mathematical studies of bond orbitals have suggested the type of bonding represented in Figure 27.1c. Where the double bond in ethylene is composed of two parts, a sigma bond and a pi bond, the isolated atoms have four electrons in the valence level (Fig. 27.2):

$$C\ 1s^2\ 2s^2 2p^1 2p^1$$

To form the bond required in ethylene three orbitals (one s orbital and two p orbitals) of each carbon combine to form three new but

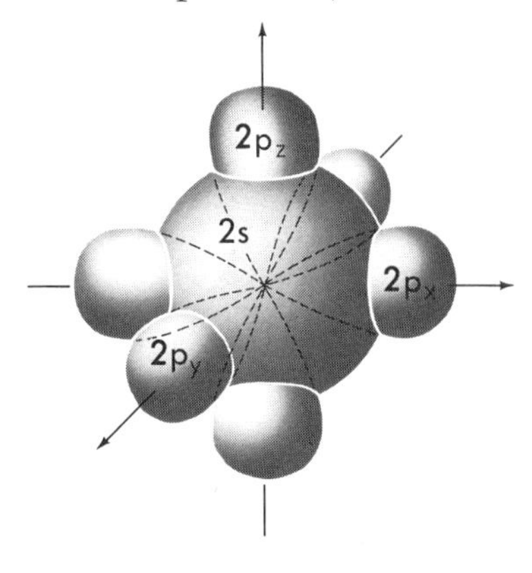

FIGURE 27.2 UNHYBRIDIZED CARBON ATOM.

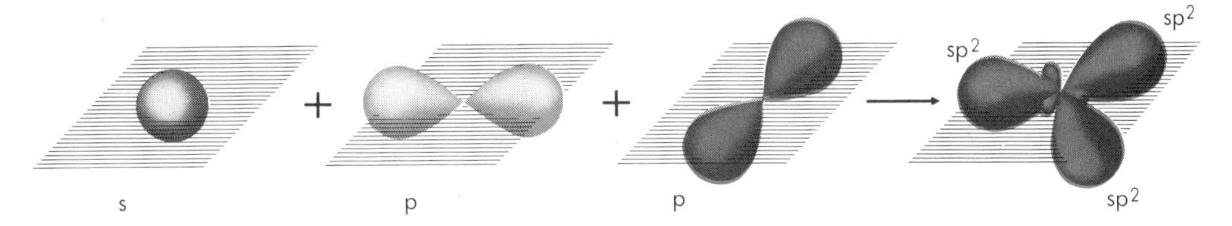

FIGURE 27.3 sp^2 HYBRIDIZED CARBON ATOM.

refer to Chapter 5 for a discussion of hybridization.

identical orbitals each called sp^2 orbitals (Fig. 27.3). The three sp^2 orbitals differ only in direction. They lie in the same plane and each points toward the apex of an equilateral triangle. The carbon atom, upon *double* bond formation, becomes

note the 120° angle between sp^2 orbitals.

$$C\ (sp^2)^1(sp^2)^1(sp^2)^1p^1$$

(The four valence electrons are placed as indicated with one in each sp^2 orbital of the hybridized atom and one in an unchanged p orbital.) A carbon atom of this type which is bonded to only three other atoms is referred to as a **trigonal carbon** atom. The ethylene molecule forms with sigma bonds between each pair of bonded atoms as shown in Figure 27.4. Each atom contributes one electron for sharing in forming each bond. Five sigma (four C to H and one C to C) bonds are formed, that is, the occupied orbitals are regularly symmetrical about the line between the two nuclei.

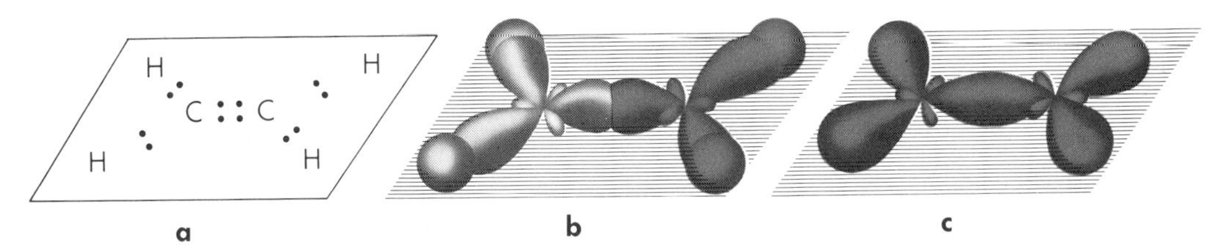

FIGURE 27.4 THE SIGMA BOND SKELETON IN ETHYLENE.

After formation of the sigma bonds there remains with each of the two carbons one p orbital perpendicular to the plane of the molecule each with one electron. The p orbitals overlap as shown in Figure 27.5, where sigma bonds are represented by straight lines, to form a pi orbital in which the two electrons are shared. The double bond in ethylene and similar compounds is therefore of two parts, a sigma bond buried within the molecule and a pi bond in which the occupying electrons are more exposed to the influence of approaching electric fields. The orbital arrangement is not unlike a hot dog. The frankfurter represents the sigma bond; the bun, the pi bond.

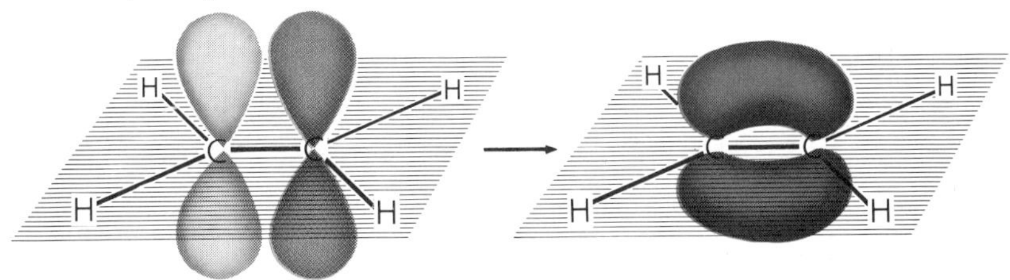

FIGURE 27.5 THE PI BOND IN ETHYLENE.

The double bond will differ in many respects from the single carbon to carbon bond as summarized in the following:

1. The carbons are held more tightly. A colliding particle requires more energy to break the molecule into halves.

2. The carbons are drawn more closely together, the distance between centers being 1.34 Å as compared to 1.54 Å for a single carbon to carbon bond.

3. The bond is *rigid.* There is no rotation about the double bond. Rotation would disrupt the pi molecular orbital and destroy the bond. In ethylene and in other compounds all atoms attached to the carbons joined by the double bond maintain the same position relative to one another. Rigidity of the double bond gives rise to another type of isomerism called **cis-trans isomerism** – discussed in Section 27.6.

4. The double bond produces a planar molecule. In more precise terms, the 2 carbons and all atoms attached to them lie in the same plane. This formula for ethylene

```
H       H
 \     /
  C==C
 /     \
H       H
```

places the atoms very nearly in their correct positions in the molecule.

5. Certain electrons (pi electrons) are more exposed than electrons of the C—C single bonds.

27.3 THE NATURE OF MULTIPLE BONDS: THE TRIPLE BOND

Acetylene (C_2H_2) contains a triple bond. Structural formulas are

$$H{:}C{:}{:}{:}C{:}H \qquad \text{or} \qquad H—C{\equiv}C—H$$

The mutual sharing of 6 electrons by the carbons is necessary if the carbons are to have the noble gas configuration. In Figure 27.6 are three representations of the acetylene molecule. The relative sizes and positions of the atoms are shown in Figure 27.6a. The illustration in Figure 27.6b shows the bonding that would result if three of the bonds from each carbon were bent together from their tetrahedral angles. The modern molecular orbitals are shown in Figure 27.6c, where the C—H bond is not altered and the triple bond leads to a linear structure for acetylene and related molecules. The C to C triple bond may be considered to be a bond of three parts. Each carbon atom of the ethyne (acetylene) which originally had the electron structure

$$C \quad 1s^22s^22p^12p^1$$

with four valence electrons, is hybridized upon formation of the

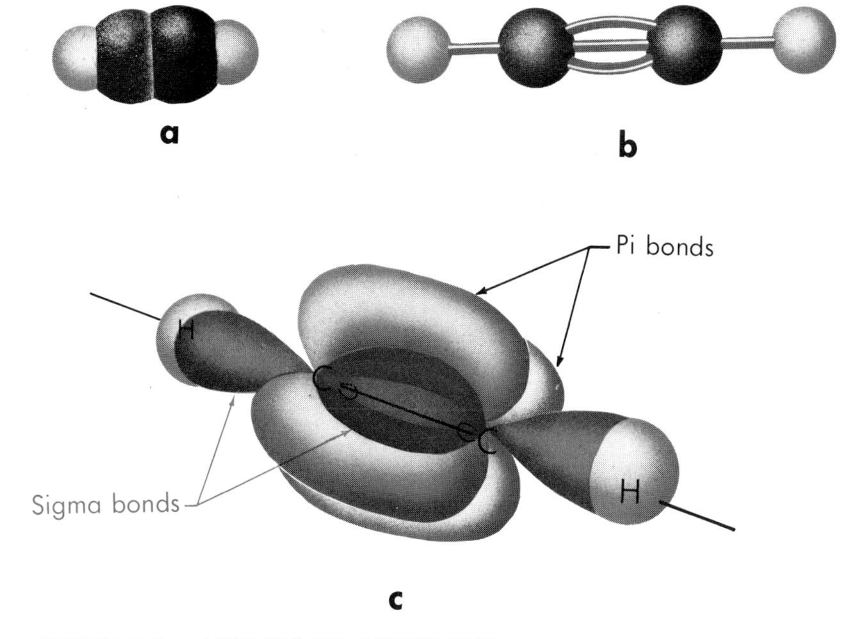

FIGURE 27.6 MODELS OF ACETYLENE.

triple bond to

$$\text{C} \quad 1s^2(2sp)^1(2sp)^1 2p^1 2p^1$$

what is the angle between the C—H and C—C bonds in acetylene?

That is, the 2s orbital and a 2p orbital combine to give two sp type orbitals (Fig. 27.7). One sp orbital forms a sigma bond with a hydrogen atom; the other with the neighboring carbon atom (Fig. 27.8). A carbon of this type which is bonded to only two other atoms is

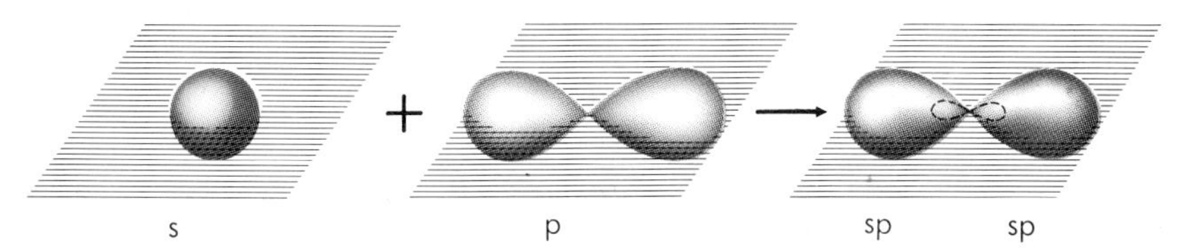

FIGURE 27.7 sp HYBRIDIZED CARBON ATOM.

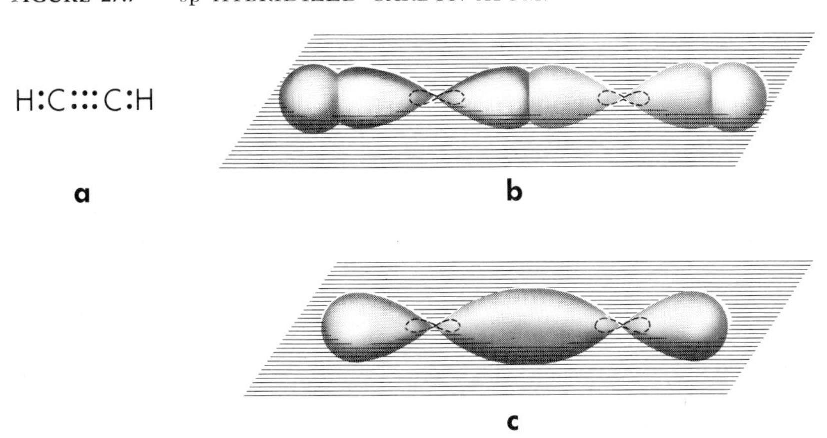

FIGURE 27.8 THE SIGMA SKELETON IN ACETYLENE.

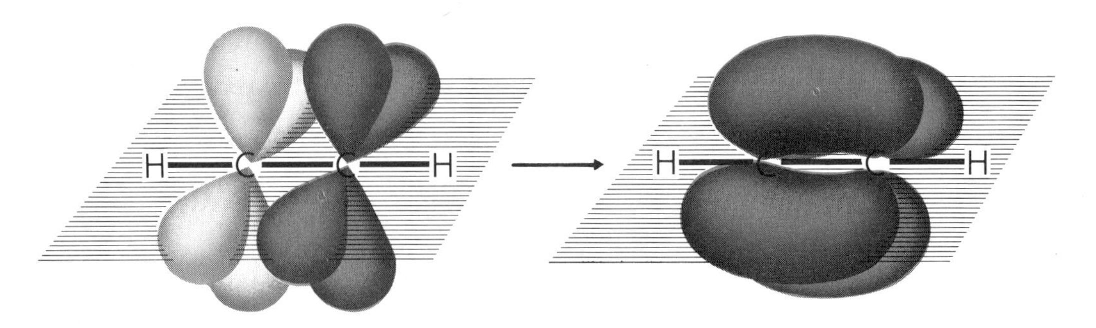

FIGURE 27.9 THE PI BONDS IN ACETYLENE.

referred to as a **digonal carbon.** In addition there remain with each carbon two electrons, each in a p orbital, with the p orbitals perpendicular to one another and perpendicular to the sigma bond between them. Each pair of p orbitals overlaps to form pi bonds (Fig. 27.9). The configuration corresponds to a hypothetical hot dog with two buns. The frankfurter is the sigma orbital and each of the buns is a pi orbital.

Compared to the single and double bonds:

C—C bond distance in acetylene is 0.13 Å shorter than that in ethylene.

1. The triple bond is a stronger bond. More energy is required to separate the bonded atoms.
2. The two carbon atoms are held more closely, the bond distance being 1.21 Å.
3. The bond is rigid; there is no rotation, but this is of no consequence, because the single bond extending from the triple bond carbon permits rotation. No cis-trans isomerism is possible.
4. The triple bond is linear. That is, the carbons and atoms attached directly to them lie in a straight line.
5. Certain electrons, the pi electrons, are more exposed than electrons of the single C to C bond. There are more of these exposed electrons in the triple bond than in the double bond (4 as compared to 2), but the degree of exposure is much the same.

27.4 NOMENCLATURE

Any open chain carbon-hydrogen compound which contains, in addition to single carbon to carbon bonds, only one double C to C bond is called an alk*ene*. Hydrocarbons with two double bonds are called alka*dienes;* with three, alka*trienes*, and so on. A cyclohydrocarbon with one double bond is a cycloalkene. Any open chain hydrocarbon which contains, in addition to single C to C bonds, only one triple C to C bond is called an alk*yne*. Compounds containing more triple bonds are called alka*diynes*, alka*triynes*, and so on. A compound containing both a double C to C bond and a triple C to C bond is an alk*eneyne*. In this text it is sufficient to give the common names for a few of the compounds, the IUPAC nomenclature system for the alkenes and the alkynes, and the systematic names for a few more complicated hydrocarbons.

Several alkenes and one alkyne are well known and carry common names. These are shown with structural formulas in Table

TABLE 27.2 NAMES AND STRUCTURAL FORMULAS OF SOME ALKENES AND ALKYNES

Formula	Common Name	IUPAC Name
$CH_2=CH_2$	ethylene	ethene
$CH_3-CH=CH_2$	propylene	propene
$CH_3-CH_2-CH=CH_2$	—	1-butene
$CH_3-CH=CH-CH_3$		2-butene
CH_3 | $CH_3-C=CH_2$	isobutylene	2-methylpropene
CH_2-CH_2 / \ CH_2 CH_2 \ / $CH=CH$	—	cyclohexene
$HC\equiv CH$	acetylene	ethyne
$CH_3-C\equiv CH$	methyl acetylene	propyne

27.2. The IUPAC rules for naming alkenes and alkynes are necessary extensions and adjustments of the rules for naming alkanes.

1. Select the longest continuous chain *containing the multiple bond.*
2. Name the chain as though it were an alkane, but change the ending to *ene* if the multiple bond is a double bond, or to *yne* if the multiple bond is a triple bond.
3. Number the carbons of the chain beginning at the end nearest the multiple bond.
4. Indicate the position of the multiple bond (if necessary) by placing the number of the first carbon to which the double bond is attached immediately preceding the stem name.
5. Complete the name, naming and indicating the position of groups as with alkanes.

is 3-methyl-2-butene a correct name for an alkene? why?

A few simple applications of the naming system are found in Table 27.2. A few more examples follow:

$$\overset{1}{CH_3}-\overset{2}{\underset{}{C}}(CH_3)=\overset{3}{CH}-\overset{4}{CHCl}-\overset{5}{CH_3}$$

4-Chloro-2-methyl-2-pentene

$$\overset{6}{CH_3}-\overset{5}{CH_2}-\overset{4}{CH}(CH_3)-\overset{3}{CH}(CH_2-CH_2-CH_3)-\overset{2}{C}\equiv\overset{1}{CH}$$

4-Methyl-3-n-propyl-1-hexyne

is 2-methyl-2-butyne a correct name for a known alkyne? why?

Note the names of a few hydrocarbons with 2 or more multiple bonds.

$$CH_3-CH=CH-CH=CH_2$$

1,3-Pentadiene

$$HC\equiv C-CH=CH_2$$

1-Butene-3-yne

$$CH_2=C(CH_3)-CH=CH_2$$

2-Methyl-1,3-butadiene

27.5 PHYSICAL PROPERTIES

The physical properties of hydrocarbons with double and triple bonds are very similar to the properties of alkanes with the same number of carbons (Table 27.3).

TABLE 27.3 PHYSICAL PROPERTIES OF SOME HYDROCARBONS WITH MULTIPLE BONDS

Name	Melting Point °C	Boiling Point °C	Density (Liquid) (g/me)
Ethene	−169.4	−102.4	0.610
Propene	−185	−47.7	0.610
1-Butene	−130	−6.5	0.626
1-Pentene	−138	30.1	0.642
1-Hexene	−98.5	63.5	0.675
Ethyne	−81.6	−83.6	0.621
Propyne	−104.7	−23.3	0.678
1,3-Butadiene	−108.9	−3	0.650
Cyclopentene	−93.3	44.2	0.772

Like the alkanes, those with 4 or fewer carbons per molecule are gases at 25°C and 1 atmosphere. Those with 5 or more carbons are liquids. All liquid alkenes or alkynes are less dense than water. Hydrocarbons with double and triple bonds are more soluble in water than the alkanes having a corresponding number of carbon atoms because the positive hydrogens of water are attracted slightly to the exposed pi electrons. Even so, the solubility is slight.

27.6 ISOMERISM

When a double or triple bond is present in a compound, additional arrangements of the atoms within the molecule are possible because multiple bonds may occupy different positions in a carbon chain. For example, with a four carbon chain a multiple bond may be between either carbons 1 and 2 or between carbons 2 and 3 as shown:

how many position isomers of an alkyne having the formula C_5H_8 are possible?

$$\underset{\text{2-Butene}}{CH_3CH{=}CHCH_3} \quad \text{or} \quad \underset{\text{1-Butene}}{CH_3CH_2CH{=}CH_2}$$

$$\underset{\text{2-Butyne}}{CH_3C{\equiv}CCH_3} \quad \text{or} \quad \underset{\text{1-Butyne}}{CH_3CH_2C{\equiv}CH}$$

Isomers of this kind which differ only in the position of the functional group are called **position isomers.**

A type of isomerism not possible among either the alkanes or the alkynes exists among the alkenes because of the rigid, nonrotating carbon to carbon double bond. For example, there are two 2-butenes – two different compounds having the structural formula, $CH_3CH{=}CHCH_3$ (Fig. 27.10). Identical groups, the two methyl or the two hydrogen groups in this example may appear on the same

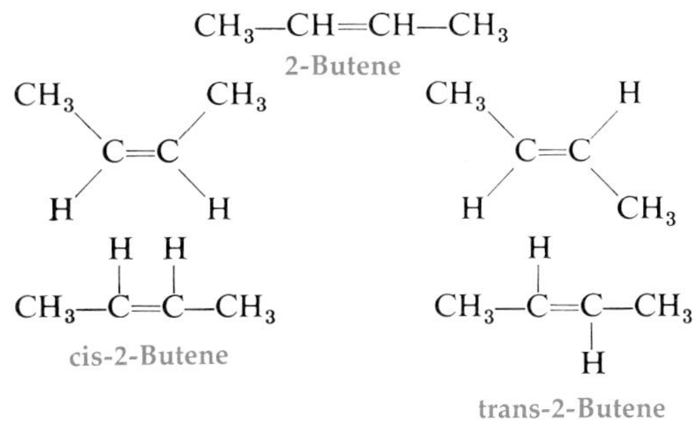

$$CH_3{-}CH{=}CH{-}CH_3$$

2-Butene

cis-2-Butene

trans-2-Butene

FIGURE 27.10 THE *CIS* AND *TRANS*- 2 BUTENES.

side of a plane which passes through the C—C double bond and perpendicular to the plane of the molecule (the *cis* compound) or may appear on opposite sides of the plane (the *trans* compound). *cis*-2-Butene and *trans*-2-butene are called **cis-trans geometrical isomers.** Because structural formulas specify only the sequence or order of the atoms in a molecule, those isomers in which the same groups are attached to the same positions of the same structure are called **stereoisomers** and are said to differ only in *configuration.* The term **configuration** is commonly used to specify the *spatial arrangement* of atoms in a molecule. Geometrical isomers are one type of stereoisomer and differ from one another in configuration.

Two stereoisomers of a compound can be isolated only if the energy required to convert one into the other is sufficiently high, as is the case with *cis-trans* isomers due to a carbon to carbon double bond. Also, *cis-trans* isomers can exist only when each of the carbons joined by the double bond has two different groups attached. No *cis-trans* isomers of 1-butene exist because two atoms (H's) attached to carbon 1 are the same, that is,

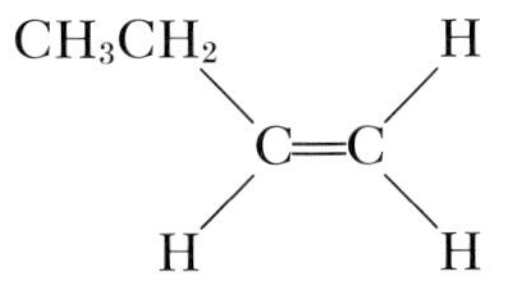

$$(CH_3CH_2)(H)C{=}C(H)(H) \quad \text{and} \quad (H)(CH_3CH_2)C{=}C(H)(H)$$

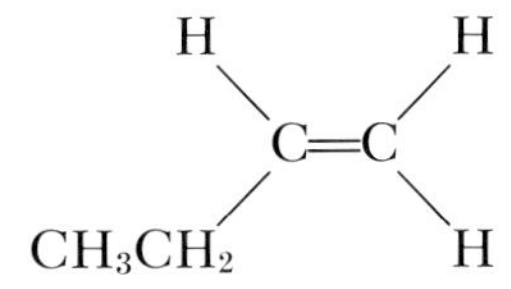

isomeric compounds cannot be superimposed upon one another.

are identical. All atoms of either structure can be matched with (superimposed upon) those of the other by a 180° rotation of the molecule out of the plane of the paper.

Geometrical isomerism could have been introduced in the discussion of the cycloalkanes; *trans*-1,3-dimethylcyclohexane obviously differs in structure from *cis*-1,3-dimethylcyclohexane (Fig. 27.11). *Cis-trans* isomerism is possible because the CH_3 group cannot rotate through the center of the ring. Rotation about the single bonds attaching the methyl bearing carbons to the ring is hindered because the angle between the single bonds must be nearly 109°28′.

The two dimethylcyclohexanes (Table 27.4) are so similar that it is difficult to distinguish one from another. *Cis* and *trans* isomers *can* however be very different both physically and chemically. Consider the *cis*- and *trans*-butendioic acids, known as maleic and fumaric acid respectively (Fig. 27.12).

◄ indicates atoms that are above the ring and . . . that are below.

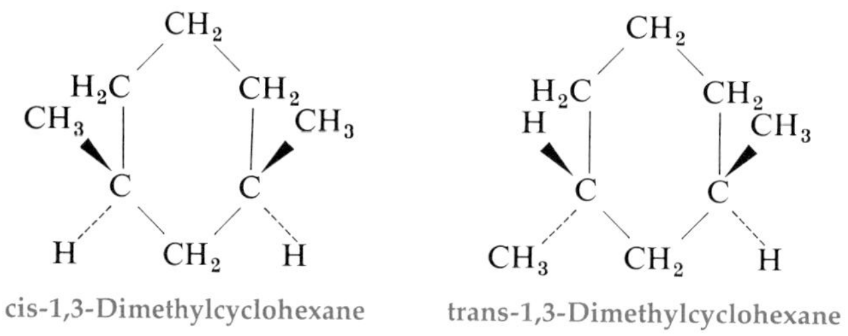

cis-1,3-Dimethylcyclohexane

trans-1,3-Dimethylcyclohexane

FIGURE 27.11 AN EXAMPLE OF *CIS-TRANS* ISOMERISM.

TABLE 27.4 PHYSICAL PROPERTIES OF SOME CIS-TRANS ISOMERS

	Melting Point °C	Boiling Point °C	Density	Solubility in Water g/100 ml of Water
cis-2-Butene	—	1.0	0.635	Insoluble
trans-2-Butene	—	2.5	0.635	Insoluble
cis-1,3-Dimethylcyclohexane	−85	121.0	1.4269	Insoluble
trans-1,3-Dimethylcyclohexane	—	119.0	1.4254	Insoluble
Maleic acid (*cis*-butendioic acid)	130	Decomposed at 135	1.590	78.8
Fumaric acid (*trans*-butendioic acid)	287	290	1.630	0.70

Physical properties of the two acids are compared in Table 27.4. The maleic acid molecule is polar because the two electron attracting —COOH groups are together on the same side, whereas fumaric acid is not polar because the groups are on opposite sides and each offsets the effect of the other. Maleic acid has a dipole moment; fumaric acid has not. Because of the difference in configuration, one chemical reaction of maleic acid is not readily available to fumaric acid. Maleic acid loses water when heated to 140°C to form an anhydride.

refer to Chapter 37 for role of fumaric acid in citric acid cycle.

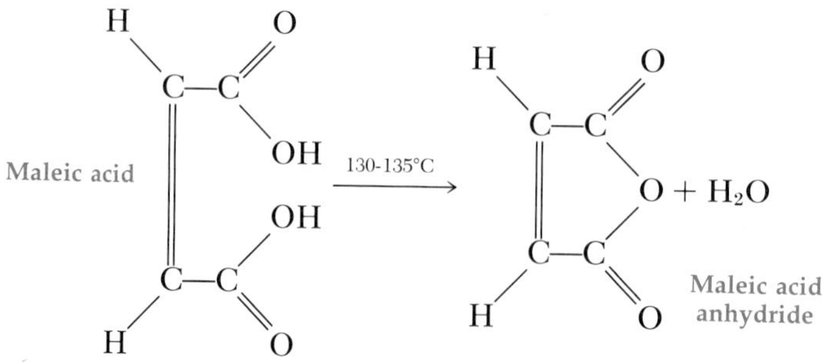

Maleic acid $\xrightarrow{130\text{-}135°C}$ Maleic acid anhydride $+ H_2O$

Heating fumaric acid does not readily split out water between the two —COOH groups because of the distance between them. Only after enough energy is added for conversion of fumaric into maleic acid does loss of water occur. **Chemical differences are emphasized**

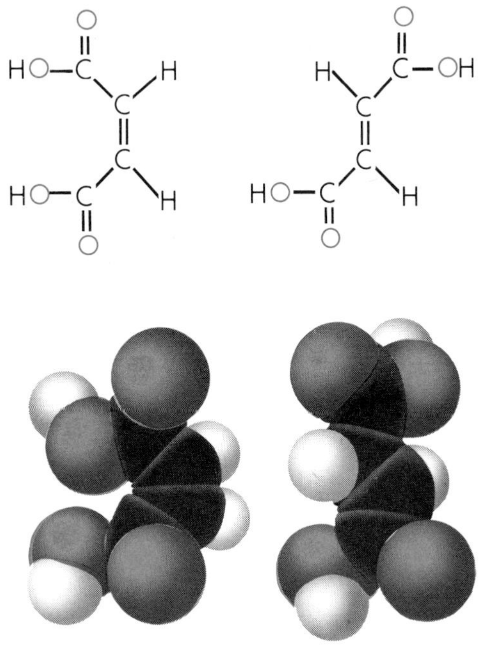

Maleic acid
(cis-butendioic acid)

Fumaric acid
(trans-butendioic acid)

FIGURE 27.12 THE BUTENDIOIC ACIDS.

by the fact that fumaric acid is a biologically important intermediate in metabolic processes whereas maleic acid is toxic to living organisms.

27.7 ADDITION REACTIONS

Many reagents add to multiple carbon to carbon bonds, that is, the adding compound splits and one portion adds to each carbon of the double bond. One product is obtained as the principal product of the reaction and the multiplicity of the bond is reduced by one when one molecule is added. For example, hydrogen bromide adds to an alkene gas bubbled through concentrated hydrobromic acid in accord with the following equations:

1. $CH_2{=}CH_2 + HBr \rightarrow CH_3CH_2Br$
2. $CH_3CH{=}CH_2 + HBr \rightarrow CH_3CH(Br)CH_3$
3. $CH_3{-}C(CH_3){=}CH_2 + HBr \rightarrow CH_3{-}C(CH_3)(Br){-}CH_3$
4. $CH_3{-}C(CH_3){=}CH(CH_3) + HBr \rightarrow CH_3{-}C(CH_3)(Br){-}CH_2CH_3$

all atoms in the reactants are combined as one unit in the product.

Only one molecule of hydrogen bromide can add to the double bond of an alkene, whereas either one or two molecules can add to the triple bond of an alkyne – the number adding being determined by the concentration of the acid, the temperature and other conditions of the reaction. The stepwise addition of two molecules of HBr to the triple bond of ethyne is shown in equations 5 and 6.

5. $CH{\equiv}CH + HBr \rightarrow CH_2{=}CHBr$
6. $CH_2{=}CHBr + HBr \rightarrow CH_3CHBr_2$

The ready addition of various reagents and their disappearance as if they were dissolving in an unsaturated solution of a multiple bond containing hydrocarbon, has gained the name "unsaturated" for alkenes and alkynes. In contrast alkanes are called saturated hydrocarbons.

In reactions 2, 3, 4 and 6 simple addition would indicate that there should be two addition products, as illustrated here for reaction 2:

$$\underset{\text{H—Br}}{CH_3CH{=}CH_2 \; + } \rightarrow CH_3CH_2CH_2Br \quad \text{and} \quad \underset{\text{Br—H}}{CH_3CH{=}CH_2 \; + } \rightarrow CH_3\underset{\underset{Br}{|}}{C}HCH_3$$

a. b.

However, the product appearing when reaction 2 is carried out is 2-bromopropane, the product of reaction b, with only traces of the other addition product, 1-bromopropane, being formed. The products written for equations 2, 3, 4 and 6 are the products obtained in the laboratory. A Russian scientist, V. Markownikoff, observed the products of addition to the double bond in many addition reactions and formulated the following rule to predict the product to be expected: "When an unsymmetrical reagent adds to an unsymmetrical multiple carbon to carbon bond, the hydrogen of the reagent adds to the carbon of the multiple bond to which a greater number of hydrogens are already attached."

Hydrogen bromide is an unsymmetrical reagent, for H is different from – and more positive than – Br. Propene, the reactant in reaction 2, is unsymmetrical; the doubly bonded C in $CH_3CH{=}$ differs from the C in $CH_2{=}$ in that the C is attached to one hydrogen and an alkyl (CH_3—) group, not to just two hydrogen atoms. The more positive part of the adding reagent (H) adds to the carbon with more hydrogens ($CH_2{=}$) to form the product 2-bromopropane.

27.8 MECHANISM OF THE ADDITION REACTION

The mechanism of the addition reaction, i.e., the proposed set of elementary reactions which constitute the addition process, assists in understanding the great reactivity of multiple bonds toward addition and accounts for the utility of Markownikoff's rule in predicting the product to be expected when more than one addition product is possible.

Addition of Hydrogen Bromide to Ethene. The addition of hydrogen bromide is believed to proceed by two steps. In step 1

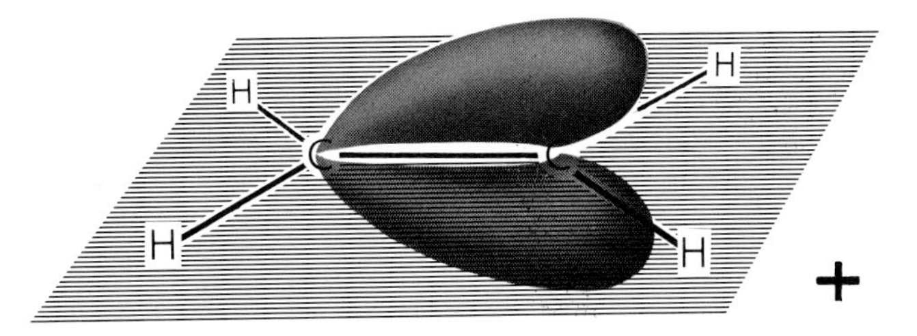

FIGURE 27.13 DISTORTION OF A PI BOND TOWARD A POSITIVE CHARGE.

the hydrogen bromide molecule approaches the multiple bond of the unsaturated compound, ethene in this case, with the positive end of the approaching dipole, the hydrogen end, directed toward the negative pi electrons. The approaching positive charge attracts the pi electrons and distorts the orbital of the pi bond in the direction of the approaching charge until it is disrupted (Fig. 27.13). The former pi electrons are then shared with the hydrogen which simultaneously releases the pair of electrons shared with bromine. In this first step, a bromide ion is formed, the hydrogen ion becomes attached to a carbon atom of the double bond by cosharing two electrons and the alkene molecule with the acceptance of the hydrogen ion becomes positively charged with the charge located on the opposite carbon of the multiple bond.

the mechanism consists of two elementary reactions (reactions 1 and 2).

$$1.\quad Br{-}H + H_2C{=}CH_2 \rightarrow Br^- + H_3C{-}CH_2^+$$

The positive ion with a charge of plus one on a carbon atom, called a **carbonium ion,** has a very brief existence. It combines with a bromide ion in the solution, step 2, to complete the addition.

$$2.\quad H_3C{-}CH_2^+ + Br^- \rightarrow H_3C{-}CH_2{-}Br$$

Reaction of Hydrogen Bromide with Propene. The mechanism for the addition of hydrogen bromide to ethene casts no light upon Markownikoff's rule because, although hydrogen bromide is an unsymmetrical addition reagent, ethene is symmetrical, and there is only one addition product. Only one product results because there is only one possible ethyl carbonium ion (ethyl cation) to combine with the bromide ion. Because the carbon which bears the charge in the ethyl carbonium ion is attached to only one other carbon atom, it is called a **primary carbonium ion.** On the other hand there are two propyl carbonium ions (propyl cations), a primary and a **secondary ion,** each one a possible product of the addition of a hydrogen ion, or proton, to a carbon of the double bond of propene.

$CH_3{-}CH^+{-}CH_3$, secondary propyl carbonium ion (secondary propyl cation), from the addition of H^+ to carbon number 1.

$CH_3{-}CH_2{-}CH_2^+$, primary propyl carbonium ion (primary propyl cation), from the addition of H^+ to carbon number 2.

Because 2-bromopropane is the product of the reaction of hydrobromic acid and propene, the second step of the mechanism of the addition is:

$$2.\quad CH_3\overset{+}{C}HCH_3 + Br^- \rightarrow CH_3CHBrCH_3$$

And the first step is the addition of an H^+ to carbon 1.

$$1.\quad HBr + CH_3CH{=}CH_2 \rightarrow CH_3\overset{+}{C}HCH_3 + Br^-$$

If the mechanism is correct, it would appear from this reaction that a secondary carbonium ion is more stable than the primary ion of the same group.

Reaction of Hydrogen Bromide with Isobutylene (2-methyl-1-propene). There are two possible butyl carbonium ions (butyl cations), each of which is a possible product of the addition of a proton to isobutylene:

$CH_3{-}\underset{\displaystyle CH_3}{\underset{|}{C}H}{-}CH_2^+$ or $(CH_3)_2CH{-}CH_2^+$, isobutyl carbonium ion (isobutyl cation), a primary ion, formed by the addition of a proton to carbon 2.

and $CH_3{-}\underset{\displaystyle CH_3}{\underset{|}{C^+}}{-}CH_3$ or $(CH_3)_3C^+$, tertiary butyl carbonium ion (tertiary butyl cation), formed by the addition of a proton to carbon 1.

2-Bromo-2-methylpropane or tertiary butyl bromide is the product of the addition of hydrogen bromide to isobutylene. Therefore, the mechanism for the reaction is:

$$1.\quad CH_3{-}\underset{\displaystyle CH_3}{\underset{|}{C}}{=}CH_2 + HBr \rightarrow CH_3{-}\underset{\displaystyle CH_3}{\underset{|}{C^+}}{-}CH_3 + Br^-$$

$$2.\quad CH_3{-}\underset{\displaystyle CH_3}{\underset{|}{C^+}}{-}CH_3 + Br^- \rightarrow CH_3{-}\overset{\displaystyle Br}{\overset{|}{\underset{\displaystyle CH_3}{\underset{|}{C}}}}{-}CH_3$$

two elementary reactions with isobutylene also occur.

According to the mechanism the tertiary carbonium ion is formed more readily than the primary ion. Similar evidence gained from a study of many reactions shows that the order of stability of carbonium ions is:

$$\text{Tertiary} > \text{Secondary} > \text{Primary}$$

The more stable carbonium ion carries the larger number of alkyl groups attached to the carbon bearing the positive charge. This may be related to the experimental observation that alkyl groups release electrons to an attached atom to a greater degree than does a hydrogen atom. The relative stability of a carbonium ion is increased as the number of "electron donating" alkyl groups bonded to the charge bearing carbon is increased. Since each ion, which has only three electrons pairs shared in sigma bonds, carries a +1 charge, the shared electron pairs are drawn closer to the positive carbon than is normal, thus shifting the positive charge in part to the three groups or atoms to which it is bound. Because alkyl groups are electron releasing, a methyl (or other alkyl) group attached to a positive carbon assumes a greater share of the positive charge than does a hydrogen atom in the same position. Therefore, the plus charge is dispersed more (is more distributed over the ion) in tertiary butyl, than in isopropyl, than in ethyl, than in methyl carbonium ion.

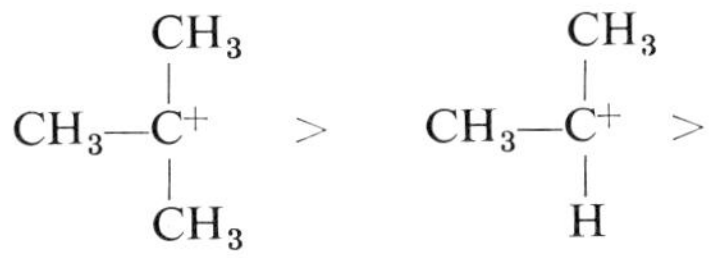

$$(CH_3)_3C^+ > (CH_3)_2CH\text{—}C^+ \text{ [i.e. } CH_3\text{—}C^+(CH_3)\text{—}H] > CH_3\text{—}CH_2^+ > CH_3^+$$

Greater dispersal of positive charge (More stable ion) — Less dispersal of positive charge (Less stable ion)

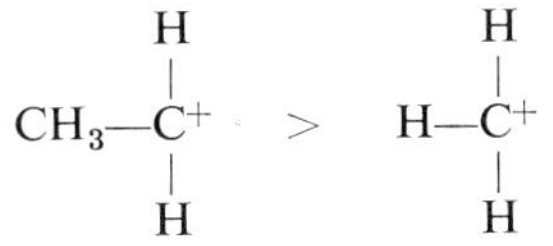

A dispersal of charge lowers the potential energy in accordance with the laws of electrostatics, thus increasing the stability of the ion concerned.

Thus, the proton adds to the carbon of an alkene having the greater number of hydrogens to give a carbonium ion in which the carbon bearing the charge is attached to the greater number of carbons. Markownikoff's rule, then, predicts the product.

In summarizing this discussion of mechanism it should be noted that in each addition reaction the multiple bond (double bond) is electron rich and provides electrons for the bonding with H^+. The H^+ is an electron seeker. In each case we have an electron rich substance reacting with an electron seeking reagent. A terminology not previously introduced is often used when referring to substances involved in these reactions. Reagents (usually positively charged ions) such as hydrogen ions, which are capable of bonding with carbon and are electron seeking, often are referred to as **electrophilic reagents** or as **electrophiles.** Electron rich reagents (usually negative ions) which are seeking positively charged centers or nuclei are designated as **nucleophilic reagents** or **nucleophiles.** Electrophiles react with nucleophiles. In each addition reaction the H^+ of the HBr and the carbonium ions formed are electrophiles (electrophilic reagents) and the alkene used and Br^- are nucleophiles.

the Lewis acids $AlCl_3$ and BF_3, being electron seekers, are electrophiles, and OH^- and CN^- are electron donors and nucleophiles.

Initial formation of carbonium ions plays an important role in many organic reactions. This will be mentioned and the terms electrophile and nucleophile employed where needed to clarify subsequent discussions.

27.9 REACTIONS OF ALKENES

Addition of Hydrohalic Acids. HCl and HI add to multiple carbon to carbon bonds in the same way as HBr. The order of reactivity is HI > HBr > HCl. Ease of addition increases as we proceed down the periodic table from the more to the less electronegative halogens.

Addition of Hydrogen Sulfate from Concentrated Sulfuric Acid. The overall reaction is:

a. $CH_2{=}CH_2 + HOSO_2OH \rightarrow CH_3CH_2OSO_2OH$

write the mechanism for the addition of $HOSO_2OH$ to propene.

The addition as expected occurs by the same two step, ionic mechanism as does HBr. This reaction is an intermediate step in the industrial preparation of alcohols, in this case, ethyl alcohol. With water the ethyl hydrogen sulfate is converted to an alcohol:

b. $CH_3CH_2OSO_2OH + H_2O \rightarrow CH_3CH_2OH + HOSO_2OH$

The addition to propene occurs as follows:

c. $CH_3CH{=}CH_2 + HOSO_2OH \rightarrow CH_3\underset{\displaystyle OSO_2OH}{\underset{|}{C}H}CH_3$

Addition of Hypohalous Acids. Although only slightly soluble in water, a halogen in solution reacts with the solvent to give an equilibrium system of halogen, hypohalous acid and hydrohalic acid as shown for a chlorine solution.

$$Cl_2 + H_2O \rightarrow \underset{\text{Hypohalous acid}}{HOCl} + HCl$$

Of the substances present in the solution the hypohalous acid is the most reactive reagent. With chlorine water, the reaction with ethylene is

$$CH_2{=}CH_2 + \underset{\text{(chlorine water)}}{HOCl} \rightarrow \underset{\displaystyle Cl}{\underset{|}{CH_2}}—\underset{\displaystyle OH}{\underset{|}{CH_2}}$$

Of the two adding fragments HO and Cl, chlorine is the more positive. Thus in agreement with Markownikoff's rule the reaction with propene is

$$CH_3CH{=}CH_2 + HOCl \rightarrow CH_3\underset{\displaystyle OH}{\underset{|}{C}H}\underset{\displaystyle Cl}{\underset{|}{CH_2}}$$

Addition of Halogens: Bromine and Chlorine. The halogens, bromine and chlorine, add rapidly to alkenes. The reaction of bromine with ethene, when the gas is bubbled through a solution of the halogen in carbon tetrachloride, is:

$$CH_2{=}CH_2 + Br_2 \rightarrow BrCH_2CH_2Br$$

As with HBr, the reaction is believed to occur in two steps:

$$1.\quad Br_2 + \begin{array}{c}H\\|\\HC\end{array}\!\!=\!\!\begin{array}{c}H\\|\\CH\end{array} \rightarrow \begin{array}{c}H\\|\\HC\\|\\Br\end{array}\!\!-\!\!\begin{array}{c}H\\|\\C^+\\|\\H\end{array} + Br^-$$

In this step at the moment of collision, bromine (an electrophile) adds as a positive bromine ion (Br^+) with 6 electrons, utilizing the two pi electrons to form a bond and give the positive $BrCH_2CH_2^+$ ion. The Br^+ is called bromonium ion. In step 2 the carbonium ion (an electrophile) combines with Br^- (a nucleophile):

$$2.\quad \begin{array}{c}H\\|\\BrC\\|\\H\end{array}\!\!-\!\!\begin{array}{c}H\\|\\C^+\\|\\H\end{array} + Br^- \rightarrow \begin{array}{c}H\\|\\BrC\\|\\H\end{array}\!\!-\!\!\begin{array}{c}H\\|\\CBr\\|\\H\end{array}$$

Evidence for this mechanism is the fact that when sodium chloride is added to the bromine solution, some 1-bromo-2-chloroethane is formed along with the dibromo compound. With Cl^- from the NaCl present, some of the carbonium ions react with this added nucleophile as well as with Br^-. With propene the overall reaction with Cl_2 is

$$CH_3CH{=}CH_2 + Cl_2 \rightarrow CH_3\!-\!\begin{array}{c}H\\|\\C\\|\\Cl\end{array}\!\!-\!\!\begin{array}{c}H\\|\\CH\\|\\Cl\end{array}$$

Water is not a suitable solvent, for hypohalous acid would be formed and would add. Carbon tetrachloride or another inert nonpolar solvent is used. Notice that because the halogen molecule is symmetrical, only one addition product is possible. Also, chlorine and bromine are the only practical halogen addends. Iodine does not react, and fluorine reacts too vigorously in both the addition to an alkene and in the substitution of F for H.

The addition of bromine (Br_2) to cyclopentene takes place in a manner similar to that observed for the addition of bromine to ethylene. The product formed is *trans*-1,2-dibromocyclopentane.

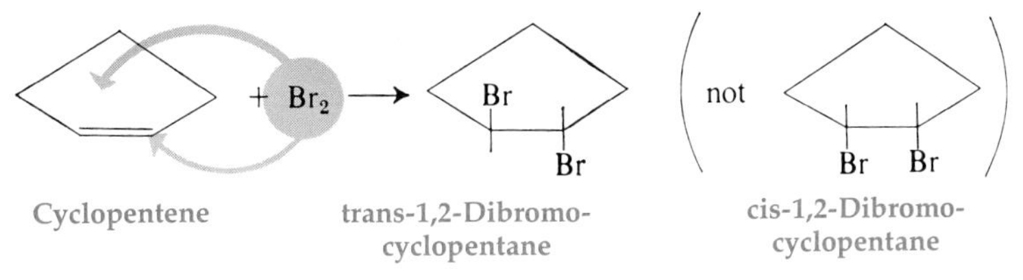

Cyclopentene — trans-1,2-Dibromo-cyclopentane — cis-1,2-Dibromo-cyclopentane

The formation of the *trans* product rather than the *cis* isomer demonstrates that in electrophilic addition to the carbon-carbon double bond the two parts of the reagent which add, in this case the two bromine atoms, add to opposite faces of the plane encompassing the carbons of the double bond and the atoms attached to it. Addition of both atoms to the same side of the molecule (*cis* addition) would have formed the *cis*-1,2-dibromocyclopentane. Conversion of one isomer into the other after formation is not possible because of the hindered rotation about the single bond between the two carbons bearing the bromine atoms. Other experimental evidence verifies the pattern of *trans* **addition** with bromine as well as other electrophilic reagents.

Since compounds containing triple bonds also add Br_2 (see next section), a convenient laboratory test for multiple bonds (unsaturated compounds) is made by shaking a hydrocarbon with a solution of bromine in carbon tetrachloride. If multiple bonds are present the reddish-brown solution is decolorized, as colorless bromoalkanes form and dissolve in the carbon tetrachloride. An alkane will decolorize a solution of bromine in CCl_4 in sunlight, but the substitution reaction is slow and HBr fumes appear.

Oxidation of Alkenes. Alkenes are oxidized rapidly by a solution of potassium permanganate (see Section 26.8). This reaction is listed because in the course of the reaction hydroxyl (—OH) groups are added to the carbons of the double bond.

complete oxidation always yields CO_2 or a carboxylic acid.

a. $$3\,CH_3CH{=}CH_2 + 2\,KMnO_4 + 4\,H_2O \rightarrow 3\,CH_3\underset{\displaystyle OH}{\underset{|}{CH}}{-}\underset{\displaystyle OH}{\underset{|}{CH_2}} + 2\,MnO_2 + 2\,KOH$$

It is difficult to arrest the oxidation of an alkene at the dihydroxy stage. However, if a *dilute* neutral solution of permanganate is used at low temperature it is sometimes possible. When solutions of $KMnO_4$ are either acidic or basic or when higher temperatures are used, the dihydroxy compound formed in this first stage is oxidized further. With propene, reaction a is followed by

$$3\,CH_3\underset{\displaystyle OH}{\underset{|}{CH}}{-}\underset{\displaystyle OH}{\underset{|}{CH_2}} + 8\,KMnO_4 \rightarrow 3\,CH_3COOK + 3\,KHCO_3 + 8\,MnO_2 + 5\,H_2O + 2\,KOH$$

The overall equation for the complete oxidation of propene is

$$3\,CH_3CH{=}CH_2 + 10\,KMnO_4 \rightarrow 3\,CH_3COOK + 3\,KHCO_3 + 10\,MnO_2 + 4\,KOH + H_2O$$

This oxidation (since alkynes react similarly) is the basis for the *Baeyer's test* for unsaturation (presence of multiple bonds). A potassium permanganate solution is purple. Manganese dioxide is a dark-brown precipitate. If a solution of potassium permanganate is bleached when shaken with a hydrocarbon, or a hydrocarbon and a brown precipitate appears, the test is positive and the original compound is unsaturated.

Hydrogenation. Hydrogen has no observable tendency to add to a multiple bond in the absence of a catalyst even at relatively high temperatures (see Section 26.10). In the presence of finely divided platinum, palladium, nickel or iron the addition proceeds smoothly at a few atmospheres pressure.

$$CH_2{=}CH_2 + H_2 \rightarrow CH_3CH_3$$
$$CH_3CH{=}CH_2 + H_2 \rightarrow CH_3CH_2CH_3$$

Obviously alkanes can be prepared from alkenes in this way. Ordinarily the method is too expensive for commercial use because the unsaturated compounds are usually more expensive than corresponding alkanes. Readily available unsaturated compounds may, however, be converted to saturated compounds by this procedure (i.e., the conversion of unsaturated oils to solid fats). Hydrogenation may also be used to evaluate the degree of unsaturation of a compound or to distinguish saturated from unsaturated compounds in the laboratory. Unsaturated hydrocarbons take up hydrogen gas, the amount being determined by their degree of unsaturation, that is by the number of double or triple bonds present in each molecule.

The addition of H_2 to the carbon-carbon double bond, in contrast with the addition of bromine to cyclopentene illustrated earlier, is an example of *cis* **addition** in which both hydrogen atoms add to the double bond from the same side of the planar double bond. The different pattern of addition in the two cases suggests different mechanisms.

27.10 REACTIONS OF ALKYNES

Addition reactions of alkynes parallel those of the alkenes. As pointed out with HBr, a triple bond has twice the capacity for adding reagents as does an alkene. Often reaction can be stopped after addition of one mole equivalent of reagent to yield substituted alkenes. Some of the more useful reactions of acetylene and related alkynes are:

a. *Addition of HBr* (outlined earlier for acetylene). The addition with propyne is in accordance with Markownikoff's rule:

$$CH_3C{\equiv}CH + HBr \rightarrow CH_3\underset{\underset{Br}{|}}{C}{=}\underset{\underset{H}{|}}{C}H$$

$$CH_3\underset{\underset{Br}{|}}{C}{=}\underset{\underset{H}{|}}{C}H + HBr \rightarrow CH_3\overset{\overset{Br}{|}}{\underset{\underset{Br}{|}}{C}}{-}\overset{\overset{H}{|}}{\underset{\underset{H}{|}}{C}}H$$

b. *Addition of Br_2 and Cl_2.* The reaction is a two stage process, since it is generally possible by proper selection of conditions to limit reaction to the first step leading to alkene formation.

$$CH_3C{\equiv}CH + Br_2 \rightarrow CH_3\underset{\underset{Br}{|}}{C}{=}\underset{\underset{Br}{|}}{C}H$$

$$CH_3\overset{\overset{Br}{|}}{C}{=}\overset{\overset{Br}{|}}{C}H + Br_2 \rightarrow CH_3{-}\overset{\overset{Br}{|}}{\underset{\underset{Br}{|}}{C}}{-}\overset{\overset{Br}{|}}{\underset{\underset{Br}{|}}{C}}H$$

c. *Oxidation of acetylenes.* Acetylenes are oxidized rapidly by solutions of potassium permanganate. The reaction as with alkenes involves addition of —OH groups to the carbons of the multiple bond.

$$3\,CH{\equiv}CH + 4\,KMnO_4 + 8\,H_2O \rightarrow 3\left[\begin{array}{c} \quad OH \quad OH \\ \quad | \qquad | \\ H{-}C{-}{-}C{-}H \\ \quad | \qquad | \\ \quad OH \quad OH \end{array}\right] + 4\,MnO_2 + 4\,KOH$$

The tetrahydroxy addition products shown cannot be isolated. Hence the brackets about the compound to indicate a *transient species*, one that is not isolated but is thought to form in the course of the reaction. The reaction proceeds further with cleavage of the carbon to carbon bond.

$$3\left[\begin{array}{c} \quad OH \quad OH \\ \quad | \qquad | \\ H{-}C{-}{-}C{-}H \\ \quad | \qquad | \\ \quad OH \quad OH \end{array}\right] + 6\,KMnO_4 \rightarrow 6\,MnO_2 + 6\,KHCO_3 + 6\,H_2O$$

The complete oxidation of acetylene is the sum of the previous two equations:

$$3\,CH{\equiv}CH + 10\,KMnO_4 + 2\,H_2O \rightarrow 6\,KHCO_3 + 10\,MnO_2 + 4\,KOH$$

d. *Hydrogenation.* As with alkenes, alkynes add hydrogen smoothly in the presence of suitable metal hydrogenation catalysts to form either an alkene *or* an alkane. Complete reaction yields an alkane.

one double bond takes up two atoms of H, a triple bond four atoms.

$$CH_3C{\equiv}CH + 2\,H_2 \rightarrow CH_3CH_2CH_3$$

With less active catalysts, palladium and nickel, the hydrogenation of an alkyne can be stopped at the first stage to give an alkene in reasonable yield.

$$CH_3C{\equiv}CH + H_2 \rightarrow CH_3CH{=}CH_2$$

e. *Addition of H_2O* in the presence of dilute sulfuric acid and mercuric sulfate.

$$CH{\equiv}CH + HOH \xrightarrow[H_2SO_4]{HgSO_4} \left[\begin{array}{c} \;H \quad OH \\ \;| \qquad | \\ HC{=}CH \end{array}\right] \rightarrow \begin{array}{c} \;H \quad H \\ \;| \qquad | \\ H{-}C{-}C{=}O \\ | \qquad \\ H \qquad \end{array}$$

the H attached to oxygen in OH moves to the adjacent carbon.

The reaction produces an unstable addition product (shown as before in brackets) that **rearranges** (undergoes a change in structure) to yield acetaldehyde, a very important compound to be discussed in a later chapter. This is a specific reaction characteristic of alkynes.

f. *Replacement of the hydrogen attached to a carbon of a triple bond.* A hydrogen attached to the end carbon of a terminal triple bond is sufficiently acidic to undergo replacement by sodium metal because electrons of the alkyne group, including those of the C—H sigma bond, are held more tightly to that group polarizing the sigma bond of the C—H linkage. When sodium is dropped into a liquid alkyne having a terminal C to C triple bond, hydrogen is released and bubbles form on the metal liquid interface and rise to the surface.

$$2\,CH_3CH_2CH_2C{\equiv}CH + 2\,Na \rightarrow 2\,CH_3CH_2CH_2C{\equiv}CNa + H_2$$

(a sodium acetylide)

No water can be present because sodium displaces hydrogen from water preferentially.

the acetylide ion, being a strong nucleophile, takes the CH_3^+ away from Br^-, forming new product just as did OH^- with CH_3Br (Section 26.7, Chapter 26).

Sodium acetylides can be used to form alkynes with longer carbon chains. The sodium compound reacts with alkyl halides.

$$CH_3CH_2CH_2C{\equiv}C^-Na^+ + CH_3Br \rightarrow CH_3CH_2CH_2C{\equiv}CCH_3 + NaBr$$

The negative acetylide ion (a nucleophile) attacks the carbon of the bromomethane, forming a bond with carbon which releases the bromide ion.

$$CH_3CH_2CH_2C{\equiv}C^- + CH_3Br \rightarrow CH_3CH_2CH_2C{\equiv}CCH_3 + Br^-$$

Sodium and bromide ions are left. Similar reactions occur with longer chain alkyl halides such as bromoethane (ethyl bromide) or 1-bromopropane (n-propyl bromide).

It is interesting to note that use of methyl bromide adds 1 carbon atom to the carbon chain of an acetylide. If an alkyne having an even number of carbon atoms is converted to its sodium salt and then reacts with methyl bromide, an alkyne having an odd number of carbon atoms results. Hydrogenation of the alkyne formed yields an alkane containing an odd number of carbon atoms. For example, a butyne can be used to prepare n-pentane.

$$2\,CH_3CH_2C{\equiv}CH + 2\,Na \rightarrow 2\,CH_3CH_2C{\equiv}CNa + H_2$$

$$CH_3CH_2C{\equiv}CNa + CH_3Br \rightarrow CH_3CH_2C{\equiv}CCH_3 + NaBr$$

$$CH_3CH_2C{\equiv}CCH_3 + 2\,H_2 \rightarrow CH_3CH_2CH_2CH_2CH_3$$

Only alkynes with terminal triple bonds react with sodium. The group ($\equiv$CH) must be present. Various acetylenes will react in the same manner with certain heavy metal ions (silver, mercury and copper) to give precipitates having characteristic colors. The metal salts are called **acetylides.** Formation of a precipitate with the proper metal ion solution may be used as a laboratory test for the presence of 1-alkynes. For example,

$$CH_3C{\equiv}CH + Ag(NH_3)_2^+ \rightarrow CH_3C{\equiv}CAg + NH_4^+ + NH_3$$

(white precipitate)

whereas

$$CH_3C{\equiv}CCH_3 + Ag(NH_3)_2^+ \rightarrow \text{No reaction}$$

and

$$CH_3CH{=}CH_2 + Ag(NH_3)_2^+ \rightarrow \text{No reaction}$$

A red precipitate of copper acetylide forms upon reaction of a terminal acetylene with ammoniacal Cu_2Cl_2. When dry the acetylides may decompose with explosive violence.

27.11 REACTION WITH OZONE (OZONOLYSIS)

When ozone (O_3) gas is passed into a solution of an alkene in carbon tetrachloride it reacts with the double bond to form an **ozonide** which has oxygen inserted into the molecule in place of the double bond.

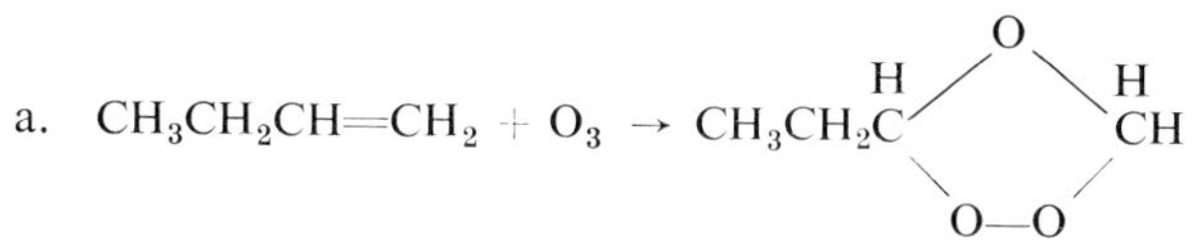

a. $CH_3CH_2CH{=}CH_2 + O_3 \rightarrow CH_3CH_2\overset{H}{C}\langle O,\ O{-}O \rangle \overset{H}{C}H$ (ozonide: CH_3CH_2CH and CH_2 joined through an O bridge and an O—O bridge)

The ozonides are unstable and explosive. Because of this ozonides are not usually isolated but are decomposed directly by treatment with water in the presence of a reducing agent, usually zinc metal. The products formed are aldehydes and ketones which can be readily identified. Reaction of the ozonide with water yields two aldehydes:

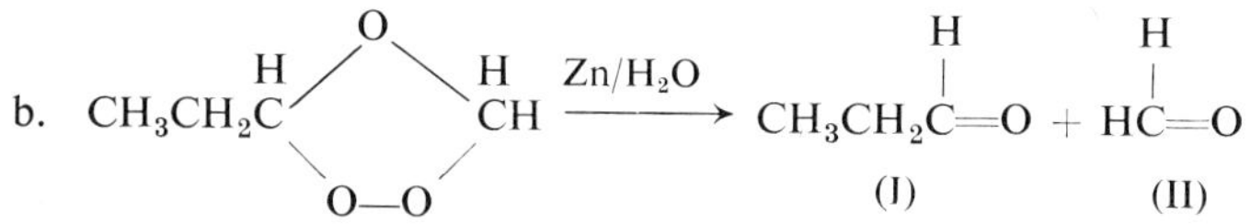

b. $CH_3CH_2\overset{H}{C}\langle O,\ O{-}O \rangle \overset{H}{C}H \xrightarrow{Zn/H_2O} CH_3CH_2\overset{H}{C}{=}O + H\overset{H}{C}{=}O$

(I) (II)

Aldehyde I is formed from the left side and II from the right side of the alkene, as shown in the equation. The degradation products show that the alkene molecule contained four carbons, three on one side and one on the other side of the carbon to carbon double bond. Knowing the numbers and arrangements of the carbon atoms in the aldehyde products, one can deduce the structure of the original alkene which reacted with ozone. For example, ozonolysis of the other isomeric C_4H_8 compounds yields different products as follows:

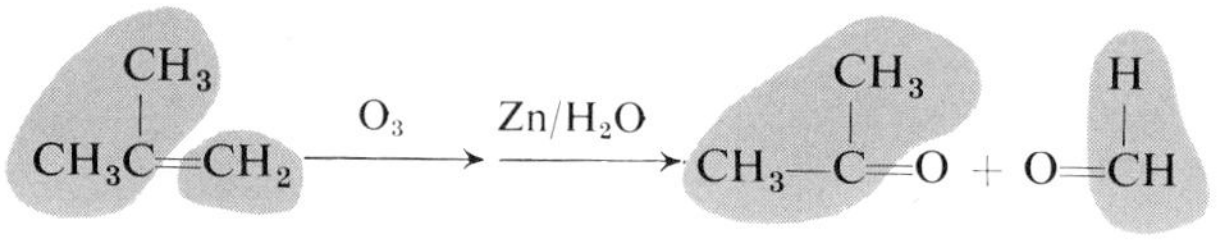

$$CH_3\overset{CH_3}{C}{=}CH_2 \xrightarrow{O_3} \xrightarrow{Zn/H_2O} CH_3{-}\overset{CH_3}{C}{=}O + O{=}\overset{H}{C}H$$

2-methyl-2-propene (a ketone)

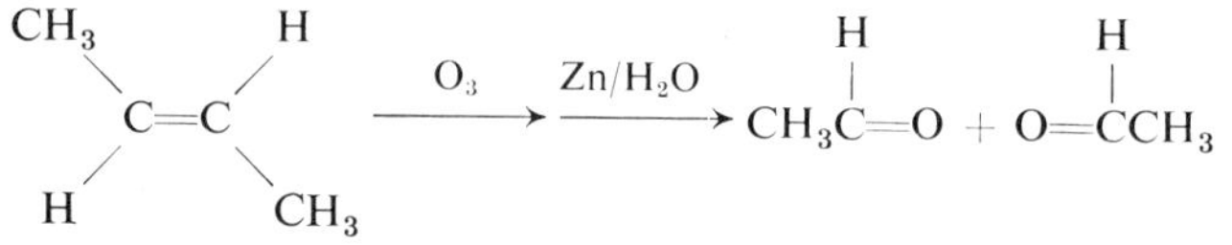

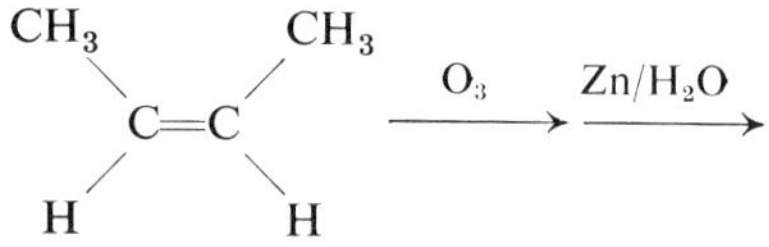

$$\overset{CH_3}{\underset{H}{}}C{=}C\overset{H}{\underset{CH_3}{}} \xrightarrow{O_3} \xrightarrow{Zn/H_2O} CH_3\overset{H}{C}{=}O + O{=}\overset{H}{C}CH_3$$

$$\overset{CH_3}{\underset{H}{}}C{=}C\overset{CH_3}{\underset{H}{}} \xrightarrow{O_3} \xrightarrow{Zn/H_2O}$$

The same products are formed with either *cis-* or *trans-*2-butene, showing that in both the double bond is between carbons 2 and 3.

Alkynes react similarly to alkenes with ozone to form ozonides which on treatment with water yield carboxylic acids rather than aldehydes or ketones. For example,

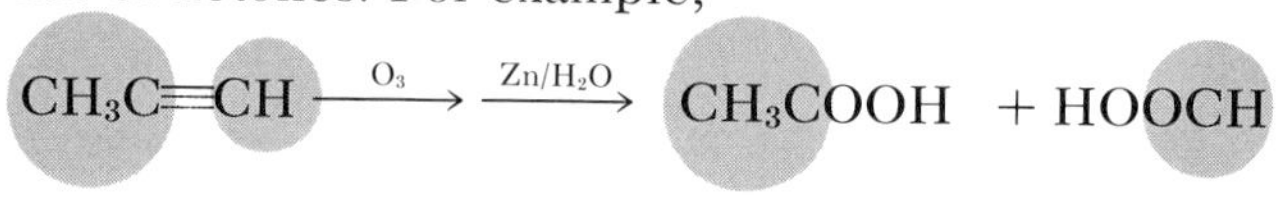

$$CH_3C{\equiv}CH \xrightarrow{O_3} \xrightarrow{Zn/H_2O} CH_3COOH + HOOCH$$

(carboxylic acids)

Simple degradation (chain segmenting) reactions of this kind are useful in structure determinations.

27.12 SOURCES OF UNSATURATED HYDROCARBONS

Unsaturated hydrocarbons are found (but only in very small percentages) in petroleum. Ethylene is often obtained when an organic substance is heated to a high temperature in the absence of air. Today, alkenes are obtained in large quantities from the "cracking" process used to obtain higher percentages of gasoline from petroleum. The smaller members, ethylene and propylene, are used extensively as raw materials for synthesis of a variety of other chemicals, of which permanent antifreeze, ethyl alcohol and acetic acid are among the most important. Chemicals produced from these and other petroleum by-products are often referred to as **petrochemicals.** The petrochemical industry is one of the largest segments of the chemical industry.

Acetylene is produced directly from inorganic chemicals: limestone, coke and water. Calcium carbide is produced by heating limestone and coke in an electric furnace.

$$\underset{\text{Limestone}}{CaCO_3} + \underset{\text{Coke}}{4\,C} \rightarrow \underset{\text{Calcium carbide}}{CaC_2} + 3\,CO$$

Calcium carbide is decomposed by water to yield acetylene:

$$\underset{\text{Calcium carbide}}{CaC_2} + 2\,H_2O \rightarrow Ca(OH)_2 + \underset{\text{Acetylene}}{C_2H_2}$$

Laboratory preparations of unsaturated hydrocarbons should be called subtraction reactions. They amount to subtracting substances from saturated compounds. They are called **elimination reactions.** Elimination of hydrogen, water or a hydrohalic acid from a compound of appropriate structure leads to formation of an unsaturated hydrocarbon.

how can you make propene from propane?

Dehydrogenation. Alkenes may be prepared directly from alkanes. At temperatures of 500 to 700°C, under the stimulus of special catalysts such as molybdenum or chromium oxides supported on aluminum oxide, ethane loses two hydrogen atoms, one from each carbon, to form ethylene:

$$CH_3CH_3 \xrightarrow{500\text{–}700°C} CH_2{=}CH_2 + H_2$$

Higher alkanes form alkenes under similar conditions, but the positions in the molecule from which the hydrogen atoms will be removed cannot be predicted. A mixture of isomeric alkenes, as well as other decomposition products, is obtained.

Dehydration. Dehydration refers to the removal of the elements of water from adjoining carbons. It is an important method for preparing alkenes. In a commercially useful process hot ethyl alcohol is passed over heated aluminum oxide (Al_2O_3) to remove water:

$$CH_3CH_2OH \xrightarrow[\text{(Hot tube 350°)}]{Al_2O_3} CH_2{=}CH_2 + H_2O$$

In the laboratory either phosphoric anhydride or concentrated sulfuric acid is used as the dehydrating agent.

$$3\,CH_3CH_2CH_2OH + P_2O_5 \rightarrow 3\,CH_3CH{=}CH_2 + 2\,H_3PO_4$$

$$CH_3CH_2OH \xrightarrow[160°]{H_2SO_4} CH_2{=}CH_2 + H_2O$$

In the dehydration of ethyl alcohol to form ethylene, sulfuric acid furnishes the H^+ which catalyzes the dehydration. The sequence of changes which comprise this dehydration (mechanism of dehydration) is as follows:

compare this with the mechanism discussed in Section 26.8, Chapter 26.

1. $CH_3CH_2OH + H^+ \rightleftarrows CH_3CH_2\overset{+}{O}H$ (with H attached to O)

2. $CH_3CH_2\overset{+}{O}H_2 \rightleftarrows CH_3CH_2^+ + H_2O$

3. $CH_3CH_2^+ \rightleftarrows CH_2{=}CH_2 + H^+$

Step 1 is the combination of the positive H^+ with CH_3CH_2OH in much the same way as it combines with water to form hydronium ion. This converts the more basic —OH into the less basic (less nucleophilic) H_2O. The attachment between $—OH_2$ and carbon is broken much more readily than that to an —OH. In step 2 a water molecule is lost to form ethyl carbonium ion (ethyl cation), which in step 3 immediately loses a proton to form ethylene. Note that, as in the case in the addition reaction, a carbonium ion is an intermediate. Here, however, the carbonium ion is formed by the loss (elimination) of a group, whereas in the addition reaction the carbonium ion forms by the addition of a proton or similar positively charged atom or group. The double arrows imply that the mechanism is reversible. Secondary alcohols with four or more carbons yield a mixture of products when subjected to dehydration. For example, 2-butanol yields both 1-butene and 2-butene.

$$CH_3CH_2\underset{\substack{|\\OH}}{\overset{\substack{H\\|}}{C}}CH_3 \xrightarrow[160°]{H_2SO_4} \underset{\textbf{(I)}}{CH_3CH{=}CHCH_3} + \underset{\textbf{(II)}}{CH_3CH_2CH{=}CH_2}$$

The amount of the various products obtained varies with the individual alcohols used. The alkene having the double bond carbons carrying the greater number of alkyl groups is usually obtained in largest yield (i.e., compound I above).

Dehydrohalogenation. When adjacent carbon atoms of a saturated compound have respectively at least one hydrogen and one halogen atom attached to them, elements of a hydrohalic acid may be removed. This elimination is called **dehydrohalogenation.** The transformation may be accomplished by using a solution of a strong base in alcohol. Potassium hydroxide in ethyl alcohol is usually employed. The reaction may be represented as follows:

$$CH_3CH_2Cl + KOH \xrightarrow{\text{Ethanol}} CH_2{=}CH_2 + KCl + H_2O$$

the strong nucleophile first removes a proton from the carbon adjacent to the carbon bearing the halogen.

The hydroxide ion, the attacking reagent, is a nucleophile. The negatively charged hydroxide ion approaches the ethyl chloride molecule from the side opposite the negative halogen and extracts a hydrogen (an electrophile) as a H^+ from the carbon adjacent to that carrying the Cl to form water. The electron pair left behind with the loss of the proton shifts into a pi bond between the carbons, and the Cl leaves with a pair of electrons as a chloride ion. This mechanism for the dehydrohalogenation is illustrated here.

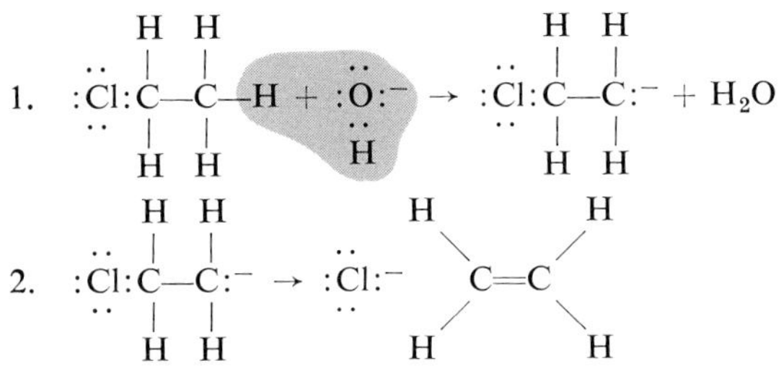

$$1.\quad :\ddot{\underset{..}{Cl}}:CH_2{-}CH_2{-}H + :\ddot{\underset{..}{O}}:^-\!{-}H \rightarrow :\ddot{\underset{..}{Cl}}:CH_2{-}\ddot{C}H_2^- + H_2O$$

$$2.\quad :\ddot{\underset{..}{Cl}}:CH_2{-}\ddot{C}H_2^- \rightarrow :\ddot{\underset{..}{Cl}}:^- \quad H_2C{=}CH_2$$

With the alkyl halides, the tendency toward dehydrohalogenation is greatest for tertiary halides (halides with Cl attached to a tertiary carbon) and least for primary halides. With a given alkyl halide, the tendency toward elimination increases with increasing concentration of the basic reagent. Concentrated alcoholic potassium hydroxide is more effective in eliminating hydrogen halide than is an aqueous solution because the hydroxide ion in water tends to attack the carbon carrying the halogen atom and to displace the halogen to form an alcohol. In essence dehydrohalogenation is the reverse of an addition reaction in the sense that the elements of HCl are removed to form an alkene which can be recovered merely by the addition of HCl to the alkene. Mixtures of products are obtained when alkyl halides having halogen attached to a secondary or tertiary carbon are used.

27.13 ALKADIENES AND RELATED COMPOUNDS

Many molecules contain two or more multiple bonds. The **alkadienes** have two double bonds. Dienes having four or more carbons may have the double bonds located in different positions with respect to each other. For example, two arrangements of the double bonds for a diene having five carbons are

$CH_2{=}CH{-}CH_2{-}CH{=}CH_2$ 1,4-pentadiene (with isolated or unconjugated double bonds)

or $CH_2{=}CH{-}CH{=}CH{-}CH_3$ 1,3-pentadiene (with conjugated double bonds)

In 1,4-pentadiene each double bond is separated from the other by a methylene group having a saturated (four bonded) carbon atom. Double bonds separated by one or more saturated carbons are said to be isolated or unconjugated. Double bonds of this type have little effect upon the chemical properties of one another, and each reacts independently.

In contrast, 1,3-pentadiene has its double bonds on adjacent carbons. It has a carbon to carbon bond sequence: double, single, double and then single. Double bonds in this arrangement (adjacent to each other) are said to be conjugated. Any compound containing a similar system of adjacent double bonds is said to contain a conjugated system (see Chapter 5).

It is not surprising that two conjugated double bonds should affect the chemical properties contributed to the molecule by each group separately, such that, when hydrogen bromide is added to 1,3-pentadiene two products are formed.

$$CH_2{=}CH{-}CH{=}CH{-}CH_3 + HBr \rightarrow \begin{cases} CH_3{-}CH{=}CH{-}\underset{Br}{CH}{-}CH_3 & \text{(1,4 addition)} \\ CH_3{-}\underset{Br}{CH}{-}CH{=}CH{-}CH_3 & \text{(1,2 addition)} \end{cases}$$

The formation of the first product, the reaction which predominates, shows that the two double bonds are not separate entities as the formula indicates, but have delocalized pi orbitals with electrons extending over the full length of the conjugated system. Both 1,4 and 1,2 addition take pi orbitals from two carbons out of the delocalized orbital system to form sigma bonds. The predominance of 1,4 addition over 1,2 addition merely indicates that a primary carbonium ion attached to a carbon atom with a double bond, $R{-}CH{=}CH{-}CH_2^+$, is generally more stable than a secondary carbonium ion in which one of the groups attached to the positively charged carbon participates in pi bonding, $R{-}C^+H{-}CH{=}CH_2$. This reaction is an example illustrating the commonly observed fact that when several functional groups are present in a molecule, one group sometimes has a significant influence on another in a neighboring position.

carbonium ions of this type are more easily formed than tertiary carbonium ions.

Many colored substances have conjugated systems of multiple bonds. Several of these are biologically important compounds. Carotene, a colored material isolated from many plants including carrots, is converted by living organisms into vitamin A. β-Carotene is converted into two molecules of vitamin A by oxidative cleavage at its central C to C bond (Fig. 27.14). Vitamin A is converted into retinene in biological systems by transforming the terminal $-CH_2OH$ (Fig. 27.14) of vitamin A into a $-CHO$ group. Retinene is an essential component of the photosensitive pigment of the eye, visual purple or rhodopsin.

Squalene (Fig. 27.15), a precursor of cholesterol and other biologically important steroids, is a hydrocarbon containing 30 carbons and 6 isolated double bonds. It is formed by the combination of six isoprene ($CH_2{=}CH{-}C(CH_3){=}CH_2$) units. Lanosterol is formed following the cyclization of squalene. It is a precursor of cholesterol.

27.14 INDUSTRIAL APPLICATIONS

Industrial utilization of the addition reactions of unsaturated hydrocarbons has been varied and extensive. Two examples are alkylation and polymerization.

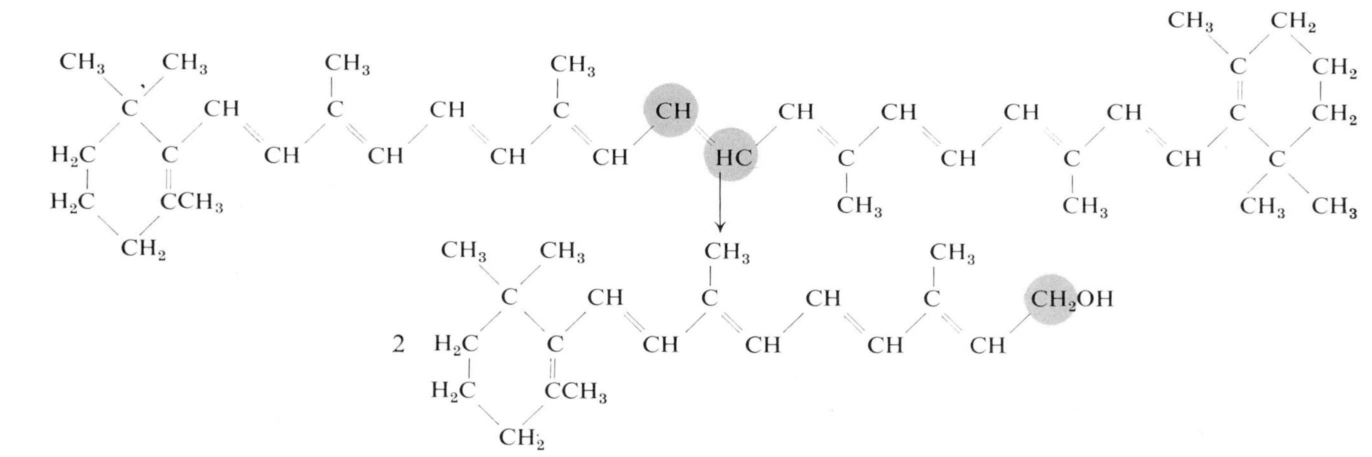

FIGURE 27.14 CLEAVAGE OF CAROTENE TO VITAMIN A.

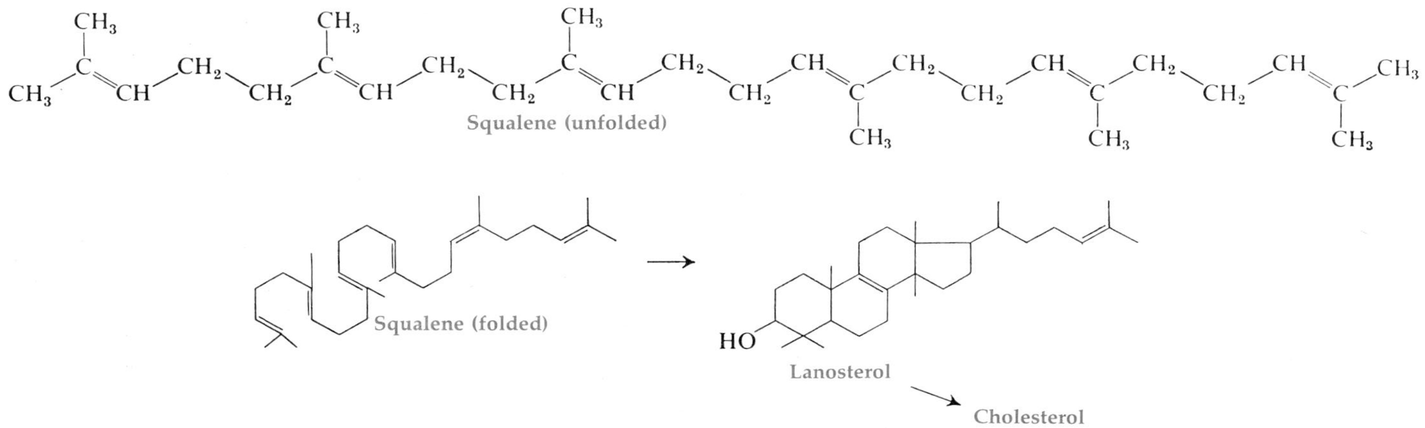

Squalene (unfolded)

Squalene (folded)

Lanosterol

Cholesterol

FIGURE 27.15 SQUALENE (SHOWING ITS RELATIONSHIP TO LANOSTEROL).

Alkylation refers to the addition of a saturated hydrocarbon to a multiple bond. Addition may take place either at high temperature and high pressure or at moderate temperature and low pressure with suitable catalysts. Isobutane may be added to ethylene at 510°C and 300 atmospheres to form isohexane and 2,2-dimethylbutane.

$$\begin{array}{c} CH_3 \\ | \\ CH_3\text{—}C\text{—}H \\ | \\ CH_3 \end{array} + CH_2\text{=}CH_2 \xrightarrow[\text{Pressure}]{\text{Heat}} \begin{array}{l} \begin{array}{c} CH_3 \\ | \\ CH_3\text{—}C\text{—}CH_2CH_3 \\ | \\ CH_3 \end{array} \\ \text{2,2-Dimethylbutane (44\%)} \\ \begin{array}{c} CH_3 \\ | \\ CH_3\text{—}CH\text{—}CH_2CH_2CH_3 \end{array} \\ \text{Isohexane (12\%)} \end{array}$$

The formulas of the major products correspond to cleavage of the alkane into hydrogen and an alkyl group, which add to the double bond.

Alkylation with one of the catalysts, aluminum chloride, sulfuric acid, or liquid hydrogen fluoride, may be carried out at temperatures between −30°C and 100°C and at pressures from 1 to 5 atmospheres, depending on the compounds. Apparently an entirely different course of reaction is followed in the presence of catalysts from that of the thermal reaction. A complex mixture of branched alkanes is obtained. The mechanism of the reaction appears to be very complicated. Nevertheless the reaction is most important as it produces highly branched alkanes for high octane fuels.

Polymerization occurs with compounds having multiple bonds. Polymerization with alkenes is merely self-addition. It provides a class of compounds referred to as **addition polymers. Polymers** are compounds of high molecular weight formed by the joining together of many smaller similar or identical units. The small unit which adds to other identical or similar units is called a **monomer.**

a large segment of the chemical industry is concerned with polymer synthesis and modification.

The self-addition of ethylene (ethene) is an excellent example. One molecule of ethylene (B) may add to another (A) to give 1-butene.

$$\underset{\text{(A)}}{CH_2\text{=}CH_2} + \underset{\text{(B)}}{CH_2\text{=}CH_2} \xrightarrow{\text{catalyst}} \underset{\text{1-Butene (a dimer)}}{CH_3CH_2CH\text{=}CH_2}$$

1-Butene is made of 2 molecules of monomer and is therefore a dimer. But 1-butene has a double bond to which a third molecule of ethylene (C) may be added.

$$\underset{\text{(A—B)}}{CH_3CH_2CH\text{=}CH_2} + \underset{\text{(C)}}{CH_2\text{=}CH_2} \xrightarrow{\text{catalyst}} \underset{\substack{\text{(A—B—C)} \\ \text{1-Hexene (a trimer)}}}{CH_3CH\text{—}CH_2CH_2\text{—}CH\text{=}CH_2}$$

Self-addition may continue indefinitely to give a polymer of many units. The product containing many units is polyethylene, the tough flexible plastic used for making cups, refrigerator dishes and similar products.

The conversion of acetylene into neoprene rubber, which is very similar in structure to natural rubber, is a good example of the polymerization process. The first step in the process is the catalytic **dimerization** of acetylene:

$$CH{\equiv}CH + CH{\equiv}CH \xrightarrow{\text{catalyst}} HC{=}CH{-}C{\equiv}CH$$

(A) (B) — (A—B) 3-Butene-1-yne

The controlled addition of HCl from concentrated hydrochloric acid gives in a second stage a substituted alkadiene:

$$CH_2{=}CH{-}C{\equiv}CH + HCl \rightarrow CH_2{=}CH{-}C(Cl){=}CH_2$$

2-Chloro-1,3-butadiene (Chloroprene)

The uniting of many molecules of chloroprene (the monomer) via 1,4-addition (because chloroprene is a conjugated diene) yields the long chain molecules of chloroprene referred to as neoprene rubber:

$$nCH_2{=}CH{-}CCl{=}CH_2 \xrightarrow{\text{catalyst}} CH_3CH{=}C(Cl)CH_2(CH_2CH{=}C(Cl)CH_2)_{n-2}CH{=}CH{-}CCl{=}CH_2$$

n, a large number, is the number of chloroprene units in the polymer. All units in this molecule except the end units are identical.

The addition reaction which comprises the polymerization step comes about as follows:

the reaction is a chain reaction.

$$A\cdot + C{::}C{:}C{::}C \rightarrow A{:}C{-}C{=}C{-}C\cdot$$

(with H, H, H, H on the carbons above, H and Cl, H below: $H_2C{::}CH{:}C(Cl){::}CH_2$ → $A{:}CH_2{-}CH{=}C(Cl){-}CH_2\cdot$)

An atom, or some other group (A·), with 1 unpaired electron (a free radical) joins with the first carbon, drawing 1 electron from the double bond and causing the electron shifts indicated. The $ACH_2CH{=}CClCH_2\cdot$ combines with the next chloroprene molecule that it encounters as did the group A· ; the electrons shift and a new fragment is ready to add to another. The initiated chain is permitted to grow in this way until the desired polymer is obtained. The effect of the combination is the loss of one double bond by addition in each unit, as is usual. The difference is that the double bond that remains, as in 1,4-additions, has been shifted to the center of the original monomer. Polymerization ceases when one of the chain termination reactions occur, i.e., combination of two growing chains, exchange of an H atom between two growing chains (**disproportionation** where one chain loses an end H atom and the other gains it), or through addition of substances that can react with ends of growing chains to terminate their growth. This is a chain process following a sequence of events related to that observed in the chlorination of methane (Chapters 25 and 26).

Natural rubber is found in colloidal suspension in the milky fluid latex of a variety of plants. The principal commercial source of this latex is the rubber tree (genus *Hevea*). Natural rubbers are linear polymers of isoprene

$$CH_2{=}CH{-}\underset{\displaystyle CH_3}{\underset{|}{C}}H{=}CH_2$$

These polymeric hydrocarbons are built up of isoprene units joined through 1,4-addition of neighboring units in a manner to give a cis-configuration (geometry) at each double bond. Natural rubber may be represented as

$$\begin{matrix} ACH_2^- & & CH_2 - & \Big(CH_2 & & CH_2 \Big) & - CH_2 & & CH_3 \\ & C{=}C & & & C{=}C & & & C{=}C & \\ CH_3 & & H & \Big(CH_3 & & H \Big)_n & CH_3 & & H \end{matrix}$$

Gum rubber is a soft pliable thermoplastic substance which when subjected to vulcanization becomes a useful product. **Vulcanization** with sulfur forms sulfur bonds (linkages) between adjacent hydrocarbon chain molecules. The result is a material having a three-dimensional polymeric network. By controlling the extent of vulcanization (and thus the extent of chain cross-linking of the rubber) products of various degrees of hardness, melting point and solubility are produced. In addition to vulcanizing agents, modern rubber technology uses a variety of other materials such as antioxidants, stiffeners, pigments, emulsion stabilizers and odorants in order to produce the multitude of today's rubber products.

many rubber-like synthetic polymers are known.

Shortages of natural rubber and a demonstrated need for rubber products of special properties led to the development of **synthetic rubbers.** Neoprene rubber, whose formation was previously discussed, is one of these. It is much more resistant to deterioration in oil than is natural rubber. Other synthetic rubbers include Buna rubber, a 1,4-addition polymer of butadiene polymerized using sodium and GRS rubber (Government Rubber Styrene), a copolymer of 75 per cent butadiene and 25 per cent styrene ($C_6H_5CH{=}CH_2$), developed when a rubber shortage was expected during World War II.

Only in recent years has it been possible to prepare truly synthetic natural rubber. It is now possible to polymerize isoprene by use of special catalysts to a polyisoprene which is almost identical with natural rubber, in which all double bonds have the cis-configuration.

Many biologically important compounds, often referred to as **natural products** since they are obtained from plants or animals, are formed from isoprene units joined in various sequences and rings. Many of these, including natural rubber, when decomposed with heat yield isoprene as a distillation product. In 1922, Ruzicka, a Swiss chemist, proposed that all these natural products are derived from one common precursor referred to as "**biological isoprene.**" This idea has been very useful in developing structures and in elucidating the possible sequence of reactions by which various biologically important compounds are formed in plants and animals. Some of the naturally occurring hydrocarbons related to isoprene and known as **terpenes** are given in Figure 27.16. Each

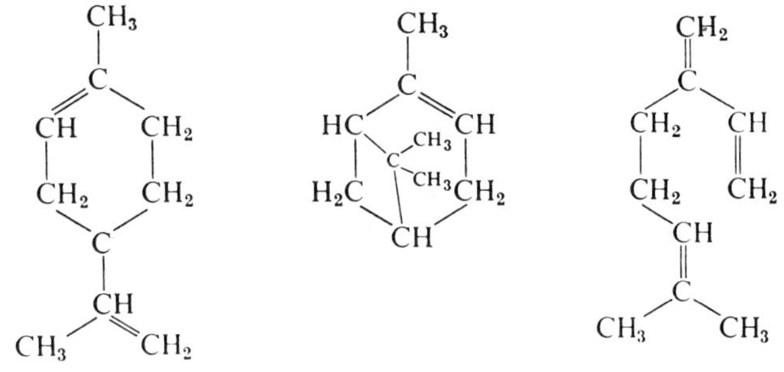

FIGURE 27.16 TERPENES.

Limonene α-Pinene Mycrene

terpene is composed of two isoprene units combined in such a manner as to give open chain or ring structures. Larger molecules containing three, four or more isoprene units are found in nature. Squalene, discussed previously in this section, is an example of one such hydrocarbon containing six isoprene units.

EXERCISES

27.1 Write the electron dot formula (valence electron structure) for (a) ethylene, (b) propylene and (c) acetylene.

27.2 Define functional group.

27.3 Write both the common names and the IUPAC names for each alkene containing no more than 4 carbons.

27.4 Write structural formulas for and name each of the isomeric pentenes by the IUPAC system.

27.5 Compare the C to C bonds in ethane, ethylene and acetylene in regard to length, rigidity and chemical reactivity.

27.6 Write structures for and name all possible compounds with the formula C_4H_8.

27.7 Name this compound:

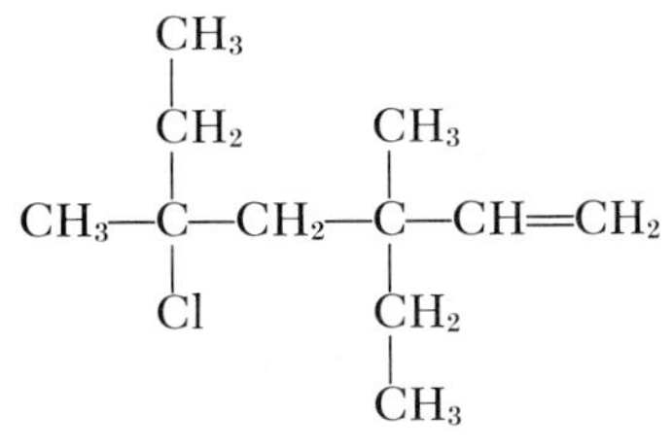

```
        CH3
        |
        CH2       CH3
        |         |
   CH3—C—CH2—C—CH=CH2
        |         |
        Cl        CH2
                  |
                  CH3
```

27.8 Define substitution reaction and addition reaction. Which is characteristic of alkenes?

27.9 Write equations for the reaction of propene with (a) concentrated

HCl, (b) chlorine water, (c) chlorine gas and (d) hydrogen gas. Show special conditions and catalysts, if any.

27.10 State Markownikoff's rule. Illustrate its application.

27.11 Write equations for a 2 step preparation of $CH_3CHOHCH_3$ (isopropyl alcohol) from propene.

27.12 Define dimer, trimer, polymer and polymerization.

27.13 Write structural formulas for and name all acetylenes having from 2 to 5 carbons.

27.14 Write equations for the stepwise addition of the following reagents to acetylene (show catalysts and special conditions on the arrow): concentrated HBr, hydrogen gas.

27.15 Write equations for 2 reactions of 1-butyne that are impossible for any of its isomers.

27.16 Write equations for the preparation of:

a. propene from 2-bromopropane
b. propene from 1-propanol ($CH_3CH_2CH_2OH$)
c. propyne from propene (2 steps)
d. isobutane from 1-butene (2 steps)
e. propene from propane (2 steps)
f. cyclohexane from bromocyclohexane
g. 2-butyne from propyne (2 steps)

27.17 Describe the procedure and write equations for:

a. two tests for unsaturation
b. a test for a terminal alkyne
c. ozonolysis of a 3-heptene

27.18 Write equations for the addition of hydrogen bromide to:

a. 2-methyl-2-pentene
b. 2,4-hexadiene
c. 1,3-butadiene
d. 2-pentene

27.19 Define and give an example of each of the following:

a. electrophile
b. carbonium ion
c. terminal alkyne
d. geometrical isomers
e. unsymmetrical double bond
f. unsymmetrical reagent
g. *cis-trans* isomers

27.20 Draw the structures and name all possible open or branched chain isomers with the formula C_5H_8. Remember to include the geometrical isomers.

27.21 Write the formula and name the tetramer of propene.

27.22 Write equations for the preparation of 1-butene from inorganic materials. (3 steps)
Hint: Prepare ethyne.

27.23 Which 1,2-dichloroethene has a dipole moment?

27.24 Which of the following molecules have structural formulas which permit *cis-trans* configurations so that *cis-trans* isomers can exist?

a. $(CH_3)_2C{=}CHCH_3$
b. $CH_3CH{=}CHCH_3$
c. $Cl_2C{=}CBr_2$
d. $HOOCHC{=}CHCOOH$
e. $ClCH_2CH{=}CHCH_3$
f. 1,2-dichlorocyclobutane
g. $ClCH{=}CHCH_3$

27.25 Of the carbonium ions possible by addition of a proton to 2-methyl-1-butene, which is the more stable? Why?

27.26 A hydrocarbon having the molecular formula C_5H_8 reacts with sodium metal to yield hydrogen gas. Write the structures of two hydrocarbons which would react in this manner.

27.27 Hydrocarbons having the following molecular formulas were allowed to react with an excess of $KMnO_4$ to give the organic products (including CO_2) listed on the right and no others. Write a structure for each alkene.

a) $C_5H_{10} \rightarrow CO_2 + (CH_3)_2CHCOOH$
b) $C_8H_{16} \rightarrow CH_3CH_2CH_2COOH$
c) $C_4H_8 \rightarrow CH_3COOH$
d) $C_4H_8 \rightarrow CO_2 + CH_3CH_2COOH$

27.28 A hydrocarbon A having the molecular formula C_6H_{12} reacts with hydrogen gas in the presence of a suitable catalyst to form C_6H_{14} and with ozone in ozonolysis to form $CH_3CH_2\overset{\overset{\displaystyle O}{\|}}{C}CH_3$ and CH_3CHO.

What is the structure of hydrocarbon A?

27.29 A compound M containing only carbon and hydrogen reacts with sulfuric acid to form a product B which reacts with water to give t-butyl alcohol, $(CH_3)_3COH$. What is the structural formula of compound A and product B?

27.30 A liquid hydrocarbon Z having the empirical formula CH_2 and a molecular weight of 126 does not react with potassium permanganate or decolorize a bromine solution. Suggest at least two structures which compound Z might have. Explain why.

27.31 An unsaturated hydrocarbon I reacts with HCl to give a good yield of a product X which was shown to be identical with one of the monochloroalkanes produced from the reaction of chlorine with isobutane. What is the structure of the hydrocarbon I?

27.32 When isoprene (2-methyl-1,3-butadiene) is allowed to react with one mole of hydrogen in the presence of a suitable catalyst, the following mixture of products is formed, in the percentages given: 12 percent 3-methyl-1-butene, 13 percent 2-methyl-1-butene, 15 percent 2-methyl-2-butene, 30 percent 2-methylbutane and 30 percent 2-methyl-1,3-butadiene. Write equations for their formation or indicate their source and suggest a reason for the presence of each.

27.33 The analysis of a compound A shows it to contain only carbon and hydrogen. The compound reacts with HI to form compound B, which reacts with Mg to form a Grignard reagent, compound C. Compound A is oxidized with $KMnO_4$ giving CO_2 and a ketone, compound D, which can be synthesized by hydration of 1-propyne using $HgSO_4$ in sulfuric acid. Write structural formulas for compounds A, B, C and D. Write equations for all reactions described.

27.34 You are given two bottles, one filled with a compound A and the other with a compound B. Both A and B have the same analysis and molecular weight. Both decolorize bromine and react slowly to dissolve in H_2SO_4. Treatment of A with $KMnO_4$ gives only CO_2 and CH_3CH_2COOH. Compound B yields CO_2 and oxalic acid (HOOC-COOH). Compound A reacts with ammoniacal CuCl; compound B does not. Write structural formulas for compounds A and B. Write equations for all reactions described.

27.35 A quantity of 0.107 gram of an olefin A with one double bond decolorized 100 ml of a bromine solution containing 0.267 g of Br_2 per milliliter. What is its molecular formula? Calculate the number of grams of H_2 with which it would combine. Suggest possible structural formulas for olefin A.

27.36 A Grignard reagent, RMgI, reacts with H_2O to produce RH and Mg(OH)I. If 0.55 g of the reagent reacts with water to form 63.5 ml of gaseous RH measured at STP, what is the molecular weight of the hydrocarbon? What is an acceptable empirical formula?

SUGGESTED READING

Allinger, N. L. and Allinger, J.: *Structures of Organic Molecules.* Prentice-Hall, Inc., Englewood Cliffs, N. J., 1965 (paperback). Chapter 4 especially.

Brown, H. C.: "New Selective Reducing Agents." J. Chem. Ed., *38*:173, 1961.

Gould, E. S.: *Mechanisms and Structure in Organic Chemistry.* Holt, Rinehart and Winston, Inc., New York, 1959. Chapter 13.

Hendrickson, J. B., Cram, D. J. and Hammond, G. S.: *Organic Chemistry,* Third Edition. McGraw-Hill Book Company, New York, 1970. Chapters 6, 14 and 15.

Jones, G.: "The Markovnikov Rule." J. Chem. Ed., *38*:297, 1961.

Morrison, R. T. and Boyd, R. N.: *Organic Chemistry,* Second Edition. Allyn and Bacon, Inc., Boston, 1966. Chapters 5 and 6.

Natta, G.: "Polymerization." Scientific American, September, 1957.

Noller, C. R.: *Chemistry of Organic Compounds,* Third Edition. W. B. Saunders Company, Philadelphia, 1965. Chapters 3 and 10.

TWENTY-EIGHT • AROMATIC HYDROCARBONS

Benzene is a familiar and widely used industrial chemical obtained as a volatile by-product of the production of coke from coal. It is the simplest of a class of organic substances known as aromatic compounds; the term aromatic refers to the characteristic fragrance of some of the first compounds of this class to be discovered. The odor, although giving a name to these compounds, has significance neither in identifying nor in defining the chemical nature of such substances.

28.1 THE BENZENE MOLECULE

Benzene, discovered in 1825 by Michael Faraday, is a hydrocarbon. Repeated analyses and molecular weight determinations show it to have the formula, C_6H_6. Because the general formula for alkanes is C_nH_{2n+2}, the formula of alkenes, C_nH_{2n}, and the formula for cycloalkenes and alkynes, C_nH_{2n-2}, benzene with the formula corresponding to C_nH_{2n-6} appears to be highly unsaturated. However, its characteristic reactions are not additions as in the case of alkenes and alkynes. It is not oxidized by neutral permanganate solution (Baeyer's test for unsaturation), nor does it decolorize bromine water. In contrast, the common effect of reagents upon benzene and benzene-like compounds is to replace or substitute groups for one or more of the hydrogens. Nitric acid reacts with benzene to give nitrobenzene:

$$C_6H_6 + HONO_2 \rightarrow C_6H_5NO_2 + H_2O$$

Sulfuric acid substitutes an —SO_3H group for a hydrogen atom:

$$C_6H_6 + HOSO_3H \rightarrow C_6H_5SO_3H + H_2O$$

Many structures were proposed for benzene but the first formulation to point toward the modern concept of the molecule was made by Kekulé in 1865, forty years after the molecular formula was first known. Kekulé represented benzene as a ring of six carbon atoms joined by alternate single and double bonds with each carbon bonded to a hydrogen (Fig. 28.1). This accounted for six carbons and six hydrogens and left each carbon sharing an octet of electrons. Such a structure might properly be called a hexatriene, i.e., 1,3-5-cyclohexatriene.

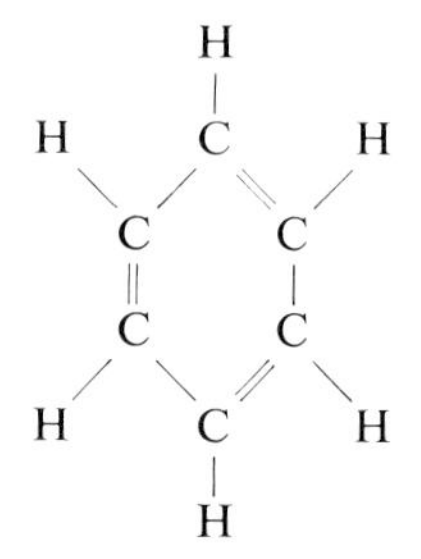

FIGURE 28.1 KEKULÉ'S CYCLOHEXATRIENE STRUCTURE FOR BENZENE.

The failure of benzene to give a positive Baeyer's test, and its tendency to undergo substitution rather than addition, implies that it is more saturated than suggested by the **Kekulé formula** with its three double bonds. On the other hand, benzene does undergo hydrogenation (adds hydrogen from hydrogen in the presence of a catalyst), taking up 3 moles for each mole of benzene in agreement with the hexatriene formulation. However, as shown in the following paragraph, the heat of that reaction with benzene is not equal to that expected of the molecule as represented by Kekulé.

Consider the hydrogenation of three six membered ring compounds: cyclohexene, C_6H_{10}, with one double bond; cyclohexadiene, C_6H_8, with two double bonds; and benzene, C_6H_6, assuming that it has three double bonds and is a triene. The molecules indicated add one, two or three molecules of hydrogen as shown in equations a, b and c. Each reaction is exothermic, giving off the number of calories shown with the equation, and each yields *cyclohexane* as the final product.

a. Cyclohexene + $H_2 \rightarrow \underset{\text{Cyclohexane}}{(C_6H_{12})}$ + 28.6 kcal ($\Delta H = -28.6$ kcal)

b. Cyclohexadiene + 2 $H_2 \rightarrow \underset{\text{Cyclohexane}}{(C_6H_{12})}$ + 55.6 kcal ($\Delta H = -55.6$ kcal)

c. Benzene + 3 $H_2 \rightarrow \underset{\text{Cyclohexane}}{(C_6H_{12})}$ + 49.8 kcal ($\Delta H = -49.8$ kcal)

The addition of 2 moles of hydrogen to the two double bonds of cyclohexadiene releases approximately twice the number of calories produced by the addition of one mole of hydrogen to the one double bond of cyclohexene.

$$2 \times (-28.6 \text{ kcal}) = -57.2 \text{ kcal (observed} = 55.6 \text{ kcal)}$$

This would indicate that the **heat of hydrogenation** of a "cyclohexatriene" in which three moles of hydrogen were added should be approximately three times that for cyclohexene.

$$3 \times (-28.6 \text{ kcal}) = -85.8 \text{ kcal}$$

The observed heat of hydrogenation of benzene, −49.8 kcal/mole, is far less than that calculated for a Kekulé "cyclohexatriene" and, in fact, is less than that released in the addition of only two moles of hydrogen to cyclohexadiene. The relationship between the observed and predicted heats of hydrogenation of the three cyclic hydrocarbons is summarized in an enthalpy diagram (heat of reaction diagram) in Figure 28.2. In accord with changes in **enthalpy** (heat of reaction, ΔH) for these reactions, the diagram shows that benzene and hydrogen have less energy by 36 kcal/mole of benzene than does a corresponding mixture of hydrogen gas and the "hexatriene," and are, therefore, more stable by 36 kcal/mole. This increased stability toward hydrogenation exhibited by benzene as compared to "cyclohexatriene" is also seen in its behavior toward other addition reactants when the heats of reaction are compared. The theory of resonance explains the source of the increased energy of stabilization of benzene.

recall the discussion of enthalpy in Chapter 26.

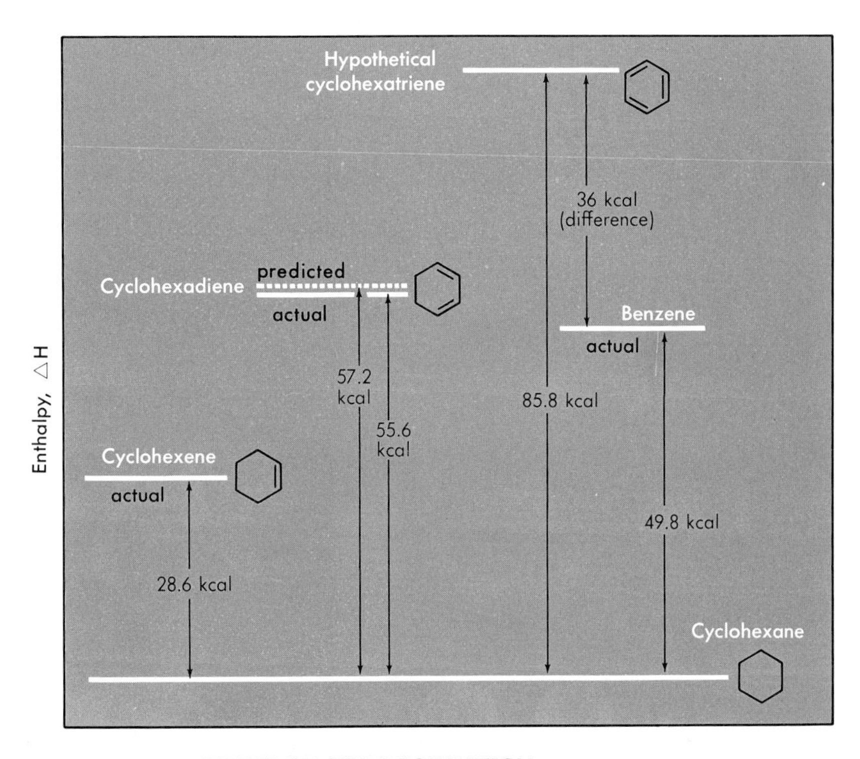

FIGURE 28.2 HEATS OF HYDROGENATION.

Benzene is a **resonance hybrid.** Many electron structures can be drawn for the molecule, some of which are

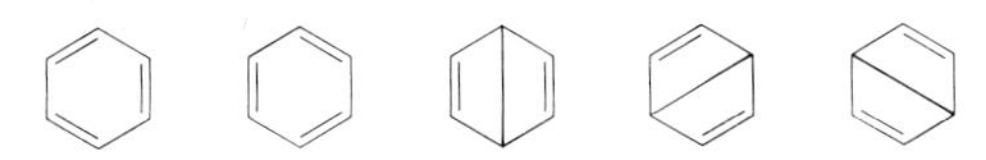

Resonance theory states that when two or more equivalent or nearly equivalent electron configurations can be drawn for the same molecule without changing the position of the atoms, the molecule is not that represented by any one of them but is a structure intermediate between the representations and is more stable than any one of them. This is true of benzene; it is more stable than any one of the possible structures we can draw using conventional notation —and is more stable than the cyclohexatriene structure by 36 kcal/mole. In other terms, the molecule is not as illustrated by the dot formula (Fig. 28.3a), but contains sigma bonds between each pair of adjacent carbons and between each carbon and a hydrogen (Fig. 28.3b). Above and below the ring of carbons and formed from a p orbital of each carbon (Fig. 28.3c) is a set of delocalized pi orbitals, one of which is shown in Figure 28.3d. These pi orbitals are occupied by the six electrons—two to an orbital—which appear as the second electron pair at the double bonds of the electron dot formula. As stated in Chapter 5, benzene, with the alternate double and single bonds and with the many possible resonance structures (or the delocalized orbitals around the complete ring), is the ultimate in delocalization and conjugation. The increased stability of the true molecule as compared to any molecule represented by any one of the resonance forms is shown in the decreased negative heat of hydrogenation cited in the last paragraph.

In recent years **x-ray and electron diffraction studies** have shown that benzene is, in fact, a completely planar molecule with the carbon atoms in a six-membered ring and with a hydrogen bound to each carbon and extending out from the ring with angles of 120° and bond lengths of 1.39 Å, as given in Chapter 5. The

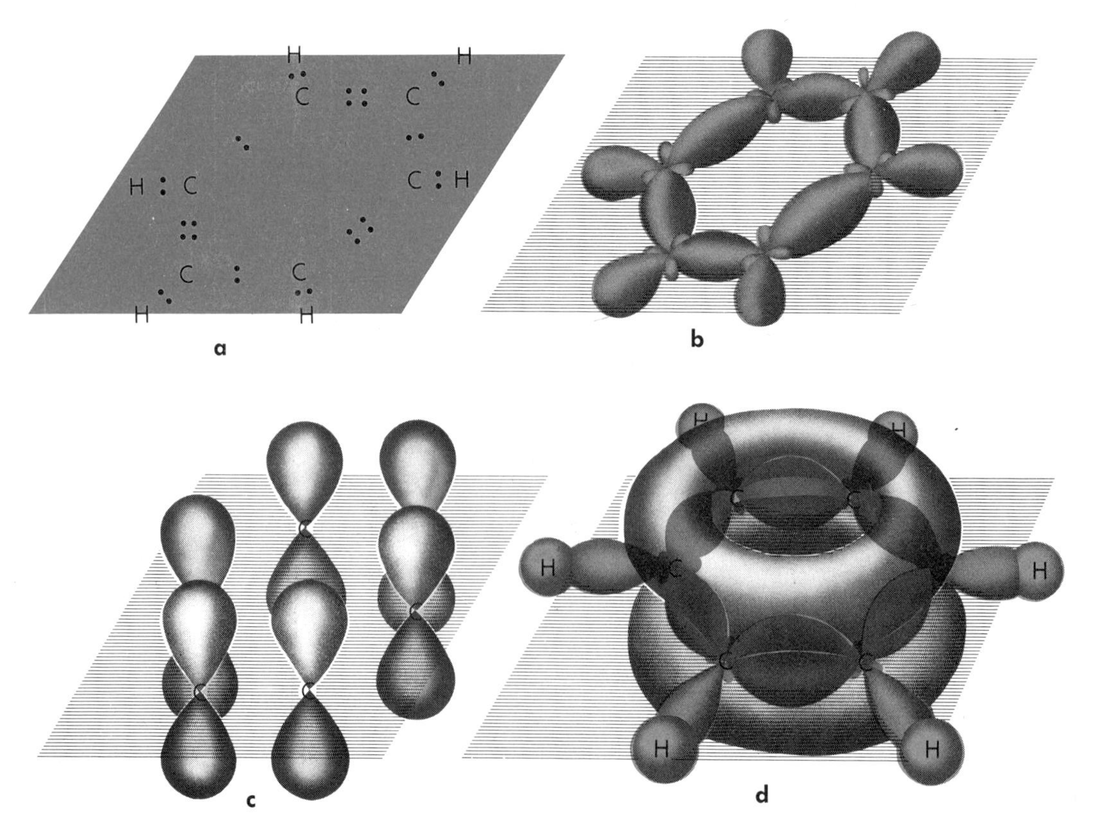

FIGURE 28.3 MOLECULAR ORBITAL REPRESENTATION OF BENZENE.

the bonding is intermediate between an alkene double bond and an alkane single bond.

carbon to carbon bonds, with bond lengths intermediate between that of a single and an isolated double bond, and with the increased stability toward addition as found in benzene-like compounds, are commonly called "**benzene**" or "**aromatic bonds.**" Because there is no good, simple way to represent those bonds and the rings in which they occur, benzene is often represented in a simplified cyclohexatriene formula as [hexagon] or, to emphasize the delocalization, as a circle inscribed in a hexagon, [hexagon with circle]. In either case, despite the inadequate symbolism, the student must be aware of the true structure and nature of benzene.

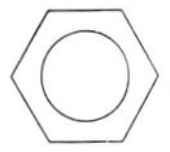

Aromatic substances may be simple derivatives of benzene (in which one or several hydrogens of benzene have been replaced with a hydrocarbon chain or other groups) or they may be more complex compounds having several benzene rings joined together, either by mutual sharing of carbons between different rings (naphthalene, Fig. 28.4) or through single covalent bonds between separate rings (diphenyl, Fig. 28.4) as shown in the following section.

28.2 NOMENCLATURE

Many aromatic compounds have names which have been adopted as official IUPAC names. A few substances are represented and named in Figure 28.4.

Each carbon atom of a ring (or rings) has a multiplicity of four in its bonding (one double and two single bonds, four single bonds, and so on), that is, a complete octet of shared electrons. Each carbon with a multiplicity of three in the ring is also attached to a hydrogen (not shown in the structural formulas, Fig. 28.4). For example, anthracene has the molecule formula, $C_{14}H_{10}$.

Other aromatic compounds are named as derivatives of the compounds given in Figure 28.4. Branches or groups attached to carbons of the rings are named in alphabetical order and according to the number of the carbon in the ring to which they are attached. The accepted pattern for numbering the rings in aromatic compounds is illustrated in Figure 28.4. Often, although not always,

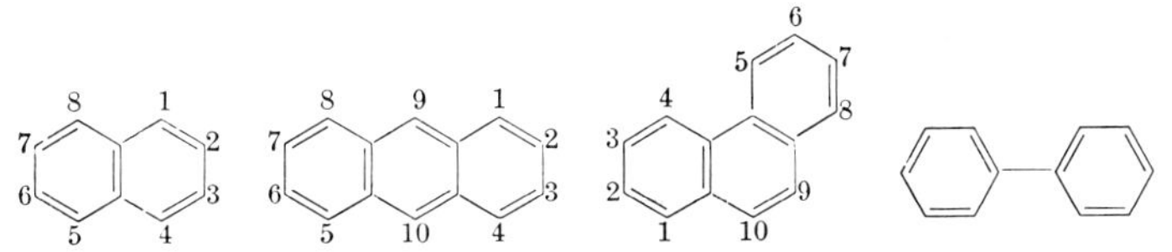

Naphthalene — Anthracene — Phenanthrene — Biphenyl (Phenyl benzene)

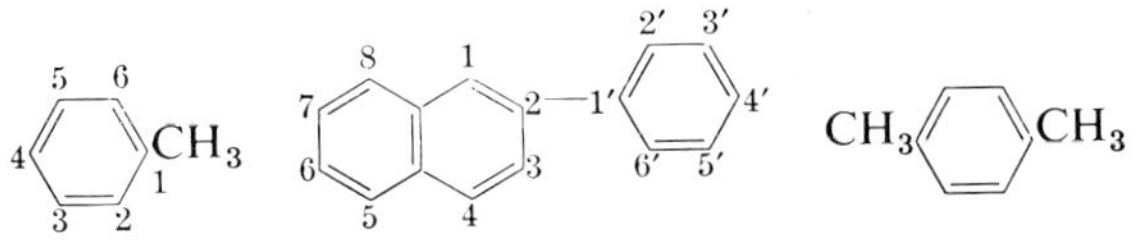

CH_3 — Methyl benzene (Toluene) — 2-Phenylnaphthalene — CH_3 CH_3 1,4-Dimethylbenzene (p-xylene)

FIGURE 28.4 SOME COMMON AROMATIC COMPOUNDS.

the numbering is done in a clockwise fashion about the complete ring system.

In naming a benzene derivative the carbons are numbered from 1 to 6, beginning with one to which a substituent is attached and continuing in the direction which gives the smallest numbers for the name. For example, the methyl group of toluene (methyl benzene) is always on carbon number 1.

Obviously these are representations of the same compound:

In each there is a methyl group on the *first carbon.*

This compound or is 1,2-dichlorobenzene. The "principle of lowest numbers" is obeyed. The compound is named 2-chloro-1,4-dimethyl benzene. No other numbering order will give such small numbers. More compounds are named in Figure 28.5.

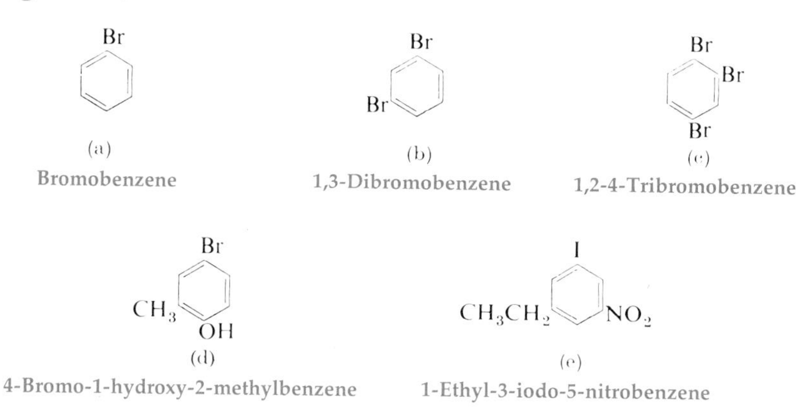

FIGURE 28.5 NAMES OF BENZENE DERIVATIVES.

Should only two substituents be attached to the ring to give a disubstituted derivative, the prefixes ortho (o-), meta (m-) and para (p-) may be used to designate the location of the two attached atoms or groups, providing no ambiguity results (Fig. 28.6). Thus when two substitutents are on adjacent (1,2) carbons (Fig. 28.6a, d) the compound is designated as the ortho isomer; if two substitutents are on alternate (1,3) carbons (Fig. 28.6b), it is a meta isomer; and if the substituents are on opposite (1,4) carbons directly across the ring (Fig. 28.6c), the compound is referred to as a para isomer.

Some compounds derived from benzene may be named also as substitution products of commonly known benzene derivatives, such as toluene or the xylenes (Fig. 28.6). For example, 2-ethyl-1-methylbenzene may be correctly named as 0-ethyltoluene or 2-ethyltoluene.

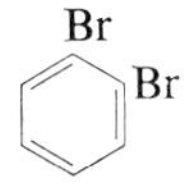

Br
Br

(a)

o-Dibromobenzene
(1,2-Dibromobenzene)

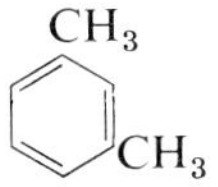

CH_3
CH_3

(b)

m-Dimethylbenzene
(1,3-Dimethylbenzene)
(m-xylene)

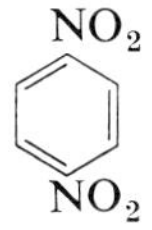

NO_2
NO_2

(c)

p-Dinitrobenzene
(1,4-Dinitrobenzene)

(d) might also be named o-nitrochlorobenzene.

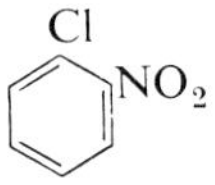

Cl
NO_2

(d)

o-Chloronitrobenzene
(1-Chloro-2-nitrobenzene)

FIGURE 28.6 NAMES OF DISUBSTITUTED BENZENES.

At times it is necessary to name an aromatic portion of a molecule as a branch. These compounds exist.

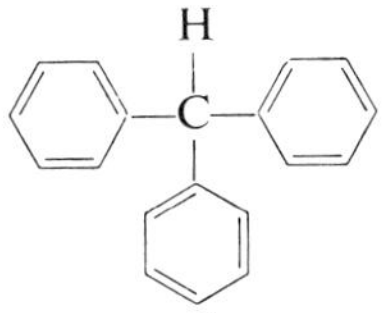

H
C

a

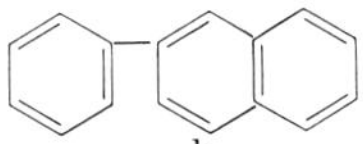

b

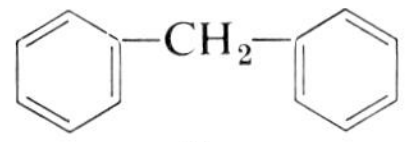

CH_2

c

CH_3

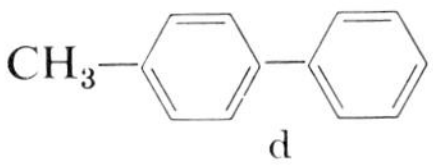

d

Names of some of the groups are:

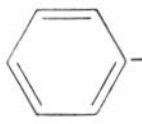

—, C_6H_5— phenyl

CH_2—, $C_6H_5CH_2$—, benzyl

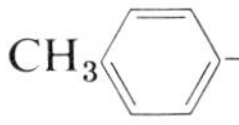

CH_3 —, p-$CH_3C_6H_4$—, p-tolyl

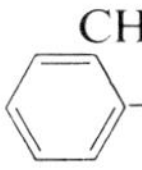

CH_3
—, o-$CH_3C_6H_6$—, o-tolyl

Compound a is triphenyl methane.
Compound b is 2-phenylnaphthalene.
Compound c is diphenylmethane or benzylbenzene.
Compound d is 1-methyl-4-phenylbenzene or p-tolylbenzene.

28.3 PHYSICAL PROPERTIES OF AROMATIC HYDROCARBONS

The physical properties of an individual aromatic hydrocarbon are very nearly equal in magnitude to the properties of other hydrocarbons containing the same number of carbons. Because no continuous chain hydrocarbon with more than 5 carbon atoms per molecule is a gas at room temperature and one atmosphere pressure and, because benzene has the fewest carbons of any aromatic hydrocarbon, no aromatic hydrocarbon is a gas at ordinary conditions. (See the normal boiling points, Table 28.1.) Benzene derivatives containing up to 10 or 12 carbons usually are liquids; those with more carbons are solids. Aromatic hydrocarbons containing several benzene rings are solids at ordinary conditions. The presence of the compact benzene ring increases the density of aromatic compounds above that of corresponding non-aromatic hydrocarbons. Even so, compounds with only one benzene ring are not as dense as water but those with more than one ring have a density greater than unity. Having no water soluble groups, the aromatic hydrocarbons, like other hydrocarbons, are insoluble in water and other polar solvents.

TABLE 28.1 PHYSICAL PROPERTIES OF AROMATIC COMPOUNDS

Name	Formula	Melting Point, °C	Boiling Point, °C	Density
Benzene	(benzene ring)	5.51	80.1	0.879
Toluene	(benzene ring)—CH_3	−95	110.6	0.867
o-Xylene	(benzene ring with CH_3, CH_3)	−29	144.4	0.880
Diphenyl	(two linked benzene rings)	70	254	1.180
Naphthalene	(two fused rings)	80.2	217	1.145
Anthracene	(three fused rings)	217	354	1.25

28.4 CHEMICAL PROPERTIES

Benzene, as was pointed out in an earlier section, has a type of bonding that enables it to undergo certain reactions of both double and single bonded compounds. It adds hydrogen and chlorine in a reaction characteristic of unsaturated hydrocarbons; on the other hand, its most typical behavior is exhibited when it reacts to exchange a hydrogen atom for another atom or group in a so-called **substitution reaction.** Catalysts are commonly used to speed up and sometimes to direct the course of the reactions which take place.

A. Addition

addition is slow because the π (pi) electrons are much less available than in ethylene.

1. *Hydrogen* adds readily to benzene when heated (180 to 200°C) with it in the presence of a nickel catalyst.

$$\text{(benzene)} + 3H_2 \xrightarrow[\text{pressure}]{\text{Ni cat./heat}} \text{cyclo-}(CH_2)_6 \quad [H_2C,\ CH_2,\ CH_2,\ CH_2,\ CH_2,\ CH_2]$$

2. The addition of a *halogen* to benzene occurs and is favored by sunlight. Benzene adds chlorine in the presence of sunlight to form 1,2,3,4,5,6-hexachlorocyclohexane, an insecticide product, which is sold commercially as Lindane or Benzenehexachloride.

$$\text{(benzene)} + 3Cl_2 \xrightarrow[\text{slowly}]{\text{sunlight}} \text{cyclo-}(CHCl)_6$$

1,2,3,4,5,6-Hexachlorocyclohexane

Bromine adds in a similar manner to form hexabromocyclohexane.

B. Substitution

Substitution, the more usual type of reaction shown by benzene and its homologs, takes place when benzene reacts with halogens, nitric acid, sulfuric acid and alkyl halides under the right conditions and in the presence of suitable catalysts.

1. Nitration. The substitution of a hydrogen atom attached to the aromatic nucleus of benzene by nitro groups is achieved through the use of concentrated nitric acid dissolved in concentrated sulfuric acid, a mixture sometimes referred to as "mixed acid." The monosubstitution product formed is nitrobenzene (oil of mirbane). The reaction is referred to as **nitration.**

$$\text{(benzene)} + HONO_2 \xrightarrow{\text{Concentrated } H_2SO_4} \text{(benzene)}\text{-}NO_2 + H_2O$$

Additional hydrogen atoms may be replaced by nitro groups if the temperature of the reaction mixture is raised somewhat and if a larger proportion of nitrating acid is used. Sulfuric acid has a dual function: It (a) acts as a water-absorbing agent and (b) reacts with the HNO_3 to form NO_2^+ (the **nitronium ion**), the species which appears to be the substituting reagent, although it is present in only very small amount in the nitric acid solution. Nitronium ion is formed by the reaction:

NO_2^+, being a strong electron attracting ion, is an electrophile. (See Section 28.5 which follows).

$$HONO_2 + 2H_2SO_4 \rightleftarrows NO_2^+ + H_3O^+ + 2HSO_4^-$$

2. Sulfonation. Benzene and related aromatic hydrocarbons react with concentrated sulfuric acid itself to form sulfonic acids. A hydrogen atom or several hydrogens are replaced with $—SO_3H$ groups with the concurrent elimination of water. The reaction is known as **sulfonation.**

$$C_6H_6 + HOSO_2OH \xrightarrow[\text{Heat}]{\text{Concentrated } H_2SO_4} C_6H_5SO_3H + H_2O$$

In this reaction sulfuric acid also has a dual function. It acts as both (a) a water-absorbing agent and (b) the source of SO_3 (sulfur trioxide), which appears to be the substituting reagent of sulfuric acid.

SO_3 is electron deficient, as is $AlCl_3$, and is a Lewis acid-type electrophile.

Raising the temperature of the reaction mixture permits the introduction of a second $—SO_3H$ group into benzene. Other reagents, such as chlorosulfonic acid ($ClSO_3H$), may be used to introduce the sulfonic acid group.

3. Halogenation. In the presence of iron or of its iron(II) salts, which act as catalysts, chlorine and bromine react readily with benzene to form chloro- or bromobenzene. Here a hydrogen atom is replaced with halogen and the reaction is known as **halogenation.** The reaction with chlorine is called chlorination; with bromine, bromination.

Cl^+ is formed as the substituting reagent when $FeCl_3$ and Cl_2 react.

$$C_6H_6 + Cl_2 \xrightarrow{FeCl_3} C_6H_5Cl + HCl$$

4. Alkylation (Friedel-Crafts Reaction). Alkyl halides in the presence of anhydrous aluminum chloride react with benzene to form alkyl substituted compounds in which an alkyl group is substituted for one or more hydrogen atoms. This reaction is spoken of as **alkylation.** According to the experimental conditions, mono-, di- or even trisubstituted benzenes may be obtained. The term **Friedel-Crafts** comes from the names of the two men, Charles Friedel and James M. Crafts, who discovered the reaction (1878). The aluminum chloride and other reagents, when used, must be dry:

CH_3^+ forms when $AlCl_3$ and CH_3Cl react.

$$C_6H_6 + CH_3Cl \xrightarrow[0^\circ]{AlCl_3} C_6H_5CH_3 + HCl$$

The same reaction carried out with normal chain alkyl halides having more than three carbons, such as n-propyl chloride, yields branched chain alkyl substituted benzenes.

$$C_6H_6 + CH_3CH_2CH_2Cl \xrightarrow{AlCl_3} C_6H_5CH(CH_3)_2 + HCl$$

Benzene n-Propyl chloride Isopropyl benzene

The attachment of the central carbon of n-propyl chloride to the benzene ring, rather than the terminal carbon which carried the halogen atom, results because the n-propyl carbonium ion, $CH_3CH_2CH_2^+$, first formed **rearranges** (undergoes a structural change) to give the more stable isopropyl carbonium ion, $CH_3\overset{+}{C}HCH_3$. This as an electrophile adds to the benzene ring to form isopropyl benzene. Alkenes also may be used to accomplish the alkylation of benzene. For example, ethylene (like ethyl chloride) yields ethyl benzene and propene, 2-phenylpropane.

$$C_6H_6 + CH_2{=}CH_2 \xrightarrow{AlCl_3} C_6H_5CH_2CH_3$$

$$C_6H_6 + CH_3CH{=}CH_2 \xrightarrow{AlCl_3} C_6H_5CH(CH_3)_2$$

Higher alkenes such as propene yield branched chain alkyl benzenes, just as the longer chain alkyl halides do.

C. Disubstitution

o-, p- products predominate when toluene and HNO_3 react.

The introduction of a single atom or group into the benzene ring involves no uncertainty in predicting the structure of the product to be formed. Since all the hydrogens which might be replaced are identical, only one monosubstitution product is possible. The positions open to a second entering substituent indicate that at least three isomers may be formed. For example, the nitration of toluene indicates the possible formation of ortho-, para- or metanitrotoluene disubstitution products.

$$C_6H_5CH_3 + HONO_2 \xrightarrow{H_2SO_4} o\text{-}CH_3C_6H_4NO_2 + p\text{-}CH_3C_6H_4NO_2 + m\text{-}CH_3C_6H_4NO_2 + H_2O$$

Likewise, when nitrobenzene is brominated the possibility is open for the formation of ortho-, para- or metanitrobromobenzene. Since three isomers are permitted

a m- product forms when nitrobenzene reacts with Br_2.

$$C_6H_5NO_2 + Br_2 \xrightarrow{FeBr_3} o\text{-}NO_2C_6H_4Br + p\text{-}NO_2C_6H_4Br + m\text{-}NO_2C_6H_4Br + HBr$$

in both of these examples, the question might be asked, "Which of the possible products can one expect when di- or polysubstitution occurs?"

Experience has shown that the **group or groups already present on the benzene ring determine the position taken by the next entering substituent.** In the nitration of toluene, a mixture of o-nitrotoluene and p-nitrotoluene is obtained, accompanied by traces of m-nitrotoluene. On the other hand, bromination of nitrobenzene forms mainly m-bromonitrobenzene, with only small amounts of the ortho and para isomers.

Such experiments show that the two groups induce the formation of differently positioned disubstitution products. This indicates that the group already attached to the ring has a marked directing effect. This directive phenomenon is referred to as **orientation** and the various types of substituents fall into two classes on the basis of their directive influence:

Class I: —OH, —NH_2, —Cl, —Br, —I, —OR, —NHR, —NR_2, —NHCOR, —R
Class II: —NH_3^+, —NO_2, —SO_3H, —COOH, —CO_2R, —CHO, —CN, —$COCH_3$

class II groups carry a + charge or have an atom with a double or triple bond to C, i.e., —C≡N.

Class I substituents direct the entering group to the ortho and para positions with respect to their location on the ring. **Class II substituents** direct the entering groups or atoms to the meta position with respect to themselves.

Application of these rules to a specific situation, the nitration of hydroxybenzene (phenol), leads one to predict that the products would be the ortho and para isomers. Experimental study of this reaction shows that actually about 40 per cent of the o-nitrophenol and about 60 per cent of p-nitrophenol are obtained.

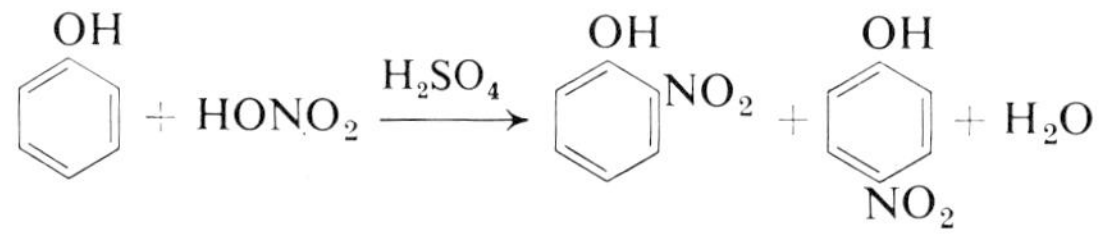

$$C_6H_5OH + HONO_2 \xrightarrow{H_2SO_4} o\text{-}NO_2C_6H_4OH + p\text{-}NO_2C_6H_4OH + H_2O$$

In addition to their directing influence, substituents on an aromatic ring modify the ease with which further substitution takes place. These groups which increase the rate of substitution are said to be "**activating.**" Groups which hinder (slow down) further substitution are said to be "**deactivating.**" Class I substituents, with the single exception of the halogens, are activating and when present make subsequent substitutions occur faster. For example, phenol, C_6H_5—OH, reacts with nitric acid much faster and more readily than does benzene. Class II substituents deactivate the aromatic ring. Nitrobenzene is more difficult to nitrate (nitrates more slowly) than benzene.

D. Polysubstitution

The orientation of additional groups introduced into the benzene nucleus after two or more have already been substituted also depends upon the groups present. Where several groups are attached to the ring, they may **reinforce** if both direct substituents to the same locations, or they **may oppose.** Class I groups take precedence over Class II groups in determining the position of the next entering substituent. Within each class, certain groups take precedence over others. The relative order of directive power for some Class I groups is —OH > —NH_2 > —OR > —Cl > —Br > —I > —CH_3. For class II groups the relative order of directive influence is: —NO_2 > —SO_3H > —CN > —COOH. The following reactions illustrate the application of these findings to particular compounds:

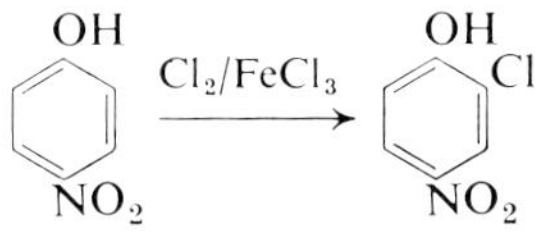

OH, NO_2 $\xrightarrow{Cl_2/FeCl_3}$ OH, Cl, NO_2 (Reinforcement)

NO_2, SO_3H $\xrightarrow{Br_2/FeBr_3}$ NO_2, Br, SO_3H (Reinforcement)

Br, CH_3 $\xrightarrow{HNO_3/H_2SO_4}$ Br, NO_2, CH_3 (Bromine has precedence over methyl)

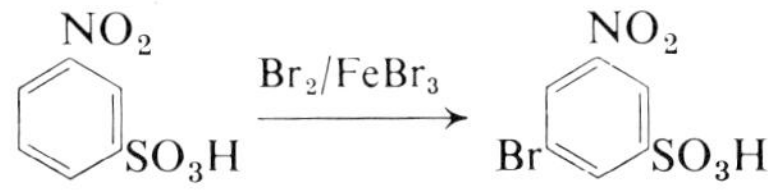

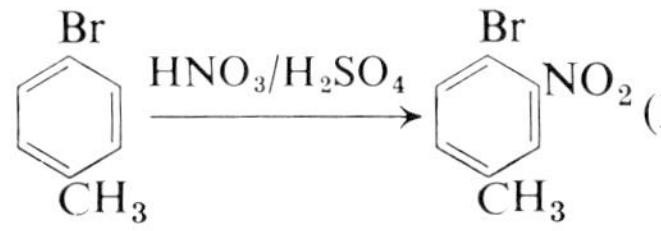

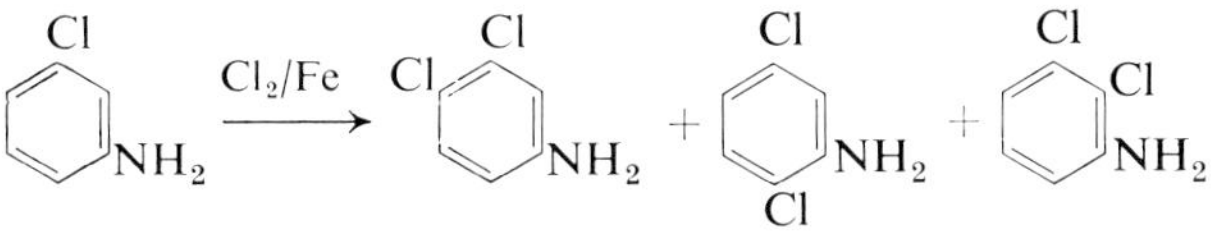

Cl, NH_2 $\xrightarrow{Cl_2/Fe}$ Cl, Cl, NH_2 + Cl, NH_2, Cl + Cl, Cl, NH_2 (Isomers) (Reinforcement)

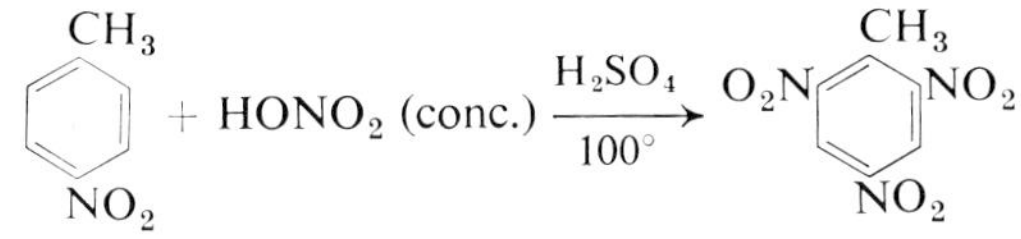

CH_3, NO_2 + $HONO_2$ (conc.) $\xrightarrow[100^\circ]{H_2SO_4}$ CH_3, O_2N, NO_2, NO_2 (Reinforcement)

OH, NO_2 + $HONO_2$ (conc.) $\xrightarrow[100^\circ]{H_2SO_4}$ OH, NO_2, NO_2

and OH, NO_2, NO_2 and OH, O_2N, NO_2

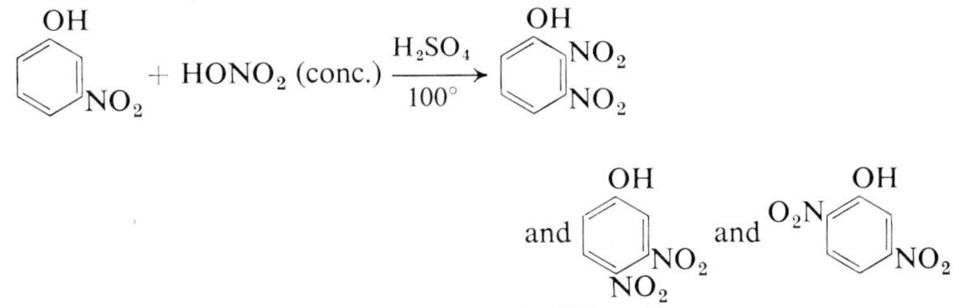

(Class I over Class II)

28.5 MECHANISM OF AROMATIC SUBSTITUTION

three elementary reactions are required; with different reagents only the electrophiles differ.

A theoretical picture for substitution of benzene and related aromatic compounds is available through an examination of the electronic structures of the compounds and the mechanism of the reaction. Substitution reactions of benzene may be pictured as a series of steps: (1) the formation of a positively charged reagent (or neutral electrophile), (2) the union (by addition) of the positively charged reagent (or neutral electrophile) with the benzene ring to form an unstable carbonium ion, which stabilizes itself by (3) loss of a proton from the carbon to which the substitution (electrophilic) reagent was added. The series of steps for the nitration of benzene is as follows:

1. $2\,H_2SO_4 + HNO_3 \rightarrow NO_2^+ + H_3O^+ + 2\,HSO_4^-$

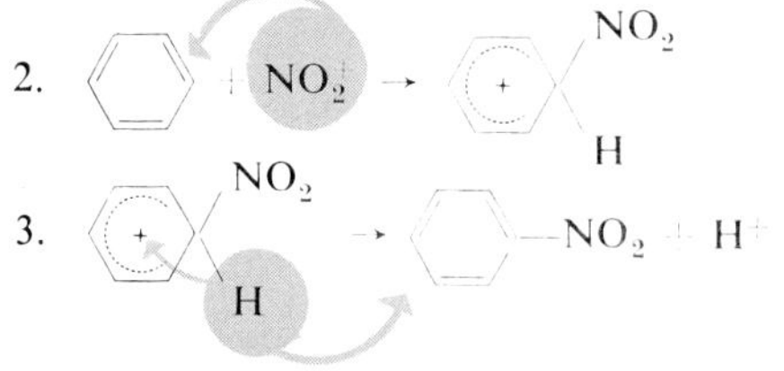

2. benzene + NO_2^+ → carbonium ion (ring with + , bearing NO_2 and H)

3. carbonium ion (ring with +, bearing NO_2 and H) → C_6H_5—NO_2 + H^+

In this nitration reaction of benzene the positive ion, the nitronium ion, is first formed by reaction of sulfuric acid and nitric acid as shown in reaction 1. The nitronium ion, the electrophile in this reaction, then approaches the benzene ring and bonds to a carbon, the result of which is withdrawal of electrons from the pi system of the ring to form a carbonium ion as shown in reaction 2. The formula for the carbonium ion cannot be shown adequately by our usual notation but is frequently represented as in reaction 2. The set of three resonance structures having electrons localized between particular pairs of carbons shows that the positive charge is not centered over a single atom but may reside equally at the

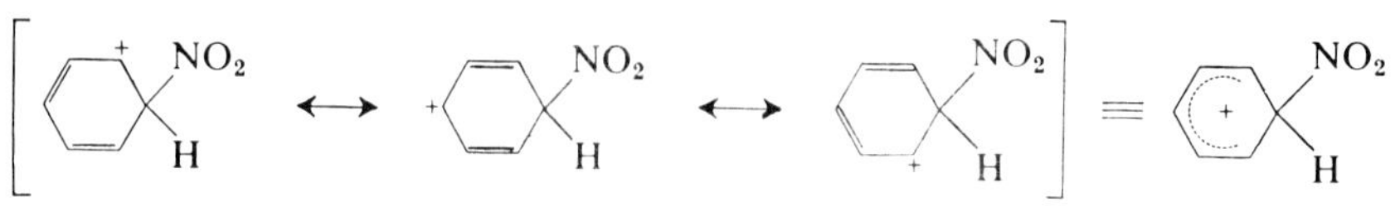

[three resonance structures of the carbonium ion bearing NO_2 and H, connected by ↔, enclosed in brackets] ≡ [ring with delocalized + bearing NO_2 and H]

ortho and the para carbons of the ring system. In the representation for the carbonium ion used in reaction 2, the carbon to which the nitro group has become attached is bonded to four different groups and is therefore an sp^3 type carbon. With no p orbitals available it is no longer included in the pi system, and in essence has been withdrawn from participation in the stabilization characteristic of the benzene ring. The positive charge brought to the ion by the entering nitronium ion is therefore distributed to the remainder of the molecule, and is associated with the five atoms which make up the pi system in the carbonium ion formed.

All five electrons available over and above those required for the sigma bonding are delocalized in pi orbitals encompassing the five sp^2 type carbons of the ring. The carbonium ion is specially stabilized due to this delocalization. It should, however, be recognized that the aromatic ring system is more stable than is the carbonium ion. Because of this the carbonium ion loses the proton which is attached to the carbon carrying the nitro group as shown in 3. With loss of the proton the nitro bearing sp^3 carbon of the carbonium ion reverts once more to an sp^2 configuration with formation of a p orbital which, as in benzene itself, once again overlaps with neighboring p orbitals to permit electron delocalization over the entire ring system. Substitution is thus accomplished with regeneration of the stable aromatic ring system. Steps 2 and 3 of the mechanism are illustrated in Figure 28.7.

Halogenation, sulfonation and alkylation are thought to follow much the same reaction pattern as outlined for nitration. The mechanism for these reactions differs only in the identity of the

Step 2

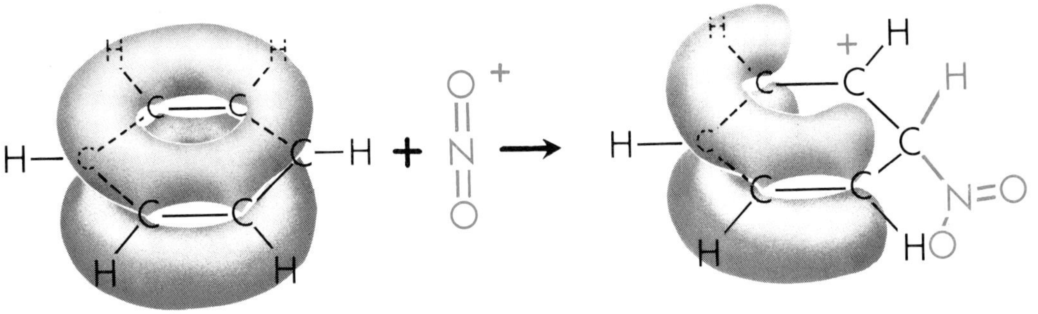

Step 3

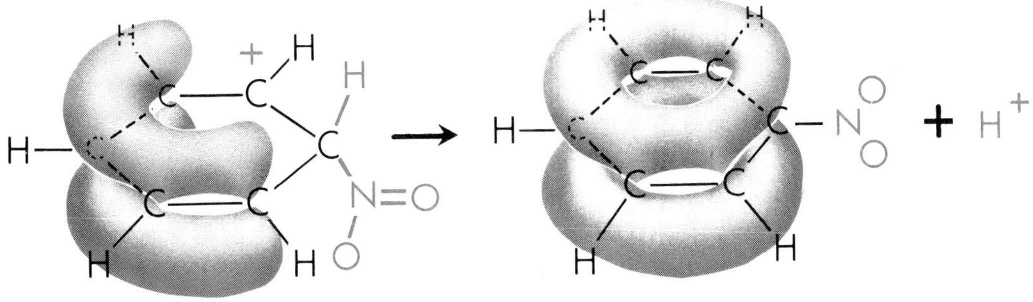

In Kekulé notation: benzene $+ NO_2^+ \longrightarrow$ (+, NO_2, H) $\longrightarrow$ NO_2 $+ H^+$

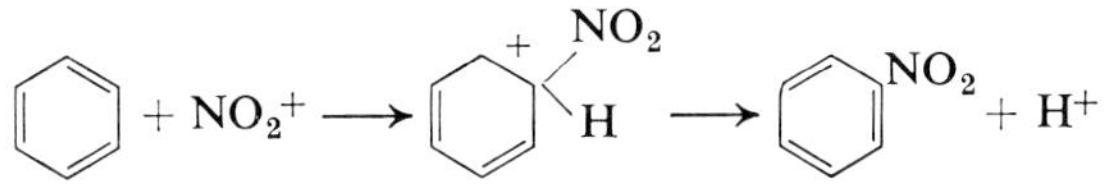

FIGURE 28.7 THE NITRATION OF BENZENE.

electrophile that attacks the ring and in the reaction that produces it. The electrophiles for these reactions are bromonium ion (for bromination), sulfur trioxide and methyl carbonium ion respectively, prepared by the following:

(1) $Br_2 + FeBr_3 \rightarrow FeBr_4^- + Br^+$

(1′) $H_2SO_4 + H_2SO_4 \rightarrow SO_3 + H_3O^+ + 1\ HSO_4^-$

(1″) $CH_3Cl + AlCl_3 \rightarrow AlCl_4^- + CH_3^+$

Notice that the iron added in a halogenation reaction to accomplish the substitution immediately reacts with the halogen present to form iron(III) halide which undergoes reaction 1 with the halogen present to yield the electrophile (i.e., bromonium or chloronium ion). Steps 2 and 3 of the alkylation proceed as follows:

(2″) benzene $+ CH_3^+ \longrightarrow$ (+, CH_3, H)

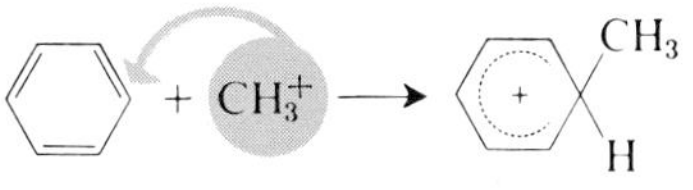

(3″) (+, CH_3, H) $\longrightarrow$ benzene$-CH_3 + H^+$

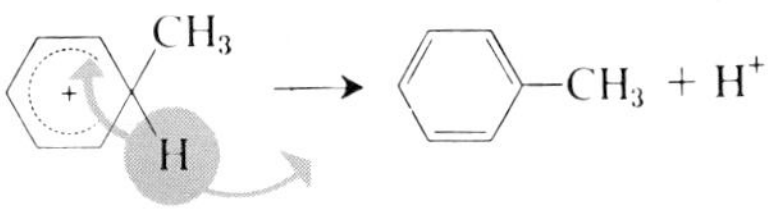

Because aromatic substitution results from the reaction of an aromatic hydrocarbon (i.e., benzene) with an electrophilic reagent, the overall process is commonly referred to as an **electrophilic aromatic substitution** reaction. All processes with mechanisms following this pattern of reactions are known as such.

28.6 ORIENTATION, ACTIVATION AND DEACTIVATION IN AROMATIC SYSTEMS

From exhaustive studies of the substitution reactions of substituted benzene compounds, it has been found that not only does the group (or groups) already on the ring direct an entering group to preferred positions on the ring, but it also increases or decreases the reactivity of (i.e., activates or deactivates) the ring toward substitution. Class I substituents, the ortho-para directing groups, activate the ring to which they are attached. The Class II substituents, meta directing groups, deactivate the ring. The substitution rates with respect to nitration of aniline and toluene with their ortho-para directing groups, NH_2 and CH_3, are compared in Table 28.2 with substitution rates of benzene and of a benzene derivative with a meta directing nitro group. It should be noted that the Cl group, although ortho-para directing, is deactivating.

TABLE 28.2 RATES OF NITRATION

Compound	Substituent Group	Class of Compound	Relative Rate of Substitution
Aniline	NH_2	I (activating)	ca. 10^6
Toluene	CH_3	I (activating)	25
Benzene	H	Standard	1
Chlorobenzene	Cl	I (deactivating)	0.033
Nitrobenzene	NO_2	II (deactivating)	ca. 10^{-6}

The activating effect of ortho-para directing groups is caused by the "feeding" of electrons into the pi system of the (aromatic) benzene ring which increases the electron density in the ring, especially at ortho and para positions. The deactivating effect of the meta directing group is due to the withdrawing of electrons from the ring, thus reducing the electron density in the ring and especially at these same, ortho and para, positions. The increased charge in the ring because of the presence of ortho-para groups attracts the electrophile more readily and especially at those positions to increase the rate of substitution. The decreased electron charge of the pi electrons on the ring bearing a meta directing group leaves lesser attraction for the approaching positive group, with the smallest attraction being at the ortho and para positions, such that the entering electrophile adds at the *lesser deactivated* meta positions.

Halo groups, illustrated by the chloro (Cl) group in Table 28.2, are exceptions to the generally observed activating effect of ortho-para directing groups. Because of their relatively large electronegativities, particularly of F and Cl, they tend to withdraw electrons from the ring, thus producing a net decrease in electron density

the two effects (electronegativity and resonance) oppose each other and net withdrawal is least at o- and p- positions.

in the aromatic ring. This withdrawal, however, is partially offset by a simultaneous "feeding in" effect to the ortho-para positions because of the resonance characteristic of these ortho-para directing groups. Of the effects the withdrawal due to electronegativity exceeds the "feeding in" due to resonance to produce an overall decrease in electron density in the ring, with the least decrease occurring at the ortho-para positions to maintain an ortho-para directing influence. The activating influence of the ortho-para directing amino group of aniline can be seen by considering three resonance structures that contribute to their resonance hybrid, the real aniline molecule,

$$\left[\bar{C}_6H_5{=}\overset{+}{N}H_2 \leftrightarrow {}^{-}C_6H_5{=}\overset{+}{N}H_2 \leftrightarrow \bar{C}_6H_5{=}\overset{+}{N}H_2 \right]$$

In these structures, the unshared pair of electrons on the nitrogen of the amino group has moved toward the ring to be shared with the carbon of the ring. These are three of the energetically most probable arrangements carrying a charge spearation in which a multiplicity of four bonds is on each carbon. The negative charge at the two ortho and the para positions in each of the structures is preserved in part in the real molecule, and any incoming electrophile is attracted more to one of these positions.

The deactivating effect of meta directing groups on the benzene ring toward substitution can be seen in three contributing structures to the resonance hybrid for the nitrobenzene molecule.

$$\left[\overset{+}{C}_6H_5{=}N\bar{O}_2 \leftrightarrow {}^{+}C_6H_5{=}N\bar{O}_2 \leftrightarrow \overset{+}{C}_6H_5{=}N\bar{O}_2 \right]$$

In these structures pi electrons associated with the nitro group are shared in a delocalized pi system which includes the pi electrons of the carbon atoms of the ring. In each of the three structures, which are the energetically most probable electron arrangements following electron withdrawal, a positive charge is placed on the ortho or para carbon. A fraction of this charge is preserved in the real molecule, which is somewhat intermediate in structure between these structures and others. The partial positive charge reduces the negative charge at those points, such that entering positive groups (electrophiles) attack the more negative meta positions.

28.7 SEVERAL REACTIONS OF ALKYL BENZENES

As shown earlier, the Friedel-Crafts reaction produces a hydrocarbon derivative of benzene, which is composed of two parts, an aromatic (aryl) and an alkyl groups. Reactions of these compounds may take place at either or both of these groups.

1. Oxidation. With toluene, which has a methyl group attached to the ring, oxidation affects the alkyl group without transforming the aromatic ring. Toluene is readily oxidized to benzoic acid with oxidizing agents such as potassium permanganate:

CH$_3$ (O) → COOH

Since alkanes are not oxidized readily under these same conditions, the aryl group must have had a modifying influence on the alkyl group, making it more subject to oxidation. When a longer alkyl group is present, oxidation also takes place with formation of benzoic acid.

CH$_2$CH$_2$CH$_3$ (O) → COOH

The ring in either case resists oxidation under these conditions, showing its inherently high degree of stability.

One application of this oxidation reaction has been in the determination of the structure of benzene derivatives. For example, if the structure of the hydrocarbon C_9H_{12} were under investigation, a study might be initiated with an experiment in which the hydrocarbon would be subjected to permanganate oxidation. As pointed out previously, such an oxidation would convert each alkyl side chain attached to the benzene ring into a —COOH group (carboxyl group). Therefore, if oxidation of C_9H_{12} gave benzene-1,3-dicarboxylic acid (see Chapter 31) as the only organic product, it could be assumed that its structure includes a benzene ring having two alkyl groups attached at positions 1 and 3 of the ring.

C_9H_{12} [0] → COOH, COOH *therefore* C_9H_{12} ≡ R, R

Benzene-1,3-dicarboxylic acid

A benzene ring without the two alkyl groups consists of the C_6H_4 unit, leaving the additional 3 carbons and 8 hydrogens of C_9H_{12} as components of the attached R groups. Since the two alkyl groups are required they must consist of one CH_3 and one CH_3CH_2 group. Because they are bonded to positions 1 and 3 of the benzene ring the original hydrocarbon C_9H_{12} is 3-ethyl-1-methylbenzene.

2. Halogenation. Halogenation of toluene is dependent upon the conditions. When toluene is chlorinated at low temperatures in the presence of a catalyst, substitution occurs on the ring as expected. At higher temperatures, when a catalyst is not used, substitution occurs in the side-chain methyl group. Mono-, di- or trisubstitution occurs, depending on the conditions and reaction time.

the reaction is a chain reaction as with CH_4 (Chapter 26).

CH$_3$ Cl_2/light/heat → CH$_2$Cl Cl_2 → CHCl$_2$ Cl_2 → CCl$_3$

Benzyl chloride — Benzal chloride — Benzotrichloride

The methyl group attached to the ring reacts with neither nitric acid nor sulfuric acid.

EXERCISES

28.1 Define: (a) aromatic character, (b) unsaturation, (c) saturation, (d) resonance, (e) orientation, (f) electrophile.

28.2 Indicate which of these molecules are planar: (a) $C_2H_2Cl_2$, (b) CH_2Cl_2, (c) C_2H_2, (d) C_2H_6, (e) $CH_2{=}CH{-}CH{=}CH_2$, (f) C_6H_6, (g) $C_{10}H_8$ (naphthalene), (h) $C_6H_6Cl_6$ (1,2,3,4,5,6-hexachlorocyclohexane).

28.3 Criticize [benzene ring drawn with three double bonds] as a symbol for benzene.

28.4 Describe the following structural features of benzene:

a. C—C—C bond angle
b. H—C—C bond angle
c. C—H bond length
d. C—C bond length

28.5 Name each of these by two correct IUPAC names:

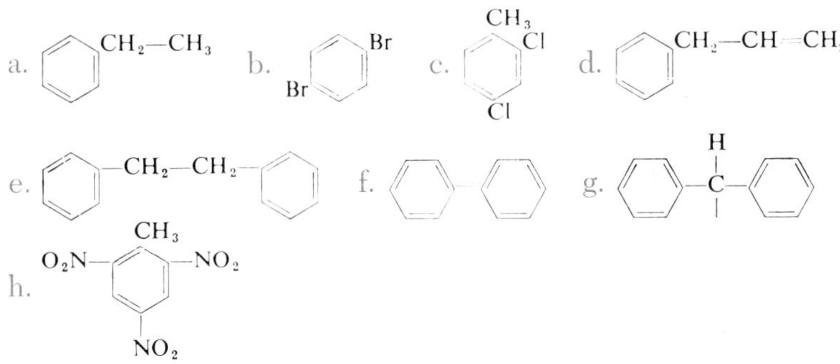

a. $C_6H_5{-}CH_2{-}CH_3$
b. benzene ring with Br, Br
c. benzene ring with CH_3, Cl, Cl
d. $C_6H_5{-}CH_2{-}CH{=}CH_2$
e. $C_6H_5{-}CH_2{-}CH_2{-}C_6H_5$
f. $C_6H_5{-}C_6H_5$
g. $C_6H_5{-}CH_2{-}C_6H_5$ (H above the central C)
h. benzene ring with CH_3, O_2N, NO_2, NO_2

28.6 Write an equation showing the conversion of benzene to cyclohexane. Show catalyst and special conditions over the arrow.

28.7 Write equations showing catalysts and special conditions for

a. the alkylation of toluene
b. the nitration of toluene
c. the sulfonation of toluene
d. the substitution of halogens on the ring of toluene

28.8 Explain the ortho-para and meta directive effect of ring substituents.

28.9 Write formulas for five meta and for six ortho-para directing groups.

28.10 Explain activation and deactivation as related to ortho-para and meta directing groups.

28.11 Write equations including over the arrow special conditions and catalysts for

a. the preparations of methylcyclohexane from benzene (two steps)
b. the preparation of m-chloronitrobenzene from benzene (two steps)
c. benzoic acid from benzene (two steps)

28.12 Explain how the ortho-para directing hydroxyl group functions.

28.13 How does the cyanide group, —CN, deactivate the benzene ring to which it is attached?

28.14 Why is the carbonium ion that is formed by adding a bromonium ion to the benzene ring more stable than the ethyl carbonium ion?

28.15 What function does aluminum chloride serve in the alkylation of benzene?

28.16 A compound, C_7H_7Cl, was oxidized with alkaline $KMnO_4$ and gave benzoic acid (C_6H_5COOH) as the only organic product. What is the structure of the compound? Explain.

28.17 An aromatic hydrocarbon, A, having a molecular weight of 106 was reacted with $Cl_2/FeCl_3$ to produce a compound having the molecular formula C_8H_9Cl. When it was vigorously oxidized with chromic acid, the only organic product isolated was 4-chlorobenzoic acid. What is the structure of compound A?

28.18 An aromatic hydrocarbon having the formula C_8H_{10} when oxidized with $KMnO_4$ gave benzene-1,2-dicarboxylic acid (o-phthalic acid) as the only organic product. What is the structure of this compound? Explain.

28.19 Compounds A and B are isomers having the formula C_8H_{10}. Treatment of A with $KMnO_4$ gives benzoic acid, C_6H_5COOH, whereas oxidation of B with chromic acid gives ortho-phthalic acid. Write structures for compounds A and B.

28.20 A compound A had the molecular formula $C_7H_5O_2NCl_2$. Oxidation of A gave 2-chloro-4-nitrobenzene-1-carboxylic acid. What is the structure of compound A? Explain.

28.21 How would you test for the presence of benzene of the following impurities, not more than one being present in any sample of benzene you are testing? (a) chlorobenzene (b) nitrobenzene (c) cyclohexane (d) 2-pentene (e) 1-pentyne.

28.22 The hydrocarbon $C_{10}H_8$ may be nitrated at a low temperature to give a compound B having the formula, $C_{10}H_7NO_2$. Vigorous oxidation of B gives compound C which contains nitrogen. 80 ml of 0.5M NaOH react with 0.2 mole of C. Reduction of B with iron and hydrochloric acid yields compound D which when oxidized gives a compound E containing no nitrogen. 80 ml of 0.5M NaOH also reacts with 0.2 mole of E. Give structural formulas for compounds B, C, D and E. Write equations for the reactions described and offer a possible explanation for the fact that B oxidizes one way and D another.

28.23 Three xylenes (dimethyl benzenes) boil at 144.1°C, 139.3°C and 138.4°C. The product boiling at 144.1°C gives two mononitrodimethylbenzenes when subjected to nitration. The xylene boiling at 139.3°C yields two mononitro derivatives and the xylene boiling at 138.4°C yields only one mononitro product. Write structures for all the mononitro products described above relating each one to the parent xylene.

28.24 Terephthalic acid is benzene-1,4-dicarboxylic acid. Which xylene yields terephthalic acid upon oxidation?

28.25 Which tribromobenzene yields three different tetrabromobenzenes when subjected to bromination? Which xylene yields only one?

SUGGESTED READING

Corwin, A. H. and Bursey, M. M.: *Elements of Organic Chemistry.* Addison-Wesley Publishing Company, Reading, Mass., 1965. Chapters 40, 41 and 42.

Duewell, H.: "Aromatic Substitution." J. Chem. Ed., *43*: 138, 1955.

Ferguson, L. N.: "The Orientation and Mechanism of Electrophilic Aromatic Substitution." J. Chem. Ed., *32*:42, 1955.

Hendrickson, J. B., Cram, D. J. and Hammond, G. S.: *Organic Chemistry,* Third Edition. McGraw-Hill Book Company, New York, 1970. Chapters 5 and 16.

Morrison, R. T. and Boyd, R. N.: *Organic Chemistry,* Second Edition. Allyn and Bacon, Inc., Boston, 1966. Chapters 10, 11 and 12.

Nelson, L. C.: "Aromatic Substitution by Free Radicals." J. Chem. Ed., *32*:606, 1955.

Newell, L. C.: "Faraday's Discovery of Benzene." J. Chem. Ed., *3*:1248, 1926.

Noller, C. R.: *Chemistry of Organic Compounds,* Third Edition. W. B. Saunders Company, Philadelphia, 1965. Chapters 20, 29 and 31.

TWENTY-NINE • ALCOHOLS, PHENOLS AND ETHERS

Oxygen is found in more organic compounds than any other element, except the ever present carbon and hydrogen. It is a constituent of many types of compounds, among which are **alcohols, phenols, ethers, aldehydes, ketones, acids, esters, carbohydrates** and **fats.** Many individual compounds of oxygen are found widely distributed in nature, many are of great biological significance and many have important industrial uses. Several chapters in this book are devoted to oxygen containing compounds. In this chapter compounds in which oxygen has two single bonds will be studied.

29.1 STRUCTURE AND COMPOSITION

Alcohols, phenols and ethers all can be viewed as derivatives of water. Alcohol is a water derivative in which an alkyl group such as methyl or ethyl has replaced one hydrogen atom (Fig. 29.1a). In a phenol a hydrogen of the water molecule is replaced by an aromatic ring (that is, the —OH group is attached directly to a carbon of an aromatic ring) (Fig. 29.1b). To form an ether both hydrogens of water are replaced by alkyl or aryl groups (Fig. 29.1c).

alcohols and phenols have one H attached to O. Ethers have only Cs attached to O.

These oxygen compounds may also be viewed as hydrocarbon derivatives formed by replacing a hydrogen of an aliphatic or aromatic hydrocarbon with an —OH (hydroxyl), an —OR (alkoxy) or an —OAr (aryloxy) group. The replacement of a hydrocarbon —H with an —OH group yields an alcohol or a phenol. Similar replacement

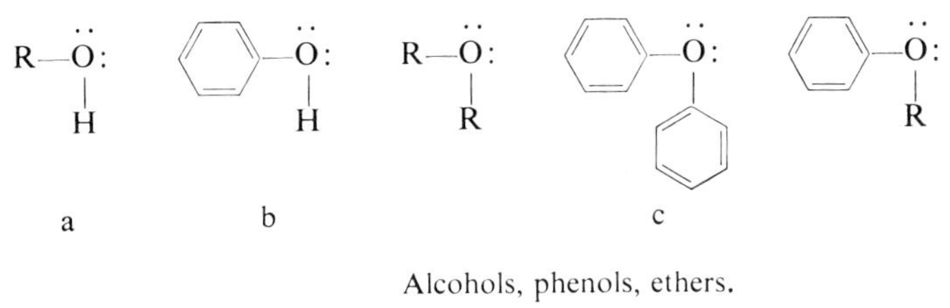

R—O—H (a) C₆H₅—O—H (b) R—O—R, C₆H₅—O—C₆H₅, C₆H₅—O—R (c)

FIGURE 29.1 ALCOHOLS(A), PHENOLS(B) AND ETHERS(C).

Alcohols, phenols, ethers.

with an alkoxy or an aryloxy group yields an ether. Note that the introduction of oxygen to form an alcohol or phenol does not alter the carbon-hydrogen ratio from that present in the parent hydrocarbon. For example, phenol, C_6H_5OH, obtained when one hydrogen of benzene, C_6H_6, is replaced with an —OH group, has the same 1:1 ratio of carbon to hydrogen as does benzene. Likewise, ethyl alcohol, C_2H_5OH, having an —OH in place of one hydrogen of ethane, C_2H_6, retains the 1:3 ratio of carbon to hydrogen found in ethane. Ethers have the carbon-hydrogen ratio of the hydrocarbon formed when the two hydrocarbon groups attached to the oxygen are united. Diethyl ether, $C_2H_5OC_2H_5$, to illustrate, has the carbon-hydrogen ratio (1:2.5) of n-butane, C_4H_{10}, obtained by joining the two ethyl groups attached to the oxygen.

Because of the structural relationship to both water and hydrocarbons the oxygen compounds bear resemblance to both. In terms of functional groups alcohols and phenols are distinguished by the presence of a hydroxyl (—OH) group, and ethers by the presence of oxygen bound to two carbons by single bonds (—O—).

29.2 STRUCTURE VERSUS PHYSICAL PROPERTIES

That the physical properties of a substance are a function of the structure of its molecules was shown in Chapter 25, in which the boiling points of the isomeric pentanes were compared. Branched chain isomers were shown to have lower boiling points than their unbranched isomers. With the introduction of a new element, oxygen, the relationship between structure and properties becomes more apparent. Not only are oxygen containing "hydrocarbons" very different in some properties from pure hydrocarbons but the various types of oxygen containing compounds differ. Al-

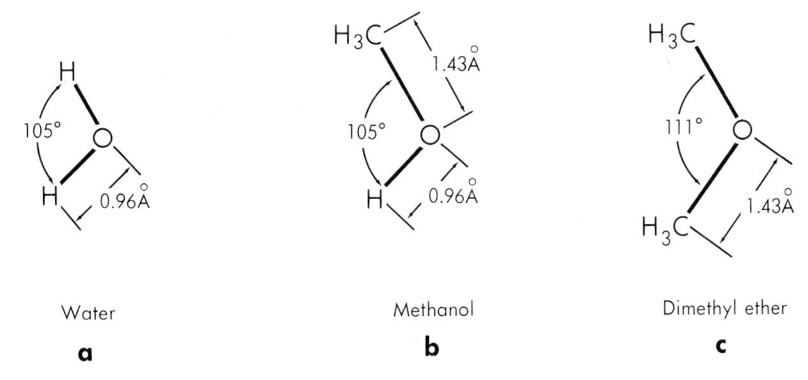

FIGURE 29.2 STRUCTURES OF WATER, METHANOL AND DIMETHYL ETHER.

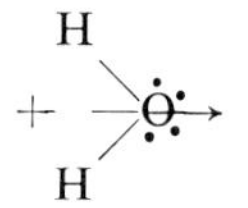

FIGURE 29.3 POLARITY OF WATER.

cohols, phenols and ethers are only three of numerous types of oxygen containing organic compounds—those types most like water. Because water is a bent molecule (Fig. 29.2a), and because oxygen has a greater electronegativity than hydrogen, the molecule is highly polar with the negative pole in the direction of the arrow in Figure 29.3. Alcohols, phenols and ethers are, like water, bent at the oxygen atom (Fig. 29.2b and c), and, because the alkyl and aryl groups have an attraction for electrons not very different from hydrogen, are also polar. The measured dipole moments in **Debye units** are given for water and several oxygen containing compounds in Table 29.1.

TABLE 29.1 DIPOLE MOMENTS OF OXYGEN CONTAINING COMPOUNDS

Compound	Formula	Dipole Moments (Debye Units)
Water	H_2O	1.84
Methyl alcohol (methanol)	CH_3OH	1.69
Ethyl alcohol (ethanol)	CH_3CH_2OH	1.69
Phenol	C_6H_5OH	1.73
Dimethyl ether	CH_3OCH_3	1.29
Diethyl ether	$CH_3CH_2OCH_2CH_3$	1.18
Methylphenyl ether	$CH_3OC_6H_5$	1.20
Diphenyl ether	$C_6H_5OC_6H_5$	1.00

Because the C to O bond length is greater than the length of the O to H bond (and because the angle is larger in ether molecules), the dipole moments decrease from water to alcohols to ethers. Because of their polarity, polar molecules in a sample of the pure substance tend to line up positive ends to negative ends and are attracted to one another and to other polar molecules with a greater attraction than exists between non-polar hydrocarbon molecules.

Similar to water molecules, molecules of these organic oxygen containing substances can participate in hydrogen bonding. A hydrogen bond (Chapter 5) is a weak attraction compared to the normal covalent bond and is a bond in which a hydrogen atom—or better, a proton—is held between the negative unshared electron pairs of two highly electronegative atoms—commonly oxygen, nitrogen and fluorine. Hydrogen bonding in water, alcohols, phenols and ethers is illustrated in Figure 29.4. Notice that each water molecule can be hydrogen bonded to four other molecules (and is, indeed, in ice) because there are two hydrogens attached to the oxygen atom in each molecule. Alcohol molecules can bond with only three other alcohol molecules on the average (if the proper arrangement in space can be attained). Ether molecules participate in hydrogen bonding only with molecules such as water, which can

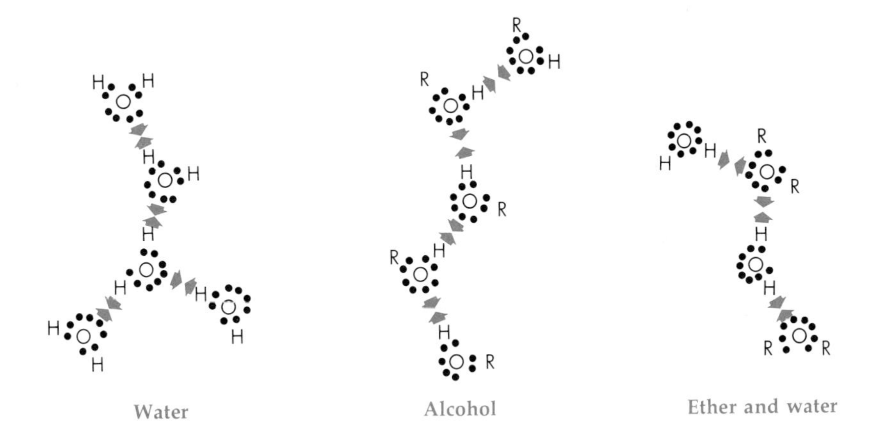

FIGURE 29.4 HYDROGEN BONDING.

furnish the needed hydrogen atom (Fig. 29.4c). Substances in which adjacent molecules are bonded together by hydrogen bonds are said to be associated. Water is highly associated, alcohols and phenols are less so and ethers are not associated in pure samples.

The lesser degree of hydrogen bonding in methanol compared to water is reflected in its boiling point, 64.9°C, which is considerably lower than the boiling point of water, even though water has a much smaller molecular weight. On the other hand, because of its hydrogen bonding, n-butanol has a boiling point of 118°C which is much greater than that of either diethyl ether, 35°C, or pentane, 36°C, although all three compounds have nearly identical molecular weights. Because of association, no alcohols are gases at room temperature and ordinary pressures, but the smaller ethers are.

association is through hydrogen bonding.

The smaller members of the oxygen containing compounds are souble in water (Table 29.2) because of association with the solvent molecules. This generalization includes the ethers, which, while unable to form hydrogen bonds with their neighbors in the pure state, participate in hydrogen bonding with water. This is illustrated by the close agreement in the solubilities in water of n-butyl alcohol and its isomer diethyl ether which are soluble to the extent of 7.9 and 7.5 grams per 100 grams of water respectively. Phenol, with a molecular weight of 94, has a solubility of 6.7 grams per 100 grams of water compared to the solubility of 0.59 grams for n-hexyl alcohol (molecular weight, 102).

These oxygen containing compounds are often thought of as having a dual nature—a Dr. Jekyll and Mr. Hyde personality. The oxy or hydroxyl group with its hydrogen bonding capability and polarity permits the compounds to dissolve in water and other polar solvents—a characteristic foreign to the non-polar, non-hydrogen bonding hydrocarbon portion of the molecule. This is true in some respects, as will be illustrated in the discussion of the reactions of the compounds. However, some properties appear not to be properties of either contrasting group but instead are an average of both. Although the oxy or hydroxyl group does not change, the hydrocarbon group or groups attached can vary greatly—from one carbon to many carbons, may be saturated or unsaturated, and may be aromatic or not—and may have differing effects on the properties of the compound.

TABLE 29.2 PHYSICAL PROPERTIES OF SOME ALCOHOLS, PHENOLS AND ETHERS

Name	Formula	Melting Point °C	Boiling Point °C	Density	Solubility g/100 g Water
Methyl alcohol	CH_3OH	−97.8	64.6	0.796	Completely
Ethyl alcohol	CH_3CH_2OH	−117	78.5	0.789	Completely
Isopropyl alcohol	$(CH_3)_2CHOH$	−88	82.3	0.785	Completely
n-Butyl alcohol	$CH_3(CH_2)_3OH$	−89	118	0.810	7.9
n-Hexyl alcohol	$CH_3(CH_2)_5OH$	−51	157.2	0.819	0.59
2,3-Hexanediol	$CH_3(CHOH)_2(CH_2)_2CH_3$	–	207	0.967	Completely
Phenol	C_6H_5OH, (benzene ring)–OH	41	182	1.07	6.7*
o-Cresol	o-$CH_3C_6H_4OH$, (benzene ring)–OH, CH_3	30	191.5	1.05	3.1
Dimethyl ether	CH_3-O-CH_3	−138.5	−24	–	Very soluble
Ethylmethyl ether	CH_3CH_2-O-CH_3	–	7.9	0.726	Soluble
Ethyl ether, Diethyl ether	CH_3CH_2-O-CH_2CH_3	−116	34.6	0.713	7.5

* 6.7 at 16°C; reasonably soluble at room temperature.

One thing is apparent: the longer the hydrocarbon chain, the greater is the molecular weight of the hydrocarbon portion of the molecule, and the less soluble the compound is in water. Addition of more oxygen containing groups tends to make the compound more soluble. It is a common rule that hydroxyl containing compounds will be soluble in water if the carbon to hydroxyl group ratio does not exceed 3/1. By this rule methyl, ethyl and propyl alcohols should be soluble, and also the two smallest ethers, dimethyl and methylethyl (if the rule may be applied to oxy groups). No phenols should be soluble unless they contain more than one hydroxyl group, for the smallest member contains six carbons. These "predictions" are in general agreement with the experimental findings.

Because of their miscibility with both hydrocarbons, owing to the alkyl group, and water, the smaller alcohols and the smallest ethers are used as solvents to bring organic and polar compounds together in the same phase. Many drugs which are insoluble in water alone are dispersed and stored in alcohol solutions called *tinctures*. Diethyl ether, with its limited solubility in water, in accord with its carbon to oxygen ratio of 4/1 and its large capability for dissolving non-polar substances, is used to extract hydrocarbon-like compounds from plant and animal materials. When shaken with mixtures, the ether dissolves the non-polar materials and collects in a layer above the aqueous layer. After the ether layer has formed and is separated from the lower layer, the volatile ether is distilled, leaving behind the extracted material.

ethers form peroxides in air and may explode when distilled to dryness.

Ethers, with the slightly heavier and slightly smaller oxygen atom in the chain with carbon and with a polarity that tends to pull the molecules more tightly together, are more dense than hydrocarbons but less dense than corresponding alcohols which have hydrogen bonding in addition. Phenols, because of the compact benzene ring, are more dense than water.

29.3 ACIDITY AND BASICITY

The marked influence of oxygen on the physical properties of alcohols, phenols and ethers extends to their chemical properties as well. A comparison of the relative acidities (or basicities) of these compounds serves to illustrate the nature and extent of this influence.

Water reacts with sulfuric acid and other strong acids by accepting a proton to form a hydronium ion by attachment of a proton to the oxygen of water as follows:

$$H_2O + H_2SO_4 \rightarrow H_3O^+ + HSO_4^-$$

In accepting a proton water behaves as a Brønsted base. In the same sense alcohols and ethers show basic properties when treated with sulfuric acid.

water, methanol and ethers with an alkyl group attached to oxygen are stronger bases than HSO_4^-.

$$CH_3OH + H_2SO_4 \rightarrow CH_3\overset{+}{O}H_2 + HSO_4^-$$

$$CH_3OCH_3 + H_2SO_4 \rightarrow CH_3\underset{\displaystyle H}{\overset{+}{O}}CH_3 + HSO_4^-$$

$$CH_3OAr + H_2SO_4 \rightarrow CH_3\underset{\displaystyle H}{\overset{+}{O}}Ar + HSO_4^-$$

The alcohol or ether in each case acts as a base by accepting a proton from sulfuric acid. The protonated alcohol or ether is called an **oxonium ion.** Because of oxonium ion formation many oxygen containing compounds readily dissolve in sulfuric acid without any other change. Solubility in concentrated sulfuric acid is used as a test to distinguish alcohols and ethers from saturated hydrocarbons. Although oxonium salts of alcohols and ethers (and of water) normally exist only in solution, the crystalline salts of certain ethers have been isolated at low temperatures. Because electron withdrawal occurs from resonance interaction of the oxygen electrons with the pi electrons of the aromatic ring, phenol and diaryl ethers exhibit little tendency to accept protons and are considerably less basic than are alcohols and other ethers. Diphenyl ether, for example, fails to form an oxonium salt in sulfuric acid. Phenol reacts predominantly as an acid through proton loss from the —OH group present.

Formation of an oxonium ion when a strong acid is present in a reaction medium is often the first stage in the reaction of an alcohol or an ether. Important reactions of alcohols, such as dehydration to form alkenes and the replacement of a hydroxyl group with a halogen atom using a hydrohalic acid (to be discussed in a later section), are dependent upon oxonium ion formation. Ether cleavage with strong acids is limited by oxonium ion formation. Alcohols and phenols, because they have a hydroxyl group, lose protons, that is, may act as Brønsted acids. Ethers, with all hydrogens bonded to carbons, do not lose protons to act as acids except under unusual circumstances.

The polarity of the hydrogen-oxygen bonds in water gives water sufficient acidic character to react readily with active metals. Water reacts rapidly with sodium to give sodium hydroxide and hydrogen. Alcohols such as ethanol react in the same way but less rapidly to give hydrogen and the sodium salt of the alcohol.

$$2\ Na + 2\ H_2O \rightarrow 2\ NaOH + H_2$$

$$2\ Na + 2\ CH_3CH_2OH \rightarrow 2\ NaOCH_2CH_3 + H_2$$

With phenol, hydrogen gas and sodium phenoxide are formed.

$$2\ Na + C_6H_5OH \rightarrow 2\ NaOC_6H_5 + H_2$$

With sodium essentially 100 per cent displacement of hydrogen occurs. The products obtained (all bases) called conjugate bases, are OH^-, $CH_3CH_2O^-$ and OPh^-. The strongest base arises from reaction of the weakest acid. Sodium ethoxide ($NaOCH_2CH_3$) and water react when mixed to form ethyl alcohol:

$CH_3CH_2O^-$ has greater attraction for a proton than does OH^- and is a stronger base and a nucleophile.

$$CH_3CH_2O^- + H_2O \rightarrow CH_3CH_2OH + OH^-$$

Water, a stronger acid (proton donor) than ethyl alcohol, releases a proton to the ethoxide ion, $CH_3CH_2O^-$. If the same reaction is performed using phenol instead of water, ethyl alcohol is obtained by a similar reaction:

$$CH_3CH_2O^- + PhOH \rightarrow CH_3CH_2OH + PhO^-$$

Phenol is a stronger acid than ethyl alcohol—sufficiently strong that, when sodium hydroxide solution is added to phenol, neutralization occurs with the formation of water and the sodium salt of phenol:

$$PhOH + OH^- \rightarrow PhO^- + H_2O$$

In this case, the hydroxide ion abstracts a proton from the phenol, showing that phenol is more acidic than water. In summary it can be concluded that the order of acidity as proton donors is

$$\text{phenol} > \text{water} > \text{ethyl alcohol}$$

This is a generally observed pattern for all alcohols and phenols. Phenols in general are more acidic than water and alcohols are less acidic, from which it follows that phenols usually are more acidic than alcohols. Further discussion of reactions will be deferred until nomenclature is reviewed in the next section.

29.4 NOMENCLATURE AND CLASSIFICATION OF ALCOHOLS

Many alcohols are best known by common names which are obtained by adding the term alcohol to the name of the alkyl group

TABLE 29.3 NAMES OF SOME ALCOHOLS

Formula	Common Name	IUPAC Name
CH_3OH	Methyl alcohol	Methanol
$CH_3—CH_2OH$	Ethyl alcohol	Ethanol
$CH_3—CH_2—CH_2OH$	n-Propyl alcohol	1-Propanol
$CH_3—CH—OH$ | CH_3	Isopropyl alcohol	2-Propanol
$CH_3CH_2CH_2CH_2OH$	n-Butyl alcohol	1-Butanol
$CH_3—CH_2—CHOH$ | CH_3	Sec-butyl alcohol	2-Butanol
$CH_3—CH—CH_2OH$ | CH_3	Isobutyl alcohol	2-Methyl-1-propanol
CH_3 | $CH_3—C—OH$ | CH_3	Tert-butyl alcohol	2-Methyl-2-propanol
$CH_3—CH_2—CH_2—CH_2—CH_2—OH$	n-Amyl alcohol (n-Pentyl alcohol)	1-Pentanol
$CH_3—CH—CH_2—CH_2—OH$ | CH_3	Isoamyl alcohol (Isopentyl alcohol)	3-Methyl-1-butanol
H_2 C / \\ H_2C CH_2 | | H H_2C C< \\ / OH C H_2	Cyclohexyl alcohol	Cyclohexanol

attached to the hydroxyl group. The common names of some alcohols appear in Table 29.3. At this point one should be familiar with all the alkyl groups named in this table, except those of the amyl alcohols, which are only specific names, and the butyl (four carbon) alcohols.

$CH_3—CH_2—CH_2—CH_2—$ n-butyl

$CH_3—CH—CH_2—$ isobutyl
|
CH_3

(n-butyl and isobutyl: primary groups)

$CH_3—CH_2—CH—$ secondary butyl (sec-butyl)
|
CH_3

CH_3
|
$CH_3—C—$ tertiary butyl (tert-butyl)
|
CH_3

Although neither n-butyl nor isobutyl is called primary butyl alcohol, both are primary alcohols. The terms **primary, secondary** and **tertiary** refer to the number of alkyl carbons to which the carbon bearing the functional group is attached; primary indicates that the carbon is attached to one other (or none other), secondary, to two other carbons, and tertiary, to three. Alcohols and other series of compounds are classified primary, secondary and tertiary by the structure of the carbon to which the functional group is attached. The distinction is useful, because these classes with the same functional group often act differently.

recall the classes of carbons and hydrogens discussed in Chapter 25.

The IUPAC rules of nomenclature may be extended to cover the naming of alcohols with one hydroxyl group (—OH) as follows:

1. Choose the longest carbon chain *which includes the carbon attached to the hydroxyl group.*
2. Name the chain as an alkane, but change the name ending by replacing *e* with *ol.* Officially, alcohols are alkanols.
3. Number the chain to give the carbon with the hydroxyl group the lowest number. The OH group takes precedence in numbering over any multiple bond that may be present.
4. Name all branches in alphabetical order and indicate the position of attachment with numbers prefixed as before. Check the IUPAC names appearing in Table 29.3 for agreement with the rules.

Other examples are:

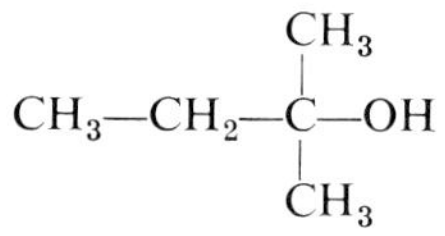

$$\begin{array}{c} \qquad\qquad CH_3 \\ \qquad\qquad | \\ CH_3—CH_2—C—OH \\ \qquad\qquad | \\ \qquad\qquad CH_3 \end{array}$$

2-methyl-2-butanol, a tertiary alcohol

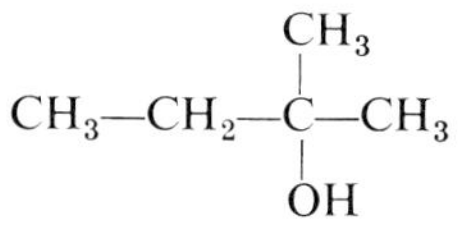

$$\begin{array}{c} \qquad\qquad CH_3 \\ \qquad\qquad | \\ CH_3—CH_2—C—CH_3 \\ \qquad\qquad | \\ \qquad\qquad OH \end{array}$$

(the same compound)

$$\begin{array}{l} \qquad\qquad\quad CH_3 \\ \qquad\qquad\quad | \\ CH_3—CH—CH—CH—OH \\ \qquad\quad | \qquad\qquad | \\ \qquad\quad Cl \qquad\quad CH_3 \end{array}$$

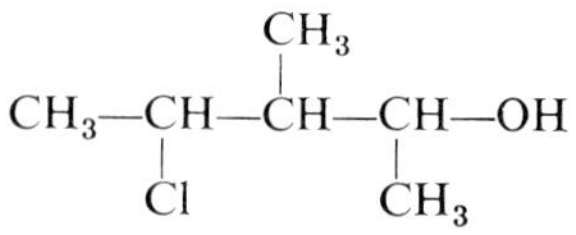

4-chloro-3-methyl-2-pentanol, a secondary alcohol

$$\begin{array}{l} \quad\; CH_3 \\ \quad\; | \\ CH_2{=}C—CH_2—CH_2OH \end{array}$$

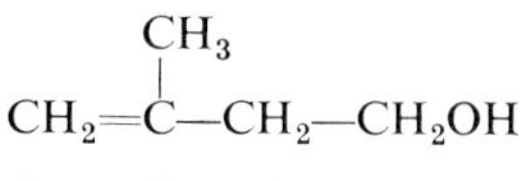

3-methyl-3-butene-1-ol, a primary alcohol

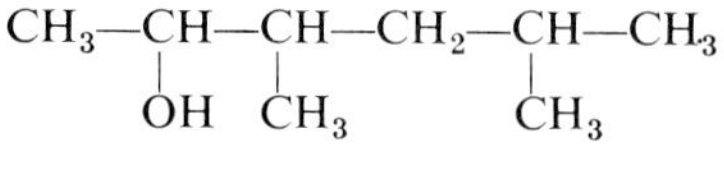

$$\begin{array}{l} CH_3—CH—CH—CH_2—CH—CH_3 \\ \qquad\quad | \qquad | \qquad\qquad\quad | \\ \qquad OH \;\; CH_3 \qquad\quad CH_3 \end{array}$$

3,5-dimethyl-2-hexanol, a secondary alcohol

Alcohols having more than one hydroxyl group are commonly known as di-, tri- or polyhydroxy alcohols. They are officially known as diols, triols or polyols. Some common polyols are:

$CH_2OH—CH_2OH$	ethylene glycol	1,2-ethanediol
$CH_3—CHOH—CH_2OH$	propylene glycol	1,2-propanediol

Both glycols (diols) named are used as permanent antifreezes.

what is meant by a permanent antifreeze?

$CH_2OH—CHOH—CH_2OH$	glycerol	1,2,3-propanetriol
$CH_2(OH)—CH(OH)—CH(OH)—CH(OH)—CH(OH)—CH_2(OH)$	mannitol	1,2,3,4,5,6-hexanehexol

29.5 REACTIONS OF ALCOHOLS

The **hydroxyl group** is responsible for the characteristic behavior of alcohols. At least four types of reactions are common to alcohols. These include reactions which result in (1) replacement of the hydrogen atom of the hydroxyl group, (2) replacement of the —OH group itself through cleavage of the alcohol carbon-oxygen bond, (3) dehydration of the alcohol and (4) oxidation of the alcohol to an aldehyde, ketone or carboxylic acid.

1. Replacement of the Hydrogen Atom of the —OH Group

which reacts faster with sodium, 2-methyl-2-pentanol or 2-methyl-3-pentanol?

The hydrogen of the hydroxyl group of alcohols reacts quite differently from other hydrogens present in the molecule. As shown in Section 29.3 of this chapter, the hydrogens attached to oxygen, in contrast to those attached to carbon, are weakly acidic. As shown in Section 29.3, the hydroxyl hydrogen is not acidic enough to react with strong bases such as sodium hydroxide, but it is replaced slowly with sodium or potassium metal liberating hydrogen.

$$2\ ROH + 2\ Na \rightarrow 2\ RONa + H_2$$

(a) Primary alcohols react most rapidly; tertiary least rapidly. The difference in the rate of reaction with sodium observed for primary, secondary and tertiary alcohols arises because they differ in acid strengths. This difference can be accounted for on the basis of the number of alkyl groups attached to the oxygen bearing carbon of each type of alcohol. An alkyl group, an "electron releasing" group (see Chapter 27), when attached to carbon induces in that carbon a more negative (less positive) character than would be the case were a hydrogen atom to occupy the same position in the molecule. The oxygen bearing carbon of the different alcohols, and therefore the oxygen atom of the alcohol itself, becomes more negative (less positive) as the number of alkyl groups attached to the carbon is increased. A primary alcohol has only one alkyl group, a secondary has two and a tertiary alcohol has three alkyl groups attached to the oxygen bearing carbon. Of the three classes of alcohols, the primary should have the least negative (most positive) and the tertiary the most negative (least positive) oxygen atom. As a result, the primary alcohol binds its proton least tightly and reacts most rapidly with sodium metal. A secondary alcohol exhibits an intermediate behavior in both acidity and reactivity.

which alkoxide reacts most rapidly with water, CH_3O^-, $(CH_3)_2CHO^-$ or $(CH_3)_3CO^-$?

The **alkoxide** formed when sodium reacts with an alcohol hydrolyzes in the presence of water re-forming the alcohol and liberating sodium hydroxide. The recommended method for the disposal of sodium wastes is to dissolve them in a secondary alcohol with which the sodium reacts relatively slowly.

(b) Alcohols react with hydroxyl containing strong acids such as HNO_3 ($HONO_2$) and H_2SO_4 to form **esters.** An ester is the other product formed when an H from the hydroxyl group of the alcohol combines with the —OH group of an acid to form water. The net result is replacement of the H of the alcohol —OH with $—NO_2$ or $—SO_3H$, depending on the inorganic acid used. Instead of an —OH, an $—ONO_2$ or $—OSO_3H$ group is attached to the R— in an ester.

note that the —OH of the water formed comes from the acid used and the H from the alcohol. How could you prove this?

$$ROH + HONO_2 \rightarrow RONO_2 + H_2O$$

$$ROH + HOSO_2OH \rightarrow ROSO_2OH + H_2O$$

The reaction of alcohols with nitric acid has wide application in the manufacture of explosives and drugs. For example, nitroglycerin, which structurally is glyceryl trinitrate, is formed when glycerol reacts with either concentrated or dilute nitric acid at low temperature

$$\begin{array}{l} CH_2OH \\ | \\ CHOH \\ | \\ CH_2OH \end{array} + 3\ HONO_2 \rightarrow \begin{array}{l} CH_2ONO_2 \\ | \\ CHONO_2 \\ | \\ CH_2ONO_2 \end{array} + 3\ H_2O$$

Glycerine — Glyceryl trinitrate (nitroglycerine)

Gun cotton, another useful explosive, results from the reaction between cellulose (in cotton), a polyhydroxy compound, and nitric acid. Esters of other inorganic acids include those of nitrous acid and phosphoric acid. Esters of phosphoric acid are of particular importance in life processes. Glucose (see Chapter 36), the sugar which is used for energy by most living organisms, is utilized by them only after its conversion into an ester of phosphoric acid.

2. Replacement of the —OH Group

(a) The replacement of —OH with a halogen atom by reaction of an alcohol with hydrohalic acids, HBr, HCl or HI, to produce an alkyl halide is a typical reaction of alcohols. The reaction between HBr and ethanol is as follows:

$$CH_3CH_2OH + HBr \rightarrow CH_3CH_2Br + H_2O$$

The same reaction with t-butyl alcohol is

$$CH_3-\overset{\displaystyle CH_3}{\underset{\displaystyle CH_3}{\overset{|}{\underset{|}{C}}}}-OH + HBr \rightarrow CH_3-\overset{\displaystyle CH_3}{\underset{\displaystyle CH_3}{\overset{|}{\underset{|}{C}}}}-Br + H_2O$$

The order of reactivity of alcohols toward replacement of the hydroxyl group with a halide by a given hydrohalic acid is R_3COH, tertiary alcohol, > R_2CHOH, secondary alcohol, > RCH_2OH, primary alcohol. The use of Lucas reagent, a mixture of concentrated HCl in zinc chloride solution, is based on this difference in reactivity. Tertiary alcohols react rapidly with the reagent, secondary alcohols react slowly and primary alcohols hardly at all, even when heated. The order of reactivity of the hydrohalic acids with respect to replacement of the —OH group is HI > HBr > HCl.

The replacement of the —OH of ethanol with a halide is accomplished by a sequence of reactions involving **displacement** or **substitution** (Chapter 26). The steps are as follows:

$$1.\quad CH_3CH_2OH + HBr \rightarrow CH_3CH_2\overset{+}{O}H(-H) + Br^- \quad \text{(fast)}$$

most primary and secondary alcohols react by an S_N2 mechanism.

$$2.\quad Br^- + CH_3CH_2\overset{+}{O}H(-H) \rightarrow \left[\overset{\delta-}{Br} \cdots\cdots C(CH_3)(H)(H) \cdots\cdots \overset{\delta-}{O}H_2 \right] \rightarrow CH_3CH_2Br + OH_2 \quad \text{(slow)}$$

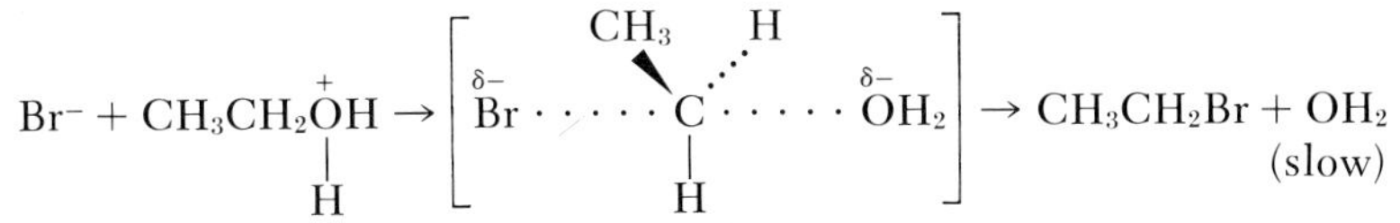

In step 1 the oxonium salt of the alcohol is formed by the addition of H^+ from the HBr to the oxygen of the alcohol, the same reaction seen in the discussion of the mechanism of the preparation of alkenes from alcohols (Chapter 27). Because alcohols are bases (proton acceptors) this reaction is usually the first reaction undergone by alcohols when an acid is present. The attachment of the proton to oxygen converts the —OH into the much less basic group, $—OH_2$. Step 2 involves attack on the carbon attached to the OH_2 group by the negative bromide ion (Br^-), a nucleophile. In a *concerted process*, as a bond between bromide ion and carbon develops, the bond between the oxygen of the OH_2 and carbon atom stretches and weakens. The reaction is consummated when the C—O bond is broken, releasing water accompanied by the simultaneous bonding of the bromide to carbon to form ethyl bromide. In essence the Br^- (the nucleophile) displaces (substitutes for) the —OH attached to the ethyl group. This displacement reaction is therefore classified as a **nucleophilic substitution** reaction. Since the rate of formation of ethyl bromide is dependent upon the presence and concentration of two substances, bromide and oxonium ion, i.e., it is a bimolecular reaction, this pathway is designated as an **S_N2 reaction.** The structure shown in the brackets represents the so-called **"transition state"** which must be attained before products can be formed in step 2.

The replacement of the —OH of t-butyl alcohol occurs more rapidly and by a pathway which differs from that followed when ethyl alcohol reacts. The series of steps by which t-butyl alcohol is transformed into t-butyl bromide is

$$1.\quad (CH_3)_3C—OH + HBr \rightarrow (CH_3)_3C—\overset{+}{O}H_2 + Br^- \quad \text{(fast)}$$

tertiary alcohols react by an S_N1 mechanism.

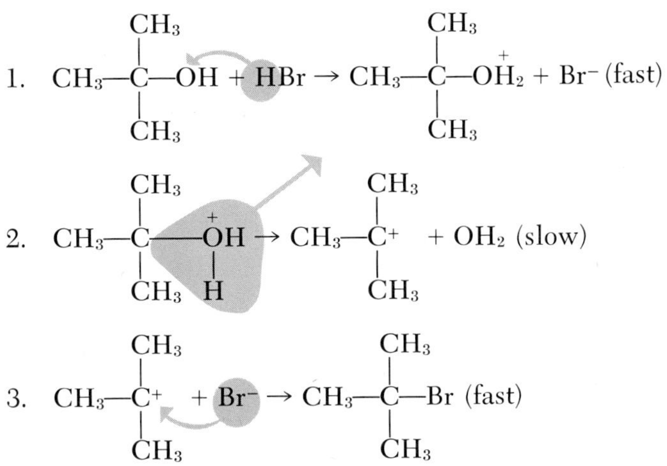

$$2.\quad (CH_3)_3C—\overset{+}{O}H(-H) \rightarrow (CH_3)_3C^+ + OH_2 \quad \text{(slow)}$$

$$3.\quad (CH_3)_3C^+ + Br^- \rightarrow (CH_3)_3C—Br \quad \text{(fast)}$$

The reaction of HBr with t-butyl alcohol to form the oxonium ion (step 1) is the same as with ethanol. In step 2 t-butyl oxonium ion undergoes ionization, losing OH_2 to form t-butyl carbonium ion prior to reaction with or attack by the bromide ion. In step 3 the t-butyl carbonium ion reacts rapidly with the Br^- to form t-butyl bromide. The ultimate result of the reaction series, as with ethanol, is replacement of the —OH with —Br. Since replacement is with a nucleophile this pathway or sequence is also referred to as a **nucleophilic substitution** reaction. Because the rate of t-butyl bromide formation is determined by the rate of ionization of the oxonium ion (one substance only—unimolecular), this series of changes is designated as an **S_N1 reaction** to distinguish it from the S_N2 reaction (two molecule—bimolecular) of ethyl alcohol. The S_N1 and S_N2 mechanisms are reaction pathways commonly encountered in organic chemistry. Other examples will be identified in subsequent chapters.

The equation for the reaction shows what happens. The mechanism provides an explanation for the manner in which the transformation is accomplished in terms of the changes which individual molecules undergo.

(b) Reaction of an alcohol with *phosphorus trichloride, tribromide* or *triiodide*, or with *phosphorus pentachloride* or *thionyl chloride* results in the replacement of an —OH group with a halogen atom. In each case an alkyl halide is formed:

each reaction follows a mechanism which is not included here because of its complexity.

$$3CH_3CH_2OH + PBr_3 \rightarrow 3CH_3CH_2Br + P(OH)_3$$

$$CH_3CH_2CH_2OH + PCl_5 \rightarrow CH_3CH_2CH_2Cl + POCl_3 + HCl$$

$$CH_3\underset{|}{\overset{CH_3}{C}}HOH + \underset{\text{Thionyl chloride}}{SOCl_2} \rightarrow CH_3\text{—}\overset{CH_3}{\overset{|}{C}}HCl + SO_2 + HCl$$

Other tertiary, secondary and primary alcohols react similarly with the phosphorus and thionyl chlorides.

3. Dehydration of Alcohols (Removal of Water)

Monohydroxyl alcohols lose water under the proper experimental conditions in the presence of dehydrating agents to yield ethers or alkenes. The dehydration of ethanol is a classical reaction which may be used to illustrate the transformations which are likely. When ethanol is heated to a relatively high temperature (160°C) in the presence of an excess of concentrated sulfuric acid, a dehydrating agent, ethylene is obtained as a result of **intramolecular** loss of water:

intramolecular loss of water leads to alkenes (Chapter 27).

$$CH_3CH_2OH \xrightarrow[\text{(Sulfuric acid)}]{160°} CH_2\text{=}CH_2 + H_2O$$

The mechanism of this reaction (an elimination reaction) is outlined in Chapter 27 in the section concerned with the sources of alkenes.

At a lower temperature and with the ratio of alcohol to sulfuric acid markedly increased, diethyl ether is obtained as the principal product.

$$2CH_3CH_2OH \xrightarrow[\text{(Sulfuric acid)}]{140°} CH_3CH_2OCH_2CH_3 + H_2O$$

Here **intermolecular** loss of water takes place. The sequence of steps by which diethyl ether is formed from ethyl alcohol under these conditions is as follows:

1. $CH_3CH_2OH + H_2SO_4 \rightleftarrows CH_3CH_2\overset{+}{O}H_2 + HSO_4^-$

2. $CH_3CH_2OH + CH_3CH_2\overset{+}{O}H_2 \rightleftarrows \left[CH_3CH_2{-}\overset{\delta+}{O}(H)\cdots C(CH_3)(H)(H)\cdots\overset{\delta+}{O}H_2\right]$

$\rightleftarrows CH_3CH_2{-}\overset{+}{O}(H){-}CH_2CH_3 + OH_2$

3. $CH_3CH_2{-}\overset{+}{O}(H){-}CH_2CH_3 \rightleftarrows CH_3CH_2{-}O{-}CH_2CH_3 + H^+$

This is another example of the S_N2 type process wherein a second molecule of alcohol in step 2 acts as a nucleophile to displace the water from the carbon of an oxonium ion formed when sulfuric acid reacts with ethanol in step 1. Step 2 parallels very closely the S_N2 reaction with bromide ion discussed in the last section. Step 3 yields the free ether by proton loss. The overall result of this sequence is intermolecular loss of water. Primary alcohols react in this way to form simple ethers—those in which the alkyl groups attached to the oxygen are identical. If a mixture of two alcohols is used a mixture of ethers is obtained from which the desired product or products must be separated.

In any case whenever alcohols are subjected to sulfuric acid both intramolecular and intermolecular loss of water occurs. Both ethers and alkenes are formed. With primary alcohols ether formation predominates. Secondary alcohols dehydrate more readily than do primary alcohols and give a higher percentage of alkene. Tertiary alcohols dehydrate intramolecularly so easily that tertiary alkyl ethers are seldom obtained. Special methods are used to obtain ethers from tertiary alcohols and from some secondary alcohols.

4. Oxidation of Alcohols

Primary and secondary alcohols are readily oxidized to **aldehydes** and **ketones:**

$$\underset{\text{1-Propanol}}{CH_3CH_2CH_2OH} + [O] \xrightarrow[OH^-]{KMnO_4} \underset{\text{An aldehyde}}{CH_3{-}CH_2{-}\overset{O}{\overset{\|}{C}}{-}H} + H_2O$$

$$\underset{\text{2-Propanol}}{CH_3{-}\underset{OH}{\overset{H}{C}}{-}CH_3} + [O] \xrightarrow[H_2SO_4]{Na_2Cr_2O_7} \underset{\text{A ketone}}{CH_3{-}\overset{O}{\overset{\|}{C}}{-}CH_3} + H_2O$$

These equations are not balanced. The symbol O is used to indicate that oxygen is gained from an **oxidizing agent.** The notation $\xrightarrow[OH^-]{KMnO_4}$ indicates that the oxidizing agent is a basic solution of potassium permanganate.

Unless the aldehyde is removed from the oxidizing solution, it is oxidized further to a carboxylic acid:

$$CH_3CH_2—\overset{\overset{O}{\|}}{C}—H + [O] \xrightarrow[OH^-]{KMnO_4} \underset{\text{An acid}}{CH_3—CH_2—\overset{\overset{O}{\|}}{C}—OH}$$

The oxidation of primary and secondary alcohols to aldehydes and ketones amounts to a **dehydrogenation:**

biological oxidations often involve a dehydrogenation in which H is transferred to the oxidizing agent.

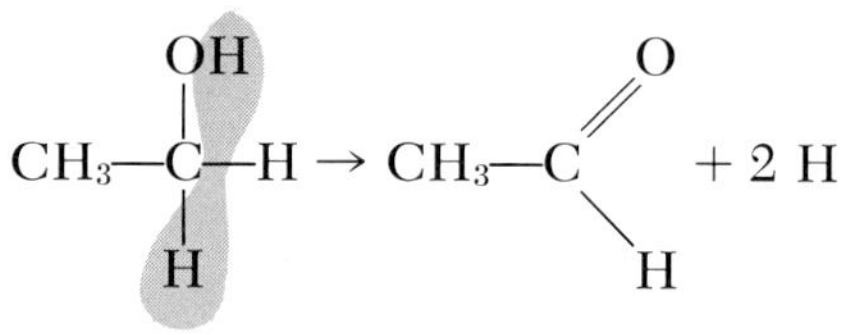

$$CH_3—\underset{\underset{H}{|}}{\overset{\overset{OH}{|}}{C}}—H \rightarrow CH_3—C\begin{matrix}\diagup O\\ \diagdown H\end{matrix} + 2\ H$$

A tertiary alcohol has no hydrogen on the carbon to which the hydroxyl group is attached and cannot be oxidized in this manner. However, under vigorous conditions tertiary alcohols undergo oxidative degradation to form compounds having fewer carbon atoms than the alcohol. Some of the products obtained when tertiary butyl alcohol is heated for several hours with an acid solution of sodium dichromate are acetic acid ($CH_3—\overset{\overset{O}{\|}}{C}—OH$), acetone ($CH_3—\overset{\overset{O}{\|}}{C}—CH_3$), carbon dioxide and water.

29.6 PREPARATION OF ALCOHOLS

Alcohols may be prepared in a variety of ways. The discussion includes methods applicable for the preparation of only one particular alcohol, as well as general methods for preparing any one of a series of alcohols.

1. Specific Methods

(a) *Methyl alcohol (methanol),* the simplest of the alcohols, is obtained commercially by the combination of hydrogen and carbon monoxide, using a specific catalytic process. In this method, water gas (a mixture of carbon monoxide and hydrogen formed by spraying hot coke with steam), with some added hydrogen, is forced at high pressure (200 atmospheres) over zinc chromite catalyst at a temperature near 400°C.

$$CO + 2\ H_2 \xrightarrow[ZnCrO_2\ catalyst]{200\ atm/400°C} CH_3OH$$

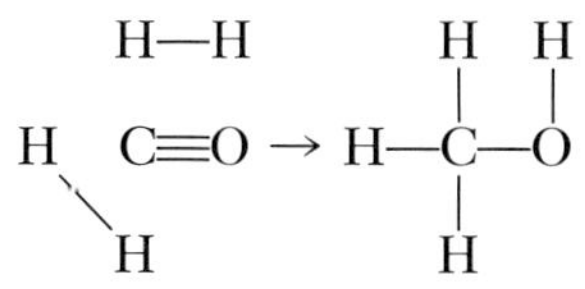

$$\begin{matrix} & H—H & \\ H & C{\equiv}O & \\ \diagdown & & \\ & H & \end{matrix} \rightarrow H—\underset{\underset{H}{|}}{\overset{\overset{H}{|}}{C}}—\overset{\overset{H}{|}}{O}$$

(b) *Ethyl alcohol (ethanol)*, another very common and widely used alcohol, is obtained by a specific procedure known as **fermentation.** In this method, sugar is transformed into ethyl alcohol and carbon dioxide under the influence of a mixture of **biological catalysts** called **enzymes** which are found in, and formed by growing yeast cells (see Chapter 37). The mixture of yeast enzymes is called "zymase." Commercially, starch and "black strap molasses" are used as starting materials. Starch is first converted into sugar by an enzyme found in freshly sprouted barley (malt):

$$\text{Starch} \xrightarrow[\text{Water}]{\text{Malt}} \underset{\text{Simple sugar}}{yC_6H_{12}O_6}$$

where y is any number.

$$yC_6H_{12}O_6 \xrightarrow{\text{Zymase}} 2yC_2H_5OH + 2yCO_2$$

Fermentation of sugar to produce ethanol is only one example of a large number of industrial processes that utilize fermentation. Many antibiotics such as penicillin and hormones such as cortisone, along with a variety of other substances, are obtained commercially by fermentation.

2. General Methods

most of the ethyl alcohol used today is made chemically.

(a) *Ethyl alcohol* is also prepared from ethylene by a strictly chemical procedure. This method involves the addition of sulfuric acid, followed by hydrolysis of the addition product with steam. Because ethylene is obtained from the cracking of petroleum, ethyl alcohol plants are often built in the vicinity of petroleum plants.

$$CH_2{=}CH_2 + H_2SO_4 \rightarrow CH_3CH_2OSO_2OH$$

$$CH_3CH_2OSO_2OH + H_2O\ (\text{steam}) \rightarrow CH_3CH_2OH + H_2SO_4$$

The reaction of sulfuric acid with alkenes other than ethylene followed by hydrolysis is a general method for the preparation of secondary alcohols. Note that the addition of sulfuric acid to alkenes places the $-OSO_2OH$ group in such a position (Markownikoff's rule) that it always, except in the case of ethylene, forms a secondary or tertiary alcohol when hydrolyzed. The formation of 2-butanol from 1-butene is accomplished as follows:

mechanism was discussed in Chapter 27.

$$CH_3CH_2\overset{H}{\overset{|}{C}}{=}\overset{H}{\overset{|}{C}}H + H_2SO_4 \rightarrow CH_3CH_2\underset{OSO_2OH}{\underset{|}{\overset{H}{\overset{|}{C}}}}CH_3$$

$$CH_3CH_2\underset{OSO_2OH}{\underset{|}{\overset{H}{\overset{|}{C}}}}CH_3 + H_2O \rightarrow CH_3CH_2\underset{OH}{\underset{|}{\overset{H}{\overset{|}{C}}}}CH_3 + H_2SO_4$$

(b) The *alkaline* hydrolysis of alkyl halides is a general laboratory method for preparing alcohols. In practice the alkyl halide (haloalkane) is heated with a concentrated aqueous solution of sodium hydroxide. The general reaction is

$$R{-}X + NaOH \xrightarrow{\text{Heat}} ROH + NaX$$

In water

where X represents any halogen and R— represents any alkyl group. The specific reaction for the preparation of isopropyl alcohol from isopropyl bromide (2-bromopropane) is:

$$CH_3{-}\underset{\substack{|\\CH_3}}{CH}Br + NaOH \xrightarrow{\text{Heat}} CH_3\underset{\substack{|\\CH_3}}{CH}OH + NaBr$$

In water

The method is best for the preparation of primary alcohols from primary alkyl halides; secondary and tertiary alkyl halides react more slowly. With tertiary alkyl halides, a considerable quantity of alkene is formed. The reason is apparent when one considers the course of the reaction mechanism discussed in Chapter 26. The attacking species is the hydroxide ion in an S_N2 type reaction. This negative ion approaches the carbon to which the negative halogen is attached from the side opposite the halogen. The hydroxide ion forms a bond with the carbon, at the same time freeing the bromide ion on the opposite side. Figure 29.5 illustrates the process with ethyl bromide and hydroxide ion. When the hydroxide ion approaches tertiary butyl bromide, it encounters difficulty in reaching the tertiary carbon imbedded in the center of the molecule. Instead, some hydroxide ions draw a charged hydrogen (a proton) from one of the CH_3— groups. The unshared pair of electrons on the CH_2 group shifts inside to form a double bond and the bromide ion is eliminated (Fig. 29.5). A reaction of this type is referred to as an **elimination reaction.**

(c) A modern, simple and convenient method for preparing alcohols is the so-called "**hydroboration-oxidation**" procedure. This

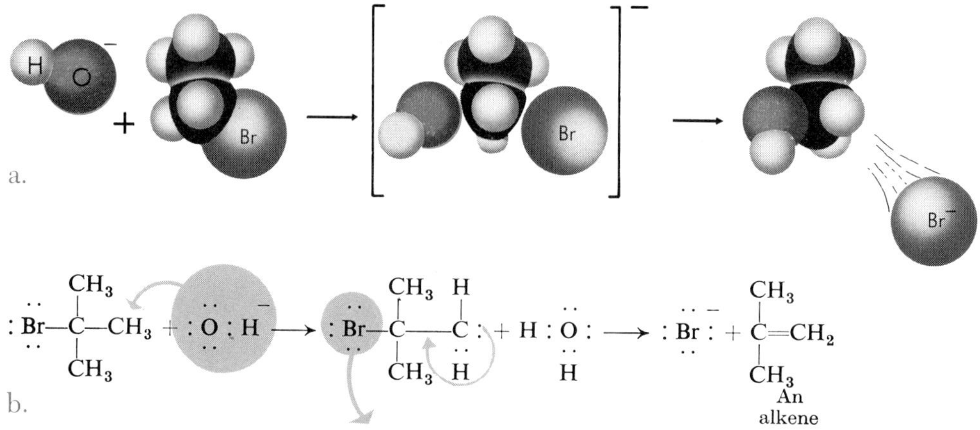

a.

b.

$$:\!\ddot{Br}{-}\underset{CH_3}{\overset{CH_3}{C}}{-}CH_3 + :\!\ddot{O}\!:\!H^{-} \longrightarrow :\!\ddot{Br}{-}\underset{CH_3}{\overset{CH_3}{C}}{-}\underset{H}{\overset{H}{C}}\!: + H\!:\!\ddot{O}\!:\!H \longrightarrow :\!\ddot{Br}\!:^{-} + \underset{CH_3}{\overset{CH_3}{C}}{=}CH_2$$

An alkene

FIGURE 29.5 COMPARISON OF REACTION OF (A) ETHYL BROMIDE AND (B) T-BUTYL BROMIDE WITH HYDROXIDE ION.

process was developed from the discovery that alkenes react with diborane, $(BH_3)_2$ or B_2H_6, to form trialkylboranes.

$$1.\quad 6\ RCH{=}CH_2 + (BH_3)_2 \xrightarrow{0°C} 2(RCH_2CH_2)_3B$$

The trialkylborane results from the addition of BH_3 to the carbon-carbon double bond of the alkene, with the boron atom acting as the electrophile and H acting as a hydride (H^-) ion, the nucleophile. The trialkylborane produced is readily oxidized, using alkaline hydrogen peroxide, to an alcohol and boric acid:

$$2.\quad (RCH_2CH_2)_3B + 3H_2O_2 \xrightarrow{OH^-} 3\ RCH_2CH_2OH + B(OH_3)$$

Reactions 1 and 2 are both carried out in the same reaction flask. The trialkylborane formed in reaction 1, without being isolated from the reaction mixture, is treated directly with alkaline hydrogen peroxide. The alcohol formed is separated from the reaction mixture and purified.

The oxidation reaction cleaves the C—B bond to each alkyl group and replaces the boron atom with an —OH group. The carbon skeleton of the alcohol obtained is that of the alkyl group attached to the boron in the trialkylborane. The reaction is useful for preparing "anti-Markownikoff" addition products from olefins, here illustrated for alcohol preparation from propene:

$$CH_3CH{=}CH_2 \xrightarrow{H_2SO_4} \xrightarrow{H_2O} \underset{\displaystyle OH}{CH_3\overset{}{C}HCH_3} \qquad \text{"Markownikoff" product}$$

$$CH_3CH{=}CH_2 \xrightarrow{(BH_3)_2} \xrightarrow{H_2O_2,\ NaOH} CH_3CH_2CH_2OH \qquad \text{"Anti-Markownikoff" product}$$

29.7 PHENOLS

A phenol is a compound containing a hydroxyl group attached to a carbon of an aromatic (benzene-like) ring. Several phenols and their common and official names are listed in Table 29.4. Phenols may be classified as mono-, di- or trihydroxyl phenols, depending on the number of —OH groups attached to aromatic rings. They are named as hydroxy derivatives of the aromatic hydrocarbon to which they are attached. Observe the names given in Table 29.4.

A. Preparation of Phenols

Two of several methods for obtaining phenols should be mentioned. Both involve the replacement of an atom already attached to the aromatic ring with a hydroxyl group.

1. Although replacement of a halogen attached directly to an aromatic ring is usually very difficult, phenol itself is prepared industrially by the *hydrolysis of chlorobenzene* under extreme conditions of temperature and pressure.

TABLE 29.4 NAMES AND STRUCTURES OF SOME PHENOLS

Name	Structure	IUPAC Name
Phenol	C_6H_5OH	Hydroxybenzene
o-Cresol	$C_6H_4(CH_3)(OH)$ (ortho)	2-Hydroxytoluene
m-Cresol	$HO-C_6H_4-CH_3$ (meta)	3-Hydroxytoluene
p-Cresol	$HO-C_6H_4-CH_3$ (para)	4-Hydroxytoluene
Resorcinol	$C_6H_4(OH)_2$ (meta)	1,3-Dihydroxybenzene
Hydroquinone	$HO-C_6H_4-OH$ (para)	1,4-Dihydroxybenzene
α-Naphthol	$C_{10}H_7OH$ (1-position)	1-Hydroxynaphthalene
β-Naphthol	$C_{10}H_7OH$ (2-position)	2-Hydroxynaphthalene

$$C_6H_5Cl + \underset{\text{Steam}}{HOH} \xrightarrow[\text{3000 psi pressure}]{\text{NaOH, 320°}} C_6H_5\bar{O}Na^+ + NaCl$$

(psi means pounds per square inch. 3000 psi = 204 atm)

The weakly acidic phenol is obtained in the highly alkaline medium as the sodium salt which is converted into phenol upon acidification.

2. The second method of fairly general application is the *fusion of the salt of the aromatic sulfonic acid* (see Chapter 31) with solid sodium hydroxide. During fusion the sulfonic acid group is replaced with a hydroxyl group. The resulting phenol is liberated by adding acid. When benzenesulfonic acid, obtained from the sulfonation of benzene, is fused with sodium hydroxide, and the fusion mixture acidified, the reaction sequence which takes place is:

$$C_6H_5SO_3H \xrightarrow[\text{Solid NaOH}]{\text{Fuse}} C_6H_5ONa \xrightarrow{\text{dil. HCl}} C_6H_5OH$$

Phenols are isolated from coal tar by extraction and distillation procedures.

3. Since 1954 phenol has been synthesized from cumene (isopropyl benzene). Cumene is obtained from petroleum or by synthesis from benzene and propene by the Friedel-Crafts synthesis. Air oxidation of cumene gives cumene hydroperoxide. This hydroperoxide on treatment with strong acid decomposes into phenol and acetone in a reaction involving hydrolysis and rearrangement of structure.

$$C_6H_5-C(CH_3)_2H \longrightarrow C_6H_5-C(CH_3)_2OOH \xrightarrow{H^+} C_6H_5-OH + (CH_3)_2C{=}O$$

B. Reactions of Phenols

Functional groups directly attached to an aromatic ring frequently exhibit unique behavior. This becomes apparent when the chemical behavior of phenols is compared with that of the alcohols. Although phenols and alcohols form many similar derivatives, they are often prepared using quite different procedures and reagents. Of the four types of alcohol reactions discussed in the previous paragraphs only the following two are of any real importance for phenols: (1) the hydrogen of the —OH of phenols is subject to replacement much as in the case of the alcohols. The replacement of the —OH group with other atoms or groups and dehydration do not occur with phenols. (2) Ready oxidation is characteristic of phenols.

Substitution of other atoms or groups in place of hydrogen atoms of the aromatic ring is also characteristic of phenols.

1. Replacement of the Hydrogen of the Phenolic —OH Group (Neutralization or Salt Formation)

The hydrogen of the —OH group in phenols is more acidic than is the corresponding hydrogen of alcohols. Phenols therefore react with sodium or potassium metal as do alcohols, but in addition are sufficiently acidic to undergo neutralization with aqueous solutions of sodium hydroxide (or other strong bases):

$$C_6H_5OH + NaOH \rightarrow C_6H_5\overset{-}{O}\overset{+}{Na} + H_2O$$

Sodium phenoxide

the net effect of delocalization is electron withdrawal from the O atom, making it less negative so a proton can leave more easily than from an alcohol.

The resulting sodium phenoxide salt is soluble in water. Because of this reaction phenols are soluble in a sodium hydroxide solution. The removal of a proton from the phenol —OH group produces the phenoxide ion, a resonance hybrid.

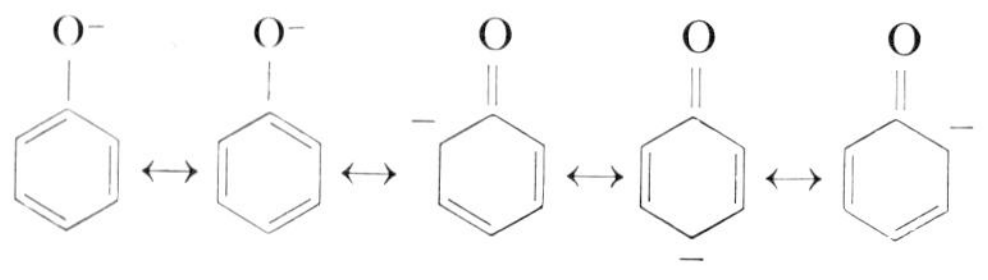

A delocalization of the negative charge occurs following removal of the proton, producing a resonance stabilization in excess of that possible for the parent phenol. The phenoxide ion is stabilized more by resonance with respect to phenol than is the alkoxide ion formed from an alcohol with respect to the parent alcohol. Thus phenol loses its proton more readily than an alcohol and is more acidic. Introduction of electron-attracting groups into the ortho or para positions of the ring enhances the acidity of phenol. Electron-donating groups decrease the acidity. The order of acidity of several substituted phenols is as follows:

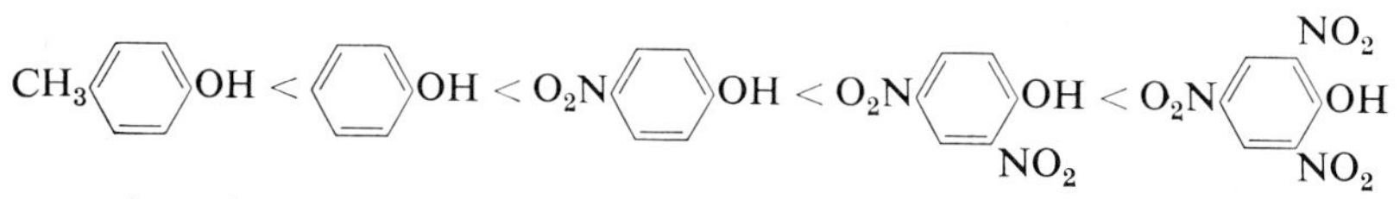

Phenols, however, are weaker acids than carbonic acid ion and do not dissolve in sodium bicarbonate solutions. The formation of soluble salts with sodium hydroxide permits the separation of phenols from alcohols and non-acidic, water soluble organic compounds.

2. Oxidation

Phenols are readily oxidized and yield a variety of complex products. Vigorous oxidation results in cleavage of the carbon ring, yielding carboxylic acids and other products. Controlled oxidation of certain polyhydric phenols is possible. For example, hydroquinone is readily transformed into p-benzoquinone.

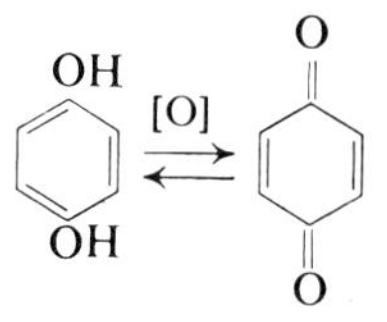

The ease with which hydroquinone undergoes oxidation makes it valuable as an **antioxidant** since it reacts readily with and destroys a variety of substances which catalyze the oxidative decomposition of organic substances.

3. Ring Substitution

Phenols undergo the usual substitution reactions characteristic of aromatic hydrocarbons. Since the hydroxyl group is a strong ortho-para directing group (a ring-activating group) phenol substitutions occur more readily than for the corresponding hydrocarbon. In fact, ease of substitution is one of the distinguishing characteristics of phenols. The rapid reaction of phenol with bromine that occurs in the absence of the usual catalyst is an excellent example. A precipitate of 2,4,6-tribromophenol appears very rapidly when phenol is added to bromine water.

ring substitution is discussed in Chapter 28.

$$C_6H_5OH + 3\,Br_2 \xrightarrow{\text{In water}} 2,4,6\text{-}Br_3C_6H_2OH + 3\,HBr$$

It is not possible to form monobromophenol by the usual methods of direct bromination.

Dilute nitric acid alone reacts readily with phenol to form a mixture of ortho- and para-nitrophenols (Chapter 28), although much oxidation occurs simultaneously. Sulfonation of phenol may be accomplished at room temperature with concentrated sulfuric acid:

$$\underset{\text{Phenol}}{C_6H_5OH} + \underset{\text{Concentrated}}{HOSO_2OH} \longrightarrow \underset{\text{o-hydroxybenzene sulfonic acid}}{o\text{-}HOC_6H_4SO_3H} \text{ and } \underset{\text{p-hydroxybenzene sulfonic acid}}{p\text{-}HOC_6H_4SO_3H} + H_2O$$

4. Reduction

Reduction of the aromatic ring of phenols is possible. Phenol itself, when reduced in the presence of a suitable catalyst, yields cyclohexanol.

$$C_6H_5OH + 3\,H_2 \xrightarrow[\text{Pressure}]{\text{Ni cat.}} \text{cyclo-}(CH(OH)CH_2CH_2CH_2CH_2CH_2)$$

Phenols and their homologs are toxic and have caustic action on skin and other tissues. Contact or vapor inhalation should be avoided. The phenols are active bactericides and insecticides. Phenol, sometimes called carbolic acid, is used as an antiseptic in hospitals and is responsible for "hospital odor." Cresol is an ingredient of creosote* and is used in sheep dip, the oily bath used to control skin infections on sheep.

29.8 ETHERS

Ethers are isomeric with alcohols of the same number of carbon atoms but lack the hydroxyl group. They have the general formula R—O—R, R—O—Ar or Ar—O—Ar in which an oxygen atom acts as a bridge between the attached R or Ar groups. Note that, as pointed out previously, the oxygen is bound to two carbons by single bonds only. Therefore, some of the physical and chemical properties of individual ethers differ markedly from those of their isomeric alcohols. These differences were outlined earlier in this chapter. Figure 29.6 shows the structure and common names of some ethers.

---

* Creosote is an oily, transparent, poisonous distillate from wood tar. It is used in the preservation of meat and as a disinfectant.

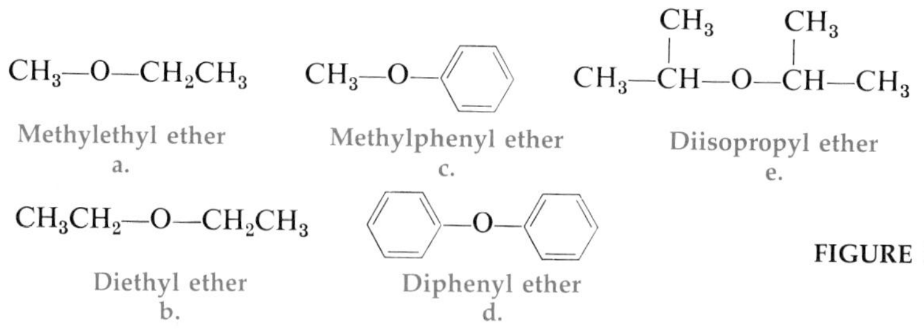

$CH_3—O—CH_2CH_3$

Methylethyl ether
a.

$CH_3—O—C_6H_5$

Methylphenyl ether
c.

$CH_3—CH(CH_3)—O—CH(CH_3)—CH_3$

Diisopropyl ether
e.

$CH_3CH_2—O—CH_2CH_3$

Diethyl ether
b.

$C_6H_5—O—C_6H_5$

Diphenyl ether
d.

FIGURE 29.6 SOME ETHERS.

Ethers are not commonly referred to by official IUPAC names but are named according to the groups on each side of the oxygen (Fig. 29.6). They also may be classified and referred to in other terms that are indicative of the structures. A simple ether has two identical groups attached to the oxygen; b, d and e in Figure 29.6 are simple ethers. Ethers with non-identical groups are mixed ethers (Fig. 29.6a, c). When both groups attached to the oxygen are alkyl groups (Fig. 29.6a, b, e), the ether may be represented by R—O—R and is called an aliphatic ether. When at least one group is attached to the oxygen through carbon of a benzene type ring, the ether is an aromatic ether (Fig. 29.6c, d).

A. Preparation of Ethers

best method for making mixed ethers.

There are two general preparations of ethers: the dehydration of alcohols with sulfuric acid (noted earlier in this chapter in the section concerned with the dehydration of alcohols) and the Williamson synthesis, in which an alkoxide (sodium salt of an alcohol) reacts with an alkyl halide. Ethyl bromide and sodium ethoxide, $NaOCH_2CH_3$, are the reactants in the preparation of diethyl ether. The equation for the reaction is:

$$CH_3CH_2ONa + CH_3CH_2Br \rightarrow CH_3CH_2OCH_2CH_3 + NaBr$$

The net change in this reaction is replacement of the Br^- of ethyl bromide with the $CH_3CH_2O^-$ group. Just as the OH^- from sodium hydroxide reacts with ethyl bromide (page 658) to displace Br^- to form an alcohol, $CH_3CH_2O^-$ reacts (with ethyl bromide) to form an ether, diethyl ether, in this case. For example, the mechanism for the synthesis of ethyl ether by the Williamson synthesis is as follows:

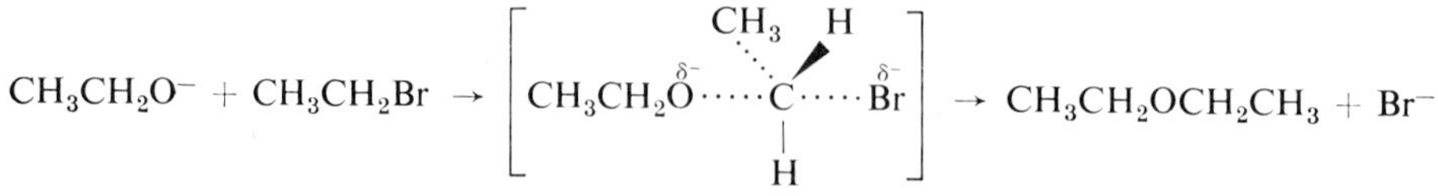

$$CH_3CH_2O^- + CH_3CH_2Br \rightarrow \left[CH_3CH_2\overset{\delta-}{O}\cdots\cdots \underset{H}{\overset{CH_3\ \ H}{C}}\cdots\cdots\overset{\delta-}{Br} \right] \rightarrow CH_3CH_2OCH_2CH_3 + Br^-$$

This is another S_N2 substitution, a concerted one-step process in which the ethoxide bonds with carbon as the bond to bromide is cleaved. Depending on the two reactants used, any one of a variety of mixed ethers can be prepared by this procedure. For example, to prepare phenetole (phenylethyl ether), sodium phenoxide is reacted with ethyl iodide,

$$C_6H_5\overset{-}{O}\overset{+}{Na} + CH_3CH_2I \xrightarrow{\text{Heat}} C_6H_5OCH_2CH_3 + NaI$$

Sodium phenoxide

Here, as before, in the substitution process the phenoxide ion bonds with the carbon holding the iodide ion and displaces it in an S_N2 type substitution to form the ether. The inverse of this reaction,

$$CH_3CH_2O^-Na^+ + C_6H_5I \not\rightarrow C_6H_5OCH_2CH_3 + NaI \quad \text{(No reaction)}$$

does not normally occur because of the unusually low reactivity of the halogen atom attached directly to an aromatic ring.

The experimental conditions for carrying out the reaction with various other phenols and alkyl halides differ widely and are determined by the reactivity of the substances employed.

B. Reactions of Ethers

solubility in H_2SO_4 is due to oxonium ion formation.

Ethers react with neither alkali metals such as sodium, nor with strong bases nor with oxidizing agents. They are similar in chemical inertness to the paraffins. However, they are soluble in concentrated sulfuric acid, a characteristic of most oxygen containing compounds, and by this test can be distinguished from hydrocarbons. Ethers are cleaved by hot concentrated mineral acids. The reaction may be written as follows:

$$CH_3OCH_2R + HI \rightarrow CH_3I + RCH_2OH$$

With an excess of HI the alcohol, RCH_2OH, is also converted into an alkyl iodide, RCH_2I, by the reaction,

$$RCH_2OH + HI \rightarrow RCH_2I + H_2O$$

The most effective acid for the cleavage is concentrated hydriodic acid.

Cleavage of ethers with HBr proceeds more slowly and requires higher temperatures. However, phenetole (phenyl ethyl ether) is cleaved with HBr to yield phenol:

$$C_6H_5OCH_2CH_3 + HBr \xrightarrow{100^\circ} C_6H_5OH + CH_3CH_2Br$$

Cleavage of an ether in the presence of a strong acid involves scission of the carbon-oxygen bond. As with most reactions of alcohols, the formation of the oxonium salt in the presence of strong acids appears to be the initial step in the process. Cleavage of the carbon-oxygen bond follows. The sequence of changes in the reaction between HI and CH_3OCH_2R appears to parallel the mechanism of the reaction of HBr with ethanol to form ethyl bromide (page 581).

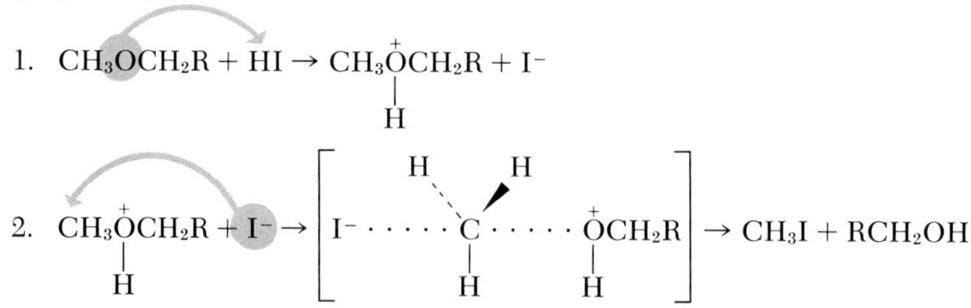

$$1.\quad CH_3OCH_2R + HI \rightarrow CH_3\overset{+}{O}(H)CH_2R + I^-$$

$$2.\quad CH_3\overset{+}{O}(H)CH_2R + I^- \rightarrow \left[I^- \cdots\cdots \overset{H\ \ H}{C}(H) \cdots\cdots \overset{+}{O}(H)CH_2R \right] \rightarrow CH_3I + RCH_2OH$$

Again with an excess of HI, RCH_2I is obtained from the alcohol formed upon ether cleavage. With HBr and HCl a similar reaction sequence is followed. Because the bromide and chloride ion are poorer nucleophiles than iodide, higher temperatures are required and the rate of reaction is slower. This is another illustration of a bimolecular nucleophilic substitution (S_N2) process (outlined first in Section 29.4 of this chapter).

Alkyl aryl ethers are sufficiently stable in the presence of acids to permit normal substitution reactions of the aromatic ring. Since the —OR group is activating and ortho-para directing, the following reaction can be expected:

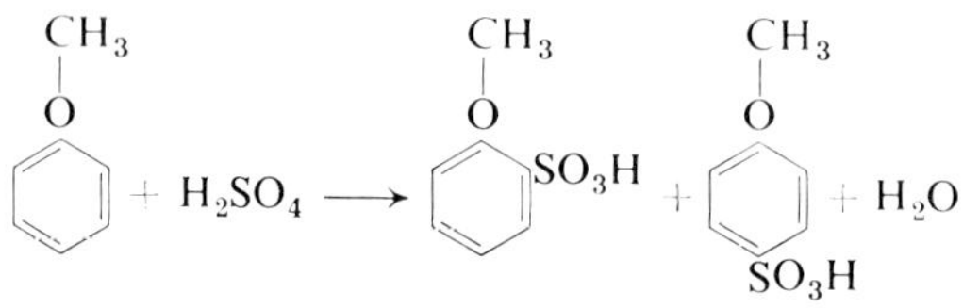

$$C_6H_5OCH_3 + H_2SO_4 \longrightarrow o\text{-}CH_3OC_6H_4SO_3H + p\text{-}CH_3OC_6H_4SO_3H + H_2O$$

The ethers are one of the less important classes of organic compounds and are used chiefly as solvents. Diethyl ether, probably the most widely used compound of this class, is employed in extracting fats from biological materials. It is also used as a general anesthetic.

C. Special Ethers – Alkene Oxides

In addition to the more conventional ethers which have just been discussed, alkene oxides should be mentioned. Ethylene oxide, $\underset{\diagdown O \diagup}{CH_2—CH_2}$, is the simplest compound of this type. **Alkene oxides** are highly reactive cyclic ethers whose behavior is said to be due to the internal strain common to three-membered ring systems. Alkene oxides react rapidly with many reagents, and in the course of these reactions the three-membered ring is opened, relieving the strain. This is illustrated by reactions of ethylene oxide with water, methanol and ammonia. Note that in each case the three-membered ring opens and addition products are obtained.

$$CH_2{-}CH_2(\text{—O—}) + HOH \rightarrow CH_2(OH){-}CH_2(OH)$$

$$CH_2{-}CH_2(\text{—O—}) + C_2H_5OH \rightarrow CH_2(OH){-}CH_2(OC_2H_5)$$

$$CH_2{-}CH_2(\text{—O—}) + HNH_2 \rightarrow CH_2(OH){-}CH_2(NH_2)$$

$$CH_2{-}CH_2(\text{—O—}) + RMgBr \rightarrow CH_2(R){-}CH_2(OMgBr)$$

The commercial importance of ethylene oxide is dependent upon this remarkable reactivity, which permits its ready conversion into many useful compounds.

EXERCISES

29.1 Which elements are necessarily found in alcohols?

29.2 Define and give an example of each by writing the formula: (a) primary alcohol, (b) secondary alcohol, (c) tertiary alcohol, (d) phenol, (e) simple ether, (f) aromatic ether, (g) alkene oxide.

29.3 Write both the correct common names and the IUPAC names for

CH_3OH, CH_3CH_2OH, $(CH_3)_2CHOH$, $(CH_3)_2CH{-}CH_2{-}OH$,

(phenol) OH and (naphthol) OH

29.4 Define the term functional group. What is the functional group in $CH_3CH_2CH(CH_3)OH$? What function does it perform when sodium metal is dropped into the anhydrous alcohol?

29.5 Give the order of reactivity of primary, secondary and tertiary alcohols and phenols with: (a) sodium metal, (b) hydrohalic acids, (c) concentrated sulfuric acid (intramolecular dehydration), (d) oxidation with basic potassium permanganate solution, (e) neutralization with aqueous sodium hydroxide.

29.6 Write balanced equations showing reaction conditions over the arrow for the reaction of bromine water: (a) with propane in the dark, (b) with propylene in the dark, (c) with phenol.

29.7 Describe a chemical method for separating the components of a solution of phenol in cyclohexanol.

29.8 Write structural formulas for each of seven compounds with the formula $C_4H_{10}O$.

29.9 In what way is the Williamson synthesis superior to dehydration with sulfuric acid in preparing mixed ethers?

29.10 Write equations for the reaction of concentrated hydriodic acid with methyl ethyl ether.

29.11 Write equations for the two-step preparation of: (a) dimethyl ether, starting with carbon monoxide and steam, (b) methyl isopropyl ether from the two alcohols, (c) isopropyl alcohol from propene.

29.12 Bottles, each containing one of the liquids 1-pentyne, 2-pentyne, secondary butyl alcohol, n-hexane and methyl n-propyl ether, are unlabeled. Describe how you would identify each chemically for labeling.

29.13 What is the first reaction which occurs when an alcohol or an ether is added to concentrated hydrochloric acid?

29.14 An ether gave methyl iodide and isobutyl iodide when allowed to react with concentrated hydriodic acid (HI). What is the probable formula of the original ether? Why?

29.15 If one uses only readily available inorganic chemicals (reagents) in addition to n-hexane, by what series of reactions may n-hexane be converted into n-hexyl alcohol?

29.16 Give a simple chemical test (including any visual observations you would expect) that would distinguish between each pair of compounds given below:

1. 1-butanol and t-butyl alcohol
2. 4-chlorophenol and 4-chloro-1-cyclohexanol
3. ethyl ether and ethanol
4. n-butanol and n-pentane

29.17 Arrange each group of compounds below in the sequence specified for each group.

1. Increasing basicity: butyl, isobutyl and t-butyl alcohols.
2. Increasing ease of proton loss: 1-pentanol, 2-pentanol, 2-methyl-2-butanol, phenol, 4-nitrophenol, diethyl ether.
3. Decreasing water solubility: ethanol, ethyl ether, 1-octanol, 2-hexene.
4. Increasing boiling point: 1-pentanol, 3-pentanol, neopentyl alcohol, 2-methyl-2-butanol, 2-butanol, 1-hexanol.

29.18 Starting with ethene and phenol, show equations for the preparation of phenyl ethyl ether.

29.19 A compound A contains carbon, hydrogen and oxygen. Sodium reacts with A to liberate hydrogen. When A is oxidized a compound B having the molecular formula $C_4H_8O_2$ is obtained. When A is permitted to react with $SOCl_2$, 1-chlorobutane is formed. What is the formula of compound A? Write equations for all these reactions.

29.20 A compound A reacts very slowly with sodium metal to form hydrogen and when treated with HBr rapidly forms a compound B. Compound B reacts readily with magnesium in dry ether to form a Grignard reagent which when allowed to react with water forms isobutane. What formula for compound A best accounts for the properties described for A? Explain.

29.21 A compound A, C_3H_8O, reacts in 10 to 15 minutes with Lucas reagent. When A is heated with concentrated H_2SO_4 a compound, C_3H_6, forms which reacts with HCl to yield compound C. What are suitable structural formulas for the compounds A, C_3H_6 and C?

29.22 An ether A is treated with an excess of HI to product two alkyl iodides which when permitted to react with KOH(aq) form products B and C. When product B is oxidized with potassium permanganate, CH_3COOH is the only organic product formed. Oxidation of compound C yields a ketone having the molecular formula C_4H_8O. What is a suitable formula for product B, product C and for the ether A? Give equations for each reaction which occurs.

29.23 Which should form an oxonium salt more readily, diphenyl ether or dicyclohexyl ether? Explain.

29.24 Two aromatic compounds A and B have the same molecular formula, C_7H_8O. Both isomers react with sodium to produce hydrogen. Compound A dissolves readily in aqueous NaOH; compound B does not. Compound B reacts readily with HBr to form C_7H_7Br; compound A does not. When subjected to hydrogenation both A and B take up three moles of hydrogen per mole of compound yielding an alcohol. Each alcohol when heated with concentrated sulfuric acid yields a single alkene. Both alkenes when treated with H_2 in the presence of palladium metal yield the same saturated cycloalkane, methyl cyclohexane. What are suitable formulas for the compounds A and B?

29.25 Compound A, $C_6H_{14}O$, reacts rapidly with Lucas reagent and rapidly loses water when warmed with acid to form compound B. Compound A reacts with sodium metal to give a gas. Oxidation of compound B with $K_2Cr_2O_7$ in acid solution gives a neutral ketone (see Chapter 30), C_4H_8O and an acid, compound D, $C_2H_4O_2$. Write structural formulas for A, B, C, and D. Write equations for each reaction described.

29.26 Give the structure of the product (or products) formed at each stage in the following sequence of reactions (the reagent to be used is shown over the arrow in each case):

$$A \xrightarrow{Na} B \xrightarrow{R'I} C \xrightarrow{HI} D \xrightarrow{KOH} E \xrightarrow{CH_3I} C_6H_5OCH_3$$

$$+\ G \xrightarrow{KOH} H \xrightarrow{PBr_3} C_2H_5Br$$

29.27 An alcohol, $C_6H_{14}O$, reacts rapidly with Lucas reagent. It also undergoes dehydration to form a hydrocarbon, C_6H_{12}, which in turn undergoes oxidation to form the products: A, C_4H_8O and B, $C_2H_4O_2$. Write possible structures for A and B and equations for the reactions described.

29.28 A reaction is carried out between $HCOOC_2H_5$ and water containing the ^{18}O isotope of oxygen. Write an equation showing the reaction of $HCOOC_2H_5$ with water which shows the location of the ^{18}O in the products formed.

SUGGESTED READING

Ferguson, L. N.: "Hydrogen Bonding and the Physical Properties of Substances." J. Chem. Ed., *33*:267, 1956.

Hendrickson, J. B., Cram, D. J. and Hammond, G. S.: *Organic Chemistry*, Third Edition. McGraw-Hill Book Company, New York, 1970. Chapters 10, 11 and 14.

Morrison, R. T. and Boyd, R. N.: *Organic Chemistry*, Second Edition. Allyn and Bacon, Inc., Boston, 1966. Chapters 15, 16, 17, 25 and 28.

Noller, C. R.: *Chemistry of Organic Compounds*, Third Edition. W. B. Saunders Company, Philadelphia, 1965. Chapters 8, 9, 26, 27 and 36.

THIRTY • ALDEHYDES AND KETONES

The study of aldehydes and ketones is the study of the carbonyl group, its chemical reactivity and the reactivity of certain neighboring atoms in the molecule.

30.1 THE CARBONYL GROUP

the carbonyl (C=O) group may bond to only two other atoms.

The **carbonyl functional group** is a carbon atom and an oxygen atom joined together by a double bond. The next neighboring atoms, which we will consider in this chapter, are either two carbons, in which case the compound is a ketone, or a carbon and a hydrogen, in which case it is an aldehyde. In one compound, formaldehyde, the simplest of carbonyl compounds, the carbonyl group is flanked by two hydrogens. Formaldehyde may be represented in several ways (Fig. 30.1). The electron configuration in (a) shows that each atom has the noble gas configuration by sharing. The bond structure which appears in (b) shows a single carbon to hydrogen bond and a double carbon to oxygen bond. In (d) the double bond appears as it would if two single bonds from each atom were drawn together. The molecular orbital representation is shown in (e). Two electrons of those linking carbon to oxygen occupy a sigma (σ) orbital directly between the two atoms. A pi (π) orbital shown above and below the sigma bond contains the second pair of electrons. As with the pi electrons of the double carbon to carbon bond, these pi electrons are more exposed and are affected to a greater extent by approaching particles than are sigma electrons. The double carbon to oxygen bond differs in one important respect from the double carbon to carbon bond. The C to O linkage is highly polar, the oxygen atom being more eleca b c d

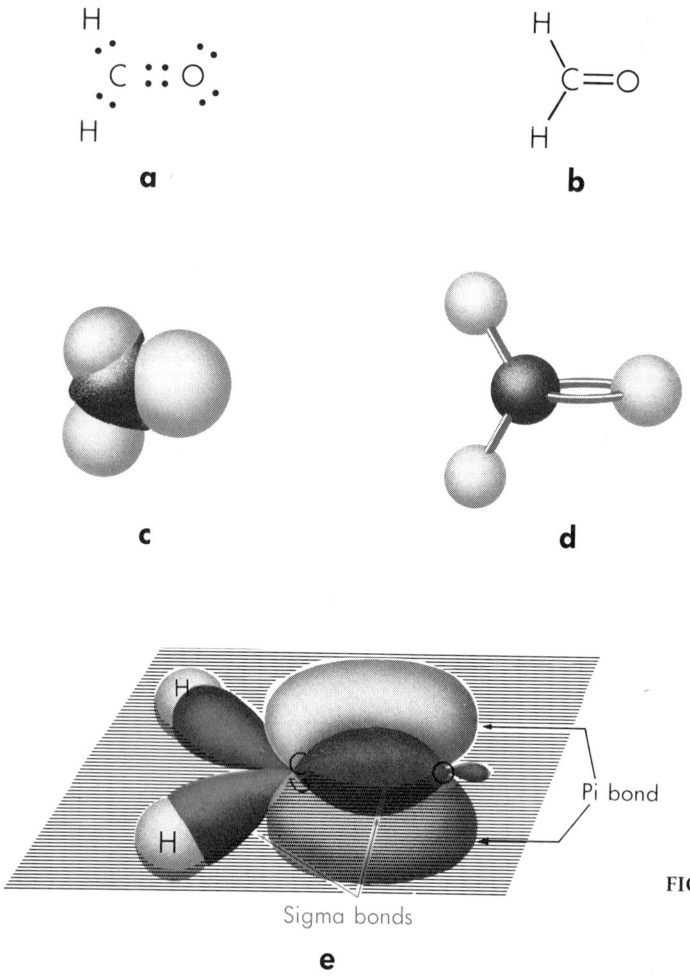

Pi bond

Sigma bonds

e

FIGURE 30.1 FORMALDEHYDE.

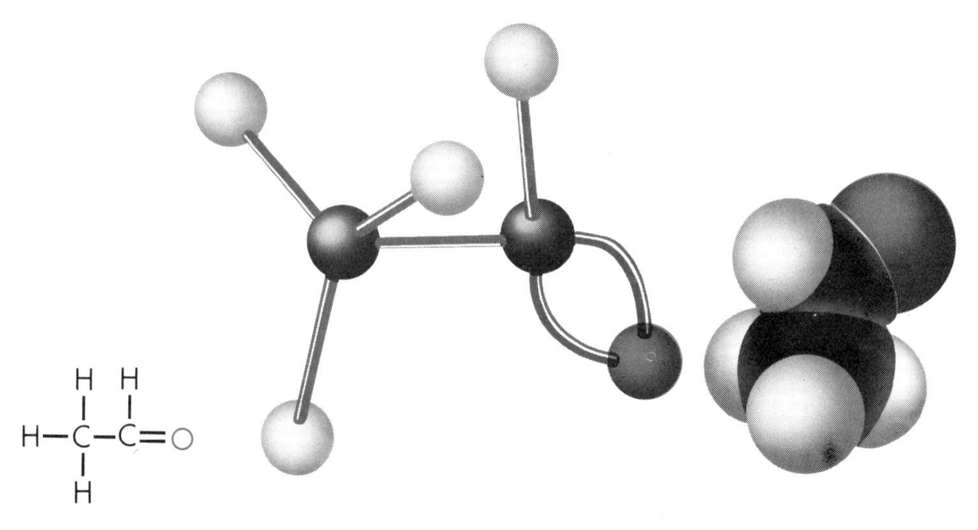

FIGURE 30.2 ACETALDEHYDE.

tronegative than the carbon. Compared to one another, carbon is positive and oxygen is negative.

Figure 30.2 shows two methods for representing the three-dimensional structure of acetaldehyde. The CH_3— group may be replaced by any other R— or Ar— group.

30.2 NOMENCLATURE

A. Aldehydes

To be attached to both another carbon and a hydrogen, the carbon of the carbonyl group of an aldehyde must be terminal (that is, it must be the end carbon). Some common aldehydes are listed in Table 30.1.

Common names of aldehydes are derived from the common names of related acids which differ only in the functional group. Formic acid has the same number of carbons as formaldehyde; acetic acid, as acetaldehyde; cinnamic acid, as cinnamaldehyde, and so on (Chapter 31). Aldehydes may be either *aromatic* or *aliphatic*. The carbonyl group of an aromatic aldehyde must be attached directly to a carbon of a benzene or benzene-like ring. Of those illustrated in Table 30.1, only benzaldehyde and vanillin

TABLE 30.1 COMMON ALDEHYDES

Formula	Common Name	IUPAC Name
$H_2C{=}O$ (CH_2O)	Formaldehyde	Methanal
$CH_3{-}CHO$ (CH_3CHO)	Acetaldehyde	Ethanal
$C_6H_5{-}CHO$ (C_6H_5CHO)	Benzaldehyde	Phenylcarboxaldehyde
$C_6H_5{-}CH{=}CH{-}CHO$	Cinnamaldehyde	3-Phenyl-2-propenal
ring with $-CHO$, $-OCH_3$, $-OH$	Vanillin	(4-Hydroxy-3-methoxyphenyl)carboxaldehyde
$CH_3{-}CH(CH_3){-}CHO$ ($(CH_3)_2CHCHO$)	Isobutyraldehyde	2-Methylpropanal

are aromatic. If the carbonyl group is not attached directly to an *aromatic* carbon, the aldehyde is *aliphatic*.

The IUPAC name ending for aldehydes is *al*, which replaces the *e* in *aliphatic* (straight chain hydrocarbon) names. For naming, the carbon of the carbonyl group must be included in the primary chain of carbons, and it becomes carbon *number one*. Because the carbonyl carbon is *number one*, its position need not be specified with a number. In other respects aldehydes are named as other organic compounds. Compare the IUPAC names in Table 30.1 with the structural formulas.

B. Ketones

Because the carbonyl group of a ketone is attached to *two* carbons, ketones may be either *aliphatic*, *aromatic* or *mixed aliphatic and aromatic*. Figure 30.3 (compare with Fig. 30.2) illustrates the three dimensional structure of acetone. Of the compounds listed in Table 30.2 only benzophenone is aromatic; acetophenone and benzylphenylketone are mixed aromatic-aliphatic; and the balance are aliphatic ketones.

TABLE 30.2 SOME KETONES

Formula	Common Name	IUPAC Name
$CH_3—C(=O)—CH_3$	Acetone, Dimethylketone	Propanone
$CH_3—C(=O)—CH_2—CH_3$	Methylethylketone	2-Butanone
$C_6H_5—C(=O)—CH_3$	Acetophenone, Phenylmethylketone	1-Phenylethanone
cyclic $C(=O)$, H_2C, CH_2, H_2C, CH_2, CH_2 ring	—	Cyclohexanone
$C_6H_5—C(=O)—C_6H_5$	Benzophenone (Diphenylketone)	Diphenylmethanone
$C_6H_5—CH_2—C(=O)—C_6H_5$	Benzylphenylketone	1,2-Diphenylethanone

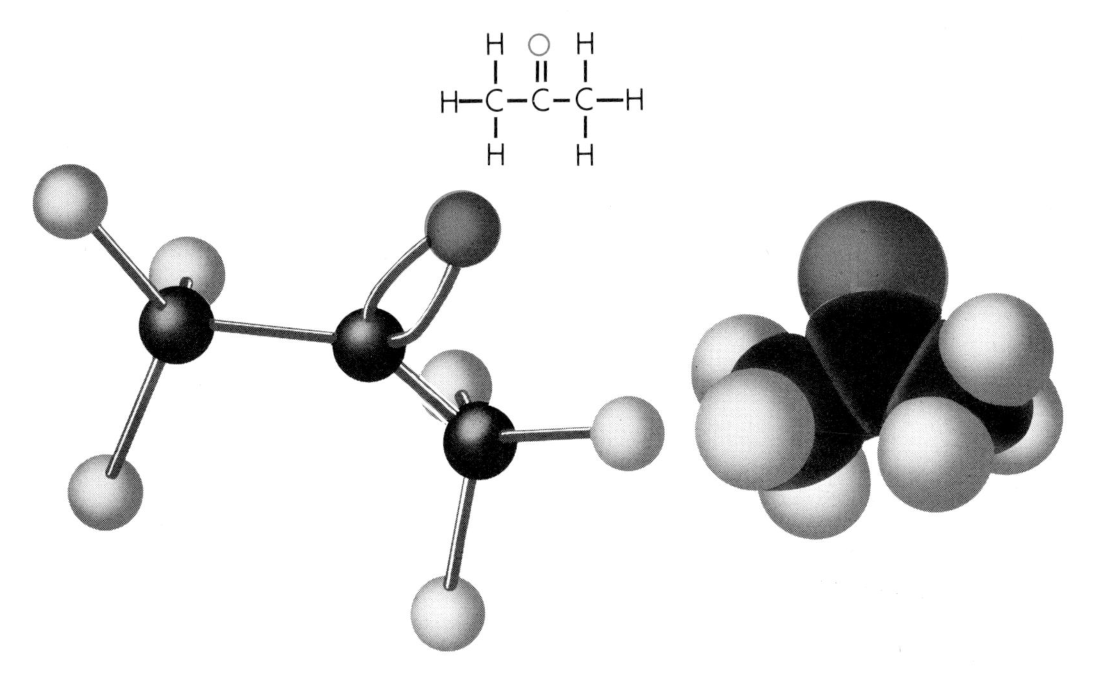

FIGURE 30.3 ACETONE.

Most common names of ketones follow a system similar to that used for naming ethers. The names of both groups attached to the carbonyl group are followed by the word ketone (Table 30.2). The IUPAC name ending for ketones is *one*, which replaces the *e* of aliphatic hydrocarbon names. The carbon of the carbonyl group must be contained in the primary carbon chain. Its position in the chain is indicated by number. In other respects, ketones are named according to the system for other organic compounds.

Ethylisopropylketone,

note that the principle of lowest numbers is followed in the IUPAC names.

$$\begin{array}{ccccccccc} & & CH_3 & & O & & & & \\ & & | & & \| & & & & \\ CH_3 & — & CH & — & C & — & CH_2 & — & CH_3, \end{array}$$

is named 2-methyl-3-pentanone. The compound

$$\begin{array}{ccccccccccc} & & & & CH_3 & & Cl & & O & & \\ & & & & | & & | & & \| & & \\ CH_3 & — & CH & = & C & — & CH & — & C & — & CH_3 \end{array}$$

(Give lowest possible number)

is named 3-chloro-4-methyl-4-hexen-2-one.

30.3 PHYSICAL PROPERTIES

The presence of the polar carbonyl group increases the attraction of molecules of carbonyl compounds for one another. As a result, the boiling points of these compounds are considerably higher than the boiling points of hydrocarbons with nearly the same molecular weight. Formaldehyde and acetaldehyde are the only gases at room temperature. The polarity of the C to O bond and the capability to hydrogen bond with a substance presenting a hydrogen bonded to either oxygen or nitrogen give them a great attraction to water molecules. The smaller carbonyl compounds, *up to 2 or 3 carbons,* are completely soluble in water. Solubility

TABLE 30.3 PHYSICAL PROPERTIES OF SOME ALDEHYDES AND KETONES

Name	Formula	Boiling Point, °C	Density	Solubility in Water. Grams per 100 g Water
Formaldehyde (Methanal)	$H{-}C({=}O){-}H$	−21	0.815	Completely soluble
Acetaldehyde (Ethanal)	CH_3CHO	20.2	0.7834	Completely soluble
Propionaldehyde (Propanal)	CH_3CH_2CHO	48.8	0.807	20
n-Butyraldehyde (Butanal)	$CH_3(CH_2)_2CHO$	75.7	0.817	3.7
Benzaldehyde (Phenylmethanal)	$C_6H_5{-}CHO$	179.6	1.050	0.33
Acetone (Propanone)	$CH_3{-}C({=}O){-}CH_3$	56.5	0.792	Completely soluble
2-Butanone	$CH_3{-}C({=}O){-}CH_2{-}CH_3$	79.6	0.805	35.3
Acetophenone (1-Phenylethanone)	$C_6H_5{-}C({=}O){-}CH_3$	202.3	1.036	Insoluble

TABLE 30.4 ODORS OF SOME COMPOUNDS

Formula	Name	Odor
$C_6H_5{-}CH{=}CH{-}C({=}O)H$	Cinnamaldehyde	Cinnamon
$CH_3{-}C(CH_3){=}CH{-}CH_2{-}CH_2{-}C(CH_3){=}CH{-}C({=}O){-}H$	Citral	Lemon
$HO{-}C_6H_3(OCH_3){-}C({=}O){-}H$	Vanillin	Vanilla
$CH_3{-}CH{<}(CH_2)(CH_2)_{12}{>}C{=}O$ (ring)	Muscone	Musk

muscone is the odoriferous chemical of the musk used in fine perfumes.

in both polar and non-polar solvents makes acetone a very useful solvent.

Many aldehydes and ketones have distinctive odors. In addition to having an acrid odor, formaldehyde vapors temporarily destroy the sense of smell. Table 30.4 gives the odors of some carbonyl compounds.

30.4 PREPARATION

Aldehydes and ketones may be prepared in several ways. Four of the most significant methods will be noted here.

A. The *oxidation of primary alcohols to yield aldehydes* and the attendant problem of further oxidation were discussed under reactions of alcohols (Chapter 29). The oxidation of *secondary alcohols to prepare ketones* was discussed in the same chapter.

A specific oxidation of a primary alcohol of industrial importance is the vapor phase oxidation of methanol with air over a hot copper catalyst.

$$2\,CH_3OH + \underset{\text{In air}}{O_2} \xrightarrow[Cu]{500°C} 2\,H{-}\overset{\overset{\large H}{|}}{C}{=}O + 2\,H_2O$$

B. *Carbonyl compounds* can be *prepared by dehydrogenation.* Using the same alcohols and the same catalysts as are used in oxidation, one can form ketones and aldehydes by dehydrogenation:

$$\underset{\text{A primary alcohol}}{CH_3CH_2CH_2OH} \xrightarrow[500°C]{Cu} \underset{\text{An aldehyde}}{CH_3CH_2\overset{\overset{\large O}{\|}}{C}{-}H} + H_2$$

$$\underset{\text{A primary alcohol}}{CH_3OH} \xrightarrow[500°C]{Cu} \underset{\text{An aldehyde}}{\underset{\underset{\large H}{|}}{\overset{\overset{\large H}{|}}{C}}{=}O} + H_2$$

$$\underset{\text{A secondary alcohol}}{CH_3{-}\underset{\underset{\large CH_3}{|}}{C}HOH} \xrightarrow[500°C]{} \underset{\text{A ketone}}{CH_3{-}\underset{\underset{\large CH_3}{|}}{C}{=}O} + H_2$$

The oxidation of methanol with air, and dehydrogenation, both with a hot copper catalyst, are the two principal methods for producing formaldehyde.

C. Aldehydes and ketones may be obtained by *hydrolysis of* **alkyl dihalides (dihaloalkanes)** in which both halogen groups are on the same carbon. Consider the hydrolysis of dichlorophenylmethane (benzal chloride) when it is heated with an aqueous solution of sodium hydroxide:

$$C_6H_5CHCl_2 + 2\,NaOH \xrightarrow{\text{Heat}} \left[C_6H_5\text{—}CH(OH)_2 \right] + 2\,NaCl$$

$$\downarrow$$

$$C_6H_5\text{—}CH{=}O + H_2O$$

Benzaldehyde

both Cl atoms are replaced with OH as a result of S_N2 attack by OH^-.

An alkyl halide may be converted to an alcohol (the halide being replaced by the hydroxyl group) by heating with an aqueous solution of a strong base. In this case replacement of the two halogen atoms leaves two hydroxyl groups on the same carbon. Such a structure, shown within the brackets in the equation, is unstable, decomposing to an aldehyde by loss of water. An aldehyde was formed in this example because the two halogens appeared on a *terminal* carbon. If both halogens were on a non-terminal carbon, hydrolysis would yield a ketone.

D. The fourth method, *addition of water* to an *acetylenic* compound, was discussed as a reaction of multiple carbon to carbon bonds. The preparation of acetaldehyde was demonstrated there. 2-Butanone can be prepared by addition of water to 2-butyne in the presence of a solution of mercuric sulfate and sulfuric acid.

$$CH_3\text{—}C{\equiv}C\text{—}CH_3 + HOH \xrightarrow[H_2SO_4]{Hg^{2+}} \left[CH_3\text{—}C(OH){=}C(H)\text{—}CH_3 \right] \rightarrow$$

$$CH_3\text{—}C(=O)\text{—}CH_2\text{—}CH_3$$

2-butanone

notice the shift of H from oxygen to carbon.

30.5 REACTIONS

The most characteristic reactions of aldehydes and ketones arise from the electronegativity of oxygen which induces an unequal distribution of electrons between the nuclei of the carbon and the oxygen atoms of the carbonyl group so that the electron density is increased at the oxygen and decreased about the carbon. Unsaturation is due to the presence between the carbon and oxygen of the pi bond which yields to addition. In view of these characteristics of the carbonyl group—polarity and unsaturation—addition should occur with the negative portion of the adding compound attaching itself to the electron-deficient carbon, the positive portion to the electron-rich oxygen. Identification of the products from a variety of addition reactions confirms this manner of addition. Furthermore, experimental studies show that the negative portion combines with the carbon of the carbonyl first. Addition to oxygen follows as a second step to complete the reaction and form the final addition product. Accordingly the addition of hydrocyanic acid (HCN) to acetaldehyde takes place as shown in Figure 30.4.

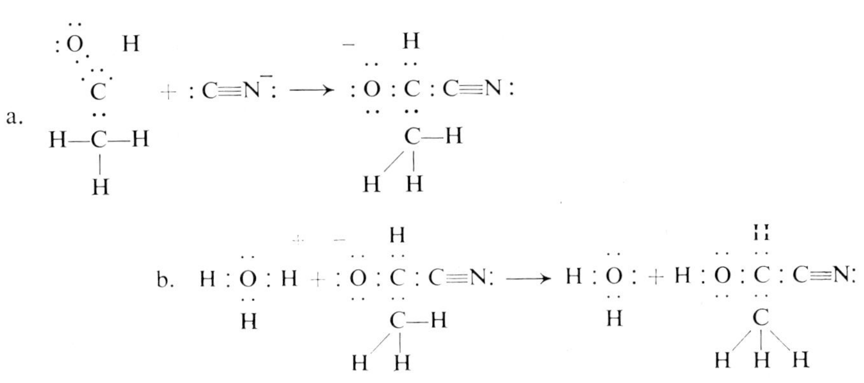

a. $CH_3CH{=}\ddot{O}$ (acetaldehyde, :O: double bonded to C with H; C bonded to CH_3) $+ :C{\equiv}N:^- \longrightarrow$ $:\ddot{O}:\ddot{C}:C{\equiv}N:^-$ (C bearing H and CH_3)

b. $H:\ddot{O}:H$ (with H) $+$ $:\ddot{O}:\ddot{C}:C{\equiv}N:^-$ (C bearing H and CH_3) $\longrightarrow$ $H:\ddot{O}:^-$ (with H) $+$ $H:\ddot{O}:\ddot{C}:C{\equiv}N:$ (C bearing H and CH_3)

FIGURE 30.4 COURSE OF ADDITION OF HYDROCYANIC ACID TO ACETALDEHYDE.

The carbonyl group, because of its character, is attacked by either polar or ionic substances. Since addition occurs first at the electron deficient carbonyl carbon, the reaction involves initially an attack by a nucleophilic group (a neutral electron-pair donor or a negative ion). Carbonyl addition reactions are therefore commonly classified as **nucleophilic addition** reactions in contrast with the electrophilic additions which are characteristic of unsaturated hydrocarbons.

Other atoms of aldehydes and ketones are affected by the presence of the carbonyl group. Among the most influenced are the *α-hydrogens,* the bond between them and the carbon being weakened by the carbonyl group.

An **α-hydrogen** is a *hydrogen attached* to an *α-carbon,* one *adjacent* to the *functional group.* The carbons of butanol and 2-hexanone are labeled with letters of the Greek alphabet:

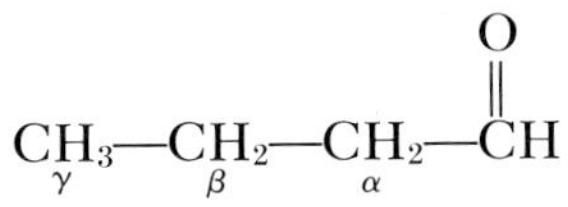

$$\underset{\gamma}{CH_3}-\underset{\beta}{CH_2}-\underset{\alpha}{CH_2}-\overset{O}{\overset{\|}{C}}H$$

Butanal

$$\underset{\alpha}{CH_3}-\overset{O}{\overset{\|}{C}}-\underset{\alpha}{CH_2}-\underset{\beta}{CH_2}-\underset{\gamma}{CH_2}-\underset{\delta}{CH_3}$$

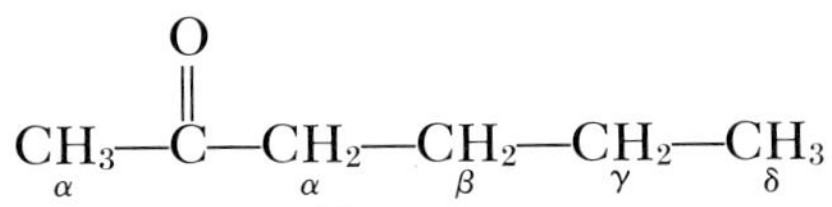

2-hexanone

Beginning at the carbon adjacent to the carbonyl group, the carbons are designated in order as alpha, beta, gamma and delta. The weakening of the α-hydrogen bond is seen in the fact that ketones and aldehydes exist in an enol form as well as in a keto form.

any one aldehyde or ketone exists in both forms in solution with the keto form in large excess.

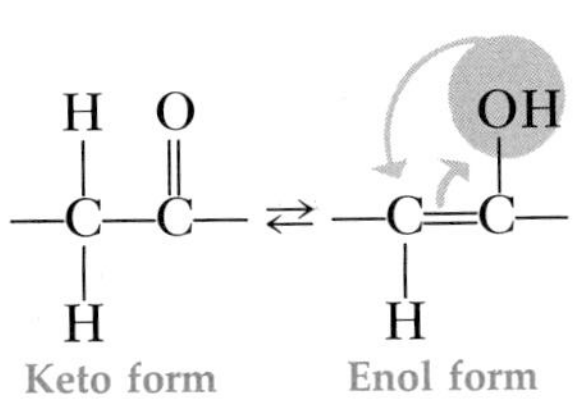

$$-\overset{H}{\overset{|}{\underset{H}{\underset{|}{C}}}}-\overset{O}{\overset{\|}{C}}- \rightleftarrows -\underset{H}{\underset{|}{C}}{=}\overset{OH}{\overset{|}{C}}-$$

Keto form — Enol form

In effect an α-hydrogen migrates to the oxygen of the carbonyl group from the α-carbon to give the enol. Actually the α-hydrogen is lost and another is gained by the oxygen. An equilibrium exists between these two forms. With most compounds the equilibrium, as illustrated, is far to the left. This phenomenon is known as **keto-enol tautomerism.** It is quite commonly observed when hydrogens are bonded to carbons adjacent to carbonyl or other electron attracting groups.

Reactions of aldehydes and ketones are largely those of addition to the carbonyl group, loss of an active hydrogen, oxidation (aldehydes), and directed substitution (aromatic aldehydes and ketones).

A. Addition Reactions

Some reagents which add to the carbonyl group are hydrocyanic acid, water, alcohols, bisulfite ion, ammonia and ammonia-like compounds. Aldehydes and ketones may even add to themselves.

1. Hydrogen Cyanide. Hydrogen cyanide (hydrocyanic acid) adds to give a cyanohydrin.

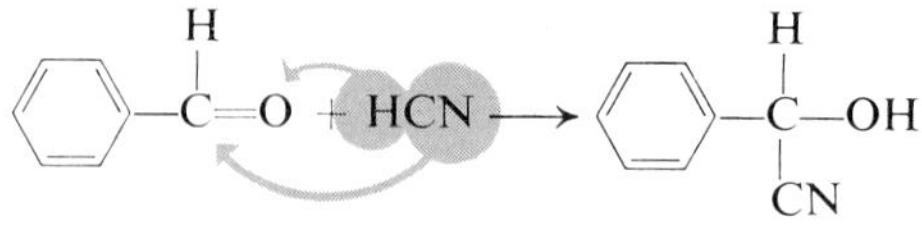

$$C_6H_5\text{—}\overset{H}{\overset{|}{C}}\text{=O} + HCN \longrightarrow C_6H_5\text{—}\overset{H}{\overset{|}{\underset{CN}{\underset{|}{C}}}}\text{—OH}$$

Benzaldehyde A cyanohydrin

2. Water. Water adds to only a few aldehydes and ketones to give stable hydrates. One preparation of carbonyl compounds, the hydrolysis of dihalides, depends on the general *instability* of the hydrates.

a. *Chloral* (2,2,2-*trichlorolethanal*) adds water to give stable chloral hydrate, the principal ingredient of "knock-out drops," a narcotic and hypnotic.

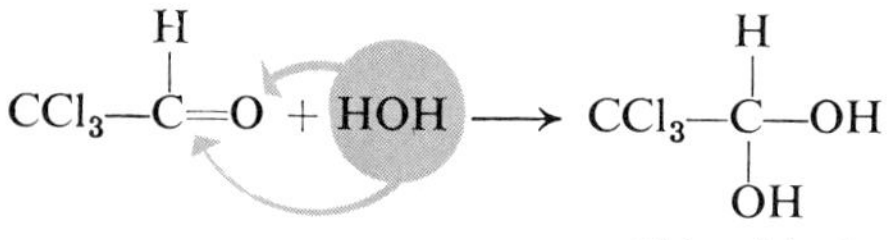

$$CCl_3\text{—}\overset{H}{\overset{|}{C}}\text{=O} + HOH \longrightarrow CCl_3\text{—}\overset{H}{\overset{|}{\underset{OH}{\underset{|}{C}}}}\text{—OH}$$

Chloral hydrate

b. *Ninhydrin,* used in the *identification* and *analysis* of proteins, is a triketone which adds water.

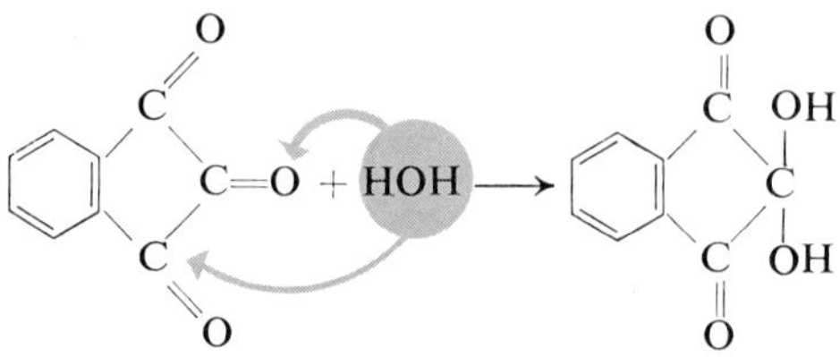

$$C_6H_4(CO)_2C\text{=O} + HOH \longrightarrow C_6H_4(CO)_2C(OH)_2$$

3. Addition of Alcohol. Many alcohols add to the carbonyl group of aldehydes in the presence of suitable acid catalysts to

monosaccharides (Chapter 36) are examples of stable hemiacetals.

form hemiacetals. Hemiacetals are unstable but may be stabilized by further reaction with alcohol to form acetals. Most ketones do not readily undergo this direct reaction. Benzaldehyde diethyl acetal (b.p. 222°C) can be prepared in 55 per cent yield by direct reaction in the presence of dry hydrogen chloride. With acetaldehyde the reaction is

$$CH_3\overset{H}{\underset{}{C}}{=}O + CH_3CH_2OH \xrightarrow{HCl} CH_3\overset{H}{\underset{OCH_2CH_3}{C}}{-}OH$$

Hemiacetal

$$CH_3\overset{H}{\underset{OCH_2CH_3}{C}}{-}OH + CH_3CH_2OH \xrightarrow{HCl} CH_3\overset{H}{\underset{OCH_2CH_3}{C}}{-}OCH_2CH_3 + H_2O$$

Acetal

4. Addition of Sodium Bisulfite. Sodium bisulfite ($NaHSO_3$) reacts with most aldehydes, methyl alkyl ketones and non-aromatic cyclic ketones to form slightly soluble solid "bisulfite addition" products. The reaction with propanal is:

$$CH_3CH_2\overset{H}{C}{=}O + NaHSO_3 \rightarrow CH_3CH_2\overset{H}{\underset{SO_3^-Na^+}{C}}{-}OH$$

Since the addition products are readily decomposed by either acid or alkali, the reaction serves as a convenient means of separating and recovering aldehydes or ketones from accompanying impurities.

note that addition of N to the carbon and H to the oxygen occurs first in each case.

5. Addition of Ammonia and Ammonia Derivatives. Ammonia and its derivatives, hydroxylamine (H_2NOH), phenyl hydrazine ($C_6H_5NHNH_2$), and semicarbazide ($H_2NCONHNH_2$), each have a free pair of electrons bound to nitrogen. These electrons are available to bond with the carbon of the carbonyl group of aldehydes and ketones. The addition products, however, are unstable reactive substances which, in most cases, lose water immediately. The compounds obtained as a result of this loss of water are stable crystalline materials that have proved to be very useful for the characterization and identification of carbonyl-containing compounds. In the case of ammonia, the dehydration products are not usually obtained because, in most cases, they polymerize to form *cyclic trimers*. Representative equations for reactions of this type follow:

$$3\,CH_3\overset{H}{C}{=}O + 3\,NH_3 \longrightarrow \left[3\,CH_3\overset{H}{\underset{NH_2}{C}}{-}OH\right] \longrightarrow \left[3\,CH_3\overset{H}{C}{=}NH\right] \longrightarrow$$

Ammonia

$$\begin{array}{c} \text{H} \\ | \\ \text{N} \\ / \quad \backslash \\ CH_3CH \quad HC{-}CH_3 \\ | \qquad\quad | \\ \text{HN} \quad \text{H} \quad \text{NH} \\ \backslash \;|\; / \\ \text{C} \\ | \\ CH_3 \end{array} + 3\,H_2O$$

$$CH_3\overset{\overset{CH_3}{|}}{C}{=}O + \underset{\text{Hydroxylamine}}{H_2NOH} \longrightarrow \left[CH_3{-}\underset{\underset{NHOH}{|}}{\overset{\overset{CH_3}{|}}{C}}{-}OH \right] \longrightarrow \underset{\text{An oxime}}{CH_3{-}\overset{\overset{CH_3}{|}}{C}{=}NOH} + H_2O$$

$$C_6H_5{-}\overset{\overset{H}{|}}{C}{=}O + \underset{\text{Phenyl hydrazine}}{H_2NNHC_6H_5} \longrightarrow \left[C_6H_5{-}\underset{\underset{NHNHC_6H_5}{|}}{\overset{\overset{H}{|}}{C}}{-}OH \right] \longrightarrow \underset{\text{A phenyl hydrazone}}{C_6H_5{-}\overset{\overset{H}{|}}{C}{=}\overset{\overset{H}{|}}{N}NC_6H_5} + H_2O$$

$$CH_3CH_2\overset{\overset{H}{|}}{C}{=}O + \underset{\text{Semicarbazide}}{H_2\overset{\overset{H}{|}}{N}\overset{\overset{O}{\|}}{C}NH_2} \longrightarrow \left[CH_3CH_2\underset{\underset{HNNHCONH_2}{|}}{\overset{\overset{H}{|}}{C}}{-}OH \right] \longrightarrow \underset{\text{A semicarbazone}}{CH_3CH_2\overset{\overset{H}{|}}{C}{=}N\overset{\overset{H}{|}}{N}CONH_2}$$

The substances given in the brackets are not usually isolated, but are the presumed reaction intermediates. The net result of these reactions is a substitution of a nitrogen compound for the oxygen of the carbonyl group.

6. Addition of Grignard Reagents. **Grignard reagents,** which have the general formula RMgX (or ArMgX) are organometallic compounds prepared by direct reaction between alkyl (or aryl) halides and magnesium metal in dry ether. These compounds, because of the relatively small electronegativity of metals, are highly polar covalent substances in which the metal atom is positive and the carbon to which it is bonded is negative. A typical Grignard reagent, $\overset{-}{R}\overset{+}{M}gX$, behaving as if it were an ionic substance, readily adds to the carbonyl group of most aldehydes and ketones. The nucleophilic R^- first adds to the carbonyl carbon and the $\overset{+}{M}gX$ associates with the oxygen to produce an addition product which is a mixed salt of divalent magnesium. This adduct, which is usually not isolated, when hydrolyzed with dilute acid yields an alcohol and the corresponding magnesium salt. The reaction has wide application for the preparation of alcohols. A typical preparation consists of three distinct operations which are carried out in sequence in the same reaction flask:

can you outline a mechanism for the reaction of an aldehyde or ketone with CH_3CH_2MgCl?

1. Formation of the Grignard reagent by reaction of an organic halide with magnesium in anhydrous ether.

$$CH_3CH_2Cl + Mg \rightarrow CH_3CH_2MgCl$$

2. Addition of a sample of anhydrous aldehyde (or ketone) to an ether solution of the Grignard reagent to form the addition product.

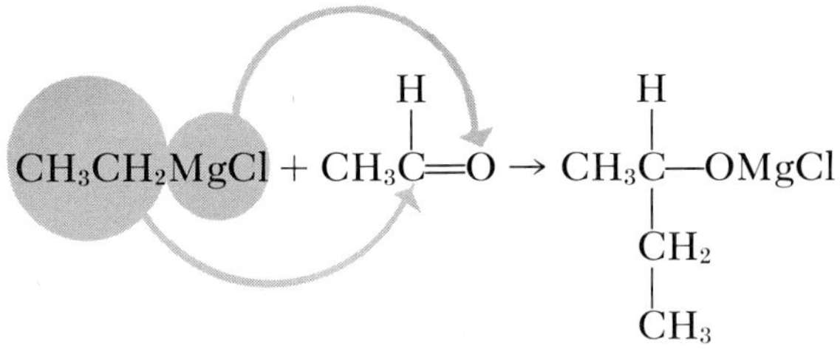

$$CH_3CH_2MgCl + CH_3\overset{\overset{H}{|}}{C}{=}O \rightarrow CH_3\overset{\overset{H}{|}}{\underset{\underset{\underset{CH_3}{|}}{CH_2}}{\underset{|}{C}}}{-}OMgCl$$

3. Hydrolysis of the addition product followed by separation and purification of the alcohol formed.

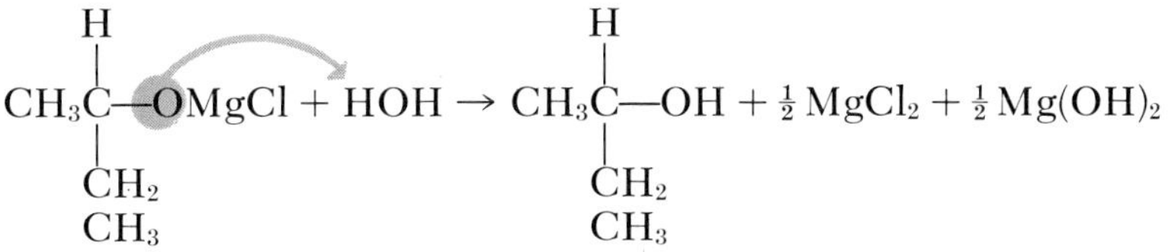

$$CH_3\overset{\overset{H}{|}}{\underset{\underset{CH_3}{CH_2}}{\underset{|}{C}}}{-}OMgCl + HOH \rightarrow CH_3\overset{\overset{H}{|}}{\underset{\underset{CH_3}{CH_2}}{\underset{|}{C}}}{-}OH + \tfrac{1}{2}MgCl_2 + \tfrac{1}{2}Mg(OH)_2$$

Reaction 2 is the addition reaction of the Grignard reagent to the carbonyl group of the compound. The overall reaction provides a method for preparing alcohols of any class, the class of alcohol obtained being determined by the particular aldehyde or ketone used. Only formaldehyde yields a primary alcohol. Other aldehydes yield secondary alcohols, and ketones yield tertiary alcohols. Examples are as follows:

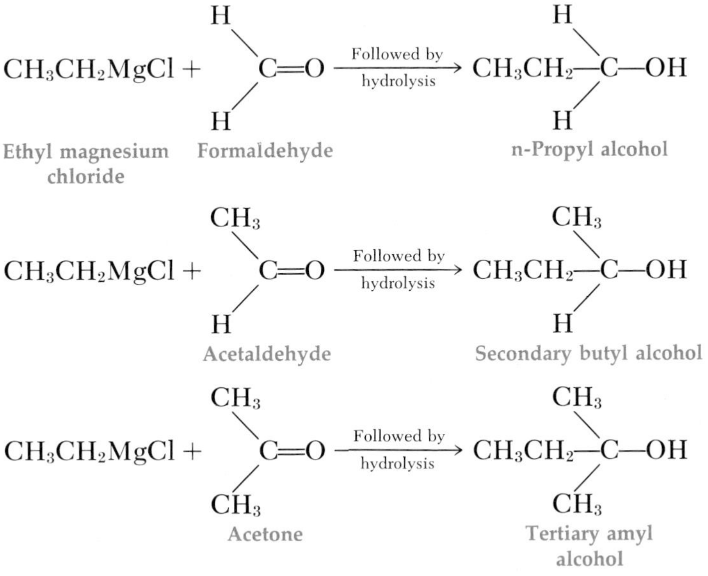

only with formaldehyde can one get a primary alcohol.

$$\underset{\text{Ethyl magnesium chloride}}{CH_3CH_2MgCl} + \underset{\text{Formaldehyde}}{(H)(H)C{=}O} \xrightarrow[\text{hydrolysis}]{\text{Followed by}} \underset{\text{n-Propyl alcohol}}{CH_3CH_2{-}C(H)(H){-}OH}$$

$$CH_3CH_2MgCl + \underset{\text{Acetaldehyde}}{(CH_3)(H)C{=}O} \xrightarrow[\text{hydrolysis}]{\text{Followed by}} \underset{\text{Secondary butyl alcohol}}{CH_3CH_2{-}C(CH_3)(H){-}OH}$$

$$CH_3CH_2MgCl + \underset{\text{Acetone}}{(CH_3)(CH_3)C{=}O} \xrightarrow[\text{hydrolysis}]{\text{Followed by}} \underset{\text{Tertiary amyl alcohol}}{CH_3CH_2{-}C(CH_3)(CH_3){-}OH}$$

The addition of Grignard reagents to carbonyl containing compounds is a versatile reaction since, by varying the structure of the alkyl halide or the carbonyl compound used, a wide variety of different alcohols—alcohols having different structures—may be prepared. For example, 3-methyl-3-hexanol can be prepared starting with either n-propyl chloride and methyl ethyl ketone or with ethyl chloride and 2-pentanone. Benzyl alcohol can be prepared using chlorobenzene and formaldehyde. In each case the alkyl or aryl halide is used to prepare a Grignard reagent which then is permitted to react with the aldehyde or ketone. Other reactions employing the Grignard reagent for synthesis will be encountered in subsequent chapters.

7. Self Addition. Self addition results from the presence of less strongly bonded α-hydrogen atoms. Self additions of aldehydes and ketones are, in general, referred to as **aldol condensations.** The term aldol condensation is taken from the name given to the specific product, aldol, formed when acetaldehyde reacts with itself in the presence of dilute base.

$$\mathrm{CH_3\overset{H}{C}{=}O + H\underset{H}{\overset{H}{C}}{-}\overset{H}{C}{=}O \xrightarrow{NaOH} CH_3\underset{OH}{\overset{H}{C}}{-}\underset{H}{\overset{H}{C}}\overset{H}{C}{=}O}$$

Aldol (an aldehyde-alcohol)

Sometimes the addition products are stable enough to be isolated. In other cases water may be lost spontaneously to form unsaturated aldehydes or ketones. Aldol itself may be isolated, but it does undergo dehydration, forming crotonaldehyde when the solution is acidified with sulfuric acid.

$$\mathrm{CH_3\underset{OH}{\overset{H}{C}}{-}\underset{H}{\overset{H}{C}}{-}\overset{H}{C}{=}O \xrightarrow{H_2SO_4} CH_3\overset{H}{C}{=}\overset{H}{C}{-}\overset{H}{C}{=}O + H_2O}$$

Ketones exhibit a smaller tendency to undergo the aldol condensation than do the aldehydes. Aromatic aldehydes possess no alpha hydrogens and can not undergo the aldol condensation. They do, however, form aldol-type condensation products with other compounds which have α-hydrogen atoms. For example, acetone reacts to form benzal acetone by the reaction:

$$\mathrm{C_6H_5\overset{H}{C}{=}O + H\underset{H}{\overset{H}{C}}{-}\overset{O}{\overset{\|}{C}}{-}CH_3 \xrightarrow{NaOH} \left[C_6H_5\underset{OH}{\overset{H}{C}}{-}\underset{H}{\overset{H}{C}}{-}\overset{O}{\overset{\|}{C}}{-}CH_3\right] \longrightarrow C_6H_5{-}\overset{H}{C}{=}\overset{H}{C}{-}\overset{O}{\overset{\|}{C}}CH_3 + H_2O}$$

Benzaldehyde acetone

Benzal acetone

Self addition does not always stop with the completion of one step. If a large number of molecules combine together as a result of a sequence of aldol condensations, viscous, water-insoluble materials known as resins or polymers are formed.

8. Reduction. Reduction, although considered separately from other addition reactions since it occurs via a different mechanism, is really the simplest of all additions. Most aldehydes and ketones add hydrogen directly to the carbonyl to form alcohols, if suitable catalysts such as platinum, palladium, nickel or copper chromite are provided.

$$CH_3CH(CH_3)CH_2CH{=}O + H_2 \xrightarrow[\text{Pressure}]{\text{Pt}} CH_3CH(CH_3)CH_2CH_2OH$$

The reduction of the carbonyl group may also be accomplished with chemical reducing agents such as sodium amalgam in water or sodium metal in ethanol.

$$C_6H_5\text{—}\overset{\overset{\displaystyle O}{\|}}{C}\text{—}C_6H_5 + 2\,Na + 2\,H_2O \longrightarrow C_6H_5\text{—}CH(OH)\text{—}C_6H_5 + 2\,NaOH$$

$$CH_3\overset{\overset{\displaystyle O}{\|}}{C}CH_2CH_2CH_2CH_2CH_3 + 2\,Na + 2\,C_2H_5OH \longrightarrow$$

$$CH_3CH(OH)CH_2CH_2CH_2CH_2CH_3 + 2\,C_2H_5\overset{-}{O}\overset{+}{Na}$$

This reaction is a reversal of the dehydrogenation of alcohols (discussed in an earlier section) as a method for preparing aldehydes or ketones.

Aldehydes and ketones may also be converted into alcohols using the more useful $\mathbf{LiAlH_4}$ and $\mathbf{NaBH_4}$. These are relatively mild reducing agents which, in contrast with reductions performed with hydrogen gas in the presence of metal catalysts, do not reduce carbon-carbon double bonds that might be present. The reaction with these hydrides proceeds through an addition process in which a hydrogen, as a **hydride ion** ($\mathbf{H^-}$), adds to the carbonyl carbon (compare with the addition of CN^-) and the Al (or B) atom attaches to the carbonyl oxygen, forming a complex which when decomposed with water yields the alcohol, $Al(OH)_3$ and LiOH. With the ketone 2-butanone (ethylmethyl ketone), the reaction proceeds in two stages forming 2-butanol.

$$4\,CH_3CH_2C(CH_3){=}O + LiAlH_4 \rightarrow (CH_3CH_2CH(CH_3)\text{—}O)_4AlLi$$

$$(CH_3CH_2CH(CH_3)\text{—}O)_4AlLi + 4\,H_2O \rightarrow 4\,CH_3CH_2CH(CH_3)\text{—}OH + Al(OH)_3 + LiOH$$

About 0.25 mole of $LiAlH_4$ is required for each mole of ketone or aldehyde used. With 2-butenal the reaction forms 2-buten-1-ol without reduction of the double bond.

$$CH_3CH{=}CHCHO \xrightarrow{0.25\ \text{mole}\ LiAlH_4} CH_3CH{=}CHCH_2OH$$

Sodium borohydride, $NaBH_4$, a milder reducing agent than $LiAlH_4$ toward other functional groups, forms with aldehydes and ketones a similar intermediate complex, which decomposes with water in the same way to yield an alcohol. Aldehydes yield primary alcohols and ketones form secondary alcohols here, just as they do with other reducing agents.

B. Oxidation

One of the important differences between aldehydes and ketones is their relative ease of oxidation. Aldehydes are oxidized very readily. Ketones undergo oxidation only with strong oxidizing agents under rather unusual conditions. The difficulty in the oxidation of ketones is understood when one realizes that the carbon to carbon bonds must be broken, degrading the compound into fragments, all of which have fewer atoms than the original ketone.

Several tests based on this difference in behavior are used to distinguish aldehydes from ketones. Very mild oxidizing agents such as *Tollens' reagent, Fehling's solution and Benedict's solution,* too mild to affect the oxidation of either multiple C to C bonds or alcohols, can be used for this purpose because they do not react with ketones. **Tollens' reagent** is an ammoniacal solution of silver nitrate. Reaction with this reagent converts an aldehyde to a salt of a carboxylic acid, and the silver ammonia complex ion is reduced to metallic silver. If the test tube in which the reaction is performed is clean, a silver mirror will form on the glass surface in the same way that mirrors can be made commercially. A simplified equation for this reaction is:

$$\underset{\text{An aldehyde}}{R\overset{H}{\overset{|}{C}}{=}O} + 2\,Ag(NH_3)_2^+ + 3\,OH^- \rightarrow R\overset{O}{\overset{\|}{C}}O^- + \underset{\substack{\text{Silver}\\ \text{mirror}}}{2\,Ag} + 2\,H_2O + 4\,NH_3$$

Fehling's and **Benedict's** reagents are alkaline solutions of copper(II) sulfate with a suitable binding reagent (called a chelating agent) added to keep the copper(II) ion in solution. Sodium citrate serves this function in Benedict's solution. The equation for the reaction may be written as:

$$R\overset{H}{\overset{|}{C}}{=}O + 2\,CuO + OH^- \xrightarrow[\text{or citrate}]{\text{tartrate}} R\overset{O}{\overset{\|}{C}}O^- + \underset{\text{Red}}{Cu_2O} + H_2O$$

The occurrence of a reaction when this test is performed is shown by the appearance of copper(I) oxide as a red precipitate. This reaction is often used to detect the presence of an amount of a "reducing" sugar in the blood or urine. The quantity of copper(I) oxide formed is a measure of the sugar concentration. Benedict's and Fehling's solutions do not oxidize aromatic aldehydes like benzaldehyde, whereas Tollens' reagent does.

With stronger oxidizing agents, aldehydes are oxidized readily to the corresponding carboxylic acids or their salts. Both aromatic and aliphatic aldehydes react smoothly with alkaline potassium dichromate, alkaline potassium permanganate, hydrogen peroxide, aqueous silver oxide, peracetic acid and oxygen. These reactions may be illustrated by the oxidation of benzaldehyde and n-butyraldehyde as follows:

$$2\, C_6H_5CHO + O_2 \longrightarrow 2\, C_6H_5COOH$$

$$3\, CH_3CH_2CH_2CH{=}O + 2\, MnO_4^- + OH^- \longrightarrow 2\, MnO_2 + 3\, CH_3CH_2CH_2COO^- + 2\, H_2O$$

C. Substitution of Alpha Hydrogens

Hydrogen atoms attached to the α-carbon of aldehydes and ketones may be readily replaced by halogen atoms in a basic solution. One or more such atoms may be substituted, depending on the reaction time and conditions under which the reaction is performed. The equation for this reaction may be written as follows:

$$CH_3CH_2CH{=}O + 2\, Cl_2 + 2\, OH^- \rightarrow CH_3CCl_2{-}CH{=}O + 2\, H_2O + 2\, Cl^-$$

$$CH_3CH_2C(=O){-}R + 2\, Br_2 + 2\, OH^- \rightarrow CH_3CBr_2{-}C(=O){-}R + 2\, H_2O + 2\, Br^-$$

When acetaldehyde reacts with chlorine in a basic solution, three α-hydrogens are replaced to form chloral (2,2,2-trichlorethanol). Substances of this type, that is, compounds having three halogens substituted on a carbon next to a carbonyl, are readily cleaved with the alkali present. The carbon-carbon bond of chloral breaks to form chloroform ($CHCl_3$). Iodoform, a yellow crystalline solid, precipitates when acetaldehyde reacts under similar conditions with iodine in an alkaline solution. This reaction with a basic solution of iodine, called the **iodoform test,** is used to detect acetaldehyde, methyl ketones or potential methyl ketones. The structure $CH_3C(=O){-}$ is needed, because the basic cleavage occurs only after $CI_3C(=O){-}$ forms. Ethanol gives a positive iodoform test, because the reagent itself oxidizes ethanol to acetaldehyde.

$$CH_3CH_2OH + I_2 + 2\,OH^- \rightarrow CH_3CHO + 2\,I^- + 2\,H_2O$$
Ethanol

Replacement of the active alpha-hydrogens with iodine follows:

$$CH_3\overset{\overset{\displaystyle H}{|}}{C}{=}O + 3\,I_2 + 3\,OH^- \rightarrow CI_3\overset{\overset{\displaystyle H}{|}}{C}{=}O + 3\,H_2O + 3\,I^-$$
Acetaldehyde

Cleavage then yields iodoform:

$$CI_3\overset{\overset{\displaystyle H}{|}}{C}{=}O + NaOH \rightarrow \underset{\text{Iodoform}}{CHI_3} + \underset{\text{Sodium formate}}{HCOONa}$$

Isopropyl alcohol and all other secondary 2-alkanols oxidize to methyl ketones and consequently give a positive iodoform test.

D. Ring Substitution Reactions

Aromatic aldehydes and ketones have the potentiality for undergoing substitution in the aromatic ring. Sulfonation, halogenation and nitration lead to ring substitution mainly in the *meta position*. The equation for a typical substitution reaction of benzaldehyde and of acetophenone is as follows

the C═O, being electron withdrawing, deactivates the aromatic ring.

$$C_6H_5\text{–}HC{=}O + HONO_2 \longrightarrow m\text{-}NO_2C_6H_4\text{–}HC{=}O + H_2O$$

$$C_6H_5\text{–}C(CH_3){=}O + HONO_2 \xrightarrow[0^\circ]{H_2SO_4} m\text{-}NO_2C_6H_4\text{–}C(CH_3){=}O$$

It should be noted that the aldehyde group is particularly sensitive to oxidation by nitric acid and halogens, and for best yields should be protected (temporarily combined with some other group) when substitution reactions using these reagents are attempted.

EXERCISES

30.1 What group of atoms is responsible for the chemical properties of: (a) alkenes, (b) alcohols, (c) aldehydes, (d) ketones?

30.2 Define: (a) functional group, (b) polar bond, (c) addition reaction, (d) electrophilic, (e) oxidation, (f) mixed ketone, (g) an aldol, (h) an α-hydrogen.

30.3 Using both common names and IUPAC names, name and write the structural formulas for three common aldehydes and three common ketones.

30.4 Use these endings correctly in IUPAC names of compounds: (a) ene, (b) ane, (c) yne, (d) one, (e) al.

30.5 Name each of these compounds correctly by the IUPAC system:

a. $(CH_3)_2CH—CHO$

b. 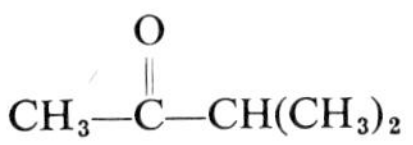 $CH_3—C(=O)—CH(CH_3)_2$

c. $CH_2Cl—CH(CH_3)—C(=O)—CH_3$

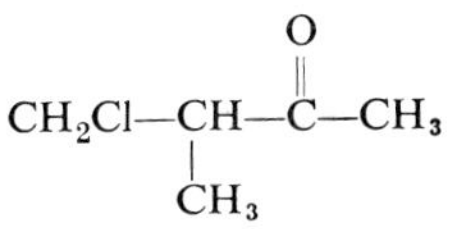

d. 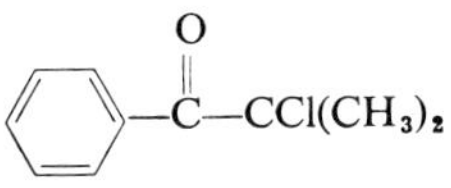 $C_6H_5—C(=O)—CCl(CH_3)_2$

30.6 Write equations showing any catalysts or special conditions for the two-step preparation of:

a. benzaldehyde (phenylcarboxaldehyde) from toluene
b. acetaldehyde from ethyl chloride
c. formaldehyde from carbon monoxide, hydrogen and oxygen
d. acetaldehyde from calcium carbide and water

30.7 Write equations for the additions of four different reagents to a carbonyl compound.

30.8 Write an equation for an aldol condensation.

30.9 Which of these compounds: (a) form a silver mirror with Tollens' reagent, (b) form a red precipitate with Fehling's solution, (c) decolorize cold basic potassium permanganate solution, (d) form iodoform when added to an aqueous solution of iodine and sodium hydroxide: (i) ethanol, (ii) isopropyl alcohol, (iii) propane, (iv) propene, (v) formaldehyde, (vi) benzaldehyde, (vii) methyl ethyl ketone?

30.10 What alcohol is formed when methyl magnesium bromide reacts with the aldehyde or ketone given below? For each alcohol, give its structure and the equations for the reactions required to form the alcohol.

1. CH_3CHO
2. 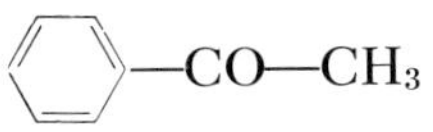 $C_6H_5—CO—CH_3$
3. CH_2O
4. $CH_3COCH_2CH_3$
5. 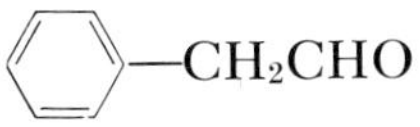 $C_6H_5—CH_2CHO$
6. 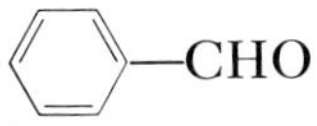 $C_6H_5—CHO$
7. Cyclohexanone
8. Propanal

30.11 Show two enol forms which might result when 3-hexanone undergoes keto-enol tautomerism.

30.12 What is unique about the structure and composition of chloral hydrate? Explain.

30.13 If a single α-hydrogen of 2-pentanone is replaced with a bromine atom, how many such monobromosubstituted 2-pentanones are possible? Why?

30.14 A compound A, C_4H_8O, reacts with methyl magnesium bromide to form a product which hydrolyzes to compound B, $C_5H_{12}O$. When B is oxidized it forms compound C, $C_5H_{10}O$. Compound C gives a positive iodoform test and forms a semicarbazone when reacted with semicarbazide hydrochloride. What is an acceptable structure for compound A? Write equations for all reactions which compound A undergoes as described above.

30.15 Compound A is one of the monochloropentanes. It reacts with magnesium to form a Grignard reagent which reacts with water to form n-pentane and with acetaldehyde to form the alcohol, 3-methyl-2-hexanol. What is a suitable structure for compound A? Why?

30.16 A compound A, $C_5H_{10}O$, reacts with hydroxyl amine to form an oxime. It does not react with Tollens' reagent but forms iodoform when reacted with $I_2 + NaOH$. Compound A may be converted by suitable reduction into n-pentane. What is an acceptable structure for compound A? Explain.

30.17 A compound A, $C_6H_{14}O$, was passed over hot Cu at 500°C to give compound B, $C_6H_{12}O$. Compound B undergoes no reaction with Fehling's solution but does form an oxime when treated with NH_2OH. Iodoform does not form when compound B is treated with NaOI, but compound A is re-formed when compound B is treated with $LiAlH_4$ at 25°C. Write a correct structure for compound A. Explain.

30.18 A volatile liquid A reacts with hydroxyl amine to yield a crystalline solid B which contains 19.18% nitrogen. Liquid A did not form a silver mirror with Tollen's reagent or Fehling's solution. The vapor density of A at STP was 29. Write structures for compounds A and B. Show how A might be prepared from an alkyne.

30.19 An aromatic oxygen containing compound A has the empirical formula C_8H_7OBr. It forms a bisulfite addition product, compound B, and gives a solver mirror test. Oxidation of A gives a compound C which has the formula, $C_7H_5O_2Br$. Compound C does not react with CH_3Cl in the presence of $AlCl_3$. Nitration of C gives only one mononitro product of two which are possible. Upon reduction, compound A, accepts one mole of H_2 to give compound D. When D is reacted with metallic sodium it yields a gas. Write structural formulas for compounds A, B, C and D. Write equations for the reactions described.

30.20 Give the structure of the product formed at each stage of the sequence of reactions outlined in the following (the reagent to be used is shown over or below the arrow in each case):

$$A \xrightarrow{CH_3MgI} B \xrightarrow{H_2O} C \xrightarrow[\text{Conc. } H_2SO_4]{\text{Heat}} D \xrightarrow{HBr} E \xrightarrow{NaOH(aq)} F \xrightarrow[500°C]{Cu} CH_3\overset{\overset{\displaystyle O}{\|}}{C}CH_3$$

SUGGESTED READING

Gould, E. S.: *Mechanism and Structure in Organic Chemistry.* Holt, Rinehart and Winston, Inc., New York, 1959. Chapters 9, 10 and 13.

Gutsche, C. D.: *The Chemistry of Carbonyl Compounds.* Prentice-Hall, Inc., Englewood Cliffs, N. J., 1966 (paperback).

Hendrickson, J. B., Cram, D. J. and Hammond, G. S.: *Organic Chemistry,* Third Edition. McGraw-Hill Book Company, New York, 1970. Chapters 12 and 13.

Morrison, R. T. and Boyd, R. N.: *Organic Chemistry,* Second Edition. Allyn and Bacon, Inc., Boston, 1966. Chapters 19, 27 and 30.

Noller, C. R.: *Chemistry of Organic Compounds,* Third Edition. W. B. Saunders Company, Philadelphia, 1965. Chapters 13, 16, 26, 27, 37 and 40.

THIRTY-ONE • ORGANIC ACIDS AND THEIR DERIVATIVES

Some compounds which are acids according to the Brønsted-Lowry definition have been encountered among the organic compounds already studied. Protons are replaced in terminal alkynes, alcohols and phenols upon reaction with sodium.

a. $2\,RC{\equiv}CH + 2\,Na \rightarrow 2\,RC{\equiv}CNa^+ + H_2$
A terminal alkyne — A sodium acetylide

b. $2\,ROH + 2\,Na \rightarrow 2\,RO^-Na^+ + H_2$
An alcohol — A sodium alkoxide

c. $2\,ArOH + 2\,Na \rightarrow 2\,ArO^-Na^+ + H_2$
A phenol — A sodium phenoxide

Of these three classes of compounds, only phenols are sufficiently acidic to react with a strong base in aqueous solution. Phenols alone are more acidic than water:

d. $ArOH + NaOH \rightarrow ArO^-Na^+ + H_2O$

The acids to be discussed in this chapter, like phenols, are more acidic than water and may therefore be neutralized by bases in aqueous solution.

31.1 ORGANIC ACIDS

Included among the variety of organic acids are the **carboxylic acids**, the **sulfonic acids**, the **alkyl hydrogen sulfates**, the **thiocar-**

boxylic acids and the **alkyl phosphoric acids.** The four most important of these acids are shown in Figure 31.1. All four are hydroxy acids, as are sulfuric acid (HO—SO_2—OH) and nitric acid (HO—NO_2), in which the active hydrogen (the one lost as a proton upon reaction with a base) is a hydrogen of a hydroxyl group. In each the oxygen losing the proton in an acid-base exchange is attached to a non-metal. Also, in each acid there is more than one oxygen to which the hydrogen might be bonded. This can be illustrated by showing the anions formed when a proton is lost from sulfuric acid, methyl sulfonic acid and ethyl hydrogen sulfate (Fig. 31.2). The negative charge carried by each ion shown in the

many synthetic detergents are salts of alkyl hydrogen sulfates.

```
    O              O                  O                 OH
    ‖              |                  |                 |
R—C—OH        R—S—OH          R—O—S—OH          R—O—P=O
                   |                  |                 |
                   O                  O                 OH
```

Carboxylic acids | Sulfonic acids | Alkyl hydrogen sulfates | Alkyl hydrogen phosphate

FIGURE 31.1 TYPES OF ORGANIC ACIDS.

figure is distributed equally over all three oxygens so that all are equivalent. This equivalence of the oxygen atoms leads to a degree of resonance stabilization (delocalization of electrons) in the ion which exceeds that in the un-ionized acid, a fact that favors ionization of the proton giving to each acid its acidic character.

For the **carboxylate ion** (the anion of a carboxylic acid) the two equivalent positions for reception of a proton by the ion are not apparent in the customary formula, CH_3COO^-, but there is resonance between the structures shown in Figure 31.3a and 31.3b. That is, the real ion is an arrangement of atoms and bonds in a structure intermediate between the two. Figure 31.3c is a better representation. Between the carbon and each oxygen is a sigma bond. A delocalized pi orbital about the carbon and the two oxygens of the carboxyl group contains the other two electrons. The extra stabilization of the carboxylate ion due to delocalization (resonance stabilization) over that of the un-ionized acid allows the hydroxyl proton of the acid to be lost more easily to a base than the proton of a hydroxyl group of an alcohol where stabilization by resonance is not involved. The carboxylic acids are, therefore, more acidic than alcohols.

delocalization of the electrons always stabilizes the species in which it occurs.

This chapter will discuss the carboxylic acids (by far the most important type of organic acids) and derivatives of carboxylic acids. Industrial sulfonic acids also will be mentioned. Some esters of phosphoric acid are shown on pages 746 and 747.

```
        ..                      ..                              ..
       :O:⁻                    :O:⁻                            :O:⁻
        ..  ..                  ..  ..                          ..  ..
H : O : S : O :         CH3—S : O :          CH3CH2—O—S : O :
        ..  ..                  ..  ..                          ..  ..
       :O:                     :O:                             :O:
        ..                      ..                              ..
```

Hydrogen sulfate ion | Alkyl sulfonate ion | Alkyl sulfate ion

FIGURE 31.2 ANIONS OF ACIDS.

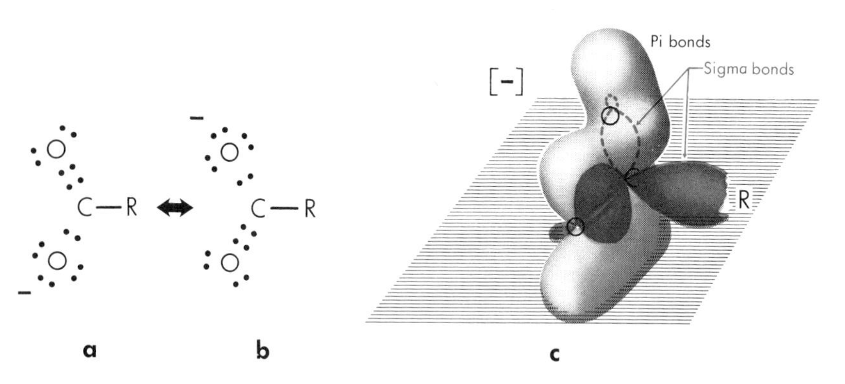

FIGURE 31.3 THE CARBOXYLATE ION.

31.2 NOMENCLATURE

Carboxylic acid may contain one or more carboxyl groups. The number of groups may be designated by the prefixes *mono-*, *di-*, and *tri-*. Acetic acid, $CH_3—\overset{\overset{\large O}{\|}}{C}—OH$, is a monocarboxylic acid. The carboxyl group may be either the end carbon of a chain or may be attached to an aromatic carbon. They are called aliphatic and aromatic carboxylic acids, respectively. Some acids are shown in Figure 31.4.

When animal fats are boiled in lye (sodium hydroxide solution) and then acidified, many monocarboxylic acids containing from 6 to 22 carbons precipitate from the solution. For this reason, **saturated monocarboxylic acids** are called **fatty acids.** Most of the R-groups of the acids obtained are unbranched alkane chains. Some long chain **unsaturated acids** (acids having one or more double carbon to carbon bonds) appear with them. These are called **unsaturated fatty acids.**

Common names are used widely for carboxylic acids. Names of the first ten members of the normal fatty acid series appear with others in Table 31.1.

common names are perhaps more frequently used than IUPAC names because of their long history of use.

$CH_3—C(=O)—OH$
Acetic acid
(an aliphatic acid)

$C_6H_5—C(=O)—OH$
Benzoic acid
(an aromatic acid)

$C_6H_4(—C(=O)—OH)_2$
o-Phthalic acid
(a benzene dicarboxylic acid)

$CH_3CH_2C(=O)—OH$
Propionic acid
(an aliphatic acid)

$HO—C(=O)—CH_2—C(=O)—OH$
Malonic acid
(an aliphatic dicarboxylic acid)

$CH_2(C(=O)OH)—COH(C(=O)OH)—CH_2(C(=O)OH)$
Citric acid
(an aliphatic tricarboxylic acid)

FIGURE 31.4 SOME CARBOXYLIC ACIDS.

TABLE 31.1 PHYSICAL PROPERTIES OF SOME ACIDS

No. of Carbons	Common* Name	Melting Point °C	Boiling Point °C	Solubility in Water g/100 g Water	IUPAC Name
1	Formic†	8.3	101	Completely	Methanoic Acid
2	Acetic	16.67	119	Completely	Ethanoic Acid
3	Propionic	−36.0	141	Completely	Propanoic Acid
4	Butyric‡	−4.7	163	5.62	Butanoic Acid
5	Valeric	−34.5	186	3.7	Pentanoic Acid
6	Caproic	−1.5	205	0.4	Hexanoic Acid
7	Enanthic	−10.5	223	—	Heptanoic Acid
8	Caprylic	16.5	237	0.25	Octanoic Acid
9	Pelargonic	12.5	254	—	Nonanoic Acid
10	Capric	31.4	269	Slightly	Decanoic Acid
16	Palmitic	63.1	268	Insoluble	Hexadecanoic Acid
18	Stearic	70.1	287	Insoluble	Octadecanoic Acid
18	Oleic	13.5	—	Insoluble	9-Octadecenoic Acid
7	Benzoic	121.0	250	0.18	Benzene-carboxylic Acid
9	Phthalic	213.0	—	0.54	1,2-Benzene-dicarboxylic Acid
2	Oxalic	101.2	Subl. 150	9.5	Ethandioic Acid
3	Malonic	135.6	Decomposes	73.5	Propandioic Acid
6	Adipic	152.0	265	1.5	Hexandioic Acid

* The term "acid" should follow each common name given.
† From *formica,* the Latin word for ants; it was first prepared by distillation of red ants.
‡ Found in rancid butter.

Systematic IUPAC names for monocarboxylic acids are obtained by replacing the ending *e* of the name of the parent hydrocarbon with the suffix *oic,* followed by the word *acid.* The carbon of the carboxyl group must be one of the chain and must be number *one.* It need not be numbered in the name. Branches and groups are named and placed by number as usual. For example,

Acetic acid $CH_3-C(=O)-OH$ is ethanoic acid.

Isobutyric acid $CH_3-CH(CH_3)-C(=O)-OH$ is methylpropanoic acid.

Phenyl acetic acid $C_6H_5-CH_2-C(=O)-OH$ is phenylethanoic acid.

Cinnamic acid $C_6H_5-CH{=}CH-C(=O)-OH$ is 3-phenyl-2-propenoic acid.

Benzoic acid $C_6H_5-C(=O)-OH$ is phenylmethanoic acid.

Dicarboxylic acids have the suffix *dioic:*

Oxalic acid $HO{-}\overset{\overset{O}{\|}}{C}{-}\overset{\overset{O}{\|}}{C}{-}OH$ is ethandioic acid.

Malonic acid $HO{-}\overset{\overset{O}{\|}}{C}{-}CH_2{-}\overset{\overset{O}{\|}}{C}{-}OH$ is propandioic acid.

Adipic acid $HO{-}\overset{\overset{O}{\|}}{C}{-}(CH_2)_4{-}\overset{\overset{O}{\|}}{C}{-}OH$ is hexandioic acid.

Acceptable systematic names may be assigned by appending the name "carboxylic acid," "dicarboxylic acid" and so forth, to the name of the compound in which the carboxyl groups replace hydrogens. For example, succinic (or butandioic*) acid,

$HO(O{=})C{-}CH_2{-}CH_2{-}C({=}O)OH$, may be called ethane-1,2-dicarboxylic acid. These names are especially useful for aromatic acids.

o-Phthalic acid, $C_6H_4(COOH)_2$ (benzene ring with two adjacent $-\overset{\overset{O}{\|}}{C}{-}OH$ groups), may be called benzene-1,2-dicarboxylic acid.

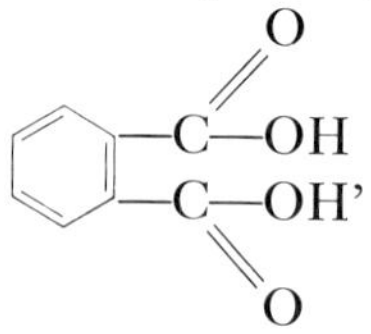

31.3 PHYSICAL PROPERTIES

Physical properties of some carboxylic acids are listed in Table 31.1. There are no carboxylic acid gases at room temperature. The smaller members of the monocarboxylic acid series are soluble (up to 4 carbons) in water. Dicarboxylic acids are solids and are about as soluble as monocarboxylic acids of the same molecular weight. This might be expected, considering the geometry and hence the polarity of the carboxyl group. The acids are more soluble in strongly basic solutions than in water because of the reaction with the base (neutralization) to form ionic compounds.

Smaller members of the monocarboxylic acid series have a sour taste and a biting odor. The odor of the second member, acetic acid, is well known. Butyric acid (4 carbons), has an acrid odor. The disagreeable odor of sour milk and rancid butter is caused largely by normal butyric acid. Caproic acid (hexanoic acid) is present in the skin and in secretions of goats. It is responsible for the "goaty" odor. Longer chain fatty acids, being much less volatile, are practically odorless.

---

* IUPAC name.

31.4 PREPARATION OF CARBOXYLIC ACIDS

Carboxylic acids may be prepared by a variety of methods, most of which involve oxidation or hydrolysis of suitable compounds.

A. Oxidation

The various oxidation reactions giving carboxylic acids as products have been discussed as reactions of other compounds. Types of compounds that may be oxidized to acids are:

1. *Unsaturated hydrocarbons.* Cleavage occurs at the multiple bond, yielding two acids unless the compound is symmetrical or unless the multiple bond is terminal (Chapter 27).
2. *Primary alcohols* can be oxidized, first to aldehydes, then to carboxylic acids (Chapter 29).
3. *Aldehydes* are oxidized to acids (Chapter 30).
4. *Side chains of aromatic compounds* are oxidized readily to acids (Chapter 28).

An interesting oxidation is that of naphthalene:

adipic acid reacts with $H_2N(CH_2)_6NH_2$, splitting out H_2O to form the polymer nylon.

$$C_{10}H_8 \text{ (naphthalene)} + [O] \xrightarrow[450\text{–}500^\circ]{air/V_2O_5} C_6H_4(COOH)_2$$

The second ring is oxidized as if it were two side chains.

The oxidation of the specific secondary alcohol, cyclohexanol, is also of interest:

$$\underset{\text{Cyclohexanol}}{\text{cyclo-}(CH_2)_5CHOH} \xrightarrow[(O)]{HNO_3} \underset{\text{Adipic Acid (Hexanedioic acid)}}{HOOC-CH_2-CH_2-CH_2-CH_2-COOH}$$

This is an intermediate reaction in the preparation of nylon. Phenol from coal tar is hydrogenated to cyclohexanol, which is oxidized by nitric acid (as just shown) to give one of the two reactants which are necessary for making nylon.

B. Hydrolysis

1. *Hydrolysis of nitriles* (cyanides) and the various types of acid derivatives yields carboxylic acids. Both alkyl and aryl carboxylic acids can be obtained in this way. For example, propionitrile yields propionic acid, and benzonitrile yields benzoic acid,

$$CH_3CH_2CN + 2\,H_2O + HCl \rightarrow CH_3CH_2COOH + NH_4Cl$$

$$C_6H_5CN + H_2O + NaOH \longrightarrow C_6H_5COONa + NH_3$$

Alkyl nitriles are readily formed from alkyl halides by the reaction:

$$CH_3CH_2Cl + NaCN \rightarrow CH_3CH_2CN + NaCl$$

This conversion makes it possible to use alkyl halides as starting materials for preparing carboxylic acids. Note that this reaction permits the conversion of commercially available alkyl halides into compounds which have one more carbon atom. Aryl nitriles, on the other hand, may be prepared from the sodium or potassium salts of aromatic sulfonic acids upon fusion with sodium cyanide.

$$C_6H_5SO_3Na + NaCN \xrightarrow{\text{fusion}} C_6H_5CN + Na_2SO_3$$

Since aromatic sulfonic acids are readily prepared by sulfonation, this reaction serves as one step in the pathway to aromatic carboxylic acids from aromatic hydrocarbons.

2. *Trihalomethyl groups are readily hydrolyzed* to carboxyl groups. This method serves as the most suitable means for obtaining some particular carboxylic acids.

$$C_6H_4(CF_3)(NO_2) + 2\,H_2O \xrightarrow{H_2SO_4} C_6H_4(COOH)(NO_2) + 3\,HF$$

3. The hydrolysis reactions of various acid derivatives are of great practical importance as a source of certain acids. These reactions will be discussed in later sections of this chapter. Many carboxylic acids are obtained by the hydrolysis of naturally occurring esters, such as the fats, oils or waxes. Esters of this class are discussed in the next chapter. The hydrolysis of proteins, amides found in all living matter, is the principal source of many amino carboxylic acids. Proteins and amino acids will be discussed in Chapter 35.

C. Carbonation of Grignard Reagents

The reaction of a Grignard reagent with carbon dioxide (**carbonation**) is an important and relatively easy laboratory method for the synthesis of carboxylic acids. An ether solution of the Grignard reagent (RMgX or ArMgX) is allowed to react with carbon dioxide, either by passing gaseous carbon dioxide into the solution or by pouring the ether solution of the Grignard reagent into a mixture of absolute ether and crushed dry ice. The reagent adds to the carbonyl group of the CO_2 to form an addition product (reaction intermediate). This readily undergoes hydrolysis upon addition of aqueous mineral acid to yield the carboxylic acid. The

CO_2 is a linear molecule with structure O=C=O and having a carbonyl group to which the RMgX adds to form $R-C(=O)-O^-MgX^+$.

following reactions are involved in the preparation of n-butyric acid:

1. $CH_3CH_2CH_2Cl + Mg \rightarrow CH_3CH_2CH_2MgCl$

2. $CH_3CH_2CH_2MgCl + CO_2 \rightarrow CH_3CH_2CH_2\overset{\overset{\displaystyle O}{\|}}{C}\text{—}OMgCl$

3. $CH_3CH_2CH_2\overset{\overset{\displaystyle O}{\|}}{C}\text{—}OMgCl + HCl(aq) \rightarrow CH_3CH_2CH_2\overset{\overset{\displaystyle O}{\|}}{C}\text{—}OH + MgCl_2$

Almost any alkyl or aryl halide may be used in place of n-propyl chloride in this reaction sequence. The reaction sequence is particularly useful for the preparation of a branched chain carboxylic acid such as 2-methylbutyric acid or 2,2-dimethylpropionic acid, starting with the appropriate alkyl halide (i.e., 2-chlorobutane or 2-chloro-2-methylpropane), often the most readily available starting material for the synthesis.

31.5 REACTIONS

The characteristic reactions of the carboxylic acids are caused by the presence of the **carboxyl functional group.** Structurally the group is a composite of a carbonyl group and a hydroxyl group, a fact clearly shown when its structure is expanded and written as

$$\text{—}\overset{\overset{\displaystyle O}{\|}}{C}\text{—}OH.$$

Although the reactions which such acids might be expected to show are those of both the carbonyl and the hydroxyl groups (studied in the two previous chapters), when combined as they are here, they do not exhibit fully the properties that might be expected. The hydroxyl influence decreases the reactivity (polarity) of the carbonyl group to such an extent that acids behave very little like aldehydes and ketones. When a nucleophile does add to the carbonyl carbon, the attached hydroxyl usually is lost.

A. Replacement of the Hydrogen of the Carboxyl Group. Acidity. Salt Formation

The characteristic chemical property of carboxylic acids is acidity. In accord with the Brønsted-Lowry definition, they are acids because each carboxyl group (due to the hydroxyl group present) has the capability of donating a proton in an acid-base exchange. The carboxyl group loses a proton readily because the carboxylate ion formed when a proton is lost is stabilized by resonance much more than is the un-ionized acid from which it arises. In an aqueous solution of any carboxylic acid, water acts as a proton acceptor (base) and an equilibrium between acid, water, carboxylate and hydronium ion is established.

$$RCOOH + H_2O \rightarrow RCOO^- + H_3O^+$$

The extent to which this reaction yields ions is determined by the degree of acidity of the acid. With stronger acids, a larger concentration of carboxylate and hydronium ions is present at equilibrium. With weaker acids, a smaller concentration is present. The equilibrium (ionization or acidity) constant for this reaction, *Ka*, given by the expression

$$K_a = \frac{(RCOO^-)(H_3O^+)}{(RCOOH)},$$

serves as a measure of the strength of the acid. The greater the magnitude of the K_a, the stronger the acid. Weak acids have relatively smaller values.

For acetic acid the acidity constant expression is

$$K_a = \frac{(CH_3COO^-)(H_3O^+)}{(CH_3COOH)}$$

The experimentally determined K_a for acetic acid is 1.8×10^{-5} (25°C). Acids with larger values than this are stronger than acetic acid, and those with smaller values are weaker. Unsubstituted monocarboxylic acids like acetic acid, having K_a values near that of acetic acid, are classified as weak acids.

Familiar substances which are acidic in the Brønsted sense when compared with acetic acid fall into this general pattern in accord with their K_a values (see Table 31.2):

$$HCl > RSO_2OH > RCOOH > C_6H_5OH > HOH > ROH > RC{\equiv}CH > NH_3.$$

Substituted carboxylic acids (see Section D which follows) having electron withdrawing groups introduced in place of one or more hydrogen atoms attached to the alpha carbon of the acid exhibit acidities greater than that of the corresponding unsubstituted acids. For example, acetic, chloroacetic, dichloroacetic and trichloroacetic acids (see Table 31.3) increase in acidity in the order named because the electronegative halogen atoms attached to the alpha carbon attract electrons away from the carboxyl group, causing it to be more positively charged.

TABLE 31.2 ACIDITY CONSTANTS OF SOME ACIDS

Name	Formula	K_a (approximate)
Hydrochloric acid	HCl	Very large
Sulfonic acids	RSO_2OH	10^{-2} to 10^{-3}
Carboxylic acids	$RCOOH$	10^{-5}
Phenols	$ArOH$	10^{-10}
Water	HOH	10^{-14}
Alcohols	ROH	10^{-16}
Acetylenes	$RC{\equiv}CH$	10^{-26}
Ammonia	NH_3	10^{-35}
Methane	CH_4	10^{-37}

notice that for CCl_3COOH more than 50 per cent of the acid is ionized, whereas with acetic acid ionization is less than 1 per cent for a 1 molar solution.

TABLE 31.3 ACIDITY CONSTANTS OF SOME SUBSTITUTED ACIDS

Name	Formula	K_a
Acetic acid	CH_3COOH	1.8×10^{-5}
Chloroacetic acid	$ClCH_2COOH$	155×10^{-5}
Dichloroacetic acid	$Cl_2CHCOOH$	5140×10^{-5}
Trichloroacetic acid	CCl_3COOH	$121{,}000 \times 10^{-5}$
2-Chlorobutyric acid	$CH_3CH_2CHClCOOH$	140×10^{-5}
3-Chlorobutyric acid	$CH_3CHClCH_2COOH$	8.8×10^{-5}
Butyric acid	$CH_3CH_2CH_2COOH$	1.5×10^{-5}

Substituents attached to carbon atoms more distant than the alpha carbon have a considerably smaller effect on proton loss than do those attached directly to the carbon adjacent to the carboxyl group. For example, 2-chlorobutyric acid exhibits about the same acidity as chloroacetic acid whereas 3-chlorobutyric acid is less acidic and is similar in acidity to butyric acid itself and to acetic acid.

Since the carboxyl —OH loses a proton much more readily than does the —OH of alcohols because of resonance stabilization, reactions of acids which correspond to those of alcohols occur more readily. Acids react with active metals and with a variety of bases, such as sodium hydroxide or sodium carbonate, to form salts. Equations for typical reactions are:

$$C_6H_5COOH + Na_2CO_3 \longrightarrow C_6H_5\overset{-}{C}OO\overset{+}{Na} + NaHCO_3$$

Benzoic acid → Sodium benzoate

$$CH_3COOH + NaOH \longrightarrow CH_3COO^-Na^+ + H_2O$$

Acetic acid → Sodium acetate

$$2\,RCOOH + Ca \longrightarrow (RCOO)_2Ca + H_2$$

Acetic acid forms soluble salts with almost all metals. Salts of larger acids are less soluble. The metal salts (particularly the sodium and potassium salts) of the long chain fatty acids, such as palmitic and stearic acids, are known as soaps. They will be discussed with the fats and oils (Chapter 32).

B. Replacement of the Hydroxyl of the Carboxyl Group

A variety of reagents react with carboxylic acids to exchange the —OH with other groups. Compounds derived from the carboxylic acids by substituting an atom or group for the —OH are known as **acid derivatives.** Typical compounds formed by replacing the carboxyl —OH include the **acid halides, acid anhydrides, acid**

amides and **esters.** These will be discussed briefly here. Further treatment will be given them later in the chapter.

1. *Alcohols react with carboxylic acids to form esters.* The classic reaction which illustrates this transformation is that of acetic acid with ethyl alcohol:

$$CH_3COOH + CH_3CH_2OH \underset{}{\overset{\text{(dry HCl)}}{\rightleftharpoons}} \underset{\text{Ethyl acetate}}{CH_3\overset{\overset{\displaystyle O}{\|}}{C}OCH_2CH_3} + H_2O$$

formation of an ester in this manner is called esterification.

To achieve this, a strong acid, such as sulfuric acid or dry hydrogen chloride gas, is usually employed as a catalyst for the reaction. An examination of the structure written for the ester shows that the —OH has been replaced with an alkoxy group (RO—).

2. *Carboxylic acids* react with *phosphorus halides* or *thionyl chloride* ($SOCl_2$) to *form acid halides,* also referred to as **acyl halides.** The acid chlorides are more often prepared than other acid halides. For this reason the equations for the reactions to illustrate this phenomenon are written for the formation of the chlorine compounds.

$$C_6H_5COOH + PCl_5 \longrightarrow \underset{\text{Benzoyl chloride}}{C_6H_5\overset{\overset{\displaystyle O}{\|}}{C}Cl} + POCl_3 + HCl$$

these reactions are analogs of the reactions exhibited by alcohols with the same reagents.

$$3\,CH_3COOH + PCl_3 \longrightarrow \underset{\text{Acetyl chloride}}{3\,CH_3COCl} + P(OH)_3$$

$$RCOOH + \underset{\text{Thionyl chloride}}{SOCl_2} \longrightarrow RCOCl + HCl + SO_2$$

3. *Acid anhydrides are not usually prepared directly from acids,* although structurally they are compounds in which the —OH of the carboxyl group is replaced with an acyl ($R\overset{\overset{O}{\|}}{C}$—) group. Alternate formulas for acetic acid anhydride are:

$$CH_3\overset{\overset{\displaystyle O}{\|}}{C}—O—\overset{\overset{\displaystyle O}{\|}}{C}CH_3 \quad \text{and} \quad \begin{matrix} CH_3—C{=}O \\ \diagdown \\ O \\ \diagup \\ CH_3—C{=}O \end{matrix}$$

4. *Likewise, acid amides, acid derivatives* in which the —OH *of the carboxyl is replaced with an* —NH_2 *group,* which would appear to be readily obtainable directly from the acid, are better prepared by other procedures. The formula for benzamide (benzoic acid amide), derived from benzoic acid, is:

$$C_6H_5\overset{\overset{\displaystyle O}{\|}}{C}—NH_2$$

C. Decarboxylation

Decarboxylation refers to the transformation which removes the carboxyl group of an acid. Direct decarboxylation can be conveniently accomplished by passing the acid through a hot tube packed with MnO as a catalyst. The equation for the reaction is:

$$\begin{matrix} \overset{\displaystyle O}{\overset{\|}{RC}}\text{—OH} \\ + \\ \underset{\displaystyle O}{\underset{\|}{RC}}\text{—OH} \end{matrix} \xrightarrow[300\text{–}400^\circ]{MnO} \begin{matrix} R \diagdown \\ \quad C{=}O \\ R \diagup \end{matrix} + H_2O + CO_2$$

More often decarboxylation is achieved by heating the sodium salt of the acid with soda-lime, a mixture of NaOH and CaO.

D. Substitution of Alpha Hydrogens

PCl_5 catalyzes the reaction by forming the acid chloride which reacts rapidly with Cl_2.

Because the highly polar carboxyl group strongly attracts the electrons of the carbon to which it is attached, the hydrogens joined to this alpha carbon are rather easily substituted. These alpha hydrogens may be displaced by halogens in the presence of a suitable catalyst.

$$CH_3CH_2COOH + Cl_2 \xrightarrow{PCl_5} CH_3\underset{\displaystyle Cl}{\underset{|}{\overset{\displaystyle H}{\overset{|}{C}}}}COOH + HCl\uparrow$$

E. Aromatic Substitution Reactions

Aromatic acids undergo the usual aromatic substitution reactions. The *carboxyl group is deactivating and meta-directing.* Bromination of benzoic acid yields meta-bromobenzoic acid by a reaction that is more difficult to achieve than the bromination of benzene.

$$C_6H_5COOH + Br_2 \xrightarrow{Fe} m\text{-}BrC_6H_4COOH + HBr$$

31.6 ACID DERIVATIVES

A. Salts

Salts are ionic compounds formed by the reaction of organic acids with bases:

$$2\,CH_3COOH + Ca(OH)_2 \longrightarrow (CH_3COO)_2Ca + 2\,H_2O$$

Calcium acetate

salts are formed by typical acid-base neutralization reactions.

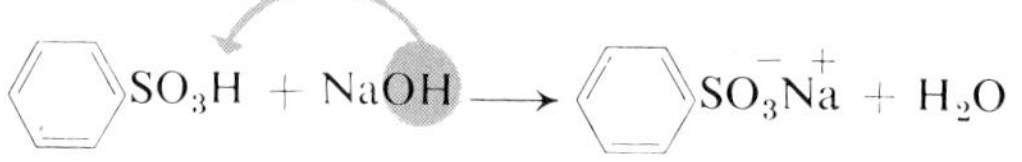

$$C_6H_5SO_3H + NaOH \longrightarrow C_6H_5SO_3^-Na^+ + H_2O$$

Benzene sulfonic acid — Sodium benzene sulfonate

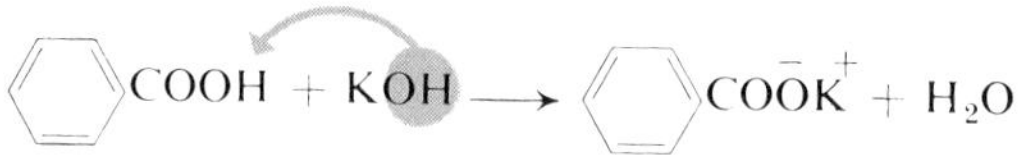

$$C_6H_5COOH + KOH \longrightarrow C_6H_5COO^-K^+ + H_2O$$

Benzoic acid — Potassium benzoate

To form the name of a salt from the acid requires two operations. First, change the *ic* ending of the name of the acid to *ate*, and then place the name of the metal in front of the term obtained. By this procedure the sodium salt of formic acid becomes sodium formate and the potassium salt of stearic acid, potassium stearate.

Particular salts of organic acids undergo characteristic thermal decompositions (decarboxylation or dehydration) which, depending on the type of salt, yield hydrocarbons, ketones, aldehydes or, with ammonium salts, amides.

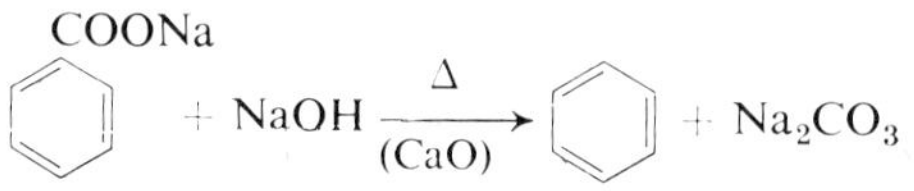

$$C_6H_5COONa + NaOH \xrightarrow[(CaO)]{\Delta} C_6H_6 + Na_2CO_3$$

Sodium benzoate — Benzene

$$(CH_3COO)_2Ca \xrightarrow{\Delta} CH_3\overset{\overset{O}{\|}}{C}CH_3 + CaCO_3$$

Calcium acetate — Acetone

$$CH_3COONH_4 \xrightarrow{\Delta} CH_3CONH_2 + H_2O$$

Ammonium acetate — Acetamide

Acid anhydrides are formed when salts react with acid halides:

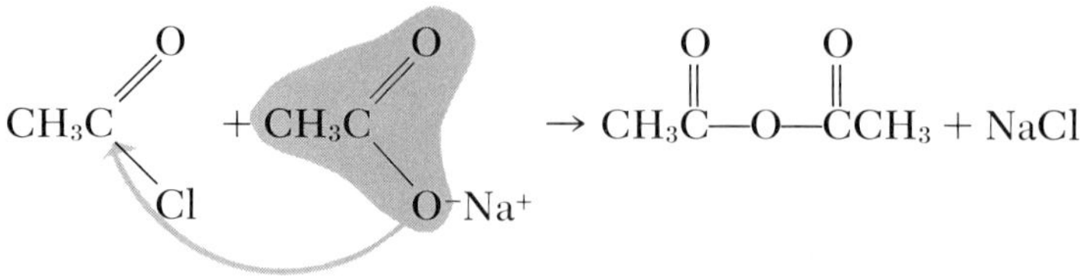

$$CH_3C(=O)Cl + CH_3C(=O)O^-Na^+ \rightarrow CH_3\overset{\overset{O}{\|}}{C}{-}O{-}\overset{\overset{O}{\|}}{C}CH_3 + NaCl$$

Acetyl chloride — Sodium acetate — Acetic anhydride

Salts also react with phosphorus halides (PCl_3), much as do the free acids, to form acid chlorides, an industrially useful reaction because the salts are obtained more cheaply than acids:

$$3\,CH_3COONa + PCl_3 \rightarrow 3\,CH_3COCl + Na_3PO_3$$

B. Acid Halides

Acid halides are formed from organic acids by replacing the —OH of the acid group with a halogen atom:

acid derivatives are formed by replacement of H or OH of the —COOH with another atom or group.

CH_3COCl — Acetyl chloride

C_6H_5COCl — Benzoyl chloride

$C_6H_5SO_2Cl$ — Benzenesulfonyl chloride

The more important acid chlorides will be employed to illustrate the preparations and the chemical transformations which such substances undergo. Acid bromides and iodides, when obtained, exhibit essentially the same reactions as the chlorides.

Common names of acid halides are formed by first replacing the *ic* ending of the name used for the acid with *yl*, forming a name to which one adds the name chloride, bromide or iodide. The acid chloride prepared from acetic acid is acet*yl chloride*. The acid chloride formed from benzene sulfonic acid becomes benzene sulfon*yl chloride* and the acid bromide of propionic acid, propion*yl bromide*.

Acid chlorides are readily prepared from either the acids or their salts. Acids react with phosphorus chlorides (PCl_3, PCl_5) or thionyl chloride ($SOCl_2$) to produce acid chlorides (see p. 702). Sodium salts also react with phosphorus trichloride (PCl_3) in much the same manner as acids to form acid chlorides (see above). It should be noted that these preparatory reactions were previously studied as reactions of acids.

Acid halides react with many compounds having reactive hydrogen atoms to introduce an $R\overset{\overset{\displaystyle O}{\|}}{C}—$ group (acyl group) in place of a hydrogen atom. The great activity of acid chlorides makes them excellent synthetic reagents. The smaller acid chlorides (acyl chlorides) of the aliphatic series react violently with water or vapor in the air to form the free acid. They are lachrymators and irritate both the eyes and throat.

$$RCOCl + HOH \rightarrow RCOOH + HCl$$

The higher members of the nonaromatic series react less rapidly, because of their decreasing solubility in water. The chlorides of aromatic acids are inherently more resistant to hydrolysis and are usually hydrolyzed only very slowly by cold water, but more rapidly with hot.

$$C_6H_5COCl + HOH \longrightarrow C_6H_5COOH + HCl \quad \text{(slowly)}$$

Benzoyl chloride

A *similar reaction with an alcohol* is referred to as **alcoholysis** and the product formed is an ester. This is a good method for preparing esters:

$$CH_3COCl + CH_3CH_2OH \rightarrow CH_3COOCH_2CH_3 + HCl$$

$$RCOCl + R'OH \rightarrow RCOOR' + HCl$$

The *corresponding reaction with ammonia is known as* **ammonolysis.** The compound obtained is an amide. This reaction is an *excellent method* for *preparing amides.* Since in each of these transformations the acyl (RCO—) group is introduced, such reactions are commonly referred to as **acylation** reactions.

$$C_6H_5SO_2Cl + NH_3 \longrightarrow C_6H_5SO_2NH_2 + HCl$$

Benzenesulfonyl chloride → Benzene sulfonamide

$$CH_3CH_2COCl + NH_3 \longrightarrow CH_3CH_2CONH_2 + HCl$$

Propionyl chloride → Propionamide

The acylation of aromatic compounds is of very practical value for the industrial synthesis of aromatic ketones. When benzene is permitted to react with an acid chloride in the presence of $AlCl_3$, an acyl group from the acid halide is attached to the ring in place of a hydrogen.

$$RCOCl + C_6H_6 \xrightarrow{AlCl_3} C_6H_5COR + HCl$$

the mechanism is a typical electrophilic aromatic substitution.

This reaction is an extension of the Friedel-Crafts reaction discussed earlier (Chapter 28), in which an acid chloride is employed rather than an alkyl chloride. An **acyl carbonium ion** (RCO^+), called an **acylium ion,** related in charge and similar in reactivity to the alkyl carbonium ion (RCH_2^+), is formed when an acid halide reacts with aluminum chloride. Addition of the acylium ion to the ring then occurs to produce an intermediate which, through loss of a proton, yields the aromatic ketone (see Chapter 28 for the mechanism). Acetophenone, an industrially important aromatic ketone, is made from acetyl chloride and benzene.

$$CH_3COCl + C_6H_6 \xrightarrow{AlCl_3} C_6H_5COCH_3 + HCl$$

Acetic anhydride, one example of the acid anhydrides to be discussed in the next section, reacts in a manner analogous to that of acetyl chloride.

$$(CH_3CO)_2O + C_6H_6 \xrightarrow{AlCl_3} C_6H_5COCH_3 + CH_3COOH$$

The similarity of the reaction with acetic anhydride, $(CH_3CO)_2O$, to that with acetyl chloride illustrates the close parallel in chemical behavior that exists between the acid halides discussed in this section and the acid anhydrides reviewed in the next section.

C. Acid Anhydrides

Structurally an acid anhydride may be regarded as the product obtained when a molecule of water is removed from two molecules of acid. If the water is removed from two molecules of the same acid, a **simple anhydride** is obtained. The anhydride formed from two different acids is known as a **mixed anhydride:**

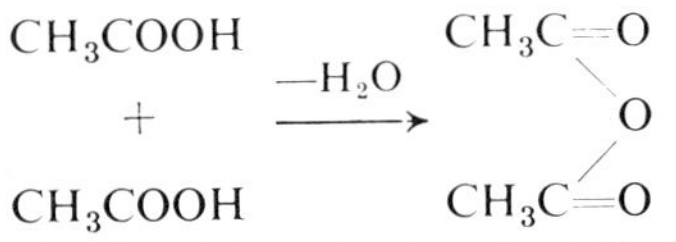

$$\begin{matrix} CH_3COOH \\ + \\ CH_3COOH \end{matrix} \xrightarrow{-H_2O} \begin{matrix} CH_3C{=}O \\ \quad\diagdown \\ \quad O \\ \quad\diagup \\ CH_3C{=}O \end{matrix}$$

Acetic acid — Acetic anhydride (a simple anhydride)

acetic anhydride is the only organic anhydride widely used in industry.

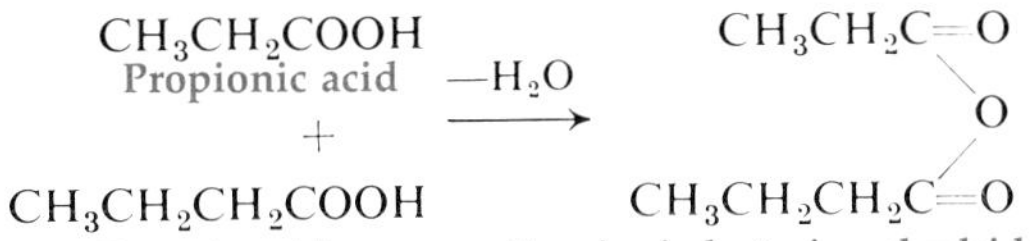

$$\begin{matrix} CH_3CH_2COOH \\ \text{Propionic acid} \\ + \\ CH_3CH_2CH_2COOH \\ \text{Butyric acid} \end{matrix} \xrightarrow{-H_2O} \begin{matrix} CH_3CH_2C{=}O \\ \quad\diagdown \\ \quad O \\ \quad\diagup \\ CH_3CH_2CH_2C{=}O \end{matrix}$$

Propionic butyric anhydride (a mixed anhydride)

The common names of acids are used to form the common names of the anhydrides. The term acid is dropped and replaced with the term anhydride in naming simple anhydrides. To name a mixed anhydride, the stem names of the two different acids are given, followed by the term anhydride.

The systematic IUPAC names are formed in a similar manner from the systematic names of the parent acids. For example, acetic anhydride becomes ethanoic anhydride and succinic anhydride, butanedioic anhydride.

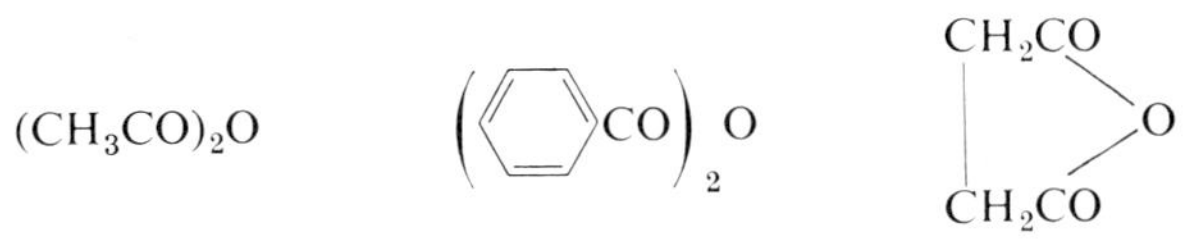

$(CH_3CO)_2O$ — Ethanoic anhydride

$(C_6H_5CO)_2O$ — Benzoic anhydride

$$\begin{matrix} CH_2CO \\ | \quad\quad \diagdown \\ | \quad\quad\quad O \\ | \quad\quad \diagup \\ CH_2CO \end{matrix}$$

Butanedioic anhydride

Although the formulation that shows the splitting out of water implies that anhydrides are formed directly from acids by dehydration, in actual practice the preparation of an acid anhydride is not that simple. The classic method (p. 704) for preparing them is the reaction of an acid halide with the sodium salt of an acid.

$$\underset{\text{Sodium propionate}}{CH_3CH_2COONa} + \underset{\text{Acetylchloride}}{Cl\overset{\overset{O}{\|}}{C}CH_3} \rightarrow \underset{\text{Acetic propionic anhydride}}{CH_3CH_2\overset{\overset{O}{\|}}{C}O\overset{\overset{O}{\|}}{C}CH_3} + NaCl$$

This reaction is particularly useful for preparing mixed anhydrides.

Anhydrides of acids may also be prepared by refluxing the acid with acetic anhydride or acetyl chloride. The acetic anhydride and chloride serve as dehydrating agents. This reaction is not a simple dehydration. The easy removal of acetic acid shifts the equilibrium far to the right:

$$2\,RCOOH + (CH_3CO)_2O \rightleftarrows (RCO)_2O + 2\,CH_3COOH$$

Anhydrides are analogous in their behavior to acid halides and show the type of reactions which were discussed in the previous section for acid chlorides. However, they are considerably less reactive.

Anhydrides undergo hydrolysis. The reaction of acetic anhydride with water is rather slow compared to that of acetyl chloride. For the higher anhydrides, it is often necessary to boil them with water for some time before they are completely hydrolyzed.

$$(CH_3CO)_2O + H_2O \rightarrow 2\,CH_3COOH$$

Alcoholysis leads to the formation of esters in a transformation entirely analogous to that discussed for acid chlorides:

$$(CH_3CO)_2O + CH_3CH_2OH \rightarrow \underset{\text{Ethylacetate}}{CH_3COOCH_2CH_3} + CH_3COOH$$

Ammonolysis forms the corresponding amide:

$$(CH_3CO)_2O + NH_3 \rightarrow \underset{\text{Acetamide}}{CH_3CONH_2} + CH_3COOH$$

Similar reactions occur with aromatic anhydrides and with the cyclic anhydrides of dibasic acids, such as succinic and o-phthalic acids:

$$\underset{\text{Benzoic anhydride}}{(C_6H_5CO)_2O} + NH_3 \longrightarrow \underset{\text{Benzamide}}{C_6H_5CONH_2} + C_6H_5COOH$$

$$\underset{\text{Phthalic anhydride}}{C_6H_4(CO)_2O} + CH_3OH \longrightarrow C_6H_4(COOCH_3)(COOH)$$

D. Amides

Amides of carboxylic acids have the general formula $RCONH_2$; those of sulfonic acids, RSO_2NH_2. Typical compounds of this class are:

CH_3CONH_2	$C_6H_5CONH_2$	$C_6H_5SO_2NH_2$
Acetamide	Benzamide	Benzenesulfonamide

urea, $H_2N\overset{O}{\overset{\|}{C}}NH_2$, an amide, is the principal form in which nitrogen is excreted in animals.

Common names are used most frequently for these compounds. The common names are formed by replacing the ending "ic acid" from the common name of the acid with "amide." The IUPAC systematic names are formed in the usual way. The names for amides of aliphatic carboxylic acids are obtained by replacing the terminal "e" of the systematic name of the alkane having the same number of carbon atoms with the term "amide." Any substituents are designated by name and located by number with the carbon of the amide group numbered as carbon 1.

$$CH_3CONH_2 \qquad CH_3CH(CH_3)CH_2CONH_2 \qquad CH_3(CH_2)_4CONH_2$$

Ethanamide — 3-methyl butanamide — Hexanamide

In previous sections of this chapter, reactions by means of which amides can be prepared have been discussed. These include their formation from ammonium salts (p. 704), from acid chlorides (p. 706) and from acid anhydrides (p. 707). They may also be prepared as a result of the reaction of esters with ammonia (p. 712).

Amides undergo hydrolysis, as do the other acid derivatives. They react extremely slowly with water alone, but they do hydrolyze when heated under reflux with excess aqueous alkali. The course of the reaction may be followed by the evolution of ammonia. Alcoholic potassium hydroxide may be used for the hydrolysis of some less soluble compounds:

$$CH_3CH_2CH_2CONH_2 + OH^- \xrightarrow{\Delta} CH_3CH_2CH_2COO^- + NH_3$$

Aqueous mineral acids act similarly:

$$CH_3CH_2CH_2CONH_2 + H_3O^+ \xrightarrow{\Delta} CH_3CH_2CH_2COOH + NH_4^+$$

Dehydration of amides leads to the formation of nitriles, whose characteristic structural feature is the presence of a —CN group. Compounds having this group are referred to as nitriles or cyanides.

$$CH_3CONH_2 \xrightarrow{\Delta/P_2O_5} CH_3CN + H_2O$$

E. Esters

Esters are formed when alcohols react with acids to split out a molecule of water. Structurally they are derived from acids by replacing the —OH group of the acid group with an alkoxy (—OR)

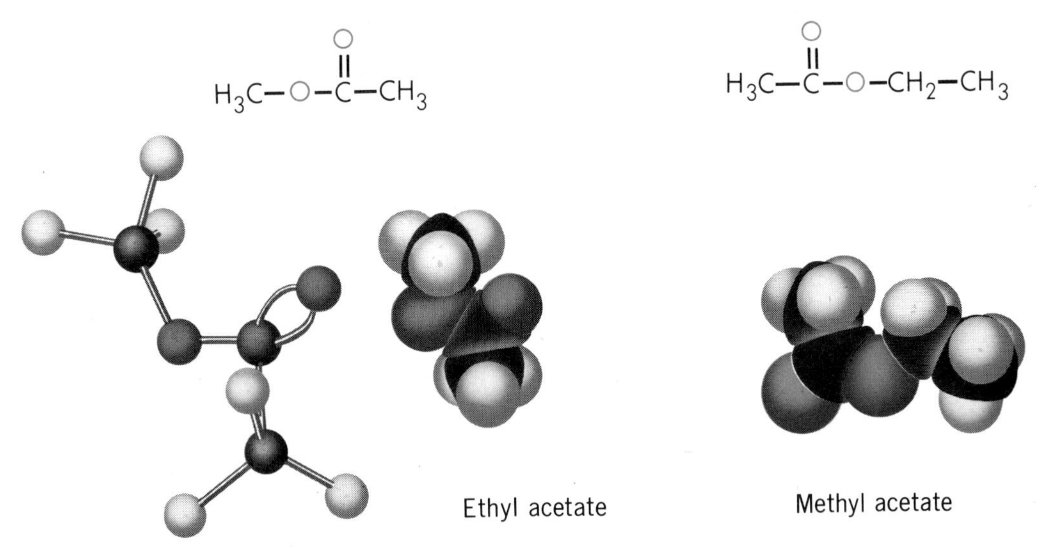

FIGURE 31.5 SIMPLE ESTERS.

group, and accordingly they are represented by the general formulas:

$$R-\overset{\overset{\displaystyle O}{\|}}{C}-OR' \qquad \text{and} \qquad Ar-\overset{\overset{\displaystyle O}{\|}}{C}-OR'$$

Three-dimensional models which show more clearly the structure of two of the simplest esters are given in Figure 31.5.

Esters are compounds of great importance. They are widely distributed in nature. The active ingredient responsible for the pleasant flavor and odor of a fruit or flower is often an ester.

Some esters and their pleasant odors are listed below:

Amyl acetate (1-pentylacetate)	$CH_3-\overset{\overset{O}{\|}}{C}-O-(CH_2)_4-CH_3$	Banana odor
Octyl acetate	$CH_3-\overset{\overset{O}{\|}}{C}-O-(CH_2)_7-CH_3$	Orange odor
Ethyl butyrate	$CH_3-CH_2-CH_2-\overset{\overset{O}{\|}}{C}-O-CH_2-CH_3$	Pineapple odor
Methyl salicylate	$C_6H_4(-OH)-\overset{\overset{O}{\|}}{C}-O-CH_3$ (benzene ring with —OH)	Oil of wintergreen
Methyl anthranilate	$C_6H_4(-NH_2)-\overset{\overset{O}{\|}}{C}-O-CH_3$ (benzene ring with —NH₂)	Intense odor of grape

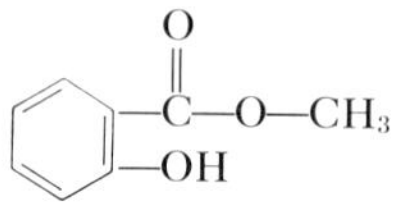

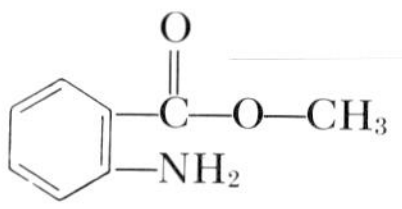

Animal and vegetable fats and oils are esters. They will be discussed more completely in the following chapter.

Esters were once thought to be salts of organic acids and therefore the common names are similar to the common names for salts. For example, the salt of sodium hydroxide and acetic acid is sodium acetate.

$$\underset{\text{Sodium hydroxide}}{NaOH} + \underset{\text{Acetic acid}}{CH_3-\overset{\overset{\displaystyle O}{\|}}{C}-OH} \rightarrow \underset{\text{Sodium acetate}}{CH-\overset{\overset{\displaystyle O}{\|}}{C}-O^-Na^+} + H_2O$$

Similarly the ester of ethyl alcohol and acetic acid is ethyl acetate.

$$\underset{\text{Ethyl alcohol}}{CH_3CH_2OH} + \underset{\text{Acetic acid}}{CH_3-\overset{\overset{\displaystyle O}{\|}}{C}-OH} \rightarrow \underset{\text{Ethyl acetate}}{\underbrace{CH_3-\overset{\overset{\displaystyle O}{\|}}{C}-O}-\underbrace{CH_2-CH_3}}$$

The common and most used name of an ester contains the alkyl group of the alcohol followed by the name of the acid with the suffix changed from *ic* to *ate*. By this procedure, the methyl ester of acet*ic* acid becomes *methyl* acet*ate*; the ethyl ester of benzo*ic* acid becomes *ethyl* benzo*ate*.

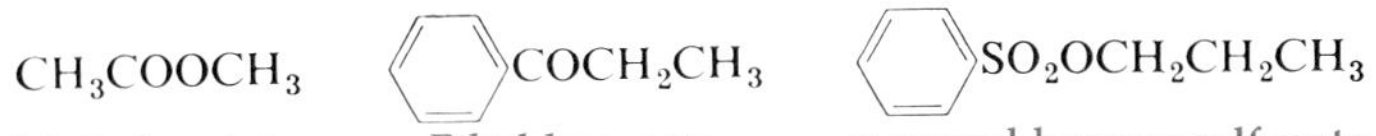

CH_3COOCH_3	$C_6H_5COCH_2CH_3$	$C_6H_5SO_2OCH_2CH_2CH_3$
Methyl acetate (methyl ethanoate)	Ethyl benzoate (ethyl benzoate)	n-propyl benzenesulfonate (n-propyl benzenesulfonate)

The IUPAC names are similar. However, the systematic names of the acids and alcohols, and not their common names, are used. For example, methyl acetate is also named *methyl* ethano*ate*; ethyl butyrate becomes *ethyl* butano*ate*.

The formation of an ester by direct reaction of an alcohol with an acid is referred to as **esterification.** This is usually effected by warming a solution of an acid in an excess of the appropriate alcohol, using as a catalyst either a little sulfuric acid, or gaseous hydrogen chloride bubbled into the alcohol-acid mixture. The esterification is presumed to occur by way of a series of reversible stages involving in sequence the following changes: (a) a protonation of the un-ionized acid, (b) a reaction of the alcohol with the carbon of the protonated acid, (c) a proton transfer from the alcohol —OH in the intermediate formed to the —OH of the acid, (d) the loss of a molecule of water, and finally (e) a loss of a proton to yield the ester. The series of changes with each transformation labeled is shown here:

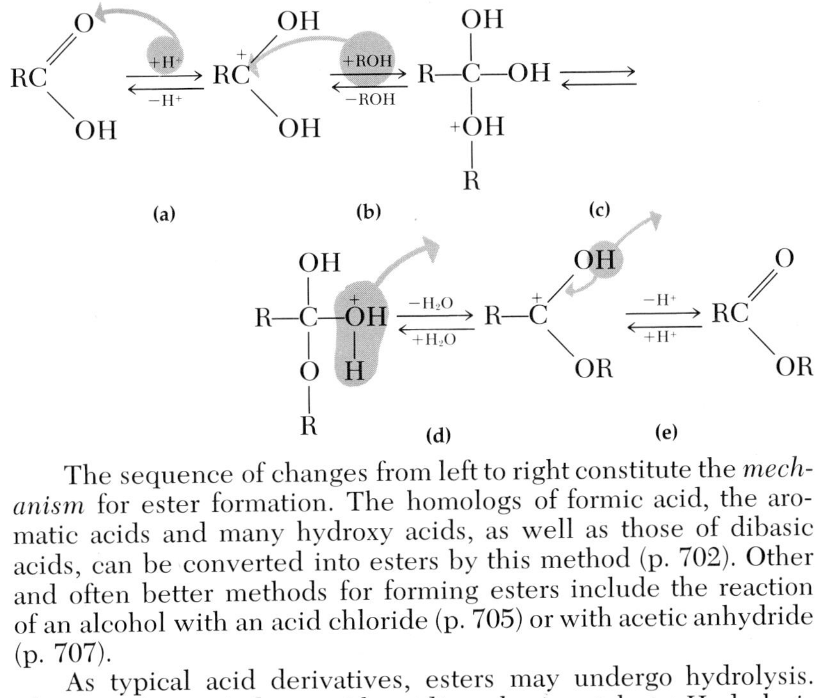

$$RC(=O)OH \underset{-H^+}{\overset{+H^+}{\rightleftharpoons}} RC^+(OH)OH \underset{-ROH}{\overset{+ROH}{\rightleftharpoons}} R{-}C(OH)(OH){-}\overset{+}{O}HR \rightleftharpoons$$

(a) (b) (c)

$$R{-}C(OH)(OR){-}\overset{+}{O}H_2 \underset{+H_2O}{\overset{-H_2O}{\rightleftharpoons}} R{-}C^+(OH)OR \underset{+H^+}{\overset{-H^+}{\rightleftharpoons}} RC(=O)OR$$

(d) (e)

The sequence of changes from left to right constitute the *mechanism* for ester formation. The homologs of formic acid, the aromatic acids and many hydroxy acids, as well as those of dibasic acids, can be converted into esters by this method (p. 702). Other and often better methods for forming esters include the reaction of an alcohol with an acid chloride (p. 705) or with acetic anhydride (p. 707).

As typical acid derivatives, esters may undergo hydrolysis. This may be carried out with acidic or basic catalysts. Hydrolysis in the presence of acid is the reverse of esterification.

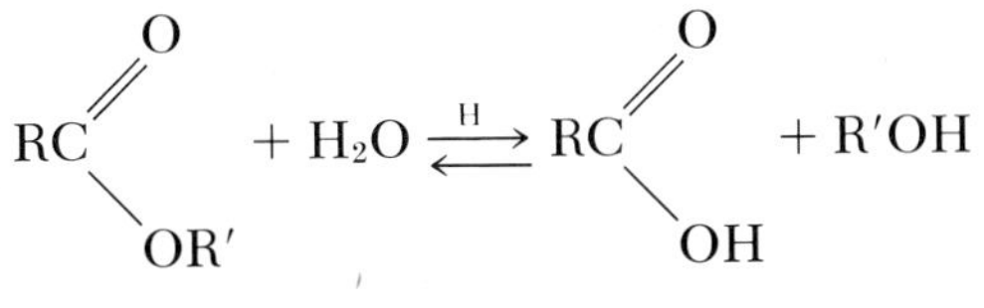

$$RC(=O)OR' + H_2O \overset{H}{\rightleftharpoons} RC(=O)OH + R'OH$$

The mechanism for acid hydrolysis is the right to left sequence of changes as just outlined for the esterification reaction. Basic hydrolysis is referred to as **saponification** and is widely used for making soaps from fats and oils. The overall equation for ester saponification is:

$$RC(=O)OR' + NaOH \rightarrow RC(=O)O^-Na^+ + R'OH$$

a salt of a long chain carboxylic acid is called a soap; hence the term saponification.

Since saponification depends on both the concentration of the ester and the hydroxide ion provided by the base, it is assumed to follow a mechanism which involves three stages: (a) an initial step in which the OH^- by nucleophilic attack bonds to the carbonyl carbon (S_N2 attack), then (b) the loss of the alkoxide ion ($R'O^-$) from the intermediate formed, followed by (c) a proton transfer from the acid to the alkoxide ion to give the carboxylate ion and the alcohol.

$$OH^- + RC(=O)OR' \underset{(a)}{\rightleftarrows} R{-}C(O^-)(OH){-}OR' \underset{(b)}{\rightleftarrows} R{-}C(=O)OH + R'O^- \underset{(c)}{\rightarrow} R{-}C(=O)O^- + R'OH$$

Esters also undergo ammonolysis to form amides; the reaction is:

$$CH_3COOCH_2CH_3 + NH_3 \rightarrow CH_3CONH_2 + CH_3CH_2OH$$

Esters may also be converted to alcohols by catalytic hydrogenation (reduction):

$$RCOOR' + H_2 \xrightarrow{Ni} RCH_2OH + R'OH$$

This reaction is important in the preparation of long chain alcohols from naturally occurring fats and oils in the industrial manufacture of synthetic detergents.

EXERCISES

31.1 List in the order of increasing acidity: water, phenol, ammonia, alcohol, sulfuric acid, sulfonic acids, carboxylic acids and alcohols.

31.2 What substances and solutions might one use to test the acidity of an unknown organic compound?

31.3 Define each term and name and write the structural formula for an example of each: (a) fatty acid, (b) diprotic (dibasic) carboxylic acid, (c) tribasic acid, (d) an aromatic acid, (e) a sulfonic acid, (f) an alkyl hydrogen sulfate, (g) an acid halide, (h) an acid anhydride, (i) an ester.

31.4 Define (a) resonance, (b) relative acidity, (c) displacement, (d) replacement, (e) substitution.

31.5 Explain why the hydrogen of the carboxyl group should be acidic. Correlate the discussion with the acidic nature of the hydrogens in sulfuric acid.

31.6 Write equations for a one-step preparation of an acid from each substance listed, showing any catalysts and special reaction conditions required: (a) ethyl benzene, (b) 2-butene, (c) ethyl acetate, (d) acetaldehyde, (e) isobutyl alcohol, (f) ethyl cyanide, (g) 1,1,1-triiodohexane.

31.7 Write equations showing all catalysts or special conditions illustrating: (a) decarboxylation, (b) salt formation, (c) esterification, (d) hydrolysis, (e) alcoholysis, (f) ammonolysis, (g) halogenation of an α-hydrogen of an acid, (h) acid chloride formation.

31.8 Write reactions for and compare the ease of reactions of acids, acid chlorides, amides, acid anhydrides and esters with: (a) water, (b) ammonia, (c) solution of NaOH, (d) alcohols.

31.9 Write equations for each step of this synthesis:

(a) 1-butyne is heated with basic potassium permanganate
(b) the liquid product is treated with thionyl chloride
(c) the reactive product formed is added to isopropyl alcohol

31.10 Define K_a and state its significance with reference to the acidity of organic substances.

31.11 Arrange the following acids in order of increasing K_a values. State the reasons why you select the order chosen.

(1) propionic acid
(2) 2,2-dichloropropanoic acid
(3) 2-aminopropanoic acid
(4) 2-nitropropionic acid

31.12 Write the expression for the K_a of chloroacetic acid.

31.13 Which exhibits the larger K_a value, phenol or p-nitrophenol? Why?

31.14 Show the reaction by which hydronium ion arises when an acidic substance is dissolved in water. To what extent do the following substances undergo this reaction: HCl, CH_3COOH, CH_3CH_2OH, phenol?

31.15 An alkyl bromide is converted to a Grignard reagent that reacts with carbon dioxide (dry ice) to produce 2,2-dimethyl propanoic acid. What is an acceptable structure for the alkyl bromide used? Explain.

31.16 A liquid of formula $C_6H_{12}O_2$ undergoes hydrolysis to yield an acid A and an alcohol B. The oxidation of compound B with hot chromic acid yields a product identical with A. What is a correct structure for $C_6H_{12}O_2$? Write equations for all reactions described for it.

31.17 Both succinic and glutaric acid yield a cyclic anhydride when heated with acetic anhydride as a dehydrating agent. When dicarboxylic acids containing seven or more carbon atoms are heated with acetic anhydride under the same conditions, polymeric anhydrides are formed. Suggest a possible reason for this difference in chemical behavior. (Hint: Refer to the Baeyer strain theory in Chapter 25).

31.18 If an alcohol containing O^{18} (i.e., $RO^{18}H$) is allowed to react with an acid, CH_3COOH, which of the products, the ester, CH_3COOR or H_2O, will contain the labeled oxygen (O^{18})? Explain your reply in terms of the mechanism proposed for esterification.

31.19 Alkenes have the general formula C_nH_{2n}. Devise a *general formula* which can be used for the class of cyclic esters (called lactones) of

```
                         CH2
                        /   \
which the compound   CH2     C=O   is an example.
                      |      |
                     CH2     O
                        \   /
                         CH2
```

31.20 A colorless liquid A containing both oxygen and chlorine gave the following analysis: carbon, 59.79%; hydrogen, 3.56%; and chlorine, 25.27%. Its vapor density at STP was 0.0063 g/ml. Reacted with NH_3 it formed compound B, which when heated with dilute NaOH gave off ammonia to form a sodium salt of a monocarboxylic acid, compound C, which had a mole weight of 122 g. Write structures for compounds A, B and C and the equations for the two reactions described.

31.21 A compound A having the empirical formula C_8H_9NO, is subjected to hydrolysis in acid solution. The hydrolysate was neutralized with dilute NaOH and extracted with ether. Evaporation of the ether extract yielded the aromatic compound B having the formula C_6H_7N, which yields N_2 when reacted with nitrous acid (HNO_2) at room temperature (see Chapter 33). Evaporation of the neutralized hydrolysate gives the sodium salt, compound C. The salt when heated with soda-lime, produces methane. Give the structural formula of compound A.

31.22 A compound A of molecular formula $C_5H_{10}O_5$ is changed to $C_{13}H_{18}O_9$ by reacting it with acetic anhydride. How many hydroxyl groups are indicated as being present in each molecule of A by this reaction?

31.23 Describe a simple chemical reaction which could be used as a test to distinguish between each pair of compounds listed (give the observations which differ for the two compounds concerned):

(a) Propionic acid and ethyl formate
(b) Benzoyl chloride and chlorobenzene
(c) Dimethyl amine and ethyl amide
(d) p-Bromobenzoic acid and benzoyl chloride
(e) Acetic anhydride and acetyl chloride

31.24 Identify the structure of compound I, the final product of the reaction sequence outlined. Write structural formulas for compounds B, C, D, E, F, G and H.

$$CH_3CH_2COOH \xrightarrow{PCl_5} B \xrightarrow{NH_3} C \xrightarrow{H_2O/H^+} D \xrightarrow{C_2H_5OH/H^+} E$$

$$E \xrightarrow{NH_3} F \xrightarrow{NaOH(Aq)} G \xrightarrow{PCl_5} H \xrightarrow{Phenol} I$$

SUGGESTED READING

Breslow, R.: *Organic Reaction Mechanisms.* W. A. Benjamin, Inc., New York, 1965 (paperback). Chapter 6.

Davidson, D.: "Acids and Bases in Organic Chemistry." J. Chem. Ed., *19*:154, 1942.

Ferguson, L. N.: "Balancing Equations for Organic Oxidation-Reduction Reactions." J. Chem. Ed., 23:550, 1946.

Hendrickson, J. B., Cram, D. J. and Hammond, G. S.: *Organic Chemistry,* Third Edition. McGraw-Hill Book Company, New York, 1970. Chapters 13, 18 and 25.

Morrison, R. T. and Boyd, R. N.: *Organic Chemistry,* Second Edition. Allyn and Bacon, Inc., Boston, 1966. Chapters 18, 20, 21, 29, 30, 31 and 37.

Noller, C. R.: *Chemistry of Organic Compounds,* Third Edition. W. B. Saunders Company, Philadelphia, 1965. Chapters 11, 12, 16, 19, 28, 31, 38, 39 and 40.

THIRTY-TWO • FATS AND OILS

The organic matter of living cells is made up largely of these types of substances: lipids, carbohydrates and proteins. Lipids are a rather heterogeneous collection of substances obtained by the exhaustive extraction of biological materials with those liquids commonly referred to as "fat solvents" (i.e., methanol, ethanol, acetone, chloroform, carbon disulfide, ether or benzene). Among the substances that are soluble in these organic solvents are *fats* and *oils*, cerebrosides (in the brain), sphingomyelins (nerve tissue), lecithins and cephalins (phosphorus containing compounds) and cholesterol (found in the blood). All are important in the chemistry of living organisms. Fats and oils comprise the bulk of the lipids.

32.1 COMPOSITION OF FATS AND OILS

The naturally occurring fats and oils—lard, beef tallow, whale blubber, cottonseed oil, palm oil, olive oil, corn oil and so on—are lipids, which are insoluble in water, but which are hydrolyzed in hot sodium hydroxide solution to give glycerol and sodium salts of long chain fatty acids called soaps.

A. Glycerol

glycerol is a trihydroxy alcohol.

Glycerol (1,2,3-propanetriol) is completely soluble in water, has a sweet taste and is non-toxic. It is used widely in medicines, cosmetics and foods. Glycerol is hygroscopic, that is, it absorbs water vapor from the atmosphere. For this reason it is used as a humectant in non-drying inks and tobacco and some is left in hand soap to keep it moist. Glycerol may be used as an antifreeze and can be added to water to prevent its freezing at low temperatures. The trihydroxy alcohol (glycerol) reacts with nitric acid to form nitroglycerin, the active constituent of dynamite (p. 652). Special precautions are required in its synthesis, however, to avoid development of excessive heat and formation of impurities which may cause premature detonation.

The glycerol obtained as a by-product in soap manufacture is not sufficient to meet the demands for dynamite production, especially during wartime. Some glycerine is made by fermentation of sugar and large quantities are made from propylene, a cracking by-product in the petroleum industry.

When heated to a high temperature, glycerol is dehydrated to acrolein. In the presence of potassium hydrogen sulfate, the reaction takes place at a lower temperature.

an overcooked steak or overheated oil gives acrolein—that burned odor in the kitchen.

$$\begin{array}{l} CH_2OH \\ | \\ CHOH \\ | \\ CH_2OH \end{array} \xrightarrow{KHSO_4} \begin{array}{l} CH_2 \\ \| \\ CH + 2\,H_2O \\ | \\ CHO \end{array}$$

Glycerol — Acrolein 2-propenal

Acrolein has a very irritating odor which may be detected in fumes from burning animal or vegetable fats.

B. Fatty Acids

When the hydrolysis mixture (resulting from the hydrolysis of a fat or an oil in a basic solution) is acidified, the acid salts are converted into an oily, insoluble mixture of carboxylic acids. Repeated analyses of the resulting acids have shown them to be mostly *long chain* acids with an *even number* of carbons. The acid chain *may* or *may not* contain *double bonds*. Only very small amounts of fatty acids with either an odd number of carbon atoms or with branched carbon chains have been found and these have come from very special sources. For example, isovaleric acid (3-methylbutanoic acid), having both an uneven number of carbons and a methyl branch, is found in dolphin and porpoise blubber.

Some of the saturated fatty acids commonly found in natural fats and oils are listed in Table 32.1. The two most commonly

TABLE 32.1 SATURATED ACIDS COMMONLY FOUND IN NATURAL FATS AND OILS

		Percentage in			
Name	Formula	Butterfat	Cottonseed Oil	Lard	Human
Myristic acid	$CH_3(CH_2)_{12}COOH$	12	1	2	2.7
Palmitic acid	$CH_3(CH_2)_{14}COOH$	25	22	29	24
Stearic acid	$CH_3(CH_2)_{16}COOH$	9	2	14	8.4
Arachidonic acid	$CH_3(CH_2)_{18}COOH$	1	2	—	1

occurring are palmitic acid (16 carbons) and stearic acid (18 carbons), with palmitic acid generally the more abundant.

Table 32.2 lists several important unsaturated fatty acids which are found in many fats and oils. It is apparent that oleic acid is the one most widely distributed in nature.

TABLE 32.2 UNSATURATED ACIDS COMMONLY FOUND IN NATURAL FATS AND OILS

Name	Formula	Percentage in Butterfat	Cottonseed Oil	Lard	Human
Palmitoleic acid	$CH_3(CH_2)_5CH{=}CH(CH_2)_7COOH$	5	1	2	5
Oleic acid	$CH_3(CH_2)_7CH{=}CH(CH_2)_7COOH$	35	28	45	47
Linoleic acid	$CH_3(CH_2)_3(CH_2CH{=}CH)_2(CH_2)_7COOH$	5	44	6	10.2
Linolenic acid	$CH_3(CH_2CH{=}CH)_3(CH_2)_7COOH$	5	—	—	1.5

C. Fats and Oils

From a consideration of their hydrolysis products, which are an alcohol and the sodium salts of acids, one must assume that *fats and oils are esters.* Analyses show that three molecules of the salt of the acid are liberated in the hydrolysis for each molecule of glycerol. Reconstructing the substance, a triester of glycerol results. The triester undergoes hydrolysis in this fashion:

$$\begin{array}{l} CH_2OC(=O)(CH_2)_{16}CH_3 \\ | \\ CHOC(=O)(CH_2)_{16}CH_3 \\ | \\ CH_2OC(=O)(CH_2)_{16}CH_3 \end{array} + 3\,NaOH \rightarrow \begin{array}{l} CH_2OH \\ | \\ CHOH \\ | \\ CH_2OH \end{array} + 3\,Na^+O^-{-}C(=O)(CH_2)_{16}CH_3$$

Stearin Glycerol A sodium salt (sodium stearate)

Stearin, the fat molecule used here to illustrate the hydrolysis of an ester, is a simple triester of glycerol, a **simple triglyceride,** with three identical acid groups. The salt of only one acid is found among the hydrolysis products. A **mixed triglyceride** is one in which there are two or three different acid groups.

Common names for simple triglycerides are formed by first naming the acid esterified with the glycerol, then dropping the ending *ic* of the name of the acid and adding the ending *in.* The triglyceride formed from palmitic acid is palmitin; from stearic acid, stearin; and from oleic acid, olein. The names of the mixed triglycerides are obtained by naming the individual acids without endings, except

$$\begin{array}{l} O \\ \| \\ CH_2OC(CH_2)_{14}CH_3 \\ |O \\ |\| \\ CHOC(CH_2)_{14}CH_3 \\ |O \\ |\| \\ CH_2OC(CH_2)_{14}CH_3 \end{array}$$

Palmitin
(glyceryltripalmitate)

$$\begin{array}{l} O \\ \| \\ CH_2OC(CH_2)_7CH{=}CH(CH_2)_7CH_3 \\ |O \\ |\| \\ CHOC(CH_2)_{14}CH_3 \\ |O \\ |\| \\ CH_2OC(CH_2)_2CH_3 \end{array}$$

α-Oleo-β-palmitobutyrin
(glyceryl-α-oleo-β-palmito-α'-butyrate)

for the last named, which is given the ending -in and then designating their location by the Greek symbols α, β and α'. The terminal positions of glycerol are designated by the α and α'. The central carbon of glycerol is specified by β. Esters of glycerol may also be named as simple esters. The names derived in this way are given in the parentheses.

Both simple and mixed triglycerides may be present in natural fats and oils. For the most part, however, they consist of a rather complicated mixture of glycerides, almost exclusively of the mixed type. Simple glycerides occur very infrequently unless one of the fatty acids happens to be present in particularly large amounts (50 per cent or more).

32.2 PHYSICAL PROPERTIES OF FATS AND OILS

butterfat and swiss cheese have a butyric acid tang!

Structurally, the fats and oils are compounds of the same type, and are distinguished from each other on the basis of their physical state at ordinary temperatures. Fats are solids and oils are liquids. It should be apparent that the distinction is often rather vague since the physical state in a particular locality depends upon the climate, weather and other not too clearly defined variables such as diet.

The physical properties of a particular fat or oil are an expression of its fatty acid composition. Correspondingly, other factors being the same, a fat or oil with a larger percentage of shorter chain (lower molecular weight) fatty acids will exhibit greater water solubility and a lower melting point than other fats with a lesser percentage. Likewise, glycerides containing larger amounts of unsaturated fatty acids will generally exhibit lower melting points than similar glycerides containing smaller quantities. Thus, if two glycerides containing the same number of carbon atoms, one of which is solid and the other liquid, are compared, the liquid glyceride invariably contains the higher percentage of unsaturated acids. Oils (liquid fats) are usually characterized by a higher content of unsaturated fatty acid groups than solid fats.

32.3 CHEMICAL PROPERTIES: REACTIONS AND PRODUCTS

A. Hydrolysis

Probably the most important chemical reaction of fats and oils is that of *hydrolysis to form fatty acids and glycerol.* Three molecules of fatty acid and one molecule of glycerol are formed per molecule of fat or oil hydrolyzed.

$$\begin{array}{l}CH_2OOCR\\ |\\ CHOOCR + 3\,HOH \rightarrow\\ |\\ CH_2OOCR\end{array}\quad\begin{array}{l}CH_2OH\\ |\\ CHOH + 3\,RCOOH\\ |\\ CH_2OH\end{array}$$

lipases function largely in the small intestines at pH's above 7.

Since this reaction takes place only very slowly in boiling water, various catalysts have been employed to increase the rate at which the transformation proceeds. Both acids and bases **catalyze** the reaction, as do certain **digestive enzymes** called **lipases.** With a base the reaction becomes essentially irreversible, since carboxylate ions (formed from the acid) have little or no tendency to recombine with the hydroxyl groups of glycerol.

When hydrolysis takes place in the presence of base, the metal salt of the acid, called a *soap,* is obtained; the process by which it is formed is called **saponification.**

Saponification is used as a rapid test to determine the average chain length of acid groups attached to the glycerol molecule. An excess of potassium hydroxide solution of known concentration is added to a sample of the fat or oil; the mixture is boiled until the reaction is complete, after which the left-over unreacted base is titrated with acid. The shorter the acid chain, the more ester molecules are contained per gram of oil, and the greater the amount of potassium hydroxide required.

$$\begin{array}{l}CH_2OOCR\\ |\\ CHOOCR + 3\,NaOH \rightarrow\\ |\\ CH_2OOCR\\ \text{Fat or oil}\end{array}\quad\begin{array}{l}CH_2OH\\ |\\ CHOH + 3\,RCOO^-Na^+\\ |\\ CH_2OH\\ \text{Glycerol}\qquad\qquad \text{Soap}\end{array}$$

The number of milligrams of potassium hydroxide required to saponify 1 g of oil is called the **saponification number.**

Soaps have been made for many years. The early American housewife saved beef tallow scraps and cooked them with lye-containing oak ashes until the reaction was complete, skimmed off the scum that formed, and poured the clear liquid into molds to harden.

The soap molecule is an interesting chemical. A characteristic soap molecule, sodium stearate, obtained from the saponification of stearin, is illustrated in Figure 32.1. It has a long hydrocarbon (petroleum-like) chain attached at one end to a negative carboxylate ion with its associated sodium ion. The molecule has a dual (Dr. Jekyll and Mr. Hyde) personality. The hydrocarbon end is oil soluble; the carboxylate end loses its sodium ion in water, is nega-

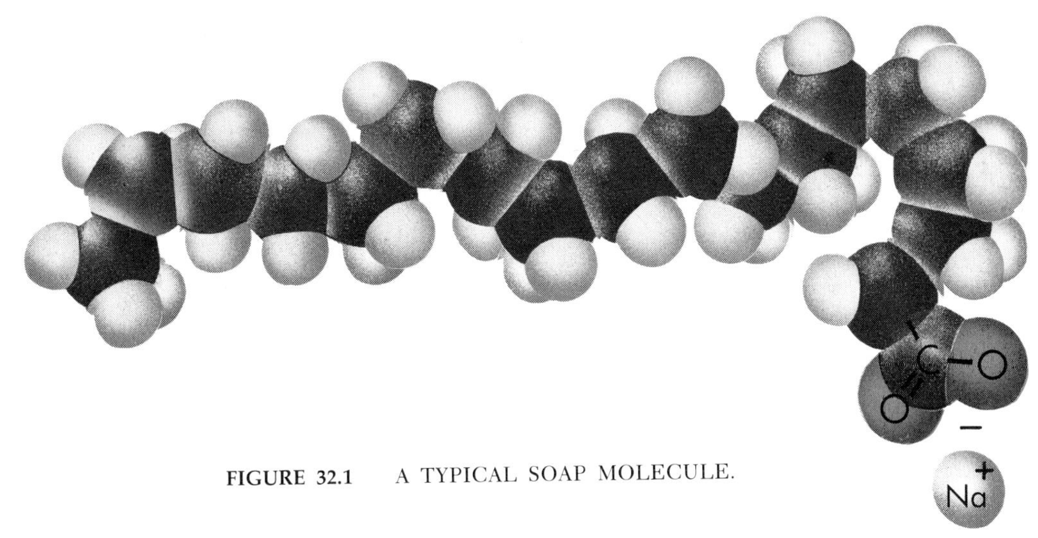

FIGURE 32.1 A TYPICAL SOAP MOLECULE.

tively charged and soluble in water. The insoluble oily ends of many molecules associate, leaving the carboxylate ions extending out into the water (Fig. 32.2). The tiny highly charged oil drop, called a micelle, is attached to polar water molecules and remains suspended in the water phase.

A soap solution is not a true solution but a **colloid.** The particles in solution are not molecules but aggregates of many molecules. The colloidal solution is quite stable, because the **charged micelles** repel one another. They can be made to precipitate only by neutralization of the charge. Dissolved calcium or magnesium salts with their doubly charged ions are attracted to the micelles, neutralize them and allow them to collect. Sodium and potassium ions remain free in solution.

refer to Chapter 24 for a discussion of colloids.

The cleansing action of soap is illustrated in Figure 32.3, which shows a soap micelle meeting an oil droplet. The oily ends of the soap molecules dissolve in the droplet, making a large micelle, still charged sufficiently to be carried off in the water. In making two insoluble phases compatible, soap acts as a detergent. All **detergents** act in a similar manner. The increased use of soap, with its ability to "cut" grease and oil and to remove bacteria imbedded in the oils, has been partially responsible for the increase in life expectancy of the human race.

B. Addition Reactions

Since most fats and oils contain unsaturated fatty acids, a variety of addition reactions is to be expected. Typical of these reactions are those with hydrogen and the halogens.

1. *Hydrogenation of oils* in the presence of a nickel catalyst results in the addition of hydrogen to carbon-carbon double bonds of the unsaturated fatty acids present. For example, olein may be converted to stearin by this means:

$$\begin{array}{l} CH_2OOC(CH_2)_7CH{=}CH(CH_2)_7CH_3 \\ | \\ CHOOC(CH_2)_7CH{=}CH(CH_2)_7CH_3 + 3\,H_2 \\ | \\ CH_2OOC(CH_2)_7CH{=}CH(CH_2)_7CH_3 \\ \text{Olein} \end{array} \xrightarrow{\text{Ni cat.}} \begin{array}{l} CH_2OOC(CH_2)_{16}CH_3 \\ | \\ CHOOC(CH_2)_{16}CH_3 \\ | \\ CH_2OOC(CH_2)_{16}CH_3 \\ \text{Stearin} \end{array}$$

recall the hydrogenation of alkenes (Chapter 26) of which this is an application.

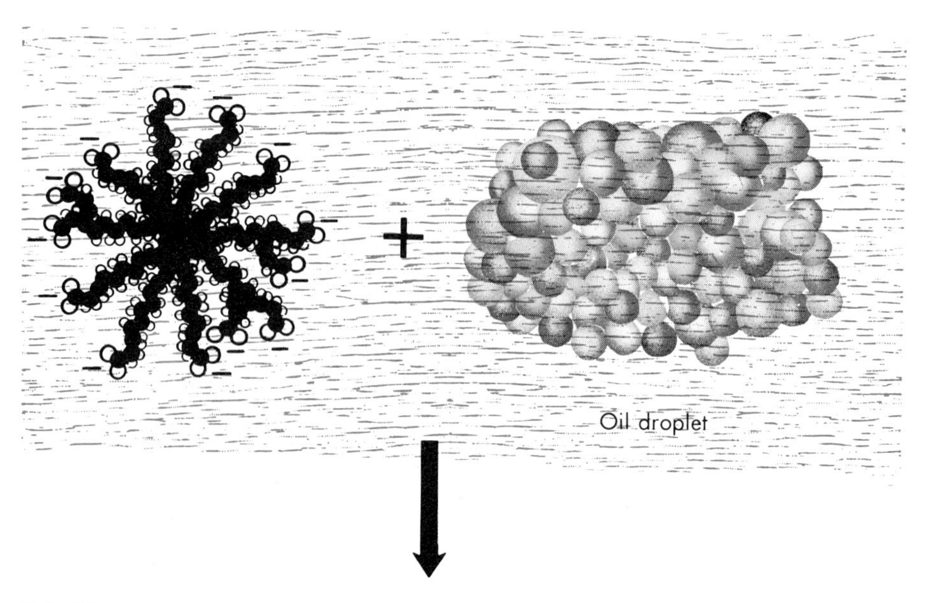

FIGURE 32.2 A SOAP MICELLE.

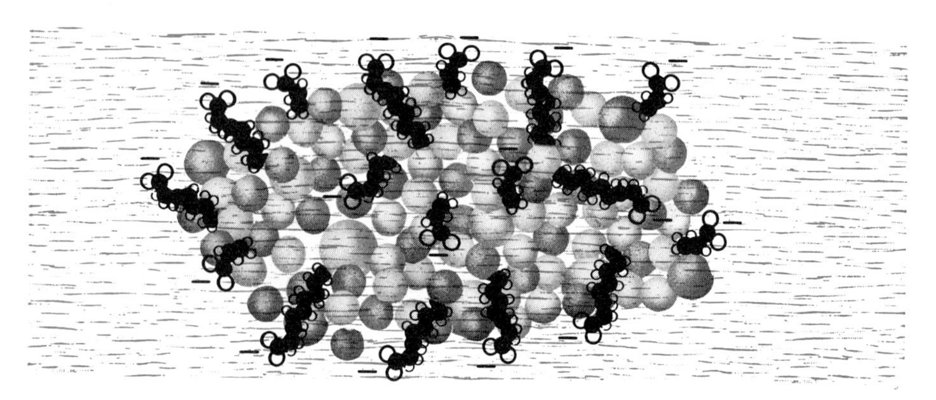

FIGURE 32.3 THE ACTION OF SOAP.

This reaction is of great industrial importance because it permits the transformation of inexpensive, unsaturated vegetable oils into fats which the American housewife finds desirable for household use. Industrially the hydrogenation is generally not continued to completion, but only until the melting point has been raised sufficiently to yield a fat of the proper consistency. Cottonseeds, soybeans and peanuts are excellent sources of oils which may be hydrogenated to give a good cooking fat. Most present day shortenings are "**hydrogenated vegetable oil.**" Butter substitutes (margarines) produced by this method are further modified by adding vitamin supplements, coloring and flavoring to simulate butter. This hydrogenation of oils to fats is often referred to as "**hardening**" of fats.

2. Among other reagents which *add readily to unsaturated bonds in fats and oils are the halogens.* Samples of oils are treated with a specially prepared iodine solution to determine their degree of unsaturation and therefore the feasibility of hardening them to make shortening. To determine how much iodine adds, an excess of this specially prepared solution of a known iodine concentration

is added to a sample of the oil; the mixture is allowed to stand for a time, and the iodine remaining is measured by titration. The grams of iodine which have been added to 100 grams of the oil are calculated. This is referred to as the oil's "**iodine number.**" A representative equation showing the addition of iodine to olein (glyceryl trioleate) follows:

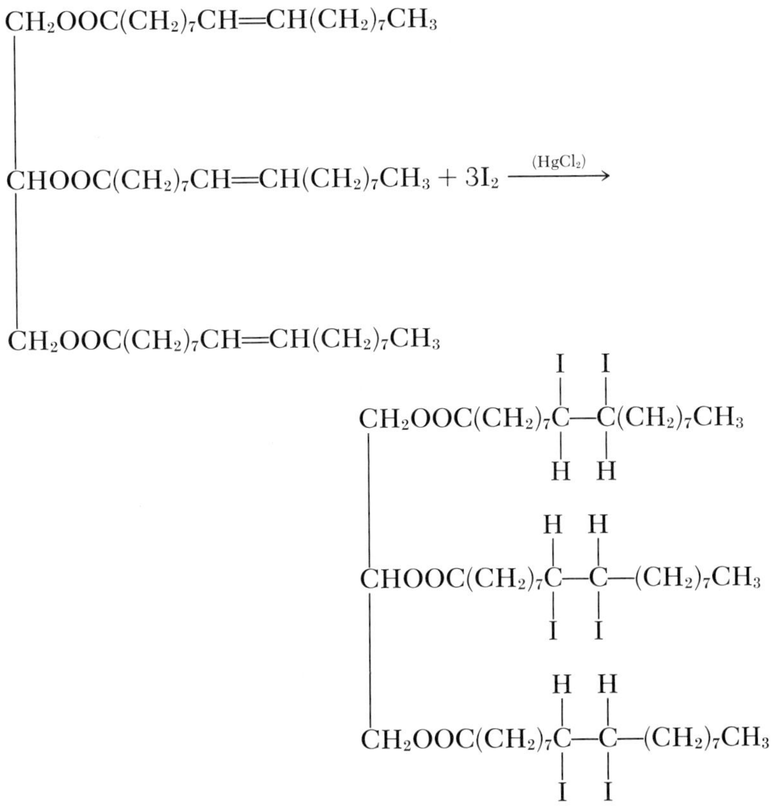

$$\begin{array}{l} CH_2OOC(CH_2)_7CH{=}CH(CH_2)_7CH_3 \\ | \\ CHOOC(CH_2)_7CH{=}CH(CH_2)_7CH_3 + 3I_2 \xrightarrow{(HgCl_2)} \\ | \\ CH_2OOC(CH_2)_7CH{=}CH(CH_2)_7CH_3 \end{array}$$

$$\begin{array}{l} CH_2OOC(CH_2)_7CI(H){-}CI(H)(CH_2)_7CH_3 \\ | \\ CHOOC(CH_2)_7CH(I){-}CH(I){-}(CH_2)_7CH_3 \\ | \\ CH_2OOC(CH_2)_7CH(I){-}CH(I){-}(CH_2)_7CH_3 \end{array}$$

3. When the *hydrogenation of a fat or oil* is carried out *using copper chromite under pressure* and at higher temperatures, in addition to adding hydrogen to the carbon-carbon double bonds present, the reduction of the carbonyl groups and the cleavage of the ester linkages are accomplished. The reaction is illustrated with α, α'-dioleo-β-palmitin.

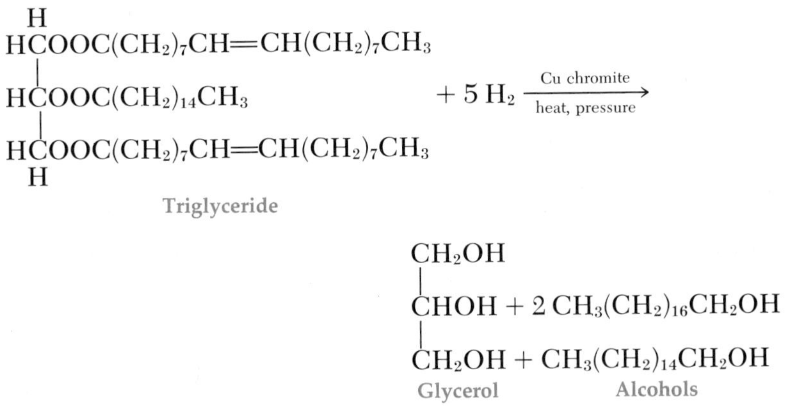

$$\begin{array}{l} H \\ HCOOC(CH_2)_7CH{=}CH(CH_2)_7CH_3 \\ | \\ HCOOC(CH_2)_{14}CH_3 \\ | \\ HCOOC(CH_2)_7CH{=}CH(CH_2)_7CH_3 \\ H \end{array} + 5\,H_2 \xrightarrow[\text{heat, pressure}]{\text{Cu chromite}}$$

Triglyceride

$$\begin{array}{l} CH_2OH \\ | \\ CHOH + 2\,CH_3(CH_2)_{16}CH_2OH \\ | \\ CH_2OH + CH_3(CH_2)_{14}CH_2OH \end{array}$$

Glycerol Alcohols

This reaction is carried out on a large scale commercially. The long chain fatty alcohols are treated with sulfuric acid to yield alkyl hydrogen sulfates.

$$\underset{\text{Alcohol}}{CH_3(CH_2)_{16}CH_2OH} + \underset{\text{Conc.}}{H_2SO_4} \rightarrow \underset{\text{An alkyl hydrogen sulfate}}{CH_3(CH_2)_{16}CH_2OSO_3H} + H_2O$$

These alkyl hydrogen sulfates are moderately strong acids; they are much stronger than carboxylic acids. Neutralization with sodium hydroxide solution yields the sodium salt.

$$CH_3(CH_2)_{17}OSO_3H + \underset{\text{Solution}}{NaOH} \rightarrow \underset{\text{Sodium alkyl sulfate}}{CH_3(CH_2)_{17}OSO_3^-Na^+}$$

Like soap, this molecule has two natures, one due to a long, oil soluble chain and the other to a water soluble ionic end. Like soap, it is a detergent, one of the synthetic detergents used for laundering clothes and washing dishes. The detergents form micelles in water and surround and carry off oil droplets much the same as soap does. These so-called "**synthetic detergents**" have advantages over soap as cleansing agents.

Magnesium and calcium ions present in "hard water" cause the soap to form a precipitate that is lighter than water. Not only is the soap unable to aid in cleansing, but the insoluble curd adheres to the fabric being washed.

$$2\ CH_3(CH_2)_{16}\overset{\overset{\Large O}{\|}}{C}—O^-Na^+ + Mg^{2+}SO_4^- \rightarrow$$

$$\underset{\text{Insoluble}}{Mg(O—\overset{\overset{\Large O}{\|}}{C}(CH_2)_{16}CH_3)_2} + Na_2SO_4$$

The magnesium and calcium salts of the alkyl hydrogen sulfates are soluble. The detergent molecules are neither precipitated nor hindered in their cleansing.

Soaps are not effective cleansing agents in weakly acid solutions, such as a solution containing fruit juices, because the soap ion gains a proton and becomes insoluble.

$$CH_3(CH_2)_{16}\overset{\overset{\Large O}{\|}}{C}—O^-Na^+ + H_3O^+ \rightarrow$$

$$\underset{\text{Insoluble}}{CH_3(CH_2)_{14}\overset{\overset{\Large O}{\|}}{C}—OH} + Na^+ + H_2O$$

Synthetic detergents do not take protons from weakly acid solutions, but remain soluble and active.

The widespread use of synthetic detergents prior to about 1965 led to a water pollution problem. The various detergents available and in use accumulated in sewage and waste waters because these detergents were not sensitive to decomposition by microorganisms as were soaps. Detergents of the types available were not **biode-**

gradable. Soaps, containing long unbranched hydrocarbon chains, are subject to bacterial decomposition by a process (described in the next section as β-oxidation) which removes carbons two at a time until the entire molecule is converted into carbon dioxide and water. Synthetic detergents, in contrast with soaps, were composed of highly branched hydrocarbon chains. Because of the branching they could not be broken down by microorganisms. In more recent years these **nonbiodegradable** ("hard") detergents have been replaced with compounds having "linear side chains," the biodegradable ("soft") detergents. Several types of compounds are in current use, the most common being the sodium salts of the alkyl benzene sulfonic acids, with the following composition:

$$R-C_6H_4-SO_3^-Na^+$$

The R represents a long chain unbranched alkyl group, attached at no specified position on the ring by use of a suitable Friedel-Crafts reaction (page 630).

32.4 BIOLOGICAL SIGNIFICANCE

It is well known that fats are stored in many different places in the human body and serve a number of important functions. The stored fat serves as (1) a protection to the body and its vital organs against bruises and shock damage, (2) an insulator to prevent excessive heat loss and (3) a support for the different organs in the body. Fats also are the most concentrated source of energy to the organism, yielding per gram over twice as many calories as do carbohydrates and proteins:

Fat → 9 cal/gram
Carbohydrate → 4 cal/gram
Protein → 4 cal/gram

Thus fat is the most efficient way of storing calories. Stored fat, referred to in man as depot fat, may be called on to furnish fuel and energy when the need arises. With the exception of certain unsaturated fatty acids, however, lipid is not essential in the diet, a fact which implies merely that most lipids can be synthesized by man from other dietary constituents at a rate adequate for normal requirements of growth and health.

Any food ingested by animals beyond normal requirements is largely converted into fat and deposited in the appropriate tissues. Evidence for this is the fat deposition observed when animals are placed on a diet rich in a carbohydrate such as starch. Because fat deposition occurs, it may be inferred that the excess of carbohydrate over actual needs can be biologically converted into fats. This process is called *fat biosynthesis*.

Studies concerned with the breakdown of starches in animals, including man, have shown that glucose, the simple carbohydrate from which starch is formed, is transformed inside cells into a very reactive two-carbon compound named *acetyl coenzyme A*. Acetyl

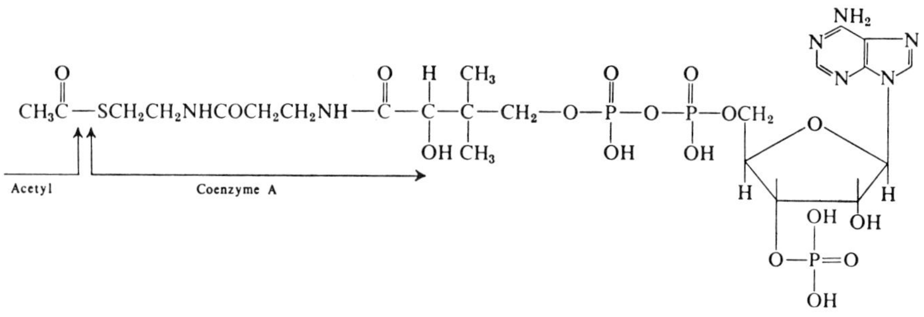

$$CH_3\overset{O}{\overset{\|}{C}}—SCH_2CH_2NHCOCH_2CH_2NH—\overset{O}{\overset{\|}{C}}—\underset{OH}{\overset{H}{C}}—\underset{CH_3}{\overset{CH_3}{C}}—CH_2—O—\underset{OH}{\overset{O}{\overset{\|}{P}}}—O—\underset{OH}{\overset{O}{\overset{\|}{P}}}—OCH_2$$

Acetyl Coenzyme A

NH_2 N N N N O H H OH OH O—P=O OH

FIGURE 32.4 ACETYL COENZYME A.

coenzyme A may be utilized by cells in a variety of ways, one being the biosynthesis of fats. Acetyl coenzyme A, whose rather complex structure (Fig. 32.4) will be represented by the formula CH_3CO—SCoA, is a thioester of acetic acid with coenzyme A, a sulfhydryl (—SH) containing compound. A fatty acid gains increased reactivity when it is converted into a coenzyme A thioester, and only those fatty acids which are combined with coenzyme A (such as acetic acid in acetyl coenzyme A) participate in fat biosynthesis. The pathway of reactions utilized for fat synthesis is outlined in Figure 32.5.

Biosynthesis of fats begins with acetyl coenzyme A (CH_3CO—SCoA) which, under the influence of a suitable enzyme catalyst, reacts with CO_2 (reaction a) to form malonyl coenzyme A ($HOOCCH_2CO$—SCoA). Malonyl coenzyme A produced in this reaction then condenses (reaction b) with any fatty acid coenzyme A thioester (RCO—SCoA), accompanied by the simultaneous loss

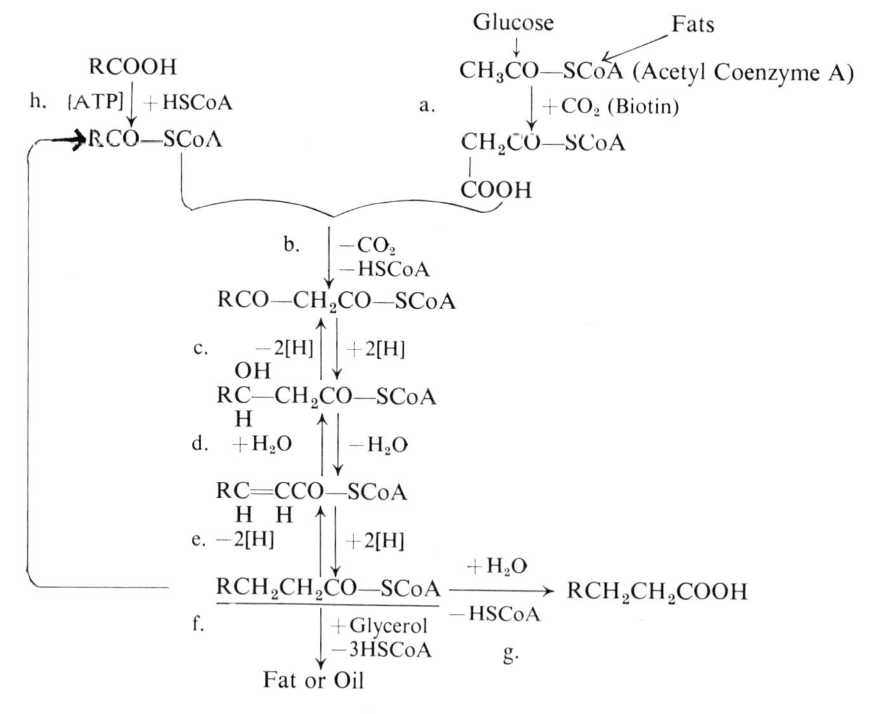

FIGURE 32.5 FATTY ACID AND FAT BIOSYNTHESIS.

of CO_2. The thioesters of the fatty acids which react with malonyl coenzyme A may be formed by reaction of free fatty acids with coenzyme A when aided by suitable catalysts. Two carbon atoms from the malonyl coenzyme A are added in reaction b to the carbon chain of the fatty acid in the reacting thioester, to form a β-ketoacid coenzyme A thioester. This ketoacid thioester then undergoes reduction (via reactions c, d and e) to yield the coenzyme A thioester of a new fatty acid having two additional carbon atoms. The newly synthesized fatty acid thioester may then react with another molecule of malonyl coenzyme A (via reaction b) to repeat the reaction sequence and attach two more carbons to the growing carbon chain. The pathway b through e may be successively repeated until sufficient pairs of carbons have been added to the carbon chain to provide a thioester of the desired fatty acid. Note that each time a fatty acid thioester traverses the reaction pattern, two carbon atoms are added to the fatty acid being formed. The fatty acid thioesters thus produced are finally combined with glycerol (change f) to yield a fat or oil, or they may be hydrolyzed to yield free fatty acids and coenzyme A.

Dietary fat synthesized as outlined (Fig. 32.5) when stored provides an important and efficient source of reserve energy for organisms because of the large amount of energy (9 cal/gram) released when fat is oxidized to CO_2 and H_2O. Whenever an organism utilizes a fat or oil for energy, the initial stage involves an enzyme catalyzed hydrolytic cleavage, releasing the free fatty acids and glycerol. Subsequent oxidation of the fatty acids obtained is the major source of the energy recovered from fats and is achieved by a process commonly called "beta oxidation." **β-Oxidation** refers to the observation that initial oxidation of a fatty acid occurs at the carbon beta to the carboxyl group of the acid. The β-oxidation of a fatty acid is achieved through the series of reactions outlined in Figure 32.6.

Just as in fat biosynthesis, the free fatty acids must first be converted into coenzyme A thioesters by reaction of the fatty acids with coenzyme A in the presence of suitable catalysts. The formula

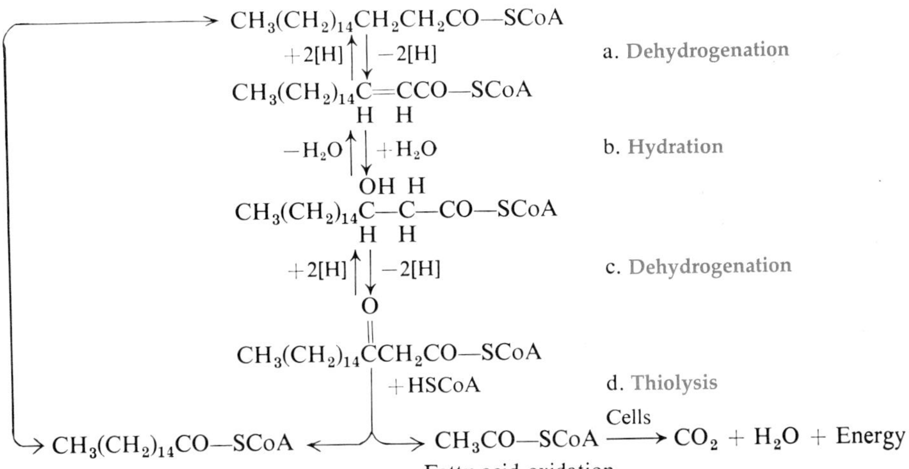

$$CH_3(CH_2)_{14}CH_2CH_2CO{-}SCoA$$

$$+2[H]\uparrow\downarrow-2[H] \qquad \text{a. Dehydrogenation}$$

$$CH_3(CH_2)_{14}\underset{H}{C}{=}\underset{H}{C}CO{-}SCoA$$

$$-H_2O\uparrow\downarrow+H_2O \qquad \text{b. Hydration}$$

$$CH_3(CH_2)_{14}\underset{H}{\overset{OH}{C}}{-}\underset{H}{\overset{H}{C}}{-}CO{-}SCoA$$

$$+2[H]\uparrow\downarrow-2[H] \qquad \text{c. Dehydrogenation}$$

$$CH_3(CH_2)_{14}\overset{O}{\overset{\|}{C}}CH_2CO{-}SCoA$$

$$+HSCoA \qquad \text{d. Thiolysis}$$

$$CH_3(CH_2)_{14}CO{-}SCoA \longleftarrow\longrightarrow CH_3CO{-}SCoA \xrightarrow{\text{Cells}} CO_2 + H_2O + \text{Energy}$$

(An arrow returns $CH_3(CH_2)_{14}CO{-}SCoA$ to the start of the cycle.)

Fatty acid oxidation.

FIGURE 32.6 FATTY ACID OXIDATION.

of the coenzyme A thioester of stearic acid is, therefore, the first compound in Figure 32.6. Because all fatty acids are oxidized as their thioesters, reactions comprising β-oxidation involve only changes in the fatty acid coenzyme A thioester and its transformation products. The thioester is first dehydrogenated (reaction a). Water is then added to the double bond of the thioester obtained to yield a β-hydroxyacid thioester (reaction b). This upon dehydrogenation (reaction c) yields the coenzyme A thioester of a β-ketoacid. The β-ketoacid thioester, being very stable, then reacts, in a process referred to as **thiolysis**, with a molecule of coenzyme A, producing one molecule each of acetyl coenzyme A and the thioester of a fatty acid (here, palmitic acid) with two fewer carbon atoms than stearic acid. The acetyl coenzyme A formed is readily oxidized and in the cell yields CO_2, H_2O and energy. Since the palmitic acid is obtained as a thioester comparable to that of stearic acid, it can itself undergo β-oxidation, just as the stearic acid thioester did, to yield another molecule of acetyl coenzyme A. This again is oxidized in the cell to CO_2, H_2O and energy. Complete degradation of stearic acid via β-oxidation would yield nine molecules of acetyl coenzyme A.

Both the biosynthesis and oxidation of fatty acids take place with the *incorporation or loss* of two carbon atoms from the fatty acid at each stage. This accounts for the natural occurrence of fatty acids with an even number of carbon atoms and the relative rarity in nature of fatty acids with an odd number of carbon atoms. The provision of separate reaction pathways for synthesis and degradation is part of the economy of cellular processes, providing independent control for better regulation of cell metabolism. The cycle-like pattern of each sequence, characteristic of most biological chemical systems, should be noted. It should also be noted that reactions c, d and e (Fig. 32.5) of biosynthesis are the reverse of reactions a, b and c (Fig. 32.6) of fatty acid degradation (oxidation). It appears that one reversible reaction is utilized in both directions in biological processes where possible.

EXERCISES

32.1 How does cottonseed oil differ structurally from petroleum oil?

32.2 What structural feature is identical in all animal and vegetable oils?

32.3 Define the following:

a. glyceride
b. ester
c. fatty acid
d. oil
e. fat
f. saponification number
g. iodine number
h. soap
i. detergent
j. "synthetic" detergent
k. "hardened" oil
l. micelle
m. lipid

32.4 Write a representative equation showing special conditions and catalysts for:

a. the saponification of a fat
b. the hydrogenation of a fat with a nickel catalyst
c. the hydrogenation of a fat with copper chromite catalyst
d. the addition of iodine to a fat

32.5 Write structural formulas for the following:

a. a fat
b. an oil
c. shortening
d. soap
e. synthetic detergent
f. glycerol
g. β-stearo-α,α'-dipalmitin
h. palmityl hydrogen sulfate
i. sodium stearyl sulfate
j. oleic acid

32.6 Write equations showing the preparation of a synthetic detergent, beginning with the natural fat palmitin. Show catalysts and other special conditions.

32.7 Write equations accounting for the failure of soap as a cleansing agent in (a) weakly acid solutions and (b) hard water.

32.8 Which type of food—protein, fat or carbohydrate—contributes the most energy per gram when oxidized completely in the body?

32.9 Explain the action of detergents.

32.10 What is the role of acetyl coenzyme A in fatty acid biosynthesis?

32.11 Outline the sequence of reactions by which butyric acid can be formed biologically from malonyl coenzyme A.

32.12 Why do most naturally occurring fatty acids contain an even number of carbon atoms?

32.13 What is beta-oxidation as applied to fatty acid breakdown?

32.14 What is the role of coenzyme A in fatty acid conversions in living cells?

32.15 A supplier of fats and oils provides a sample of linseed oil (Iodine No. = 202) which the customer suspects was diluted with cottonseed oil (Iodine No. = 103). What chemical property could the customer examine in order to settle the dispute? Describe how this examination would resolve the matter.

32.16 What volume of H_2 (measured at STP) would be required to convert 500 g of cottonseed oil having an iodine number of 110 into a saturated fat?

32.17 One mole of glycerol on treatment with conc. H_2SO_4 loses two moles of water. What is a likely structure for the product formed? Why?

32.18 0.416 g of a carboxylic acid having a mole weight of 104 g reacts with 80 ml of 0.1N NaOH in a titration using phenolphthalein as an indicator. How many carboxyl groups are present in this acid and what must be its formula?

32.19 The names of a series of α-substituted acetic acid derivatives are as follows with their pK_a values in parentheses ($pK_a = -\log K_a$): hydroxyacetic acid (3.83); methoxyacetic acid (3.48); chloroacetic acid (2.80); cyanoacetic acid (2.44); thiohydroxyacetic acid (3.55); iodoacetic acid (3.13); phenylacetic acid (4.31). Using the pK_a values, arrange the substituted groups, OH, OCH_3, CN, SH, I, and phenyl (C_6H_5) in order of decreasing attraction for electrons.

SUGGESTED READING

Green, D. E.: "Metabolism of Fats." Scientific American, January, 1954.
Mahler, H. R. and Cordes, E. H.: *Basic Biological Chemistry.* Harper and Row, Inc., New York, 1968. Chapter 16.
Morrison, R. T. and Boyd, R. N.: *Organic Chemistry,* Second Edition. Allyn and Bacon, Boston, 1966. Chapters 15 and 20.
Noller, C. R.: *Chemistry of Organic Compounds,* Third Edition, W. B. Saunders Company, Philadelphia, 1965. Chapter 12.
Snell, F. D.: "Soap and Glycerol." J. Chem. Ed., *19*:172, 1942.
Snell, F. D. and Snell, C. D.: "Syndets and Surfactants." J. Chem. Ed., *35*:271, 1958.
White, A., Handler, P. and Smith, E. L.: *Principles of Biochemistry,* Fourth Edition. McGraw-Hill Book Company, New York, 1968. Chapters 4, 21 and 22.

THIRTY-THREE • ORGANIC COMPOUNDS OF NITROGEN, PHOSPHORUS AND SULFUR

Organic compounds containing nitrogen, phosphorus and sulfur are relatively abundant and are widely distributed in nature in a great variety of substances. Many of these substances are key compounds in living organisms. They induce important physiological responses in plants and animals. Proteins (nitrogen containing compounds; Chapter 35) mediate the rates of almost all biological reactions. Organic phosphorus compounds are present in all living cells and actively participate in the dynamic processes by which food is converted into usable energy. Many substances must be phosphorylated (combined with phosphoric acid) before undergoing changes in living systems. The amino acid cystine (Chapter 35), a nitrogen-sulfur-organic compound, is the main structural unit in hair, fingernails and animal hoofs. The polymer nylon contains nitrogen.

Many organic compounds of these elements are either synthesized or isolated for use in medicine and agriculture.

33.1 COMPOUNDS OF NITROGEN

Of the many types of organic nitrogen containing compounds, amines and related compounds are the most important and will be

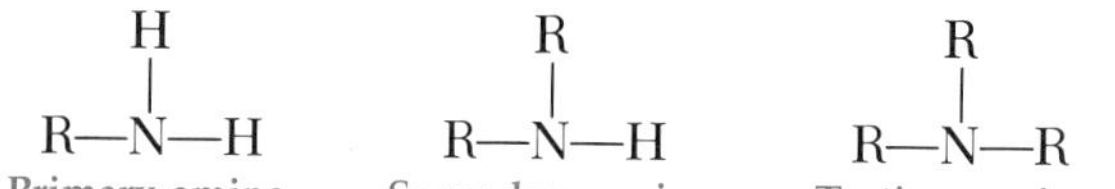

$$\underset{\text{Primary amine}}{R{-}\overset{\displaystyle H \atop |}{N}{-}H} \qquad \underset{\text{Secondary amine}}{R{-}\overset{\displaystyle R \atop |}{N}{-}H} \qquad \underset{\text{Tertiary amine}}{R{-}\overset{\displaystyle R \atop |}{N}{-}R}$$

FIGURE 33.1 CLASSES OF AMINES.

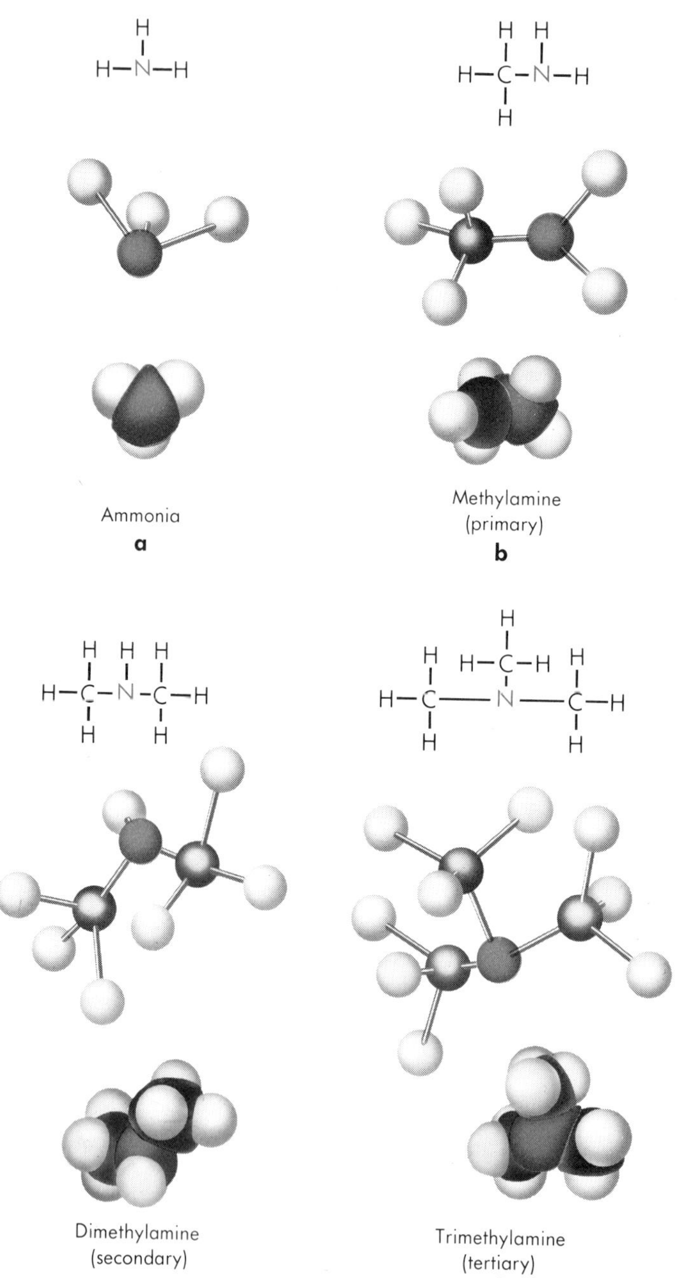

FIGURE 33.2 STRUCTURES OF THE METHYL AMINES.

note the new way in which the terms primary, secondary and tertiary are used when specifying amines.

discussed in detail. Amines are organic bases closely related to ammonia in structure and in chemical behavior. Amines may be considered as formed from ammonia by replacing *one or more* hydrogens with alkyl or aryl groups.

Depending upon the number of alkyl or aryl groups introduced, amines are classified as **primary, secondary** or **tertiary** (Fig. 33.1). Specific examples of primary, secondary and tertiary amines are given in Figure 33.2. Ammonia is included to emphasize the close structural relationship. Primary amines may be either aliphatic or aromatic. In accordance with the practice followed in the classification of ethers, secondary and tertiary amines are referred to as simple or mixed, depending on the identity of the organic groups attached to the nitrogen atom. Diphenylamine is a simple aromatic amine. Ethylmethylamine is a mixed aliphatic amine. Other examples are given in Table 33.1.

TABLE 33.1 AMINES

Formula	Classification	Name
CH_3NH_2	Primary; aliphatic	Methylamine
$C_6H_5NH_2$	Primary; aromatic	Aniline (Phenylamine)
$(CH_3CH_2)_2NH$	Secondary; simple aliphatic	Diethylamine
$(C_6H_5)_2NH$	Secondary; simple aromatic	Diphenylamine
$(CH_3CH_2CH_2)_3N$	Tertiary; simple aliphatic	Tri-*n*-propylamine
$(C_6H_5)_3N$	Tertiary; simple aromatic	Triphenylamine
$CH_3CH_2N(H)CH_3$	Secondary; mixed aliphatic	Ethylmethylamine
$O_2N{-}C_6H_4{-}N(H){-}C_6H_5$	Secondary; mixed aromatic	p-Nitrophenylphenylamine
$C_6H_5N(CH_3)_2$	Tertiary; mixed aromatic-aliphatic	Dimethylaniline (Dimethylphenylamine)

33.2 NOMENCLATURE

Common names for amines are obtained by naming the groups attached to the nitrogen atom, followed by the term "amine." For example, $CH_3CH_2NH_2$ is ethylamine. $(CH_3)NH(CH_2CH_3)$ is ethylmethylamine and $(C_6H_5)_3N$ is triphenylamine. Table 33.1 provides examples of other names. The similarity of this system to that used for naming ethers and ketones should be apparent. Names of amines formed in this manner are also approved systematic IUPAC names.

Aromatic amines related to aniline (phenylamine), particularly the mixed amines, are usually named as aniline derivatives. Thus the compound which has a methyl group substituted for one of the hydrogen atoms attached to the nitrogen of aniline is named monomethylaniline. If two methyl groups are present in place of hydrogens, the compound is named dimethylaniline. Common names of ring substituted aromatic amines are formed in a similar way. Some aromatic amines have special names. meta-Toluidine, for example, is a name often used for meta-methylaniline.

Systematic IUPAC names may be formed also by considering amines as amino (—NH_2) or substituted amino (—NHR or —NR_2) hydrocarbons. Systematic names for several amines derived in this way are listed in Table 33.2.

TABLE 33.2 AMINES—SYSTEMATIC NAMES

Structure	Name
$CH_3CH_2CH_2NH_2$	1-Aminopropane
$CH_3CH(NH_2)CH_2CH_2CH_3$	2-Aminopentane
$CH_3CH_2NHCH_2CH_2CH_3$	1-(N-Ethylamino)propane
$C_6H_5NH_2$	Aminobenzene (aniline)
$C_6H_5NHCH_2CH_3$	N-Ethylaminobenzene

33.3 PHYSICAL PROPERTIES

Methylamine, dimethylamine and trimethylamine are gases which have a strong ammoniacal, or fishy, odor and are flammable. Higher amines of the aliphatic series are liquids or colorless solids and exhibit decreasing solubility with increasing molecular weight. On the other hand, the aromatic amines all are liquids or solids. They are colorless, when freshly prepared, but become dark brown or red on standing. They have a peculiar non-ammoniacal odor and are, in general, much less soluble in water than aliphatic amines. *Methylamine* is the simplest of the alkyl amines, and *aniline* is the simplest of the aromatic amines. As with the alcohols and aldehydes, to be aromatic, the key atom or group (in this case, nitrogen) must be joined directly to a benzene or benzene-like ring. The structures of these amines follow, with a brief description of each.

$H_3C—NH_2$

Methylamine

$C_6H_5—NH_2$

Aniline

Properties of Specific Compounds

1. Methylamine. Methylamine occurs in fish (herring) brine, in the urine of dogs and other animals fed on a high protein diet, and in certain plants. In most respects it has properties very similar to ammonia. For example, it is very soluble in water. One volume of water dissolves 959 volumes of the amine at 25°C. It is soluble in alcohol and completely miscible with ether. The primary use for methylamine is in the tanning of leather and in organic synthesis and it is manufactured by the ton for these uses.

2. Aniline. Aniline (phenylamine) was first obtained in 1826 by the dry distillation of indigo. It is an oily, colorless liquid which rapidly darkens on exposure to air and light. It is flammable and volatile with steam (b.p. 184 to 186°C; m.p. −0.6°C). The compound is much less soluble in water than is ammonia or methylamine (1 g dissolves in 28.6 ml of water), but it is miscible with alcohol, benzene and most organic solvents. Aniline is toxic; serious poisoning may result from the ingestion of as little as 0.25 cc.

33.4 PREPARATION

Two reactions used for the preparation of amines will be considered, **alkylation** (of ammonia with alkyl halides) and **reduction** (of nitro compounds or nitriles). Special reactions for preparing specific amines will not be described.

1. Alkylation

Alkylation, the most direct method for preparing amines, refers to the reaction between ammonia or amines and an alkyl halide, introducing alkyl groups directly into the molecule in place of one or more hydrogens. The reaction between ammonia and ethyl iodide yields ethyl ammonium iodide as the initial product.

$$NH_3 + CH_3CH_2I \rightarrow CH_3CH_2\overset{+}{N}H_3\overset{-}{I}$$

$$H:\underset{\underset{H}{..}}{\overset{\overset{H}{..}}{N}}: + H{-}\underset{\underset{H}{|}}{\overset{\overset{CH_3}{|}}{C}}{-}I \rightarrow \left[H:\underset{\underset{H}{..}}{\overset{\overset{H}{..}}{N}}:\underset{\underset{H}{|}}{\overset{\overset{CH_3}{|}}{C}}{-}H\right]^+ I^-$$

alkylation results from S_N2 attack of NH_3 on CH_3CH_2I.

In this reaction, since halide ions generally are weak bases and rather reluctant to share their electrons, iodide is readily displaced by the stronger base, ammonia. The reaction proceeds in a manner entirely analogous to the displacement of the halide ion from alkyl halides by hydroxide, in the preparation of alcohols. Ethylamine is liberated from the salt with excess ammonia or, even more readily and completely, with sodium hydroxide.

$$CH_3CH_2\overset{+}{N}H_3\overset{-}{I} + \underset{(NH_3)}{OH^-} \rightarrow CH_3CH_2NH_2 + H_2O + \underset{(NH_4I)}{I^-}$$

Primary and secondary amines are subject to further alkylation. The reaction between ethyl iodide and ammonia does not stop with replacement of only one hydrogen. Ethylamine formed in this first phase reacts with a second molecule of ethyl iodide, again displacing the iodide, to form diethylammonium iodide. Diethylamine is liberated with the excess base.

$$CH_3CH_2I + CH_3CH_2NH_2 \longrightarrow (CH_3CH_2)_2\overset{+}{N}H_2\overset{-}{I} \xrightarrow{NH_3} (CH_3CH_2)_2NH + NH_4I$$

Alkylation of diethylamine yields the tertiary amine, triethylamine, which, when subjected to further alkylation, forms the expected quaternary ammonium salt, tetraethylammonium iodide.

In all cases, the reaction of an alkyl halide with ammonia gives a mixture of three types of amines, plus a quaternary salt; the amount of each product is dependent on the relative quantities of ammonia and alkyl halide used. With an excess of the halide, a larger proportion of the tertiary amine and quaternary salt will be obtained. When an excess of ammonia is used, a larger percentage of the mixture will be primary and secondary amines. The mixture, however, is usually separable by fractional distillation of the free amines.

Aryl groups cannot be directly introduced in this way without employing high temperatures and pressures, conditions suitable for industrial synthesis but not often employed in the laboratory. Just as phenol is prepared industrially from chlorobenzene and steam (Chapter 29), using unusual conditions, aniline is produced by a reaction between chlorobenzene and ammonia at a high temperature (300°C) and pressure (500 psi).

2. Reduction of Nitro Compounds

The reduction of nitro compounds is one of the most useful reactions for preparing amines. Nitro compounds are characterized by the presence of the *nitro* (—NO_2) functional group. Nitrobenzene (Chapter 28) and nitromethane (Chapter 25) are typical examples of compounds of this type. Nitro groups of both aromatic (nitrobenzene) and aliphatic (nitromethane) compounds may be reduced to give primary amines.

$$C_6H_5NO_2 + 3\,H_2 \xrightarrow{Fe/HCl} C_6H_5NH_2 + 2\,H_2O$$

$$CH_3NO_2 + 3\,H_2 \xrightarrow{Pd} CH_3NH_2 + 2\,H_2O$$

Since aromatic nitro compounds are readily prepared by direct nitration (Chapter 28), this procedure is the most common means of introducing an amino group into an aromatic ring. Aliphatic nitro compounds (Chapter 25) are less readily obtainable and are less useful.

3. Reduction of Nitriles (Cyanides)

Organic compounds which have an alkyl or aryl group attached to a cyano (—CN) group are known as **nitriles or cyanides** (Chapter 31). Their general formula is RCN, with R representing either an alkyl or an aryl group. Nitriles are readily prepared by replacing the halogen of an alkyl halide with the cyano group.

$$CH_3CH_2I + KCN \rightarrow CH_3CH_2CN$$

Nitriles, once obtained, are readily reduced by catalytic hydrogenation to give primary amines, the cyano group being converted into an aminomethyl ($—CH_2NH_2$) group.

$$CH_3CH_2CN + 2\ H_2 \xrightarrow[\text{Pressure}]{\text{Ni}} CH_3CH_2CH_2NH_2$$

Examination of these two reactions shows that, in the formation of the nitrile from the alkyl iodide, one carbon is added to the R group which, when the amine is formed, is converted into a methylene ($—CH_2—$) group. The reaction is used to increase the length of a carbon chain.

33.5 REACTIONS

1. Basicity

Basicity is the characteristic property of amines. It determines the types of reactions which they undergo and influences their physical properties. As in ammonia, the basicity of amines is caused by an unshared (free) pair of electrons on the nitrogen atom. The chemical behavior of amines is, therefore, quite analogous to that of ammonia. Amines, because of a free pair of electrons, are electron-rich molecules in the same sense as ammonia or the hydroxide ion. Correspondingly, amines seek the relatively most positive site of other molecules with which to react chemically. For example, methylamine reacts with hydrogen chloride, forming a hydrochloride salt.

this is a base displacement of Cl^- from a proton with the stronger base CH_3NH_2.

$$CH_3NH_2 + HCl \rightarrow \begin{matrix} & H & & H & \\ H : & \ddot{C} & : & \ddot{N}^+ & : H \\ & \ddot{H} & & \ddot{H} & \end{matrix} \quad Cl^- \quad (\text{or } CH_3NH_2{\cdot}HCl)$$

The driving force for this addition resides in the tendency of nitrogen to share its free electron pair with the positively charged proton of the hydrogen chloride. If the proton is combined with chloride when collision occurs, the chloride ion is displaced. If it is present in solution as the hydrogen ion, the positive ion merely becomes attached to the amine. The salt is named methylammonium chloride or methylamine hydrochloride; the name methylammonium chloride is indicative of its similarity to ammonium chloride.

In a related reaction with boron trifluoride (BF_3), which is an electron-deficient substance, methylamine, by providing both electrons for the bond, yields an amine-boron trifluoride analog of the salt formed when ammonia reacts with BF_3.

$$CH_3NH_2 + BF_3 \rightarrow H : \underset{\ddot{H}}{\overset{H}{\ddot{C}}} : \underset{\ddot{H}}{\overset{H^+}{\ddot{N}}} : \underset{\ddot{F}}{\overset{F^-}{\ddot{B}}} : F$$

Secondary and tertiary amines undergo similar reactions with acids to form stable salts.

$$C_6H_5NH(CH_3) + HCl \longrightarrow C_6H_5\overset{+}{N}H_2(CH_3)\overset{-}{Cl}$$

$$(CH_3)_2NH + HCl \longrightarrow (CH_3)_2\overset{+}{N}H_2\overset{-}{Cl}$$

$$C_6H_5N(CH_3)_2 + HCl \longrightarrow C_6H_5\overset{H+}{N}(CH_3)_2\overset{-}{Cl}$$

$$(CH_3CH_2)_3N + HCl \longrightarrow (CH_3CH_2)_3\overset{+}{N}H\overset{-}{Cl}$$

The readiness with which an amine accepts protons or reacts with other electron-deficient reagents depends on the availability of the unshared electrons, since these electrons are necessary for bond formation. An increase in the electron density about the nitrogen atom makes the unshared pair more available; a decrease makes these same electrons less available. Alkyl amines, in general, are stronger bases than ammonia. This fact is attributed to the nature of alkyl groups which are more electron-releasing than hydrogen. Alkyl groups induce an increased electron density on the nitrogen of amines above that of ammonia. The free pair on the nitrogen becomes more available, giving aliphatic amines their greater basicity.

On the other hand, aromatic amides are less basic than ammonia. Aryl groups generally are electron-withdrawing and decrease the electron density about the nitrogen atom. The basicity of several amines is listed in terms of ionization constants in Table 33.3.

Quaternary Ammonium Compounds. Closely related to and derived directly from the amines are the quaternary ammonium salts and bases. Quaternary salts are formed when tertiary amines react with alkyl halides. Tetramethylammonium iodide, a typical quaternary ammonium salt, has the formula shown in Figure 33.3.

acetyl choline $CH_3COOCH_2CH_2\overset{+}{N}(CH_3)_3\overset{-}{O}H^-$, which is liberated at nerve endings, is a quarternary ammonium compound of great biological importance.

Structurally, this compound is related to ammonium iodide. It is a substituted ammonium iodide in which four hydrogens have been replaced with methyl groups. The properties of tetramethylammonium iodide verify this relationship. It is a solid with a high melting point (above 200°C). The compound is a true ionic salt and dissolves in water to give a solution which conducts an electric current. Other quaternary ammonium salts are similar. If one of the R groups is a long chain hydrocarbon, the salt takes on the properties of a detergent. Detergents of this class are known as **invert soaps** because the water soluble polar end is a positive rather than a negative ion. It might be recalled that the water soluble part of

TABLE 33.3 BASICITY OF AMINES

Name	Formula	K_b
Ammonia	NH_3	2×10^{-5}
Methylamine	CH_3NH_2	4×10^{-4}
Aniline	$C_6H_5NH_2$	5.4×10^{-10}
Dimethylamine	$(CH_3)_2NH$	5.4×10^{-4}
Methylaniline	$C_6H_5NH(CH_3)$	2.5×10^{-10}
Trimethylamine	$(CH_3)_3N$	5.9×10^{-5}
Dimethylaniline	$C_6H_5N(CH_3)_2$	2.4×10^{-10}
Triphenylamine	$(C_6H_5)_3N$	less than 10^{-10}

$$K_b = \frac{[RNH_3^+][OH^-]}{[RNH_2]}$$

soap and the usual detergents is a negative carboxylate or sulfonate group.

Just as ammonium iodide can be converted into ammonium hydroxide by treatment with silver hydroxide (silver iodide precipitates), so tetramethylammonium iodide can be converted into tetramethylammonium hydroxide.

$$\left[\begin{array}{c} CH_3 \\ | \\ H_3C\text{—}N\text{—}CH_3 \\ | \\ CH_3 \end{array}\right]^+ I^-$$

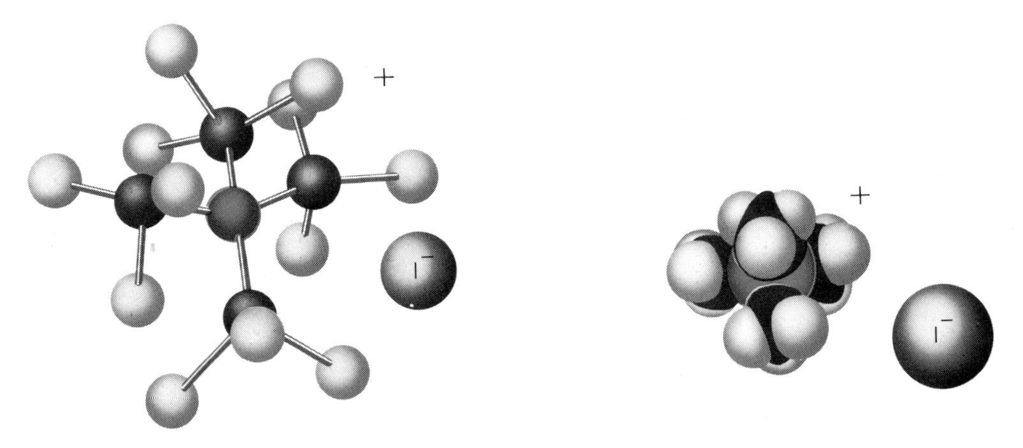

FIGURE 33.3 TETRAMETHYLAMMONIUM IODIDE.

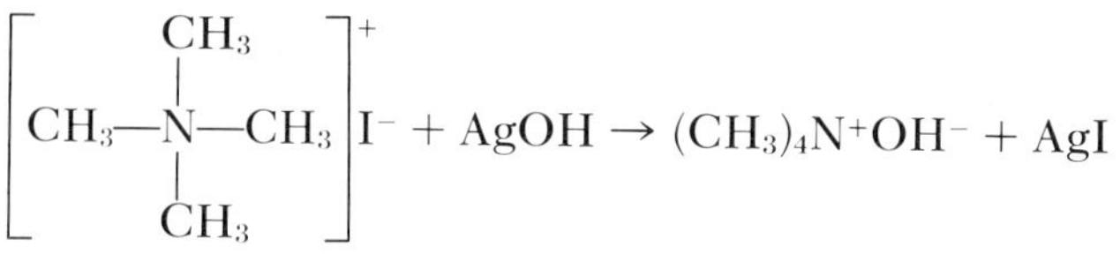

$$\left[\begin{array}{c} CH_3 \\ | \\ CH_3{-}N{-}CH_3 \\ | \\ CH_3 \end{array}\right]^{+} I^- + AgOH \rightarrow (CH_3)_4N^+OH^- + AgI$$

Tetramethylammonium hydroxide is a strong base. It readily absorbs carbon dioxide from air, just as sodium or potassium hydroxide does. An aqueous solution of this base is colorless, odorless, bitter, caustic and strongly alkaline in the same sense as are solutions of alkali metal hydroxides. Like sodium hydroxide, tetramethylammonium hydroxide imparts a soapy feel to the skin. A quaternary salt having an aryl group is trimethylphenylammonium iodide.

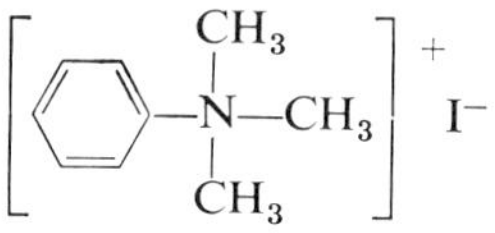

$$\left[\begin{array}{c} CH_3 \\ | \\ C_6H_5{-}N{-}CH_3 \\ | \\ CH_3 \end{array}\right]^{+} I^-$$

2. Alkylation

Alkylation of amines is a displacement reaction discussed previously (pp. 735–736).

3. Acylation

The reaction between a primary or secondary amine and an acid derivative (acid halide, acid anhydride or ester; see Chapter 31) yields an amine derivative in which an acyl ($R{-}\overset{O}{\overset{\|}{C}}{-}$) group has been substituted for a hydrogen attached to the amine nitrogen. The term **acylation** is used because an **acyl group** is introduced into the molecule.

Tertiary amines which have no hydrogen attached to nitrogen do not undergo this reaction.

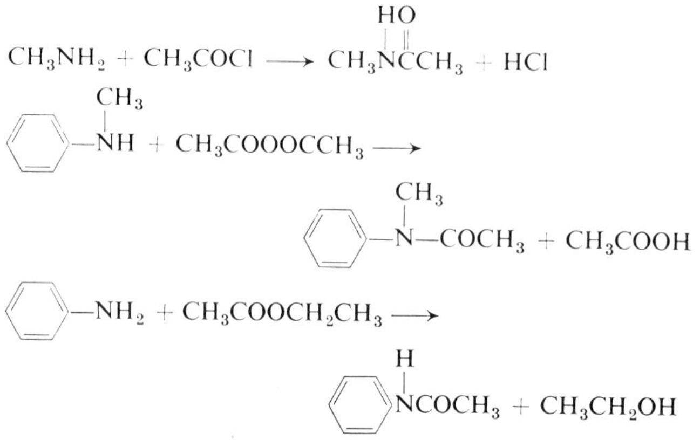

$$CH_3NH_2 + CH_3COCl \longrightarrow CH_3\overset{H}{\overset{|}{N}}\overset{O}{\overset{\|}{C}}CH_3 + HCl$$

$$C_6H_5{-}\overset{CH_3}{\overset{|}{N}}H + CH_3COOOCCH_3 \longrightarrow C_6H_5{-}\overset{CH_3}{\overset{|}{N}}{-}COCH_3 + CH_3COOH$$

$$C_6H_5{-}NH_2 + CH_3COOCH_2CH_3 \longrightarrow C_6H_5\overset{H}{\overset{|}{N}}COCH_3 + CH_3CH_2OH$$

Acylation may be considered to be an acid-base reaction. The amine (a nucleophile and base) attacks the carbonyl carbon (positive electrophilic center) of the acid derivative to displace a negative ion (i.e., chloride from an acid chloride). As a result of this loss of a negative ion, a positively charged complex remains which immediately adjusts (stabilizes) itself through loss of a proton to give the stable acylated amine (amide). The reaction reflects the electron-rich (basic) nature of amines and emphasizes the electron-deficient character of the carbonyl carbon atom. Having the most positive carbonyl carbon, acid halides react most readily, whereas esters, with the least positive carbonyl carbon, react much more slowly.

Close scrutiny of the formula of the product shows that two organic groups are united by an $-\overset{H}{\overset{|}{N}}-\overset{O}{\overset{\|}{C}}-$ group of atoms. This is called an **amide linkage.** It is present in acid amides (Chapter 31) and gets its name from these compounds. It is an important type of bonding commonly encountered in biologically interesting substances, particularly the proteins (Chapter 35).

4. Reactions with Nitrous Acid

The products formed when amines react with nitrous acid depend on the amine used. Each class of amine exhibits different behavior.

Primary aliphatic amines liberate gaseous nitrogen, the evolution of which serves as a useful test for amines of this class. The reaction is used also for the quantitative determination of free amino groups in proteins and amino acids by measuring the volume of nitrogen gas evolved. Aliphatic diazonium compounds (see page 742), that decompose immediately are formed as intermediates in this reaction.

$$CH_3NH_2 + HONO \rightarrow CH_3OH + N_2 + H_2O$$

Secondary aliphatic and *aromatic amines* form nitrosoamines. Nitrosoamines are usually yellow, neutral compounds, insoluble in dilute mineral acids. When formed in aqueous solutions, they separate as an oily layer, a behavior used as a test for secondary amines.

reaction occurs because of the activation of the phenyl ring by the dimethylamino group.

$$(CH_3CH_2)(CH_3)NH + HONO \rightarrow (CH_3CH_2)(CH_3)N{-}N{=}O + H_2O$$

Tertiary aliphatic amines yield nitrous acid salts that dissolve in the aqueous medium. Substitution of the ring occurs with aromatic tertiary amines.

$$C_6H_5N(CH_3)_2 + HONO \longrightarrow O{=}N{-}C_6H_4{-}N(CH_3)_2 + H_2O$$

The reaction of nitrous acid with primary, secondary and tertiary aliphatic amines serves as a test for distinguishing between these three types of compounds because the different products formed can readily be detected.

The reaction of primary aromatic amines with nitrous acid is one of the most important reactions in organic chemistry. Primary aromatic amines react to form **diazonium compounds** (salts) which are stable at low temperatures in aqueous solution. The nitrous acid for this reaction is formed *in situ* by adding hydrochloric acid to a cold solution of sodium nitrite. Aniline and nitrous acid form benzene diazonium chloride:

$$C_6H_5NH_2 + HONO + HCl \longrightarrow C_6H_5\overset{+}{N}{\equiv}N\overset{-}{Cl} + 2\,H_2O$$

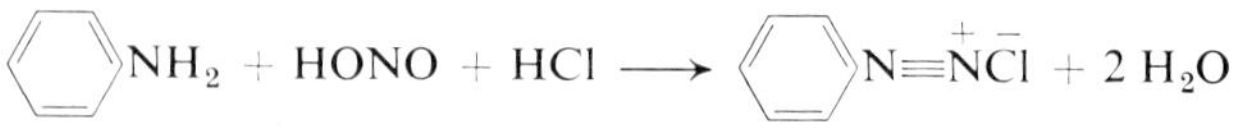

The product formed is extremely useful because of the unique behavior of the **diazonium** (—N≡N$^+$) **group.** Diazonium salts are versatile. The diazonium group may readily be replaced with a variety of other atoms or groups, or it may undergo a coupling reaction to form diazo compounds.

Replacement of the diazonium group is an excellent general method for introducing F, Cl, I, CN, OH and H into aromatic (benzene-like) rings. The equation for the introduction of iodine is:

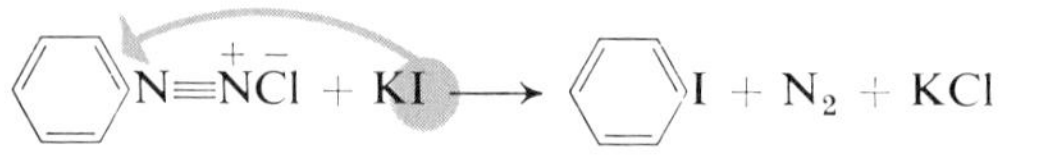

$$C_6H_5\overset{+}{N}{\equiv}N\overset{-}{Cl} + KI \longrightarrow C_6H_5I + N_2 + KCl$$

Coupling takes place with phenols and aromatic amines. **Azo compounds**, recognized by the *azo* (—N=N—) group, are formed. Benzene diazonium chloride reacts with phenol to give p-hydroxyazobenzene:

coupling takes place only with highly activated aromatic rings.

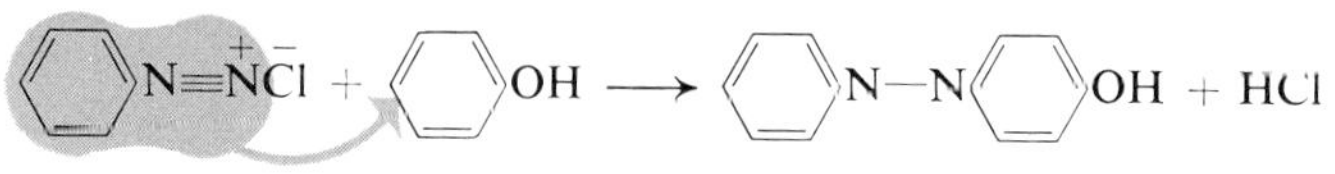

$$C_6H_5\overset{+}{N}{\equiv}N\overset{-}{Cl} + C_6H_5OH \longrightarrow C_6H_5N{=}NC_6H_4OH + HCl$$

Most azo compounds are colored. Many are useful dyes or biological stains. About half the dyes in industrial use today are azo dyes. Methyl orange is an azo compound which you may have used as an acid-base indicator (Fig. 33.4).

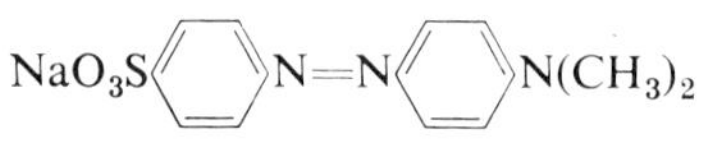

$$NaO_3SC_6H_4N{=}NC_6H_4N(CH_3)_2$$

FIGURE 33.4 METHYL ORANGE.

5. Dyes and Dyeing

Man's eye is color sensitive only to the small range of the **electromagnetic spectrum** lying between wavelengths 4000 Å and 7500 Å, commonly known as the visible spectrum. The colors violet, green, yellow, orange and red are within this range. Perception of color is a physiological response associated both with the wavelengths of light and the intensities of the different wavelengths striking the retina of the eye. Color can be observed when a light source emits light of a wavelength in the visible region. The common sodium vapor lamp emits yellow light. Color is also observed when a substance absorbs light of wavelengths in the visible range from white light to which it is exposed. The color complementary to that removed by the absorption is seen. For example, if violet light (4100 Å) is absorbed, the absorbing substance appears yellow to the eye; if red light (7000 Å) is absorbed, the absorber assumes a green color. Absorption of radiation outside the range of visible light, 4000 Å to 7500 Å, has no effect on the observed color.

All organic molecules absorb electromagnetic radiation (light). Absorption occurs when the energy provided by the light is taken up by (transferred to) electrons present in the molecule. Each type of electron absorbs light of only a certain energy content. When energy is absorbed by an electron, the electron is promoted from an orbital of lower energy to an orbital of higher energy. Less firmly bound (higher energy) electrons, such as pi electrons, require less energy to be promoted to higher orbitals than do more tightly bound (lower energy) electrons. Thus, pi electrons, when present, absorb light of lower energy and longer wavelength. More tightly bound electrons, such as electrons in sigma orbitals, absorb light of higher energy or shorter wavelengths. Since the energies of the electrons present reflect the structure of a molecule, the spectrum of the light absorbed indicates the molecular structure of a compound and the structural features present in the compound.

Many substances are colored because they absorb light in the visible region. All of these, however, cannot be classified as dyes. A **dye** is a material that absorbs light strongly in the visible region, and also can be firmly attached to a surface or fiber by chemical or physical bonding. To serve as a dye, a substance must contain a

TABLE 33.4 CHROMOPHORES

Name	Structure
Ethylenic group	$-\overset{\vert}{C}=\overset{\vert}{C}-$
Azo group	$-N=N-$
Nitro group	$-\underset{\substack{\vert \\ O}}{N}=O$
Thiocarbonyl group	$>C=S$
Carbonyl group	$>C=O$
Conjugated system	$-\overset{\vert}{C}=\overset{\vert}{C}-\overset{\vert}{C}=\overset{\vert}{C}-$
Quinoid group	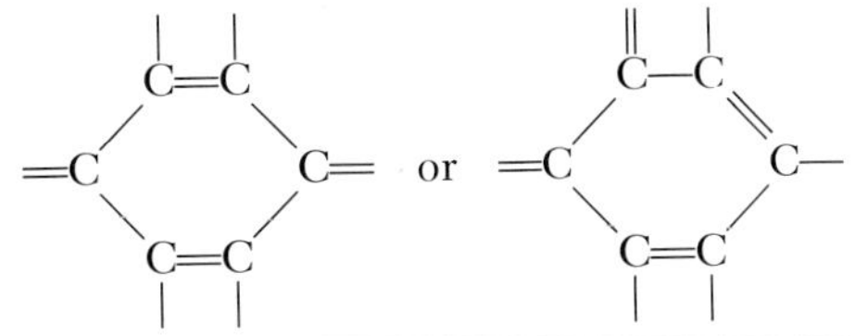 see structures below

```
        |   |                      ||  |
        C===C                      C---C
       /     \                    /     \\
   ==C         C==      or    ==C         C--
       \     /                    \     /
        C===C                      C===C
        |   |                      |   |
```

structural feature that provides electrons which can absorb wavelengths of light in the visible region, and it must also have a group or groups that permit chemical or physical combination with the material to be dyed. Azobenzene is a colored substance with a deep red coloration from the azo (—N=N—) group. Such structures, permitting the absorption of light causing the compound to be colored, are referred to as **chromophores.** Every colored compound and dye contains at least one chromophore. Some of the common chromophores are given in Table 33.4. Note that each structure shown in Table 33.4 has pi electrons in unsaturated linkages between atoms. Azobenzene is not a dye because it does not adhere (bond) readily to fabrics.

In contrast to azobenzene, p-hydroxyazobenzene, having a hydroxyl group in place of a para hydrogen of azobenzene, is a dye because it contains both the azo group that produces the color and a hydroxyl group enabling the compound to adhere to fabrics of certain types. An anchoring group such as the hydroxyl acts as an **auxochrome,** a type of group that, when conjugated to chromophores, deepens the color. Each permanent dye must contain both a chromophore and an auxochrome that is resistant to washing procedures. Table 33.5 includes a variety of groups classed as auxochromes. They are divided into two categories according to their acidic or basic character.

TABLE 33.5 AUXOCHROMES

Name	Structure
Basic groups	
Amino	$—NH_2$
Alkyl amino	—NHR
Dialkyl amino	$—NR_2$
Acidic groups	
Phenolic hydroxyl	—OH
Carboxyl	—COOH
Sulfonic acid	$—SO_3H$

A group whose presence causes light absorption of a parent compound to move to longer wavelengths is called a **bathochromic group.** Most of these have unshared electron pairs. Most auxochromes are also bathochromic groups. **Hypsochromic groups** have the opposite effect to that of the bathochromic groups and cause absorption to shift to shorter wavelengths.

Dyes may be classified either by chemical structure or by differences in their mode of application. Some chemically important classes of dyes are the azo dyes, the triphenylmethane dyes, the anthraquinone dyes and the nitrophenol dyes. **Azo dyes** contain the azo (—N=N—) as the main chromophore. Methyl orange (Fig. 33.4), a common dye and laboratory indicator, is an example of this class of dye. Three phenyl groups joined to one central carbon atom, with one of the rings having a quinoid structure, characterize the **triphenylmethane dye** structure. Pararosaniline chloride (Fig. 33.5) is a typical compound of this class.

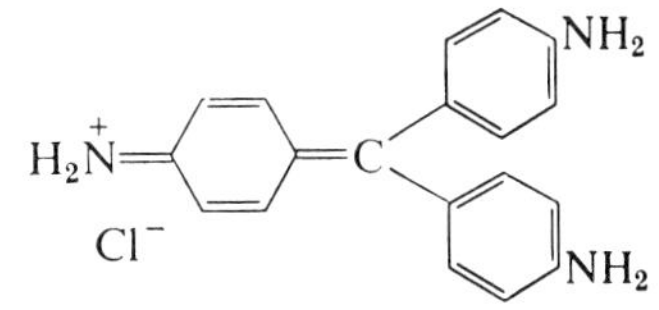

FIGURE 33.5 PARAROSANILINE.

The anthraquinone ring system is the distinguishing feature of **anthraquinone** dyes. Alizarin (Fig. 33.6) is a useful dye of this type.

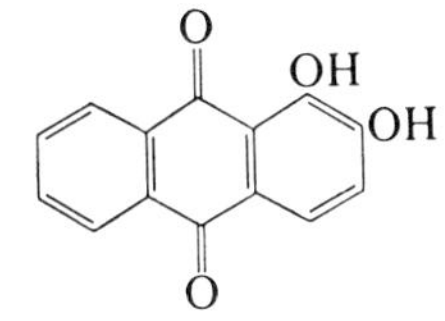

FIGURE 33.6 ALIZARIN.

A **nitrophenol** dye, such as acid yellow I (naphthol yellow S) (Fig. 33.7), contains both nitro groups and a phenolic group. The sulfonic acid group present increases its water solubility.

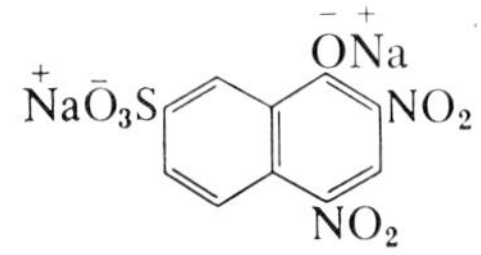

FIGURE 33.7 ACID YELLOW I.

In accordance with their method of application to materials there are such dyes as the direct or substantive dyes, the ingrain dyes, and the vat dyes, to name three classes.

Direct dyes are applied directly to the fiber or cloth and bind firmly to the material without the assistance of other chemicals, through direct reaction of an acidic or basic group of the dye with a group of the opposite character on the fabric. **Ingrain** dyes are formed chemically within the fiber itself by reaction between the necessary ingredients after application to the material. **Vat dyeing** also consists of dye formation directly in the fabric. Vat dyes are water insoluble but are rendered water soluble by reduction in alkaline solution. Cotton absorbs this colorless soluble reduced form into its fibers, and then the less soluble dye is produced by direct oxidation of the soluble colorless form trapped in the fabric. Numerous other methods of applying dyes are presently used, the details of which are readily available in special monographs on the subject.

33.6 PHOSPHORUS COMPOUNDS

Many organic phosphorus compounds are important. Two types of particular interest are **alkyl phosphines** and the organic esters of phosphoric acid. Alkyl phosphines are derived from phosphine (PH_3) and resemble amines in structure. They are less basic and more subject to oxidation than amines. Phosphine itself burns spontaneously in air. The basicity of phosphines increases with the introduction of alkyl groups. Table 33.6 lists the known types of phosphine compounds.

TABLE 33.6 ALKYL PHOSPHINES

Formula	Classification
RPH_2	Primary
R_2PH	Secondary
R_3P	Tertiary
R_4P^+	Quaternary

Esters are perhaps the most important, and certainly the most frequently encountered, organic phosphorus compounds. Esters of phosphorus, phosphoric and pyrophosphoric acids are known. *Tributyl phosphate* and tricresyl phosphate (TCP), widely used plasticizers, are triesters of phosphoric acid. Monophosphate esters of phosphoric acid with glucose, fructose, and other carbohydrates (Chapters 36 and 37) are significant biological compounds. Glucose-1-phosphate, fructose-1,6-diphosphate and 2-phosphoglyceric acid are three of the recognized metabolites formed during the fermentation of glucose and also in its transformation in muscle. The structures of several of these compounds are as follows:

$(CH_3CH_2CH_2CH_2O)_3PO$

Tri-n-butyl phosphate

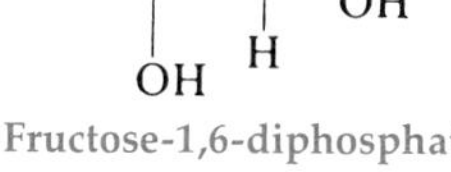

$(CH_3C_6H_4O)_3PO$

Tricresyl phosphate (TCP)

Glucose-1-phosphate

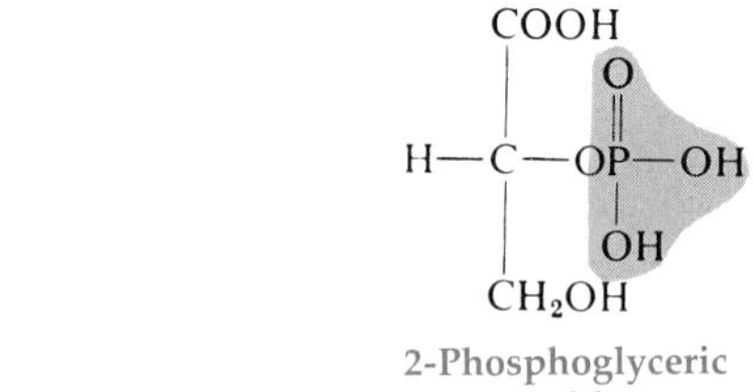

Fructose-1,6-diphosphate

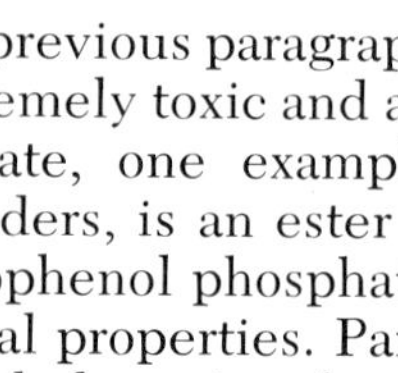

2-Phosphoglyceric acid

In contrast to the metabolites mentioned in the previous paragraph, certain organic phosphorus compounds are extremely toxic and are used as insecticides. Tetraethyl pyrophosphate, one example, useful for control of aphids, mites and red spiders, is an ester of pyrophosphoric acid. Para-oxon (diethyl p-nitrophenol phosphate) is a phosphoric acid ester with good insecticidal properties. Parathion, the corresponding thiophosphate (diethyl p-nitrophenol thiophosphate), is also an effective insecticide and is more widely used than para-oxon. The formulas for these compounds are as follows:

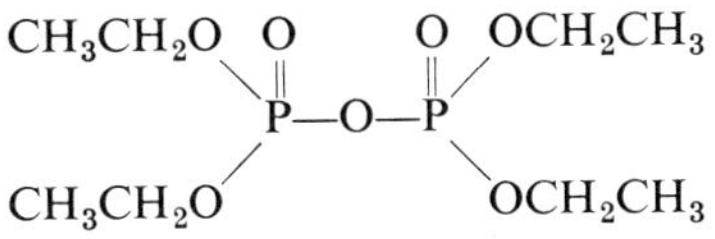

$$(CH_3CH_2O)_2P(=O)-O-P(=O)(OCH_2CH_3)_2$$

Tetraethyl pyrophosphate

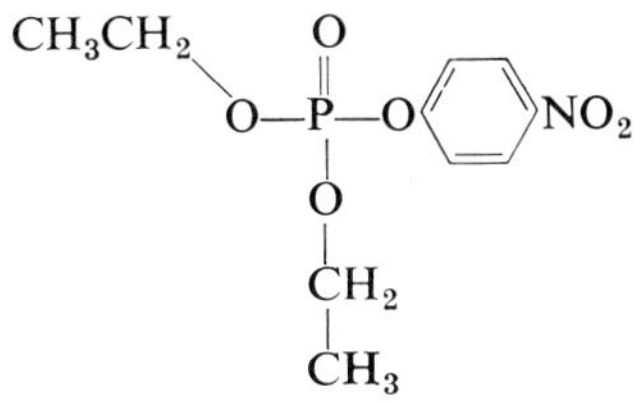

$$CH_3CH_2-O-P(=O)(OCH_2CH_3)-O-C_6H_4-NO_2$$

Para-oxon

$$CH_3CH_2O-P(=S)(OCH_2CH_3)-O-C_6H_4-NO_2$$

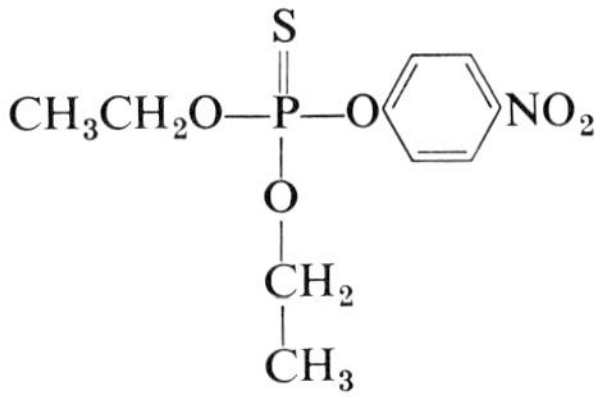

Parathion

33.7 SULFUR COMPOUNDS

Sulfur compounds are important, particularly in the field of dyes and medicinals. They are closely parallel in structure to the compounds of oxygen discussed in earlier chapters of this text. The similarity is shown in Table 33.7, listing the important oxygen compounds and their corresponding analogs.

TABLE 33.7 SULFUR COMPOUNDS

Oxygen Compounds		Sulfur Compounds	
Type Name	Type Formula	Type Formula	Type Name
Alcohol	ROH	RSH	Thioalcohol (Mercaptan)
Aldehyde	RCHO	RCHS	Thioaldehyde
Carbamic acid	H_2NCOOH	H_2NSO_3H	Sulfamic acid
Ether	ROR	RSR	Thioether (Sulfide)
Peroxide	ROOR	RSSR	Disulfide
Phenol	 C_6H_5OH	 C_6H_5SH	Thiophenol

Many other types of compounds might be compared, but the compounds listed in Table 33.7 are sufficient for illustrating the point that sulfur occupies the same position in sulfur compounds that oxygen does in similar oxygen containing substances. Although, in accord with their structural similarity, many reactions of sulfur compounds are similar to those of oxygen compounds, notable differences appear because of the larger size of the sulfur atom and its smaller electronegativity. For example, whereas peroxides are relatively rare and are unstable, the corresponding disulfides (RSSR) are numerous and stable. Correspondingly, ethers may form explosive peroxides when oxidized, whereas the oxidation of thioethers (sulfides) yields stable compounds called *sulfoxides* (RSOR) or *sulfones* (RSO_2R). Some specific compounds illustrative of the classes listed in Table 33.7 include butyl mercaptan, methionine and cystine. Their structures are as follows:

$$CH_3CH_2CH_2CH_2SH$$

n-Butyl mercaptan

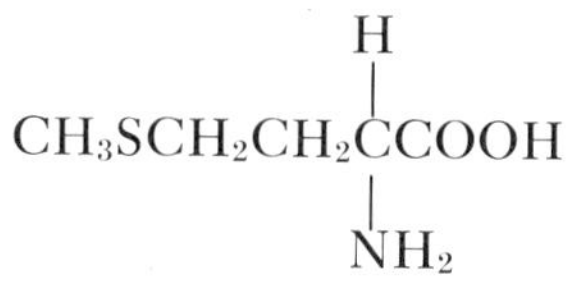

$$CH_3SCH_2CH_2\overset{\displaystyle H}{\underset{\displaystyle NH_2}{C}}COOH$$

Methionine

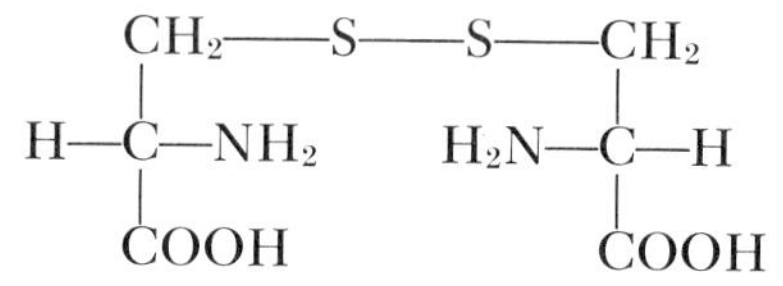

$$\begin{array}{ccc} CH_2-S & -S- & CH_2 \\ | & & | \\ H-C-NH_2 & & H_2N-C-H \\ | & & | \\ COOH & & COOH \end{array}$$

Cystine

Sulfathiazole is a typical example of a medicinal compound containing sulfur. It is a sulfonamide-type drug. It should be noted that nitrogen is also present in this drug.

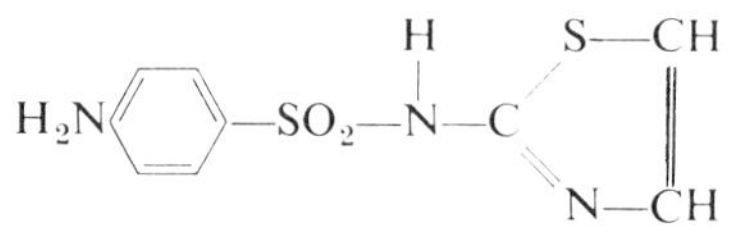

$$H_2N-C_6H_4-SO_2-\overset{H}{N}-C\begin{array}{l} S-CH \\ \| \\ N-CH \end{array}$$

Sulfathiazole

Many sulfur dyes are known. Some of them are obtained by heating different organic substances with sodium polysulfide solutions. These sulfur dyes have complex structures about which little is known. They can be used for dyeing cotton.

33.8 HETEROCYCLIC COMPOUNDS

Many organic compounds are cyclic in structure. Some of these compounds, such as benzene or cyclohexane, are classed as carbocyclic compounds because all atoms of the ring (or cyclic) structures present are carbon. Other compounds have ring structures in which one or more non-carbon atoms are present along with carbon. For example, in the six-atom ring of the cyclic compound pyridine, five atoms are carbon and one is nitrogen (Fig. 33.8). A cyclic compound such as pyridine is classified as a "**heterocyclic**" compound and the non-carbon ring atom is referred to as a "**hetero**" atom. Nitrogen, oxygen and sulfur are the hetero atoms most commonly present. For this reason heterocyclic compounds are introduced at this point in the text. Although only a brief discussion can be provided here, these compounds, because of their great abundance in nature, do comprise a very large segment of the total field of organic chemistry. Moreover, they are important because of the

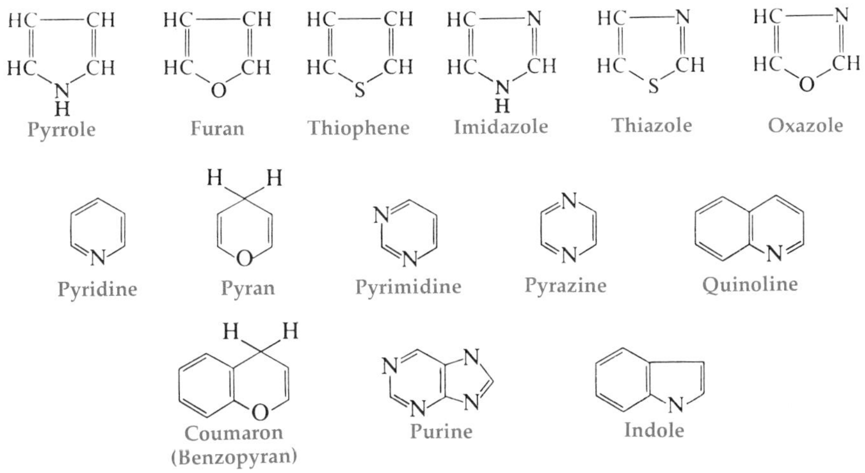

Pyrrole Furan Thiophene Imidazole Thiazole Oxazole

Pyridine Pyran Pyrimidine Pyrazine Quinoline

Coumaron (Benzopyran) Purine Indole

FIGURE 33.8 SOME HETEROCYCLIC RINGS AND RING SYSTEMS.

potent biological effects that many of them have on living organisms. Many serve critical functions in cellular processes and reactions; others have significant medicinal uses. Some heterocyclic compounds present relatively simple structures; others are extremely complex. Some contain a single hetero atom; others contain several. Some contain only a single type of heterocyclic ring; others contain several or many. Examples of some of the less complex heterocyclic rings and ring systems are given in Figure 33.8.

An almost endless number and variety of substances occurring in nature could be cited as illustrations of important compounds having heterocyclic structures, but only a few compounds will be presented to illustrate inclusion of the more common types of heterocyclic structures shown in Figure 33.8.

Many vitamins contain heterocyclic rings as part of their structure. Vitamin C (ascorbic acid) contains a furan-type ring (Fig. 33.9). It may be isolated from citrus fruits, tomatoes and other fresh vegetables. Biologically it appears to function in oxidation-reduction processes in cells, where it undergoes reversible oxidation-reduction in performing its function.

FIGURE 33.9 ASCORBIC ACID.

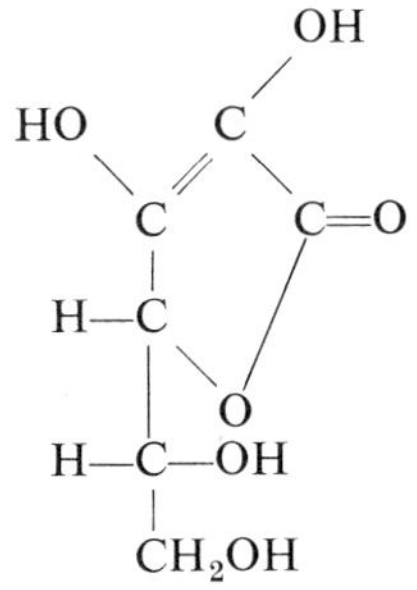

Vitamin E is a mixture of several compounds classed as **tocopherols.** The structure of α-tocopherol (5,7,8-trimethyltocol) is given in Figure 33.10. It contains a benzopyran ring structure with an oxygen hetero atom. It is obtained from lettuce, milk and hard-grain cereals. It appears to function as an antioxidant in cells.

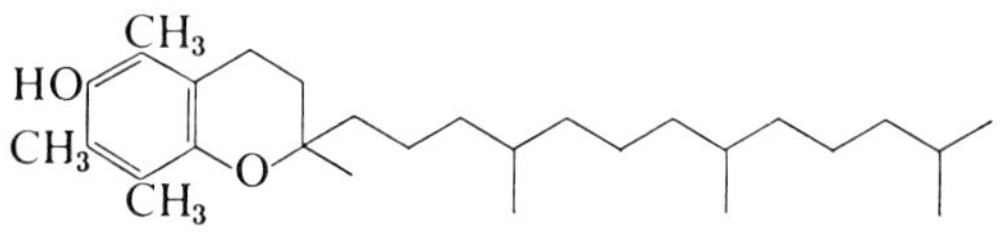

FIGURE 33.10 α-TOCOPHEROL. Each bend corresponds to a single carbon, as does each end of a line.

Thiamine (vitamin B_1), with the structure shown in Figure 33.11, contains two heterocyclic rings. Both the pyrimidine ring (on the left) and the thiazole ring (on the right) are prominent structures. Thiamine is prevalent in brown polishings of rice and plays a significant role in the oxidation of and loss of carbon dioxide from amino acids and other compounds in cells. It is usually found in cells as its pyrophosphate derivative in which

$$-\underset{\underset{OH}{|}}{\overset{\overset{O}{\|}}{P}}-O-\underset{\underset{OH}{|}}{\overset{\overset{O}{\|}}{P}}-OH$$

has replaced the H of the OH of the CH_2CH_2OH group.

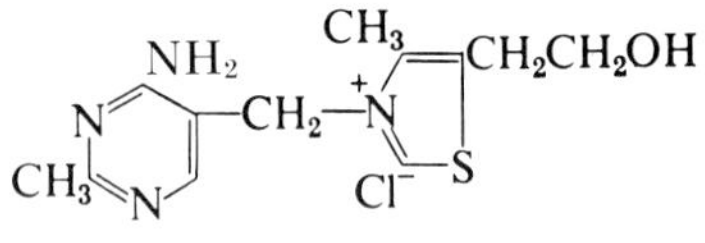

FIGURE 33.11 THIAMINE CHLORIDE.

Nicotinic acid amide (or niacin), the antipellagra vitamin, is a relatively simple heterocyclic derivative of pyridine. It is the amide having the structure shown in Figure 33.12. This vitamin is a component of one of the most important oxidation-reduction catalysts found in cells, and the pyridine ring undergoes reversible oxidation-reduction in its functions. This is its only known physiological role. It is widely distributed in plants and animals, meat products being the most important sources.

FIGURE 33.12 NIACIN.

One of the components of the vitamin B_6 complex, **pyridoxal,** also contains the pyridine ring system (Fig. 33.13). Pyridoxal, found in cells, is combined with phosphate as pyridoxal phosphate. The phosphate group, $-PO_3H_2$, replaces the H of the CH_2OH group of pyridoxal. Pyridoxal phosphate plays a significant role in transformations of amino acids in cells.

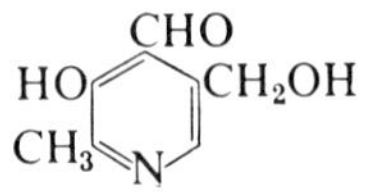

FIGURE 33.13 PYRIDOXAL.

Certain of the essential amino acids (Chapter 35) have structures containing heterocyclic rings. **Tryptophan** (Fig. 33.14) contains the indole ring system. **Histidine** (Fig. 33.15), which seems to have an important role in the action of many enzyme catalysts in cells, contains the imidazole ring system.

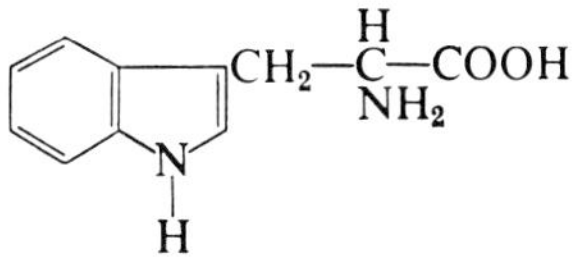

FIGURE 33.14 TRYPTOPHAN.

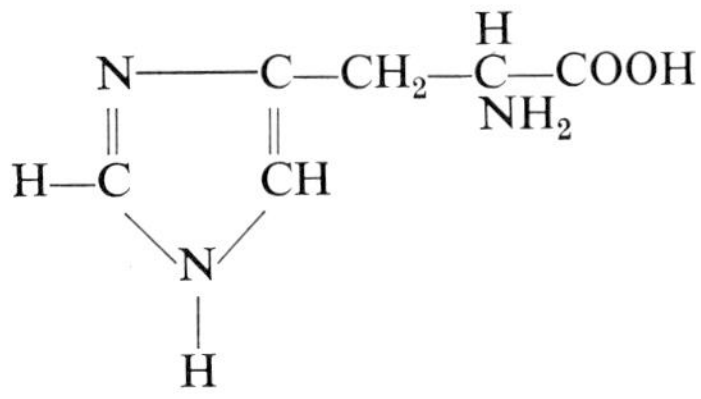

FIGURE 33.15 HISTIDINE.

Heterocyclic compounds containing either the **pyrimidine or purine ring systems** are important constituents of the genes and chromosomes. Examples are shown in Figure 33.16. These substances, along with phosphoric acid and pentose sugars correctly combined, form a class of substances referred to as **nucleic acids.** They are designated as **ribose nucleic acid** (RNA) or **deoxyribose nucleic acid** (DNA) according to the pentose sugar they contain. A nucleic acid combined with a suitable protein yields a nucleoprotein.

Uracil Cytosine Adenine Guanine

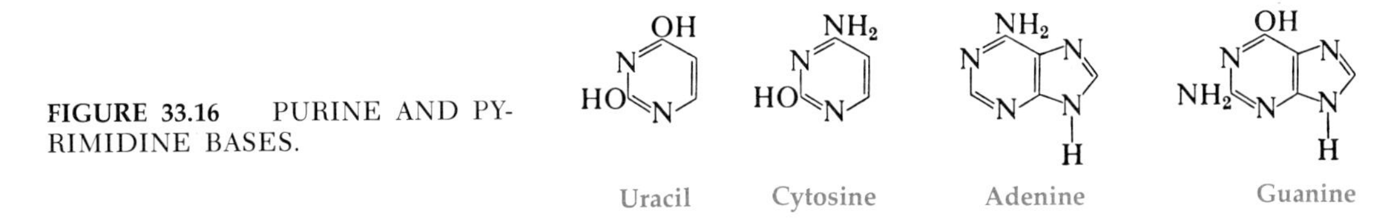

FIGURE 33.16 PURINE AND PYRIMIDINE BASES.

Heme, a component of hemoglobin (Fig. 33.17), and chlorophyll (Fig. 33.18) provide examples of one class of complex heterocyclic compounds of great biological importance. Both heme and chlorophyll have a complicated structure consisting of four pyrrole rings joined into a single larger ring system by methine (=CH—) groups. Various substituents are attached to each pyrrole ring in place of hydrogens. The large 4-pyrrole ring structure present in these two compounds is called the **porphyrin ring system.**

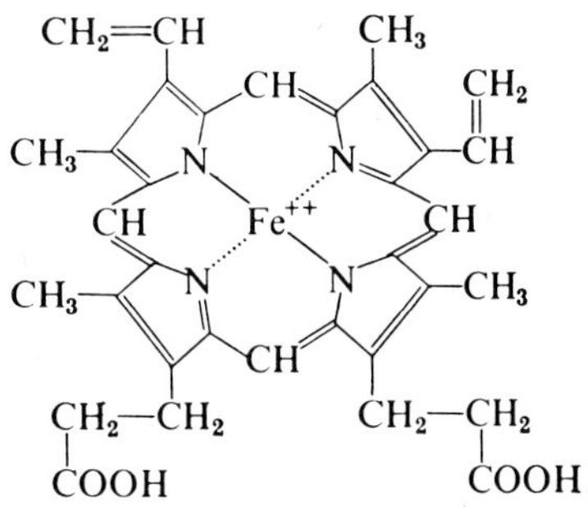

FIGURE 33.17 HEME.

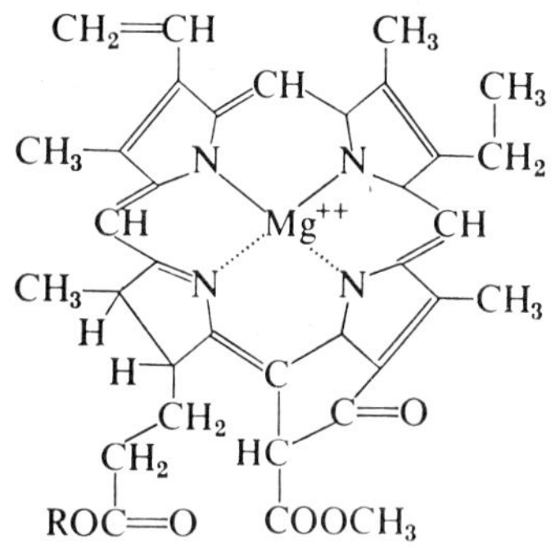

FIGURE 33.18 CHLOROPHYLL A.

Chlorophyll is the photosynthesizing pigment that effects the conversion of light energy into chemical energy which is then used for the conversion of CO_2 into carbohydrate in plants. Heme, when joined to the protein (see Chapter 35), gives the compound hemoglobin. Heme is the colored component of hemoglobin. Heme and its close relatives structurally are capable of catalyzing many different biological reactions. Hemoglobin transports oxygen in the blood. Catalase, the enzyme which catalyzes the decomposition of hydrogen peroxide, contains this iron porphyrin system, with the iron in the +3 oxidation state. The cytochromes, which are catalysts for cellular oxidations, have the same iron porphyrin as catalase. Note that magnesium is the central metal atom in chlorophyll, in contrast to iron in the other compounds mentioned. These compounds are colored because of the extended conjugated pi electron systems, which permit light absorption in the visible range or wavelengths. A detailed study of heterocyclic compound chemistry is of great value to anyone interested in an understanding of biological phenomena.

EXERCISES

33.1 Why is nitrogen an important element in organic compounds?

33.2 Define the following and give an example for each italicized term:

a. an *amine*
b. a *primary amine*
c. a *secondary amine*
d. a *mixed secondary amine*
e. an *alkyl aryl amine*
f. *alkylation of an amine*
g. reduction
h. basicity
i. a *tertiary amine*
j. a *quaternary ammonium salt*
k. an *invert soap*
l. acylation
m. *nitrosoamine*
n. amide linkage

33.3 Write two names for each:

a. $CH_3—CH_2—NH_2$

b. $(CH_3CH_2)_2NH$

c. 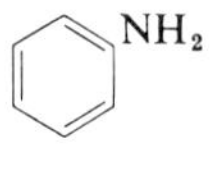 NH_2

d. $(CH_3)_4\overset{+}{N}I^-$ (one name)

e. $CH_3—\overset{O}{\overset{\|}{C}}—NH—CH_3$ (one name)

f. 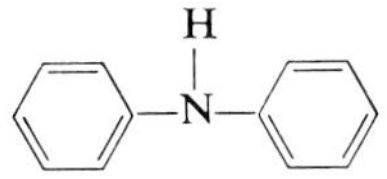 $C_6H_5—\overset{H}{\overset{|}{N}}—C_6H_5$

g. 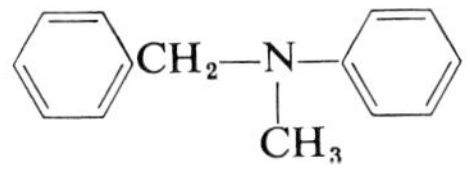 $C_6H_5CH_2—\underset{CH_3}{\underset{|}{N}}—C_6H_5$

33.4 Write equations for the preparation of aniline, starting with benzene.

33.5 Write structural formulas for all the organic products formed when methyl iodide reacts with ammonia.

33.6 List in the order of increasing basicity: (a) ammonia, (b) sodium hydroxide, (c) water, (d) ethyl alcohol, (e) aniline, (f) hydrochloric acid, (g) propylamine.

33.7 Write an equation for the reaction between aniline and benzoyl chloride.

33.8 What reagent might one use to distinguish methylamine, dimethylamine and trimethylamine? Write reactions.

33.9 What is a heterocyclic compound? Give an example of such a compound having more than one kind of hetero atom.

33.10 How many heterocyclic rings are present in the compound adenosine triphosphate (ATP) whose structure is shown here? To which rings of Figure 33.8 are they similar in structure?

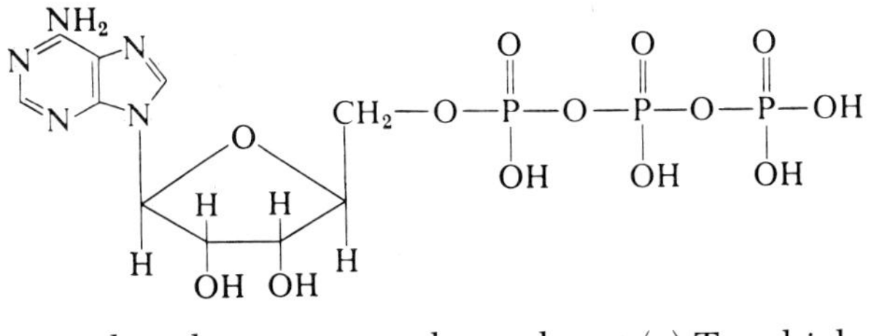

NH₂ N N N N O H H H H OH OH CH₂—O—P—O—P—O—P—OH O O O OH OH OH

33.11 Sucrose sugar has the structure shown here. (a) To which structure of Figure 33.8 is the six membered ring most closely related? (b) To which structure given in Figure 33.8 is the five membered ring most closely related? Why?

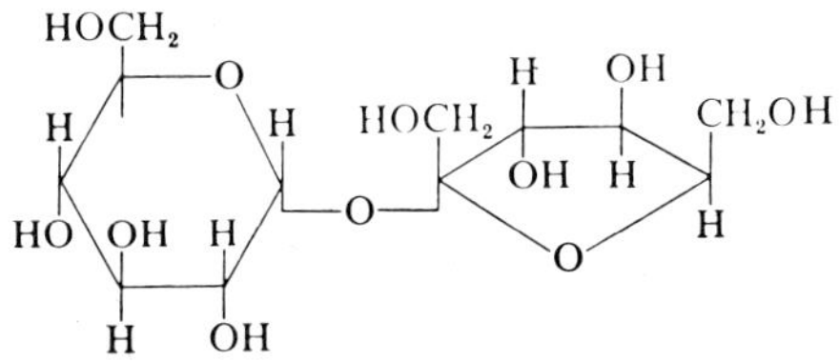

HOCH₂ O H H HO OH H H OH O H OH HOCH₂ OH H CH₂OH H O

33.12 What is a dye?

33.13 Name the chromophores present in the dyes shown here:

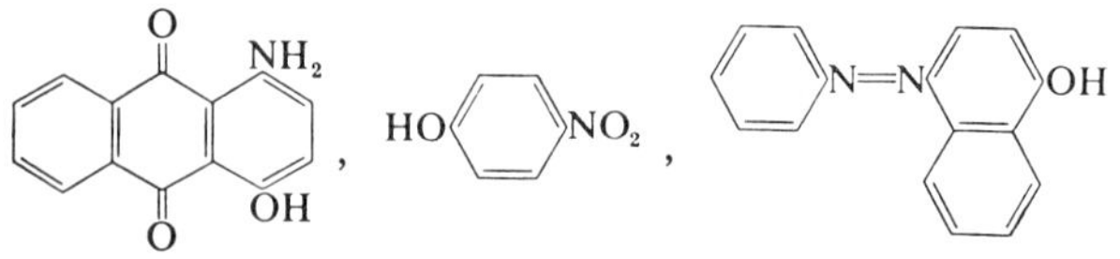

33.14 Refer to Exercise 33.13. What auxochromes occur in these dyes?

33.15 Give the name and formula of a colored compound not classified as a dye. What must be done to convert this compound into a dye?

33.16 Name a vitamin containing each of the following type heterocyclic rings in its structure: (a) pyridine, (b) pyrimidine, (c) thiazole.

33.17 In what way are chlorophyll A and heme similar in their structures? What is one significant difference?

33.18 Name a type of heterocyclic ring or ring system which is part of the structure of:

(a) Heme
(b) Tryptophan
(c) α-Tocopherol
(d) Sulfathiazole
(e) Acetyl coenzyme A (see Chapter 32)
(f) Guanine
(g) Cytosine
(h) Pyridoxal

33.19 Name two compounds which are closely related to heme in structure and composition. State their biological role in living systems.

33.20 Describe a chemical reaction by which one could distinguish between the two compounds listed in each item to follow (describe the significant observations which make the test useful):

(a) Triethyl amine and n-heptane
(b) Diethyl amine and n-butyl amine
(c) Aniline and n-hexyl amine
(d) Ethyl alcohol and ethyl amine
(e) CH_3CH_2SH and CH_3CH_2OH

SUGGESTED READING

Barron, F., Jarvik, M. E. and Bunnell, S. Jr.: "The Hallucinogenic Drugs." Scientific American, April, 1964.
Gates, M.: "Analgesic Drugs." Scientific American, November, 1966.
Gillis, R. G.: "The Mechanism of Diazotization." J. Chem. Ed., *31*:344, 1954.
Hendrickson, J. B., Cram, D. J. and Hammond, G. S.: *Organic Chemistry*, Third Edition. McGraw-Hill Book Company, New York, 1970. Chapters 12, 14, 16, 19, 24, 25, 26 and 27.
Morrison, R. T. and Boyd, R. N.: *Organic Chemistry*, Second Edition. Allyn and Bacon, Inc., Boston, 1966. Chapters 21, 22, 23, 24, 36 and 37.
Noller, C. R.: *Chemistry of Organic Compounds*, Third Edition. W. B. Saunders Company, Philadelphia, 1965. Chapters 14, 15, 16, 22, 23, 24, 25, 26, 27 and 31.
Paquette, L. A.: *Modern Heterocyclic Chemistry*. W. A. Benjamin, Inc., New York, 1968.
Robinson, T.: "Alkaloids." Scientific American, July, 1959.
Webb, H. A.: "Dyes and Dyeing." J. Chem. Ed., *19*:460, 1942.

THIRTY-FOUR • STEREO-ISOMERISM

34.1 ISOMERISM

The thread of isomerism runs throughout the study of organic chemistry. There are many sets of two or three or more compounds with the same molecular formula. The existence of isomers greatly multiplies the number, the complexity and the variety of organic compounds found in nature. The several types of isomerism discussed in previous sections of the text will be enumerated and illustrated by examples before optical isomerism is introduced.

Chain Isomers. Chain isomers are found among the simplest of all organic compounds, the alkanes, and the simplest isomers are n-butane, $CH_3CH_2CH_2CH_3$, and isobutane, $(CH_3)_2CHCH_3$, both with the molecular formula, C_4H_{10}. Because the number of carbons in a saturated hydrocarbon may be great, 30 or 40 or more, and because any carbon atom in the molecule may be bonded to one, two, three or four other carbon atoms, many chain isomers are possible for long chain hydrocarbons.

Relatively small differences exist between the physical properties of members of a set of chain isomers. As cited in Chapter 25, more highly branched—more compact—isomers tend to have slightly lower boiling points and correspondingly smaller heats of vaporization. Chemical differences are largely centered in the numbers of possible substitution products obtainable upon reaction with a particular reagent, varying with accessibility or lack of accessibility of the hydrogens toward attacking substances.

Functional Group Isomers. Ethyl alcohol, CH_3CH_2OH, and dimethyl ether, CH_3OCH_3, are isomers with the common molecular formula C_2H_6O. They differ in that ethyl alcohol has the —OH functional group and dimethyl ether has the ether functional unit,

—O—. Acetone, CH_3COCH_3, and propionaldehyde, CH_3CH_2CHO, have a common formula, C_3H_6O, as do 1-butyne, $CH_3CH_2C{\equiv}CH$, and butadiene, $CH_2{=}CH{-}CH{=}CH_2$. Because all sets of functional group isomers have the same elements, they have some common properties, but they can also differ greatly. An alcohol, with the active hydrogen of the hydroxyl group and with its capability for hydrogen bonding, differs from an ether; an aldehyde, with its instability toward oxidation, is not similar in this respect to a ketone; and, although both will add many reagents in common, a terminal alkyne with its active hydrogen has reactions impossible for an alkadiene.

differences between functional isomers are more apparent than their similarities.

Position Isomers. Position isomers have the same chain or the same ring system and the same functional groups but differ in the position or positions to which those groups (or that group) are attached; 1- and 2-chlorobutane differ only in the position occupied by the chloro "group." These position isomers are a special case of chain isomers with branching in the alkyl radical. For example, 2-chloro-2-methylpropane is a chain isomer of the two position isomers, 1- and 2-chlorobutane, just cited.

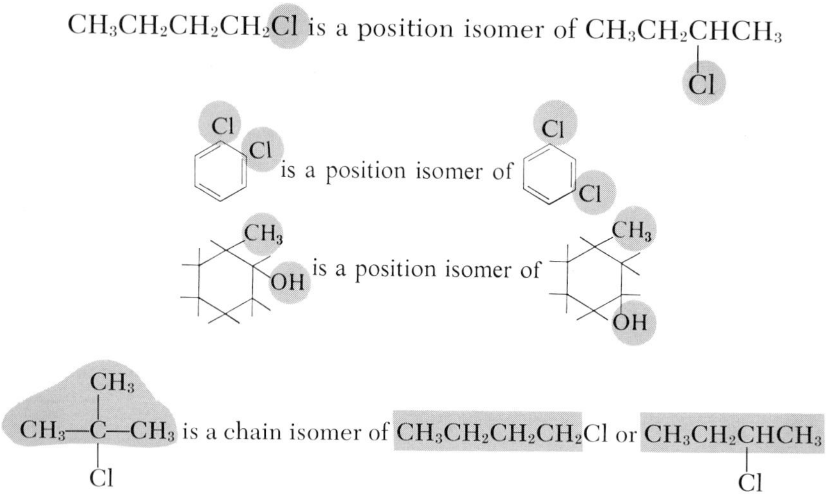

$CH_3CH_2CH_2CH_2Cl$ is a position isomer of $CH_3CH_2CH(Cl)CH_3$

(1,2-dichlorobenzene: Cl, Cl) is a position isomer of (1,3-dichlorobenzene: Cl, Cl)

(cyclohexane ring with CH_3 and OH on adjacent carbons) is a position isomer of (cyclohexane ring with CH_3 and OH)

$CH_3{-}C(CH_3)(Cl){-}CH_3$ is a chain isomer of $CH_3CH_2CH_2CH_2Cl$ or $CH_3CH_2CH(Cl)CH_3$

These three isomers are quite similar but are different in that the alkyl halide is primary, secondary or tertiary respectively, depending on the number of carbon atoms attached to the carbon bearing the halide. These compounds, because of that difference, are known to act differently toward an approaching nucleophile such as hydroxide ion.

1-Butyne and 2-butyne, and 1,3-hexadiene and 1,4-hexadiene are pairs of position isomers. The terminal alkyne with the active hydrogen has reactions not possible for the 2-alkyne, and the conjugated alkadiene with resonance or delocalization behaves differently from the diene with isolated multiple bonds.

The ortho-, meta- and parabenzene dicarboxylic acids have the same molecular formula, with the same two functional groups attached to the same ring, but they are different compounds. Only ortho-benzenedicarboxylic acid (o-phthalic acid) can lose water to a dehydrating agent to form an intramolecular anhydride.

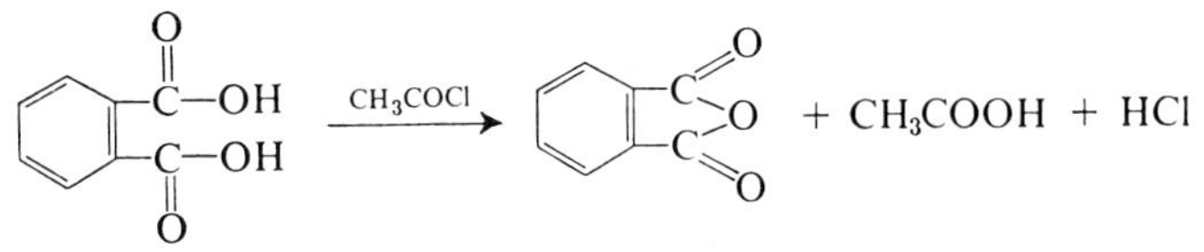

$$\xrightarrow{CH_3COCl} \quad + CH_3COOH + HCl$$

The heterocyclic compounds, pyrimidine and pyrazine (Fig. 33.8), are positional isomers in respect to the relative positions of the nitrogen atoms in the ring.

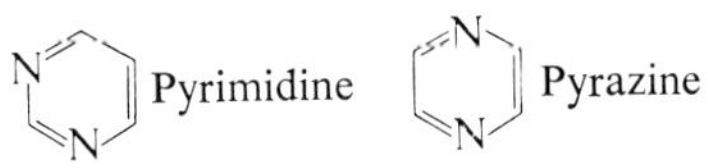

Pyrimidine Pyrazine

Chain Ring Isomerism. Certain continuous and branched chain compounds are isomeric with closed ring compounds. 1-Hexene and 4-methyl-1-pentene are isomers of cyclohexane. Many other examples could be cited. There is considerable difference between the chemical properties of the ring compounds compared to the chain compounds, but, of course, the formula requires that a functional group, the double carbon to carbon bond, be present in the chain compound.

Geometrical (cis-trans) Isomerism. *cis-trans* Isomerism appears when a double bond or some other structural feature prevents rotation about a bond between two atoms, each of which is attached to two different groups. **Geometrical isomers** contain the same structural features, the same atomic groups, the same bond angles, and are attached to the same positions on identical rings or chain frameworks. They differ only in the arrangement of some of the groups in space. Isomers with these subtle differences are called **stereoisomers**; *cis-* and *trans*-butene are the simplest of the geometrical isomers.

all geometrical isomers are stereoisomers, but not all stereoisomers are geometrical isomers.

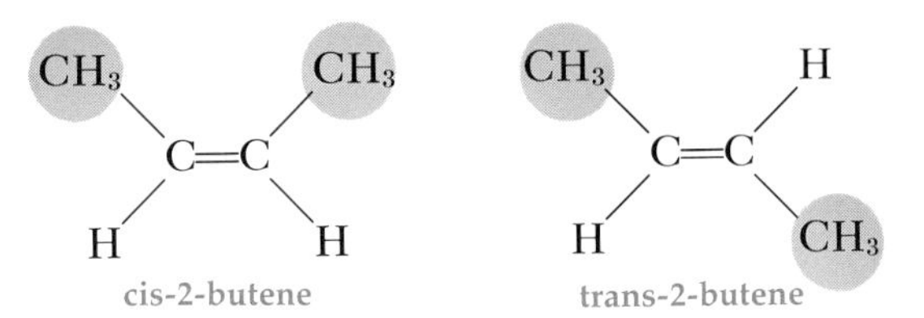

cis-2-butene trans-2-butene

There are also *cis* and *trans* forms of 1,2-dibromocyclohexane.

trans-1,2-dibromocyclohexane cis-1,2-dibromocyclohexane

While *cis* and *trans* isomers are often similar in both chemical and physical properties, great differences can exist. If the attached groups are highly electron attracting, the *cis* compound may be highly polar while the *trans* compound may have no dipole moment. Furthermore, some reactions are possible to one isomer and not the other, such as the intramolecular dehydration of maleic acid as cited in Chapter 27, a reaction not possible with the *trans* isomer, fumaric acid.

34.2 OPTICAL ISOMERISM

Optical isomers have the same skeletal ring or chain system and contain the same groups. Furthermore, two optical isomers may exhibit identical melting and boiling points and have the same solubilities in the same solvents. Yet the difference that does exist may be so important that one isomer is necessary for the growth of an organism whereas the other is poisonous. An understanding of optical isomerism is necessary for an understanding of carbohydrates and proteins, both necessary for life, for both are composed of units each of which is an optical isomer exhibiting optical rotation.

A. The Lactic Acids

lactic acid may build up in muscles during strenuous exercise.

In 1780 an acidic substance was isolated from sour milk by Scheele. The substance, called lactic acid, was a product of the fermentation of milk sugar. In 1807 Berzelius discovered a *similar* substance, an extract from muscle tissue. During the next half century the acid found in sour milk and the muscle extract were found to be remarkably alike. *Both substances:*

1. have the same molecular formula, $C_3H_6O_3$;
2. have the same melting point (26°C) and the same boiling point (122°C [15 mm]);
3. are hygroscopic (i.e., they absorb water from the atmosphere);
4. are very soluble in both water and organic solvents;
5. are decomposed by hot sulfuric acid to acetaldehyde and formic acid;
6. are oxidized to acetic acid and carbon dioxide;
7. have the same functional groups—a hydroxy group and a carboxyl group.

In fact, the functional groups occupy the same position. Each substance has the same structural formula,

$$CH_3-\overset{\displaystyle H}{\underset{\displaystyle OH}{\overset{|}{\underset{|}{C}}}}-COOH$$

and the same name, α-hydroxypropionic acid (2-hydroxypropanoic acid). The evidence is overwhelming. With much less data, the compounds would have been "proved" to be identical, were it not for the fact that the plane of plane polarized light is rotated clockwise when passed through a solution of the muscle extract, but is unaffected when passed through a solution of the compound (as then prepared) from sour milk.

Polarization and the Polarimeter. Plane polarized light is light that has passed through a polarizer. A polarizer is a crystal or a film that allows only that component of light vibrating in one direction to pass through. When ordinary light, vibrating in all directions perpendicular to the direction it is traveling, falls upon the polarizer, only half passes through. Figure 34.1 shows light vibrating in three directions approaching the polarizer. That parallel to the crystal passes through unchanged; that perpendicular is stopped completely; and only the parallel projection of that vibrating on a 45° angle (to the crystal) passes through.

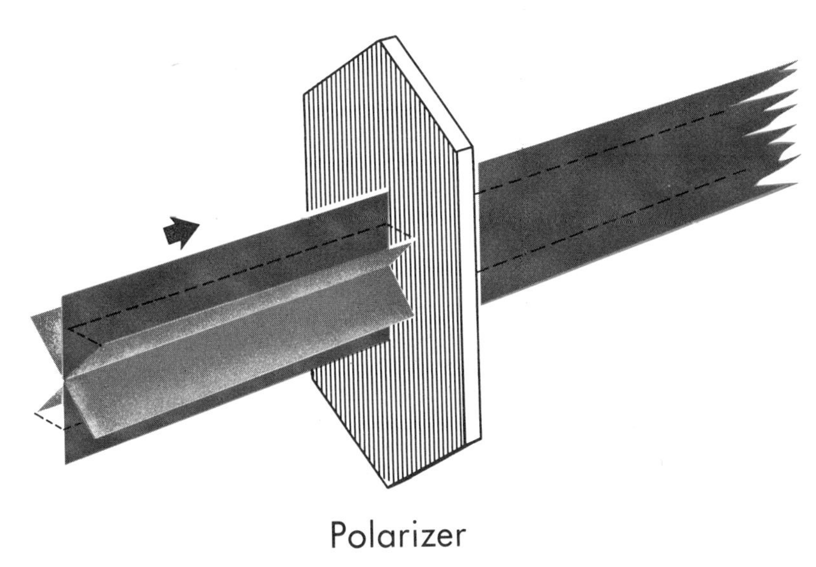

FIGURE 34.1 PLANE POLARIZATION OF LIGHT.

This is true also for a second polarizer (Fig. 34.2a) when placed behind and 90° to the first (Fig. 34.2b); the second polarizer lets no light pass through. Only when they are exactly "crossed" is the light excluded completely.

The polarimeter used to measure the optical rotation of solutions (Fig. 34.3) is arranged so that light passes through a polarizer, through the solution to be tested and through a second polarizer (analyzer) that can be rotated. The scale measuring rotation reads zero when the two polarizers are perpendicular to one another. If the light is excluded when the second polarizer is at zero (perpendicular to the first), the solution has no optical rotation, and is optically inactive. If the second polarizer must be rotated clockwise to exclude all the light, the solution is optically active; it is said to rotate light to the right, or in the positive (+) direction. If the second polarizer must be turned counterclockwise, the solution rotates light to the left, or in the minus (−) direction.

The number of degrees (magnitude) of rotation depends on the nature of the substance dissolved, the distance (length of the path) the light travels through the solution, and the concentration of the solution. For the purpose of this course, it shall be sufficient to know that some solutions rotate light positively (+); some rotate light negatively (−).

In the study of the two seemingly identical compounds, investigators eventually came to the conclusion that the muscle extract was a pure substance, and that the acid removed from sour milk

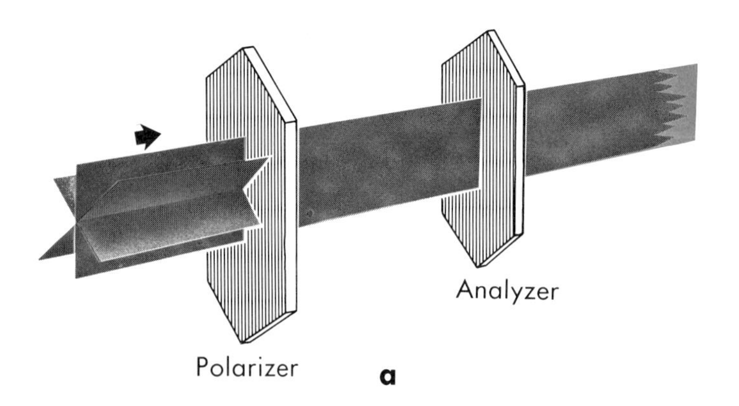

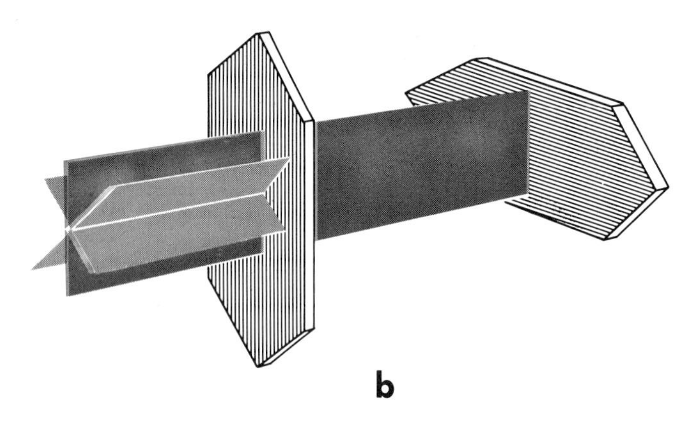

FIGURE 34.2 TOTAL EXCLUSION OF LIGHT BY CROSSED POLARIZERS.

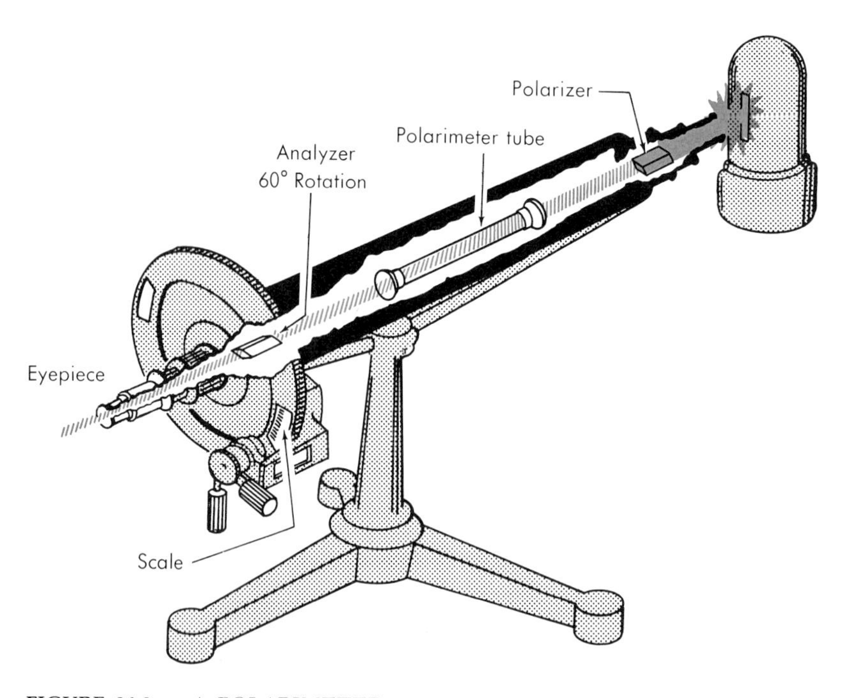

FIGURE 34.3 A POLARIMETER.

was a mixture of this pure substance and an isomer. This conclusion was reached when another acid, tartaric acid, also optically inactive, was separated by Louis Pasteur into two equal portions, each of which rotated light equally but in opposite directions. From this work of Pasteur it came to be expected that if one optically active substance existed, another did also with an equal and opposite optical activity. The acid in muscle tissue is a pure isomer, and therefore rotates the vibrational plane of light. The acid obtained from sour milk contains an equal amount of both isomers. Each isomer rotates light but in an opposite direction, making it appear that no rotation has taken place.

B. Asymmetric Molecules

Pasteur's separation of seemingly optically inactive tartaric acid is most ingenious. His observations gave the key to possible differences in the isomers. He found that the sodium ammonium salt of tartaric acid, when crystallized from a solution, formed distinctive hemihedral crystals. Furthermore, two kinds of crystals formed. The crystals were almost identical, differing only in the sense that the *left hand* differs from the *right hand.* Pasteur separated the two types of crystals with a pair of tweezers. Solutions of each of these were optically active, exhibiting rotations of equal magnitude but of opposite direction.

In 1874 Jacobus Hendricus van't Hoff and Joseph Achille LeBel independently presented the theory of optical activity accepted today. They saw that just as some crystals could be to one another as right and left hands, so could some molecules. A molecule that is **asymmetric** (that has no *plane of symmetry*) will be optically active, and will be one of a pair of isomers. The other isomer will be its *mirror image* and will differ as the right hand differs from the left hand, and its solutions will rotate light equally in the opposite direction. Any molecule having only one carbon atom present attached to four *different* groups is asymmetric (Fig. 34.4). A carbon atom of this type is called an *asymmetric carbon atom.* One cannot place his or her hands together in any way to have them exactly the same. Because a hand has four different aspects, the back, the palm, the fingers and the thumb, it is not symmetrical; it is asymmetric and there exists another arrangement, its mirror image, which is not the same. Were it symmetrical, were there a plane of symmetry in the molecule so that one half of it was a mirror image of the other half, a non-identical mirror image of the whole would be impossible.

Now study the model in Figure 34.4. Observe that four different groups are attached to the central atom. Just as a left hand appears in a mirror as a right hand, this molecule appears as its mirror image. Again, because the four groups are different, the mirror image has a different arrangement from that of the optical isomer positioned before the mirror.

A model of the lactic acid molecule is shown in front of a mirror (Fig. 34.5). Only one carbon atom is asymmetric; the carbon with two bonds to oxygen is not an asymmetric atom. The mirror image is another arrangement, the only other possible arrangement of the four groups. Both compounds were in the acid from sour milk. One compound is the acid extracted from muscles.

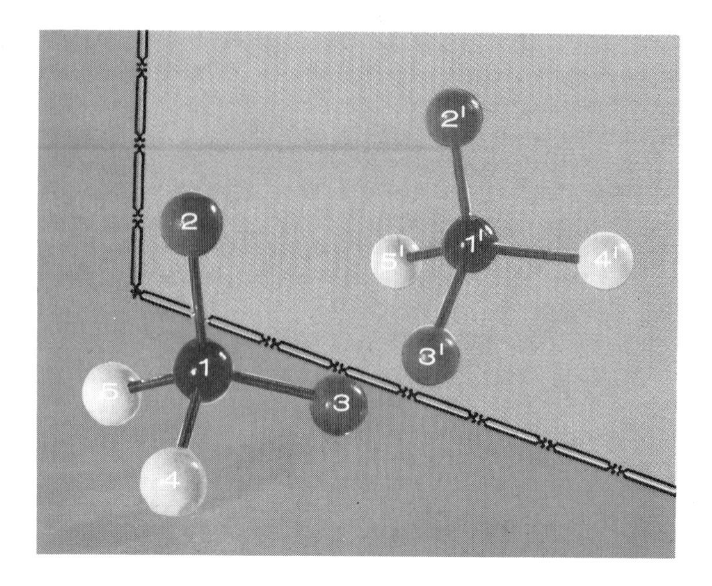

FIGURE 34.4 AN ASYMMETRIC MOLECULE AND ITS MIRROR IMAGE.

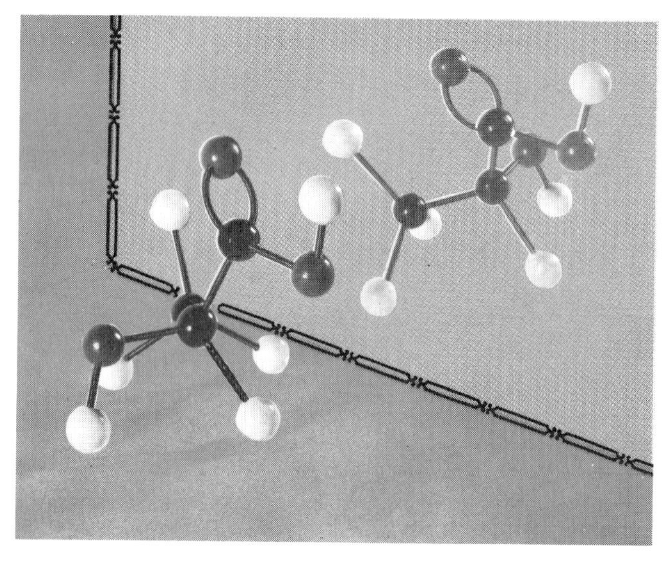

FIGURE 34.5 STRUCTURES OF THE ENANTIOMERS OF LACTIC ACID.

Any two isomers (mirror images) related to each other as are the (+)-rotating and the (−)-rotating lactic acids are called **enantiomers.** It should be noted that all molecules have mirror images. If the mirror images are identical (superimposable), as are those of methane (CH_4) or any other symmetrical molecule, they are not enantiomers. Enantiomers are non-identical (non-superimposable) mirror images. Each is the enantiomer of the other. A molecule may be an enantiomer of only one other molecule, its non-identical mirror image.

nonsuperimposability of mirror image isomers is the test for nonidentity. Superimposability is a test for identity.

C. Physical Properties of Optical Isomers

It is now obvious that the two isomers of lactic acid, and, in fact, any pair of enantiomers, must have the same physical properties except for optical rotation. It should also be apparent that if a solution of one enantiomer rotates light clockwise (+), a solution of the other will rotate light equally and counterclockwise (−).

D. Formulas and Nomenclature of Isomers

Because organic compounds are three dimensional, and because the asymmetry of optical isomers is not apparent from a two-dimensional formula, a convention must be adopted to indicate this third dimension.

When writing or observing structural formulas, each asymmetric carbon, assumed to be a tetrahedron, is written so that one edge is forward and horizontal with another edge behind and vertical. The asymmetric carbon occupies the central position (the tetrahedron) with groups above, below, to the right and to the left. The horizontal groups (those to right and left) occupy the forward edge and are in a place above the paper, whereas the vertical groups reside in the same plane as the paper. To illustrate this, the formula for lactic acid is shown in Figure 34.6. The formula shown in Figure 34.6 has the structure represented by the tetrahedron (Fig. 34.6b).

```
   COOH            COOH              COOH             COOH
    |                                  |
H—C—OH        H ◇ OH           HO—C—H         HO ◇ H
    |                                  |
   CH3             CH3               CH3              CH3
    a.              b.                c.               d.
     One enantiomer                     One enantiomer
```

FIGURE 34.6 ENANTIOMERS OF LACTIC ACID.

Figure 34.6c and d represent one and the same molecule, the enantiomer of the molecule represented in Figure 34.6a and b. Other compounds having one asymmetric carbon may be compared to lactic acid if their structure is written on the page in the same way. It should be obvious that in rewriting the formulas one must be careful to check their identity or non-identity. Two molecules are superimposable if like groups can be shown to occupy identical positions in space when the molecules are compared.

In naming optically active compounds, the direction of optical rotation is signified by a (+) or a (−), for clockwise and counterclockwise, respectively. A capital D- or L- is used to indicate that the molecule is related in structure to either D- or L-glyceraldehyde. For example:

```
   HC=O            HC=O
    |
H—C—OH   or   H ◇ OH       is D-(+)-glyceraldehyde
    |                         (rotates light clockwise)
   CH2OH           CH2OH
```

whereas

```
   HC=O            HC=O
    |
HO—C—H   or   HO ◇ H       is L-(−)-glyceraldehyde
    |                         (rotates light counterclockwise)
   CH2OH           CH2OH
```

The substance:

O ‖ C—OH | H—C—OH | CH_3 or H, OH (tetrahedral drawing with C—OH above and CH_3 below)

is D-(—)-lactic acid (Fig. 34.6a)

The lactic acid found in muscle extract has a (—) rotation and is D-lactic acid (Fig. 34.6abcd) for it can be related to D-glyceraldehyde. L-Glyceraldehyde, the other enantiomer, yields L-lactic acid. Each transformation from the glyceraldehyde to the acid involves several steps. L-Alanine,

O ‖ C—OH | H_2N—C—H , | CH_3

is related to L-glyceraldehyde and L-lactic acid. The name as written does not reveal the direction of optical rotation. The rotation of L-alanine is +14.5° (in 6 N HCl). The complete name would be L-(+)-alanine. Note that the D- or L- prefix does not specify the direction of rotation, but only the structural similarity to D- or L-glyceraldehyde.

E. Diastereomers

A molecule may possess several asymmetric carbons. The four-carbon sugars are examples of compounds with two asymmetric carbons. The optical isomers all have common names but all are 2,3,4-trihydroxybutanal.

$HOCH_2$—C*(H)(OH)—C*(H)(OH)—C(=O)H

disastereomers are neither mirror images nor superimposable.

There are four optical isomers of 2,3,4-trihydroxybutanal (all known) and these are shown in Figure 34.7 by two-dimensional drawings, called Fisher projection formulas, that imply the actual three-dimensional character of the molecules. When testing for superimposability of isomers represented by such formulas, each formula may be rotated any number of degrees, but only while the formula remains in the plane of the paper; that is, only rotations on the flat surface of the page are permitted. A formula may not be taken out of the plane and flipped over.

Two pairs of enantiomers are shown in Figure 34.7. D-Threose and L-threose are mirror images and have the same chemical and

D-Erythrose	L-Erythrose	D-Threose	L-Threose
CHO	CHO	CHO	CHO
H—C—OH	HO—C—H	HO—C—H	H—C—OH
H—C—OH	HO—C—H	H—C—OH	HO—C—H
CH_2OH	CH_2OH	CH_2OH	CH_2OH

FIGURE 34.7 THE OPTICAL ISOMERS (STEREOISOMERS) OF 2,3,4-TRIHYDROXY-BUTANAL.

physical properties except for the direction in which they rotate the plane of plane polarized light. They rotate polarized light equally, but in opposite directions. D-Erythrose and L-erythrose, likewise, are mirror images; they have the same properties except for the rotation of plane polarized light. Again, each rotates polarized light equally, but in opposite directions. An erythrose and a threose, in contrast, however, are not mirror images nor do they necessarily have physical and chemical properties that are identical. Stereoisomers such as these, which are not mirror images, are called **diastereomers.** D-Erythrose is a diastereomer of both D-threose and L-threose. Each of the four isomeric 2,3,4-trihydroxy-butanals (tetroses—see Chapter 36) is an enantiomer of one of the other three optical isomers and a diastereomer of two.

As the number of asymmetric carbon atoms present in a molecule becomes greater, the number of optical isomers possible increases. The number may be calculated using the formula, $N = 2^n$, where N equals the number of optical isomers possible, and n equals the number of asymmetric carbon atoms in the molecule. The number of optical isomers which actually exists is, however, sometimes less than this because of the existence of meso forms. Meso forms, to be discussed in the next section, are symmetrical compounds—thus not optically active—which have present in their structure two or more asymmetric carbon atoms. When meso forms exist, the calculated number of optical isomers is decreased by two for each such form or compound known. No meso forms of the 2,3,4-trihydroxybutanals are possible.

F. Meso Forms

Oxidation of an erythrose with nitric acid yields an α,β-dihydroxysuccinic acid (a tartaric acid) with two asymmetric carbon atoms as indicated by the stars (*).

CHO		COOH
H—C*—OH	(O) →	H—C*—OH
H—C*—OH		H—C*—OH
CH_2OH		COOH

Although the acid is formed from an optically active compound, and although it contains two asymmetric carbons, solutions of the

* Asymmetric carbon atom.

acid exhibit no optical activity. The molecule is not asymmetric. Because the bottom half (the lower two carbons as written) is the mirror image of the top half (the upper two-carbon structure), there is a plane and center of symmetry in this molecule, and the molecule is identical with its mirror image.

		180°	
COOH	COOH		COOH
H—C—OH	HO—C—H		H—C—OH
H—C—OH	HO—C—H		H—C—OH
COOH	COOH		COOH

the mirror images of the tartaric acid shown here are superimposable.

A symmetrical molecule such as this dihydroxysuccinic acid, containing several asymmetric carbons, is called a *meso* compound. The molecule is said to be internally compensated to optical rotation, since the mirror image halves of the molecule influence the rotation of polarized light equally but in opposite directions. The presence of symmetry eliminates the possibility of optical rotation because only asymmetry, irrespective of its source, leads to optical activity. This compound is named meso-α,β-dihydroxysuccinic acid (or more commonly, meso-tartaric acid).

The oxidation product of D-threose, also a dihydroxysuccinic acid (and a tartaric acid), is a stereoisomer but not an enantiomer nor an optical isomer of the meso-dihydroxysuccinic acid.

CHO		COOH
HO—C—H	(O) ⟶	HO—C—H
H—C—OH		H—C—OH
CH_2OH		COOH

This acid, with two asymmetric carbon atoms, in contrast with the meso-acid, is asymmetric, has an enantiomer—a non-identical mirror image—and its solutions rotate the plane of plane polarized light.

COOH	COOH
HO—C—H	H—C—OH
H—C—OH	HO—C—H
COOH	COOH

D- and L-α,β-Dihydroxysuccinic acid (D- and L-tartaric acid).

the mirror images of the tartaric acid shown here are not superimposable.

For the acid of structural formula, HOOC—CHOH—CHOH—COOH, three stereoisomers exist. Two are enantiomers and their solutions rotate equally and in opposite directions. One is a meso structure and exhibits no optical activity. For each compound having a structure permitting the existence of optical isomerism, that structure must be examined for possible arrangements in space that permit symmetry to exist, in order to determine the possibility of the existence of meso forms. Meso forms, being diastereomers of the optically active forms, can usually be readily separated from the other isomers because of differences in properties.

G. Racemates (Racemic Forms)

A 50–50 mixture of enantiomers, such as that obtained by Scheele when he isolated lactic acid from sour milk, is called a **racemic mixture.** Ordinary chemical synthesis of optically active compounds, because of an equal probability of forming either enantiomer, yields racemic mixtures. To obtain the individual isomers —the (+)-rotating and (−)-rotating forms—this mixture must be separated. The process of separating a racemic mixture into the two enantiomers of which it is composed is called **resolution.**

Resolution may be accomplished in a variety of ways. Pasteur accomplished the resolution of his sodium ammonium tartrate salts by physically picking out the different shaped crystals of the two enantiomers with a pair of tweezers. This procedure, because such differences in crystal forms are rarely observed, is very seldom useful. The more common method is to react a racemic mixture with an optically active reagent. For example, if a racemate of an optically active acid is permitted to react with one enantiomer of an optically active base, the following occurs:

Mixture of enantiomers (Racemate)		Mixture of diastereomers (Separable)
$\left\{\begin{array}{l}(+)\text{-acid}\\(-)\text{-acid}\end{array}\right\}$	$+ (+)\text{-base} \rightarrow$	$\left\{\begin{array}{l}(+)\text{-acid}\cdot(+)\text{-base}\\(-)\text{-acid}\cdot(+)\text{-base}\end{array}\right\}$

This device is useful because it converts the mixture of enantiomers that are not readily separable into a pair of new optically active compounds that are diastereomers of each other. These diastereomers have different physical properties and may be separated by crystallization, distillation or by some other technique based on a difference in physical properties. After the diastereomers are separated, the original enantiomers are recovered by decomposing the now purified diastereomers into starting materials:

$$(+)\text{-acid}\cdot(+)\text{-base (a salt)} + HCl \rightarrow (+)\text{-acid} + (+)\text{-base}\cdot HCl$$

and in a separate experiment:

$$(-)\text{-acid}\cdot(+)\text{-base (a salt)} + HCl \rightarrow (-)\text{-acid} + (+)\text{-base}\cdot HCl$$

Isolation of the enantiomer (i.e., the (+)-acid or the (−)-acid) from the optically active reagent (here the (+)-base·HCl) yields the pure enantiomer. Resolution—recovery of the pure enantiomers—is achieved. A detailed procedure must be developed for each racemate that one desires to separate.

Biochemical resolution of a racemic mixture is sometime possible because certain microorganisms can utilize one isomer and not the other of an enantiomeric pair. When allowed to "feed" upon a racemic mixture, one isomer is consumed and destroyed leaving the other unchanged. Purification of the remaining isomer makes this one available. Biochemical resolution, however, has limited preparative application. Often it is difficult to find a suitable microorganism and often the isomer destroyed is the one desired in pure form. In any case 50 per cent of the racemate is lost.

In many cases one can start with an optically active pure substance and prepare another without destroying the optical rotation caused by racemate formation. For example, biosynthesis (synthesis in living organisms) is usually stereospecific, leading to the formation of one or the other of the enantiomers, but not both. Another example is that of "pure" S_N2 reactions at asymmetric centers.

34.3 ASYMMETRIC MOLECULES WITHOUT ASYMMETRIC CARBON ATOMS

Many organic compounds without an asymmetric carbon atom show optical activity. The only requirement is that the mirror image not be superimposable. 2,3-Pentadiene, a compound of the allene type (Fig. 34.8), is one of the least complex structures in the open chain series of compounds exhibiting optical activity with no asymmetric carbon atom present. Asymmetry arises because the groups attached to each of the terminal carbons are not in the same plane. In the structure drawn for 2,3-pentadiene the darkened bonds (▬ or ►) project upward from the page; the bonds shown on dotted lines (----) are behind the page. The single unthickened undotted lines (—) represent bonds in the same plane as the page. Manipulation of either structure (each the mirror image of the other) shows that they are not superimposable. In this case, in contrast with the situation with the planar structures discussed earlier, three-dimensionality is included in the drawing and the models may be removed from the paper for manipulation. The molecule represented is asymmetric and exhibits optical activity. The asymmetry is a property of the overall structure and is not attributable to any one asymmetric center.

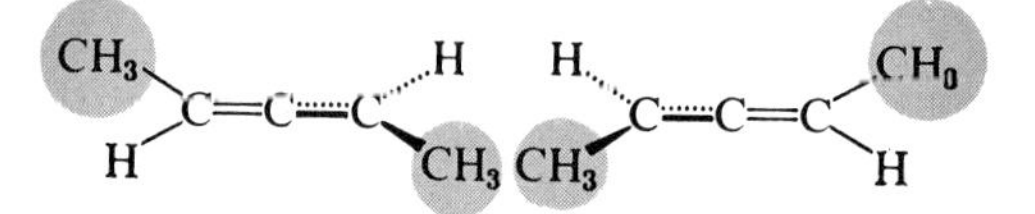

FIGURE 34.8 2,3 PENTADIENE.

the three carbons joined by double bonds provide a rigid fixed structure to which the H and CH_3 groups are bonded.

Another type of asymmetry appears in the case of ortho substitution derivatives of biphenyl. In the case of biphenyl (Fig. 34.9), the two benzene rings are coaxial, are free to rotate about the single bond joining the two rings and may assume a symmetrical coplanar structure.

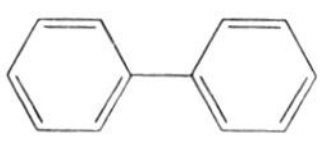

FIGURE 34.9 BIPHENYL.

In the optically active biphenyl derivatives, such as the o,o′-dinitrodiphenic acids (Fig. 34.10), the two coaxial rings, instead of being free to rotate completely, are held in approximately perpendicular planes by the repulsions of the ortho substituents. The two configurations shown for o,o′-dinitrodiphenic acid are mirror images that are non-superimposable. These two optical isomers have been separated (resolved), demonstrating that asymmetry has been produced by the hindrance to free rotation. The parent diphenic acid, with H atoms in place of the nitro groups, has not been resolved, thus indicating the absence of asymmetry. Every optically active biphenyl derivative must have bulky groups present in the ortho positions of both rings to prevent free rotation, otherwise the molecule can become planar so that it possesses a plane of symmetry. On the other hand, if one of the groups attached to the ring is asymmetric itself because of the presence of an asymmetric carbon atom, the substance may be optically active even if the rings can achieve coplanarity. Many other compounds analogous to the allenes and biphenyl derivatives are known and have been well studied.

COOH NO_2 NO_2 COOH

NO_2 COOH COOH NO_2

FIGURE 34.10 o,o′-DINITRODIPHENIC ACIDS.

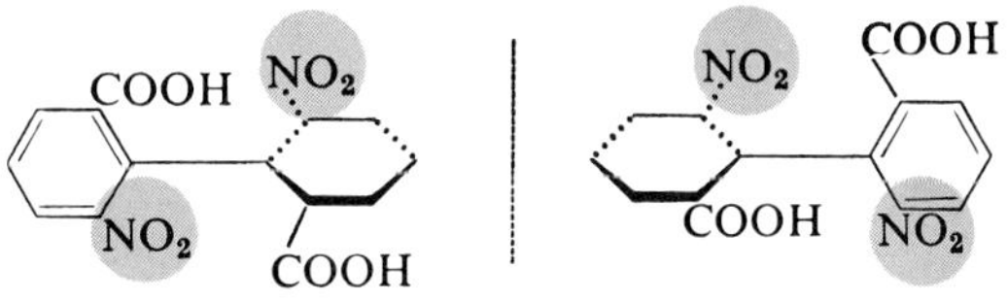

34.4 ABSOLUTE AND RELATIVE CONFIGURATION

Absolute configuration refers to the actual existing arrangement in space of the atoms of a stereoisomer. Prior to 1951 the absolute configuration of no optical isomer was known. Therefore, before 1951 the configuration assigned to a compound was a relative configuration arrived at by demonstrating a structural relationship between the compound and a reference compound selected as a standard. The reference was assigned a configuration and all compounds shown to be structurally related to it were assumed to have that same arrangement of atoms. Glyceraldehyde was selected as the standard reference compound. Dextro-rotatory glyceraldehyde was assigned the configuration given on page 763 and also shown in Figure 34.11, and it was named D-(+)-glyceraldehyde. Its enantiomer was assigned the mirror image structure and named L-(−)-glyceraldehyde.

CHO | H—C—OH | CH_2OH ; CHO | HO—C—H | CH_2OH ; or H—C—OH ; HO—C—H

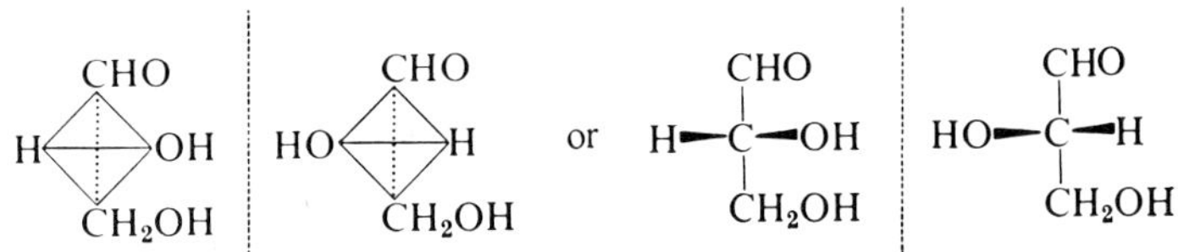

D-(+)-Glyceraldehyde L-(−)-Glyceraldehyde

FIGURE 34.11 D-(+)- AND L-(−)-GLYCERALDEHYDE.

The capital D- and L-, as indicated earlier, designate the configuration; (+) and (−) are used to specify the direction of optical rotation. When a compound was shown by suitable chemical transformations to be structurally like D-glyceraldehyde it was designated as belonging to the D-series of compounds; those related to L-glyceraldehyde were assigned to an L-series. For example, simple sugars are classified as D-sugars because the asymmetric carbon used for classifying them has the same configuration as D-glyceraldehyde. On the other hand, amino acids obtained from proteins, because the alpha asymmetric carbon has the same configuration as L-glyceraldehyde, are members of the L-series.

THIRTY-FIVE • AMINO ACIDS; PEPTIDES; PROTEINS

The name protein is a term applied to a widely distributed class of natural polymers noted particularly for their unique association with living things. Proteins make up the bulk of the solid matter of all living cells, and for these units of living matter they serve either as functional participants in the many activities that take place, or as components of the structural framework. Proteins chemically are very complex substances. Their study presents the chemist with many difficulties because of their extreme sensitivity to heat and chemical reagents.

Amino acids and peptides are simple, nitrogen containing compounds obtained by the hydrolysis of proteins. Since their structures are relatively simple and because their reactions are more easily understood, we shall begin by examining first the structure and behavior of the alpha amino acids, fundamental building blocks from which all proteins are constructed. The simpler compounds of the amino acids, the peptides, will then be discussed, after which proteins will be considered.

35.1 AMINO ACIDS

A. General Aspects

An **amino acid,** as the name implies, is a difunctional compound containing both an amino and an acidic group. The acid group may be one of several types, the most commonly encoun-

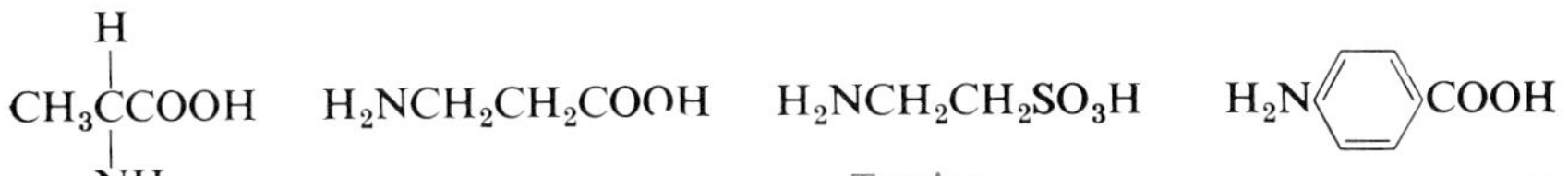

$$CH_3\underset{NH_2}{\overset{H}{C}}COOH \qquad H_2NCH_2CH_2COOH \qquad H_2NCH_2CH_2SO_3H \qquad H_2N{-}C_6H_4{-}COOH$$

Alanine — β-Alanine — Taurine (β-aminoethane sulfonic acid) — p-Aminobenzoic acid

FIGURE 35.1 TYPES OF AMINO ACIDS.

tered being the carboxyl. Amino sulfonic acids are known, but they are encountered less frequently. Several biologically important amino acids that serve to illustrate the various types of known amino acids are given in Figure 35.1.

taurine is found combined as an amide with cholic acid in human bile.

Taurine is an amino sulfonic acid. Alanine is an **amino carboxylic acid.** More specifically, alanine is an example of an alpha amino acid, a distinction based on terminology designating the carbon next to a carboxyl group as an alpha carbon atom. The β- in β-alanine, an amino acid found in nature, signifies that the amino group of the molecule is attached to a carbon located in a beta position with respect to the carboxyl group. Alanine and β-alanine are positional structural isomers. Para-aminobenzoic acid is an example of an aromatic amino acid. Formerly thought to be a vitamin in its own right, p-aminobenzoic acid is now recognized to be important mainly because it is a component of another vitamin, folic acid.

All but two of the acids that have been isolated from protein hydrolysates are α-amino acids. Two imino acids, proline and hydroxyproline, which are secondary amines having the imino $(\overset{H}{\overset{|}{-N-}})$ group, have also been obtained from protein hydrolysates, the reaction mixture obtained when a protein is hydrolyzed. They are not true alpha amino acids, since the nitrogen present is tied up in a cyclic structure. The formulas for proline and hydroxyproline are given in Table 35.1, which gives the names, formulas and different classifications for 21 amino acids whose presence in proteins has been verified repeatedly through isolation from protein hydrolysates. It should be noted that the names given are common names, more widely employed than the systematic official names. This means that a certain number of them will have to be committed to memory. The starred amino acids are those commonly designated as the essential amino acids. Essential amino acids are no more essential to the animal body than other amino acids, but they are not manufactured by animal organisms and must be provided in the diet.

in proline and hydroxyproline the R group is attached directly to the —COOH of the amino acid.

The general formula used for α-amino acids is:

$$R{-}\underset{NH_2}{\overset{H}{\underset{|}{\overset{|}{C}}}}{-}COOH$$

Figure 35.2 shows the model of a simple amino acid, alanine, with each atom labeled to show more clearly the actual structure. It illustrates that in alanine, as in all α-amino acids, both a carboxyl group and an amino group are attached to the same carbon. Differ-

TABLE 35.1 PROTEIN AMINO ACIDS

Amino Acid	Structural Formula (R group)	Classification
Alanine	$CH_3—$	Neutral, aliphatic
Arginine*	$H_2N\overset{\overset{NH}{\|}}{C}NHCH_2CH_2CH_2—$	Basic, aliphatic
Aspartic acid	$HOOCCH_2—$	Acidic, aliphatic
Cysteine	$HSCH_2—$	Neutral, aliphatic, sulfur-containing
Cystine	$SCH_2—$ | $SCH_2—$	Neutral, aliphatic, sulfur-containing
Glutamic acid	$HOOCCH_2CH_2—$	Acidic, aliphatic
Glycine	$H—$	Neutral, aliphatic
Histidine*	HC=C—CH_2— (ring: HC, N, C(H), NH, C)	Basic, heterocyclic
Hydroxylysine	$H_2NCH_2\underset{\underset{OH}{\|}}{C}HCH_2CH_2—$	Basic, aliphatic, hydroxyl-containing
Hydroxyproline	HOCH—CH_2; CH_2, CH—; N—H (ring)	Neutral, imino, heterocyclic, hydroxyl-containing
Isoleucine*	$CH_3CH_2\overset{\overset{CH_3}{\|}}{C}H—$	Neutral, aliphatic

*Essential amino acids.

ences between the individual amino acids reside in variations in the structure of the R group, often called the side chain. The properties characteristic of particular amino acids are determined largely by the nature of this group.

B. Classification of Amino Acids

Amino acids may be classified in various ways according to the nature of the R group. For example, if the R (side chain) is a hydrocarbon group, as is the case for alanine or phenylalanine, the amino

TABLE 35.1 PROTEIN AMINO ACIDS (CONTINUED)

Amino Acid	Structural Formula (R group)	Classification
Leucine*	$(CH_3)_2CHCH_2$—	Neutral, aliphatic
Lysine*	$H_2NCH_2CH_2CH_2CH_2$—	Basic, aliphatic
Methionine*	$CH_3SCH_2CH_2$—	Neutral, aliphatic, sulfur-containing
Phenylalanine*	C_6H_5 ring —CH_2—	Neutral, aromatic
Proline	H_2C——CH_2 | | H_2C CH— N | H	Neutral, imino heterocyclic
Serine	$HOCH_2$—	Neutral, aliphatic, hydroxyl-containing
Threonine*	CH_3CH— | OH	Neutral, aliphatic hydroxyl-containing
Tryptophan*	CH_2— N | H	Neutral, aromatic heterocyclic
Tyrosine	HO— ring —CH_2—	Neutral, aromatic, phenolic
Valine*	$(CH_3)_2CH$—	Neutral, aliphatic

* Essential amino acids.

FIGURE 35.2 ALANINE.

acid will contain one basic group (the amino group) and one acidic group (the carboxyl group), and the amino acid is classified as a **neutral amino acid.** Should the side chain itself contain a carboxyl group, as is true for aspartic and glutamic acids, the amino acid will have two carboxyl groups and only one amino group and is an **acidic amino acid.** On the other hand, if the radical, R-, carries a basic group, the number of basic groups present will be larger than the number of acidic groups and the compound is a **basic amino acid.**

what type of amino acid is lysine? why?

Other terms are used at times to indicate further structural features of the R- group. If the radical is an alkyl group, the acid is an **aliphatic amino acid.** When a benzene ring is part of the side chain, the compound is considered to be an **aromatic amino acid.** Other amino acids, of which proline is an example, are designated as heterocyclic to specify the presence in the side chain of a cyclic structure in which atoms other than carbon are part of the ring. Amino acids containing hydroxyl groups or sulfur atoms are often designated as hydroxyl containing or sulfur containing amino acids. The various classifications of several amino acids are given in Table 35.1.

C. Physical Properties of Amino Acids

Some physical properties of amino acids differ rather noticeably from properties usually associated with organic compounds. Because of their unusually high melting points, often above 250°C and usually with decomposition, and because they are generally more soluble in water than in non-polar solvents, so that organic solvents may be added to precipitate them from aqueous solutions, investigators predicted and subsequently verified experimentally that amino acids exist as dipolar ions in water solutions. This comes about through an internal acid-base reaction in which the proton of the carboxyl group (either from the same amino acid or from a neighbor) shifts to the amino group:

$$R-\underset{NH_2}{\underset{|}{\overset{H}{\overset{|}{C}}}}-COOH \rightarrow R-\underset{NH_3^+}{\underset{|}{\overset{H}{\overset{|}{C}}}}-COO^-$$

notice that the base of an amino acid is the $—COO^-$ group and the acid is the $^+NH_3$ group.

This reaction yields an ion that, although neutral as a whole, has a positively charged ammonium and a negative carboxylate group. Experimental evidence indicates that dipolar ions are present in the crystals of solid amino acids.

With the acidic ammonium group, $—NH_3^+$, and the basic carboxylate group, $—COO^-$, amino acids are amphoteric, i.e., may act as either acids or bases. The reactions of glycine with hydronium ion in an acidic (low pH) solution or with hydroxide ion in a basic (high pH) solution are:

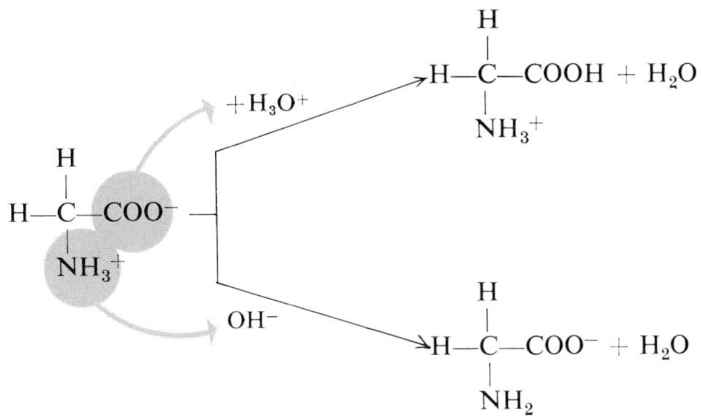

$$H{-}\overset{\overset{\large H}{|}}{\underset{\underset{\large NH_3^+}{|}}{C}}{-}COO^- \begin{cases} \xrightarrow{+H_3O^+} H{-}\overset{\overset{\large H}{|}}{\underset{\underset{\large NH_3^+}{|}}{C}}{-}COOH + H_2O \\ \xrightarrow{OH^-} H{-}\overset{\overset{\large H}{|}}{\underset{\underset{\large NH_2}{|}}{C}}{-}COO^- + H_2O \end{cases}$$

In an acidic solution the carboxylate ion accepts a proton and the amino acid carries the resultant positive charge. In a basic solution the ammonium group loses a proton and the amino acid carries only a negative charge.

It is apparent that the electrical character of the amino acid molecule in a solution is dependent upon the acidity or pH. If an amino acid solution is placed in an electrolysis cell, the amino acid migrates either to the anode (positive pole) or the cathode (negative pole), depending upon the pH of the solution. The pH at which the dipolar (neutral) form of the amino acid predominates, so that the amino acid does not migrate in an electrical field, is defined as its isoelectric point. Only at the isoelectric point (or isoelectric pH) does the amino acid, attracted equally from both directions, remain at rest. The isoelectric point, because the acidic and basic groups present are usually not of equal strength, is generally not at pH 7. Each amino acid, however, has a characteristic isoelectric point. Isoelectric points above 7 are characteristic of basic amino acids; those much below 7, of acidic amino acids. For other amino acids the isoelectric point is near but slightly below 7.

the isoelectric point of an amino acid lysine is 9.74. What charge does it carry at pH 6?

D. Reactions of Amino Acids

Reactions of amino acids may be grouped in two categories: (1) those which all amino acids undergo and (2) those which are peculiar to individual amino acids. For example, all amino acids show reactions attributable to the presence of carboxyl or amino groups. Tyrosine, on the other hand, although taking part in reactions expected for carboxyl and amino groups, in addition exhibits certain reactions characteristic of phenols, since the aromatic ring of the side chain has a phenolic hydroxyl attached to it. Cysteine, the only amino acid with a free —SH (sulfhydryl) group, undergoes a series of reactions attributable to this group.

the amino group reacts as does a primary amine.

(1) Reactions of category (1).

$$\text{a. } CH_3{-}\overset{\overset{\large CH_3}{|}}{CH}{-}\overset{\overset{\large H}{|}}{\underset{\underset{\large NH_2}{|}}{C}}{-}COOH + HONO \longrightarrow CH_3{-}\overset{\overset{\large CH_3}{|}}{CH}{-}\overset{\overset{\large H}{|}}{\underset{\underset{\large OH}{|}}{C}}{-}COOH + N_2 + H_2O$$

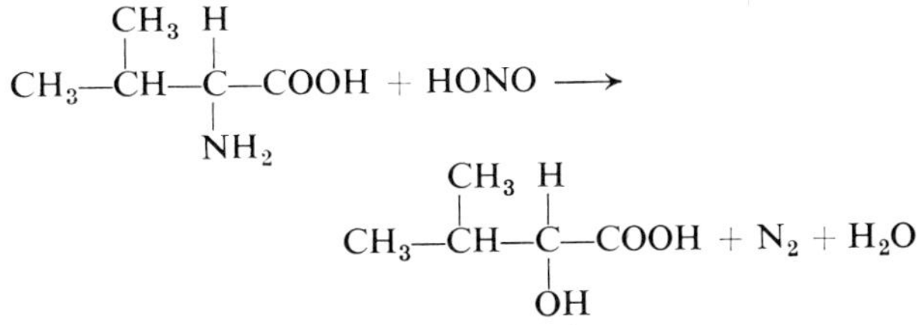

b. $C_6H_5COCl + H_2N{-}CH_2COOH + NaOH \longrightarrow C_6H_5C(=O){-}NH{-}CH_2COOH + NaCl + H_2O$

c. $HCHO + H_2NC(CH_3)(H)COOH \xrightarrow{(NaOH)} HOCH_2N(H)C(CH_3)(H)COOH$

d. $RCH(NH_2)COOH + 2\ \text{(ninhydrin)} \longrightarrow RCHO + CO_2 + \text{(indanedione)}C{=}N{-}CH\text{(indanedione)}$

The reactions of amino acids with acids and bases were discussed in the previous section. The amino group is responsible for the reactions that take place between amino acids and nitrous acid, with acylating reagents such as benzoyl chloride, and with formaldehyde and ninhydrin. Equations for the reactions of particular amino acids with these reagents are shown on page 740.

the —COOH undergoes esterification as does any carboxylic acid.

Esterification illustrates a type of reaction for which the carboxyl group is responsible.

$$C_6H_5CH_2CH(NH_2)COOH + CH_3CH_2OH \xrightarrow{(dry\ HCl)} C_6H_5CH_2CH(NH_2)COOCH_2CH_3 + H_2O$$

The special reactions (category (2)) which individual amino acids show will not be discussed at this point. They will, however, be alluded to in the discussion of color tests for the presence of particular amino acids in proteins in Section 35.3*c*.

35.2 PEPTIDES

Peptides are condensation products of amino acids. Structurally they occupy an intermediate position between the relatively simple α-amino acids and the considerably more complex proteins. The formation of a peptide by condensation is shown in Figure 35.3.

If two amino acids condense (eliminate a molecule of water) between the carboxyl group of one acid and amino radical of the

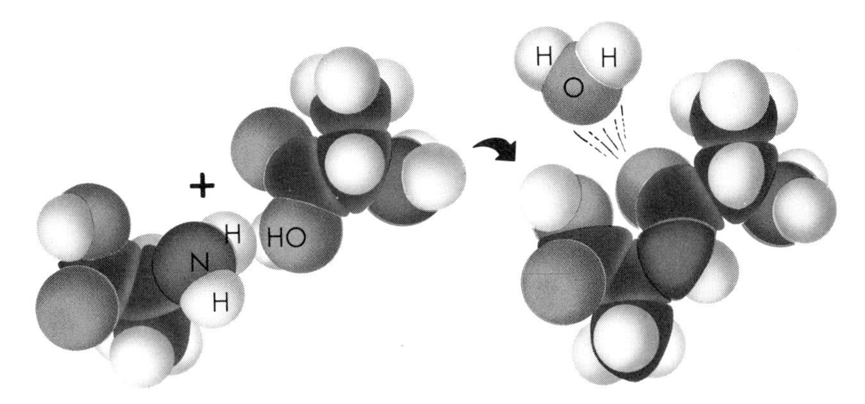

FIGURE 35.3 PEPTIDE FORMATION.

other, a substituted amide of a particular type, commonly referred to as a **peptide,** is formed (Fig. 35.4). More precisely, a two amino acid peptide of this kind is designated as a **dipeptide,** whereas the amide linkage (—CO—NH—) that serves to bond together the two α-amino acids in Figure 35.4 is called a **peptide bond.** The constituent amino acids of peptides and proteins are linked to each other through peptide bonds.

FIGURE 35.4 A DIPEPTIDE.

```
       CH3 O  H  H
       |   ‖  |  |
H2N—C———C—N—C—COOH
       |        |
       H        CH2OH
```

L-Alanyl-L-serine

A three amino acid peptide is called a **tripeptide,** a four amino acid compound, a **tetrapeptide,** and so on. Compounds in which a large number of amino acids are linked to each other through peptide bonds are known as **polypeptides.** Proteins are polypeptides having very high molecular weights, ranging from about 10,000 to many millions. Polypeptides with molecular weights less than 10,000 are considerably more stable than proteins and usually do not undergo so readily the degradation, or changes in physical properties, so easily induced in proteins by heat and chemical reagents.

To form the name of a peptide, give the names of the constituent amino acids in the order they appear in the molecule. The endings of all (except the amino acid with the free terminal carboxyl group) are changed from *ine* to *yl.* Typical peptides named in this way are:

```
  HOHCH3
  | ‖| |
H2NCCNCCOOH
   |   |
   H   H
```

Glycylalanine

```
                              C6H5
      CH3 OH   H  OH  CH2
      |   ‖|   |  ‖|  |
H2N—C———CN—C—CN—C—COOH
      |        |      |
      H        H      H
```

Alanylglycyltyrosine

```
    OH              SH
    |               |
    CH2     GH3     CH2
    |       |       |
H2NC—CONHCCONHC—COOH
    |       |       |
    H       H       H
```

Serylalanylcysteine

Glutathione, a peptide occurring in nearly all living cells, has the name γ-L-glutamyl-L-cysteinylglycine. The γ-notation signifies that the glutamic acid is linked through its γ-carboxyl group to the amino group of cysteine.

Peptides of various types are widely distributed in nature. Some are characterized by the presence of rather unusual amino acids. For example, the D-isomer of phenylalanine is found to be a constituent of certain peptides with antibiotic properties. An amino acid not found in proteins, L-ornithine, the five carbon analog of lysine, is obtained from certain natural peptides. Gramicidin S, a cyclic decapeptide containing five different amino acids, contains both D-phenylalanine and L-ornithine, along with L-valine, L-leucine and L-proline, α-amino acids commonly found in proteins. L-Ornithine is much better known for its important role as a catalyst in connection with the formation of urea in the liver of mammals.

Hydrolysis is the most characteristic reaction of peptides. Hydrolysis signifies the introduction of the elements of water between the carbon and nitrogen of the peptide linkage to liberate a free carboxyl and a free amino group. Complete hydrolysis of a peptide results in the liberation of all its constituent amino acids. The kinds and amounts of each kind of amino acid can then be determined by suitable analytical procedures now available. The hydrolysis of glutathione is shown in the following:

$$\mathrm{HOOC{-}\underset{\displaystyle NH_2}{\overset{\displaystyle H}{C}}{-}CH_2CH_2\overset{\displaystyle O}{\overset{\|}{C}}{-}\overset{\displaystyle H}{N}{-}\underset{\displaystyle H}{\overset{\displaystyle \overset{SH}{CH_2}}{C}}{-}CONHCH_2COOH + 3\,HOH \xrightarrow{H^+}}$$

$$\begin{array}{c} \mathrm{COOH} \\ | \\ \mathrm{CH_2} \\ | \\ \mathrm{CH_2} \\ | \\ \mathrm{HC{-}NH_2} \\ | \\ \mathrm{COOH} \end{array} + \begin{array}{c} \mathrm{SH} \\ | \\ \mathrm{CH_2} \\ | \\ \mathrm{HC{-}NH_2} \\ | \\ \mathrm{COOH} \end{array} + \begin{array}{c} \mathrm{CH_2COOH} \\ | \\ \mathrm{NH_2} \end{array}$$

A complete hydrolysis of a peptide such as glutathione provides information as to the relative amounts and kinds of amino acids present in a peptide but gives no information concerning the order or position that the individual amino acids occupy in the peptide chain. The procedure for determining the amino acid sequence of any peptide utilizes hydrolysis as a significant reaction but requires other more sophisticated techniques and is more time consuming than a simple determination of the amino acid composition.

Elucidation of the amino acid sequence begins with a determination of the terminal (end) amino acids of the peptide (**end group analysis**). Detection of the terminal amino acids is based on the fact that terminal residues are different from all others present, the **N-terminal residue** containing a free amino group and the **C-terminal residue,** a free carboxyl group. The N-terminal residue is usually identified by treating the peptide with 2,4-dinitrofluorobenzene (DNFB). This reacts with the free amino group, giving a yellow derivative with the 2,4-dinitrophenyl (DNP) group substituted into the amino group in place of a hydrogen. The reaction with glutathione is:

$$\text{2,4-}(O_2N)_2C_6H_3F + H_2N\underset{}{\overset{COOH}{C}}HCH_2CH_2CONH\overset{CH_2SH}{C}HCONHCH_2COOH \xrightarrow{OH^-}$$

$$\text{2,4-}(O_2N)_2C_6H_3\text{—}\underset{H}{N}\overset{COOH}{C}HCH_2CH_2CONH\overset{CH_2SH}{C}HCONHCH_2COOH$$

When this peptide derivative, the DNP peptide, is subjected to complete hydrolysis and all the peptide linkages are broken, the N-terminal amino acid remains attached to the DNP, forming a yellow derivative. Identification of the DNP amino acid obtained reveals the N-terminal amino acid.

The C-terminal residue can be identified by subjecting the peptide to enzyme catalyzed hydrolysis, using the enzyme carboxypeptidase. Carboxypeptidase induces cleavage of peptide bonds adjacent to amino acid residues having a free alpha carboxyl group. By controlling the concentration of enzyme and the time of reaction, the reaction can be limited so that the C-terminal amino acid, the one initially carrying the free carboxyl group, is separated. Identification of the liberated amino acid reveals the C-terminal amino acid. With glutathione this would be glycine.

When the above procedures are applied to glutathione as indicated, the DNP amino acid obtained upon hydrolysis of the DNP glutathione is DNP glutamic acid. The amino acid released first upon treatment of glutathione with carboxypeptidase is glycine. A glutamic acid residue is therefore present at the N-terminal position of glutathione; glycine at the C-terminal end of the peptide chain. Once both terminal amino acid residues are known for glutathione, the complete sequence of amino acids is recognized, as only three amino acid units are present in the molecule. The amino acids, in order, are glutamic acid, cysteine and glycine.

With more complex longer chain peptides, once the terminal amino acids have been determined, a series of partial, but selective, hydrolyses of the peptide must be carried out. Different conditions and catalysts must be used to break the large peptide into a number of shorter peptide units having fewer amino acid residues. The sequence in each of the peptide units is elucidated in a manner similar to that illustrated with glutathione. Once these peptides have been obtained and identified (usually by chromatographic techniques), they can be correlated or matched together in such a way as to reconstruct the complete sequence of amino acids in the order in which they appear in the peptide.

The various identified hydrolysis products (peptides or amino acids) resulting from different modes of hydrolysis may be said to correspond to pieces of a jigsaw puzzle. The enzyme catalysts act as the "saws" to provide the variety of small sized hydrolysis products formed. Just as the pieces of a jigsaw puzzle may be fitted together to reconstruct the picture from which they were originally obtained, so may the small peptides formed on hydrolysis be matched to reproduce the complete peptide sequence. This may be illustrated for a decapeptide whose composition obtained by complete hydrolysis is given in Table 35.2. "End group analysis" of this decapeptide by the methods just described gave DNP tyrosine and phenylalanine. Selective hydrolysis of different samples using different enzyme catalysts gave the tripeptides listed in Table 35.3, each of whose composition and sequence was determined by a procedure such as that outlined for glutathione.

TABLE 35.2 AMINO ACID COMPOSITION OF A DECAPEPTIDE

Amino Acid	Moles/mole of peptide hydrolyzed
Arginine (Arg)	1
Cysteine (CyST)	1
Glutamic acid (Glu)	1
Glycine (Gly)	2
Leucine (Leu)	1
Phenylalanine (Phe)	2
Tyrosine (Tyr)	1
Valine (Val)	1

TABLE 35.3 PEPTIDES

1. Gly-Phe-Phe
2. Tyr-Leu-Val
3. Arg-Glu-Gly
4. Gly-Arg-Glu
5. Val-CyST-Gly

By matching these peptides in the manner indicated here, the amino acid sequence of the decapeptide becomes evident:

```
N-terminal  Tyr
            Tyr-Leu-Val 2.
                    Val-CyST-Gly 5.
                             Gly-Arg-Glu 4.
                                 Arg-Glu-Gly 3.
                                         Gly-Phe-Phe 1.
                                                 Phe     C-terminal
            ------------------------------------------
            Tyr-Leu-Val-CyST-Gly-Arg-Glu-Gly-Phe-Phe
```

The peptide is tyrosylleucylvalylcysteinylglycylarginylglutamylglycylphenylalanylphenylalanine. For still larger peptides and proteins the effort requires many more hydrolyses, is much more time consuming and the matching becomes a considerably more complicated problem because many more peptides are obtained. Some of these, although coming from widely separated parts of a long chain, may be identical.

Synthesis, in contrast to hydrolysis, most often has as its aim the preparation of substances identical with biologically important, naturally occurring peptides whose structures have been worked out with techniques similar to those just outlined for the decapeptide. A variety of synthetic procedures have been developed, many in recent years, which permit the bonding together of different optically active amino acids to form peptide chains of a predetermined length and sequence. Formation of peptides containing different amino acids requires reaction between the carboxyl group of one and the amino group of another. Interaction between the carboxyl of one amino acid and an amino group of another molecule of the same acid must be avoided. Such a reaction can be prevented by attaching to the amino group of the acid furnishing the carboxyl group a substituent that makes the amino group unreactive, and also a substituent that may be readily removed so as to regenerate the —NH_2 after peptide bond formation with another amino acid has been accomplished. The group attached is called a "protecting group" because it protects the amino from undesired reactions. The most common protecting group, of many now available, is the carbobenzoxy group ($C_6H_5CH_2OCO$). It is readily removed by reduction with H_2 in the presence of palladium to regenerate the original amino group. A peptide synthesis utilizing the carbobenzoxy group as a protecting group may be illustrated by the formation of glycylalanine. Synthesis of this simple dipeptide proceeds through the following stages:

1. $C_6H_5CH_2OCOCl + NH_2CH_2COOH \rightarrow C_6H_5CH_2OCONHCH_2COOH$

2. $C_6H_5CH_2OCONHCH_2COOH + SOCl_2 \rightarrow C_6H_5CH_2OCONHCH_2COCl + SO_2 + HCl$

3. $C_6H_5CH_2OCONHCH_2COCl + H_2NCH(CH_3)COOH \rightarrow C_6H_5CH_2OCONHCH_2CONHCH(CH_3)COOH + HCl$

4. $C_6H_5CH_2OCONHCH_2\mathbf{CONH}CH(CH_3)COOH + H_2 \xrightarrow{Pd} C_6H_5CH_3 + CO_2 + H_2NCH_2\mathbf{CONH}CH(CH_3)COOH$

Reaction 1 introduces the carbobenzoxy group into the amino group of glycine to make it unreactive ("protect" it). In reaction 2 the carboxyl group (—COOH) is converted into the highly reactive acid chloride (—COCl) group, which is more reactive than the free carboxyl group. The acyl chloride reacts rapidly with the amino group, as shown in reaction 3, to form a peptide linkage between the carbobenzoxy substituted glycine and alanine. Step 4 accomplishes the regeneration of the amino group of glycine to form the desired peptide. The carbobenzoxy group is removed in reaction 4 by reduction with hydrogen gas in the presence of palladium metal. Toluene and carbon dioxide are formed along with the peptide. The advantage of the reduction procedure is that it does not affect the peptide bond.

Through utilization of the variety of techniques and procedures developed in recent years, known peptides of many amino acids have been synthesized. The peptides that have been prepared synthetically include the hormones vasopressin (an octapeptide) and oxytocin (an octapeptide), ACTH, the adrenocortical hormone (containing 39 amino acid residues), insulin (which contains 51 amino acids) and growth hormone with a mole weight of 21,500 and 189 amino acids.

The method outlined for determining amino acid sequences of peptides has been applied widely in the study of proteins, as have the techniques of peptide synthesis.

35.3 PROTEINS

A. Composition and Classification

As pointed out earlier in this chapter, proteins are polymers of amino acids. More specific structural studies have shown them to be high molecular weight polypeptides that because of their giant size possess certain unique properties. Figure 35.5a and b show a portion of a typical protein molecule.

In terms of elementary composition, proteins contain all the elements found in the various amino acids: carbon, hydrogen, oxygen, nitrogen and sulfur. Smaller amounts of other elements, such as phosphorus, and certain metals, such as iron, may be present. There is no known limitation to the variety of elements. Nitrogen, however, is considered to be the characteristic element, and its average percentage composition in proteins is about 16 per cent.

Concerning the amino acid composition of proteins, it was pointed out in the section on amino acids that some 20 amino acids have been obtained from proteins. The variety of amino acids in any one protein may be large and will sometimes consist of most, if not all, of these 20. Casein, the milk protein, is an example of a polypeptide containing most of the common amino acids. In other proteins, however, a few amino acids may make up the bulk of the constituent amino acids. For example, about 50 per cent of the amino acid molecules liberated when silk fibroin is hydrolyzed are glycine. Keratin is another example of a protein in which a single amino acid makes up a rather large share of the units. Cystine comprises about 14 per cent of the amino acid residues.

Several not entirely satisfactory methods are used to classify

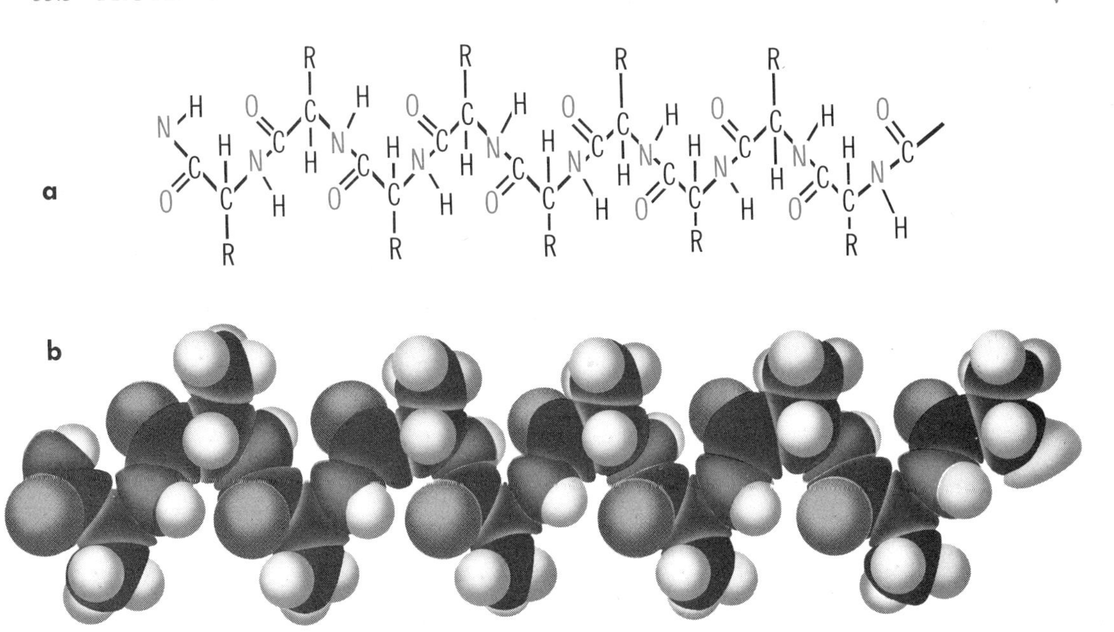

FIGURE 35.5 A PROTEIN FRAGMENT.

proteins. On the basis of gross structure, proteins are divided into **fibrous** and **globular** proteins. Fibrous proteins are not soluble nor dispersible in water. Keratin (hair), collagen (connective tissue) and silk fibroin (silk) are examples of proteins of this type. X-ray studies have shown them to consist of a more or less elongated peptide chain or group of such chains.

In contrast, globular proteins are generally soluble or dispersible in water. They appear to consist of a rather compactly folded or coiled chain or of a group of such chains. Most readily crystallized proteins are of this type. Insulin, a pancreatic hormone, and hemoglobin are considered to be globular proteins.

Proteins may also be classified as *simple, conjugated* or *derived* proteins. If a protein, when completely hydrolyzed, yields only α-amino acids, it is classed as a **simple protein.** Egg albumin (egg white) and the γ-globulins of blood proteins are usually considered to be simple proteins.

A **conjugated protein,** on the other hand, is composed of a simple protein combined with a characteristic non-amino acid substance. When the non-amino acid substance is an organic compound of some type, it is usually referred to as a **prosthetic group.** Vitamins often are components of these prosthetic groups. Familiar conjugated proteins are hemoglobin, casein, tobacco mosaic virus, polio virus, enzymes and mucin.

The class of **derived proteins** encompasses a rather heterogeneous group of materials that form when proteins are altered by chemical or physical treatment. Coagulated egg white is an example of a substance of this type.

Solubility differences between proteins may be used to further subdivide the compounds within a given category. For example, albumins may be distinguished from globulins on this basis. Albumins are soluble in both water and salt solutions, whereas globulins are sparingly soluble in water.

For more details concerning protein classification, the student should consult a standard textbook of biochemistry.

B. Physical Properties of Proteins

The physical properties of proteins reflects their large size. Most proteins do not diffuse through membranes such as cellophane or cell walls. They exist in solution as colloidal particles and show properties associated with the colloidal state of matter. For example, despite their large size they do not settle from suspension unless placed in an ultracentrifuge and rotated at speeds producing centrifugal forces many times that of gravity.

Solid proteins are usually amorphous, although some have been obtained as crystalline solids. Being composed of optically active amino acids, they are also optically active. Their solubility varies widely, depending both upon their composition and the nature and pH of the solvent used to dissolve them. Proteins decompose at relatively low temperatures without melting, and they are very sensitive to changes in the environment. Exposure to heat or extremes of pH usually produces rather drastic nonhydrolytic changes in the structure of the protein. These changes often produce marked alterations in physical properties. Heating often leads to a marked decreased in solubility. This may result in the precipitation of the protein as a result of its coagulation. If the action of heat, pressure, irradiation, strong acids or other chemical agents makes such a profound change in the protein that its structure cannot be restored to its original form, the protein is said to have been *denatured*. Denaturation converts certain proteins (egg albumin) highly resistant to digestive enzymes into products that are easily digested.

recall what occurs when an egg is fried or boiled.

Proteins are also amphoteric in behavior, a result of the presence of free basic and free acidic groups. Most of these groups come from the acidic and basic amino acids that are components of the polypeptide chain. A free amino group will also be located at the N-terminal and a free carboxyl group at the C-terminal end of the protein chain. As with amino acids, protons shift easily from carboxyl groups to amino groups, forming negative carboxylate and positive ammonium ions. Proteins, as a result, carry many negative and positive charges and contain a large number of dipoles. This makes them highly polar.

The net charge carried by a free native protein at a certain pH depends mainly upon its amino acid composition. Should a protein contain a larger percentage of acidic than basic groups, it will be an acidic protein and usually carry a negative charge at rather high pH's. A protein with a preponderance of basic amino acids will normally carry a positive charge even at rather low pH's and will be classed as a basic protein.

In solution, however, the charge carried by the protein will also depend upon and vary with the pH, just as the charge on an amino acid varies with the acidity of the solution. Proteins react with both acids and bases. Protons attach themselves to the negative carboxylate groups, and with bases the positive ammonium ion gives up a proton. In solutions of low pH a protein takes on a net positive charge and will migrate to the cathode in an electrolysis cell. In solutions of high pH, a protein assumes a net negative charge and migrates to the anode. At some intermediate pH value the number of negative charges in the protein will equal the number of positive charges and it will carry a net charge of zero. At this pH the protein does not migrate in an electric field. This pH is referred to as its **isoelectric point.** Just as the isoelectric point is a specific property of individual amino acids, it is a unique property also of individual proteins. Many properties of proteins show a minimum or maximum value at the isoelectric point. For example, most proteins are least soluble at their isoelectric point and often are least stable to heat and chemical transformation at this pH.

the isoelectric point of albumin is 4.7. What charge does it carry in the blood (pH = 7.35)?

C. Chemical Properties and Tests

Many proteins precipitate from aqueous solution when organic solvents such as alcohol or acetone are added. This addition favors the attractive forces of adjacent protein molecules for each other by tying up many of the water molecules to the added solvent, aiding their precipitation. Neutral salts are also effective in precipitating certain proteins in the same way. Ammonium sulfate is widely used in procedures for separating proteins by selective precipitation.

Heavy metal salts of lead, mercury or copper are usually insoluble. Solutions of these ions usually precipitate proteins. The positive metal ions react with the negative charges on the protein to form the salts. Likewise, acidic reagents, such as trichloroacetic acid, react with proteins to form insoluble salts. The negative trichloroacetate ion reacts with the positive groups on the protein, forming salts that precipitate. Egg white is used as an antidote for mercury or lead poisoning because it reacts with and traps these metal ions until the stomach can be pumped to remove the metal ions containing protein.

A variety of color tests or reactions have been used in the study of proteins. The **biuret test,** for example, is a general reaction used to show the presence and amount of proteins in solution. On the other hand, other tests are more specific and are used to detect certain functional groups, which, when shown to be present, are evidence for the presence of certain amino acids. Some of these tests are discussed here.

Biuret Test. Substances containing two or more peptide linkages produce a reddish-violet color with dilute copper sulfate in strong alkali. The color results from the formation of a complex between the copper(II) ion and several of the peptide bonds in the protein. This complex is shown as follows:

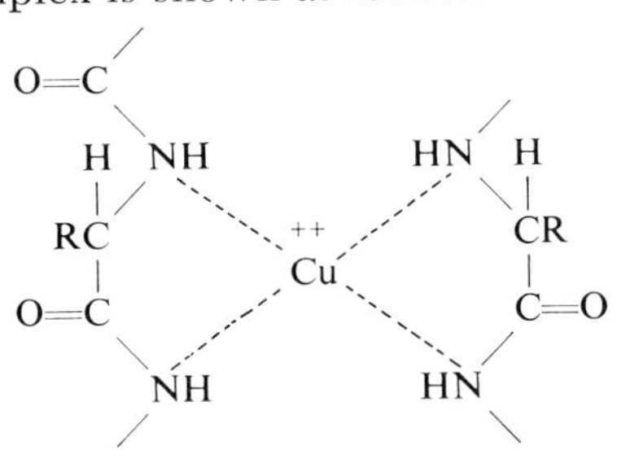

```
          /
      O=C
          \                     /
        H  NH               HN   H
        | /   .           .   \  |
       RC       .   ++  .       CR
        |         Cu            |
      O=C       .    .          C=O
          \   .        .      /
           NH            HN
          /                \
```

The biuret test is generally positive for proteins and polypeptides. It is called the biuret test because the compound biuret, formed when urea is heated exhibits the reddish-violet color characteristic of substances having two or more peptide linkages.

Xanthoproteic Test. When proteins containing aromatic groups are heated with concentrated nitric acid, they turn yellow. The chemical reaction responsible for the color is the nitration of aromatic compounds. Easily nitrated aromatic compounds, such as tyrosine and tryptophan, undergo this reaction. The yellow nitro compounds formed are converted to orange colored salts by base. Phenylalanine does not respond. Skin stains yellow when nitric acid contacts it, because aromatic amino acid units occur in the proteins. Subsequent addition of a base turns the proteins reddish-orange.

Millon's Test. Millon's reagent, a mixture of mercury(II) and mercury(I) nitrates in concentrated nitric acid, forms a red color with compounds containing phenolic groups. Tyrosine, having a phenolic ring in its R-group, undergoes this reaction whether free or combined in protein.

mercury (II) ions bind firmly to the free —COO^- present in the protein.

Hopkins-Cole Test. In the presence of concentrated sulfuric acid, glyoxylic acid condenses with indole compounds to form brightly colored compounds. The phenomenon can be observed when glyoxylic acid is added to certain protein solutions and concentrated sulfuric acid is poured slowly down the container so that it forms a layer on the bottom. If indole is present in the protein, a violet ring is formed at the interface between the two solutions. This test is specific for the presence of tryptophan in proteins since tryptophan is the only amino acid having an indole ring.

Nitroprusside Reaction. The presence of free sulfhydryl groups is shown by the formation of a red color when sodium nitroprusside in an ammoniacal solution is added to a solution of a compound. Proteins that give a red coloration on addition of ammoniacal sodium nitroprusside solution contain cysteine units with their free sulfhydryl groups.

D. Biological Significance of Proteins

The ability to synthesize proteins is a unique characteristic of living organisms. Proteins are intimately connected with almost all the chemical and physical activities that constitute life; in fact, there is hardly an important physiological function in which proteins do not participate.

Some proteins, called **enzymes,** serve as catalysts to speed up the variety of chemical transformations that living cells are able to accomplish. Other proteins function as **hormones** whose principal activity is to control processes occurring in living things. The growth hormone produced by the pituitary gland is an example of such a hormone. Likewise, hemoglobin, a protein in the blood, serves as an **oxygen carrier** to transport oxygen from the lungs to the various tissues. Still another protein, actomyosin, participates in muscular contraction. Nucleoproteins, typical conjugated proteins composed of substances called nucleic acids combined with simple proteins, are the main constituents of genes, the carriers of heredity. In addition, **antibodies,** which are proteins produced in response to the introduction of foreign proteins into the blood, function to protect man against disease. Less spectacular are the proteins such as hair and fingernails that serve as important structural elements of the body.

In concluding this chapter some attention should be given to protein structure, since biological activity is a reflection of the very specific and often complex structures that characterize proteins. Proteins made by living organisms are composed of L-amino acids, that is, amino acids related in their configuration to L-glyceraldehyde discussed in the previous chapter. This contrasts with carbohydrates (see next chapter), which are related in their configuration to D-glyceraldehyde. Architecturally each protein has a primary, secondary, tertiary and often a quaternary structure.

Primary structure has reference to the amino acid sequence. Every molecule of any one protein displays a sequence of amino acids that evidence indicates is genetically controlled. Any small change, even in only one amino acid in this sequence, may have a marked effect on the biological activity of the protein. Hemoglobin serves as a good example of a protein in which such an influence is observed. Structurally the hemoglobin molecule is composed of a porphyrin (heme) and a protein (globin). The protein is made up of four polypeptide chains—two referred to as α-chains and two referred to as β-chains. Each chain (whether of the α- or β-type) is present in an α-helical folded arrangement. These folded chains, in turn, are held together, forming a still more highly complex three-dimensional structure. The particular role of hemoglobin biologically is the transport of oxygen and carbon dioxide in the blood.

Many hemoglobins are known. In man are found normal adult hemoglobin, hemoglobin A(HbA) and a number of variants, some of which are connected with pathological conditions. One of these pathological conditions is called sickle-cell anemia. This blood disease is the result of the presence in the blood of red cells containing an abnormal hemoglobin referred to as hemoglobin S(HbS), which when deoxygenated becomes considerably less soluble than ordinary HbA. Red blood cells containing this hemoglobin when deoxygenated take on elongated and sickle shapes. When enough of these types of cells are present, they precipitate and clog the small blood vessels (capillaries) and cause the symptoms that characterize the disease.

A comparison of the two types of hemoglobin, HbA and HbS, shows that the two differ from each other in primary structure in only one amino acid. Normal hemoglobin, HbA, has the following sequence of amino acids at the N-terminal end of the β-chain:

Val-His-Leu-Thr-Pro-Glu-Glu-Lys----------
1 2 3 4 5 6 7 8 ....

Abnormal hemoglobin, present in sickle-cell anemia, in this same segment of its β-chain has this sequence:

Val-His-Leu-Thr-Pro-Val-Glu-Lys--------
1 2 3 4 5 6 7 8 ....

The difference is at position 6. At position 6 in HbS valine is found in place of a glutamic acid at this position in normal adult hemoglobin, HbA. This single exchange of valine for glutamic acid transforms the −1 charge carried by HbA owing to the COO^- group at this position in the β-chain into a no charge condition at this position in HbS. The exchange causes the protein to be less soluble than HbA, reducing its effectiveness as an oxygen carrier under conditions of oxygen deficiency.

Since the discovery of the structural difference between HbS in persons with sickle-cell disease and HbA, a large number of abnormal hemoglobins (by definition those differing from HbA) have been identified in the blood of individuals suffering other blood diseases. Hemoglobin C, HbC, for example is another variant, in which lysine is exchanged for the glutamic acid at position 6 of HbA giving it the β-chain N-terminal sequence:

Val-His-Leu-Thr-Pro-Lys-Glu-Lys---------
1 2 3 4 5 6 7 8 ....

Other hemoglobins are known in which exchange has occurred at other positions in both the α- and β-chains. Symptons appear only when the capability of the hemoglobin to carry out its function is significantly altered. Other modifications in normal hemoglobin not involving the protein chain are known, but discussion here is limited to modification in the primary structure of the polypeptide chains concerned.

Secondary structure refers to the folding which the polypeptide chains characteristically undergo because of hydrogen bonding between amino acids at adjacent points along the peptide chain. A protein normally assumes an α-helical type of secondary structure. Each polypeptide chain of hemoglobin is present in an α-helical type of arrangement.

Hemoglobin, as stated previously, has a still more complex three-dimensional structure. Its **tertiary structure** consists of a three-dimensional bundle of four polypeptides as α-helices arranged (α-, β-, α-, β-) about the porphyrin ring containing the iron (Fe^{2+}) of hemoglobin. Proteins also sometimes have a **quaternary structure** in which the tertiary bundles are packed together into still larger groupings. For hemoglobin this consists of four of the tertiary units linked together by physical or chemical bonding into a still more complex architectural arrangement. Hemoglobin is normally present in its quaternary structure when it is performing its biological function in blood. Many other proteins have similar levels of complexity exhibited in their overall structure, and for most the loss of this highly complex tertiary or quaternary arrangement results in loss of ability to carry on their normal biological function. Denaturation is commonly assumed to involve a breaking up of this protein structure under the stimulus of denaturing agents. Proteins are so important biologically that life as we know it would cease under conditions in which protein synthesis could not be carried forward or under conditions in which proteins could not exist.

modification of the primary structure may effect the stability of the secondary structure.

EXERCISES

35.1 Define:

a. amino acid
b. α-amino acid
c. dipeptide
d. polypeptide
e. peptide bond
f. acidic amino acids
g. difunctional compounds
h. condensation product
i. hydrolysates
j. essential amino acids
k. aromatic amino acids
l. isoelectric point
m. amphoteric
n. cathode
o. polymer
p. fibrous protein
q. globular protein
r. simple protein
s. conjugated protein
t. derived protein
u. denaturation

35.2 What elements are invariably found in proteins? What other elements are sometimes found in proteins?

35.3 Explain, using structural formulas, the existence of amino acids in solution as dipolar ions. Show how an amino acid in solution may be given either a positive or a negative charge by changing the pH.

35.4 Compare the physical properties of fats and proteins.

35.5 Write equations showing special conditions and catalysts for the reaction of glycine with: (a) acetyl chloride, (b) formaldehyde, (c) ethyl alcohol.

35.6 Write an equation for the hydrolysis of the tripeptide of alanine.

35.7 Give the reagent(s), the procedure, and the observation necessary for a positive test and the group detected by the:

a. biuret test
b. xanthoproteic test
c. Millon's test
d. Hopkins-Cole test
e. nitroprusside reaction

35.8 What is the primary structure of the protein named egg albumin?

35.9 Which level of protein structure is most complex?

35.10 How many amino acid differences are required before proteins exhibit different properties?

35.11 In what way does the primary structure of HbC differ from that of HbS?

35.12 What have the authors defined as an abnormal hemoglobin?

35.13 Why is hydrolysis not used to remove the carbobenzoxy group to regenerate the free —NH_2 group in the preparation of glycylalanine?

35.14 What probably occurs when a protein undergoes denaturation without a change in amino acid composition?

35.15 What is the name of an octapeptide hormone that has been synthesized in the laboratory?

35.16 Write equations outlining how valylphenylalanine can be synthesized?

35.17 What comprises the tertiary structure of hemoglobin?

35.18 If the amino acid valine is exchanged for the glutamic acid at position 6 of HbA, what change in the solubility of HbA would occur?

35.19 How many different dipeptides could be obtained if two amino acids, A and B, are combined in all possible sequences? How many tripeptides if three amino acids, A, B and C, are combined in all possible sequences? How many tetrapeptides are possible when four amino acids, A, B, C and D, are combined in all possible sequences? Compare the number of peptides found possible in the three cases cited above and, using these results as your guide, suggest a method for calculating the number of isomeric decapeptides that should be possible were ten different amino acids combined in all possible sequences.

35.20 Assume that the following compounds have been isolated from the reaction mixture obtained when a small peptide was hydrolyzed: alanylglycine, tyrosylcysteinylphenylalanine, phenylalanylleucyltryptophane, cysteinylphenylalanylleucine, valyltyrosylcysteine, glycylvaline and glycylvalyltyrosine. Assume also that only one amino acid of each kind is present in the peptide subjected to hydrolysis. Determine the sequence and number of amino acids present in the peptide from the data presented. Write the name of the peptide.

35.21 What charge would the tetrapeptide, glutaminylglycylaspartylalanine, have at a pH of 1? at a pH of 12? How many different dipeptides are possible were it hydrolyzed in a manner such as to give only dipeptides?

35.22 Write the structure for all the tripeptides which could be formed from the three L-amino acids: alanine, leucine and methionine.

35.23 A heptapeptide(A) when subjected to end-group analysis using 2,4-dinitrofluorobenzene gave 2,4-dinitrophenylglycine. Partial hydrolysis of A gave a variety of peptides containing two or three amino acids. Each peptide was completely hydrolyzed and the amino acids present identified. Sequences were not determined directly. The peptide number with its amino acid composition is given in the following:

Peptide Number	*Amino acid content*
1	alanine, glycine
2	cysteine, glutamic acid
3	alanine, cysteine
4	glutamic acid, methionine, alanine

Peptide Number	*Amino acid content*
5	leucine, alanine
6	alanine, leucine, methionine

Suggest an amino acid sequence for heptapeptide A which is in agreement with the data.

SUGGESTED READING

Doty, P.: "Proteins." Scientific American, September, 1957.

Hendrickson, J. B., Cram, D. J. and Hammond, G. S.: *Organic Chemistry*, Third Edition. McGraw-Hill Book Company, New York, 1970. Chapter 25.

Ingram, V. B.: *Biosynthesis of Macromolecules*. W. A. Benjamin, Inc., New York, 1966 (paperback).

Morrison, R. T. and Boyd, R. N.: *Organic Chemistry*, Second Edition. Allyn and Bacon, Inc., Boston, 1966. Chapter 37.

Noller, C. R.: *Chemistry of Organic Compounds*, Third Edition. W. B. Saunders Company, Philadelphia, 1965. Chapter 19.

Perutz, M. F.: "The Hemoglobin Molecule." Scientific American, November, 1964.

Sanger, F.: "The Chemistry of Insulin" (Nobel Lecture). Chemistry and Industry, *104*, 1959.

White, A., Handler, P. and Smith, E. L.: *Principles of Biochemistry*, Fourth Edition. McGraw-Hill Book Company, New York, 1968. Chapters 5, 6 and 7.

THIRTY-SIX • CARBOHYDRATES

The term carbohydrate is used to identify a rather diverse group of materials found in nature and known to have important functions as constituents of both plants and animals. They provide a structural framework for plants and serve as a source of energy for both plants and animals. Table sugar, wood, cotton, milk sugar, potato starch and honey are carbohydrates familiar to most of us. Chitin, the hard substance forming the shell of insects and crustacea, is a less familiar carbohydrate.

Carbohydrates, as the name implies, are composed mainly of carbon, hydrogen and oxygen, although other elements have been found in a few compounds. For example, chitin contains nitrogen. The observation that many of the simpler compounds had carbon, hydrogen and oxygen in the ratio CH_2O ($C_nH_{2n}O_n$) or $C_n(H_2O)_n$ gave rise to the name carbohydrate (hydrates of carbon), although they are in no sense hydrates.

Structurally, carbohydrates are polyfunctional compounds. They contain two kinds of functional groups, the hydroxyl group and the carbonyl group. They are polyhydroxy aldehydes or ketones, or are more complex materials that yield such aldehydes or ketones when hydrolyzed. Since the carbonyl group, because of intramolecular cyclization with a hydroxyl group, may be part of a ring structure, some carbohydrates are said to contain potential aldehyde or ketone groups.

36.1 CLASSIFICATION

The behavior of carbohydrate materials when subjected to acid hydrolysis provides the basis for an initial separation into groups. The simplest of these do not hydrolyze into smaller units. Substances of this kind are the least complex of the carbohydrates and are known as **monosaccharides.*** Compounds, on the other hand, which undergo hydrolysis to liberate two or more, but fewer than eleven, monosaccharide molecules are designated as **oligosaccharides.** Many of these lower molecular weight carbohydrates have a sweet taste. Because of this they are often referred to as "sugars."

If more than ten molecules of monosaccharide result when hydrolysis occurs, the compound is referred to as a **polysaccharide.** Thus, using this response of different carbohydrates to one type of chemical reaction as a basis, three classes of carbohydrates may be distinguished: monosaccharides, oligosaccharides and polysaccharides.

Subdivisions within each group specified here will be discussed in the section dealing with each class of carbohydrate. Structural considerations are used to make finer distinctions and a further designation is employed to show more fully the constituents of each carbohydrate.

36.2 MONOSACCHARIDES

emphasis is placed on common names.

Names most used for carbohydrates are the common names and these will be employed in this chapter. The characteristic ending used in naming monosaccharides is *ose*. This is attached to a word element that may reflect either the source or the historical origin of the carbohydrate. Systematic names can be obtained and will be given occasionally, along with the common names, to illustrate the way such names for carbohydrates are formed.

A. Types of Monosaccharides

tobacco mosaic virus contains the pentose ribose as a constituent.

Monosaccharides are divided into subgroups on the basis of the number of carbon atoms in each molecule. A two-carbon monosaccharide is referred to as a **biose,** a three-carbon compound as a **triose** and one with four carbons as a **tetrose.** The **pentoses,** with five carbons, are important constituents of viruses and related substances. **Hexose** refers to the six-carbon monosaccharides. They along with the pentoses, are the most important monosaccharides from the standpoint of abundance and distribution in living things.

The monosaccharides, with the exception of the bioses, may be either polyhydroxy aldehydes or polyhydroxy ketones. Monosaccharides that have an aldehyde or potential aldehyde group are referred to as "aldoses." Similarly, those with a keto or a potential keto group are designated as "ketoses." Within each category of monosaccharides, except for the biose, we find both aldoses and ketoses.

Table 36.1 lists the various types of monosaccharides along with several of the more important compounds of each kind.

Examination of the table reveals several important facts. Only aldoses are listed for the biose and tetrose classifications. Only one biose, an aldehyde (glycol aldehyde), is possible. The ketotetroses

* The Greek word *sakcharon* means sugar. Saccharides is a general term for carbohydrates.

TABLE 36.1 MONOSACCHARIDES

Classification		Example	Optical Rotation $[\alpha]_D^{20}$
Biose	Aldose	Glycol aldehyde	Inactive
Triose	Aldose	D-Glyceraldehyde	+21.2°
	Ketose	Dihydroxy acetone	Inactive
Tetrose	Aldose	D-Erythrose D-Threose	−14.5° −12.3°
Pentose	Aldose	D-Ribose D-Arabinose D-Xylose D-Lyxose	−25.0° +105.1° +18.6° −14.0°
	Ketose	D-Xylulose	
Hexose	Aldose	D-Glucose D-Galactose D-Mannose D-Gulose D-Allose	+52.5° +79.3° +14.2° −20.4° +14.4°
	Ketose	D-Fructose	−92.0°

are not particularly important, so they are not included in the table. It should be observed that the name of each optically active sugar listed in the table is preceded by a capital D. This notation is used to designate a structural relationship between the compound and that stereoisomer of glyceraldehyde which was assigned the name D-glyceraldehyde. The name of a monosaccharide that can be converted into, or can be made from, D-glyceraldehyde by suitable chemical reactions will be preceded by a capital D to show this relationship. Monosaccharides related to L-glyceraldehyde in the same way are designated with a capital L. The relationship described here is referred to as the configuration of the molecule. The optically active compounds listed in the table all have the D configuration. Again, to emphasize the point, this relationship is structural and is not directly related to the direction of optical rotation of the compound. Optical rotation is a specific physical property whose value can be determined by measurement but which bears no simple relationship to configurations. Only D-monosaccharides are listed, because the naturally occurring and biologically important monosaccharides are of this configuration. L-Monosaccharides are rarely encountered and are relatively insignificant.

B. The Hexoses

The hexoses as a class are by far the most important of the monosaccharides. **Glucose,** the most important hexose, warrants special consideration. From a chemical standpoint, studies of the chemistry and properties of glucose have provided much of what is now known concerning the structure and behavior of carbohydrates in general. Furthermore, glucose is the central carbohydrate of living organisms of all types and is the major source of the energy that humans require for their day-to-day activities. It is widely distributed in nature as the monosaccharide in grape sugar and honey, as a component of the disaccharides, lactose, maltose, cellobiose and sucrose, and as the building unit from which the polysaccharides, starch, cellulose and glycogen, are formed.

1. Glucose

The structural formula for glucose was arrived at from a consideration of facts and conclusions such as the following:

a. Quantitative analysis establishes the empirical formula CH_2O. This type of formula shows only the kinds of, and ratio between, atoms present in a compound.

b. The molecular weight of glucose, determined from a study of the freezing point of glucose solutions, shows a value of 180. When this is compared with the empirical formula weight (30), the conclusion is reached that the molecular formula is $6(CH_2O)$, or $C_6H_{12}O_6$.

recall that n-hexane has six carbons in a continuous chain.

c. Glucose, when subjected to reduction under the proper conditions, is converted into n-hexane. This is possible only if the carbon atoms of glucose are united with each other to form an open chain of six carbon atoms.

d. Glucose reacts readily with reagents commonly employed to show the presence of a carbonyl group (Chapter 30). For example, glucose adds HCN, reacts with hydroxylamine and phenylhydrazine. The compounds obtained from glucose are those one would expect if an aldehyde (—CHO) group were present. It may be concluded, therefore, that glucose has an aldehyde functional group.

e. The presence and the number of hydroxyl groups are shown by reacting glucose with acetyl chloride. Five acetyl groups are introduced into the molecule to form pentaacetyl glucose. Five hydroxyl groups are therefore assumed to be present, and since two hydroxyl groups attached to one carbon are usually unstable, it is further concluded that the five hydroxyl groups are on separate carbon atoms.

f. Oxidation reactions are in agreement with the above facts. Mild oxidation gives a six-carbon monocarboxylic acid. This is characteristic of an aldehyde. More drastic oxidation forms a six-carbon dicarboxylic acid in which carbons 1 and 6 are carboxyl carbons.

Combining these facts to design a structural formula leads to this *tentative* structure for glucose:

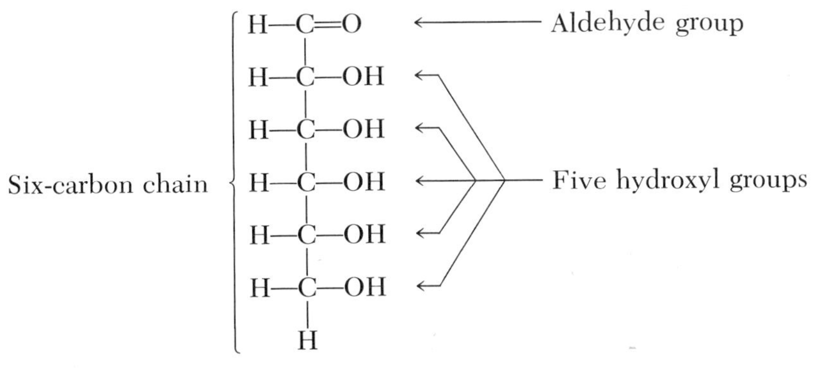

```
                    H—C=O    <———— Aldehyde group
                      |
                    H—C—OH   <
                      |
                    H—C—OH   <
                      |
Six-carbon chain    H—C—OH   <———— Five hydroxyl groups
                      |
                    H—C—OH   <
                      |
                    H—C—OH   <
                      |
                      H
```

The formula shows this carbohydrate to be a polyhydroxy aldehyde. Glucose, however, is optically active and further examination shows the presence of four asymmetric carbon atoms, which, if the carbonyl carbon is numbered 1, are carbons number 2, 3, 4 and 5. The presence of four asymmetric carbon atoms indicates the existence of 16 optical isomers of glucose. The number may be calculated from the van't Hoff rule: *The number of optical isomers equals 2^n, where n equals the number of asymmetric carbon atoms present.* $2^4 = 16$. All 16 have been synthesized and identified.

CONFIGURATION OF GLUCOSE. Since all the optical isomers of glucose have the structural features shown in the formula given above, the various isomers must differ from each other in the spatial (stereo) arrangements of the hydroxyl and hydrogen atoms about the carbons present. To establish the arrangement of the groups around each asymmetric carbon atom involves a rather lengthy and complicated series of reactions whose completion required some fifty years of study but which can be verified today by a few brief but conclusive experiments. The structural formula assigned to glucose on the basis of these studies is shown in Figure 36.1. This structure for glucose was firmly established with the achievement of its synthesis by Emil Fischer in 1890. Although the structure explained all the reactions known up to that time, new reactions were to be found shortly that would require an extension of this picture.

```
 O    H
  \\ /
   C
   |
H—C—OH
   |
HO—C—H
   |
H—C—OH
   |
H—C—OH
   |
H—C—OH
   |
   H
```

FIGURE 36.1 D-(+)-GLUCOSE.

A French chemist by the name of Tanret in 1895 observed the phenomenon of **mutarotation** with glucose, an observation that led to the proposal and establishment of the presently accepted cyclic structure for glucose and other monosaccharides. If ordinary D-glucose is dissolved in dilute acetic acid and allowed to crystallize, the material obtained has a specific rotation of +113°. However, if another portion of this same ordinary D-glucose is dissolved in water, acetic acid is added, and the solution is heated to 100° and rapidly cooled, a crystalline material is obtained that exhibits a specific rotation of +19°. If either of these crystalline materials is dissolved in water and allowed to stand, the specific rotation of the solution prepared changes gradually until a final value of +52.5° is obtained. Tanret could explain this gradual change in optical rotation (mutarotation) only if there were two forms of D-glucose that could change from one to another rather easily. Each method of crystallization yields one form. Once crystals form, no reaction can occur. As soon as the glucose is dissolved, conversion takes place and an equilibrium mixture with a rotation of +52.5° is obtained.

Two kinds of D-glucose could exist if the aldehyde group of the sugar were to react with an alcohol group of the same molecule to form an *intra*molecular **hemiacetal** (Chapter 30).

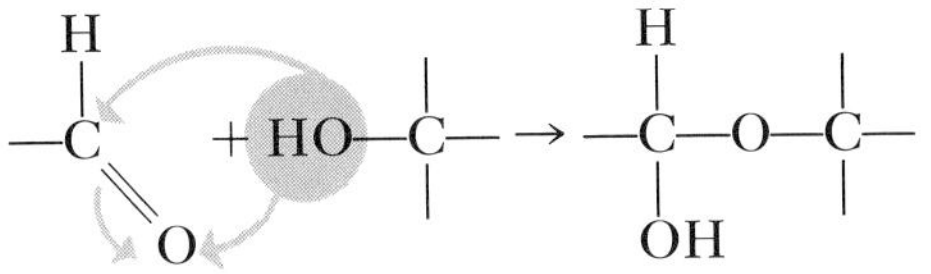

attention was called to this in Chapter 30.

H—C(=O)— + HO—C— → —C(H)(OH)—O—C—

Such an intramolecular reaction would result in formation of a ring (Fig. 36.2).

Because the normal bond angle of the sp^3 carbon present in carbohydrates, as in alkanes, is 109° 28′, strainless 5- or 6-membered ring formation is favored in this cyclization. Note that reaction of the OH on carbon number 5 with the aldehyde of carbon number 1 gives the 6-membered ring.

Carbon number 1 becomes asymmetric as a result of the cyclization reaction. It is now attached to 4 different groups, making more isomers possible. This means simply that when the —H of the —OH on carbon 5 adds to the O of the planar aldehyde group, the —OH group formed may move either left or right. The left position, as indicated in Figure 36.2, gives **β-D-glucose** and the right position, **α-D-glucose.**

It is apparent that the cyclic glucoses (Fig. 36.2) differ from one another and that they both differ from the open chain structure in the center. All three exist in equilibrium (after mutarotation) in a solution of glucose, although only traces of the free aldehyde form are present in the mixtures.

Note that the compounds on the right and left have the —OH groups situated on opposite sides of carbon 1. The isomer having the hydroxyl to the right of this carbon is designated as α-D-glucose, the one having the hydroxyl on the left, as β-D-glucose. Because ring closure forms a fifth asymmetric carbon, each of the 16 isomers of glucose predicted using the open chain structure, when cyclized, will form two different substances known as **anomers.** A grand total of 32 isomers of glucose is therefore possible. Again, all are known.

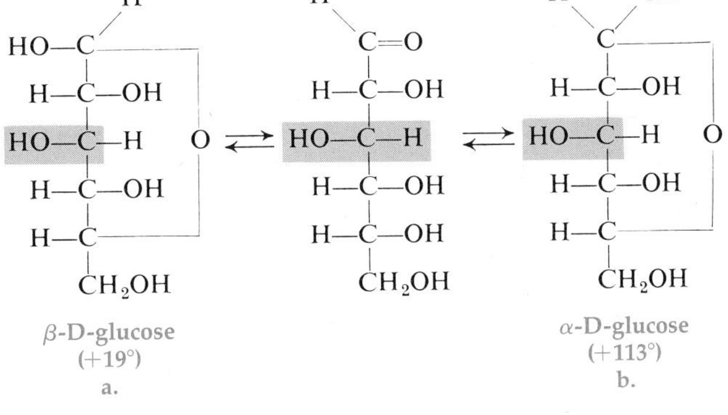

FIGURE 36.2 MUTAROTATION.

β-D-glucose (+19°) a. ⇌ open chain ⇌ α-D-glucose (+113°) b.

Mutarotation is now explained. The initial rotations of +113° or +19° are caused by the presence of α-D-glucose (+113°) or β-D-glucose (+19°), respectively. Since the two anomers are interchangeable through conversion into the open chain structure, the gradual change in specific rotation is attributed to the establishment of an equilibrium between the forms. The equilibrium mixture, 37 per cent α- and 63 per cent β-, has a specific rotation of +52.5°.

The linear cyclic structures shown (Fig. 36.2) do not accurately depict the true shape of this molecule. A formulation suggested by **Haworth,** in which rings are written as more or less regular hexagons, is more correct.

When the carbon-oxygen ring is written this way, the lower edge of this ring is assumed to be nearest the reader. Groups projecting to the right or left in the linear formula are shown as projecting above or below the plane of this ring.

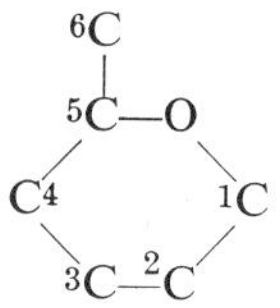

6C — 5C—O — C4 — 1C — 3C—2C   or more simply   (hexagon with O)

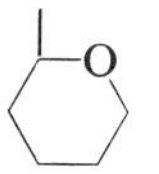

The transition from the open chain to the Haworth formula, in which the open chain is first folded and the primary alcohol shown in its true steric relation to other groups, is as follows:

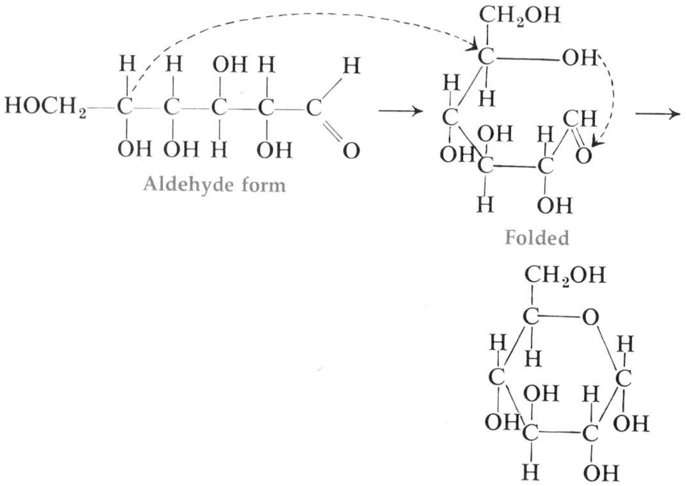

$HOCH_2$—C—C—C—C—C(H)=O   →   Folded   →

Aldehyde form

Folded

α-D-glucose

Using this notation, α-D-glucose and β-D-glucose are written as:

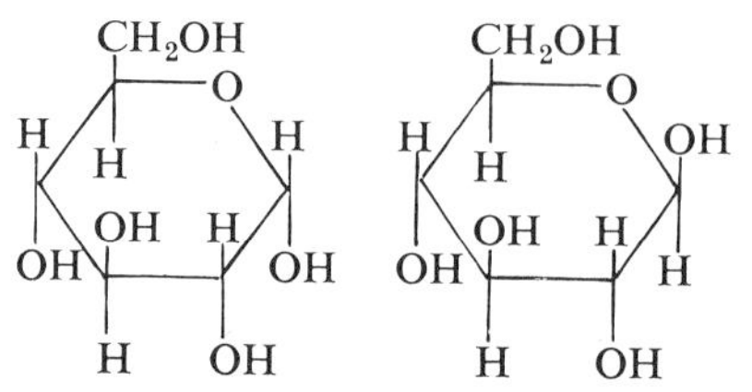

α-D-glucose (α-D-glucopyranose)   β-D-glucose (β-D-glucopyranose)

These cyclic formulas are particularly useful for showing the structures of disaccharides and polysaccharides.

The cyclic structures usually written for α- and β- anomeric forms are compared in Figure 36.3 with the three-dimensional cyclic models constructed of these same compounds, which give a more realistic picture of their actual shape. These should be compared with those in Figure 36.2.

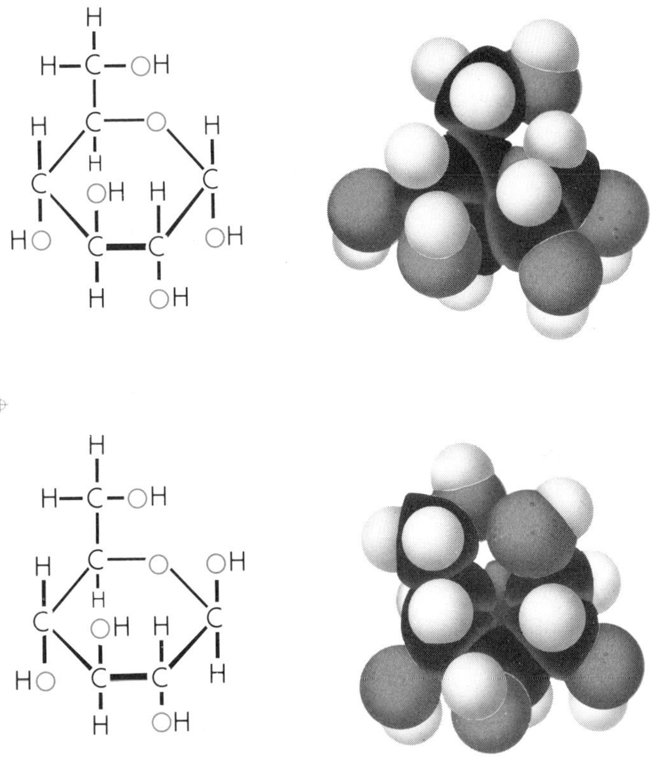

FIGURE 36.3 GLUCOSE STRUCTURES.

In writing equations for the reactions of glucose, both the open chain and the cyclic structures will be used. The particular formula that appears to be the most convenient will be employed.

Table 36.2 shows the open chain structures for the optically active aldose monosaccharides in the D-series.

A corresponding series of L-monosaccharides could be written. For each of the monosaccharides forming cyclic structures, two isomeric compounds exist and are designated as α- and β- anomers.

The structural relationship of all D-sugars to D-glyceraldehyde is clearly seen from the table, if one compares the configuration of the hydroxyl group attached to the starred carbon atom of D-glyceraldehyde with the configuration of the hydroxyl groups on the starred carbon of D-erythrose, D-arabinose and D-glucose. In each the hydroxyl is to the right.

When the aldehyde or keto group is written at the top, if the hydroxyl on this asymmetric carbon farthest from the carbonyl is to the right, the sugar is a D-sugar. If this hydroxyl is on the left,

TABLE 36.2 D-ALDOSES

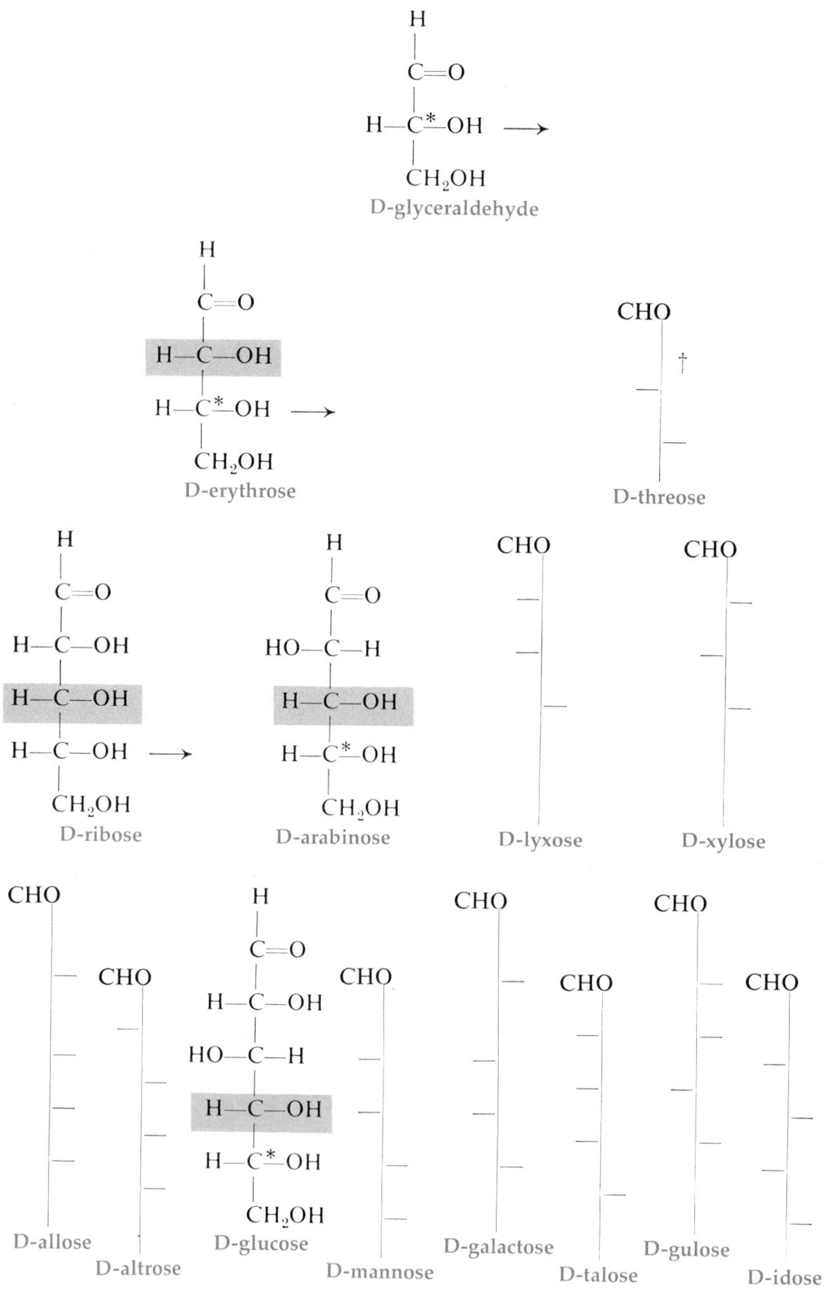

H
|
C=O
|
H—C*—OH →
|
CH_2OH
D-glyceraldehyde

H
|
C=O
|
H—C—OH
|
H—C*—OH →
|
CH_2OH
D-erythrose

CHO †
D-threose

H
|
C=O
|
H—C—OH
|
H—C—OH
|
H—C—OH →
|
CH_2OH
D-ribose

H
|
C=O
|
HO—C—H
|
H—C—OH
|
H—C*—OH
|
CH_2OH
D-arabinose

CHO
D-lyxose

CHO
D-xylose

CHO
D-allose

CHO
D-altrose

H
|
C=O
|
H—C—OH
|
HO—C—H
|
H—C—OH
|
H—C*—OH
|
CH_2OH
D-glucose

CHO
D-mannose

CHO
D-galactose

CHO
D-talose

CHO
D-gulose

CHO
D-idose

* Represents the D-glyceraldehyde asymmetric carbon.
† Horizontal lines represent an —OH group.

it is an L-sugar. The formulas for L-glyceraldehyde and L-glucose are:

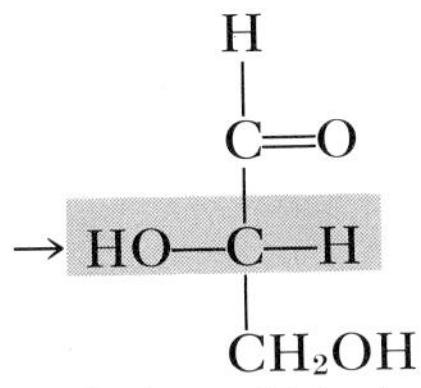

```
      H                          H
      |                          |
      C=O                        C=O
      |                          |
→ HO—C—H                     HO—C—H
      |                          |
      CH2OH                   H—C—OH
L-glyceraldehyde                 |
                             HO—C—H
                                 |
                          → HO—C—H
                                 |
                                 CH2OH
                              L-glucose
```

L-glyceraldehyde

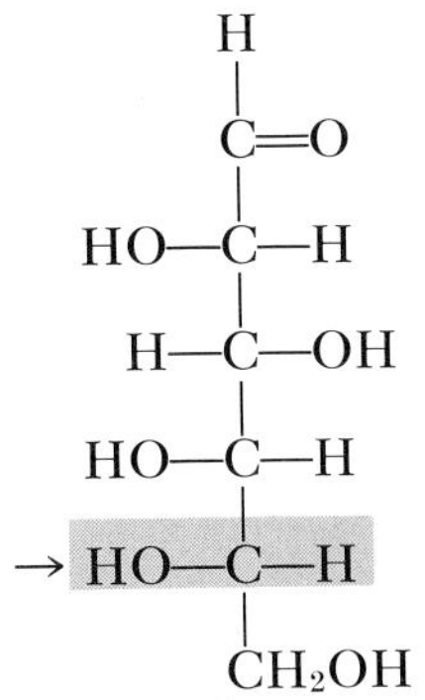

L-glucose

The D-configuration is shown in the cyclic structure by the —CH_2OH being up if the ring is written with the oxygen of the ring located at the upper right, as in the cyclic structure shown for glucose. For a cyclic structure written in this way, if the —OH attached to carbon 1 projects below the ring, the sugar is designated an *alpha* sugar; if this —OH projects up (above the ring), the sugar will be a *beta* form. Formulas for α-D-galactose and β-L-mannose are:

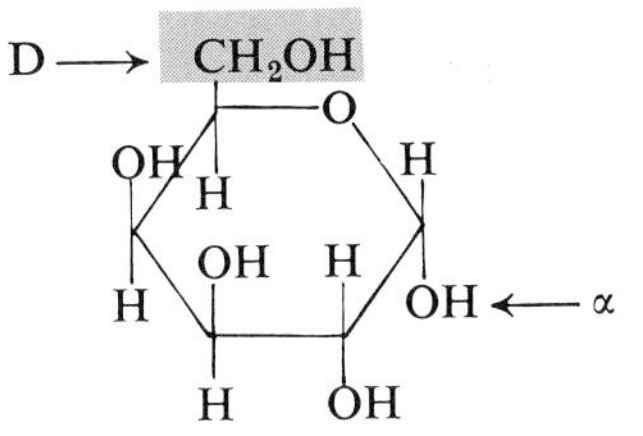

D ⟶ CH_2OH L ⟶ H

OH H O H   OH CH_2OH O OH ⟵ β

OH H H OH ⟵ α   H H H H

H OH   OH OH

α-D-galactose 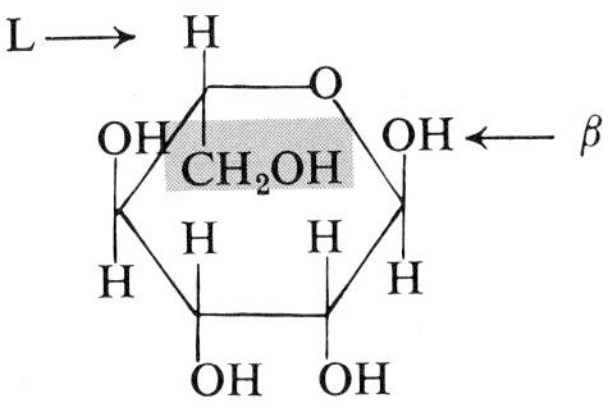   β-L-mannose

Glucose, also known as dextrose, is widely distributed in nature. It is used extensively as a food and sweetening agent. Large quantities are made commercially from starch each year. Glucose is the primary carbohydrate utilized by the body tissues, is the common sugar of blood and other body fluids, and is used for intravenous feeding of medical patients unable to take nourishment in other ways. Glucose appears in the blood in excessive amounts in "sugar" diabetes (diabetes mellitus).

2. Fructose

Fructose is the ketohexose closely related to glucose. It is the most common of the ketoses and is the monosaccharide obtained along with glucose when the disaccharide, sucrose, undergoes hydrolysis. In the course of its degradation in cells and tissues, glucose is converted into fructose derivatives. Fructose is sometimes given the name levulose because it rotates light to the left (levo-). Note that, although it is a levorotatory sugar, it has the

D-configuration and is classed as a D-sugar. Fructose has the following structures:

$$\begin{array}{c} CH_2OH \\ | \\ C{=}O \\ | \\ HO{-}C{-}H \\ | \\ H{-}C{-}OH \\ | \\ H{-}C{-}OH \\ | \\ CH_2OH \end{array}$$

D-Fructose

or

α-D-Fructose (Pyranose ring)

or

α-D-Fructose (Furanose ring)

The cyclic formula for fructose formed by reaction of the hydroxyl on carbon 6 with the carbonyl group gives the usual six-membered ring. However, fructose with the five-membered ring is found in nature. This results from reaction of the hydroxyl on carbon 5 with the carbonyl group.

Figure 36.4 illustrates the open chain and the Haworth formulas of fructose along with the corresponding molecular models.

To distinguish the two types of fructose, the one with a six-atom ring structure is named α-D-fructopyranose, and the molecule having the five-atom ring is β-D-fructofuranose. **Pyranose** comes

$$\begin{array}{c} H \\ | \\ H{-}C{-}OH \\ | \\ C{=}O \\ | \\ HO{-}C{-}H \\ | \\ H{-}C{-}OH \\ | \\ H{-}C{-}OH \\ | \\ H{-}C{-}OH \\ | \\ H \end{array}$$

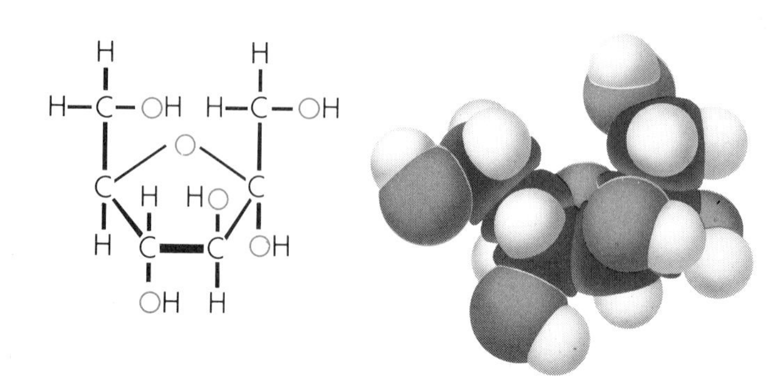

FIGURE 36.4 STRUCTURE OF FRUCTOSE.

from the word pyran, the name for the six-atom ring having five carbons and one oxygen (Chapter 33). Furan has a ring with four carbons and one oxygen (Chapter 33).

Pyran Furan

Note that the configurations of carbons 3, 4, 5 and 6 of fructose are exactly the same as those in glucose.

3. Mannose

Mannose is the monosaccharide obtained from the polysaccharide mannan. It is considered here only to point up another structural relationship that exists between monosaccharides. The structure of mannose is:

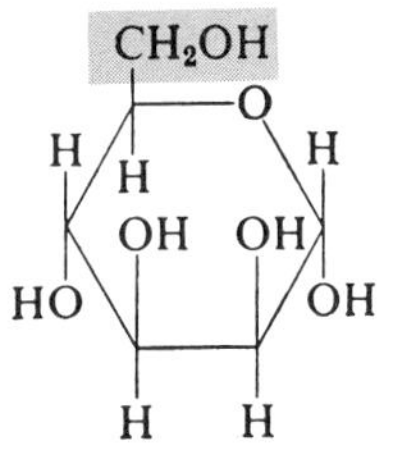

α-D-mannose 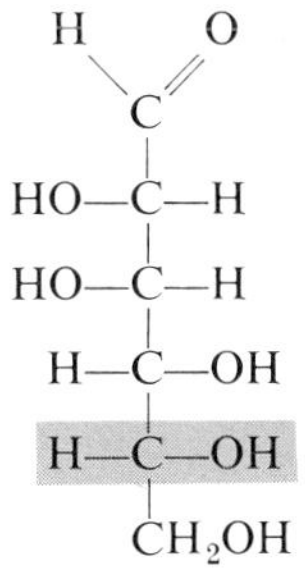 D-mannose

An examination of the structure shows that it differs from glucose only in the configuration at carbon 2. The position of the hydroxyl on carbon 2 in mannose is the reverse of its position in D-glucose. Two sugars which differ in this way are said to be **epimers.**

4. Galactose

Galactose is important because it is a component of the disaccharide, lactose (milk sugar). It is also a constituent of the fatty substances of brain and nerve tissue. It has the following structure:

what monosaccharide is an epimer of D-galactose?

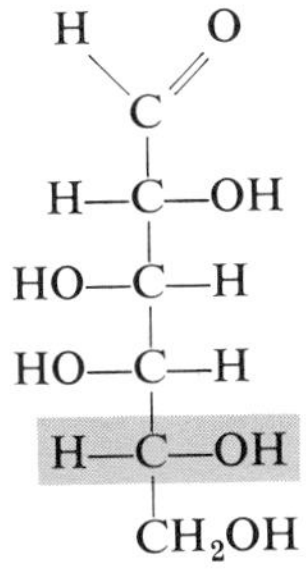

D-galactose 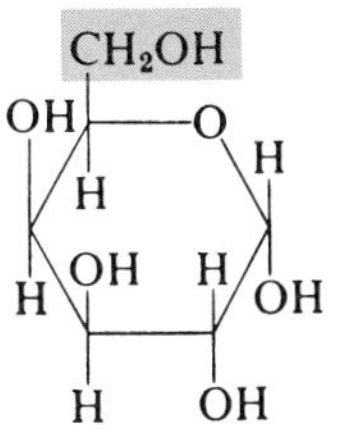 α-D-galactose

C. Reactions of Monosaccharides

Monosaccharides, because of their polyfunctional nature, can be expected to show reactions of aldehydes, ketones or alcohols.

1. Oxidation. Monosaccharides generally are oxidized very readily by such mild oxidizing agents as Tollens', Fehling's and Benedict's reagents. All monosaccharides, whether open chain structures or cyclic structures, are easily oxidized. The free hydroxyl group attached to carbon 1 in the cyclic form activates the molecule toward oxidation. Sugars that react with mild oxidizing agents are referred to as **reducing sugars.** Products of the reaction with Benedict's or Fehling's solution (Chapter 30) are red copper(I) oxide and the monocarboxylic acid of the aldehyde. Glucose gives D-gluconic acid and red copper(II) oxide. Conversion of the aldehyde of glucose into a carboxyl yields D-gluconic acid.

Nitric acid oxidizes aldohexoses to saccharic acids. Saccharic acids are terminal dicarboxylic acids. D-galactose and lactose (see next section) yields the saccharic acid, mucic acid. It is an insoluble crystalline compound and its formation is used to indicate the presence of galactose.

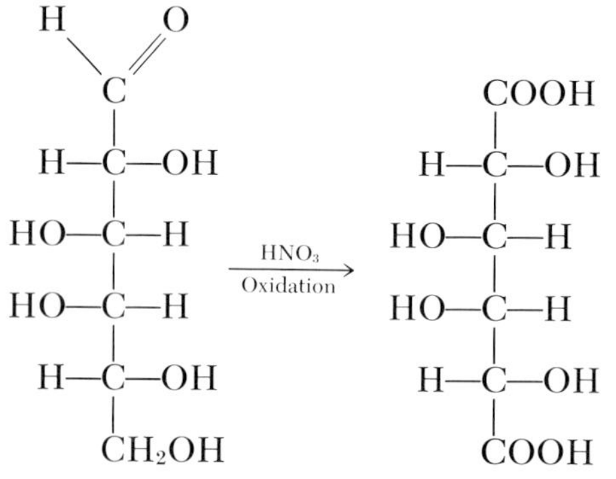

```
  H     O
   \   //
     C                         COOH
     |                          |
  H—C—OH                     H—C—OH
     |                          |
 HO—C—H        HNO3          HO—C—H
     |      ---------->         |
 HO—C—H      Oxidation       HO—C—H
     |                          |
  H—C—OH                     H—C—OH
     |                          |
    CH2OH                      COOH
```

Galactose

Mucic acid (optically inactive)

is mucic acid an optically active or a meso acid?

2. Osazone Formation. Phenylhydrazine reacts with aldoses to form phenylhydrazones (Chapter 30), which then react further to finally give compounds referred to as phenylosazones. Osazones are usually easily crystallized and may be used to show the presence of particular sugars by the characteristic crystals which appear or by the melting points observed for the products obtained. The sequence of reactions for glucose may be written as follows:

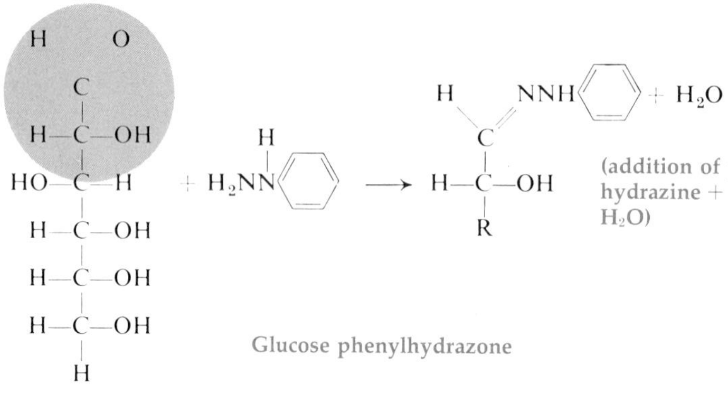

```
  H    O
    C                                  H      NNH—C6H5  + H2O
    |                                    \   //
 H—C—OH             H                      C
    |               |                      |
HO—C—H   +  H2N—N—C6H5   ---->          H—C—OH
    |                                      |
 H—C—OH                                    R
    |
 H—C—OH
    |
 H—C—OH
    |
    H
```

(addition of phenylhydrazine + loss of H_2O)

Glucose phenylhydrazone

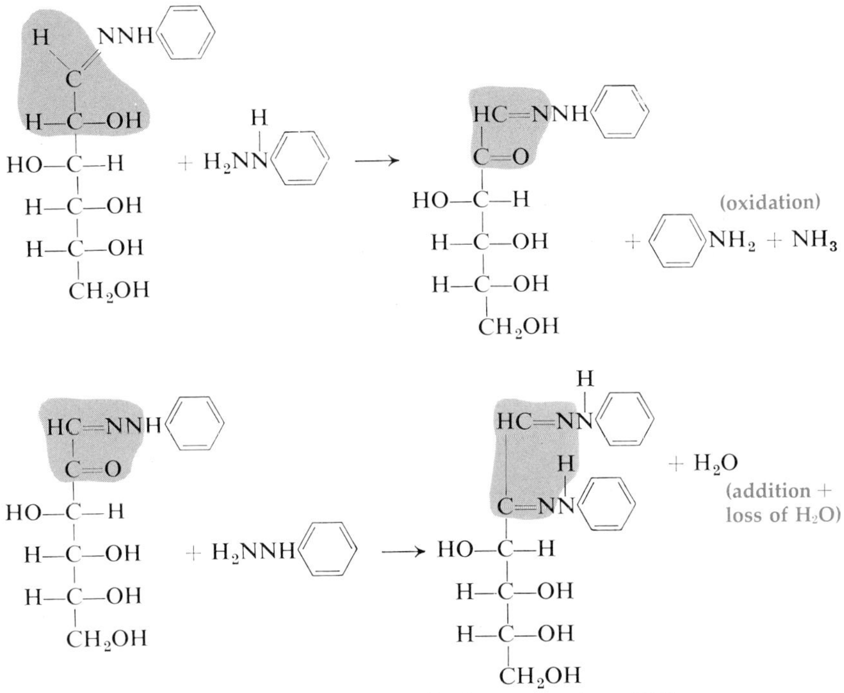

H NNH–C_6H_5
C
H—C—OH
HO—C—H + $H_2NN(H)C_6H_5$ ⟶ HC=NNH–C_6H_5
H—C—OH C=O
H—C—OH HO—C—H
CH_2OH H—C—OH + $C_6H_5NH_2$ + NH_3 (oxidation)
H—C—OH
CH_2OH

note that the H—C—OH undergoes dehydrogenation to form a C=O group at carbon 2.

HC=NNH–C_6H_5
C=O
HO—C—H
H—C—OH + H_2NNH–C_6H_5 ⟶ HC=NNH–C_6H_5
H—C—OH C=NNH–C_6H_5 + H_2O (addition + loss of H_2O)
CH_2OH HO—C—H
H—C—OH
H—C—OH
CH_2OH

D-glucosazone (m.p. 208°)

Examination of this sequence of reactions shows that only the top two carbons (carbons 1 and 2) of the monosaccharide take part in this reaction. The rest of the glucose molecule remains unchanged. This observation was of considerable significance in arriving at the structures of the isomers of glucose. Fischer found that D-mannose and D-fructose both gave the same osazone as did D-glucose. From this he concluded correctly that the arrangement of the atoms around carbons 3, 4, 5 and 6 of these three was the same.

Examine the structures of these sugars given in Table 36.2 and on page 803 to verify this.

3. Glycoside Formation. The formation of mixed acetals by the interaction of monosaccharides with alcohols in the presence of traces of mineral acids is a significant reaction. The reaction of glucose with methanol (or ROH) is:

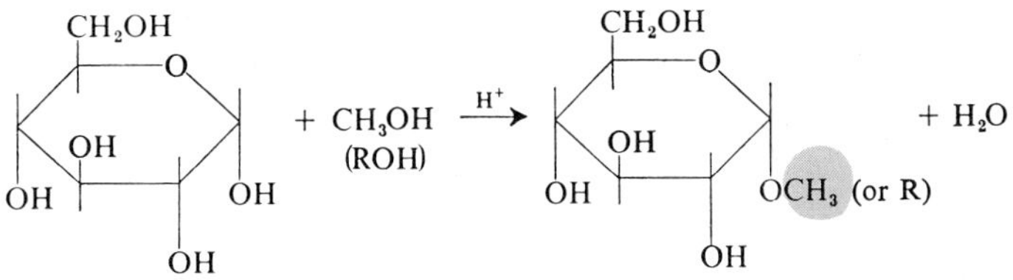

CH_2OH (ring: O, OH, OH, OH, OH) + CH_3OH (ROH) $\xrightarrow{H^+}$ CH_2OH (ring: O, OH, OH, OH, OCH_3 (or R)) + H_2O

Observe that this reaction takes place between the hydroxyl attached to carbon 1 of glucose (the hemiacetal hydroxyl) and that the hydrogen of the hydroxyl is replaced with a methyl (or R) group. Derivatives of carbohydrates formed when hydrogens of this kind are replaced with R groups are referred to as *glycosides,* and the bond between the radical and the monosaccharide is called a *glycoside linkage.* A compound obtained from glucose would be designated more specifically as a **glucoside** because it is a derivative of glucose. In this instance, reaction with CH_3OH, it is a methyl glucoside, and, more specifically, a methyl α-D-glucoside. If the methanol reacted with β-D-glucose, methyl β-D-glucoside would be obtained. Both alpha and beta glucosides are found in nature.

write the structure of methyl β-D-glucoside.

This type of combination is of interest because the disaccharides to be discussed next are joined to each other through glycoside linkages of this type. Disaccharides are glycosides in which the R group is another monosaccharide.

36.3 OLIGOSACCHARIDES

The most common and most important oligosaccharides are the disaccharides. Disaccharides are all glycosides of a particular kind in which the R- group is another monosaccharide. Four of these will be discussed: maltose, cellobiose, lactose and sucrose. Of these, sucrose is the most common.

A. Maltose

Maltose hydrolyzes to yield only glucose. Structurally it is a glucoside in which the R group is another glucose molecule. It consists of two glucose units joined together by an alpha glycosidic linkage between carbon 1 of one unit and carbon 4 of another. Its structure is:

Maltose

The presence of a free hydroxyl group attached to carbon 1 on the unit on the right makes this a potential aldehyde group. Maltose is, therefore, a reducing sugar which would be expected to show mutarotation and to react with Tollens', Fehling's and Benedict's reagents, which it does. Like glucose, it forms a phenylosazone of characteristic crystalline appearance. Both monosaccharides have the D-configuration.

Maltose is used in baby foods and malted milk. It is particularly important as an intermediate in the manufacture of ethyl alcohol from starch. The first step in the digestion of starch is its hydrolysis to maltose. The enzyme maltase in the small intestine completes its conversion into glucose.

B. Cellobiose

Cellobiose is a beta glucoside. The R-group is a second molecule of glucose. It consists of two glucose units joined by a beta glycosidic linkage from carbon 1 of one unit to carbon 4 of the second. Its formula is:

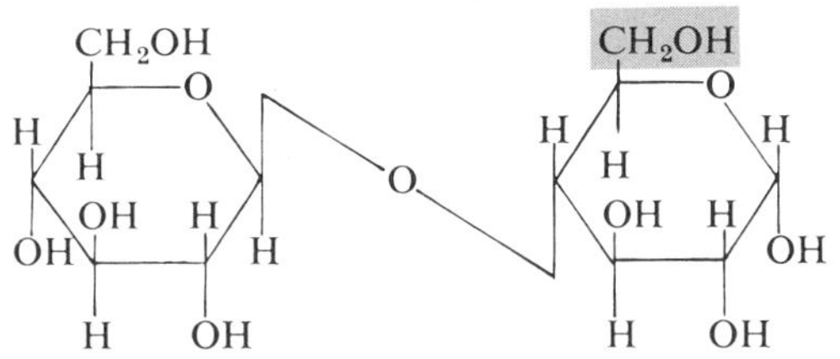

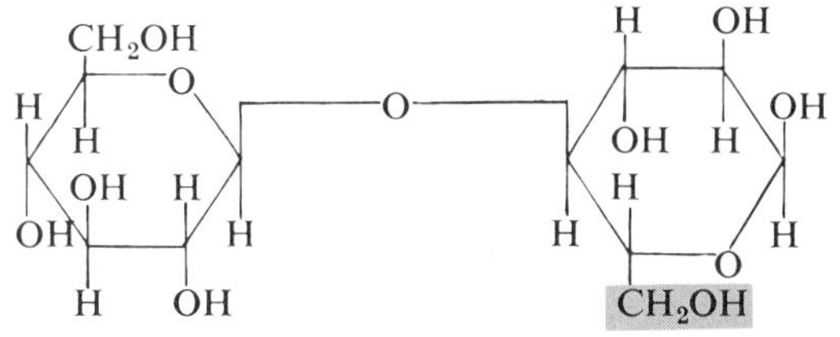

the β-glycosidic linkage is more difficult to hydrolyze than the α-linkage.

The second cellobiose formula was obtained by rotating the glucose unit on the right in the first formula through an angle of 180°, to give a better picture of the actual molecule. Since cellobiose has a free hydroxyl attached to carbon 1 of the unit to the right, it is a reducing sugar and shows the corresponding reactions. It is obtained by careful hydrolysis of cellulose. The distinctive difference between cellobiose and maltose, the type of glycosidic linkage present, is important in determining the structure and nature of the polysaccharides formed from glucose. Both monosaccharides in cellobiose have the D-configuration.

C. Lactose

Lactose, a reducing sugar, is the important carbohydrate in milk. It is commonly referred to as milk sugar. Human milk contains 5 to 8 per cent and cow's milk 4 to 6 per cent. Hydrolysis of lactose yields both glucose and galactose. In contrast to maltose and cellobiose, lactose is a beta galactoside, in which the two component monosaccharides are joined through a galactosidic linkage between carbon 1 of the galactose and carbon 4 of the glucose unit. Its formula is:

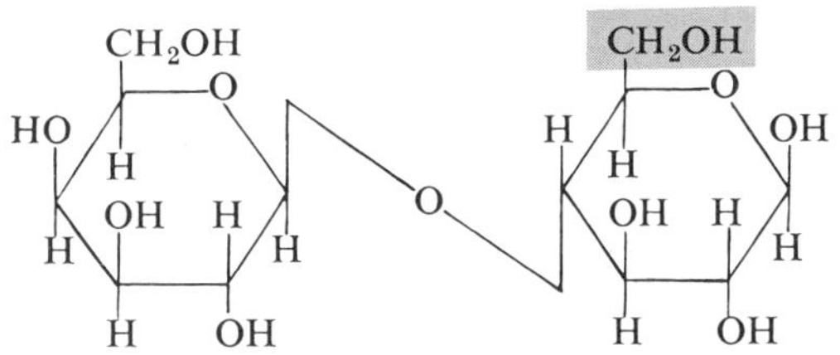

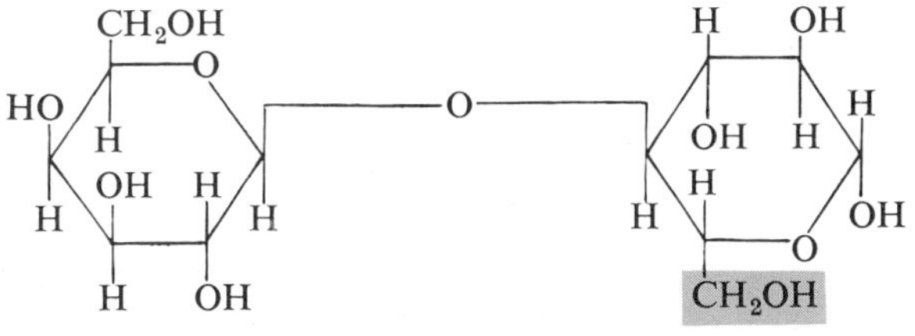

sandy ice cream results when hard α-lactose hydrate crystals form during its preparation.

Lactose forms an osazone having characteristic crystals that are of great value in indicating the presence of this sugar. The sugar is readily fermented to lactic and butyric acids—a reaction that occurs in the making of cheese. Both glucose and galactose have the D-configuration in lactose.

D. Sucrose

Sucrose (table sugar) is produced in larger quantities today than any other organic chemical manufactured as a pure compound. It is obtained commercially from sugar cane and sugar beets. Although its structure has been known for many years, its chemical synthesis was not accomplished until recently (1957), and then only milligram quantities were prepared. Sucrose is used principally as a food. Hydrolysis of sucrose yields glucose and fructose in equal molar quantities, a mixture often referred to as invert sugar. Structurally, sucrose is both a fructoside and a glucoside, since the two monosaccharide units are joined from carbon 2 of the fructose to carbon 1 of the glucose unit. The fructose component of this disaccharide is beta fructose, and the glucose component is alpha glucose. The structures of sucrose and fructose are shown below:

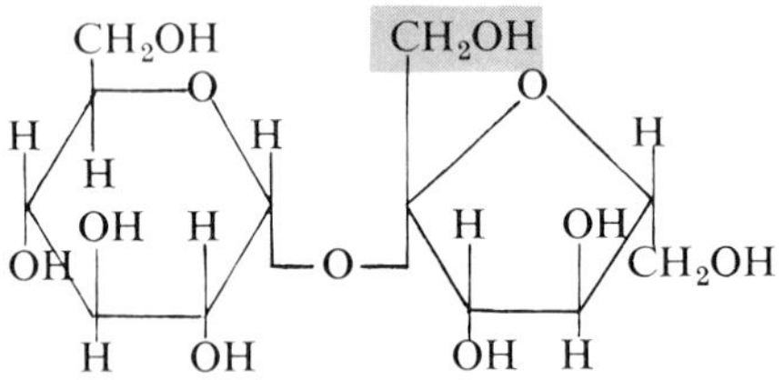

Sucrose

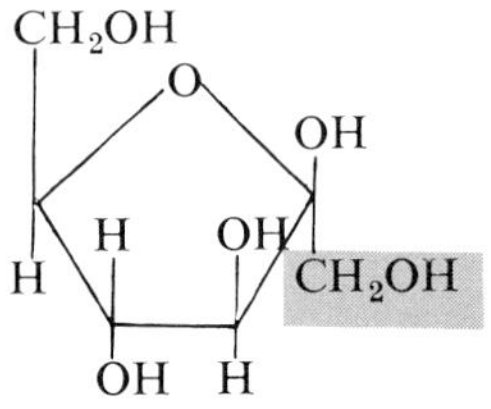

β-D-Fructofuranose

Note that to place the beta hydroxyl on carbon 1 of the fructose next to the alpha hydroxyl attached to carbon 1 of glucose, it is necessary to rotate the fructose vertically through 180°, a transformation placing all the groups that are *up* in the formula on the right *down* in the structure written for sucrose.

Hydrolysis of sucrose is often referred to as an inversion. The specific rotation of pure sucrose is +66°. The specific rotation of the equal molar mixture of fructose and glucose is −20°. This conversion of a dextro (+) rotation solution into one having a levo (−) rotation is called inversion. The −20° is an average of the +52° of the glucose and the −92° of the fructose dissolved in the solution.

36.4 POLYSACCHARIDES

Polysaccharides are polymers in which one finds the component monosaccharide units joined to each other by glycosidic linkages. Polysaccharides found in nature are molecules of high molecular weight, and correspondingly consist of large numbers of monosaccharide units. In many of these, D-glucose is the sole, or major, component. However, polysaccharides containing D-mannose, D- and L-galactose, D-xylose and L-arabinose are known.

Also, polysaccharides formed from sugar derivatives such as glucuronic acid or glucosamine have been found. Chitin, mentioned earlier, is a polymer of acetylated glucosamine. Polysaccharides differ structurally as well as in their monosaccharide components. The most important polysaccharides are starch, glycogen and cellulose. Others of interest include inulin, gums and pectins.

A. Cellulose

Cellulose is the first polysaccharide discussed because it is the most abundant and contains 50 per cent or more of all the carbon in vegetation. The source of the purest cellulose is cotton. Complete hydrolysis of cellulose yields only D-glucose and, as pointed out in an earlier section, its careful partial hydrolysis yields cellobiose. The formation of cellobiose shows that the glucose units in cellulose are united through beta linkages. Cellulose is therefore a linear polymer of a large number of β-D-glucose units joined from carbon 1 of one unit to carbon 4 of the next unit. The structure may be written in the following way:

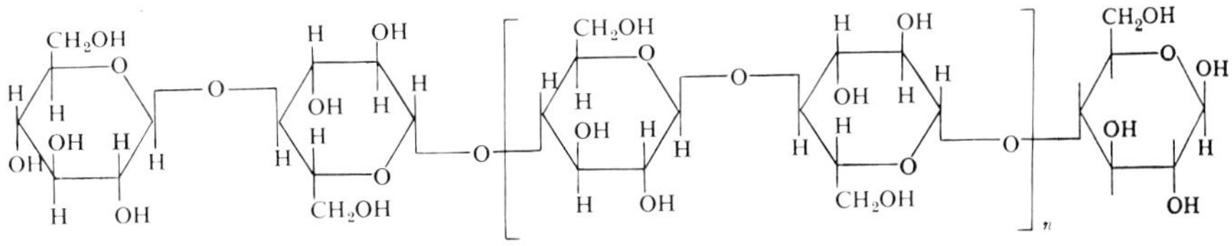

a fortune awaits one who can economically hydrolyze cellulose into glucose.

Cellulose

The molecular weights of celluloses obtained from different sources and by various methods have been estimated to be between 50,000 and 500,000.

cellulose acetate is one example.

The average numbers of glucose units in the different samples of cellulose, therefore, range between about 300 and 2500 units per molecule. In the structure given here, n will be equal to some number between 150 and 1250. Celluloses with the lower molecular weights are probably degradation products of the larger polymers. It is unlikely, however, that all molecules of native cellulose are of the same weight.

Cellulose is insoluble in water and in all organic solvents. It does not react with Tollens', Fehling's or Benedict's reagents, nor does it form a phenylhydrazone. It dissolves in a variety of special reagents, such as an ammoniacal solution of copper(II) sulfate (Schweitzer's reagent), from which cellulose may be precipitated to produce rayon fiber. Many types of rayon and other derivatives have been prepared by chemical modification of cellulose.

B. Starch

The polysaccharide starch serves as the nutritional reserve carbohydrate in many plants. It is chiefly concentrated in the seeds and roots. On complete hydrolysis, starch yields glucose as the only monosaccharide. Enzymatic hydrolysis gives relatively large amounts of maltose, thus indicating that the glucose units in starch are joined through alpha glucosidic linkages. Starch is clearly a polymer of α-D-glucose.

Starch in its native state is observed as microscopically visible granules. The starch grains of plants differ in size and shape; these differences are apparent when seen under the microscope and their appearance may be used to identify the source.

Starches found in nature contain in general two types of compounds, **amylose** and **amylopectin,** which are separable from each other. On heating starch with water, a soluble portion (amylose) is obtained, and an insoluble pasty material (amylopectin) remains. Since both of these substances are composed of α-D-glucose, they must differ structurally.

C. Amylose

The polysaccharide amylose is believed to be composed of long, unbranched chains of glucose units. As a linear polymer it resembles cellulose; however, it has a lower average molecular weight, ranging between 10,000 and 50,000. This indicates that there are 60 to 300 units in each molecule of this component of starch. The structure of amylose may be written as follows:

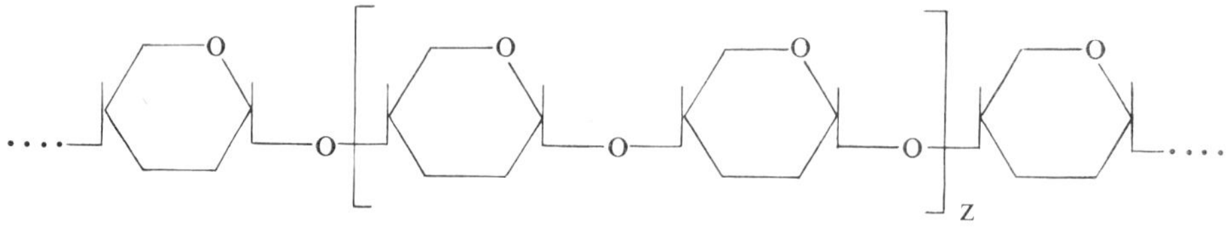

Maltose unit

Here the repeating unit is maltose, in contrast to the repeating cellobiose units in cellulose. Amyloses give a deep-blue color with iodine (in contrast to the red-purple color formed with amylopectins). Potato starch contains about 20 per cent of the amylose fraction, a proportion found also in many other starches. The remainder is amylopectin. The exact per cent of each varies from starch to starch.

D. Amylopectin

Amylopectins are composed of molecules larger than those of the amyloses. They differ from amylose in having a branched chain structure. Amylopectins have one chain-terminal glucose for every 24 to 30 glucose units in the molecule. Molecular weights up to 300,000 or more have been found. This corresponds to molecules having around 1800 glucose residues per molecule. With branching occurring often enough so that each branch has only 20 to 30 units, a molecular weight of 300,000 would correspond to a molecule with over 60 branching points.

Branching results from the linking of glucose units to a polymeric chain through one or more of the hydroxyl groups not utilized in forming the linear structure. Branches appear to be the result of the joining of certain glucose units through carbon 1 of the first unit to carbon 6 of the second unit. A disaccharide of the following formula illustrates this:

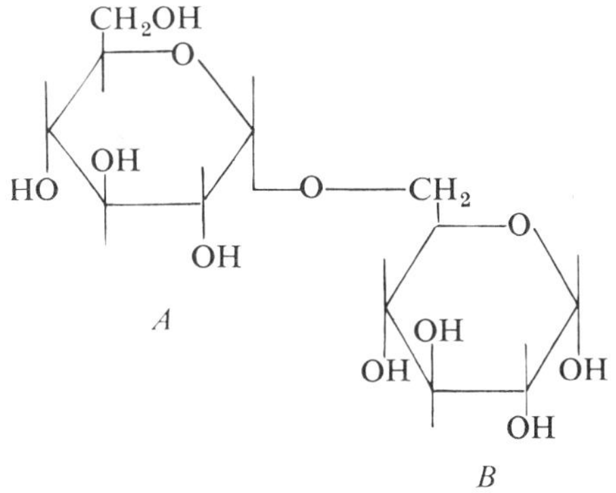

This disaccharide is known as *isomaltose*. Two glucose molecules may be joined in this way, even if the glucose unit B is already attached to neighboring glucose units through alpha glucosidic linkages between carbons 1 and 4. Whenever a glycosidic linkage from carbon 1 to carbon 6 occurs along a glucose chain, a branch appears. The following formula shows this:

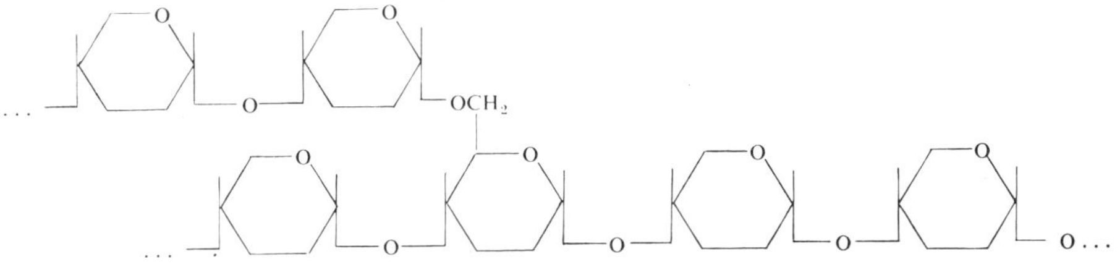

In simpler language, amylopectin is a treelike molecule in contrast to the linear arrangement in amylose. This highly ramified structure permits the formation and existence of the bigger insoluble molecules.

E. Glycogen

glycogen has been called "animal" starch.

In the animal body the polysaccharide glucose is stored in both the liver and the muscles as glycogen. Structurally, glycogen is a branched chain polysaccharide resembling amylopectin. Glycogen consists largely of α-D-glucose units 1,4-linked through glucosidic bonds with considerable branching involving 1,6-linkages. Glycogen differs from amylopectin in being more highly branched, having one chain-terminal glucose for every 12 to 20 glucose units. Glycogens have molecular weights ranging from about 250,000 to several millions, thus requiring the presence of 1500 to 20,000 residues per molecule. Molecular sizes vary widely even within samples obtained from a single source.

Glycogen occurs in animal cells in smaller particles than are normally observed for starch granules. It gives a violet-red color with iodine and is quite stable with base, but it undergoes acid hydrolysis quite readily. It is precipitated from solution with alcohol.

EXERCISES

36.1 Define:

a. carbohydrate
b. aldohexose
c. ketohexose
d. monosaccharide
e. oligosaccharide
f. polysaccharide
g. mutarotation
h. hemiacetal
i. alpha sugar
j. dextrose
k. epimers
l. glycoside
m. α-glucoside
n. an invert sugar
o. blood sugar

36.2 Name four hexoses that occur in nature either as monosaccharides or as units in other molecules.

36.3 Why are there no ketobioses?

36.4 Why is glucose considered the most important hexose?

36.5 Explain the significance of each fragment of the name D-(+)-glucose.

36.6 Give some experimental evidence for each of the following facts: Glucose

a. has the formula $(CH_2O)_6$
b. is a straight chain compound
c. is an aldehyde
d. has 5 hydroxyl groups

36.7 Use structural formulas to explain the mutarotation of D-glucose.

36.8 Write open chain formulas for

a. D-glucose
b. D-mannose
c. D-fructose
d. D-galactose

36.9 Explain the structural differences between cellobiose, sucrose, maltose and lactose.

36.10 Explain the structural similarities and differences between (a) starch and cellulose, (b) amylose and amylopectin, (c) amylopectin and glycogen.

36.11 What color is formed with iodine and each of these: (a) starch, (b) amylose, (c) amylopectin, (d) glycogen?

36.12 Compare the relative molecular weights of (a) glucose, (b) sucrose, (c) cellobiose, (d) dextrins, (e) amylose, (f) amylopectin, (g) cellulose, (h) starch, (i) glycogen.

36.13 What structural characteristics are necessary if a disaccharide is to be a reducing sugar?

36.14 What are Tollens' reagent, Benedict's solution, and Fehling's solution?

36.15 What kind of a product is formed when an aldohexose is oxidized mildly? When it is oxidized with hot nitric acid?

36.16 Compare the polymers: (a) starch, (b) cellulose, (c) silk, (d) polyethylene.

36.17 D-Galactose yields an optically inactive diacid, mucic acid, when oxidized with nitric acid. Why is mucic acid optically inactive?

36.18 Which hexose is an epimer of gulose?

36.19 Which hexose will produce the same osazone as altrose?

36.20 Write the structural formula for a disaccharide made from glucose and gulose, with no reducing properties.

36.21 Two aldotetroses, A and B, give the same osazone. When oxidized with concentrated HNO_3, tetrose A gives a 4 carbon dicarboxylic acid which is optically inactive. Aldotetrose B when oxidized in the same way gives an optically active 4 carbon dicarboxylic acid. Write structural formulas for compounds A and B.

SUGGESTED READING

Claude, F.: "The Basic Work of Fischer and van't Hoff in Carbohydrate Chemistry." J. Chem. Ed., *30*:120, 1953.

Hendrickson, J. B., Cram, D. J. and Hammond, G. S.: *Organic Chemistry*, Third Edition. McGraw-Hill Book Company, New York, 1970. Chapters 25 and 27.

Hudson, C. S.: "Emil Fischer's Discovery of the Configuration of Glucose." J. Chem. Ed., *18*:353, 1941.

Hudson, C. S.: "Symposium on Wood as a Chemical Raw Material." J. Chem. Ed., *35*:482, 1958.

Morrison, R. T. and Boyd, R. N.: *Organic Chemistry*, Second Edition. Allyn and Bacon, Inc., Boston, 1966. Chapters 33 and 34.

Noller, C. R.: *Chemistry of Organic Compounds*, Third Edition. W. B. Saunders Company, Philadelphia, 1965. Chapter 18.

Schoch, T. J.: "Cellulose, Glycogen and Starch." J. Chem. Ed., *25*:626, 1948.

White, A., Handler, P. and Smith, E. L.: *Principles of Biochemistry*, Fourth Edition, McGraw-Hill Book Company, New York, 1968. Chapters 2, 3, 18, 19 and 20.

THIRTY-SEVEN • BIOCHEMISTRY, AN INTRODUCTION

37.1 INTRODUCTION

When the study of the chemistry of the compounds of carbon (Chapter 25) was initiated, the point was made that the title "Organic Chemistry" came into wide use because organic compounds were first obtained only from plants and animals. These compounds appeared to have a unique association with life. Organic chemistry has no such limitations today. Many organic compounds do, however, have important functions in and are obtainable generally from living plants, animals or microorganisms. Proteins, carbohydrates and fats are examples of such substances. Proteins, in particular, have an association with life processes so clear and unique that many consider the ability to synthesize protein, essentially a chemical process, to be the one property most characteristic of living organisms and of life itself. Irrespective of how one may define life, from the chemical standpoint, muscle contraction, vision, absorption of foods and nerve conduction, vital processes, are maintained only through a complex set of chemical and physical changes so adjusted as to provide both the energy and the chemical compounds required.

Because of the close tie between chemistry and biological processes, a field of study referred to as **biochemistry** has evolved and grown into a major discipline. Knowledge of the subject matter of this field contributes greatly to the understanding of many principles of zoology, botany and the other fields of biology. Modern courses in biology lean heavily on chemistry in explaining essential biological phenomena.

Biochemistry seeks to discover, describe and explain the variety of chemical and physical processes which take place in living organisms. For the chemist it is the field which focuses on the chemical events that maintain life. Others consider it to be one of the life sciences, a point of view based on the fact that the subject of its study is the living organism. In essence, the **biochemist** uses chemical methods to study biological phenomena and from his observations seeks to understand cellular changes through use of recognized chemical and physical laws. One important objective of biochemistry is to bridge the gap, large as it is, that exists between the highly integrated activity of the living cell and the properties of the individual components that make up the cell. The study of biochemistry will be introduced in this chapter by an examination of several important chemical reactions and processes occurring within cells. Several essentially biological phenomena will be described and some significant factors in achieving biological reactions will be discussed. Earlier consideration of this topic has been delayed because a basic knowledge of chemistry, and particularly of organic reactions, is fundamental for understanding the various events that characterize biological processes.

37.2 METABOLISM

Wherever one examines the world of living things in detail, chemical changes are found to be taking place. Within all living organisms foodstuffs undergo profound chemical changes. The totality of these chemical transformations that substances experience is called **metabolism.** Included are all changes a nutrient undergoes, from the moment of its entry into the organism until that moment when a product formed is used for the synthesis of other compounds or is discharged as waste. Based on the observation that a variety of intermediate products ("intermediates") are formed during the conversion of most nutrients into ultimate products, metabolism is frequently referred to as **intermediary metabolism.** Two other terms, **anabolism** and **catabolism,** descriptive of different aspects of metabolism, are also used. For example, green plants use solar energy to make a variety of complex organic compounds from simple starting materials such as water, carbon dioxide and small amounts of other inorganic compounds such as nitrates and phosphates. This formation of larger, sometimes rather complex molecules from simple starting materials in living organisms is called *anabolism.* Such reactions or processes require energy. Living organisms also have the ability to decompose or degrade larger molecules into smaller ones. Reactions or processes of this kind provide a net yield of energy and constitute the phase of metabolism called *catabolism.* Catabolism permits cells to exploit, for their own use, the chemical energy locked up within the different compounds that reach them. In this way cells obtain all the energy that must be expended to accomplish their various anabolic reactions and for carrying on bodily functions.

Metabolism serves as an example of an essentially biological phenomenon. It accomplishes two things: (1) it provides the energy required of all organisms for maintaining life, and (2) it achieves

Glycogen
↓↑
Glucose-1-phosphate
↓↑
Glucose → Glucose-6-phosphate
↓↑
Fructose-6-phosphate
↓
Fructose-1,6-diphosphate
↓↑
Glyceraldehyde-3-phosphate ⇆ Dihydroxyacetone phosphate
↓↑
1,3-Diphosphoglyceric acid
↓
3-Phosphoglyceric acid
↑↓
2-Phosphoglyceric acid
↑↓
Phosphoenolpyruvic acid
↓↑
Pyruvic acid (⇌ Lactic acid)

FIGURE 37.1 EMBDEN-MEYERHOF PATHWAY FOR METABOLISM OF GLUCOSE.

the conversion of dietary constituents into cellular components. Catabolism and anabolism, two opposing aspects of metabolism, constitute a dynamic system in which both are taking place simultaneously and continuously. When a proper balance exists between them, the organism experiences normal growth, development and function.

Chemically, metabolism consists of a sequence of related interconnected chemical reactions. The sequence of equations outlining in order the series of reactions by which a particular substance is transformed within a cell is referred to as a **metabolic pathway.** One such sequence is the Embden-Meyerhof pathway* for the conversion of glucose or glycogen into pyruvic acid, a sequence outlined in Figure 37.1. This series of reactions provides a limited amount of energy for cellular processes; but, more importantly, it accomplishes the conversion of a 6-carbon sugar into two molecules of a three-carbon compound that can be readily oxidized to provide additional energy—and, as a matter of fact, to provide the bulk of the energy recovered by the cell from the degradation of glucose. Each reaction shows only the important metabolic change occurring and does not include all the reactants required. For example, a phosphorus compound, adenosine triphosphate, commonly abbreviated ATP, is required and provides the phosphate group for the formation of glucose-6-phosphate. The structure of

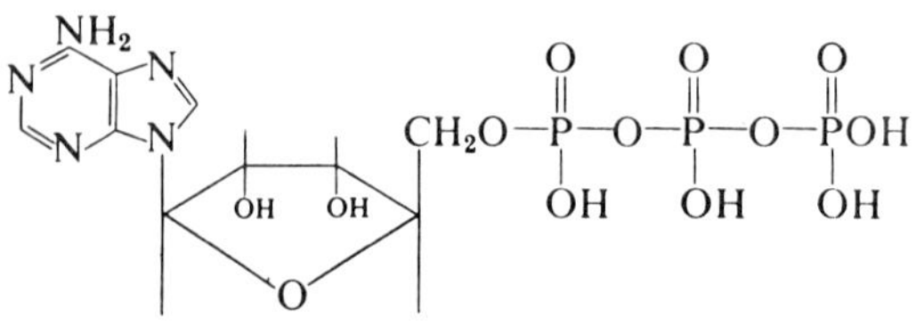

FIGURE 37.2 ADENOSINE TRIPHOSPHATE (ATP).

* Named after G. Embden and O. Meyerhof, whose studies led to the discovery of this pathway.

ATP is given in Figure 37.2. ATP is composed of the heterocyclic base, adenine (Chapter 33), the sugar, ribose (Chapter 36), and three phosphoric acid residues. It reacts with glucose in the presence of enzyme hexokinase (Table 37.2) as follows:

note that a $—PO_3H_2$ group is exchanged for the H of the CH_2OH of glucose, i.e., transferred from ATP to glucose.

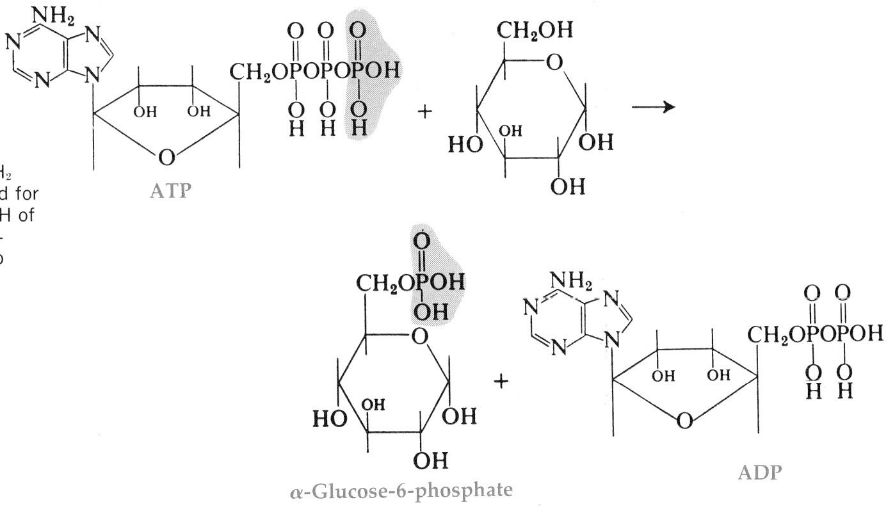

ATP + glucose → α-Glucose-6-phosphate + ADP

ATP is a type of biological substance referred to as a "high energy" compound, and the transfer of phosphate to glucose is accompanied by the release of energy. ATP is capable of transferring phosphate onto alcoholic hydroxyl groups (as illustrated here), onto acyl groups or amide groups, when appropriate enzymes are present. ATP is perhaps the most significant source of chemical energy for driving biological processes in living systems. Oxidations in living systems are accompanied by the formation of large amounts of ATP. The compound obtained through loss of one phosphate residue is named adenosine diphosphate, ADP; that resulting from the loss of two residues is adenosine monophosphate, AMP.

AMP is one of the building units (nucleotides) from which nucleic acids are formed (Chapter 33).

Each reaction of the pathway is referred to as a **metabolic** or **biochemical reaction.** Both terms will be used when referring to individual reactions. Any one of the transformations shown in Figure 37.1 serves as an example of such a reaction. For example:

$$\text{Glucose-1-phosphate} \rightleftarrows \text{Glucose-6-phosphate}$$

or

$$\text{Fructose-6-phosphate} \rightleftarrows \text{Fructose-1,6-diphosphate}$$

Biochemical reaction is thus a descriptive term used to designate a certain chemical transformation that a substance experiences in a living cell.

As is the case with all chemical reactions, biochemical transformations are controlled by the concentrations of reacting substances and the energy changes involved. However, they are also subject to other controlling factors and influences, to which most organic reactions performed outside cells are not. Included is physiological control by the nervous system and by hormones produced in the glands of internal secretion (endocrine glands) such as the pituitary, adrenal or thyroid. These, however complicated, are chemical in nature, obeying the same sort of laws of chemical behavior discussed throughout the text.

If metabolic reactions are compared in other ways with non-biologically controlled organic reactions, one is immediately struck by the high rate with which they proceed in cells under the mild conditions of temperature and pH of the physiological environment, and by the highly specific nature (specificity) of the changes that take place. For example, an important intracellular ("in vivo") reaction, hydration of fumaric acid, may be compared with its chemical hydration outside the cell in a test tube ("in vitro"):

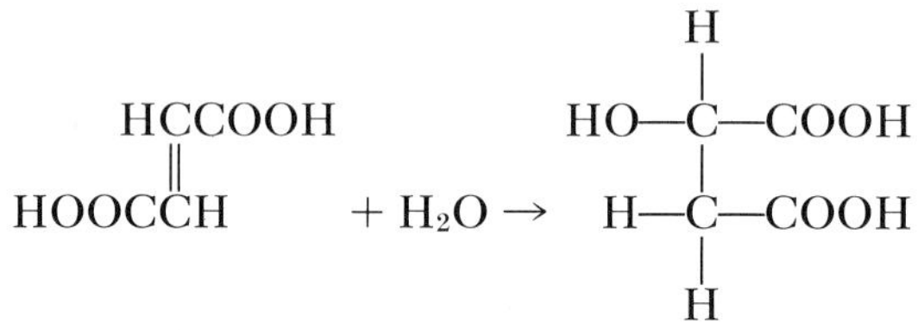

$$\underset{\text{Fumaric acid}}{\begin{matrix} \text{HCCOOH} \\ \| \\ \text{HOOCCH} \end{matrix}} + H_2O \rightarrow \underset{\text{Malic acid}}{\begin{matrix} & \text{H} & \\ & | & \\ \text{HO—} & \text{C} & \text{—COOH} \\ & | & \\ \text{H—} & \text{C} & \text{—COOH} \\ & | & \\ & \text{H} & \end{matrix}}$$

Under physiological conditions (temperature 37°C; pH 7.34) this reaction takes place rapidly and smoothly while producing only one product, L-(—)-malic acid. In contrast, chemical hydration with water outside the cell occurs only under much more drastic conditions (temperature 150–160°C in a closed tube), and then not too readily, to produce a non-specific racemic mixture of both optical isomers of malic acid. One is equally impressed with the sequential and predominantly cyclic pattern found for many of the processes used to achieve rather complex transformations.

the citric acid cycle (Figure 37.5) is an important example.

37.3 ENZYMES

The rapid rates of metabolic reactions have been shown to result from the presence in cells of special catalysts called *enzymes*. Catalysts, as you may recall, are substances that speed up the rates of chemical reactions when present in extremely small amounts, without undergoing any permanent change or being consumed in the process. Enzymes, in addition to catalyzing reactions, also determine the type of reaction and, thus, the nature of the product that will be obtained from a given reactant. This may be referred to as their **directive effect.** Enzymes, because of these capabilities, are of great importance and will be considered in more detail.

Chemically, enzymes are proteins (see Chapter 35) and are, themselves, products of cellular metabolism. Being proteins, they are chemical compounds susceptible to the same chemical reactions and modifications as all other proteins. For example, most enzymes are unstable to heat, a characteristic of most proteins. Enzymes, when kept at an elevated temperature, may undergo structural changes (denaturation), often including the loss of the ability to function.

The characteristic property and function of an enzyme is the ability to catalyze chemical reactions. Catalysis is accomplished through a set of reactions (see Fig. 37.3) including, first, a rapid

$$E + S \rightleftarrows ES$$
$$ES \rightarrow P + E$$

FIGURE 37.3 MECHANISM OF ENZYME CATALYSIS.

combination of the enzyme (E) and a reactant referred to as the substrate (S) to form an enzyme-substrate compound (or complex) (ES). This complex (ES) in a second step—a somewhat slower reaction—decomposes or separates into products (P) and enzyme (E). The enzyme is recovered unchanged and is available for recombination with another molecule of substrate, permitting a repetition of the entire process. Such repetition allows the two-step process to occur many times with the same enzyme.

This picture for the catalytic function of an enzyme is referred to as the Michaelis-Menten hypothesis.* The idea rests on a firm experimental foundation, although the nature of the ES compound has been directly demonstrated for only a few enzymes. The lifetime of the complex is extremely short because of rapid decomposition into products, making detection of the complex difficult. According to this concept one enzyme molecule can achieve conversion of many molecules of substrate. Therefore, only small amounts of enzyme are required. The combination of enzyme and substrate (to be discussed later) aids in explaining the specificity of enzymes.

An enzyme contributes no energy to an overall reaction. Each reaction that is catalyzed must be energetically possible without the enzyme. Enzymes merely increase the rate of formation of products. The final yield (equilibrium concentration) of product is the same with or without the enzyme. Reaction in the absence of the enzyme takes much longer and sometimes is so slow as to be almost undetectable during a reasonable time of observation. Table 37.1 gives some relative rates for the decomposition of H_2O_2 into water and oxygen, $2H_2O_2 \rightarrow 2H_2O + O_2$, when conducted under the same conditions with and without catalysts present.

* Named after the investigators, L. Michaelis and M. L. Menten, who offered the first mathematical treatment of the concept.

TABLE 37.1 RELATIVE RATES OF ENZYME CATALYZED REACTIONS

Reaction	Catalyst	Rates (0°C)*	Rate catalyzed/Rate of Non-catalyzed
Decomposition of H_2O_2	None	3.75	–
	MnO_2	8.50	2.3
	Colloidal Pt	11.90	3.1
	Liver catalase	36.7	9.8
Hydrolysis (inversion) of sucrose	H_3O^+ (acid)	0.87	–
	Yeast invertase	12.3	14.1

* Compared to rate of reaction at 30°C set equal to 100.

Note that the enzyme catalyzed decomposition of H_2O_2 proceeds about ten times faster than the reaction without a catalyst and three to four times as fast as its decomposition in the presence of the inorganic catalysts, MnO_2 or Pt. Sucrose hydrolyzes about 14 times as rapidly in the presence of invertase as in acid solution.

invertase catalyzes the inversion (hydrolysis) of sucrose; hence the name.

As stated, metabolic reactions are highly specific with regard to both the type of substrate involved and the particular reaction that the substrate undergoes. Each reaction proceeds in a manner such as to produce a specific change in the compound concerned. For example, glucose-6-phosphate may be transformed directly into any one of several different products, namely, glucose-1-phosphate, fructose-6-phosphate, glucose or 6-phosphogluconolactone.

1. Glucose-6-phosphate $\leftrightarrows$ Glucose-1-phosphate
2. Glucose-6-phosphate $\leftrightarrows$ Fructose-6-phosphate
3. Glucose-6-phosphate + H_2O $\leftrightarrows$ Glucose + H_3PO_4
4. Glucose-6-phosphate $\leftrightarrows$ 6-Phosphogluconolactone + 2 H

usually only one or a few closely similar reactions are speeded up by one enzyme.

Each reaction yields a different product. The product formed within the cell is determined by the enzyme present at the site of reaction. Each enzyme has the capability of selecting and catalyzing one from among all the possible reactions. Glucose-6-phosphate is converted into glucose-1-phosphate if the enzyme phosphoglucomutase is present. With phosphoglucoisomerase, reaction 2 takes place to give fructose-6-phosphate. Should the enzyme phosphatase be present, glucose-6-phosphate is hydrolyzed to glucose and phosphoric acid as shown in reaction 3. 6-Phosphogluconolactone is formed when glucose-6-phosphate is oxidized under the influence of the enzyme glucose-6-phosphate dehydrogenase as shown in reaction 4. Only if all four enzymes mentioned are present can all four products be expected.

Each enzyme carries an inherent or built-in capability for catalyzing a particular reaction with certain substrates. This capability is referred to as its *specificity*. Enzymes are highly specific catalysts. Catalytic activity, as just postulated, is dependent upon and operates through a combination of enzyme and substrate. Specificity is not some mysterious property of an enzyme, but a phenomenon resulting from a structural kinship between substrate and enzyme. It is a reflection of both (1) the way in which the enzyme unites with the substrate, and (2) how well this attachment of substrate to enzyme takes place. For example, since four different products are obtainable from glucose-6-phosphate, depending upon which of four different enzymes are present, it may be assumed that each enzyme unites with glucose-6-phosphate in a different way. The type of combination occurring predetermines the reaction that will take place and the products that will form. The rate of appearance of product, however, will be dependent upon how well the combination takes place. Specificity reflects an overall relationship between the structural features of the enzyme and substrate. The relationship either facilitates or hinders satisfactory combination. If a close reciprocal structural relationship or fit between enzyme and substrate exists, they combine firmly in such a way that a particular reaction occurs. Without a fit the enzyme-substrate complex does not form and no reaction takes place. A partial combination with a structurally similar compound may occur, in which case reaction sometimes takes place but usually at a greatly reduced rate.

the combination referred to is that of a particular structural fit.

Also, combination of the enzyme with a foreign unreactive substance can prevent reaction by excluding substrate from the enzyme surface or active site. Substances which combine with the enzyme and prevent it from functioning as a catalyst are called *inhibitors.*

The lock and key relationship is a simple analogy employed to visualize this idea or concept of structural fit. Only those keys (substrates) which fit properly into the lock ("active site of the enzyme") induce the lock (enzyme) to open up (catalyze the reaction). The general idea is illustrated in Figure 37.4.

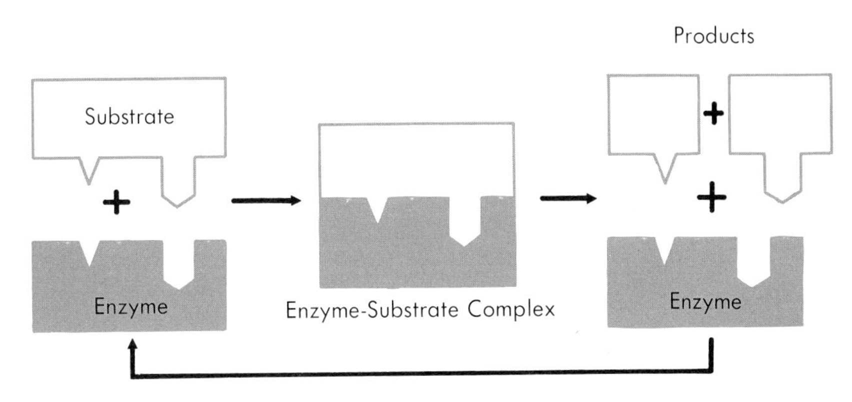

FIGURE 37.4 ENZYME-SUBSTRATE COMBINATION.

Although our present view concerning how enzymes function, when examined in detail, is much more sophisticated than the lock and key analogy, the idea serves reasonably well as an illustration of the concept that specificity depends on and is determined by the degree of structural matching between enzyme and substrate. Since such matching would permit only certain reactions to occur (at reasonable rates) and particular products to form, the idea can account for the two most characteristic features of enzyme function—an ability to catalyze reactions and specificity with regard to types of reactions catalyzed and substrates transformed.

The real question, "What actually happens when the enzyme-substrate complex forms to cause the substrate to divide into products upon separating from the enzyme?" remains unanswered. It could be reasoned that structural and conformational adjustments are required of enzyme and substrate molecules to come together in a perfect fit. The two molecules might merely have to unwind and line up, or it might be necessary for some bonds to stretch or for a chain to twist to bring mutually attractive groups together. The partial immobilization of the substrate in the complex, the readjustments in bond energies with the stretching and twisting of the molecule or the bringing of active groups in the substrate into the complex formation might account for the increased rate of reaction.

37.4 METABOLIC CYCLES

Let us now return to a consideration of the sequential and cyclic nature of metabolic processes. Each reaction of a metabolic pathway is catalyzed by a specific enzyme. A different enzyme is required for each reaction. Any pathway, therefore, becomes a **multienzyme system** composed of a sequence of enzyme-catalyzed reactions requiring careful control and coordination. This illuminates the necessity for the variety of controls, neural, hormonal and enzymatic, that influence cellular processes. The pattern of reactions of one pathway, the Embden-Meyerhof scheme shown in Figure 37.1, has already been outlined. Let us now examine this pathway in more detail.

The formation of pyruvic acid from glucose involves a series of ten reactions, each of which accomplishes only a small, though important, change in the structure of the product of the preceding reaction. As an initial stage, either glucose or glycogen, depending upon the cell studied, is converted into a phosphomonoester of glucose. This appears to be necessary because the enzymes concerned with glucose metabolism by this pathway act only on phosphorylated derivatives. Note that each of the intermediates in the sequence of reactions is a phosphate derivative. Not included as intermediates are glucose or glycogen and the final product, pyruvic acid. Combination with phosphate seems to activate the bond of the derivatives, so the reactions outlined take place readily.

phosphate is biologically important for uses other than bone formation.

For each reaction, the name of the enzyme catalyzing the reaction and the type of change achieved by the reaction are summarized in Table 37.2. Only one step in this sequence involves an oxidation. Both oxidation and phosphorylation are required for the conversion of 3-phosphoglyceraldehyde into 1,3-diphosphoglyceric acid.

TABLE 37.2 REACTION AND ENZYMES OF THE EMBDEN-MEYERHOF PATHWAY OF GLUCOSE METABOLISM

Reaction	Enzyme	Type of change
Glycogen ⇄ Glucose-1-phosphate	Phosphorylase	Phosphorylation
Glucose-1-phosphate ⇄ Glucose-6-phosphate	Phosphoglucomutase	Phosphate interchange
Glucose ⇄ Glucose-6-phosphate	Hexokinase	Phosphorylation
Glucose-6-phosphate ⇄ Fructose-6-phosphate	Phosphoglucoisomerase	Isomerization
Fructose-6-phosphate ⇄ Fructose-1,6-diphosphate	Phosphofructokinase	Phosphorylation
Fructose-1,6-diphosphate ⇄ Glyceraldehyde-3-phosphate	Aldolase	Cleavage of C—C bond
Dihydroxyacetonephosphate ⇄ Glyceraldehyde-3-phosphate ⇅ Dihydroxyacetone phosphate	Triosephosphate isomerase	Isomerization
Glyceraldehyde-3-phosphate ⇄ 1,3-diphosphoglyceric acid	Phosphoglyceraldehyde dehydrogenase	Oxidation and Phosphorylation
1,3-diphosphoglyceric acid ⇄ 3-Phosphoglyceric acid	Phosphoglyceryl kinase	Loss of phosphate
3-Phosphoglyceric acid ⇄ 2-Phosphoglyceric acid	Phosphoglyceromutase	Phosphate interchange
2-Phosphoglyceric acid ⇄ Phosphoenolpyruvic acid	Enolase	Loss of water
Phosphenolpyruvic acid ⇄ Pyruvic acid	Pyruvic kinase	Transfer of phosphate

The net result of the sequence of reactions from glucose to pyruvic acid is the oxidation of glucose accompanied by cleavage into two molecules of pyruvic acid:

$$C_6H_{12}O_6 + O_2 \rightarrow 2\ CH_3COCOOH + 2\ H_2O$$

Note that for each glucose (or glucose-6-phosphate) molecule transformed by way of this pathway, two molecules of pyruvic acid are obtained. This follows from the fact that the dihydroxy acetone phosphate is directly converted into 3-phosphoglyceraldehyde and is thus ultimately transformed into pyruvic acid.

Pyruvic acid, however, does not usually accumulate in cells because of further reaction. The further reaction that it experiences depends upon the type of cell in which it is metabolized. Glucose, when **fermented** (yeast cells), is initially converted into pyruvic acid through the sequence of reactions outlined in Figure 37.1. However, to complete the fermentation process yielding alcohol, pyruvic acid undergoes a loss of carbon dioxide and reduction to give ethanol. On the other hand, in muscle cells the pyruvic acid is reduced directly to lactic acid without loss of carbon dioxide. In muscle the source of glucose is glycogen stored in muscle cells, and the sequence of reactions by which lactic acid is formed from glycogen is referred to as **glycolysis.** The reactions that produce pyruvic acid from glucose-6-phosphate in both fermentation and glycolysis are the same. Fermentation and glycolysis differ only in the source of the glucose-6-phosphate utilized and in the disposition of the pyruvic acid formed. In fermentation glucose-6-phosphate is obtained by direct phosphorylation of glucose, whereas in muscle cells it arises from glucose-1-phosphate that is formed directly from muscle glycogen. Fermentation and glycolysis can both be accomplished anaerobically—without oxygen—if the oxidation of 3-phosphoglyceraldehyde to 1,3-diphosphoglyceric acid is coupled with the reduction of pyruvic acid to ethanol or lactic acid.

The degradation of glucose, however, is more often carried out with oxygen readily available within the yeast or muscle cell. Normally, pyruvic acid is oxidized to carbon dioxide and water without formation of either alcohol or lactic acid. Lactic acid accumulates in muscle only when the supply of oxygen is not sufficient to permit oxidation of all the pyruvic acid formed. Oxidation of pyruvic acid is accomplished through a combination of metabolic reactions (a cycle of reactions) referred to as the TCA* or **citric acid cycle.** The citric acid cycle is one of the most important metabolic pathways found in cells and is presented here (see Fig. 37.5) to illustrate the cyclic nature of many biological processes. The sequence of reactions may be pictured as starting and terminating with the same compound, *oxaloacetic acid.*

In Figure 37.5 pyruvic acid is shown to enter the TCA cycle following its conversion into "active" acetate through a process in which it undergoes oxidation and loss of carbon dioxide (oxidative decarboxylation), in accordance with the equation:

$$CH_3COCOOH + \tfrac{1}{2}O_2 \rightarrow \text{“}CH_3COOH\text{”} + CO_2$$

The equation for the overall reaction for the oxidation of pyruvic acid and the term "active" acetate are employed to simplify the series of events (or reactions) involved in the introduction of pyruvic acid into the tricarboxylic acid (Krebs) cycle. In reality, however, "active" acetate is the compound acetyl coenzyme A (see Chapter 32), formed from pyruvic acid through a series of at least four reactions involving the substances thiamine pyrophos-

intermediates are present only momentarily and in low concentrations due to their rapid conversion into the next intermediate.

* TCA = Tricarboxylic acid cycle.

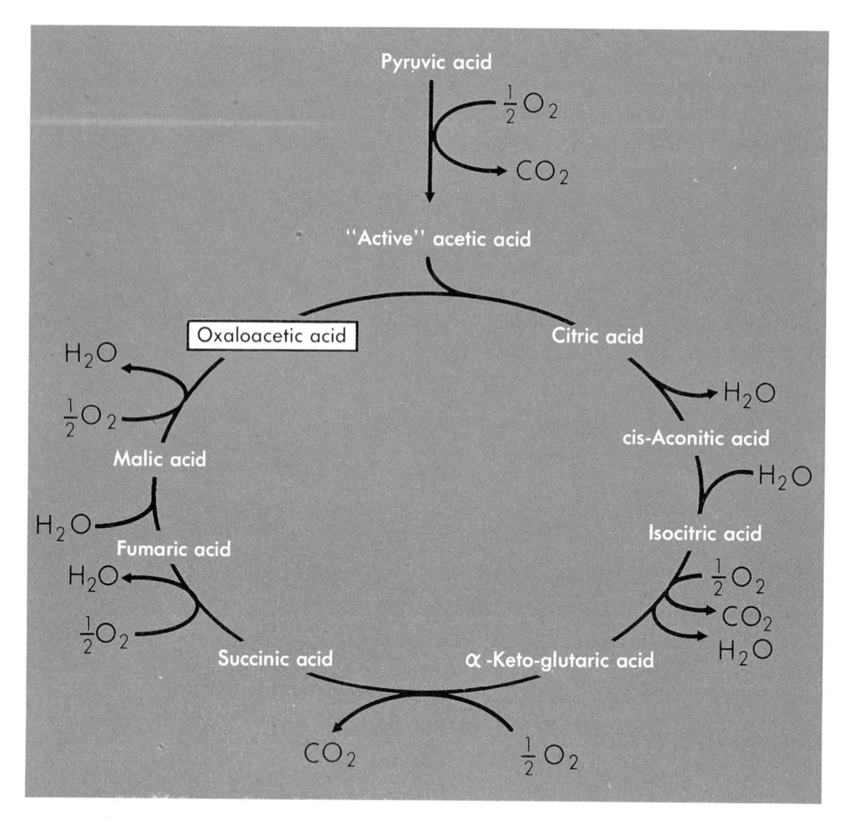

FIGURE 37.5 CITRIC ACID CYCLE.

phate (Chapter 33), lipoic acid, coenzyme A (Chapter 32), diphosphopyridine nucleotide DPN+) and, of course, pyruvic acid. Thiamine pyrophosphate, lipoic acid, coenzyme A and DPN+, whose formulas are shown in Figure 37.6, are commonly referred to as coenzymes since they are required for enzymatic conversion of pyruvic acid into acetyl coenzyme A but are never used up permanently in the process. In the sense that they are present at the start of the reaction, are used in accomplishing the reaction and then may be recovered unchanged after it has been completed, they may be considered to be functioning as catalysts.

many vitamins are essential components of coenzymes.

The series of biochemical reactions by means of which pyruvic acid is converted into acetyl coenzyme A is outlined in Figure 37.7.

Following its formation "active" acetate (acetyl coenzyme A) enters the cycle by reacting with oxaloacetic acid (accompanied by the release of coenzyme A) to form citric acid, which then experiences the sequence of changes outlined in Figure 37.5, the end result of which is the re-formation of oxaloacetic acid.

The regenerated oxaloacetic acid can again react with "active" acetic acid and repeat the cycle. One trip through the cycle of reactions, oxaloacetic acid to oxaloacetic acid, accomplishes the conversion of one "active" acetic acid into carbon dioxide and water. The net result, starting with pyruvic acid, is the transformation of one molecule of pyruvic acid into CO_2 and water.

$$CH_3COCOOH + \frac{5}{2}\,O_2 \rightarrow 3\,CO_2 + 2\,H_2O$$

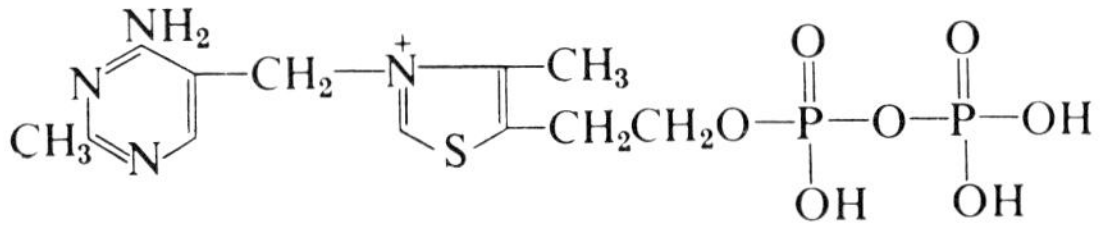

NH2 N CH3 N —CH2—N —CH3 S —CH2CH2O—P—O—P—OH O O OH OH

Thiamine pyrophosphate (TPP)

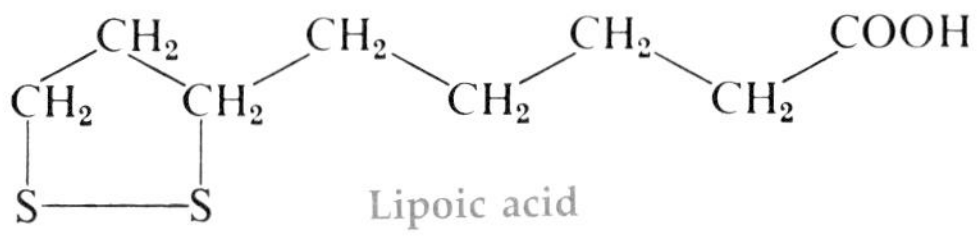

CH2 CH2 CH2 CH2 CH2 CH2 CH2 COOH S—S

Lipoic acid

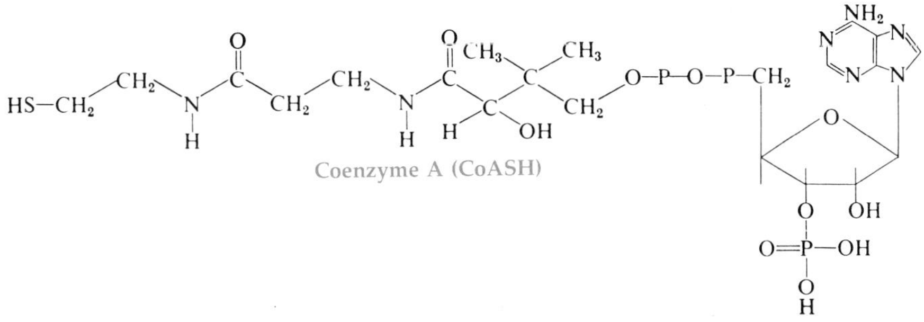

HS—CH2 CH2 N H C O CH2 CH2 N H C O CH3 C CH3 CH2 C H OH O—P—O—P—CH2 NH2 N N N N O O OH O=P—OH O H

Coenzyme A (CoASH)

O C—NH2 N CH2O—P—O—P—OCH2 O O OH OH NH2 N N N N O OH OH O OH OH

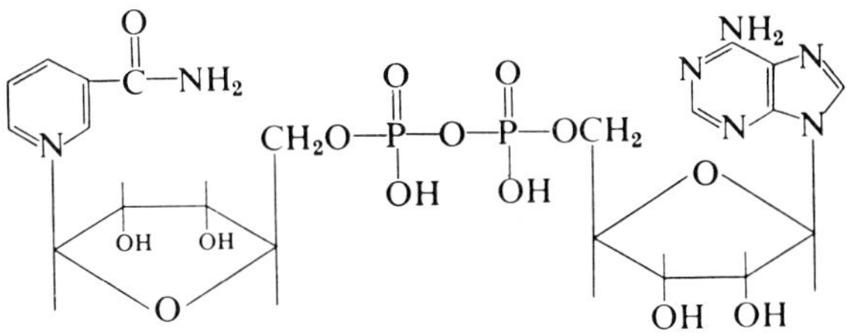

Diphosphopyridine nucleotide (DPN^+)
Nicotinamide adenine dinucleotide (NAD^+)

FIGURE 37.6 SOME COENZYMES.

The oxidation of the two molecules of pyruvic acid obtained from glucose in fermentation or glycolysis requires that two molecules traverse the cycle. The equation for the oxidation of two molecules is:

$$2\ CH_3COCOOH + 5\ O_2 \rightarrow 6\ CO_2 + 4\ H_2O$$

In summary, if one molecule of glucose is subjected to the Embden-Meyerhof pathway and the citric acid cycle in turn, the chemical change accomplished is the complete degradation of glucose to carbon dioxide and water. The Embden-Meyerhof phase achieves the conversion of one glucose into two molecules of pyruvic acid.

the energy released in this process is captured and retained as ATP.

$$\text{a.}\quad C_6H_{12}O_6 + O_2 \rightarrow 2\ CH_3COCOOH + 2\ H_2O$$

The pyruvic acid is then completely oxidized to CO_2 and water

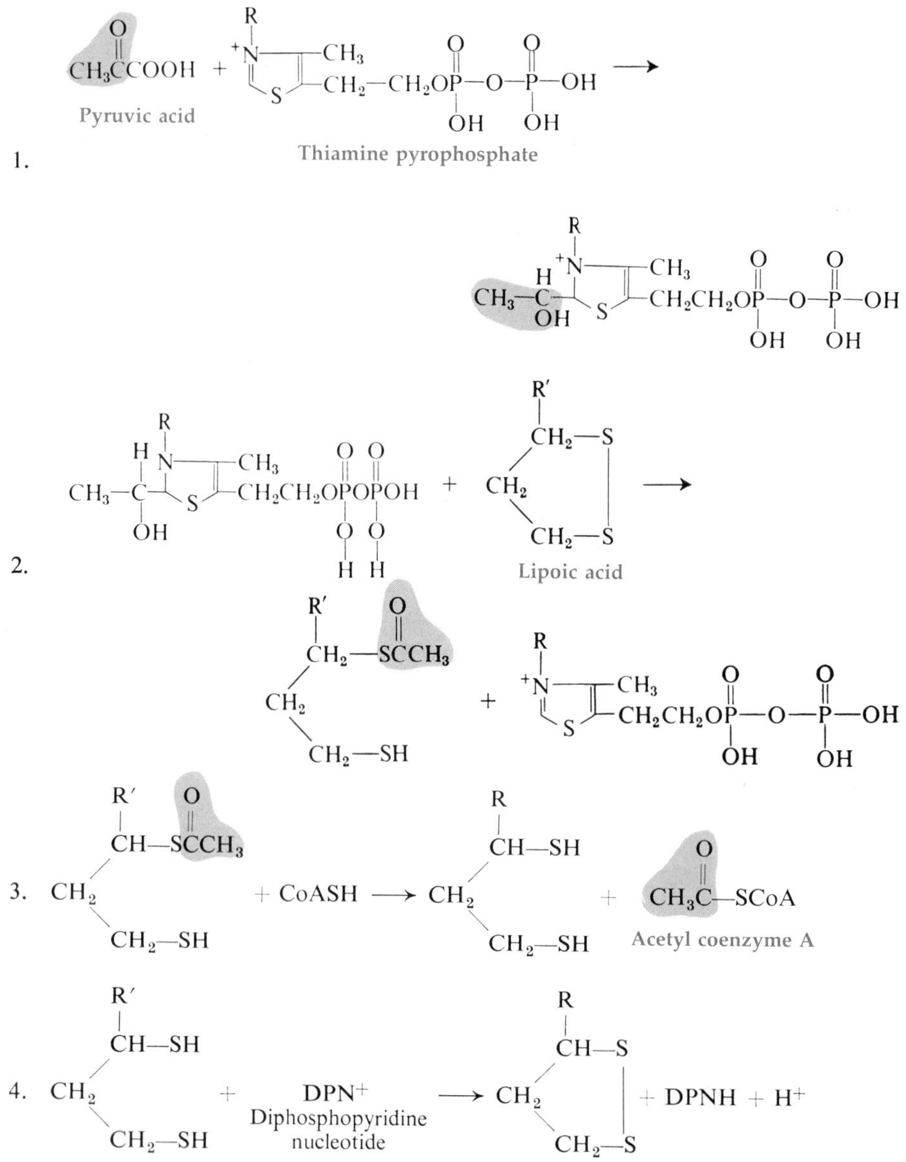

1. $CH_3COCOOH$ (Pyruvic acid) + Thiamine pyrophosphate ($R-N^+$ thiazolium ring with $-CH_3$ and $-CH_2-CH_2O-PO(OH)-O-PO(OH)-OH$) →

$CH_3-CH(OH)-$ (H) attached to C-2 of the thiazolium ring ($R-N^+$, $-CH_3$, $-CH_2CH_2OP(O)(OH)-O-P(O)(OH)-OH$)

2. $CH_3-C(H)(OH)-$ thiazoline ring (R, N, S, $-CH_3$, $-CH_2CH_2OPOPOH$ with two $=O$ and two $-OH$) + Lipoic acid ($R'-CH_2-S$, CH_2, CH_2-S; cyclic disulfide) →

R', CH_2-SCCH_3 (with $C=O$), CH_2, CH_2-SH + Thiamine pyrophosphate ($R-N^+$ thiazolium ring, $-CH_3$, $-CH_2CH_2OP(O)(OH)-O-P(O)(OH)-OH$)

3. $R'-CH(-SCOCH_3)-CH_2-CH_2-SH$ + CoASH → $R-CH(-SH)-CH_2-CH_2-SH$ + $CH_3C(=O)-SCoA$

Acetyl coenzyme A

4. $R'-CH(-SH)-CH_2-CH_2-SH$ + DPN^+ (Diphosphopyridine nucleotide) → $R-CH(-S)-CH_2-CH_2-S$ (cyclic disulfide) + DPNH + H^+

5. $DPNH + H^+ + \frac{1}{2}O_2 \xrightarrow[\text{enzymatic process}]{\text{Via}} DPN^+ + H_2O$

Overall: $CH_3C(=O)COOH + \frac{1}{2}O_2 + CoASH \rightarrow CH_3COSCoA + CO_2 + H_2O$

(Coenzyme A is recovered when acetyl CoA reacts with oxaloacetic acid)

Note: R = 2-methyl-4-amino-pyrimidine ring (NH_2, N, N, CH_3-) $-CH_2-$ and $R'=HOOC(CH_2)_3CH_2-$

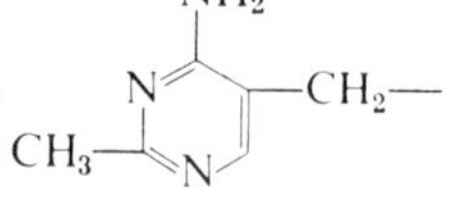

FIGURE 37.7 FORMATION OF "ACTIVE" ACETATE (ACETYL COENZYME A).

11.19 pages 236 and 238–239

11.21 page 241

11.23 The electrons are a greater distance from the same charge in the larger atom.

11.25 (a) 60 g, (b) 16.9 L, (c) 2.06 L

EXERCISES Page 264

12.1 (a) same properties and composition throughout, (b) more than one chemical substance together, (c) homogeneous mixture, (d) a substance in solution dispersed in the other, (e) the continuous phase in a solution

12.3 (a) air, (b) salt water, (c) jewelry gold

12.5 the proportion of the solute in the solution

12.7 1/3

12.9 Most (those with a positive heat of solution) are more soluble at higher temperatures.

12.11 freezing point depression, vapor pressure lowering, boiling point elevation, osmotic pressure

12.13 page 255

12.15 (a) 0.60 mole, (b) 0.94 mole, (c) 0.49 mole, (d) 2.0 moles

12.17 (a) 1800 ml, (b) 2300 ml

12.19 320 ml

12.21 1 M H_2SO_4

12.23 solution (a)

12.25 (a) 23.4 mm, 24.4 atm, −1.86°C, 100.52°C
(b) 23.0 mm, 48.8 atm, −3.72°C, 101.04°C
(c) 23.2 mm, 37.4 atm, −2.85°C, 100.79°C
(d) 23.4 mm, 23.4 atm, −1.78°C, 100.50°C

12.27 (a) 50 g, (b) 746.3 mm, (c) −1.86°C, (d) 24.4 atm

12.29 23.6 moles % alcohol

12.31 (a) or (b)

EXERCISES – SET I Page 293

13.1 page 267

13.3 non-electrolyte, weak electrolyte and strong electrolyte

13.5 pages 267–269

13.7 Test the electrical conductivity and the colligative properties of the solution and a series of dilutions.

13.9 the reactants and products as they really are in the solution

13.11 an ion that accompanies a reacting ion

13.13 pink, colorless

13.15 because the products conduct an electric current

13.17 oxalic acid, phosphoric acid and sulfuric acid

13.19 It includes acid-base reactions taking place outside water solutions.

EXERCISES – SET II Page 294

13.21 (a) $H_3O^+ + OH^- \rightarrow 2\ H_2O$
(b) $Ca(OH)_2 + 2\ H_3O^+ \rightarrow Ca^{2+} + 4\ H_2O$
(c) $2\ HOAc + OH^- \rightarrow 2\ OAc^- + 2\ H_2O$
(i) $Ba^{2+} + SO_4^{2-} \rightarrow BaSO_4$
(l) $Cl_2 + 2\ Br^- \rightarrow Br_2 + 2\ Cl^-$
(m) $AgCl + 2\ NH_3 \rightarrow Ag(NH_3)_2^+ + Cl^-$

PROBLEMS Page 295

13.23 (a) −1.40°C, (b) 100.39°C, (c) 350.3 mm, (d) 18.3 atm

13.25 (a) 36.5 g, (b) 60 g, (c) 32.7 g, (d) 45 g, (e) 40 g, (f) 85.5 g, (g) 37 g, (h) 26 g, (i) 17 g

13.27 (a) 220 g, (b) 3.6 g, (c) 27.8 g, (d) 102.4 g, (e) 175 g

13.29 (a) 2.4 M, (b) 0.19 M, (c) 12.1 M, (d) 6 M, (e) 0.015 M

13.31 normality = n × molarity, where n is a small number

13.33 (a) 3 N, (b) 9.4×10^{-4} N, (c) 4.28 N, (d) 0.085 N, (e) 5.0 N, 1.72 N, (f) 0.193 N

EXERCISES – SET I Page 318

14.1 (a) loss of electrons, (b) gain of electrons, (c) page 190, (d) substance reduced, (e) substance oxidized

14.3 (a) Mn, (b) reduced, (c) MnO_4^- and H_2O in becoming MnO_2 and OH^-

14.5 Zn, Cu, Fe

14.7 See Exercise 14.16.

14.9 negative

14.11 $Cu + 2\ Ag_2SO_4 \rightarrow CuSO_4 + 2\ Ag$ (overall)
$Cu \rightarrow Cu^{2+} + 2\ e^-$ (anode)
$2\ Ag^+ + 2\ e^- \rightarrow 2\ Ag$ (cathode)
$Cu + 2\ Ag^+ \rightarrow Cu^{2+} + 2\ Ag$ (net ionic)

14.13 No, zinc is more active than H_2.

14.15 $PbSO_4 + H^+ + 2\ e^- \rightarrow HSO_4^- + Pb$
$PbSO_4 + 2\ H_2O \rightarrow HSO_4^- + PbO_2 + 3\ H^+ + 2\ e^-$

EXERCISES – SET II Page 319

14.17 59.2 L

14.19 113 g

21.11 Ti^{3+} has an unpaired electron, whereas Ti^{4+} does not.

21.13 sp^2d, square planar; sp, linear; sp^3d^2, octahedral

21.15 dichlorotetraaquochromium(III) chloride, aquopentaamino-cobalt(II) sulfate

21.17 the higher the ion charge of the same atom, the tighter it holds on to the electron pair of the ligand, here on the pair on oxygen. The hydrogen atom then has a smaller share of the electron pair.

EXERCISES Page 510

22.1 Al, Sn, Pb

22.3 SiO_2, $CaSiO_3$, Na_2SiO_3, $MgSiO_3$, $BaSiO_3$

22.5 multiple bonding and ring formation

22.7 because of the three-dimensional covalent bonding coupled with great electron attraction

22.9 There is no regular bonding between the graphite layers; they are far apart, giving a low density to graphite.

22.11 120°

22.13 The carbon dioxide exhaled by humans must equal the carbon dioxide taken up by the plants or there will be an imbalance.

EXERCISES Page 566

25.3 all are satisfactory

25.5 methyl, ethyl, propyl and isopropyl

25.7 $(CH_3)_4C$, $(CH_3)_2CHCH_3$, $CH_3(CH_2)_3—CH(CH_3)_2$ and $(CH_3)_4C$

25.9 The insolubility of alkanes and their low boiling points result from their non-polar character: non-polar molecules have little attraction for each other.

25.13 Ethane, propane, butane, isobutane, pentane, isopentane, hexane, isohexane and 2,3-dimethylbutane.

25.15 (a) 1-bromo-3-methylpentane, (b) 2,4-dimethyl-3-nitro-hexane, (c) 2-bromo-3,3,5-trimethylheptane, (d) 2,3-di-bromo-5-chloro-4-isopropyl-3-methylheptane, (e) methyl-cyclopropane, (f) 1,2-dimethylcyclobutane

25.17 n-hexane, three; isohexane, five; neohexane, three; 2,3-dimethylbutane, two.

25.21 The longest continuous chain was not selected in (a), (d) and (g). The chain was numbered from the wrong end in (b) and (c). For (e) the numbering around the ring was incorrectly chosen. The error in (f) was that no numbers are required.

25.23 The sample is a mixture of two isomeric hydrocarbons.

25.25 RI is ethyl iodide. R'I is 1-iodopropane

25.27 C_6H_{14}; $CH_3CH_2CH(CH_3)CH_2CH_3$

25.29 (a) $(CH_3)_3CCH_2CH_3$, $(CH_3)_2CHCH(CH_3)_2$ or $(CH_3)_3CCH_2CH_3$; (b) $(CH_3)_2CHCH_2CH_2CH_3$; (c) all structures give more than two monochloro substitution products.

EXERCISES Page 584

26.3 $\Delta G = \Delta H - T\Delta S$. Large increases in the entropy (ΔS) increase the $T\Delta S$ term of this mathematical expression, which in turn leads to a more negative value for ΔG. Negative ΔG indicates an energetically favorable reaction.

26.5 Increase in entropy occurs. The structure in the liquid state is generally more random than in the solid state, indicating a larger entropy.

26.7 The catalyst could be added to increase the rates of what can occur but it would have no influence on the yield of product to be expected.

26.9 $\Delta G = -57{,}800 \text{ cal} - 298° \times -10.6 \text{ cal/°C} = -57{,}800 + 3160 = -54{,}640 \text{ cal}$

EXERCISES Page 617

27.3 ethylene (ethene); propylene (propene); -butylene (1-butene); -butylene (2-butene); isobutylene (2-methylpropene)

27.5 ethane, 1.54 Å, free rotation, slow substitution; ethylene, 1.34 Å, rigid, rapidly adds a variety of reagents; acetylene, 1.21 Å, rigid, rapidly adds a variety of reagents

27.7 5-chloro-3-ethyl-3, 5-dimethyl-1-heptene

27.11 $CH_3CH{=}CH_2 + H_2SO_4 \rightarrow CH_3CH(OSO_3H)CH_3$; $CH_3CH(OSO_3H)CH_3 + H_2O \rightarrow CH_3CH(OH)CH_3 + H_2SO_4$

27.13 acetylene (ethyne), propyne, 1-butyne, 2-butyne, 1-pentyne, 2-pentyne and 3-methyl-1-butyne

27.15 The reactions of 1-butyne not possible for its isomer are the reactions with sodium metal and with the ammoniacal $AgNO_3$ and Cu_2Cl_2.

27.17 (a) the decolorization of a Br_2 or $KMnO_4$ solution; (b) the formation of a precipitate with ammoniacal $AgNO_3$ or Cu_2Cl_2; (c) ozone adds, causing cleavage of the molecule at the double bond, when reacted with water to yield aldehydes

27.21 The tetramer of propene is $ACH(CH_3)CH_2CH(CH_3)CH_2CH(CH_3)CH_2CH(CH_3){=}CH_2$. A is the radical or ion which initiates chain formation.

27.23 cis-1,2-dichloroethene

27.25 $(CH_3)_{2+}CCH_2CH_3$, the tertiary carbonium ion, is more stable owing to easier accommodation of the positive charge than is the case with either a secondary or primary ion.

27.27 (a) $CH_3CH(CH_3)CH{=}CH_3$,
(b) $CH_3(CH_2)_2CH{=}CH(CH_2)_2CH_3$, (c) $CH_3CH{=}CHCH_3$,
(d) $CH_3CH_2CH{=}CH_2$

27.29 Compound M is $(CH_3)_2C{=}CH_2$; compound B is $(CH_3)_3COSO_2OH$.

27.31 Hydrocarbon I is $(CH_3)_2C{=}CH_2$.

27.33 Compound A is $CH_3CH(OH)CH_3$; compound B is CH_3CHICH_3; compound C is $CH_3CH(MgI)CH_3$; compound D is $CH_3C(CH_3){=}O$ (acetone)

27.35 Calculations show that 64 g of olefin combine with one mole of Br_2. The nearest molecular formula for an olefin such as this is C_5H_8. The 0.107 g of olefin combines with 0.0033 g of H_2 gas. A possible structure would be cyclopentene.

EXERCISES Page 639

28.3 The symbol implies more unsaturation, that two ortho isomers exist and that there is a higher energy content for the molecule than is found for benzene.

28.5 ethylbenzene (phenylethane); p-dibromobenzene (1,4-dibromobenzene); 2,4-dichlorotoluene (1-methyl-2,4-dichlorobenzene); allyl benzene (3-penylpropene); 1,2-diphenylethane (2-phenylethylbenzene); diphenylmethyl group; 2,4,6-trinitrotoluene (1-methyl-2,4,6-trinitrobenzene)

28.9 meta directing groups: CN, CHO, COOH, NO_2, NH_4^+; ortho-para directing groups: Cl, CH_3, OH, NH_2, OCH_3, $NHCH_3$

28.13 The CN, being strongly electron attracting, pulls electron density out of the ring, decreasing its reactivity toward electrophiles and in electrophilic substitution.

28.15 Aluminum chloride ($AlCl_3$) is an electron attracting reagent (Lewis acid) which highly polarizes the C—Cl bond in the RCl to which it bonds, increasing the electrophilic character of the alkyl halide.

28.17 Compound A is ethyl benzene.

28.19 Compound A is ethyl benzene. Compound B is o-dimethyl benzene.

28.21 (a) Perform a sodium fusion and test for Cl^-; (b) perform a sodium fusion and test for CN^-; (c) test for the solubility in concentrated H_2SO_4—cyclohexane is insoluble; (d) test for a 1-alkyne with ammoniacal $AgNO_3$ or Cu_2Cl_2.

28.23 o-xylene (b.p. 144.1°C) yields 1,3-dimethyl-2-nitrobenzene and 1-nitro-2,4-dimethylbenzene; p-xylene (b.p. 138°C) yields 1,4-dimethyl-2-nitrobenzene; m-xylene (b.p. 138°C) yields 1,3-dimethyl-2-nitrobenzene, 1,3-dimethyl-4-nitrobenzene and 1,3-dimethyl-5-nitrobenzene.

EXERCISES Page 667

29.1 carbon, hydrogen and oxygen

29.3 Methyl alcohol (methanol), ethyl alcohol (ethanol), isopropyl alcohol (2-propanol), isobutyl alcohol (2-methyl-1-propanol), phenol (phenol) and -naphthol (1-naphthol).

29.5 (a) phenol 1° 2° 3°; (b) 3° 2° 1° phenols; (c) same as (b); (d) same as (a); (e) phenols react readily; alcohols do not react

29.7 Phenol and cyclohexanol can be separated with dilute NaOH. Phenol dissolves in the dilute NaOH and cyclohexanol does not.

29.9 The Williamson synthesis gives one product only when it reacts.

29.13 The first reaction is formation of the oxonium salt, ROH_2^+.

29.15 Hexane with Cl_2 yields 1-chlorohexane. 1-Chlorohexane reacts with $NaOH_{(aq)}$ to give n-hexyl alcohol (1-hexanol).

29.17 (1) n-butyl, isobutyl, t-butyl; (2) diethyl ether, 2-methyl-2-butanol, 2-pentanol, 1-pentanol, phenol, 4-nitrophenol; (3) ethanol, 1-octanol, 2-hexene, ethyl ether; (4) 2-butanol, 2-methyl-2-butanol, neopentyl alcohol, 3-pentanol, 1-pentanol and 1-hexanol.

29.19 Compound A is $CH_3CH_2CH_2CH_2OH$.

29.21 Compound A is $CH_3CH_2CH_2OH$. Compound C is $CH_3CH(Cl)CH_3$.

29.23 Dicyclohexyl ether. The oxygen is more basic than in diphenyl ether.

29.25 Compound A is $CH_3CH_2C(CH_3)(OH)CH_2CH_3$. Compound B is 3-methyl-2-pentene. Compound C is methylethyl ketone. Compound D is acetic acid.

29.27 The compound $C_6H_{14}O$ is 3-methyl-3-pentanol. Compound A is $CH_3CH_2C(=O)CH_3$; compound B is CH_3COOH.

EXERCISES Page 688

30.1 (a) alkenes, C=C double bond, (b) alcohol, OH group (hydroxyl), (c) aldehydes, CHO group (aldehyde group), (d) ketones, C=O group (carbonyl).

30.3 acetaldehyde (ethanal), benzaldehyde (phenylcarboxaldehyde), propionaldehyde (propanal), acetone (propanone), acetophenone (1-phenylethanone) and methylethyl ketone (2-butanone).

30.5 a. 2-methylpropanal, b. 3-methylbutanone, c. 4-chloro-3-methylbutanone, d. 2-chloro-2-methyl-1-phenylpropanone.

30.9 (a) formaldehyde and benzaldehyde; (b) formaldehyde; (c) ethanol and isopropyl alcohol (slow), propene and formaldehyde (fast) and benzaldehyde; (d) ethanol, isopropyl alcohol and methylethyl ketone.

30.11 The two enols are $CH_3CH{=}C(OH)CH_2CH_2CH_3$ and $CH_3CH_2C(OH){=}CHCH_2CH_3$.

30.13 Two are possible because there are two different kinds of alpha H's in the molecule.

30.15 Compound A has the structure $CH_3CH_2CH_2CH_2CH(Cl)CH_3$.

30.17 Compound A may be either of two structures, $CH_3CH_2CH_2CH(OH)CH_2CH_3$ or $(CH_3)_2CHCH(OH)CH_2CH_3$.

30.19 Compound A is p-bromophenylethanal. Compound B is the bisulfite addition compound of compound A. Compound C is p-bromobenzoic acid. Compound D is 2-(p-bromophenyl) ethanol.

EXERCISES Page 712

31.1 Ammonia, water, alcohol, phenol, carboxylic acids, sulfonic acids, sulfuric acid.

31.5 The hydrogen atoms of the carboxyl group of carboxylic acids should be acidic because the ion is stabilized more by resonance than is the unionized molecule.

31.7 (a) $C_6H_5COOH + NaOH(CaO)$ yields $C_6H_6 + Na_2CO_3$
(c) $CH_3COOH + CH_3CH_2OH$ yields $CH_3COOCH_2CH_3 + H_2O$
(f) $CH_3CH_2COOCH_2CH_3 + NH_3$ yields $CH_3CH_2CONH_2 + CH_3CH_2OH$

31.9 (b) 3 CH_3CH_2COOH + 3 $SOCl_2$ yields 3 CH_3CH_2COCl + 3 SO_2 + 3 HCl

31.11 propionic acid, 2-aminopropanoic acid, 2,2-dichloropropanoic acid, 2-nitropropanoic acid.

31.13 p-Nitrophenol exhibits the larger K_a. This is due to electron withdrawal, causing a decreased electron density on the phenyl ring which aids proton loss from the OH.

31.15 The alkyl bromide is t-butyl bromide, $(CH_3)_3CBr$. Only this alkyl bromide will yield 2,2-dimethyl propanoic acid using the Grignard reaction as outlined.

31.17 Baeyer strain theory suggests that only smaller five- and six-membered rings will form readily since five- and six-membered rings appeared to be most stable. If reagents used could form only smaller or larger rings, linear polymers would be favored over ring formation.

31.19 The general formula would be $C_nH_{2n-2}O_2$.

31.21 Compound A is acetanilide, $C_6H_5NHCOCH_3$. Compound B is aniline, $C_6H_5NH_2$. Compound C is acetic acid, CH_3COOH.

31.23 (a) Sour taste, neutralization of dilute NaOH demonstrate presence of propionic acid.

(b) Add aqueous silver nitrate and AgCl would form with benzoyl chloride.
(c) Dimethyl amine is soluble in dilute HCl.
(d) Same test as in (b) above.
(e) Same test as in (b) above. Note that other tests not described can be used also.

EXERCISES Page 728

32.1 Cottonseed oil is a mixture of mixed triglycerides, whereas petroleum oil is a mixture of long chain alkanes (saturated hydrocarbons).

32.3 (a) A glyceride is an ester of a fatty acid with glycerol.
(c) A fatty acid is a saturated monocarboxylic acid.
(e) A fat is a solid triglyceride.
(g) the number of grams of I_2 which, if added, could add to 100 grams of fat or oil
(i) Detergents are compounds having a water soluble and a fat soluble group within the same molecule, so they are partially soluble in both polar and non-polar materials.
(k) an oil which has undergone complete or partial hydrogenation
(m) a substance which is soluble in the usual fat solvents (benzene, ether, CCl_4)

32.7 (a) In acid solution fatty acids separate: $RCOONa + H_3O^+$ yields $RCOOH + Na^+ + H_2O$.
(b) In hard water Ca^{2+} and Mg^{2+} salts precipitate: $2\,RCOONa + Ca^{2+}$ yields $(RCOO)_2Ca + 2\,Na^+$

32.9 Detergents dissolve materials by forming micelles, which have insoluble fat and oil at the center surrounded by hydrophilic (water soluble) groups at the surface, rendering the micelle soluble.

32.11 outlined in detail beginning on page 726 of the text

32.13 Beta-oxidation is the oxidative process at the beta-carbon (number 3 carbon) of the carbon chain of a carboxylic acid.

32.15 Determine the iodine number; adulteration would yield a value that is much too low.

32.17 acrolein, $CH_2{=}CHCHO$

32.19 C_6H_5, OH, SH, OMe, I, Cl, CN

EXERCISES Page 752

33.1 Nitrogen is a key element in compounds important for living systems (proteins, DNA).

33.3 ethyl amine (aminoethane), diethyl amine (ethylaminoethane), aniline (aminobenzene), tetramethyl ammonium iodide, acetylmethyl amine (N-methylethanamide), diphenyl amine (N-phenylaminobenzene), and benzylmethylphenylamine (N-benzyl-N-methylaminobenzene)

33.5 CH_3NH_3I, $(CH_3)_2NH_2I$, $(CH_3)_3NHI$ and $(CH_3)_4NI$ (all salts)

33.7 $C_6H_5NH_2 + C_6H_5COCl$ yields $C_6H_5NHCOC_6H_5 + HCl$

33.9 A ring (cyclic) compound which has one or more atoms of an element other than carbon present as a ring atom.

33.11 (a) pyran, (b) furan

33.13 A quinone structure, a nitro group and an azo group.

33.15 Azobenzene. Add a group that will permit it to attach to a fiber.

33.17 Both are porphyrin derivatives. A significant difference is the central metal atom complexed in the porphyrin unit; Mg^{2+} is in chlorophyll and Fe^{2+} in hemoglobin.

33.19 Cytochrome c, used in cellular electron transport; catalase, an enzyme catalyzing the decomposition of H_2O_2.

EXERCISES Page 770

34.1 a. compounds having the same molecular formulas but different structural formulas
c. isomers having the same functional groups but which differ only in the chain sequence of the carbon atoms present
e. isomers having the same structural formula and geometry but which differ in the three-dimensional spatial (stereo) distribution of the atoms present
g. isomers which are capable of rotating the plane of plane polarized light
i. carbon atoms which are bonded to four different groups or atoms
k. an equal molar mixture of a pair of enantiomers
m. light which has its plane of vibration confined to only one plane
o. the actual (real) spatial (three-dimensional) distribution of the atoms present

34.7 *cis*-1,3-cyclohexane dicarboxylic acid and *trans*-1,3-cyclohexane dicarboxylic acid. Many *cis-trans* pairs differ in properties as much or more than do maleic and fumaric acids.

34.9 $CH_3CH_2CH(CH_3)COOH$

34.11 All alpha amino acids except glycine, $CH_2(NH_2)COOH$, have one or more asymmetric carbon atoms.

34.13 Three; two are optically active, one is a meso compound.

34.15 five

EXERCISES Page 791

35.1 a. a difunctional compound containing both an amino and an acidic group (COOH)
c. a compound formed by loss of water between an amino group and a carboxyl group of two amino acids

e. the —NHCO— group which bonds two amino acids of a peptide or a protein
g. compounds containing two like or unlike functional groups
i. the mixture of products obtained when a hydrolysis reaction has been completed
k. amino acids having a benzene or benzene-like ring present in their structures
m. a substance which reacts both as an acid or a base, depending on the conditions
o. a large molecular weight compound formed from a large number of identical or very similar structures by suitable chemical reaction (polypeptides)
q. an amino acid polymer which is globular in shape and soluble or dispersible in water
u. the transformation of a protein from its state in nature into a structure which is not readily restored to its original condition

35.3 Amino acids are in solution as dipolar ions of the form: $H_3\overset{+}{N}CH(R)COO^-$. In acidic solution it becomes $H_3^+NCH(R)COOH$, and in basic solution, $H_2NCH(R)COO^-$.

35.5 (a) $H_2NCH_2COOH + CH_3COCl$ yields $CH_3CONHCH_2COOH + HCl$
(b) $H_2NCH_2COOH + CH_2O$ yields $HOCH_2NHCH_2COOH$
(c) $H_2NCH_2COOH + CH_3CH_2OH$ yields $H_2NCH_2COOCH_2CH_3 + H_2O$

35.7 See pages 787–788 of the text.

35.9 quaternary structure

35.11 HbC and HbS differ at position 6 of the alpha chain (lysine is present in HbC and valine in HbS).

35.13 Hydrolysis would cause cleavage of the peptide bond.

35.15 oxytocin; vasopressin.

35.17 The three-dimensional bundle of four polypeptide as arranged about the porphyrin ring.

35.19 Two; six; twenty-four; use the formula n!, where n is the number of amino acids.

35.21 The charge at pH 1 is +1. The charge at pH 12 is −3. Four peptides are possible.

35.23 Glycylalanylcysteinylglutaminylmethionylalanylleucine

EXERCISES Page 814

36.1 a. a polyhydroxy aldehyde or ketone or a compound which yields them on hydrolysis
c. a six carbon monosaccharide which has a non-terminal carbonyl group
e. a carbohydrate which yields from two to ten monosaccharides when hydrolyzed

g. a change in the optical rotation during the standing of an optically active solution
k. aldoses which differ only in the configuration about carbon 2 of the carbon chain
n. an equimolar mixture of glucose and fructose obtained upon hydrolysis of sucrose

36.3 For a two carbon chain the carbonyl group cannot be on a non-terminal carbon.

36.5 D designates the configuration; (+) designates the direction of optical rotation.

36.9 Maltose is an alpha-1,4-glucoside; cellobiose is a beta-1,4-glucoside; lactose is a beta-1,4-galactoside; sucrose is both a glucoside and a fructoside.

36.11 (a) starch (blue); (b) amylose (blue); (c) amylopectin (reddish-violet); (d) glycogen (reddish-violet).

36.13 the presence of an aldehyde group or an unsubstituted OH on carbon 1 in the cyclic (Haworth) structure

36.15 Mild oxidation yields a monocarboxylic acid, referred to as an -onic acid. Oxidation with nitric acid forms a saccharic acid which is a dicarboxylic acid.

36.17 Mucic acid is a meso acid. The molecule is symmetrical.

36.19 D-Allose.

36.21 Compound A is meso-tartaric acid. Compound B is a racemate of D- and L-tartaric acids.

EXERCISES Page 829

37.1 The goal is to discover, describe and explain the variety of chemical and physical processes which occur in living systems.

37.3 any product of a metabolic reaction which is further transformed is an intermediate

37.5 It is a useful process (1) for providing energy and (2) for converting dietary materials into useful and essential biological compounds.

37.7 any reaction which takes place in a living system

37.9 The determining factor is the type of enzyme available at the site of reaction.

37.11 An important function of the tricarboxylic acid cycle is the conversion of pyruvic acid, obtained upon the cleavage of carbohydrate, into CO_2 and H_2O to obtain energy.

37.13 This is the Embden-Meyerhof pathway for transformation of glucose.

37.15 The inversion of sucrose takes place 14 times faster in the presence of invertase.

37.17 phosphoglucomutase; phosphoglucoisomerase; phosphoglyceromutase; pyruvic kinase

37.19 non-protein entities associated with enzymes and necessary for the reactions they catalyze to occur; diphosphopyridine nucleotide (DPN) and coenzyme A

37.21 ATP is a phosphate donor and energy source.

37.23 one molecule

37.25 Lipoic acid is a substituted carboxylic acid containing S atoms on carbons 6 and 8.

INDEX

Note: Page numbers in *italics* indicate an illustration; page numbers followed by the letter "t" indicate tabular material. A page number followed by the letter "m" indicates material in a marginal note.

F

G

I

K

L

Q

R

S

T

U

V

TABLE OF ATOMIC WEIGHTS

(Based on Carbon-12)

	Symbol	*Atomic No.*	*Atomic Weight*
Actinium	Ac	89	227
Aluminum	Al	13	26.9815
Americium	Am	95	[243]*
Antimony	Sb	51	121.75
Argon	Ar	18	39.948
Arsenic	As	33	74.9216
Astatine	At	85	[210]
Barium	Ba	56	137.34
Berkelium	Bk	97	[249]
Beryllium	Be	4	9.0122
Bismuth	Bi	83	208.980
Boron	B	5	10.811
Bromine	Br	35	79.909
Cadmium	Cd	48	112.40
Calcium	Ca	20	40.08
Californium	Cf	98	[251]
Carbon	C	6	12.01115
Cerium	Ce	58	140.12
Cesium	Cs	55	132.905
Chlorine	Cl	17	35.453
Chromium	Cr	24	51.996
Cobalt	Co	27	58.9332
Copper	Cu	29	63.54
Curium	Cm	96	[247]
Dysprosium	Dy	66	162.50
Einsteinium	Es	99	[254]
Erbium	Er	68	167.26
Europium	Eu	63	151.96
Fermium	Fm	100	[253]
Fluorine	F	9	18.9984
Francium	Fr	87	[223]
Gadolinium	Gd	64	157.25
Gallium	Ga	31	69.72
Germanium	Ge	32	72.59
Gold	Au	79	196.967
Hafnium	Hf	72	178.49
Helium	He	2	4.0026
Holmium	Ho	67	164.930
Hydrogen	H	1	1.00797
Indium	In	49	114.82
Iodine	I	53	126.9044
Iridium	Ir	77	192.2
Iron	Fe	26	55,847
Krypton	Kr	36	83.80
Lanthanum	La	57	138.91
Lawrencium	Lw	103	[257]
Lead	Pb	82	207.19
Lithium	Li	3	6.939
Lutetium	Lu	71	174.97
Magnesium	Mg	12	24.312
Manganese	Mn	25	54.9380
Mendelevium	Md	101	[256]
Mercury	Hg	80	200.59
Molybdenum	Mo	42	95.94
Neodymium	Nd	60	144.24
Neon	Ne	10	20.183
Neptunium	Np	93	[237]
Nickel	Ni	28	58.71
Niobium	Nb	41	92.906
Nitrogen	N	7	14.0067
Nobelium	No	102	[253]
Osmium	Os	76	190.2
Oxygen	O	8	15.9994
Palladium	Pd	46	106.4
Phosphorus	P	15	30.9738
Platinum	Pt	78	195.09
Plutonium	Pu	94	[242]
Polonium	Po	84	210
Potassium	K	19	39.102
Praseodymium	Pr	59	140.907
Promethium	Pm	61	[145]
Protactinium	Pa	91	231
Radium	Ra	88	226.05
Radon	Rn	86	222
Rhenium	Re	75	186.2
Rhodium	Rh	45	102.905
Rubidium	Rb	37	85.47
Ruthenium	Ru	44	101.07
Samarium	Sm	62	150.35
Scandium	Sc	21	44.956
Selenium	Se	34	78.96
Silicon	Si	14	28.086
Silver	Ag	47	107.870
Sodium	Na	11	22.9898
Strontium	Sr	38	87.62
Sulfur	S	16	32.064
Tantalum	Ta	73	180.948
Technetium	Tc	43	[99]
Tellurium	Te	52	127.60
Terbium	Tb	65	158.924
Thallium	Tl	81	204.37
Thorium	Th	90	232.038
Thulium	Tm	69	168.934
Tin	Sn	50	118.69
Titanium	Ti	22	47.90
Tungsten	W	74	183.85
Uranium	U	92	238.03
Vanadium	V	23	50.942
Xenon	Xe	54	131.30
Ytterbium	Yb	70	173.04
Yttrium	Y	39	88.905
Zinc	Zn	30	65.37
Zirconium	Zr	40	91.22

*A value given in brackets denotes the mass number of the longest-lived or best-known isotope.